Category of measurement	Name of SI unit	Symbol	Expression in terms of Other units	Base units
area	square meter	m^2		m^2
volume	cubic meter	m^3		m^3
velocity	meter per second	$m\ s^{-1}$		$m\ s^{-1}$
flow	cubic meter per second	$m^3\ s^{-1}$		$m^3\ s^{-1}$
density	kilogram per cubic meter	$kg\ m^{-3}$		$kg\ m^{-1}$
force	newton	N		$kg\ m\ s^{-2}$
pressure	pascal	Pa	$N\ m^{-2}$	$kg\ m^{-1}s^{-2}$
energy, heat	joule	J	N m	$kg\ m^2\ s^{-2}$
power	watt	W	$J\ s^{-1}$	$kg\ m^2\ s^{-1}$

There are also a number of other units that are not part of SI but nonetheless are widely used and thus acceptable: the liter (l), a unit of volume equal to one one-thousandth of a cubic meter; the are (a), a unit of area equal to 100 square meters, and the more often used hectare (ha), equal to 100 a and 10,000 m^2; minute, hour, and day, familiar units of time. The calorie (cal) a familiar unit of heat energy, is not derived from SI base units and should be abandoned in favor of the joule (=0.239 cal), even though ecologists continue to use the cal and kcal.

Multiples of 10 of the SI units are given special names by adding the following prefixes to the unit name:

Factor	Prefix	Symbol		Name
10^9	giga	G	1 000 000 000	billion[a]
10^6	mega	M	1 000 000	million
10^3	kilo	k	1 000	thousand
10^2	hecto	h	100	hundred
10^1	deka	da	10	ten
10^{-1}	deci	d	0.1	tenth
10^{-2}	centi	c	0.01	hundredth
10^{-3}	milli	m	0.001	thousandth
10^{-6}	micro	μ	0.000 001	millionth
10^{-9}	nano	n	0.000 000 001	billionth

[a] Milliard in the United Kingdom; the British billion is a million million (10^{12}), equivalent to the U.S. trillion, commonly used only in astronomy and the Federal budget.

Hence a thousand meters becomes a kilometer (km) and one thousandth of a meter, a millimeter (mm); a thousand joules is a kilojoule (kj) and a million watts is a megawatt (MW).

ECOLOGY

ECOLOGY

THIRD EDITION

Robert E. Ricklefs

University of Pennsylvania

W. H. Freeman and Company

New York

The cover illustration is a slightly cropped reproduction from Henri Rousseau (French, 1844–1910) EXOTIC LANDSCAPE, 1910 oil on canvas, reproduced with the permission of THE NORTON SIMON FOUNDATION. F.71.3.P.

Library of Congress Cataloging-in-Publication Data

Ricklefs, Robert E.
Ecology / Robert E. Ricklefs.—3rd ed.
p. cm.
Includes bibliographical references.
ISBN 0-7167-2077-9: $45.00
1. Ecology. I. Title.
QH541.R53 1990 89-17046
574.5—dc20 CIP

Printed in the United States of America

2 3 4 5 6 7 8 9 0 HL 9 0

Contents

IV

Population Ecology

V

Population Interactions

VI

Evolutionary Ecology

VII

Community Ecology

Preface

A decade has passed since publication of the second edition of *Ecology*. The discipline of ecology has grown tremendously, and publication of ecological literature has increased perhaps twofold over publication in the previous decade. Several new scientific journals have been introduced to absorb this increased productivity. These journals reflect the broadening interests of ecologists, with ever greater attention being paid to genetic and evolutionary aspects of populations and communities, to connections between behavior and ecology, to physiological and morphological bases of organism adaptations, and to the dynamics of ecosystems. The third edition of *Ecology* reflects these broader interests.

Notwithstanding *Ecology*'s changed appearance, a major reorganization of chapters, considerable additions, and some deletions, I have tried to retain the style, the equal emphasis on phenomena and theory, and the balance of examples from marine, freshwater, and terrestrial habitats that characterized preceding editions. I continue to put high value on three didactic attributes:

First, a solid grounding in natural history, which is the wellspring of all study of ecology The more we know about habitats and their resident organisms, the better we are able to generalize, which is the ultimate goal of the study of ecology.

Second, an appreciation of the organism as the fundamental unit of ecology The structure and dynamics of populations, communities, and ecosystems express the activities of, and interactions among, the organisms they comprise. That a population of insects increases to outbreak proportions depends upon the fecundity and survival of individuals in the population, and the fecundity and survival reflect, in turn, the interaction of the individual with resources, predators, and physical conditions of the environment. Similarly, the regeneration of nutrients within an ecosystem depends in large part on the activities of individual microorganisms that make waste products of their own feeding and metabolism available as resources to plants.

Third, the central position of evolutionary thinking in the study of ecology The qualities of all ecological systems express the evolutionary adaptations of their component species. It is impossible to understand how ecological systems develop, function, and respond to perturbation without understanding the evolutionary dynamics of populations.

Readers familiar with the second edition of *Ecology* will note, in particular, several changes in organization and topics. The discussion of ecosystems now appears immediately after the discussion of organisms in the physical environment, rather than at the end of the book. This makes sense to me because ecosystem studies are concerned with physical and chemical characterizations of energy flux and the cycling of elements. In this section I have considered more thoroughly the roles of microorganisms in nutrient cycling and the mechanisms of regulation of ecosystem function.

I have deleted the larger portion of several chapters that dealt explicitly with the history of evolution. Despite the intrinsic interest of this material to ecologists, little of it is integrated into courses in ecology. I recommend that every student of ecology learn the fundamentals of evolutionary biology, but at the same time I recognize that the limits of space in a comprehensive ecology text no longer afford the luxury of including such a treatment of evolution. Even so, the reader should note that in Part VI, titled "Evolutionary Ecology," I have considerably expanded discussion of the evolutionary interpretation of the ecology and behavior of organisms. I have also brought the discussion of competition and predation, formerly divided between parts on organism and population biology, into a single part called "Population Interactions," and I have emphasized evolutionary as well as ecological aspects of these interactions.

The third edition of *Ecology*, like its predecessors, emphasizes mathematical approaches to the study of ecology, including mathematical analysis and modeling. I have tried where practical to employ a step-by-step development of mathematical treatment so as to make it accessible to readers with only modest backgrounds in mathematics. Elsewhere, much of the mathematics may be skipped without loss of the biological thread of discussion.

Amidst all that is new in this edition as well as what is constant in the ideas that have guided me in writing, I hope I have conveyed the intriguing aspects of natural history and the excitement of the study of ecology. If I have conveyed my own fascination with ecology, I shall consider this book a success.

Preface to the Second Edition

Almost six years will have passed between the first and second editions of *Ecology.* These years have seen substantial progress in many aspects of ecology, among them the application of evolutionary principles to interpreting life history patterns and the development of theory about community organizations. In revising *Ecology,* I have tried to reflect this progress and similar progress in other studies and to bring coverage of the literature up to date for ecology as a whole. I have also tried to provide more balance in examples and in application of principles, giving equal emphasis to aquatic and terrestrial systems and to studies of plants, invertebrates, and vertebrates.

Many of the changes the reader will find in the book have resulted from changes in my own perception of ecology. Six years of teaching, reading, and research have broadened my outlook and given me time to think about the relationships among the many subdisciplines of ecology and between ecology and allied fields of genetics, evolution, behavior, and comparative anatomy and physiology. But my belief remains firm in the principles that evolution is the most important unifying concept in ecology and that ecological systems are best understood in terms of interactions among their components. These principles are revealed most clearly at the organism and population levels of organization, yet one sees a strong movement to apply them to studies of communities and ecosystems.

In appearance, organization, and tone, *Ecology* has changed little. It is still intended for courses in general ecology at the college level. I have continued to try to convey to the reader both the important findings and theories of ecology and the fascinating diversity of patterns in the natural world, which ecologists seek to explain. The coverage remains broad, although I have alloted less space to some topics, notably ethology, while enlarging the discussion of others, particularly life history patterns, quantitative genetics, community ecology, and nutrient cycling. In addition, I have written new chapters on soils, extinction, and niche theory.

For me, *Ecology* is a continuing experience and a significant part of my professional life. I have been immensely pleased by the response of students and colleagues to the book, and I am grateful for the many comments and suggestions provided by its readers. I hope that *Ecology* will continue to be as readable, as stimulating to students, and as helpful to teachers as it has been gratifying to me.

1979

Preface to the First Edition

The array of complex ecological interactions — among individuals, among populations, and between life forms and the physical and chemical environment — may be understood in terms of a few basic principles that govern these interactions. I have written this book primarily to show how fundamental ecological principles elucidate the meaning of patterns in nature.

We study ecology for two reasons: to gain the intellectual gratification that comes from understanding natural patterns and processes, and to apply that understanding to environmental problems that confront mankind. The recent emphasis that ecologists have placed on practical applications is timely and appropriate, but it also tends to narrow our view of ecology as a science. The kinds of knowledge that we draw upon to solve environmental problems represent only a portion of the science of ecology. This book tries to bring into balance the inevitable tendency toward specialization in science, and it tries to keep in view the sensible relationship between ecology the science and ecology the solver of environmental problems. My aim is to impart understanding of, and provoke curiosity about, ecology the science. Where applications demonstrate a particular point better than do natural ecosystems, I will discuss the applications. Readers will quickly appreciate the links between natural systems and practical problems. Thorough understanding of principles of structure and function in natural communities is, I think, the best preparation for tackling those problems.

The scope of *Ecology* is wider than that of most other texts; I have included sections on the physiology and behavior of organisms, on genetics, and on evolutionary biology. I feel that such breadth is important because the structure and function of biological communities are based on the attributes of organisms that form the communities. These attributes, in turn, are evolved characteristics that are determined by the interaction of the organism with its environment. An understanding of evolution and adaptation is essential to, and inseparable from, the study of all ecological relationships.

I have written this book primarily for readers beginning their study of ecology at the college level, but the book will also serve as a source of material and ideas for those whose study is more advanced and for those engaged in the practice of ecology as a profession. The reader does not need a strong background in biology or other sciences to understand what appears here. The first part of the book is not meant to be rigorous and demanding, and it leans heavily on detailed examples to make its points. As the narrative progresses, subjects are presented more quantitatively, and more diversified knowledge is summarized and synthesized. The organization of the book, though clearly proceeding from organism level to population and community levels of ecological interaction, is not rigid; from time to time, related subjects are discussed in detail to show that ecology is intimately tied to all science and that boundaries should not be stringently drawn around disciplines.

Although this book is intended as an introduction, I have not shied away from presenting a mathematical treatment of population processes. I have tried to explain the mathematics thoroughly, including those algebraic steps that are not always evident, and I hope that readers who feel uncomfortable with mathematics will try to work through the mathematical presentations to their own satisfaction. None of the derivations in the book is difficult. I have included the mathematical treatment to demonstrate ways in which ideas can be described and organized quantitatively, and to demonstrate that one may gain insights that are not always intuitively obvious. Some acquaintance with mathematical description and mathematical modeling of ecological processes is required to evaluate contemporary thought in ecology.

I hope this book will provide a broad but sound introduction to a fascinating and essential area of human thought. If it encourages critical evaluation of our thinking about ecology and stimulates some readers to contribute further to our knowledge and understanding of ecology, I will be all the more gratified.

1973

ECOLOGY

I

Introduction

1

The Environment and Its Inhabitants

The English word "ecology" is taken from the Greek *oikos,* meaning house, the immediate human environment. In 1870, the German zoologist Ernst Haeckel first gave the word its broader meaning, the study of the natural environment and of the relations of organisms to each other and to their surroundings. General use of the word came only in the late 1800s, when European and American scientists began to call themselves ecologists. The first societies and journals explicity devoted to ecology appeared in the early decades of this century. Since that time, ecology has undergone immense growth and diversification, so much so that persons devoting their professional lives to ecology now number in the tens of thousands. With the dual crises of rapid growth of human population and accelerating deterioration of the earth's environment, ecology has taken on the utmost importance to everyone. Management of biotic resources in a way that sustains a reasonable quality of human life depends upon wise ecological principles, not merely to solve or prevent environmental problems but to inform our economic, political, and social thought and practice.

Nitrogen cycles in forests, energy budgets of pond organisms, stability of marine food chains, pollination systems of plants, mating calls of toads, predator-prey oscillations — all fall within the realm of ecology. This diversity is well illustrated by the complexities that meet our effort toward control and management of populations, whether to maintain resources of food or fuel, to reduce the incidence of disease, or to preserve species of aesthetic or cultural worth. Management of populations requires us to understand the causes of population change and the consequences of population change in any natural system. Several years ago, the Nile perch was introduced into Lake Victoria in East Africa. This was done with the well-intentioned purpose of providing additional food for the peoples living in the area and additional income from export of the surplus catch. Because simple ecological principles were ignored, the result was virtual destruction of the lake's entire fishery (Barel et al. 1985). Until the introduction of the Nile perch, Lake Victoria supported endemic fishing for a variety of species of cichlids, which themselves fed primarily on detritus and plants. The large Nile perch feeds upon other fish, the smaller cichlids in this instance. Because energy is lost with each step in the food chain, a predatory fish cannot be harvested at as high a rate as its herbivorous prey. Furthermore, the perch was alien to Lake Victoria, and evolution of the local cichlids did not include behavior to escape predation. Inevitably, the perch annihilated the cichlid populations, destroying the native fishery and all but eliminating its own food, thereby, in turn, hastening its own demise as an exploitable fish. To be sure, the native fishery was already precariously overexploited owing to growth of the local human population and recent application of less primitive fishing technology, but the appropriate solution to these problems would have been better management of the cichlids, not introduction of an efficient predator upon them.

Still other results of the introduction of the alien fish turn the story into a tragic comedy of errors. The flesh of the Nile perch is not particularly liked by the peoples living near the shores of Lake Victoria. They prefer the accustomed texture and flavor of the native fishes. Moreover, the oily meat of the Nile perch must be preserved by smoking rather than sun-drying, and local forests are being cut rapidly for firewood. Because the larger Nile perch requires larger and more elaborate nets, local subsistence fishermen cannot compete with more prosperous outsiders who are equipped for commercial fishing. The lesson is painful yet simple: Man is an integral part of the ecology of the Lake Victoria area. Traditional local fishing had been sustained for thousands of years until the pressure of population and the perception of an opportunity for an export fishery led to an ecologically unsound decision and to an economic and social disaster.

Half a world away, efforts to save the California sea otter illustrate the intricate intermingling of ecology and other human concerns (Booth 1988). The sea otter was once widely distributed around the North Pacific Rim from Japan to Baja California, but in the 1700s and 1800s intense hunting for fine otter pelts reduced the population to near extinction. The fur hunting industry, of course, collapsed from the overexploitation. Subsequent protection enabled the California sea otter population to rise to more than 1500 individuals by the 1980s. The sea otter's success, alas, has irked California fishermen, who claim that the otters — which do not need commercial fishing licenses — drastically reduce stocks of valuable abalone, sea urchins, and spiny lobsters. The matter has deteriorated to the marine equivalent of a range war between fishermen and conservationists, with the otter, often fatally, caught in the line of fire. Ironically, the otters benefit a different commercial marine enterprise, the harvesting of kelp, a large seaweed used in making fertilizer. The kelp grows in shallow waters in stands called kelp forests, which provide refuge and feeding grounds for several larval fish. The kelp is also grazed by sea urchins which, like goats on land, can, given the right conditions, denude an area of vegetation. One of those conditions is the absence of a principal sea urchin predator, the sea otter (Van Blaricom and Estes 1988). When the expanding sea otter population has spread into new areas, urchin populations have been controlled, allowing kelp forests to regrow. Human involvement in the local ecosystem requires that management of otter populations and various commercial exploitation be balanced in an economically and socially acceptable manner, an impossibility without understanding the complex ecological role of the otter in the system.

Although the plight of endangered species may arouse us emotionally, there is a beginning realization that the only means of preserving and using natural resources is through the conservation of entire ecological systems and of broad-scale ecological processes (Ricklefs et al. 1984). Individual species, including the ones that humans rely upon for food and other products, themselves depend upon the maintenance of a healthy ecological support system. We have already seen how such predators as the Nile perch and sea otter may assume key roles in maintaining the natural properties of a system or in dramatically altering its properties. On land, pollinators and seed-dispersers are necessary to the long-term maintenance of plant formations. Therefore, the management of forests without attention to maintaining populations of insects, birds, and mammals, upon which the long life of the forest depends, is ecologically naïve. Natural systems are diverse and highly evolved in the sense that each species is adapted to its particular role within the system and the system depends to some degree on each player acting out its role. It turns out that any reduction in biodiversity can upset the balance of a system and alter its function (Wilson 1988).

Most natural systems sustain cycles of disturbance (fire, hurricanes, disease) and rejuvenation. Natural restoration depends on there being patches of intact habitat from which species may recolonize disturbed areas. But as habitats become more and more fragmented, as are forests by the spread of agriculture and urban development, disturbances can so thoroughly destroy an area as to leave little chance of complete recovery, even with substantial help from man (Janzen 1988). Disturbance cycles, whatever their spatial and temporal scale, are as much a part of the natural landscape as are key species within a local system and must be taken equally into account in any effort to maintain the system.

Many of the now rapid changes in the world bring a seeming improvement in the quality of human life, but those very changes increasingly strain the earth's ecological support systems and in reality bring decline in the quality of life. On any scale on which we view the surface of the earth, the deterioration of the environment is alarming. Partially enclosed systems, such as the estuaries of major rivers, allow us to isolate the causes of much of the change. In San Francisco Bay, for example, the pressures of population and agriculture have gradually but profoundly altered it to the point that it would be unrecognizable to nineteenth-century inhibitants of the region (Nichols et al. 1986). Diking of wetlands to create farmland, filling to create land for urban construction, pollution from sewage and from agricultural and industrial waste, and introduction of exotic species have transformed a healthy, productive, and economically valuable system into an ecological disaster, whose greatly diminished value lies entirely in recreation, itself of ever less value, and in commercial shipping, not in its once generous biotic resources. The present human population of the San Francisco Bay area may well be incompatible with a pristine estuary, but much of the change came about through state subsidy of unprofitable agricultural ventures, absence of ecologically and economically sensible city planning, careless introduction of pest species, use of the bay as a cheap sewer, and uncontrolled harvesting of fish and shellfish.

The multiplication of technology and population has brought ecological crisis to far larger regions of the earth than San Francisco Bay. One of the most critical is the Sahel in Africa (Sinclair and Fryxell 1985). The Sahel is a vast band of semiarid habitat lying at the southern edge of the Saharan Desert and extending from Gambia in the west to Ethiopia in the east. Population pressure, expressed through intensified grazing and collection of firewood, has caused extensive deforestation of the entire region (Sai 1984, Eckholm et al. 1984). As humans and domesticated animals clear vegetation over such vast areas, climate is affected. Rainfall decreases because less water vapor is recirculated to the atmosphere and heating and cooling patterns of the land surface are changed (Otterman 1974, Otterman et al. 1975). Even when plentiful rains result in good yields from crops, the abundance may be snatched away by plagues of grasshoppers and locusts, whose populations grow rapidly under the moister conditions (Walsh 1986). The Sahel is at present truly devastated.

In wetter parts of the world, tropical rainforests are being cleared at an alarming rate for forest products and largely unsuitable agriculture, bringing profit to a few in the short term and yet another disaster in the long term. The destruction of rainforests dismantles an intricate, self-sustaining system that cannot be rebuilt. Species, most of them unknown to science and many of them with potential commercial value, are being driven to extinction (Lewin 1986, Vietmeyer 1986, Wilson 1988). Burning of cut vegetation adds greatly to the already dangerous level of carbon dioxide in the atmosphere. In this instance, the ecological principles that should govern human use of the rainforest are generally understood. As Michael Robinson, director of the National Zoo, Washington, D.C., has pointed out, "The problems are not [the result of] ignorance and stupidity. The problems derive from the poverty of the poor and the greed of the rich" (Lewin 1986).

The destruction of rainforests is scarcely alone in its far-reaching effect. Automobile emissions reduce tree and crop production over large areas (Reich and Amundson 1985). Sulfurous gases from burning coal have acidified the rain and snow over large areas of eastern North America and Europe, killing forests and degrading freshwater ecosystems (Schindler 1988). Chlorofluorocarbons (CFCs) used to pressurize spray cans and as refrigerants have reduced the ozone concentration in the upper atmosphere, increasing the amount of dangerous ultraviolet radiation that reaches the earth's surface.

As a result of industrial growth and deforestation, levels of carbon dioxide, methane, and CFCs have increased in the atmosphere so much that they are changing the earth's climate. In the so-called greenhouse effect, long-wave radiation from the earth's surface is absorbed and retained by the atmosphere. As a result the temperature of the earth is increasing and is expected to rise by 2 to 6 degrees Celsius during the next century (Schneider 1989). Such change will cause a rise in sea level as the polar ice caps melt, it will disrupt agriculture and shift ecological habitats across the landscape as temperature and moisture patterns change, and it will perhaps cause widespread extinction of forms of life unable to adjust to the change and its effect (Roberts 1989).

Our ability to wreak havoc upon ourselves and the world in which we live is virtually unlimited. The most appalling evidence of this ability is our means of waging global nuclear war, wherein destruction of ecological systems and extinction of species, possibly including ourselves, would be virtually without parallel in the history of the earth (Turco et al. 1983, Ehrlich et al. 1983). Nuclear war remains unthinkable even if we ignore the ecological consequencies. What we are gradually accomplishing in the degradation of a livable planet may end by being just as punishing. Merely to understand ecology will not solve our environmental problems in all their political, economic, and social dimensions. But as we contemplate the need for global management of natural systems, we are doomed to fail if we do not understand their structure and function, an understanding that depends upon the principles of ecology.

2

The Order of the Natural World

At first, we are bewildered by the diversity and complexity of nature. Every species plays a different role in the ecological play. Some connections make themselves apparent to us from the start. The caterpillar depends upon the tree whose leaves it eats. The tree sends its roots deeply in the soil, from which it obtains water and mineral nutrients needed for growth and maintenance. Feeding relationships unite all species into a system. But other connections also govern the structure and function of ecological systems. What, for example, determines how many species live in a given patch of forest? Do their mutual interactions somehow limit the kinds of organisms that can coexist? What is responsible for the particular form and behavior of each species? How do the forest, the grassland, and the oyster bed replenish the resources they take from the environment?

To understand ecology, one must first see patterns in the natural world.

What better place to experience nature than in the tropical rainforest, where life is most luxurious and diverse? Nowhere else is one so acutely aware of nature. At night, especially, the multitude of active creatures make the life of the greatest cities seem paltry by comparison. In the Panamanian rainforest, the delicate first light of a November day is shattered by the explosive cries of a nearby troop of howler monkeys. The howling subsides, but it is presently answered by that of another troop farther away, and then another and another. We can easily locate the nearby troops of monkeys and so, we would guess, can the howlers themselves. Are we witnessing a morning proclamation of territory by each group?

It is now light enough so that we can begin to wend our way along the narrow forest trails, being careful not to brush against the long sharp spines that viciously ornament the trunk of the black palm. We make our way

Figure 2–1
Opening of the fruit of *Rinorea sylvatica* (violet family), showing how the seeds are shot out as the pod dries.

around the giant buttresses that spread out from the bases of trees over the forest floor. There is a faint snapping sound, and our eyes quickly search for some small animal moving in the bushes. Another snap. A small object catapults in front of us and strikes the ground nearby. Finally we locate the source — a small bush whose fruits have opened and are shooting their seeds over the forest floor. As the fruit, which resembles a three-sided pea pod, dries out, the edges of the pod come together, squeezing the seeds with greater and greater force until the stalk of the seed finally breaks and the seed is ejected from the pod (Figure 2–1).

The drab browns and greens of the inner rainforest are occasionally broken by the flashing of iridescent butterfly wings. A blue morpho butterfly streaks by us, just out of reach. How striking it is that these butterflies should be so conspicuous. How different from the moths that are attracted at night to the lights around buildings; they possess every conceivable device for being unobtrusive. Their brown and gray colors look like dead leaves and bark. Many of them conceal their legs, normally a sure betrayal of any animal, beneath their wings; others benefit from legs modified to resemble bark or lichens. One species characteristically protrudes one leg or another from underneath its regular outline to break up the symmetry so characteristic of animal forms. Others have wings that, although intact, often have the broken and contorted appearance of dead leaves. The hind edge of the wing of another species bears the unmistakable picture of a rolled-up leaf complete with its shadow. How perfectly these moths blend with their backgrounds when they are resting on leaves or the bark of trees. How have these moths come to be that way, and why are other kinds of butterflies and moths so conspicuous?

We are constantly impressed by a sense of purpose in the forms of animals and plants we encounter. As we walk through the forest we wonder about the purpose of nature. Whose purpose? What purpose? Our thoughts are suddenly interrupted as we step, as if going from one room to another through a door, into a large clearing where we are blinded momentarily by the sun. A recent treefall caused this break in the forest canopy. The giant did not exit gracefully, but took with it at least a dozen other trees directly in its path or bound to it by vines. The clearing gives us our first glimpse of the sky in several hours. A long, graceful ribbon of hawks silently glides southward many hundreds of feet above. Their flight seems effortless—none of the birds moves its wings. The narrow ribbon is caught by a small updraft, sending it spiraling upward in wide, lazy circles until it breaks away again toward South America, seemingly impatient to be gone. Are these the same birds that we saw earlier in the fall, gliding along the ridges of the Appalachians in Pennsylvania? What has sent them south? Why are they going all the way to Argentina? What cue will they experience there that tells them spring is coming once again in the north and it is time to make the return journey?

Our own journey through the forest is coming to an end. By the time we return to our house, our eyes and ears have experienced more than our minds can sort out in a day, and we are tired after the long walk. But a shower and a good meal start us on an evening of conversation that takes us back over the day's journey to recount our observations and to ask questions—mostly to ask questions—far into the night.

The natural world is diverse and complex.

The naturalist stands in awe of the diversity of forms in nature and of the intricate interactions of these forms with one another and with their environments. In our walk through the tropical forest, we passed several hundred varieties of trees, and yet to untrained eyes they all seemed to be "doing" about the same thing. They were using the energy of sunlight assimilated by their green leaves to convert carbon dioxide from the air, and water and minerals from the soil, into the organic molecules that make up their structure. Why should there be so many kinds of plants when one seemingly could perform the same functions as all the others? Indeed, we know that vast forests consisting of one or, at most, a very few species of coniferous trees stretch across the middle latitudes of Canada.

There are hundreds of species of butterflies and moths in the forest, and thousands of species of other insects. Their appearance varies so much that one wonders if there are as many different environments within one forest as there are kinds of insects. In spite of how diverse the butterflies and moths look to us and, presumably, to their natural predators, they are remarkably similar in their feeding habits. As adults, they all have long tubular snouts, which normally lie curled beneath their heads, but which can extend to the depths of tubular flowers for feeding. And as larvae—caterpillars—they all eat green vegetation. Most of the species are picky eaters and can be found on only one or a few species of plants. A few feed very widely from the forest's smorgasbord. How do adults know which are suitable food plants to lay their eggs on? Why are some specialists and some generalists? Adult butterflies and moths, as well as many other insects, birds, and bats, perform a vital function in the forest by carrying pollen from flower to flower, thus ensuring that the plants will set seed. How do they decide which flowers to visit? Why is it that some species of moths—the silk

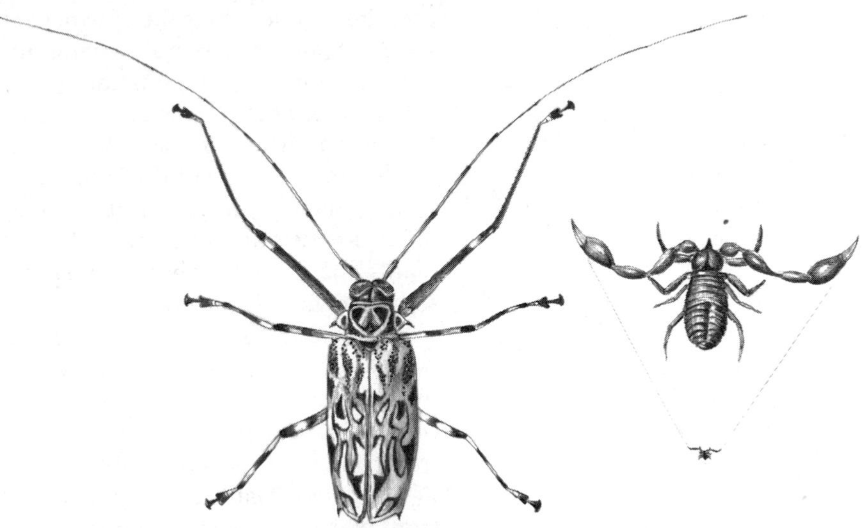

Figure 2–2
A harlequin beetle (about half natural size) and its pseudoscorpion traveling companion (much enlarged).

moths — do not feed as adults and die within a few days of emerging from the pupa?

Many species are mutually bound to others for their survival. The delicately colored harlequin beetle, for example, carries on its back a small community of mites and pseudoscorpions that feed upon the mites (Figure 2–2). This particular kind of pseudoscorpion is found nowhere else in nature. The mites perform a beneficial function for the beetle by scouring fungi from its delicate membranous hind wings (fungi seem to grow on almost everything in the tropics). The pseudoscorpions take advantage of this ready food supply, but the mites are at least partly protected when they hide in the numerous small pits that dot the forewings of the beetle. Examples of organisms living in intimate, mutually interdependent relationships abound. Lichens are a combination of alga and fungus — the first makes carbohydrates by photosynthesis, the second obtains water and minerals from the rock or tree trunk on which the lichen lives. Much of the digestion that occurs in the guts of animals is carried out by specialized bacteria and protozoa living there. Plants are aided in the uptake of minerals from the soil by specialized fungi intimately associated with their roots. Insects pollinate their flowers. Birds and mammals disperse their seeds.

The natural world is dynamic, but it is also stable and self-replenishing.

The large tree that falls in a forest may be several hundred years old when it finally topples; its life is exceptionally long compared to most living things. Death comes in many ways to organisms. Some fall victim to predators and parasites, while others die of exposure to the physical environment. A cold snap in spring or a pond's drying up late in the summer casts death's net very broadly. The natural community, whose presence seems so stable, actually undergoes constant turnover, with replacement by new individuals, just as the organic structure of our body is continually replaced during our lifetime. In spite of this dynamic aspect, the natural world is also measurably stable — an equilibrium is maintained. The forest we walk through today is very much as it was 5, 10, or even 100 years ago. The same kinds of plants and animals persist to the present, though most of these are not the same individuals that were here earlier.

The dead bodies of organisms and the wastes of biological processes do not pile up. They are broken down and their component parts are recycled by the community. The dead leaves rustling under our feet cover the decomposing remains of other leaves in the soil beneath them. Soil organisms transform their elegant shapes into an amorphous mass of decaying and decomposing plant tissues, finally reducing them to the mineral elements from which they were once, in part, synthesized.

Populations of organisms also are continually replaced. An insect may lay thousands of eggs each year, and some marine organisms shed millions of eggs into the water — more than necessary to compensate losses of individuals from populations. Yet in spite of tremendous potential for growth, various checks and balances keep most populations within rather narrow limits.

One must not assume, however, that the world is unvarying. Quite the opposite. Variation and equilibrium are in constant tension. All systems suffer disturbance — from weather, fire, tree falls, even a cow pat creates a major disturbance for some organisms — from which they continually recover. Patches of disturbance may range from a few millimeters, as an earthworm eats its way through the soil, to large portions of the earth, as global weather patterns shift. Much variation is imposed by the physical environment — climate, ocean currents, landscape evolution — but much is also generated by the biological world. Many organisms create disturbances for others as they forage and generally move about. The dynamics of population interactions can establish cycles of population change that send ripples throughout the biological community. Variation must be considered a part of the equilibrium of the natural world.

The natural world is organized by physical and biological processes.

We constantly recognize patterns, organization, and interrelationships in the complexity of our surroundings. We are able to perceive these patterns because of the predictability of their elements. We can anticipate nature in various ways, but most efficiently by the generalization of past experience. For example, experience with the movement of objects thrown into the air enables us to predict their trajectories with accuracy. A good outfielder can predict where a baseball will land long before it begins to drop.

Our lives are organized around patterns of our environment. Only if nature is predictable can we respond properly to it. Birds live in the woods, fish inhabit the sea. Without ever having been in a particular forest, we could expect to find birds rather than fish simply on the basis of past experience with birds and fish and forests and seas. By experiencing the unnaturalness of surrealism and dreams, we realize how completely our minds are bound up in the various patterns we recognize in nature. As ecologist G. Evelyn Hutchinson once pointed out, "[if] we imagine ourselves encountering in the middle of a desert a rock crystal carving of a sewing machine associated with a dead fish to which postage stamps are stuck, we may suspect that we have entered a region of the imagination in which ordinary concepts have become completely disordered."

Patterns have two sources of predictability, one achieved through observation, and the second by understanding the mechanism that produces the pattern. In the first case, predictions are based on extrapolating observations to new but similar situations. We do this when we predict the flight path of a ball. But by applying the laws of motion, we could have predicted the trajectory of the ball without any previous experience with the phenomenon, knowing only its initial speed and direction. Similarly, by understanding the principles governing the form and functioning of animals, we would know that we would not be likely to find an organism resembling a fish living in a forest.

In the development of science, empirically observed patterns almost always preceded the discovery of the causative principles that produce the patterns. After detailed observations, the German astronomer Johannes Kepler discovered that the time required for a planet to revolve around the sun is inversely related to its distance from the sun. Only later did the English physicist Isaac Newton formulate laws of motion whose predictive powers are so great that they made possible the detection of unseen planets on the basis of their gravitational effects on the motion of some of the known planets. Alfred Wegener did not understand the underlying mechanism when in 1915 he proposed, from observations of geological and geographical relationships of the continents, that the major land masses slowly drift over the surface of the earth. In fact, for many years Wegener was ridiculed and his ideas rejected for lack of a plausible mechanism. (In the last two decades, however, irrefutable evidence for continental drift has been discovered, its rate has been measured directly by satellite observations, and plausible mechanisms have been proposed.) The same has been true in the biological sciences. Early naturalists classified organisms into a regular hierarchy of species and other taxonomic groupings based on similarities. But only after Charles Darwin proposed his theory of evolution was the basis for these patterns understood.

Perception limits our understanding of nature.

Our world view is very much a product of the information we receive. Overwhelmed by the sights and sounds in our surroundings, we often forget that other organisms perceive the world differently than we do. We mostly see and hear the world around us. Smells drift by largely unnoticed. Our sense of taste is dull. We also become accustomed to interpreting natural phenomena in terms of surroundings that we are familiar with. Islanders know water better than natives of Kansas, for example.

Because of our size, mobility, and the pace of our lives, we are sensitive to particular scales of variation in time and space. Our perception of distance differs from that of the collembolan, which finds a jungle in a pinch of soil, or the water flea, which perceives a drop of pond water as we would the entire pond. For us, a few tens of years are a lifetime. But our lives are instants in the clock of evolution and ecosystem development, and eternities to microorganisms.

Our limited perception of nature can encumber our ability to understand natural phenomena. For years, most ecologists were trained and worked in Europe and temperate North America, where seasonal fluctuations in temperature are a major aspect of the environment. When naturalists visited the tropics, they were impressed by the constant year-round temperature and they assumed that biological communities are less variable in tropical regions than in temperate regions. Only recently have ecologists bothered to count individuals of tropical species over long periods. Surprisingly, they have found that populations undergo marked seasonal fluctuations in the tropics, and additionally that they can vary considerably from year to year. Temperature defines the seasons in a temperate climate and forms a familiar pattern. But the patterns are different in the tropics, where the seasons are marked by wet and dry periods, and where rainfall is notoriously unpredictable at many times of the year.

Our interpretation of the natural world is also plagued by the way that size determines pertinent scales of time and space. A flea can jump a hundred times its length. We immediately react by translating that distance to a familiar scale—comparing flea lengths to human lengths. The comparable human jump would clear two football fields. How incredibly strong fleas are! But the flea does not actually perform athletic miracles. As size changes, the relationships between distance, power, and time change accordingly. Consider the wing beat of flying organisms. Try to move your arms up and down as rapidly as possible. How fast? Two or three times per second? Most large birds can flap their wings between 2 and 20 times per second, and some small insects can do so up to 500 times per second. This may seem amazing, but as the size scale changes, so does the relevant time scale. You can demonstrate this very simply for yourself. Tie a weight to a string, and start it swinging like a pendulum. As you shorten the length of the string, you will notice that the weight swings back and forth much faster. In fact, the rate of swinging varies inversely with the length of the string. Halve the length of the pendulum, and its swinging frequency doubles. The wings of most small insects are less than a centimeter long, so we should not be surprised that an insect can beat its wings more than a hundred times faster than we can flap our arms.

We study nature by observation, theory, and experiment.

Scientists look at the natural world from many different viewpoints, depending on their training and temperament, the problems they study, and the systems with which they work. All perspectives and approaches are valid to the extent that they can help us to understand the natural world. But in spite of their differences, all scientists employ similar methods of study. A question is the starting point of any inquiry. Phenomena—patterns in nature—prompt us to inquire how or why a pattern came to be. We form several hypotheses to answer our question, and then test each hypothesis by suitable experiments or further observations. If the results are consistent with a hypothesis, we may begin to generalize our understanding of a phenomenon and make valid predictions based on this understanding. More importantly, if the results are not consistent with a hypothesis, we may reject that hypothesis as flawed and thus narrow our search for the "truth" of nature. Experimental results—tests of ideas—provide new observations, which in turn may prompt us to ask new questions or rephrase old ones. Scientific inquiry is thus self-perpetuating; once a new inquiry is begun, it snowballs, and questions lead to new questions.

The principal features of scientific inquiry—the detection of pattern in nature, the proposal of ideas to explain patterns (theories), and the testing of theories by evaluating predictions they permit us to make about previously unappreciated phenomena—will be illustrated over and over again in this book. Although there are accepted guidelines for developing and testing theory, there is no single best way to approach science. Scientific inquiry is as much an art form as painting or the composition of music. While our culture dictates certain aesthetic standards, individual compositions or paintings are unique products of the human spirit and imagination. So it is with science. Differences in intuition and approach among individual practitioners make science exciting and guarantee its progress. Much of this variation arises from different experiences with the diversity of nature, as we shall see in Chapter 3.

3

Natural History

The basic information of ecology concerns the lives of organisms and the environments in which they live. In its raw form, fresh from the field, such information is called natural history. The observations of natural history provide a basis upon which scientific inquiry begins to establish patterns and determine the mechanisms responsible for those patterns. It should be obvious that the direction of scientific inquiry, and to some degree its ultimate success, depends on the quality of natural history observation. Many of ecology's greatest theoreticians, from Charles Darwin on, were great naturalists. Their insights sprang from close observation of organisms and habitats.

In this chapter we shall see how natural history colors our perception of ecological relationships by considering the lives of three species, each living in a different environment and having to find solutions to different problems. The first, the giant red velvet mite, lives in the harsh environment of the Mojave Desert of southern California; its activities are closely geared to physical factors in its environment, particularly to temperature and rainfall. The next species, the chestnut-headed oropendola, a large relative of North American blackbirds and grackles, lives in the lowland tropical forest of Panama. The oropendola is partner to a complicated arrangement among a diverse group of interacting species. Unlike the harsh desert environment, the climate in wet tropical areas is highly favorable for the development of life; therefore biological factors become more conspicuous parts of the environment, and physical factors are relegated to less important roles as influences on organism function and agents of natural selection. Our third example, that of the baboon on the savannas of Africa, is striking in its complex social life. As for all animals and plants, physical and biological aspects of the baboon's environment have laid the foundations of its adaptations and guide its daily life, but the rich social environment of the baboon has additionally shaped the behavior of the species.

Figure 3–1
Adult giant red velvet mites on the surface of the ground *(left)* and in a vertical burrow (*right*, a cutaway view of the ventral side of the mite). (Photographs by Philip L. Boyd, Deep Canyon Research Station, courtesy of Lloyd Tevis, Jr.)

In the harsh desert environment, the giant red velvet mite spends only a few hours aboveground each year.

The Mojave Desert of southern California is an area of little rain, searing summer heat, and chilling winter cold. The desert appears to be nearly devoid of life for most of the year. But the desert's silence and apparent sterility are occasionally broken during the milder days of winter by swarms of insects and other creatures that appear on the surface or fly above it for a few hours, and then disappear as mysteriously as they came. One of the more conspicuous of these creatures is the giant red velvet mite, *Dinothrombium pandorae* (Figure 3–1). The mite's generic name paints a vivid picture of this close relative of spiders. *Dinothrombium* is derived from the Greek *deinos,* meaning terrible, and *thrombos,* a lump or a clot, describes the mite's resemblance to a clot of blood.

Several decades ago, biologists Lloyd Tevis and Irwin Newell (1962) began a study of the behavior of the mite in relation to the physical conditions of its environment. They found that the mites spend most of the year in burrows dug in the sand. The particular conditions that favor the emergence of mites occur infrequently in the Mohave Desert. During 4 years of observation, adults appear aboveground only 10 times, always during the cooler months of December, January, or February, when they can tolerate the temperatures on the desert's surface. Tevis and Newell could predict from their observations that an emergence would occur on the first sunny day after a rain of more than three-tenths of an inch, provided that air temperatures were moderate. An individual mite appeared only once each year.

On the day of a major emergence, the mites came out of their burrows between 9:00 and 10:00 A.M., and by late morning one could find thousands of mites scurrying across the desert sands in all directions. At midday, between 11:30 and 12:30, the mites dug back into the sand, waiting until the following year before emerging again.

These "terrible clots" do not leave their burrows to terrorize unsuspecting desert travelers. During their 2- or 3-hour stay aboveground each year, each mite must perform two important functions: feeding and mating. On the same day the mites emerge, large swarms of termites appear flying over the desert sand, their own emergence presumably triggered by the same physical factors that cause the mites to leave their burrows. It is upon these termites that the mites feed. A flying termite cannot, of course, be caught by the earthbound mites; a mite must locate its prey after the termite drops to the ground and sheds its wings, but before it burrows into the sand. All this happens very quickly, giving the mites about an hour to find their prey.

Because the mites are solitary in their burrows, they must mate during their brief period aboveground each year. Courtship of the giant red velvet mite is similar to that of spiders and their relatives. The males walk nervously around and over a feeding female, tapping and stroking her, and cover the sand around her with loosely spun webs. Males court *feeding* females for two reasons. First, the females have ravenous appetites (they have not eaten for a year, after all) and would just as likely devour a male mite as a termite. Second, and perhaps more important, females can produce eggs only if they have had a meal. Thus, by mating with a feeding female, the male guarantees that his efforts to reproduce will not be wasted.

About midday, after the mites have fed and mated, they congregate in troughs on the windward sides of sand dunes, where surface temperatures and the size of the sand particles are just right (less than one-half millimeter in diameter), and reenter the sand almost simultaneously, as if urged on by the same unseen hand. Mites continue to dig their new burrows on the first day until the coolness of the late winter afternoon slows their activity. Burrowing continues on subsequent days when the sand becomes warm enough, until the burrows are completed. During the rest of the year, the adult mite spends its time moving up and down in the burrow to follow the movement of its preferred temperature zone as the surface of the sand heats and cools each day.

Females lay eggs during the early spring. The eggs soon hatch, and the young mites crawl to the surface of the desert to search for hosts, usually grasshoppers, to which they attach themselves. While it is growing, the young mite remains with its host, obtaining its nourishment from the body fluids. When it is fully grown, the young mite drops off its unwilling host and seeks a suitable spot to dig its own burrow in the sand, thus renewing the life cycle of the giant red velvet mite.

Oropendolas and their cowbird brood parasites are part of a complex interrelationship of species in the humid tropics.

The tropical surroundings of the oropendola in the Republic of Panama are a far cry from the rigorous environment of the giant red velvet mite. In the tropics, air temperatures vary little during the year, and abundant rainfall maintains the lush vegetation. Life abounds; diverse animals and plants are intricately interwoven into a rich fabric of biological interactions.

The situation we are going to examine involves two birds, the chestnut-headed oropendola and its brood parasite the giant cowbird (Figure 3–2). As we shall see, two insects—a bot fly and a wasp—also play an integral part in the interaction between them. This story was discovered by Neal

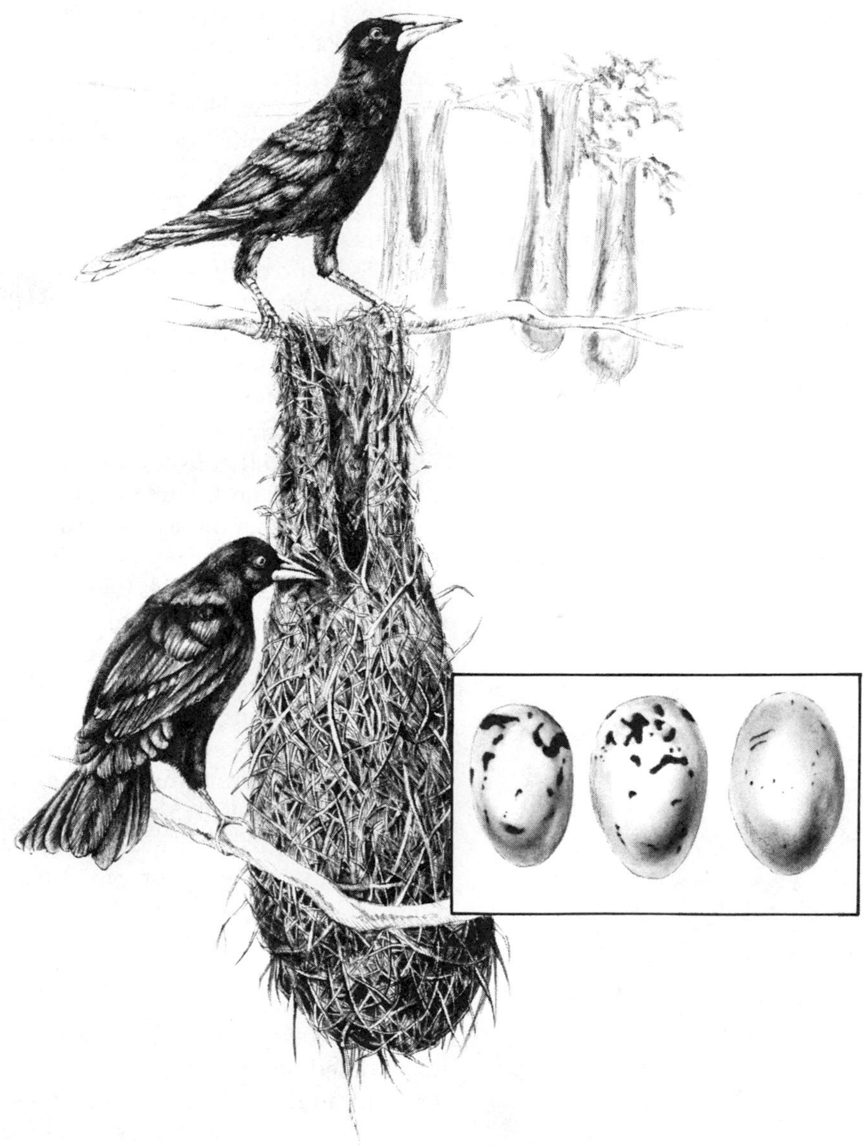

Figure 3–2
Female chestnut-headed oropendola at her nest *(above)* and giant cowbird *(below)*. In the inset, the oropendola egg *(left)* is compared with a mimetic cowbird egg *(center)* and a nonmimetic cowbird egg *(right)*.

Smith (1968), a staff biologist at the Smithsonian Tropical Research Institute in the Republic of Panama.

Brood parasites (cowbirds and cuckoos are familiar ones) have been known for a long time. They are so-named because the female lays her eggs in the nest of another species, the host, and the young are raised by the foster parents (Payne 1977). The presence of a brood parasite in a nest usually reduces survival of the host young because parasitic young compete for the food brought by the host parents. In many cases, the brood parasite removes one or more host eggs before laying her own. Many potential hosts can detect parasite eggs in their nests and eject them (Rothstein 1971, 1975). To counter this defense, many species of brood parasite have evolved an elaborate egg mimicry to fool the host species into accepting the alien egg as one of its own (Wickler 1968).

Chestnut-headed oropendolas breed in colonies, which include anywhere from 10 to 100 of their long, hanging nests, placed along the branches of large, isolated trees. From the observations of other naturalists, Smith knew that the oropendolas were often parasitized by the giant cowbird. Eggshells found on the ground underneath the colonies, thrown out of nests by females after the young had hatched, included those of both oropendolas and cowbirds. But Smith noted that in some colonies the eggshells of the cowbirds were distinctly different from those of the oropendolas, whereas in other colonies the eggshells of the two species so closely resembled each other that they could be distinguished only on the basis of shell thickness. Thus it appeared that in some colonies the cowbirds had evolved to mimic the eggs of their host, while in others they had not.

On the basis of these observations, Smith set out to determine whether the presence or absence of egg mimicry in the cowbirds elicited different behavior of the oropendolas toward foreign eggs in their nests. A number of objects — including mimetic cowbird eggs, nonmimetic cowbird eggs, other kinds of eggs, and a variety of objects only remotely, if at all, resembling eggs — were put into nests in both kinds of oropendola colonies. Smith found, as he had suspected, that in oropendola colonies where the cowbird eggs closely mimicked those of their hosts, the oropendolas removed virtually everything from their nests except their own eggs and very closely matching cowbird eggs. These oropendolas were discriminators; they tried, often in vain, to discover and eject the cowbird eggs. In oropendola colonies where cowbirds were poor egg mimics, the oropendolas were willing to accept all sorts of alien objects.

In what other ways did these two types of oropendola colonies differ? An extensive survey of the oropendolas in central Panama revealed that in nondiscriminator colonies, young oropendolas were often infested with the larvae of a species of bot fly. These parasites sometimes killed the nestling oropendolas, and frequently so weakened them that their chances of surviving to adulthood were slim. In discriminator colonies these parasites were rarely present. Here was a major difference between the two types of colonies. Was it possible that the role of the cowbird in the colonies was linked to bot-fly parasitism?

Smith examined oropendola young in nondiscriminator colonies (susceptible to bot-fly parasitism), and discovered that the incidence of bot flies was higher in nests that did not contain cowbirds than in nests that did, as shown by the following data:

	Number of nestling oropendolas in nests	
	With cowbirds	Without cowbirds
With bot-fly parasites	57	382
With no parasites	619	42

Further observations showed that the nestling cowbirds would snap at anything small that moved within the nest, including adult bot flies and, furthermore, they would remove bot-fly larvae from the skin of the nestling oropendolas. This behavior on the part of the young cowbird benefited the oropendola and could account for the acceptance of the brood parasites in nondiscriminator colonies.

The cowbird young are well suited to groom their nest mates. They hatch 5 to 7 days before the oropendola young, and develop precociously. Their eyes are open within 48 hours after hatching, whereas the eyes of oropendola nestlings open 6 to 9 days after hatching. Also, cowbirds are born with a thick covering of down, absent in the oropendola young, which presumably deters bot flies from laying their eggs on the skin of the young cowbirds. By the time the oropendolas hatch, the cowbirds are sufficiently developed to groom them. In discriminator colonies, which are not troubled by bot-fly parasitism, cowbird young perform no such useful function for the oropendolas; and because they compete with the oropendola young for food, cowbirds reduce the productivity of the colony.

The role of the cowbird in this story is now evident, but we have not yet determined why bot flies are present in some colonies and absent from others. Smith noted that all the discriminator colonies, and none of the nondiscriminator colonies, were built near the nests of wasps or bees, whose occupants swarm in large numbers around their nests and virtually fill the air throughout the oropendola colony. Wasps and bees presumably prevent the bot flies from entering the colonies. At any rate, rolls of flypaper hung in the two types of colonies revealed that adult bot flies rarely enter the area around discriminator nests.

Oropendolas and cowbirds are not bothered by the same wasps and bees that viciously attack other intruders at their nests and hives. The birds could benefit the wasps inasmuch as their defenses against nest predators may also protect the wasp nests. In addition, the oropendolas have a characteristic, strong odor that may act as a chemical signal to the insects to suppress their normal defensive reactions. Clearly, the oropendola story may someday chronicle subtle mutual interdependences of which we now have only tantalizing hints.

The nesting success of the oropendolas in nondiscriminator colonies shows the advantage of having cowbirds as nest mates (Table 3–1). Relatively few young are raised from the nest of nondiscriminating oropendolas —less than one-half of a young per nest on the average—owing to preda-

Table 3-1
Fledging success of oropendolas in discriminator and nondiscriminator colonies tabulated according to the presence or absence of cowbird nest mates

Number of young in nest		Fledging success*	
Oropendola	Cowbird	Discriminator colonies	Nondiscriminator colonies
2	0	0.53	0.19
3	0	0.55	0.19
2	1	0.28	0.53
2	2	0.20	0.43

* Fledging success is expressed as number of oropendola offspring leaving the nest.
Source: Smith 1968.

tion, starvation, and abandonment of nests. Nondiscriminator broods with cowbirds nonetheless produce more than twice as many oropendola young as nest without parasites. The difference in success between nondiscriminator nests and discriminator nests, in the absence of cowbirds, may be attributed to the detrimental effects of bot-fly parasitism in the nondiscriminator nests.

We might expect that in the presence of cowbirds, discriminator and nondiscriminator nests would produce similar numbers of oropendola young. Neither is beset by bot-fly parasitism, and both have the same number of mouths to feed. But, in fact, cowbirds reduce success more in discriminator colonies, where they are disadvantageous to the host, than in nondiscriminator colonies, where they confer some advantage. In discriminator colonies, a cowbird chick reduces the number of host young fledged by about one-half. In nondiscriminator colonies, however, a brood parasite more than doubles the nesting success of the host. These facts suggest that if there were no bot-fly parasitism, nondiscriminator colonies would be the more successful, perhaps because they can select colony sites without regard to the presence or absence of wasp nests. Furthermore, nondiscriminator colonies breed earlier in the year than discriminator colonies because they do not have to wait for wasp nests to become active. More food may be available earlier in the year.

For the oropendola, the presence or absence of wasps in the vicinity of the colony completely alters the role of the cowbird as a factor of the environment. Accordingly, the behavior of the oropendolas toward cowbirds and their eggs also varies between the two types of colonies, and in turn affects the environment that molds the behavior of the cowbird. In discriminator colonies, adult oropendolas not only eject the eggs of cowbirds from their nests when they can distinguish them, but they also chase adult cowbirds out of the colony. The oropendolas in nondiscriminating colonies are indifferent to cowbirds. The behavior of the cowbird in the two types of colonies reflects this difference. In discriminator colonies, female cowbirds are cautious and always enter the colony singly. Behavioral adaptation has gone so far that the cowbirds mimic the behavior of the discriminator oropendolas; they often gather nest-building materials and act as if they are beginning to build a nest, a most uncharacteristic behavior of brood parasites. Conversely, cowbirds that parasitize nondiscriminator colonies are often gregarious and enter the colonies in small groups. They behave aggressively toward the oropendolas and sometimes even chase them from their nests.

It is not sufficient for a successful mimic merely to produce an egg that is indistinguishable from host eggs. It must also lay the egg at the proper time. A female oropendola normally lays her 2 eggs on consecutive days. She is sensitive to the appearance of eggs in the nest before or after her laying period, and will frequently desert her nest if it contains more than 3 eggs. Even when confronted by perfect egg mimicry, a discriminating oropendola can be fooled only if a single egg is laid by the cowbird soon after the first oropendola egg has been laid. If the cowbird lays its egg a day too early or too late, it reveals its presence and the oropendola will abandon the nest and start another. In nondiscriminator colonies, however, neither the number of eggs laid by the cowbird nor their appearance matters to the oropendola. Commonly, a female cowbird lays 2 to 5 eggs over several days in the nest of a single nondiscriminating oropendola. Smith referred to these birds as dumpers.

The interaction between the oropendola and the cowbird shows how the pattern of the environment determines the adaptations of the organism and how two or more kinds of organisms can be major factors in each other's environments (Table 3–2). The oropendola, the cowbird, the bot fly, and the wasp are mutually important and affect each other's evolution. Such biotic relationships differ sharply from the relationship to the physical environment of an organism that neither evolves nor responds by adaptation to changes in the biotic environment. Whereas the physical environment is passive, the biotic environment is responsive. Each species continually readjusts to evolutionary changes in others with which it interacts. We might expect evolution in a biotically dominated environment to differ from evolution in a physically dominated environment. Different populations,

Table 3-2
Attributes of discriminator and nondiscriminator colonies of the chestnut-headed oropendola

	Colony type	
	Discriminator	Nondiscriminator
Wasp nests	Present	Absent
Possibility of bot-fly parasitism	Slight or absent	Heavy
Effect of cowbird on oropendola	Disadvantageous	Advantageous
Foreign objects in nest	Rejected	Accepted
Cowbird eggs	Mimetic	Nonmimetic
Cowbird eggs per nest	One	Several
Cowbird behavior in colony	Timid	Aggressive
Colony structure	Compact	Open
Nesting season	Late	Early

Source: Smith 1968.

like the oropendola and the cowbird, may coevolve with respect to each other to form a mutually beneficial relationship that could not be achieved between a population and its physical environment.

Behavioral interactions between individuals define both the social environment of an organism and the behavioral patterns that promote and cope with sociality. These patterns are clearly expressed in the social behavior of baboons, which confers another layer of pattern to our observations of nature.

The behavior of the baboon has evolved in a social context.

Social behavior has developed to different degrees in various animals, but nowhere has its study excited so much interest as in the subhuman primates, the closest relatives of man (Hinde 1983, Smuts et al. 1987). The social group consists of individuals of the same species, many of whom are closely related, upon whom the individual depends in part for its survival. The extent to which some animals have evolved in a social context is demonstrated by their utter dependence upon the group for survival. Such is the case among the baboons of East Africa. In groups, these animals are avoided by all predators except lions; individual baboons, despite their strength and fearsome teeth, have very poor prospects. As a consequence of living in a social group, the individual adapts to an environment determined largely by the behavior of its group members. Whereas individual antagonism and avoidance are the general rules among nonsocial animals, group cohesiveness is absolutely necessary for the survival of such social animals as the baboon. This cohesiveness is promoted by the evolution of a high level of organization within the baboon troop.

Anthropologists S. L. Washburn and Irven DeVore (1961) studied baboons intensively at two localities in Kenya: a small park near Nairobi and the Amboseli Reserve at the foot of Mount Kilimanjaro. Months of careful observation revealed details of the social interactions among the baboons. For example, the troop, moving in an open savanna, has a spatial structure that reflects the roles of individuals (Figure 3–3). When resting in trees, baboons are normally safe from predators, but when they move across open grasslands without the immediate safety of trees, members of a troop assume an organized, defensive pattern. The less dominant adult males and some of the older juveniles occupy the periphery of the group. Females without infants and some of the older juveniles stay closer to the center. Females with infants and young juveniles, the most vulnerable members of the troop, move at the center with the dominant males. When the troop is threatened by a predator, the dominant males move into position between the center of the troop and the threat, while the rest of the baboons seek safety behind them. The importance of trees as potential escape routes is so great that baboons are not often found in areas lacking them, even if the habitat is otherwise perfectly suitable.

Interactions among individuals within the troop are many and diverse.

Figure 3–3
A moving troop of baboons. Females with infants riding on their backs are near the center, close to some large males. Small juveniles and young males are closer to the periphery of the troop.

The baboon has a strong tendency to stay with other baboons. Even within the troop, individuals form small, persistent friendship groups, within which there is much grooming and playing. Most baboons, particularly adult females, spend many hours each day grooming others that present themselves in the proper manner. To groom, one baboon parts the hair of the other with its hands and removes dirt, lice, ticks, and other objects with its hands and mouth. Grooming and play additionally strengthen the social bonds within the troop and promote its stability and cohesiveness.

Care of infant baboons is left to their mothers. The fathers' attentions are directed toward the well-being of the troop as a whole. After an infant has been weaned, its strong bond with its mother is replaced by looser associations within a play group of other juveniles. As young baboons reach adulthood, they enter the dominance hierarchy of the troop in which their position is determined by fighting, bluffing, and the relative rank of those with whom they associate most frequently. Several males may band together and enhance each other's position in the hierarchy, although individually they hold lower ranks. The position of an individual in the social hierarchy determines to some degree the amount of grooming it receives and its precedence over other baboons at a food and water source. Rank also extends privileges in mating with females.

The female is sexually receptive for about 1 week each month. At the beginning of her estrus cycle, a female often mates with the older juveniles and some of the subordinate males. But when she becomes fully fertile she mates only with the dominant males, who consequently sire most of the offspring in the troop. During the mating period, a male and female form what is called a consort pair for a few hours to several days; all other social functions are disrupted for them. Paired baboons generally move at the periphery of the troop, and the male can be extremely aggressive toward others. Baboons appear not to form long-lasting pair bonds, although lack of close association within the troop need not imply a weak social bond.

The baboon must cope with a wide range of behavioral interactions during its lifetime. The relationship between mother and infant, among individuals in a friendship or play group, among males in a dominance hierarchy or among those individuals that cooperate to defend their dominance rank, between the male and female in a consort pair—all produce a complex social environment.

Social adaptation does not occur independently of the physical and biotic environments (Crook 1970, Kummer 1971). Baboon troops in different habitats exhibit considerable variation in their social organization (Stammbach 1987). For example, in the open savannas, an average-sized baboon troop, perhaps 40 individuals near Nairobi and 80 individuals in the Amboseli, may occupy a home range commonly of 2 to 6 square miles, but sometimes as large as 15 square miles (DeVore and Washburn 1964). In Uganda, the baboons inhabit forest or mixed forest-grassland where troops of 40 to 80 usually range over less than 2 square miles (Rowell 1967). In the Ugandan forests, the preferred food of the baboon, which is fruit, is much more plentiful than in the open savannas of Kenya. Thus the troops can satisfy their food needs in a smaller area.

The social organization within the troops of the two areas also differs in several important aspects. In the open savannas of Kenya, antagonistic interactions between individuals are more frequent than in Uganda. In the forest, an individual can go out of sight behind a bush until its antagonist "forgets" about the dispute between them. No such social escape exists on the open savannas. Forest baboons can fill their stomachs in an hour's feeding each day, whereas baboons of the open grasslands spend most of their time in search of food. Whereas grassland troops are often on the move, forest baboons are frequently engaged in grooming, playing, sitting, and drowsing. Because grassland baboons forage almost continuously, individuals interact less socially, and when they do they are more often antagonistic than are forest dwellers.

Hamadryas baboons, inhabiting the arid grassland at the southern edge of the Danakil Desert in Ethiopia, are organized at three social levels: (1) single males with several females and their offspring; (2) persistent bands of several of these single-male family groups; and (3) a larger troop composed of many bands (Kummer 1968, Dunbar 1983, Abegglen 1984; Figure 3–4). During the day, the hamadryas disperse to feed as small bands or single-male groups (Sigg and Stolba 1981). At dusk, they congregate in large numbers to sleep on cliffs, where they are protected from nocturnal predators. Before breaking up in the morning, the troop congregates near the roosting cliffs for a morning social hour, filled with grooming and, among infant and juvenile males, play. The gelada baboon, a cliff-roosting species of the Ethiopian highlands, forages in large groups when food is abundant but disperses into single-male groups during seasons of scarcity (Crook 1966, Dunbar 1983). In contrast to hamadryas, geladas do not appear to have distinct bands within the troop. The patas monkey of the dry grasslands of Uganda has an entirely different social structure, based solely on single-male groups that seldom meet and are rarely on friendly terms (Hall 1965). These studies, and others describing additional variations on the basic single-male groups, multi-male bands, and the larger troops of

Figure 3–4
Above: A troop of hamadryas baboons departing from the roosting cliff. Smaller bands are not evident during this early morning procession. *Below:* A single-male group resting at mid/morning after splitting from the rest of the troop to feed. (After Kummer 1971; courtesy of H. Kummer.)

savanna-dwelling species, emphasize the role of the environment in determining social organization. The particular social structure itself influences adaptations of behavior that strengthen social bonds and both organize and facilitate social interactions.

Natural history observations are not limited to details of the daily lives of organisms. They also include bacterial degradation of plant litter, population cycles of fur-bearing mammals, community development on bared patches of rocky shore, and variation in the diversity of biological communities over the surface of the earth. With this knowledge, ecologists begin to perceive patterns in the natural world and propose explanations for these patterns. These explanations are put to the test by experimentation and further observation as we seek to understand the basic processes responsible for the structure and function of ecological systems. Each of the parts of this book adresses the observations of a different part of ecology, from the relations of organisms to physical factors in the environment to the regulation of community structure. Natural history provides the basic theme for every chapter, from the most descriptive to the most theoretical. We shall begin by considering the physical environment, the setting within which ecological systems develop.

4

Life and the Physical Environment

We often contrast the living and the nonliving as opposites: biological versus physical and chemical, organic versus inorganic, biotic versus abiotic, animate versus inanimate. While these two great realms of the natural world are almost always distinguishable and separable, they do not exist in isolation from each other. Life depends upon the physical world. Living beings also affect the physical world. Their impact is often subtle, but soils, the atmosphere, lakes and oceans, and many sedimentary rocks owe their characteristics in part to the activities of plants and animals.

Living systems require the purposeful expenditure of energy to keep the organism out of equilibrium with and distinct from its physical surroundings. In this chapter, we shall explore how organisms maintain their integrity as open systems, continually exchanging materials and energy with the physical environment.

Life has unique properties not shared by physical systems.

Motion and reproduction are the two most evident of the properties that distinguish living organisms from inanimate objects. Motion expresses a fundamental property of life, the ability to perform work directed toward a predetermined goal; biological reproduction derives its need from the mortality of the individual and ensures the continuation of life. Although distinct from physical systems, living beings nonetheless function within constraints set by physical laws. Like internal combustion engines, they transform energy to perform work. The automobile engine's burning of gasoline is chemical; its transmission of power from the cylinder to the tires is mechanical. The organism's metabolism of carbohydrates and its movement of appendages follow related chemical and physical principles. The biological world is therefore not an alternative to the physical world, but an extension of it.

While biological systems operate on the same principles as physical systems, there is an important difference. In physical systems, energy transformations act to even out differences in energy level throughout the system, always following the path of least resistance. But in biological systems, the organism purposefully transforms energy to keep itself out of equilibrium with the physical forces of gravity, heat flow, diffusion, and chemical reaction. The goal of keeping itself distinct from the physical world applies whether the organism is pursuing prey, producing seeds, or maintaining basic body functions.

In a sense, the organism's use of energy is the secret of life. A boulder rolling down a steep slope releases energy during its descent, but it performs no useful work. The source of the energy—in this case, gravity—is external, and as soon as the boulder comes to rest in the valley below, it is once more in equilibrium with the forces in its physical environment. A bird in flight, on the other hand, constantly expends energy to maintain itself aloft against the pull of gravity. The bird's source of energy—the food it has assimilated—is internal, and the bird uses that energy to perform useful work—to pursue prey, to escape from predators, or to migrate.

The ability to act against external physical forces is the one common property of all living forms, the source of animation that distinguishes the living from the nonliving. Bird flight supremely expresses this property, but plants just as surely perform work to counter physical forces when they absorb soil minerals into their roots and synthesize the highly complex carbohydrates and proteins that make up their structure.

The ultimate source of energy for life is light from the sun. Pigments in the green tissues of plants absorb light and capture its energy; that energy is then converted to food energy through the manufacture of carbohydrates from simple inorganic compounds—carbon dioxide and water. This energy-trapping process is called photosynthesis—literally, a putting together with light. Energy locked up in the chemical bonds of sugars—and thence of proteins, fats, and other organic compounds—is used by plants and by animals, which either eat plants or eat other animals that eat plants, and so on, to fire the engines of life.

The biological and physical worlds are interdependent.

Life depends totally on the physical world. Not only do organisms ultimately receive their energy from sunlight, they must also tolerate the extremes of temperature, moisture, salinity, and other physical factors of their surroundings. The heat and dryness of deserts exclude most life forms, just as the bitter cold of polar regions turns back all but the most hardy. The form and function of plants and animals must obey the rules of the physical world. The viscosity and density of water require that fish be streamlined according to restrictive hydrodynamic design principles if they are to be both efficient and swift. The concentrations of oxygen in the atmosphere and in lakes, streams, and oceans place upper bounds on the metabolic rates of animals and microorganisms. Similarly, the limited ability of plants and animals to dissipate body heat—accomplished by the purely physical means of evaporative cooling, thermal conduction, and radiation of heat from the body surface to the surroundings—determines their rate of activity and their safe exposure to direct sunlight.

While organisms depend totally on the physical world, they also affect the physical world, sometimes in a profound manner. The oxygen we take for granted with every breath is the by-product of photosynthesis by bacteria that lived eons before the appearance of most forms of life (Cloud 1968, 1974; Schopf 1983; Van Valen 1971). Before photosynthetic bacteria evolved in primitive seas, metabolism was accomplished slowly by anaerobic fermentation of organic molecules. The atmosphere of the earth consisted of strongly or mildly reduced gases, perhaps methane (CH_4) or carbon dioxide (CO_2), ammonia (NH_3) or molecular nitrogen (N_2), water vapor (H_2O), and hydrogen (H_2); geologic evidence points quite strongly to the absence of free oxygen (O_2) (Berkner and Marshall 1965). As bacteria began to utilize sunlight as a source of energy, photosynthesis liberated oxygen, some of which escaped from the oceans and accumulated in the atmosphere. Over the past $3\frac{1}{2}$ billion years, the known span of life, photosynthetic bacteria and plants have assimilated carbon and nitrogen; they have been partly replaced in the atmosphere by oxygen produced during photosynthesis.

Figure 4–1
Mima mounds in the vicinity of San Diego, California. Pocket gophers form mounds around their burrow entrances as excavated dirt accumulates (Cox and Allen 1987); they locate new entrances on established mounds because the lower areas between mounds are flooded during spring periods of high rainfall. (Courtesy of G. W. Cox; from Cox 1984.)

Plants and microorganisms play an equally influential role in the development of soil from rock (Crocker 1952). Plant roots invade tiny crevices and pulverize rock as they grow and expand. The "rotting" of plant detritus by bacteria and fungi produces organic acids, which dissolve minerals out of rock, thereby weakening the rock's crystalline structure and speeding its weathering. Fragments of detritus eventually alter the physical structure of the soil. Certain bacteria and cyanobacteria are responsible for the biological fixation of nitrogen from the atmosphere (Broda 1975). Animals play a part in the development of soil by burrowing, trampling, and defecating. Burrowing rodents, for example, create a striking mound topography where drainage is poor and the climate is seasonally dry (Cox 1984; Figure 4–1).

The Dust Bowl that developed in the American Midwest during the 1930s provided a vivid example of what can happen when the environmental roles of plants and animals are disrupted. The Dust Bowl region is normally dry and windy, but the roots of the native perennial grasses were extensive enough to hold the soil in place. When the prairies were brought under the plow, annual crops having less extensive root systems replaced the perennial grasses. A series of dry years reduced crop growth and turned the soil surface to fine dust. The result, illustrated in Figure 4–2, still haunts the memories of local old-timers.

Plants also influence the movement of water. Rain does not accumulate where it falls. If it did, New York State would be under 60 meters of water within a lifetime. Some water flows over the soil or through the underlying earth to enter rivers, lakes, and, eventually, the ocean. The remainder escapes by evaporation from the ground surface and by transpiration from vegetation. In such places as the eastern United States, where the leaves of

Figure 4–2
The Dust Bowl of the midwestern United States. Wind erosion begins when soils are plowed but too little rain falls to support crop growth. *Left:* A winter-wheat crop failure in Finney County, Kansas, has resulted in soil blowing (March 1954). *Right:* A dust storm, composed of wind-blown soil particles, approaches Springfield, Colorado, in May 1937, during the height of the Dust Bowl tragedy. (Courtesy of the U.S. Soil Conservation Service.)

Figure 4–3
Overgrazing by sheep has obliterated most of the vegetation in this area of northern Kenya. (Courtesy of the World Wildlife Fund/IUCN.)

deciduous trees have about four times the surface area of the ground underneath, vegetation provides the major pathway for water escaping from the soil to the atmosphere. When a forest is cut, much of the water that would have evaporated from the leaves flows instead into rivers. Without provision for extensive replanting, deforestation causes flooding, increased erosion, downstream silt deposition, and removal of mineral nutrients from the denuded soil. In the humid tropics, where vegetation holds most of the nutrients in the system, deforestation can decrease soil fertility tragically.

Vegetation absorbs sunlight more efficiently than does bare ground. Because it also has a humidifying effect on the atmosphere, vegetation alters local heat budgets and fosters precipitation, thereby contributing to local weather. In vast regions of Africa just south of the Sahara Desert, the removal of most native vegetation—a consequence of overgrazing and the collection of wood for fuel—greatly intensified the devastating drought of the 1980s (Figure 4–3).

Life forms can increase their energy levels by thermodynamically improbable transformations.

Physical and chemical processes usually lead to transformations that release energy; those that require energy are highly improbable. As a result, the energy level of any bounded physical or chemical system decreases with time as the system loses energy to its surroundings; in other words, such a system spontaneously changes from a higher to a lower energy state. The oxidation of a carbohydrate—for example, the burning of a piece of paper—releases energy in the form of light and heat, and the products of this oxidation (carbon dioxide and water, in this case) contain less energy than the reactants (oxygen and carbohydrate).

Physical systems also dissipate energy. A swinging pendulum contains a certain amount of energy that is periodically transformed between the kinetic energy of the weight moving at the bottom of its swing and the potential energy stored at its highest point. In a frictionless environment, a pendulum would continue to swing forever without loss of energy. But in an atmosphere, the weight sets molecules of oxygen and nitrogen in motion and thereby transfers some of its energy to its surroundings. As it loses energy, the weight swings through smaller and smaller arcs. In this way, energy initially residing in the pendulum becomes more evenly distributed throughout the larger system.

If we could perceive energy density as values of light, organisms would appear to us as beacons against the dim background of the physical world. Animals and plants represent immense concentrations of energy—energy derived from the brightest light in the surroundings, the sun. To prevent physical and chemical processes from dispersing its energy more evenly throughout the system in which it exists, the organism performs work on the system in a manner designed to maintain its own integrity. By way of analogy, imagine yourself as a high mound of sand piled steeply on a flat landscape. Little avalanches of grains tumble down from your sides, the wind blows others away, and rain erodes your stature and carries your substance off in milky rivulets. To maintain your prominence, you continually scoop up nearby sand and pile it on top of your head. You may even reach across to another pile, where sand is easier to get, to maintain or perhaps add to your own substance. If you are clever, you build walls at your base to help retain your sand. In many ways living forms elaborate this theme, with the sand representing their energy and substance.

While each organism maintains a highly improbable concentration of energy in its body, it continuously expends energy to preserve its integrity. The energy expended must be balanced by energy gained, either by the assimilation of sunlight directly or by the assimilation of food. The energy released upon the metabolism of carbohydrates to carbon dioxide and water equals the net energy a plant assimilates to manufacture that amount of carbohydrates from raw inorganic materials. This assimilation of energy runs counter to the tendencies of physical and chemical processes, and it requires the expenditure of energy to build and maintain the individual's structure. Plants and animals also need additional energy, over and above this amount, because biochemical transformations are not perfectly efficient. The metabolism of carbohydrates is like a slow burn, but the organism cannot harness all the energy released; much is lost as heat, a form of energy that neither plants nor animals can use. Through the many biochemical transformations of the living organism, and because of the inefficiency of these transformations, all assimilated energy eventually is dissipated to the physical system—that is, energy is only transient in biological systems.

In a thermodynamic steady state, energy gain balances energy loss.

A uniform sphere of metal in space provides a simple illustration of a physical thermodynamic system. Completely lifeless, the sphere passively intercepts light emanating from the sun, or from some more distant star, and radiates energy into the black depths of space (Figure 4–4). Energy is assimilated, transformed, and lost from the system. When the sphere absorbs light, its temperature rises as molecules are caused to move more rapidly; the energy in the light is transformed to heat energy. But the hotter an object, the more rapidly it loses energy to its surroundings, and so the sphere radiates more and more energy to empty space. Eventually, the sphere heats up so much that it loses energy as rapidly as it absorbs it, and at this point the temperature of the sphere comes into equilibrium with its environment. This equilibrium is a dynamic steady state, in which the thermal characteristics of the system remain constant but the system sustains a continual flux of energy.

Simple physical principles allow us to characterize the steady state quantitatively and predict how the equilibrium temperature will vary with the conditions of the physical environment and the qualities of the sphere. The rate of change in the sphere's heat energy content (H) is the difference between its rate of absorption of radiant energy from the sun (A) and its rate of radiation of energy from the sphere to the environment (R). Thus, $dH/dt = A - R$.

Absorption depends on

F, the radiation flux through the sphere's environment, expressed as energy per unit of time per unit of area perpendicular to the source

C, the cross-sectional area of the sphere, which determines how much energy is intercepted

a, the absorptivity of the surface, which determines how much of the intercepted light is absorbed rather than reflected

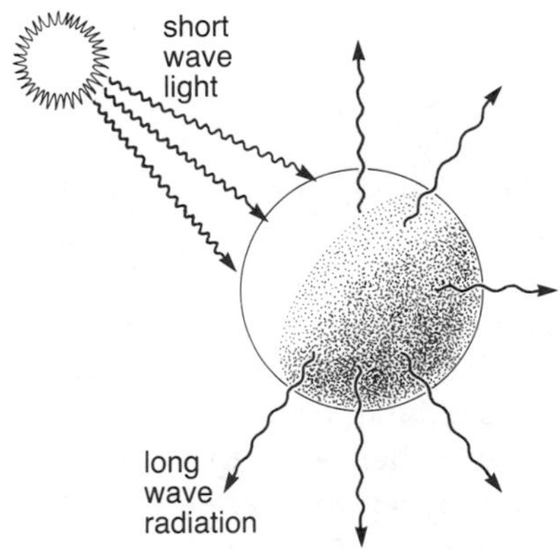

Figure 4–4
Thermal model of a solid sphere, which receives energy as it absorbs sunlight and loses energy by radiation to space.

In this book, we shall express energy in units called joules (J) and rate of energy flux in units of joules per second, or watts (W), the common electrical power rating (see the endpapers for the International System of Units; for comparison, 1 joule = 0.239 calories). The intensity of the energy flux through the environment can be expressed as watts per square meter ($W\ m^{-2}$). (The flux of radiant energy from the sun at the distance of the earth is about 1350 $W\ m^{-2}$.) The cross-sectional area of a sphere can be expressed in square meters (m^2). Absorptivity is a dimensionless proportion of the total incident radiation. Hence, multiplying the flux density by the cross-sectional area by the absorptivity gives the rate of absorption in watts ($W\ m^{-2}$)(m^2) = W; that is, $A = aFC$.

Radiation varies in direct relation to

T^4, the temperature of the system, measured in degrees Kelvin (°K) raised to the fourth power (°K = °C + 273; that is, absolute zero, 0°K, is −273°C)

S, the area of the surface of the sphere (m^{-2})

e, the emissivity of the surface, a property akin to absorptivity

In fact, good absorbers are also good emitters. Because radiation is measured in watts per unit of surface area, emissivity must have units of watts per area per fourth root of thermodynamic temperature ($W\ m^{-2}\ °K^{-4}$) to make radiation come out in watts.

Absorption and radiation can now be combined in a single equation of the form $dH/dt = A - R$; namely,

$$\frac{dH}{dt} = aFC - eST^4$$

The system comes into equilibrium when $dH/dt = 0$; that is, when

$$eST^4 = aFC$$

radiation equals absorbance (Figure 4–5). The temperature of the sphere at its steady state can be found by rearranging the condition for equilibrium to obtain

$$\hat{T} = \left(\frac{aCF}{eS}\right)^{1/4}$$

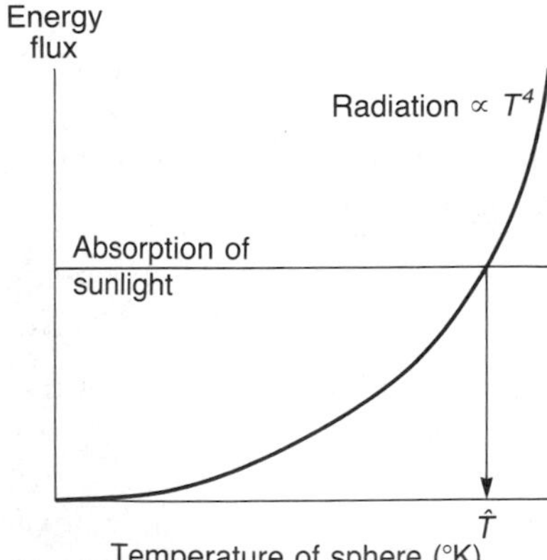

Figure 4–5
Relationship of radiation to temperature of the radiating surface. In the example of the sphere in space, when temperature increases to the point that loss of heat by radiation equals gain from the sun, the temperature has reached its equilibrium point ($\hat{T}$).

Because the surface area of a sphere is four times its cross-sectional area, the ratio C/S in this expression can be replaced by 4. Thus, equilibrium temperature (T) depends only on the absorptivity, the emissivity, and the radiation flux. Spheres having the same surface qualities but different sizes equilibrate at the same temperature; those closer to the sun (higher F) achieve higher temperatures than those more distant. All other things being equal, radiation flux varies inversely with the square of the distance from a heat source (that is, $F \propto D^{-2}$). Equilibrium temperature therefore varies inversely with the square root of the distance from the source ($\hat{T} \propto (D^{-2})^{1/4} = D^{-1/2}$). Temperature may be adjusted by changing the surface properties of absorptivity and emissivity—for example, by covering a satellite with a reflective surface to prevent the interior from overheating in the full sunlight of space.

Organisms can control the flux of energy or material between their internal and external environments.

The flux of energy or material across the surface of an organism can be described in general terms

$$\text{flux} = \text{surface area} \times \text{gradient} \times \text{conductance}$$

The gradient is the difference between the levels or concentrations of energy or material inside the organism and those in its surroundings. Conductance is the ease with which the energy or material crosses the surface barrier. In electrical terms, which are often used in analogies of ecological systems or to construct analog models, flux is current (amperes), gradient is electrical potential (volts), and conductance is the inverse of the electrical resistance (ohms per square meter of surface or cross-sectional area of wire).

Organisms can control flux by altering any of its three components. A mammal's thick winter fur reduces the conductance of heat (thermal conductance) from its body to its surroundings. The hard, waxy cuticles covering the exoskeletons of insects increase the resistance of the body's surface to water flux. Because pale colors reflect more light than do dark colors, desert organisms frequently adopt them to reduce radiational heat load. Biological surfaces may also actively transport materials either from inside out or from outside in. Many ions are moved into and out of cells in this way by membrane-bound molecular "pumps" that maintain suitable concentrations of physiologically important ions (for example, sodium or potassium). Such active transport requires the expenditure of energy.

Surface area can be manipulated in many ways. Organisms can reduce exchange areas by sealing off some portion of the surface, thus restricting flux to smaller regions that may be highly specialized. The exterior surface of a leaf is protected from water loss by a wax cuticle not unlike the surface of an insect. Gas exchange form the interior of the leaf takes place through stomates, tiny holes whose size can be controlled precisely in accordance with concentrations of water within the leaf. Surfaces can also be increased by elaborate folding, as in the external gills of salamander larvae or in the interior lungs of mammals.

The gradient between an organism and its surroundings depends, of course, on the internal and external environments. The internal environment is dictated in part by the ranges of body temperature, ion concentrations, and other conditions suitable for life processes. Moreover, the biochemical transformations often exaggerate gradients and thereby maintain high fluxes. For example, as our tissues consume oxygen, internal concentrations are reduced, enhancing the outside-inside gradient and encouraging oxygen flow across the lung surface. Conversely, metabolism produces carbon dioxide and, as its concentration builds up within tissues and the blood, its outward flux increases.

The gradient between internal and external environments may also change when organisms move through a heterogeneous environment.

Shady and sunny spots present dramatically different radiative environments to an organism concerned with regulating its temperature. When humidity, ion strength, pH, soil nutrients, dissolved oxygen, prey abundance, disease organisms, and other factors vary over distances that are small compared to an individual's daily range of movement, choice of environment has a direct bearing on flux. The responses of plants and animals to many of these factors will be considered in the following chapters, but all may be understood in terms of the components of flux across the organism-environment boundary.

Finally, we must distinguish between passive (physical) and active (biological) flux. The first is the natural thermodynamic tendency of material or energy to move from areas of high concentration to areas of low concentration. Passive fluxes across a surface occur in direct proportion to the gradient or difference across the surface. Given time, and in the absence of other processes, passive flux continually reduces a gradient until flux drops to zero. Thus, in the absence of an energy input, a hot sphere gradually cools to the temperature of its surroundings.

Active flux allows biological systems to accumulate substance against a physical gradient as, earlier, we were accumulating sand into unstable piles of self. Active flux requires the expenditure of energy because it opposes the energy-releasing tendencies of physical fluxes. Therefore, in terms of the flux equation, active flux has a negative conductance.

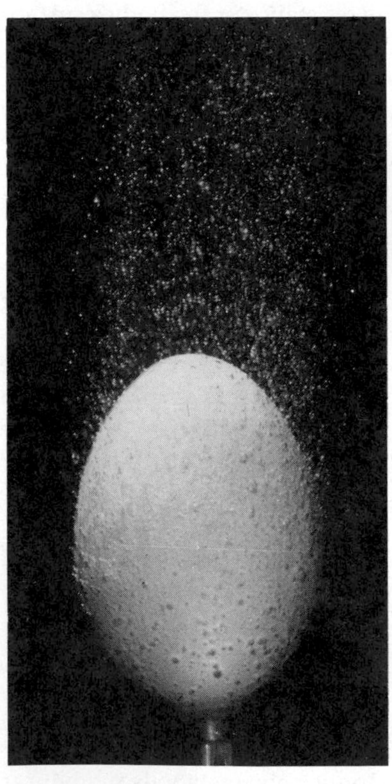

Figure 4–6
An eggshell, which is filled with pressurized air, emits thousands of tiny bubbles through minute pores distributed over its surface. (Courtesy of C. Carey and H. Rahn.)

The avian egg illustrates the role of surface properties in determining the organism-environment flux.

A bird's egg contains all the nutrients that the embryo needs to grow to hatching. The incubating parent generally maintains the egg's temperature within narrow limits, about 34 to 38°C. Its only exchanges with the outside world during the period of embryonic development are the influx of oxygen for metabolism and the effluxes of carbon dioxide, produced by respiration, and water. These exchanges are essential for growth, for if the egg surface were completely sealed, the embryo would suffocate. Exchange occurs through tiny pores that penetrate the eggshell and bring the surrounding atmosphere into contact with the respiratory membranes of the embryo. Gases pass freely through these pores by physical diffusion (Figure 4–6). But while the egg must gain oxygen from the environment and rid itself of carbon dioxide, loss of water by the same route is undesirable because it might lead to dehydration. The egg therefore is challenged to accomplish necessary gas exchange while limiting water loss to a tolerable level.

Gas exchange is a flux, determined by surface, gradient, and conductance according to the general relationship just outlined. In the early 1970s, O. Douglas Wangensteen et al. (1970–1971) discovered that the movement of gas through the pores of an eggshell obeyed simple laws of physical diffusion. Hermann Rahn and C. V. Paganelli (1981), in whose laboratory at the State University of New York at Buffalo much of this research was conducted, tell how the investigation got started:

The beginning of our interest in the gas exchange of avian eggs can be quite clearly documented. It occurred in 1968 after the arrival of O. Douglas Wangensteen as a Postdoctoral Fellow from the University of Minnesota. One day he asked us how eggs breathe. Since none of us had ever thought about this problem, we suggested that he might find out. Using shell fragments and an O_2 electrode he established that gas transport across the eggshell is essentially limited to gas phase diffusion and obeys Fick's law in contrast to gas exchange of man and other animals, where transport is dominated by convection.

Fick's law concerns the diffusion of a gas across a constant concentration gradient (Schmidt-Nielsen 1975). Accordingly, the rate of diffusion (V) of gas across an eggshell, measured in cubic centimeters per second ($cm^3\ s^{-1}$), is the product of

1. pore area, A (cm^2)
2. the diffusion coefficient of the gas, D($cm^3\ cm^{-1}\ s^{-1}$)
3. the inverse of the length (L) of the pore (cm), that is, shell thickness
4. the difference in the concentration of gas (C) between the egg and the surrounding air ($cm^3\ cm^{-3}$)

(Paganelli 1980)

Because gases occur in an aqueous phase inside the egg, the difference in vapor pressure, P (torr; see Table 4–1), between the inside and outside of the egg can be substituted for the concentration difference (C) with an appropriate conversion factor. Therefore, gas exchange may be described by the equation $V = (A/L)DP$. Experimentally determined values of D($cm^3\ cm^{-1}\ s^{-1}$) in air at 38°C (the incubation temperature of chicken eggs) are 0.27 for water, 0.23 for oxygen, and 0.18 for carbon dioxide; the heavier gas molecules of CO_2 diffuse more slowly than the lighter molecules of O_2 and H_2O.

Table 4–1
Units of pressure

Pressure is a force (weight) per unit area. In the English system of measurement, the usual unit of pressure is the pound per square inch (lb in^{-2}). The pressure of the atmosphere at sea level—that is, the total weight of atmosphere above a square inch of surface—is 14.7 lbs in^{-2}. This value is often expressed in terms of the height of a column of mercury (Hg) having the same weight (29.9 in or 760 mm). A pressure of 1 mm Hg is sometimes referred to as 1 torr. In the International System of Units (see the endpapers), the unit of pressure is the pascal (Pa), which is the weight of 1 newton (N) per square meter.

Expressed in these various units, sea-level atmospheric pressure is equal to:

1	atmosphere of pressure (atm)
14.7	pounds per square inch (lbs in^{-2})
29.9	inches of mercury (in Hg)
760	millimeters of mercury (mm Hg)
760	torr
101,325	pascals (Pa)
0.101	megapascals (MPa)
1.01	bar (1 bar = 100,000 Pa)

The idealized chicken egg has a pore area of 0.023 cm^2 and a shell thickness of 0.030 cm. The vapor pressure of water at 38°C is 50 torr. That is, the tendency of water to evaporate from a liquid surface at 38°C is exactly balanced by the tendency of water to move from the gas to the liquid phase when the pressure of water vapor at the surface is 50 torr, or about 7 per cent of the total pressure of the atmosphere at sea level (760 torr).

Rahn et al. (1977) used a simple device, called an "egg hygrometer," to determine the vapor pressure in the immediate environment of the egg under the incubating parent. An egg hygrometer is constructed in the following manner. First, the conductance of water vapor across an eggshell is determined by placing the egg in a dessicator, an airtight container in which the water vapor pressure is kept close to zero by using silica gel to absorb water from the air. Under these conditions, the external water vapor pressure is zero and the internal pressure is a simple function of temperature. Therefore, the conductance per torr of a known vapor pressure gradient can be calculated from the weight loss of the egg, which can be measured simply by change in weight. Then the contents of the egg are removed through a small hole and replaced with silica gel, thereby reducing the vapor pressure of water *within* the egg hygrometer close to zero. The hole is sealed and the egg is replaced in a nest with the rest of the clutch. Under these circumstances, the weight of water *gained* by the egg indicates the vapor pressure of water in the nest environment. For example, suppose a chicken egg kept in a dessicator at 25°C (vapor pressure of water equal to 24 torr) loses 240 milligrams of water per day. This is equivalent to 10 mg d^{-1} $torr^{-1}$. If the same egg filled with silica gel and placed in a nest gained 200 mg d^{-1}, the vapor pressure of the nest would be 20 torr.

By placing calibrated egg hygrometers in the nests of many species of bird, Rahn and his co-workers determined that the vapor pressure of water in the nests of most species is maintained between 18 and 26 torr regardless of the temperature and water content of the surrounding air. This regulation is achieved by the parent's adjusting the water permeability of the nest and the parent's temporal pattern of sitting.

Because the pressure gradients of oxygen and carbon dioxide are controlled primarily by the metabolic activity of the embryo, they therefore change during the course of incubation. The pressure of oxygen in sea-level atmosphere is about 150 torr, and that of carbon dioxide is close to zero. In an unincubated chicken egg, the partial pressures of oxygen and carbon dioxide equilibrate with the air surrounding the egg. As the embryo approaches hatching, however, it metabolizes energy rapidly and partial pressures of gases within the egg decrease to 104 torr for oxygen and increase to 37 torr for carbon dioxide (Rahn et al. 1974). Note that as the requirement for oxygen increases, the pressure gradient also increases, enabling oxygen to enter the egg more rapidly. Conversely, as the embryo produces carbon dioxide more rapidly, the concentration of the gas within the egg increases, causing a rise in CO_2 flux from the egg. But gas conductances ultimately do place an upper limit on the embryo's rate of metabolism; above this limit, levels of carbon dioxide would become toxic for the embryo. Increasing the gas conductance of the shell to alleviate this problem would result in excessive water loss.

The different environments in which birds breed place different demands on gas exchange. At high altitude, atmospheric pressure is reduced; at 3800 meters, for example, it is 60 per cent of sea-level value. Atmospheric oxygen pressure drops from 150 torr at sea level to 90 torr at 3800 m; within the egg, the partial pressure of oxygen drops accordingly to far below the level in the egg at sea level. Wangensteen et al. (1974) found that chickens kept at 3800 m elevation adapt to the reduced oxygen concentration in the egg by reducing embryonic metabolism and prolonging the incubation period slightly. Although one might think that the problem of delivering oxygen to the embryo could be solved by increasing pore area, the limitation at high altitude is not the rate at which oxygen moves across the shell; rather, it is the low absolute level of oxygen. In fact, gas conductances increase with altitude because gas molecules move through thinner air and therefore experience fewer collisions. Because air pressure (hence density) at 3800 m is 60 per cent lower than at sea level, gas conductances are 60 per cent higher. To prevent excessive water loss at high altitude, bird eggs actually have reduced pore area (Rahn et al. 1977).

Wedge-tailed shearwaters, seabirds that breed on remote oceanic islands, have reduced the pore area of their eggs to prevent excessive water loss resulting not from high altitude but from a greatly prolonged incubation period (Ricklefs 1984). Although the egg of the wedge-tailed shearwater is about the same size as a chicken egg, its incubation, for some unknown reason, takes 52 days as compared with 21 days for the chicken. If the shearwater egg lost water at the same daily rate as the chicken egg, the embryo would die of dehydration before the end of the incubation period. Because water vapor pressures in the nests of the shearwater and the chicken are similar, the only solution to this problem of excessive water loss is to reduce the pore area, and hence the gas conductance, of the eggshell, and this the shearwater has done.

The water vapor conductance of the shearwater egg is 6.2 mg d^{-1} $torr^{-1}$, compared with 14.4 mg d^{-1} $torr^{-1}$ for the chicken egg (Ackerman et al., 1980). As Ar and Rahn (1980) have shown, total gas conductance in bird eggs is adjusted so that the amount of water lost per gram of egg over the entire incubation period is approximately the same for all species. For the 60-gram chicken egg with a water vapor conductance of 14.4 mg d^{-1} $torr^{-1}$ over 21 days, this value is 5.0 mg g^{-1} $torr^{-1}$; for the 60-gram shearwater egg, it is 6.2 mg d^{-1} $torr^{-1}$ $\times$ 52 days/60 g = 5.4 mg g^{-1} $torr^{-1}$.

One problem creates another, however, as the reduced pore area also reduces the embryo's supply of oxygen. To solve this new problem, the shearwater embryo pips (that is, breaks a tiny hole through the eggshell) several days before it hatches and before its oxygen requirements reach their maximum. It then begins to breathe air directly (Ackerman et al. 1980).

One final example involving avian eggs will further illustrate the consistency and inevitability of the responses of organisms to their physical environments. Brush turkeys *(Alectura)* belong to a chickenlike group of Australian birds known as the megapodes. The brush turkey is unique among birds in that incubation takes place within mounds of vegetation built by the male. The eggs are warmed by the heat of the vegetation as it decomposes. This unusual style of incubation creates unusual problems, for within the nest mound, oxygen is scarce and both carbon dioxide and water vapor increase to high levels. Seymour and Rahn (1978) measured partial pressures of 48 torr for water vapor, 100 torr for oxygen, and 62 torr for carbon dioxide in brush turkey nests. Water loss is not a problem because the vapor pressures inside and outside the egg are similar. Oxygen is almost as scarce in the mound as it is at 3000 meters altitude. The most critical problem is that to get rid of carbon dioxide, the embryo must tolerate levels of carbon dioxide in the egg in excess of 62 torr. To increase the flux of carbon dioxide without increasing the concentration of the gas within the egg, brush turkeys have greatly increased the gas conductance of their eggs, to 2.6 times as much as that of an egg of similar size of a species that builds a more typical nest.

Compromise dominates adaptations of life forms.

Compromise is a consistent theme in the relationship of organisms to their environments. As the avian egg illustrates, terrestrial organisms cannot reduce water loss without reducing access to oxygen or carbon dioxide in the atmosphere. The same thick coat of fur that promotes the conservation of body heat in cold surroundings prevents the dissipation of excess heat when the environment warms up. Modifications of the legs and feet of horses that enable them to run swiftly (Figure 4–7) also produce a built-in stiffness that makes the limbs useless for scratching and swatting flies. Of course, horses have found ways around this problem. They have long tails to swish flies off their hindquarters, and as for scratching, horses love nothing so much as rolling in the dust.

Still, every adaptation has its costs. No organism has unlimited time, resources, or body tissues. What it allocates to one function, it must take from another. Nothing is free. In the absence of any benefit, even small costs become apparent. The eye, so important to humans, is useless to cave-dwelling fish that live in total darkness; the cost of producing eyes and their associated muscles and nerves apparently is sufficiently great that many species of cave organisms have reduced them to tiny, rudimentary structures (Figure 4–8).

Although the examples in this chapter have concentrated on physical factors in the struggle for existence, organisms must also contend with biological aspects of their environment: predators, prey, pathogens, and even collaborators. These factors, too, impose certain requirements of structure and function for successful living, and they also create conflicts of allocation. Time taken to watch for predators is time taken from feeding. Carbohydrates that a plant devotes to spines as a defense against herbivores are carbohydrates that cannot be packaged in seeds. The bargain that each type of organism strikes among these conflicting needs is determined by the pressures it faces in its environment.

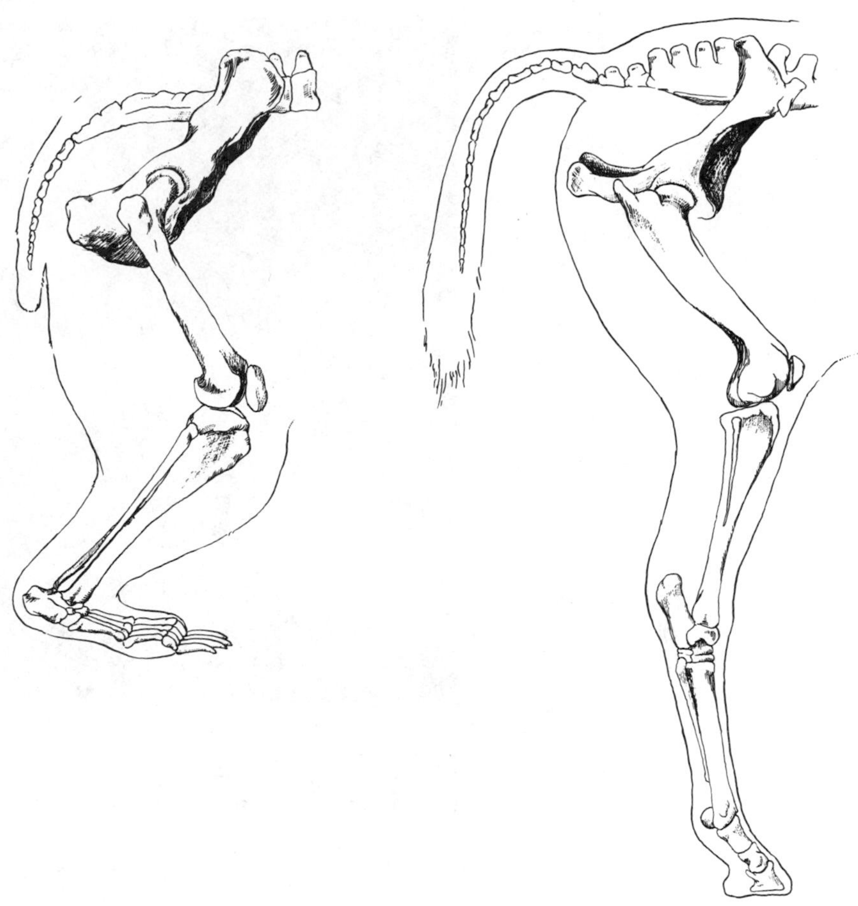

Figure 4–7
Hind limbs of a bear *(left)* and a horse, showing the elongated and simplified structure of the horse's foot for rapid locomotion. (After Dunbar 1960.)

The activities of organisms transform energy and materials, linking the physical and biological worlds into ecosystems.

In the course of maintenance, growth, and reproduction, organisms transform materials and energy, often against the thermodynamic grain. By doing so, they to some extent transform the physical world. Also, many physical and chemical processes transform the environment independently of life; others undo what living forms have done. These transformations shape the physical world in which organisms live, particularly by determining the conditions for life and by influencing the availability of nutrients. The idea that life processes and physical processes are intimately related is the basis of the ecosystem concept in ecology, which encompasses both the

Figure 4–8
The northern cavefish *(Amblyopsis spelaea)*, an inhabitant of subterranean streams in Indiana and Kentucky, has rudimentary eyes and is blind. (Courtesy of T. C. Barr; from Barr 1968.)

biological and physical realms. Materials cycle through both organic and inorganic forms, and their flux through these cycles is determined by biological and by physical transformations. Just as the egg and its developing embryo are only a small segment of the total life cycle of the bird, the exchanges of gases across the shell between the egg and its immediate surroundings are themselves only a small part of a vast machine—the ecosystem, occupying a thin, wet film on the surface of the earth.

SUMMARY

1. Life is an extension of the physical world but differs from that world in its purposeful expenditure of energy for self-perpetuation.

2. Life is totally dependent on the physical world for energy and materials, and living organisms must tolerate the physical conditions of their surroundings—among them, temperature, moisture, and salinity.

3. Organisms also affect the physical world by transpiring water vapor, recycling elements, abetting soil development, precipitating minerals in aquatic environments, and altering terrestrial topography.

4. Spontaneous physical and chemical processes result in transformations that release energy. Organisms differ from physical systems because they can increase their energy levels by thermodynamically improbable transformations. The ultimate source of energy for these transformations is light energy from the sun harnessed by photosynthesis.

5. Organisms approach thermodynamic steady states in which energy assimilated approximately balances energy lost to the environment. Fluxes of materials into and out of living systems are also approximately balanced over the long term.

6. Fluxes of materials and energy across the surfaces of organisms are equal to the product of the surface area, the gradient between the interior and exterior environments, and the conductance. Values of conductance are material- and surface-specific. Active uptake or export against a gradient can be represented by a negative conductance, indicating an energy-requiring process.

7. Organisms can alter, within limits, all three components of flux by modifying surface areas and surface properties and by choosing suitable environments.

8. Principles pertaining to fluxes across surfaces are illustrated by gas and water vapor conductances across the avian egg; these conductances are controlled by the size, number, and length of pores that penetrate the shell. These variables respond adaptively to concentrations of oxygen, carbon dioxide, and water vapor in the atmosphere surrounding the egg, and to the developing embryo's needs to consume oxygen and to rid itself of carbon dioxide.

9. Adaptive modification of organisms to several factors in the environment often results in compromise to balance the conflicting costs and benefits of such modifications.

10. Energy transformations and material fluxes couple organisms to physical processes in a single functional unit referred to as ecosystems.

5

Conditions of the Physical Environment

The physiologist Lawrence J. Henderson remarked in his book *The Fitness of the Environment* (1913) that "Darwinian fitness is compounded of a mutual relationship between the organism and the environment. Of this, fitness of environment is quite as essential a component as the fitness which arises in the process of organic evolution; and in fundamental characteristics the actual environment is the fittest possible abode of life." It is difficult to imagine life evolving under conditions other than those found at the surface of the earth. Indeed, could we imagine life in any form other than our own?

Biologists before and after Henderson have argued that certain physical and chemical conditions and properties must be met for any sort of life to evolve and that only a limited range of these suitable conditions exist on the earth (Westheimer 1987). Any imaginable form of life must have a liquid basis; gases are too diffuse and structureless, solids too rigid. The basic medium of life on the earth is water; thus, life processes occur only at temperatures between the melting and boiling points of water, those that prevail at the surface. Henderson argued that of all compounds only water has all the qualities necessary to support life. In this chapter, we shall discuss the many serendipitous properties of water and the other features of the environment important to life: temperature, carbon and oxygen, inorganic nutrients, salts, and light.

Water is essential to life.

All organisms are composed mostly of water, whether they dwell in oceans, lakes, or rivers, or on land. Water is generally abundant on the earth's surface and, within the temperature range usually encountered, it is liquid.

Water has many thermal and solvent properties favorable to life. One must add or remove a large amount of heat energy to change its tempera-

Table 5–1
Thermal properties of water

Specific heat is the quantity of heat energy required to raise the temperature of 1 g of water 1°C: 1 calorie (cal) or 4.2 joules (J).

Heat of melting is the quantity of heat energy that must be added to ice to melt 1 g of water at 0°C: 80 cal or 335 J.

Heat of vaporization is the quantity of heat energy that must be added to evaporate 1 g of water: 597 cal or 2498 J at 0°C, 536 cal or 2243 J at 100°C.

Thermal conductivity is the flux of heat through a 1 cm^2 cross-section at a gradient of 1°C cm^{-1} (units are J cm^{-1} sec^{-1} $°C^{-1}$): 0.0055 at 0°C, 0.0060 at 20°C, 0.0063 at 40°C, and 0.022 for ice at 0°C.

Density is the mass per unit volume:

water at 30°C = 0.99565 g cm^{-1}
20°C = 0.99821
10°C = 0.99970
4°C = 0.99997 (maximum density)
0°C = 0.99984
ice at 0°C = 0.917

ture, and because water conducts heat rapidly the temperatures of organisms and aquatic environments tend to be relatively constant and homogeneous. Water also resists change of state between solid (ice), liquid, and gaseous (water vapor) phases. To evaporate a quantity of water requires an input of over 500 times more energy than that needed to raise its temperature by 1 degree Celsius. Freezing requires the removal of 80 times as much heat as that needed to lower temperature by 1 degree (Table 5–1). Another curious thermal property of water is that whereas most substances become more dense when they are cooled, water becomes less dense as it cools below 4°C. Consequently ice floats (Figure 5–1), not only making ice-skating possible but also keeping the bottoms of lakes and oceans from freezing, thus providing refuge for aquatic plants and animals during winter.

Water has an immense capacity to dissolve inorganic compounds, making them accessible to living systems and providing a substrate within which they can react to form new compounds. The formidable solvent properties of water derive from the strong attraction of water molecules for other compounds. Molecules are composed of electrically charged atoms or groups of atoms called ions. Common table salt, sodium chloride (NaCl), is made up of a positively charged sodium atom (Na^+) and a negatively charged chlorine atom (Cl^-). When salt is placed in water, the attraction of the water molecules for the charged sodium and chlorine atoms is so great, compared with the bonds that hold the molecule together, that the salt molecule readily dissociates onto its component atoms—another way of saying that the salt dissolves. The dissociation of sodium chloride into its component ions may be written

$$NaCl \rightleftharpoons Na^+ + Cl^-$$

or, portraying the role of the water molecules as a solvent, as

$$NaCl + H_2O \rightleftharpoons H_2ONa^+ + Cl^-$$

Figure 5-1
Water becomes less dense as it freezes, and ice therefore floats. Because the density of ice is 0.92 g cm^{-2} more than 90 per cent of the bulk of this antarctic iceberg lies below the surface.

The arrows indicate that even in solution ions continually rejoin as well as dissociate. The absolute magnitudes of the dissociation and association rates determine how fast a substance dissolves. The magnitudes of these rates relative to each other determine the solubility of the substance at equilibrium.

The ability of soil to retain water is related to the size of soil particles.

Most terrestrial plants obtain the water they need from the soil. The amount of water that soil holds and its availability to plants vary according to the physical structure of the soil particles (Brady 1974). Soil consists of grains of clay, silt, and sand, and particles of organic detritus. Grains of clay, produced by the weathering of minerals in certain types of bedrock, are the smallest; grains of sand, which are the quartz crystals that remain after minerals more susceptible to weathering are removed from rock, tend to be the largest; silt particles are intermediate. Collectively, these particles are known as the soil skeleton; as the name implies, these comprise a stable component that influences the physical structure of the soil and its water-holding ability, but do not play a major role in its chemical transformations.

Water is sticky. The capacity of water molecules to cling to each other and to surfaces they touch accounts for the familiar phenomena of surface tension and the rise of water against gravity in capillary tubes. Water clings

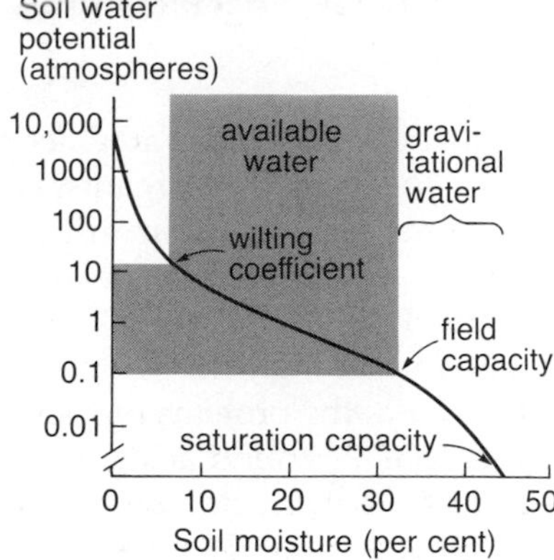

Figure 5-2
Relationship between water content of a loam and the average force of attraction of the water to soil particles (soil water suction). The difference between the soil water content at field capacity (0.1 atmosphere) and the wilting coefficient (15 atm) is the water available to plants. (After Brady 1974.)

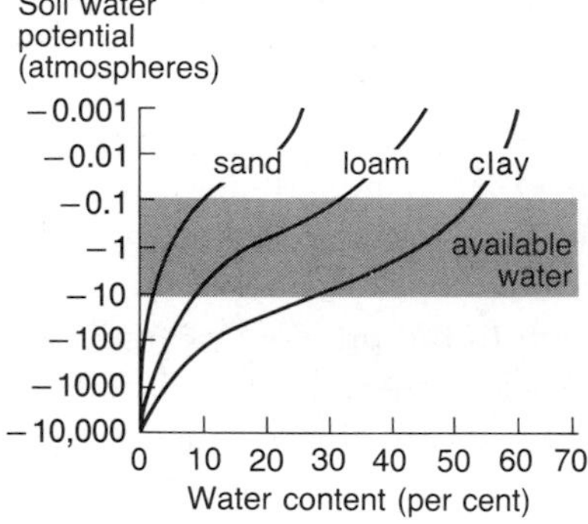

Figure 5-3
Relationship between soil water content and soil water suction for sand, loam, and clay soils. Loams have the highest water availability (field capacity minus wilting coefficient). Sandy soils have large pore spaces that allow free drainage of water; the small interstices between particles in clay soils hold water tightly. (After Brady 1974.)

tightly to surfaces of the soil skeleton. Because the total surface area of particles increases as their size decreases, silty soils hold more water than coarse sands, through which water drains quickly.

Water capacity is not equivalent to water availability. Plant roots easily take up water that clings loosely to soil particles by surface tension, but water near the surface of sand and silt particles is bound tightly by stronger forces. These forces are called the water potential of the soil. Soil scientists measure soil water potential and the strength with which the cells of root hairs can absorb water from the soil by equivalents of atmospheric pressure (see Table 4-1). Capillary attraction holds water in the soil with a force equivalent to a pressure of $\frac{1}{10}$ to $\frac{1}{5}$ atmosphere (76-152 torr or 0.01-0.02 megapascals [1 MPa = 1,000,000 Pa]). Water attracted to soil particles with less force than $\frac{1}{10}$ atmosphere (water in the middle of interstices between large soil particles, hence at great distance from their surfaces) drains out of the soil under the pull of gravity and joins the groundwater in the crevices of the bedrock below. The amount of water held against gravity by forces of attraction greater than $\frac{1}{10}$ to $\frac{1}{5}$ atmosphere is called the field capacity of the soil.

A force equivalent to $\frac{1}{10}$ atmosphere is sufficient to raise a column of water nearly 1 meter. We know that plant roots can exert a much greater pull on water in the soil, because water rises in the tallest trees to leaves more than 100 meters above the ground. In fact, plants can exert a pull of about 15 atmospheres (1.5 MPa) on soil water. Once plants under drought stress have taken up all the water in the soil held by forces weaker than 15 atmospheres, they can no longer obtain water and they wilt, even though water remains in the soil. Thus, a soil water potential of 15 atmospheres is referred to as the wilting coefficient, or wilting point, of the soil (Slatyer 1967, Meidner and Sheriff 1976).

As soil water is depleted, the remainder is held by increasingly stronger forces, on average, because a greater proportion of the water is situated close to the surfaces of soil particles. The relationship between water content and water potential is shown in Figure 5-2 for a typical soil with a more or less even distribution of soil particle sizes from clay (up to 0.002 mm) through silt (0.002-0.05 mm) to sand (0.05-2.0 mm). Such soils are called loams. When saturated, a loam holds about 45 grams of water per 100 grams of dry soil (45 per cent water). The field capacity is about 32 per cent, and the wilting coefficient about 7 per cent. The difference between the field capacity and the wilting coefficient, about 25 per cent in this case, measures the water available to plants. Of course, plants obtain water most readily when the soil moisture is close to the field capacity.

In soils with predominantly smaller particles the soil skeleton has a relatively large surface area; such soils hold a larger amount of water at both the wilting coefficient and the field capacity, and a correspondingly larger proportion of soil water is held by forces greater than 15 atmospheres. Soils with predominantly larger skeletal particles have less surface area and larger interstices between particles. More of the soil water is held loosely and is thus available to plants, but such soils have lower field capacities. Plants can obtain the most water from soils having a variety of particle sizes between sand and clay (Figure 5-3).

Life processes occur optimally within a narrow range of temperature.

Life processes, as we know them, are restricted to the temperatures at which water is liquid: 0 to 100°C at the earth's surface. Relatively few organisms can survive body temperatures above 45°C. Some bacteria occur in hot springs close to the boiling point of water, and the photosynthetic cyanobacteria tolerate temperatures as hot as 75°C (Brock 1970, 1985; Brock and Darland 1970). The properties that permit existence at high temperatures are not well understood. Compared to most bacteria, the proteins of thermophilic bacteria have subtly different proportions of amino acids; as a result, the structure of these proteins remains stable at temperatures up to 95°C (Singleton and Amelunxen 1973, Hochachka and Somero 1984).

While temperatures on the earth rarely exceed 50°C, except in hot springs and at the soil surface in hot deserts, temperatures below the freezing point of water are common over large portions of the surface. When living cells freeze, the crystal structure of ice disrupts most life processes and may damage delicate cell structures, leading rapidly to death. Many species successfully cope with freezing temperatures either by maintaining their body temperatures above the freezing point of water, or by activating mechanisms to resist freezing or to tolerate its effect (Somme 1964, Baust 1973). The freezing point of water may be depressed by dissolved substances that interfere with the formation of ice. For example, the freezing point of seawater, which contains about three-and-a-half per cent dissolved salts, is −1.9°C.

The blood and body tissues of most vertebrates contain less than half the salt content of seawater and thus may freeze at a higher temperature than the freezing point of the ocean. Saltier blood could enable vertebrates to live in polar seas, but protein structure and function are too sensitive to salt concentration to make this a practical solution. Many organisms reduce the freezing points of their body fluids with large quantities (up to 30 per cent in some terrestrial invertebrates) of glycerol and glycoproteins, which act like antifreeze; their presence in the blood and tissues allows antarctic fish, for example, to remain active in seawater that is colder than the normal freezing point of the blood of fish in temperate or tropical seas (DeVries 1980, 1982).

Supercooling provides a second solution to the problem of freezing. Under certain circumstances, fluids can fall below the freezing point without ice crystals forming. Ice generally forms around an object, called a seed, which can be a small ice crystal or some other particle. In the absence of seeds, pure water may be cooled more than 20°C below its melting point. Supercooling has been recorded to −8°C in reptiles (Lowe et al. 1971) and −18°C in invertebrates (Baust and Morrissey 1975).

Finally, some organisms can tolerate freezing of most or all of the water in their bodies. Such organisms employ some unknown mechanism to restrict ice formation to the spaces between cells rather than within them; hence ice does not destroy cell structure (Kanwisher 1959). But because salts are excluded from ice and are therefore concentrated in the liquid water within cells, freezing-tolerant organisms must also cope with extremely high salt levels in their tissues during the winter.

Temperature has several opposing effects on life processes. First, heat increases the kinetic energy of molecules and thereby accelerates chemical reactions; the rate of biological processes commonly increases between two and four times for each 10°C rise in temperature throughout the physiological range (Schmidt-Nielson 1983; Hochachka and Somero 1973, 1984). Second, enzymes and other proteins become less stable and may not function properly or retain their structure at high temperatures. Third, the level of heat energy in the cell influences the conformations of proteins, which are balanced between the natural kinetic motions induced by heat and the forces of chemical attraction between different parts of the molecule. The physical properties of fats, which are important components of cell membranes and are accumulated by many animals as a reserve of food energy, also depend on temperature. When too cold, fats become stiff like the fat on a piece of meat taken from the refrigerator; when warm, they become fluid (Singer and Nicholson 1972, Sinensky 1974).

Enzymes function well only when they assume the proper shape. Too hot, the molecule may open its structure and tend to unfold; too cold, the molecule may close up so that substrates do not fit properly to active sites (Hochachka and Somero 1984). The combination of these factors results in an optimum temperature range for the occurrence of biological systems. The structures of enzymes and other molecules enable them to function best within the normal range of body temperature of the organism.

Most biological energy transformations are based on the chemistry of carbon and oxygen.

Plants and animals consist of many elements joined together into organic molecules that form the structure of the individual. Such organic compounds also provide the energy needed to maintain the organism in the form of chemical bonds between atoms and molecules. These energy-containing bonds arise from chemical changes in the atoms of various elements. In biological systems, one of the most prevalent of these transformations is the chemical reduction of carbon, accomplished when electrons are added to the atom. An oxidized form of carbon is carbon dioxide (CO_2). During photosynthesis, plants reduce the carbon atom in carbon dioxide. This altered atom forms new compounds, such as the carbohydrate glucose ($C_6H_{12}O_6$), within which its energy level is greatly increased. The added energy comes from light, of course. To release this stored energy for other purposes, both plants and animals undo the results of photosynthesis by oxidizing carbon back to carbon dioxide. During this transformation energy is released, a portion of which organisms harness; the rest escapes as heat.

Photosynthesis and respiration involve the complementary reduction and oxidation of carbon and oxygen. Oxygen's common oxidized state is molecular oxygen (O_2), which occurs as a gas in the atmosphere and dissolved in water. In a reduced state, oxygen readily forms water molecules (H_2O). Thus, as carbon is reduced during photosynthesis, oxygen is oxidized from its form in water to its molecular form. During respiration, inhaled or absorbed oxygen is reduced to its form in water as carbon is oxidized to its form in carbon dioxide. Why, then, does the coupling of an oxidation reaction to a reduction reaction result in a net release of energy? Because the reduction of oxygen is thermodynamically more favorable (requires less energy input) than the reduction of carbon, the oxidation of carbon releases more energy than the reduction of oxygen requires. (This is why oxygen is such a good oxidizer.)

Plants reduce more carbon than they oxidize (otherwise they would not grow), and they therefore require an external source of carbon. The only practical source of inorganic carbon, carbon dioxide, has an extremely low concentration in the atmosphere (about 0.03 per cent, or a partial pressure at sea level of 0.2 torr). As a result, gradients of CO_2 concentration between the atmosphere and the interior of plant cells are very low, certainly much lower than gradients of water vapor pressure between the plant and the surrounding atmosphere. This creates special problems for water conservation by plants, especially in arid environments, and accounts for the fact that plants transpire 500 grams of water, more or less, for every gram of carbon assimilated.

Carbon availability poses less of a problem for aquatic plants than for terrestrial plants because of the high solubility of carbon dioxide in water. When carbon dioxide dissolves, some of the molecules react with water to form carbonic acid (H_2CO_3) and associated compounds, which provide a reservoir of inorganic carbon. The concentration of carbon dioxide in the atmosphere is about 0.0003 $cm^3\ cm^{-3}$; its solubility in fresh water and under ideal conditions is nearly the same, about 0.0003 $cm^3\ cm^{-3}$. Depending on the acidity of the water, carbonic acid molecules dissociate into bicarbonate (HCO_3^-) and carbonate (CO_3^{2-}) ions; within the range of acidity of most natural waters (pH 6–9) bicarbonate ion is the most common form (Hutchinson 1957, Stumm and Morgan 1981). Bicarbonate ion dissolves readily in water (69 g of $NaHCO_3$ per liter of water, for example). As a result, seawater normally contains concentrations of bicarbonate ion equivalent to 0.03 to 0.06 cm^3 of carbon dioxide gas per cm^3 of water, over 100 times the concentration of the dissolved gas (Nicol 1967).

Whereas carbon dioxide poses difficulties for plants in terrestrial environments, it is oxygen that limits animals in aquatic habitats. Compared to its concentration of 0.21 $cm^3\ cm^{-3}$ in the atmosphere, the maximum solubility of oxygen (at 0°C in fresh water) is 0.01 $cm^3\ cm^{-3}$—only 14 parts per million (ppm) by weight. Furthermore, below the limit of light penetration in deep bodies of water and in waterlogged sediments and soils, where aquatic plants are absent, and no oxygen is produced by photosynthesis, animal and microbial respiration may severely deplete dissolved oxygen. Deeper layers of water in lakes and the mucky sediments of marshes frequently become totally deprived of oxygen (anaerobic, anoxic). Similar conditions in waterlogged soils of swamps pose problems for terrestrial plants, whose roots need oxygen for respiration just as animals do.

Life requires inorganic nutrients.

Organisms assimilate a wide variety of chemical elements. After hydrogen, carbon, and oxygen, the elements required in greatest amount are nitrogen, phosphorus, sulfur, potassium, calcium, magnesium, and iron. Their primary functions are summarized in Table 5–2. Many other elements, such as boron and selenium, are known to be required in smaller quantity even though their physiological functions are not well understood (Treshow 1970). As techniques for measuring quantities of elements in the parts-per-million and parts-per-billion ranges develop, the importance of these trace elements to ecological relationships of plants and animals will emerge more clearly.

Plants acquire mineral nutrients other than oxygen, carbon, and some nitrogen from water. They obtain nitrogen in the form of ammonia ion (NH_4^+) or nitrate ion (NO_3^-), phosphorus in the form of phosphate ion (PO_4^{3-}), calcium and potassium in the form of their elemental ions (Ca^{2+}, K^+), and so on. The availability of each of these elements varies with their chemical form in the soil, and with the temperature, acidity, and presence of other ions in the soil water. A comparison of the amounts of elements in the soil and accumulated by vegetation suggests that certain elements are likely to be scarcer than others (Table 5–3). Phosphorus, in particular, often limits plant production because even when abundant most of the compounds it forms in the soil do not dissolve easily.

All natural waters contain some dissolved substances. Although nearly pure, rainwater acquires some dissolved minerals from dust particles and droplets of ocean spray in the atmosphere (Ingham 1953, Likens et al. 1977). Most lakes and rivers contain 0.01 to 0.02 per cent dissolved minerals and roughly $\frac{1}{20}$ to $\frac{1}{40}$ the average salt concentration of the oceans (3.4 per cent), in which salts and other minerals have accumulated over the millennia (Hutchinson 1957).

Dissolved minerals in freshwater and saltwater differ in composition as well as in quantity (Table 5–4). Seawater abounds in sodium and chlorine, with significant amounts of magnesium and sulfate. Fresh water contains a

Table 5–2
Major nutrients required by organisms, with some of their primary functions

Element	Function
Nitrogen (N)	Structural component of proteins and nucleic acids
Phosphorus (P)	Structural component of nucleic acids, phospholipids, and bone
Sulfur (S)	Structural component of many proteins
Potassium (K)	Major solute in animal cells
Calcium (Ca)	Structural component of bone and of material between woody plant cells; regulator of cell permeability
Magnesium (Mg)	Structural component of chlorophyll; involved in function of many enzymes
Iron (Fe)	Structural component of hemoglobin and many enzymes
Sodium (Na)	Major solute in extracellular fluids of animals

Table 5-3
Typical concentrations of elements in soils and annual uptake by plants

Element	Soil content (weight %)*	Annual plant uptake (kg ha^{-1} yr^{-1})	Soil content/ annual plant uptake (years)†
Silicon (Si)	33	20	21,000
Aluminum (Al)	7	0.5	180,000
Iron (Fe)	4	1	52,000
Calcium (Ca)	1	50	260
Potassium (K)	1	30	430
Sodium (Na)	0.7	2	4600
Magnesium (Mg)	0.6	4	2000
Titanium (Ti)	0.5	0.08	62,000
Nitrogen (N)	0.1	30	40
Phosphorus (P)	0.08	7	150
Manganese (Mn)	0.08	1	1000
Sulfur (S)	0.05	2	320
Fluorine (F)	0.02	0.01	26,000
Chlorine (Cl)	0.01	0.06	220
Zinc (Zn)	0.005	0.01	6500
Copper (Cu)	0.002	0.006	4200
Boron (B)	0.001	0.03	400
Molybdenum (Mo)	0.0003	0.0003	13,000
Selenium (Se)	0.0000001	0.0003	40

* Carbon, oxygen, hydrogen, and some additional trace elements make up the remaining percentage.
† Soil content (g m^{-2}) divided by annual uptake (g m^{-2} yr^{-1}) yields a ratio, the time to soil depletion in the absence of replenishment, in years.
Source: Bohn et al. 1979.

more even distribution of diverse ions, but calcium is usually the most abundant cation (an ion carrying a positive charge) and carbonate and sulfate the most abundant anions (those carrying a negative charge). The composition of freshwater and saltwater differs owing to the different rates

Table 5-4
Percentage composition of dissolved minerals in rivers (freshwater), in seawater, and in the blood plasma and cells of frogs

Mineral ion	Delaware River	Rio Grande River	Seawater	Frog plasma	Frog cells
Sodium (Na^+)	6.7	14.8	30.4	35.4	1.3
Potassium (K^+)	1.5	0.9	1.1	1.3	77.7
Calcium (Ca^{2+})	17.5	13.7	1.2	1.2	3.1
Magnesium (Mg^{2+})	4.8	3.0	3.7	0.4	5.3
Chlorine (Cl^-)	4.2	21.7	55.2	39.0	0.8
Sulfate (SO_4^{2-})	17.5	30.1	7.7	—	—
Carbonate (CO_3^{2-})	33.0	11.6	0.4	22.7	11.7

Note: The percentages of the negatively charged ions (anions) exceed those of the positively charged ioins (cations) because, ion for ion, anions are much the heavier; the numbers of positive and negative ions are approximately equal. The sums of all columns do not equal 100 because all dissolved substances are not included.
Source: Reid 1961, Gordon 1968.

of solution and solubilities of substances. Few compounds reach their maximum solubilities in fresh water; their concentrations reflect the composition and rates of solution of materials in the rock and soil that the water contacts. Limestone consists primarily of calcium carbonate, which dissolves quickly; thus, water in limestone areas contains abundant calcium ion, making it "hard." Granite contains such minerals as quartz and feldspar, which do not contain calcium and which dissolve slowly; water flowing through granitic areas contains few dissolved substances and is "soft."

The oceans are like large stills, concentrating minerals as pure water evaporates from the surface and nutrient-laden water arrives via streams and rivers. Here the concentrations of some minerals, particularly calcium carbonate, are limited by their maximum solubilities. Calcium carbonate dissolves only to the extent of 0.000014 grams per gram (ca. 1 cm^3) of water. Its concentration in the oceans reached this level eons ago, and the excess calcium ion entering oceans each year from streams and rivers precipitates to form limestone sediments. At the other extreme, the solubility of sodium chloride (0.36 g per g of water) far exceeds its concentration in seawater; most of the sodium chloride washing into ocean basins remains dissolved.

Substances dissolved in the aqueous environment pose osmotic problems.

Organisms obtain nutrients from the soil, water, or their food. Often these are much more concentrated in the tissues than in the surroundings, and organisms must therefore assimilate these nutrients against the prevailing gradient. But organisms also must exclude from their bodies many abundant substances in the environment that are metabolically useless or even toxic at high concentration. When a surface permits the flux of desirable substances, it can keep out others only by selective permeability (if wanted and unwanted substances differ greatly in size or electrical charge) or by actively pumping unwanted substances out across the surface.

Left to their own devices, ions diffuse across the surfaces of organisms from regions of high to low concentration, thereby equalizing their concentrations. Water also moves across permeable membranes (the process is called osmosis) toward regions of high ion concentrations (that is, low water concentration), tending to equalize concentrations of dissolved substances on both sides of the membrane. This tendency of a solution to attract water is known as its osmotic potential. The osmotic potential of seawater is high, that of freshwater is low, and that of the body fluids of vertebrate animals is intermediate. Gradients of osmotic potential pose different problems for freshwater fish, which tend to gain water and lose solutes, and saltwater fish, which tend to gain salt and lose water. Most organisms solve their osmotic problems by pumping ions in one direction or the other, expending considerable energy in the process, across various body surfaces (skin, kidney tubules, and gills).

Certain environments pose special salt and osmotic problems. Aquatic environments with salt concentrations greater than that of seawater occur in some landlocked basins, particularly in arid zones where evaporation is great. The Great Salt Lake (20 per cent salt) in Utah and the Dead Sea (23 per cent salt), lying between Israel and Jordan, are well-known examples. The osmotic potential of such environments would suck the water from most animals and plants; but a few aquatic creatures, such as brine shrimp *(Artemia),* can survive in saltwater concentrated to the point of crystallization (300 g per liter, or 30 per cent). Brine shrimp excrete salt at a prodigious rate to maintain their body fluids hypotonic to (less concentrated than) their surroundings (Croghan 1958, Gilles 1975).

The small copepod *Tigriopus* (Figure 5–4) lives in pools high in the splash zone along rocky coasts. The pools receive fresh seawater infrequently and, as the water evaporates, the salt concentration rises to high levels. Unlike *Artemia, Tigriopus* solves its water loss problem by increasing the osmotic potential of its body fluids. It accomplishes this by synthesizing certain amino acids abundantly (Burton and Feldman 1982). These small molecules increase the osmotic potential of the body to match that of the habitat without the deleterious physiological effects of high levels of salt (Yancey et al. 1982).

Terrestrial plants living at the edge of the sea have special salt problems. As we have seen, plants transpire hundreds of grams of water for every gram of dry-matter production. Salts in the water that move from the roots to the leaves stay behind in the leaves as water evaporates from their surfaces. Although this poses relatively little problem for plants when the source of water in the soil is fresh, mangrove trees and salt marsh grasses

Figure 5–4
The copepod *Tigriopus californicus.* Adults are 1–2 mm long. (Courtesy of R. S. Burton.)

Figure 5–5
Left: The roots of mangrove vegetation immersed in saltwater. Some species excrete excess salt from their roots. Specialized glands in the leaves of the button mangrove *Conocarpus erecta (right),* excrete salt, which precipitates on their outer surfaces.

actively excrete most of the salt they take in with water (Waisel 1972, Tomlinson 1986) (Figure 5–5).

Light energy is the ultimate driving force of life processes.

Light is the primary source of energy for the ecosystem. Green plants absorb light and assimilate its energy by photosynthesis. But not all light striking the earth's surface is useful in photosynthesis. Rainbows and prisms show that light consists of a spectrum of wave lengths that we perceive as different colors. Wave lengths of light are generally expressed in micrometers (μm) (one-millionth of a meter [10^{-6} m]) or nanometers (nm) (one-billionth of a meter [10^{-9} m]). The visible spectrum extends between wave lengths of about 400 nm (violet) and 700 nm (red). The energy content of light varies with wave length and hence with color; short wave length blue light has a higher energy level than longer wave length red light.

Light that reaches the upper part of the earth's atmosphere from the sun extends far beyond the visible range: through the ultraviolet region

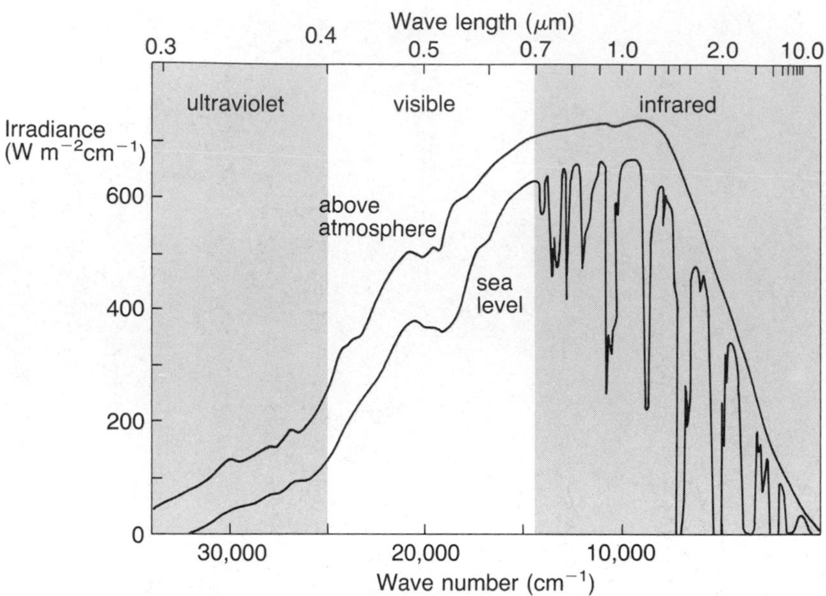

Figure 5-6
Spectral distribution of direct sunlight above the atmosphere and at sea level. The difference between the two, which is proportionately greatest in the ultraviolet region of the spectrum, represents light absorbed by the atmosphere. Wave number is the inverse of the wave length in centimeters and represents the number of oscillations a wave goes through while traveling 1 centimeter. (After Gates 1980.)

toward the short wave length, high-energy X-rays at one end of the spectrum, and through the infrared region to such extremely long wave length, low-energy radiation as radio waves at the other end of the spectrum (Gates 1980). Because of its high energy level, ultraviolet light can damage exposed cells and tissues. As light passes through the atmosphere, however, most of its ultraviolet components are absorbed, primarily by a molecular form of oxygen known as ozone (O_3), which occurs in the upper atmosphere. The atmosphere thus shields life at the earth's surface from the most damaging wave lengths of light (Figure 5-6).

Vision and the photochemical conversion of light energy to chemical energy by plants occur primarily within that portion of the solar spectrum at the earth's surface containing the greatest amount of energy. Absorption of radiant energy depends on the nature of the absorbing substance. Water only weakly absorbs light whose characteristic wave lengths fall in the visible region of the spectrum of energies; as a result, a glass of water appears colorless. Dyes and pigments are strong light absorbers of some wave lengths in the visible region and reflect or transmit light of definite colors that become identifying characteristics. Plant leaves contain several kinds of pigments, particularly chlorophylls (green) and carotenoids (yellow), which absorb light and harness its energy. Carotenoids, which give carrots their orange color, absorb primarily blue and green (Figure 5-7) and reflect light in the yellow and orange regions of the spectrum. Chlorophyll

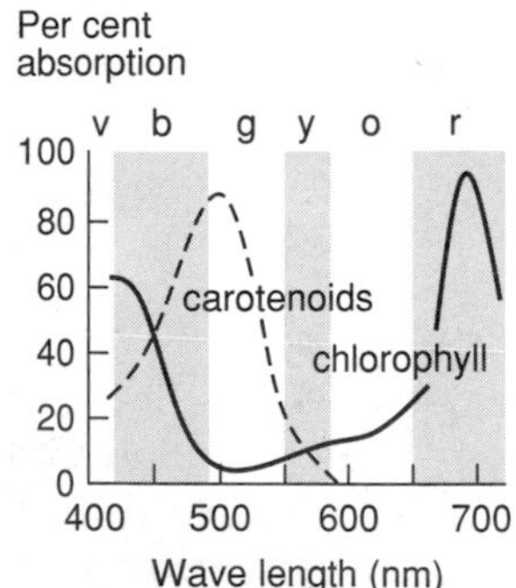

Figure 5-7
Absorption of light of different wave lengths by two groups of plant pigments —chlorophylls and carotenoids—that capture light energy for photosynthesis. The colors of the spectrum are violet (v), blue (b), green (g), yellow (y), orange (o), and red (r). (After Emerson and Lewis 1942.)

absorbs red and violet light while reflecting green and blue. When chlorophylls and carotenoids occur together in a plant, green light is absorbed least, hence the leaves appear green to us.

The attenuation of light in water limits photosynthesis in aquatic environments.

Water absorbs or scatters enough light to limit the depth of the sunlit zone of the sea. The transparency of a glass of water is deceptive. In pure seawater, the energy content of light in the visible part of the spectrum diminishes to 50 per cent of its surface value within 10 meters depth, and to less than 7 per cent within 100 meters depth (Sverdrup 1945). Furthermore, water absorbs longer wave lengths more strongly than shorter ones; virtually all infrared radiation disappears within the topmost meter of water (Weisskopf 1968). Short waves of light (violet and blue) tend to be scattered by water molecules and thus also do not penetrate deeply. As a consequence of the absorption and scattering of light by water green light tends to predominate with increasing depth. The photosynthetic pigments of aquatic plants are adapted to this spectral shift. Plants near the surface of the oceans, such as the green alga *Ulva* (sea lettuce), have pigments resembling those of terrestrial plants and best absorb light of blue and red color. The deep-water red alga *Porphyra* has additional pigments that enable it to utilize green light more effectively in photosynthesis (Figure 5–8).

Because photosynthesis requires light, the depth at which plants exist in the oceans is limited by the penetration of light to a fairly narrow zone close to the surface, in which photosynthesis exceeds plant respiration. This range of depths is called the euphotic zone. The lower limit of the euphotic zone, where photosynthesis just balances respiration, is called the compensation point. It may be defined by either depth or light level. When algae in the phytoplankton sink below the compensation point or are carried below it by currents, and do not soon return to the surface on upwelling currents, they die.

In some exceptionally clear ocean and lake waters, the compensation point may lie 100 meters below the surface. But this is a rare condition. In productive waters with dense phytoplankton, or in water turbid with suspended silt particles, the euphotic zone may be as shallow as 1 meter. In some polluted rivers, little light penetrates beyond a few centimeters.

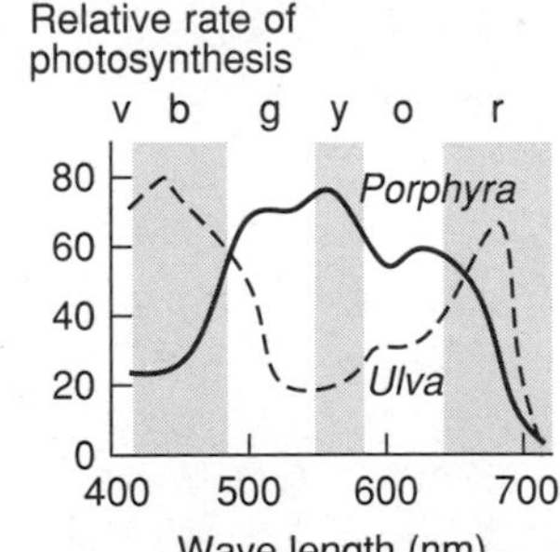

Figure 5–8
Relative rates of photosynthesis by the green alga *Ulva* and the red alga *Porphyra* as a function of the color of light. (After Haxo and Blinks 1950.)

Radiation, conduction, and convection define the thermal environments of terrestrial organisms.

Much of the solar radiation absorbed by objects is converted to heat. The earth warms each day and cools by night. As the days lengthen and the sun rises higher in the sky toward summer, the surroundings become warmer; more heat is added each day than is lost. The warmth of the sun heating the atmosphere drives the winds. Light absorbed by water provides the major source of heat for evaporation.

Each object and each organism continually exchanges heat with its environment (Gates 1980, Porter and Gates 1969, Tracy 1982). When the temperature of the environment exceeds that of the organism, the organism gains heat and becomes warmer. When the environment is cooler, the organism loses heat and cools. The heat budget of an organism includes avenues of heat gain and avenues of heat loss (Figure 5–9). When temperature reaches an equilibrium, gains equal losses. When gains exceed losses, energy is stored or accumulated in the body and temperature rises. When losses exceed gains, temperature drops.

The major categories of heat transfer between an organism and its environment may be described as follows:

Radiation (gain or loss) is the absorption or emission of electromagnetic radiation. Sources of radiation in the environment are the sun, the sky (scattered light), and the landscape, including vegetation. At night, although we cannot see the infrared radiation, objects that have warmed up in the sunlight radiate their stored heat to colder parts of the environment and, eventually, to space. The bodies of organisms, especially warm-blooded birds and mammals, often are the brightest objects in the night. Because radiation increases with the fourth power of thermodynamic temperature (°K), we radiate tremendous quantities of energy to the clear, black night sky. We can also receive radiation from atmospheric water vapor and from vegetation, which balances much of our radiation loss.

Conduction (gain or loss) is the transfer of the kinetic energy of heat between substances in contact. Thermal conductance (k) is expressed in

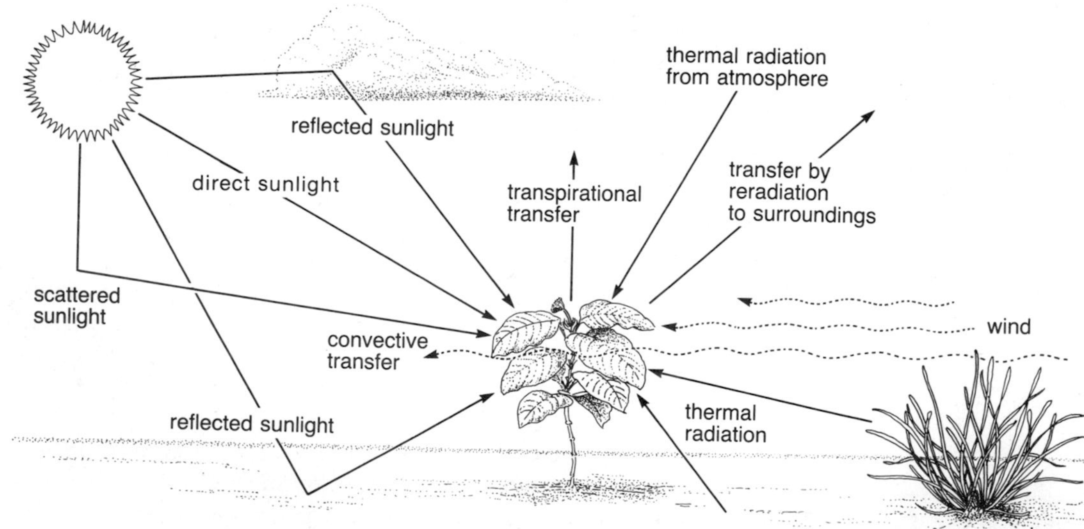

Figure 5–9
Pathways of exchange of energy between a plant and its environment by radiation, convection, and transpiration. (After Gates 1980.)

watts (joules per second), normalized by the cross-sectional area (cm^2), the inverse of the distance traversed (cm^{-1}), and the temperature gradient (°C). Thus, the units of conductance are W cm^{-2} cm $°C^{-1}$, or W cm^{-1} $°C^{-1}$. No heat is conducted to a vacuum ($k = 0$); water ($k = 0.006$) conducts heat better than air (0.00026), owing to its greater density. Some metals, such as silver (4.3) and copper (4.0), conduct heat very rapidly. The rate of conductance between two objects, or between the inside and the outside of an organism, depends on insulative properties of the surface (the resistance to heat transfer, a function of k and surface thickness), surface area, and temperature gradient. An organism may either gain or lose heat depending on its temperature relative to that of the environment.

Convection (gain or loss) is the movement of liquids and gases of different temperatures, particularly over surfaces across which heat is transferred by conduction. Air conducts heat poorly. In still air, a boundary layer of air forms over a surface. A warm body tends to warm this boundary layer to its own temperature, effectively insulating itself against heat loss. A current of air flowing past a surface tends to disrupt the boundary layer and increase the rate of heat exchange by conduction. This convection of heat away from the body surface is the basis of the "wind-chill factor" heard on the evening weather report.

Evaporation (loss) requires heat. The evaporation of 1 gram of water from the body surface removes 2.43 kJ of heat at 30°C. As plants and animals exchange gases with the environment, some water evaporates from respiratory surfaces. Rate of evaporative heat loss depends on the amount of water exposed on the surface of the organism, the relative temperatures of the surface and of the air, and the vapor pressure of the atmosphere. Like heat, moisture can be trapped in the boundary layer of air that forms around bodies, and so convection increases evaporative as well as conductive heat loss. Warm air holds more water than cold air (51 g m^{-3} at 40°C, 17.3 g m^{-3} at 20°C, and 4.8 g m^{-3} at 0°C); hence, it has greater potential for evaporating water than does cold air. Where water is plentiful in hot climates, animals evaporate water from their skins and respiratory surfaces to cool themselves. For warm-blooded animals in cold climates, evaporation can become an unavoidable problem as cold air containing little water is warmed in contact with the body surface. We visualize such water loss on winter days when water evaporated from the warm surfaces of the lungs condenses as our breath mingles with the cold atmosphere.

Aquatic and terrestrial environments present different opportunities and constraints.

Life arose in the sea. Conditions in shallow coastal waters were ideal for the development and diversification of the first plants and animals. Temperature and salinity varied little; sunlight, dissolved gases, and minerals were abundant. Water itself is buoyant and supports both delicate structures and massive bodies with equal ease.

Several hundred million years elapsed between the time modern life began to flourish in the sea and the appearance of life on land. Yet in spite of the harshness of some terrestrial environments, life has generally attained a higher degree of organic diversity and productivity on the land.

To appreciate fully the distinction between aquatic and terrestrial environments, we should contrast the properties of water and air rather than those of water and earth. The density of water (about 800 times that of air) and its ability to dissolve gases and minerals largely determine the form and functioning of aquatic organisms. Water provides a complete medium for life. In contrast, both the atmosphere and the land make essential contributions to the environment of terrestrial life: air provides oxygen for respiration and carbon dioxide for photosynthesis, while soil is the source of water and minerals. Air offers less resistance to motion than does water, and thus constrains movement less, but it also offers less support against the pull of gravity.

Water and air differ in buoyancy and viscosity.

Because water is dense, it provides considerable support for organisms that, after all, are themselves mostly water. But organisms also contain bone, proteins, dissolved salts, and other materials that are more dense than salt or freshwater. These would cause organisms to sink were it not for a variety of mechanisms that reduce their density or retard their rate of sinking. Many fish have a swim bladder, a small gas-filled structure whose size can be adjusted to make the density of the body equal to that of the surrounding water (Denton 1960). Some large kelps, a type of seaweed found in shallow waters, have analogous gas-filled organs. The kelps are attached to the bottom by holdfasts, and gas-filled bulbs float their leaves to the sunlit surface waters.

Fats and oils have densities of between 0.90 and 0.93 g cm^{-2} (90 to 93 per cent of the density of pure water). Many microscopic, unicellular plants (phytoplankton), which float in great numbers in the surface waters of lakes and oceans, contain droplets of oil that compensate for the natural tendency of cells to sink (Gross and Zeuthen 1948, Steen 1970). Fish and other large marine organisms also accumulate lipids to provide buoyancy (Hochachka and Somero 1973: 304).

Aquatic organisms further lighten their bodies by reducing skeleton, musculature, and perhaps even the salt concentration of body fluids. It has been argued that aquatic vertebrates maintain low osmotic concentrations in their blood and body fluids (about one-third to one-half that of seawater) to reduce density.

The high viscosity of water lends a hand to some organisms that would otherwise sink more rapidly, but hampers the movement of others. Tiny marine animals often have long, filamentous appendages that retard sinking, just as a parachute slows the fall of a body through air. The wings of maple seeds, the spider's silk thread, and the tufts on dandelion and milk-

Figure 5–10
The streamlined shapes of young mackerel reduce the drag of water on the body and allow the fish to swim rapidly with minimum energy expenditure. (Courtesy of the U.S. Bureau of Commercial Fisheries.)

weed seeds provide a similar function and increase the dispersal range of terrestrial species. But to reduce the drag encountered in moving through a medium as dense and viscous as water, fast-moving animals must assume streamlined shapes. Mackerel and other swift fish of the open ocean closely approach the hydrodynamicist's body of ideal proportions (Figure 5–10). Of course, air offers far less resistance to movement, having less than 1/50th the viscosity of water. But the atmosphere provides little buoyancy. To provide lift against the pull of gravity, birds and other flying organisms expend prodigious energy. The mechanics and aerodynamics of animal flight, involving the conversion of the movement of the muscles and appendages to lift and forward thrust, are described by Pennycuick (1975).

Form and function change allometrically with body size.

The relationships of organisms to their environments change with the size of the organism owing to nonproportional scaling of various physical processes with respect to size (Calder 1984, Schmidt-Nielsen 1984). In ecology, relationships between rates of processes and dimensions of objects are often described by the allometric relationship

$$Y = aX^b$$

where Y is being compared to X and a and b are constants pertaining to the relationship. Y, for example, might be the frequency of the heartbeat and X the body mass. The constant b is referred to as the allometric constant. When b is 1, Y is directly proportional to X. When b is greater than 1, Y increases proportionately more rapidly than X, and so the ratio of Y to X increases with larger X. When b is less than 1, the ratio of Y to X decreases with larger X; when b is less than 0, the absolute value of Y decreases with larger X. Because the equation describing the allometric relationship is a power function, it is often transformed to its logarithmic form

$$\log(Y) = \log(a) + b\log(X)$$

which is an equation describing a straight line.

Figure 5–11 shows an example: the relationship between heart rate and body mass for mammals ranging over many orders of magnitude in size. The equation for the line that best fits the points has an allometric constant less than 0 ($b = -0.23$), in accordance with the slower heart rates of larger species. In contrast, the resting metabolic rates (RMR) of mammals increase with larger size, but less rapidly than mass itself; the allometric constant of the relationship between RMR and body mass is 0.73 (Figure 5–12). Because allometric constants usually differ from 1, the properties of organisms of different sizes are not directly comparable. Thus, the life span of an albatross may be 10 times that of a sparrow, but this difference is related in part to physiological consequences of size over and above differences in their environments or adaptations. But environment is important, too—albatrosses live longer than geese of similar size—and its effect appears as variation above and below the allometric line of relationship for all species together.

The allometry of certain relationships arises from simple physical or geometrical considerations. For example, the volume of a sphere increases in proportion to the cube of its diameter, but surface area increases in proportion to only the square of the diameter. Hence the relationship be-

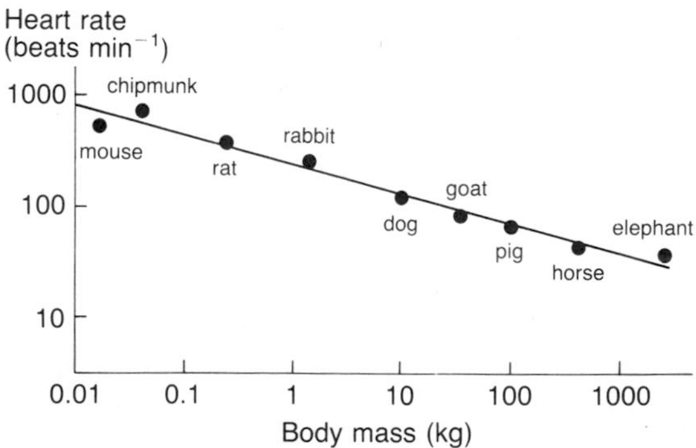

Figure 5–11
Allometric relationship between heart rate and body mass in a variety of mammals ranging in size from mouse to elephant. (Data from Altman and Dittmer 1964.)

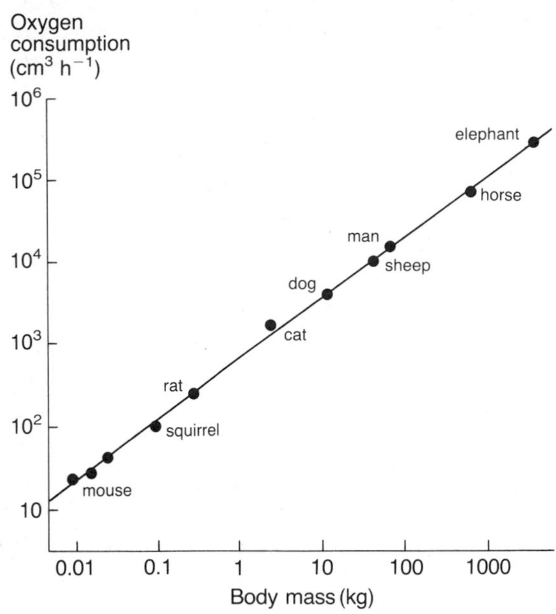

Figure 5–12
Allometric relationship between resting metabolic rate and body mass in mammals. (Data from Altman and Dittmer 1964.)

tween surface (S) and volume (V) has an allometric constant of $\frac{2}{3}$, or

$$S = aV^{0.67}$$

Therefore, larger organisms have relatively smaller surfaces compared to the bulk of their bodies. A sphere of radius r has a surface area of $4\pi r^2$ and a volume of $4\pi r^3/3$; the ratio of surface to volume is $3/r$. Thus a sphere having a diameter of 1 millimeter, comparable in size to a small water flea, has 1000 times as much surface per unit of volume as a bear-sized sphere 1 meter in diameter. This makes the uptake of oxygen relatively easy for the smaller organism but heat and water loss pose severe problems. Larger organisms elaborate larger interior surfaces (lungs) to obtain enough oxygen, and employ circulatory systems to move things around inside; conversely, heat and water conservation pose less severe problems. In general, because large animals have great bulk per unit of surface compared to smaller animals, they have greater capacity to resist change imposed by changes in the environment.

Size changes everything in ecology. In aquatic habitats, drag on the body depends on size and speed. Whereas a whale can coast on its momentum, a copepod stops dead in the water as soon as its power stroke ends; its momentum isn't sufficient to break through the viscosity of water (Vogel 1981, 1988). On land, size scales the movement of appendages and greatly affects locomotion (Schmidt-Nielsen 1984, McMahon 1984). All these considerations influence the manner in which plants and animals adapt to the conditions of the environment.

SUMMARY

1. Water is the basic medium of life. It has the density of solids and the fluidity of gases. It is liquid within the range of temperatures encountered over most of the earth. It has an immense capacity to dissolve inorganic compounds. These properties, and its abundance at the earth's surface, make water an ideal medium for living systems.

2. Water is scarce in some terrestrial environments. Because water clings tightly to soil surfaces, it is less readily available in clay soils where small particle size results in larger surface area.

3. Most kinds of organisms cannot survive temperatures greater than 45°C, but thermophilic bacteria grow in hot springs at up to 95°C; they apparently tolerate such temperatures by increasing the intramolecular forces that hold proteins together.

4. Organisms in cold environments withstand freezing temperatures by metabolically maintaining elevated body temperatures; by lowering the freezing point of their body fluids with salts, glycerol, or glycoproteins; by supercooling their body fluids; or by tolerating freezing.

5. Biological energy transformations are based mostly on the chemistry of carbon and oxygen. Energy is assimilated during photosynthesis as carbon is reduced and, in a coupled reaction, oxygen is oxidized from its form in water to molecular oxygen. The energy stored in carbohydrates is released by the oxidation of carbon (respiration).

6. Carbon dioxide is scarce in the atmosphere (0.03 per cent) but is more abundantly distributed in aquatic systems, where it forms soluble carbonates. Oxygen, abundant in the atmosphere, is relatively scarce in water, where its solubility and rate of diffusion are low, and may be depleted by bacterial respiration of organic matter (producing anoxic conditions) in deep, stagnant layers.

7. Organisms assimilate many elements necessary to life processes and biological structure. The availability of these elements relative to their requirement varies tremendously among habitats. Nitrogen and phosphorus often limit plant growth because of their scarcity relative to need.

8. In aquatic habitats, differences in the concentrations of dissolved salts establish osmotic gradients between the organism and its surroundings. Hyperosmotic organisms, which have greater salt concentrations, tend to lose salt and gain water; hypo-osmotic organisms gain salt and lose water. Organisms achieve osmotic balance by altering the osmolarity of body fluids and actively pumping salts across membranes.

9. All the energy for life ultimately comes from sunlight. Solar radiation varies over a spectrum of wave lengths. Plants extract energy primarily in the high-intensity, short wave length portion of the spectrum, which roughly coincides with visible light.

10. Light is attenuated by water. The depth of the euphotic zone, at the bottom of which photosynthesis balances respiration (the compensation point), varies from 100 meters in clear, unproductive waters, to a few tens of centimeters in turbid or polluted water.

11. The thermal environment of organisms, especially in terrestrial habitats, is determined by radiation, conduction, convection, and evaporation.

12. Water is more dense than air and provides more buoyancy; but it is also more dense and viscous and therefore impedes movement.

13. Many factors bearing on the ecology of organisms scale disproportionately with respect to overall body size. Such relationships are described by the slope (allometric constant) of the line relating the logarithm of a measurement of some structure or function to the logarithm of body size. Surface area scales to the 0.67 power of body size when shape is held constant; hence larger organisms have proportionately smaller surface-to-volume ratios. This and other allometric relationships have important consequences for organisms of different sizes.

6

Adaptation to the Physical Environment

The physical environment includes many factors important to the well-being of the organism. Some of these are consumed and transformed, others influence the internal environment. To accumulate resources while maintaining suitable internal conditions poses a major challenge. Plants and animals can control fluxes across their surfaces by seeking appropriate environments, by adjusting the internal environment with respect to conditions of the surroundings, and by modifying the quality and area of the body surface. We have seen how the number of pores in the shell of a bird's egg may be modified to satisfy the embryo's requirement for gas exchange, and in response to concentrations of oxygen, carbon dioxide, and water vapor in the surrounding air. Such modifications, which better suit the organism to its particular environment, are called adaptations.

Adaptations may arise in two ways. First, they may be evolved properties of the species, genetically inherited by each individual from its parents. Such evolutionary responses to new or changed environmental conditions result from many generations of natural selection. A second mechanism of adaptation is by behavioral, physiological, or developmental response of the individual. Both evolutionary and individual adaptation modify the structure or function of the organism so as to improve its relationship to the environment.

In this chapter, we shall explore a variety of mechanisms by which animals and plants adapt to several prominent environmental factors. A common theme is manipulation of the components of flux—surface area, gradient, and conductance— but this theme is manifested differently in adaptations to each type of environmental factor. Furthermore, the relationship of an organism to any one factor depends on its relationship to all others. This point is illustrated by the coupled problems of regulating salt and water balance, with which we shall begin our discussion.

Salt and water balance go hand in hand.

Left to their own devices, ions diffuse across cell membranes from regions of high to low concentration, thereby tending to equalize their concentrations. Water also moves across membranes (osmosis) toward regions of high ion concentrations, tending to dilute dissolved substances. Maintaining an ionic imbalance between the organism and the surrounding environment (osmoregulation) against the physical forces of diffusion and osmosis requires energy and often is accomplished by organs specialized for salt retention or excretion (Potts and Parry 1964).

Ion retention is critical to terrestrial and freshwater organisms (Burton 1973). Freshwater fish, for example, continually gain water by osmosis through the mouth and gills, which are the most permeable tissues exposed to the surroundings. (The skin is relatively impermeable.) To counter this influx, individuals continually eliminate water in the urine. But, of course, if fish do not also selectively retain dissolved ions, they would soon become lifeless bags of water. A fish's kidneys retain salts by actively removing ions from the urine and feeding them back into the bloodstream. In addition, its gills can selectively absorb ions from the surrounding water and secrete them into the bloodstream (Schmidt-Nielsen 1983). Terrestrial animals acquire mineral ions in the water they drink and the food they eat, although sodium deficiencies in some areas force animals to obtain salt directly from mineral sources—salt licks (Botkin et al. 1973, Hebert and Cowan 1971, Weeks and Kirkpatrick 1976); plants absorb ions dissolved in soil water.

Keeping ions out of the body poses as big a problem for saltwater fish as the retention of ions does for freshwater species. In contrast to their freshwater relatives, the gills and kidneys of saltwater fish actively excrete ions. Marine fish also drink seawater to replenish water lost in urine and by osmosis across the surfaces of the gills and, to a lesser extent, the skin.

Some sharks and rays have achieved a rather elegant solution to the problem of osmotic balance. Sharks retain urea [$CO(NH_2)_2$]—a common nitrogenous waste product of metabolism in vertebrates—in the bloodstream instead of excreting it in the urine. The urea raises the ionic concentration of the blood to the level of seawater without having to increase the concentration of sodium chloride. Urea does tend to destabilize proteins, but this problem is overcome by high blood levels of another compound, trimethylamine oxide [$NO(CH_3)_3$] (Yancey and Somero 1980, Yancey et al. 1982). Although sharks and rays must regulate the diffusion of specific ions in and out of their bodies, high levels of urea in the blood effectively cancel the tendency of water to leave by osmosis. And inasmuch as sharks do not have to drink water to replace osmotic losses of water, they also do not ingest large quantities of salt. The fact that the few freshwater sharks and rays do not accumulate urea in their blood (Thorson et al. 1967) emphasizes the osmoregulatory role of urea in marine species.

Salt excretion is not limited to fishes among marine vertebrates. Oceanic birds and reptiles ingest more salt in the food than their kidneys can excrete. Many species have additional regulatory organs known as salt glands, usually located near the eyes or nasal passages, which help to rid the body of excess salt while conserving water (Peaker and Linzell 1975).

Water and salt balance are intimately related in terrestrial as well as aquatic organisms. As we have seen before, plants inevitably take up salts along with the water that passes into their roots. In hypersaline environments, excess salts are excreted back into the soil by the roots, which therefore function as the plant's kidneys (Scholander 1968), or extruded by the leaves (Reimold and Queen 1974). Animals, especially meat eaters, obtain salts in their food in excess of their requirements. Where water is abundant, such animals can drink large quantities of water to flush out salts that would otherwise accumulate in the body. Where water is scarce, however, animals must produce a concentrated urine to conserve water. And so, as one would expect, desert animals have champion kidneys (Schmidt-Nielsen 1964). For example, whereas humans can concentrate salt ions in their urine to about 4 times the level in blood plasma, the kangaroo rat's kidneys produce urine with a salt concentration as high as 14 times that of its blood (Schmidt-Nielsen 1964), and the Australian hopping mouse, a desert-adapted species, 25 times that of its blood (MacMillen and Lee 1967).

Excretion of nitrogenous wastes presents terrestrial animals with special problems.

Most carnivores, regardless of whether they eat crustacea, fish, insects, or mammals, consume excess nitrogen, as well as excess salts, in their diets. This nitrogen, ingested in the form of proteins and nucleic acids, must be eliminated from the body when these compounds are metabolized (Campbell 1970, Schmidt-Nielsen 1983). Animals lack the biochemical mechanisms possessed by some microorganisms for producing molecular nitrogen (N_2) and, consequently, they cannot dispose of nitrogen as a gas that would escape from the blood through the lungs or gills. Many oxidized, inorganic forms of nitrogen — nitrates, for one — are highly poisonous and cannot be produced in quantity without toxic effects. To solve the problem of nitrogen excretion, most aquatic organisms produce the simple metabolic by-product, ammonia (NH_3). Although ammonia is mildly poisonous to tissues, aquatic organisms can eliminate it rapidly in a copious, dilute urine before it reaches a dangerous concentration.

Terrestrial animals cannot afford to use large quantities of water to excrete nitrogen. To circumvent this problem, they produce protein metabolites that are less toxic than ammonia and therefore can be concentrated in the blood and urine without dangerous effect (Hochachka and Somero 1973). In mammals, this waste product is urea, the same substance produced and retained by sharks and rays to achieve osmotic balance in marine environments (Mommsen and Walsh 1989). Because urea dissolves in water, its excretion requires some urinary water loss, the amount depending on the concentrating power of the kidneys. In birds and reptiles, adaptation to terrestrial life has been carried one step further: they excrete nitrogen in the form of a double-ringed compound, uric acid ($C_5H_4N_4O_3$). Uric acid bestows the distinct advantage in desert environments of being crystallized out of solution and thereby greatly concentrated in the urine. Uric acid excretion enhances water conservation by birds and reptiles, but it is perhaps most clearly adapted to nitrogen excretion by the embryos of birds and reptiles (Needham 1931). Because no liquid water leaves the egg across the shell, nitrogenous waste produced by the embryos must be retained within the egg. These are sequestered harmlessly as uric acid crystals in special membranes external to the embryo, which are discarded at hatching (Clark and Fisher 1957).

For terrestrial organisms, conservation of water becomes more difficult with increasing temperature.

Avoiding heat stress is critical to an individual's survival (Louw and Seely 1982). Warm-blooded animals maintain their body temperatures only 6°C, or thereabouts, below the upper lethal maximum. In cool environments, plants and animals dissipate excess heat, generated by activity or absorbed from the sun, by conduction and radiation to their surroundings. But when air temperature approaches or exceeds body temperature, individuals can dissipate heat only by evaporating water from their skin and respiratory surfaces. In deserts, scarce water makes evaporative heat loss a costly mechanism; in hot, humid climates, the high water vapor pressure of the air slows evaporation from body surfaces and reduces the effectiveness of evaporative heat loss. In either case, to balance heat budgets in hot environments, organisms are often forced to reduce heat-generating activity, use cool microclimates, or undertake seasonal migrations to cooler regions (Figure 6–1); many desert plants orient their leaves to avoid the direct rays of the sun (Ehleringer and Forseth 1980), or they may shed their leaves and become inactive during periods of combined heat and water stress.

Even in the absence of heat stress, hot, dry environments rob the body of water simply because the vapor pressure of water at body surfaces increases with temperature, roughly doubling with each 10°C rise. Where water is scarce, one often sees adaptations to reduce water loss. Among mammals, the kangaroo rat is well adapted to life in a nearly waterless environment (Schmidt-Nielsen and Schmidt-Nielsen 1952, 1953). The large intestine of the kangaroo rat's digestive tract resorbs water from waste material so efficiently as to produce virtually dry feces.

Kangaroo rats recover much of the water that evaporates from the lungs by condensation in enlarged nasal passages (Schmidt-Nielsen et al. 1970). When the kangaroo rat inhales dry air, moisture in its nasal passages evaporates, cooling the nose and saturating the inhaled air with water (Figure 6–2). When moist air is exhaled from the lungs, much of its water condenses on the cool nasal surfaces. By alternating condensation with evaporation during breathing, the kangaroo rat minimizes its respiratory water loss.

Figure 6–1
A jackrabbit seeking refuge from the hot sun of southern Arizona in the shade of a mesquite tree. The large ears and long legs of desert-inhabiting jackrabbits effectively radiate heat when the environment is cooler than the body. (Courtesy of the U.S. Fish and Wildlife Service.)

In extremely hot temperatures, organisms can cool themselves only by evaporation. So unless water is freely available, individuals must reduce their activity or seek cool microenvironments to balance both their heat and water budgets. The kangaroo rat does this in part by restricting its feeding to the cooler night-time periods, and spending the daylight hours below ground in relatively cool, humid burrows. In sharp contrast, ground squirrels remain active during the day. They conserve water by restricting evaporative cooling. As a consequence, their body temperatures rise when they are aboveground and exposed to the hot sun. But before their body tempera-

Figure 6–2
Water conservation in the respiratory passages of the kangaroo rat. (Photograph courtesy of the U.S. Fish and Wildlife Service.) Inhaled air is warmed and humidified; at the same time the nasal passages are cooled by the evaporation of water from the surface membranes *(upper left).* Water-saturated air cools as it exhales through the nose, leaving behind condensed water *(right).* Cross-sections of the nasal passages *(lower left)* at depths of 3 and 9 millimeters, respectively, from the external opening of the nose exhibit convolutions that increase the surface area of the nasal membranes. (After Schmidt-Nielsen et al. 1970.)

tures become dangerously high, they return to their cool burrows, where they dissipate their heat load by conduction and radiation, rather than by evaporation. In this manner, ground squirrels extend their activity into the heat of the day and pay a relatively small price in water loss (Hudson 1962).

Plants obtain water from dry soils by increasing the water potential of their roots.

As we have seen in Chapter 5, water is held in the soil by a variety of forces, whose strengths may be expressed in terms of water potentials, or pressures. When water sticks tightly to the surfaces of soil particles it is referred to as hygroscopic water and is held with a pressure of at least −50 bars (or −5 megapascals). (The bar, which is equal to 0.1 MPa, is approximately equal to sea-level atmospheric pressure. By convention, water potentials, which are the converse of osmotic potentials, are given negative values. The larger the negative value, the greater the attraction for water.) Water within the interstices between soil particles is held by capillary attraction of water molecules to each other by forces as low as about −0.15 bars. Water in large pores in the soil and at great distance from the surfaces of soil particles, held with pressures less than −0.15 bars, usually drains through the soil under the pull of gravity. Plants tap the soil water held with the least pressure first; it is easiest to extract. But as soils dry out, the remaining water is held with greater and greater water potential, eventually causing water stress (Hsiao 1973, Comstock and Ehleringer 1984). As a general rule of thumb, plants cannot obtain water held with pressures more negative than −15 bars, the so-called permanent wilting point of the soil.

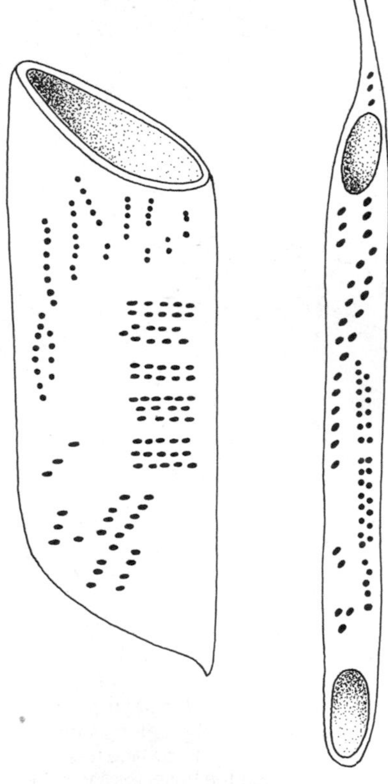

Figure 6–3
Narrow vessel members in the xylem of oak trees conduct water through the stem. Individual vessel elements are the walls of wood cells from which the living matter has disappeared. The cells are arranged vertically in the wood, and water flows from cell to cell through openings in the ends. (After Esau 1960.)

Plants obtain water from the soil by osmosis and so the ability of the roots to take up water depends upon their osmotic potential (Osonubi and Davies 1978). Osmotic potential, in turn, is a function of the concentration of dissolved molecules (including sugars) and ions in the cell sap. A 1 molar solution, for example, has an osmotic potential of about 22 bars (that is, a water potential of −22 bars). By manipulating the osmotic potential of root cells, plants can alter their ability to remove water from soil. Plants growing in dry areas have been shown to increase the water potential of their roots to as much as −60 bars, although they pay a high metabolic price to maintain such concentrations of dissolved substances.

Plants conduct water to their leaves through xylem elements, the empty remains of cells connected end-to-end, which function pretty much as pipes are used to conduct water over long distances (Figure 6–3). If water is to move from the root to the leaves, the water potential in the xylem must be more negative than that in the surrounding epidermal and cortex cells of the roots into which water enters from the soil. Similarly, the water potential of the leaves must sufficiently exceed that of the roots to draw water upward against the pull of gravity and the resistance of the xylem.

Leaves generate water potential by transpiring water to the atmosphere. Dry air has a water potential of as much as −1000 bars. R. O. Slatyer

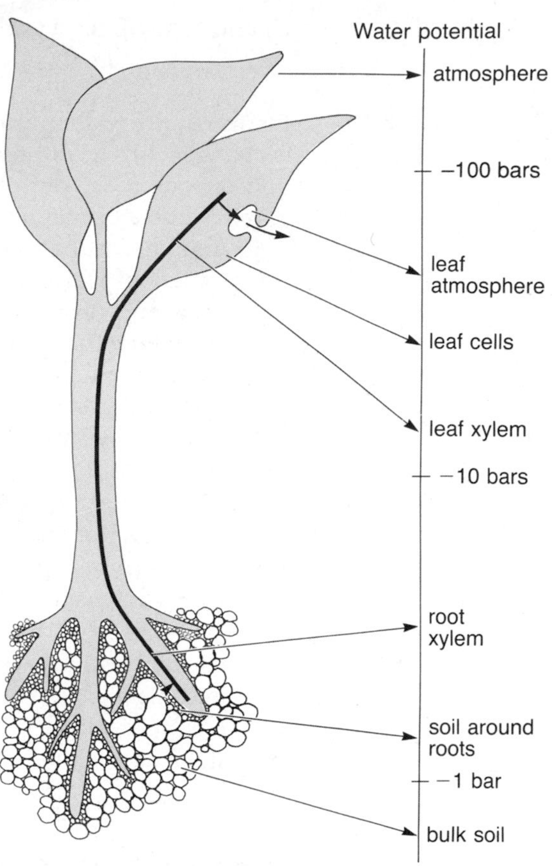

Figure 6–4
Water potentials and resistances to the flow of water between the soil and the leaves of a representative plant. (After Ray 1972.)

(1967) was the first to appreciate that the transduction of this potential to the root tissue depends on the resistances to water flow encountered in the intervening plant tissues—stomatal resistance determining the conductance of water from the interior of the leaf to the atmosphere, mesophyll resistance within the leaf, and the xylem resistance between the leaf and the root (Larcher 1980, Kramer 1983, Jarvis and McNaughton 1986, Figure 6–4).

Resistance of movement of water from the soil into the root depends partly on resistance in the epidermis and cortex of the root and partly on the rate of flow of water through the soil. In moist soils, water moves rapidly by percolation and capillary attraction into the zone immediately surrounding the fine roots. In dry soils, water is bound more closely to soil particles and does not move rapidly through the soil. Thus, in drought-adapted species, some fine roots die as local soil water is depleted, and others grow into regions of the soil not yet exploited; in dry soils, roots grow faster than water moves.

C_4 and CAM photosynthesis increase water use efficiency.

Plants require water primarily to offset losses incurred during periods of gas exchange with the atmosphere. Because the atmosphere contains little carbon dioxide (0.03 per cent), the concentration gradient for water loss is several orders of magnitude greater than that for carbon dioxide assimilation. And because the vapor pressure of water increases with temperature, the problem of water loss is magnified in hot environments.

Heat- and drought-adapted plants exhibit modifications of anatomy and physiology that reduce transpiration across plant surfaces, reduce heat loads, and enable plants to tolerate high temperatures (Hadley 1970, Berry and Bjorkman 1980). When plants absorb sunlight, they heat up. Plants can minimize overheating by increasing their surface area for heat dissipation and protecting their surfaces from direct sunlight with dense hairs and spines (Ehleringer 1984, Figure 6–5). Spines and hairs also produce a still boundary layer of air that traps moisture and reduces evaporation. Plants further reduce transpiration by covering their surfaces with a thick, waxy cuticle impervious to water and by recessing the stomata in deep pits, often themselves filled with hairs (Figure 6–6).

The biochemical mechanism of carbon assimilation has also been modified in some plants to conserve water. Most species in mesic environments (those with adequate water) assimilate the carbon in carbon dioxide into an organic molecule in a single biochemical step in a pathway known as the

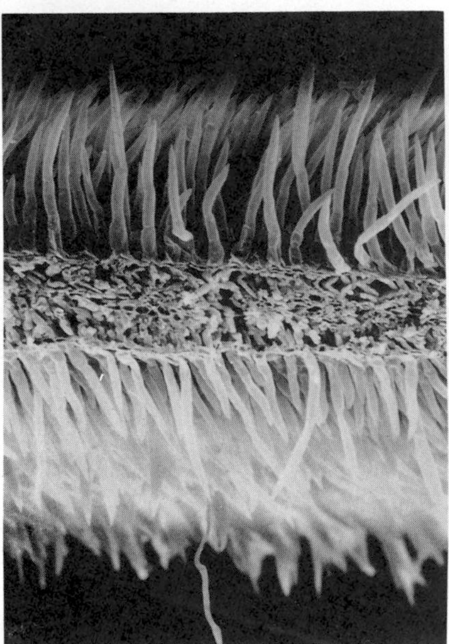

Figure 6–5
Cross-section *(left)* and surface view *(right)* of the pubescent leaf of the desert perennial herb *Enceliopsis argophylla.* (Courtesy of J. Ehleringer; from Ehleringer 1984.)

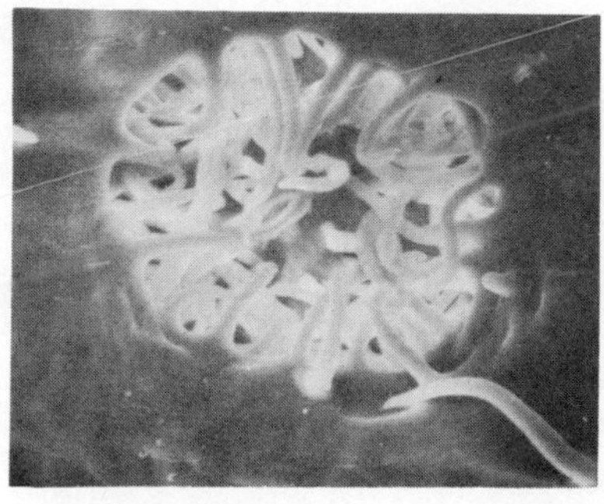

Figure 6-6
In oleander, a drought-resistant plant, stomata lie deep within hair-filled pits on the leaf's undersurface (magnified about 500 times). The hairs reduce water loss by slowing air movement and trapping water. (Courtesy of M. V. Parthasarathy.)

Calvin cycle (Figure 6-7). This step may be represented as

$$CO_2 + \text{RuBP} \longrightarrow 2\ \text{PGA}$$

where RuBP (ribulose bisphosphate) is a 5-carbon organic compound and PGA (phosphoglycerate) is a 3-carbon compound. (Because the immediate product of carbon assimilation is a 3-carbon compound, this mechanism is referred to as C_3 photosynthesis.) Several biochemical steps beyond the production of PGA, 1 molecule of RuBP is regenerated to complete the Calvin cycle while 1 carbon atom is made available for the synthesis of glucose. Inasmuch as glucose is a 6-carbon compound, each molecule of glucose produced requires six turns of the Calvin cycle.

The enzyme responsible for assimilation of carbon—RuBP carboxylase—has a very low affinity for carbon dioxide. As a result, at the concentration of CO_2 in the atmosphere and its resulting partial pressure in the plant mesophyll cell, plants assimilate carbon very inefficiently. In order to achieve high rates of assimilation, plants must pack their cells with large amounts of RuBP carboxylase. An additional problem with this enzyme is that it facilitates the oxidation of RuBP in the presence of high oxygen and low carbon dioxide concentrations, especially at elevated leaf temperatures. Oxidation of RuBP undoes what RuBP carboxylase accomplishes when it reduces carbon, thus making photosynthesis inefficient. In fact, carbon assimilation tends to be self-inhibiting unless CO_2 is maintained at a high level in the cell. This can be accomplished only when the stomatal resistance remains low—which, of course, leads to high water loss. It is not clear why plants did not develop a better enzyme for such a crucial step.

Many plants in hot climates exhibit a modification of C_3 photosynthesis that involves an additional step in the assimilation of carbon dioxide and the spatial separation of the initial assimilation step and Calvin cycle pathways within the leaf. This modification is called C_4 photosynthesis because the assimilation of CO_2 initially results in a 4-carbon compound:

$$CO_2 + \text{PEP} \longrightarrow \text{OAA}$$

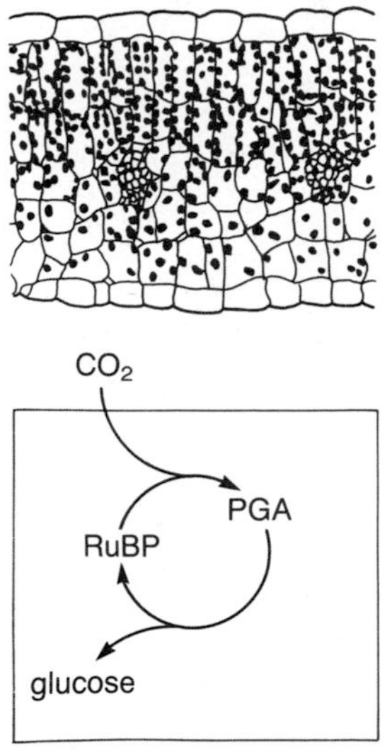

Figure 6-7
Above: An idealized cross-section of a leaf of a C_3 plant, illustrating the general dispersion of chloroplasts throughout the mesophyll. *Below:* The major steps of the Calvin cycle of carbon dioxide assimilation. (Courtesy of J. Ehleringer.)

where PEP (phosphoenol pyruvate) is a 3-carbon compound and OAA (oxaloacetic acid) is a 4-carbon product (Hatch and Slack 1970, Bjorkman and Berry 1973, Edwards and Walker 1983). The assimilatory reaction is catalyzed by PEP carboxylase, which, in contrast to RuBP carboxylase, has a high affinity for CO_2. Assimilation occurs in the mesophyll cells of the leaf, but in most C_4 plants, photosynthesis, including the Calvin cycle, is localized in specialized cells surrounding the leaf veins (Figure 6-8). These are called bundle sheath cells. Oxaloacetic acid, or another 4-carbon derivative such as malate, diffuses into the bundle sheath cells, where it is metabolized to produce CO_2 plus pyruvate, a 3-carbon compound. The pyruvate moves back into the mesophyll cells, where enzymes convert it to PEP to complete the carbon assimilation cycle.

The CO_2 released by metabolism of OAA in the bundle sheath cells enters the Calvin cycle, as it does in C_3 plants. C_4 photosynthesis confers an advantage because CO_2 can be concentrated within the bundle sheath cells

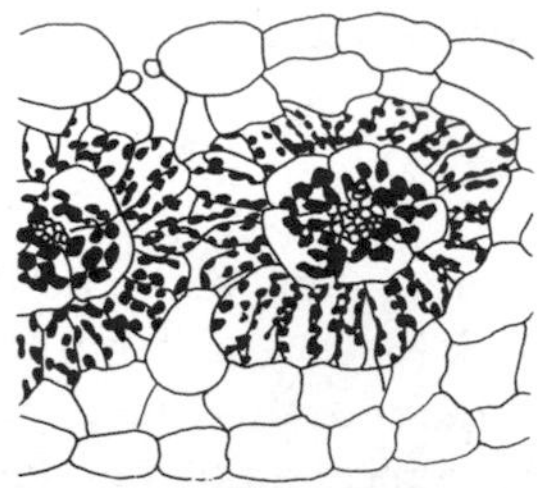

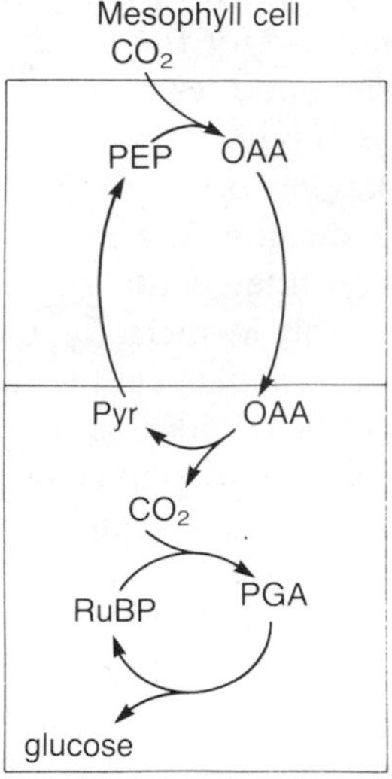

Figure 6–8
Above: An idealized cross-section of a leaf of a C_4 plant, illustrating the concentration of chloroplasts in cells of the bundle sheath (Kranz anatomy). *Below:* The major steps of carbon dioxide assimilation, including the transport of carbon from the mesophyll to the bundle sheath. (Courtesy of J. Ehleringer.)

to a level, determined by the concentration of OAA or malate, that far exceeds its equilibrium established by diffusion from the atmosphere. At this higher concentration, the Calvin cycle operates more efficiently. Furthermore, because the enzyme PEP carboxylase has a high affinity for CO_2, it can bind CO_2 at a lower concentration in the cell, thereby allowing the plant to increase stomatal resistance and reduce water loss.

C_4 photosynthesis increases the rate of carbon assimilation in hot environments with abundant solar radiation, and increases water use efficiency in dry habitats. Why, then, don't all plants use the C_4 pathway? The answer appears to have two parts. First, C_4 plants must expend energy to regenerate PEP from pyruvate, and so C_4 photosynthesis is energetically less efficient than C_3 photosynthesis. Second, at temperatures below 25°C RuBP carboxylase operates relatively efficiently because it has less tendency to behave as an oxidase. Hence the advantage of the C_4 pathway diminishes in cool environments, particularly when the availability of light, rather than that of carbon dioxide, limits photosynthesis (Figure 6–9). Maximum photosynthesis is achieved by C_4 plants at about 45°C, close to the maximum tolerable temperature, and by C_3 plants at between 20 and 30°C. As a result, C_4 plants predominate in hot climates, and C_3 plants in cool climates (Teeri and Stowe 1976, Pearcy and Ehleringer 1984).

C_4 plants segregate initial carbon assimilation and the Calvin cycle into different tissues. Certain succulent plants in desert environments utilize the same biochemical pathways as C_4 plants, but segregate assimilation and the Calvin cycle temporally between day and night. Because this arrangement was first recognized in plants of the family Crassulaceae (stonecrop family;

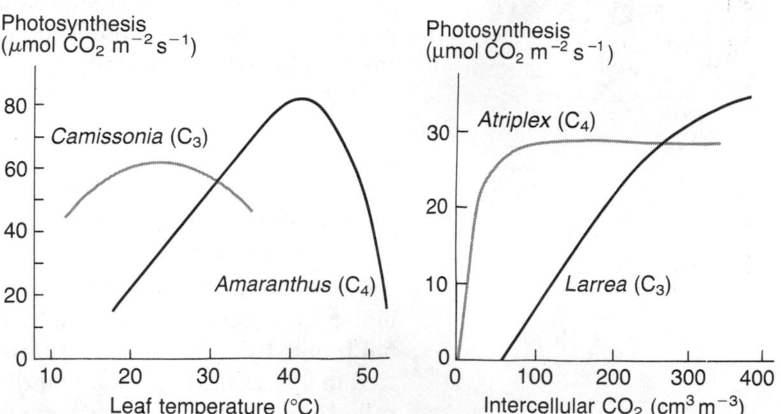

Figure 6–9
Left: Relationship of photosynthesis (measured in micromoles of carbon dioxide assimilated per square meter of leaf area per second) to leaf temperature in the winter-active C_3 desert herb *Camissonia claviformis* and the summer-active C_4 desert herb *Amaranthus palmeri*. *Right:* Response of net photosynthesis to intercellular carbon dioxide concentration in two desert shrubs, *Atriplex hymenelytra* (C_4) and *Larrea divaricata* (C_3) under high light intensity and leaf temperatures of 30°C. Under these conditions, the C_4 plant utilizes CO_2 more efficiently at low concentrations. (After Pearcy and Ehleringer 1984.)

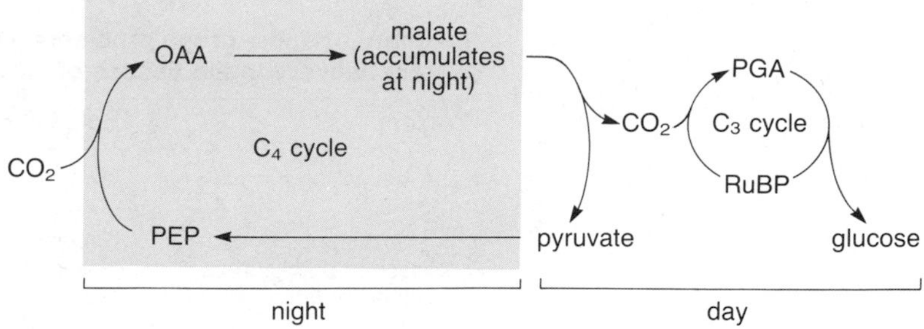

Figure 6–10
Outline of the photosynthetic pathway in CAM plants. (After Harbourne 1982.)

Sedum is one example), and because it entails the storage of certain 4-carbon organic acids (malic acid and OAA), it is referred to as crassulacean acid metabolism, or CAM (Osmond 1978, Kluge and Ting 1978).

CAM plants open their stomates for gas exchange during the cool desert night, at which time transpiration of water is minimal. CAM plants initially assimilate CO_2 in the form of 4-carbon OAA and malic acid, which the leaf tissues store in high concentration (Figure 6–10). During the day, the stomates of the leaves close and the stored organic acids are gradually recycled to release CO_2 to the Calvin cycle. Assimilation of CO_2 and regeneration of phosphoenol pyruvate are regulated by enzymes having different temperature optima. CAM photosynthesis results in extremely high water use efficiencies and enables some types of plants to exist in habitats too hot and dry for other, more conventional species.

Adaptations to procure oxygen are critical to animals, particularly in aquatic habitats.

The chemical energy in organic compounds is made available to living systems primarily by oxidative metabolism. The biochemical pathways involved, which are the converse of photosynthesis in many respects, are collectively referred to as respiration. Because oxygen plays such an important role in releasing energy, its availability in the environment can restrict metabolic activity, particularly in aquatic habitats where low solubility and slow diffusion limit oxygen concentrations. Even for terrestrial organisms, which breathe an atmosphere containing abundant oxygen, delivery of oxygen to the tissues where it is needed requires its solution and transport through the aqueous medium of the body. The various methods by which organisms procure oxygen demonstrate the influence that physical properties of the environment have on their design (Table 6–1).

Small aquatic organisms usually obtain oxygen by diffusion from the surrounding environment into their tissues. Carbon dioxide produced by

Table 6–1
Summary of some of the problems overcome in the evolution of high rates of oxygen delivery to the tissues of large organisms

Problem	Solution	Biological occurrence
None in small or inactive organisms	Oxygen is obtained by diffusion through cells	Protozoa, sponges, cnidarians
Diffusion distance from surface to body core is too great in large organisms	Circulatory system pumps fluids from surface to core	Widespread: pumping by body muscles in roundworms; open system without capillaries in arthropods and many mollusks; closed capillary systems in vertebrates
The solubility of oxygen in water limits oxygen transport by circulating fluids	Incorporation of oxygen-binding proteins (e.g., hemoglobin) in blood	Hemoglobin is widespread in vertebrates, but sporadic in lower groups where other oxygen-binding pigments may be found
High concentrations of proteins increase osmotic level of blood	Respiratory proteins are tightly packed in red blood cells	All vertebrates, some mollusks and echinoderms

respiration escapes the body by traveling the same route in the opposite direction. When the concentration of oxygen in an organism's tissues is lower than that of the surrounding medium, oxygen diffuses in; because respiratory metabolism consumes oxygen, it keeps the concentration of oxygen in the body low.

Diffusion can satisfy the oxygen needs of tiny aquatic organisms (and the leaves of terrestrial plants), but distances between the external environment and the centers of large organisms are too great for diffusion to ensure a rapid supply of oxygen. In fact, diffusion is ineffective at distances greater than about 1 millimeter (Harvey 1928, Schmidt-Nielsen 1975). Plants and insects have solved this problem by conduit systems that carry air directly to the tissues. Other animals have blood circulatory systems to distribute oxygen from the respiratory surface throughout the body.

Blood itself cannot carry enough dissolved oxygen (a maximum of 1 per cent by volume or about 14 parts per million by weight) to support a high rate of activity. To increase oxygen-carrying capacity, the blood of most groups of animals contains complex protein molecules, such as hemoglobin, to which oxygen molecules readily attach and are thereby taken out of solution (Hochachka and Somero 1984). When oxygen binds to hemoglobin — 4 molecules of oxygen for each of hemoglobin — dissolved oxygen is taken out of solution, making room for the diffusion of more into the blood across the blood-gas boundary. The binding process is reversible so that oxygen can be released to the tissues. While plasma itself carries only limited oxygen in solution, whole blood transports up to 50 times more oxygen, which is bound to oxygen-carrying molecules. Hemoglobin binds oxygen most effectively when present in very high concentration, close to the point of crystallization. If hemoglobin occurred in the plasma, the osmotic level of the blood would be too great for proper physiological function, particularly for proper salt balance; instead hemoglobin is packaged inside red blood cells (erythrocytes), each of which may contain upward of a quarter billion hemoglobin molecules (Schmidt-Nielsen and Taylor 1968). The close association of hemoglobin molecules within red blood cells may alter their interaction among each other and with the surrounding blood environment in such a way as to facilitate oxygen binding and release. An analogous arrangement is found in several worms and snails, which have hemoglobin but lack special blood cells. In these species, hemoglobin is aggregated into groups of 20 to 50 molecules (Schmidt-Nielsen 1983).

Countercurrent circulation increases the flux of oxygen across gill surfaces.

A recurrent adaptation that enhances the uptake of dissolved oxygen from water is countercurrent circulation, an arrangement of the structure of gills whereby water and blood flow in opposite directions (Figure 6–11). In a countercurrent system, as blood picks up oxygen from the water flowing past, it comes into contact with water having progressively greater oxygen concentration. This is possible because the water has flowed past a progressively shorter distance of the gill lamella (Figure 6–12). With this arrangement, the oxygen concentration of the blood plasma can approach very nearly the concentration in the surrounding water. If blood and water were to flow together through the gill, and equilibrium oxygen concentration would eventually be established with equal, intermediate levels in the blood and water flowing past the gill. The countercurrent system keeps the blood and water out of equilibrium and maintains a constant gradient across which oxygen can flow.

The countercurrent principle appears frequently in adaptations to increase flux of heat or materials across surfaces. Among terrestrial organisms,

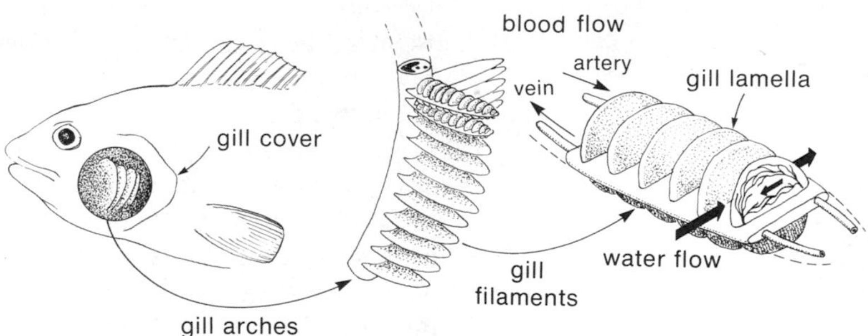

Figure 6–11
A fish's gill consists of several gill arches, each of which carries two rows of filaments. The filaments bear thin lamellae oriented in the direction of the flow of water through the gill. Within the lamellae, blood flows in a countercurrent direction to the movement of water past the surface. (From Randall 1968.)

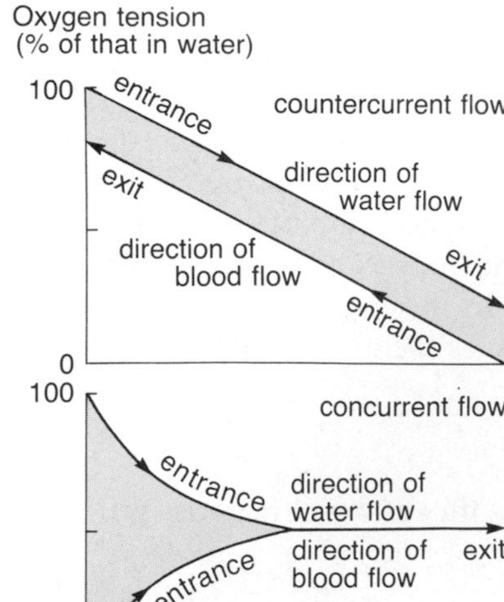

Figure 6–12
Changes in oxygen tension of blood and water in counterconcurrent and concurrent systems. The former maintains a constant gradient for oxygen to diffuse across; as a result, the oxygen tension of the blood leaving the system approaches that of the incoming water. (After Schmidt-Nielsen 1975.)

birds have a unique lung structure, which, unlike the lungs of mammals, results in a one-way flow of air (Schmidt-Nielsen 1972, 1983). This adaptation allows birds to deliver the high rates of oxygen required by their active lives with lungs whose weight and volume are compatible with flight.

The oxygen-binding capacity of hemoglobin balances oxygen availability and oxygen requirement.

Active organisms require an abundant supply of oxygen for cell respiration. Hemoglobin, and other oxygen-carrying pigments such as hemocyanin, act as go-betweens for the uptake of oxygen from the surrounding environment and its release to the cells. The two functions conflict, however, for the hemoglobin molecule that readily binds oxygen in the lungs or gills also holds oxygen tenaciously when it must be released to active tissues.

The compromise between the oxygen-uptake and oxygen-release functions of hemoglobin, or other blood pigments, expresses itself in the oxygen dissociation curve (Figure 6–13). The dissociation curve portrays the amount of oxygen bound to hemoglobin, expressed as a per cent of the total possible (saturation), as a function of the concentration of oxygen in the blood plasma. Oxygen binds to hemoglobin reversibly, the per cent of bound and unbound oxygen reaching an equilibrium described by the expression $Hb + O_2 \rightleftharpoons HbO_2$. As oxygen diffuses into the blood plasma in the gills or lungs, the equilibrium is shifted to the right and additional oxygen is bound to hemoglobin. As active tissues deplete the blood of oxygen, the equilibrium shifts toward the left, and oxygen is released from the oxyhemoglobin complex. In general, the proportion of hemoglobin with bound oxygen increases with greater concentration of dissolved oxygen in the blood plasma until the hemoglobin becomes saturated.

The concentration of oxygen, or any other gas, dissolved in a liquid is commonly expressed in terms of oxygen tension or partial pressure, whose units correspond to the familiar scale of a barometer: millimeters of mercury, or torr. At every water-air interface, gas molecules continually enter into solution and escape back into the air. For a given gas under a given pressure, these comings and goings achieve an equilibrium with no net movement of gas molecules either into or out of solution. Under these conditions, the concentration of the gas dissolved in water can be expressed in terms of the pressure of the gas in air. Physiologists have adopted this convention because the movement of a gas between air and water (or blood plasma) depends on the difference between oxygen tensions rather than absolute concentrations.

Sea-level atmospheric pressure is 760 torr. Because approximately one-fifth of the atmosphere is oxygen, the partial pressure of oxygen is about one-fifth of 760, or 159 torr. Thus, aqueous dissolved oxygen in equilibrium with sea-level atmosphere has a partial pressure of 159 torr. By comparison, the absolute concentration of oxygen in the atmosphere is 209 cm^3 per liter, but dissolved in water it is only between 5 and 10 cm^3 per liter at equilibrium, depending on temperature and salinity.

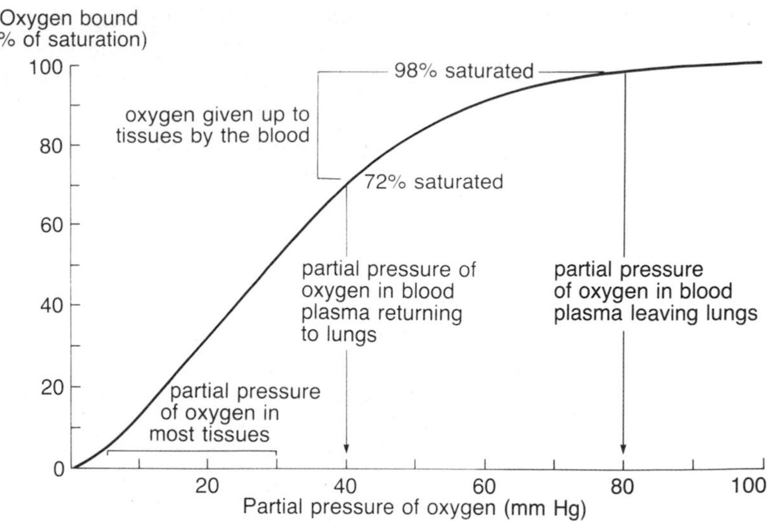

Figure 6–13
The oxygen dissociation curve for human hemoglobin. The blood releases about 25 to 30 per cent of the oxygen bound to hemoglobin to the tissues, which have oxygen tensions of 5 to 30 torr.

In humans, oxygen enters the lungs at a partial pressure of 159 torr and diffuses rapidly into the bloodstream of capillaries surrounding the air sacs. When the blood leaves the lungs, oxygen dissolved in the plasma has a partial pressure of 80 torr, and the hemoglobin is about 98 per cent saturated with oxygen. (Saturated blood carries more than 200 cm^3 of oxygen gas per liter of blood, almost 50 times the amount of oxygen that can be dissolved in water at the normal temperature of the human body and very nearly the absolute concentration of oxygen in the atmosphere.) The partial pressure of oxygen in various tissues of the body ranges from 5 to 30 torr, and oxygen therefore diffuses readily from the blood plasma into the tissues.

Tissues receive oxygen at a rate proportional to the difference between the partial pressures of the oxygen in the tissues and in the blood: the higher the gradient, the higher the rate of diffusion. When dissolved oxygen leaves the bloodstream to enter the tissues, the partial pressure of oxygen in the plasma decreases. As the difference between oxygen concentrations in the blood and in the tissues decreases, diffusion slows. But as the partial pressure of oxygen in the blood plasma decreases, hemoglobin releases some bound oxygen, thereby tending to restore the oxygen concentration. Thus oxygen drawn from the blood plasma into the tissues is partially replenished by that bound to hemoglobin.

As blood travels through the human body, hemoglobin loses 25 to 30 per cent of its bound oxygen. The blood returns to the lungs carrying about 144 cm^3 of oxygen per liter. The partial pressure of oxygen in the blood plasma at this point (40 torr) barely exceeds the partial pressure in many tissues. Such depleted blood can, however, continue to release oxygen to tissues because of a phenomenon known as the Bohr effect (Riggs 1960). As carbon dioxide enters the blood from active tissues, where it is produced during cellular respiration, the blood plasma becomes increasingly acid owing to the formation of carbonic acid. Increased acidity shifts the oxygen dissociation curves of most hemoglobins to the right, facilitating the release of bound oxygen from hemoglobin and raising the partial pressure of oxygen in the blood plasma.

Small changes in amino acid sequence may modify the oxygen-binding capacity of the hemoglobin molecule with respect to both the availability of oxygen in the environment and its requirement by the organism. Because small animals have high oxygen demands, their oxygen dissociation curves are generally shifted to the right for easier unloading of oxygen. The dissociation curve of a mammalian embryo is shifted to the left of that of its mother because its oxygen supply derives from the mother's blood, which has a lower partial pressure of oxygen than the air she breaths.

The total amount of hemoglobin in the bloodstream also is subject to adjustment. For example, the blood of the sedentary goosefish has a total oxygen capacity of 5 per cent by volume; that of the more active mackerel is 16 per cent (Laglar et al. 1962). This difference reflects the hemoglobin concentration in the blood and parallels the sizes of the gills in the two species, the surface area of those of the mackerel being 50 times that of the goosefish relative to body weight (Gray 1954).

High-altitude acclimation involves the gas-exchange characteristics of the blood. When humans from sea-level spent several weeks at an elevation of 5000 meters, the oxygen-carrying capacity of their blood increased from 21 to 25 per cent by volume (210 to 250 cm^3 per liter); it did not, however, reach the 30 per cent capacity of the local high-altitude residents (Prosser and Brown 1961). Again, these differences were largely related to the concentration of hemoglobin in the blood. Increased lung size, breathing rate and volume, heart size, rate and stroke volume of the heartbeat, and capillary density are also important adaptations that influence the overall rate of activity that can be sustained at high altitude (Frisancho 1975).

Adaptations for procuring oxygen illustrates a set of solutions to the problems organisms must confront at the interface between themselves and their environments. The design of the hemoglobin molecule—influencing simultaneously its oxygen-binding and oxygen-releasing properties—further emphasizes the fact that adaptation often requires compromise.

Plants resort to active uptake when soil nutrients are scarce.

Plants acquire mineral nutrients—nitrogen, phosphorus, potassium, calcium, and others—from dissolved forms of these elements in soil water (Chapin 1980, Harley and Russell 1979, Haynes and Goh 1978). When an element is abundant and highly mobile in the soil solution—as in the case of calcium and magnesium ions (Ca^{2+} and Mg^{2+})—uptake is limited primarily by the absorptive capacity of the root (Nye 1977). Plants compensate for reduced levels of a limiting nutrient in the soil by active uptake of the nutrient and by increasing the extent of the root system, that is, the absorptive surface of the roots. In laboratory experiments, barley and beet roots passively took up phosphorus by diffusion when its concentration in the water surrounding the root was greater than 0.2 to 0.5 millimolar (Barber 1972). Under these conditions, concentrations of phosphorus in root tissues and in the water conducted to the leaves were less than that in the soil solution. At lower external concentrations than 0.2 to 0.5 millimolar, roots actively transported phosphorus across their surfaces and concentrated the element within the root cortex. This active absorption requires the expenditure of energy as the root tissue moves ions against a concentration gradient.

Plants may also respond to decreased soil nutrient availability by increasing the allocation of photosynthate to root growth at the expense of shoot growth. This brings the nutrient requirement of the plant into line with availability by reducing the nutrient demand created by the leaves, by increasing the absorptive surface area of the root system, and by growing roots into new areas of soil whose scarce minerals have not been depleted (Chapin 1980). Such modifications cannot completely compensate for decreasing soil fertility, however, and both the concentration and total amount of a limiting nutrient in a plant vary in direct relationship to the availability of the nutrient in the soil.

Crop plants and wild species growing on fertile soils have a great capacity to absorb nutrients across their root surfaces and vary their growth rates in response to variation in soil nutrient levels, but species adapted to nutrient-poor soils are much more conservative (Chapin 1980). They cope with low nutrient status by allocating a large fraction of their biomass to roots; by establishing symbiotic relationships with fungi, which enhance mineral absorption; and by growing slowly and retaining leaves for long periods, thereby reducing nutrient demand. Such species typically cannot respond to artificially increased nutrient levels by increasing growth rate. In response to a nutrient flush, their roots absorb more nutrients than the plant requires and store them for subsequent utilization when the nutrient status of the soil decreases (Grime 1979, Grime and Hunt 1975).

Plants and animals can adapt their temperature optima to match the temperature of the environment.

Unlike carbon dioxide, oxygen, and soil nutrients, temperature is not a resource consumed by plants and animals. Rather, temperature influences organisms thermodynamically through its effects on rates of physical and biochemical processes and on conformations of such biologically important molecules as proteins and lipids (Hochachka and Somero 1973, 1984). As a rule, within the temperature range between the freezing point of water (0°C) and the upper limit for most life forms (40–50°C), higher temperature quickens the pace of life by increasing the kinetic energy of the organism and its surroundings. Temperature affects such physical processes as diffusion and evaporation as well as biochemical reactions. For example, increased temperature speeds the diffusion of gases through the shell of a bird's egg or of mineral ions through the soil solution.

Temperature influences the shapes of enzymes and other proteins because kinetic energy imparted to such molecules can break the weak forces holding parts of the structure together. Too much kinetic energy (high temperature) and the molecules tend to unravel; too little and they tend to fold tightly upon themselves. In many cases, the activity of an enzyme depends on its ability to bind closely to the substrate or substrates upon which it acts; binding strength depends on a close fit between the shapes of the enzyme and substrate. Consequently, each enzyme functions best over a narrow range of temperature. That range is determined by the amino acid sequences of protein chains, particularly as they affect the number and strength of bonds that hold the protein structure together. These sequences are, of course, subject to evolutionary modification.

Temperature adaptation is strikingly revealed by comparing organisms taken from environments with different characteristic temperatures. When we examine the performances of such organisms over a range of temperatures, each typically exhibits its maximum activity at a temperature corresponding to that of its normal environment. Many fish in the freezing oceans surrounding Antarctica swim just as actively as fish living among tropical coral reefs. A graph of the relationship between oxygen consumption of fish and water temperature (Figure 6–14) shows that the metabolic rates of cold-water fish are adjusted such that their activity in cold water is on a par with that of warm-water fish at the higher temperatures characteristic of their usual environments. Put a tropical fish in cold water, however, and it becomes sluggish and soon dies; conversely, antarctic fish cannot tolerate temperatures warmer than 5 to 10°C.

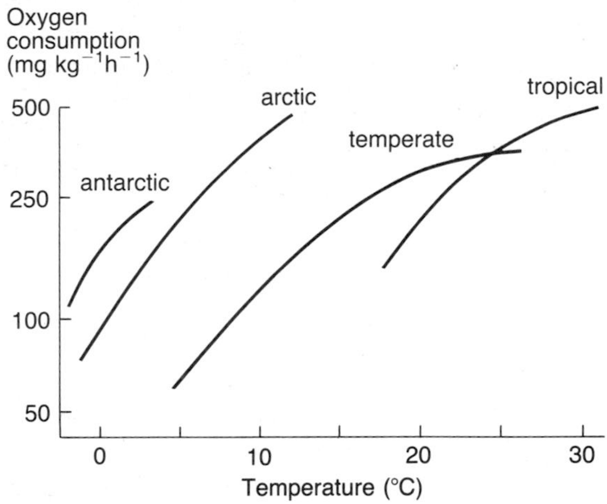

Figure 6–14
Temperature compensation in the rates of oxygen consumption of fishes from different thermal regimes. Although metabolism in each increases with temperature, the curves are offset so that fishes from different thermal environments have similar metabolic rates at the prevailing temperatures of their native habitats. (After Hochachka and Somero 1984.)

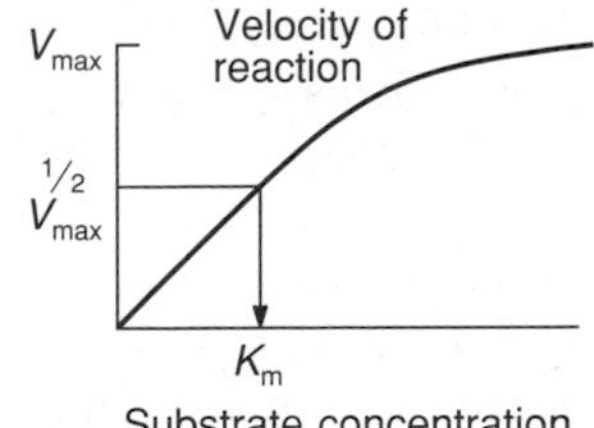

Figure 6–15
Velocity of an enzymatic reaction as a function of substrate concentration. The reaction rate levels off when the enzyme is saturated (V_{max}). The K_m is the concentration of the substrate when the reaction proceeds at half its maximum rate.

How can fish from cold environments be as active metabolically as fish from the tropics? Metabolism consists of a series of biochemical transformations, most of which are catalyzed by enzymes. Because a given transformation occurs more rapidly at high temperature than at low temperature, the compensation observed in cold-adapted organisms must involve either a quantitative increase in the amount of substrate or the amount of enzyme that catalyzes each step, or a qualitative change in the enzyme itself (Somero, 1978). Many enzymes occur in slightly different forms whose characteristics reflect the substitution of certain amino acids in the protein chain. These differences usually result from small changes in the gene that encodes the structure of the protein. Not only do the structures of enzymes sometimes differ among species and among populations of the same species, more than one form of an enzyme may occur within the same population (in which case it is polymorphic for the enzyme) or the same individual (a different variant of the enzyme inherited from the father and the mother, in which case the individual is heterozygous for the enzyme).

Laboratory studies of the function of isolated enzymes have shown, in many cases, that the different variants have different catalytic properties when tested over ranges of temperature, pH, salt concentration, and level of substrate. One measure of catalytic ability is the facility with which the enzyme binds with its substrate. This usually is expressed as the Michaelis-Menten constant (K_m), which is the concentration of the substrate at which the velocity of the reaction is one-half of its maximum. Optimally adapted enzymes are thought to have K_ms in the range of substrate concentrations normally encountered within tissues (Graves et al. 1983). In this way, the rate of the catalyzed reaction is most sensitive to changes in substrate concentration (Figure 6–15). K_m values for lactate dehydrogenase enzymes (LDH) in three species of barracudas *(Sphyraena)* illustrate compensation: within the range of temperatures normally encountered by each of the species, the K_m values of their particular LDHs are similar (Figure 6–16).

This picture of metabolic compensation is greatly simplified, of course, Adapting to changes in the environment requires a complete retuning of metabolic pathways, which may involve changes in enzyme structure or concentration, or the use of alternative metabolic pathways. The design of metabolic systems is not fully understood; it is certainly beyond the scope of this discussion. Clearly, however, plants and animals employ a wide range of structural and functional adaptations to fit themselves as well as possible to their environments (Hochachka and Somero 1984, Prosser 1986).

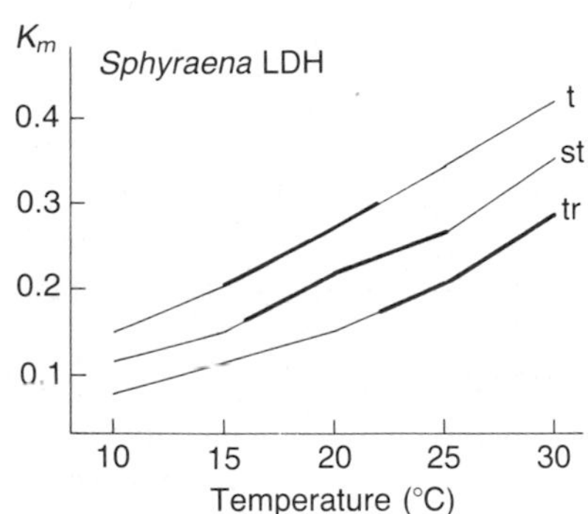

Figure 6–16
Relationship between K_m and temperature for the LDH enzymes of three species of barracudas *(Sphyraena)* from temperate (t), subtropical (st), and tropical (tr) waters. Because of temperature compensation, the K_ms of the three species are nearly identical within the normal temperature ranges of their environments. (After Hochachka and Somero 1984.)

SUMMARY

Adaptations to the physical environment involve modifications of the components of flux across the organism's surface and adjustment of metabolic pathways with respect to conditions of the internal environment.

1. To maintain salt and water balance, marine organisms whose internal environment is hypo-osmotic actively exclude salts; freshwater orga-

nisms, all of which are hyperosmotic, retain salts while excreting the water that continually diffuses into the body; terrestrial organisms minimize water loss, in part by concentrating salts and nitrogenous waste products in their urine.

2. Nitrogenous waste products of protein metabolism are excreted as ammonia by most aquatic organisms, as urea by mammals, as uric acid by birds and reptiles. Because uric acid crystallizes out of solution, it may be excreted at high concentrations and therefore offers considerable water economy.

3. Water stress increases with temperature. In dry environments, animals seek cool microclimates and plants increase the stomatal resistance of their leaves. Such responses uniformly reduce productivity in return for enhanced survival.

4. Plants draw water from the soil by osmotic potential in their roots, and from the roots to the leaves by water potential generated by evaporation of water from leaves. Resistance to water conductance from the soil, through the root and xylem, to the leaves must balance an adequate supply of water against the risk of causing leaves to wilt.

5. During photosynthesis, plants assimilate carbon through a reaction (the C_3 pathway) catalyzed by the enzyme RuBP carboxylase. This enzyme has a low affinity for carbon dioxide and brings about photo-oxidation at high temperatures, resulting in low efficiency. Plants adapted to high temperature interpose a more efficient (C_4) carbon-assimilation step, which is spatially separated from the C_3 reactions (Calvin cycle) in the leaf. In desert environments, some plants separate carbon assimilation and the Calvin cycle reactions temporally between nighttime and daytime phases (CAM photosynthesis). Both C_4 and CAM increase water use efficiency at high temperatures, but are energetically less efficient than C_3 photosynthesis, which prevails in cool, moist environments.

6. Oxygen diffuses through tissues too slowly to meet the metabolic needs of large animals. This problem is overcome either by conducting air directly to the tissues by a multibranched tracheal system (as in insects) or by transporting oxygen dissolved in circulating fluids. The low solubility of oxygen in water is compensated by oxygen-binding proteins, such as hemoglobin, that take oxygen out of solution and transport it at high concentrations.

7. Uptake of the oxygen by aquatic organisms is greatly facilitated by countercurrent circulation of blood through the gills in a direction opposite to that of water flowing over the gill surfaces. In this way, countercurrent circulation maintains high gradients of oxygen concentration, and the blood can achieve nearly the oxygen concentration of the surrounding water.

8. Oxygen transport proteins optimize the balance between uptake of oxygen in lungs and gills and release of oxygen to tissues. Binding characteristics of these proteins may be modified with respect to the availability of oxygen in the environment and its demand by tissues.

9. Plants obtain nutrients in dissolved form from soil water by passive diffusion when nutrients are abundant and by active uptake when they are scarce. When the concentrations of soil nutrients are low, plants may in-

crease total root surface area at the expense of shoot growth. Plants also may store nutrients when they are temporally or locally abundant.

10. Although most physical and biological processes are accelerated at higher temperatures, temperature optima of metabolic processes may be adjusted to match the characteristic temperature of the environment by altering the structure and quantity of key enzymes.

Overall, adaptation to the physical environment depends upon the adjustment of compromises between opposing functions to maximize the survival and productivity of the individual in a particular environment.

7

Response to Variation in the Environment

Change pervades an organism's surroundings — the annual cycle of the seasons, daily periods of light and dark, frequent unpredictable turns of climate. The survival of each individual depends on its ability to cope with variation in the environment. Humans can be aware of their own responses to change. When we step from a warm room into the outdoors on a cold day, we shiver to generate heat. A few weeks on the beach and our skin darkens to block damaging radiation from the sun. We respond to environmental change in order to maintain our internal conditions at optimum levels for proper functioning. But what determines the best internal condition? At what rate should the individual function? How can one respond to environmental change most effectively?

Responses must be analyzed, like a problem in economics, in terms of costs and benefits. When the environment cools, shivering increases the probability of survival of an organism that must maintain a high body temperature to function properly. But shivering also requires energy to produce body heat, which in turn may deplete fat reserves and render life more precarious in the face of a sudden food shortage. The organism may reduce the cost of such temperature regulation by lowering the regulated temperature, just as we turn down the thermostat to save fuel. But turning down the fire of an organism's life also reduces its rate of activity and hence its food-gathering and predator-avoiding abilities.

When we consider the many factors affecting costs and benefits of a particular response, the optimum becomes a subtle concept. In this chapter we shall explore some facets of response to environmental change in order to understand why different organisms regulate their internal conditions at different levels, or not at all, and why they employ different means of response to environmental change.

Homeostasis depends upon negative feedbacks.

Homeostasis is the ability of the individual to maintain constant internal conditions in the face of a varying external environment. All organisms exhibit homeostasis to some degree with respect to some environmental conditions, although the occurrence and effectiveness of homeostatic mechanisms varies.

Most mammals and birds closely maintain their body temperature between 36 and 41°C, even though the temperature of their surroundings may vary from −50 to +50°C. Such regulation, referred to as homeothermy, guarantees that biochemical processes within cells can occur under constant temperature (homeothermic) conditions. The internal environments of cold-blooded (poikilothermic) organisms, such as frogs and grasshoppers, conform to external temperature. Of course, frogs cannot function at either high or low temperature extremes, and so they can be active only within a narrow part of the range of environmental conditions over which mammals and birds thrive.

How is body temperature regulated? Most homeotherms have a sensitive thermostat in their brain (Hammel 1968, Calder and King 1974). This thermostat responds to changes in the temperature of the blood by secreting hormones into the bloodstream to slow down or accelerate the generation of heat in body tissues. In addition, most homeotherms partly regulate body temperature by altering gains and losses of heat from the environment. For example, humans put on heavy clothes in cold weather and avoid standing in the sun when it is hot; birds fluff up their feathers to provide greater insulation against the cold.

Because the so-called cold-blooded organisms cannot generate heat unless they are very active, many adjust their heat balance behaviorally by simply moving into or out of the shade, or by orienting the body with respect to the sun (Bogert 1949, Heatwole 1976, McGinnis and Dickson 1967, Stevenson 1985). When cold, horned lizards increase the profile of their bodies exposed to the sun by lying flat against the ground; when hot, they decrease their exposure by standing erect upon their legs (Heath 1965). By lying flat against the ground, horned lizards also can gain heat by conduction from the sun-warmed surface. Such behavior, widespread among reptiles, effectively regulates body temperature within a narrow range, which is elevated considerably above the temperature of the surrounding air (Cowles and Bogert 1944, Hammel et al. 1967, Huey 1974).

Regardless of the particular mechanism of regulation, all homeostasis exhibits properties of a negative feedback system, exemplified by the working of a thermostat. When a room becomes too hot, a temperature-sensitive switch turns off the heater; when the temperature drops too low, the switch turns the heater on. When we walk from a dark room into bright sunlight, the pupils of our eyes rapidly contract, which restricts the amount of light entering the eye. A sudden exposure to heat brings on sweating, which increases evaporative heat loss from the skin and helps to maintain body temperature at its normal level.

Such patterns of response represent negative feedbacks, meaning that

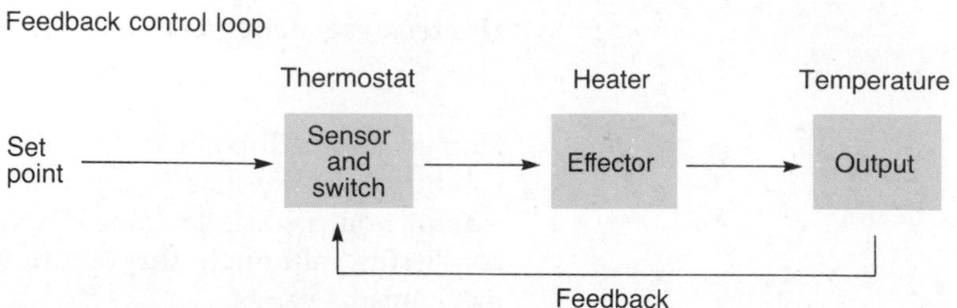

Figure 7–1
The major attributes of a negative feedback system, with particular components indicated for the regulation of body temperature.

when external influences alter a system from its norm, or desired state, internal response mechanisms act to restore that state. The essential elements of a negative feedback system are (1) a mechanism of sensing the internal condition of the organism; (2) a means of comparing the actual internal state with the desired state; and (3) an apparatus to alter the internal condition in the direction of the desired condition. A schematized and simplified negative feedback mechanism for regulating body temperature is illustrated in Figure 7–1.

Homeostasis requires the expenditure of energy.

To maintain internal conditions significantly different from the external environment requires work and energy. Regulation of constant high body temperatures by birds and mammals in cold environments exemplifies the metabolic cost of homeostasis. As air temperatures decrease, the gradient between the internal and external environments increases, and the body surface loses heat proportionately more rapidly. An animal that maintains its body temperature at 40°C loses heat twice as fast at an ambient (that is, surrounding) temperature of 20°C (a gradient of 20°C) as at an ambient temperature of 30°C (a gradient of 10°C). To maintain a constant body temperature, an organism must replace heat that is lost by generating heat metabolically. Thus, the rate of metabolism required to maintain body temperature increases in direct proportion to the difference between body and ambient temperature, all other things being equal (Figure 7–2).

If the only purpose of metabolism were temperature regulation, the individual would require no metabolic heat production when body temperature equaled ambient temperature, at which point no heat would flow between it and its surroundings. But organisms release energy unrelated to temperature regulation to support such functions as heartbeat, breathing, muscle tone, and kidney function, regardless of the ambient temperature. The metabolic rate of an organism that is resting quietly and has not eaten recently is called basal or resting metabolic rate (BMR or RMR). This is the

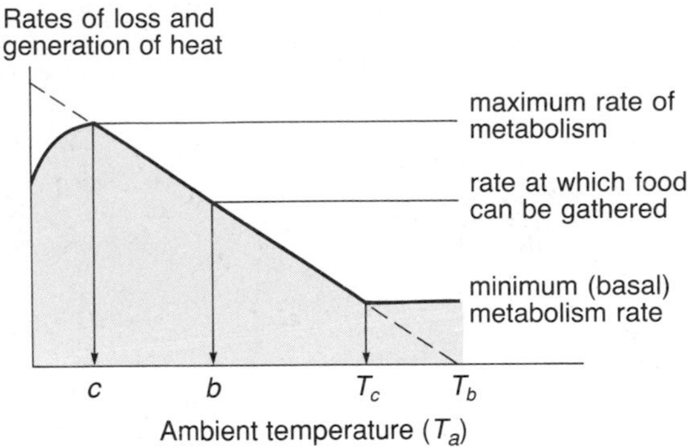

Figure 7-2
Relationship between energy metabolism and ambient temperature for a homeothermic bird or mammal whose body temperature is maintained at T_b. T_c is the lower critical temperature, below which metabolism must increase to maintain body temperature. Points *c* and *b* are the lower lethal temperature and the lowest temperature at which the organism can maintain itself indefinitely.

lowest level of energy release under normal conditions. At this basal level, the individual produces sufficient heat to maintain its body temperature when ambient temperatures exceed a certain level, the lower critical temperature (T_c). This temperature depends on BMR and thermal conductance. As body size increases among species, basal metabolic rate increases more rapidly (allometric slope 0.75) than body surface area (0.67), and thicker fur and feathers tend to reduce thermal conductances. As a result, T_c decreases with increasing size, from about 30°C in sparrow-sized birds to below 0°C in penguins and other large species (Calder and King 1974).

An organism's ability to sustain a high body temperature while exposed to very low ambient temperatures is limited over the short term by its physiological capacity to generate heat, and over the long term by its ability to gather food to satisfy the metabolic requirements of generating heat. The maximum rate at which an organism can perform work generally is no more than 10 to 15 times its basal metabolism (Kleiber 1961, King 1974). Over the course of a day, few organisms expend energy at a rate exceeding four times BMR (Drent and Daan 1980). When the environment becomes so cold that heat loss exceeds the organism's ability to produce heat (point *c* in Figure 7-2), body temperature begins to drop, a condition fatal to most homeotherms.

The lowest temperatures that homeotherms can survive for long periods often depend on their ability to gather food (point *b*) rather than their ability to assimilate and metabolize the energy in food. Animals can quite literally starve to death at low temperatures because they metabolize food energy more rapidly than they can gather food. In temperate and arctic climates, the coldest temperatures are often associated with storms and snow accumulation, thereby taxing both the physiological and ecological sources of heat energy.

Points *c* and *b* in Figure 7-2 might appropriately be called the lower

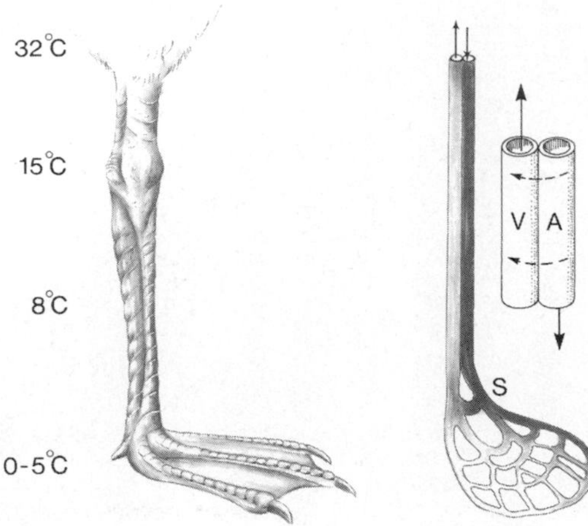

Figure 7–3
Skin temperatures of the leg and foot of a gull standing on ice. The anatomical arrangement of blood vessels and countercurrent heat exchange between arterial blood (A) and venous blood (V) are diagrammed at right. Arrows indicate direction of blood flow, and dashed arrows indicate heat transfer. A shunt at point S allows the gull to constrict the blood vessels in its feet, thereby reducing blood flow and heat loss further, without having to increase its blood pressure. (After Irving 1966, and Schmidt-Nielsen 1983.)

critical physiological temperature and the lower critical ecological temperature. Below *c*, organisms die if exposed too long. Between *c* and *b*, they survive but only for relatively brief periods and with a negative energy balance. Above *b*, energy balance is positive and not only can organisms survive indefinitely, but they also can engage in energy-consuming activities besides foraging.

When homeostasis costs more than the individual can afford, certain economy measures are available. For example, one may lower the regulated temperature of portions of the body, thereby reducing the temperature difference between air and body. Because the legs and feet of birds are not feathered, they would be a major avenue of heat loss in cold regions if they were not kept at a lower temperature than the rest of the body (Figure 7–3). Gulls accomplish this by a countercurrent heat exchanger in which warm blood in arteries leading to the feet cools as it passes close to the veins that return cold blood to the body (Scholander 1955, Scholander and Schevill 1955). In this way, heat is transferred from arterial to venous blood and transported back into the body rather than lost to the environment.

Because of their small size, hummingbirds have a large surface area relative to their weight and consequently lose heat rapidly compared with their ability to produce heat. As a result, hummingbirds require very high metabolic rates to maintain their at-rest body temperature near 40°C. Species that inhabit cool climates would risk starving to death overnight if they did not become torpid, a condition of lowered body temperature and inactivity resembling hibernation. The West Indian hummingbird, *Eulampis jugularis,* drops its temperature to between 18 and 20°C when resting at night. It does not cease to regulate body temperature, but merely changes

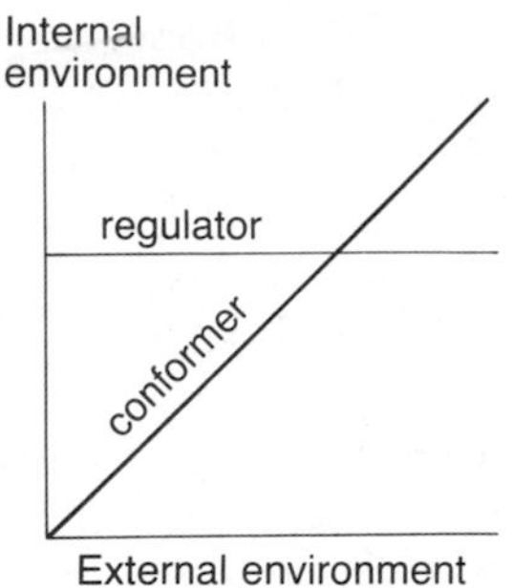

Figure 7–4
Relationship between internal and external environments in idealized regulating and conforming organisms. Regulators maintain constant internal environments with homeostatic mechanisms whereas conformers allow their internal environments to follow changes in the external environment.

the setting on its thermostat to reduce the difference between ambient and body temperature (Hainsworth and Wolf 1970, Hainsworth et al. 1977).

The level of regulation balances costs and benefits.

Animals that maintain constant internal environments are usually referred to as regulators; those that allow their internal environments to follow external changes are called conformers (Figure 7–4). Few organisms are ideal conformers or regulators. Frogs regulate the salt concentration of their blood but conform to external temperature. Even warm-blooded animals are partial temperature conformers: in cold weather, our hands, feet, noses, and ears—our exposed extremities—become noticeably cool.

Organisms sometimes regulate their internal environments over moderate ranges of external conditions, but conform under extremes. Small aquatic amphipods of the genus *Gammarus* regulate the salt concentrations of their body fluids when placed in water with less concentrated salt than their blood, but not in water with more concentrated salt (Figure 7–5). The freshwater species *G. fasciatus* regulates the salt concentration of its blood at a lower level than the saltwater species *G. oceanicus,* and thus begins to conform to concentrated salt solutions at a lower level. In their natural habitat, however, neither the freshwater species nor the saltwater species encounters salt more concentrated than that in its blood. Among animals that inhabit salt lakes and brine pools, however, the salt concentration of the blood is actively kept below that of the surrounding water. You will recall that the brine shrimp *Artemia* maintains an internal salt concentration below 3 per cent even when placed in a 30 per cent salt solution (Croghan 1958).

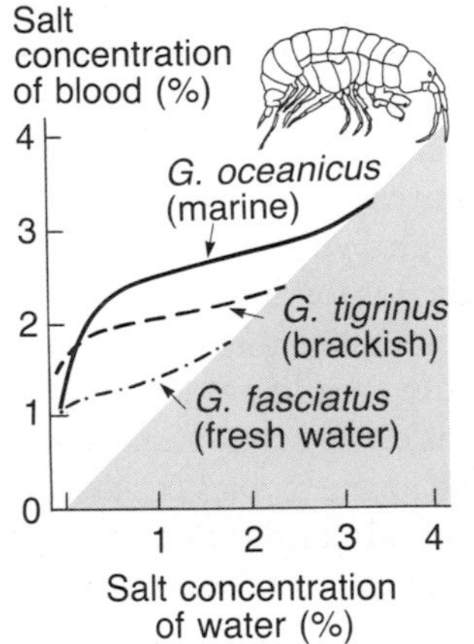

Figure 7–5
Salt concentration in the blood of three gammarid crustaceans from different habitats as a function of the salt concentration of their external environment. The normal salt concentration of seawater is 3.5 per cent. (After Prosser and Brown 1961.)

Whereas most fully aquatic invertebrates cannot regulate the salt concentration of their blood below that of the surrounding water, land crabs and many intertidal invertebrates that are periodically exposed to air do possess this capability (Prosser 1973). This seems puzzling until one recalls that land animals must sometimes tolerate the loss of body water. When blood volume decreases, salt concentration increases—dangerously so unless the organism can excrete salt. In addition, impermeable outer skins or shells, which restrict water loss when the individual is exposed to air, also reduce the movement of salt into the body when submersed. Thus, biological solutions to problems of water and salt balance imposed by terrestrial environments enable organisms to regulate the salt concentrations of their body fluids to levels below those in their environment.

Body temperatures further demonstrate the phenomenon of partial regulation. Only birds and mammals are referred to as endotherms in that they generate heat metabolically to regulate body temperature. But pythons maintain high body temperatures while incubating eggs (Gans and Pough 1982, Hutchison et al. 1966, Vinegar et al. 1970). Some large fish, such as the tuna, use a countercurrent arrangement of blood vessels to maintain temperatures up to 40°C in the center of their muscle masses (Carey et al. 1971);

swordfish employ specialized metabolic heaters, derived from muscle tissue, to keep their brains hot (Carey 1982, Block 1987). Large moths and bees often require a preflight warm-up period during which flight muscles shiver to generate heat (Heinrich 1979, Heinrich and Bartholomew 1971). Even among plants, temperature regulation based upon metabolic heat production has been discovered in the floral structures of philodendron and skunk cabbage (Knutson 1974, Nagy et al. 1972).

Clearly, organisms other than birds and mammals are physiologically capable of generating heat to maintain elevated body temperatures, and many do so under certain conditions. Why, then, is the distribution of endothermy throughout the animal and plant kingdoms so limited? Part of the answer certainly lies in a consideration of body size. Birds and mammals are relatively large as animals go. As body size increases, volume increases relatively more rapidly than surface area, across which heat leaves the body. In general, the lower the ratio of surface to volume, the more comprehensive and precise regulation can be made.

Although body size may explain why mammals are endotherms and insects generally are not—large moths that exhibit preflight warm-up approach the size of small mammals—large fish and reptiles also generally have not made the shift to homeothermy. For most fish, low availability of oxygen and high rate of heat loss in the aquatic environment, owing to the high thermal conductance of water, preclude the high metabolic rates necessary for temperature regulation in all but the most active species. The metabolic rates of resting birds and mammals may be 10 times those of fish, amphibians, and reptiles of similar size (Schmidt-Nielsen 1984).

Why have reptiles not evolved temperature regulation? After all, the body sizes of contemporary reptiles are comparable to those of small mammals, and they breathe air. The major innovation leading to endothermy appears to have been the insulation provided by fur and feathers. Birds and mammals evolved from reptile lineages but did not displace their ancestors completely, probably owing to advantages of ectothermy under certain environmental conditions; reptiles lack the high activity levels of "warm-blooded" animals, but their low energy requirements are well suited to highly seasonal and erratic food supplies (Pough 1980).

Most homeotherms maintain their body temperature considerably above that of the surrounding air. Logic tells us that to minimize the energetic cost of temperature regulation, temperature should be regulated close to the average ambient temperature. Why do birds and mammals maintain such high body temperatures whether they live in the tropics or close to the poles (McNab 1966)? The advantages of high body temperatures are several. First, increased temperature raises the level of sustained activity and may increase alertness, both contributing to capture of food and avoidance of predators. Second, elevated body temperature reduces the need for frequent use of evaporative cooling to dissipate body heat. For terrestrial organisms, potential savings in water may partly justify the energetic cost of elevated temperature. Third, a large gradient in temperature between body and environment may allow more rapid and precise adjustment of heat loss in response to changes in rate of activity than is possible with a small gradient.

These considerations illuminate only one facet of the optimization of physiological functions. Clearly, however, organisms have developed a variety of physiological mechanisms, morphological devices, and behavioral ploys to lessen the tension in their relationships to the physical environment. The adaptations of desert birds and mammals to their stressful environment illustrate this versatility particularly well.

Temperature regulation and water balance in hot deserts require diverse homeostatic adaptations.

Among vertebrates, birds are perhaps the most successful inhabitants of the desert. Certainly, they remain active in the heat long after other animals have sought refuge. The success of birds derives from their low excretory water loss (recall that birds excrete nitrogen as crystallized uric acid rather than as urea) and from feeding on insects, from which they obtain some free water. Even some seed-eating species can persist without water in the desert, provided they avoid both full sun and shade temperatures above 35°C.

But the behavior of the cactus wren, a desert insectivore (Figure 7–6), shows that it too must respect the physiological demands of its climate. In cool air, wrens lose 2 to 3 cm^3 of water each day in the air they exhale. Water loss increases rapidly above 30 to 35°C, to over 20 cm^3 per day at 45°C; active birds might use water at five times that rate to dissipate their heat load. The wren's body contains about 25 cm^3 of water. In the cool temperatures of the early morning, wrens forage throughout most of the environment, actively searching for food among foliage and on the ground (Ricklefs and Hainsworth 1968). As the day brings warmer temperatures, they select cooler parts of their habitat, particularly the shade of small trees and large shrubs, always managing to avoid feeding where the temperature of the microhabitat exceeds 35°C (Figure 7–7). When the minimum temperature in the environment rises above 35°C, wrens become less active; they even feed their young less frequently during hot periods.

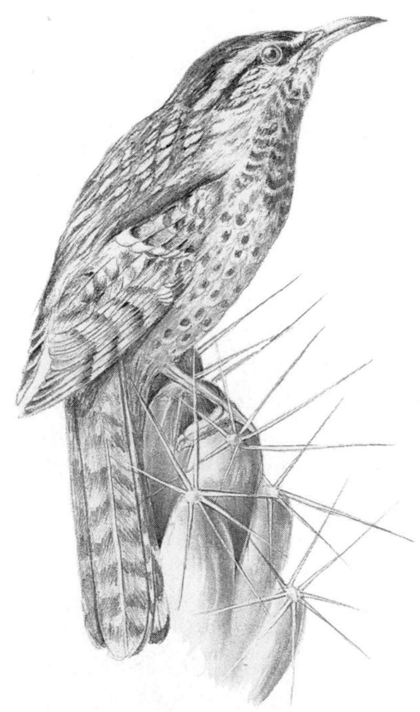

Figure 7–6
The cactus wren *(Campylorhynchus brunneicapillus)*, a conspicuous resident of deserts in the southwestern United States and northern Mexico.

Many desert birds build enclosed nests or place their nests in holes in the stems of large cacti, where the young are protected from the sun and from extremes of temperature. Cactus wrens build an untidy nest, a bulky and somewhat haphazard ball of grass, with a side entrance. Once a pair of wrens have built their nest, they cannot change its position or orientation. For a month and a half, from the laying of the first egg until the young leave the nest, the nest must provide a suitable environment day and night, in hot and cool weather. Cactus wrens usually rear several broods of young during the breeding period of March through September in southern Arizona. They build early nests so that the entrances face away from the direction of the cold winds of early spring; during the hot summer months, nests are oriented to face prevailing afternoon breezes, which circulate air through the nest and facilitate heat loss (Ricklefs and Hainsworth 1969; Figure 7–8). It makes a difference! Nests oriented properly for the season are consistently

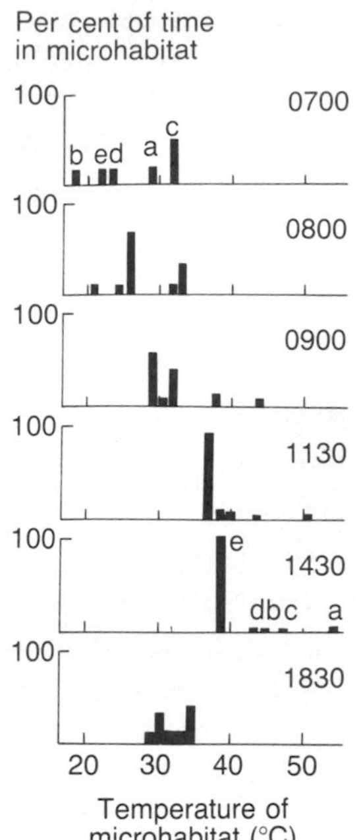

Figure 7-7
Microhabitat use by cactus wrens in southeastern Arizona during the course of a day in late spring. Microhabitats vary in degree of thermal stress between exposed ground (a) and the deep shade of trees (e). Wrens distribute their activity among all microhabitats in the cool hours of early morning (0700 h) but restrict their activity to cool shade (e) during the hottest part of the day (1430 h) when other microhabitats are above 40°C. (From Ricklefs and Hainsworth 1968.)

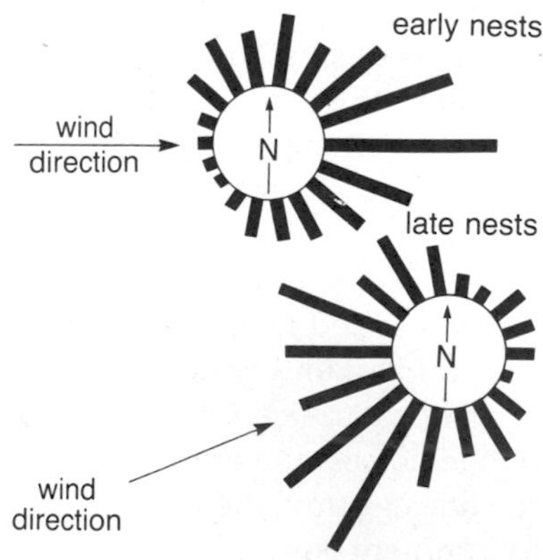

Figure 7-8
Orientation of nest entrances of the verdin *(Auriparus flaviceps)* during the early (cool) and late (hot) part of the breeding season in Nevada and Arizona. Orientation of cactus wren nest entrances is similar. (Data courtesy of G. T. Austin.)

more successful, 82 per cent, than nests facing the wrong direction, 45 per cent (Austin 1974).

Cactus wrens use water so conservatively that they do not waste even the water in the feces of their young. The fecal sacs, which adults remove in order to keep the nest sanitary during cool parts of the year, are left in the nest during hot weather. The evaporation of water from the fecal sacs presumably helps to cool the air in the nest, and the increased humidity reduces respiratory water loss.

The reuse of fecal and excretory water is common in desert organisms. In the desert iguana, secretions from salt glands, located in the orbits of the eyes, run down to small pits at the entrance of the nasal passages where the water evaporates into the inhaled air, thereby reducing respiratory water loss (Dunson 1976). Some storks defecate on their legs during periods of hot weather, thus benefiting from evaporative cooling of their fecal water. Adult roadrunners eat the fecal sacs of their young; the kidneys of adults apparently have greater urine-concentrating abilities than the kidneys of the young, so adults can extract water from the ingested fecal sacs (Calder 1968). These examples merely emphasize the diversity of adaptations—morphological, physiological, and behavioral—that are brought to bear as solutions to a single problem: heat dissipation and water balance in desert organisms.

Graininess describes the scale of spatial and temporal variation in the environment.

The suitability of the environment for each organism varies over time and space. At any given moment, the cactus wren may feed within any of several parts of the habitat, each with different thermal characteristics and access to food. These distinctive patches of microhabitat shift continuously throughout the daily cycle of solar radiation and temperature, and seasonally as both the physical environment and populations of suitable prey organisms vary. The wren must adjust behaviorally and physiologically to these ever-changing conditions.

Appropriate responses to temporal changes in environmental conditions differ depending on the duration of the change. The choices presented to organisms by spatial variation depend on the distance between patches of different habitat type. But what seems long or brief, and near or far, depends upon the life span, response time, and mobility of the individual. Humans do not distinguish upper and lower surfaces of a leaf as we push aside the branch of a tree, but these present different worlds to the aphid that sits on the undersurface and sucks plant juices. A tropical storm or passing cold front may mean little more to us than spoiling plans for a picnic, but may completely encompass the adult life of a mayfly during which it must mate and lay eggs.

These different perspectives on the scale of environmental variation were summarized by Richard Levins (1968) in his concept of grain. Imagine for a moment that environmental alternatives consist of patches of different conditions distributed as a mosaic in time and space, resembling a patchwork quilt or the pattern made by alternating fields of different crops. Patches have different sizes, and patterns of various kinds of patches can be superimposed. For example, patches of air temperature tend to be larger than patches of soil moisture because topographic influences subdivide the soil moisture landscape more finely.

The concept of grain relates the size of a patch to the activity space of the organism. We define a coarse-grained environment as one in which the patches are relatively so large that the individual can choose among them. In a fine-grained environment, patches are so small the individual cannot usefully distinguish among them, and the environment appears essentially uniform. To anyone other than a trained botanist, a field appears as a uniform carpet of plants—a type of patch readily distinguished from patches of forest, marsh, beach, and so on. But to the caterpillar, a single plant within the field may be its home for the duration of its larval life. Grasshoppers, which can fly from plant to plant, choose carefully among the various species in the field, for some provide better feeding than others. Therefore, whereas individual plants in a field appear as fine grains to us, they are coarse grains to small insects. Needless to say, grain also depends on activity. If we set out in the field to pick flowers of a particular kind we, like grasshoppers, would perceive individual plants as coarse grains.

Patches may be thought of as occurring in time as well as space. At one extreme, conditions that fluctuate through a daily cycle or over a shorter period may be thought of as fine-grained to most organisms, inasmuch as such rapid changes offer too little time for animals and plants to undergo major adjustments in response. At the other extreme, seasonal changes and longer trends are decidedly coarse-grained for most organisms. Given persistent changes of several months duration, plants and animals can undertake morphological changes, such as dropping their leaves and increasing the thickness of their fur, corresponding changes in physiological mechanisms, and, in the case of animals, migrations to areas with more favorable conditions.

An organism's choice of environment patches defines its activity space.

Animals actively move among patches of conditions in a coarse-grained environment as the environment changes temporally. Conditions within each spatially defined patch vary over diurnal and seasonal cycles; by moving among them animals can remain within ranges of conditions close to their optima. Although plants cannot uproot themselves and move, most regulate their activity according to the suitability of conditions at a particular time. Simply by closing the stomates on their leaves, plants can shut themselves off from some unfavorable conditions.

Most plants cannot choose where they grow; they survive in suitable sites, elsewhere they die. British plant ecologist John Harper and his coworkers have shown that for proper germination, seeds require quite specific combinations of light, temperature, and moisture, which vary even among closely related species (Harper et al. 1965). Irregularities in the surfaces of natural soils provide the variety of conditions needed to allow the germination of many species, but Harper dramatized the differences between species by creating an artificially heterogeneous soil environment. Plantains are common lawn and roadside weeds in the genus *Plantago.* Three species sown in seedbeds responded differently to the modifications in environment produced by slight depressions, by squares of glass placed on the soil surface, and by vertical walls of glass or wood (Figure 7–9). Relatively few seeds germinated on the smooth surfaces of soil that had not been disturbed experimentally.

Unlike plants, animals can choose where they live. Such choices may encompass major categories of habitat, such as forest, grassland, or marsh, within which an individual may spend all its time (Hilden 1965; Wecker 1963, 1964; Cody 1985). Or they may allow the individual to follow the most suitable conditions within a habitat, as we have seen the cactus wren do. As the conditions within different patches change, the activity space also changes.

The diurnal behavior cycle of lizards is geared to the varying temperatures of habitat patches (Heatwole 1970, Gans and Pough 1982). Although few lizards generate heat metabolically for temperature regulation, they do take advantage of solar radiation and warm surfaces to maintain their body temperatures within the optimum range. At night, these sources of heat are

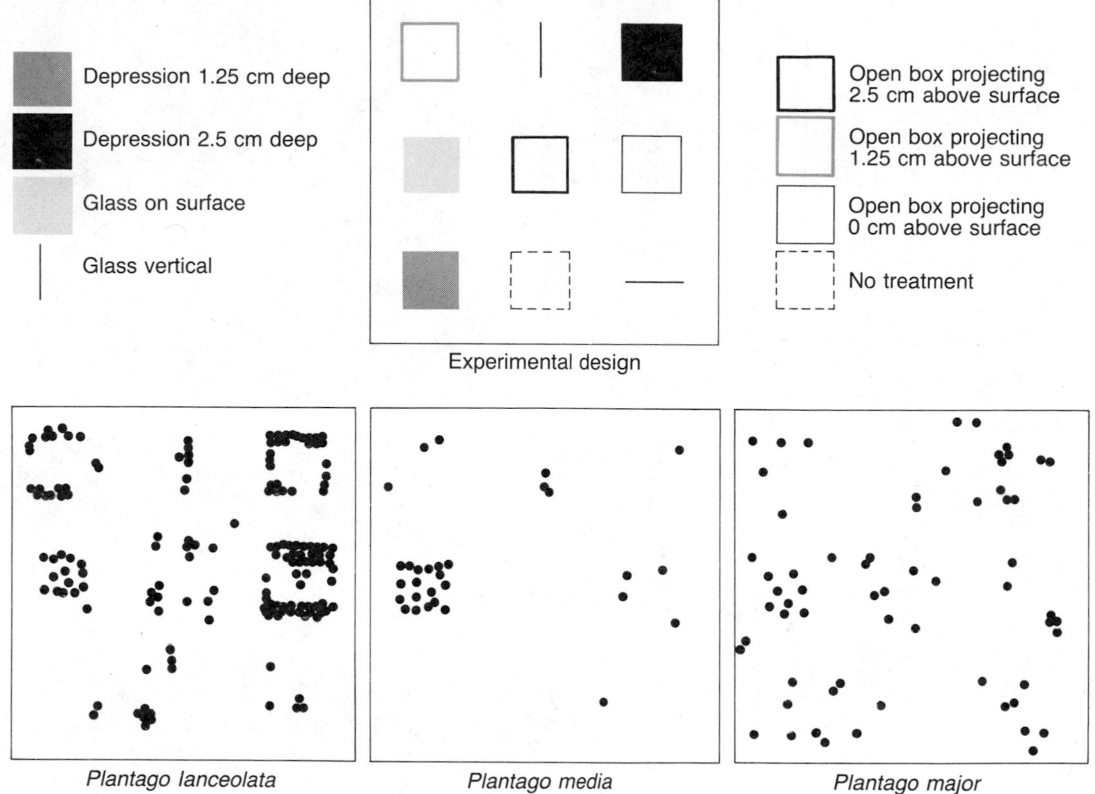

Figure 7–9
Germination of seedlings of three species of plantains (genus *Plantago*) with respect to artificially produced variation in the soil surface. (After Harper et al. 1965.)

not available, and the lizard's body temperature gradually drops to that of the surrounding air. The mallee dragon *(Amphibolurus fordi),* an agamid lizard of Australia, is fully active only when its body temperature falls between 33 and 39°C (Cogger 1974). In the early morning, before its body temperature has risen above 25°C and when it still moves sluggishly, the mallee dragon basks within large clumps of the grass of the genus *Triodia,* within which it finds protection from predators (Figure 7–10). In fact, *Amphibolurus* depends so much upon these grass clumps that it occurs only where *Triodia* grows. When the temperature of an individual rises above 25°C, it leaves the *Triodia* clump and basks in the sunshine nearby, with its head and body in direct contact with the ground surface, from which it absorbs additional heat. When body temperature enters the range for normal activity (33 to 39°C), *Amphibolurus* ventures farther from *Triodia* clumps to forage, its head and body normally raised above the ground as it moves. When body temperature exceeds 39°C, it moves less rapidly and seeks the shade of small *Triodia* clumps; above 41°C, it reenters large *Triodia* clumps, at whose centers it finds cooler temperatures and deeper shade. It

Figure 7–10
The mallee dragon *(Amphibolurus fordi)* at different times during its activity cycle: *top right*, early morning basking in *Triodia* grass clump; *left*, midmorning basking on ground (note body flattened against surface to increase exposed profile and contact with warm soil); *bottom right*, normal foraging attitude. (Courtesy of H. Cogger, from Cogger 1974.)

may also pant to dissipate heat by evaporation. If heat stress is not avoided, *Amphibolurus* loses locomotor ability above 44°C, and will die if body temperature exceeds 46°C.

On a typical summer day, during which air temperature varies from about 23°C at dawn to 34°C at midday, the mallee dragon does not begin to forage until about 0830 hours. By 1130, the habitat has become too hot for normal activity and most individuals seek shade and become inactive. By 1430, the habitat has cooled off enough for the dragons to resume foraging, but by 1800, it has cooled so much that they must retreat back into *Triodia* clumps and their bodies rapidly cool. Individuals remaining in the open after this time of day are sluggish and easily caught by warm-blooded predators.

The desert iguana *(Dipsosaurus dorsalis)* of the southwestern United States faces a more severe environment, with greater annual fluctuation,

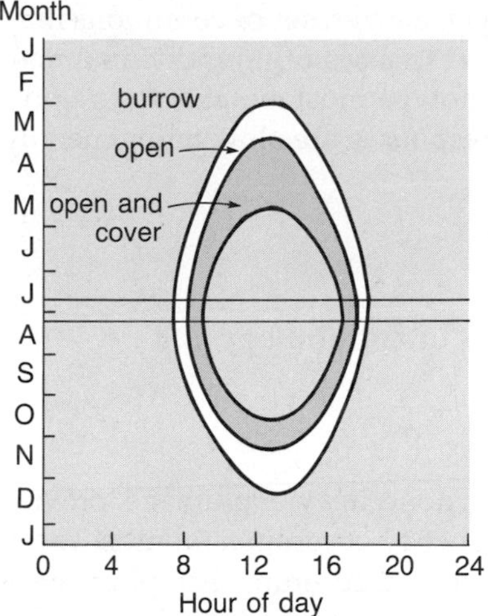

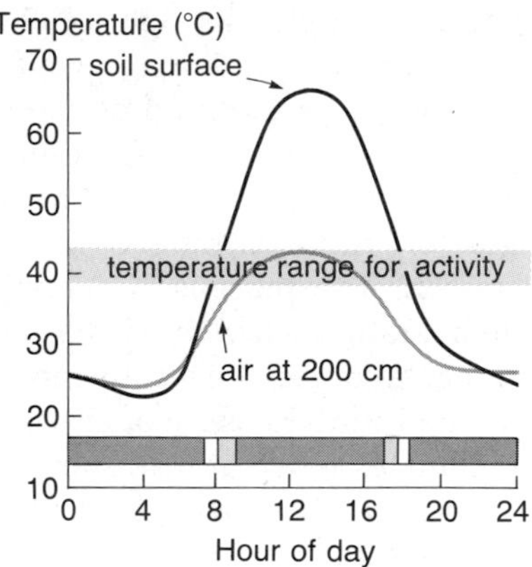

Figure 7–11
Seasonal activity space of the desert iguana *(Dipsosaurus dorsalis)* in southern California. *Top:* the daily activity budget for an entire seasonal cycle. *Bottom:* the activity budget for July 15 is shown with the time course of environmental temperature. (After Beckman et al. 1973.)

than the mallee dragon (Beckman et al. 1973, DeWitt 1967, Norris 1953, Porter et al. 1973). Shade temperatures can reach 45°C in summer and plunge below freezing in winter. During mid-July, the thermal environment courses so rapidly between extremes that the desert iguana can be normally active within its preferred body temperature range of 39 to 43°C for only about 45 minutes in mid-morning and a similar period in the early evening (Figure 7–11). During the remainder of the day, it seeks the shade of plants or the coolness of its burrow, where the temperature rarely rises above the preferred range. At night, the desert iguana enters its burrow, where it is safe from predators; at dawn the burrow is warmer than the desert surface, and so the early morning warm-up period is correspondingly brief.

Whereas in summer the desert iguana restricts its activity to two brief bouts separated by inactivity through midday to avoid heat stress, it finds more favorable temperatures for activity in spring. The thermal environment in May does not exceed the preferred range of *Dipsosaurus,* and individuals forage actively aboveground from 0900 to 1700, only occasionally seeking the cool shade of plants. Winter cold restricts *Dipsosaurus* to brief periods of activity in the middle of the day when body temperature rises to the point that individuals can come aboveground and forage. Between early December and the end of February, most days are so cold that the desert iguana cannot venture from its burrow.

For the lizard, patches of different environmental temperature are coarse-grained and vary both in time and space. Its response to this environment, particularly its movement between patches, allows it to exploit this mosaic of conditions to best advantage.

Homeostatic responses vary in their time courses.

The time course of a response to changing conditions must be substantially shorter than the period of environmental change. Otherwise today's form and function may reflect yesterday's conditions. For each type of response, some types of environmental fluctuations are coarse-grained, others are fine-grained. Responses fall naturally into three general categories: regulatory, acclimatory, and developmental responses. Regulatory responses are accomplished most rapidly, developmental responses most slowly.

Regulatory responses involve both changes in the rate of physiological processes and changes under behavioral control. Examples are the mallee dragon's metabolic response to cold stress and its shade-seeking behavior under heat stress. These responses do not require modification of existing morphology or biochemical pathways.

Acclimatory responses involve more substantial change, such as thickening of fur in winter, increase in number of red cells in the blood at high altitude, and production of enzymes with different temperature optima. These changes may be thought of as shifts in the ranges of the regulatory responses of the individual. Both regulatory and acclimatory responses are reversible, as they must be to follow the ups and downs of the environment.

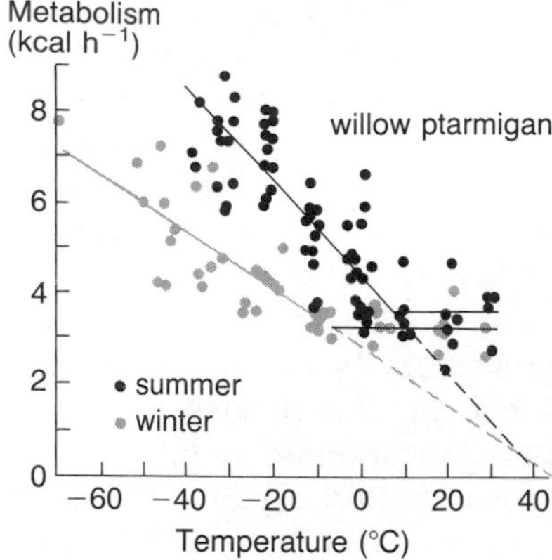

Figure 7–12
Metabolic responses of willow ptarmigan acclimated to summer and winter temperatures. Winter-acclimated birds have thicker plumages, providing better insulation than summer birds. Hence their metabolic rates are lower at any given temperature, and their lower critical temperatures are also lower. (After West 1972.)

When the environment changes slowly, a given set of conditions may persist during the adult life span and the responses of an individual may alter its development to produce the phenotype most suitable to the prevailing conditions. Such developmental responses are slow and generally are not reversible.

Acclimation involves reversible changes in morphology and physiology.

During the cold winter months, many birds don a heavier plumage, providing greater insulation, than they have during the hot summer months. These species replace their body feathers in spring and autumn; each plumage is suited to the typical conditions of the environment between each molt. The willow ptarmigan, a ground-feeding arctic bird, trades its lightweight, brown summer plumage in the fall for a thick, white winter plumage, which provides both insulation and camouflage against a background of snow. With increased insulation, ptarmigan expend less energy to maintain their body temperatures during the winter (Figure 7–12). Seasonal change in plumage thickness effectively shifts the regulatory response range to match the prevalent temperature range of the season. The ptarmigan needs a similar metabolic rate to maintain its body temperature when the surroundings are −40°C in winter and −10°C in summer (a conceivable temperature in its arctic home). Although winter-acclimated individuals would seem well adapted for both winter and summer climates when at rest, summer activity combined with a winter plumage would quickly produce heat prostration. Adjusting insulation to enhance heat conservation in winter and to facilitate heat dissipation in summer maintains constant body temperature at the least possible cost.

Temperature-conforming animals and plants also acclimate to seasonal changes in their environments (Schmidt-Nielsen 1983, Hochachka and Somero 1984). By switching between enzymes and other biochemical systems with different temperature optima, cold-blooded animals adjust their tolerance ranges in response to prevalent environmental conditions. The relationship of the swimming speed of fish to water temperature shows at once the capabilities and limitations of acclimation (Fry and Hart 1948). Goldfish swim most rapidly when acclimated to 25°C and placed in water between 25 and 30°C, conditions that closely resemble their natural habitat (Figure 7–13). Lowering the acclimation temperature to 15°C increases the swimming speed at 15°C but reduces it at 25°C. (Increased tolerance of one extreme often brings reduced tolerance of the other.) Reducing acclimation temperature further, to 5°C, well beyond the normal lower range experienced by goldfish in nature, does not noticeably increase swimming speed, but does reduce swimming speed at moderate temperature.

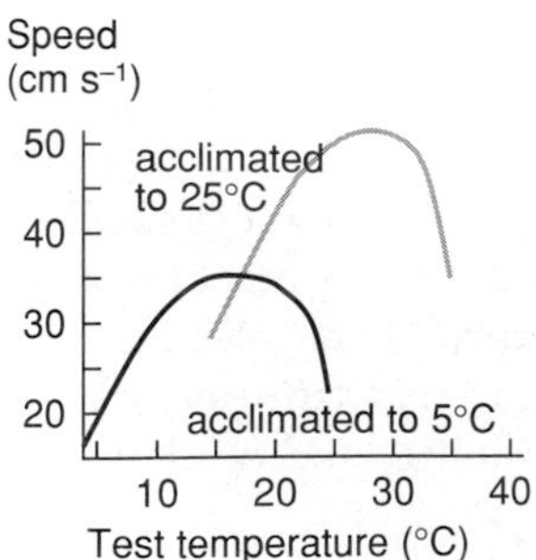

Figure 7–13
Swimming speed of the goldfish as a function of temperature. Separate curves are graphed for individuals acclimated to 5°C and 25°C. (After Fry and Hart 1948.)

When goldfish are placed in environments whose temperatures approach their upper lethal limits, they exhibit progressive stages of behavioral abnormality, passing from hyperexcitability to equilibrium loss and

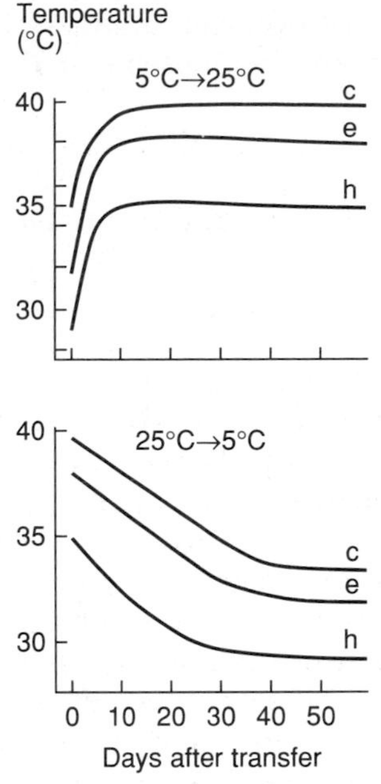

Figure 7–14
Time courses of acclimation for three behavioral indices to high temperature stress—hyperexcitability (h), loss of equilibrium (e), and coma (c)—in goldfish. Individuals were first acclimated to either 5°C *(top)* or 25°C *(bottom)* and then transferred to the other temperature. (From Cossins et al. 1977.)

coma. The temperatures at which these syndromes appear can be shifted by acclimation over periods of 2 to 6 weeks, as shown in Figure 7–14. A. R. Cossins and his co-workers (1977) found that these neurological traits were strongly correlated with the fluidity of lipids in the membranes of nerve synapses, apparently affecting the transmission of nerve impulses to the muscles. Acclimation depends on turnover of lipid components of the membrane, which, as one would expect, requires longer at 5°C than at the higher acclimation temperature of 25°C.

Acclimation of photosynthetic rate to temperature shows that the capacity for acclimation is often linked to the range of temperatures experienced in the natural environment (Figure 7–15). *Atriplex glabriuscula,* a species of saltbush native to cool coastal regions of California, does not increase its photosynthetic rate at high temperature when acclimated at 40°C; plants acclimated to 16°C are uniformly more productive at all temperatures. The thermophilic species *Tidestromia oblongifolia* cannot acclimate to cool temperatures. Two species that inhabit interior deserts, but are photosynthetically active during the cool winters as well as the hot summers (*Atriplex hymenelytra* and *Larrea divaricata* [creosote bush]), show the classic shift in temperature optima characteristic of thermal acclimation. The basis

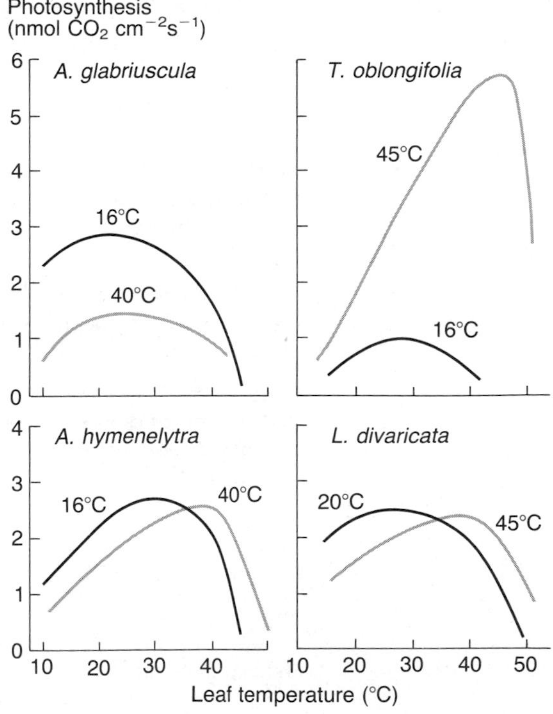

Figure 7–15
Light-saturated photosynthetic rate as a function of leaf temperature in four species of plants (genera *Atriplex, Tidestromia,* and *Larrea*) grown under moderate and hot temperatures. (From Hochachka and Somero 1984, after Bjorkman et al. 1980.)

for this acclimation appears to be associated with changes in viscosity of membranes directly related to photosynthetic pathways (Raison et al. 1980).

Developmental responses are conspicuous in plants and in animals with several generations per year.

In many organisms, growth and differentiation are sensitive to environmental variation. Such developmental responses are generally not reversible; once fixed during development, they remain unchanged for the remainder of the organism's life. Developmental responses cannot accommodate short-term environmental changes owing to their long response time and irreversibility; as a general rule, therefore, plants and animals exhibit developmental flexibility only in environments with persistent variation for the individual. When environmental changes occur slowly compared to the life span of an organism, as do seasonal conditions for short-lived animals, developmental responses are often appropriate. Also, for plant species whose seeds may settle in many different kinds of habitats, the strategy of developmental flexibility makes good sense.

Light intensity is one of many factors that can influence the course of development in plants. Loblolly pine seedlings grown in shade have smaller root systems and more foliage than seedlings grown in full sunlight (Bormann 1958). Because the shaded environment taxes the plant's water economy less, shade-grown seedlings can allocate more of their production to stem and needles. Sun-grown seedlings must develop more extensive root systems to obtain sufficient water. The larger proportion of foliage of the shade-grown seedlings results in a higher rate of photosynthesis per plant under given light conditions, particularly under low light intensity (Table 7–1).

A more complicated example of developmental response involves wing development of water striders, which are freshwater bugs of the genus

Table 7–1
Distribution of dry matter between roots, needles, and stems, and rates of photosynthesis, in loblolly pine *(Pinus taeda)* seedlings grown under shade and full sunlight

	Shade-grown	Sun-grown
Per cent of dry weight in*		
Roots	35	52
Needles	47	37
Stem	18	11
Photosynthetic rate ($mgCO_2\ hr^{-1}\ gm^{-1}$)†		
Low light intensity	1.9	1.0
Moderate light intensity	4.6	4.0
High light intensity	7.2	6.6

*Six-month-old seedlings.
†Four-month old seedlings; light intensities were 500-, 1500-, and 4500-foot candles.
Source: After Bormann 1958.

Table 7–2
Wing lengths of water striders *(Gerris)* inhabiting bodies of freshwater with different levels of permanence and predictability

Characteristic of habitat	Characteristic wing length*	Mechanism of determination
Permanent	Short	Genetic
Fairly persistent but unpredictable	Both short and long	Genetic dimorphism
Seasonal	Seasonally dimorphic (summer)	Developmental switch
Very unpredictable	Long	Genetic

*Short wings are not functional and prevent dispersal.
Source: After Vepsalainen 1974.

Gerris (Brinkhurst 1959; Vepsalainen 1973, 1974a, 1974b, 1974c). European species fall into four categories of wing length depending on the place in which they live (Table 7–2). At one extreme, species inhabiting large, permanent lakes have short wings, or none at all, and do not disperse between lakes. At the other extreme, species living in temporary ponds usually have long, functional wings and disperse to find suitable sites for breeding each year. Between these extremes, species characteristic of small ponds, which are more or less persistent from year to year but tend to dry up during the summer, frequently have both long-winged and short-winged forms.

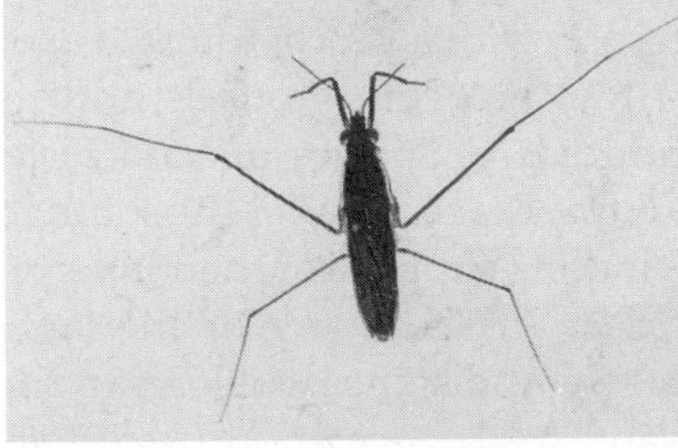

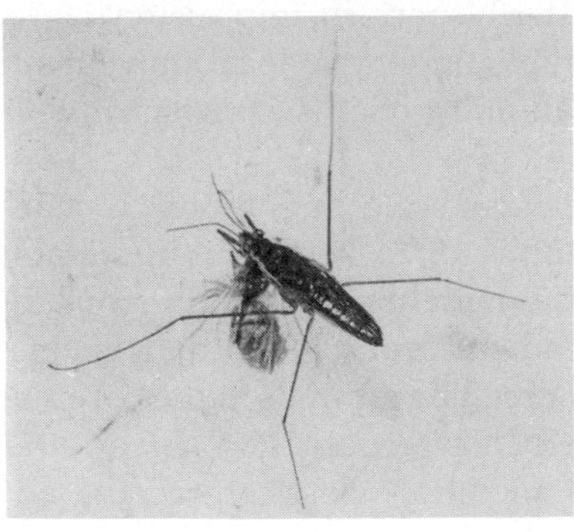

Figure 7–16
Alary polymorphism in the water strider *Gerris odontogaster*. The macropterous (long-winged) form is above and the micropterous (short-winged) form is below. (Courtesy of K. Vepsalainen.)

The life cycle of most *Gerris* species in central Europe, England, and southern Scandinavia includes two generations per year. The first (summer) generation hatches during the spring, reproduces during the summer, and then dies. The second, hatched from eggs laid by females of the summer generation, develops to the adult stage during late summer, then overwinters before breeding in early spring the following year. In species that inhabit seasonal ponds, the summer generation is dimorphic, having both long-winged and short-winged forms (Figure 7–16). All the individuals in the winter generation have long wings and can fly. They leave the pond in late summer and move into nearby woodlands for the winter. In the spring, they return to small bodies of water to lay eggs.

Dimorphism in the summer generation reflects two extreme strategies. The long-winged forms can fly to other habitats if their pond dries, especially when this happens early in the season. The short-winged forms gamble on their pond persisting. What do water striders gain by producing short wings rather than functional ones? It has been suggested that the short-winged forms convert nutrients that would have become wings and flight muscles into eggs, but present evidence is equivocal.

Because all of the winter generation have long wings, seasonal dimorphism must be controlled during development. K. Vepsalainen (1971) found that wing length is determined primarily by day length. When length of the day increases continually during the larval period and, in southern Finland, exceeds 18 hours during the last nymphal (immature) stage prior to adulthood, individuals grow short wings. When day length begins to decrease before the end of the nymphal stage (as it does when larval development extends beyond June 21, the summer solstice), long-winged adults are produced. Thus summer-generation individuals become short-winged or long-winged depending upon hatching date and rate of nymphal development. The switch between long and short wings also is influenced by temperature (Vepsalainen 1974); high temperatures, which cause ponds to dry quickly, favor the development of long-winged forms. The kind of developmental response described by Vepsalainen for wing length of water striders is only one of a large number of similar responses in various insects with short generations (for example, see Fraser Rowell 1970, Lamb and Pointing 1972, Lees 1966, Steffan 1973, Young 1965).

Migration, storage, and dormancy allow organisms to tolerate seasonally unsuitable conditions.

In many parts of the world, freezing temperatures, extreme drought, low light, and other adverse conditions prevent animals and plants from pursuing their usual activities. Under such conditions, organisms resort to a number of extreme responses. These include moving to another region where conditions are more suitable, relying on resources stored during bountiful seasons for use during lean periods, or becoming inactive.

Many animals, particularly among those that fly or swim, undertake extensive migrations (Baker 1978, Gathreaux 1981). The arctic tern probably holds the record for long-distance migration with a yearly round trip of 30,000 kilometers between its North Atlantic breeding grounds and its antarctic wintering grounds. Each fall hundreds of species of land birds leave temperate and arctic North America, Europe, and Asia for the south in anticipation of cold winter weather and dwindling supplies of their invertebrate food (Keast and Morton 1980). In East Africa, many of the large ungulates, such as the wildebeest, migrate long distances following the geographical pattern of seasonal rainfall and fresh vegetation (Maddock 1979). A few insects, like the monarch butterfly, perform impressive migratory movements each year, and some others undertake local movements (Dingle 1978), but most species of temperate and arctic insects overwinter in dormant stages as eggs or pupae.

Some marine organisms undertake large-scale migrations to reach spawning grounds, to follow a food supply, or to keep within suitable temperature ranges for development. The migration of salmon from the ocean to their spawning grounds at the headwaters of rivers and the reverse migration of adult freshwater eels to their breeding grounds in the Sargasso Sea are striking examples (Jones 1968, McCleave et al. 1984, Gross et al. 1988). Many whales undertake seasonal migrations between feeding and breeding areas (for example, see Jones et al. 1984).

Some populations exhibit irregular or sporadic movements that are tied to food scarcity during particular years rather than to seasonal conditions (Svardson 1957). The occasional failure of cone crops in coniferous forests of Canada and the mountains of the western United States forces large numbers of birds that rely on seeds to move to lower elevations or latitudes.

Birds of prey that normally feed on rodents disperse widely when their prey populations decline sharply (Gross 1947, Snyder 1947). Even insects are subject to irregular movements. Outbreaks of migratory locusts, from areas of high local density where food has been depleted, can reach immense proportions and cause extensive crop damage over wide areas (Gunn 1960, Waloff 1966; Figure 7–17). This behavior is a developmental response to population density (Uvarov 1961). When the locusts grow up in sparse populations, they become solitary and sedentary as adults. In dense populations, frequent contact with others stimulates young individuals to develop a gregarious, highly mobile behavior, often leading to mass emigration following local depletion of food resources.

Although homeostasis and migration help maintain function in the face of a changing physical environment, environmental changes often plunge organisms from feast into famine. When an environment marginally supports life, even small fluctuations in food or water supply can be critical; to prevent disaster in such circumstances, many plants and animals store resources during periods of abundance for use in times of scarcity. During infrequent rainy periods desert cacti store water in their succulent stems. Plants growing on infertile soils absorb more nutrients than they require in times of abundance and use them when soil nutrients are depleted. Many

Figure 7–17
A dense swarm of migratory locusts in Somalia, Africa, in 1962. (Courtesy of the U.S. Dept. of Agriculture.)

temperate and arctic animals accumulate fat during mild weather in winter as a reserve of energy for periods when snow and ice make food sources inaccessible. Some winter-active mammals (beavers and squirrels) and birds (acorn woodpeckers and jays) cache food supplies underground or under the bark of trees for later retrieval (Ritter 1938, Swanberg 1951, Anderson and Krebs 1978, Tomback 1980, Sherry 1984). In habitats that frequently burn severely—as in the chaparral of southern California—perennial plants store food reserves in fire-resistant root crowns, which sprout and send up new shoots shortly after a fire has passed (Figure 7–18).

The environment sometimes becomes so extremely cold, dry, or nutrient-poor that animals and plants can no longer function normally; under such circumstances, those incapable of migration enter physiologically dormant states. Many tropical and subtropical trees shed their leaves during seasonal periods of drought; temperate and arctic broad-leaved trees shed theirs in the fall because they cannot obtain from frozen soil the moisture needed to maintain them (Vegis 1964). For many small invertebrates and cold-blooded vertebrates, freezing temperatures directly curtail activity and lead to dormancy. Many mammals hibernate because they lack food, not because they are physiologically unable to cope with the physical environment.

In most species, conditions requiring dormancy are anticipated by a series of physiological changes (for example, production of antifreezes, dehydration, fat storage) that prepare the organism for a complete shutdown (Prosser 1973, Mrosovsky 1976, Clutter 1978, Gregory 1982, Lyman 1982).

Prior to winter, insects enter into a resting state known as diapause, in which water is chemically bound or reduced in quantity to prevent freezing, and metabolism drops to near nought (Hochachka and Somero 1984, Lee et al 1987). In summer diapause, drought-resistant insects either allow their

Figure 7–18
Root-crown sprouting by chamise *Adenostoma fasciculatum)* following a fire in the chaparral habitat of southern California. The photograph on the left was taken on May 4, 1939, 6 months after the burn. The photograph on the right showing extensive regeneration, was taken on July 16, 1940. (Courtesy of the U.S. Forest Service.)

bodies to dry out and tolerate desiccation, or they secrete an impermeable outer covering to prevent drying. Plant seeds and the spores of bacteria and fungi have similar dormancy mechanisms (for example, see Koller 1969, Wareing 1966).

Regardless of the mechanism, dormancy serves the single purpose of reducing exchange between the organism and its environment. In this way animals and plants ride out unfavorable conditions and await better ones before resuming an active and interactive state.

Proximate cues enable organisms to anticipate predictable environmental change.

What stimulus indicates to birds wintering in the tropics that spring approaches in northern forests, or urges salmon to leave the seas and migrate upstream to their spawning grounds? How do aquatic invertebrates in the arctic sense that if they delay entering diapause a quick freeze may catch them unprepared for winter? J. R. Baker (1938) made the important distinction between proximate factors — cues by which organisms can assess the state of the environment — and ultimate factors — features of the environment that bear directly upon the well-being of the organism. Virtually all plants and animals sense the length of the day (photoperiod) as a proximate factor that indicates season, and many can distinguish periods of lengthening and shortening days (Bunning 1967, Lofts 1970, Vince-Prue 1975, Murton and Westwood 1977, Beck 1980). For example, the oriental fruit moth enters diapause in the early fall when days become less than 13 hours long (Figure 7–19), corresponding to the date of early frosts (Dickson 1949).

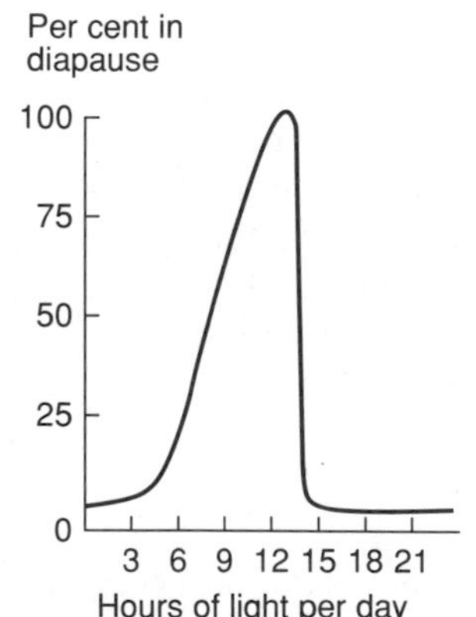

Figure 7–19
The influence of day length upon entrance into diapause by larvae of the oriental fruit moth *(Grapholitha molesta)*. Note that the sharpest break in the response occurs between 13 and 14 hours of day length, corresponding to early fall, which is biologically the most critical point for the inception of diapause. (After Dickson 1949.)

Similar organisms may differ strikingly in their responses to photoperiod in different areas. Under controlled cycles of light and dark, southern populations (30°N) of side oats gramma grass flowered when day length was 13 hours, whereas more northerly populations (47°N) flowered only when the light period exceeded 16 hours each day (Olmsted 1944). The longer period of light suppressed flowering in the southern populations. At 45°N in Michigan, populations of small freshwater crustaceans, known as water fleas *(Daphnia)*, form diapausing broods at photoperiods of 12 hours of light (mid-September) or less (Stross and Hill 1965). In Alaska, at 71°N, related species enter diapause when the light period decreases to fewer than 20 hours per day, which corresponds to mid-August (Stross 1969). Warm temperature and low population density tend to shorten the day length necessary for diapause (and hence delay the inception of diapause in the fall), suggesting that these portend more favorable environmental conditions for *Daphnia*.

When day length does not accurately predict sporadically changing conditions, animals and plants must take their cues more directly from changes in ultimate factors in the environment. Annual cycles in equatorial regions, where day length is nearly constant, follow upon seasonal cycles of rainfall and their effects on humidity and vegetation. In such highly unpredictable environments as deserts, many organisms adopt a conservative strategy of readiness during the entire period in which sporadic rains are likely to occur. In some desert birds, photoperiod stimulates the development of reproductive organs to a point just short of breeding. The gonads maintain this state of readiness throughout the time when rains might occur, but rainfall itself finally kicks off the completion of their physiological development and the initiation of breeding (Marshall and Disney 1957, Immelmann 1971).

Ecotypic differentiation reflects adaptation to local conditions.

Botanists have long recognized that individuals of a species grown in different habitats may exhibit various forms corresponding to the conditions under which they are grown. In many cases, these differences among habitats result from developmental responses, but experiments on some species have revealed genetic adaptations to local conditions; the phenotypes of individuals are fixed, but vary among individuals from place to place. In the hawk-weed *Hieracium umbellatum,* for example, woodland plants generally have an erect habit; those from sandy fields are prostrate; and those from sand dunes are intermediate in form. Leaves of the woodland ecotype are broadest, those of the dunes ecotype are narrowest, and those of sandy fields intermediate. Plants from sandy fields are covered with fine hairs, a trait the others lack. When individual plants of the same species are taken from different habitats and grown in a garden under identical conditions, differences in growth form persist generation after generation.

About 65 years ago, the Swedish botanist Göte Turesson collected seeds of several species of plants that occurred in a wide variety of habitats and grew them in his garden. He found that even when grown under identical conditions, many of the plants exhibited different forms depending upon their habitats of origin. Turesson (1922) called these forms ecotypes, a name that persists to the present, and suggested that ecotypes represented genetically differentiated strains of a population, each restricted to specific habitats. Because Turesson grew these plants under identical conditions, he realized that the differences between the ecotypes must have had a genetic basis, and that they must have resulted from evolutionary differentiation within the species according to habitat.

Jens Clausen and co-workers David Keck and William Hiesey (1948) conducted similar experiments on a species of yarrow, *Achillea millefolium,* in California. *Achillea,* a member of the sunflower family, grows in a wide variety of habitats ranging from sea level to more than 3000 meters elevation. Clausen collected seed from plants at various points along the altitude gradient and planted them at Stanford, California, near sea level. Although the plants were grown under identical conditions for several generations, individuals from montane populations retained their distinctively small size and low seed production (Figure 7–20), thereby demonstrating ecotypic

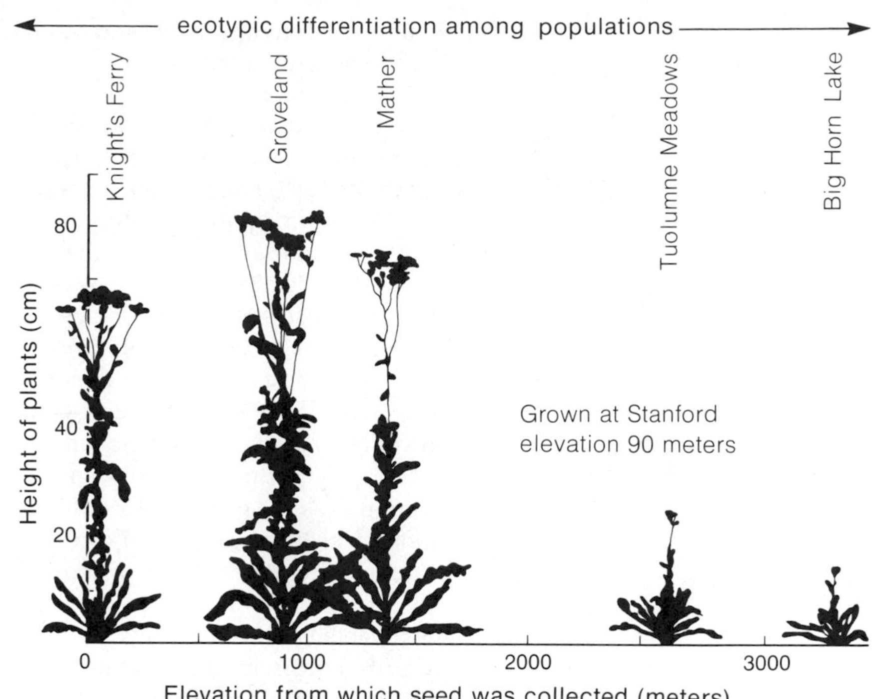

Figure 7-20
Ecotypic differentiation in populations of the yarrow, *Achillea millefolium*, demonstrated by raising plants derived from different elevations under identical conditions in the same garden. (After Clausen et al. 1948.)

differentiation within the population. Such region and habitat differences in adaptations undoubtedly broaden the ecological tolerance ranges of many species by dividing them into smaller subpopulations, each differently adapted to consistent local environmental conditions (Antonovics 1971, Bjorkman 1968, Coyne et al. 1983, Eickmeier et al. 1975, Hiesey and Milner 1965, Kruckenberg 1951, McMillan 1959, McNaughton 1973, Mooney and Billings 1961, Schemske 1984, Silander 1985).

SUMMARY

In varying environments organisms must be able to respond to changing external conditions in order to maintain suitable internal environments. Responses include a variety of physiological changes and morphological modifications, each having a characteristic time period.

1. Maintenance of constant internal conditions, called homeostasis, depends upon negative-feedback responses. Organisms sense changes in

their internal environments and respond in such a manner as to return those conditions toward their optimum.

2. Homeostasis requires energy when a gradient must be maintained between internal and external conditions. This is illustrated by birds and mammals, which maintain elevated body temperatures by generating heat metabolically to balance loss of heat to their cooler surroundings.

3. The magnitude of the body-environment gradient and the constancy of internal conditions that organisms adopt balance the costs of homeostasis (maintaining the physiological apparatus and sustaining gradients) against the benefits of closely regulated internal conditions.

4. A major component of homeostasis is the selection of microhabitats that minimize the body-environment gradient. This is illustrated by temperature-dependent foraging behavior of desert birds and the construction of nests by such species to ameliorate the environments of their chicks.

5. The pattern of distribution of patches of habitat and microhabitat is referred to as the grain of the environment. Coarse habitat grains are ones among which the individual can choose; it cannot distinguish fine grains and therefore utilizes them in proportion to their occurrence.

6. The activity space of an individual, its selection of habitat and microhabitat grains, depends on the suitability of conditions and availability of resources in each patch at each time. This was illustrated by the influence of changing thermal conditions on habitat utilization by desert lizards.

7. Behavioral selection of habitats cannot fully compensate for environmental change, and organisms must bring a variety of homeostatic responses into play. The most rapid of these are regulatory responses, which involve changes in rates of processes and the attitude of the organism, but not change in morphological structure or biochemical pathway. Shivering in response to cold stress and contraction of the pupil in response to bright light are examples. Regulatory responses are rapid and reversible.

8. Acclimatory responses involve reversible changes in structure (for example, fur thickness) and biochemical pathway (induction or changes in the amounts of enzymes and their products), which require longer periods (usually days to weeks) than regulatory responses. Acclimation plays a prominent role in response to seasonal change by long-lived organisms.

9. Developmental responses express the interaction between an organism and its environment during the growth period. Different environments lead to characteristic structures and appearances that generally are not reversible. Responses to sun versus shade by plant seedlings, and the presence or absence of wings in water striders depending on the permanence of their habitat, are examples of developmental responses.

10. When conditions exceed ranges of tolerance, organisms may migrate elsewhere, rely on materials stored during periods of abundance, or enter inactive states (for example, torpor, hibernation, diapause).

11. In many cases, to respond appropriately to environmental change, the individual must anticipate changes in conditions. Organisms rely on proximate cues, such as day length and seasonal changes in climate, to predict changes in ultimate factors, such as food supply and temperature, that directly affect their well-being.

12. When populations become subdivided in habitats with different conditions, each subpopulation may evolve genetic adaptations to the local environment. Such differentiated forms are called ecotypes. Well-known examples come from plants that, because of their low mobility, exhibit evolutionary isolation of subpopulations on small geographic scales.

8

Biological Factors in the Environment

Many factors in the environment affect each individual's success—its ability to survive and reproduce. The form and function of plants and animals evolve in response to these factors to best match the organism to its surroundings. This adaptation of the organism proceeds by selection of those individuals genetically best-suited to the environment at a particular time and place. The winners of the evolutionary contest are those that leave the most progeny.

Each set of physical conditions presumably has its corresponding combination of anatomical, physiological, and behavioral traits that best suits each particular life form, whether an herbaceous plant or a beetle, to local conditions. Once this ideal has been achieved, and providing the environment does not change, the engines of evolution, to the degree they are fueled by the exigencies of physical factors, slow to a halt. Furthermore, because organisms existing side-by-side experience similar conditions, selection by physical factors promotes uniformity of appearance among them. Indeed, plants and animals do arrive at common forms dictated by factors in the environment. For example, in arid habitats around the world, plants form small, leathery leaves as adaptations to reduce water loss. Yet, from the earliest explorations of natural history, variety—not uniformity—has most profoundly held our attention. The diverse forms of animals and plants suggest the application of selective forces over and above those of the physical environment.

The markings that blend moths into their resting places during the day clearly allow them to escape the notice of most predators. By their insistent colors and fragrances flowers present themselves to our attention and to the notice of insects and birds that carry pollen from one flower to the next and effect fertilization. The agents that have selected these adaptations are biological and their effects differ from those of physical factors in two ways. First, biological factors elicit interactive traits: the predator shapes its prey's adaptations for escape, but its own adaptations for pursuit and capture are just as surely shaped by the prey. Second, biological factors tend to diversify adaptations rather than promote convergence. Organisms are specialists with respect to biological factors in their environments; each pursues a different combination of prey, strives to avoid a different combination of predators and disease organisms, and engages in cooperative arrangements with a unique set of pollinators, seed-dispersers, or gut flora.

The physical environment within a habitat offers a variety of places—called microenvironments or microhabitats—each presenting a distinctive set of conditions to the individual. The number of these microhabitats increases with the presence of plants and animals, owing to their effects on local physical conditions. For a small insect, the physical conditions in the forest canopy and in the leaf litter on the ground are worlds apart. This variety allows for some specialization with respect to physical factors, but very little, really, compared to the diverse adaptations of organisms to biological factors in their environments. But because each population of animal, plant, and microbe has a separate evolutionary history and future, and because the interaction of each pair of organisms creates its own special "biological microenvironment," the possibilities for diversification are limited only by the number of kinds of organisms.

Relative size influences the relationship between predator and prey.

When we think of predator and prey, we usually think of lynx and rabbit, of bird and beetle—predators that pursue, capture, and eat individual prey. Such prey are usually smaller than the predator, but large enough to be worth pursuing. Other predators consume minute organisms in vast numbers. The blue whale weighs many tons but eats small, shrimp-like krill, fish fry, and the like. On a smaller scale, clams and mussels pump water through tiny filtering devices that trap minute plankton; many protozoa, sponges, and rotifers filter bacteria and other microorganisms from the water.

As the size of prey increases in relation to predator, prey become more difficult to capture, and predators become specialized for pursuing and subduing their prey (Figure 8–1). Beyond a critical size ratio, predators lack sufficient strength and swiftness to capture potential prey. Lions will attack animals their own size or a little larger, but they are no match for fully grown elephants (Schaller 1972). A few species, such as wolves, hyenas, and army ants, hunt cooperatively and thus can run down and subdue larger prey (Kruuk 1972, Mech 1970, Rettenmeyer 1963).

At the other end of the relative size spectrum we find the "live-on" and "live-in" predators, or parasites—the myriad viruses, bacteria, protozoa, worms, and others that invade the body of the host and feed on its tissues or blood, or on the partially digested food in its intestine. Parasitism differs utterly from predation or filter feeding because the survival of most parasites depends upon the survival of its host rather than its death; the parasite must not bite the hand that feeds it.

Depending upon what parts of a plant they eat, herbivores resemble either predators or parasites. Parasites live off the productivity of a host

Figure 8–1
African lions taking a midday break in Amboseli Park, Kenya. With their powerful legs and jaws, lions can subdue prey somewhat larger than themselves. But because they cannot maintain speed over long distances, successful hunting relies on stealth and surprise.

organism without killing it; therefore, a deer, browsing on trees and shrubs, is functionally a parasite. A sheep that consumes an entire plant, pulling it up by the roots and macerating it into lifeless shreds, is functionally a predator. The beetle larva that develops within a seed, thereby destroying the embryonic plant organism it contains, also is a predator.

Predator adaptations demonstrate the importance of the biotic environment as an agent of natural selection.

The modifications of predators for feeding are too widespread to discuss in detail here, but the following examples illustrate a number of common types.

The nature of the diet often profoundly affects the evolution of teeth. Herbivores, especially those that eat grass, have teeth with large grinding surfaces to break down tough, fibrous plant materials. The teeth of predators have cutting and biting surfaces that both immobilize the prey in the mouth and cut it into pieces small enough to swallow. Seemingly simple differences in dentition reflect important ecological differences. The upper and lower incisors of horses, for example, are strongly opposed so that they

Figure 8–2
Some species of snake have enlarged their gape as much as 20 per cent by shifting the articulation of the jaw with the skull from the quadrate bone (black) to the supratemporal (stippled). (After Gans 1974.)

can cut the fibrous stems of grasses. Other ungulates, such as cows, sheep, and deer, lack upper incisors; their lower teeth press against the upper jaw at an angle for gripping and pulling plant material (Gwynne and Bell 1968).

Many predators use their forelegs to help tear their food into small morsels. Among birds, for example, the hawks, eagles, owls, and parrots use their powerful, sharp-clawed feet and hooked beaks for this purpose. Diving birds often eat large fish, but must swallow them whole because their hind legs are specialized for swimming and diving rather than for grasping and dismantling prey. Some species of snake compensate for their lack of grasping appendages with distensible jaws that enable them to swallow large prey whole (Figure 8–2).

Quality of diet influences adaptations of the predator's digestive and excretory systems as well as structures directly related to procuring food. Because plants contain cellulose and lignin, which make vegetation more difficult to digest than the high-protein diets of carnivores, the digestive tracts of herbivorous animals are often greatly elongated. In addition, many contain sac-like offshoots (caecae of rabbits; rumens of cows) that, like fermentation vats, house bacteria and protozoa which aid digestion (Swenson 1977). With a larger volume of intestines, herbivores can keep meals in the digestive tract longer and digest them more completely. Balancing this seemingly happy situation, herbivores must carry quantities of undigested food in their bellies, adding weight and reducing mobility.

To locate and capture food, predators possess senses commensurate with their habitats, feeding tactics, and their prey's ability to avoid detection (Gordon 1968, Prosser 1973, Lythgoe 1979). We ourselves primarily use vision to locate food, particularly as it is now displayed on the shelves of supermarkets. Yet our sight is pitiable compared to that of hawks and falcons (Fox et al. 1976), and most insects perceive ultraviolet light invisible to us (Figure 8–3). Insects also can detect rapid movement, such as that of wings beating 300 times per second; we see such movement as only a blur.

Among the more unusual sensory organs of predators are the pit organs of pit vipers, a group that includes the rattlesnake. The pit organs, located on each side of the head in front of the eyes, detect the infrared (heat) radiation given off by the warm bodies of potential prey—a sort of "seeing in the dark" (Grinnell 1968; Figure 8–4). Pit vipers are so sensitive to infrared radiation that they can detect a small rodent several feet away in less than a second. Moreover, because the pits are directionally sensitive, vipers can locate warm objects precisely enough to strike them.

A few aquatic animals have developed the sensory ability to detect electrical fields (Bullock 1982). Some species of "electric fish" produce continuous discharges of electricity from specialized muscle organs, creating a weak electric field around them. Nearby objects distort the field, and these changes are picked up by receptors on the surface of the electric fish (Machin and Lissman 1960). Some species use electric signals to communicate between individuals. The specialized electric ray *Torpedo* uses powerful electrical currents, up to 50 volts at several amperes, to defend itself and to kill prey (Keynes and Martins-Ferreira 1953). As one might expect, the production and sensation of electric fields are most highly developed in fish inhabiting murky waters. In other habitats with poor visibility, bottom-

Figure 8–3
The appearance of flowers to the human eye *(left)* and to eyes that are sensitive to ultraviolet light *(right). Above:* marsh marigolds; *below:* five species of yellow-petaled Compositae from central Florida. (Courtesy of T. Eisner; from Eisner et al. 1969.)

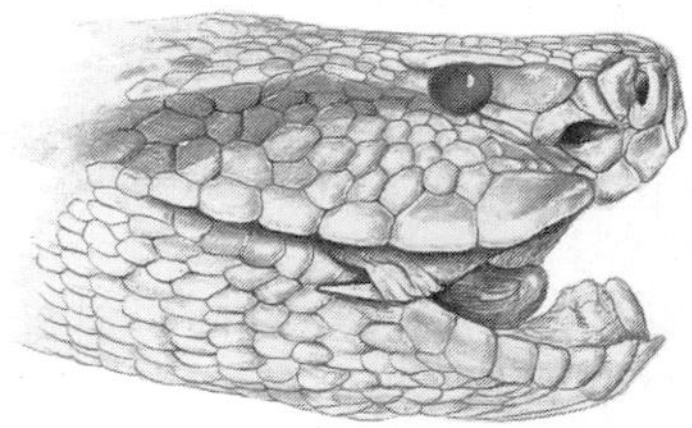

Figure 8–4
Head of the western rattlesnake *(Crotalis viridis)*, showing the location of the infrared-sensitive pit between, and slightly lower than, the eye and the nostril. (After a photograph in Burkhardt et al. 1967.)

dwelling species such as catfish use elongated fins and barbels around the mouth as sensitive touch and taste receptors.

In contrast to the magnificent senses of many predators, others perceive their surroundings only dimly and rely upon chance to bump into prey. (Of course, their prey must be equally oblivious for this tactic to work.) But even such "blind" predators adopt searching patterns that increase the chances of encountering prey. For example, the predatory larvae of the ladybird beetle feed on mites and aphids that infest the leaves of certain plants, and they must physically contact their prey to recognize them (Figure 8–5). Their movements on the leaves are not oriented toward the prey, but neither are they random. The veins and rims of leaves make up less than 15 per cent of the leaf surface, yet larvae spend most of their time searching on these areas — which they recognize by touch — where almost 90 per cent of aphids are distributed (Dixon 1959).

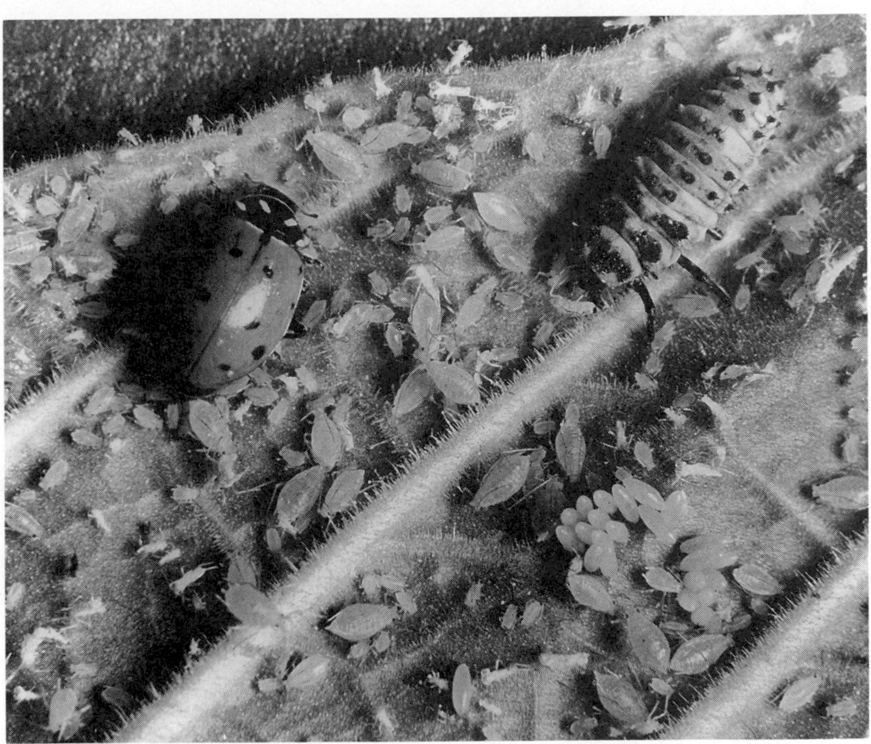

Figure 8-5
Adult and larval ladybird beetles (family Coccinellidae) feeding on aphids in a laboratory culture. Note the hairs on the veins of the leaf, which deter the aphids from penetrating the plant and sucking its juices. (Courtesy of the U.S. Dept. of Agriculture.)

Prey can hide, flee, or fight to escape predation.

The adaptations of prey organisms to avoid predators are as diverse as the adaptations of predators. While organisms that are extremely small compared to their predators—those captured by filter feeders, for example—exhibit few adaptations to avoid being caught, larger quarry may either hide, fight, or flee. Many circumstances of the predator-prey relationship determine which strategy, or combination of strategies, minimizes the chance of being caught. Grassland offers no hiding places for large ungulates, and so escape depends on early detection of predators and on swiftness (Figure 8-6). Plants cannot flee and must rely on thorns and toxic chemicals to fend off herbivores.

Protective defenses rarely involve physical combat because few prey can match their predators, and predators carefully avoid those that can. Instead, seemingly defenseless organisms produce foul-smelling or stinging chemical secretions to dissuade predators (Eisner and Meinwald 1966). Whip scorpions and bombardier beetles direct sprays of noxious liquids at threatening animals. Many plants and animals are inedible because they

Figure 8–6
These wildebeest in the Masai Mara region of Kenya rely on early detection and rapid flight to escape predators.

contain toxic substances (Whittaker and Feeny 1975). Slow-moving animals, such as the porcupine and armadillo, protect themselves with spines or armored body coverings. Such defenses cost time, energy, and materials, which may be limited in supply. That organisms accept this burden of antipredator adaptations indicates the major influence of predation on fitness, and the strength of predators as agents of natural selection.

Crypsis, warning coloration, and mimicry demonstrate the complexity of antipredator adaptations.

The perfection of cryptic appearances and positions to avoid detection by predators fascinates observers of the natural world, and bears testament to the force and pervasiveness of natural selection (Cott 1940, Wickler 1968).

Many organisms achieve crypsis by matching the color and pattern of the background upon which they rest. Various animals resemble inedible objects: sticks, leaves, flower parts, even bird droppings. The great pains that evolution has taken to conceal their heads, antennae, and legs underscores the importance of these cues to predators. In the stick-mimicking phasmids (stick insects) and leaf-mimicking katydids, legs are often concealed in the resting position either by being folded back upon themselves or upon the body, or by being protruded in a stiff, unnatural fashion (Robinson

Figure 8–7
This Central American mantis of the genus *Acanthops* resembles a dead, curled-up leaf and thus escapes the notice of most predators.

1969). The dead-leaf-mimicking mantis *Acanthops* partially conceals its head under its folded front legs (Figure 8–7). Asymmetry is also a good cover for animals, but it is difficult to achieve. The leaf-mimicking moth *Hyperchiria nausica* produces the appearance of an asymmetrical midvein by folding one forewing over the other (Figure 8–8). Moths sometimes rest with a leg protruding to one side, but not to the other, or with the abdomen twisted to one side to break their symmetry.

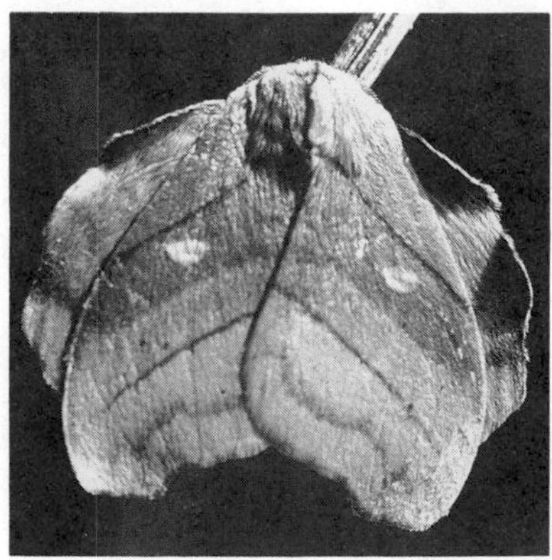

Figure 8–8
The moth *Hyperchiria nausica* partly obliterates its symmetry by folding one wing over the other.

The behavior of cryptic organisms must correspond to their appearance. A leaf-mimicking insect resting on bark, or a stick insect moving rapidly along a branch, will not likely fool many predators. Mimicry involves behavior as well as form, often expressed as subtle choice of appropriate backgrounds (Sargent 1966, 1969; Endler 1984). The behavior of cryptic insects also negates the characteristics to which their predators are most sensitive. For example, many katydids and mantids apparently avoid detection either by moving very slowly or by rocking from side to side while they move, mimicking natural leaf movements (Robinson 1969).

When predators discover cryptic organisms, they may be confronted with a variety of second-line defenses, including startle-and-bluff displays and various attack-and-escape mechanisms (Blest 1964). The green caterpillar of the hawkmoth *Leucorampha omatus* normally assumes a cryptic position. When disturbed, however, it puffs up its head and thorax, looking for all the world like the head of a small poisonous snake, complete with a false pair of large shiny eyes; the caterpillar consummates this display by

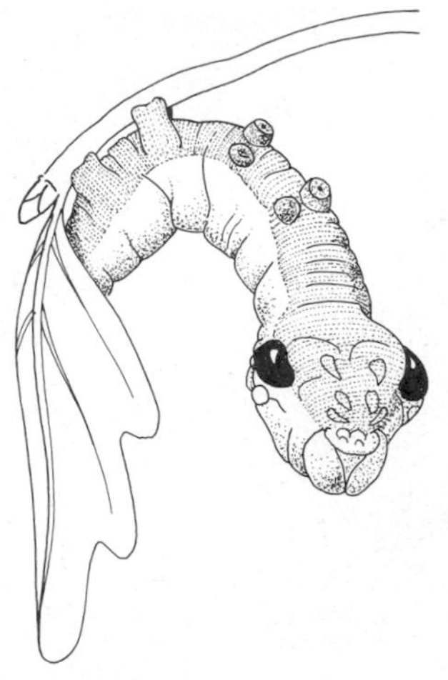

Figure 8–9
Snake display by the caterpillar of the sphingid moth *Leucorampha omatus*. (From Robinson 1969.)

weaving back and forth, while hissing like a serpent (Figure 8–9). The eyespots displayed by many moths and other insects when disturbed (Figure 8–10) frighten their predators, because they resemble the eyes of large birds of prey (Blest 1957).

The appropriate line of defense sometimes depends on features of the environment other than predators. Large-bodied moths at low elevations in the tropics rarely exhibit such protective displays as eyespots or flash coloration. During the day, the air temperature rises to within 6°C of the optimum working temperature of the flight muscles (Heinrich and Bartholomew 1971), and moths may escape from predators by flight. At higher elevations (1500 meters), moths require a lengthy preflight warm-up period because air temperature may be 15°C lower than the working temperature of the flight muscles. Because the cold precludes escape, a large proportion of moth species at these elevations do have special protective displays (Blest 1963).

Crypsis is a strategy of palatable animals. Others have rejected crypsis and taken a bolder approach to predator defense: they produce noxious chemicals, or accumulate them from food plants, and advertise the fact with conspicuous color patterns. Predators learn quickly to avoid such conspicuous markings as the black and orange stripes of the monarch butterfly; its taste causes loss of appetite for monarchs for some time to come (Brower 1969). It is not a coincidence that many aposematic (warningly colored) forms adopt similar patterns of coloration: black and either red or yellow stripes characterize such diverse animals as yellow-jacket wasps and coral snakes. Predators appear to have a generalized instinctive aversion for such patterns (Smith 1975, 1979).

Distasteful animals and plants that display warning coloration often serve as models for the evolution of mimicking color patterns by palatable forms (Wickler 1968, Rettenmeyer 1970). Some potential prey may even

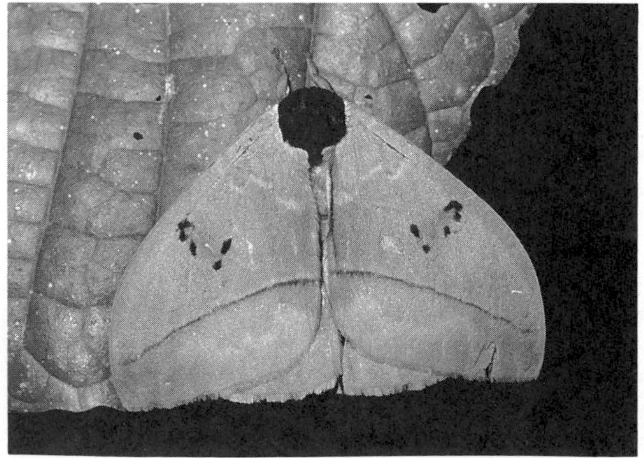

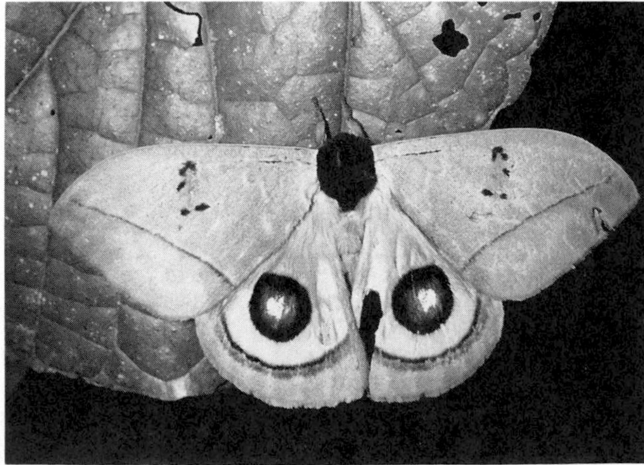

Figure 8–10
The eyespot display of an automerid moth from Panama: *left*, normal resting attitude; *right*, reaction when touched.

Figure 8–11
The wing markings of the tephritid fly *Rhagoletis pomonella* closely resemble the forelegs and pedipalps of jumping spiders (Salticidae). (Courtesy of the U.S. Dept. of Agriculture.)

resemble their predators to avoid predation (Figure 8–11; Mather and Roitberg 1987, Green et al. 1987). These relationships are collectively referred to as Batesian mimicry, named after its discoverer, the nineteenth-century English naturalist Henry Bates. In his journeys to the Amazon region of South America, Bates found numerous cases of palatable insects that had forsaken the cryptic patterns of close relatives and had come to resemble brightly colored, distasteful species. Bates rightly guessed their deception.

Experimental studies have subsequently demonstrated that mimicry does confer advantage to the mimic. Jane and Lincoln Brower (1962) showed that toads fed live bees thereafter avoided the palatable drone fly, which mimics bees; when toads were fed only dead bees from which the stings had been removed, they relished the drone fly mimics. Similar results were obtained with blue jays as predators; the distasteful monarch butterflies were the models and their viceroy butterfly mimics were the experimental subjects (Brower 1958).

The study of crypsis provided one of the first demonstrations of evolutionary change caused by natural selection, a confirmation of evolutionary theory that bears repeating here in order to emphasize the responsiveness of the phenotype to changes in its environment.

Studies on industrial melanism demonstrated natural selection in action.

The English have always been avid butterfly and moth collectors, and such enthusiasts look carefully for rare variant forms. Early in the nineteenth century, occasional dark (or melanistic) specimens of the common peppered moth *(Biston betularia)* were collected. Over the succeeding 100 years, the dark form, referred to as *carbonaria,* became increasingly common in industrial areas until, at present, it makes up nearly 100 per cent of some populations. The phenomenon aroused considerable interest among geneticists, who showed by cross-mating light and dark forms of the moth that melanism is an inherited Mendelian trait.

In the early 1950s, H. B. D. Kettlewell, an English physician who had been practicing medicine for 15 years and an amateur butterfly and moth collector, changed the course of his life to pursue the study of industrial melanism (Kettlewell 1959). Several facts about melanism were known before Kettlewell began his studies: (1) the melanistic trait is an inherited characteristic, hence its spread reflected genetic changes in populations; (2) the earliest records of *carbonaria* were from forests near heavily industrialized regions of England; (3) where there is relatively little industrialization, the light form of the moth still prevails. It was also known that melanism is not unique to the peppered moth; dark forms have appeared in many other moths and other insects (Bishop and Cook 1980).

The peppered moth inhabits dense woods and rests on tree trunks during the day. Kettlewell reasoned that where melanistic individuals had become common, the environment must have been altered in some way to give the dark form a greater survival advantage than the light form. Could natural selection have led to the replacement of the "typical" light form by *carbonaria?* To test this hypothesis, Kettlewell had to find some measure of fitness other than the relative evolutionary success of the two forms.

To determine whether *carbonaria* had greater fitness than typical peppered moths in areas where melanism occurred, Kettlewell chose the mark-release-recapture method. Adult moths of both forms were marked with a dot of cellulose paint and then released. The mark was placed on the underside of the wing so that it would not call the attention of predators to a moth resting on a tree trunk. Moths were recaptured by attracting them to a mercury vapor lamp in the center of the woods and to caged virgin females at the edge of the woods. (Only males could be used in the study because females are attracted neither to lights nor to virgin females.)

In one experiment, Kettlewell (1955, 1956) released 201 typicals and 601 *carbonaria* in a wooded area near industrial Birmingham, where the tree bark was darkened by pollution. The results were as follows:

Number of moths recaptured	Typicals	Melanics
Number of moths released	201	601
Number of moths recaptured	34	205
Per cent recaptured	16.0	34.1

These figures indicated that more of the dark form survived over the course of the experiment.

Although consistent with Kettlewell's original hypothesis, the results could be interpreted otherwise: as differential attraction of the two forms to the traps or as differential dispersion of the two forms away from the point of release. Variables besides differential mortality had to be accounted for.

To test the hypothesis of natural selection unequivocally, Kettlewell ran a control experiment in an unpolluted forest near Dorset, where the tree bark was lighter, with the following results: of 496 marked typicals released, 62 (12.5 per cent) were recaptured; of 473 marked melanics, 30 (6.3 per cent) were recaptured. Thus, in the unpolluted forest, light adults had a higher recapture rate than dark adults. If typicals and melanics were differently attracted to light traps or dispersed from a release point at different rates, the level of pollution would not have influenced the results. In fact, only differential survival could account for a reversal in the relative rates of recapture. This confirmed Kettlewell's hypothesis and established natural selection as responsible for the high frequency of *carbonaria* in industrial areas.

The specific agent of selection was easily identified. Kettlewell reasoned that in industrial areas pollution had darkened the trunks of trees so much that "typical" moths stood out against them and were readily found by predators. Any aberrant dark forms would be better camouflaged against the darkened tree trunks, and their coloration would confer survival value (Figure 8–12). Eventually, differential survival of dark and light forms would lead to changes in their relative frequency in a population. To test this idea, Kettlewell placed equal numbers of the light and dark forms on tree trunks in polluted and unpolluted woods and watched them carefully at some distance from behind a blind. (A blind is a tentlike structure intended to conceal an observer from his subjects, more appropriately called a "hide" in England.) He quickly discovered that several species of birds regularly searched the tree trunks for moths and other insects and that these birds more readily found a moth which contrasted with its background than one which blended with the bark. Kettlewell tabulated the following in-

Figure 8–12
Typical and melanistic forms of the peppered moth at rest on a lichen-covered tree trunk in an unpolluted countryside *(left)* and on a soot-covered tree trunk near Birmingham, England *(right)*. (From the experiments of H. B. D. Kettlewell.)

stances of predation:

	Individuals taken by birds	
	Typicals	Melanics
Unpolluted woods	26	164
Polluted woods	43	15

These data are fully consistent with the results of the mark-release-recapture experiments. Together they clearly demonstrate the operation of natural selection, which over a long period resulted in genetic changes in populations of the peppered moth in polluted areas. Many decades were required for the replacement of one form by the other. The agents of selection were insectivorous birds whose ability to find the moths depended on the coloration of the moth with respect to its background. Kettlewell's study shows clearly how the interaction between the organism and its environment determines its fitness.

Host specificity and complex life cycles characterize many parasites.

Parasites usually are much smaller than their prey, or hosts, living either on their surfaces (ectoparasites) or inside their bodies (endoparasites). Both types demonstrate characteristic adaptations to their way of life. First, because parasites generally live inside of, or in close association with, a larger organism, they expend little effort to maintain their own internal environments. Thus, endoparasites often have little more than food-processing and egg-producing capacities. Second, parasites must disperse through the hostile environment between hosts. Many accomplish this through complicated life cycles, one or more stages of which can cope with the external environment. These principles apply equally to such parasitic plants as mistletoes (Kuijt 1969, Lamont 1983).

Ascaris, an intestinal roundworm that parasitizes humans, has a relatively simple life cycle. A female *Ascaris* may lay tens of thousands of eggs per day, which pass out of the host body in the feces. Where sanitation is poor or where human excrement is applied to farmland as fertilizer, the eggs may be inadvertently ingested. The egg is the only stage during the life cycle of *Ascaris* that occurs outside the host, and it is well protected by a sturdy, impermeable outer covering. The parasite relies on the host to consume these eggs and complete the life cycle.

Schistosoma, a trematode worm (blood fluke) that commonly infects humans and other mammals in tropical regions, has a more complicated life cycle that involves a freshwater snail as an intermediate host. Male-female pairs of adult worms live in the blood vessels that line the human intestine or the bladder, depending on the species of *Schistosoma.* The eggs pass out of the host body in the feces or urine. When the eggs are deposited in water, they develop into a free-swimming larval form (the miracidium), which burrows into a snail within 24 hours. In the snail, the miracidium produces cells that eventually develop into free-swimming cercariae, which leave the snail and can penetrate the skin of the next host in the cycle. Once inside the body of a human or other host, the cercariae travel a circuitous route through blood vessels until they become lodged in an appropriate place, where they metamorphose into adult worms (Jordan and Webbe 1969).

A snail infected with one miracidium may liberate from 500 to 2000 cercariae per day over a period of a month (McClelland 1965). During their lifetimes, which average between 4 and 5 years in human hosts, adults may produce as many as several hundred eggs per day. Relatively little is known about the ecology and population dynamics of *Schistosoma* (Hairston 1965, Barbour 1982), although the parasite, which causes the incurable disease schistosomiasis, probably affects 200 million humans (Maldonado 1967) and countless domestic animals (Soulsby 1968).

The life cycle of the protozoan parasite *Plasmodium,* which causes malaria, resembles that of *Schistosoma* in that it involves two hosts, a mosquito and man or some other mammal, bird, or reptile (Mattingly 1969). But whereas the sexual phase of the schistosome life cycle occurs in man, that phase of the malaria parasite's life cycle takes place within the mosquito (Figure 8–13). When an infected mosquito bites a human, cells called spor-

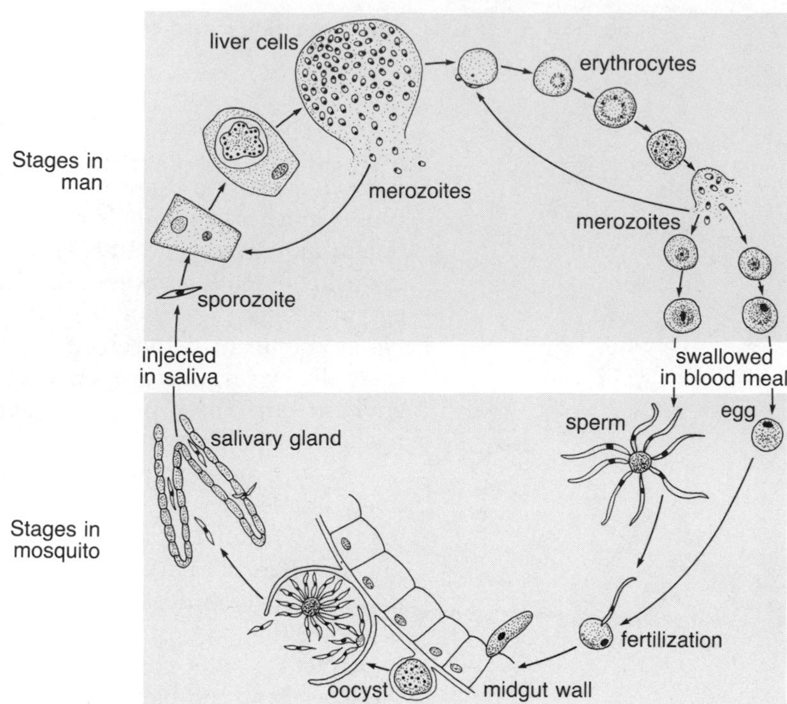

Figure 8–13
Stages in the life cycle of the malaria parasite *Plasmodium.* (After Buchsbaum 1948 and Sleigh 1973.)

ozoites are injected into the bloodstream with the mosquito's saliva. The sporozoites enter the red blood cells and feed upon hemoglobin. When the malaria cell becomes large enough, it undergoes a series of divisions (asexual reproduction), and the daughter cells break out of the red blood cell. Each daughter cell enters a new red blood cell, grows, and repeats the cycle, which takes about 48 hours. (When the infection has built up to a high level, the emergence of daughter cells corresponds to periods of high fever.) After several of these cycles, some of the cells that enter red blood corpuscles change into sexual forms. If these are swallowed by a mosquito along with a meal of blood, the sexual cells are transformed into eggs and sperm, and fertilization (sexual reproduction) takes place. The fertilized egg then divides and produces sporozoites, which work their way into the salivary glands of the mosquito, from which they may enter a new intermediate host, thereby completing the life cycle.

Parasites usually are host-specific, each species being restricted to one host organism, or to a few closely related ones, in each stage of its life cycle. Furthermore, most parasite-host relationships have evolved a fine balance; parasites rarely impair the health of their host. Parasitic organisms so virulent as to kill their hosts also kill themselves. As a general rule, therefore, selection favors benign infections, provided that different family lines of the parasite do not compete in the same host (see Ewald 1983). This delicate balance depends in part upon the adaptation of the parasite to its host environment. Disease or other stresses can change the internal environment of the host and upset the normal host-parasite balance, often leading to the death of the host (Latham 1975). The famous black plagues of medieval Europe are thought to have followed periods of widespread famine, which generally increased susceptibility to all disease (Cartwright and Biddis 1972).

The balance between parasite and host also depends on the immune responses and other defenses of the host. Parasites introduced to human populations that have not developed the proper mechanisms of immunity can increase rapidly, with disastrous results. Time and time again, foreigners have introduced parasites to susceptible native populations with the result that normally benign disease organisms become virulent.

The mutual accommodation of parasites and their hosts raises many questions (Jackson et al. 1969, 1970). How, for example, does the intestinal worm avoid being digested? How do organisms that invade the blood stream of vertebrates avoid being destroyed by immunity mechanisms which usually stimulate the formation of antibodies that specifically attack foreign proteins?

Successful parasites have found several ways to circumvent the immune mechanism (Bloom 1979). Some microscopic disease organisms produce chemical factors that suppress the immune system of the host (Schwab 1975). Others have surface proteins that mimic host antigens and thus escape notice by the immune system (Damien 1964). Some schistosomes are known to excite an immune response when they enter the host but do not succumb to antibody attack because they coat themselves with proteins of the host before antibodies become numerous (Smithers et al., 1969). As a consequence, parasites that subsequently infect a host face a barrage of antibodies stimulated by the earlier entrance of now-entrenched parasite individuals. Gerhard Schad (1966) has suggested that parasites could use the immune response of their hosts in this way to exclude competing parasites. When this response affects closely related species of parasite, it is known as cross-resistance (Cohen 1973, Kazacos and Thorson 1975). For example, most of the predominantly human forms of schistosomiasis disobey the common rule and are extremely virulent. But when a person has previously been infected by other schistosome organisms — some of which have little effect on man — the impact of the parasite is moderated considerably (Lewert 1970).

Plants use structural and chemical defenses against herbivores.

The conflict between herbivores and plants resembles that between parasite and host in that both are waged primarily on biochemical battlegrounds. Plant defenses against herbivores include the inherently low nutritional value of most plant tissues and the toxic properties of so-called secondary substances produced and sequestered for defense (Whittaker and Feeny

Figure 8–14
Spines protect the stems and leaves of many plants: an agave (century plant) from Baja California *(top left)*, a *Parkinsonia* (bean family) from the Galapagos Islands *(top right)*, and a cholla cactus *(Opuntia)* from Arizona *(bottom)*.

1971, Robbins and Moen 1975, Seigler and Price 1976). Sessile marine organisms, including both plants and animals, also employ a variety of chemical defenses (Bakus et al. 1986). Structural defenses, such as spines, hairs, and tough seed coats, are important as well (Figure 8–14).

The nutritional quality and digestibility of plant foods is critical to herbivores (Scriber and Slansky 1981, Scriber 1984, Moran and Hamilton 1980). Because young animals have a high protein requirement for growth, the reproductive success of grazing and browsing mammals depends upon the protein content of their food. Herbivores usually select plant food according to its nutrient content (Gwynne and Bell 1968). Young leaves and flowers are chosen frequently because of their low cellulose content; fruits and seeds are particularly nutritious compared to leaves, stems, and buds (Short 1971). Working with saplings of 46 species of canopy-forming trees on Barro Colorado Island, Panama, Elizabeth Coley (1983) analyzed rates of grazing on leaves with respect to their nutritional value and physical structure. In comparisons among species, she found that increased levels of cellulose and increased leaf toughness significantly reduced herbivory, but that the presence of hairs on the leaf surface and high levels of water and nitrogen seemed to encourage feeding (Table 8–1).

Many plants use chemicals to reduce the availability of their proteins to herbivores. For example, tannins sequestered in vacuoles in the leaves of oaks and other plants combine with leaf proteins and digestive enzymes in a

Table 8-1
Correlations between rates of herbivory and defenses of leaves of saplings among 46 species of trees in lowland rain forest habitat in central Panama

Leaf attribute	Range of values	Correlation coefficient *(r)*
Chemical		
Total phenols (% dry mass)	1.7–22.6	−0.10
Cellulose (% dry mass)	10.2–30.4	−0.47**
Lignin (% dry mass)	3.3–20.8	−0.23
Physical		
Toughness (Newtons)	2.5–11.6	−0.52**
Undersurface hairs (number mm^{-2})	0–18	0.64**
Nutritional		
Water (%)	49–82	0.51**
Nitrogen (% dry mass)	1.7–3.1	0.29*

Statistical significance: *$P < 0.05$; **$P < 0.01$.
Source: Coley 1983.

herbivore's gut, thereby inhibiting protein digestion. Thus, tannins considerably slow the growth of caterpillars and other herbivores, reducing the quality of tannin-laden hosts as food plants (Feeny 1969). With the buildup of tannins in oak leaves during the summer, fewer and fewer leaves are attacked by herbivores (Feeny 1968, 1970, Feeny and Bostock 1968). Insects, for their part, can reduce the inhibitory effects of tannins by detergent-like surfactants in their gut fluids, which tend to disperse tannin-protein complexes (Martin and Martin 1984, Martin et al. 1985).

Whereas tannins exhibit a generalized reaction with proteins of all types, many secondary substances of plants (that is, those not used for metabolism but for other purposes, chiefly defense) interfere with specific metabolic pathways or physiological processes of herbivores. But because the sites of action of such substances are localized biochemically, herbivores may counter their toxic effects by modifying their own physiology and biochemistry. Detoxification may involve one or several biochemical steps, including oxidation, reduction, or hydrolysis of the toxic substance, or its conjugation with another compound (Smith 1962).

Several early studies of the chemical give-and-take between plants and herbivores focused upon the larvae of bruchid beetles, many of which infest the seeds of legumes (pea family). Adult bruchids lay their eggs on developing seed pods. The larvae then hatch and burrow into the seeds, which they consume as they grow. To counter this attack, legumes have mounted a variety of defenses (Janzen 1969, Center and Johnson 1974), including the evolution of tiny seeds. Each larva feeds on only one seed. To pupate successfully and metamorphose into an adult, the larva must attain a certain size, ultimately limited by the amount of food in the seed. The small seeds of some species of legumes contain too little food to support the growth of a single bruchid larva (Janzen 1969).

Most legumes also contain substances that inhibit proteolytic enzymes produced in the herbivore's digestive organs. While these toxins provide an effective biochemical defense against mosts insects, many bruchid beetles

```
H     H   H H H    O
 \    |   | | |   //
  N-C-N-O-C-C-C-C
 /    ||  | | |   \
H     N   H H N    OH
      |      / \
      H     H   H
```

canavanine

```
H     H H H H H    O
 \    | | | | |   //
  N-C-N-C-C-C-C-C
 /    ||  | | | |  \
H     N   H H H N   OH
      |        / \
      H       H   H
```

arginine

Figure 8–15
The toxic non-protein amino acid (L-canavanine) and its protein amino-acid analogue (arginine). The shaded area highlights the difference between the two molecules. (From Harborne 1982.)

have metabolic pathways that either bypass or are insensitive to them (Applebaum 1964, Applebaum et al. 1965). Among legume species, however, soybeans stand out as being resistant to attack even by most bruchid species. When bruchid lay their eggs on soybeans, the first instar larvae die soon after burrowing beneath the seed coat; chemicals isolated from soybeans have been shown to inhibit the development of bruchid larvae in experimental situations. Seeds of the tropical leguminous tree *Dioclea megacarpa* contain 13 per cent by dry weight of L-canavanine, a non-protein amino acid toxic to most insects because it interferes with the incorporation into proteins of arginine, which it closely resembles (Figure 8–15). One species of bruchid, *Caryedes brasiliensis,* possesses enzymes that discriminate L-canavanine and arginine during protein formation and additional enzymes that degrade L-canavanine to forms that can be utilized as a source of nitrogen (Rosenthal et al. 1976).

The tobacco hornworm (larval stage of the moth *Manduca sexta;* Figure 8–16) can tolerate nicotine concentrations in its food far in excess of levels that kill other insects. Nicotine disrupts normal functioning of the nervous system by preventing the transmission of impulses from nerve to nerve. The hornworm has circumvented this defense by excluding nicotine from the nerve at the cell membrane; in other species of moths, nicotine readily diffuses into nerve cells (Yang and Guthrie 1969). Resistance to nicotine enables *M. sexta* to feed on tobacco *(Nicotiana tabacum),* a member of the tomato family (Solanaceae), but some other species of *Nicotiana* produce other alkaloid toxins that the tobacco hornworm cannot tolerate. When tobacco hornworms were grown on 44 species of *Nicotiana* in greenhouse experiments, the larvae grew normally on 25 species, but were retarded or stopped completely on the others. In addition, 15 of the species caused moderate to severe incidences of mortality (Parr and Thurston 1968).

Figure 8–16
Larva of the tobacco hornworm *Manduca sexta* (Sphingidae).

Most plants produce toxic, defensive compounds (Beck 1965; Fraenkel 1959, 1969; Harborne 1982; Levin 1976; Rosenthal and Janzen 1979). Many of these are important sources of pesticides (which is how plants use them) and drugs (some of their pharmacological effects are beneficial in small doses) (Balandrin et al. 1985). Secondary plant substances, or compounds, can be divided into three major classes of chemical structure: nitrogen

Table 8-2
Secondary plant compounds involved in plant-animal interactions

Class	Approximate number of structures	Distribution	Physiological activity
NITROGEN COMPOUNDS			
Alkaloids	5500	Widely in angiosperms, especially in root, leaf, and fruit	Many toxic and bitter-tasting
Amines	100	Widely in angiosperms, often in flowers	Many repellent-smelling; some hallucinogenic
Amino acids (non-protein)	400	Especially in seeds of legumes but relatively widespread	Many toxic
Cyanogenic glycosides	30	Sporadic, especially in fruit and leaf	Poisonous (as HCN)
Glucosinolates	75	Cruciferae and 10 other families	Acrid and bitter
TERPENOIDS			
Monoterpenes	1000	Widely, in essential oils	Pleasant smells
Sesquiterpene lactones	600	Mainly in Compositae, but increasingly found in other angiosperms	Some bitter and toxic, also allergenic
Diterpenoids	1000	Widely, especially in latex and plant resins	Some toxic
Saponins	500	In over 70 plant families	Haemolyse blood cells
Limonoids	100	Mainly in Rutaceae, Meliaceae, and Simaroubaceae	Bitter-tasting
Cucurbitacins	50	Mainly in Cucurbitaceae	Bitter-tasting and toxic
Cardenolides	150	Especially common in Apocynaceae, Asclepiadaceae, and Scrophulariaceae	Toxic and bitter
Carotenoids	350	Universal in leaf, often in flower and fruit	Colored
PHENOLICS			
Simple phenols	200	Universal in leaf, often in other tissues as well	Antimicrobial
Flavonoids	1000	Universal in angiosperms, gymnosperms, and ferns	Often colored
Quinones	500	Widely, especially in Rhamnaceae	Colored
OTHER			
Polyacetylenes	650	Mainly in Compositae and Umbelliferae	Some toxic

Source: Harborne 1982.

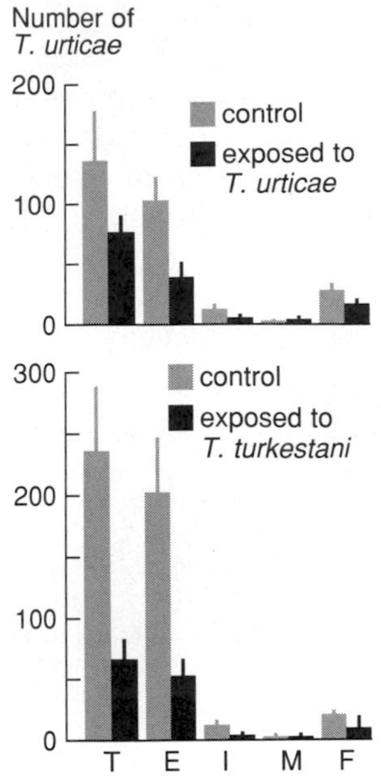

Figure 8-17
Mean number of the mite *Tetranychus urticae* on cotton plants previously exposed to mites and on controls with no previous exposure. The effect is similar whether the initial exposure is to *T. urticae (top)* or to the related species *T. turkestani (bottom)*. (T = total population, E = eggs, I = immatures, M = adult males, F = adult females.) (From Karban and Carey 1984.)

compounds ultimately derived from amino acids, terpenoids, and phenolics (Table 8-2). Among the nitrogen-based substances are lignin, a highly condensed polymer that resists digestion; alkaloids, such as morphine (derived from poppies), and atropine and nicotine (from various members of the tomato family Solanaceae); non-protein amino acids, such as L-canavanine; and cyanogenic glycosides, which produce cyanide (HCN). Terpenoids include essential oils, latex, and plant resins; among the phenolics, many simple phenols have antimicrobial properties.

Where herbivory is more intense, plants have more varied and concentrated toxins (for example, Dolinger et al. 1973). Where plant defenses are strong, adaptations of herbivores to detoxify poisonous substances proliferate. This war between plants and herbivores promotes biochemical specialization of herbivores to certain restricted groups of plants with similar toxins. Associations of plants and herbivores in groups based on plant chemistry and structure have been referred to as plant defense guilds (Atsatt and O'Dowd 1976).

Recent investigations have shown that plant defenses may be induced by herbivore damage. Rhoades (1979) found that alkaloids, phenolics, N-oxidases, and proanthocyanins, all of which are linked to antiherbivore defenses, increased dramatically in many plants following defoliation by herbivores or clipping of leaves by the investigators. Other studies have shown that plant responses to grazing can substantially reduce subsequent herbivory (Bryant and Kuropat 1980, Fowler and MacGarvin 1986, Haujioka 1980, Karban and Carey 1984; Figure 8-17). Inducibility suggests that chemical defenses are often too costly to maintain economically under light grazing pressure. Undoubtedly, the offensive biochemical tactics of herbivores also are expensive.

SUMMARY

The environments of organisms include biological factors—food, predators, diseases, other individuals of the same species—which interact with the physical environment in molding adaptations. Unlike physical factors, biological factors themselves evolve and diversify. The relationship between predators and their prey illustrates the mutual influence between organisms and biological factors in the environment.

1. Depending on the relative sizes of predators and their prey, predatory behavior includes active pursuit of individual large prey and filtering of tiny organisms from large quantities of water.
2. Predators are well-adapted to pursue, capture, and eat particular types of prey. Carnivorous and herbivorous mammals, for example, differ in the conformation of their teeth and the size and design of their digestive systems.
3. Organisms escape predation by avoiding detection; by chemical, structural, and behavioral defenses; and by escape. Crypsis and warning

coloration are examples of contrasting defenses of edible and unpalatable organisms.

4. Studies on industrial melanism of moths demonstrated that crypsis evolves because predators detect and capture individuals that do not match their surroundings more readily than those that blend in.

5. Host-parasite relations represent a specialized interaction characterized by complex life cycles of the parasite, made necessary by the difficulties of locating and infecting new hosts. In addition, parasite and host often evolve a delicate balance because the parasite's well-being may depend on the survival of its host.

6. Plants have evolved numerous structural and chemical defenses to deter herbivores. These include factors that influence the nutritional quality and digestibility of plant parts, and specialized chemicals—secondary compounds—that have toxic effects on animals and microorganisms. Most of these are nitrogen-containing compounds, terpenoids, or phenolics derived by modification and elaboration of normal metabolic pathways.

7. Herbivores themselves have evolved means of detoxifying many secondary chemicals of plants, enabling them to specialize on plant hosts that are poisonous to most other species.

9

Climate, Topography, and the Diversity of Natural Communities

No single type of animal can tolerate all the conditions found on the earth. Each thrives only within relatively narrow ranges of temperature, precipitation, soil conditions, salinity, and other physical factors. Biotic factors also influence the ability of populations to maintain themselves under such conditions. The preferences and tolerances of each species differ from those of every other; thus, no two experience exactly the same spectrum of ecological conditions.

Because all species are adapted to their surroundings by modifications of form and function, differences in physical conditions from place to place often reveal themselves as differences in the forms of plants and animals. Variation in the physical conditions derive from two major classes of factors: climate and topography. For terrestrial organisms, climate includes the characteristic temperature and precipitation of a region. Climate interacts with topography and other features of the land to create local variation. For aquatic organisms, temperature and salinity serve as climatic factors, and the land underlying the oceans, streams, and lakes creates topographic diversity to which aquatic organisms respond by way of their adaptations and the distributions of their populations.

The earth is a giant heat machine.

Clues to the origin of diversity in the biological world can be found in the study of variation in the physical world. The surface of the earth, its waters, and the atmosphere above it behave as a giant heat machine, obeying the laws of thermodynamics. Climate is determined by the absorption of the energy in sunlight and its redistribution over the globe. As the surface varies from bare rock to forested soil, open ocean, and frozen lake, its ability to absorb sunlight varies as well, thus creating differential heating and cooling.

The heat energy absorbed by the earth is eventually radiated back into space, but not before undergoing further transformations that perform the work of evaporating water and contributing to the circulation of the atmosphere and oceans. All these factors have created a great variety of physical conditions, which, in turn, have fostered the diversification of ecosystems.

Variation in solar radiation with latitude creates major global patterns in temperature and rainfall.

The earth's climate tends to be cold and dry toward the poles and hot and wet toward the equator. Although this oversimplification has many exceptions, climate nonetheless does exhibit broadly defined patterns (Barry and Chorley 1970, Flohn 1968, Lowry 1969, Trewartha 1954).

Global variation in climate is determined largely by the position of the sun relative to the surface of the earth. The sun warms the atmosphere, oceans, and land most when it is directly overhead. Not only does a beam of sunlight spread over a greater area when the sun is close to the horizon, it also travels a longer path through the atmosphere, where much of its energy is either reflected or absorbed by the atmosphere and reradiated into space as heat. The sun's highest position each day varies from directly overhead in the tropics to near the horizon in polar regions; hence the warming effect of the sun increases from the poles to the equator.

Warming air expands, becomes less dense, and thus tends to rise. Its ability to hold water vapor increases, and evaporation accelerates; the rate of evaporation from a wet surface nearly doubles with each rise in temperature of 10°C.

Because warm tropical air can hold much more water than the cooler temperate and arctic air, annual precipitation is greatest in tropical regions (Figure 9–1). The tropics are wet not because more water occurs in tropical latitudes than elsewhere but because water cycles more rapidly through the tropical atmosphere. The heating effect of the sun causes water to evaporate; energy input, not the quantity of water, primarily determines latitudinal patterns in rainfall. The distribution of continental land masses exerts a secondary effect. At a given latitude, rain falls more plentifully in the southern hemisphere because oceans and lakes cover a greater proportion of its surface (81 per cent, compared with 61 per cent of the northern hemisphere, List 1966). Water evaporates more readily from exposed surfaces of water than from soil and vegetation.

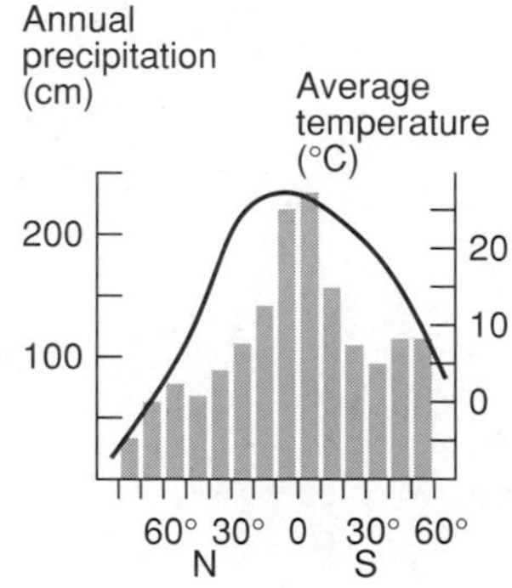

Figure 9-1
Average annual precipitation (vertical bars) and temperature (line) for 10° latitudinal belts within continental land masses. The figure presents averages for many localities, which obscures the great variation within each latitudinal belt. (From data in Clayton 1944.)

Energy from the sun drives the winds, distributing water vapor through the atmosphere. Indeed, wind patterns strongly influence precipitation. The mass of warm air that rises in the tropics eventually spreads to the north and south in the upper layers of the atmosphere. It is replaced from below by surface-level air from subtropical latitudes. The tropical air mass that rises under the warming sun cools as it radiates heat back into space. By the time this air has extended to about 30° north and south of the solar equator (the parallel of latitude that lies directly under the sun), the cooled air mass

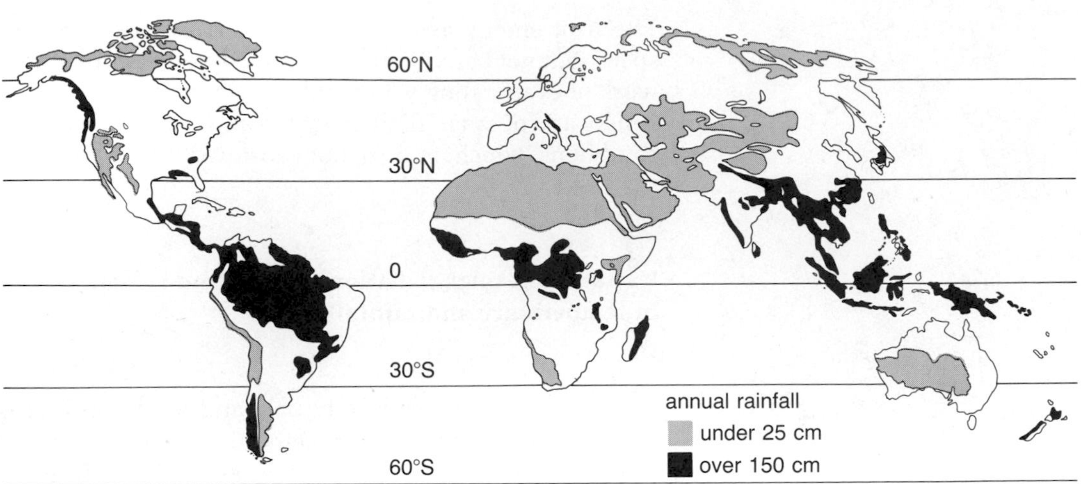

Figure 9-2
Distribution of the major deserts (regions with less than 25 centimeters [10 inches] annual precipitation) and wet areas (more than 150 centimeters [80 inches] annual precipitation). (After Espenshade 1971.)

becomes more dense and begins to sink back to the surface. Condensation has already removed much of its water; its capacity to evaporate and hold water increases further as it sinks and warms. As the air mass strikes the ground in subtropical latitudes and spreads to the north and south, it draws moisture from the land, creating zones of arid climate centered about 30° latitude north and south of the equator (Figure 9–2). The great deserts—

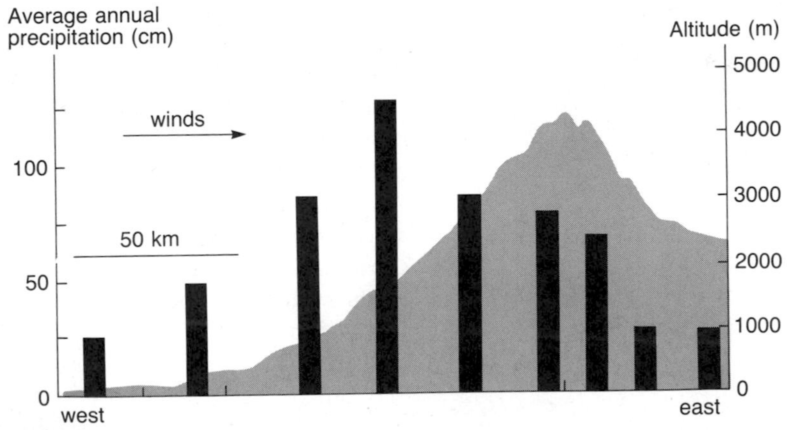

Figure 9-3
Influence of the Sierra Nevada mountain range on local precipitation and in causing a rain shadow to the east. Weather comes predominantly from the west across the central valley of California. As moisture-laden air is deflected upward by the mountains, it cools and its moisture condenses, resulting in heavy precipitation on the western slope of the mountains. As the air rushes down the eastern slope, it warms and begins to pick up moisture, creating arid conditions in the Great Basin. (After Pianka 1988.)

the Arabian, Sahara, Kalahari, and Namib of Africa, the Atacama of South America, the Mohave, Sonoran, and Chihuahuan of North America, and the Australian—all belong to regions within these belts.

Major land masses produce some exceptions to this pattern. Mountains force air upward, causing it to cool and lose its moisture as precipitation on the windward side of the range. As the air descends the leeward slopes and travels across the lowlands beyond, it picks up moisture and creates arid environments called rain shadows (Figure 9–3). The Great Basin deserts of the western United States and the Gobi Desert of Asia lie in the rain shadows of extensive mountain ranges.

The interior of a continent is usually drier than its coasts simply because the interior is farther removed from the major site of water evaporation, the surface of the ocean. Furthermore, coastal (maritime) climates vary less than interior (continental) climates because the great heat-storage capacity of water reduces temperature fluctuations. For example, the hottest and coldest mean monthly temperature near the Pacific coast of the United States at Portland, Oregon, differ by 16°C. Farther inland, this range increases to 18°C at Spokane, Washington; 26°C at Helena, Montana; and 33°C at Bismarck, North Dakota.

Ocean currents play a major role in transferring heat over the surface of the earth. In large ocean basins, cold water tends to move toward the tropics

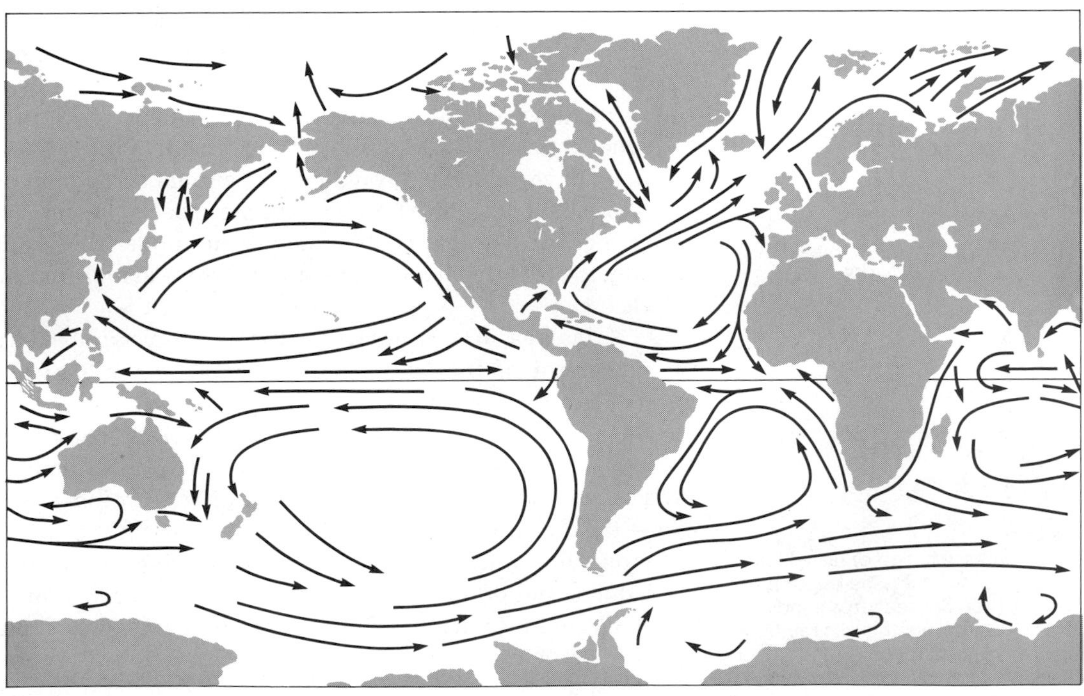

Figure 9-4
The major ocean currents. Water movement generally proceeds clockwise in the northern hemisphere and counterclockwise in the southern hemisphere. (After Duxbury 1971.)

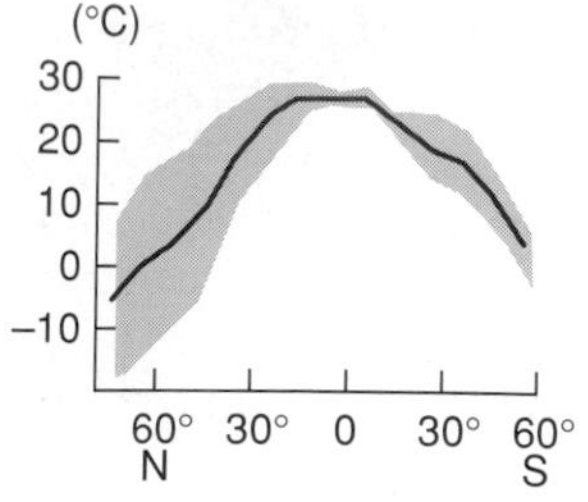

Figure 9-5
Annual range of mean monthly temperatures (shaded area) as a function of latitude. The mean annual temperature is indicated by the line. (After data in Clayton and Clayton 1947.)

along the western coasts of the continents, and warm water tends to move toward temperate latitudes along the eastern coasts of the continents (Figure 9–4). The cold Humboldt Current moving north from the Antarctic Ocean along the coasts of Chile and Peru causes the establishment of cool deserts along the west coast of South America right to the equator. Conversely, the warm Gulf Stream, emanating from the Gulf of Mexico, carries a mild climate far to the north into Western Europe and the British Isles.

The seasons bring predictable changes in the environment.

Patterns of change in climate influence plants and animals as much as long-term averages of temperature and precipitation (Wolfe 1979). Periodic cycles in climate follow astronomical cycles: the rotation of the earth upon its axis causes daily periodicity; the revolution of the moon around the earth determines the tides; the revolution of the earth about the sun brings seasonal change.

The equator is tilted slightly with respect to the path the earth follows in its orbit around the sun. As a result, the northern hemisphere receives more solar energy than the southern hemisphere during the northern summer, less during the northern winter. The seasonal change in temperature increases with distance from the equator (Figure 9–5). At high latitudes in the northern hemisphere, mean monthly temperatures vary by an average of 30°C, with extremes of more than 50°C annually; the mean temperatures of the warmest and coldest months in the tropics differ by as little as 2 or 3°.

Latitudinal patterns in rainfall seasonality result in part from the seasonal northward and southward movement of belts of wet and dry climate. Rainfall varies most throughout the annual cycle in broad latitudinal belts lying about 20° north and south of the equator (Figure 9–6). As the seasons change, these regions alternately come under the influence of the solar equator, bringing heavy rains, and subtropical high-pressure belts, bringing clear skies.

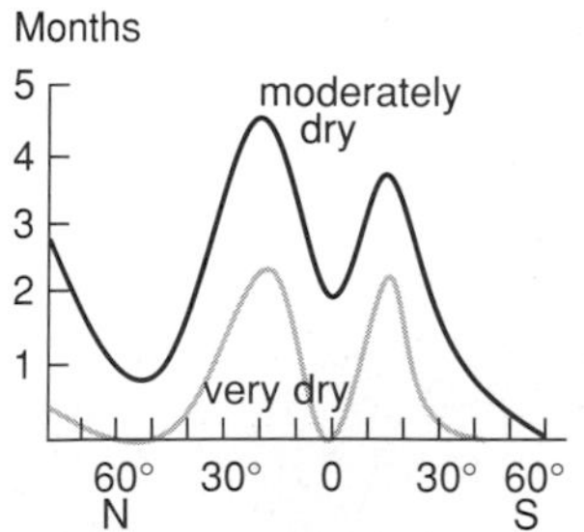

Figure 9-6
Seasonal distribution of rainfall as a function of latitude. The length and severity of the dry season are indicated by the number of months with rainfall less than a certain fraction ($\frac{1}{30}$ and $\frac{1}{100}$) of the annual total. If rainfall were evenly distributed throughout the year, $\frac{1}{12}$ of the total would fall during each month. (Data compiled from Clayton and Clayton 1947.)

Panama, at 9°N, lies within the wet tropics, but even there the seasonal movement of the solar equator profoundly influences the climate. The major tropical belt of high rainfall remains south of Panama during most of the northern winter, but it lies directly overhead during the northern summer. Hence the winter is dry and windy, the summer humid and rainy. Panama's climate is wetter on the northern (Caribbean) side of the isthmus —the direction of prevailing winds—than on the southern (Pacific) side; mountains intercept moisture coming from the Caribbean side of the isthmus and produce a rain shadow. The Pacific lowlands are so dry during the winter months that most trees lose their leaves. Tinder-dry forests and bare branches contrast sharply with the wet, lush, more typically tropical forest during the wet season (Figure 9–7).

Farther to the north, at 30°N in central Mexico, rainfall comes only during the summer when the solar equator reaches its most northward limit

Figure 9-7
Kiawe forest on the island of Maui, Hawaii, during the peak of the dry season. The trees have dropped their leaves and the grasses of the forest understory are tinder-dry. (Courtesy of the U.S. Forest Service.)

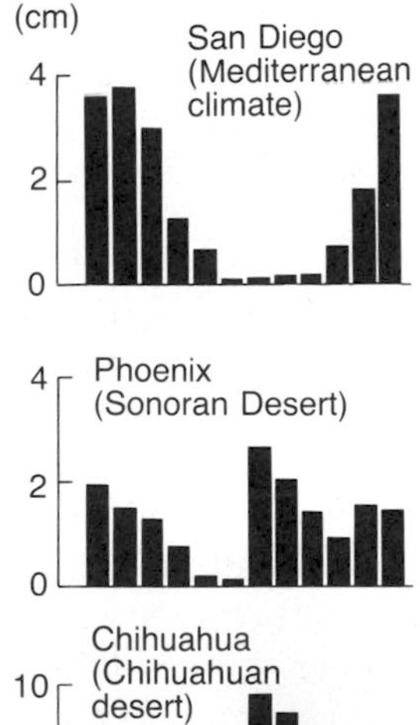

Figure 9–8
Seasonal occurrence of rainfall at three localities in western North America: the summer rainy season of the Chihuahuan Desert, the winter rain–summer drought of the Pacific Coast (Mediterranean climate type), and the combined climate pattern of the Sonoran Desert. (Rainfall data from Clayton 1944.)

(Figure 9–8). During the rest of the year this region falls within the dry, subtropical high-pressure belt. The influence of the solar equator, bringing summer rainfall, extends into the Sonoran Desert of southern Arizona and New Mexico. This area also receives moisture during the winter from the Pacific Ocean, carried by the southwesterly winds emanating from the subtropical high-pressure belt farther south. Southern California lies beyond the summer-rainfall belt and has a winter-rainfall, summer-drought climate, often referred to as a Mediterranean climate.

The sun warms the seas just as it does the continents and atmosphere, but the ocean's great mass of water acts like a heat sink to dampen daily and seasonal fluctuations in temperature. Where ocean temperature does change seasonally, it more often reflects seasonal movements of water masses of different temperature than local heating and cooling. During the Panamanian dry season, roughly January to April, steady winds blowing in a southwesterly direction create strong upwelling currents in the Pacific Ocean along the southern and western coasts of Central America. During the upwelling period, winds blow warm surface water away from the coast, where cooler water moves upward from deeper regions to replace it. As a result, water temperature on the Pacific coast of Panama varies annually three times as much as on the Caribbean coast (Rubinoff 1968).

Small temperate-zone lakes are more sensitive than oceans to the changing seasons. Temperature cycles shape the nutrient budgets of lakes because changes in the gradient of temperature from surface to bottom lead to vertical mixing of the water twice each year, during the spring and the autumn (Hutchinson 1957). In winter, the lake has an inverted temperature profile; that is, the coldest water (0 °C) lies at the surface, just beneath the ice. (Because the density of water increases between freezing and 4 °C, the warmer water within this range sinks, and temperature increases up to 4 °C with depth.) In early spring, the sun warms the lake surface gradually. But until surface temperature exceeds 4 °C, the sun-warmed surface water tends to sink into the cooler layers immediately below. This minor vertical mixing creates a uniform temperature distribution throughout the water column. Without thermal layering to impede mixing, winds cause deep vertical movement of water in early spring (spring overturn), bringing nutrients to the surface from regions of nutrient release in bottom sediments, and oxygen from the surface to the depths.

Later in spring and early summer, as the sun rises higher each day and the air above the lake warms, surface layers of water gain heat faster than deeper layers, creating a sharp zone of temperature change at intermediate depth, called the thermocline, across which water does not mix. Now, at temperatures exceeding 4 °C, the warmer surface water literally floats on the cooler water below. The depth of the thermocline varies with local winds and with the depth and turbidity of the lake. It may occur anywhere between 5 and 20 meters below the surface; lakes less than 5 meters deep usually lack stratification.

The thermocline demarcates an upper layer of warm water (the epilimnion) and a deep layer of cold water (the hypolimnion). Most of the primary production of the lake occurs in the epilimnion, where sunlight is intense. Photosynthesis supplements mixing of oxygen at the lake surface to keep the epilimnion well aerated and thus suitable for animal life, but plants often deplete dissolved mineral nutrients and thereby curtail their own production. The hypolimnion is cut off from the surface of the lake and its animals and bacteria, remaining mostly below the euphotic zone of photosynthesis, deplete the water of oxygen, creating anaerobic conditions.

During the autumn, surface layers of the lake cool more rapidly than deeper layers and, becoming heavier than the underlying water, begin to sink. This vertical mixing (fall overturn) persists into late fall, until the temperature at the lake surface drops below 4 °C, and winter stratification ensues. Fall overturn causes greater vertical mixing of water than spring overturn because temperature differences in the lake during summer stratification exceed those during winter stratification. Fall overturn speeds the movement of oxygen to deep waters and rushes nutrients to the surface. Where the hypolimnion becomes fairly warm in midsummer, deep vertical mixing may take place in late summer when temperatures remain favorable for plant growth. Infusion of nutrients into surface waters at this time often causes a burst of phytoplankton population growth—the fall bloom. In deep, cold lakes, vertical mixing does not penetrate to all depths until late fall or early winter, when water temperatures are too cold to support plant growth.

Irregular fluctuations in the environment are superimposed on periodic cycles.

Everyone knows that weather is difficult to predict far in advance. We often remark that a year was particularly dry or cold compared to others. Almost all aspects of climate are unpredictable to some extent. Rainfall is most variable where it is sparsest. At a given locality, mean monthly precipitation varies most during the driest season. Year-to-year variation in temperature on a particular date is greatest where temperature fluctuates most during the year. The most extreme conditions are also generally the rarest, but these may exert disproportionate selection upon the adaptations of organisms, which must prepare for all possible contingencies.

Some events—earthquakes, tornadoes, volcanic eruptions, and hurricanes—create conditions exceeding the capacities of organisms to respond, and their general effect on the biological community is devastating.

Figure 9–9
Nesting colony of Peruvian boobies on an island off the coast of Peru. The dense population depends on the anchovy stocks in the rich Humboldt Current. (Negative number 327672; photograph by R. C. Murphy, courtesy of the Department of Library Services, American Museum of Natural History.)

The rich Peruvian fishing industry, as well as some of the world's largest seabird colonies, thrives on the abundant fish in the rich waters of the Humboldt (Peru) Current, a mass of cold water that flows up the western coast of South America and finally veers offshore at Ecuador, where warm tropical inshore waters prevail (Figure 9–9). Each year a warm countercurrent known as El Niño ("little boy" in Spanish and in this use referring to the Christ child because it appears about Christmastime) moves down the coast of Peru, sometimes strongly enough to force the cold Humboldt Current offshore, taking with it the food supply of millions of birds. The ornithologist Robert Cushman Murphy (1936: 102) described the result:

> The immediate result of an advance of El Niño is to raise the temperature of the littoral ocean water by five or more degrees Centigrade. The normal plankton of the cool Humboldt Current waters next succumbs, perhaps because of the increased temperature, perhaps in part because of a different composition of salts in the water. The common schooling fish leave the region or die, and less familiar species, such as flying fish, dolphins, and other tropical types, invade the shore waters and even enter harbors. Later, if the incursion of tropical waters is marked and widespread, disease attacks the population of cormorants, boobies, pelicans, and other guano birds belonging to the normal Humboldt Current fauna. Carcasses drift ashore in vast numbers, and the survivors of such species are driven southward.

During "normal" years between El Niño events, a steady wind blows across the equatorial central Pacific Ocean from an area of high atmospheric pressure centered over Tahiti to an area of low pressure centered over Darwin, Australia. An El Niño event appears to be triggered by a reversal of

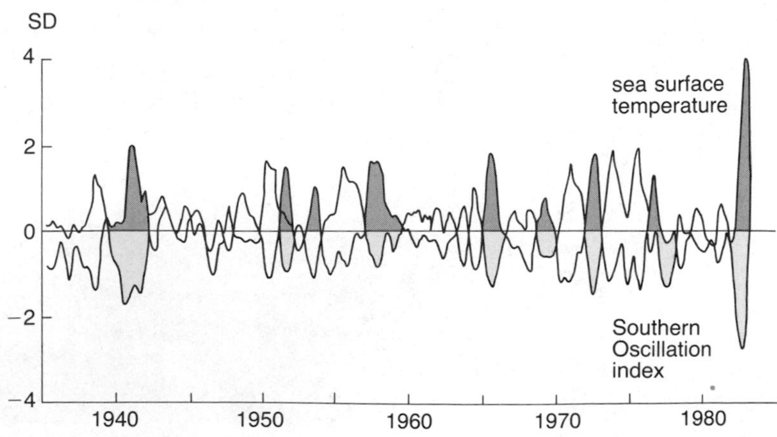

Figure 9–10
Major occurrences of the El Niño–Southern Oscillation are marked by strong negative swings in the Southern Oscillation Index and large positive sea-surface temperature anomalies in South American coastal waters. (From Rasmussen 1985.)

these pressure areas (the so-called Southern Oscillation) and of the winds that flow between them (Graham and White 1988). As a result, the westward-flowing equatorial currents stop or even reverse, upwelling off the coast of South America weakens or ceases, and warm water—the El Niño current—piles up along the coast of South America. Historical records of atmospheric pressure at Tahiti and Darwin, and of sea-surface temperature on the Peruvian coast, reveal pronounced ENSO (El Niño–Southern Oscillation) events at irregular intervals of between 2 and 10 years (Figure 9–10).

The climatic and oceanographic effects of an ENSO event extend over much of the world, affecting weather and biotic communities in such distant areas as India, South Africa, Brazil, and western Canada (Barber and Chavez 1983, Cane 1983, Glynn 1988, Rasmussen 1985, Rasmussen and Wallace 1983). A major ENSO event in 1982–1983 disrupted fisheries and destroyed kelp beds in California (McGowan 1984, Dayton and Tegner 1984), caused reproductive failure of seabirds in the central Pacific Ocean (Schreiber and Schreiber 1984), and resulted in widespread mortality of coral in Panama (Glynn 1984).

Topographic and geologic influences superimpose local variation on global patterns.

Variation in topography and geology can create variation in the environment within regions of uniform climate. In hilly areas, the slope of the land and its exposure to the sun influence the temperature and moisture content of the soil. Soils on steep slopes drain well, often causing moisture stress for plants at the same time that the soils of nearby lowlands are saturated with water. In arid regions, stream bottomlands and seasonally dry riverbeds may support well-developed riparian forests, which accentuate the bleakness of the surrounding desert. In the northern hemisphere, south-facing slopes directly face the sun, whose warmth and drying power limit vegetation to shrubby, drought-resistant (xeric) forms. The adjacent north-facing slopes remain relatively cool and wet and harbor moisture-requiring (mesic) vegetation (Figure 9–11).

Air temperature decreases with altitude by about 6°C for each 1000-meter increase in elevation. Even in the tropics, if one climbs high enough, one eventually encounters freezing temperatures and perpetual snow. Where the temperature at sea level is 30°C, freezing temperatures are reached at about 5,000 meters, the approximate altitude of the snow line on tropical mountains.

In north-temperate latitudes, a 6°C drop in temperature corresponds to the temperature change over an 800-kilometer increase in latitude. In many respects, the climate and vegetation of high altitudes resemble those of sea-level localities at higher latitudes (Billings and Mooney 1968). But despite their similarities, alpine environments usually vary less from season to season than their low-elevation counterparts at higher latitudes. Temperatures in tropical montane environments remain nearly constant and frost-

Figure 9–11
The influence of exposure on the vegetation of a series of mountain ridges near Aspen, Colorado. The north-facing (left-facing) slopes are cool and moist, permitting the development of spruce forest. Shrubby, drought-resistant vegetation grows on the south-facing slopes.

free over the year, allowing many tropical plants and animals to live in the cool environments found there.

In the mountains of the southwestern United States, changes in plant communities with elevation result in more or less distinct belts of vegetation, referred to as life zones by the nineteenth-century naturalist C. H. Merriam (1894). Merriam's scheme of classification included five broad zones, which he named, from low to high elevation (or from north to south): Lower Sonoran, Upper Sonoran, Transition, Canadian (or Hudsonian), and Alpine (or Arctic-Alpine).

At low elevations in the Southwest, one encounters a cactus and desert-shrub association characteristic of the Sonoran Desert of northern Mexico and southern Arizona (Figure 9–12). In the woodlands along stream beds, plants and animals have a distinctly tropical flavor. Many hummingbirds and flycatchers, ring-tailed cats, jaguars, and peccaries make their only temperate-zone appearances in this area. In the Alpine zone, 2500 meters higher, one finds a landscape resembling the tundra of northern Canada and Alaska. Thus, by climbing 2500 meters, one experiences changes in climate and vegetation that would require a journey to the north of 2000 kilometers, or more, at sea level.

Local variation in the bedrock underlying a region promotes the differentiation of soil types and enhances biotic heterogeneity. In the northern Appalachian Mountains and in mountains near the Pacific coast of the United States, outcrops of serpentine (a kind of igneous rock) weather to form soils having so much magnesium that species of plants characteristic of

Hudsonian Zone
Elevation 2500 meters

Alpine Zone
Elevation 3500 meters

Upper Sonoran Zone
Elevation 1500 meters

Transition Zone
Elevation 2000 meters

Lower Sonoran Zone
Elevation 900 meters

Upper Sonoran Zone
Elevation 1200 meters

Figure 9–12
Vegetation at different elevations in the mountains of southeastern Arizona. The Lower Sonoran Zone supports mostly saguaro cactus, small desert trees such as paloverde and mesquite, numerous annual and perennial herbs, and small succulent cacti. Agave, ocotillo, and grasses are conspicuous elements of the Upper Sonoran Zone, with oaks appearing toward its upper edge. Large trees predominate at higher elevations: ponderosa pine in the Transition Zone, spruce and fir in the Hudsonian Zone. These gradually give way to bushes, willows, herbs, and lichens in the Alpine Zone above the treeline. (Courtesy of the U.S. Soil Conservation Service, the U.S. Forest Service, W. J. Smith, and R. H. Whittaker, from Whittaker and Niering 1965.)

surrounding soil types cannot grow (Whittaker 1954, Walker 1954, Proctor and Woodell 1975). Serpentine barrens, as they are called, usually support little more than a sparse covering of grasses and herbs, many of which are distinct endemics (species found nowhere else) that have evolved a high tolerance of magnesium (Figure 9–13). Depending on the composition of

Figure 9–13
A small serpentine barren in eastern Pennsylvania. The soils surrounding the barren support oak-hickory-beech forest.

the bedrock and the rate of weathering, granite, shale, and sandstone also can produce a barren type of vegetation. The extensive pine barrens of southern New Jersey, where mature trees attain no more than waist height in some areas, occur on a large outcrop of sand, which produces a dry, acid, infertile soil (McPhee 1968, McCormick 1970, Forman 1979).

Physical characteristics of the soil and of the underlying rock also influence drainage and the ability of the soil to hold moisture. For example, the extensive pine forests of the coastal plain of the southeastern United States grow on sandy soils that drain too well for most broad-leaved trees. Climate, too, plays an important role in the weathering of rock and the formation of soils; in temperate and arctic regions of the northern hemisphere, glaciation during the last 100,000 years has influenced soil characteristics over vast areas, scraping some areas to bare bedrock, while burying others under tens of meters of wind-blown glacial dust, or loess (Bunting 1967).

Integrated descriptions of climate emphasize the interaction of temperature and availability of water.

We find it easier to dissect climate into its component properties of temperature, humidity, precipitation, wind, and solar radiation than to appreciate at once all the implications of these factors for the ecosystem. We must, however, understand interactions among climate factors in their effect on life.

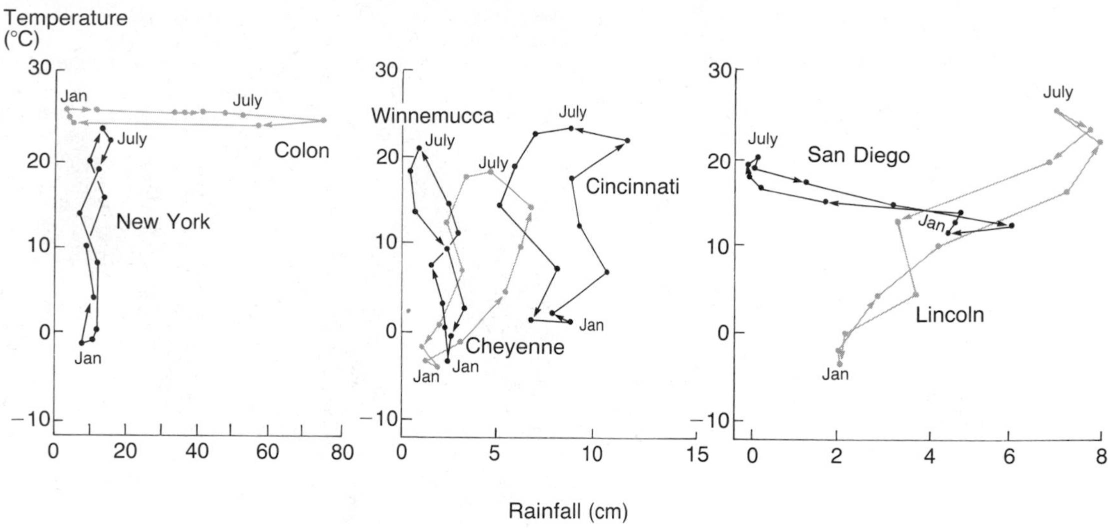

Figure 9–14
Climographs of representative localities in North and Central America. Each point represents the mean temperature and precipitation for one month. Lines connecting the months at each locality indicate seasonal change in climate. (Data from Clayton and Clayton 1947.)

For example, seasonal rainfall promotes plant growth more strongly during warm months than during cold months. Wind and solar radiation interact with temperature to determine thermal stress; temperature and humidity together influence water balance.

The climograph portrays seasonal changes in temperature and precipitation simultaneously, permitting a visual comparison of climates at different localities (Figure 9–14). The seasons at Colon, on the Caribbean coast of Panama, bring marked variation in rainfall but little change in temperature, just the reverse of the situation in New York City. Only during July and August in New York and April in Colon are the climates of the two localities similar. Moving east to west across the United States from Cincinnati, Ohio, to Winnemucca, Nevada, climate becomes more arid but temperatures remain within the same range. The change in vegetation from deciduous, broad-leaved forest in Ohio to short-grass prairie in Wyoming and desert shrubs in Nevada thus depends on the water relations of different plant forms rather than on temperature tolerance.

Although San Diego's weather in January resembles that of Lincoln, Nebraska, in April, the overall climates differ as much as their vegetation. San Diego's Mediterranean-type climate, with hot, dry summers and cool, moist winters, favors slow-growing, drought-resistant shrubs (chaparral), while Lincoln's wet summers and cold, dry winters favor the development of tall-grass prairie.

Although the climograph provides a useful basis for comparing localities, it fails to combine the effects of temperature and precipitation in any biologically meaningful way, and it does not reveal cumulative effects of weather. For example, during dry seasons both evaporation and transpira-

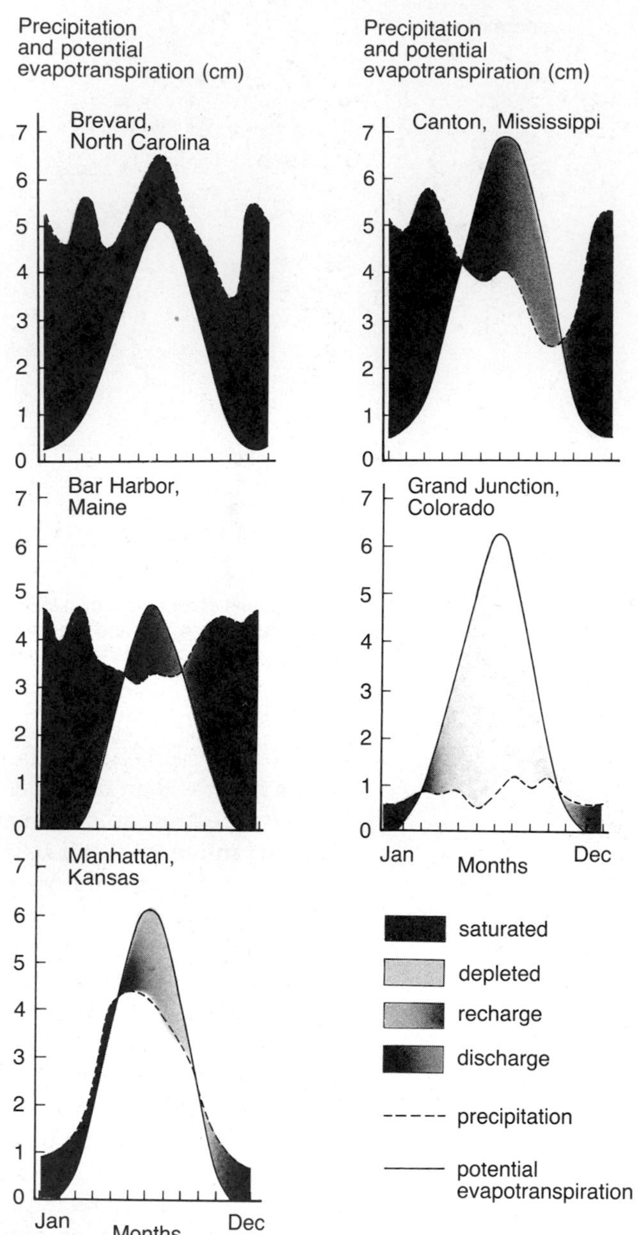

Figure 9–15
The relationship of precipitation and potential evapotranspiration to changes in the availability of soil moisture. When evapotranspiration exceeds precipitation, water is withdrawn from the soil until the deficit exceeds 10 centimeters (4 inches), the average amount of moisture that soils can hold. (After Thornthwaite 1948.)

tion (evaporation of water from leaves) remove water from the soil. When rainfall does not balance evaporation and transpiration losses, the water deficit in the soil steadily increases, perhaps for months at a time; soil water reflects last month's conditions as well as more recent ones.

In 1948, the geographer C. W. Thornthwaite published a method for utilizing climate data to estimate the seasonal availability of water in the soil. He compared the rate at which water is drawn from the soil by plants and by direct evaporation with the rate at which it is restored by precipitation. The sum of evaporation and transpiration is the total evapotranspiration of the habitat. Evaporation and transpiration increase with temperature by a factor of nearly two for each 10°C rise in temperature, other things being equal, although the character of the soil and vegetation cover also influence water loss.

In natural environments, evapotranspiration is at times limited by the availability of water in the soil. Potential evapotranspiration, the amount of water that would be drawn from soil if moisture were unlimited, can be calculated from temperature and precipitation (Ward 1967). Because potential evapotranspiration increases with temperature, temperature and water stress go hand in hand. Thus, boreal regions receiving 25 to 50 centimeters of precipitation each year have more favorable water budgets for plant production than tropical regions with similar levels of precipitation.

When precipitation input to the soil (rainfall minus surface runoff) exceeds potential evapotranspiration at all seasons, the soil remains saturated with water throughout the year, as at Brevard, North Carolina (Figure 9–15). At Bar Harbor, Maine, precipitation falls below potential evapotranspiration during the warm summer months, and so the soil is depleted of water during late summer and early fall. Canton, Mississippi, receives rainfall similar in amount to that of Bar Harbor, but the hotter climate increases the potential evaporation of water, resulting in serious water deficits during the summer months. Manhattan, Kansas, receives much less rain than Canton, but because most rain falls during the summer period of maximum potential evapotranspiration, soils do not develop serious water deficits. In the dry climate of Grand Junction, Colorado, soils are depleted of water most of the year and become saturated only briefly, after heavy rains. Plants are correspondingly unproductive.

Broad climate patterns interact with local topography and geologic influences to define the area over which physical conditions are suitable for a particular species. The geographical range of the species may be more narrowly limited than this area owing to barriers to dispersal and the population effects of competition, predation, and disease, which can make otherwise suitable areas uninhabitable. Nevertheless, the distributional limits of plants and animals frequently correspond to particular climate conditions.

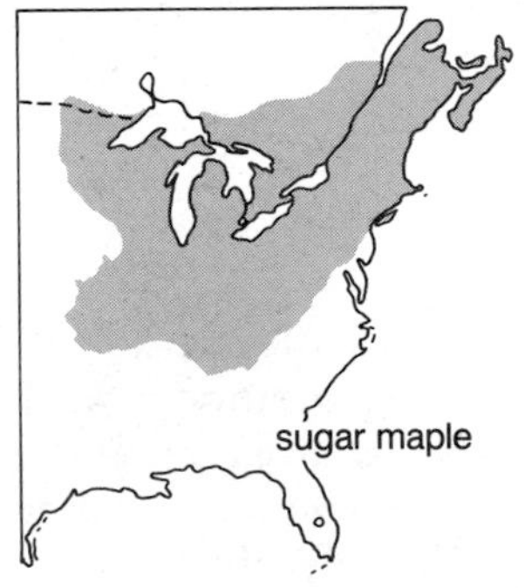

Figure 9–16
The range of the sugar maple in eastern North America. (After Fowells 1965.)

The geographical distributions of plants are related to climate patterns.

The range of the sugar maple, a common forest tree in the northeastern United States and southern Canada, is limited by cold winter temperatures

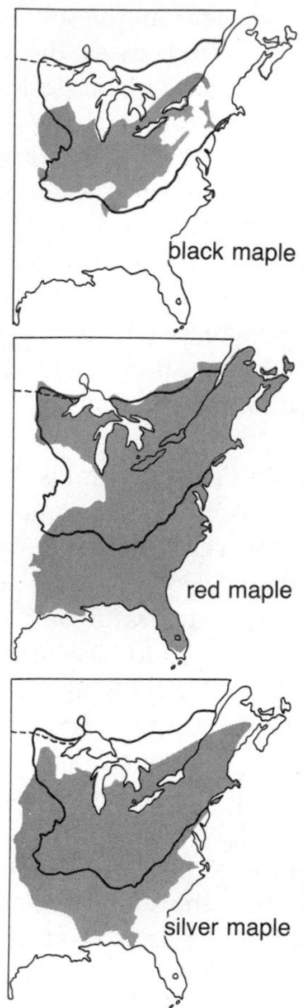

Figure 9–17
The ranges of black, red, and silver maples in eastern North America. The range of the sugar maple is outlined on each map to show the area of overlap. (After Fowells 1965.)

to the north, hot summer temperatures to the south, and summer drought to the west (Figure 9–16). Attempts to grow sugar maples outside their normal range have shown that they cannot tolerate average monthly high temperatures above (24 to 27°C) or below about −18°C (Fowells 1965). The western limit of the sugar maple, determined by dryness, coincides with the western limit of forest-type vegetation in general. Because temperature and rainfall interact to determine availability of moisture, sugar maples tolerate lower annual precipitation at the northern edge of their range (about 50 cm) than at the southern edge (about 100 cm). To the east, the range of the sugar maple stops abruptly at the Atlantic Ocean. Within its geographical range, sugar maple is more abundant in northern forests, where it sometimes forms single-species stands, than in the more diverse forests to the south. Sugar maple occurs most frequently on moist, slightly acid soils.

Differences in the distributions of the sugar maple and other tree-sized species of maples—black, red, and silver—suggest differences in ecological tolerances (Figure 9–17). Where their ranges overlap, maples exhibit distinct preferences for local environmental conditions created by differences in soil and topography. Black maple frequently occurs together with the closely related sugar maple, but prefers drier, better-drained soils with higher calcium content (therefore less acidic). Silver maple is widely distributed but prefers the moist, well-drained soils of the Ohio and Mississippi river basins. Red maple is peculiar in preferring either wet, swampy conditions or dry, poorly developed soils.

The local distribution of plants is determined largely by topography and soils.

The distribution of plants manifests the effects of different factors on different scales of distance. Climate, topography, soil chemistry, and soil texture exert progressively finer influences on geographical distribution. Elevation, slope, exposure, and underlying bedrock—factors that greatly influence the plant environment—vary most in mountainous regions, and ecologists frequently turn to the varied habitats of mountains to study plant distribution.

Along the coast of northern California, mountains create conditions for a variety of plant communities, ranging from dry coastal chaparral to tall forests of Douglas fir and redwood (Waring and Major 1964). When localities are ranked on scales of available moisture, the distribution of each species among the localities exhibits a distinct optimum (Figure 9–18). The coast redwood dominates the central portion of the moisture gradient and frequently forms pure stands. Cedar, Douglas fir, and two broad-leaved evergreen species with small thick leaves—manzanita and madrone—occur at the drier end of the moisture gradient. Three deciduous species—alder, big-leaf maple, and black cottonwood—occupy the wetter end.

Change in one environmental condition usually brings about changes in others. Increasing soil moisture alters the availability of nutrients. Varia-

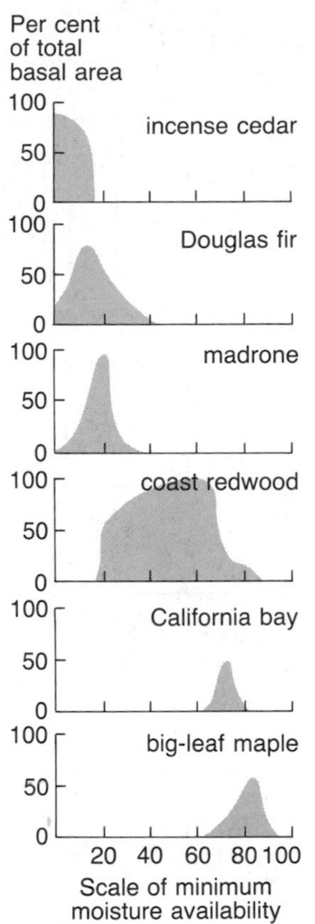

Figure 9–18
The distribution of species of trees along a gradient of minimum available soil moisture in the northern coastal region of California. (After Waring and Major 1964.)

tion in the amount and source of organic matter in the soil creates parallel gradients of acidity, soil moisture, and available nitrogen. Such factors often interact in complex ways to determine the distributions of plants. Figure 9–19 relates the distributions of some forest-floor shrubs, seedlings, and herbs in woodlands of eastern Indiana to levels of organic matter and calcium in the soil (Beals and Cope 1964). These soils contain between 2 and 8 per cent organic matter and between 2 and 6 per cent exchangeable calcium. Within the range of soil conditions in these woodlands, each species shows different preferences. Black cherry seedlings occur only within a narrow range of calcium but tolerate variation in the percentage of organic matter. Bloodroot is narrowly restricted by the per cent of organic matter in the soil but is insensitive to variation in calcium. The distributions of yellow violets and cream violets extend more broadly over levels of organic matter and calcium in the soil, but the two species do not overlap. Cream violets prefer relatively higher calcium and lower organic-matter content than yellow violets; where one occurs, the other usually does not.

The suitability of the environment determines the success of a population.

Organisms can live only under favorable combinations of conditions. For the Mediterranean fruit fly optimum temperatures for survival and reproduction lie between 16 and 32°C, with preferred relative humidities between 65 and 75 per cent. Populations thrive under these conditions provided, of course, that there is food. Outside these optimum ranges, fruit flies can barely maintain populations under conditions between about 10 and 35°C, and 60 to 90 per cent relative humidity; individuals can persist at temperatures as low as 2°C and relative humidity as low as 40 per cent, but

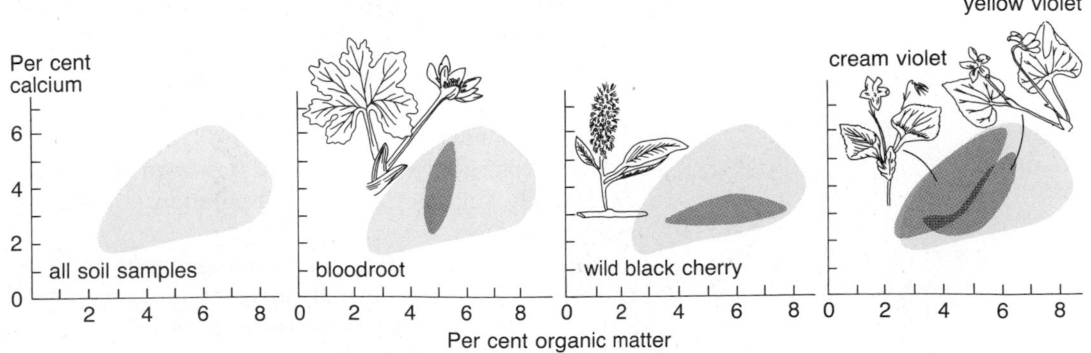

Figure 9–19
The occurrence of four forest-floor plants with respect to the calcium and organic-matter contents of the soil in woodlands of eastern Indiana. (After Beals and Cope 1964.)

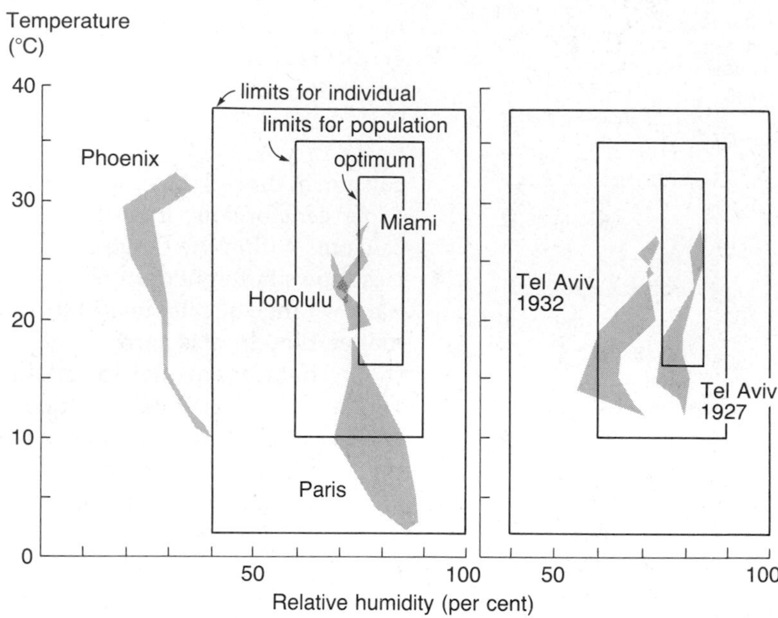

Figure 9–20
The seasonal course of air temperature and relative humidity at selected locations in relation to conditions favorable for the Mediterranean fruit fly. The inner rectangle encloses conditions optimum for growth, the middle rectangle encloses conditions suitable for development, and the outer rectangle delimits the extreme tolerance range. (After Bodenheimer, in Allee et al. 1949.)

they do not reproduce under these conditions; more extreme conditions are usually lethal (Figure 9–20).

The Mediterranean fruit fly is a major agricultural pest in many parts of the world, but populations reach outbreak proportions only where conditions remain within the biological optimum most of the year and rarely exceed the limits of tolerance. Thus Tel Aviv, Israel, where mean monthly temperatures vary between 7 and 31°C, and humidity ranges between 59 and 73 per cent, frequently experiences plagues of the fly and requiring extensive control measures, although conditions favor outbreaks more in some years (1927, for example) than others (1932). The climate of Paris, France, is generally too cold for fly populations to reach damaging levels, and that of Phoenix, Arizona, is too dry. The climates of Honolulu, Hawaii, and Miami, Florida, where the pest has been accidentally introduced, are unfortunately suitable for rapid population growth and so the need for controls continues.

Distributions of plants over heterogeneous soils have revealed many factors that limit plant growth, survival, and reproduction. In the coastal ranges of northern California, several species of pines *(Pinus)* and cypresses *(Cupressus)* are restricted to serpentine soils, while others occur only on extremely acid soils (Kruckelberg 1954, McMillan 1956). When grown on soils from different localities, seedlings of these endemics often do best when planted in soil from the native habitat. Thus, lodgepole pine grows

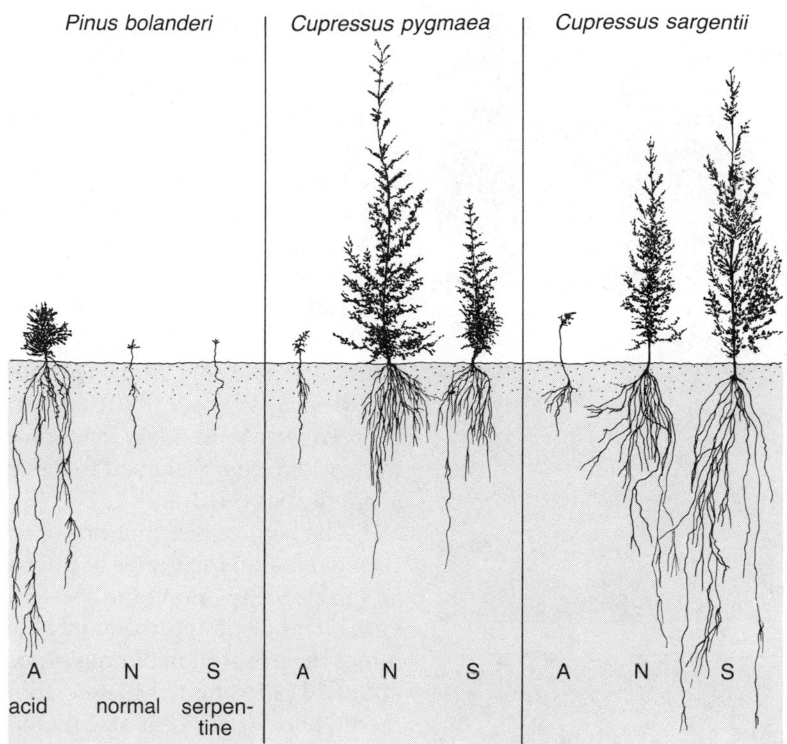

Figure 9–21
Seedling growth of lodgepole pine, pygmy cypress, and Sargent cypress in acid *(left)*, "normal" *(center)*, and serpentine *(right)* soils. Acidity (pH) of the three soils was 4.5, 6.8, and 7.1; exchangeable calcium (millequivalents per 100 g) was 0.7, 12.9, and 3.4; exchangeable magnesium (me per 100 g) was 0.05, 8.9, and 17.0. (After McMillan 1956.)

only in acid soils and Sargent cypress, a serpentine endemic, grows somewhat better on serpentine soil than on "normal" soil, and not at all on acid soil (Figure 9–21). But not all endemics perform best on their home soil. When given the chance in an experimental garden, pygmy cypress, normally restricted to acid soils, grows much better on "normal" and serpentine soils. Where does it derive its tolerance for serpentine soils? What factors exclude pygmy cypress from soils on which its seedlings grow vigorously? Clearly, factors other than edaphic conditions (those inherent in the soil), likely including competitors and pathogens, influence the distributions of these species.

Adaptations of plants and animals match the conditions within their environments.

The adaptations of an organism—its form, physiology, and behavior—cannot easily be separated from the environment in which it lives. Insect larvae from stagnant aquatic environments in ditches and sloughs can survive longer without oxygen than related species from well-aerated streams and rivers; species of marine snails that occur high in the intertidal zone, where they are frequently exposed to air, tolerate desiccation better than species from lower levels.

Compare the leaves of deciduous forest trees with those of desert species. The former are typically broad and thin, providing a large surface area for light absorption and unavoidably, therefore, for water loss. Desert trees have small, finely divided leaves—or sometimes none at all (Figure 9–22). Leaves heat up in the desert sun. Structures lose heat by convection most rapidly at their edges, where wind currents disrupt the insulating boundary layers of still air. The more edges, the cooler the leaf and the lower the water loss (Vogel 1970), and small size means a large portion of each leaf is given over to its edge. Even on a single plant, leaves exposed to full sun may be differently shaped to dissipate heat and conserve water better than shade leaves (Figure 9–23).

The vertical distributions of species of algae within the intertidal range closely parallel their rates of photosynthesis in water and when exposed to air (Table 9–1). Along the coast of central California, the tide varies over a vertical range of approximately 2.5 meters. Height within this range determines the proportion of time exposed to air on an annual basis. Exposure is about 10 per cent at 0 meters above designated sea level, 40 per cent at 1 meter, and 90 per cent at 2 meters. Algae from the lower part of the tide range (*Prionitus* and *Ulva*) reduce their photosynthetic rate while exposed; algae from the middle and upper parts of the tidal range elevate their rates of photosynthesis in air (Johnson et al. 1974). The higher of the two mass-specific rates of photosynthesis, those obtained while exposed or submerged, varies directly with the surface area of the alga per unit of dry mass. The rate

Figure 9–22
Leaves of some desert plants from Arizona. Mesquite *(Prosopis)* leaves *(top)* are subdivided into numerous small leaflets, which facilitate the dissipation of heat when exposed to sunlight. The palo verde *(Cercidium)* carries this adaptation even further *(center)*; its leaves are tiny and the thick stems, which contain chlorophyll, are responsible for much of the plant's photosynthesis (hence the name paloverde, which is Spanish for green stick). (Cacti rely entirely on their stems for photosynthesis; their leaves are modified into thorns for protection.) Unlike most desert plants, limberbush *(Jatropha) (bottom)* has broad succulent leaves, which it produces for only a few weeks during the summer rainy season in the Sonoran Desert. (Photographs about one-quarter size.)

Table 9–1
Characteristics of intertidal species of algae from various height zones

Characteristic	Species				
	Prionitis	*Ulva*	*Iridaea*	*Porphyra*	*Fucus*
Mean height above sea level (ft)	−1.0	+0.5	+1.0	+3.0	+3.0
Time exposed to air (per cent)	5	15	25	40	40
Leaf area per unit weight ($dm^2\ g^{-1}$ dry weight)*	0.4	3.3	1.3	1.9	0.5
Photosynthesis rate ($mgCO_2\ g^{-1}\ hr^{-1}$)					
(a) in air	1.1	12.2	5.2	17.7	5.8
(b) in water	1.2	16.7	1.7	6.3	0.9
(c) air to water ratio	0.9	0.7	2.9	2.8	6.6
Photosynthesis drops to one-half submerged rate at					
(a) water loss (per cent)	10	24	32	47	60
(b) hours of exposure	0.4	0.6	1.3	3.6	5.3
Rate of water loss ($\%/hr^{-1}$)	25	40	25	13	11

*dm = decimeter; 1 dm^2 = 100 cm^2.
Source: Johnson et al. 1974.

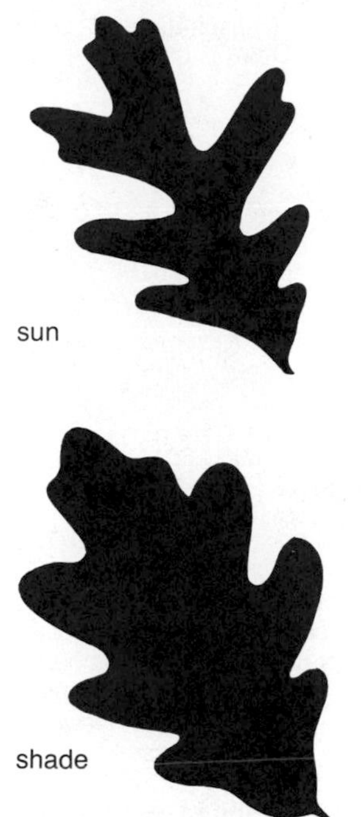

Figure 9–23
Silhouettes of sun and shade leaves of white oak. The sun leaves have more edge per unit of surface area and therefore dissipate heat more rapidly. (After Vogel 1970.)

of photosynthesis in air relative to that while submerged depends on tolerance of desiccation. In algae from the lower portion of the intertidal, photosynthesis decreases after exposure to air compared to photosynthesis in plants immersed in water. After 0.4 to 0.6 hours of exposure to air, plants have lost 10 to 24 per cent of water (at a rate of 25 to 40 per cent per hour). In algae from the middle and upper intertidal zone, photosynthesis first increases upon exposure. It drops to one-half the value while submerged only after 1.3 to 5.3 hours and a 32 to 60 per cent loss of water (at a rate of 11 to 25 per cent per hour). Thus, the adaptations of these algae—low rates of water loss and relative insensitivity to desiccation—suit them to the long periods of exposure to air in the upper parts of the tidal range.

The water relations of coastal sage and chaparral plants in southern California demonstrate divergent courses of adaptation (Harrison et al. 1971, Mooney and Dunn 1970). Chaparral plants generally occur at higher elevations than the coastal sage, and thus experience cooler and moister conditions. Both vegetation types are exposed to prolonged summer drought, but soils have greater water deficits in the sage habitat. Plants of the coastal habitat typically have shallow roots and small, delicate deciduous leaves (Figure 9–24). Chaparral species have deep roots, often extending through tiny cracks and fissures far into the bedrock; their leaves are typically thick and have a waxy outer covering (cuticle) that reduces water loss. Most coastal sage species shed their delicate leaves during the summer drought period; the tougher leaves of chaparral plants persist.

Leaf morphology influences photosynthetic rate in tandem with its influence on transpiration (Table 9–2). The thin leaves of coastal sage species lose water rapidly, but also assimilate carbon rapidly when water is available to replace transpiration losses. To demonstrate this relationship, scientists clipped leaves from plants and placed them in a chamber within which they could monitor transpiration and photosynthesis. Both functions declined as the leaves dried out and their stomata closed to prevent further water loss. Rates of both photosynthesis and transpiration of such coastal sage species as the black sage were high initially, but shut down quickly

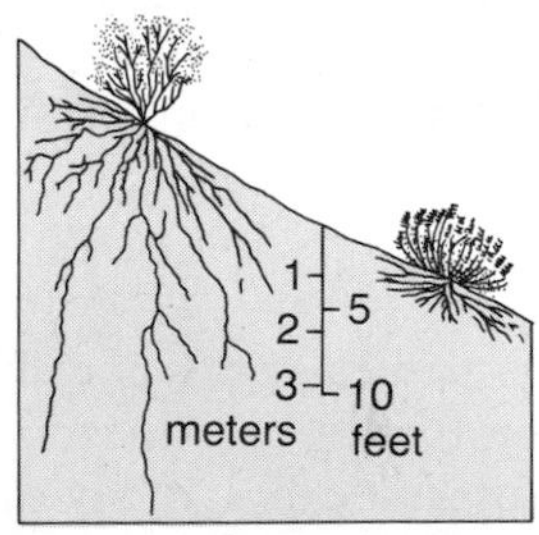

Figure 9–24
Profiles of the root systems of chamise *(Adenostoma fasciculatum)*, a chaparral species *(left)*, and black sage *(Salvia mellifera)*, a member of the coastal sage community. (After Hellmers et al. 1955.)

Table 9–2
Characteristics of chaparral and coastal sage vegetation in southern California

Characteristics	Vegetation type: Chaparral	Vegetation type: Coastal sage
Roots	Deep	Shallow
Leaves	Evergreen	Summer deciduous
Average leaf duration (months)	12	6
Average leaf size (cm^2)	12.6	4.5
Leaf weight (g dry wt dm^{-2})*	1.8	1.0
Maximum transpiration (g H_2O dm^{-2} hr^{-1})	0.34	0.94
Maximum photosynthetic rate (mg C dm^{-2} hr^{-1})	3.9	8.3
Relative annual CO_2 fixation	49.8	46.8

*dm = decimeter; 1 dm^2 = 100 cm^2.
Source: Harrison et al. 1971, Mooney and Dunn 1970.

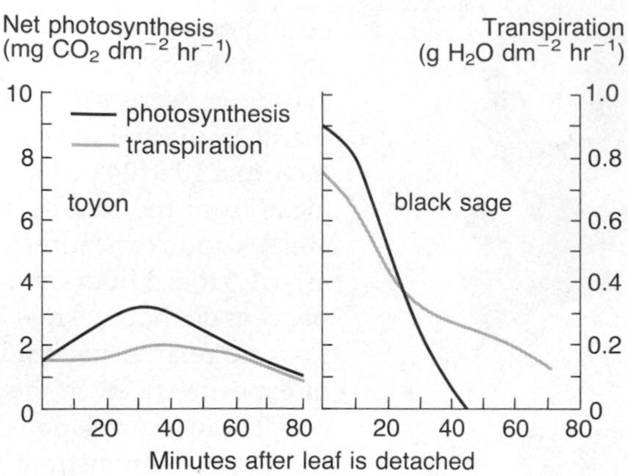

Figure 9–25
Time courses for photosynthesis and transpiration under standard drying conditions: *left,* a chaparral species (toyon, *Heteromeles arbutifolia*); *right,* a coastal-sage species (black sage, *Salvia mellifera*). Note that transpiration continues well after photosynthesis has been shut off; hence leaf dormancy is an ineffective long-term solution to drought. (After Harrison et al. 1971.)

owing to rapid water loss (Figure 9–25); photosynthetic rates of such chaparral species as the toyon (rose family) were only one-fourth to one-third those of coastal sage species at most, but the leaves resist desiccation better and continued to be active under drying conditions for longer periods.

When chaparral and coastal sage species grow together near the overlapping edges of each other's ranges, they exploit different parts of the environment: deep, perennial sources of water versus shallow, ephemeral sources of water. In spite of these differences and the corresponding adaptations of leaf morphology and drought response, they are equally productive at intermediate levels of water availability. In drier habitats, the prolonged seasonal absence of deep water tips the balance in favor of the deciduous coastal sage vegetation. Increasing availability of deep water at higher elevations favors the evergreen chaparral vegetation.

Classifications of plant associations based on plant form correspond closely to climate.

Natural history, and later ecology, grew out of schemes of classification—systems by which animals and plants are given names based upon their similarities. European botanists had described most local plant species by the end of the last century; they then began to develop systems of classification for entire communities of plants. They based most of these schemes on structure—height of vegetation, leaf or needle structure, deciduousness, and dominant plant form. Because these properties adapt plants to the physical environment in which they live, vegetation zones and climate correspond closely (Figure 9–26).

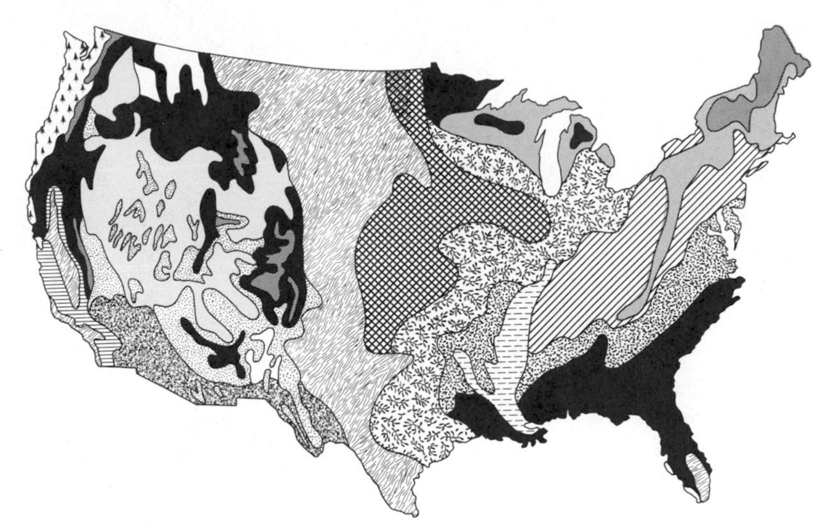

Figure 9–26
Vegetation map of the United States based primarily upon forest types. (After Fowells 1965.)

The earliest classifications of vegetation described the most important plants of each major association (see Schimwell 1971, Mueller-Dombois and Ellenberg 1974). These classifications embraced complete analysis by name of the species composition of communities (floristic analysis), on one hand, and description of plant forms, regardless of species, on the other. Floristic analysis proved useful in Europe, where botanists knew all the species and where minor differences between communities involved the replacement of species by similar ones with slightly different ecological requirements. But floristic analysis proved unworkable on a global scale because biogeographical barriers restrict the distributions of individual species, rendering floristic comparisons ecologically meaningless. Forests in Europe and the United States, shrublands in California and Australia, grasslands in Africa and South America, while structurally similar, have few species in common.

Worldwide classification of vegetation depends on an analysis of form and function of plants rather than their scientific names. Botanists have derived various sets of symbols to describe such characteristics as plant size, life form, leaf shape, size, and texture, and per cent of ground coverage. A. W. Kuchler (1949) worked out a system of letter and number symbols that he combined into formulas describing the structural characteristics of diverse plant formations. Thus M6iCXE5cD3i6H2pL1c represents an oak-yew woodland and E4hcD2rGH2rL1c(b), a madrone-holly scrub. A similar symbolic method of description, devised by Pierre Dansereau (1957), portrays vegetation formations by using lollipop- and ice-cream-cone-shaped figures with internal shading and symbols that vary according to the nature of the plant (Figure 9–27). Although they emphasize plant form rather than floristics, the classifications of Kuchler and Dansereau are primarily descriptive. Too complex for hierarchical systems, they have been useful when

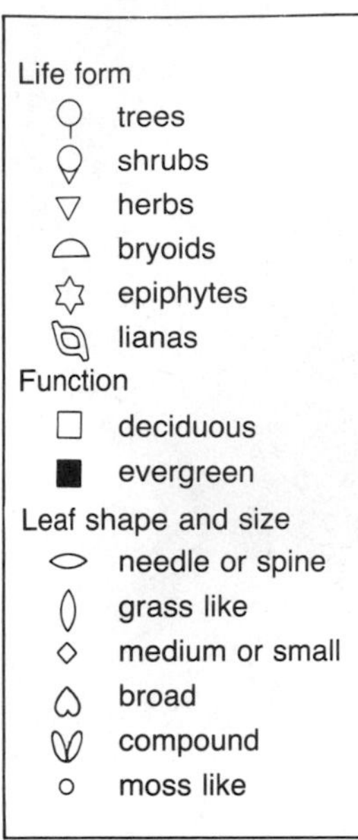

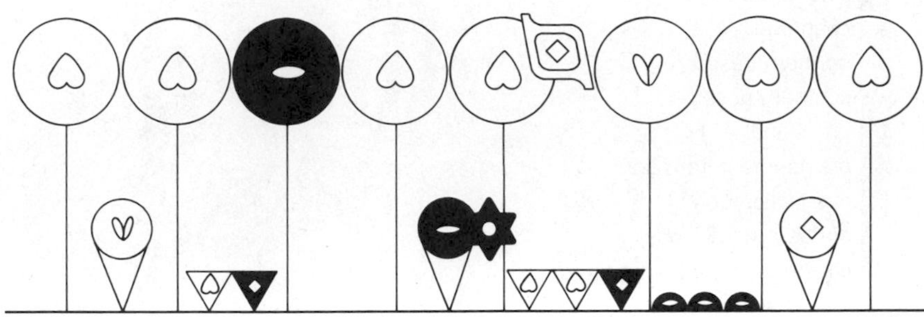

Figure 9–27
Symbolic representation of a forest by the Dansereau method of vegetation classification. (After Dansereau 1957.)

researchers wish to classify only the predominant features of a plant formation.

In 1903, the Danish botanist Christen Raunkiaer proposed to classify plants according to the position of their buds (regenerating parts), and found that the occurrence of his major categories corresponded closely to climatic conditions (Raunkiaer 1934). He distinguished five principal life forms (Figure 9–28). Phanerophytes (from the Greek *phaneros,* visible) carry their buds on the tips of branches, exposed to extremes of climate. Most trees and large shrubs are phanerophytes. As one might expect, this plant form predominates in moist, warm environments where buds require little protection. Chamaephytes (from the Greek *chamai,* on the ground, dwarf) comprise small shrubs and herbs that grow close to the ground (prostrate life form). Proximity to the soil protects the bud. In winter, snow cover often protects the buds from extreme cold. Chamaephytes occur most frequently in cool, dry climates. Hemicryptophytes (from the Greek *kryptos,* hidden) persist through the extreme environmental conditions of the winter months by dying back to ground level, where the regenerating bud is protected by soil and withered leaves. This growth form is characteristic of cold, moist zones. Cryptophytes are further protected from freezing and desiccation because their buds are completely buried beneath the soil. The bulbs of irises and daffodils are the regenerating buds of cryptophyte plants. Like hemicryptophytes, cryptophytes are found in cold, moist climates. Therophytes (from the Greek *theros,* summer) die during the unfavorable season of the year and do not have persistent buds. Therophytes are regenerated

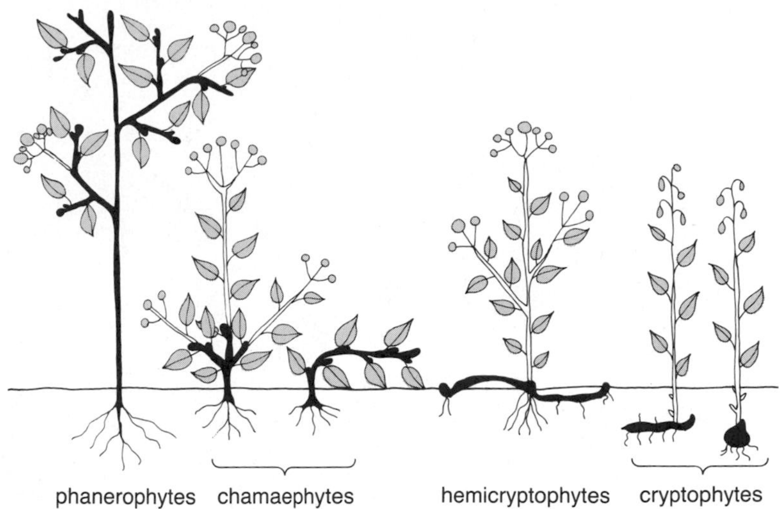

Figure 9–28
Diagrammatic representation of Raunkiaer's life forms. Lightly shaded parts of the plant die back during unfavorable seasons, while the solid black portions persist and give rise to the following year's growth. Proceeding from left to right, the buds are progressively better protected. Therophytes, whose persistent parts are seeds, are not illustrated. (After Raunkiaer 1937.)

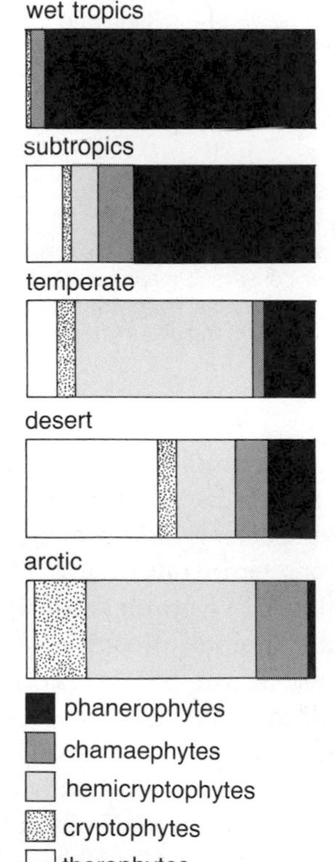

Figure 9–29
Proportion of plant life forms, classified according to Raunkiaer (1934, 1937), in various climatic regions. (After compilations of Richards 1952, Dansereau 1957, and Daubenmire 1968.)

solely by seeds, which resist extreme cold and drought. The therophyte form includes most annual plants and occurs most abundantly in deserts and grasslands. Raunkiaer's life forms and climate go closely together (Figure 9–29).

The classification schemes of Holdridge and Whittaker stress the interaction of temperature and precipitation.

Botanist L. R. Holdridge (1967) proposed a classification of the world's plant formations based solely on climate (Figure 9–30). Holdridge considered temperature and rainfall to prevail over other environmental factors in determining vegetation, although soils and exposure may strongly influence plant communities within each climate zone.

Holdridge's scheme, while not widely used, classifies climate according to the biological effects of temperature and rainfall on vegetation. As in Thornthwaite's analysis of climate, temperature and rainfall are seen as interacting to define humidity provinces separated by critical ratios of potential evapotranspiration to precipitation. Potential evapotranspiration (PE) estimates the total evaporation and transpiration from a habitat in the presence of continuously abundant water. Thus the ratio of PE to actual evapotranspiration provides an index to aridity.

Humidity provinces relate temperature and rainfall to the water rela-

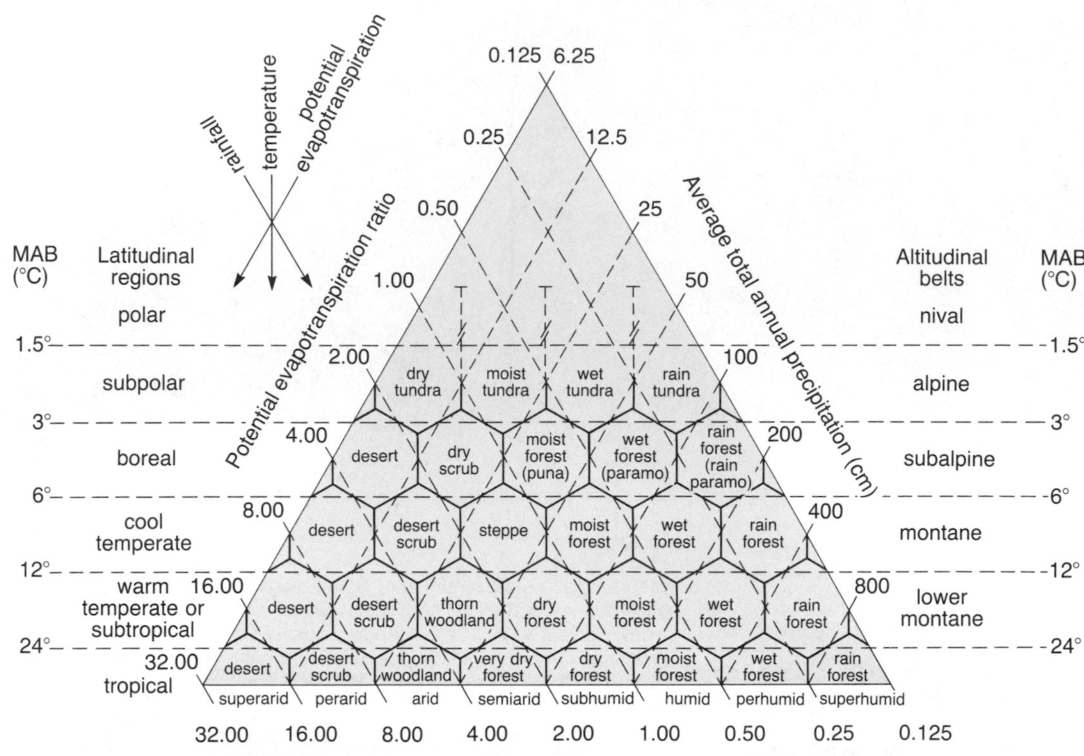

Figure 9–30
The Holdridge scheme for the classification of plant formations. Mean annual biotemperature (MAB) is calculated from monthly mean temperatures after converting means below freezing to 0°C. The potential evapotranspiration ratio is the potential evapotranspiration divided by the precipitation; the ratio increases from humid to arid regions. (After Holdridge 1967.)

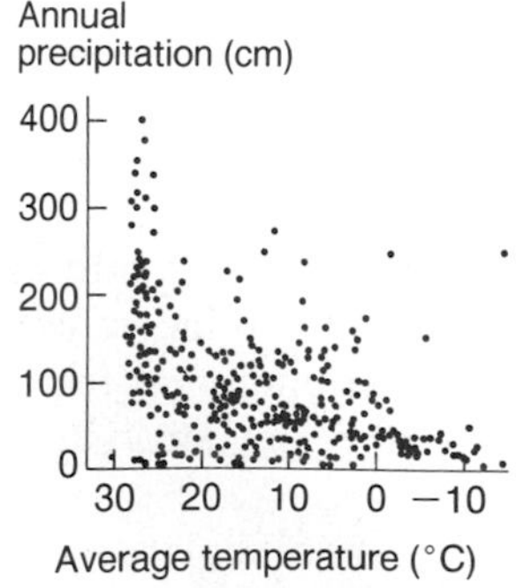

Figure 9–31
Average annual temperature and precipitation for a sample of localities more or less evenly distributed over the land area of the earth. Most of the points occur within a triangular region that includes the full range of climates, excluding those of high mountains. (Data from Clayton 1944).

tions of plants. Holdridge's formula indicates, for example, that moisture is equally available to plants in wet tundra, with an annual precipitation of 25 centimeters and an average temperature near freezing, and wet tropical forest, with 400 cm precipitation and an average temperature of 27°C.

When we plot a sample of terrestrial localities on a graph according to their mean annual temperature and rainfall, most points fall within a triangular area whose three corners are warm-moist, warm-dry, and cool-dry (Figure 9–31). Cold regions with high rainfall are conspicuously absent; water does not evaporate rapidly at low temperature, and the atmosphere in cold regions holds little water vapor. Plant ecologist R. H. Whittaker (1975) combined several structural classifications of plant communities with Holdridge's idea of humidity provinces into a single scheme, which he has transposed onto this graph of temperature and rainfall (Figure 9–32).

Within the tropical and subtropical realms, with mean temperatures between 20 and 30°C, vegetation types grade from true rain forest, which is wet throughout the year, to desert. Intermediate climates support seasonal forests, in which some or all trees lose their leaves during the dry season, and short, dry forests or scrublands with many thorntrees.

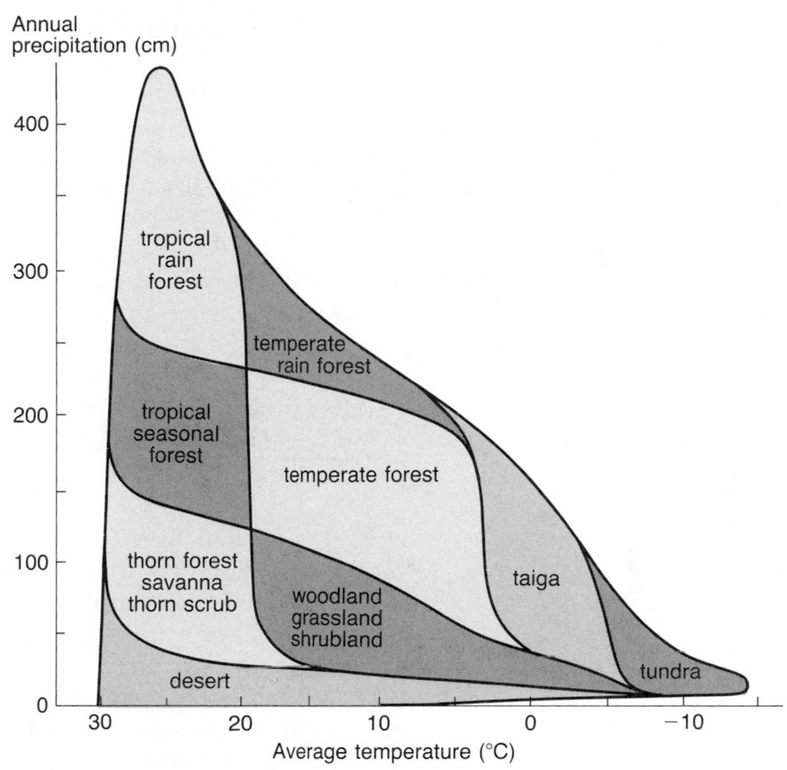

Figure 9–32
Whittaker's classification of vegetation types superimposed upon the range of terrestrial climates. In climates intermediate between those of forested and desert regions, fire, soil, and climate seasonality determine whether woodland, grassland, or shrubland develops. (From Whittaker 1975).

Plant communities in temperate areas follow the pattern of tropical communities, with the same vegetation types distinguishable in both. In colder climates, however, precipitation varies so little from one locality to another that vegetation types are poorly differentiated on the basis of climate. Where mean annual temperatures are below −5°C, Whittaker lumped all plant associations into one type: tundra.

Toward the drier end of the rainfall spectrum within each temperature range, fire plays a distinct role in shaping the form of plant communities (Borchert 1950, Daubenmire 1968). For example, in the African savannas and midwestern American prairies, frequent fires kill the seedlings of trees and prevent the establishment of tall forests, for which favorable conditions otherwise exist. Burning favors perennial grasses with extensive root systems that can survive fire. After an area has burned over, grass roots send up fresh shoots and quickly revegetate the surface. In the absence of frequent fires, tree seedlings can become established and eventually shade out prairie vegetation.

As in all classifications, exceptions appear frequently. Boundaries between vegetation types are at best fuzzy. Moreover, all plant forms do not respond to climate in the same way. For example, some species of Australian eucalyptus trees form forests under climate conditions that support only shrubland or grassland on other continents. Finally, plant communities reflect factors other than temperature and rainfall. Topography, soils, fire, and seasonal variation in climate all leave their mark, further emphasizing the adaptation of life forms to the diversity of environments on the earth.

SUMMARY

Plants and animals are specialized to exist within narrow ranges of environmental conditions. Climate therefore determines both the distributions of organisms and the characteristic life forms of plants within each region.

1. Global patterns of temperature and rainfall are determined by the local interception of solar radiation and the redistribution of heat energy with winds and ocean currents. Prominent features of terrestrial climates include a band of warm, moist climate over the equator and bands of dry climate at about 30° north and south latitude.
2. Seasonality is caused by the annual progression of the sun's path northward and southward, and the latitudinal movement of associated belts of wind and precipitation.
3. Seasonal warming and cooling profoundly changes the characteristics of temperate-zone lakes. During summer, such lakes are stratified, with a warm surface layer (epilimnion) separated from a cold bottom layer (hypolimnion) by a sharp thermocline. In spring and autumn, the profile of temperature with depth becomes more uniform, allowing vertical mixing.
4. Irregular and unpredictable variations in climate, such as El Niño–Southern Oscillation events, may cause major disruptions of biological communities on global scales.
5. Topography and geology impose local variation in environmental conditions over more general climate patterns. Mountains intercept rainfall, creating arid rain shadows in their lees. Conditions at higher altitudes resemble conditions at higher latitudes. Soil characteristics reflect the quality of the underlying bedrock and sometimes foster specialized floras, such as those of serpentine barrens.
6. Descriptions of climate, such as the climograph and Thornthwaite's analysis of seasonal water availability, integrate the effects of both temperature and precipitation.
7. The geographical distributions of plants are determined primarily by climate, whereas local distributions within regions vary according to topography and soils.
8. Climate profoundly affects adaptations of plants and animals. As a consequence, each climate region has characteristic vegetation forms differing in general habit, leaf morphology, and phenology.
9. Although early schemes of classifying plant associations were based on species (floristics), those based on plant form have proven more

broadly applicable. Raunkiaer classified plants according to the degree of protection of their buds and showed that proportions of these plant types within a local flora varied systematically with climate.

10. Recognizing that plant form was directly related to climate through adaptation, Holdridge and Whittaker designated major regions of vegetation on the basis of local temperature and precipitation. All such schemes emphasize the interaction between temperature and water availability, and further recognize the modifying effects of climate seasonality, soils, and fire.

III

Energy and Materials in the Ecosystem

10

The Ecosystem Concept

As the diverse information gathered by naturalists during the last century coalesced into a unified picture of nature two concepts emerged, pushing the study of ecology in new directions. The first was the realization that species of plants and animals formed natural associations, each with distinctive members. Just as morphological data had allowed systematists to assign species to a hierarchy of taxonomic groups, detailed studies of the ecological distributions of plants led to the classification of biological communities (Shimwell 1971).

A second new concept was the realization that organisms are linked both directly and indirectly by means of their feeding relationships. Humans have appreciated since their beginnings that one organism both preys upon and is preyed upon by others, but that these feeding relations linked species into a functional unit was a novel idea at the turn of the present century.

The analogy of the organism has been applied to biological communities.

From its inception, thinking about systems of interacting populations has been divided over a major issue: whether the whole exceeds the sum of its parts. On one hand, we may believe that the system as a whole has attributes that cannot be understood in terms of the workings of its parts. On the other, the system may be viewed merely as a collection of independently functioning populations.

Many ecologists believe that there is determinism (and perhaps purpose or causality) at the level of the multispecies system analogous to the determinism imparted at the organism level by natural selection. No matter how much one knows about the anatomy of elephants, one cannot explain the special relationship between the whole of an elephant and its surroundings. Is it also true, then, that all possible knowledge of grass and crickets and mice and lizards cannot unlock the secrets of the prairie?

The frequency with which ecologists have compared associations of species living together (biological communities) to organisms attests to a strong undercurrent of holistic sentiment. Functional similarities between associations and organisms—primary production and feeding, predation and metabolism, species and organs, the regular succession of stages from fallow field to mature forest and the development of the individual—are obvious. The influential American plant ecologist F. E. Clements (1916, 1936) extrapolated these similarities into a concept of biological communities as discrete vegetation types (climaxes), each occurring in a particular region defined by climate and soil and having a characteristic sequence of developmental stages (the sere) leading from bare or cleared ground to the mature state. Clements viewed plant communities as superorganisms. He had little need to question the analogy between organism and superorganism because his work was primarily descriptive and the organism concept adequately embraced what he saw in nature.

Clements's alter ego was also an American plant ecologist, H. A. Gleason (1926, 1939), for whom the local community, far from being a distinct unit like an organism, was merely the fortuitous association of species whose adaptations enabled them to live in a particular place. While recognizing that species do interact (all animals must eat!), Gleason argued that the presence or absence of any one species is independent of all others. In his view, we may define an association for convenience, but it does not represent a natural unit and it has no functional significance beyond the roles played by each of its members.

Charles Elton described communities in terms of feeding relationships.

Following the tradition of floristic analysis in plant community ecology, Clements and Gleason were concerned with the species composition of communities. By the mid-1920s, however, other ecologists had begun to consider functional patterns within communities. Foremost among the proponents of this viewpoint was the English ecologist Charles Elton. During his student days at Oxford, Elton accompanied an ecological expedition to Spitsbergen Island in the North Atlantic Ocean, where, in collaboration with the botanist V. S. Summerhayes, he worked out the feeding relationships among inhabitants of a simple tundra community (Figure 10–1).

By the time Elton was twenty-six, he had developed a new concept of communities organized by the feeding relationships within them. In his book *Animal Ecology* (1927), which was to become a landmark for modern ecology, he wrote:

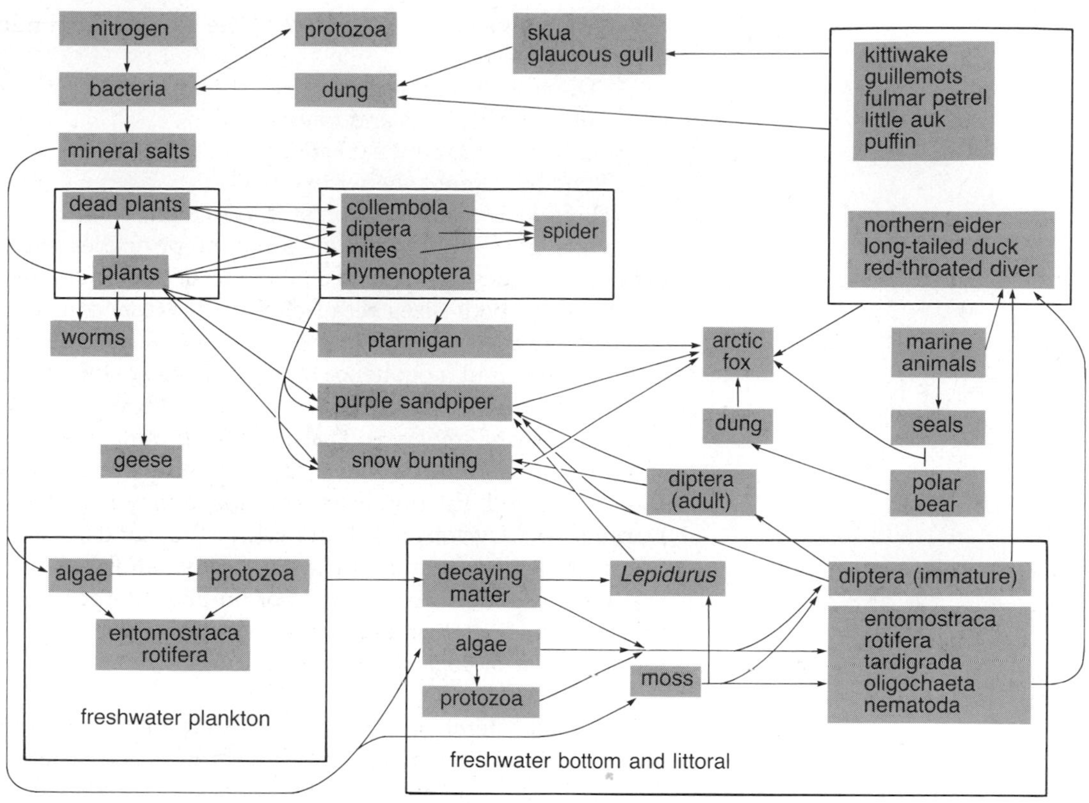

Figure 10-1
Summerhayes and Elton's 1923 depiction of the feeding relationships among animals on Bear Island, near Spitsbergen. (From Elton 1927.)

> Food is the burning question in animal society, and the whole structure and activities of the community are dependent upon questions of food-supply . . . animals have to depend ultimately upon plants for their supplies of energy, since plants alone are able to turn raw sunlight and chemicals into a form edible to animals. Consequently herbivores are the basic class in animal society. . . . The herbivores are usually preyed upon by carnivores, which get the energy of the sunlight at third-hand, and these again may be preyed upon by other carnivores, and so on, until we reach an animal which has no enemies, and which forms, as it were, a terminus on this food-cycle. There are, in fact, chains of animals linked together by food, and all dependent in the long run upon plants. We refer to these as "food-chains," and to all the food-chains in a community as the "food-cycle."

"Food-cycle" became "food web" over the years, but Elton's basic concept has otherwise survived unchanged.

Elton prefaced his Chapter V ("The Animal Community") with three Chinese proverbs:

"The large fish eat the small fish; the small fish eat the water insects; the water insects eat plants and mud."

"Large fowl cannot eat small grain."

"One hill cannot shelter two tigers."

The first proverb is Elton's "food-chain." The second and third are cornerstones of another of Elton's general principles, the pyramid of numbers. As one goes up the food chain, one also ascends a more or less regular procession of body sizes because most predators consume prey somewhat smaller than themselves. Progressively larger animals require progressively more space to find food; hence their numbers are lower. Elton noted that an oak wood harbors "vast numbers of small herbivorous insects like aphids, a large number of spiders and carnivorous ground beetles, a fair number of small warblers (insectivorous birds), and only one or two hawks. Similarly in a small pond, the numbers of protozoa may run into millions, those of Daphnia and Cyclops into hundreds of thousands, while there will be far fewer beetle larvae, and only a very few small fish."

Elton explained this pyramid of numbers, large at the base of the food chain and small at the top, by using arguments from population biology and the scaling of biological functions to body size.

> The small herbivorous animals which form the key-industries in the community are able to increase at a very high rate (chiefly by virtue of their small size), and are therefore able to provide a large margin of numbers over and above that which would be necessary to maintain their population in the absence of enemies. This margin supports a set of carnivores, which are larger in size and fewer in numbers. These carnivores in turn can only provide a still smaller margin, owing to their large size which makes them increase more slowly, and to their smaller numbers.

And so on to the tiger on the hill.

During the 1930s, the idea of the community as an association of interacting species more and more became the focus of ecological thinking, but was far from universally accepted.

A. G. Tansley coined the term "ecosystem" in 1935.

By the mid-1930s, the superorganism concept of the community was still very much alive. Clements remained influential. In a three-part article, "Succession, Development, the Climax and the Complex Organism: An Analysis of Concepts," English ecologist John Phillips (1934, 1935) elaborated the Clementsian view that the community has unique properties of function and organization, a "newness springing from the interaction, interrelation, integration and organisation of qualities . . . (which) could not be predicted from the sum of the particular qualities or kinds of qualities concerned: integration of the qualities thus results in the development of a whole different from, unpredictable from, their mere summation." But Phillips also took a strong philosophical stance on causation in ecology by attributing "operative causes" and "inherent, dynamic characteristics" to communities; he stressed the "fundamental nature of the *factor of holism* innate in the very being of community, a *factor of cause.*" He further explained:

> At different levels the whole reacts upon habitat, changing (ameliorating) this for higher level wholes: the reaction of a whole, taken into account with its particular habitat and with the interrelations existing among its constituent organisms, shows as emergent changes in the habitat that are different from the sum of the changes that the constituent organisms would undergo were these not in communal association.

The English plant ecologist A. G. Tansley (1936) rejected the superorganism notion of Clements and Phillips, preferring to regard the animals and plants in associations, together with the physical factors of their surroundings, simply as systems. He wrote that

> The more fundamental conception is, as it seems to me, the whole *system* (in the sense of physics), including not only the organism-complex, but also the whole complex of physical factors forming what we call the environment of the biome — the habitat factors in the widest sense. Though the organisms may claim our primary interest, when we are trying to think fundamentally we cannot separate them from their special environment, with which they form one physical system.

This Tansley called the ecosystem.

A. J. Lotka espoused a thermodynamic view of the ecosystem.

The controversy over the superorganism analogy went unnoticed by A. J. Lotka, a chemist by training, whose different vision of biological systems still influences us through the legacy of his book *The Elements of Physical Biology,* published in 1925. Lotka was the first to treat populations and communities as thermodynamic systems. In principle, he said, each system can be represented by a set of equations that govern transformations of mass among its components. Such transformations include the assimilation of carbon dioxide into organic carbon compounds by green plants and the consumption of plants by herbivores and animals by carnivores.

Lotka believed that the size of a system and the rate of transformations within it were determined according to certain thermodynamic principles. In the same sense that heavy machines and fast machines require more fuel to operate than their lighter and slower counterparts, and efficient machines require less fuel than inefficient ones, the energy transformations of ecosys-

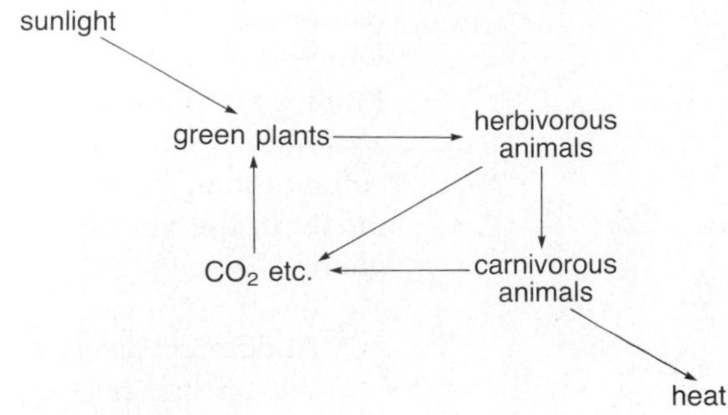

Figure 10–2
Lotka's simple diagram of the mill-wheel of life. (From Lotka 1925.)

tems grow in direct relation to their size (roughly the total masses of their constituent organisms), productivity (rate of transformations), and inefficiency. Lotka regarded the ecosystem as a part of the world machine responsible for the transformation of the energy of sunlight reaching the surface of the earth. Not all the energy enters biological pathways of transformations; in fact, most of it drives the circulation of winds and ocean currents and the evaporation of water. But the portion that plants assimilate by photosynthesis drives a part of the world machine, what Lotka referred to, by way of analogy, as the mill-wheel of life (Figure 10–2).

When Lotka published his book, most ecologists were preoccupied with the problem of species associations and missed the implications of a thermodynamic characterization of the natural world (Kingsland 1985). Tansley never referred to *The Elements of Physical Biology,* even though Lotka had provided the mechanical analogy—thus a tangible model—for his concept of the ecosystem.

Raymond Lindeman developed the trophic-dynamic concept.

The idea of the ecosystem as an energy-transforming system was brought to the attention of many ecologists for the first time in a paper published in 1942 by Raymond Lindeman, a young aquatic ecologist from the University of Minnesota. The full story behind this historic paper has been recounted by Robert Cook (1977): the journal *Ecology* first rejected the manuscript on the advice of reviewers who felt the treatment was too theoretical; but strong advocacy by Yale ecologist G. E. Hutchinson finally led to its publication.

Lindeman's framework for understanding ecological succession based on sound thermodynamic principles made a deep impression. He adopted Tansley's notion of the ecosystem as the fundamental unit in ecology and

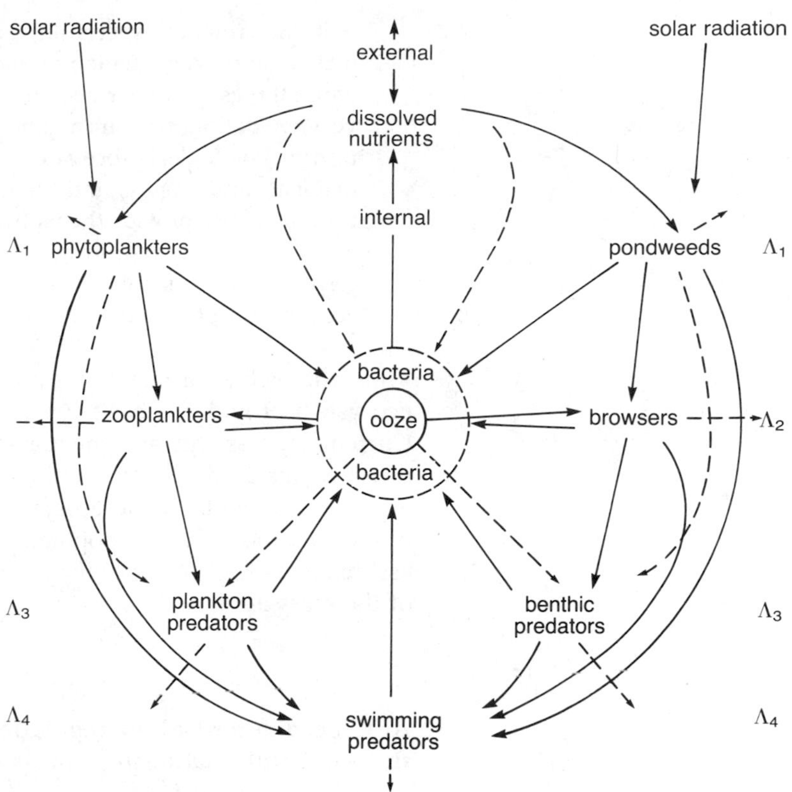

Figure 10-3
Lindeman's diagram of the generalized "food-cycle" relationships within a temperate-zone lake. (After Lindeman 1942.)

Elton's concept of the food web, including inorganic nutrients at the base, as the most useful expression of ecosystem structure (Figure 10-3).

The food chain consisted of steps—primary producer, herbivore, carnivore—that Lindeman referred to as trophic levels. But rather than seek regularity in a trophic pyramid of numbers, as Elton had, he visualized a pyramid of energy transformation. He argued that less energy is available to each higher trophic level owing to the work performed and to the inefficiency of biological energy transformations on the next lower trophic level. Thus, of the light impinging on a lake (Λ_0), plants assimilate only a fraction (Λ_1), the primary production of the system. Herbivores assimilate less energy (Λ_2) than do plants because plants utilize some of their own production to maintain themselves before herbivores consume them. The ratio of production on one trophic level to that on the level below it (for example, Λ_2/Λ_1) is the biological efficiency of that link in the food chain.

By the 1950s, the ecosystem concept had fully pervaded ecological thinking and provided the beginnings for a new branch of ecology. University of Michigan ecologist Francis C. Evans (1956) summarized the essential points in a brief essay:

> In its fundamental aspects, an ecosystem involves the circulation, transformation, and accumulation of energy and matter through the medium of living things and their activities. Photosynthesis, decomposition, herbivory, predation, parasitism, and other symbiotic activities are among the principal biological processes responsible for the transport and storage of materials and energy, and the interactions of the organisms engaged in these activities provide the pathways of distribution. The food-chain is an example of such a pathway. . . . The ecologist, then, is primarily concerned with the quantities of matter and energy that pass through a given ecosystem and with the rates at which they do so.

Thus, the cycling of matter and the associated flux of energy through the ecosystem provided a basis for characterizing its structure and function. Currencies of energy and the masses of such elements as carbon allowed direct comparison of plants, animals, microbes, and abiotic sources of energy and elements in the ecosystem. Taxonomic lists of species and the numbers of individuals in populations gave way to measurements of energy assimilation and energetic efficiencies in this new thermodynamic concept of the ecosystem.

A. J. Lotka described the regulation of ecosystem function in terms of the ecological relationships of its component populations.

Although Evans did not explicitly address the issue of the ecosystem as a superorganism, his writing expressed a mechanical notion of the function of the ecosystem and its regulation. Each population was an individual machine and the work it performed added to that performed by all the others to yield the work performed by the ecosystem as a whole. In Evans's view, probably the prevalent one in the 1950s, energy transformation by each population depended on factors in its environment that directly affected the activities of individuals in the population.

> Ecosystems are further characterized by a multiplicity of regulatory mechanisms, which, in limiting the numbers of organisms present and in influencing their physiology and behavior, control the quantities and rates of movement of both matter and energy. Processes of growth and reproduction, agencies of mortality (physical as well as biological), patterns of immigration and emigration, and habits of adaptive significance are among the more important groups of regulatory mechanisms.

Lotka (1925) had formalized such relationships in a system of equations, one for each population. Each equation described the rate of change in the numbers of individuals (one could substitute biomass or its energy equivalent) in terms of physical and biological factors in the environment, including the species it fed upon and which fed upon it. Formally, Lotka considered a system with components X_1 through X_n: "In general the rate of growth dX/dt of any one of these components will depend upon, will be a function (F) of the topography, climate, etc. If these latter features are defined in terms of a set of parameters $P_1 P_2 \ldots P_j$, we may write, . . .

$$\frac{dX_1}{dt} = F_1(X_1, X_2, \ldots X_n; P_1, P_2, \ldots P_j)$$

. . . In general there will be n such equations, one for each of the n components." When Xs are expressed in terms of energy, the sum of the X_is is the total energy content of the system at a given time and the sum of the dX_i/dts is the overall rate of change in the structure of the system. Energy transformations are dictated by individual terms of the functions F, which include all the increments of gain and loss to X.

In Raymond Lindeman's symbolism, Λ_n is the total energy content of trophic level n and the rate of change in Λ_n ($d\Lambda_n/dt$) is the sum of the gains (λ_n) and losses (λ_n') of energy from the trophic level; hence,

$$\frac{d\Lambda_n}{dt} = \lambda_n + \lambda_n'$$

"where λ_n is by definition positive and represents the rate of contribution of energy from Λ_{n-1} (the previous level) to Λ_n, while λ_n' is negative and represents the sum of the rate of energy dissipated from Λ_n and the rate of energy content handed on to the following level Λ_{n-1}. The more interesting quantity is λ_n which is defined as the true productivity of level Λ_n" (Lindeman 1942).

According to Lotka's equations, the dynamics of a single link in the food web of the ecosystem—for example, between components (species, physical sources) i and j, are defined by terms in F_i and F_j. The dynamical properties of the entire system are governed by all the terms of the individual functions F_i. The whole is equal to, and can be understood in terms of, the sum of its parts.

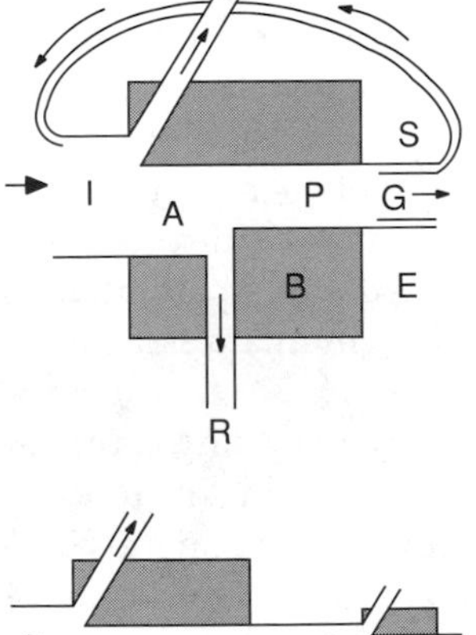

Figure 10–4
E. P. Odum's "universal" model of ecological energy flow, which can be applied to any organism; I = ingestion, A = assimilation, P = production, NU = not utilized, R = respiration, G = growth, E = egesta, S = storage, as in the form of fat, for future utilization, and B = biomass. (After Odum 1968.)

Eugene P. Odum popularized the study of ecosystem energetics.

With a clear conceptual framework for the ecosystem and a currency of energy to describe its structure, ecologists began to measure energy flow and the cycling of nutrients in the ecosystem. One of the strongest proponents of this approach has been Eugene P. Odum of the University of Georgia, whose text *Fundamentals of Ecology,* first published in 1953, influenced a generation of ecologists.

Odum depicted ecosystems as simple energy flow diagrams (Figure 10–4). For any one trophic level, such a diagram consisted of a box representing the biomass (or its energy equivalent) at any given time and pathways through the box representing the flow of energy. Feeding relationships linked energy flow diagrams into a food web.

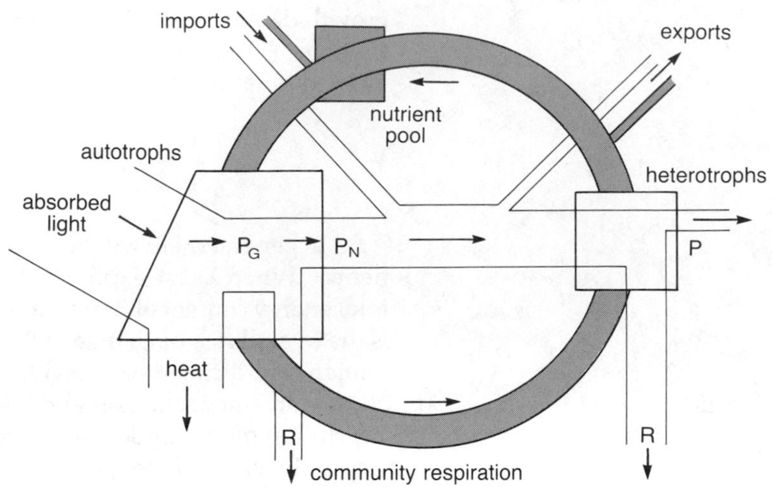

Figure 10–5
E. P. Odum's flow diagram of an ecosystem showing the one-way flow of energy and the recycling of materials; P_G = gross production, P_N = net production, P = heterotrophic production, and R = respiration. (After Odum 1960.)

Energy flow diagrams were elaborated to include the cycling of elements (Figure 10–5). Unlike energy, which ultimately comes from sunlight and leaves the ecosystem as heat, nutrients are regenerated and retained within the system. In the development of ecosystem studies, the cycling of elements has assumed nearly equal standing with the flow of energy. One reason for this prominence is that the amounts of elements and their movement between components can provide a convenient index to the flow of energy, which is difficult to measure directly. Carbon, in particular, bears a close relationship to energy content because of its intimate association with the assimilation of energy by photosynthesis.

A second reason for the prominence of nutrient cycling is the fact that levels of certain nutrients regulate primary production. In deserts, plant growth reflects amount of water rather than sunlight or minerals in the soil in most areas. By contrast, the open oceans are deserts by virtue of their scarce nutrients. Understanding how elements cycle between components of the ecosystem seems crucial to understanding the regulation of ecosystem structure and function.

The ecosystem concept has fostered a systems approach to ecology.

General acceptance of the ecosystem viewpoint provided yet another point of departure for interpreting biological systems. On one hand, many ecologists feel that understanding an ecosystem requires analysis of the interactions of most of its individual components. On the other, a quite distinct group of ecologists feels that the properties of ecosystems transcend the behavior of their individual components and can be understood only at the system level. Each component has its particular behavior, of course, just as each organ of the body has its particular function, but in a holistic concept these behaviors are subjugated to the behavior of the system as a whole.

The individualistic viewpoint, which has been referred to as the ecosystem viewpoint or *systems ecology*, relies heavily on mathematical analysis to handle the complexity of natural systems. George Van Dyne (1966) summarized the approach and its difficulties.

> The tools and processes required for systems ecology are different from those for conventional phases of ecology because of the complexity of the total ecosystem as compared with a segment of it. When we consider the totality of interactions of populations with one another and with their physical environs—i.e., ecosystem ecology—we face a new degree of complexity. . . . One of the major problems in systems ecology is that of analyzing and understanding interactions. Events in nature are seldom, if ever, caused by a single factor. They are due to multiple factors which are integrated by the organism or the ecosystem to produce an effect which we observe. To further complicate the matter, various combinations of factors and their interactions may be interpreted and integrated by the ecosystem to produce the same result.

According to Van Dyne, the goal of the ecosystem approach is to provide a model of the ecosystem capable of mimicking its dynamical properties. Disturbances, such as pollution, grazing, fire, introduced pests, hunting, and severe climate conditions, alter Lotka's system of equations dX_i/dt for certain species by changing the conditions of the environment (P) or the abundances of interacting species (X_j). By their effect on X_i, such disturbances indirectly affect other species or components (X_k) to which i is linked in the system. Only a detailed model that faithfully depicts the interactions among all components can assess the impact of a stress or other disturbance on the structure and function of the ecosystem as a whole (Van Dyne 1969, Watt 1968).

Organismic features have been attributed to ecosystems.

Systems ecologists understand ecosystems by taking them apart and seeing how each piece functions in the whole. Their philosophy asserts that just as one cannot understand a house without detailed study of the blueprints and direct nail-by-nail observation of its construction, one cannot understand an organism without a map of its genes and step-by-step knowledge of its development and functioning.

But other ecologists view large systems as having a purposeful structure that cannot be comprehended by detailed examination of the parts. An understanding of how an organism functions in its particular environment differs from an understanding of the biochemistry of cell processes or the control of gene expression during development. Darwin did not have a detailed, or even correct, knowledge of genetics and yet he was able to appreciate the mechanism of evolution and the purpose in adaptation. An architect designs the structure of a house with purpose in mind. The detailed arrangement of the bricks and plumbing are merely the means to achieve this design; they do not offer insight into the purpose of the house as a complete, integrated structure that serves the purposes of human habitation.

It is only natural that organismic thinking should influence our concepts of ecosystems, as earlier in this century it influenced our concepts concerning associations of species. Odum (1959) expressed the beginnings of such a viewpoint:

> Some attributes, obviously, become more complex and variable as we proceed from [the cell to the ecosystem level of organization], but *it is an often overlooked fact that other attributes become less complex and less variable as we go from the small to the large unit.* Because homeostatic mechanisms, that is, checks and balances, forces and counter forces, operate all along the line, a certain amount of integration occurs as smaller units function within larger units. . . . *When we consider the unique characteristics which develop at each level,* there is no reason to suppose that any level is any more difficult or any easier to study quantitatively. . . . Furthermore, the findings at any one level *aid in the study of another level, but never completely explain the phenomena occurring at that level.*

In a paper entitled "The Strategy of Ecosystem Development," Odum (1969) developed this theme more explicitly, suggesting that evolution of components (species) with respect to each other has resulted in (perhaps, even, is directed toward) increased integrity and stability of the system:

> [Ecological succession] culminates in a stabilized ecosystem in which maximum biomass (or high information content) and symbiotic function between organisms are maintained per unit of available energy flow. In a word, the "strategy" of succession as a short-term process is basically the same as the "strategy" of long-term evolutionary development of the biosphere—namely increased control of, or homeostasis with, the physical environment in the sense of achieving maximum protection from its perturbations.

Information content in the ecosystem has been compared to the genetic information that directs the development of the organism and specifies the physiological apparatus that maintains its integrity. In ecosystems as in organisms, information content, specificity of ecosystem responses, and stability in the face of perturbation are interrelated (also see Margalef [1963], who considered biological diversity as a measure of the information in a system and Odum 1988).

Odum went on to describe how the adaptations of species to others in a system could contribute to overall stability:

> The time involved in an uninterrupted succession allows for increasingly intimate associations and reciprocal adaptations between plants and animals, which lead to the development of many mechanisms that reduce grazing—such as the development of indigestible supporting tissues (cellulose, lignin, and so on), feedback control between plants and herbivores, and increasing predatory pressure on herbivores. Such mechanisms enable the biological community to maintain the large and complex organic structure that mitigates perturbations of the physical environment. . . . there can be little doubt that the net result of community actions is symbiosis, nutrient conservation, stability, a decrease in entropy, and an increase in information. . . . The overall strategy is, as I stated at the beginning of this article, directed toward achieving as large and diverse an organic structure as is possible within the limits set by the available energy input and the prevailing physical conditions of existence (soil, water, climate, and so on).

If there was still any doubt about Odum's attraction to the organism analogy, he then posed the "intriguing question": "Do mature ecosystems age, as organisms do? In other words, after a long period of relative stability of 'adulthood,' do ecosystems again develop unbalanced metabolism and become more vulnerable to diseases and other perturbations?"

The resolution of the systems-versus-organism dichotomy poses a formidable challenge for several reasons. First, it is difficult to phrase an organismic theory of the ecosystems so as to make predictions that can be falsified by observation or experiment. The behavior of an organism certainly suits the purpose of the organism; but we may also describe that behavior mechanically in terms of the actions and reactions of the organism's component parts. At a higher level of organization, observed efficiencies of energy transfer between trophic levels may maximize the power output of the ecosystem, as H. T. Odum and R. C. Pinkerton (1955) have suggested, but they may also follow naturally from adaptations of organisms to maximize their power output individually, provided that some connection links power output and evolutionary fitness. At what level of system—organism or ecosystem—is purpose being served? How can we know (Loehle and Pechmann 1988)?

G. M. Weinberg (1975) suggested that ecosystems pose conceptual difficulties because they fall in the range of "middle-number systems" that cannot be treated appropriately either by differential equations, as Lotka proposed, or by statistical approaches whereby we may characterize the properties of a collection of components by calculating means and variances. Simple physical systems with few components, such as the gravitational attraction among the nine planets of our solar system and contrived laboratory microcosms with few species, are "small-number simple systems." "Large-number simple systems" consist of immense numbers of similar items—a quantity of gas consisting of billions and billions of atoms, for example. The temperature and pressure of the gas depend on the motion and collisions of individual atoms, but averages of these properties adequately describe the whole system.

Ecosystems are too complex to model as small-number systems of differential equations. Also, their parts differ so in function that simple averages, even within trophic levels, would obscure essential features of ecosystem structure. Thus, the ecosystem is what Weinberg calls a middle-number system—its parts are too numerous to describe fully, and too few and diversified to average meaningfully.

New and uncertain concepts of ecological systems are emerging.

A few biologists have recently attempted to find new ways of conceptualizing ecosystems. We cannot at present judge the success of these endeavors; certainly their proponents have yet to attract large followings. These approaches suffer from borrowing analogies from inappropriate branches of study—physics, psychology, sociology, linguistics, and communication—and they may provide little more than a new set of jargon in the place of old conceptual frustrations.

Proponents of these new concepts attribute independent levels of causation to each level in the hierarchy of organization of systems, from organism to ecosystem. T. F. H. Allen and Thomas B. Starr (1982), of the University of Wisconsin, elaborated the notion of hierarchical structure in ecological systems, borrowing heavily from Arthur Koestler's 1967 book *The Ghost in the Machine:*

> The salient feature of an entity is that it is an integration of all its parts. Koestler uses the image of a doorway between the parts of the structure and the rest of the universe. The entity has a duality in that it looks inward at the parts and outward at an integration of its environment; it is at once a whole and a part. At every level in a hierarchy there are these entities, and they have this dual structure. As in taxonomy, where each level in the hierarchical classification is called a "taxon," Koestler call his two-faced entities "holons." . . . [Quoting Koestler:] "Every holon has dual tendency to preserve and assert its individuality as a quasi autonomous whole; and to function as an integrated part of (an existing or evolving) larger whole. . . . The self-assertive tendencies are the dynamic expression of holon wholeness, the integrative tendencies of its partness."

Thus, ecosystems can be viewed as parts within parts, each level of the hierarchy having its own scale of time and space. The more disparate the scales of any two levels, the less likely they are to be dynamically coupled. According to this insight, if the behavior of large systems with dynamics of long period is purposeful, it can be understood only in terms of response to physical processes of similar scale. To attribute structural and functional "adaptations" of ecosystems to processes that vary over periods of days or even seasons makes little sense; the dynamics of individuals and populations are far closer in scale.

This is new ground for ecology. Perhaps it will be infertile; at worst, some "concepts" may turn out to be means of legitimizing a mysticism that has no place in science. Caution against the quick conclusion was raised, perhaps in self-defense, by Bernard Patton (1982), one of the few strong advocates of these new approaches:

> In constructing environ theory to its present form, the various plateaus were marked as they were reached with new words, to hold the place, and make it possible to explore further avenues and then return. The new terms tend to be resisted, which impedes acceptance of the theory. However, they are not ready to be discarded because they do express key ideas. The reader is asked to be tolerant; ignore the terms if necessary but hold the thoughts. The words in question are holon, creaon, genon, taxon, and environ.

Then, seemingly having given himself license, Patton one-ups the organismic concept with a more extreme analogy to the cell:

> In addition, one further lexical liberty is taken for didactic reasons. The environ center will be referred to as its nucleus and the remainder as enviroplasm.

We shall not plumb the meaning of "environ" here. Rather our first task will be to characterize the ecosystem in terms of energy flow, in the next chapter, and cycles of elements, in Chapters 12 and 13. Then, in Chapter 14 we shall return to the regulation of ecosystem structure and function.

SUMMARY

The ecosystem is the whole complex of organisms and the physical environments they inhabit. Although this concept is of central importance to present-day ecology, its development was slow and unsteady.

1. A consistent theme in ecology has been the dichotomy between one concept of large systems having properties that result from the interactions of their smaller components, and a second concept of systems expressing properties of organization and regulation that cannot be deduced from their parts. F. E. Clements favored the second, organismic viewpoint early in this century in his writings about plant associations, while H. A. Gleason felt that the properties of these associations were only the sums of the properties of their parts.
2. Charles Elton described communities in terms of feeding relationships and emphasized the pyramid of numbers as a dominant organizing principle in community structure.
3. In 1935, A. G. Tansley coined the term "ecosystem" to include organisms and all the abiotic factors in the habitat.

4. A. J. Lotka, in 1925, provided a thermodynamic perspective on ecosystem function, whereby the movement and transformations of mass and energy conform to thermodynamic laws. He also showed how the behavior of whole systems could be described mathematically in terms of the interactions of their components.

5. Raymond Lindeman, in 1942, popularized the idea of the ecosystem as an energy-transforming system, providing a formal notation for the energy flux in trophic levels and for ecological efficiency.

6. The study of ecosystem energetics dominated ecology during the 1950s and 1960s, largely due to the influence of Eugene P. Odum, who championed energy as a common currency for describing ecosystem structure and function.

7. Systems ecology, based upon Lotka's idea that ecosystems could be represented by a set of equations describing the dynamic interrelationships of components, grew to prominence in the 1960s and 1970s with the availability of the computer. Systems ecologists portrayed ecosystems in ever-increasing complexity with the objective of understanding the dynamical behavior of the whole system from the behavior of mathematical analogies.

8. Frustration with the complexity of the systems approach has led to a resurgence of the organism analogy in the study of ecosystems, with no resolution of these opposing viewpoints in sight as yet.

11

Energy Flow in the Ecosystem

Organisms dissipate energy in two ways. First, they perform work on the system in which they live. Thus, for example, as an animal runs it transfers kinetic energy to the atmosphere and to the ground. Second, the biochemical transformations required for movement, biosynthesis, secretion, and cell maintenance are inefficient. Organisms lose energy in the form of heat at each biochemical step.

In spite of these losses, the energy budget of the ecosystem is balanced because plants assimilate energy from the sun. Plants then use this energy, stored in the chemical bonds of carbohydrates, for their metabolic needs. But herbivores also take a portion, and thus start some of the energy assimilated by plants on its way up the food chain. In this chapter, we shall see how the flux of energy through the ecosystem depends on rates at which plants assimilate energy, rates of consumption at each trophic level, and energetic efficiencies of transforming food into biomass. A logical place to start is at the beginning of the food chain with plant production.

Plants assimilate energy by photosynthesis.

Photosynthesis is the process by which plants capture light energy and transform it into the energy of chemical bonds in carbohydrates. Glucose and other organic compounds (starch and oils, for example) may be stored conveniently, and their energy later released in respiratory metabolic pathways. Photosynthesis chemically unites two common inorganic compounds, carbon dioxide (CO_2) and water (H_2O), to form glucose ($C_6H_{12}O_6$), with the release of oxygen (O_2). The overall stoichiometry (chemical balance) of photosynthesis is

$$6CO_2 + 12H_2O \longrightarrow C_6H_{12}O_6 + 6O_2 + 6H_2O$$

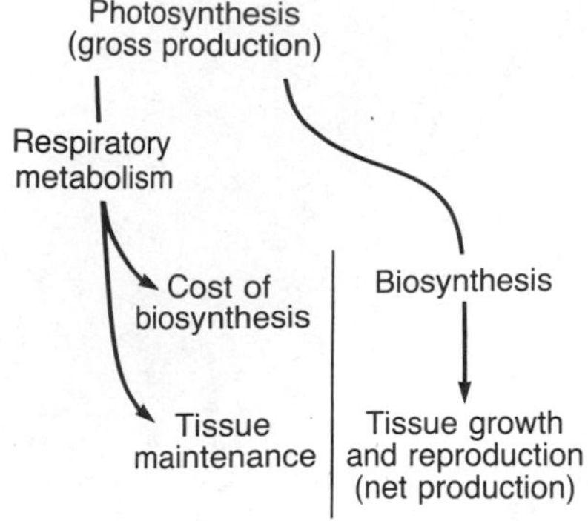

Figure 11–1
A diagram of the partitioning of energy by plants.

Photosynthesis transforms carbon from an oxidized state in CO_2 to a reduced state in carbohydrate. Because work is performed on carbon atoms, photosynthesis requires energy, which is provided by visible light. For each gram of carbon assimilated, the plant gains 39 kilojoules (kJ) of energy. But because of inefficiencies in the various biochemical steps of photosynthesis, no more than 34 per cent, and usually much less, of the light energy absorbed by photosynthetic pigments eventually appears in carbohydrate molecules (Rabinowitch and Govindjee 1969).

Photosynthesis supplies the carbohydrate building blocks and energy the plant needs to synthesize tissues and grow. Rearranged and joined together, glucose becomes fats, oils, and cellulose. Combined with nitrogen, phosphorus, sulfur, and magnesium, simple carbohydrates derived ultimately from glucose produce an array of proteins, nucleic acids, and pigments. Plants cannot grow unless they have all these basic building materials (Clarkson and Hanson 1980). Chlorophyll, for example, contains an atom of magnesium; even though all other necessary elements might be present in abundance, a plant lacking magnesium cannot produce chlorophyll and thus cannot grow.

Plants build and maintain tissues by complex, energy-requiring biochemical transformations. Because plants utilize much of the energy assimilated by photosynthesis to supply these needs, they always contain in their tissues substantially less energy than the total assimilated. Therefore, ecologists must distinguish two measures of assimilated energy: gross production, the total energy assimilated by photosynthesis, and net production, the accumulation of energy in plant biomass; that is to say, plant growth and reproduction. Because plants are the first link in the food chain, ecologists refer to these measures as gross or net primary production. The difference between them is the energy of respiration, that utilized for maintenance and biosynthesis (Figure 11–1).

Methods of measuring plant production vary with habitat and growth form.

Because primary production involves fluxes of carbon dioxide, oxygen, minerals, and water, on one hand, and the accumulation of plant biomass, on the other, the rates of any of these should provide an index to the overall rate of plant production. In practice, the appropriate measure depends on habitat and growth form. Furthermore, one uses different measures to estimate gross and net production, or the production of an entire system and that of a small part of a single plant.

Net production can be expressed conveniently as grams of carbon assimilated, dry weight of plant tissues, or their energy equivalents. Ecologists use such indices interchangeably because they have found a high degree of correlation among them. The energy content of an organic compound depends primarily on its carbon content (assimilated at a cost of 39 kJ per gram) and the energy added or subtracted during various transformations. For example, glucose contains 40 per cent carbon by mass, and should therefore contain an equivalent of $0.40 \times 39 = 15.6$ kJ g^{-1}. The observed value is about 17.6 kJ g^{-1}, as determined by burning samples in devices called bomb calorimeters (Paine 1971).

A calorimeter has a small chamber where the sample is burned, in a fast chemical version of cellular respiration, to oxidize it completely to carbon dioxide and water. Oxygen is forced under high pressure into the chamber to ensure complete combustion. A water jacket surrounds the chamber and absorbs the heat produced. The increase in temperature of a known amount of water in the jacket provides a direct estimate (4.2 kJ $°C^{-1}$ kg^{-1}) of the heat energy released by combustion. Measured in this way, amounts of energy released in complete oxidation average approximately 17.6 kJ g^{-1} of carbohydrate, 23.8 kJ g^{-1} of protein, and 39.7 kJ g^{-1} of fat (King and Farner 1961, Kleiber 1961). Thus, by measuring the composition of plant tissues, one can estimate their energy content.

In terrestrial ecosystems, ecologists usually estimate plant production by the annual increase in plant biomass, although measurements of CO_2 flux also are practical (Newbould 1967, Milner and Hughes 1968). In areas of seasonal production, annual growth is determined by cutting, drying, and weighing the plants at the end of the growing season (Odum 1960). Root growth usually is ignored because roots are difficult to remove from most soils; thus, harvesting measures the annual aboveground net productivity (AANP), the most common basis for comparing terrestrial communities.

Because the atmosphere contains little carbon dioxide (0.03 per cent), uptake by plants can measurably reduce its concentration in enclosed chambers within short periods. Ecologists have used this principle to measure production in terrestrial habitats, beginning with the classic study of H. N. Transeau (1926) on field corn. The most convenient application of the method is to enclose samples of vegetation (whole herbaceous plants or branches of trees) in clear chambers (light must penetrate for photosynthesis) and to measure the change in concentration of CO_2 in air passed through the chamber. Nowadays ecologists measure the absorbance of infrared light by the airstream inasmuch as CO_2 strongly absorbs light in that part of the spectrum (Mooney and Billings 1961). CO_2 uptake per gram of dry weight or square centimeter of surface area of leaves in the chamber can then be extrapolated to the entire tree or forest.

Carbon dioxide flux during the light period of the day includes both assimilation (uptake) and respiration (output) and thus measures net production. Respiration can be estimated separately by carbon dioxide production during the night, when photosynthesis shuts down. Daniel Botkin et al. (1970) measured CO_2 uptake (grams per square meter of ground area per day) of the three most important species of trees (two oaks and a pine) in the Brookhaven National Laboratory forest over the course of a single growing season. During daylight hours plants assimilated 4336 g m^{-2} d^{-1}. Over 24 hours the value was reduced by nighttime respiration to 3702 g m^{-2} d^{-1}, a difference of 635 g m^{-2} d^{-1}. Because the average chemical composition of plant (tree) carbohydrate is $C_6H_{10}O_5$, one gram of CO_2 assimilated is equivalent to 0.614 g of carbohydrate produced. Using this ratio, Botkin and his colleagues estimated the net dry matter production of the forest to be

$3702 \times 0.614 = 2273$ g m^{-2} d^{-1}, which agreed well with the harvest technique applied to the same forest (Whittaker and Woodwell 1969).

The radioactive isotope carbon-14 (^{14}C) provides a useful variation on the gas exchange method of measuring productivity. When one adds a known amount of ^{14}C-carbon dioxide to an airtight enclosure, plants assimilate the radioactive carbon atoms in the same proportion as they occur in the air inside the chamber. Thus, one may calculate the rate of carbon fixation by dividing the amount of ^{14}C in the plant by the proportion of ^{14}C in the chamber at the beginning of the experiment. For example, if a plant assimilated 10 milligrams of ^{14}C in an hour, and the proportion in the chamber was 0.05, we could calculate that the plant assimilated carbon at a rate of 200 mg h^{-1} (10 divided by 0.05).

Gas exchange provides the best estimate of primary production in aquatic habitats.

Although production of large aquatic plants, such as kelps, can be estimated by harvesting (Penfound 1956, Mann 1973), small size and rapid turnover of phytoplankton preclude the method as a general approach to aquatic production. But whereas high levels of atmospheric oxygen preclude using the production of oxygen by photosynthesis as a measure in terrestrial habitats, the low natural concentration of oxygen dissolved in water makes the measurement of small changes in oxygen concentration practical in most aquatic systems (Strickland 1960, Strickland and Parsons 1968).

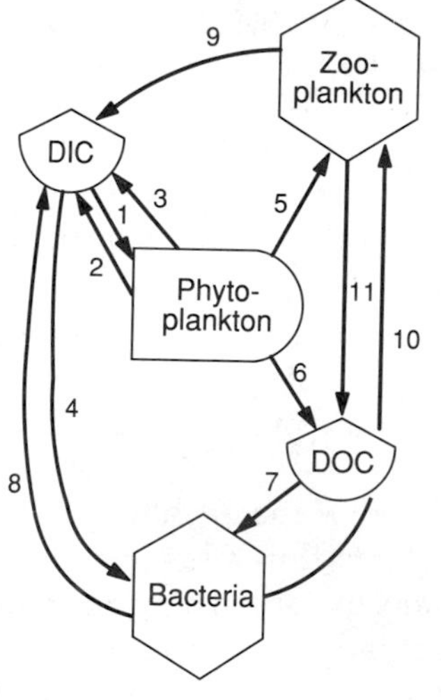

Figure 11–2
Carbon pools and pathways influencing ^{14}C-CO_2 uptake experiments with natural plankton communities. DIC = dissolved inorganic carbon (carbon dioxide and carbonates); DOC = dissolved organic carbon; numbers indicate 11 pathways of carbon flux that can affect the measurement and interpretation of phytoplankton production. (From Peterson 1980.)

To measure production, samples of water containing phytoplankton are suspended in pairs of sealed bottles at desired depths beneath the surface of a lake or the ocean; one of the pair (a "light bottle") is clear and allows sunlight to enter; the other (a "dark bottle") is opaque. In the light bottles, photosynthesis and respiration occur together, and part of the oxygen produced by the first process is consumed by the second. In the dark bottles, respiration consumes oxygen without its being replenished by photosynthesis. Thus, to estimate gross production one adds the change in oxygen concentration in the light bottle (photosynthesis—respiration) to the amount consumed in the dark bottle.

In unproductive waters, such as those of deep lakes and the open ocean, changes in oxygen concentration due to photosynthesis and respiration are small compared to the amount dissolved, and light and dark bottle measurements are not practical. Steeman Nielsen (1951) pioneered the measurement of ^{14}C uptake to estimate production under such conditions. The principle of the ^{14}C method is the same in aquatic systems as in terrestrial ones, except that the isotope is usually provided in the form of hydrogen carbonate ion (HCO_3^-). In practice, however, the interpretation of ^{14}C uptake values is plagued by problems, including photorespiration at high light intensities and the uptake and release of carbon in both organic and inorganic forms by bacteria and zooplankton (Peterson 1980; Figure 11–2).

A final method for estimating plant production in aquatic habitats is based on the idea that the concentration of chlorophyll sets an upper limit to

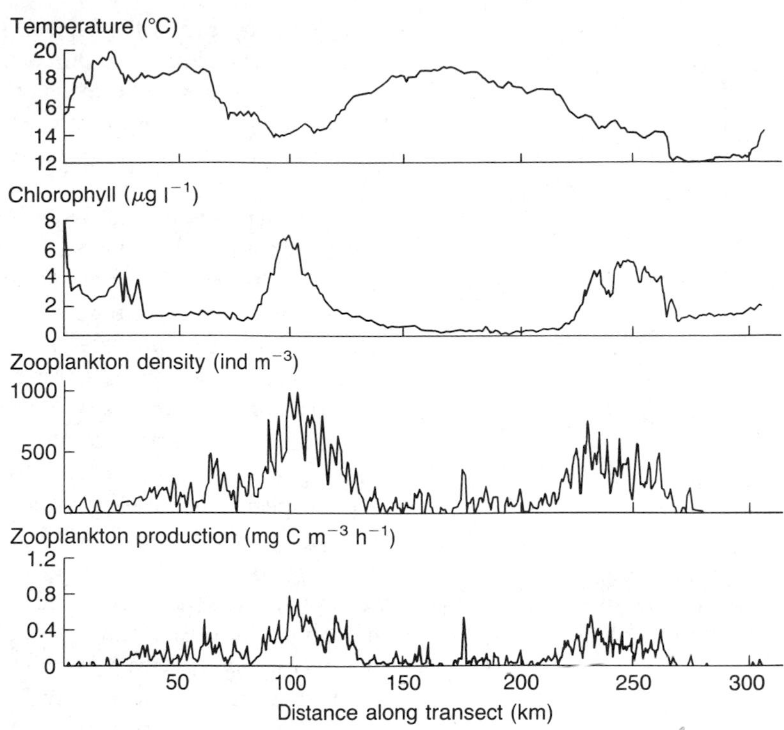

Figure 11–3
Temperature, chlorophyll concentration, zooplankton density, and estimated zooplankton production at a depth of 4 meters along a transect between Portland, Maine, and Yarmouth, Nova Scotia, during the summer of 1979. (From Huntley and Boyd 1984.)

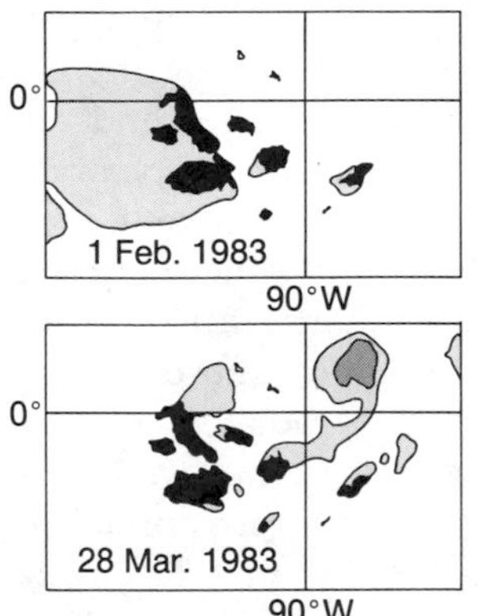

Figure 11–4
Phytoplankton pigment concentrations around the Galápagos Islands obtained from satellite imagery on 1 February 1983 *(top)* and 28 March 1983 *(bottom)*. The light shading indicates 1–3 mg chlorophyll m^{-3}; dark shading, 3–10 mg chlorophyll m^{-3}. In general, concentrations throughout the region were less than 0.2 mg m^{-3}. Note the shift in ocean current direction from westerly to northeasterly with the development of El Niño conditions during March 1983. (After Feldman 1984.)

the rate of photosynthesis at high light intensities. J. H. Ryther and C. S. Yentsch (1957), who advocated application of the method, determined that marine algae assimilate a maximum of 3.7 grams of carbon per gram of chlorophyll per hour (with variation between 2.1 and 5.7 over several studies). With experimentally determined relationships between light intensity and rate of photosynthesis by known concentrations of phytoplankton (Ryther 1956, Steele 1962), and measurements of chlorophyll concentrations and light penetration down through the water column, one can estimate total production per unit of surface area. What the chlorophyll method lacks in precision it makes up in speed and simplicity. It has been used with transect surveys (Figure 11–3) and satellite observations of surface water (Figure 11–4) to map productivity and follow its temporal change over vast areas.

Rate of photosynthesis varies in relation to light and temperature.

The rate of photosynthesis varies in direct proportion to light at low intensity, usually less than one-quarter that of full sunlight (Ryther 1956, Kramer

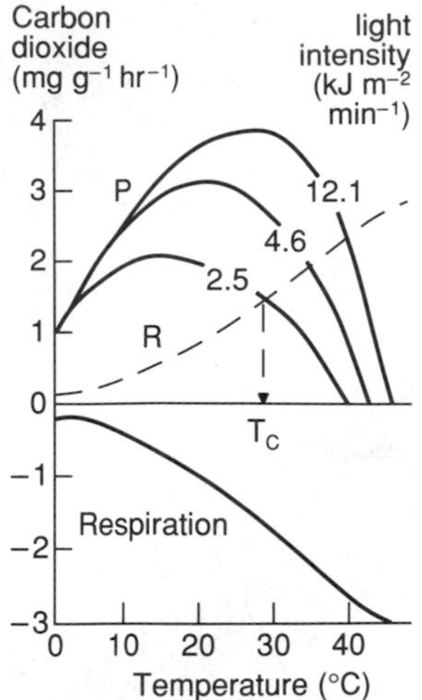

Figure 11-5
Net photosynthetic rate as a function of leaf temperature and light intensity in the heath *Loiseleuria.* (After Larcher et al. 1975.)

1957, Mooney and Billings 1961, Loach 1967). Brighter light saturates the photosynthetic pigments, and as intensity increases rate of photosynthesis increases more slowly or levels off (Berry 1975). In many plants, extremely bright light impairs photosynthesis owing to deactivation of photosynthetic reactions and enhanced photorespiration (Downes and Hesketh 1968, Ogren 1984).

The response of photosynthesis to light intensity has two reference points. The first, called the compensation point, is the level of light intensity at which photosynthetic assimilation of energy just balances respiration. Above the compensation point, the energy balance of the plant is positive; below it, its energy balance is negative. The second reference point is the saturation point, above which the rate of photosynthesis no longer responds to increasing light intensity. Among terrestrial plants, the compensation points of species that normally grow in full sunlight (approximately 500 watts per square meter) occur between 1 and 2 $W\,m^{-2}$. The saturation points of such species usually are reached between 30 and 40 $W\,m^{-2}$, less than a tenth of the energy level of bright, direct sunlight.

Like most other physiological processes, photosynthesis proceeds most rapidly within a narrow range of temperature; the optimum varies with environment, from about 16°C in many temperate species to as high as 38°C in tropical species. Optimum temperature also varies with light intensity in some species, such as the alpine heath *Loiseleuria* of Austria (Figure 11-5). Net production depends on the rate of respiration as well as that of photosynthesis, and respiration generally increases with increasing leaf temperature (Kramer 1958).

Photosynthetic efficiency, which is the per cent of the energy in incident radiation converted to net primary production during the growing season, provides a useful index to rates of primary production under natural conditions. Where water and nutrients do not severely limit plant production, photosynthetic efficiency varies between 1 and 2 per cent.

What happens to the remaining 98 to 99 per cent of the energy? Leaves and other surfaces reflect anywhere from one-quarter to three-quarters. Molecules other than photosynthetically active pigments absorb most of the remainder, which is converted to heat and either radiated or conducted across the leaf surface or dissipated by the evaporation of water from the leaf (transpiration). For example, in a study of an oak forest, annual incident photosynthetically active radiation was 1.9×10^6 $kJ\,m^{-2}$ (an average of 60 $W\,m^{-2}$). Of this, 56 per cent was absorbed by leaves. Annual transpiration was approximately 0.5×10^6 $g\,m^{-2}$. At 2.24 $kJ\,g^{-1}$ of water evaporated (the heat of vaporization), transpiration accounted for 1.1×10^6 $kJ\,m^{-2}$, which was approximately the total light energy absorbed (Loucks 1977).

Transpiration efficiency relates net production to water availability.

The tiny openings (stomates) through which leaves exchange carbon dioxide and oxygen with the atmosphere also allow passage of water vapor

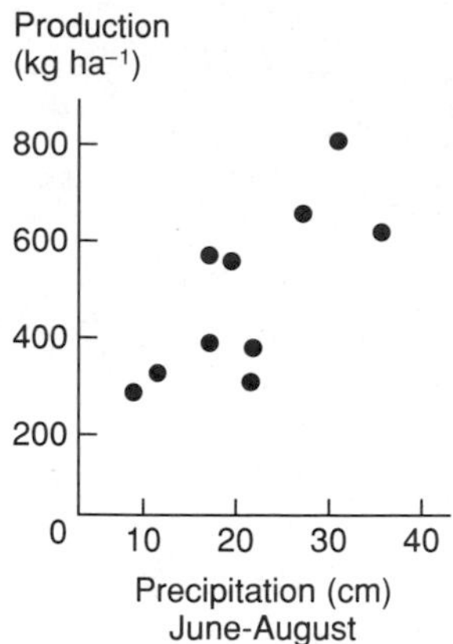

Figure 11–6
Relationship between production and summer precipitation on perennial grassland in southern Arizona. (After Cable 1975.)

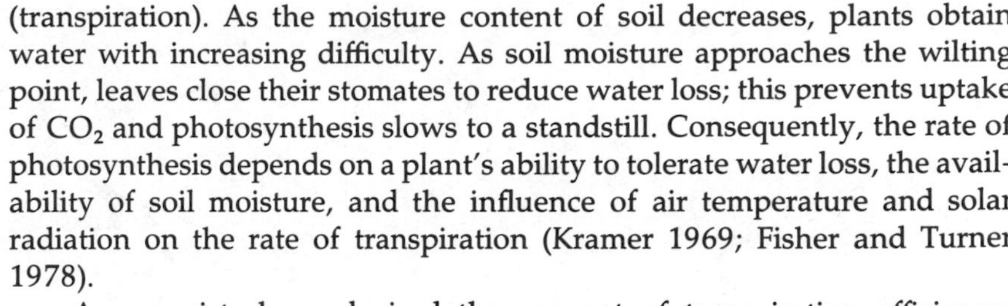
(transpiration). As the moisture content of soil decreases, plants obtain water with increasing difficulty. As soil moisture approaches the wilting point, leaves close their stomates to reduce water loss; this prevents uptake of CO_2 and photosynthesis slows to a standstill. Consequently, the rate of photosynthesis depends on a plant's ability to tolerate water loss, the availability of soil moisture, and the influence of air temperature and solar radiation on the rate of transpiration (Kramer 1969; Fisher and Turner 1978).

Agronomists have devised the concept of transpiration efficiency, which is the ratio of net production to transpiration, as an index to the drought resistance of plants. Transpiration efficiencies are less than 2 grams of production per kilogram of water transpired in most plants, but they may be as high as 4 g kg^{-1} in drought-tolerant crops (Odum 1971). Because the transpiration efficiency varies within narrow limits across a wide variety of plants, production is directly related to water availability in the environment (Webb et al. 1978). For example, annual production of perennial grasses in southern Arizona varies in direct relation to summer rainfall during the year of production (Figure 11–6).

Nutrients limit plant production in many habitats.

Fertilizers stimulate plant growth in most habitats (Albrektson et al. 1977, Bengtson 1979, Miller 1981, Thomas 1955, Tilman 1984, Treshow 1970). For example, when G. S. McMaster et al. (1982) applied nitrogen and phosphorus fertilizers singly and in combination to chaparral habitat in southern California, most species responded by increased leaf and stem production to the application of nitrogen, but not of phosphorus, indicating that the availability of nitrogen limited production (Figure 11–7). In contrast, production of the California lilac *(Ceanothus greggii)*, which harbors nitrogen-fixing bacteria in root nodules, responded to the application of phosphorus, but not of nitrogen. The productivity of annual plants (forbs and grasses) in the same habitat increased when nitrogen was applied (biomass production of 66 g m^{-2} versus 30 g m^{-2} for the control), but was depressed somewhat by the application of phosphorus alone (to 16 g m^{-2}). When the same amounts of nitrogen and phosphorus were added to a single plot, however, production soared to 252 g m^{-2}, indicating that the plants could take advantage of increased phosphorus only in the presence of high levels of nitrogen.

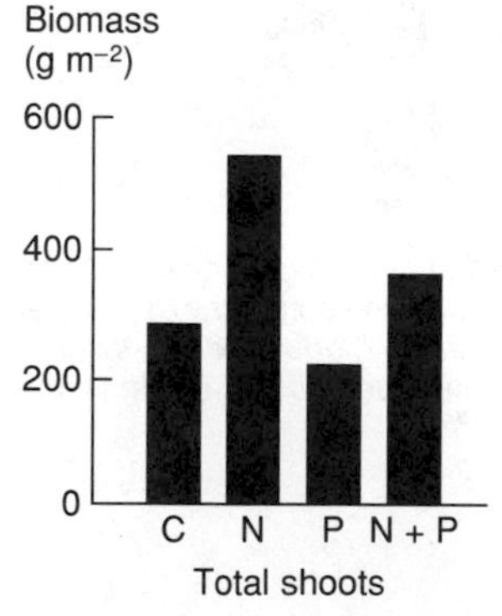

Figure 11–7
Response of the chaparral shrub *Adenostema* to nitrogen (N) and phosphorus (P) fertilization (C = control). (After McMaster et al. 1982.)

Nutrient limitation generally is felt most strongly in aquatic habitats, particularly the open ocean where scarcity of dissolved minerals reduces production far below terrestrial levels (Howarth 1988). Even in shallow coastal waters, where vertical mixing, upwelling currents, and runoff from the land maintain nutrients at high concentrations, the addition of fertilizer (often inadvertently through water pollution) may greatly enhance aquatic production. In a study along the southern coast of Long Island, New York, Ryther and Dunstan (1971) found a close correlation between phytoplank-

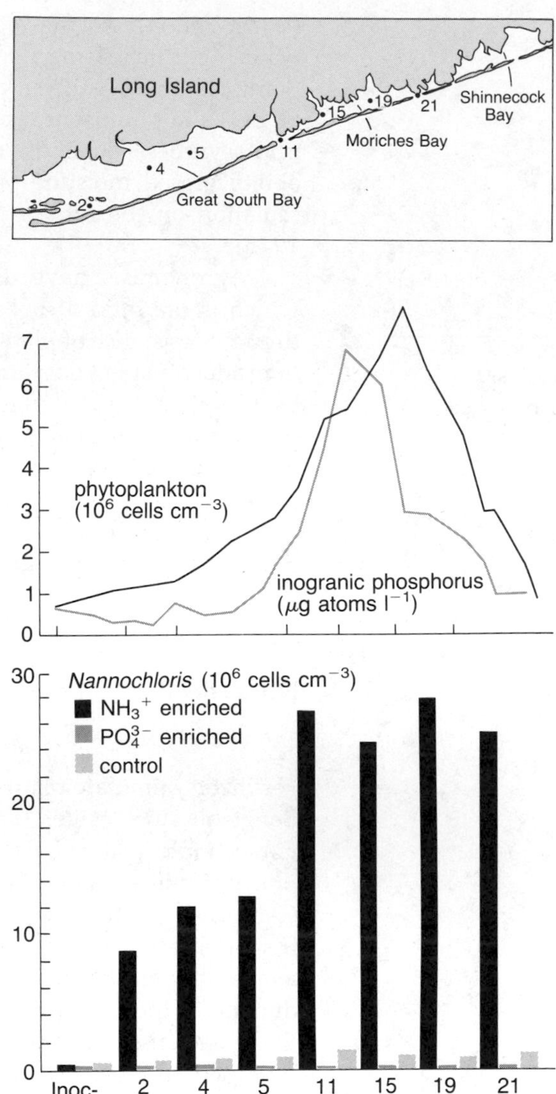

Figure 11–8
Center: phytoplankton and inorganic phosphorus concentrations in water collected from stations in Great South Bay and Moriches Bay, Long Island, in the summer of 1982 (see map *above*). *Bottom:* growth of standard cultures of the alga *Nannochloris* in water, taken from each of the stations, to which ammonium or phosphate had been added. (From Ryther and Dunstan 1971.)

ton abundance and the level of inorganic phosphorus, the latter being a general indicator of pollution. But the addition of nutrients to standard cultures of the alga *Nannochloris* in water samples taken from different areas demonstrated that nitrogen rather than phosphorus limited primary production in both polluted and relatively clean waters (Figure 11–8).

Variation in the productivity of terrestrial ecosystems is related primarily to light, temperature, and rainfall.

The favorable combination of intense sunlight, warm temperature, and abundant rainfall in the humid tropics results in the highest productivity on the earth. Low winter temperatures and long winter nights curtail production in temperate and arctic ecosystems. Within a given latitude belt, where light and temperature do not vary appreciably from one locality to the next, net production is directly related to annual precipitation. W. Webb et al. (1978) have shown that above a certain threshhold of water availability, production increases by 0.4 grams per kilogram of water in hot deserts and 1.1 g kg^{-1} in short-grass prairies and cold (Great Basin) deserts. Forest ecosystems, whose water use efficiencies are on the order of 0.9 to 1.8 g kg^{-1}, are relatively insensitive to variation in water availability, but do not develop unless precipitation exceeds at least 50 cm annually in the cooler regions of the United States and perhaps 100 cm in the warmer eastern and southeastern areas.

Ecologists Robert Whittaker and Gene Likens (1973) estimated the net primary production for representative terrestrial and aquatic ecosystems (Table 11–1). Their values were based on the results of many studies employing a wide variety of techniques, but what these lack in strict comparability does not override the general patterns they reveal. Whittaker and Likens's summary shows that production of terrestrial vegetation is greatest in the wet tropics and least in tundra and desert habitats. Swamp and marsh

Table 11–1
Average net primary production and related dimensions of the earth's major habitats

Habitat	Net primary production ($g\ m^{-2}\ yr^{-1}$)	Biomass ($kg\ m^{-2}$)	Chlorophyll ($g\ m^{-2}$)	Leaf surface area ($m^2\ m^{-2}$)
Terrestrial				
Tropical forest	1800	42	2.8	7
Temperate forest	1250	32	2.6	8
Boreal forest	800	20	3.0	12
Shrubland	600	6	1.6	4
Savanna	700	4	1.5	4
Temperate grassland	500	1.5	1.3	4
Tundra and alpine	140	0.6	0.5	2
Desert scrub	70	0.7	0.5	1
Cultivated land	650	1	1.5	4
Swamp and marsh	2500	15	3.0	7
Aquatic				
Open ocean	125	0.003	0.03	—
Continental shelf	360	0.01	0.2	—
Algal beds and reefs	2000	2	2.0	—
Estuaries	1800	1	1.0	—
Lakes and streams	500	0.02	0.2	—

Source: Whittaker and Likens 1973.

ecosystems, which occupy the interface between terrestrial and aquatic habitats, can be as productive as tropical forests.

Production in aquatic ecosystems is limited primarily by nutrients.

The open ocean is a virtual desert, where scarcity of mineral nutrients limits productivity to a tenth or less that of temperate forests (See Table 11–1). Upwelling zones (where nutrients are brought to the surface from deeper waters) and continental-shelf areas (where exchange between shallow, nutrient-rich bottom sediments and surface waters is well developed) support greater production. In shallow estuaries, coral reefs, and coastal algal beds, production approaches that of adjacent terrestrial habitats (Barnes and Mann 1980, Bunt 1973, Mann 1973, McLusky 1981). Primary production in freshwater habitats is comparable to that of marine habitats, being greatest in rivers, shallow lakes, and ponds, and least in clear streams and deep lakes.

Ecological efficiencies characterize the movement of energy along the food chain.

Plants manufacture their own "food" from raw inorganic materials. Hence they are referred to as autotrophs (literally, "self-nourishers"). Animals and most microorganisms, which obtain their energy and most of their nutrients by eating plants, animals, or their dead remains, are called heterotrophs (literally, "nourished from others"). The dual roles of living forms as food producers and food consumers give the ecosystem a trophic structure, determined by feeding relationships, through which energy flows and nutrients cycle. The food chain from grass to caterpillar to sparrow to snake to hawk delineates one particular path of energy through the trophic structure. But with each link in the food chain, much energy is dissipated before it can be consumed by organisms feeding on the next higher trophic level. All the grass in Africa piled together would dwarf a mound of all the grasshoppers, gazelles, zebras, wildebeests, and other animals that eat grass. But that mound of herbivores would be overwhelming beside the pitiful heap of all the lions, hyenas, and other carnivores that feed on them.

As Lindeman (1942) pointed out, the amount of energy reaching each trophic level is determined by the net primary production and the efficiencies with which food energy is converted to biomass energy within each trophic step. Of the light energy assimilated by plants, 15 to 70 per cent is used for maintenance and therefore is unavailable to consumers. Most herbivores and carnivores are more active than plants and expend correspondingly more of their assimilated energy on maintenance. As a result, the productivity of each trophic level is usually no more than 5 to 20 per cent

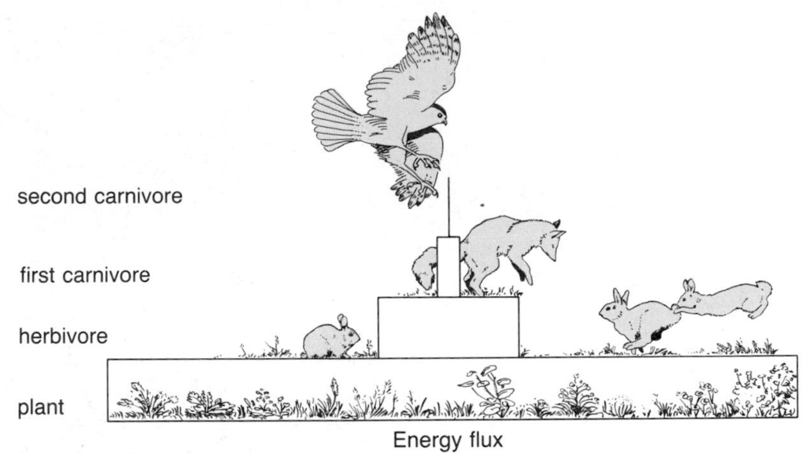

Figure 11–9
An ecological pyramid in which the breadth of each bar represents the net productivity of each trophic level in the ecosystem. For this particular system, ecological efficiencies are 20, 15, and 10 per cent between trophic levels, but these values vary widely between communities.

that of the level below it (Figure 11–9). The percentage transfer of energy from one trophic level to the next is called both the ecological efficiency and the food chain efficiency.

The individual link in the food chain is the basic unit of trophic structure.

Once food is eaten, its energy follows a variety of paths through the organism (Figure 11–10). Regardless of an organism's source of food, what it digests and absorbs is referred to as assimilated energy, which supports maintenance, builds tissues, or is excreted in unusable metabolic by-products. The energy used to fulfill metabolic needs, most of which is lost as heat, is known as respired energy. A smaller fraction of assimilated energy is excreted in the form of organic, nitrogen-containing wastes (primarily ammonia, urea, or uric acid) produced when the diet contains an excess of nitrogen. Assimilated energy retained by the individual organism is available for the synthesis of new biomass (production) through growth and reproduction, which may then be consumed by herbivores, carnivores, and detritivores.

Not all food can be fully assimilated; hair, feathers, insect exoskeletons, cartilage and bone in animal foods, and cellulose and lignin in plant foods resist digestion by most animals. These materials are egested either by defecation or by regurgitation of pellets of undigested remains. Some egested wastes are substances that have been relatively unaltered chemically during their passage through an organism, but nearly all have been

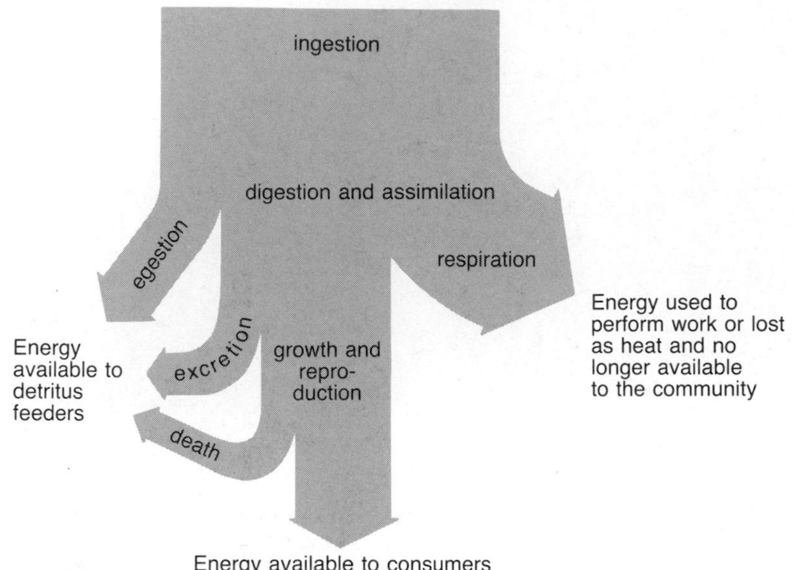

Figure 11–10
Partitioning of energy within a link of the food chain.

mechanically broken up into fragments by chewing and by contractions of the stomach and intestines, which makes them more readily usable by detritus feeders (Figure 11–11).

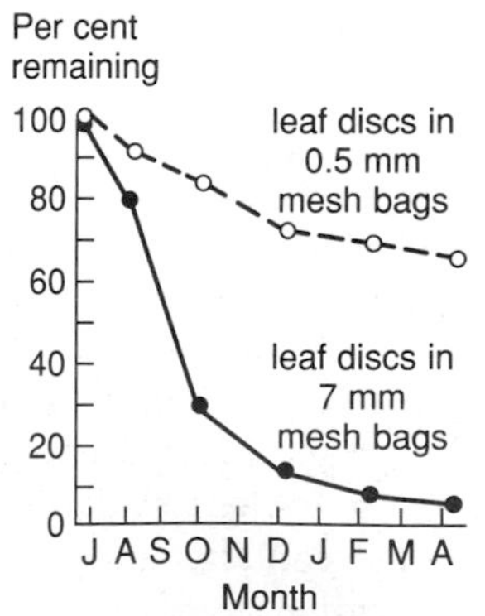

Figure 11–11
The influence of mechanical breakdown of plant litter by large detritus feeders on its overall consumption. Leaf discs were enclosed in mesh bags with either large (7 mm) or small (0.5 mm) openings. The small openings admitted bacteria, fungi, and small arthropods, but excluded such large species as earthworms and millipedes. (After Edwards and Heath 1963, in Phillipson 1966.)

Assimilation and production efficiencies determine ecological efficiency.

Ecological efficiency is the product of efficiencies with which organisms exploit their food resources and convert them into biomass available to the next higher trophic level. Because most biological production is consumed, exploitation efficiency is 100 per cent overall, and ecological efficiency depends on two factors: the proportion of consumed energy assimilated, and the proportion of assimilated energy incorporated in growth, storage, and reproduction. The first proportion is called the assimilation efficiency and the second, the net production efficiency (Table 11–2). The product of the assimilation and net production efficiencies is the gross production efficiency: the proportion of food energy that is converted to consumer biomass energy.

For plants, net production efficiency is defined as the ratio of net to gross production. This index has been found to vary between 30 and 85 per cent, depending on habitat and growth form. Rapidly growing plants in temperate zones, whether trees, old-field herbs, crop species, or aquatic

Table 11-2
Definitions of several energetic efficiencies

$$\text{(1) Exploitation efficiency} = \frac{\text{Ingestion of food}}{\text{Prey production}}$$

$$\text{(2) Assimilation efficiency} = \frac{\text{Assimilation}}{\text{Ingestion}}$$

$$\text{(3) Net production efficiency} = \frac{\text{Production (growth and reproduction)}}{\text{Assimilation}}$$

$$\text{(4) Gross production efficiency} = (2) \times (3) = \frac{\text{Production}}{\text{Ingestion}}$$

$$\text{(5) Ecological efficiency} = (1) \times (2) \times (3) = \frac{\text{Consumer production}}{\text{Prey production}}$$

plants, have uniformly high net production efficiencies (75 to 85 per cent). Similar vegetation types in the tropics exhibit lower net production efficiencies, perhaps 40 to 60 per cent. Thus, respiration increases relative to photosynthesis at low latitudes.

Animal food is more easily digested than plant food; assimilation efficiencies of predatory species vary from 60 to 90 per cent. Vertebrate prey are digested more efficiently than insect prey because the indigestible exoskeletons of insects constitute a larger proportion of the body than the hair, feathers, and scales of vertebrates. Assimilation efficiencies of insectivores vary between 70 and 80 per cent, whereas those of most carnivores are about 90 per cent.

The nutritional value of plant foods depends upon the amount of cellulose, lignin, and other indigestible materials present (Grodzinski and Wunder 1975). Herbivores assimilate as much as 80 per cent of the energy in seeds, and 60 to 70 per cent of that in young vegetation (Chew and Chew 1970). Most grazers and browsers (elephants, cattle, grasshoppers) assimilate 30 to 40 per cent of the energy in their food. Millipedes, which eat decaying wood composed mostly of cellulose and lignin (and the microorganisms that occur in decaying wood), assimilate only 15 per cent (O'Neill, 1968).

The effect of food quality on assimilation can be seen in the efficiencies with which a single animal retains energy from different portions of its diet. In studies with mountain hares *(Lepus timidus)*, Ake Pehrson (1983) determined that the efficiency of energy assimilation on a diet of small willow twigs was 39 per cent, of which 5 per cent was lost through urinary nitrogen excretion. Assimilation efficiencies were lower on larger twigs (31 per cent), presumably because of their thick, less digestible bark, and on birch twigs (23 to 35 per cent), on which hares could not maintain a constant weight. By measuring the content of fiber (cellulose and lignin) in the food and feces, Pehrson found that the digestibility of fiber was between only 15 and 25 per cent. Assimilation of nutritional components of the diet, estimated by similar input-output measurements (for example, on willow twigs 3 to 5 mm in diameter), varied between 9 per cent for phosphorus and 81 per cent for magnesium (Figure 11-12).

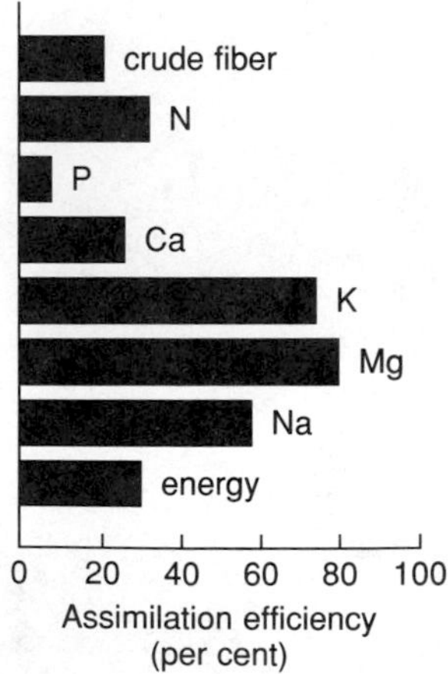

Figure 11-12
Proportion of dietary components assimilated by caged mountain hares *(Lepus timidus)* during the winter. (After Pehrson 1983.)

Maintenance, movement, and, in warm-blooded animals, heat production require energy that otherwise could be utilized for growth and reproduction. Active, warm-blooded animals exhibit low net production efficiencies: birds less than 1 per cent, small mammals with high reproductive rates up to 6 per cent. More sedentary, cold-blooded animals, particularly aquatic species, channel as much as 75 per cent of their assimilated energy into growth and reproduction (Welch 1968). The extreme high value approaches the biochemical efficiency of egg production and tissue growth, between 70 and 80 per cent in domesticated animals (Kielanowski 1964, Ricklefs 1974).

The efficiency of biomass production within a trophic level (gross production efficiency) is the product of assimilation efficiency and net production efficiency. Gross production efficiencies of warm-blooded, terrestrial animals rarely exceed 5 per cent, and those of some birds and large mammals fall below 1 per cent (Turner 1970). Gross production efficiencies of insects lie within the range of 5 to 15 per cent, and those of some aquatic animals exceed 30 per cent (Figure 11–13).

Terrestrial plants, especially woody species, allocate much of their production to structures that are difficult to ingest, let alone digest. As a result, most terrestrial plant production is consumed as detritus by organisms specialized to attack wood and leaf litter. This partitioning between herbivory and detritus feeding establishes two food chains in terrestrial communities. The first originates as relatively large animals feed on leafy vegetation, fruits, and seeds; the second originates with relatively small animals

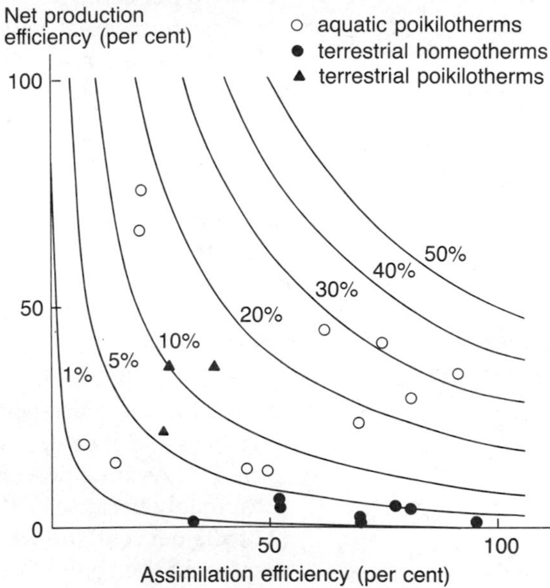

Figure 11–13
Relationships between assimilation efficiency and net production efficiency for a variety of animals. Gross production efficiencies are indicated by the curved lines on the graph. Homeotherms are warm-blooded animals; poikilotherms have variable body temperature, that is, are cold-blooded animals. (From Ricklefs 1979.)

and microorganisms consuming detritus in the litter and soil layer. These separate food chains sometimes mingle considerably at higher trophic levels, but the energy of detritus tends to move into the food chain much more slowly than the energy consumed by herbivores.

The relative importance of predatory-based and detritus-based food chains also varies greatly among communities. Predators predominate in plankton communities, detritus feeders in terrestrial communities. The proportion of net production that enters herbivore-predator strands of the food web depends on the relative allocation of plant tissue between structural and supportive functions, on one hand, and growth and photosynthetic functions, on the other. Herbivores consume 1.5 to 2.5 per cent of the net production of temperate deciduous forests, 12 per cent of that in old-field habitats, and 60 to 99 per cent in plankton communities (Wiegert and Owen 1971).

How long does energy take to flow through the community?

Food chain efficiencies indicate the amount of energy that eventually reaches each trophic level of the community. A second component of the energy dynamics of the ecosystem is the rate of transfer of energy or, inversely, its residence time in each trophic level. For a given rate of production, the residence (or transit) time of energy in the community and the storage of energy in living biomass and detritus are directly related: the longer the residence time, the greater the accumulation.

The average residence time in a particular link in the food chain is equal to the energy stored divided by the rate of flux, or

$$\text{residence time (yr)} = \frac{\text{biomass (kJ m}^{-2}\text{)}}{\text{net productivity (kJ m}^{-2}\text{ yr}^{-1}\text{)}}$$

(the residence time defined by this equation is sometimes calculated in terms of mass rather than energy and is then referred to as the biomass accumulation ratio). According to Whittaker and Likens (1973), wet tropical forests produce an average of 2000 grams of dry matter per square meter per year and have an average living biomass of 45,000 grams per square meter. Inserting these values into the equation for average residence time, we obtain 22.5 years (45,000/2000). Average residence times of energy in plants in representative ecosystems (Table 11–3) vary from more than 20 years in forested terrestrial environments to less than 20 days in aquatic plankton-based communities. These figures underestimate the total residence time of energy in primary production, however, because they do not include the accumulation of dead organic matter in the litter.

The residence time of energy in accumulated litter can be estimated by an equation analogous to that for the biomass accumulation ratio

$$\text{residence time (yr)} = \frac{\text{litter accumulation (g m}^{-2}\text{)}}{\text{rate of litter fall (g m}^{-1}\text{)}}$$

Table 11-3
Average transit time of energy in living plant biomass (biomass/net primary production) for representative ecosystems

System	Net primary production (g m^{-2} yr^{-1})	Biomass (g m^{-2})	Transit time (yrs)
Tropical rain forest	2,000	45,000	22.5
Temperate deciduous forest	1,200	30,000	25.0
Boreal forest	800	20,000	25.0
Temperate grassland	500	1,500	3.0
Desert scrub	70	700	10.0
Swamp and marsh	2,500	15,000	6.0
Lake and stream	500	20	0.04*
Algal beds and reefs	2,000	2,000	1.0
Open ocean	125	3	0.024†

* 15 days.
† 9 days.
Source: From data in Whittaker and Likens 1973.

In forests, the average residence time for energy varies from 3 months in the wet tropics to 1 to 2 years in dry and montane tropical habitats, 4 to 16 years in the southeastern United States, and more than 100 years in temperate mountains and boreal regions (Olson 1963). Warm temperature and abundance of moisture in lowland tropical regions create optimum conditions for rapid decomposition.

Energy transfer through the food chain has been followed by radioactive tracers.

Organic compounds containing energy can be labeled with a radioactive element, and their movement followed. In radioactive tracer studies, investigators label plants (or water in an aquatic system) with a radioactive isotope, usually phosphorus-32 (^{32}P) applied in a phosphate solution. They then collect consumer species at intervals after the initial labeling and examine them with a radiation counter.

Eugene P. Odum and his co-workers at the University of Georgia used this technique to follow the movement of ^{32}P through components of an old-field community (Odum 1962, Odum and Kuenzler 1963). They labeled a common plant—telegraph weed *(Heterotheca)*—with drops of ^{32}P-phosphate solution placed directly on leaves, and then collected insects, snails, and spiders at intervals of several days for about 5 weeks. Certain herbivores, notably crickets and ants, began to accumulate ^{32}P within a few days of its initial application and attained peak amounts of the tracer at 2 to 3 weeks. Ground-living, detritus-feeding insects (carabid and tenebrionid beetles, and gryllid crickets), and predatory spiders did not accumulate peak amounts of tracer until 3 weeks after the start of the experiment, Thus, the ^{32}P label appeared first in herbivores and later in detritivores and predators. Moreover, even though the average residence time of primary production in temperate grasslands is on the order of 3 years, these experiments show that a portion begins to move up the food chain very quickly.

A similar study on a small trout stream in Michigan gave results comparable to those of Odum's (Ball and Hooper 1963): ^{32}P-phosphate was added to the stream at one point and its accumulation in plants and animals downstream from the release site was monitored for 2 months. The median time for each population to accumulate its maximum concentration of ^{32}P varied from a few days for aquatic plants to 1 to 2 weeks for filter feeders and other herbivores, 3 to 4 weeks for omnivores, 4 to 5 weeks for detritus feeders, and 4 weeks to more than 2 months for most predators. The results suggest that most of the energy assimilated in aquatic ecosystems is dissipated within a few weeks, although a small portion may linger for months in the predator food chain, and perhaps for years in organic sediments on the bottoms of streams and lakes (King and Ball 1967, Minshall 1967, Fisher and Likens 1972).

Energy transfer and accumulation have been used to describe the structure and function of ecosystems.

Flux of energy and its efficiency of transfer summarize certain aspects of the structure of an ecosystem: the number of trophic levels, the relative importance of detritus and predatory feeding, the steady-state values for biomass and accumulated detritus, and the turnover rates of organic matter in the community (Weigert 1988). The importance of these measures was argued by Lindeman (1942), who constructed the first energy budget for an entire biological community—that of Cedar Bog Lake in Minnesota. The proliferation of energy flow studies during the 1950s and 1960s more clearly reflected energy's value as a universal currency, a common denominator to which all populations and their acts of consumption could be reduced.

The overall energy budget of the ecosystem reflects the balance between income and expenditure, just as it does in a bank account. The ecosystem gains energy through photosynthetic assimilation of light by green plants and the transport of organic matter into the system from outside. Sources of the latter are referred to as allochthonous inputs (from the Greek *chthonos,* of the earth, and *allos,* other); local production is referred to as autochthonous.

In Root Spring, near Concord, Massachusetts, the energy flux of herbivores is 0.31 W m^{-2}, but the net productivity of aquatic plants is only 0.09 W m^{-2}, the balance being transported into the spring by leaf fall from nearby trees (Teal 1957). G. W. Minshall (1978) surveyed the relative amounts of autochthonous and allochthonous inputs in aquatic systems: large rivers, lakes, and most marine ecosystems are dominated by autochthonous production; allochthonous imports were most important in small streams and springs under the closed canopies of forests; life in caves and abyssal depths of the oceans, to which no light penetrates, subsists entirely on energy transported in from outside.

Lindeman constructed the Cedar Bog Lake energy budget from measurements of the harvestable net production on each of three trophic levels (all carnivores were lumped together) and from laboratory determinations of respiration and assimilation efficiencies. The animals and plants collected at the end of the growing season constituted the net production of the trophic levels to which each species was assigned.

Trophic level	Harvestable production ($kJ\ m^{-2}\ y^{-1}$)	($W\ m^{-2}$)
Primary producers (green plants)	2944	0.0934
Primary consumers (herbivores)	293	0.0093
Secondary consumers (carnivores)	54	0.0017

Lindeman estimated the energy dissipated by respiration from ratios of respiratory metabolism to production measured in the laboratory: 0.33 for aquatic plants, 0.63 for herbivores, and 1.4 for the more active carnivores in the lake. He calculated the gross production of carnivores as the sum of their harvestable production (54 $kJ\ m^{-2}\ y^{-1}$) and respiration ($54 \times 1.4 = 76$ $kJ\ m^{-2}\ y^{-1}$), a total of 130 $kJ\ m^{-2}\ y^{-1}$ (Table 11–4). He then estimated the gross production of primary consumers as the sum of their harvestable production, respiration (production $\times$ 0.63), and the consumption of primary consumers by secondary consumers. In making the last calculation, Lindeman assumed that the assimilation efficiencies of secondary consumers were 90 per cent. By backtracking another step in this manner, he calculated the production of primary producers as well.

Lindeman's energy budget is somewhat startling in that organisms at

Table 11–4
An energy flow model for Cedar Lake Bog, Minnesota

Energy ($kcal\ m^{-2}\ yr^{-1}$)	Trophic level: Primary producers	Primary consumers	Secondary consumers
Harvestable production*	704	70	13
Respiration	234	44	18
Removal by consumers			
assimilated	148	31	0
unassimilated	28	3	0
Gross production (totals)	1114	148	31

* Does not include net production removed by consumers. Actual net production, including removal by consumers, was 879 $kcal\ m^{-2}\ yr^{-1}$ for primary producers, 104 $kcal\ m^{-2}\ yr^{-1}$ for primary consumers, and 13 $kcal\ m^{-2}\ yr^{-1}$ for secondary consumers.
Source: Lindeman 1942.

the next higher trophic level failed to consume 83 per cent of the net primary production and 70 per cent of the net secondary production. This surplus production is transported out of the system by sedimentation; the lake is filling with layers of organic detritus.

Even with sedimentation, the overall ecological efficiency of energy transfer between trophic levels in Cedar Bog Lake was about 12 per cent. After comparing similar analyses for five aquatic communities, D. G. Kozlovski (1968) concluded that (1) assimilation efficiency increases at higher trophic levels; (2) net production efficiency decreases at higher levels; (3) gross production efficiency also decreases; and (4) ecological efficiency (assimilation or gross production at level n divided by that at level $n - 1$) remains constant between trophic levels at about 10 per cent.

The length of food chains is limited by ecological efficiencies.

Kozlovski's 10 per cent generalization is not a fixed law of ecological thermodynamics. In Silver Springs, Florida, Odum (1957) measured ecological efficiencies of 17 per cent between the producer and herbivore level but only 5 per cent between herbivores and secondary consumers. Ecological efficiencies are usually lower in terrestrial habitats, and a useful rule of thumb states that the top carnivores in terrestrial communities can feed no higher than the third trophic level on average, whereas aquatic carnivores may feed as high as the fourth or fifth levels (Fenchel 1988). This is not to say that there can be no more than three links in a terrestrial food chain; a tiny fraction of the total energy may travel through a dozen links before it is dissipated by respiration. Such high trophic levels do not, however, contain enough energy to fully support a single predator population.

We can estimate the average length of food chains in a community from net primary production, average ecological efficiency, and average energy flux of a top predator population. The energy available [$E(n)$] to a predator at a given trophic level [n (plants being level 1)] is equal to the product of the net primary production (*NNP*) and the intervening ecological efficiencies (*Eff*). Thus

$$E(n) = NNP\ Eff^{n-1}$$

where *Eff* is the geometric mean of the efficiencies of transfer between each level. This expression can be rearranged to give

$$n = 1 + \frac{\log[E(n)] - \log(NPP)}{\log(Eff)}$$

Using this equation and some rough estimates for the values on its right-hand side, we can calculate the average number of trophic levels to be about 7 for marine plankton-based systems, 5 for inshore aquatic communities, 4 for grasslands, and 3 for wet tropical forests (Table 11–5). These estimates

Table 11-5
Average number of trophic levels in various ecosystems calculated from primary production, consumer energy flux, and ecological efficiencies

Community	Net primary production (kcal m^{-2} yr^{-1})	Predator ingestion (kcal m^{-2} yr^{-1})	Ecological efficiency (per cent)	Number of trophic levels
Open ocean	500	0.1	25	7.1
Coastal marine	8000	10.0	20	5.1
Temperate grassland	2000	1.0	10	4.3
Tropical forest	8000	10.0	5	3.2

Values are approximations based on many studies.

should be taken with a grain of salt, to be sure, but they do indicate the general size of the pyramid of energy built upon a base of primary production within an ecosystem.

SUMMARY

The ecosystem is a giant thermodynamic machine that continually dissipates energy in the form of heat. This energy initially enters the biological realm of the ecosystem by photosynthesis and plant production, which is the ultimate source of energy for all animals and nonphotosynthetic microorganisms.

1. Gross primary production is the total energy fixed by photosynthesis. Net primary production is the accumulation of energy in plant biomass; hence, it is the difference between gross production and plant respiration.
2. Primary production is measured by one or some combination of a variety of methods: harvest, gas exchange (carbon dioxide in terrestrial habitats, oxygen in aquatic habitats), assimilation of radioactive carbon (^{14}C), and production indices based on chlorophyll content.
3. Rate of photosynthesis varies in direct relation to light intensity up to the saturation point (usually 30–40 W m^{-2}), above which it levels off or decreases (bright sunlight is approximately 500 W m^{-2}). The compensation point, above which photosynthesis exceeds respiration, occurs at light intensities of 1–2 W m^{-2} in terrestrial vegetation. The efficiency of photosynthesis (gross production/total incident light energy) is 1 to 2 per cent in most habitats.
4. Because plants lose water in direct proportion to the amount of carbon dioxide assimilated, plant production in dry environments is limited by, and varies in direct proportion to, availability of water. Transpiration efficiency, the ratio of production (grams dry mass) to water transpired (kilograms), is typically 1 to 2, and rarely as high as 4 in drought-adapted species.

5. Production in both terrestrial and aquatic environments can be enhanced by application of various nutrients, especially nitrogen and phosphorus, indicating that nutrient availability limits production. This effect is greatest in the open ocean and deep lakes, where nutrient inputs from sediments and runoff are least.

6. The movement of energy and materials through the food chain can be characterized by efficiencies of assimilation (assimilation/digestion) and net production (production/assimilation). Unassimilated material enters detritus-based food chains.

7. Assimilation efficiency depends on the quality of the diet, particularly the amounts of digestion-resistant structural material (cellulose, lignin, chitin, keratin), and varies from about 15 to 90 per cent. Net production efficiency is lowest in animals whose costs of maintenance and activity are greatest, especially warm-blooded vertebrates, for which values of 1 to 5 per cent contrast with values of 15 to 45 per cent typical of invertebrates.

8. Gross production efficiency (production/ingestion) varies between about 5 and 20 per cent in most studies.

9. The average residence time of energy or biomass in a single link of the food chain is the ratio of biomass to rate of net production. Residence times for primary production vary from 20 years in some forests to 20 days or less in aquatic, plankton-based communities.

10. Radioactive tracer studies show that a portion of energy moves into food chains almost immediately after it is assimilated by plants. Conversely, much energy is tied up in accumulated litter and organic sediments with long residence times, or may be transported out of the system.

11. Considerations of energy flux and ecological efficiencies suggest that the highest trophic level at which a consumer population can be maintained ranges from the third level in terrestrial food chains to the seventh level in plankton-based communities of the open ocean.

12

Pathways of Elements in the Ecosystem

Nutrients, unlike energy, are retained within the ecosystem, where they are continually recycled between organisms and the physical environment. Most of these nutrients originate in rocks of the earth's crust. But because they are made available very slowly, high rates of organic production require that materials assimilated by one organism can be reused by others.

All the energy assimilated by green plants is "new" energy received from outside the ecosystem. In contrast, most of the nutrients assimilated by plants have been used before, many of them quite recently. Nitrates absorbed from the soil by roots might have been released from decaying leaves on the forest floor during the same day. The carbon dioxide assimilated by every terrestrial green plant was produced by animal, plant, or microbial respiration. On average, a molecule of carbon dioxide resides in the atmosphere for about 8 years before being reassimilated, although carbon dioxide can be recycled almost instantaneously (for example, within the closed canopy of a forest) before it escapes to the general atmospheric circulation.

Each element follows a unique route, determined by its particular biochemical transformations, in its cycle through the ecosystem. Living systems transform elements and their compounds in order to provide nutrients for building structure and to mobilize the energy required by all life processes. Over long periods, processes that transform elements from one form to a second form must be balanced by processes that restore the first.

Elemental cycles sometimes do become unbalanced, and elements accumulate in or are removed from the system. For example, during periods of coal and peat formation, dead organic materials accumulate in the sediments of lakes, marshes, and shallow seas where anaerobic conditions slow their decomposition by microorganisms. In other instances, under intensive cultivation or following removal of natural vegetation, erosion can wash away nutrient-laden layers of soil that took centuries to develop. By and large, however, most systems exist in a steady state in which export of elements is approximately balanced by import. Furthermore, gains and losses usually are small compared to the rate at which nutrients are cycled within the system.

This chapter is devoted to describing the pathways of elements in relation to the chemical and biochemical transformations responsible for their cycling and the conditions of the environment that influence these transformations. Most of these pathways are discussed in detail in texts on microbial ecology and environmental chemistry, of which I recommend for further reference Bohn et al. (1979), Fenchel and Blackburn (1979), Grant and Long (1981), Griffin (1981), Moat (1979), Rheinheimer (1980), and Stumm and Morgan (1981).

The movement of many elements parallels energy flow through the community.

Transformations that result in the production of organic forms of a particular element are referred to as assimilatory processes. Photosynthesis, by which "inorganic" (oxidized) carbon (carbon dioxide) is reduced to the "organic" carbon of carbohydrates, is the most obvious assimilatory transformation of an element, although others involving nitrogen and sulfur, for example, are equally important. In the overall cycling of carbon, photosynthesis is balanced by respiration, a complementary dissimilatory process that involves the oxidation of organic carbon with the accompanying release of energy and the return of carbon to its available inorganic form.

Not all transformations of elements in the ecosystem are biological, nor do all involve the net assimilation or release of useful quantities of energy. Many chemical reactions take place in the air, soil, and water. Some of these, like the weathering of bedrock, release certain elements (potassium, phosphorus, and silicon, for example) to the ecosystem. Other physical and chemical process, such as the sedimentation of calcium carbonate in the oceans, remove elements from ecosystem circulation to rocks in the earth's crust, where they can remain locked up for eons.

Most energy transformations are associated with biochemical oxidation and reduction of carbon, oxygen, nitrogen, phosphorus, and sulfur. In each case, an energy-releasing transformation is coupled with an energy-requiring transformation, so that energy is transferred from the reactants in the first to the products in the second (Figure 12–1). When more energy is released by the first than required by the second, the balance is lost as heat—hence the thermodynamic inefficiency of life processes. A typical coupling between transformations might involve oxidation of carbon in carbohydrate (perhaps glucose), with the release of energy, and reduction of nitrate-nitrogen to amino-nitrogen, which requires energy. These particular transformations are not directly linked; many intermediate steps of the type portrayed in Figure 12–1 are required for the transfer of energy between them.

The initial input of energy into the ecosystem is accomplished by an assimilatory transformation—the reduction of carbon—in which the

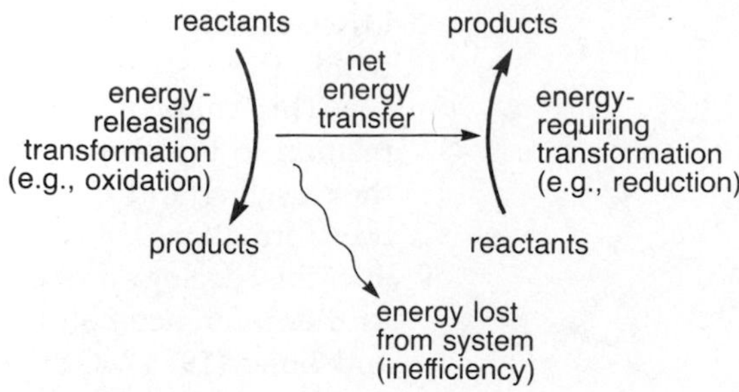

Figure 12–1
The coupling of energy-releasing and energy-requiring transformations is the basis of energy flow in the ecosystem.

source of energy is light rather than a coupled dissimilatory process. A portion of that energy is lost with each subsequent transformation. Some of these involve assimilation of other elements required for growth and reproduction; most involve biochemical transformations required for maintaining the cell environment and for movement. The cycling of elements between living and physical parts of the ecosystem is related to energy flow by the coupling of the dissimilatory part of one cycle to the assimilatory part of another (Figure 12–2).

Compartment models portray the cycling of elements in the ecosystem.

Each form of an element can be thought of as occupying a separate compartment in the ecosystem, like a room in a house. In this analogy, biochemical transformations are fluxes of the elements among the compartments;

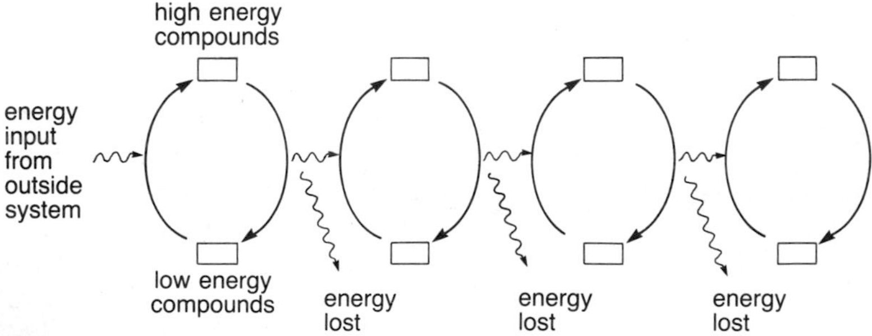

Figure 12–2
As energy flows through the ecosystem, elements alternate between assimilatory and dissimilatory transformations, thus going through cycles.

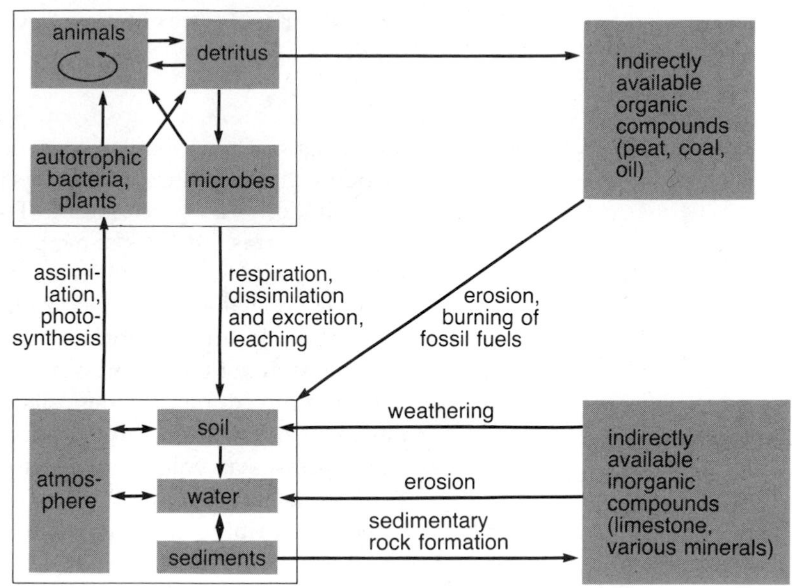

Figure 12–3
A generalized compartment model of the ecosystem.

that is, their movement between rooms (Figure 12–3). Carbon occurs in the forms of carbon dioxide both in the atmosphere and dissolved in water, carbonate and bicarbonate ions dissolved in water, calcium carbonate (limestone) sediments, and all organic molecules. Photosynthesis moves carbon from the carbon dioxide compartment to that containing organic forms of carbon (assimilation); respiration brings it back (dissimilation). The compartment of organic carbon has many subcompartments: animals, plants, microorganisms, and detritus. Herbivory, predation, and detritus feeding move carbon between subcompartments.

Some transformations involve changes in energy state. Photosynthesis adds energy to carbon, which may be thought of as lifting the element to the second floor of a house. In descending the respiration "staircase," carbon releases this stored chemical energy, which can then be used for other purposes.

Both organic and inorganic forms of elements may be removed from rapid circulation within the ecosystem to compartments that are not readily accessible to transforming agents. For example, coal, oil, and peat contain vast quantities of carbon removed from the ecosystem. Sedimentation removes inorganic (oxidized) carbon from ecosystem circulation primarily by sedimentation as calcium carbonate, which forms thick layers of limestone over large areas presently or formerly covered by seas. These forms of carbon are returned to the rapid cycling compartments only by the slow geological processes of uplift and erosion. Of course, man has greatly accelerated one component of this flux by burning fossil fuels.

The water cycle provides a physical model for element cycling in the ecosystem.

Although water is chemically involved in photosynthesis, most water flux through the ecosystem occurs by the physical processes of evaporation, transpiration, and precipitation (Figure 12–4). Light energy absorbed by water performs the work of evaporation. The condensation of atmospheric water vapor to form clouds releases the potential energy in water vapor as heat. Thus, evaporation and condensation resemble photosynthesis and respiration thermodynamically.

More than 90 per cent of all water is locked up in rocks in the earth's core and in sedimentary deposits near the surface. Water from this compartment enters the hydrological cycles of ecosystems through such geological processes as volcanic outpourings of steam; most of the water at the earth's surface originated in this way, although the great reservoirs in the interior contribute little to contemporary water flux in the ecosystem.

Over land surfaces, precipitation (23 per cent of the global total) exceeds evaporation and transpiration (16 per cent). Evaporation of water from the oceans exceeds replenishment from precipitation. Much of the water vapor that winds transport from the oceans to the land condenses and falls as precipitation over mountainous regions and where rapid heating of the land surface creates vertical air currents. The net flow of atmospheric water vapor from ocean to land is balanced by runoff from the land into ocean basins.

The energy that drives the hydrological cycle approximately equals the global evaporation of water (378×10^{18} g yr^{-1}) times the energy required to evaporate 1 gram of water (2.24 kJ). The product, $8.5 \times d10^{20}$ kJ yr^{-1},

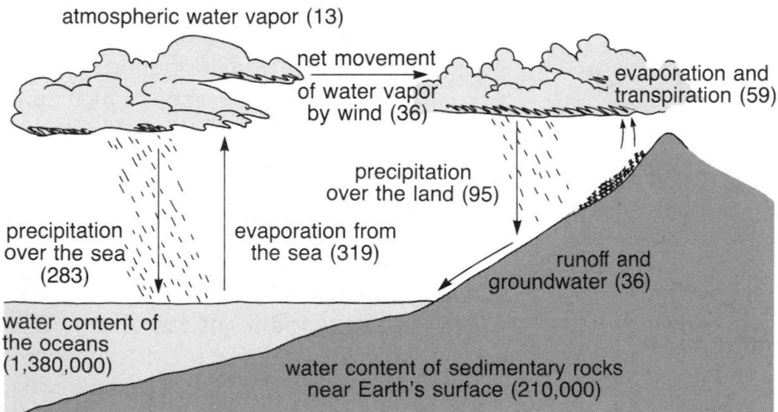

Figure 12–4
The water cycle, with its major components expressed on a global scale. All pools and transfer values (shown in parentheses) are expressed as billion billion (10^{18}) grams and billion billion grams per year. (After Hutchinson 1957, Barry and Chorley 1970.)

represents about one-fifth of the total energy income from the sun's radiation striking the earth. The remaining energy is either reflected, or absorbed and reradiated as heat. Of the total radiation striking temperate forests, for example, plants absorb about 40 per cent, of which they dissipate two-thirds through evapotranspiration (Stanhill 1970). Only a few per cent, at most, are assimilated by photosynthesis.

Evaporation, not precipitation, determines the flux of water through the ecosystem. The absorption of radiant energy by liquid water couples an energy source to the hydrological cycle. Evaporation and precipitation are closely linked because the atmosphere has a limited capacity to hold water vapor; any increase in evaporation of water into the atmosphere creates an excess of vapor and results in an equal increase in precipitation.

Water vapor in the atmosphere at any one time (the compartment size) corresponds to an average of 2.5 centimeters (1 inch) of water spread evenly over the surface of the earth. An average of 65 cm (26 in) of rain or snow falls each year (the water flux), which is 26 times the average amount of water vapor. The steady-state content of water in the atmosphere—the atmospheric pool—is therefore replaced 26 times each year on average. (Conversely, water has an average transit time of $\frac{1}{26}$ of a year, or 2 weeks.) The water content of soils, rivers, lakes, and oceans exceeds 100,000 times that of the atmosphere. Fluxes through both pools are the same, however, because evaporation balances precipitation. Thus, the average transit time of water in its liquid form at the earth's surface (about 3650 years) is 100,000 times longer than in the atmosphere.

The oxidation-reduction (redox) potential of a system indicates its energy level.

Water gains energy as it evaporates; hence, the energy content of water vapor and its capacity to release energy exceed those of liquid water. The chemical energy of a compound is indicated by its ability to reduce other compounds and, conversely, by its ability to become oxidized. In chemical reactions, an atom is oxidized when it donates an electron to another atom, which becomes reduced by accepting the electron. An oxidizing agent is therefore one that accepts electrons. The more readily a substance accepts electrons, the greater its oxidizing potential. A reducing agent donates electrons, and its strength depends on how readily it does so. Giving and taking electrons are opposite sides of the same coin; each substance may serve as an oxidizing or a reducing agent. But because strong oxidizers hold onto electrons tightly, they are always weak reducers, and vice versa. Put another way, a strong oxidizer can always oxidize (be reduced by) a weaker oxidizer.

Every oxidation-reduction (redox) reaction can be characterized by a pair of half-reactions, one for the reduction (electron-accepting) step and the other for the oxidation (electron-donating) step. For example, when molecular oxygen acts as an oxidizing agent, each atom accepts 2 electrons; i.e., $O_2 + 4e^- = 2O^{2-}$. In this reduced form, oxygen readily combines with any of a variety of positive ions, such as H^+, giving

$$O_2 + 4e^- + 4H^+ = 2H_2O$$

or C^{4+}, giving

$$O_2 + 4e^- + C^{4+} = CO_2$$

These half-reactions include only the reduction step. The electrons must come from an oxidation half-reaction; for example,

$$CH_2O = C^{4+} + H_2O + 4e^-$$

In which case, the carbon atom donates electrons (it is oxidized) and the electrons are available to reduce an oxidizer. The overall reaction combining the last two half-reactions has the form

$$CH_2O + O_2 = CO_2 + H_2O$$

which is the familiar equation for respiration.

The relative strengths of oxidizers and reducers are quantified by their electrical potentials (Eh) measured in volts (V) (Table 12–1). (An electrical current represents the movement of electrons; a chemical battery generates an electrical potential by a pair of redox reactions, one at each electrode.) Because oxygen has a high redox potential (0.81 volts at pH 7 and 25 °C), it is an excellent oxidizer. So are nitrate (NO_3^-), ferric iron (Fe^{3+}), and, to a lesser extent, sulfate (SO_4^{2-}). Near the bottom of the redox scale are carbon dioxide (CO_2) and molecular nitrogen (N_2). The electrical potentials of the half-reactions reducing CO_2-carbon (C^{4+}) to organic carbon (C^0, −0.42 V) and

Table 12–1
Redox potentials of selected half-reactions at 25 °C and neutral pH

Reaction	Eh (V)
$O_2 + 4H^+ + 4e^- = 2H_2O$	0.81
$NO_3^- + 6H^+ + 6e^- = \frac{1}{2}N_2 + 3H_2O$	0.75
$NO_3^- + 2H^+ + e^- = NO_2^- + H_2O$	0.42
$NO_3^- + 10H^+ + 8e^- + NH_4 + 3H_2O$	0.36
$Fe^{3+} + e^- = Fe^{2+}$	0.36
$NO_2^- + 8H^+ + 6e^- = NH_4^- + 2H_2O$	0.34
$CH_3OH + 2H^+ + 2e^- = CH_4 + H_2O$	0.17
$CH_2O + 2H^+ + 2e^- = CH_3OH$	−0.18
$SO_4^{2-} + 8H^+ + 6e^- = S + 4H_2O$	−0.20
$SO_4^{2-} + 10H^+ + 8e^- = H_2S + 4H_2O$	−0.21
$CO_2 + 8H^+ + 8e^- = CH_4 + 2H_2O$	−0.24
$N_2 + 8H^+ + 6e^- = 2NH_4^+$	−0.28
$H^+ + e^- = \frac{1}{2}H_2$	−0.41
$CO_2 + 4H^+ + 4e^- = \frac{1}{6}C_6H_{12}O_6$ (glucose) $+ H_2O$	−0.43
$CO_2 + 4H^+ + 4e^- = CH_2O + H_2O$	−0.48
$Fe^{2+} + 2e^- = Fe$	−0.85

Source: Stumm and Morgan 1981.

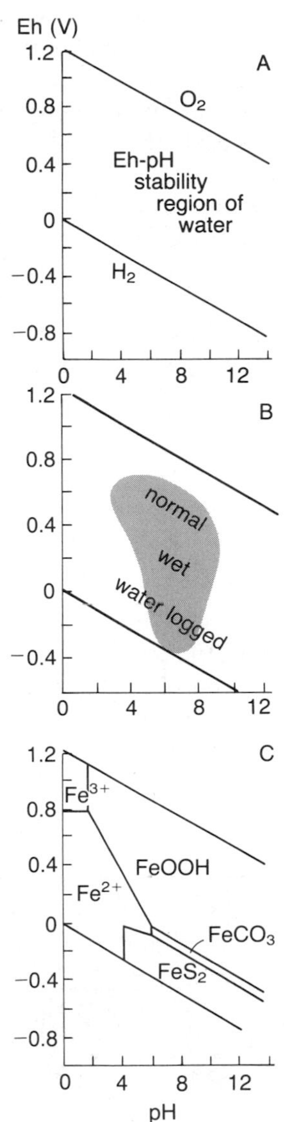

Figure 12–5
Distribution of pH and electrical potentials of soils and the ion compounds associated with these conditions. *A*: The stability region for water sets outer limits to soil conditions. *B*: Measured values of pH and Eh in soils fall within the shaded region; waterlogged soils tend to have lower (more reduced) electrical potentials. *C*: Stability regions for various iron ions and compounds. Iron tends to precipitate at neutral and alkaline pHs. (After Bohn et al. 1979.)

N_2 to organic (ammonia)-nitrogen (N^{3-}; −0.25 V) are unfavorable thermodynamically.

Any oxidization reaction having a high redox potential coupled with a reduction reaction having a low redox potential can proceed with the release of energy. Thus the oxidation of carbon by oxygen (O_2) releases energy because oxygen is a stronger oxidizing agent than carbon dioxide. For the reaction to proceed in the opposite direction (for example, fixation of carbon by photosynthesis), energy must be added.

Oxygen and carbon are not the only electron acceptors and donors of importance to biological reactions. Nitrate is a powerful oxidizer; hydrogen sulfide (H_2S) and ammonium (NH_3) are good reducing agents. Indeed, the synthesis and metabolism of many organic compounds, including amino acids, involve assimilatory and dissimilatory redox reactions of nitrogen and sulfur. In general, oxygen is the oxidizer of choice owing to its ubiquity and the fact that its common reduced form (H_2O) is innocuous. In contrast, the reduction of nitrate produces nitrite (NO_2^-), which most organisms can tolerate only in minute quantities. Oxidizers such as nitrate and ferric iron can nonetheless become important in soils and aquatic sediments when oxygen has been depleted (anaerobic conditions), as we shall see below.

When all other oxidizers have been exhausted (extremely reduced conditions), hydrogen ion can be used as an electron acceptor: $2H^+ + 2e^- = H_2$. This reaction is, however, thermodynamically very unfavorable (redox potential = −0.41 V) and therefore requires considerable energy. Organic carbon also can be an electron acceptor when other oxidizing agents are absent:

$$CH_2O + 2H^+ + 4e^- = CH_4 + O^{2-}$$

This reaction occurs in anaerobic fermentations in some oxygen-depleted sediments and, notably, in the rumens of ungulates, which lack inorganic electron acceptors (Wolin 1979). Methane, or natural gas (CH_4), is one of the end-products of this set of reactions.

Hydrogen can be a powerful electron donor, by the half-reaction $H_2 = 2H^+ + 2e^-$, depending on the hydrogen ion concentration of the environment, or pH. At lower pH (hydrogen ions more abundant, more acid conditions), hydrogen gives up electrons less readily and the electrical potential of the redox reaction increases (that is, hydrogen is a poorer reducing agent). As the pH increases, the supply of hydrogen ions (H^+) decreases, the equilibrium of the reaction $H_2 = 2H^+ + 2e^-$ shifts to the right, and electrode potentials decrease.

With respect to oxidation-reduction reactions, an environment may be characterized with respect to pH and electrical potential (Eh). The first measures the availability of hydrogen ions, the second the availability of electrons (Figure 12–5). Because hydrogen and oxygen are so abundant, their reactions limit naturally occurring values of pH and Eh. For a given pH, electrical potential has a practical upper bound set by the reaction $O_2 + 2e^- = 2O^{2-}$ (Eh = 0.81 V at pH 7) and a lower bound set by the reaction $2H^+ + 2e^- = H_2$ (Eh = −0.41 V). When a powerful oxidizer raises the Eh above 0.81 V, it oxidizes the O_2 in water, liberating oxygen (O^{2-}), until the oxidizing agent is fully consumed (reduced) and Eh drops to that of O_2.

When a powerful reducing agent decreases the Eh below −0.41 V, it reduces hydrogen ions to molecular hydrogen (H_2) until the reducing agent is fully oxidized. In general, high values of Eh occur in well-oxygenated soils; lower values occur in waterlogged soils and the deep waters and sediments of lakes and estuaries, into which oxygen diffuses from the atmosphere much more slowly than it is consumed by the respiration of organisms.

Each element that undergoes redox reactions assumes a predominant form that may vary depending upon the Eh and pH of the environment. For example, iron occurs in its ferric form (Fe^{3+}) under oxidizing conditions but in its ferrous form (Fe^{2+}) under reducing conditions. Under acidic conditions, both forms tend to occur as dissociated ions, but as the hydrogen ion concentration decreases, they more readily form insoluble compounds (hydroxides, carbonates, and sulfides) (Figure 12–5). As we shall see below, conditions of pH and Eh have important consequences for the cycling of many elements.

The carbon cycle is most closely associated with energy flow in the ecosystem.

Three major classes of process are involved in the cycling of carbon in aquatic and terrestrial systems (Figure 12–6). The first includes the assimilatory and dissimilatory redox reactions of carbon in photosynthesis and respiration. These are the major energy-transforming reactions of life. Approximately 10^{17} grams (10^{11} metric tons) of carbon enter into such reactions worldwide each year.

The second class of process is the physical exchange of carbon dioxide

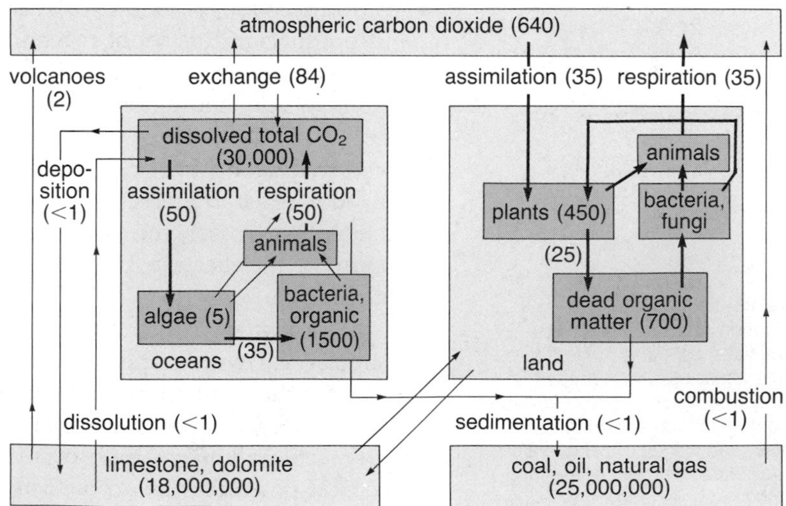

Figure 12–6
The global carbon cycle. Pools and fluxes (shown in parentheses) are in billions of metric tons (10^{15} g). (After Fenchel and Blackburn 1979, Grant and Long 1981.)

between the atmosphere and oceans, lakes, and streams. Carbon dioxide dissolves readily in water; indeed, the oceans contain about 50 times as much CO_2 as the atmosphere. Exchange across the air-water boundary links the carbon cycles of terrestrial and aquatic systems, but the two may be considered virtually independent because production balances consumption of CO_2 in both, and dissolved and atmospheric CO_2 achieve an approximate equilibrium globally.

The third class of process is the solution and precipitation (deposition) of carbonate compounds as sediments, particularly limestone and dolomite. On a global scale, these processes are approximately balanced, although certain conditions favoring precipitation have led to the deposition of extensive layers of calcium carbonate-rich sediments in the past. In aquatic systems, dissolution and deposition occur about two orders of magnitude more slowly than assimilation and dissimilation. Thus, exchange between sediments and the water column is relatively unimportant to the short-term cycling of carbon in the ecosystem. Locally, and over long periods, it can assume much greater importance; in fact, most of the ecosystem's carbon resides in sedimentary rocks.

When carbon dioxide dissolves in water, it forms carbonic acid

$$CO_2 + H_2O \rightleftharpoons H_2CO_3$$

which readily dissociates into bicarbonate and carbonate ions,

$$H_2CO_3 \rightleftharpoons H^+ + HCO_3^-$$

$$HCO_3^- \rightleftharpoons H^+ + CO_3^{2-}$$

At low pH, abundant hydrogen ions in the environment drive these reactions to the left (Figure 12–7). When calcium is present, it also equilibrates with the carbonate and bicarbonate ions

$$CaCO_3 \rightleftharpoons Ca^{2+} + CO_3^{2-}$$

Calcium carbonate ($CaCO_3$) has low solubility under most conditions and readily precipitates out of the water column.

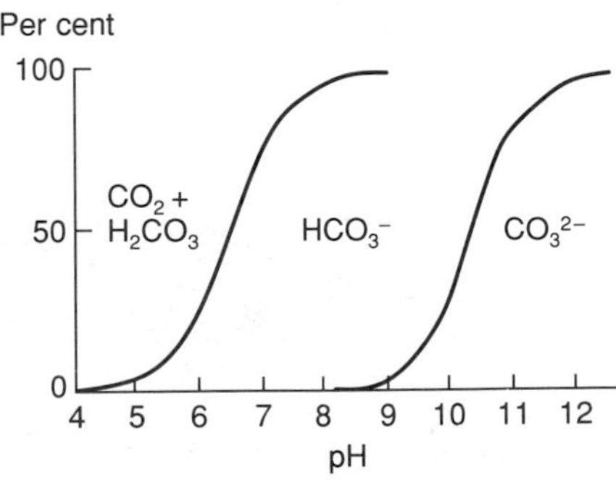

Figure 12–7
Proportion of carbon as carbonic acid, bicarbonate, and carbonate in solution as a function of pH.

Under acid conditions, carbonate ion is removed from the system as the equilibrium $H^+ + CO_3^{2-} \rightleftharpoons HCO_3^-$ shifts to the right and the dissociation of $CaCO_3$ increases. This reduces the amount of dissolved calcium carbonate in the water, which, in turn, enhances the dissolution of calcium carbonate rocks and sediments that the water passes over. By this process, slightly acid streams and groundwater dissolve limestone sediments (and acid rain erodes marble statuary). As these calcium-laden fresh waters enter the oceans and mix with water of nearly neutral pH, the equilibria shift back toward undissociated forms of carbonate, and calcium carbonate precipitates out of the water column to form sediments.

Dissolution and dissociation may be affected locally by the activities of organisms. In the marine system, under approximately neutral conditions, the carbonate system has the overall equilibrium

$$CaCO_3 \text{ (insoluble)} + H_2O + CO_2 \rightleftharpoons Ca(HCO_3)_2 \text{ (soluble)}$$

Removal of CO_2 by photosynthesis shifts the equilibrium to the left, resulting in the formation and precipitation of calcium carbonate. Many algae excrete this calcium carbonate to surrounding water, but reef-building algae and coralline algae incorporate the substance as hard structure. In the system as a whole, when photosynthesis exceeds respiration, as it does during algal blooms, calcium tends to be precipitated out of the system.

Nitrogen assumes many oxidation states in its paths through the ecosystem.

The ultimate source of nitrogen to the ecosystem is molecular nitrogen in the atmosphere. This dissolves to some extent in water but none is found in native rock. So, were it not for biological processes, under the present oxidizing conditions at the earth's surface virtually all nitrogen would occur in its molecular form.

Molecular nitrogen enters biological pathways of the nitrogen cycle (Figure 12–8) owing to assimilation (nitrogen fixation) by certain microorganisms. Although this pathway ($N_2 \rightarrow NH_3$) constitutes a small fraction of the earth's annual nitrogen flux, all biologically cycled nitrogen can be

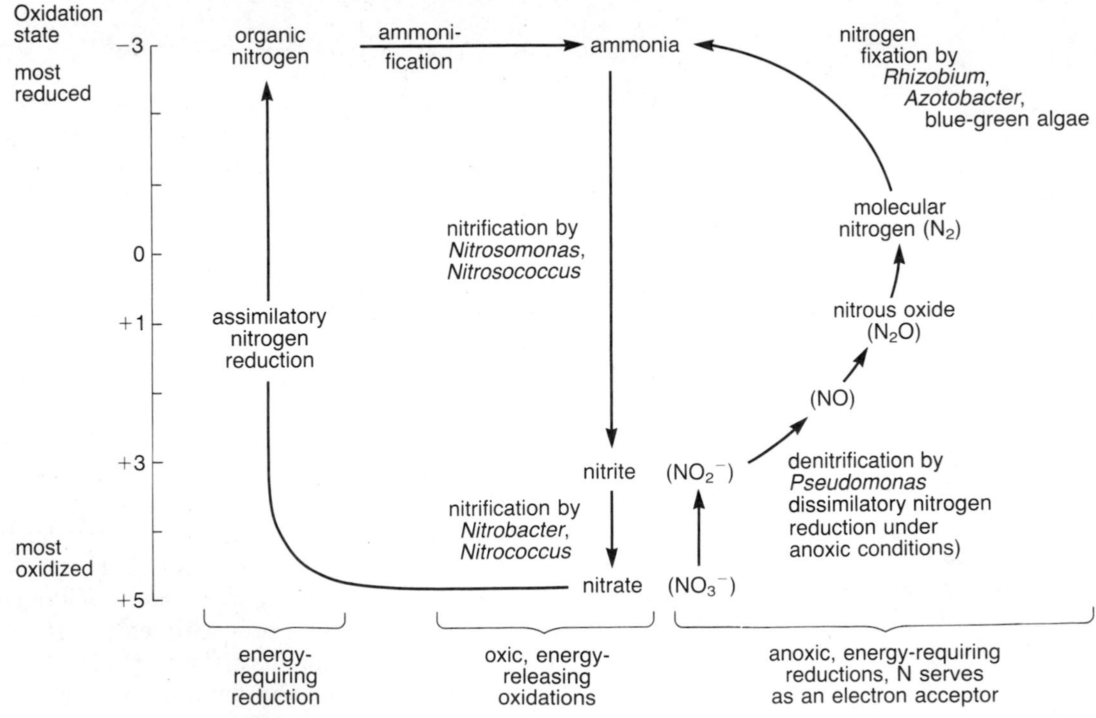

Figure 12–8
Schematic diagram of transformations and oxidation states of compounds in the nitrogen cycle.

Table 12–2
Biochemical processes involved in the ecological cycling of nitrogen and several other elements

Process	Organism	Yield (kJ/mole)
Respiration		
$C_6H_{12}O_6 + 6O_2 \longrightarrow 6CO_2 + 6H_2O$	Virtually universal	2870
Denitrification		
$C_6H_{12}O_6 + 6KNO_3 \longrightarrow 6CO_2 + 3H_2O + 6KOH + 3N_2O$	*Pseudomonas denitrificans*	2280
$5C_6H_{12}O_6 + 24\ KNO_3 \longrightarrow 30CO_2 + 18H_2O\ 24KOH + 12N_2$	*Pseudomonas denitrificans*	2385
$5S + 6KNO_3 + 2CaCO_3 \longrightarrow 3K_2SO_4^+\ 2CO_2 + 3N_2$	Anaerobic sulfur bacteria	552
Ammonification		
$C_2H_5NO_2 + 1\frac{1}{2}O_2 \longrightarrow 2CO_2 + H_2O + NH_3$	Many bacteria; most plants and animals	736
Nitrification		
$NH_3 + 1\frac{1}{2}\ O_2 \longrightarrow HNO_2 + H_2O$	*Nitrosomonas* bacteria	276
$KNO_2 + \frac{1}{2}O_2 \longrightarrow KNO_3$	*Nitrobacter*	73
Nitrogen fixation		
$2N_2 + 3H_2 \longrightarrow 2NH_3$	Some cyanobacteria, *Azotobacter*	−616
Oxidation of sulfur		
$2H_2S + O_2 \longrightarrow S_2 + 2H_2O$		335
$S_2 + 3O_2 + 2H_2O \longrightarrow 2H_2SO_4$		1004
Oxidation of iron		
$Fe^{2+} \longrightarrow Fe^{3+}$		48

$C_6H_{12}O_6$ = glucose; CO_2 = carbon dioxide; $C_2H_5NO_2$ = glycine (an amino acid); $CaSO_4$ = calcium sulfate; $CaCO_3$ = calcium carbonate; HNO_2 = nitrous acid; H_2S = hydrogen sulfide; H_2SO_4 = sulfuric acid; KNO_2 = potassium nitrite; KNO_3 = potassium nitrate; KOH = potassium hydroxide; NH_3 = ammonia; N_2O = nitrous oxide; S = sulfur.
Source: Delwiche 1970, Rheinheimer 1980.

traced back to nitrogen fixation. Once in the biological realm, the cycling of nitrogen is much more complicated than that of carbon owing to its numerous oxidation states. Beginning with reduced (organic) nitrogen, the first step in the nitrogen cycle is ammonification: the hydrolysis of protein and oxidation of amino acids, which is accomplished by all organisms. During the initial breakdown of amino acids, some carbon is oxidized, releasing energy (Table 12–2), but the oxidation state of nitrogen does not change (valence remains at −3).

Nitrification involves the oxidation of nitrogen, first from ammonia to nitrite, then from nitrite to nitrate (Brown et al. 1974, Fochte and Verstaete 1977). Both steps are carried out only by specialized bacteria: $NH_3 \rightarrow NO_2^-$ ($N^{3-} \rightarrow N^{3+}$) by *Nitrosomonas* in the soil and by *Nitrosococcus* in marine systems; $NO_2^- \rightarrow NO_3^-$ (N^{5+}) by *Nitrobacter* in the soil and *Nitrococcus* in the seas.

Since nitrification steps are oxidations, they require oxic conditions under which oxygen can act as an electron acceptor. Under the anoxic conditions of waterlogged soils and sediments, nitrate and nitrite can act as electron acceptors (oxidizers) and the nitrification reactions reverse (Delwiche and Bryan 1976). This denitrification occurs in soil when the redox potential is less than 0.2 V, and is accomplished by such bacteria as *Pseudomonas denitrificans.* Additional physical reactions can result in the production of molecular nitrogen; that is,

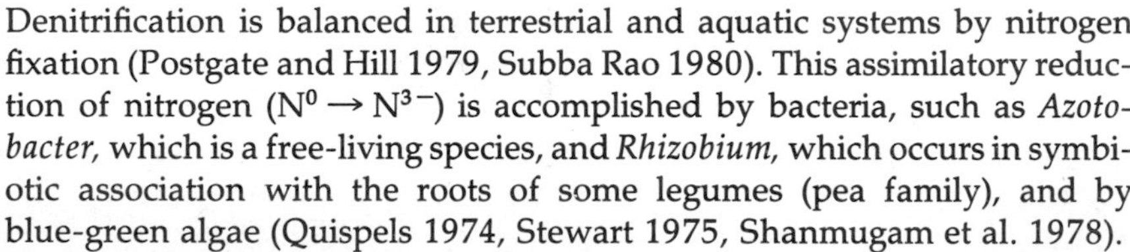

$$NO_3^- \longrightarrow NO_2^- \longrightarrow NO \longrightarrow N_2O \longrightarrow N_2$$

with the consequent loss of nitrogen from biological circulation. In soils, for example, nitrate produced by nitrification in oxygenated surface layers may be washed to deeper, anaerobic layers where denitrification may take place primarily by physical reactions.

Nitrogen-fixation is an energy-requiring reduction.

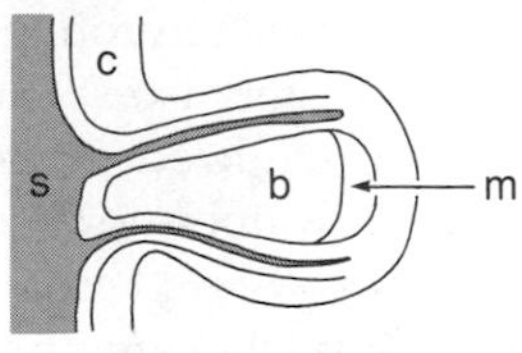

Figure 12–9
The root system of an Austrian winter pea plant showing the clusters of nodules that harbor symbiotic nitrogen-fixing bacteria. The diagram shows the arrangement of tissues within a nodule: s = vascular stele, c = root cortex, b = bacteroid-containing tissue, m = nodule meristem. (Courtesy of the U.S. Soil Conservation Service, and from Grant and Long 1981.)

Denitrification is balanced in terrestrial and aquatic systems by nitrogen fixation (Postgate and Hill 1979, Subba Rao 1980). This assimilatory reduction of nitrogen ($N^0 \rightarrow N^{3-}$) is accomplished by bacteria, such as *Azotobacter,* which is a free-living species, and *Rhizobium,* which occurs in symbiotic association with the roots of some legumes (pea family), and by blue-green algae (Quispels 1974, Stewart 1975, Shanmugam et al. 1978).

Nitrogen fixation is energetically expensive (Table 12–2) although no more so than the conversion of an equivalent amount of nitrate to ammonia by plants (Hardy and Havelka 1975). The reduction of an atom of nitrogen from chemical valence 0 (N_2) to -3 (NH_3) requires approximately the amount of energy released by the oxidation of an atom of carbon from valence 0 (glucose, $C_6H_{12}O_6$) to $+4$ (CO_2).

Nitrogen-fixing microorganisms obtain the energy and reducing power (organic carbon) they need to reduce N_2 to NH_3 by oxidizing sugars or other organic compounds. Free-living bacteria must obtain these resources by metabolizing organic detritus in the soil, sediments, or water column. More abundant supplies of energy are available to bacteria that enter into symbiotic relationships with plants, which provide them with photosynthate. The best known of these associations is the infection of root nodules of leguminous plants (peas, alfalfa, and their relatives) by the nitrogen-fixing bacterium *Rhizobium* (Quispel 1974, Broughton 1983). The nodules are specialized structures of the root cortex, whose development is stimulated by the *Rhizobium* (Figure 12–9). The nodules provide an optimum environment of abundant photosynthate and low oxygen for nitrogen fixation. Nitrogen-fixing symbioses are by no means limited to legumes, and such associations of bacteria or blue-green algae have been identified in other terrestrial plants (alders, *Casaurina*), some marine algae (Wiebe 1975), lichens (Milbank and Kershaw 1969), and shipworms (Carpenter and Culliney 1975).

On a global scale, nitrogen fixation approximately balances the production of N_2 by denitrification. The fluxes are about 2 per cent of the total cycling of nitrogen through the ecosystem (Hardy and Havelka 1975). On a local scale, nitrogen fixation can assume much greater importance, especially in nitrogen-poor habitats. When land is exposed to colonization by plants (for example, areas left bare by receding glaciers) such species with nitrogen-fixing capabilities as the alder *Alnus* dominate newly established vegetation.

The phosphorus cycle is closely tied to the acidity and the redox potential of the environment.

Ecologists have studied the role of phosphorus in the ecosystem intensively because organisms require phosphorus at a high level (about one-tenth that of nitrogen) as a major constituent of nucleic acids, cell membranes, energy-transfer systems, bones, and teeth. Phosphorus is thought to limit plant productivity in many aquatic habitats, and the influx of phosphorus to rivers and lakes, in the form of sewage and runoff from fertilized agricultural lands, artificially stimulates production in aquatic habitats with undesirable consequences.

The phosphorus cycle has fewer steps than the nitrogen cycle: plants assimilate phosphorus as phosphate (PO_4^{3-}) directly from the soil or water and incorporate it into various organic compounds as phosphate esters. Animals eliminate excess organic phosphorus in their diets by excreting phosphorus salts in urine; phosphatizing bacteria also convert organic phosphorus in detritus to phosphate ion. Phosphorus does not enter the atmosphere in any form other than dust; thus, the phosphorus cycle involves only the soil and aquatic compartments of the ecosystem.

Phosphorus occurs in the environment as orthophosphate (PO_4^{3-}, valence +5) and it does not undergo oxidation-reduction reactions in its cycling through the ecosystem except in very limited microbial transformations (Cosgrave 1977). The availability, and consequently the uptake and assimilation, of phosphorus is greatly affected by the pH of the environment. In general, at low pH (acid conditions), phosphorus binds tightly to clay particles in soil and forms relatively insoluble compounds with ferric iron [strengite, $Fe(OH)_2H_2PO_4$] and aluminum [variscite, $Al(OH)_2H_2PO_4$]. At high pH, other insoluble compounds are formed; for example, with calcium [hydroxyapatite, $Ca_{10}(PO_4)_6(OH)_2$]. These solubility relationships can be portrayed in a solubility diagram (Figure 12–10), in which the concentration of dissolved $H_2PO_4^-$ in equilibrium with insoluble compounds is shown as a function of the acidity of the soil. At low pH and in the presence of ferric iron or aluminum (which are present in virtually all soils, sediments, and waters), the equilibrium is driven to a very low level of dissolved $H_2PO_4^-$. Similarly, in alkaline soils in the presence of calcium, the equilibrium also is very low. When both ferric iron or aluminum and calcium are present, the highest concentration of dissolved phosphate—that is, the greatest availability of phosphorus—occurs at a pH of between 6 and 7. Under anoxic conditions in aquatic sediments, phosphorus is solubilized when iron is reduced from Fe^{3+} to Fe^{2+} because the iron then forms sulfides rather than phosphate compounds.

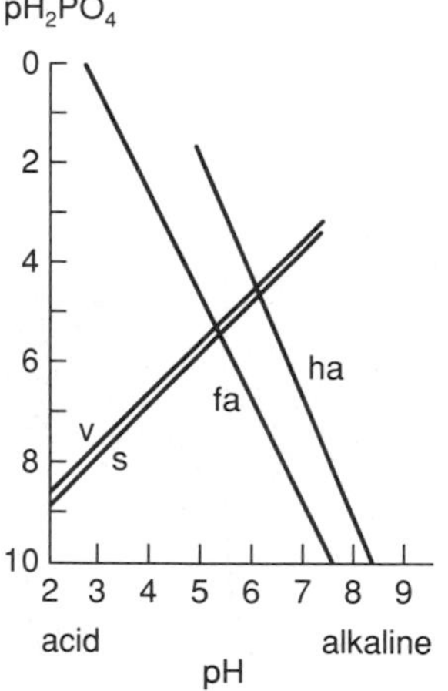

Figure 12–10
The solubility of phosphate ion in equilibrium with several insoluble forms of phosphorus as a function of pH of the soil; v = variscite, s = stengite, fa = flouroapatite, ha = hydroxyapatite. (After Lindsay and Moreno 1960.)

Sulfur takes part in many redox reactions.

Sulfur is a component of the amino acids cysteine and methionine; hence, its requirement by plants and animals. But the importance of sulfur in the ecosystem goes far beyond its assimilation by plants. Like nitrogen, sulfur has many oxidation states. Like nitrogen, therefore, it follows complex chemical pathways and affects the cycling of other elements (Trudinger 1969; Figure 12–11).

The most oxidized form of sulfur is sulfate (SO_4^{2-}, valence +6); the most reduced are sulfide (S^{2-}) and the organic (thiol) form of sulfur (also S^{2-}). Under oxic conditions, assimilatory sulfur reduction ($SO_4^{2-} \rightarrow$ organic S) is balanced by the oxidation of organic sulfur back to sulfate, either directly or with SO_3^{2-} (sulfite, S^{4+}) as an intermediate step. This oxidation is accomplished by most animals when they excrete excess dietary organic sulfur and by microorganisms when they decompose plant and animal detritus.

Under anoxic conditions (Eh < 0, dependent upon pH), sulfate may function as an oxidizer (Table 12–1). The bacteria *Desulfovibrio* and *Desulfomonas* both couple dissimilatory sulfate reduction ($SO_4^{2-} \rightarrow S^{2-}$) to the oxidation of organic carbon in order to make energy available (Le Gall and Postgate 1973). The reduced S^{2-} may then be used as a reducing agent ($S^{2-} \rightarrow S^0 + 2e^-$) by photoautotrophic bacteria, which assimilate carbon by pathways analogous to photosynthesis in green plants. Sulfur takes the place of the oxygen atom in water as an electron donor. The elemental sulfur

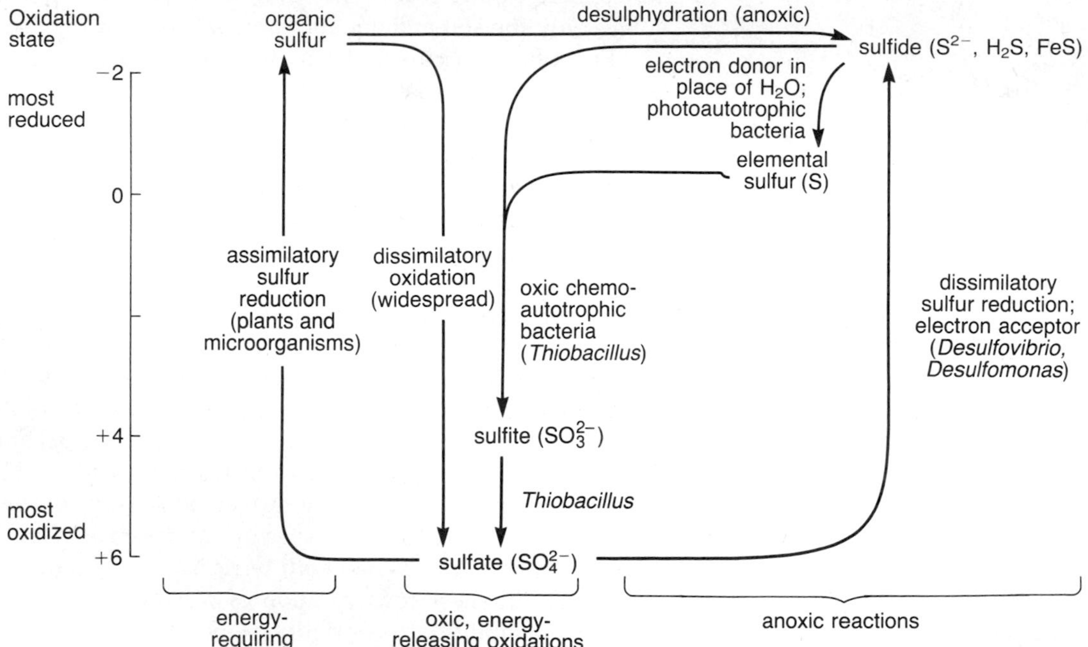

Figure 12–11
Schematic diagram of transformations and oxidation states of compounds in the sulfur cycle.

(S^0) accumulates unless the sediments are exposed to aeration or oxygenated water, at which point the sulfur may be further oxidized to SO_3^{2-} and SO_4^{2-}.

The fate of S^{2-} produced under oxic conditions depends on the availability of positive ions. Frequently hydrogen sulfide forms; it escapes from shallow sediments and mucky soils as a gas having the familiar smell of rotten eggs. Because oxic conditions generally favor the reduction of ferric iron (Fe^{3+}) to ferrous iron (Fe^{2+}), the presence of iron in sediments leads to the formation of iron sulfide (FeS). Sulfides are commonly associated with coal and oil. When exposed to the atmosphere in mine wastes or burned for energy, the reduced sulfur is oxidized, with the help of *Thiobacillus* bacteria in mine wastes, and the oxidized forms combine with water to form the sulfuric acid of acid mine drainage and acid rain.

Microorganisms assume special roles in element cycles.

Many of the transformations discussed in this chapter are accomplished mainly or entirely by microorganisms (Coleman et al. 1983). In fact, if it were not for the activities of bacteria and fungi, many element cycles would be drastically altered and the productivity of the ecosystem much reduced. For example, wood is broken down for the most part only by microorganisms, especially fungi (Kirk et al. 1977, Crawford and Crawford 1980). In their absence, organic carbon would accumulate as cellulose and lignin rather than passing through detritus-based food chains. Over long periods, primary production would drop as levels of carbon dioxide in the atmosphere decreased. For a second example, without the capacity of some microbes to utilize sulfur and iron as electron acceptors, little decomposition would occur in anoxic organic sediments, whose resulting accumulation would further reduce the amount of inorganic carbon in the ecosystem.

Nitrogen fixation, the sole source of biological nitrogen, is carried out only by microorganisms. Without them, life could not exist. Plants probably can assimilate ammonia-nitrogen as easily as, and at less energetic expense than, nitrate-nitrogen. Thus, nitrifying bacteria may actually reduce the primary productivity of terrestrial and aquatic plants by competing for nitrogen substrates in the soil and water.

Many of the transformations carried out by microorganisms, such as the metabolism of sugars and other organic molecules, are accomplished in similar ways by plants and animals. The bacteria and cyanobacteria (blue-green algae) are distinguished physiologically primarily by the ability of many species to metabolize substrates under anoxic conditions and to use substrates other than organic carbon as energy sources (Moat 1979, Grant and Long 1981). The requirements of fungi for inorganic substrates as sources of nitrogen, phosphorus, potassium, sulfur, and other elements are similar to those of higher plants (Griffin 1981). What distinguishes the fungi from the bacteria is their greater ability to break down such complex polysaccharides as cellulose and lignin, which make up a large proportion of terrestrial plant litter (Grant and Long 1981).

Every organism needs above all a source of carbon for building organic structure and a source of energy to fuel the life processes. Organisms can be distinguished according to their source of carbon. Heterotrophs, or organotrophs, obtain carbon in reduced (organic) form by consuming other organisms or organic detritus. All animals and fungi, and many bacteria, are heterotrophs. Autotrophs assimilate carbon as carbon dioxide and expend energy to reduce it to an organic form. Photoautotrophs utilize sunlight as their source of energy for photosynthesis. These include all green plants and the cyanobacteria, which use H_2O as an electron donor ($O^{2-} \rightarrow O^0 + 2e^-$) and are aerobic (Wolk 1973), and the purple and green bacteria, which have different light-absorbing pigments from green plants, use H_2S or organic compounds as electron donors, and are anaerobic.

Organisms may obtain energy from three sources: sunlight (photoautotrophs), the oxidation of organic compounds (all heterotrophs), and the oxidation of inorganic compounds (chemoautotrophs). Oxidation of glucose, represented as $CH_2O + O_2 \rightarrow CO_2 + H_2O$, incorporates an oxidation reaction that releases energy, in this case $C^0 \rightarrow C^{4+} + 4e^-$, and an electron-accepting (reduction) reaction, $2O^0 + 4e^- \rightarrow 2O^{2-}$. Heterotrophs all use C^0 as reduced substrate for oxidation. Under oxic conditions, O_2 is used as an electron acceptor. But many bacteria (sulfate-reducing bacteria, methanogens) are either facultatively or obligately capable of utilizing $SO_4{}^{2-}$, Fe^{3+}, or CO_2 as electron acceptors under anoxic conditions.

Chemoautotrophs all use CO_2 as a carbon source, but obtain energy for its reduction by the oxic oxidation of inorganic substrates: methane (for example, *Methanosomonas, Methylomonas*), hydrogen *(Hydrogenomonas, Micrococcus)*, ammonia (nitrifying bacteria *Nitrosomonas, Nitrosococcus*), nitrite (nitrifying bacteria *Nitrobacter, Nitrococcus*), hydrogen sulfide, sulfur, and sulfite *(Thiobacillus)*, and ferrous iron salts *(Ferrobacillus, Gallionella)*. The chemoautotrophs are almost exclusively bacteria, which apparently are the only type of organism that can become so specialized biochemically as to make efficient use of inorganic substrates and efficiently dispose of the products of chemoautotrophic metabolism.

In this chapter we have examined the cycling of several important elements from the standpoint of their chemical and biochemical reactions. Each type of habitat presents a different chemical environment, however, and it stands to reason that the patterns of element fluxes differ greatly among them. In the next chapter, we shall contrast element cycling in aquatic and terrestrial habitats by focusing on how some of the unique physical features of each of these environments affect chemical and biochemical transformations involved in organic production and recycling of elements.

SUMMARY

Unlike energy, nutrients are retained within the ecosystem and are cycled between its physical and biotic components. The paths of elements through the ecosystem depend upon chemical and biological transformations, which are functions of the chemical characteristics of each element and their requirement by organisms.

1. The movement of energy through the ecosystem parallels the paths of several elements, particularly carbon, whose transformations either require or release energy.
2. The cycling of each element may be thought of as movement between compartments of the ecosystem, the major ones being living organisms, organic detritus, immediately available inorganic forms, and unavailable organic and inorganic forms, generally in sediments.
3. The water cycle provides a physical analogy for element cycling. Energy is required to transform water from its liquid phase to atmospheric vapor. Upon condensation and precipitation, that energy is released as heat.
4. Energy transformations in biological systems occur primarily during oxidation-reduction (redox) reactions. An oxidizer is a substance that readily accepts electrons (O_2, NO_3^-); a reducer is one that readily donates electrons (H_2, organic-C). Upon being reduced, oxidizers gain energy; upon being oxidized, reducing agents release energy. Elements in organic compounds tend to be in reduced forms; hence, biological assimilation of many elements (carbon, nitrogen, sulfur) requires energy.
5. All organisms require organic carbon as the primary substance of life. It also is the major source of energy for most animals and microorganisms. Carbon shuttles between living forms and the carbon dioxide compartment of the ecosystem by way of photosynthesis and respiration. Precipitation of calcium carbonate in the oceans accounts for the accumulation of limestone sediments.
6. Nitrogen has many oxidation states and consequently follows many pathways through the ecosystem. Quantitatively, most of the flux follows the cycle nitrate → organic-N → ammonia → nitrite → nitrate. The last two steps, referred to as nitrification, are accomplished by certain bacteria under oxic conditions. Under anoxic conditions in soils and sediments, nitrate replaces oxygen as an electron acceptor (denitrification) and the reactions reverse: nitrate → nitrite → (eventually) molecular nitrogen (N_2). This loss of nitrogen from the general biological cycling is balanced by nitrogen fixation by some microorganisms.
7. Phosphorus does not change oxidation state. It is assimilated by plants in the form of phosphate (PO_4^{2-}), whose availability is largely determined by the acidity and oxidation level of the soil or water.
8. Sulfur is an important redox element in anoxic habitats, where it may serve as an electron acceptor in the form of sulfate (SO_4^{2-}) or an electron donor (for photoautotrophic bacteria) in the forms of elemental sulfur (S^0) and sulfide (S^{2-}).
9. Many elemental transformations, particularly under anoxic conditions, are accomplished by biochemically specialized microorganisms (bacteria, cyanobacteria). The activities of these organisms therefore assume important roles in the cycling of elements through the ecosystem.

13

Nutrient Regeneration in Terrestrial and Aquatic Ecosystems

The paths of elements through the ecosystem are directed by their chemical properties, which in turn determine chemical and biochemical reactions in the ecosystem. This much we discussed in detail in Chapter 12. We also have seen that utilizable forms of elements must be regenerated within the system to maintain high productivity. Elements enter the rapid-cycle compartments from external sources — for example, by nitrogen fixation and weathering of bedrock — too slowly to support the energy fluxes characteristic of most habitats. Thus the regeneration of assimilable forms of nutrients is a key to understanding the regulation of ecosystem function.

Processes of regeneration differ between terrestrial and aquatic systems. The chemical and biochemical transformations involved are basically the same; oxidation of carbohydrates, nitrification, chemoautotrophic oxidation of sulfur, among many others, result from similar biochemical mechanisms in both habitats. It is the material basis for nutrient regeneration that distinguishes terrestrial and aquatic habitats. In terrestrial systems, most nutrients cycle through detritus at the soil surface. In most aquatic systems, lakes and oceans in particular, sediments are the ultimate source of regenerated nutrients. Small streams have functional affinities with terrestrial habitats to the extent that production is based on the decomposition of terrestrial detritus, often under well-aerated conditions (Cummins 1974, Petersen and Cummins 1974, Anderson and Sedell 1979). To be sure, both aquatic and terrestrial systems exhibit great variation in element cycles. Our goal in this chapter will be a general understanding of nutrient regeneration in each of these systems.

Regenerative processes in terrestrial ecosystems occur in the soil.

While soil is difficult to define, it can be described as whatever overlies unaltered rock at the surface of the earth. It includes minerals derived from the parent rock, altered minerals formed anew within the zone of weathering, organic material contributed by plants, air and water within the pores of the soil, living roots of plants, microorganisms, and the larger worms and arthropods that make the soil their home. Five factors largely determine the characteristics of soils: climate, parent material (underlying rock), vegetation, local topography, and, to some extent, age (Jenny 1941, 1980, Brady 1974).

Soils are in a dynamic state, changing as they develop on newly exposed rock material. But even after soils achieve stable properties, they remain in a constant state of flux. Groundwater removes some material; other material enters the soil from vegetation, in precipitation, and as dust from above and by the weathering of rock from below. Where little rain falls, the parent material weathers slowly and plant production adds little organic detritus to the soil. Thus arid regions typically have shallow soils, with bedrock lying close to the surface (Figure 13–1). Soils may not form at all where weathered material and detritus erode as rapidly as they form. Soil development also stops short on alluvial deposits, where fresh layers of silt deposited each year by flood waters bury weathered material. At the other extreme, weathering proceeds rapidly in parts of the humid tropics, where chemical alteration of parent material may extend to depths of 100 meters

Figure 13–1
Profile of a poorly developed soil in Logan County, Kansas, illustrating shallow soil depth and absence of soil zonation. (Courtesy of the U.S. Soil Conservation Service.)

Figure 13–2
Soil profiles from the central United States illustrating distinct horizons. *Left:* This profile, from eastern Colorado, is weathered to a depth of about 2 feet where the subsoil contacts the original parental material, consisting of loosely aggregated, calcium-rich, wind-deposited sediments (loess). The A_1 and A_2 horizons are not clearly distinguished except that the latter is somewhat lighter-colored. The B horizon contains a dark band of redeposited organic materials that were leached from the uppermost layers of the soil. The C horizon is light-colored and has been leached of much of its calcium. Some of the calcium has been redeposited at the base of the C horizon and at greater depths in the parent material. *Right:* The profile is of a typical prairie soil from Nebraska. Rainfall is sufficient to leach readily soluble ions completely from the soil. Hence there are no B layers of redeposition, as in the drier Colorado soil, and the profile is more homogeneous. The A horizon is weakly subdivided into a darker upper layer and lighter lower layer. The weathered soil lies upon a parent material composed of loess, the wind-blown remnants of glacial activity. The depth scale, in feet, at right, applies to both profiles; the soil horizons, at left, apply only to the left-hand profile. (Courtesy of the U.S. Soil Conservation Service.)

(Bunting 1967, Eyre 1968). Most soils of temperate zones are intermediate in depth, extending to about 1 meter, as a rough average.

Soil horizons reflect the changing influence of soil-forming factors with depth.

Where a recent roadcut or excavation exposes soil in cross-section, one often notices distinct layers, called horizons (Figure 13–2). These horizons have been described with complex and sometimes conflicting terminology by soil classifiers (see, for example, Buol et al. 1973). A generalized, and somewhat simplified, soil profile has four major divisions: O, A, B, and C horizons, with two subdivisions of the A horizon. Arrayed in descending order from the surface of the soil, the horizons and their predominant characteristics are:

O Primarily dead organic litter. Most soil organisms are found in this layer.

A_1 A layer rich in humus, consisting of partly decomposed organic material mixed with mineral soil.

A_2 A region of extensive leaching (or eluviation) of minerals from the soil. Because minerals are dissolved by water (mobilized) in this layer, plant roots are concentrated here.

B A region of little organic material whose chemical composition resembles that of the underlying rock. Clay minerals and oxides of aluminum and iron leached out of the overlying A_2 horizon are sometimes deposited here (illuviation).

C Primarily weakly weathered material, similar to the parent rock. Calcium and magnesium carbonates accumulate in this layer, especially in dry regions, sometimes forming hard, impenetrable layers.

Soil horizons demonstrate the decreasing influence of climate and biotic factors with increasing depth. Critical to soil formation is the movement of mineral elements upward and downward through the soil profile. But before considering these processes in detail, we shall examine the initial weathering of the bedrock and how it influences soil characteristics.

Weathering is the physical and chemical breakdown of rock material near the earth's surface.

Weathering of rock is fostered by the action of both physical and chemical agents. It occurs only where surface water penetrates. Repeated freezing and thawing of water in crevices breaks up rock into smaller pieces and exposes greater surface area to chemical action. Initial chemical alteration of the rock occurs when water dissolves some of the more soluble minerals, especially sodium chloride (NaCl) and calcium sulfate ($CaSO_4$). Other materials, particularly the oxides of titanium, aluminum, iron, and silicon, dissolve less readily.

The weathering of granite exemplifies some basic processes of soil formation. Granite, an igneous rock, forms when less dense molten material deep within the earth rises to the surface, cools, and crystallizes. The minerals making up the grainy texture of granite—feldspar, mica, and quartz—consist of various combinations of oxides of aluminum, iron, silicon, magnesium, calcium, and potassium, along with other, less abundant compounds. The key to weathering is the displacement of certain elements in these minerals—notably calcium, magnesium, sodium, and potassium—by hydrogen ions, followed by the reorganization of the remaining oxides of aluminum, iron, and silicon into new minerals. For example, in the weathering of bedrock in the Hubbard Brook Experimental Forest of New Hampshire, virtually all the calcium is displaced and leached from the developing soil, as are 16 to 24 per cent of the sodium, magnesium, and potassium; in contrast, only 2 per cent of the aluminum and silicon are lost (Likens et al. 1977).

Feldspar, which consists of aluminosilicates of potassium, weathers rapidly owing to the displacement of potassium (K) by hydrogen ions according to

$$KAlSi_3O_8 + 4H_2O + 4H^+ \longrightarrow K^+ + Al^{3+} + 3Si(OH)_4$$

(Bohn et al. 1979). The silicon hydroxide is soluble and subject to leaching under these conditions, but new insoluble materials, particularly clay particles, normally form and most of the aluminum and silicon stay in the soil:

$$2Al^{3+} + 2Si(OH)_4 + H_2O \longrightarrow 5H^+ + Al_2Si_2O_5(OH)_4$$

This new mineral is called kaolinite; along with other clay minerals, it helps retain leachable ions (for example, Ca^{2+}, K^+) in the soil, as we shall see below.

Mica grains consist of aluminosilicates of potassium, magnesium, and iron. As in feldspar, the potassium and magnesium are displaced readily during weathering, and the remaining iron, aluminum, and silicon form various kinds of clay particles. Quartz, a form of silica (SiO_2), is relatively insoluble and therefore remains more or less unaltered in the soil as grains of sand. Changes in chemical composition as granite weathers from rock to soil in different climates show that weathering is most severe under tropical conditions of high temperature and rainfall (Figure 13–3).

An important factor in the initial weathering of parent material, regardless of the chemical nature of the rock, is the presence of hydrogen ions (H^+) in the water that percolates to the bedrock. These are derived from two sources. All precipitation contains dissolved carbon dioxide, which, as we

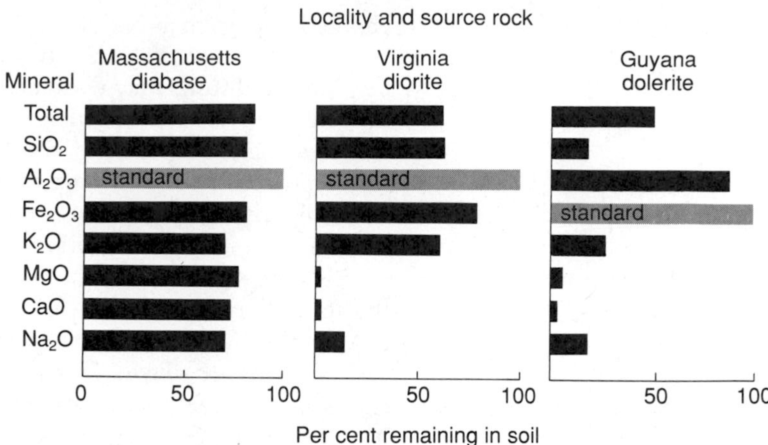

Figure 13–3
Differential removal of minerals from granitic rocks as a result of weathering in Massachusetts, Virginia, and British Guiana. Values are compared to either aluminum or iron oxides (standards = 100 per cent), which are assumed to be the most stable components of the mineral soil. (After Russell 1961.)

have seen, forms carbonic acid, some of which dissociates to H^+ and HCO_3^-. In regions not affected by pollution-caused acidification, concentrations of hydrogen ions in rainwater are between 1 and 2×10^{-5} moles per liter, equivalent to a pH of about 5 (Galloway et al. 1984).

The hydrogen ions in rainwater are supplemented by others generated by the oxidation of organic material in the soil. For example, nitrification of ammonia acidifies the soil by the reaction

$$NH_4^+ + 2O_2 \longrightarrow 2H^+ + H_2O + NO_3^-$$

Similarly, the oxidation of carbohydrate results in the production of CO_2 of the generation of additional H^+ by the dissociation of carbonic acid. In the Hubbard Brook Forest, internal processes account for about 30 per cent of the hydrogen ions needed for weathering of bedrock, but internal sources may be much more important in the tropics, leading to more rapid weathering (Johnson et al. 1975).

The clay and humus content of soil determines its cation exchange capacity.

Plants obtain mineral nutrients from the soil in the form of dissolved ions, which are electrically charged atoms or compounds, whose solubility is determined by their electrostatic attraction to water molecules. Positively

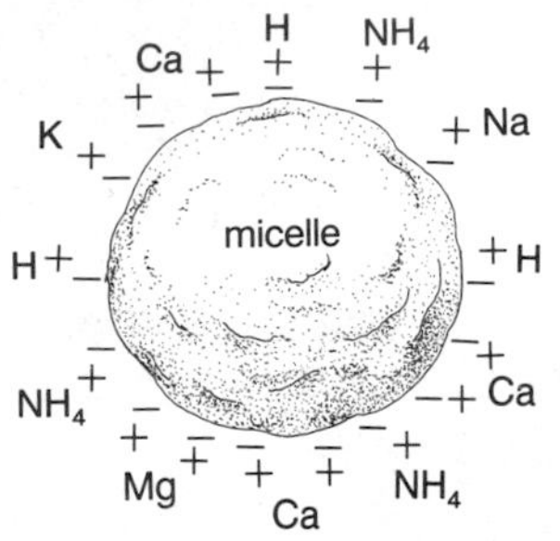

Figure 13–4
Schematic representation of a clay or humus particle (micelle) with hydrogen ions and mineral ions attracted by negative charges at its surface. (After Eyre 1968.)

charged ions, such as Na^+, are called cations; negatively charged ones (Cl^-), anions. Because ions are dissolved in water, those not immediately taken up by plants and fungi may wash out of the soil profile if they are not strongly attracted to stable soil particles. Clay and humus particles, separately or associated in complexes, are large enough to form a stable component of the soil. These particles and complexes, known as micelles, have negative electrical charges at their surfaces that are critical to maintaining the smaller, more mobile ions in the soil (Figure 13–4).

The negative charges of micelles arise by two processes. First, during the formation of clay particles by crystallization, an atom of magnesium might substitute in the crystal lattice for one of aluminum, or one of aluminum for one of silicon. In each of these substitutions, a less positively charged atom replaces a more positively charged atom (that is, Mg^{2+} for Al^{3+}, or Al^{3+} for Si^{4+}). The result is a net negative charge on the clay particle. The second process is the ionization of several types of functional groups; for example, hydroxide (—OH) in clay and the carboxyl (—COOH), amine ($—NH_2$), or phenolic ($—C_6H_4OH$) groups in organic compounds. The dissociation of such groups (for example, $—COOH \rightleftharpoons —COO^- + H^+$) exposes negative charges that can then hold other ions in the soil. Charges resulting from the substitution of metal atoms in clay particles are called permanent charges because they are insensitive to the concentration of hydrogen ions (pH) in the soil. In contrast, the dissociation of functional groups varies inversely with hydrogen ion concentration and therefore is pH dependent.

The bonds between soil particles and such ions as K^+ and Ca^{2+} are relatively weak, so they constantly break and reform. When a potassium ion (K^+) dissociates from a micelle, its place may be taken by any other positive ion (cation) close by. Some ions cling more strongly to micelles than others: in order of decreasing tenacity, hydrogen (H^+), calcium (Ca^{2+}, magnesium (Mg^{2+}), potassium (K^+), and sodium (Na^+). Hydrogen ions thus tend to displace calcium and all other cations in the soil. If cations were not added to or removed from the soil, the relative proportions of the various cations associated with clay-humus particles would assume a steady state. But carbonic acid in rainwater and organic acids produced by the decomposition of organic detritus continually add hydrogen ions to the upper layers of the soil; these readily displace other cations, which are then washed out of the soil and into the groundwater. The influx of hydrogen ions in water percolating through the soil largely causes the mobility of ions in the soil and the differentiation of layers in the soil profile.

Negative ions important to plant nutrition, such as nitrate, phosphate, and sulfate, can be adsorbed onto clay particles by means of ion "bridges." These bridges form under more acid conditions by the association of an additional hydrogen ion with a functional group; for example, $—OH + H^+ \rightarrow —OH_2^+$, which makes possible $—OH_2^+ \cdots NO_3^-$.

It should be clear from the cation-binding properties of clay and humus particles that the potential long-term fertility of soil—its capacity for storing nutrients—depends in large part on its clay content. Furthermore, the realized nutrient storage and the immediate availability of ions to plants depends to a large degree on the acidity of the soil.

Podsolization and laterization reduce soil clay content and long-term soil fertility.

Under mild, temperate conditions of temperature and rainfall, sand grains and clay particles resist weathering and form stable components of the soil skeleton. In acid soils, however, clay particles break down in the A horizon of the soil profile, and their soluble ions are transported downward and deposited in lower horizons. This process, known as podsolization, reduces the ion-exchange capacity and, therefore, the fertility of the upper layers of the soil.

Acid soils occur primarily in cold regions where coniferous trees dominate the forests. The slow decomposition of plant litter produces organic acids. In addition, in regions of podsolization rainfall usually exceeds evaporation. Under these moist conditions, water continually moves downward through the soil profile, so little clay-forming material is transported upward from the weathered bedrock below.

In North America, podsolization advances farthest under spruce and fir forests in New England and the Great Lakes region and across a wide belt of southern and western Canada. A typical profile of a podsolized soil (Figure 13–5) reveals striking bands corresponding to regions of leaching (eluviated horizons) and redeposition (illuviated horizons). The topmost layers of the profile (O and A_1) are dark and rich in organic matter. These are

Figure 13–5
Profile of a podsolized soil in Plymouth County, Massachusetts. The light-colored, eluviated A_2 horizon and the dark-colored, illuviated B_1 horizon immediately below it form distinct bands. Note the general absence of roots in the A_2 horizon compared with the lower B_1 horizon. (Courtesy of the U.S. Soil Conservation Service.)

underlain by a light-colored horizon (A_2), which has been leached of most of its clay content. As a result, A_2 consists mainly of sandy skeletal material that holds neither water nor nutrients well. One usually finds a dark band of deposition immediately under the eluviated A_2 horizon. This is the uppermost layer of the B horizon, where iron and aluminum oxides are redeposited. Other, more mobile minerals may accumulate to some extent in lower parts of the B horizon, which then grades almost imperceptibly into a C horizon and the parent material.

Because of the warm, wet climate, tropical soils tend to be deeply weathered, and the deeper the ultimate source of nutrients in the unaltered bedrock the poorer the surface layers tend to be. Rich soils do develop in many tropical regions, particularly in mountainous areas where erosion continuously removes nutrient-depleted surface layers of the soil. But the soils of other areas, especially those in low-lying regions (the Amazon Basin, for example) and those that develop on parent material deficient in quartz (SiO_2) but rich in iron and magnesium (basalt, for example), contain little clay and therefore do not hold nutrients well. Clay fails to form in any abundance because of a lack of silicon. Instead, iron and aluminum oxides predominate in the soil horizon. This type of weathering is known as laterization; oxides give lateritic soils (latisols) their typical red color.

The development of soil and its steady-state conditions have been studied by measuring cation budgets of large areas.

Natural weathering of bedrock occurs under deep layers of soil where it is inaccessible to direct investigation. The process may be studied indirectly, however, by measuring the net efflux of certain elements from the system. When a soil is in a steady state, as it is in undisturbed areas, the efflux of an element equals the weathering of that element from the parent material plus any influx from other sources, such as precipitation. The basic cations—calcium, potassium, sodium, and magnesium—are good candidates for studying such balances because they are readily leached from the soil profile and leave the system in stream water as dissolved ions, which are easily measured. Elements that form less soluble compounds—silicon, iron, and aluminum—leave the system as solid particles ranging in size from suspended silt to large boulders carried down stream beds during exceptionally heavy runoff. These are more difficult to monitor.

Cation budgets of large areas are most easily studied in a small watershed whose entire drainage can be sampled at a single point along the stream that drains it. Ideally, the watershed should be uniform in bedrock, soil, and vegetation so that measurements reflect processes in only a single type of habitat. In addition, the watershed must lie upon impervious bedrock so that groundwater cannot move into and out of the area except by streams.

Many such watersheds have been studied in detail, particularly in temperate regions (Binkley and Richter 1987). The best-known is the Hub-

Figure 13-6
Rain gauges installed in a ponderosa pine stand in California to intercept precipitation falling through the canopy of the forest and running down the trunks of trees. Analysis of the nutrient content of water collected in sampling programs like this helps determine the overall cation budget of the forest and the specific routes of mineral cycles. (Courtesy of the U.S. Forest Service.)

bard Brook Forest, where G. E. Likens, F. H. Bormann, and their colleagues have investigated patterns of water and nutrient cycles since the early 1960s (Likens et al. 1977). Detailed cation budgets were obtained for several small watersheds in the area by measuring the inputs in rainwater collected at various locations in the watershed (Figure 13-6) and outputs in water leaving the watershed by way of the stream that drains it (Figure 13-7).

At Hubbard Brook, during the period 1963-1974, annual input of calcium in precipitation averaged 2.2 kg ha^{-1} while loss of dissolved Ca^{2+} in stream flow was 13.7 kg ha^{-1}; therefore, the net loss to the system was 11.5 kg ha^{-1}. The investigators also determined that living and dead biomass increased in the watershed during this period, owing to the fact that the forest was recovering from earlier clearing; net assimilation of calcium in vegetation and detritus brought its overall removal from the mineral soil to 21.1 kg ha^{-1} yr^{-1}. Because calcium constitutes about 1.4 per cent of the

Figure 13–7
A stream gauge at the lower end of a watershed at the Coweeta Hydrological Laboratory, North Carolina. The V-shaped notch regulates the flow of water through the weir so that the flow rate is proportional to the water level in the basin. (Courtesy of the U.S. Forest Service.)

weight of the bedrock in the area, its annual loss equaled the weathering of 1500 kg (21.1/0.014) of bedrock per hectare, or approximately 1 millimeter of depth per year. The same calculations based on elements other than calcium gave quite different (typically lower) estimates of weathering. Nevertheless, such studies agree in concluding that weathering in cool, temperate regions proceeds very slowly compared to the annual uptake of soil cations by plants and their release by leaching and decomposition of detritus within the system (Table 13–1).

Table 13–1
Cation budgets for representative temperate forest ecosystems

	Precipitation input	Stream outflow	Net loss	Uptake by vegetation
Calcium	2–8	8–26	3–18	25–201
Potassium	1–8	2–13	1–5	5–99
Magnesium	1–11	3–13	2–4	2–24
Sodium	1–58	6–62	4–21	—

Values are reported as ranges, in kg ha^{-1} yr^{-1}; 1 hectare (ha) equals 10,000 square meters, or 2.47 acres.
Source: Carlisle et al 1966; Likens et al. 1967, Duvigneaud and Denayer-de-Smith 1970.

Most nutrients in terrestrial systems cycle through detritus.

Plants assimilate elements from the soil far more rapidly than they are generated by weathering of the parent material. Most basic cations (Ca^{2+}, Mg^{2+}, K^+, Na^+) do not figure prominently in biochemical transformations and, for the most part, are merely taken up with the water that plants need in such quantity. Excess uptake of these elements is either sequestered or excreted, and their depletion from the soil probably has little direct effect on plant production. Not so with supplies of such important nutrients as nitrogen, phosphorus, and sulfur, which are poorly represented in parent material. Igneous rocks, such as granite and basalt, contain no nitrogen, only 0.3 per cent of phosphate (P_2O_5), and 0.1 per cent of sulfate (SO_3) by mass (Bohn et al. 1979). Hence, weathering adds little of these nutrients to the soil; inputs from precipitation and nitrogen fixation also are small. Plant production therefore depends on the rapid regeneration of nutrients from detritus.

Organic detritus occurs everywhere, most conspicuously in terrestrial systems where the resistance of woody plant parts to herbivory results in the accumulation of abundant plant remains (Vogt et al. 1986, Harmon et al. 1986). Regardless of the habitat, however, this reservoir of nutrients is regenerated by the activities of a wide variety of worms, snails, insects, mites, bacteria, and fungi that consume detritus for food—their primary source of carbon and energy (Birch and Clark 1953, Griffin 1972, Harley 1972, Petersen and Cummins 1974, Anderson and Sedell 1979, Coleman et al. 1983).

Of all the detritus-based communities, the organisms that consume the litter of leaves and branches on the forest floor are probably best known (Witkamp 1966, Minderman 1968, Dickinson and Pugh 1974, Hayes 1979). The breakdown of leaf litter occurs in three ways: (1) leaching of soluble minerals and small organic compounds by water; (2) consumption by large detritus-feeding organisms (millipedes, earthworms, woodlice, and other invertebrates); and (3) further attack by fungi and eventual mineralization of phosphorus, nitrogen, and sulfur by bacteria. Between 10 and 30 per cent of the substances in newly fallen leaves dissolve in cold water; leaching rapidly removes most of these (salts, sugars, amino acids) from the litter (Daubenmire and Prusso 1963), leaving behind complex carbohydrates and other organic compounds. Although large detritus feeders assimilate no more than 30 to 45 per cent of the energy available in leaf litter, and even less from wood, they nonetheless speed the decay of litter because they macerate leaves in their digestive tracts, and the finer particles in their egested wastes expose new surfaces to microbial feeding (Hartenstein 1986).

Leaves of different species of trees decompose at different rates, depending on their composition (Witkamp and Van der Drift 1981, Taylor et al. 1989). For example, in eastern Tennessee, weight loss of leaves during the first year after leaf fall varies from 64 per cent for mulberry, to 39 per cent for oak, 32 per cent for sugar maple, and 21 per cent for beech (Shanks and Olson 1961). The needles of pines and other conifers also decompose

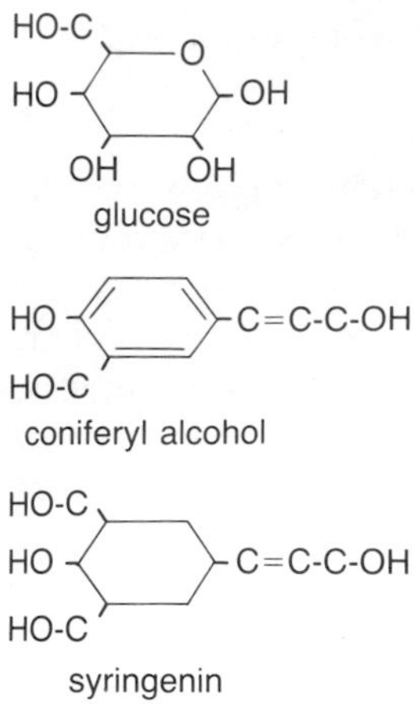

HO-C
HO O OH
OH OH
glucose

HO C=C-C-OH
HO-C
coniferyl alcohol

HO-C
HO C=C-C-OH
HO-C
syringenin

Figure 13–8
Chemical structure of the subunits of cellulose *(top)* and lignin *(center and bottom).*

slowly. Differences between species depend to a large extent upon the lignin content of the leaves. Lignins are a heterogeneous class of phenolic polymers (long chains of phenolic subunits; Figure 13–8). They lend wood many of its structural qualities, and are even more difficult to digest than cellulose (Swain 1979). In fact, only the "white rot" fungi can break down lignin (Grant and Long 1981, Griffin 1981).

The resistance of some types of litter to degradation points up the unique role of fungi in the regeneration of nutrients (Harley 1972). The familiar mushrooms and shelf fungi are merely fruiting structures produced by the mass of the fungal organism deep within the litter or wood (Figure 13–9). Most fungi consist of a network, or mycelium, of hyphae, threadlike elements that can penetrate the woody cells of plant litter that bacteria cannot reach. Like bacteria, fungi secrete enzymes into the substrate itself and absorb the simple sugar and amino-acid breakdown products of this exocellular digestion. Fungi differ from bacteria in being able to digest cellulose (which a few bacteria, protozoa, and snails also can accomplish) and, especially, lignin. Cellulose digestion begins by hydrolyzing polysac-

Figure 13–9
Shelf fungi speed the decomposition of a fallen log. The brackets are fruiting structures produced by the fungal hyphae, together called the mycelium, that grow throughout the interior of the log, slowly digesting its structure. (Courtesy of the U.S. National Park Service.)

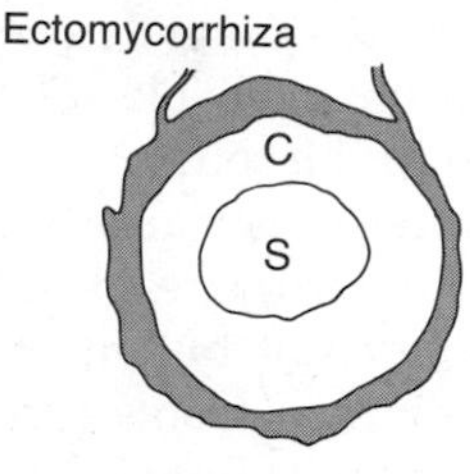

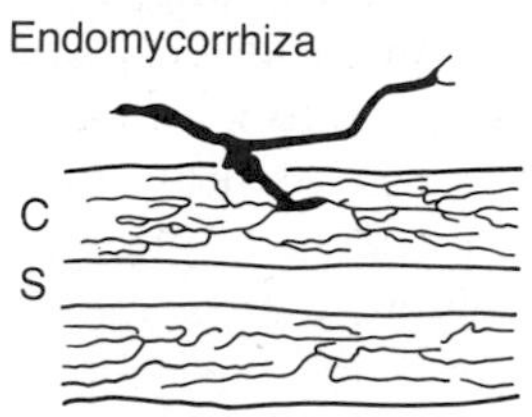

Figure 13–10
Generalized structure of ectomycorrhiza and endomycorrhiza; C = root cortex, S = vascular tissue of stele. (After Grant and Long 1981.)

charides (long chains of sugar subunits) into simple sugars that can be absorbed into the hyphae. The breakdown of lignin by fungi apparently is initiated by an oxidation that cleaves the aromatic (phenolic) ring structure (Kirk et al. 1977).

Mycorrhizae are a symbiotic association of fungi with plant roots.

In addition to their role in decomposing detritus, some kinds of fungi grow on the surface of or inside the roots of many types of plants, especially woody species. This association, called a mycorrhiza (literally, fungus root), enhances the plant's ability to extract mineral nutrients from the soil. Although many forms of the association are recognized, they are classified as endomycorrhizae when the fungus penetrates the root tissue and ectomychorrhizae when it forms a sheath over the root surface (Figure 13–10). Mycorrhizae, whose fundamental importance to production and element cycling is gaining increasing appreciation, have become the subject of an extensive literature. General accounts may be found in Rovira (1965), Wilde (1968), Marks and Kozlowski (1973), Smith (1980), and Harley and Smith (1983).

Mycorrhizae occur everywhere, but their importance is best demonstrated experimentally by growing plants on sterile soil to which spores of mycorrhizae-forming fungi are either added or not added. For example, in an experiment with *Pinus strobus* seedlings, ectomycorrhizae increased by two to three times the uptake of nitrogen, phosphorus, and potassium per unit of root mass and greatly improved the growth of the plant (Bowen 1973). The infections seem to be more successful in soils that are relatively depleted of nutrients (Table 13–2).

Mycorrhizae increase a plant's uptake of minerals by penetrating a greater volume of soil and increasing the total surface area for nutrient assimilation. Mycorrhizae, especially external forms, may also protect the plant root from infection by pathogens by physically excluding them or

Table 13–2
The effect of phosphate fertilizer and inoculation with the mycorrhizal fungus *Endogone macrocarpa* on the growth of tomato *Lycopersicon esculentum*

Phosphate level (mEq per plant)	Infection with *Endogone*	Dry weight of foliage (g)
0.6	yes	0.78
	no	0.14
1.2	yes	0.96
	no	0.86
2.4	yes	1.14
	no	1.01

Source: Harley and Smith 1983.

producing antibiotics (antibacterial toxins). What do the fungi derive from the association? The main advantage appears to be a reliable source of carbon in the form of photosynthates transported to the roots.

The commonest type of endomycorrhiza is the vesicular-arbuscular type (Harley and Smith 1983), so-called because of structures developed within the host tissue. These do not grow freely in the soil, but infect plant roots from spores left behind from dead, previously infected roots. (Presumably spores blow into virgin soils along with other dust.) Fungi that form vesicular-arbuscular mycorrhizae derive their carbon exclusively from the plant root. Mineral nutrients are derived from the soil, into which the fungi send long hyphae. They apparently can use sources of phosphorus not available to plants, such as the highly insoluble "rock phosphorus" $Ca_3(PO_4)_2$ (Murdoch et al. 1967); plants can use only the more soluble forms $CaHPO_4$ and KH_2PO_4. Perhaps the hyphae excrete hydrogen ions or organic acids, or grow in association with phosphate-solubilizing bacteria (Smith 1980).

Ectomycorrhizae are very widespread, especially among trees and shrubs. In addition to the hyphae, which extend out into the soil, the fungus forms a tough sheath around the root that may account for as much as 40 per cent of the weight of the mycorrhizal association. The sheath stores large quantities of soil-derived nutrients and carbon compounds, which may be one of the chief advantages of the mycorrhizae to the plant. Ectomycorrhizae are found predominantly in association with vegetation experiencing seasonal or intermittent growth, where they may serve to extend the productive period. Regardless of how they function, they do account for a major part of energy and nutrient budgets in some habitats. K. A. Vogt et al. (1982) estimated that the carbon assimilated by mycorrhizae in a fir *(Abies)* forest accounted for 15 per cent of the total net primary production. The mycorrhizal fungus is undoubtedly responsible for a much greater share of the mineral nutrient uptake.

Nutrients are regenerated more rapidly in tropical than in temperate forests.

Nutrient cycling in tropical and temperate ecosystems differs because of the effects of climate on weathering and regeneration of detritus. The deep weathering and poor cation exchange capacity of many tropical soils result in relatively low fertility. But rapid regeneration of nutrients from detritus under warm, humid conditions and efficient nutrient retention supports high productivity. Ecologists have concluded that in the "typical" tropical system, most of the nutrients occur in the living biomass and elements are regenerated and assimilated very rapidly. This has important implications for tropical agriculture and conservation (Jordon 1985).

Many rich and fertile tropical soils are well-suited to the removal-type agriculture practiced commonly in temperate regions. Even with clearing of the forests and frequent removal of crops, such soils maintain their fertility.

Figure 13–11
An area of about two hectares on a steep slope in Costa Rica that has been cut and burned for shifting agriculture. One can see small corn plants emerging among the debris in the closeup. This practice exposes soil to erosion and promotes leaching of nutrients and laterization. Such clearings produce crops for two or three years at most; decades of forest regeneration are required to restore the fertility of the soil. (Courtesy of D. H. Janzen.)

But over extensive regions within the tropics, planting such crops as corn on clear-cut land has had disastrous consequences (Figure 13–11). The practice of cutting and burning vegetation releases many mineral nutrients to the soil. But while these may support a year or two of crop growth, they are readily leached without the natural tropical vegetation to assimilate them, and soil fertility declines rapidly. Furthermore, as the exposed soil dries, upward movement of water draws iron and aluminum oxides to the surface where they form a concrete-like substance called laterite; surface runoff over the impenetrable laterite accelerates erosion, further depleting nutrients and choking streams with sediment.

The ecological lesson taught by such experience is that vegetation is critical to the development and maintenance of soil fertility. Even in temperate zones, removal of vegetation points up its important role in retention of soil nutrients (Figure 13–12). Clear-cutting of small watersheds in the Hubbard Brook Forest increased stream flow several times owing to removal of transpiring leaf surfaces; losses of cations increased 3 to 20 times over comparable undisturbed systems (Likens et al. 1977). The nitrogen budgets of the cutover watersheds sustained the most striking change. Plants assimilate available soil nitrogen so rapidly that undisturbed forest gained nitrogen at the rate of 1 to 3 kg ha^{-1} yr^{-1}. In the clear-cut watershed net loss of nitrogen as nitrate soared to 54 kg ha^{-1} yr^{-1}, a value comparable to the

Figure 13–12
Clear-cut watershed at the Coweeta Hydrological Laboratory, North Carolina, employed in studies of evapotranspiration and runoff in forest ecosystems. (Courtesy of the U.S. Forest Service.)

annual assimilation of nitrogen by vegetation, and many times the precipitation input (7 kg ha^{-1} hr^{-1}). The loss of nitrate resulted from nitrification of organic nitrogen at the normal annual rate by soil microorganisms without simultaneous rapid uptake by plants.

Comparative studies of temperate and tropical forests shed further light on their differences. Of the total biomass of vegetation and detritus, litter on the forest floor comprises an average of about 20 per cent in temperate coniferous forests, 5 per cent in temperate hardwood forests, and only 1 to 2 per cent in tropical rain forests (Ovington 1965). The ratio of litter to the biomass of living leaves is between 5 and 10 in temperate forests, but less than 1 in tropical forests. Of the total organic carbon, more than 50 per cent occurs in the soil and litter in northern forests, but less than 25 per cent in tropical rain forests (Kira and Shidei 1967). Clearly, dead organic material decomposes rapidly in the tropics and does not form a substantial nutrient reservoir, as it does in temperate regions.

Fewer data are available on the relative proportions of nutrients in the soil compared to the living vegetation. The distribution of potassium, phosphorus, and nitrogen in a temperate and a tropical forest with similar living biomass are compared in Table 13–3. Although these data may not be typical, two points stand out. First, the accumulation of nutrients in vegetation, on a weight for weight basis, is somewhat greater in the tropical forest. For example, the total dry weight of living vegetation in the Belgian ash-oak forest exceeds that of the tropical deciduous forest in Ghana by 14 per cent, but the accumulation of the three elements per gram of dried vegetation is 32 to 38 per cent lower in the temperate forest. Second, the ratio of each element in soil to its level in biomass is much lower in the tropics; more than 90 per cent of the phosphorus in the tropical system resides in the living biomass.

C. F. Jordan and R. Herrera (1981) recognized the general nutrient poverty of many tropical soils but also distinguished nutrient-rich and nutrient-poor soils within the tropics. The former, which they called eutrophic (literally "well-nourished"), develop in geologically active areas where natural erosion is high and soils are relatively young. With the bedrock closer to the surface, weathering adds nutrients more rapidly and soils retain nutrients more effectively. In the Western Hemisphere, such eutrophic soils occur widely in the Andes Mountains, in Central America, and in the Caribbean. By contrast, nutrient-poor (oligotrophic) soils develop in old, geologically stable areas, particularly on sandy alluvial deposits (as in much of the Amazon Basin), where intense weathering removes clay and reduces nutrient retention.

Table 13–3
Distribution of mineral nutrients in the soil and living biomass of two temperate and one tropical ecosystems

Forest (and locality)	Biomass (tons ha^{-1})	Nutrients (kg ha^{-1})		
		Potassium	Phosphorus	Nitrogen
Ash and oak (Belgium)	380			
Living		624	95	1260
Soil		767	2200	14,000
Soil/living		1.2	23.1	11.1
Oak and beech (Belgium)	156			
Living		342	44	533
Soil		157	900	4500
Soil/living		0.5	20.5	8.4
Tropical deciduous (Ghana)	333			
Living		808	124	1794
Soil (30 cm)		649	13	4587
Soil/living		0.8	0.1	2.0

Source: Ovington 1965, Duvigneaud and Denayer-de-Smet 1970, Greenland and Kowal 1960.

Table 13–4
Standing crops and fluxes of dry biomass and calcium in a nutrient-rich and a nutrient-poor tropical rain forest

		Units	Eutrophic*	Oligotrophic*
Soil	Exchangeable calcium†	kg ha^{-1}	1900	306
Standing crop	Living vegetation	T ha^{-1}	263	298
Production	Living vegetation	g m^{-2} yr^{-1}	1033	1012
Calcium standing crop	Living vegetation	kg ha^{-1}	760	529
Calcium flux	Precipitation	kg ha^{-1} yr^{-1}	21.8	16.0
	Subsurface runoff	kg ha^{-1} yr^{-1}	43.1	13.2
	Net change	kg ha^{-1} yr^{-1}	−21.3	13.2

*Eutrophic: montane tropical rain forest, Puerto Rico; oligotrophic: Amazonian rain forest, Venezuela.
†Soil calcium measured to a depth of 40 cm.
Source: Jordan and Herrera 1981, after various sources.

Comparison of an oligotrophic (Amazon Basin) and a eutrophic forest (Puerto Rico) illustrates that although production and nutrient flux are similar, the distributions of nutrients (calcium, in this case) in the oligotrophic forest shift from the soil to biomass relative to the eutrophic forest (Table 13–4). Moreover, nutrients are held more tightly; that is, loss of calcium from the eutrophic forest in Puerto Rico is equivalent to about half the annual flux through vegetation, whereas the oligotrophic system appears to gain calcium through precipitation input.

Especially in nutrient-poor areas, nutrient retention by vegetation is the key to the productivity of tropical ecosystems. This is accomplished by a very dense mat of roots (and associated fungi) close to the surface (Stark and Jordan 1978) and even extending up the trunks of trees to intercept nutrients washing down from the canopy (Sanford 1987). Data from Africa revealed between 68 and 85 per cent of the root biomass of forests concentrated within the top 25 to 30 centimeters of the soil (Greenland and Kowal 1960). Application of isotopically labeled compounds showed that nutrients regenerated by the leaching and decomposition of detritus are intercepted by the root mat before they can penetrate into the mineral soil and be washed out of the system (Luse 1970; Odum 1970).

In aquatic ecosystems, production depends upon the rapid assimilation of regenerated nutrients within the photic zone.

Because most chemical and biochemical processes involved in the cycling of elements take place in an aqueous medium, the processes themselves do not differ markedly in terrestrial and aquatic systems. What is unique and distinctive about most rivers, lakes, and oceans is the sedimentation of nutrients into bottom deposits from which they are regenerated and returned to zones of productivity very slowly.

The sediments in aquatic systems are comparable to the detritus layer in terrestrial systems. But terrestrial detritus differs from sediments in two important ways. First, the regeneration of nutrients from terrestrial detritus takes place near plant roots. Aquatic plants assimilate nutrients directly from the water column in the uppermost sunlit (photic) zones, often far removed from sediments at the bottom. Second, decomposition of terrestrial detritus is accomplished, for the most part, aerobically, hence relatively rapidly. Aquatic sediments often become anoxic, greatly slowing most biochemical transformations.

The maintenance of high aquatic production depends in part on the proximity of bottom sediments to the photic zone at the surface, or the presence of upwelling currents bringing nutrients regenerated from sediments back to the surface. A map of productivity of the oceans (Figure 13–13) shows that the rate of carbon fixation is greatest in shallow seas, in both the tropics (for example, Coral Sea and waters surrounding Indonesia) and high latitudes (Baltic Sea, Sea of Japan), and in zones of upwelling. The latter occur along the western coasts of Africa and the Americas, where winds blow surface waters away from shore, thus establishing a vertical current to replace them (Boje and Tomczak 1978).

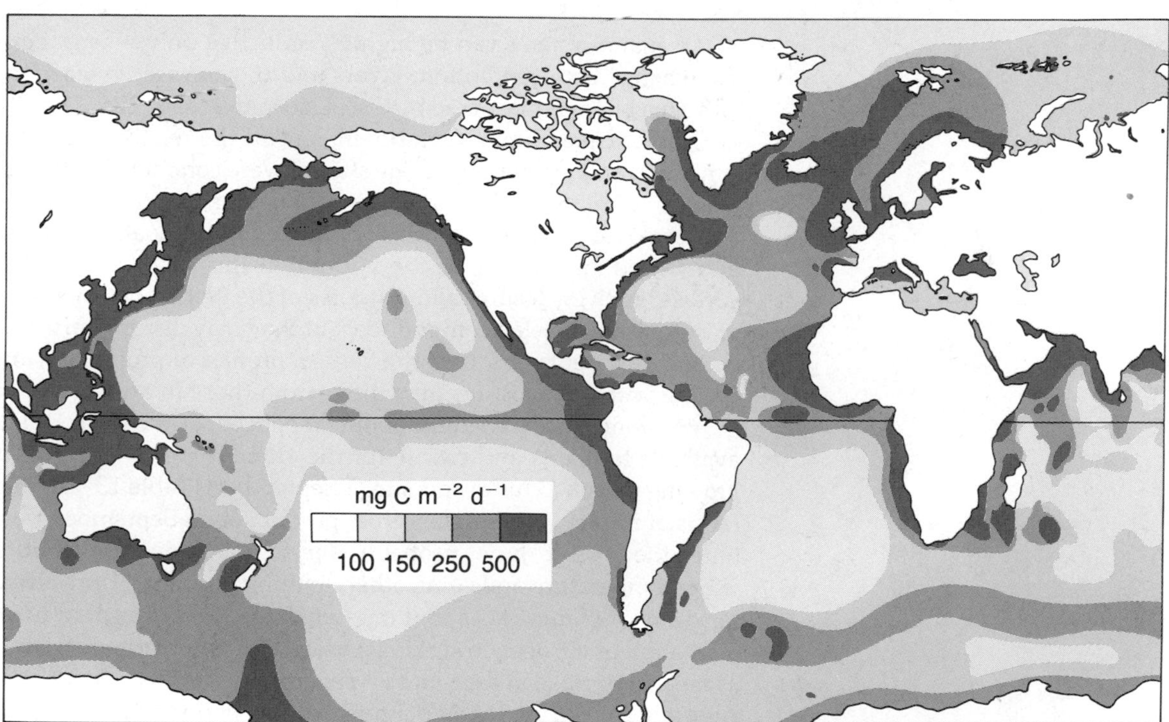

Figure 13–13
Primary production in the world's oceans in milligrams of carbon fixed per square meter per day. Productivity is greatest on the continental shelves and regions of upwelling on the west coasts of Africa and South and Central America. (After Barnes and Mann 1980.)

Regeneration of nutrients in the water column occurs by excretion and microbial decomposition, just as it does in terrestrial systems. For some elements, rates of regeneration are highly correlated with rates of production. For example, in a study of the nitrification of ammonia ($NH_4^+ \rightarrow NO_3^-$) off the western coast of North America, W. G. Harrison (1978) found that phytoplankton assimilated about half the ammonia directly and bacteria nitrified the other half, regardless of the overall flux. Studies in temperate freshwater systems indicate turnover rates of organic nitrogen on the order of 0.2 to 2 days; in temperate marine systems, turnover varies between 1 and 10 days, suggesting lower productivity overall (Harrison 1978).

Plants assimilate regenerated nitrogen very rapidly, especially in nutrient-depleted waters (Axler et al. 1981, Lehman 1980, Sterner 1986). J. J. McCarthy and J. C. Goldman (1979) demonstrated that phytoplankton can take up more nitrogen than they need for growth. This ability to store nitrogen enables algae to take advantage of local "pockets" of nutrients made available by discrete bursts of excretion of decomposition. While we think of water as homogeneous, the concentration of nutrients in organisms makes the distribution of elements such as nitrogen and phosphorus extremely heterogeneous (Lehman and Scavia 1982). Excretion and decomposition produce transient local abundances of nutrients that may be the primary practical source of nutrients. As McCarthy and Goldman point out, once turbulent mixing disperses them, they may become too sparse to be assimilated.

Aquatic systems can be highly productive only where there is strong interchange between bottom layers and the surface, at least on occasion. Nitrogen budgets measured by C. F. Liao and D. R. S. Lean (1978) in the Bay of Quinte, Lake Huron, Ontario, illustrated the relative magnitudes of assimilation and regeneration. The studies were conducted within columns of water in "limnocorrals," which are triangular or circular in cross-section, and enclosed by sheets of plastic suspended by floats at the surface and entrenched in sediment at the bottom, in this case 4 meters beneath the surface. Such enclosures allow studies of the fluxes of elements by addition of isotopically labeled compounds, but they may disrupt normal patterns of mixing of water layers; thus the vertical profiles of production and nutrient cycling within enclosures may differ from those in adjacent open water.

Ammonia accounted for about 90 per cent of the regenerated nitrogen available to plants and nitrate for the other 10 per cent, both early in the growing season (5 June) and late (4 September) (Table 13–5). But although levels of nitrate were similar, gross production in September was nearly 10 times the level in June, probably due to a combination of differences in water temperature and some other, limiting nutrient. Short-term uptake of nitrogen by plants was about one-tenth the level of carbon fixation. Sedimentation of nitrogen in sinking particulate matter amounted to 14 per cent as much as uptake in June and 28 per cent as much as uptake in September. Accordingly, during the September sample period when the total nitrogen in the system (NH_4^+, NO_3^-, and particulate) was 586 g l^{-1} and sedimentation was 36 g l^{-1} d^{-1}, physical removal of nitrogen from the system could deplete the resource quickly. But sedimentation of particulate nitrogen was approximately balanced throughout the season by return of ammonia, be-

Table 13-5
Estimates of release and uptake of nitrogen (μg N 1^{-1} da^{-1}) in limnocorrals in the Bay of Quinte, Ontario

	5 June 1974	4 September 1974
Concentration		
Ammonia (NH_4^+)	120	176
Nitrate (NO_3^-)	84	72
Primary production (μg C 1^{-1} da^{-1})		
Gross	185	1281
Net	−139	807
Uptake		
Ammonia	20	117
Nitrate	4.5	4.5
Nitrogen fixation	1.2	2.7
Total	26	124
Release		
Zooplankton grazing	9.2	27
Sedimentation	3.7	63
Total	13	90

Source: Liao and Lean 1978.

cause total nitrogen in the water column varied little relative to internal cycling.

Thermal stratification impedes vertical mixing in deep aquatic systems.

The vertical mixing of water requires an input of energy to accelerate water masses and keep them moving. Winds supply a part of this energy, causing turbulent mixing of shallow water. Mixing can be prevented in such systems when freshwater floats over more dense saltwater, as in the bay of a large river (Dyer 1973; Taft et al. 1980), or when the sun heats surface water to establish a warmer layer of less density overlying a cooler layer of greater density. Other processes promote vertical mixing. In marine systems, when evaporation exceeds freshwater input, surface layers become more saline, hence denser, and literally fall through the lighter water below. Similarly, when surface layers of water cool during the autumn, the now denser surface water tends to fall through the warmer, hence less dense, layers below.

Vertical mixing of water affects production in two opposing ways. On one hand, mixing can bring nutrient-rich water from the depths to the sunlit surface and thereby promote production. On the other hand, mixing can carry phytoplankton below the zone of photosynthesis and thereby reduce production. Indeed, if vertical mixing extends far below the photic zone, the

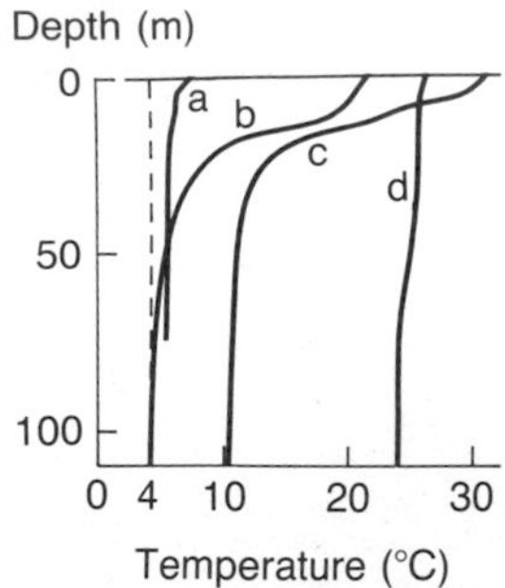

Figure 13–14
Temperature profiles of lakes at the height of summer stratification; a = Flakevatn, Norway; b = Cayuga, New York; c = Ikedako, Japan; d = Edward, Uganda. (From Hutchinson 1957.)

algae of the phytoplankton may not be able to maintain themselves, much less reproduce. Under such conditions, primary production may shut down altogether, resulting in the seeming contradiction of nutrient-rich water without primary production (Sverdrup 1953, Barnes and Mann 1980).

The more typical situation in many temperate-zone freshwater systems is one in which thermal stratification during the summer prevents vertical mixing; then, as sedimentation removes nutrients from the surface layers, production decreases. We may think of the cycle as beginning during the winter, when the temperature of the entire water column of a temperate lake is uniformly cold. There is little or no production owing to low light, particularly when the lake is frozen over. In early spring, the ice melts and the sun rises higher in the sky each day. But strong winds keep the water column thoroughly mixed and production, at least in deep lakes, remains low.

As the surface layers of the lake begin to warm, however, thermal stratification ensues, and a layer of nutrient-rich water becomes "trapped" at the surface, thus creating optimal conditions for high production. The resulting spurt of algal growth is often refered to as the spring bloom of phytoplankton production. When stratification continues into the summer, nutrients are depleted in the surface layers and production decreases as phytoplankton are consumed by larger zooplankton and the excreta and dead remains of these zooplankton sediment out of the water (Lampert et al. 1986). Nutrients may be regenerated in the deeper layers of the lake, but these cannot reach the surface.

Thermal stratification develops only weakly, if at all, at both high and low latitudes (Figure 13–14). In arctic and subarctic regions, the heat input into lakes is not sufficient to counter turbulent mixing and the water column heats up uniformly to the extent that water temperature rises at all. In the tropics, the lack of a pronounced seasonal temperature cycle reduces the sharpness of thermal stratification, although surface layers of water may be warmed by the sun during the day.

In temperate zones, as temperatures begin to decrease in the fall, the surface layers of water become more dense as they cool and therefore begin to fall through the water column, commencing a period of vertical mixing that is accelerated by wind-driven turbulence. At this time, bottom water with regenerated nutrients rises to the surface, where it may support a new rush of phytoplankton growth, refered to as the fall bloom. Further cooling of the lake and the darkening days of the oncoming winter complete the annual cycle.

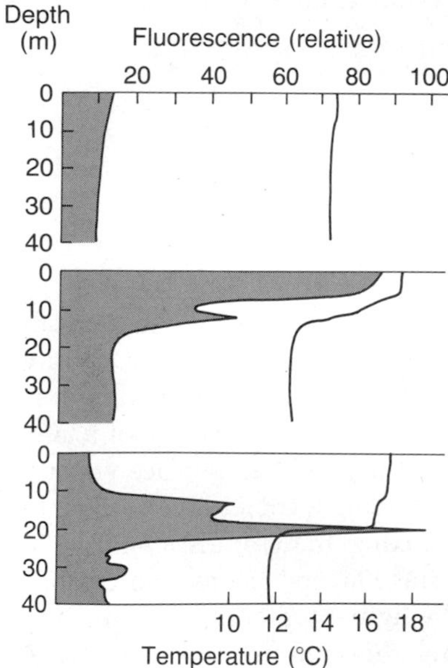

Figure 13–15
Vertical profiles of chlorophyll concentration (flourescence) and temperature in the western English Channel in July 1975: *top*, a well-mixed water mass; *bottom*, a stratified water mass; *center*, the "front" in the region of mixing between the two water masses. (From Barnes and Mann 1980.)

In marine systems, two very different water masses may meet at a "front" where intermixing creates special conditions for high production (Walanski and Hamner 1988). Sometimes at the boundary of a shallow-water system and a deep-water system, mixed (deep) and stratified (shallow) water masses are brought together felicitously. On the mixed side, nutrients may be abundant but phytoplankton do not remain within the photic zone. On the stratified side, nutrients may have been depleted from the surface waters. Where the two meet, some of the nutrient-laden mixed water enters the stratified layer, creating ideal conditions for photosynthesis and nutrient assimilation (Figure 13–15).

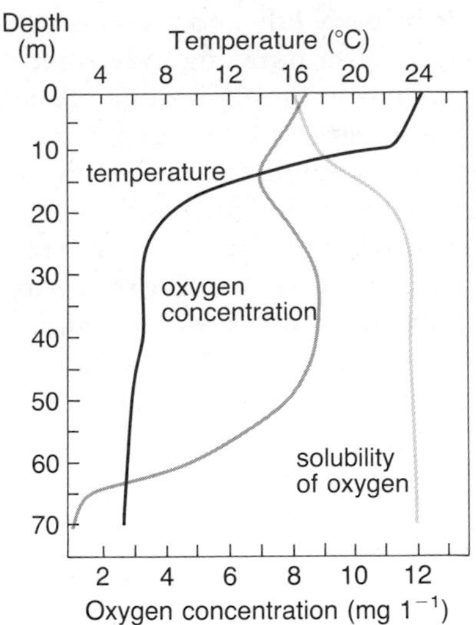

Figure 13–16
Profiles of temperature and oxygen concentration in Green Lake, Wisconsin, illustrating oxygen depletion in the hypolimnion during summer stratification. (From Hutchinson 1957.)

Regeneration of nutrients in the hypolimnion depends upon oxygen concentration.

Lake ecologists call the region above the sharp change (thermocline) in a stratified temperature profile the epilimnion; the zone below the thermocline is the hypolimnion. During periods of stratification, bacterial respiration in the hypolimnion depletes the oxygen supply and the water may become anoxic if stratification is prolonged and there is abundant organic matter to oxidize (Figure 13–16).

In the low redox environment in bottom sediments and waters immediately over them, the shift of such redox elements as iron, manganese, and nitrogen from oxidized to reduced forms greatly affects their solubility. In particular, as ferric iron (Fe^{3+}) is reduced to ferrous iron (Fe^{2+}), insoluble iron-phosphate complexes become solubilized and both elements may move into the water column. This was first demonstrated in C. H. Mortimer's classic (1941, 1942) studies on the exchange of dissolved substances between the mud and water of lakes, and reconfirmed in many more recent studies (for example, Holdren and Armstrong 1980, Riley and Prepas 1984).

Changes in the water chemistry of the hypolimnion of an English lake, Esthwaite Water, during the course of a single season show the effects of anoxic conditions (Figure 13–17). After stratification, the level of oxygen at the deepest level of the lake decreased gradually while dissolved carbon

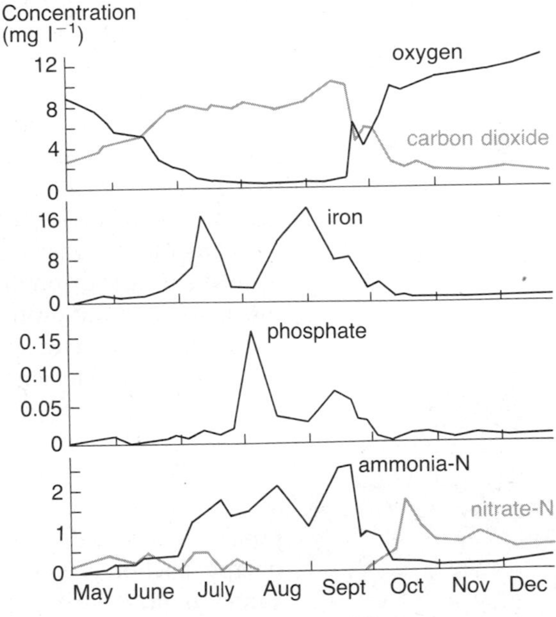

Figure 13–17
Seasonal course of water chemistry of the hypolimnion of Esthwaite Water, England, showing the solubilizing of reduced phosphorus and iron compounds during the period of summer anoxia. (From Hutchinson 1957.)

dioxide increased. The water became anoxic by early July and remained so until the end of stratification and the onset of vertical mixing in late September. During the period of anoxia, levels of ferrous iron, phosphate, and ammonia increased dramatically as they were released by reduction processes in the sediments and at the sediment-water boundary. The return of oxidizing conditions in the fall completely altered the chemistry of the bottom water, initially because of the replacement of bottom with surface water but ultimately because oxidized forms of several redox elements formed insoluble compounds, which precipitated out of the water column. Nitrogen was a conspicuous exception; under oxic conditions nitrifying bacteria convert ammonia to nitrate, which generally remains in solution.

Phosphorus is a major factor in eutrophication.

Phosphorus is often scarce in the well-oxygenated surface waters of lakes, and low levels of phosphorus limit the production of an aquatic system (Dillon and Rigler 1974, Prepas and Trew 1983). In small lakes on the Canadian Shield, productivity increased dramatically in response to the addition of phosphorus, but not nitrogen or carbon (in the form of sucrose) (Figure 13–18). Natural lakes exhibit a wide range of productivity, depending on inputs of nutrients from outside (external loading: rainfall, streams) and the regeneration of nutrients within the lake (internal loading). In shallow lakes lacking a hypolimnion internal loading occurs continuously through resuspension of bottom sediments (Reynoldson and Hamilton 1982). In somewhat deeper lakes, where the thermocline is weakly developed, vertical mixing may occur periodically as a result of occasional strong winds or cooling periods during the summer. Such mixing returns regenerated nutrients to the surface and stimulates production (Larsen et al. 1981). In very deep lakes, bottom waters rarely mix with the surface and production depends almost entirely on external loading.

Aquatic ecologists classify lakes on a continuum ranging from poorly nourished (oligotrophic) to well-nourished (eutrophic) depending on their nutrient status and production (Beeton 1965, National Academy of Sciences 1969). Naturally eutrophic lakes have characteristic temporal patterns of production and nutrient cycling that maintain the system in a steady state. Artificial addition of nutrients in sewage and drainage from fertilized agricultural lands can cause inappropriate nutrient loading and greatly alter natural cycles (Hasler 1947, Likens 1972).

Increased production is not bad in and of itself; indeed, many lakes and ponds are artificially fertilized to increase commercial fish production. But overproduction can lead to imbalance when natural regeneration processes cannot handle the increased demands on cycling. Heavy organic pollution, such as that which results from dumping raw sewage into rivers and lakes, creates what is called biological oxygen demand (BOD) due to oxidative breakdown of the detritus by microorganisms. Inorganic nutrients stimulate the production of organic detritus, adding to the BOD. In its worst manifes-

Figure 13–18
Experimental lake demonstrating the crucial role of phosphorus in eutrophication. The near basin, fertilized with carbon (in sucrose) and nitrogen (in nitrates), exhibited no change in organic production. The far basin, separated from the first by a plastic curtain, received phosphate in addition to carbon and nitrogen and was covered by a heavy bloom of blue-green algae within 2 months. (Courtesy of D. W. Schindler, from Schindler 1974.)

tations, this type of pollution can deplete the surface water of its oxygen, leading to the suffocation of fish and other obligately aerobic organisms.

In spite of what can be devastating effects of external nutrient loading, culturally eutrophied lakes can recover their original condition if inputs are shut off. Eventually, oxic conditions return, phosphorus precipitates out of the water column, and the normal cycles of assimilation and regeneration are restored. A spectacular and convincing demonstration of this occurred after diversion of sewage from Lake Washington, in Seattle, where an advanced and rather ugly case of eutrophication was quickly reversed (Edmondson 1970).

Estuaries and marshes may provide net inputs of energy and nutrients to marine systems.

Shallow estuaries, which are semi-enclosed coastal regions subject to both freshwater inputs from rivers and tidal inputs from the sea, are among the most productive ecosystems on the earth (McClusky 1981). Salt marshes, intertidal areas with emergent vegetation (Figure 13 – 19), combine the most favorable attributes of aquatic and terrestrial systems, resulting in similarly high production (Odum 1988). But in addition to these local attributes of estuaries and coastal marshes, their significance for marine systems extends

Figure 13 – 19
Salt marshes are a common feature of protected bays along most temperate coasts.

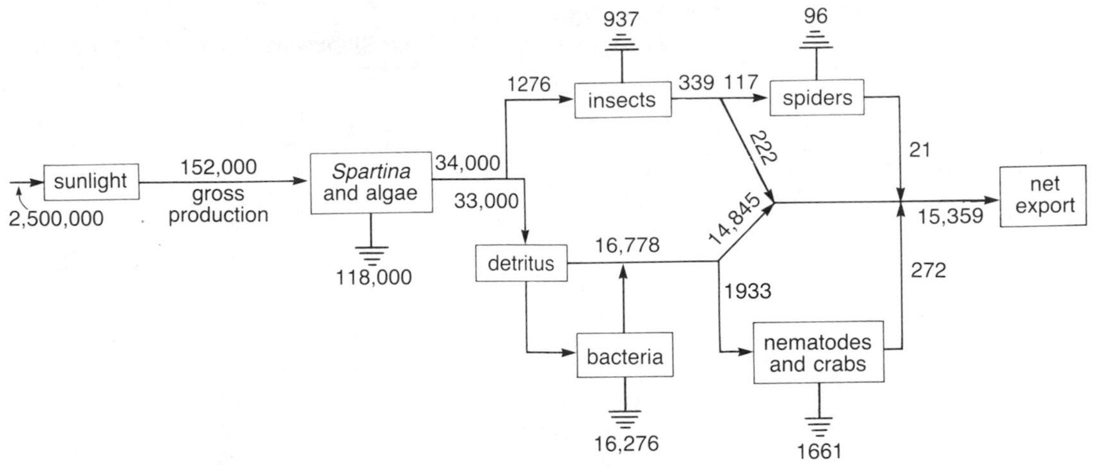

Figure 13–20
Energy-flow diagram for a Georgia salt marsh; units are in kJ m^{-2} yr^{-1}. (From McClusky 1981; after Teal 1962.)

seaward through net export of production. From a Georgia salt marsh, nearly 10 per cent of the gross primary production and almost half the net primary production is exported into surrounding marine systems in the form of organisms, particulate detritus, and dissolved organic material carried out with the tides (Teal 1962; Figure 13–20). Because of their high productivity and, owing to their structural complexity, the hiding places offered prey organisms, coastal marshes and estuaries are important feeding areas for the larvae and immature stages of many fish and invertebrates that later complete their life cycles in the sea.

The high production of coastal systems must be supported by high nutrient levels as well. Since a large fraction of the production of the coastal habitat is carried out to sea, exported nutrients must be replaced by imports to maintain a balance. The nitrogen budget of Great Sippewissett Marsh on Cape Cod reveals inputs to the marsh through precipitation (minor), groundwater flow from surrounding terrestrial systems, and local fixation of atmospheric nitrogen, which are approximately balanced by losses through denitrification, local accumulation of sediments, and net export in tidal water (Table 13–6). The high rate of denitrification, which occurred primarily in the creek bottom draining the salt marsh, underscores the role of anoxic processes in salt marsh metabolism. The rich organic sediments are mostly anoxic just below their surfaces owing to high rates of microbial decomposition of organic matter. As a result, oxidations based upon denitrification ($N^{5+} \rightarrow N^{0}$) and sulfate reduction become important for the regeneration of nutrients (Howarth 1984).

As we have seen, the basic chemical and biochemical transformations of the element cycles are uniquely molded by each type of terrestrial and aquatic system according to physical and chemical conditions created in

Table 13–6
Nitrogen budget for Great Sippewissett Marsh (kg N yr^{-1})

	Input	Output	Net exchange
Precipitation			
Nitrate	110		
Ammonia	70		
Dissolved organic	190		
Particulate	15		
Total	390		390
Groundwater flow			
Nitrate	2920		
Ammonia	460		
Dissolved organic	2710		
Total	6120		6120
Nitrogen fixation	3280		3280
Tidal water exchange			
Nitrate	390	1210	
Nitrite	150	170	
Ammonia	2620	3540	
Dissolved organic	16,300	18,500	
Particulate	6740	8200	
Total	26,200	31,600	−5350
Denitrifation		6940	−6940
Sedimentation		1295	−1295
Totals	35,990	39,835	−3845

Source: Valiela and Teal 1979.

these environments. Paths of elements discussed in this chapter describe patterns of nutrient cycling. In Chapter 14, we shall examine the regulation of the fluxes of nutrients through their cycles; hence, the overall productivity of the ecosystem.

SUMMARY

Nutrient cycles in terrestrial and aquatic systems are based upon similar chemical and biochemical reactions, but differ according to the physical configurations of the habitats.

1. Nutrient regeneration in terrestrial systems takes place in the soil. The characteristics of soil reflect the influences of the bedrock below and the climate and vegetation above. Weathering of bedrock results in the breakdown of some native minerals (feldspar and mica) and their reformation into clay particles, which mix with organic detritus entering the soil from the

surface. These vertically graded processes usually result in distinct soil horizons.

2. The clay and humus content of the soil determines its ability to retain nutrients required by plants. Clay and humus particles (micelles) have negative charges on their surfaces that attract cations (Ca^{2+}, K^+, NH_4^+) directly and anions (PO_4^{3-}, NO_3^-) indirectly under acid conditions. Hydrogen ions tend to displace other cations on micelles and thereby reduce soil fertility.

3. In acid temperate-zone (podsolized) soils and deeply weathered (laterized) tropical soils, clay particles break down and the fertility of the soil is much reduced.

4. The rate of weathering of bedrock and the associated release of new nutrients proceed very slowly compared to the assimilation of nutrients from the soil by plants. Hence, the productivity of vegetation depends on the regeneration of nutrients from plant litter.

5. Nutrients are regenerated from litter by leaching of soluble substances, consumption by large detritus-feeders (millipedes, earthworms), further attack by fungi to break down cellulose and lignin, and eventual mineralization of phosphorus, nitrogen, and sulfur by bacteria.

6. Mycorrhizae are a symbiotic association of certain types of fungi with the roots of plants. The fungi, which may either penetrate the root tissue or form a dense sheath around the root, enhance the uptake of soil nutrients by plants. They do this primarily by enlarging the volume of soil accessed by the roots. In return, they obtain a reliable source of carbon from the plant.

7. In many lowland tropical habitats, soils are deeply weathered and retain nutrients poorly. In such habitats, regeneration and assimilation of nutrients are rapid and most of the nutrients, especially phosphorus, occur in the living vegetation. Such soils clear-cut for agriculture soon lose their fertility as a consequence of nutrients being removed along with the native vegetation.

8. In aquatic systems, the sediments at the bottoms of lakes and oceans are analogous to the detritus of terrestrial systems, but differ in two important respects. First, aquatic sediments are spatially removed from the site of assimilation by aquatic plants. Second, sediments often develop anoxic conditions that retard the regeneration of some nutrients.

9. The productivity of aquatic systems is maintained either by the transport of nutrients from bottom sediments to the surface, as in shallow waters and areas of upwelling, or by the tight recycling of nutrients regenerated within the photic zone.

10. Vertical mixing is inhibited by formation of layers of water of different density (stratification) based upon differences in temperature or salinity. Stratification enhances aquatic production by retaining phytoplankton within the photic zone, but diminishes production to the extent that algae are carried below the level of sufficient light for photosynthesis.

11. The annual cycle of temperate-zone lakes is characterized by a period of temperature stratification during the summer between the spring and fall periods of vertical mixing. Nutrients are brought to the photic zone

near the surface during the periods of mixing. During the summer, nutrients are depleted by the sedimentation of organic material.

12. Nutrients are regenerated in aquatic sediments by bacterial decomposition. Anaerobic conditions that develop beneath the thermocline, due to the consumption of oxygen by bacteria, result in the reduction of iron and magnesium and the solubilizing of their phosphate compounds. Thus phosphorus is released from anoxic sediments and may be transported to the photic zone by vertical mixing.

13. Because phosphorus is readily precipitated by iron under the oxic conditions of surface waters, it frequently is in short supply and limits aquatic production. Artificial inputs of phosphorus through the pollution of water can greatly alter natural patterns of production and nutrient cycling and upset natural balances in aquatic ecosystems. One consequence of increased production, augmented by sewage inputs of organic compounds, is depletion of oxygen from all levels of the lake when bacterial consumption exceeds production of oxygen by photosynthesis and diffusion of oxygen from the atmosphere. Fish and other animals may suffocate under such conditions.

14. Shallow-water communities, particularly estuaries and salt marshes, are extremely productive owing to rapid and local regeneration of nutrients and the external loading of additional nutrients from nearby terrestrial habitats. Studies of nutrient budgets indicate that marshes and estuaries are major exporters of both organic carbon and mineral nutrients to surrounding marine systems, hence an indispensable component of marine production in some areas.

14

Regulation of Ecosystem Function

Primary production indicates the magnitude of energy flux and cycling of elements within ecosystems. As we have seen, the most productive systems are tropical rain forests, coral reefs, and estuaries, where favorable combinations of high temperature, abundant water, and intense sunlight promote rapid photosynthesis and assimilation. But plant growth also depends on the regeneration of nutrients through biological processes, which themselves are sensitive to temperature, moisture, and other conditions. How, then, can we untangle the factors responsible for the regulation of ecosystem function? The correlation of production with physical factors does not provide a cause-effect explanation, because physical factors may exert their influence on production through a variety of pathways.

Discerning how ecosystem function is regulated would be relatively simple if, for example, a single resource necessary for production were used up; whatever controlled the supply of that resource would control ecosystem function, the way an accelerator pedal controls the speed of a car. But under the most favorable circumstances plants assimilate only 1 to 2 per cent of the light energy striking the earth's surface. Not all the water entering the system leaves by transpiration from plant surfaces. Furthermore, plants do not exhaust nutrient elements from the soil.

It is not surprising that ecologists have yet to resolve the control of productivity in most ecosystems. Their data allow broad comparisons of function, but simple correlations between resources, physical conditions, and production show only that external factors are regulators, not how or where they act. A second approach to understanding ecosystem function is experimentation. But experiments are difficult to conduct on the scale of ecosystems. How, for example, could one eliminate nitrifying bacteria from soil without changing related conditions and resources? Where experiments have been performed (for example, by adding water or nutrients to a system), it is difficult to ascertain which step or steps in the cycling of critical elements are affected or whether observed responses mimic natural changes in systems.

A third approach, after comparison and experiment, is to model ecosystem function by studying the behavior of mathematical or electrical analogies of systems in response to alteration of symbolic or electrical equivalents of resources and conditions. This approach requires detailed knowledge of all processes critical to regulating ecosystem function (which could be interpreted as everything that happens in the system!), including validation of each part of the model by field observation and experiment.

The systems modeling approach was a major focus of American activities in the International Biological Program (IBP) during the 1960s and early 1970s (Worthington 1975), particularly in the Tundra and Grassland Biome Projects (Miller et al. 1976; Innis 1975, 1978; Reuss and Innis 1977; Van Dyne and Anway 1976). But the models became so complex as to diminish their general value. Even those developed to mirror specific ecosystems depended upon the sometimes erratic behavior of particular processes. To make a model resemble the complexities of nature, it must to be adjusted with respect to a long list of unverifiable variables. The unfulfilled promise of the IBP systems models soured many ecologists on the value of modeling for general applications, although the approach maintains a strong following and in relation to certain types of problems (see especially Jeffers 1978, Innis and O'Neill 1979, Kitching 1983, Odum 1983).

In this chapter, we shall develop simple systems models with the limited aim of obtaining some insights into the control of ecosystem function. We shall then use these insights to interpret particular observations and experimental results.

Systems models incorporate external forcing functions and internal control feedbacks.

Ecosystems include populations that interact through predation, pollination, seed dispersal, mutualism, and competition, and physical components that exchange material with other physical components through chemical transformations and with biological components through assimilation and regeneration. Furthermore, ecosystems function in the context of factors such as light, temperature, precipitation, bedrock, and atmosphere, which are external in the sense that they exist independently of the development of the ecosystem.

In systems terminology, understanding ecosystem function requires knowledge of how external forcing functions and internal control feedbacks are integrated. External forcing functions simply are material inputs from outside the system and physical conditions of the environment that influence the system's structure and function. Consider the following analogy. The rate at which a bucket fills with water depends on an external forcing function—the flow of water from the tap. If the bucket has a hole in its bottom, the hole, as a property of the system, could be considered an internal control feedback.

Applied to ecosystems, external forcing functions include light, temperature, weathering and precipitation inputs, salinity, and other kindred factors. Internal control feedbacks result from the chemical behavior of elements in the physical part of the system and the responses of organisms to the physical environment and to each other. As we have seen in Chapter 13, ecosystem production depends on the regeneration of nutrients by microorganisms and others; their activities form an internal feedback control in the sense that they determine the availability of nutrients to plants. Because external forcing functions can act on both assimilatory and regenerative phases of ecosystem processes, one cannot understand the control of ecosystem function without a detailed appreciation of the influence of external forcing functions on all parts of the feedback control loops within the system.

There are two issues here. The first is the degree to which variation in ecosystem function is driven by variation in external forcing variables as opposed to unique internal properties of ecosystems — essentially the difference between variation in the flow of water coming from the tap as opposed to variation in the number and size of holes in the bucket. The second issue concerns the particular means by which external forcing variables exert their influence on internal feedback controls.

Primary production is highly correlated with external forcing functions.

Ecologists are reasonably agreed that basic aspects of ecosystem structure and function are determined by external forcing variables. Maps of worldwide terrestrial production have been drafted by using predictive equations, based on a small number of studies, to translate temperature and rainfall into grams of dry matter (for example, Lieth 1973). As early as 1862, Justus Liebig speculated that levels of production could be generalized from local studies to global patterns: "If we think of the surface of the earth as being entirely covered with a green meadow yielding annually 5000 kg/ha, the total CO_2 content of the atmosphere would be used up within 21 – 22 years if the CO_2 were not replaced." Remarkably, Liebig's estimate of world production, 230 – 240 X 10^9 metric tons of CO_2 assimilated annually, is very close to present estimates and better than any other estimate in the intervening century and a quarter.

Liebig's was a lucky guess; continents are not covered uniformly with green meadows. But as more measurements of local production have become available, it has been possible to test the relationship between production and external forcing variables (particularly precipitation and temperature) on a global scale. H. Lieth (1973) summarized annual total dry matter production (g m^{-2}), average temperature, and annual precipitation for 53 localities distributed throughout the world. Estimates of production varied 50-fold, from 70 to over 3500 g m^{-2}; precipitation varied 48-fold, from 94 to 4500 mm annually; and temperature ranged from -14.2 to 27.1°C.

We can use Lieth's tabulation to develop a predictive equation relating production to temperature and precipitation. Because we are interested in factorial rather than absolute differences between sites, we first transform the values for productivity and precipitation to logarithms. We cannot treat temperatures in the same way because some of the values are negative (average temperatures below zero Celcius!), and one cannot take the logarithm of a negative number. So we must leave them untransformed. Now we fit a simple algebraic model to the data:

$$P = a + bR + cR^2 + dT + eT^2 + fRT$$

where P is the logarithm of productivity, R is the logarithm of precipitation, and T is the temperature. If the squared (quadratic) terms were significant, they would indicate nonlinearity in the relationship. A significant R-times-T interaction would indicate any synergism between temperature and precipitation. The results of the analysis provide two types of information: one concerning the contribution of variation in temperature and rainfall to variation in production; the other concerning the amount of variation not explained by the relationship.

Variation in terrestrial production clearly is strongly related to the physical environment. All the terms of the predictive model make significant, unique contributions to variation in production in spite of the fact that temperature and rainfall are, themselves, strongly correlated (Table 14–1). The signs of the coefficients b and c show that productivity increases in direct relation to precipitation over the lower part of the range (b positive) and then levels off (c negative), presumably because some other factor becomes more limiting. When variation in precipitation is factored out statistically, we find that higher temperatures lead to decreased production (d and e negative), confirming the notion that water goes farther in cold climates than in warm climates. But there is also a positive synergism between temperature and rainfall (f positive) that partially balances this effect. That is, hot, humid climates have higher productivity than one would predict from the separate relationships of production to precipitation and temperature.

The analysis shows that external forcing variables are in some way responsible for much of the variation in terrestrial production. But how much? Statistically, their effect is estimated by the coefficient of determination (R^2), which is the proportion of the variation in production that is

Table 14–1
Coefficients of multiple regression equation relating dry matter productivity to temperature and rainfall

Coefficent	Independent variable	Value	Standard error
b	$\log_{10}$(precipitation, mm)	4.51	1.23
c	$[\log_{10}$(precipitation, mm)$]^2$	−0.82	0.26
d	temperature (°C)	−0.11	0.04
e	(temperature, °C)^2	−0.0015	0.0005
f	temperature X $\log_{10}$(precipitation)	0.053	0.019

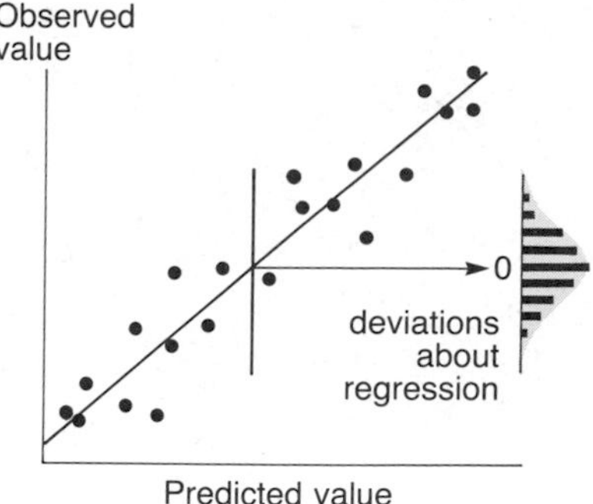

Figure 14–1
An example of a regression of observed values (for example, of production) upon values predicted by several independent variables (for example, temperature and rainfall). The vertical deviationsof values about the regression line are summarized by the histogram at right.

related to variation in temperature and precipitation. For Lieth's set of data, $R^2 = 0.73$; that is, 73 per cent of the variation in productivity is explained by the model in Table 14–1, and 27 per cent is not. Another way to consider the unexplained variation is to quantify the deviations of individual data points from the predictive model (Figure 14–1). In the present example, these have a standard deviation of 0.20 $\log_{10}$ units, which means that two-thirds of the observations fall within 0.2 units of the line and 95 per cent fall within 0.4 units of the line. A range of 0.4 units (0.2 above and 0.2 below) corresponds to a factor of 2.5, a range of 0.8 units to a factor of 6.3.

Temperature and precipitation clearly "control" terrestrial production to a large extent. The unexplained variation could be attributable to measurement error, external factors not included in the model (seasonal variation in temperature and precipitation, solar radiation, and nutrient status of the soil, to name a few), and internal control feedbacks unique to each locality. The latter may derive from particular species, perhaps herbivores, present in one locality but not others. My intuition tells me that additional external factors and inconsistent measurements account for most of the unexplained variation. It is evident, however, that one index to ecosystem function—net aboveground primary productivity—is responsive to external forcing functions.

To what degree are energy and element fluxes concordant in ecosystem function?

Energy is the most generalizable currency of ecosystem function, but availability of energy—the external forcing function, light—may have little to do with differences between systems. In this chapter, our interest focuses upon factors that regulate ecosystem function and cause its diversification. But how does one identify the component or components of the system upon which regulators act?

Peter Vitousek, a plant ecologist at Stanford University, has attempted to find these components by comparing the flux of individual elements with the flux of energy through the system as a whole. One might suppose that the element whose flux shows the strongest correlation with primary production exercises predominant control. In his search, Vitousek (1982) brought together data from studies on forests throughout the world, from subarctic to tropical latitudes.

Because energy and nutrient fluxes are difficult to measure directly, Vitousek substituted indirect but nonetheless reasonable indices. When a forest achieves a steady state, net aboveground primary production is approximately equal to litter production. Similarly, the flux of each element is approximately equal to its amount in the litter. Vitousek's analyses compared the flux of total dry matter in the litter (proportional to carbon, hence to energy content) to the flux of several elements. The data for nitrogen and phosphorus are shown in Figure 14–2 to illustrate Vitousek's analysis and his most important conclusion.

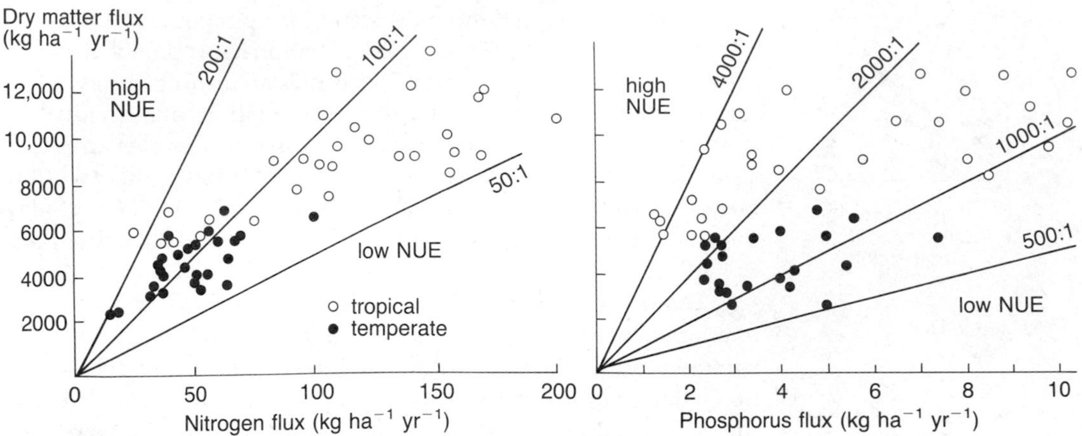

Figure 14–2
The relationship of dry matter flux in litter fall to nitrogen and phosphorus fluxes in temperate and tropical forests; NUE = nutrient use efficiency. (After Vitousek 1982.)

As you can see, production more closely parallels the flux of nitrogen than that of phosphorus (or similarly, calcium.) Vitousek concluded, quite reasonably it would seem, that factors regulating the cycling of nitrogen in forests predominantly controlled primary production.

Vitousek's study also showed that production is not directly proportional to nitrogen flux. Rather, forests with low fluxes have relatively greater production per unit of nitrogen cycled (nutrient use efficiency) than those with high fluxes. Two factors result in increased nutrient use efficiency: trees assimilate more energy per unit of nutrient assimilated; alternatively, trees retain nutrients for reuse by drawing them back into the stem before leaves are dropped. It is not possible to distinguish between these hypotheses from the data in Figure 14–2. Nonetheless nutrient use efficiency clearly is greater where nutrient flux is lower and, for phosphorus, it is greater in the tropics than in temperate latitudes.

Because nutrient use efficiency partly expresses certain adaptations of the plant, and partly may be a fortuitous consequence of nutrient availability, it is difficult to ascribe a regulatory role to an element based on the relationships in Figure 14–2. Just as nutrient use efficiency varied so much for nitrogen, whose flux supposedly limits production, the patterns for phosphorus could also reflect different adaptations of plants under different conditions of phosphorus limitation. Lacking a mechanistic model of ecosystem function, these patterns remain ambiguous.

This ambiguity is emphasized by another analysis based only on tropical forests (Vitousek 1984). In this study, Vitousek first related litter production (dry weight) to temperature and rainfall, much as we did earlier with Lieth's data on primary production. As reported in other studies, such as those of Meentemeyer et al. (1982) and Spain (1984), these external forcing functions account for a large part of the variance in litter production of tropical forests (42 per cent). Vitousek then compared deviations from

Table 14–2
Elemental composition as a percentage of dry weight of leaves of broad-leaved trees

	Number of species	N	P	Ca	K
Living leaves					
Southern Ontario	34	2.08 (0.41)*	0.119 (0.068)	0.721 (0.317)	0.737 (0.269)
Panama	26	1.87 (0.70)	0.104 (0.046)	1.090 (0.53)	1.233 (0.74)
Fresh leaf litter					
Ghana (mixed species)		2.10	0.087	2.02	1.00
Eastern United States	14	0.68	0.140	2.07	0.65

* Number in parentheses are standard deviations among species.
Source: Ricklefs and Matthew 1982, Ricklefs and Stevens unpubl., Nye 1961, McHargue and Roy 1932.

values of production predicted from rainfall and temperature to the nutrient use efficiencies of phosphorus and nitrogen; the residuals were significantly related to the nutrient use efficiency of phosphorus, but not to that of nitrogen. Does this suggest that phosphorus regulates production?

Vitousek's analyses are made more difficult to interpret by the effect of nutrient retention on nutrient use efficiency. The nitrogen and phosphorus contents of living vegetation are not so well known as the corresponding contents of litter. But several studies indicate little difference between tropical and temperate species or between forests within a latitude belt (Table 14–2). The lowest nutrient use efficiencies calculated from litter correspond to the nitrogen and phosphorus contents of living vegetation, suggesting that higher nutrient use efficiencies result from greater retention of nutrients in the vegetation rather than lower nutrient requirements for production.

In order to sort out the factors responsible for the regulation of ecosystem function, we shall develop simple systems models that incorporate expressions for the effects of these factors. Then we shall ask whether variation in these factors produce diagnostic, measurable variation in ecosystem properties. First, however, some basics.

A systems approach can reveal factors regulating ecosystem function.

Figure 14–3
Representation of a single compartment in an ecosystem model, with input (J_0), output (J_1), and compartment size, or pool (X_1).

Each form of a nutrient or energy within a system is a distinct compartment, which we shall designate X_i for the ith form (organic-N, ammonia, nitrate, for example), and each compartment has inflows and outflows, which we shall designate by Js. A schematic diagram of a single compartment (X_1) with one input (J_0) and one output (J_1) is shown in Figure 14–3. If the compartment is the water in our bucket, then J_0 is the flow from the tap and J_1 is what leaks out through the hole in the bottom. The laws of conservation of matter

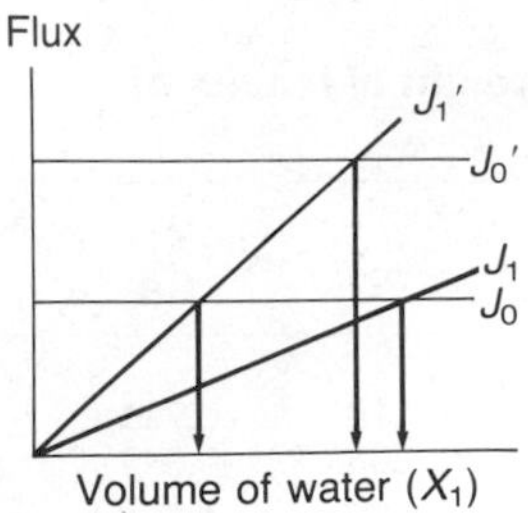

Figure 14–4
The effect of changing input (J_0) and output (J_1) rates on the volume of water (X_1) in a bucket. Because water leaves the bucket through a hole in the bottom, the rate of output is proportional to the water pressure at the bottom, and thus to the depth and volume of water.

and energy dictate that the rate of change in the amount of water in the bucket (dX_1/dt) is equal to the difference between the input and the output —algebraically

$$\frac{dX_1}{dT} = J_0 - J_1$$

The amount of water in the bucket is in a steady state (that is, $dX_1/dt = 0$) when inflow equals outflow ($J_1 = J_0$).

Each flux (J) may be a constant or it may vary depending on the state of other factors, including the value of X and the values of external forcing functions. We may express such variable fluxes in symbolic notation as $J = f(X; S; P)$, where f denotes that J is a function of the values inside the parentheses; S represents the state of the system (number and size of holes in the bucket, leaf area available for photosynthesis, population density of nitrifying bacteria); and P stands for various parameters that are the external forcing functions.

In the bucket example, suppose that J_0 is a constant but that J_1 increases in direct proportion to the volume of water in the bucket (that is, $J_1 = k_1X_1$). (As the bucket fills, the water pressure at the bottom increases the flow of water through the hole. In practice, this will not be a linear function, but portraying it as such will serve our purpose.) The amount of water in the bucket (X_1) reaches a steady state (providing the bucket is large enough) when $J_0 = k_1X_1$. This may be rearranged to show that the steady state amount of water is equal to J_0/k_1. Hence when the value of the forcing function (J_0) increases, water reaches a higher level in the bucket. Reducing the size of the hole at the bottom (the state variable k_1) has the same effect (Figure 14–4).

Single compartment models may be used to describe organisms or populations, but at least two compartments are required to depict the internal cycling of elements within ecosystems, and realistic systems models can become much more complex. Consider the simplest case, two compartments that cycle an element between them (Figure 14–5). Compartment sizes are X_1 and X_2, and fluxes are J_1 and J_2. As we have seen above, the Js represent functions. These may be of zero order, in which case J is a constant ($J = c$); first order (linear), in which case J_1 is a function of X_1; or second order (quadratic), J_1 being a function of both X_1 and X_2.

Little can be said of the functioning of a particular system without knowing the details of flux functions. Simple models do, however, lead to a general understanding of some broad features. The global water cycle, for example, can be thought of as two compartments, vapor (X_1) and liquid (X_2). Knowledge of the physics of state change of water allows us to describe functions for the fluxes in very general terms. Precipitation (J_1) is a first order equation depending only on X_1 and various external forcing functions such as air temperature. The amount of water at the earth's surface (X_2) probably has little direct effect on precipitation. Air has a limited capacity to hold water vapor at a given temperature, and so condensation increases disproportionately as water vapor increases toward this limit.

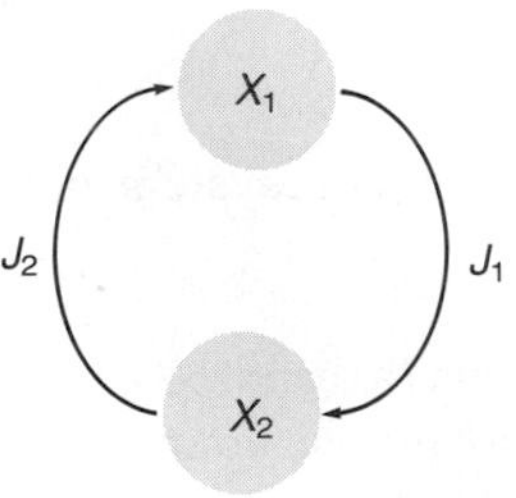

Figure 14–5
Diagram of a closed, two-compartment system.

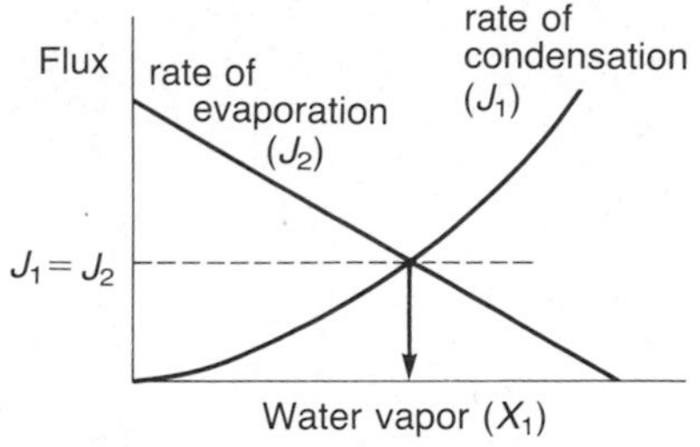

Figure 14–6
Rate of evaporation decreases and rate of condensation increases with increasing water vapor in the atmosphere. This creates an equilibrium level of water vapor in the atmosphere, at which point evaporation equals condensation.

Evaporation (J_2) is a second order equation depending on the surface area of water (some function of X_2), vapor pressure of water in the atmosphere (proportional to X_1), and the external forcing functions, temperature, and insolation (intensity of sunlight). On a global scale, precipitation probably has a smaller effect on the surface area of water than evaporation has on atmospheric water vapor, so it might be possible to ignore X_2 in the function J_2. Thus we may portray the general features of the hydrological cycle on a graph relating J_1 and J_2 to X_1 (Figure 14–6). The model shows that the system assumes a steady state with $J_1 = J_2$ at $\hat{X}_1$ and that changes in the forcing functions affecting J_1 or J_2 (change in temperature, for example) adjust the equilibrium point for the entire system. The model can't be used to predict the local weather.

A. J. Lotka developed the first model of ecosystem function.

In his book *The Elements of Physical Biology* (1925), Lotka investigated the behavior of biological systems using the insights of thermodynamics and the tools of mathematical modeling borrowed from the study of chemical equilibria and other physical phenomena. In the space of a few pages, he outlined the application of systems modeling to the interpretation of nutrient cycling in ecosystems. Lotka treated the specific case of three compartments X_1, X_2, X_3 through which some element or other material cycles with fluxes J_1, J_2, J_3 (Figure 14–7).

Change in any compartment—for example, dX_1/dt—equals the difference between the fluxes into and out of the compartment (J_3 and J_1 in the case of compartment 1). To illustrate the behavior of such a system, Lotka described each flux as a first order term $f_i(X_i)$, arbitrarily having the form g_iX_i; g_i is the rate at which material in compartment i is transferred to the next compartment. Thus $dX_1/dt = g_3X_3 - g_1X_1$. When the system is in a steady state, all the fluxes are equal; hence, $J_1 = J_2 = J_3$, or $g_1X_1 = g_2X_2 = g_3X_3$. Because $X_i = J_i/g_i$ and all the Js are equal, the sizes of the compartments are in the relative proportions.

$$X_1{:}X_2{:}X_3 = \frac{1}{g_1}{:}\frac{1}{g_2}{:}\frac{1}{g_3}$$

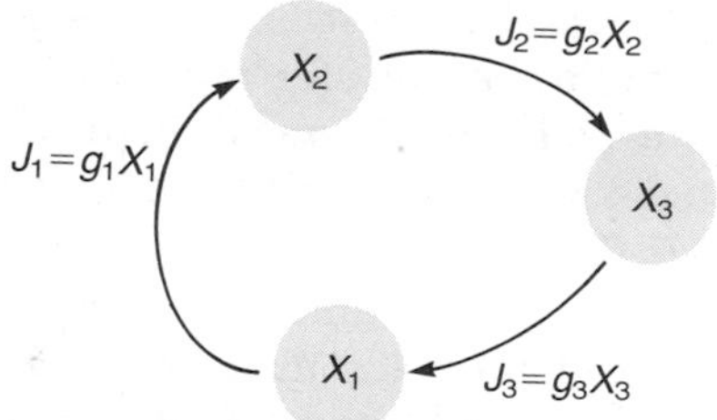

Figure 14–7
A diagram of Lotka's system of three compartments in which the flux between one compartment and the next is a function only of the first compartment.

Now, if all the compartments of the system contain a total M ($= X_1 + X_2 + X_3$) of the cycling material, the amount in any one compartment is the proportion in that compartment (P_i) times the total (that is, P_iM). The proportion depends strictly on the functions g_i because

$$P_i = \frac{X_i}{X_1 + X_2 + X_3} = \frac{1/g_i}{1/g_1 + 1/g_2 + 1/g_3}$$

A little algebraic rearrangement provides the following expressions for compartment size and flux:

$$X_1 = M\left(\frac{g_2 g_3}{g_1 + g_2 + g_3}\right)$$

and

$$J = g_1 X_1 = M\left(\frac{g_1 g_2 g_3}{g_1 + g_2 + g_3}\right)$$

These equations tell us that the structure (Xs) and function (*J*s) of the system are defined by the transfer functions (*g*s), which incorporate the external forcing variables and internal feedback controls.

The derivative of the equation for J with respect to g_i tells us how sensitive J is to changes in that expression. That derivative, which has the same form for all the *g*s in the model above, is

$$\frac{dJ}{dg_1} = M\left[\frac{g_2 g_3 (g_2 + g_3)}{(g_1 + g_2 + g_3)^2}\right]$$

Can we determine whether changes in rate (g) of one transformation influences overall cycling (J) more than change in another? Because g_1 appears only in the denominator of the expression dJ/dg_1, the larger the value of g_1 relative to g_2 and g_3, the smaller the sensitivity of J to g_1. Alternatively, as g_1 becomes very small, the sensitivity of J to changes in g_1 approaches a maximum of $Mg_2g_3/(g_2 + g_3)$. As Lotka pointed out, the relative sensitivities of J to g_1 and g_2, for example, are determined by the ratio

$$\frac{dJ/dg_1}{dJ/dg_2} = \frac{g_2 g_3 (g_2 + g_3)}{g_1 g_3 (g_1 + g_3)}$$

Therefore, if $g_1 < g_2$, then $dJ/dg_1 > dJ/dg_2$.

This model tells us that differences in structure and function between systems likely derive from differences in the lowest transfer rates. Low transfer rates place bottlenecks in the path of material flow through a system leading to the accumulation of material in the immediately preceding compartment. Therefore, the underlying cause of variation in flux through a system should be apparent in shifts of materials among compartments in the system. For example, if an increase in J is accompanied by a shift of material from compartment 1 to compartments 2 and 3, we may infer that an increase in the function g_1 was responsible. Distinguishing the roles of state variables, internal control feedbacks, and external forcing functions in this change would require additional study of the system, but research efforts are greatly focused by heeding the insights of the systems model.

Relative transfer rates can be estimated from compartment sizes.

The cycling of nitrogen within a water column by and large follows the path (ammonia → nitrate) → particulate -N → dissolved organic nitrogen (DON) → (ammonia → nitrate) (algae may assimilate either ammonia or nitrate; nitrite is converted so readily to nitrate that it can be ignored here). In the study of Liao and Lean (1978) on nitrogen transformations in the water column of the Bay of Quinte, Ontario, the sizes of some compartments changed dramatically with season (Figure 14–8). Between winter and summer, particulate and dissolved organic nitrogen increased while nitrate decreased. This shift implies that the primary difference between the cycling of nitrogen in summer and winter is the rate of uptake of nitrate by phytoplankton. A simple first order systems model will illustrate how we can elaborate this idea.

When the system in Figure 14–8 has achieved a steady state, fluxes into and out of each compartment must be balanced. For example, $J_2 = J_3$, $J_3 = J_4 + J_5$, and so on. Hence, under steady state conditions,

$$g_2X_2 = g_3X_3 = (g_4 + g_5)X_4 = (g_1X_1 + g_4X_4)$$

From the last two quantities, we can show algebraically that $X_1/X_4 = g_5/g_1$, and by making the appropriate substitutions, we obtain the relationship

$$X_1:X_2:X_3:X_4 = \frac{g_5}{g_1(g_4 + g_5)}:\frac{1}{g_2}:\frac{1}{g_3}:\frac{1}{(g_4 + g_5)}$$

Setting each X_i equal to its proportion of the total nitrogen allows us to estimate relative values for the g_is during the winter and summer (Table 14–3). These indicate quite dramatically that the ratio g_1/g_5 is an order of magnitude lower during the winter than it is during the summer. The sum

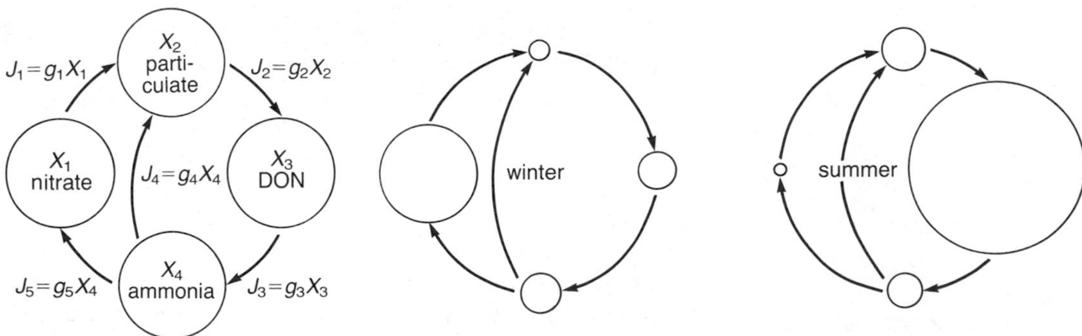

Figure 14–8
A first-order systems model of nitrogen cycling in the Bay of Quinte, Ontario, showing the changes in compartment sizes between summer and winter.

Table 14–3
Transfer rates (g_i) for nitrogen in the Bay of Quinte, Ontario

		Winter	Summer
g_2	Production of dissolved organic nitrogen (DON) by grazers, excretion, and leakage	8.0	4.1
g_3	Ammonification	4.4	2.0
$g_4 + g_5$	Assimilation of ammonium plus nitrification	4.2	4.7
$\frac{g_1}{g_5}(g_4 + g_5)$		2.4	24
$\frac{g_1}{g_5}$	Ratio of nitrate assimilation to nitrification	0.6	5.1

Source: From data in Liao and Lean 1978.

$g_4 + g_5$ differs little between the seasons, relative to g_2 and g_3, so it is evidently the value of g_1 that decreases so much between summer and winter (g_1 describes the rate of assimilation of nitrate-N by plants). During the winter it is clearly much reduced compared to nitrification (g_5) and ammonification (g_3), suggesting that some factor, such as light, might limit production (hence assimilation).

The preceding model may oversimplify or even misrepresent nitrogen transformations in the water column. It was meant to illustrate an approach. Most systems models are much more complicated, often including dozens of compartments and equations of much higher order than one. In our simple model, for example, fluxes J_1 and J_4 almost certainly should have been represented by second order equations involving X_2 (the population of the phytoplankton which accomplish the assimilation; none would be assimilated in the absence of plants). More realistic equations could be developed for particular systems, but simplified models, such as the one that follows, may also serve to direct inquiry into more general comparisons of function between ecosystems.

A model for terrestrial ecosystems incorporates compartments for soil, plant biomass, and detritus.

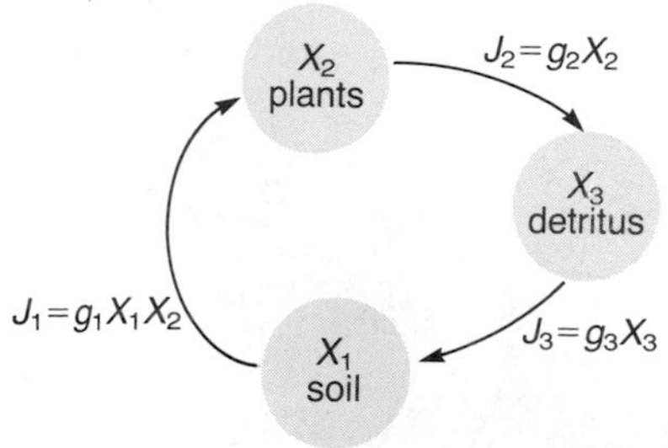

Figure 14–9
A second order systems model of nitrogen cycling in a forest ecosystem. Notice that the assimilation of nitrogen by plants (J_1) is a function both of the availability of nitrogen in the soil and the amount of living plant biomass.

In broad comparisons among terrestrial ecosystems, measurements of net primary productivity (g dry matter m^{-2} yr^{-1}) vary almost 30-fold between desert shrub (70) and tropical rain forest (2000). This variation clearly is related to climate, yet ecologists have not determined where external forcing variables exert their influence within the system. A systems approach can provide clues to fruitful avenues of study in this case.

Let us consider the cycling of nitrogen again. We shall represent an ecosystem as three compartments: mineral soil (X_1), living plant biomass (X_2), and organic detritus (X_3), with fluxes J_1, J_2, and J_3 (Figure 14–9). (Animals are downgraded in status to that of trivial hangers-on; the activities of microorganisms are implicit in the flux J_3). J_2 (the annual dropping of leaves and other detritus) and J_3 (the mineralization of detritus by microorganisms) can be considered as first order processes. Assimilation must be modeled as a second order process because its rate depends both on nutrient availability and plant abundance. Under steady-state conditions, therefore, we obtain the relationships

$$g_1X_1X_2 = g_2X_2 = g_3X_3$$

and

$$X_1X_2{:}X_2{:}X_3 = \frac{1}{g_1}{:}\frac{1}{g_2}{:}\frac{1}{g_3}$$

This simple model offers some surprises. For example, the level of inorganic nitrogen in the soil, mostly nitrate (X_1), is equal to the ratio of g_2 (rate of detritus production) to g_1 (rate of nitrogen assimilation), and is independent of g_3 (rate of microbial regeneration of inorganic nitrogen). This is not to say that nitrogen flux is independent of g_3. The equation expressing flux as a function of the rates g_i is a bit complex, but with some algebra it can be shown to be

$$J = \left[\frac{g_2g_3(g_1 - g_2)}{g_1(g_2 + g_3)}\right]M$$

Because fluxes of elements are difficult to measure directly, we have to estimate relative values of g_i from the sizes of compartments. Differences in the relative rates of transfers between compartments in different systems can, however, provide insights into points of feedback control.

D. J. Ovington (1962) tabulated estimates of compartment sizes of various elements in forest systems. For illustration, we shall compare the cycles of nitrogen and phosphorus in a 47-year-old Scots pine (*Pinus sylvestris*) plantation in England and a 50-year-old mixed tropical forest in Ghana. Compartment sizes and relative values of g are shown in Figure 14–10. The differences are striking. In England, for both nitrogen and phosphorus, values of g_1, g_2, and g_3 are of the same magnitude. In the nitrogen cycle in Ghana, g_1 and g_2 are similar, but g_3 is almost two orders of magnitude greater. Thus the regeneration of mineral nutrients apparently proceeds much more rapidly in the tropics than in temperate areas, which is confirmed by direct measurement of litter decomposition (Olson 1963). Relative transfer rates of phosphorus behave similarly, but reveal another difference between the tropical and temperate forests. In Ghana, the assimilation coefficient (g_1) is relatively much greater than the rate at which vegetation gives up nutrients (g_2), compared to the more nearly equal values in England. These values of g imply that, compared with *Pinus sylvestris*, tropical forest trees assimilate phosphorus more efficiently and hold onto it more tightly (for example, by withdrawing phosphorus into the stem before shedding leaves).

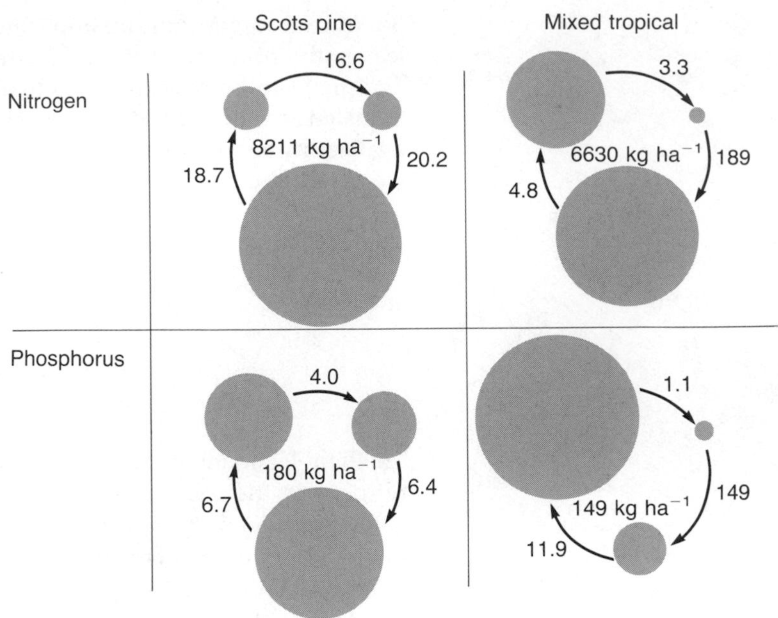

Figure 14–10
Compartment sizes and values of g_i for nitrogen and phosphorus estimated from compartment sizes in a Scots pine plantation in England and a tropical forest in Ghana. (After data in Ovington 1962, and Greenland and Kowal 1960.)

The values estimated for gs in Figure 14–10 are adjusted so that fluxes (J) are equal to 1.0 when compartments are expressed as fractions of the total. Therefore, although the gs are internally consistent, they can be compared between localities only when one of the fluxes is known. The flux in either nitrogen or phosphorus in Ghana is probably no more than double that in England and the total amounts of the two elements are similar in the two areas. Therefore, for nitrogen, both g_1 and g_2 are probably lower in the tropical locality than the temperate locality. For phosphorus, g_1 is probably much higher and g_2 somewhat lower. Unquestionably, however, much of the difference between temperate and tropical localities in overall flux is due to the tremendous increase in rate of decomposition of litter.

The regulation of production in temperate and tropical forests has yet to be resolved.

Our understanding of ecosystem processes should be judged by how convincingly we can explain variation in ecosystem structure and function. Attempts to do so for forest production on a global scale suggest that we are

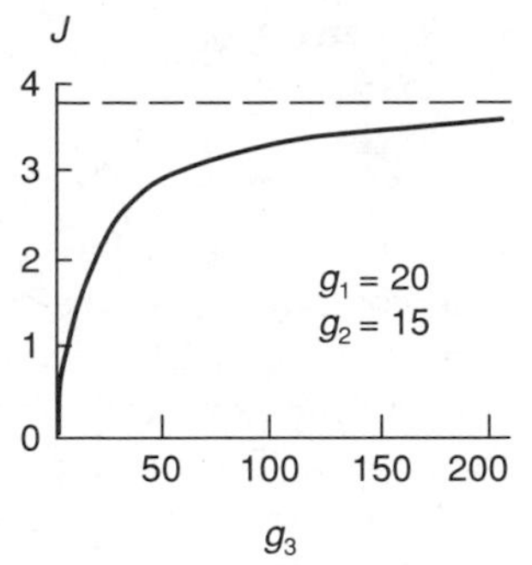

Figure 14–11
Relationship between flux and the rate constant for decomposition of detritus (g_3) in a second order systems model of a forest. Flux is relatively insensitive to variation in g_3 at high values relative to g_1 and g_2.

far from understanding one of the most intensively studied systems on the earth.

Long-term fertilization and irrigation of forests demonstrate that both biomass (X_2) and production per unit of biomass ($g_1 X_1$) can be increased by adding nitrogen, phosphorus, or water (X_1) (Albrektson et al. 1977, Waring 1983). Because different treatments in such experiments involve the same species of trees growing on the same soils, g_1 probably does not differ between treatments, and variation in production and biomass results from increased inputs to one or more compartments (X_1) that limit production. Because nutrient addition experiments involve increases in X_1, they can reveal little about differences between forests unless precipitation or weathering inputs contribute importantly to variation. More probably, differences in production result from the effects of external forcing functions or internal control feedbacks on transfer functions (g_i), which cannot be manipulated easily.

Measurements of rates of litter decomposition (Olson 1963) suggest that the effect of external forcing variables on g_3 can explain part of the variation in production between temperate and tropical forests. This effect may be seen in the forest model by varying transfer rates. For example, when one sets $g_1 = 20$ and $g_2 = 15$, production (J) increases rapidly as g_3 increases up to about 20 and then increases much more slowly with further increase in g_3 (Figure 14–11). Production doubles between $g_3 = 10$ and $g_3 = 60$, at which point J is 80 per cent of its maximum value, given the values defined for g_1 and g_2.

If, in the tropics, g_3 is uniformly high, production would be relatively insensitive to variation in g_3. How then can we account for variation in production among tropical forests? Vitousek showed that almost half the variation in production is associated with variation in temperature and rainfall. But where do these forcing functions act? When g_3 is very large compared to g_2, which is probably generally true in the lowland tropics, the equation for flux may be simplified to

$$J = \frac{g_2}{g_1}(g_1 - g_2)$$

because $g_3/(g_2 + g_3)$ is approximately 1. With g_3 out of the picture, J varies with g_1 and g_2 only, according to the relationships $dJ/dg_1 = (g_2/g_1)^2$ and $dJ/dg_2 = (g_1 - 2g_2)/g_1$. Note that g_1 depends on physical conditions in the soil, the adaptations of plants for nutrient uptake and carbon fixation, and interactions between roots, bacteria, and fungi; g_2 depends on the life expectancy of plant parts and mechanisms of nutrient retention. Moreover, adaptations of trees that affect g_1 and g_2 may be functionally interrelated in ways that constrain their evolutionary adjustment.

All these factors are beyond our current understanding. Systems models, even the simple ones developed in this chapter, can suggest control points in ecosystem processes, but particular feedback controls can be understood only by additional observation and experiment. And, of course, natural systems are potentially much more complex than such simple models suggest.

Information and feedback controls play a key role in systems analysis.

The models described above represent the compartments and fluxes of materials or energy within the system. Fluxes and compartments are both regulated by transfer functions (the gs in previous equations). The transfer functions depend in part on the conditions of the environment and in part upon the state of other compartments within the system. The terms of the transfer functions (for example, $g_3 = g$ [temperature, moisture, pH, litter chemistry, population sizes of microbes and soil invertebrates, fungus-bacteria interactions, and so on]) are sometimes thought of as the flow of information through the system (Margalef 1958; Kitching 1983). In schematic systems models, information and material fluxes are sometimes portrayed differently, as in the Forrester diagram (Forrester 1961) shown in Figure 14–12.

Although "information" is given almost mystical qualities by some systems ecologists, it in fact couples the many material and energy flow circuits within the same system. Consider, for example, the flow of energy and the cycling of nitrogen. These can be represented as separate compartment diagrams, yet they are intimately connected by processes of assimilation and dissimilation. The transfer function governing carbon (energy)

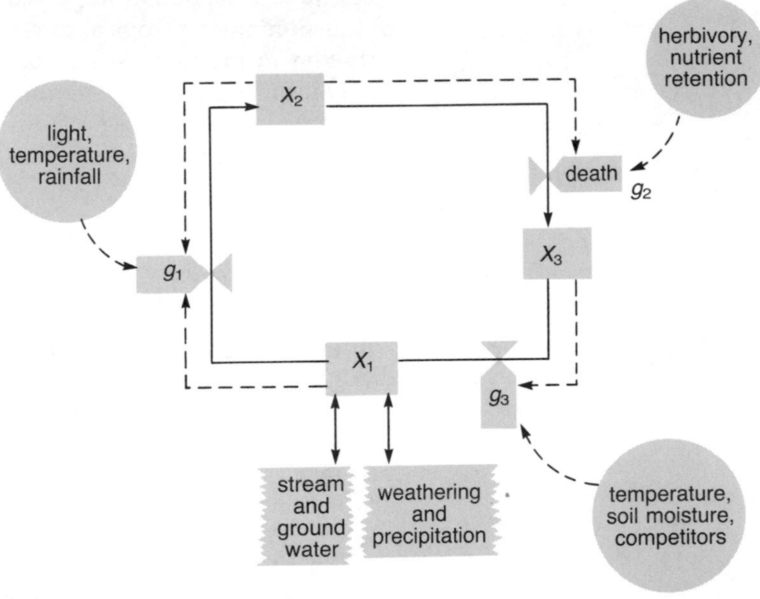

Figure 14–12
A systems model distinguishing between material fluxes (solid lines) and "information" fluxes (dashed lines). (From Forrester 1961.)

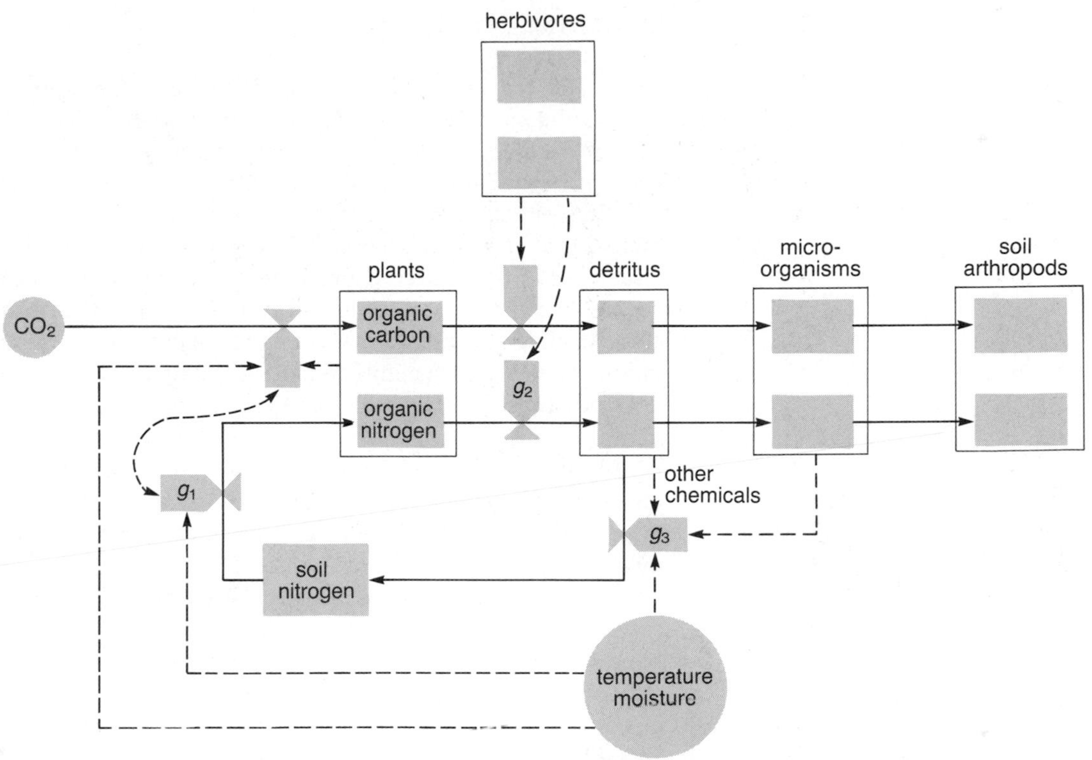

Figure 14–13
A model illustrating the parallel and coupled fluxes of nitrogen and carbon (energy) within the system.

fixation must be related to the transfer function governing nitrogen assimilation because both are required by plants in an approximately fixed ratio (Figure 14–13). What some systems ecologists call information flow is equivalent to the functional integration of the processes of organisms (carbon and nitrogen are assimilated together), the relationship of these processes to conditions in the environment (production increases with soil moisture), and the interactions between organisms (trees compete for soil nutrients; caterpillars eat leaves).

Energy and materials flow through the ecosystem in large part because of the activities of organisms. Understanding the regulation of ecosystem function therefore depends in large part on how organisms respond to physical conditions and how they interact with others of the same and different species. These interactions form the basis for the study of population processes, which are the subject of Part IV. Although population and ecosystem functions are different expressions of many common processes, the studies of ecosystem and population processes have yet to be fully integrated by ecologists.

SUMMARY

1. Ecologists use three methods to study the regulation of ecosystem function: comparison, experiment, and modeling. The results of comparisons and experiments are most interpretable when compared to the predictions of models based on mechanisms of ecosystem function.

2. Systems models incorporate external forcing functions, which are factors that affect the system from outside, and internal control feedbacks, which result from interactions among the components of a system.

3. Examination of terrestrial systems shows that primary production is closely related to the external forcing functions of temperature and precipitation. Hence, ecosystem function is sensitive to its physical context and does not appear to depend upon the presence or absence of unique components.

4. Peter Vitousek's attempt to understand the regulation of terrestrial production by the correlation between production and element cycling is discussed in detail. The results are difficult to interpret because no particular model of ecosystem function is tested.

5. Systems models based upon a particular element or upon energy consist of compartments (*X*s) and fluxes (*J*s) between them. When compartments are in steady state, input equals output. For a cycling element in steady state, fluxes through each segment of the cycle are equal. Each flux out of a compartment is equal to a transfer rate (g) times the compartment size. Each transfer rate may be a function of the size of one or more compartments, other properties of the system, and factors external to the system.

6. A. J. Lotka was the first to apply systems models to ecological systems. He showed that flux of a cycling element was most sensitive to variation in the smallest transfer rate.

7. When a system is in steady state, relative transfer rates can be estimated from compartment sizes. For a simple model of nitrogen cycling within the water column of a lake, seasonal changes in nitrogen compartments indicated that the rate of assimilation of nitrate-N decreased dramatically during the winter.

8. A three-compartment model (soil-plant-detritus) was developed for forest ecosystems. Comparisons of compartment sizes in temperate and tropical forests indicate that nitrogen and phosphorus are regenerated from litter much more rapidly in the tropics and that tropical trees withdraw phosphorus from leaves about to be dropped to a greater extent than do temperate trees. Furthermore, within the tropics, production is probably insensitive to variation in litter decomposition rate, which is uniformly high. Such conclusions are only tentative, however, because of the simplicity of the models on which they are based.

9. Compartment models for each element and for energy are linked together by their participation in the same processes of assimilation and regeneration, and by interactions between various compartments within the system. These interactions include herbivory, predation, competition, and various mutualisms, the study of which falls within the realm of population ecology and is the subject of Part IV.

IV

Population Ecology

15

Population Structure

A population comprises the individuals of a species within a prescribed area. A population's boundaries may be the natural ones imposed by geographical limits of suitable habitat or they may be defined arbitrarily at the convenience of the investigator. Population processes — births, deaths, and movements of individuals — express the varied interactions of organisms with their environments. Moreover, evolution by natural selection and the regulation of community structure and ecosystem function transcend the individual organism and can be comprehended only in terms of the dynamics of populations. Hence the focus of much of ecology at the population level.

Individuals within populations resemble each other in many respects. Thus for many purposes one may determine average birth and death rates for small samples of individuals and multiply them by the total population size to characterize the overall dynamics of a population. But populations also are heterogeneous. Genetic variation in the interaction of individuals with their environments results in evolution. Variation in the suitability of habitat leads to parallel variation in population density and may determine the net movement of individuals within the population. Population characteristics also vary with respect to sex, social position, and the accumulated effects of accident and chance.

The population has temporal continuity because individuals alive at one time are the descendants of others alive at an earlier time. The population also has spatial continuity because the individuals in different parts of the distribution are descended from common ancestors; generally the greater the distance between two individuals, the more remote in time is their common ancestor. Individuals within a population derive their genes from a common pool and thus share a common history of adaptation to the environment. This is why it is valid to consider the traits of individuals in a population as approximately uniform, particularly compared to differences in those traits between populations.

Individuals interact with one another ecologically, either directly through social contact or indirectly by utilizing shared resources. Within its lifetime, an individual affects, and is affected by, a relatively small fraction of the total population. Each individual's sphere of ecological influence depends in part on the distance it travels. But regardless of how much individuals move within a population, the dynamics of one local population may be partially independent of those of others. When dispersal is infrequent and selective factors differ, subpopulations may also evolve independently, leading to the formation of genetically distinctive ecological races, or ecotypes.

Understanding the dynamics of a population requires that one characterize both local processes affecting birth and death and the movement of individuals between localities. In principle, the tasks of population biology are straightforward. In practice, population variables are difficult to estimate; much of this chapter is devoted to the techniques of population study.

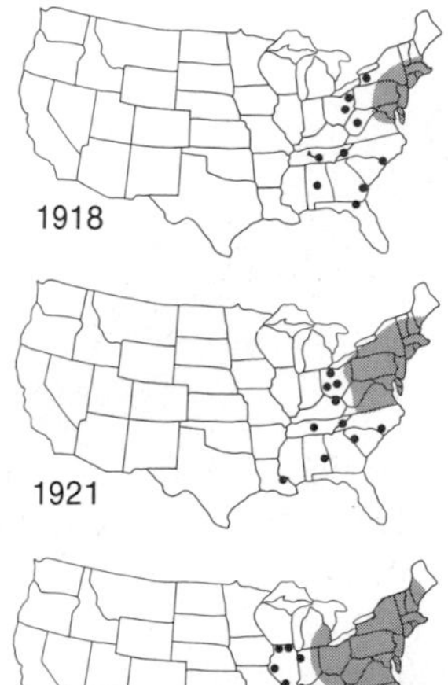

1918

1921

1926

1932

1941

1949

Figure 15-1
Western expansion of the range of the European starling (*Sturnus vulgaris*) in the United States. The shaded areas represent the breeding range; dots indicate records of birds in preceding winters. The population now inhabits the entire country. (After Kessel 1953.)

The distribution of a population is determined by ecologically suitable habitat.

The spatial structure of a population has three main properties: distribution, dispersion, and density. The distribution of population is its geographical and ecological range, determined primarily by the presence or absence of suitable habitat conditions. Hence the distribution of the sugar maple in the United States and Canada is limited to the east by the Atlantic Ocean, to the west by low precipitation, to the north by cold winters, and to the south by hot summers. Undoubtedly much suitable habitat exists throughout the world, especially in Europe and Asia, where sugar maple does not occur because it has not had the opportunity to disperse to such areas.

The distributional limits imposed by barriers to long-distance dispersal are revealed dramatically when introduced species expand successfully into newly colonized regions. In 1890 and 1891, 160 European starlings were released in Central Park, New York City. Within 60 years, the population had expanded to cover more than 3 million square miles and stretched from coast to coast (Figure 15-1). The habitats occupied by starlings in North America, principally city parks, farms, and pastures, resemble preferred habitats in Europe.

Not all introductions succeed (Long 1981, Diamond and Case 1986, Moulton and Pimm 1986). Subtle ecological factors, intrinsic qualities of a species that determine whether it is a good or poor colonizer, and chance determine the outcome of introductions. Even the prolific starling was introduced at least three times to the United States, in Pennsylvania, Ohio, and Oregon, before it finally took hold in New York.

Within the geographical range of a population, individuals will be found only where the habitat is suitable. Sugar maples do no grow in marshes, on serpentine barrens, on newly formed sand dunes, in recently burned areas, or in a variety of other habitats that are simply outside their range of ecological tolerance. Hence the geographical range of the sugar maple is a patchwork of occupied and unoccupied areas, just as it is for most other species.

For example, climate, topography, soil chemistry, and soil texture exert progressively finer influence on the geographical distribution of the perennial shrub *Clematis fremontii* (Figure 15-2). Climate and perhaps interactions with ecologically related species restrict this species of *Clematis* to a small part of the midwestern United States. The variety *riehlii* occurs only in Jefferson County, Missouri. Within its geographical range, *Clematis fremontii* is restricted to dry, rocky soils on outcroppings of dolomite. Small varia-

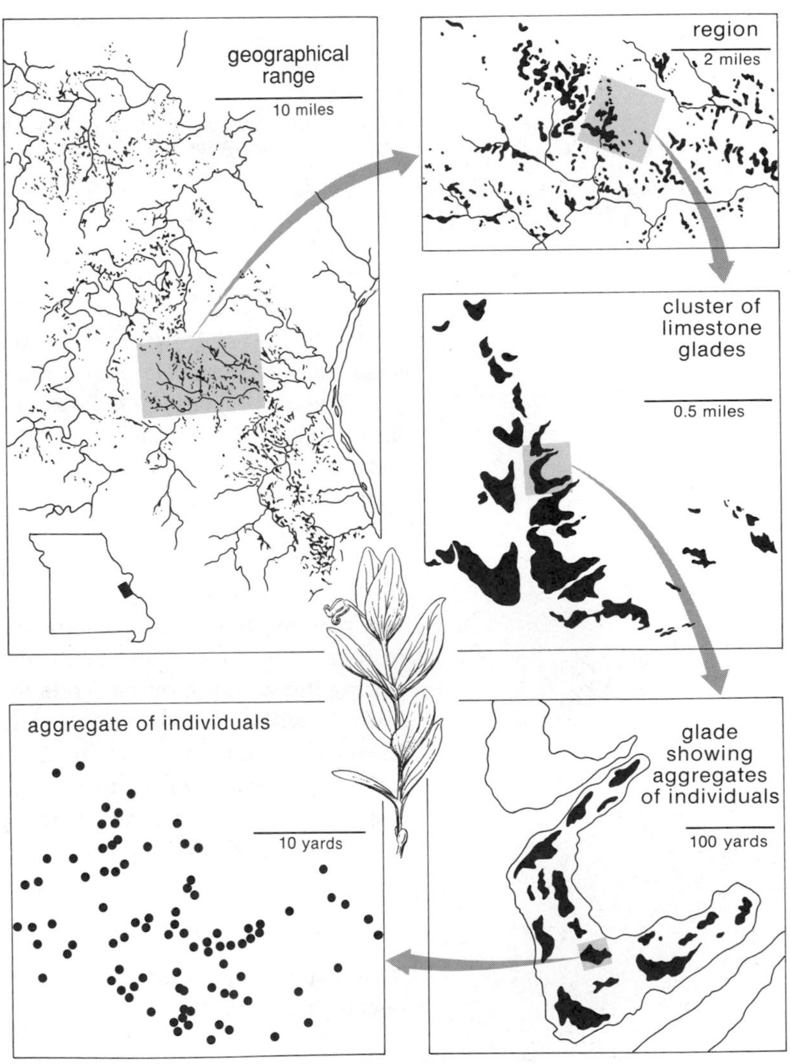

Figure 15-2
Hierarchy of patterns of geographical distribution of *Clematis fremontii*, variety *riehlii*, in Missouri. (After Erickson 1945.)

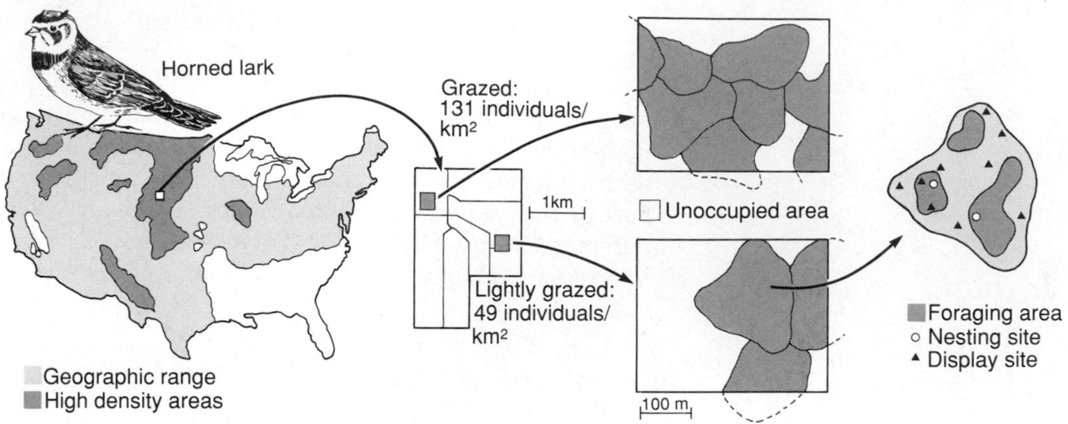

Figure 15-3
Hierarchy of patterns of distribution of the horned lark (*Eremophila alpestris*) during the breeding season. (After Wiens 1973.)

tions in relief and soil quality further restrict the distribution of *Clematis* within each dolomite glade to sites with suitable conditions of moisture, nutrients, and soil structure. Local aggregations occurring on each of these sites consist of many more or less evenly distributed individuals.

The horned lark *(Eremophila alpestris),* a small songbird of short grasslands, occurs more widely in North America than *Clematis fremontii* but its local distribution reflects subtle variation in habitat; even some parts of an individual's territory are utilized more than others (Figure 15-3).

The distributions of populations include all the areas they occupy during their life cycle. Thus the geographical range of salmon includes not only the rivers that serve as spawning grounds but also vast areas of the sea where individuals grow to maturity before making the long migration back to their birthplace. Many birds make annual migrations to warm climates during the winter months. Thus the range of the blackburnian warbler lies entirely within the northeastern United States and southeastern Canada during the summer, but within Central America and northern South America during the winter (Figure 15-4). The size and year-to-year variations of the population of such migratory species are determined by interactions of individuals with their environments throughout the year.

The dispersion of individuals reflects habitat heterogeneity and social interactions.

Within the distribution of a population, dispersion characterizes the spacing of individuals with respect to one another, forming patterns that vary from

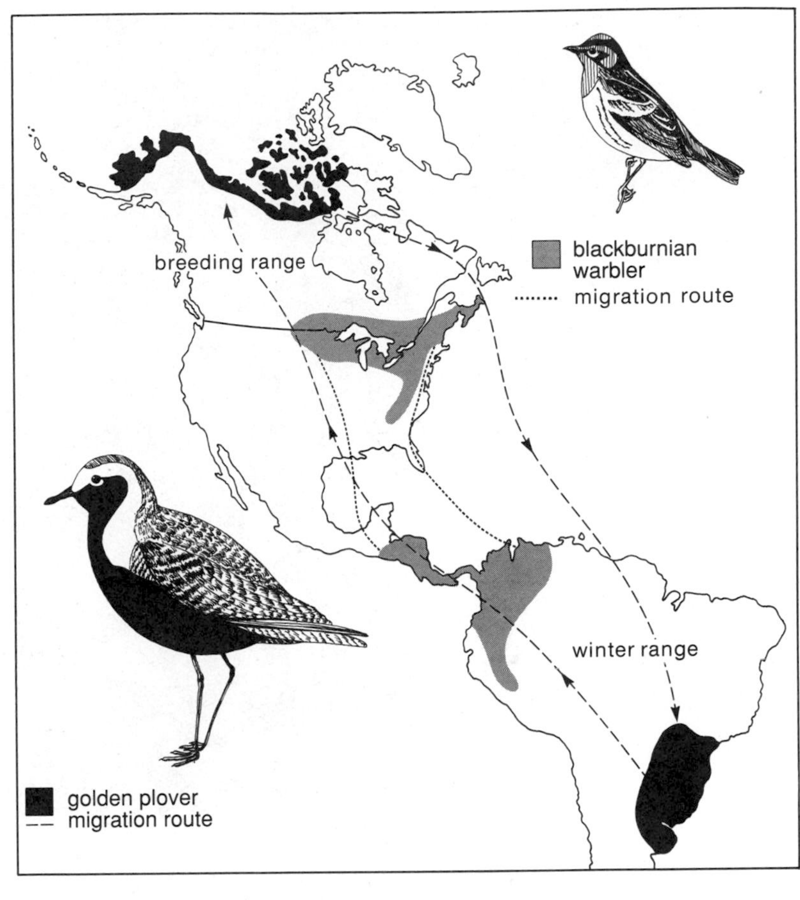

Figure 15-4
Breeding and wintering ranges of the golden plover (black) and blackburnian warbler (gray). Migration routes are indicated by dashed lines.

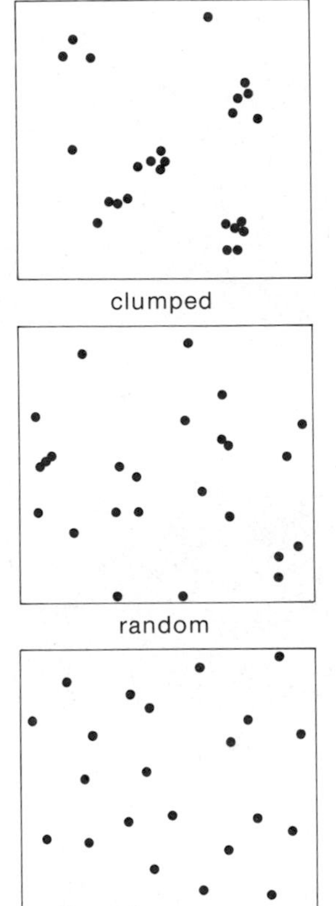

Figure 15-5
Diagrammatic representation of the spatial distributions of individuals in clumped, random, and evenly spaced populations.

tight aggregation in discrete clumps to even, or regular, distribution (Figure 15-5). Between these extremes is the pattern of random dispersion, in which individuals are distributed throughout a homogeneous area without regard to the presence of others.

Regular and clumped distribution patterns derive from different processes. Even spacing—sometimes called overdispersion—commonly arises from direct interactions between individuals. The maintenance of a minimum distance between each individual and its nearest neighbor results in even spacing; for example, in their crowded collonies, seabirds place their nests just beyond their neighbors' reach (Figure 15-6). Plants situated too close to larger neighbors often suffer the effects of shading and root competition; as close neighbors die, spacing of individuals becomes more even.

Clumping, or aggregation, results from social tendencies of individuals

Figure 15-6
A nesting colony of cape gannets on an island off the coast of South Africa. The densely packed birds space their nests more or less evenly, the distance being determined by behavioral interactions between individuals. Along the entire length of the South African coast, however, seabird populations are clumped during the breeding season on a few offshore islands with suitable nesting sites.

to form groups, the clumped distributions of resources, and the tendency of progeny to remain in the vicinity of their parents. Salamanders that prefer to live under logs exhibit clumped distributions corresponding to the patterns of fallen dead wood. Birds that travel in large flocks often aggregate to find safety in numbers.

Statistical tests of dispersion compare observed distributions to random patterns.

Mathematical measures of dispersion can be compared to the values expected of random spatial distributions (Pielou 1977: 116–165, Poole 1974: 101–125). They may be calculated from either the distributions of individuals among sample plots or the distances between individuals within a population (plotless sampling).

Suppose you establish a sampling area divided into plots or other units of equal area, and count the number of individuals in each of the plots. Some of the plots will have one individual, some many, and others none, depending on the density of the population and the size of the plots. Consider, for example, the number of red mites counted on each of 150 apple leaves (Table 15-1): 70 of the leaves had no mites, 38 had one, 17 had 2, and so on; no leaf had more than 7 mites. Unless we can determine whether individuals are more clumped or more spaced than one would expect if they were randomly distributed, we can say little about this pattern other than to describe it.

What would a random distribution of mites over the leaves be? Suppose that each mite in the population had an equal probability of finding itself on each of the leaves (1/150, or 0.067, per leaf) irrespective of the number of other mites present. When mites are distributed according to a random process such as this, the expected proportion of leaves (P) with a given number of mites (x) is the Poisson variate

$$P(x) = \frac{M^x e^{-M}}{x!}$$

where M is the mean number of individuals per sampling unit, and $x!$ is the

Table 15-1
Observed and expected frequencies of adult female red mites on 150 apple leaves. The expected frequencies are based on the Poisson distribution

Mites per leaf	Number of leaves observed	Poisson distribution	Number of leaves expected
0	70	0.3177	47.65
1	38	0.3643	54.64
2	17	0.2089	31.33
3	10	0.0798	11.98
4	9	0.0229	3.43
5	3	0.0052	0.79
6	2	0.0010	0.15
7	1	0.0002	0.02
8 or more	0	0.0000	0.00
total	150	1.0000	149.99

Source: Poole 1974.

factorial of x, or $x(x-1)(x-2)$. . . . In the example, 172 mites were recorded from 150 leaves, and so $M = 172/150 = 1.1467$ mites per leaf. To calculate the expected number of leaves with 3 mites, for example, substitute $M = 1.1467$ and $x = 3$ into the equation for $P(x)$; this gives a value of 0.080, or 8 per cent of the 150 leaves (12 leaves).

The expected distribution of number of mites per leaf is shown in Table 15-1. As you can see, fewer leaves than expected had 1 or 2 mites and more leaves had 0 and 4 or more mites, indicating a clumped distribution. The statistical significance of this difference from the expected random distribution can be determined by a chi-square or other test.

According to the Poisson distribution, the variance in the number of individuals per plot (in our example, mites per leaf) is equal to the mean. Hence the ratio of the variance to the mean (V/M) provides a convenient index to the degree of dispersion within a population. With even spacing, most plots will have the same number of individuals and the variance in number of individuals per plot will be considerably less than M (hence $V/M < 1$). With clumping, there will be many plots with no individuals and many with a large number, and the variance will considerably exceed the mean ($V/M > 1$). The variance is estimated by

$$V = \frac{1}{N-1} \sum_{x=0}^{\infty} F(x)x^2 - \frac{[F(x)x]^2}{N}$$

where $F(x)$ is the frequency of plots with x individuals and N is the total number of plots. In the mite example, the variance is 2.44 and the ratio of the variance to the mean is 2.1, showing clumping. Because the sampling distribution of this ratio is known, we can determine whether a given value represents a significant departure from 1, the expectation of a random distribution (see Pielou 1977: 125).

When organisms are distributed among discrete units of habitat, such as leaves on a tree, individual plants, and small ponds, it is reasonable to compare the frequency distributions of number of individuals per unit against a Poisson distribution. When individuals are distributed over a more or less uniform area, one may either arbitrarily divide the area into subplots and treat each as a sampling unit, as we did above, or use a plotless sample method in which the degree of aggregation within a population is estimated by distances between individuals. P. J. Clark and F. C. Evans (1954) devised an index to dispersion based on the distance between nearest neighbors within a population. When individuals are distributed at random with respect to each other, the expectation of the distance between an individual and its nearest neighbor [$E(r)$] is one-half of the square root of π divided by the density of individuals (provided, of course, that distance and area are measured in the same units). For example, at a density of 4 individuals per square meter the expected (average) nearest-neighbor distance is 0.44 meters [$\sqrt{(\pi/4)}/2$]. The ratio of the observed average nearest neighbor distance to the expected value provides an index to dispersion. Ratios less than unity indicate clumping; those greater than unity indicate even spacing. Other nearest neighbor methods are discussed by Pielou (1977), Diggle (1979), and Warren and Batchelor (1979; in Cormack and Ord 1979).

The estimation of population size is critical to the study of population dynamics.

The dynamics of populations describe changes in the number of individuals over time. In order to measure change from one period to the next, one must be able to count individuals. Rarely can one enumerate an entire population. One such case is the whooping crane, whose population numbers fewer than 200 individuals; it winters in a small area of the Aransas National Wildlife Refuge, where taking a census is easy.

Where counting entire populations is impractical, either because of numbers or area of distribution, biologists estimate densities within defined areas as indices to population size. Density itself is an important property of a population because it indicates the potential intensity of interaction among individuals. General methods for estimating density may be found in many manuals and texts, among them Cormack, Patil, and Robson (1979), Grieg-Smith (1964), Mueller-Dombois and Ellenberg (1974), Ralph and Scott (1981), and Southwood (1978).

When individuals are not mobile—plants and sessile marine invertebrates, for example—their densities may be estimated by counting individuals within plots of known area. When individuals move between plots faster than the investigator can count the number within a plot, an alternative is the mark-recapture method. Suppose that M specially marked individuals are released into a population of size N. If the marked individuals mix thoroughly with the unmarked ones, a subsequent sample of the population should contain proportion M/N marked individuals. Therefore, in a sample of n from the population, the number of marked individuals recaptured (x) should be

$$x = nM/N$$

Because we know the quantities x, n, and M, this equation can be rearranged to solve for population size

$$N = nM/x$$

This expression is known as the Lincoln Index to population size (Overton 1969, Caughley 1977). The standard error of N, $N\sqrt{[(N-M)(N-n)/nM(N-1)]}$, indicates the accuracy of the estimate of population size. By way of illustration, suppose we capture 25 sunfish in a small pond, mark them by a small, distinctive notch in the tail fin, and then put them back into the pond. Two days later, by which time we suppose the marked fish will have mixed thoroughly with the unmarked ones, we take another sample of 35 fish, of which 14 are marked. The Lincoln Index estimate of the population size is $35 \times 25/14 = 62.5$ individuals with a standard error of 8.7. Because we can be 95 per cent certain that the true number of fish lies between two standard errors above and two standard errors below the estimate, the probable number of fish in the pond is between 45 and 80 individuals.

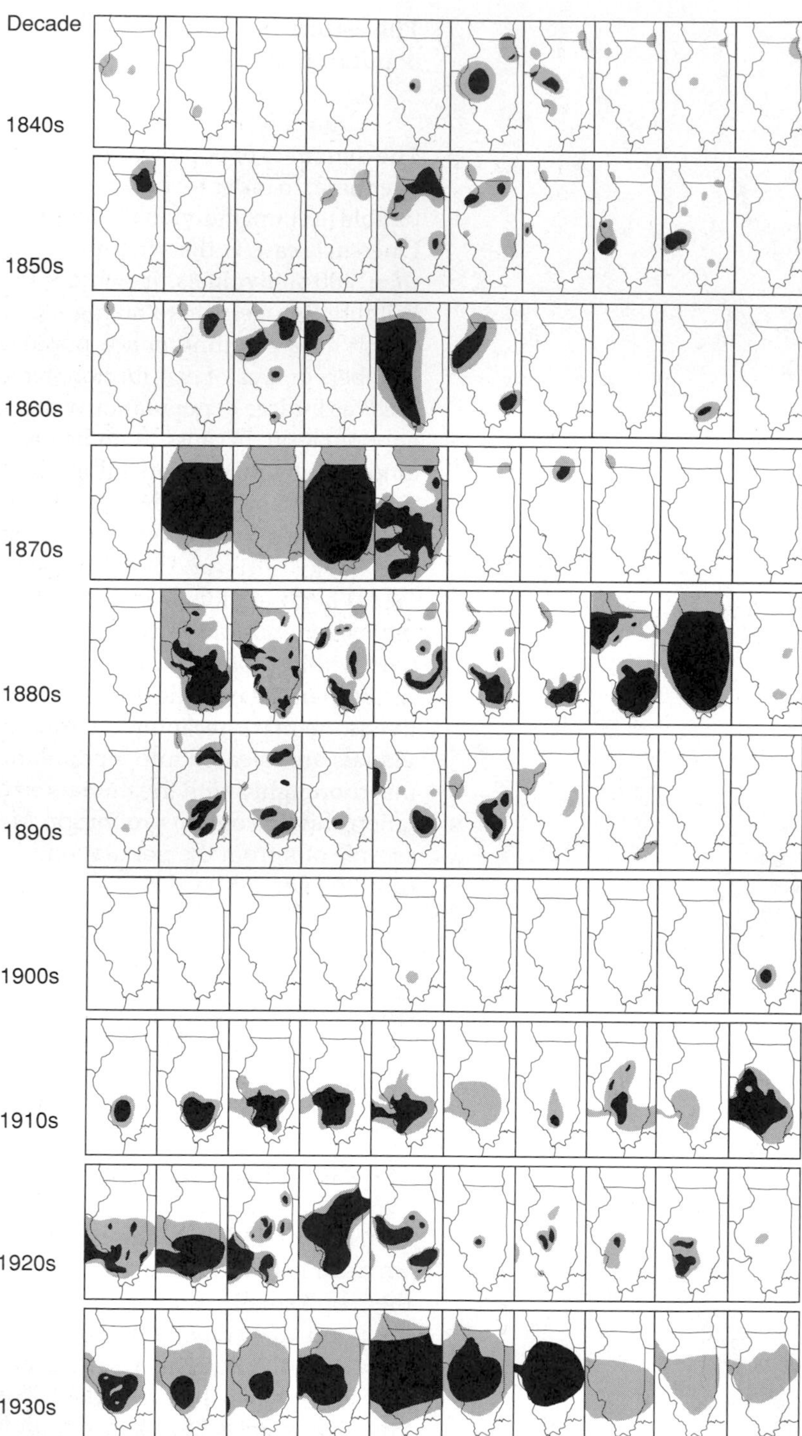

Figure 15-7
Distribution of crop damage caused by the chinch bug (*Blissus leucopterus*) in Illinois between 1840 and 1939. (From Shelford and Flint 1943.)

As this preceding example illustrates, small samples do not provide accurate estimates. Furthermore, violation of the assumptions of the model, that marked fish mix thoroughly with the population and that marked and unmarked fish are caught with equal probability, may bias the estimate, and it is difficult to ascertain the validity of these assumptions. For example, suppose that, instead of trying to estimate numbers of a confined population, like those of fish in a small pond, we wish to estimate the density of flies in an open grassland. Because the grassland poses no practical limits to the dispersal of flies, the total population that the marked individuals mix with increases as the flies disperse from the point of release. Hence our estimate of the fly population would increase with time, a very undesirable result. The mark-recapture technique has been elaborated greatly to deal with some of these problems, so that analyses are now very complex (see Cormack 1979, 1981, Jolly 1965, 1979, Seber 1965, 1973). Practical methods of marking various types of organisms are discussed in Giles (1969) and Southwood (1978).

Population density varies in time and space.

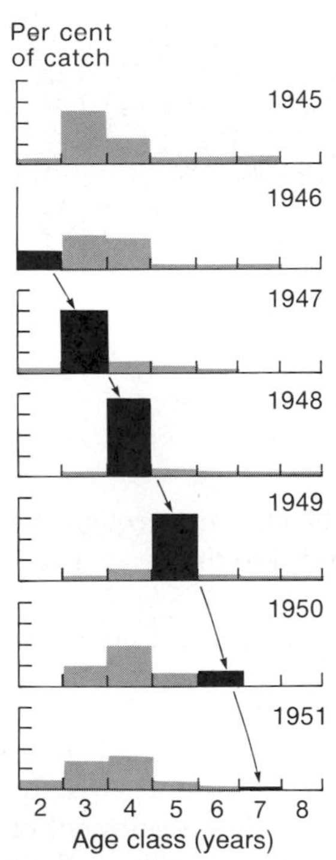

Figure 15-8
Age composition of samples from the commercial whitefish catch from Lake Erie between 1945 and 1951. Note the prominent 1944 class (solid bars). (From Lawler 1965.)

Dealing with populations is complicated by the fact that numbers change over time and space. A goal of population biology is to provide explanations for these patterns, but the initial challenge is to describe the patterns themselves. There probably is no such thing as a characteristic population of a species; one's perception is inevitably colored by where and when one looks. Records of the population of the chinch bug *(Blissus leucopterus)* in Illinois between 1840 and 1939 illustrate this point (Figure 15-7). These records exist because chinch bugs damage cereal crops; the Illinois State Entomologist's Office and later the State Natural History Survey Division determined the importance of monitoring the population. The populations were estimated from county reports of crop damage, which have been shown to be highly correlated with chinch bug density.

To illustrate the numbers involved, during 1873, when crop damage was serious over most of the state, a ballpark estimate for the population would have been 1000 chinch bugs per square meter over an area of 300,000 square kilometers, or a total of 3×10^{14} (300 trillion, more or less). In 1870 and 1875, hardly any damage was reported. Severe outbreaks were often restricted to small portions of the state, sometimes in the north, sometimes in the south. Continuity between years is evident in Figure 15-7 in the waxing and waning of infestations. This example illustrates the tremendous variations possible in numbers and spatial distribution of a population.

Temporal variation in population dynamics within a locality may also be perceived in the age structure of a population; that is, the relative frequencies of individuals of each age. The age composition of samples from the Lake Erie commercial whitefish catch for the years 1945–1951 shows that during 1947, 1948, and 1949, most of the individuals caught belonged to the 1944 year class (Figure 15-8). Age can be estimated from growth rings

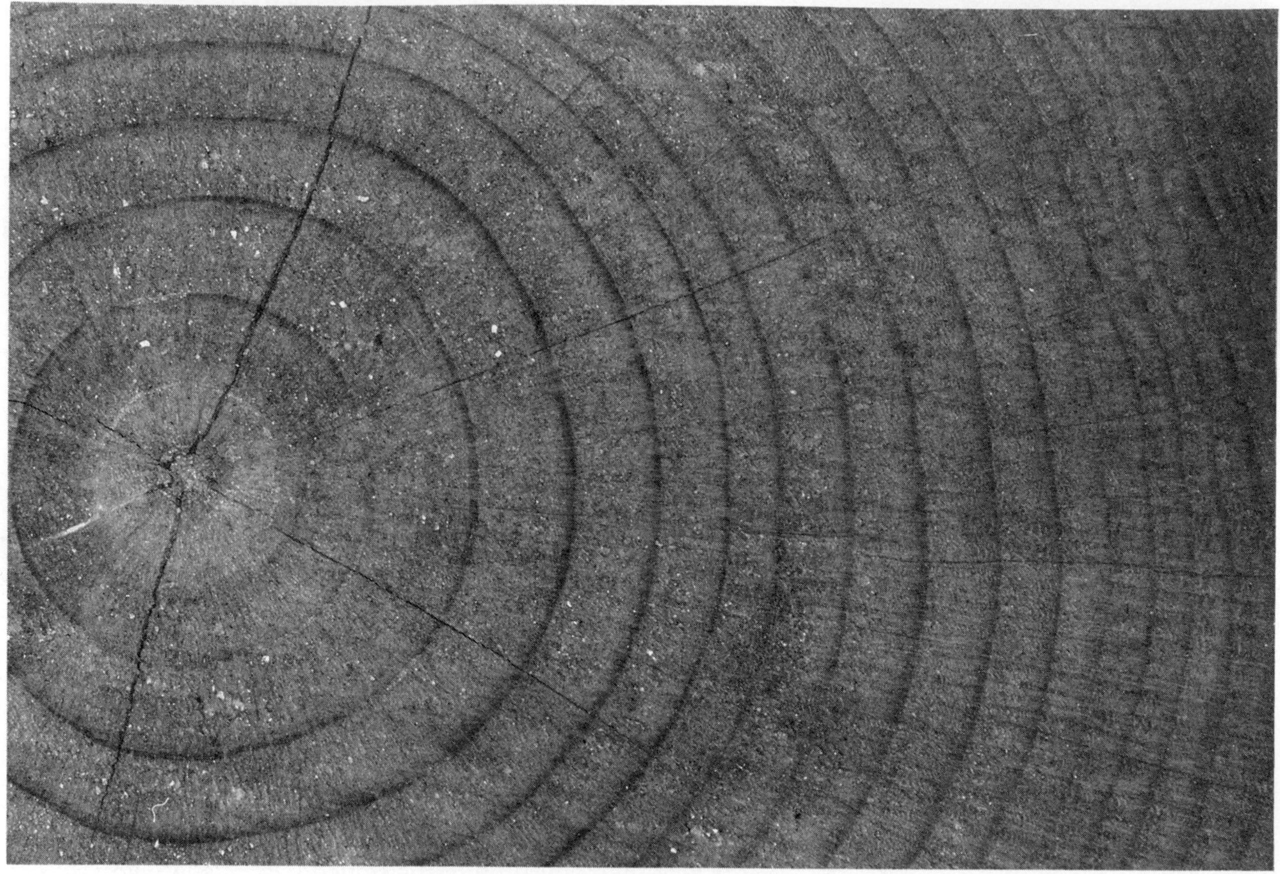

Figure 15-9
Cross-section of the trunk of a Monterey pine showing annual rings of winter (dark) and summer (light) growth.

on the scales, and the data suggest that 1944 was an excellent year for spawning and recruitment, particularly compared to the several years that followed.

Variation in the annual entrance of individuals into a population has also been revealed by the age structure of stands of trees. Age may be estimated from the growth rings in the woody tissue of the trunk of the tree, one ring being added each year under normal circumstances (Figure 15-9). The pattern in a virgin stand of timber surveyed near Hearts Content, Pennsylvania, in 1928, is typical and shows that the recruitment of most species was heterogeneous over the nearly 400-year span of the record (Figure 15-10). Many white pines became established between 1650 and 1710, undoubtedly following a major disturbance, perhaps associated with the serious drought and fire year of 1644. Fire can open a forest enough to permit the establishment of white pine seedlings, which do not tolerate deep shade. In contrast, the age distribution of beech, a species that can

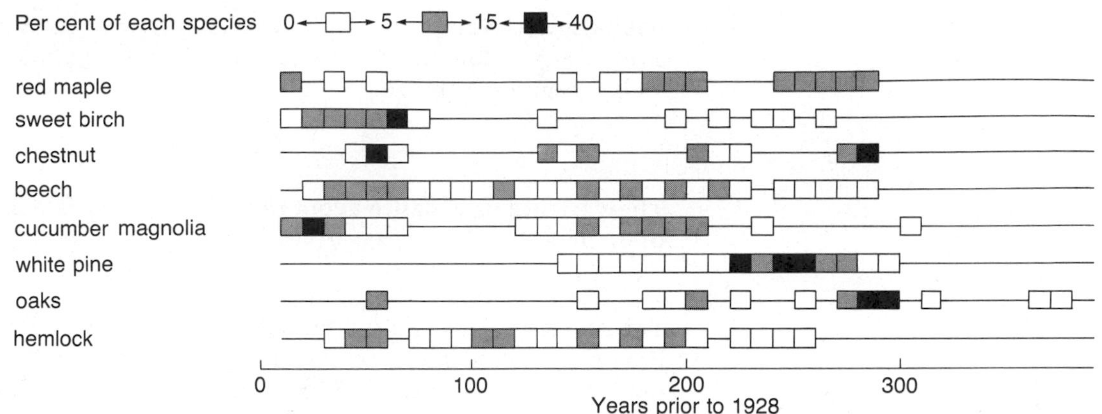

Figure 15-10
Age distribution of forest trees near Hearts Content, Pennsylvania, in 1928. (After Hough and Forbes 1943.)

become established under the canopy of a closed forest, showed a relatively even age distribution.

The spatial coherence of population processes is determined by the dispersal of individuals.

Local population changes may be due to the movement of individuals between populations. Such movement is frequently referred to as dispersal or, when discussed with respect to a particular population, as emigration and immigration.

Dispersal, particularly over long distances, is difficult to measure because its detection requires the recapture of marked individuals. Most attempts to estimate dispersal come from studies of population genetics in which investigators wish to determine the genetic structure of a population and estimate both the effective population size for evolutionary processes and the amount of gene flow between populations. The first attempts to measure dispersal in natural populations were the experiments of Theodosius Dobzhansky and Sewall Wright (1943, 1947), who described the movement away from a release point of fruit flies (*Drosophila*) distinguished by a visible mutation. Traps at various distances from the release point captured some of the marked flies as they dispersed through the population.

Dispersal is described by a number of mathematical indices, each making different assumptions about the structure of the population. The principal distinction in these models hinges on whether individuals are continuously distributed over space, as in a population inhabiting a spatially homogeneous environment (Wright 1943, 1969: 290–344), or individuals are aggregated in discrete subpopulations, perhaps because their preferred habitat is discontinuously distributed. The latter are referred to as stepping-stone models (Kimura and Weiss 1964). Here we will consider only one simple continuous model.

The variance of distance about the release point is a measure of dispersal.

The simplest model of dispersal is that of random movement, analogous to Brownian motion in physics. With respect to a fixed point of release, some random movements will take an individual farther away and some will bring it closer, although on average distance will tend to increase with time. The position of any individual at any moment is the sum of many random increments of distance; the probability of finding an individual at a given distance from the release site is described by the "normal" frequency destribution (Figure 15-11). This is a bell-shaped curve, whose peak coincides with the release site (that is, distance = 0) and whose breadth is characterized by a single variable, the standard deviation (s). The value of s provides a convenient index to dispersal distance.

The standard deviation (s) of a normal distribution is the square root of its variance (s^2). The variance is estimated by the expression

$$s^2 = \frac{1}{N} \sum_{i=1}^{N} (x_i - \bar{x})^2$$

where x_i is the position of individual i, $\bar{x}$ is the position of the release site, and N is the number of individuals in the sample. (The larger the sample of dispersing individuals the better the estimate of s.) When we define the

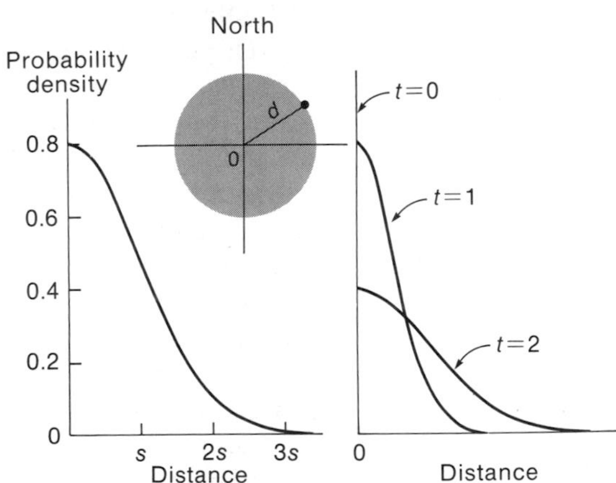

Figure 15-11
Probability density of individuals around a release point assumes a normal distribution when movement occurs at random. The distance dispersed may be characterized by the standard deviation of the curve, which increases in direct proportion to time t since release.

release site as distance 0 and measure the position of individual i by its distance from the release site d_i, the equation may be written

$$s^2 = \frac{1}{N}\sum_{i=1}^{N} d_i^2$$

D. W. Crumpacker and J. S. Williams (1973) measured the rate of dispersal of fruit flies (*Drosophila pseudoobscura*) in grasslands in central Colorado. Flies were marked with minute fluorescent dust particles; individuals were caught at traps placed at regular intervals in 8 directions of the compass at distances of up to 351 meters from the release point. After 1 day of dispersal, the estimated values of s were 139 and 171 meters at two localities.

A felicitous property of the standard deviation as a measure of dispersal is that the variances (s^2) of the distances add over time. Thus when s describes the distribution of dispersal distances after 1 day, the variance in distances after 2 days will be $2s^2$, and after t days it will be ts^2. Thus if adults of *D. pseudoobscura* live an average of 23 days and the average dispersal parameter (s) per day is 151 meters, s (23 days) would be the square root of 23×151^2, or 743 meters.

Sewall Wright (1946) defined neighborhood size within a population as the number of individuals within a circle of radius $2s(t)$, where t is the average reproductive life span of the individual. Although Wright devised this parameter to describe the genetic structure of a population, it also provides an index to the number of individuals in a population that are potentially coupled by strong interactions. In the case of *D. pseudoobscura,* Crumpacker and Williams (1973) estimated densities to be about 0.38 flies per 100 m^2. The area of a circle of radius $2s(t)$ is $4\pi[s(t)]^2$, which would have included an estimated 26,387 individuals.

George Barrowclough (1980) used recaptures of birds banded as nestlings to estimate $s(t)$ for several small songbirds in the United States. Because juvenile birds usually move much farther from their birthplaces to breed than adults move from one year to the next, Barrowclough distinguished $s_{juvenile}$ and s_{adult} and combined them according to the rule of adding variances; that is, $s^2(t+1) = s^2_{juv} + s_{ad}{}^2t$, where t is the expected life span of an adult bird (the juvenile stage lasts about 1 year, from hatching to first breeding). For 8 species $s(t+1)$ varied between 344 and 1681 meters, densities varied between 16 and 480 individuals per square kilometer, and neighborhood size varied between 151 and 7679 individuals.

J. J. D. Greenwood (1974) estimated dispersal parameters for three populations of the land snail *Cepaea nemoralis* in Europe, finding the following values:

s (meters) after 1 year	Individuals per square meter	Neighborhood size
5.5	20	7603
9.7	2.0	2365
ca. 10	1.4	1759

Notice that the neighborhood sizes of these slowly dispersing snails are on the same order as those of small birds. From mark-recapture data gathered by Frank Blair (1960) on the rusty lizard (*Sceloporus olivaceous*) near Austin, Texas, H. W. Kerster (1964) estimated $s(t)$ to be 89 meters and neighborhood size between 225 and 270 individuals.

In most studies of dispersal, the principal source of information has been the movement of individuals away from a point of release. Two other kinds of observations are pertinent, however. The first of these concerns the spread of introduced populations, which can only occur by the movements of individuals. The second concerns the genetic differentiation of local populations, which is hampered by movements of individuals.

The European starling spread almost 2500 miles across the United States in 60 years, an average rate of about 40 miles per year. This figure is much higher than Barrowclough's estimate for dispersal distance within populations of small songbirds. Adult starlings tend to nest in the same area year after year and so most dispersal is accomplished by young birds. Furthermore, the westward spread of the starling was characterized by frequent sightings of nonbreeders before breeding populations were established in an area. It is unlikely that such long-distance movements of juveniles could be detected within an established population, particularly because of the limited area of intensive population study.

Whereas a few long-distance movements may be sufficient to expand the border of a population, they may have less effect on the dynamics of widely separated, but established subpopulations. Given the propensity of populations for increase, a small number of colonizing individuals can grow to a large population in a short period. But those same individuals immigrating to an established population may have a negligible impact.

Examples of local genetic differentiation of populations suggest that the movement of individuals between populations is small compared to the forces of selection, mutation, and random change (genetic drift). Strong

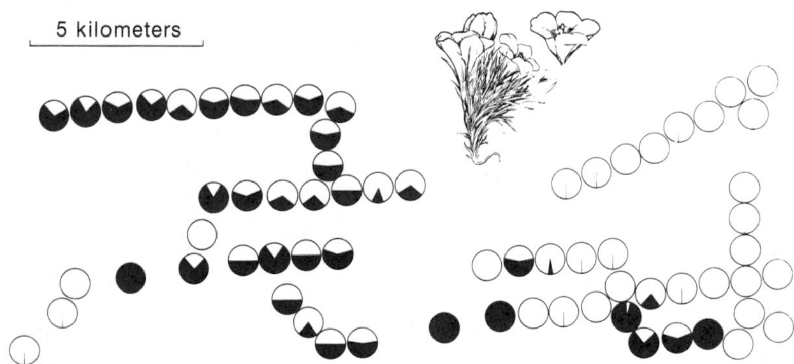

Figure 15-12
Distribution of blue flowers (whose frequency is indicated by the black portion of the circle) in populations of *Linanthus parryae* from a small region of southern California. Sampling localities were located along roads, which accounts for their distribution. (After Epling and Dobzhansky 1942.)

selection can lead to divergence between populations in the face of considerable gene flow (Endler 1977), as we have seen in the establishment of distinct ecotypes of plants in different habitats. The presence of apparently nonadaptive differences between close populations provide stronger evidence of limited gene flow. Such a situation occurs in populations of the small annual plant *Linanthus parryae.* In one small area in southern California, blue flowers—the normal flower color is white—are haphazardly distributed without apparent relation to environmental factors (Figure 15-12). Some samples of individuals separated by less than 1 kilometer differed in flower color by 100 per cent. When populations can maintain such genetic distinction, their dynamics cannot be closely linked.

Life tables summarize the survival and fecundity of individuals within populations.

The accurate projection of population growth requires a knowledge of the age structure of the population and the survival and fecundity of individuals of each age. These variables are determined in part by genotype, sex, social rank, and the effects of chance accident. Because these factors are difficult to measure, however, most models of population growth assume that individuals are uniform with respect to all traits except age. As a result, ecologists have devoted much effort to determining the age structure of populations and the age-specific birth rates and probabilities of death of individuals. These statistics collectively are known as the life table.

Demographic records have been kept for the human populations of some countries for centuries, well before the development of population models that could use the data (Lotka 1907). With the introduction of such models during the 1920s, life tables became a popular tool for both laboratory and field biologists. The first life table assembled for a nonhuman species was that of Raymond Pearl and S. L. Parker (1921) for a laboratory population of the fruit fly *Drosophila melanogaster.*

The life table sets out the fecundities and probabilities of survival for each age class of individuals in a population. Because of the difficulties of ascertaining paternity in many species, life tables usually are based entirely upon females. For some populations with highly skewed sex ratios or unusual mating systems, this can pose difficulties, but in most cases a female-based life table provides a workable population model.

Age is designated in a life table by the symbol x, and age-specific variables are indicated by the subscript x. When reproduction occurs during a brief breeding season each year, each age class is composed of a discrete group of individuals born at approximately the same time. When reproduction is continuous, as it is in the human population, each age class x is designated arbitrarily as comprising individuals between ages $x - \frac{1}{2}$ and $x + \frac{1}{2}$.

The fecundity of females, expressed in terms of female offspring produced per breeding season or age interval, is designated by b_x. (Think of b for births; *m* [for maternity] is used in some treatments, but this can be confused with *m* used here for mortality.)

Life tables portray the statistics of mortality in several ways. The fundamental measure is the probability of survival (s_x) between ages x and $x+1$, or its alternative, the probability of death m_x, where m is mortality (those who use m for fecundity often use q for mortality). Because an individual either lives or dies, $m_x = 1 - s_x$. The probabilities of survival over many age intervals are summarized by the survivorship to age x (l_x), which is the probability that a newborn individual will be alive at age x. Because by definition all newborns are alive at age 0, $l_0 = 1$. The proportion of newborns alive at age 1 is the probability of surviving from age 0 to age 1, hence $l_1 = s_0$. Similarly, $l_2 = s_0 s_1$ and, by extension, $l_x = s_0 s_1 s_2 \cdots s_{x-1}$. In product notation,

$$l_x = \prod_{i=0}^{x-1} s_i$$

(the large pi indicates the product of the terms s_i for values of i from 0 to $x-1$). Because the probability of death may be considered as a rate applied over time, some life tables include the exponential mortality rate k_x (think of k for killing power; Haldane 1949). According to the law of exponential decrease, survival over t units of time follows the expression $s(t) = e^{-kt}$ and therefore $k = -\log_e s(t)/t$. When one wishes to calculate the exponential mortality rate during each age interval, $t = 1$ and $k_x = -\log_e s_x$. Exponential mortality rate and survivorship are related in the following way: $\log_e l_x = \log_e s_0 + \log_e s_1 + \cdots + \log_e s_{x-1} = -k_0 - k_1 - \cdots - k_{x-1}$. In summation notation,

$$\log_e l_x = -\sum_{i=0}^{x-1} k_i$$

One additional measure is sometimes included in the life table: the expectation of further life (e_x) of an individual of age x, which is derived in the following manner. Individuals that die between ages i and $i+1$ only live to age i. These are fraction $l_i - l_{i+1}$ of all individuals and fraction $(l_i - l_{i+1})/l_x$ of individuals alive at age x. An individual dying between ages i and $i+1$ survives $i - x$ time units beyond age x. Now, the expectation of further life is simply the weighted average of these survival periods; that is,

$$e_x = \frac{1}{l_x} \sum_{i=x}^{\infty} (i - x)(l_i - l_{i+1})$$

The life table variables are summarized in Table 15-2, and further explained by example in the life table of the annual meadow grass *Poa annua* (Table 15-3). This life table follows the survival and fecundity of a planting of the grass under experimental conditions for 2 years, by which time the last individual had died. Age is tabulated in units of 3 months. Because *Poa* is hermaphroditic, sexes are not distinguished. Of 843 plants alive at time 0 (germination) 722, or 85.7 per cent, were alive at 3 months ($t = 1$). Hence

Table 15-2
Summary of life table variables

l_x	Survival of newborn individuals to age *x*
b_x	Fecundity at age *x*
m_x	Proportion of individuals of age *x* dying by age *x* + 1
s_x	Proportion of individuals of age *x* surviving to age *x* + 1
e_x	Expectation of further life of individuals of age *x*
k_x	$-\log_e s_x$, the exponential mortality rate between age *x* and *x* + 1

Table 15-3
Life table of the grass *Poa annua*

Age (x)*	Number alive	Survivorship (l_x)	Mortality rate (m_x)	Survival rate (s_x)	Expectation of life (e_x)	Fecundity (b_x)
0	843	1.000	0.143	0.857	2.114	0
1	722	0.857	0.271	0.729	1.467	300
2	527	0.625	0.400	0.600	1.011	620
3	316	0.375	0.544	0.456	0.685	430
4	144	0.171	0.626	0.374	0.503	210
5	54	0.064	0.722	0.278	0.344	60
6	15	0.018	0.800	0.200	0.222	30
7	3	0.004	1.000	0.000	0.000	10
8	0	0.000				

*Number of 3-month periods; i.e., 3 = 9 months.
Source: Begon and Mortimer, after Law 1975.

$s_0 = 0.857$, $m_0 = 1 - 0.857 = 0.143$, $l_1 = 0.857$, and $k_0 = -\log_e(0.857) = 0.154$ per 3 months (or 0.051 per month). The life table shows that the probability of dying increased with increasing age. As a consequence, the expectation of further life decreased with advancing age. Fecundity rose to a peak of 620 seeds per 3-month period at 6 months of age and then declined.

The estimation of survival schedules in natural populations employs several sampling techniques.

Survivorship can be estimated from four kinds of information: (1) survival of individuals to a particular age, (2) survival of individuals in each age class from one time period to the next, (3) ages at death within a population, and (4) the age structure of a population. The first kind of data forms the basis of the "cohort life table" or "dynamic life table," which follows the fate of a group of individuals born at the same time from birth to the death of the last individual. The life table for *Poa annua* presented in Table 15-3 is of this type. This method is most readily applied to populations of plants and sessile animals in which marked individuals can be continually resampled over the course of their life spans. Herein lies one of the disadvantages of this method: it can take a long time to collect the data, particularly if one studies redwood trees. It is also difficult to apply to highly mobile animals.

The second method, the "time-specific" life table, sidesteps the time problem by considering the survival of individuals of known age during a single time interval; say, 1 year. Thus each age-specific survival value is estimated independently for each subset of the population during the same period (hence the "time-specific" label). A limitation of this method is that one must know the ages of individuals, so it is practical only when age can be estimated by growth rings, tooth wear, or other reliable index.

The distribution of ages at death was first used to construct a life table for a nonhuman population by Edward Deevey (1947), who worked from Olaus Murie's (1944) data on the ages of death of Dall mountain sheep in Mount McKinley (now Denali) National Park, Alaska. Murie estimated age at death from the size of the horns, which grow continuously during the lifetime of the sheep (Figure 15-13). In all, he found 608 skeletal remains:

Figure 15-13
A group of Dall mountain sheep in Alaska. The size of the horns increases with age. (Courtesy of the American Museum of Natural History.)

Table 15-4
Life table for the Dall mountain sheep *(Ovis dalli)* constructed from the age at death of 608 sheep in Mount McKinley (now Denali) National Park

Age interval (years)	Number dying during age interval	Number surviving at beginning of age interval	Number surviving as a fraction of newborn (l_s)
0–1	121	608	1.000
1–2	7	487	0.801
2–3	8	480	0.789
3–4	7	472	0.776
4–5	18	465	0.764
5–6	28	447	0.734
6–7	29	419	0.688
7–8	42	390	0.640
8–9	80	348	0.571
9–10	114	268	0.439
10–11	95	154	0.252
11–12	55	59	0.096
12–13	2	4	0.006
13–14	2	2	0.003
14–15	0	0	0.000

Source: Based on data in Murie, 1944, quoted by Deevey 1947.

121 sheep were judged to have been less than 1 year old at death; 7 between 1 and 2 years, 8 between 2 and 3 years, and so on, as shown in Table 15-4. Deevey reasoned that all 608 dead sheep must have been alive at birth; all but 121 that died during the first year must have been alive at the age of 1 year (608 − 121 = 487), all but 128 (121 dying during the first year and 7 dying during the second) must have been alive at the end of the second year (608 − 128 = 480), and so on. He built his life table in this fashion until the oldest individuals had died, during their fourteenth year.

Deevey's age-at-death method is frought with potential biases. In particular, it assumes that an equal number of newborn forms the basis for each age class of deaths. This assumption is violated in expanding and declining populations, in which younger and older age classes, respectively, are over-represented.

Life tables constructed from the age structure of a population at a particular time suffer the same problem. The number of individuals of age x alive at time i is simply the number of newborn x years in the past times the survivorship to age x. Hence survivorship can be estimated by $l_x = n_x(i)/n_0(i - x)$. If one is willing to live dangerously and assume that equal numbers of offspring are born each year, then $n_0(i - x) = n_0(i)$ and $l_x = n_x(i)/n_0(i)$.

The role of life tables in modeling population dynamics will be explained in more detail in the following chapters. For more information on life tables themselves, consult Begon and Mortimer (1981) for a general account; Deevey (1947) for numerous examples; Caughley (1977), Eberhardt (1969), and Brownie et al. (1978) for wildlife populations, mostly birds and mammals; Ricker (1958) for fish; Varley and Gradwell (1970) and Varley, Gradwell, and Hassell (1975) for insects; and Harper (1977) for plants.

Four classes of population models are distinguished by discreteness of breeding episodes and overlap of generations.

As we shall see in the next chapter, the life table provides the foundation upon which all population models are built. But the particular form of a model depends on two aspects of the age structure of a population, which, taken together, result in four possible combinations. First, reproduction may be either discrete (that is, restricted to a narrow portion of each year) or continuous. Second, generations may overlap or not. The four combinations of these traits are summarized in Table 15-5 with examples of each type of population.

The dynamics of each type of population are modeled in a slightly different way. Those with nonoverlapping generations are the simplest because they may be described by only two variables—survival from birth to reproduction (s_0) and fecundity (b_x, where x is the age at reproduction). With overlapping generations, population models become more complex because each age class contributes differently to the dynamics of the population, and each individual moves through progressively older age classes during its lifetime.

In populations with discrete breeding seasons, age classes correspond to episodes of life. When breeding is continuous, one may either divide the life span into age classes arbitrarily or treat birth and death as instantaneous rates that change continuously over time. While the latter is attractive to theorists because such populations may be modeled by differential equations, it usually is more practical to gather and analyze data by age group. As we shall see, ecologists have used one or the other model as their purposes have required.

Table 15-5
Four types of population models

	Seasonal (discrete) reproduction	Continuous reproduction
Nonoverlapping generations	Annual plants and insects	Bacteria
Overlapping generations	Most seasonally reproducing vertebrates and higher plants	Humans, fruit flies

SUMMARY

1. Ecologists define populations either by their geographical limits of distribution or by an arbitrary boundary around a smaller area of intensive study. The primary objectives of population studies are to describe the spatial structure and temporal behavior of populations and to gather data relevant to understanding their dynamics. Populations achieve a genetical cohesion through their breeding structure and a dynamical cohesion through the interactions and movements of individuals.

2. The spatial structure of a population may be defined in terms of its density and pattern of dispersion. Density is scaled in terms of the number of individuals per unit of area. Density may be estimated either directly by counts of individuals within plots of known size or indirectly by such methods as the mark-recapture technique.

3. Dispersion describes the relationship of individuals to the position of others in the population. Clumped distributions may result from independent aggregation of individuals in suitable habitats, spatial proximity of direct descendants, or tendencies to form social groups. Evenly spaced distributions may result from antagonistic interactions between individuals. Indices of dispersion, derived either from the frequency distribution of number of individuals per plot or from the distribution of nearest neighbor distances, most often reveal clumped distributions.

4. The age structure of populations often indicates temporal heterogeneity of recruitment of individuals. For example, certain seedlings tend to become established in forests primarily following such major disturbances as fire, drought, or storms. Thus population processes may be sporadic rather than uniform over time.

5. The movement of individuals within populations may be characterized by the variance in distance from a point of release. Neighborhood size, defined as the number of individuals within a circle of area 4π times the variance of lifetime dispersal distance, is an index to the number of interacting individuals within a population. For several species, neighborhood sizes have been estimated as on the order of 10^2 to 10^4 individuals.

6. The life table sets out the fecundities (b_x) and probabilities of survival (s_x) of individuals by age class (x). These are the principal variables in models of the dynamics of populations. Survival rates may be estimated from the fates of individuals born at the same time (cohort or dynamic life table), the survival of individuals of known age during a single time period (static or time-specific life table), the age structure of a population at a particular time, or the age distribution of deaths.

7. The age structure of populations may be divided into four classes depending on whether reproduction is continuous or occurs during discrete seasons and whether generations overlap or do not. Models for each of these populations differ in their formulation and, as we shall see in later chapters, in their dynamical properties.

16
Population Growth

The immense capacity of populations for increase cannot be better illustrated than by that of the human population, which has grown prolifically, at times doubling each quarter-century. From the earliest time that the rapid increase in the number of humans could be appreciated, population growth has caused concern. The English economist Thomas Malthus, in his essay on population pulished in 1798, warned of the dangers of unchecked population growth. He began, "I think I may fairly make two postulata. First, that food is necessary to the existence of man. Secondly, that the passion between the sexes is necessary and will remain nearly in its present state." He went on to note that "Population, when unchecked, increases in a geometrical ratio: Subsistence increases only in an arithmetical ratio. A slight acquaintance with numbers will shew the immensity of the first power in comparison of the second."

By contrasting geometric growth and arithmetic growth, Malthus demonstrated an appreciation of the most basic property of population growth; exponential increase has been the foundation of all efforts to derive formulae for the projection of population size through time.

A number that increases arithmetically is added to by a constant amount during each interval of time. When a worker shovels coal into a bunker, the coal therein increases at a constant rate, say 1 ton per hour, regardless of the amount accumulated. That is, the growth of the coal pile is determined, not by the amount either already shoveled or still remaining, but by the pace of the worker. Accumulation rate is constant. Growth is arithmetic.

In contrast, geometric (or exponential) increase occurs in proportion to the amount accumulated. The larger the population the more rapidly it grows, because increase in numbers depends upon reproduction by individuals in the population. For coal, this would require some attraction of the chunks in the bunker for those remaining in the pile outside. For populations of plants or animals, Malthus's "passion between the sexes" is quite sufficient. Greatly impressed by Malthus's arguments on population,

Charles Darwin wrote, in *On the Origin of Species,* "There is no exception to the rule that every organic being naturally increases at so high a rate, that, if not destroyed, the earth would soon be covered by the progeny of a single pair." To make his case as forcefully as possible, Darwin put forth a conservative example:

> The elephant is reckoned the slowest breeder of all known animals, and I have taken some pains to estimate its probable minimum rate of natural increase; it will be safest to assume that it begins breeding when thirty years old, and goes on breeding till ninety years old, bringing forth six young in the interval, and surviving till one hundred years old; if this be so, after a period of from 740 to 750 years there would be nearly nineteen million elephants alive, descended from the first pair.

Because baby elephants grow up, mature, and themselves have babies, the elephant population grows exponentially. Imagine, however, a population in which only one female reproduces; such a population, like the coal pile, would increase arithmetically, and very slowly at that as "a slight acquaintance with numbers will shew."

Populations grow exponentially.

While Darwin took great pains to calculate the growth of an elephant population, we now use convenient equations to project the course of a population through time. These equations are the fruits of modern demography, the study of how birth and death processes determine the age structures and growth rates of populations. A population growing exponentially increases according to

$$N(t) = N(0)e^{rt} \tag{16-1}$$

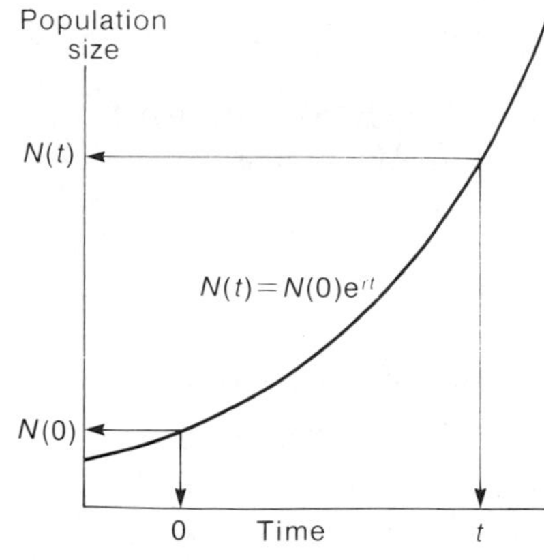

Figure 16–1
The curve of exponential growth for a population growing at rate r between time 0 and t. During this period the number of individuals increases from $N(0)$ to $N(t)$.

where $N(t)$ is the number of individuals in the population after t units of time, $N(0)$ is the initial population size ($t = 0$), and r is the exponential growth rate. The constant e is the base of the natural logarithms, having a value of approximately 2.72. Exponential growth results in a continuous curve of increase (or decrease) whose slope varies in direct relation to the size of the population (Figure 16–1).The term e^r is the factor by which the population increases during each time unit, often written as the lower case Greek lamda (λ). That is, when $t = 1$, $N(t) = N(0)e^r$, or $N(t) = N(0)\lambda$. We shall have more to say about λ later.

Equation (16–1) would have helped Darwin with his elephant problem. Although he omitted the details of his calculations, we can reasonably estimate the variables needed to solve the equation for exponential growth. Of the 6 offspring, 3 would likely be female, and the average age of the mother at birthing might be 60 years. Hence each female would replace herself by 3 female progeny ($e^r = 3$) at an average generation time of 60 years ($t = 1$). Darwin's 750 years are 12.5 generations ($t = 12.5$), and we assume a starting population $N(0)$ of 1 female (with a male, of course, to provide the passion between the sexes). Thus evaluated, equation (16–1) is $N(12.5) = 1 \times 3^{12.5} = 920{,}483$. Even doubled to account for the males produced, this number falls short of Darwin's calculation. Consider, however, that baby elephants born to a mother early in her life might mature and produce babies even while their mother is still in her productive years. As we shall see below, this overlapping of generations in the population effectively reduces the generation time to less than the average age of mothers at birth. For the elephants, a generation time of 51 years is short enough for the progeny of a single pair to swell to 19,000,000 in 750 years. (Like us, Darwin must have been a bit unsure of his own calculations. In contrast to the sixth edition [1872] of *On the Origin of Species,* quoted above, in the first edition [1859] he brings forth 15 million elephants in 500 years.) Regardless of the details, however, the important point is simply that elephants produce more elephants.

The equation for exponential growth is derived by calculus.

Populations grow like an interest-bearing bank account. Each time the account is compounded (that is, the interest is added to the principle), the amount bearing interest increases. As each baby in a population matures, the number of reproducing individuals similarly increases. According to a bank's practices, an account might be compounded annually, quarterly, daily, or perhaps continuously. The first three—periodic compounding—are examples of geometric growth, with increases posted at discrete intervals. As we shall see below, geometric models of growth have broad application in ecology. The last case, that of continuous compounding, is exponential growth.

Exponential growth is just geometric growth with the interval between growth increments reduced to zero. Consider the following equation for geometric growth with interest rate i,

$$N(t) = N(0)(1 + i)^t \qquad (16\text{–}2)$$

For a \$1000 account bearing interest at a rate of 10 per cent per year compounded annually, equation (16–2) may be evaluated as

$$\begin{aligned} N(t) &= 1{,}000 \times (1 + 0.10)^1 \\ &= 1100 \end{aligned}$$

after 1 year ($t = 1$). When an account is compounded q times per year, equation (16–2) may be written

$$N(t) = N(0)(1 + i/q)^{qt}$$

The above account compounded quarterly ($q = 4$) would yield \$1,103.81 and, compounded daily ($q = 365$), would yield \$1,105.16. To determine the yield with instantaneous compounding ($q =$ infinity) requires taking the limit of the equation as q goes to infinity, or $1/q$ goes to zero. First, we rearrange the equation by substituting x for i/q, so that

$$N(t) = N(0)(1 + x)^{it/x} \tag{16–3}$$

From calculus we know that

$$\lim_{x \to 0}(1 + x)^{1/x} = \mathrm{e}$$

where e is the base of natural logarithms (e = 2.7183, approximately). Substituting this limit into equation (16–3), we obtain

$$N(t) = N(0)\mathrm{e}^{it}$$

When discussing the growth of populations, the exponential rate of increase (i, above) is always designated as r. And because it is a rate, r has units of 1/time.

The differential form of the exponential equation expresses the rate of population increase or decrease.

The rate of increase of a population undergoing exponential growth at a particular instant in time—the instantaneous rate of increase—is the derivative of the exponential equation; that is,

$$\frac{\mathrm{d}N}{\mathrm{d}t} = rN \tag{16–4}$$

This equation expresses two principles: first, the rate of increase ($\mathrm{d}N/\mathrm{d}t$) is directly proportional to the size of the population (N) and, second, the exponential growth rate (r) expresses population increase (or decrease) on a per individual basis. In words, equation (16–4) could read

$$\begin{bmatrix}\text{The change in}\\ \text{population size}\\ \text{between the present}\\ \text{and some time}\\ \text{in the future}\end{bmatrix} = \begin{bmatrix}\text{The contribution}\\ \text{of each individual}\\ \text{to the population}\\ \text{at some time}\\ \text{in the future}\end{bmatrix} \begin{bmatrix}\text{The number of}\\ \text{individuals}\\ \text{presently in}\\ \text{the population}\end{bmatrix}$$

In the exponential description of growth, reproduction occurs continuously in the population. Therefore, "some time in the future" is arbitrarily soon and the change in population size is an instantaneous rate ($\mathrm{d}N/\mathrm{d}t$).

The individual, or per capita, contribution to population growth is the difference between the birth rate (b) and the death rate (d) calculated on a per capita basis. Rates of birth and death are abstractions and have little meaning for the individual. An elephant dies only once, so it can't have a rate of death. Babies are produced in discrete litters separated by intervals required for gestation, not at constant rates, such as 0.05 offspring per day. But when births and deaths are averaged over the population as a whole, they take on meaning as rates of demographic events in the population. If 1000 individuals were to produce 10,000 progeny in a year, it would be reasonable to assign a per capita birth rate of 10 per year and to assume that a population of 1 million would produce 10 million progeny, still 10 per individual, under the same conditions. If, of the ground hogs alive on their day in one year, only half survived to February 2 of the next, we would ascribe a death rate of 50 per cent per year to the population even though some of the ground hogs had died completely and others hadn't died at all.

A ground hog might become very nervous on learning that mortality rates in populations can exceed 100 per cent annually. Of course, no individual dies twice. But the probability of death during a short interval—say, a day—may be high enough that most of the individuals in a population pass on during a period considerably shorter than a year. With a probability of death of 1 per cent per day (0.01 day^{-1})—365 per cent on an annual basis—99 per cent of individuals survive each day. Those surviving both today and tomorrow are the fraction $0.99 \times 0.99 = 0.99^2 = 0.98$ of the initial population. Those surviving a full year would constitute only $0.99^{365} = 0.0255$ (2.55 per cent) of the original population.

When death occurs at exponential rate $-d$, the number of individuals in a population having no births changes over time according to $N(t) = N(0)e^{-dt}$, and $dN/dt = -dN$. The minus sign must be included because death removes individuals from the population. A pure birth process (no deaths) adds individuals to the population at rate $dN/dt = bN$. Combining birth and death in the same equation,

$$\frac{dN}{dt} = bN - dN$$

$$= (b - d)N$$

Furthermore, because r is the difference between birth rate and death rate $(b - d)$, $dN/dt = rN$. When the death rate exceeds the birth rate, r is negative and the population declines.

Geometric equations describe discrete growth processes.

The human population grows continuously because babies are born and added to the population at all seasons of the year. But this situation is unusual in natural populations, most of which restrict reproduction to a

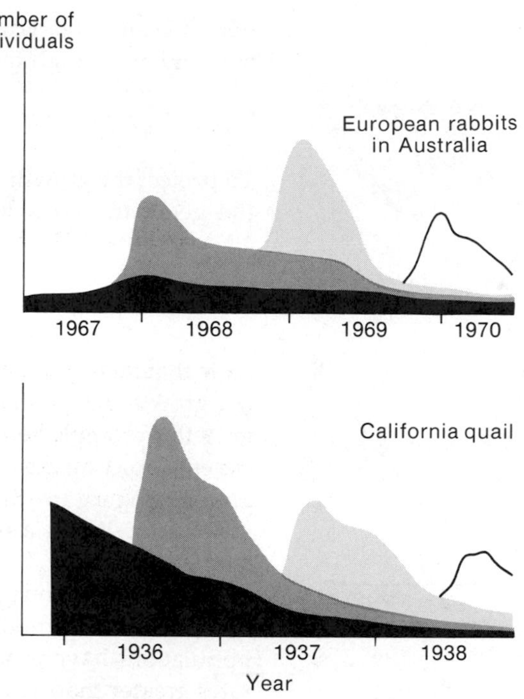

Figure 16–2
Growth of populations with discrete (seasonal) reproduction: *above*, rabbits in a subalpine population in New South Wales, Australia (after Myers 1970); *below*, California quail (after Emlen 1940).

particular time of year. The population grows during the breeding season, and then declines between one breeding season and the next (Figure 16–2). In the case of the California quail, the number of individuals doubles or triples each summer as adults produce their broods of chicks, but then dwindles by nearly the same amount during the fall, winter, and spring. Within each year, population growth rate varies tremendously depending on seasonal changes in the balance of birth and death processes.

Growth in populations with discrete breeding seasons is treated somewhat differently than growth in continuously breeding populations. The size of populations with discrete breeding seasons must be measured at a particular time of year to have any meaning. If one were interested in projecting population growth, it would be pointless to compare numbers in August, recently augmented by the chicks of the year, with those of May, after the winter has taken its toll. One must count individuals at the same time each year, with each count separated by the same cycle of birth and death processes.

When populations are enumerated periodically, growth rate is most conveniently expressed as the ratio of the population in one year [for example, $N(1)$] to that in the preceding year [$N(0)$ in this case]. Demographers have assigned the symbol λ to this ratio; hence $\lambda = N(1)/N(0)$. This defini-

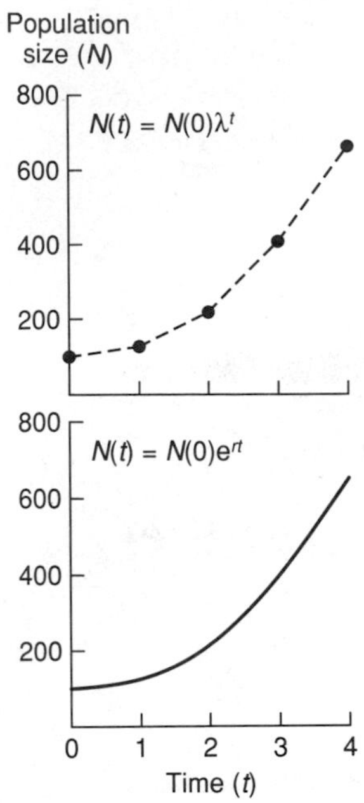

Figure 16–3
Increase in the number of individuals in populations undergoing geometric growth *(top)* and exponential growth *(bottom)* at equivalent rates ($\lambda = 1.6$, $r = 0.47$).

tion of λ can be rearranged to provide a formula for projecting the size of the population through time,

$$N(t + 1) = N(t)\lambda \qquad (16\text{–}5)$$

To project the growth of a population, the original number is multiplied by the geometric growth rate λ once for each unit of time passed. Hence, $N(1) = N(0)\lambda$, $N(2) = N(0)\lambda^2$, $N(3) = (0)\lambda^3$, and

$$N(t) = N(0)\lambda^t \qquad (16\text{–}6)$$

Note that this equation for geometric growth is identical to that for exponential growth except that λ takes the place of e^r, the amount of exponential growth accomplished in one time period. Because of this relationship, curves depicting the two models of growth can be superimposed upon one another (Figure 16–3), and there is a direct correspondence between values of λ and r. When a population's size remains constant, $r = 0$ and $\lambda = 1$ ($r = \log_e \lambda$, and $\log_e 1 = 0$). Decreasing populations have negative exponential growth rates and geometric growth rates less than 1 (but greater than 0; a real population cannot have a negative number of individuals). Increasing populations have positive exponential growth rates and geometric growth rates greater than 1.

A population's growth rate depends on its age structure.

When birth and death rates are uniform over all members of the population, the total population size (N) is the proper basis for projecting population growth into the future. But when birth and death rates vary with respect to age, the contribution of younger and older individuals to population growth must be figured separately; two populations with identical birth and death rates at corresponding ages, but with different proportions of individuals in each age class, will grow at different rates. A population composed wholly of prereproductive adolescents and oldsters too feeble to breed will not increase at all in the near future. While this example may be somewhat ridiculous, smaller variations in age distribution can also have a profound influence on population growth rate.

The growth rate of an age-structured population is the sum of the growth rates owing to individuals in each age class weighted by their proportion in the population. For exponential growth,

$$\frac{dN}{dt} = N \int_{x=0}^{\infty} r_x c_x dx$$

where r_x is the exponential growth rate attributable to age x and c_x is the proportion of individuals at age x. Even with continuously breeding populations, one cannot determine birth and death rates at every age because an individual remains a particular age only for an instant. Demographers get around this problem by dividing populations into age classes, to which they can assign probabilities of giving birth and dying. This practice is particularly appropriate for populations having discrete breeding seasons and, therefore, discrete age classes of individuals.

Consider a population that produces young in a single burst each year. At the time of breeding (the present, t) each age class contains $n_x(t)$ individuals. An individual of age x gives birth to b_x babies on average. Following reproduction, individuals of age x survive to the next breeding season with probability s_x. To project the population to the next breeding season ($t+1$), we first calculate the number of births, which will then make up the 0 age class in the population:

$$n_0(t) = n_1(t)b_1 + n_2(t)b_2 + n_3(t)b_3 + \ldots$$

continuing the sum through the oldest age class in the population.

From one breeding season to the next, individuals in age class x become a year older and move into age class $x+1$. Because they suffer mortality during this period, we must multiply $n_x(t)$ by the survival rate s_x to obtain $n_{x+1}(t+1)$. That is,

$$n_1(t+1) = n_0(t)s_0,$$

$$n_2(t+1) = n_1(t)s_1,$$

$$n_3(t+1) = n_2(t)s_2,$$

and so on to the last age class.

The number of newborn in the population at time $t+1$ is

$$n_0(t+1) = n_1(t+1)b_1 + n_2(t+1)b_2 + n_3(t+1)b_3 + \ldots$$

Remembering that $n_x(t+1) = n_{x-1}(t)s_{x-1}$, $n_0(t+1)$ can be expressed in terms of the age classes at time t, as

$$n_0(t+1) = n_0(t)s_0b_1 + n_1(t)s_1b_2 + n_2(t)s_2b_3 + \ldots$$

The total size of a population is equal to the sum of its age classes: in summation notation, $N(t+1) = \Sigma n_x(t+1)$. Thus we may express the population in year $x+1$ in terms of the population during year x as two sums, one of the survival terms and the second of the birth terms,

$$N(t+1) = \sum_{x=0}^{\infty} n_x(t)s_x + \sum_{x=0}^{\infty} n_x(t)s_xb_{x+1}$$

$$= \sum_{x=0}^{\infty} n_x(t)s_x(1 + b_{x+1})$$

This equation allows us to project the growth of a population through time. As it stands, however, it merely formalizes the pencil and paper figuring

Darwin used to calculate the increase in elephants. The size of each age class must be calculated from the size of the previous age class during the previous year to determine the total population and its growth from year to year. But as we shall see below, populations growing for long periods with constant values of b_x and s_x assume a characteristic age distribution and constant rate of growth, which can be calculated directly from the life table.

A population with a fixed life table assumes a stable-age distribution and grows at a constant exponential rate.

When age-specific birth and survival rates remain unchanged for a sufficient period, a population will assume a stable age distribution. Under such conditions each age class in the population grows or declines at the same rate and so, therefore, does the total size of the population. This was first shown by A. J. Lotka in 1922. A little pencil and paper figuring with a contrived population will demonstrate this result. Imagine a population of 100 individuals having the following characteristics:

Survival	Birth rate	Number of individuals ($t = 0$)
$s_0 = 0.5$	$b_0 = 0$	$n_0(0) = 20$
$s_1 = 0.8$	$b_1 = 1$	$n_1(0) = 10$
$s_2 = 0.5$	$b_2 = 3$	$n_2(0) = 40$
$s_3 = 0.0$	$b_3 = 2$	$n_3(0) = 30$

Because all 3-year-olds die, there are no 4-year-olds in the population; newborns have a fecundity of zero, as seems biologically reasonable. Applying the equations in the previous section to the number of individuals in each age class at time 0, and then to the numbers at time 1, and so on, gives the following picture of population increase:

	Year									
	0	1	2	3	4	5	6	7	8	% of column 8
n_0	20	74	69	132	175	274	399	599	889	(63.4)
n_1	10	10	37	34	61	87	137	199	299	(21.3)
n_2	40	8	8	30	28	53	70	110	160	(11.4)
n_3	30	20	4	4	15	14	26	35	55	(3.9)
N	100	112	118	200	279	428	632	943	1403	
λ		1.12	1.05	1.69	1.40	1.53	1.48	1.49	1.49	

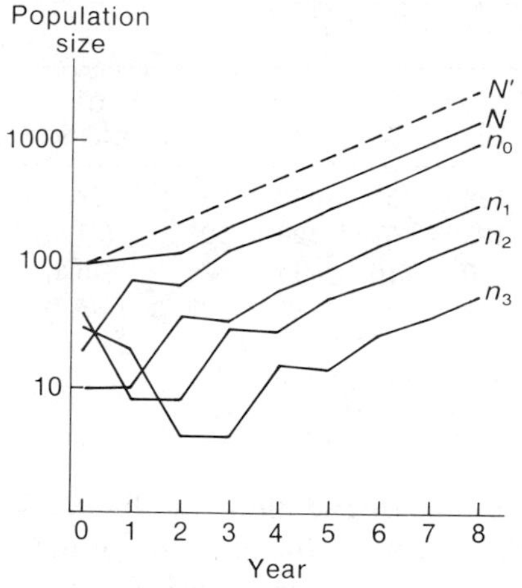

Figure 16–4
Growth of age classes as a population achieves its stable age distribution.

In this example, the growth rate of the population is at first very erratic, fluctuating between $\lambda = 1.05$ and $\lambda = 1.69$, before settling down to a constant value of 1.49. At this point, the population has achieved a stable age distribution; the percentage of individuals in each age class at $t = 8$ is shown at the right of the table. Even by the end of the fourth year, the population approached its stable age distribution, with proportions in each of the age classes of 62.7, 21.8, 10.0, and 5.4 per cent. Under stable-age conditions, each age class grows at the same rate from year to year (Figure 16–4).

Had we initiated this population with a stable age distribution, its growth rate (λ) would have been constant from the beginning, as shown by the following example.

	Year									
	0	1	2	3	4	5	6	7	8	% of column 8
n_0	63	94	138	207	306	453	674	1002	1491	(63.4)
n_1	21	31	47	69	103	153	226	337	501	(21.3)
n_2	12	17	25	38	55	82	122	181	270	(11.5)
n_3	4	6	8	12	19	27	41	61	90	(3.8)
N	100	148	218	326	483	715	1063	1581	2352	
λ		1.48	1.47	1.50	1.48	1.48	1.49	1.49	1.49	

(The small variations in growth rate in the earlier years are due to rounding individuals to whole numbers when the population is small.) As you can see, the population grows at the same rate ($\lambda = 1.49$) as that achieved in the first example after several rounds of birth and death. Also the age structure of the second population remains unchanged.

The stable age distribution and growth rate depend upon birth and survival values. Any change in the life table alters the stable age distribution and results in a new rate of population growth. Consider the following example, in which survival and fecundity of our imagined population are reduced to

Survival	Birth rate	Number of individuals
$s_0 = 0.3$	$b_0 = 0$	$n_0(0) = 1491$
$s_1 = 0.6$	$b_1 = 1$	$n_1(0) = 501$
$s_2 = 0.3$	$b_2 = 2$	$n_2(0) = 270$
$s_3 = 0$	$b_3 = 1$	$n_3(0) = 90$

We start with the population left at the end of the previous example. Because of the change in the survival and birth rates, births no longer exceed deaths and the population declines:

	Year									
	0	1	2	3	4	5	6	7	8	% of column 8
n_0	1491	1130	965	776	642	525	431	352	286	(57.7)
n_1	501	447	339	289	233	193	157	129	106	(21.4)
n_2	270	301	268	203	174	140	116	94	76	(15.3)
n_3	90	81	90	81	61	52	42	35	28	(5.7)
N	2352	1959	1662	1349	1110	910	746	610	496	
λ		0.83	0.85	0.81	0.82	0.82	0.82	0.82	0.81	

As often happens in a declining population, the distribution of individuals shifts toward the older age classes. In this example, the shift to the new stable age distribution involves small changes and occurs quickly, after which the population achieves a growth rate of $\lambda = 0.82$. Early on, however, some age classes briefly increase (n_2 between $t = 0$ and $t = 1$) as the age distribution of the population readjusts to the new life table values.

The intrinsic rate of increase of a population is determined by its life table values.

The intrinsic rate of increase is the exponential or geometrical rate assumed by a population with a stable age distribution. When changing environmental conditions alter life table values, the age structure of the population continually readjusts to the new schedule of birth and death rates. In practice, therefore, populations rarely achieve stable age distributions and grow at their intrinsic rates of increase. Because the growth performance of a population depends as much on past conditions, which determine its age structure, as on the present life table values, the intrinsic rate of increase is more useful than observed growth rates for assessing the effect of environmental conditions or individual attributes on population growth.

Each life table has a single intrinsic rate of increase. Lotka (1907, 1922) first showed how one could derive the intrinsic rate of increase from birth and death rates. The derivation follows upon the fact that in a stable age population each age class grows at the same exponential rate, as we have seen in the examples presented above (Figure 16–4) and as Lotka (1922) proved mathematically. The steps in the derivation below follow Lotka's, although the symbols have been changed to maintain consistency throughout this book.

The number of newborn individuals (n_0) at time t is the sum of offspring born to individuals of each age class (x) in the population:

$$n_0(t) = n_x(t)b_x \qquad (16\text{–}7)$$

Because the number of individuals of age x alive at time t [$n_x(t)$] is equal to the number of newborns x years ago times their survival to the present (l_x), we may substitute the expression

$$n_0(t-x)l_x = n_x(t)$$

in equation (16–7), giving

$$n_0(t) = \sum_{x=0}^{\infty} n_0(t-x)l_x b_x \qquad (16\text{–}8)$$

Remember that in a stable age population each age class grows exponentially according to the relation $n_x(t) = n_x(0)e^{rt}$. One may also calculate the size of an age class at some point in the past ($t-x$) by reversing the sign of r (going backwards in time), hence

$$n_0(t-x) = n_0(t)e^{-rx}$$

Substituting this expression into equation (16–8), we obtain

$$n_0(t) = n_0(t)e^{-rx}l_x b_x$$

Finally, dividing both sides of this equation by $n_0(t)$ gives

$$1 = \sum_{x=0}^{\infty} e^{-rx}l_x b_x \qquad (16\text{–}9)$$

This equation and the equivalent expression for geometric growth

$$1 = \sum_{x=0}^{\infty} \lambda^{-x}l_x b_x \qquad (16\text{–}10)$$

are known as Euler's (pronounced oiler's) equation, or the characteristic equation of a population.

The characteristic equation is a high order polynomial of the form $1 = a\lambda^{-1} + b^{-2} + c\lambda^{-3} + \ldots q\lambda^{-k}$, where k is the oldest ages class (note that the term for $x = 0$ is not included because b_0 is always 0; additional terms are left out when reproduction is delayed beyond age 1, as it is in humans). Multiplying all the terms in this polynomial by λ^{k+1} and rearranging slightly puts the equation in a more convenient form:

$$\lambda^{k+1} - a\lambda^{k} - b\lambda^{k-1} - c\lambda^{k-2} - \ldots - q = 0 \qquad (16\text{–}11)$$

From your experience in algebra factoring quadratic equations to determine their roots, you can appreciate the difficulty of trying to solve this equation when the number of age classes is greater than two or three. In practice, the roots of the equation must be found by trial and error. We must be grateful, however, for one property of equation (16–11); because there is only one change of sign in the coefficients of the terms, the characteristic equation has only one real positive root, which is the intrinsic rate of increase.

To illustrate the estimation of λ, let's use the simple life table example introduced earlier in the chapter:

x	s_x	l_x	b_x	$l_x b_x$	$x l_x b_x$
0	0.5	1.0	0	0	0
1	0.8	0.5	1	0.5	0.5
2	0.5	0.4	3	1.2	2.4
3	0.0	0.2	2	0.4	1.2
Sum				2.1	4.1

We wish to chose a value of λ or r for which the terms of the characteristic equation sum to 1 or, equivalently, the terms in equation (16–11) sum to 0. To obtain a reasonable initial estimate of r (r_a), Caughley (1977) recommends using

$$r_a = \frac{\sum l_x b_x \log_e \sum l_x b_x}{\sum x l_x b_x}$$

From the sums in the preceding table, we calculate r_a to be 0.38; this is equivalent to $\lambda = 1.46$, close to the observed value of about 1.48 after the population achieved a stable age distribution. When we substitute the value $r_a = 0.38$ into the right-hand terms of the characteristic equation (16–9), these sum to 1.032. In order to reduce the sum to 1, we must increase the value of r. When evaluated for $r = 0.39$ and $r = 0.40$, the terms add to 1.013 and 0.995, and we know that the positive root of the equation must lie between these values. A linear interpolation gives an estimate of $r = 0.397$, or $\lambda = 1.487$, making the sum of the right-hand term equal to 1.000, which is close enough. Indeed, for most ecological work, particularly because comparisons between populations usually are more important than the precise characterization of a single population, estimates such as r_a are adequate. For example, for the life table of red deer on the Isle of Rhum, Great Britain, prior to 1957 (Table 16–1), $l_x b_x = 1.316$, $x l_x b_x = 7.720$, and $r_a = 0.47$. After 1957, $l_x b_x = 0.923$, $x l_x b_x = 4.307$, and $r_a = -0.017$, showing that management practices implemented in 1957 reduced the intrinsic growth rate to below zero.

The Lewis-Leslie matrix formalizes the projection of population growth for computer analysis.

Economic forecasters and actuarians require precise predictions of population change and have developed complicated mathematical programs for turning data on birth rates, death rates, and age structure into projections of

Table 16-1
Static life table of the red deer on the Island of Rhum constructed from cross-sectional data on ages at death and fecundity during a short period

x	a_x*	l_x	m_x	s_x	b_x
1	129	1.000	0.116	0.884	0
2	114	0.884	0.009	0.991	0
3	113	0.876	0.055	0.945	0.311
4	81	0.625	0.037	0.963	0.278
5	78	0.605	0.245	0.755	0.302
6	59	0.457	—†	—	0.400
7	65	0.504	0.155	0.845	0.476
8	55	0.426	0.545	0.455	0.358
9	25	0.194	0.639	0.361	0.447
10	9	0.070	0.114	0.886	0.289
11	8	0.062	0.129	0.871	0.283
12	7	0.054	0.704	0.296	0.285
13	2	0.016	0.500	0.500	0.283
14	1	0.008	—	—	0.282
15	4	0.031	0.484	0.516	0.285
16	2	0.016	1.000	0.000	0.284
17	0	0.000			

* a_x = number of individuals alive at a particular age (x).
† Mortality and survival rates cannot be calculated because the number of individuals *increased* from this age class to the next.
Source: Lowe 1969.

population growth. Demographers subdivide human populations into many age classes, generally year by year, because the bureaucratic cornucopia of actuarial data allows such fine characterization. The equations for projecting population growth are similarly long. Furthermore, human birth and death rates are blown about by the winds of war, epidemics, and economic vagaries, resulting in complicated age structures (Figure 16-5). To handle these data, demographers have used the shorthand of matrix algebra, which allows one to portray sets of related equations as single variables. First developed by Lewis (1942) and Leslie (1945, 1948), the population matrix not only provides a convenient method for the projection of population growth, but also has played a prominent role in developing demographic theory. (Leslie's motivation to formulate demographic equations in matrix terms stemmed from his own work with laboratory populations of the vole *Microtis agrestis;* Leslie and Ranson 1940).

Recall the system of equations for projecting population growth written earlier in this chapter:

$$n_0(t+1) = n_0(t)s_0b_1 + n_1(t)s_1b_2 + n_2(t)s_2b_3 + n_3(t)s_3b_4 + \ldots$$

$$n_1(t+1) = n_0(t)s_0$$

$$n_2(t+1) = n_1(t)s_1$$

$$n_3(t+1) = n_2(t)s_2$$

$$\ldots$$

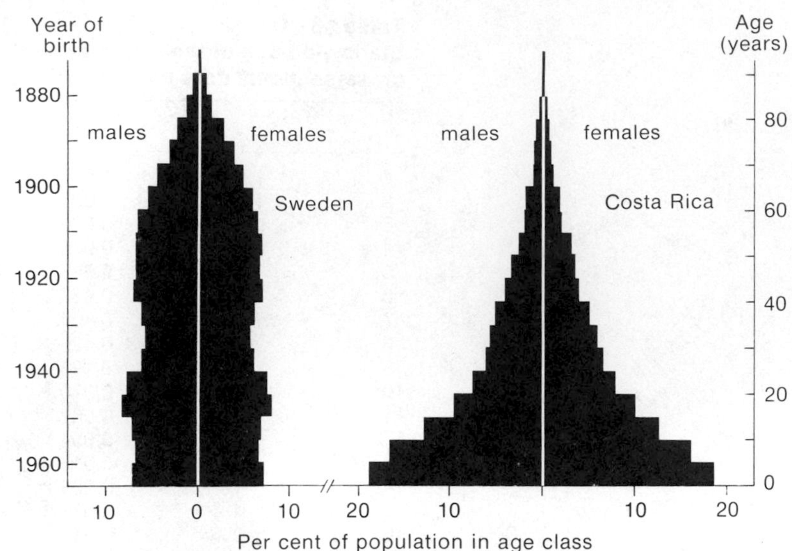

Figure 16–5
Population age structure of Sweden in 1965 and of Costa Rica in 1963. Because Sweden's population has grown slowly, its age structure is distributed toward older ages. Declining birth rates during the Depression and the baby boom that followed World War II are responsible for irregularities in the age structure. Costa Rica's rapid population growth, caused by a high birth rate, has resulted in a bottom-heavy age structure. (After data in Keyfitz and Flieger 1968.)

These equations may be written somewhat differently to show that they belong to the same set:

$$n_0(t+1) = n_0(t)s_0b_1 + n_1(t)s_1b_2 + n_2(t)s_2b_3 + n_3(t)s_3b_4 + \dots$$
$$n_1(t+1) = n_0(t)s_0 + n_1(t) \times 0 + n_2(t) \times 0 + n_3(t) \times 0 + \dots$$
$$n_2(t+1) = n_0(t) \times 0 + n_1(t)s_1 + n_2(t) \times 0 + n_3(t) \times 0 + \dots$$
$$n_3(t+1) = n_0(t) \times 0 + n_1(t) \times 0 + n_2(t)s_2 + n_3(t) \times 0 + \dots$$
$$\dots$$

Because each column in the right-hand set of terms has the same coefficient in $n_x(t)$, these coefficients may be factored out of the equations as a single column of entries. We are then left with a matrix of s_x and b_x terms and two columns, one of $n_x(t+1)$ and the other of $n_x(t)$. In matrix notation, the set of equations depicted above is expressed as

$$\begin{bmatrix} f_0 & f_1 & f_2 & f_3 & \dots \\ s_0 & 0 & 0 & 0 & \dots \\ 0 & s_1 & 0 & 0 & \dots \\ 0 & 0 & s_2 & 0 & \dots \\ . & . & . & . & \dots \end{bmatrix} \begin{bmatrix} n_0(t) \\ n_1(t) \\ n_2(t) \\ n_3(t) \\ \dots \end{bmatrix} = \begin{bmatrix} n_0(t+1) \\ n_1(t+1) \\ n_2(t+1) \\ n_3(t+1) \\ \dots \end{bmatrix}$$

The square matrix of coefficients on the left is the population matrix, designated in capital and boldface as **A.** Note that, for convenience, the terms s_0b_1, s_1b_2, . . . , are replaced by f_0, f_1, . . . The **A** matrix has the same number ($k+1$) of rows as columns, where k is the oldest age class in the population. The columns of $n_x(t)$ and $n_x(t+1)$, called column vectors and always named in lower-case bold letters, are called $\mathbf{n}(t)$ and $\mathbf{n}(t+1)$. Therefore, the projection of the population from time t to $t+1$ can be expressed by the matrix multiplication

$$\mathbf{An}(t) = \mathbf{n}(t+1)$$

For the simple life table given earlier, with four age classes $x = 0, 1, 2, 3$, the terms f_x are 0.5, 2.4, 1.0, and 0. Starting with a population structure $[n_x(t)]$ of 20, 10, 40, and 30 individuals, the matrix multiplication for the first unit of time, t to $t+1$, is

$$\begin{bmatrix} 0.5 & 2.4 & 1.0 & 0 \\ 0.5 & 0 & 0 & 0 \\ 0 & 0.8 & 0 & 0 \\ 0 & 0 & 0.5 & 0 \end{bmatrix} \begin{bmatrix} 20 \\ 10 \\ 40 \\ 30 \end{bmatrix} = \begin{bmatrix} 74 \\ 10 \\ 8 \\ 20 \end{bmatrix}$$

So far, the math is no different than in the pencil and paper example we worked earlier. For many time units of growth, the matrix equation is written

$$\mathbf{A}^t\mathbf{n}(0) = \mathbf{n}(t)$$

and here is where computer programs designed to handle matrix manipulations can greatly simplify demographic calculations.

The matrix approach also provides a simple proof of the existence of a stable age distribution. Because **A** is a square matrix, there is some column vector $\mathbf{n}(s)$ that satisfies the equation

$$\mathbf{An}(s) = \lambda\mathbf{n}(s) \qquad (16-12)$$

The constant λ is the factor by which each element of the vector of $n_x(t)$ increases as a result of a single matrix multiplication, that is, the geometric growth rate. The solution to equation (16–12) has as many roots (values of λ) as there are rows or columns but, as we have seen above, only one of these is a real positive number and therefore has biological meaning. Similarly, only one vector $\mathbf{n}(s)$ will be used. In matrix algebra, the roots of a square matrix and their corresponding vectors **n** are called eigenvalues (λ) and eigenvectors (after the German *eigen,* which may be translated as "characteristic").

The use of matrix algebra in population analysis is discussed by Begon and Mortimer (1981: 41 ff), Elseth and Baumgardner (1981: 285 ff), Jeffers (1978: 48 ff), and Pielou (1969: 33 ff, 1977: 41 ff), among others. Pielou shows how the eigenvalues of matrix **A** can be determined by finding the roots of the polynomial equation

$$\lambda^{k+1} - f_0\lambda^k - s_0f_1\lambda^{k-1} - \ldots - (s_0s_1 \ldots s_{k-1})f_k = 0$$

which is identical to equation (16–11) above. In the case of the life table presented earlier, this equation is $\lambda^4 - 0.5\lambda^3 - 1.2\lambda^2 - 0.4\lambda - 0 = 0$. Its one positive root, determined by computer (many hand calculators can now handle the job), is $\lambda = 1.4875$, precisely the same as our earlier trial-and-error estimate of 1.4874. The second pencil and paper life table, that depicting a declining population, has an eigenvalue of $\lambda = 0.8196$.

Each eigenvalue of **A** has an infinite number of eigenvectors because when each element of **n**(s) is multiplied by the same constant, equation (16–12) remains satisfied. This makes sense in that as a stable-age population grows, each of its age classes grows at the same rate; only the relative proportions of individuals in the age classes remain constant. For this reason, one must arbitrarily fix the size of one of the age classes or the total number in the population to calculate the stable-age eigenvector. For example, setting the numbers in the oldest age class (k) equal to 1, the elements of **n**(s) are

$$n_0 = \lambda^k/(s_0 s_1 s_2 s_3 \ldots s_{k-1})$$

$$n_1 = \lambda^{k-1}/(s_1 s_2 s_3 \ldots s_{k-1})$$

$$n_2 = \lambda^{k-2}/(s_2 s_3 \ldots s_{k-1})$$

and more generally

$$n_x = \lambda^{k-x}/(\Pi s_i)$$

$$n_{k-1} = \lambda/s_{k-1}$$

$$n_k = 1$$

The proportions of individuals in each age class can be determined by dividing the number by the total in the population, as shown below:

	Population 1		Population 2	
x	$n_x(s)$	c_x	$n_x(s)$	c_x
0	16.4566	0.634	10.1956	0.577
1	5.5316	0.213	3.7319	0.211
2	2.9750	0.115	2.7320	0.155
3	1.000	0.039	1.0000	0.057
Sum	25.9632		17.6595	

These numbers should be compared with the pencil and paper populations that we worked with earlier.

Stochastic population models reflect the probabalistic nature of births and deaths.

In characterizing population growth thus far, we have described the realized growth of a population as determined by the birth and death rates of the population. Particular values for birth and death rates **(A)** and initial population structure [**n**(0)] always will result in the same course of population growth or decline. Such models, whose outcomes are certain, are called deterministic. But when conditions remain constant, the numbers of births and deaths in two identical populations over the same period might be expected to vary just by chance alone. For example, when two children each flip 10 pennies, although the probability of a head on each toss is one-half, one might turn up 6 heads and the other, 3. If the process were repeated, the average outcome would be 5 heads and 5 tails, but many trials would turn up 4 or 6 heads, somewhat fewer would yield 3 or 7 heads, and runs with all heads or all tails would occur once in 1024 trials, on average.

Penny-flipping provides a useful analogy for population growth. Thus if the probability of birth is 0.5 per year, a population of 10 individuals has an expected number of offspring of 5, but the realized number may vary considerably from that average. To what extent does such variability reduce the usefulness of deterministic models? Perhaps the best way to answer this question is to construct models of population growth that take into account the probabilistic nature of birth and death. These are called stochastic models.

The simplest stochastic model is that of a pure birth process having exponential rate b in a continuously growing population (Pielou 1969, 1977). The probability of giving birth in a small interval Δt equals $b\Delta t$, and the probability of not giving birth is $(1 - b\Delta t)$. Now, a little calculus gives us the probability $[P_N(t)]$ of the population changing in size from $N(0)$ to a particular value of $N(t)$ over t units of time:

$$P_N(t) = \binom{N(t) - 1}{N(0) - 1} e^{-N(0)bt}(1 - e^{-bt})^{[N(t)-N(0)]}$$

Figure 16–6 illustrates the outcome of a stochastic birth process for a popu-

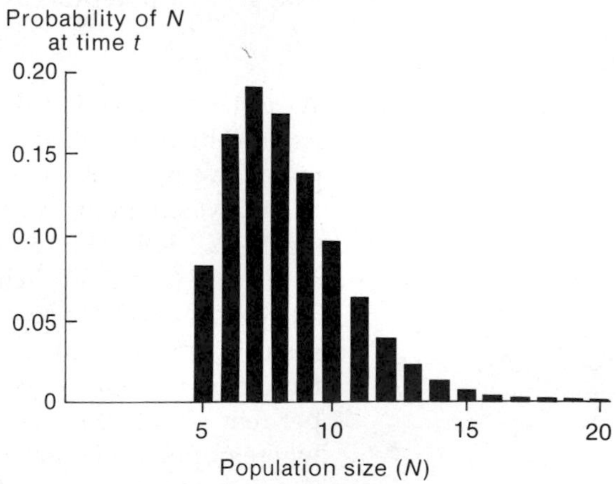

Figure 16–6
The probability distribution of the number of individuals at time t in a population undergoing a pure birth process. (After Pielou 1977.)

lation with an initial size $N(0)$ of 5 and for $bt = 0.5$. Notice that birth rate and time occur in the equation only as their product. Hence a high reproductive rate over a shorter period will result in the same probability distribution of sizes as a low reproductive rate over a longer period, so long as the product bt is the same. The average, or expected, size of the population is $N(t) = N(0)e^{bt}$, just as in deterministic exponential growth. For the present example, the expected value is therefore 8.24. As you can see, however, no births will have occurred $[N(t) = 5]$ with a fair probability, and it is a reasonable possibility that the population will have increased to 15 or more individuals.

The pure birth process described above was first proposed by G. U. Yule (1924) to characterize the evolution of new species within genera, the process of speciation being to species what birth is to individuals. Since then, stochastic models have become much more complex and far-reaching, even extending to the population matrix. Clearly each population process has probabilistic qualities. Number of births per season may be fairly uniform among individuals within a population if impregnation is certain and the cost of producing offspring is small. In this case, stochastic models may be less appropriate than deterministic characterizations. Death almost always falls capriciously on individuals, which are constantly at risk of being caught by predators or suffering other accidents. Therefore, death is probabilistic and should be treated stochastically.

In this book, however, we shall use the simpler deterministic models to describe population growth. Stochastic modeling in population biology is still young enough and difficult enough that only a small body of results has accumulated, and these do not present a picture that differs substantially from a deterministic view of life. With respect to some ecological problems, however, stochastic models have played a major role in our thinking and these will be discussed where appropriate.

The growth potential of populations is great.

We can best appreciate the capacity of a population for growth by following its rapid increase when introduced to a new region with a suitable environment. Two male and 6 female ring-necked pheasants released on Protecton Island, Washington, in 1937 increased to 1325 adults within 5 years (Einarsen 1942, 1949). The 166-fold increase represents a 180 per cent annual rate of increase ($\lambda = 2.80$). When domestic sheep were introduced to Tasmania, a large island off the coast of Australia, the population increased from less than 200,000 in 1820 to more than 2 million in 1850 (Davidson 1938). The tenfold increase in 30 years is equivalent to an annual rate of increase of 8 per cent ($\lambda = 1.08$). Even such an unlikely creature as the elephant seal, whose population had been all but obliterated by hunting during the nineteenth century, increased from 20 individuals in 1890 to 30,000 in 1970 ($\lambda = 1.096$; Bonnell and Selander 1974). If you are unimpressed, consider than another century of unrestrained growth would find 27,000,000 elephant seals crowding surfers and sunbathers off southern California beaches. Before the end of the next century, the shore lines of the Western Hemisphere would give lodging to a trillion of the beasts.

Elephant seal populations do not hold any growth records. Life tables of populations maintained under optimum conditions in the laboratory have exhibited potential annual growth rates (λ) as great as 24 for the field vole (Leslie and Ranson 1941), 10 billion (10^{10}) for flour beetles (Leslie and Park 1949), and 10^{30} for the water flea *Daphnia* (Marshall 1962).

Rapid growth rates may be conveniently expressed in terms of the time required for the population to double in number. The relationship between geometric growth rate (λ) and doubling time (t_2), derived from the equation for geometric growth, is $t_2 = \log_e 2/\log_e \lambda$; that is, $0.69/\log_e \lambda$. Hence for the field vole ($\lambda = 24$), $t_2 = 0.69/\log_e(24)$, which is 0.22 years or 79 days. Corresponding doubling times are 7.6 years for the elephant seal, 8 months for the pheasant, 10 days for the flour beetle, and less than 3 days for the water flea. Populations of microorganisms and many unicellular plants and animals can double in a day or a few hours.

The intrinsic growth rate of a population varies with environmental conditions and population density.

The intrinsic growth rate of a population is strictly determined by the life table values, which express the interaction between the individual and its environment. As a consequence of this interaction, the life table (hence the intrinsic growth rate) varies with respect to the conditions of the environment.

These conditions vary spatially and temporally, creating differences in population dynamics from place to place and leading to changing population dynamics over time. Such effects of the environment are best revealed in experimental studies in which groups of individuals from the same population are provided with different conditions or combinations of conditions. For example, L. C. Birch (1953) determined the intrinsic rate of increase (λ) of populations of two species of grain beetles, *Calandra oryzae* and *Rhizopertha dominica,* over a wide range of temperature and moisture (Figure 16–7). Neither species performed well at low temperatures and humidities; moreover, the optimum conditions for growth differed between the species, *Rhizopertha* populations growing most rapidly at somewhat warmer temperatures. Of the two, *Rhizopertha* has the more tropical distribution in nature.

Climate and other environmental conditions have been shown to be important determinants of life table values of populations in their natural settings, as one would expect. For example, the European rabbit, introduced to Australia in the last century and now widespread throughout many habitats, survives longer but produces fewer offspring per year in arid regions than in the more mesic Mediterranean climate regions (Table 16–2).

The rabbit example raises the important issue of the regulation of population size. Continuous exponential growth leads, with time, to an

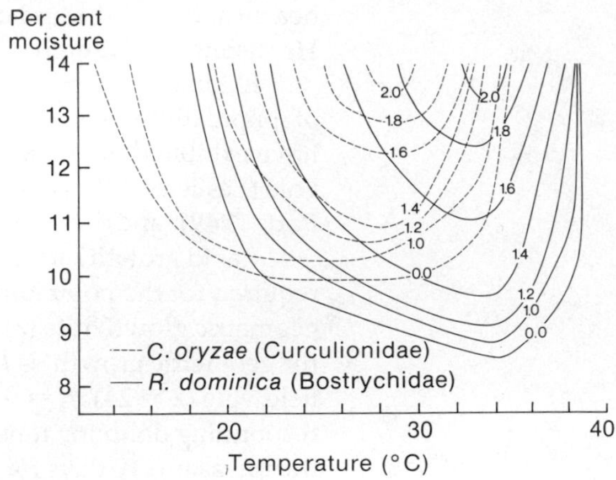

Figure 16-7
The influence of temperature and the moisture content of grain on the geometric rate of increase of populations of the grain beetles *Calandra oryzae* and *Rhizopertha dominica* living in wheat. Rates of increase are indicated by contour lines that describe conditions with identical values of λ. (After Birch 1953a.)

inconceivable number. As Darwin (1872) put it, "Even slow-breeding man has doubled in twenty-five years, and at this rate, in less than a thousand years, there would literally not be standing-room for his progeny." The growth potential of populations is driven home by the example of rabbits in Australia. In 1859, 12 pairs were released on a ranch in Victoria to provide sport for hunters. Within 6 years, the population had increased so rapidly

Table 16-2
Condensed life tables for European rabbits in arid and Mediterranean climates of Australia

Age (months)	Pivotal age* (months)	Survival (l_x)†	Proportion of female population	Females pregnant at any one time	Litter size (embryos)	b_x	$l_x b_x$	$x l_x b_x$
Mediterranean								
3-6	4.5	0.222	0.177	0.23	4.4	1.8	0.198	0.891
6-12	9.0	0.160	0.359	0.43	5.6	8.7	0.696	6.264
12-18	15.0	0.075	0.303	0.53	5.9	9.4	0.357	5.358
18-24	21.0	0.028	0.107	0.45	5.9	2.8	0.039	0.823
>24	37.5	0.006	0.053	0.55	6.2	1.8	0.005	0.203
Total							1.295	13.539
Arid								
3-6	4.5	0.570	0.140	0.05	3.8	0.2	0.057	0.257
6-12	9.0	0.457	0.265	0.15	4.7	1.8	0.411	3.702
12-18	15.0	0.302	0.217	0.32	4.6	3.1	0.468	7.022
18-24	21.0	0.186	0.192	0.37	4.3	3.1	0.288	6.054
>24	37.5	0.061	0.187	0.34	4.5	2.9	0.089	3.317
Total							1.313	20.352

* Midpoint of the age interval, frequently used to calculate generation time when reproduction is more or less continuous, as it is in rabbits in Australia.
† Based on rabbits aged 0-3 months; hence $l_{1.5} = 1$.
Source: Myers 1970.

that 20,000 rabbits were killed in a single hunting drive. Even by conservative estimate, the population must have increased by a factor of at least 10,000 in 6 years, an exponential rate (r) of about 1.5 per year. Yet the life tables of present-day populations (Table 16–2) suggest growth rates (r_a) of 0.30 and 0.21. Considering the difficulty of estimating survival of the young to reproduction and the statistical errors involved in such studies, these values probably do not differ from $r = 0$. Furthermore, under environmental conditions so different as to produce up to twofold or threefold differences in survival rates, the intrinsic growth rates of the populations are nearly identical. To achieve this, differences in rates of fecundity must balance the differences in survival. Where survival is higher, fecundity is lower.

How can one reconcile the initial rapid growth with the eventual stabilization of the rabbit population? Either birth rates decreased, death rates increased, or both when the population became more numerous. When there are more rabbits, there is less food for each; with fewer resources, fewer offspring can be nourished and these survive less well. Such density dependence of the life table values underlies the regulation of population size, the subject of the next chapter.

SUMMARY

The economic and social consequences of the rapid increase of the human population stimulated efforts to develop mathematical models to predict the future course of growth. These models were quickly applied to the growth of natural populations in the 1920s and have since provided a quantitative framework for understanding population dynamics. In this chapter we have characterized population growth mathematically as a basis for considering factors that regulate population growth, including interactions with other populations, in subsequent chapters.

1. Population growth can be described by the exponential rate of increase (r) in the expression

$$N(t) = N(0)e^{rt}$$

The factor by which a population increases in one unit of time (e^r) is the geometric growth rate of the population (λ). λ and e^r are interchangeable in population equations.

2. The exponential growth rate is the difference between the birth and death rates averaged over individuals (per capita) in the population (that is, $r = b - d$).

3. The instantaneous rate of increase of an exponentially growing population is $dN/dt = rN$.

4. Populations with discrete breeding seasons increase geometrically by periodic increments according to the relation $N(t+1) = N(t)\lambda$.

5. When birth and death rates vary according to the age of the individual (summarized by the life table), one must additionally know the proportions of individuals in each age class in order to project population growth.

6. A population with a fixed life table assumes a stable age distribution in which the numbers in each age class, as well as the population as a whole, increase at the same exponential or geometric rate known as the intrinsic rate of increase of the population.

7. The intrinsic rate of increase for any given life table is the positive root of a polynomial equation of the form $1 = \Sigma\lambda^{-x}l_x b_x$

8. Matrix algebra provides a shorthand method of handling discrete-time population equations.

9. Birth and death are probabilistic processes and are more realistically treated by stochastic, rather than deterministic, population models. But although stochastic models show that realized population growth is variable and depends on chance events, deterministic models appear to be adequate for most purposes, particularly when initial population size is large; they have been adopted widely by ecologists for both descriptive and theoretical purposes.

10. Finally, the life table of a population may be expected to vary with the conditions of the environment and the density of the population.

17

Population Regulation

Even the most slowly reproducing species would cover the earth in a short time if its population growth were unrestrained. Malthus (1798) understood that this "implies a strong and constantly operating check on population from the difficulty of subsistence." Pushing the point further, he wrote:

> Through the animal and vegetable kingdoms, nature has scattered the seeds of life abroad with the most profuse and liberal hand. She has been comparatively sparing in the room and the nourishment necessary to rear them. The germs of existence contained in this spot of earth, with ample food, and ample room to expand in, would fill millions of worlds in the course of a few thousand years. Necessity, that emperious all pervading law of nature, restrains them within the prescribed bounds. The race of plants, and the race of animals shrink under this great restrictive law.

Darwin echoed this view in *On the Origin of Species*: . . . as more individuals are produced than can possibly survive, there must in every case be a struggle of existence, either one individual with another of the same species, or with the individuals of distinct species, or with the physical conditions of life. It is the doctrine of Malthus applied with manifold force to the whole animal and vegetable kingdoms; for in this case there can be no artificial increase of food, and no prudential restraint from marriage. Although some species may be now increasing, more or less rapidly, in numbers, all cannot do so, for the world would not hold them.

At about the same time Herbert Spencer, an English philosopher and contemporary of Darwin, presented a more rigorous, quantitative concept of population regulation in his book *First Principles* (1863):

> Every species of plant and animal is perpetually undergoing a rhythmical variation in number—now from abundance of food and absence of enemies rising above its average, and then by a consequent scarcity of food and abundance of enemies being depressed below its average. . . . [A]mid

these oscillations produced by their conflict, lies that average number of the species at which its expansive tendency is in equilibrium with surrounding repressive tendencies. Now can it be questioned that this balancing of the preservative and destructive forces which we see going on in every race must necessarily go on. Since increase of numbers cannot but continue until increase of mortality stops it; and decrease of number cannot but continue until it is either arrested by fertility or extinguishes the race entirely.

This essentially modern view of the regulation of populations grew out of an awareness of the immense capacity of populations for exponential increase. In a sense, a population's growth potential and the relative constancy of its numbers cannot be logically reconciled otherwise. However, to say that faltering birth rates and increasing death rates check a population as its numbers grow is to do little more than elaborate the definition of a regulated population. Yet during the 1920s, when Lotka and others gave population biology its quantitative foundations, the projection of growth in a regulated population was one of the first problems to be tackled.

The logistic equation describes the growth of a regulated population.

In 1920, Raymond Pearl and L. J. Reed, at the Institute for Biological Research of Johns Hopkins University, published a paper in the *Proceedings of the National Academy of Sciences (U.S.)* entitled "On the Rate of Growth of the Population of the United States Since 1790 and its Mathematical Representation." Thorough and accurate population data had been gathered even in colonial times. Indeed, the phenomenal population growth of the American colonies had greatly impressed upon Malthus how rapidly humans could multiply; this was not so evident in the more crowded European countries of his time. Pearl and Reed wished to project the future growth of the population, which they supposed must eventually reach a limit. Data for the population to 1910, the latest census then available, had revealed a decline in the exponential rate of growth (Figure 17–1). Pearl and Reed reasoned that if the decline followed a regular pattern that could be described mathematically, then one could predict the future course of the population, so long as the decline in the exponential growth rate continued. They also reasoned that changes in the exponential rate of growth must be related to the size of the population rather than to time, because the time scale is arbitrary with respect to any particular population. And so, in place of the constant r in the differential equation for unrestrained population growth ($dN/dt = rN$), Pearl and Reed made r a function of N; that is, $dN/dt = r(N)N$. Specifically, they defined

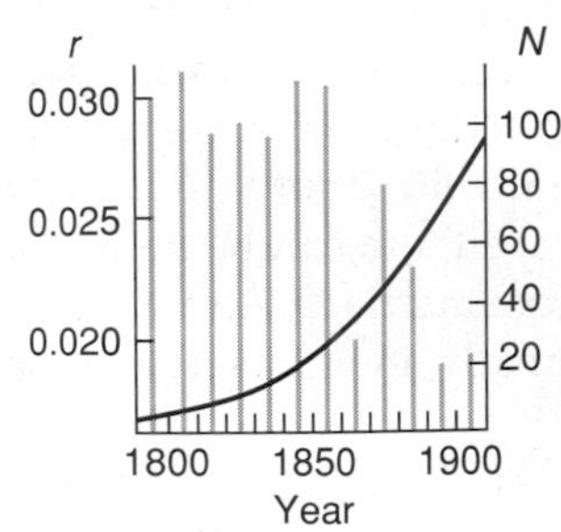

Figure 17–1
Increase in the population of the United States between 1790 and 1910 (continuous line) and the exponential rate of increase during each 10-year period (vertical bars). (From data in Pearl and Reed 1920.)

$$r(N) = r(1 - N/K)$$

Thus their differential equation describing restricted population growth

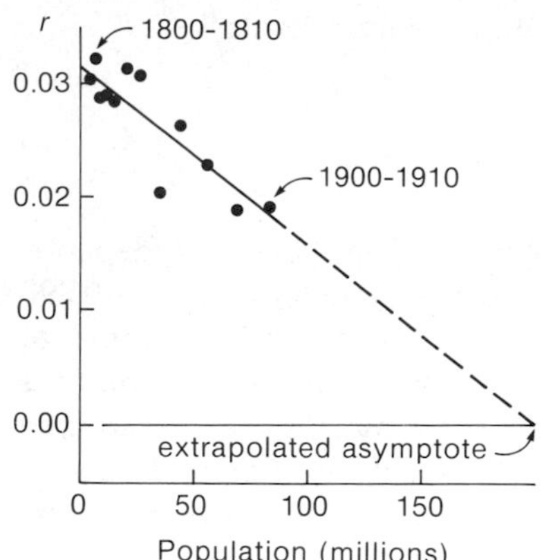

Figure 17–2
Exponential rates of population increase in the United States during each decade between 1790 and 1910 plotted as a function of the population size during that decade (the geometric mean of the beginning and ending numbers). (Data from Figure 17–1.)

became

$$\frac{dN}{dt} = rN\left(1 - \frac{N}{K}\right) \tag{17-1}$$

According to this equation, which is called the logistic equation, the exponential rate of increase decreases as a linear function of the size of the population. This reasonably approximated the data for the population of the United States (Figure 17–2). So long as population size (N) does not exceed the constant K—that is, N/K is less than 1—the population continues to increase, albeit at a slowing rate. When N exceeds the value of K, the ratio N/K exceeds 1, the term in parentheses $(1 - N/K)$ becomes negative, and the population decreases. Because populations below K increase and those above K decrease, K is the eventual equilibrium size of a population growing according to equation (17–1). The resulting relationships between N, $r(N)$, and dN/dt are shown together in Figure 17–3. The curve for dN/dt is a parabola, having a maximum at intermediate population size, specifically when $N = K/2$, and falling to 0 as N approaches either 0 or K. The humped shape of this curve arises from the circumstances that dN/dt is the product of two linear terms, one of which (N) increases with population size while the other [$r(N)$] decreases.

The course of population growth according to Pearl and Reed's equation can be found by integrating the differential. This is done by rearranging equation (17-1) to

$$\frac{dN}{N(1 - N/K)} = rdt$$

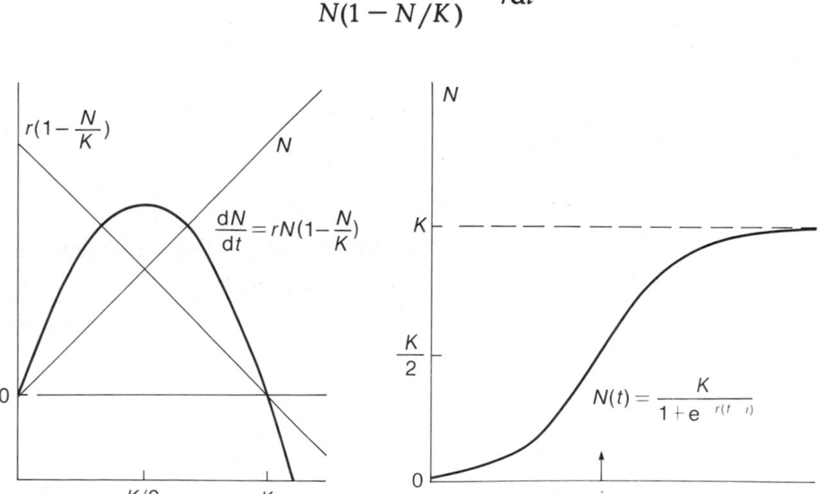

Figure 17–3
Various representations of the logistic curve of population growth. *Left:* The absolute rate of growth (dN/dt), which is the product of population size (N) and the expression $r(1 - N/K)$, reaches a maximum when population size is one-half the asymptote ((K). *Right:* The increase in numbers over time follows an S-shaped curve that is symmetrical about the inflection point.

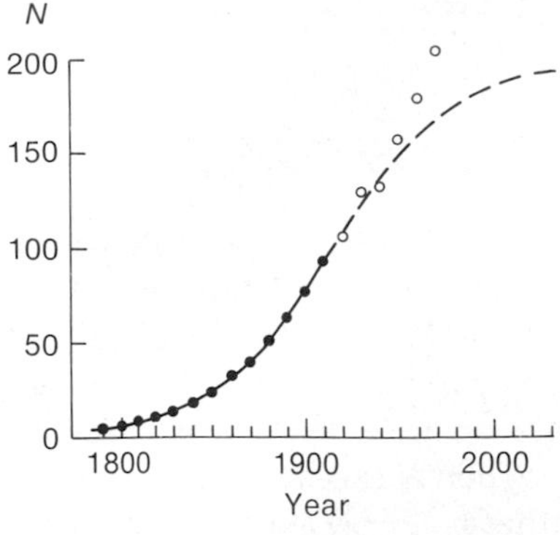

Figure 17–4
A logistic curve fitted to the population of the United States between 1790 and 1910 (solid dots). Subsequent censuses (open dots) have departed from the population curve projected by Pearl and Reed's 1920 equation.

and integrating. The steps are a bit messy but the final result is

$$N(t) = \frac{K}{1 + be^{-rt}} \tag{17-2}$$

where b is a constant equal to $[K - N(0)]/N(0)$. The value of b depends arbitrarily on the size of the population at the designated time zero—1790 in the case of Pearl and Reed's data. Equation (17-2) describes a sigmoid, or S-shaped, curve: the population grows slowly initially, then more rapidly as the number of individuals increases, and finally more slowly, gradually approaching the equilibrium number K. Applied to the growth of the population of United States from 1790 to 1910, this curve is illustrated in Figure 17–4.

Pearl and Reed obtained the best fit of their equation to the population data when the value of K was 197,273,000 and that of r was 0.03134. Thus even though the population in 1910 was only 91,972,000, they extrapolated its future growth to twice the 1910 level, based upon earlier growth performance. As Lotka (1925) remarked, however, "Such a forecast as this, based on a rather heroic extrapolation, and made in ignorance of the physical factors that impose the limit, must, of course, be accepted with reserve." Certainly Lotka was correct in his reservation. The U.S. population reached 197,000,000 between 1960 and 1970, when it was still growing vigorously, and although a leveling off in the mid-200,000,000s can now be predicted on the basis of a much reduced birth rate in recent years, all this could easily change.

Lotka's comment upon ignorance of the factors that limit population size was telling. We shall say more about such factors later, but first a brief history of Pearl and Reed's equation (Lloyd 1967, Hutchinson 1978) will provide a little diversion and some insight into the tortuous but inexorable path of progress in science.

The history of the logistic equation begins with Verhulst in the nineteenth century.

Malthus's arguments about population had stimulated considerable interest in mathematical formulations of population growth, and many scholars had taken into account that imperious, all pervading, restrictive law of necessity which restrains populations within prescribed bounds. The most successful of these attempts was by the mathematician Pierre-François Verhulst, who, in 1838, published an equation identical to that developed by Pearl and Reed 82 years later. In a second paper (1845), Verhulst named this formula the logistic equation (French, *logistique*) because of its logarithmic-exponential form. Many scientists of the day discussed the equation and compared it with less successful alternatives, but Verhulst's contribution to population biology was submerged by a wave of other scientific developments, particularly, and somewhat ironically, that of Darwinism; the logistic equation disappeared from view by mid-century.

The early twentieth century saw increasing interest in mathematical biology and the accumulation of data to which calculations could be applied. In 1908, T. B. Robertson independently derived Verhulst's equation to describe the growth of organisms. Robertson coined the name "autocatakinetic" for his equation, capturing the essence of his idea that populations of cells, like populations of individuals (an analogy he did not make), increase by self-replication (that is, exponentially) but eventually come under the influence of growth restraints. In 1911, A. G. M'Kendrick and M. K. Pai, working at the Pasteur Institute of Southern India, again rederived the logistic equation to describe the growth of populations of the gut bacterium *Escherichia coli* cultured in artificial medium.

In 1913, the Swedish biologist Tor Carlson devised some wonderful equations to describe the growth of cultured yeast cells. One of these had the form

$$dN/dt = rK^n e^{qt}(1 - N/K)^n$$

which made $r(N)$ both an exponential function of time (t) and a function of population size (N). The equation had a variable, n, to rescale the effect of population size on $r(N)$ in a nonlinear fashion. But toward the end of his paper, Carlson recognized the virtures of simplicity and adequacy in M'Kendrick and Pai's equation and applied it to his yeast cells.

Then in 1920, Pearl and Reed, apparently unaware of these earlier papers, wrote their equation for the growth of populations. The following year, Pearl acknowledged Verhulst's work and his name "logistic" for their equation.

The logistic, finally made generally known by Pearl, was the first equation in ecology to generate research that would provide the data to test a mathematical model. Pearl reviewed the growth of populations in an article published in the *Quarterly Review of Biology* in 1927. Besides data on human populations, and Carlson's excellent data on yeast, which Pearl gave a prominent place, the only experimental work to which he could refer was his own on the fruit fly *Drosophila* (Pearl 1925). This was followed quickly, however, by similar work on flour beetles by Chapman (1928), water fleas by Tera and Tenaka (1928), and the protozoan *Paramecium* by Gause (1931); they all attempted to determine whether the growth of populations fit the shape of the logistic curve. By and large they were successful, and the logistic equation took a permanent place in population biology.

Lotka showed the relationship of the logistic equation to Taylor's polynomial series.

With his characteristic mathematical insight, Lotka (1925) pointed out that the logistic equation is merely the smallest number of terms of a Taylor

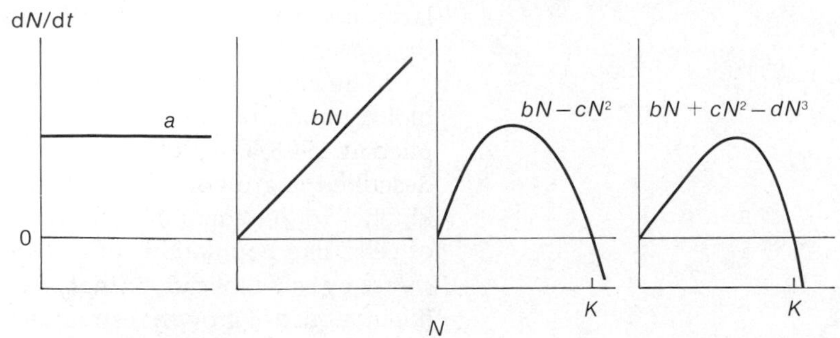

Figure 17–5
The absolute rate of growth of a population (d*N*/d*t*) as a function of density (*N*) modeled by zero-, first-, second-, and third-order polynomial equations.

expansion that can satisfactorily characterize restrained population growth. Taylor's Theorem states that any equation $f(X)$ in a single variable X can be approximated by a series of terms of the form $a + bX + cX^2 + dX^3 + \ldots$. Lotka pointed out that any equation describing the growth rate of a population in terms of its size (that is, $dN/dt = f(N)$) can therefore be expressed

$$dN/dt = a + bN + cN^2 + dN^3 + \ \ldots$$

Applied to the problem of describing population growth, the first term (a) must be deleted because as population size goes to zero the realities of life require that population growth rate also goes to zero. The second term is simply exponential growth, with $r = b$; hence it alone is not sufficient to describe regulated population growth. The second and third terms of Taylor's series together are, however, adequate if b is positive and c is negative; hence $dN/dt = bN - cN^2$ (Figure 17–5). Rearranged, this equation becomes $dN/dt = bN(1 - [c/b]N)$. Substituting r for b and K for b/c gives the familiar logistic equation. There is no reason why additional terms of Taylor's series should not be added to an equation for population growth. One example is shown graphically in Figure 17–5, and many others are possible. But, as Carlson realized, the logistic cannot be beat for simplicity and adequacy.

Populations are regulated by the effects of density-dependent factors.

Lotka was quick to point out that blind application of equations like the logistic was meaningless and potentially misleading unless one understood the biological processes responsible for the behavior that one could describe mathematically. The logistic equation describes a pattern of population growth; its terms do not explicitly describe underlying processes. Clearly, biologists needed to delve more deeply into the factors that caused logistic-like population growth curves to crop up time after time.

In their monograph on the growth and control of the gypsy moth population in the United States, L. O. Howard and W. F. Fiske (1911) made the useful distinction between factors whose effect on the population depended on density and those whose effect did not.

> In order that this balance (between growth potential and limits to population size) may exist it is necessary that among the factors which work together in restricting the multiplication of the species there shall be at least one, if not more, which is what is here termed facultative (for want of a better name), and which, by exerting a restraining influence which is relatively more effective when other conditions favor undue increase, serves to prevent it . . . .
>
> A very large proportion of the controlling agencies, such as the destruction wrought by storm, low or high temperature, or other climatic conditions, is to be classed as catastrophic, since they are wholly independent in their activities upon whether the insect which accidentally suffers is rare or abundant. The storm which destroys 10 caterpillars out of 50 which chance to be upon a tree would doubtless have destroyed 20 had there been 100 present, or 100 had there been 500 present. The average percentage of destruction remains the same, no matter how abundant or how near to extinction the insect may have become.

Howard and Fiske's "facultative" and "catastrophic" factors were called "density-dependent" and "density-independent" factors by H. S. Smith (1935) and these names have stuck. Many things influence rate of population growth, but only density-dependent factors, whose effect varies in relation to crowding, can possibly bring the population under control. Of prime importance among these must be limitation by food supply and places to live, and predators, parasites, and diseases whose effects are more strongly felt in crowded than in sparse populations. Other factors, such as temperature and moisture, influence birth and death rates without regard to numbers of individuals in a population; they therefore contribute only to the linear term of Taylor's series, that of exponential growth, while not influencing the quadratic term, which expresses the force of regulation.

Pearl's (1925, 1927) experimental studies on *Drosophila* provided the earliest indication of how density affects population processes. When confined to a bottle with a fixed supply of food, the descendants of a single pair of flies increase in number rapidly at first, but soon reach a limit. To measure the effects of density on birth and death rates, Pearl introduced different numbers of pairs of flies to otherwise identical culture bottles and recorded deaths of adults and number of progeny produced per day. Pearl found that the number of progeny raised per pair decreased as the density of flies in the bottle increased (Figure 17–6). Subsequent work has shown that this effect expresses competition among the larvae for food, which results in high mortality in dense cultures. Pearl also showed that adult life was cut short at high densities, well above the levels that affected the survival of larvae.

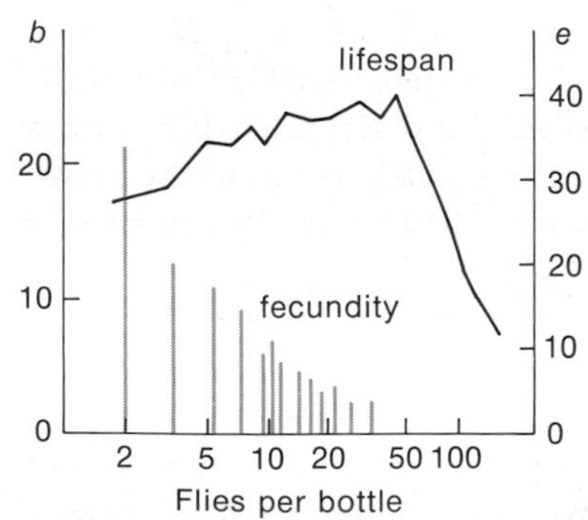

Figure 17–6
The influence of density on lifespan (*e*, days) and fecundity (*b*, progeny per day) in laboratory populations of the fruit fly *Drosophila*. (After Pearl 1927.)

Larvae of grain beetles compete for resources by direct confrontation.

Pearl's work provided the beginnings of what has become a vast body of research on the detailed relations of population processes to the density of individuals. Indicative of the early progress in this field was the work of A. C. Crombie (1944), of the University of Cambridge, on the grain beetle *Rhizopertha dominica. Rhizopertha* is a tiny bettle that completes its larval development within a single grain (of wheat, for example), living off the kernel of the seed. Chapman and others had already determined that grain beetle populations follow logistic growth curves, implying density-dependent regulation.

Females lay their eggs on the surfaces of seeds. Immediately after hatching the larva bores its way into the seed, where it commences development; later, the mature larva bores its way out of the same seed to pupate. Tough seed coats reduce the success of first instar *Rhizopertha* larvae entering grains of wheat, and mortality at this stage may reach 50 per cent. Once inside, the larva enjoys security so long as it is alone; a kernel of wheat cannot support more than one beetle. Yet, in a dense population, females may lay many eggs on the same grain.

Crombie discovered that females lay eggs without regard to the presence or absence of other eggs on a grain of wheat. He then sought to determine whether the larvae bored into grains at random or avoided those already occupied. To do so, he placed 150 wheat grains in a dish and introduced 100 first instar larvae. Twenty-four hours later, he dissected the grains to determine how many larvae had bored into each. If larvae had distributed themselves at random over the grains, oblivious of the presence or absence of other larvae, the proportion of grains infested by $x = 0, 1, 2, 3, \ldots$ larvae would have followed the Poisson distribution,

$$P(x) = \frac{M^x e^{-x}}{x!}$$

where M is the mean number of larvae per seed, in this case 100/150 or 0.67. The results of the experiment were statistically indistinguishable from a random distribution (Table 17–1); thus density played no part in the population process at this stage. In all probability, first instar larvae cannot determine whether a grain is already occupied.

What happens when two larvae meet in the same grain is another story, however. As Crombie described it:

> When two larvae in the first, second, third or fourth instars were put together into a small hole drilled in a wheat grain and watched under a binocular microscope, they were often seen to attack each other with their mandibles, and eventually either one or both left the hole. When a larva entered such a hole it always went to the bottom and turned round so as to face outwards. Other larvae trying to enter the hole were fiercely attacked. Sometimes such combats resulted in the body wall of one of the antagonists becoming punctured and its bleeding to death. In their tunnels in

Table 17–1
Distribution of numbers of larvae of grain beetles per seed compared to the Poisson distribution

	Number of grains containing *x* larvae				
	0	1	2	3	χ^2
Expected (Poisson distribution)	77	52	17	4	
Rhizopertha	74	59	14	3	1.84*
Sitotraga	71	63	12	4	2.42*
Mixed					
Rhizopertha		29	6	2	
Sitotraga		32	6	1	
Total	74	61	12	3	3.40*

Note: Each trial was based on 100 larvae introduced to 150 grains.
* Not significant at the 5% level.
Source: Crombie 1944.

> wheat grains larvae of all instars were always found curled up with the head facing towards the way they had entered. Furthermore, in all grains dissected during the experiments to be described, whenever two larvae were found in the same tunnel at least one of them was always dead.

The experiments Crombie referred to were as follows. He infested a single wheat grain with from 1 to 8 larvae, giving them 6 hours to become established in the grain. He then transferred the infested grain to a dish with 5 other fresh grains and left them together for 48 hours. At the end of the experiment, he dissected all the grains to determine how many larvae had been killed and how many had migrated to a different grain. The results (Figure 17–7) showed conclusively the strong effects of density on the survival and movement of larvae even within the first 2 days.

Crombie's beetles contested each grain of wheat through direct confrontation. After all, sole possession of a single grain is a larva's ticket to success. In this population, density dependence came into play abruptly as the number of larvae approached the number of grains of wheat. Above that level, regardless of the number of larvae relative to wheat grains the end result was always the same: one grain, one larva.

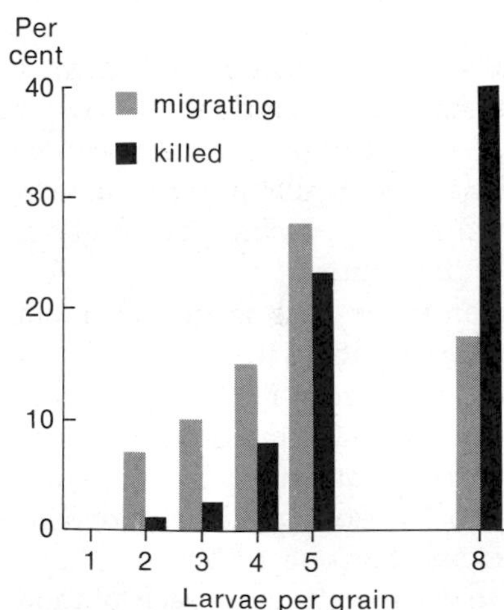

Figure 17–7
Effects of density (number of larvae per grain) on migration and mortality in the grain beetle *Rhizopertha.* (From data in Crombie 1944.)

Water fleas compete by reducing levels of their resources.

Unlike grain beetles, water fleas exert a less direct influence on each other through their consumption of resources. Each water flea consumes millions of prey, the single-celled green algae and diatoms of the plankton. As prey are consumed, the availability of food to other water fleas decreases gradually, leading to a graded response of birth and death rates to prey density. To demonstrate this effect, P. W. Frank, C. D. Boll, and R. W. Kelly (1957) cultured water fleas *(Daphnia pulex)* in small beakers in which populations were maintained at densities between 1 and 32 individuals per cubic centi-

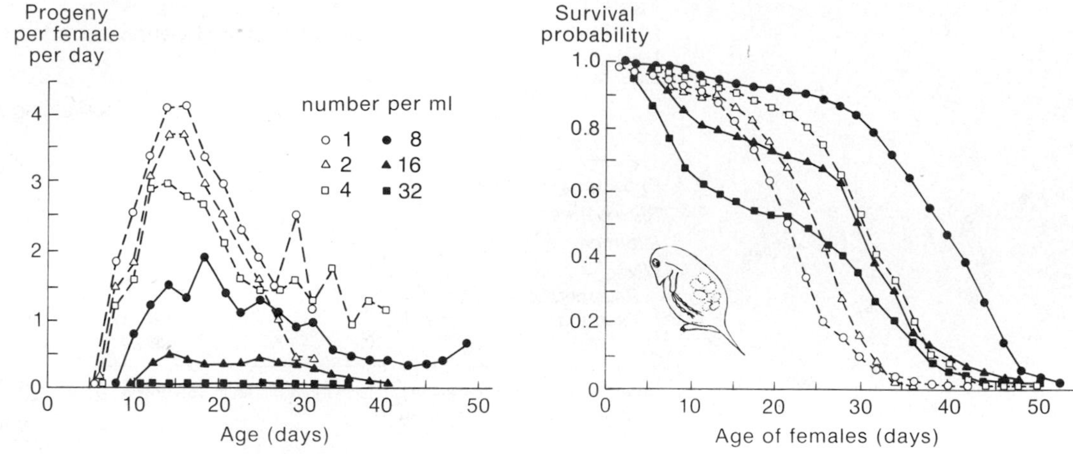

Figure 17–8
Fecundity and survival in laboratory populations of *Daphnia pulex* at different densities. (From Frank, Boll, and Kelly 1957.)

meter; they added green algae periodically to each culture, and noted the survival and fecundity of females for 2 months after the beginning of the experiment. Life tables for populations at each density revealed that fecundity decreased markedly with increasing population density (Figure 17–8). Somewhat unexpectedly, survival increased at densities up to 8 individuals cm^{-3} before decreasing at higher densities. At densities of 8 individual cm^{-3} and above, stunted body growth suggested that depletion of food resources between periodic replenishment of the alga stock culture limited birth rates and survival; the effect was clearly density-dependent.

The geometric rate of population growth (λ) calculated from the water flea life tables decreases linearly with increasing density and falls below 1.0 at a density of about 20 cm^{-3} (Figure 17–9). Therefore, under the conditions of temperature, light, water quality, and food availability provided in the laboratory, *Daphnia* populations would attain a stable size of about 20 cm^{-3}, regardless of the initial density of the culture.

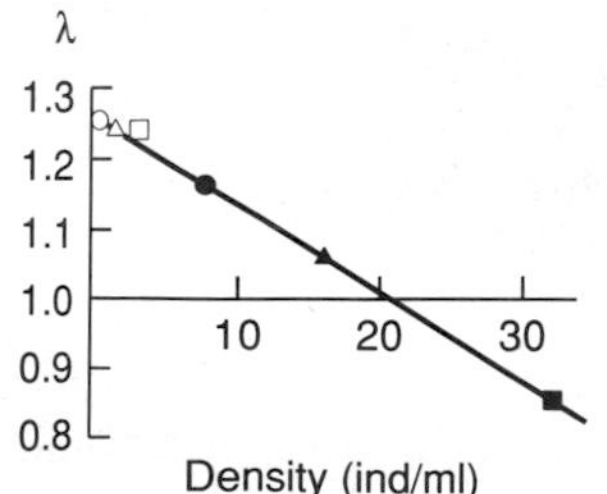

Figure 17–9
Values of λ calculated from the life-table data for *Daphnia pulex* portrayed in Figure 17–8. Population growth rate decreases as a function of density. (After Laughlin 1965.)

The data of Frank and his colleagues on water fleas relates birth and death rates to population density, but does not elucidate the mechanism of density dependence. Depletion of food seems the most likely explanation for reduced survival and fecundity at increased density, but excreted waste products, such as ammonia—which may accumulate in confined populations to the point that they act as poisons—could also cause density-dependent effects (Hammond 1938, 1939, Peters and Barbosa 1977).

Stephen Carpenter (1983) attempted to disentangle the effects of food and waste products in an experimental study of density dependence in the mosquito *Aedes triseriatus*, a species native to North America that carries human encephalitis in some areas. In the upper midwestern states, water-filled holes in trees provide particularly favorable breeding sites; mosquito larvae graze the protozoa, bacteria, and fungi that grow upon decaying leaves and other plant detritus in the tree holes (Fish and Carpenter 1982). Carpenter reproduced the essential character of these breeding sites in the laboratory by partly filling plastic petri dishes with water retrieved from the trunks of trees during rainstorms and providing measured quantities of beech leaves as a natural substrate for the microorganisms the mosquito larvae would feed upon (Figure 17–10). To distinguish the effects of food supply and waste products, he established cultures of larvae at all 27 combinations of three population sizes (10, 20, and 40 individuals), three amounts of leaves (25, 50, and 100 mg per larva), and three volumes of water (150, 300, and 600 cm³). To detect the biological effects of these treatments, Carpenter measured the survival of the larvae and the mass of the pupae; the latter partly determines the fecundity of the adult female. If food alone caused density dependence these traits would respond only to the ration of leaves per individual independently of the number of individuals or the volume of water. If excreted waste products exerted a strong influence, survival and pupal mass would respond to the number of individuals per volume of water.

The concentration of ammonia at the end of the experiment (71 days)

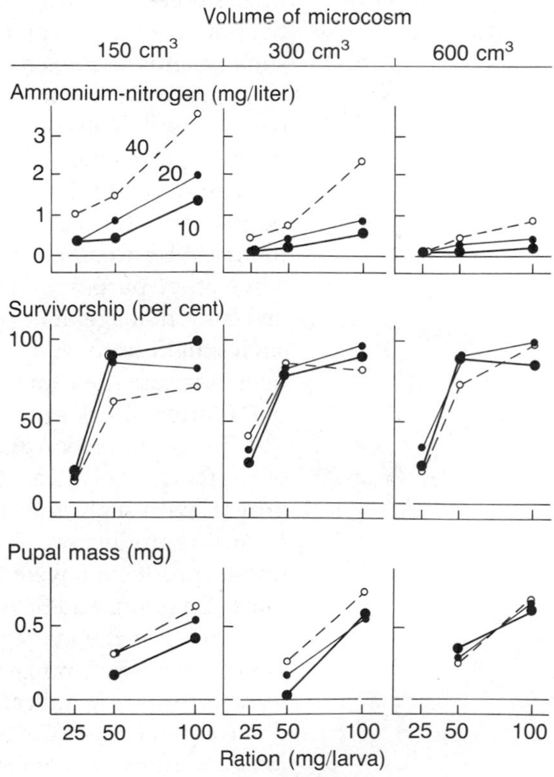

Figure 17–10
Ammonium concentration, larval survival, and pupal mass of mosquito larvae grown in microcosms of three different volumes, at three different rations of food, and at three different numbers of larvae per microcosm. (After Carpenter 1983.)

varied as predicted with respect to each of the factors (top row of boxes in Figure 17–10): greatest at the lowest volume of water, at the highest number of individuals within each volume, and at the highest ration for each number within each volume. The effect of ration on ammonia suggests that the metabolism of the mosquito larvae varies in direct relation to the amount of leaves, presumably a consequence of higher feeding rates.

The response of the mosquitoes to each of the factors clearly demonstrates the overriding importance of food ration on survival and pupal mass, and equally clearly shows that the levels of excretory products obtained in the experiment were not high enough to inhibit growth and survival. Within each volume treatment, and for each number of individuals per treatment, both survival and mass increased markedly with food ration. At the lowest food levels, growth was so slow that few of the surviving larvae achieved the minimum size required for pupation.

Density dependence influences reproduction in populations of deer.

Most studies of density dependence have centered upon laboratory populations where factors can be controlled experimentally. But the simplicity of such systems also leads one to question the relevance of laboratory findings to populations in more complex natural surroundings—where physical conditions change continuously, and food and predation are not controlled by the experimenter. Ideally one would like to conduct the same experiment in nature as in the laboratory; that is, to alter the density of individuals in the population while keeping all else constant. In practice, this difficult experiment can be accomplished only with populations managed intensively for some other purpose. Game animals are sometimes maintained at altered levels by management practices, and ecologists have taken advantage of such situations to study population processes. Studies of various species of deer illustrate the expression of density-dependent factors (see, for example, Clutton-Brock et al. 1985, Skogland 1985).

The reproduction and survival of deer depend directly upon the quality of the food supply. Deer browse leaves, and require large quantities of new growth with high nutritional content to maintain high growth rates and normal reproduction. The direct physiological effects of food availability upon reproduction were shown in a study of white-tailed deer in New York State (Chaetum and Severinghaus 1950), where the proportion of females pregnant and the average number of embryos per pregnant female were directly related to range conditions (Table 17–2). The number of *corpora lutea* in the ovary indicates the number of eggs ovulated, hence the reproductive potential of the female. The difference between number of *corpora lutea* and number of embryos showed that poor range condition, resulting in poor nutrition of pregnant females, caused embryo death and resorption. In the central Adirondack area, where the habitat for deer was very bad, even ovulation was greatly reduced.

Table 17–2
Reproductive parameters of white-tailed deer (*Odocoileus virginianus*) in five regions of New York State, 1939–1949

Region*	Per cent of females pregnant	Embryos per female	*Corpora lutea* per ovary
Western (best range)	94	1.71	1.97
Catskill periphery	92	1.48	1.72
Catskill central	87	1.37	1.72
Adirondack periphery	86	1.29	1.71
Adirondack center (worst range)	79	1.06	1.11

* Arranged by decreasing suitability of range.
Source: Chaetum and Severinghaus 1950.

Many studies on deer have shown that high populations can lead to deterioration of the habitat over time through overgrazing (Klein 1968). In one population in Ohio, whose size had increased following protection from hunting, deteriorating range condition caused by overgrazing led to decrease in growth rate, adult body weight, ovulation rate, and production of fawns (Nixon 1965).

Range deterioration caused by overgrazing often can be reversed by selective hunting to thin dense populations. During the course of Chaetum and Severinghaus's study of the white-tailed deer in New York, one of the worst ranges in the state, located in the Adirondack Mountains, was opened to hunting. Two study areas, DeBar Mountain and Moose River, had been investigated between 1939 and 1943, just prior to the open hunting seasons of 1943 and 1945, and then again in 1947. Because Moose River is relatively remote, few deer were shot; but in the more readily accessible DeBar Mountain area, 250 deer were removed from the 8000-acre range in 1945 alone, and reproduction improved dramatically over its level before the hunting and in the relatively unchanged Moose River area (Table 17–3).

Table 17–3
Reproductive parameters of white-tailed deer (*Odocoileus virginianus*) in the Adirondack Mountains of New York State prior to and after hunting

Region	Per cent of females pregnant	Embryos per female	*Corpora lutea* per ovary
DeBar Mountain			
1939–1943 (prehunting)	57	0.71	0.60
1947 (after heavy hunting)	100	1.78	1.86
Moose River			
1939–1943 (prehunting)	91	1.00	0.98
1947 (after light hunting)	69	1.00	1.13

Source: Cheatum and Severinghaus 1950.

In crowded populations, plants suffer high mortality, slowed growth, and reduced fecundity.

The sensitivity of life table values of plants to density has been shown in many studies, summarized in detail by the British plant ecologist John Harper (1977). Like animals, plants experience increased mortality and reduced fecundity at high density. Unlike many animals, however, plants respond to density by slowed growth, along with its consequences for fecundity and, to a lesser extent, survival. Starve an animal and most often you will kill it; starve a plant and it won't grow. This difference arises in part from the modular construction of plants. The repeated units of leaves and buds that make up the plant body exist more or less independently of each other; that a branch fails to grow has little effect on the well-being of other branches on the tree. The more complicated arrangement and interdependence of the parts of animals dictate greater constraint on size and shape.

Flexibility of plant growth is revealed in the size of flax plants *(Linum usitatissimum)* grown to maturity at different densities (Figure 17–11). When sown sparsely at a density of 60 seeds per square meter, the modal dry weight of individuals was between 0.5 and 1 gram, and many plants attained weights exceeding 1.5 g. When sown at densities of 1440 and 3600 seeds m^{-2}, most of the individuals weighed less than 0.5 g, and few grew to large size. The variation in size within a planting arises due to chance factors early in the seedling stage, particularly the date of germination and quality of the site in which the seedling grows. Early germination in a favorable spot gives the plant an initial growth advantage over others, which becomes accentuated as the larger plants grow and crowd their smaller neighbors (Weiner 1988).

The flexibility of plant growth does not preclude mortality in crowded situations. When horseweed (*Erigeron*) seed was sown at a density of 100,000 m^{-2} (equivalent to about 10 seeds in the area of your thumbnail),

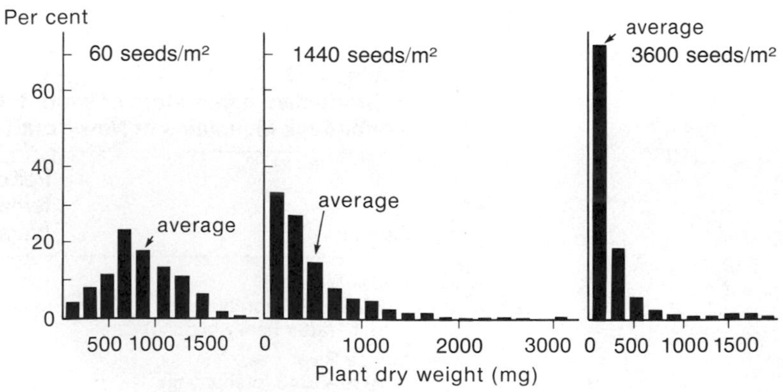

Figure 17–11
The distribution of dry weights of individuals in populations of flax plants sown at different densities. (After Harper 1967.)

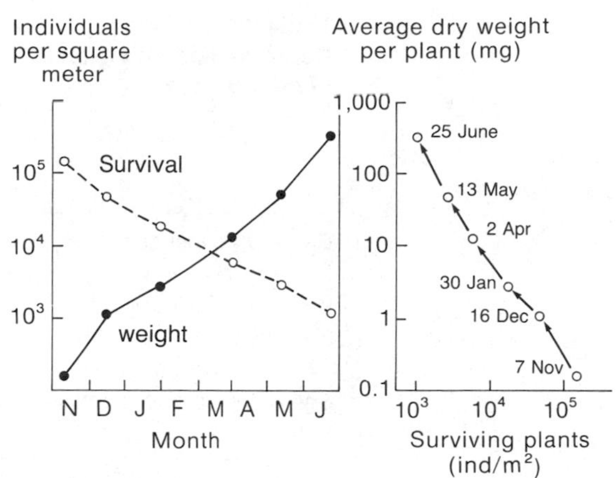

Figure 17–12
Progressive change in plant weight and population density in an experimental planting of horseweed *(Erigeron canadensis)* sown at a density of 100,000 seeds per square meter (10 cm^{-2}). The relationship between plant density and plant weight as the season progressed is shown at right. (After Harper 1967.)

the young plants competed vigorously. As the seedlings grew, many died and the density of surviving seedlings decreased (Figure 17–12). But at the same time, the growth of surviving plants exceeded the decline of the population, and the total weight of the planting increased. Over the entire growing season, the hundredfold decrease in population density was more than balanced by a thousandfold increase in the average weight of each plant.

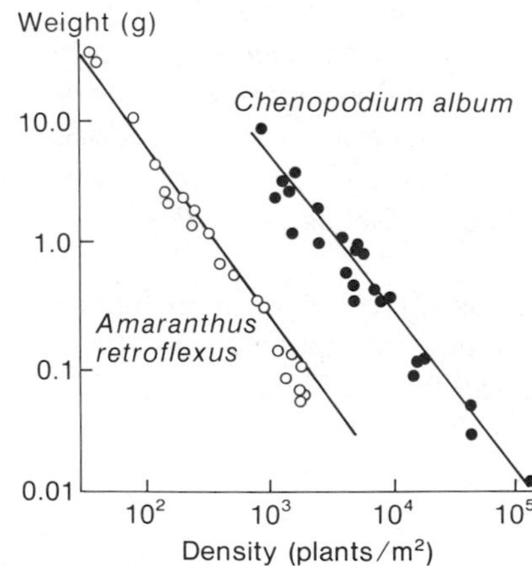

Figure 17–13
Change in plant density and mean plant weight with time for plantings of *Amaranthus* and *Chenopodium.* The two variables are related by a slope of $-3/2$ (-1.5), which has been extremely consistent in experiments of this type. (After Harper 1977.)

When one plots the logarithm of average plant weight as a function of the logarithm of density, data points recorded during the growing season fall on a line with a slope of approximately $-3/2$ (Yoda et al. 1963; Figure 17–13). Plant ecologists call this relationship between average plant weight and density a self-thinning curve (Weller 1987, Osawa and Sugita 1989). Such is the regularity of this relationship that many have referred to it as "the $-3/2$ power law" (Harper 1977), although its biological basis is not understood. The self-thinning curve, whose position on the graph varies from species to species, appears to represent the maximum average dry weight for each density of a population. When seeds are planted at a moderate density, so that the beginning combination of density and average dry weight lies below the self-thinning curve, plants grow without appreciable mortality until the population reaches its self-thinning curve. After that point, the intense crowding associated with further increase in average plant size causes the death of smaller individuals.

Stunted growth under crowded conditions dramatically alters the life tables of plants (Table 17-4). I. Palmblad (1968) germinated seeds of the plantain *Plantago major* at densities of 1 to 200 per pot (55 to 11,000 m^{-2}). Although only a quarter of the plants in the densest treatment died, many of the individuals that flowered failed to produce seeds, evidently lacking the resources to do so. Although the total dry weight of all the plants in each pot

Table 17–4
Reproduction and growth in experimental plantings of the plantain *Plantago major*

	Sowing density (seeds per pot)				
	1	5	50	100	200
Germination (per cent)	100	100	93	91	90
Mortality (per cent)	0	7	6	10	24
Per cent reproducing	100	93	72	52	34
Per cent vegetative	0	0	15	29	32
Dry weight per pot (g)	8.05	11.09	13.06	13.74	12.57
Seeds per reproductive plant	11,980	2733	228	126	65
Seeds per pot	11,980	12,760	8208	6552	4420

Source: Palmblad 1968.

taken together was similar across densities, the total number of seeds produced decreased from about 12,000 per pot at low density to 4420 at the highest density.

Andrewartha and Birch challenged density-dependent regulation of population size.

The year 1954 brought the publication of two important books on population biology, each remarkable on its own merits, but even more striking together in the utter opposition of their points of view. The English ornithologist David Lack, in *The Natural Regulation of Animal Numbers,* vigorously advocated the regulation of population size by density-dependent factors, principally food, predators, and disease. His arguments and evidence added to the momentum of belief in density dependence and the regulation of populations about stable equilibria gained by Lotka, Gause, A. J. Nicholson, and others.

The same year, two Australian entomologists, H. G. Andrewartha and L. C. Birch, published *The Distribution and Abundance of Animals,* in which they argued that most populations, particularly those of insects and other small invertebrates, are influenced primarily by density-independent factors, and that periods of favorable environmental conditions for population growth ultimately control the size of the population:

> The numbers of animals in a natural population may be limited in three ways: (a) by shortage material resources, such as food, places in which to make nests, etc.; (b) by inaccessibility of these material resources relative to the animals' capacity for dispersal and searching; and (c) by shortage of time when the rate of increase r is positive. Of these three ways, the first is probably the least, and the last is probably the most, important in nature. Concerning c, the fluctuations in the value of r may be caused by weather,

predators, or any other component of environment which influences the rate of increase.

Although Andrewartha and Birch marshaled abundant evidence to support their view that exponential growth rates reflect density-independent factors, they were greatly influenced in their thinking by the study of J. Davidson and Andrewartha (1948) on populations of the tiny insect pest *Thrips imaginis* near Adelaide, South Australia. The thrip, which infests roses and other cultivated plants, undergoes regular periods of increase when seasonally favorable conditions prevail, followed by rapid decline when conditions are less suitable (Figure 17–14). Adelaide has a Mediterranean climate; winters are cool and rainy, summers hot and dry. The spring (October through December in the Southern Hemisphere) brings an ideal combination of moisture, warmth, and plant flowering. Thrips subsist mainly on plant pollen, whose abundance varies seasonally with the production of flowers. But Adelaide has a sufficiently mild climate that some flowers are always available, and thrips remain active all year. During the winter, however, the depressing effect of cool temperature on development rate and fecundity causes the thrip population to decline to low levels. Population growth is further checked by high mortality of immature stages: thrips take so long to mature during the winter that before many can complete their development, the flowers in which they live have withered and fallen to the ground—carrying the thrips with them to their deaths. The warm weather of spring speeds development and increases the daily fecundity of females. Under these conditions, the thrips rapidly increase to infestation levels; their populations are not checked until the heat and dryness of the austral summer bring about a sharp rise in adult mortality.

Not only did the number of thrips mirror seasonal changes in weather, but also the peak density of the population varied from year to year, in

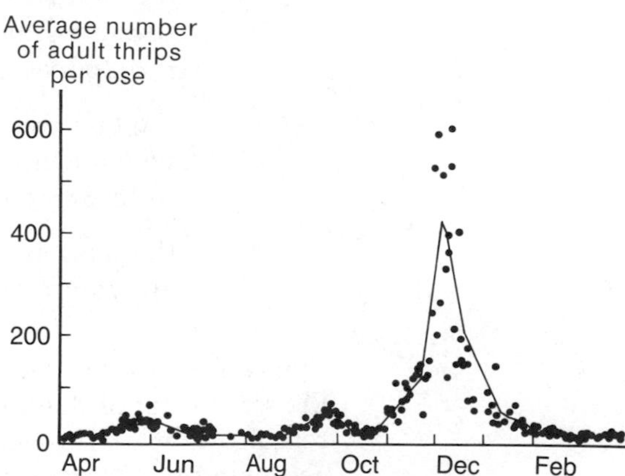

Figure 17–14
Number of *Thrips imaginis* per rose from April 1932 through March 1933 near Adelaide, Australia. Dots indicate daily records; the curve is a moving average for 15 days. (After Davidson and Andrewartha 1948.)

apparent relation to climate, during the course of Davidson and Andrewartha's study. Some years were warmer or moister, others cooler or drier. Food apparently had little to do with the vagaries of the population because, as Andrewartha and Birch (1954) explained:

> Even when the thrips are most numerous, the flowers in which they are breeding do not appear to be overcrowded except perhaps locally or temporarily: while the thrips are multiplying, the flowers increase even more rapidly. Then, when the population begins to decline, the flowers become less crowded still. . . . Considerations of this sort led to the hypothesis that the numbers achieved by the thrips during each year were determined largely by the duration of the period that was favorable for their multiplication. When this period was prolonged, the thrips would ultimately reach higher numbers; when it was briefer, the decline would set in while the numbers were still relatively low.

To determine the relationship between thrip populations and weather, Davidson and Andrewartha (1948) compared the peak population abundance each year between 1932 and 1945 to four weather variables, subjecting the lot to a statistical analysis (multiple regression). Specifically, the dependent (response) variable was

Y the average of the logarithm of the number of thrips per flower over the 30 days before the population peak.

The dependent, or regressor, variables were

X_1 the effective degree days from the first rains of the winter season to August 31, which might determine the growth of the annual plants on whose pollen the thrips would feed later in the season

X_2 the rainfall during September and October, the spring rains sustaining the thrips' food plants and promoting the survival of thrip pupae in the soil

X_3 the effective degree days for September and October, a time when temperatures are becoming marginally adequate for thrips reproduction, and, finally

X_4 the same as X_1, only for the previous year to take into account any carryover of thrips or seeds from one year to the next.

The variables were chosen to represent conditions that could affect either the well-being of the thrips or the production of their food. The analysis revealed that year-to-year variation in the four climate variables together accounted for 78 per cent of the year-to-year variation in the population (Figure 17–15). Individually, the warmth of the fall and winter (X_1) made the largest contribution to the prediction of the population, followed by spring rainfall (X_2) and the warmth of the previous winter (X_4). Variation in spring temperature appeared not to be important.

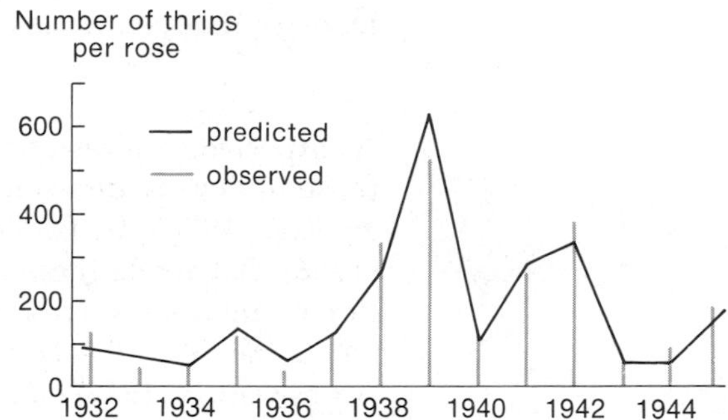

Figure 17–15
Observed peak numbers of thrips per rose between 1932 and 1945 (vertical lines) compared to prediction of a regression equation based upon four climate variables (solid line). (After Davidson and Andrewartha 1948.)

Andrewartha and Birch made two observations concerning the thrips example. First, the thrips were at no time numerous enough to consume more than a small fraction of the food available. Second, variation in the population can be accounted for satisfactorily by the physical conditions of the environment, which they presumed to affect the thrips independently of the density of their population. Regarding the last point about the 78 per cent of variance explained, they went so far as to assert the following:

> This left virtually no chance of finding any other systematic cause for variation, because 22 per cent is a rather small residuum to be left as due to random sampling errors. All the variation in maximal numbers from year to year may therefore be attributed to causes that are not related to density: not only did we fail to find a "density-dependent factor," but we also showed that there was no room for one.

How did Andrewartha and Birch address the abundant evidence of density dependence? First, they claimed, most of the evidence came from the special circumstances of the simplified and controlled laboratory populations, and had dubious application to natural populations. Second, the demonstration of density dependence does not necessarily imply regulation of the population by density-dependent factors. One must show under natural circumstances that such factors cause dense populations to decrease and allow sparse populations to increase. And what about the argument that in the absence of density-dependent factors populations would either decrease to extinction or increase without bound? Andrewartha and Birch pointed out that thrip populations do, indeed, die out every year in many local situations where the climate has been particularly severe. Such areas presumably are recolonized from populations persisting in more favorable places. As to unlimited increase, the thrip population simply does not have enough time each year to reach limits imposed by food or other resource. Each favorable season of the year, which is the spring, is invariably followed by an unfavorable period in the summer.

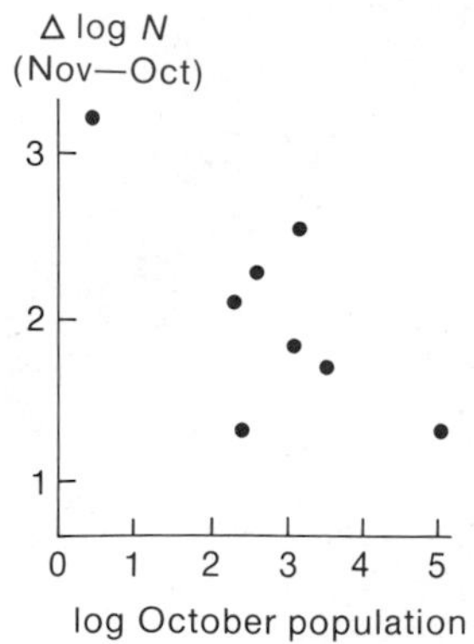

Figure 17–16
Increase in the size of the thrip population between October and November as a function of the size of the population in October. (From data in Davidson and Andrewartha 1948, after Smith 1961.)

Density independence created a major controversy among ecologists.

As expected, Andrewartha and Birch's challenge prompted a vigorous defense of density dependence, notably by M. E. Solomon (1957), A. J. Nicholson (1958), G. C. Varley (1963), all entomologists, and David Lack (1966). But while these authors criticized the logic of Andrewartha and Birch's arguments, Frederick E. Smith, then at the University of Michigan, went straight to the *Thrips* data. Smith (1961) argued that Davidson and Andrewartha (1948) had analyzed their data in such a way that they could not have detected density dependence. He pointed out that arguments about density dependence do not address the absolute size of a population at a particular time during the season; rather, they relate changes in population size to the initial size of the population.

Accepting the premise of exponential growth (that is, changes in population size [N] are properly expressed as increments in the logarithm of N), the presence of density dependence can be determined by comparing the increase in $\log N$ from one time to the next [$\Delta \log N(t) = \log N(t+1) - \log N(t)$] to $\log N$ at the beginning of the period. Smith made several comparisons based on Davidson and Andrewartha's data (for example, Figure 17–16). As you can see, ΔN is negatively related to N, as required by density dependence.

Smith posed a second argument based on the variation from year to year in thrip numbers during each month. Davidson and Andrewartha (1948) had tabulated the average number of thrips per rose for 81 consecutive months between April 1932 and December 1938 (Figure 17–17). From these, Smith calculated the average and the variance for each month. The variance is the sum of squared deviations of each value from the average of the values for the month; hence it is a measure of variability. Smith noticed that the variance was relatively low during the fall and winter, but rose to a high level coinciding with the period of most rapid population growth in October, after which it fell rapidly.

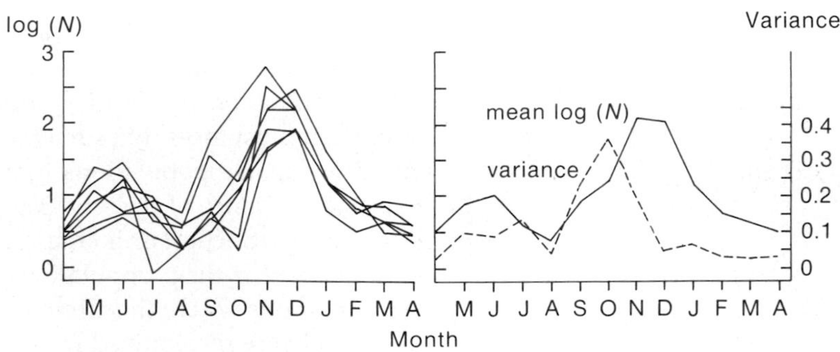

Figure 17–17
Left: Monthly mean number of thrips per rose over nearly 7 years; *Right:* the mean and variance for each month of the year. (From data in Davidson and Andrewartha 1948, after Smith 1961.)

From his training in statistics, Smith knew that when two numbers vary independently of each other, the variance of their sum is equal to the sum of their variances; that is, $\mathrm{Var}(X+Y)=\mathrm{Var}(X)+\mathrm{Var}(Y)$. Populations grow according to the expression $\log N(t+1)=\log N(t)+\Delta\log N(t)$. Therefore, if $\Delta\log N$ is independent of $\log N$—hence density-independent growth—the variance in $\log N$ must increase, on average, with each increment of population change, whether positive or negative (the variance is always positive). But the variance in $\log N$ declines between October and December, indicating a lack of independence between $\Delta\log N$ and $\log N$.

The degree of relationship between two random variables X and Y can be described by their covariance, where $\mathrm{Cov}(X,Y)$ is the sum of the product of the deviations in X and Y from their means. If Y tends to be below average when X is above average, as in Figure 17–17, $\mathrm{Cov}(X,Y)$ will contain products mostly of positive and negative deviations and will therefore be negative. When two variables are related, $\mathrm{Var}(X+Y)=\mathrm{Var}(X)+\mathrm{Var}(Y)+2\mathrm{Cov}(X,Y)$. Applied to the thrip population, Smith pointed out that for the variance of $\log N$ to decrease, $\mathrm{Cov}(\log N,\Delta\log N)$ must be a larger negative number than $-0.5\mathrm{Var}(\Delta\log N)$, which implies strong density dependence.

Andrewartha was not convinced by Smith's arguments and said so in a response printed in *Ecology* the following year (1963). Smith's further response, appended to the end of Andrewartha's note, opens with an expression of the frustrating gulf between their views: "In an effort, not to continue an argument, but to leave it as free of error as possible, . . ."

Density-dependent and density-independent viewpoints may be reconciled by considering how physical and biological factors interact to determine population growth rate.

Those who believed that density-dependent factors regulate populations discounted not the influence of density-independent factors on population size but that such factors alone could enable a population to persist. They based their case on logic strengthened by laboratory experiment. Those who could not accept the role of density-dependent factors failed to see evidence of such factors in natural populations. But as so often happens in scientific controversy, the two sides differed more over semantics than substantive issues; a particular problem arose from different interpretations of the terms density dependence and density independence.

Andrewartha could not accept density dependence in the thrip population because he could see no evidence for it—at least for what he considered to be density dependence. Climate, which, according to Howard and Fiske (1911) and Nicholson (1933), had only a density-independent effect, appeared to exert such an overriding influence on the thrip population that there was little room for the action of such density-dependent factors as food and predators. Ironically, Davidson and Andrewartha (1948) had provided a resolution of the problem. Referring to the thrips in their discussion, they wrote:

> At the end of spring most of the population occurred in situations which readily became unfavourable with the approach of summer. Thus, while the population was large, a small increase in the severity of the weather destroyed a relatively large proportion of the population. As the numbers were reduced the insects were restricted to an increasing extent to the more favourable situations and relatively large increases in the severity of the weather destroyed relatively fewer of the total population. Indeed no conceivable variation in the weather is likely to destroy all the insects in these situations in the time available: by the time spring comes the population ceases to decline and begins to increase once again in response to the changing physical environment.
>
> In this special sense the physical environment (as determined by the weather) becomes one of the "density-dependent" components of the environment, and the facts observed with *T. imaginis* during the period of the year when the population was declining can be fitted into the general theory. But this interpretation requires a restatement of the theory with respect to detail since Nicholson (1933, pp. 135–6) clearly excludes climate from the list of possible "density-dependent factors."

With his characteristic insight, Darwin (1859) had appreciated this point nearly a century earlier:

> Climate plays an important part in determining the average number of species, and periodical seasons of extreme cold or drought seem to be the most effective of all checks. . . . The action of climate seems at first sight to be quite independent of the struggle for existence; but in so far as climate chiefly acts in reducing food, it brings on the most severe struggle between the individuals, whether of the same or of distinct species, which subsist on the same kind of food. Even when climate, for instance, extreme cold, acts directly, it will be the least vigorous individuals, or those which have got least food through the advancing winter, which will suffer most.

We see now that the controversy did not arise between density dependence and density independence, but rather between weather and other physical factors, on one hand, and food, predators, and diseases, on the other. The heterogeneity of the environment dictates that some sites afford more protection than others from the ravages of climate during population decline, or provide more favorable circumstances during periods of population growth. The larger the population, the fewer favorable sites are available per individual, hence $\Delta\log N$ and $\log N$ will be inversely correlated, which is the proper definition of density dependence.

SUMMARY

The contrast between the potential of all populations for rapid growth and the observation of relative constancy of populations over long periods led naturally to the concept of density dependence of population processes. Population growth could be checked only if birth rate and survival decreased as populations grew. Limitation of food and increasing pressure of predators and disease were suspected of exerting their effects on population processes in such a density-dependent manner. The mathematical characterization of density-dependent population growth spawned laboratory and, eventually, field studies to test the idea. And although the idea generated controversy at times as it matured, density dependence has withstood severe tests and is now firmly established as a tenet of population ecology.

1. In 1920, Raymond Pearl and L. J. Reed derived an equation with density dependence to describe the growth of the population of the United States. In fact, they had rediscovered an equation formulated by Pierre-François Verhulst in 1838, which he called the logistic, and having the differential form

$$\frac{dN}{dt} = rN\left(1 - \frac{N}{K}\right)$$

with its integral

$$N(t) = \frac{K}{(1 + be^{-rt})}$$

2. Laboratory and field studies of animal and plant populations, starting with Pearl's work on the fruit fly *Drosophila,* showed in detail the expression of density-dependent factors in population processes. Experimental laboratory systems were developed with fruit flies, protozoa, flour and grain beetles, and water fleas during the 1930s. Similar studies of plants and of animals in natural habitats came later, based mostly upon such economically important species as crop plants, weeds, insect pests, and game species.

3. In 1954, Australian entomologists H. G. Andrewartha and L. C. Birch challenged the notion of density-dependent regulation of population size, pointing out that most of the variation in insect populations could be accounted for by density-independent effects of climate. They argued that populations were regulated by the length of the period favorable for population growth, the increase always being checked each year by the return of seasonally unfavorable conditions.

4. In 1961, F. E. Smith established two criteria for identifying density dependence in populations, both of which were met by the thrip population studied by Andrewartha and Birch. First, the change in population size between one period and the next [$\log N(t+1) - \log N(t)$, or $\Delta \log N(t)$] must be negatively correlated with the initial population size [$\log N(t)$] when the change is caused by density-dependent factors. Second, when seasonal changes in population size are repeated from year to year, the variance in population size must decrease when strong density-dependent factors act. Specifically, $\text{Var}[\log N(t+1)] = \text{Var}[\Delta \log N(t)] + 2\text{Cov}[\log N(t), \log \Delta N(t)]$. Thus the variance in population size can decrease only when $\log N(t)$ and $\Delta \log N(t)$ are negatively correlated.

5. Andrewartha and Birch's views can be reconciled with density-dependent population regulation when one accepts that climate can exert a density-dependent effect on population processes. In a heterogeneous habitat, some sites are more favorable than others. Individuals in poor sites are killed sooner than those in more favorable locations. Therefore, as a population decreases under the pressure of adverse climate conditions, an increasing proportion of remaining individuals occupy favorable sites and the death rate decreases, giving rise to the density dependence necessary for population regulation.

18

Population Fluctuations and Cycles

A. J. Lotka perceived populations as continually moving toward equilibria determined by the qualities of organisms and the conditions of their environments. But the environment varies, and so do populations. Moreover, patterns of variation derive not only from variability of the environment but also from the intrinsic dynamics of population responses. The study of variation in population size has its roots in two kinds of phenomena. The first includes the responses of populations to perceptible variation in the environment. The second includes regular cycles of variation in numbers unrelated to obvious periodic changes in the conditions of the environment.

Fluctuation is the rule for natural populations.

The degree of variation in the size of a population depends both on the magnitude of fluctuation in the environment and on the inherent stability of the population. After sheep became established on Tasmania, their population varied irregularly between 1,230,000 and 2,250,000—less than a two-fold range—over nearly a century (Figure 18-1). Davidson (1938) explained much of the variation in terms of changes in grazing practices, markets for wool and meat, and pasture management, which could be considered as factors in the environment of the sheep industry, if not the sheep themselves.

In sharp contrast to sheep, populations of small, short-lived organisms may fluctuate wildly over many orders of magnitude within short periods. The combined populations of species of green algae and diatoms that made up the phytoplankton of Lake Erie during 1962 illustrate tremendous population increases and crashes over periods of a few weeks (Figure 18-2); short-period fluctuations may be superimposed over changes with longer periods, perhaps occurring on a seasonal basis. Sheep and algae differ in

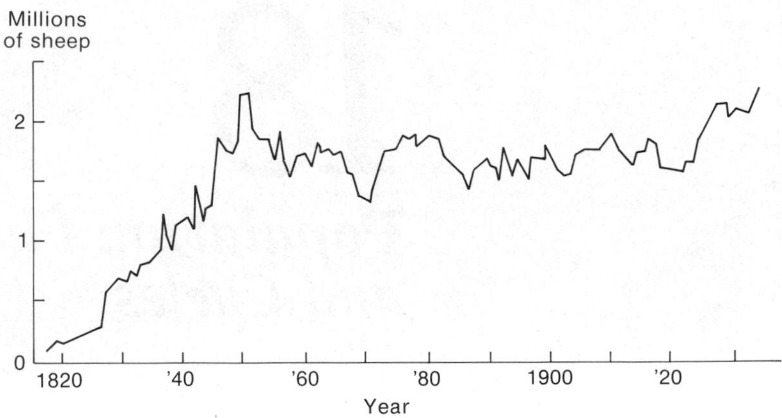

Figure 18-1
Number of sheep on the island of Tasmania since their introduction in the early 1800s. (After Davidson 1938.)

their sensitivities to environmental change and the response times of their populations. The larger sheep have greater capacities for homeostasis and therefore better resist the effects of environmental change. Furthermore, because sheep live for several years, the population at any given time includes individuals born over a long period, thereby evening out the effects of short-term fluctuations in birth rate on population size. The lives of single-celled algae span only a few days; populations turn over rapidly and thus bear the full impact of a capricious environment.

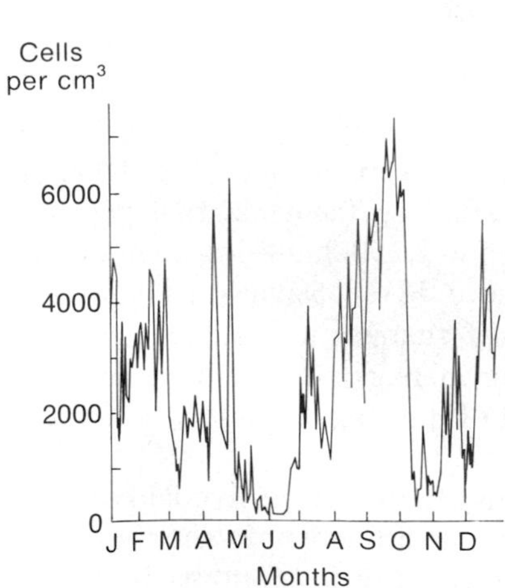

Figure 18-2
Variation in the density of phytoplankton in samples of water from Lake Erie during 1962. (After Davis 1964.)

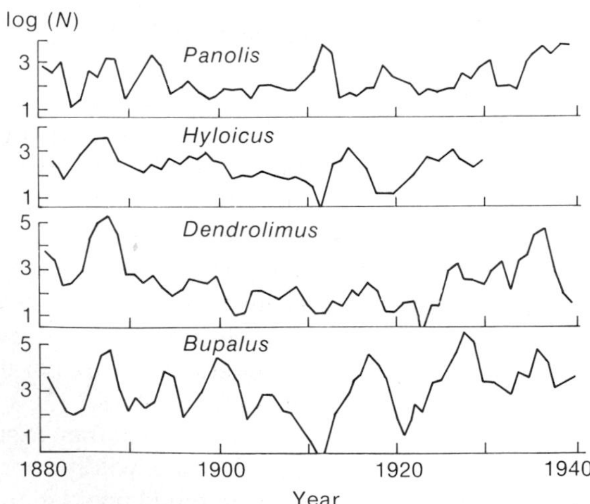

Figure 18-3
Fluctuations in the number of pupae of four species of moth (hibernating larvae in *Dendrolimus*) in a managed pine forest in Germany over 60 consecutive midwinter counts. (After Varley 1949.)

Populations of similar species in the same place may respond to different environmental factors. For example, the densities of four species of moths, whose larvae feed upon pine needles, fluctuated more or less independently over a 60-year period, as shown in Figure 18-3. The populations were sampled by counting the number of pupae (or hibernating larvae in *Dendrolimus*) per square meter of forest floor in a managed pine forest in Germany. The populations varied over three to five orders of magnitude with irregular periods of a few years. Furthermore, the highs and lows of the populations did not coincide closely, suggesting that even though the species fed on the same resource in the same forest, their populations were governed independently by different factors. Weather and food supply seem unlikely agents, as these should have affected each species similarly; population regulation may possibly have involved specialized predators or parasites.

Some populations of birds and mammals undergo periodic cycles of abundance.

Population cycles have contributed to the lore of population ecology since Charles Elton's paper "Periodic Fluctuations in the Numbers of Animals: Their Causes and Effects," published in the *British Journal of Experimental Biology* in 1924. Most of Elton's data concerned fur-bearing mammals in the Canadian boreal forests and tundra, where the Hudson's Bay Company had kept detailed records of the numbers of furs brought in by trappers each year. Data for the lynx and its principal prey, the snowshoe hare, revealed regular fluctuations of great magnitude (Figure 18-4). Each cycle lasts ap-

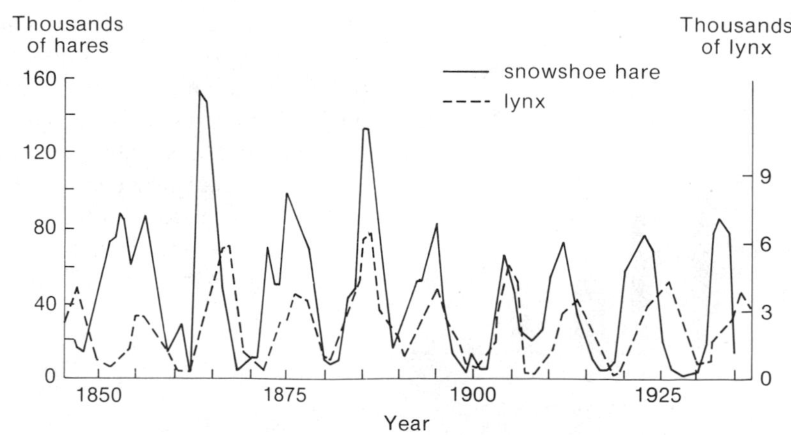

Figure 18-4
Population cycles of the lynx (*Lynx canadensis*) and the snowshoe hare (*Lepus americanus*; see photograph) in the Hudson's Bay region of Canada, as indicated by fur returns to the Hudson's Bay Company. (After MacLulich 1937; photograph courtesy of U.S. Fish and Wildlife Service.)

proximately 10 years and cycles of the two species are highly synchronized, with peaks in lynx abundance tending to trail those in hare abundance by a year or two.

The periods of population cycles vary from species to species, and even within a species. In Canada, most cycles have periods of either 9 to 10 years or of 4 years. Although colored foxes have a ten-year cycle over most of their range, they exhibit a pronounced 4-year cycle in Labrador and on the Ungava Peninsula, where they prey mostly on lemmings, which have 4-year cycles as well (Elton 1942). In general, small herbivores, such as voles and lemmings, have 4-year cycles; large herbivores—snowshoe hares, muskrat, ruffed grouse, and ptarmigan—have 9- to 10-year cycles. Predators that feed on short-period herbivores (arctic fox, rough-legged hawk, snowy owl) themselves have short population cycles. Predators of larger herbivores (red fox, lynx, marten, mink, goshawk, horned owl) have longer cycles (Dymold 1947, Keith 1963). The length of the cycle also appears related to habitat, with longer periods in forest-dwelling species and shorter ones in tundra-dwelling species.

Populations may cycle quite regularly. For example, a string of seven 4-year periods in the colored fox returns to the Moravian missions between 1847 and 1880 prompted Elton (1942) to remark that an Eskimo hunter "might have reflected that his good luck and his bad luck chased each other with sufficient regularity to amount to a natural law." But frequent variations in the periods between peaks as well as irregularities in peak and low numbers convinced some observers that population fluctuations reflect random processes, not regular cycles (see Finerty [1980] for a review). Lamont Cole (1951) dispelled this notion by showing that the distribution of periods between population peaks differed significantly from that expected for series of random numbers. To derive the random distribution, Cole reasoned that turning points in a series of random increments—that is, either a high point or a low point—occur with a probability of one-half for each time interval. For example, an increase in one interval is followed by another increase with probability 0.5, and by a decrease (defining a peak between the two) with probability 0.5. Furthermore, one-half the turning points are maxima. Using probability theory, Cole calculated the expected proportions of periods between peaks of two, three, four, . . . intervals. The expected mean interval of the random series is exactly 3 years. The average interval between peaks in fluctuating mammal populations is characteristically greater than 3 and the distributions of intervals differ markedly from the expectation for a random series (Figure 18-5). Therefore, population cycles cannot be dismissed simply as the result of random processes.

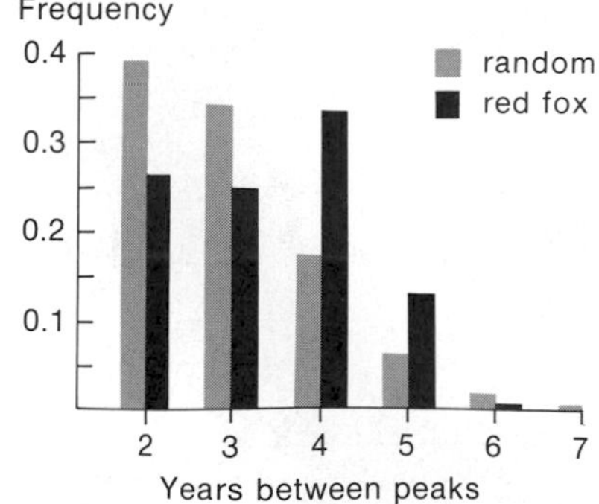

Figure 18-5
Frequencies of periods between peaks of populations of the red fox and a series of random numbers. (After Cole 1951.)

"Key-factor" analysis seeks to identify factors responsible for population change.

In the wake of the controversy created by Andrewartha and Birch, most ecologists realized that populations vary in response to both density-dependent and density-independent factors. Subsequent research focused on identification of the most important factors and on learning how their effects were expressed through population processes. Even before Smith's critique of Andrewartha's analysis of the thrip population, Canadian and English entomologists concerned with the control of insect pests had begun to apply new methods of analysis to identify factors that caused populations of insects to vary. In the introduction to his paper "Single-factor Analysis in Population Dynamics," R. F. Morris (1959), of the New Brunswick Forest Biology Laboratory, explained his purpose:

> A preliminary examination of rather extensive life-table data for the spruce budworm . . . suggested that the factors affecting this species in any one place are of two types—those that cause a relatively constant mortality from year to year and contribute little to population variation, and those that cause a variable, though perhaps much smaller, mortality and appear to be largely responsible for the observed changes in population. . . . A factor of the latter type will here be called a "key factor," meaning simply that changes in population density from generation to generation are closely related to the degree of mortality caused by this factor, which therefore has predictive value.

Morris conceived of changes in insect populations by a simple model,

$$N(t + 1) = N(t)FS_1S_2S_3 \ . \ . \ .$$

in which the factor of population increase from year t to $t + 1$ is the product of fecundity (F) and survival with respect to potential mortality factors (S_i). In logarithmic form,

$$\log N(t + 1) = \log N(t) + \log F + \log S_1 + \log S_2 + \ . \ . \ .$$

He wished to determine if variation in $\log N(t + 1)$ followed upon variation in one or a few of the values of F and S_i. He reasoned that variation in one or more of these variables reduced the correlation between $N(t + 1)$ and $N(t)$. But if S_2, for example, were responsible for changes in N from year to year, $N(t + 1)$ would be more strongly correlated with $N(t)S_2$ than with $N(t)$ alone.

Morris applied his analysis to data for the black-headed budworm (Figure 18-6). The budworm belongs to the moth family Tortricidae (leaf-rollers), native to eastern Canada, and is a major defoliator of fir trees in New Brunswick. One generation of adults appears each year; the population overwinters in the egg stage. Twelve years of observations on the density of budworm populations and the percentage of parasitism of the larvae by wasps and flies revealed a fluctuation in numbers of over two orders of magnitude with a period of about 9 years. The survival of larvae at risk to parasites (S) followed a similar course, being high during the population buildup and low during periods of decline. In this example, the correlation (r) between $N(t + 1)$ and $N(t)$ was 0.67, indicating that 45 per cent (r^2) of the variation in population size was related to variation in population size during the previous year. When $N(t)$ was multiplied by S, the correlation

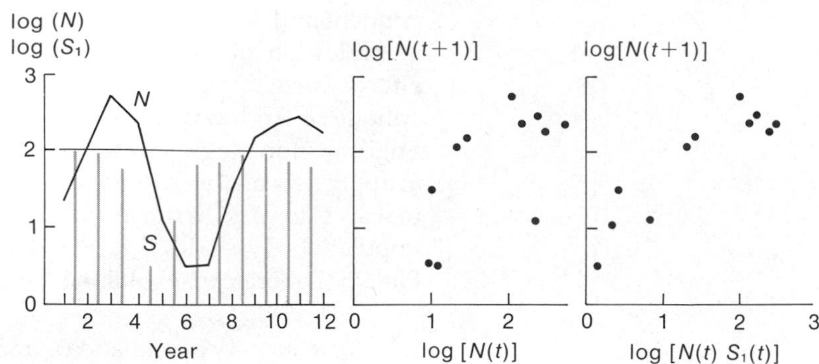

Figure 18-6
Left: Population size (*N*, solid line) and larval survival (*S*, vertical bars) for the black-headed budworm; *center:* the relationship between population size in one year and its size in the previous year was weak. *Right:* The relationship between population size in one year and the number of larvae surviving parasitism from the previous year was strong. (After Morris 1959.)

between $N(t + 1)$ and $N(t)S$ was 0.93, and the proportion of variation explained was 86 per cent. From this analysis, Morris concluded that larval parasitism was a key factor in the population processes of the black-headed budworm.

English entomologists G. C. Varley and G. R. Gradwell (1960) extended Morris's analysis to examine the effects of many factors and determine statistically which contributed to variation in population size. Following the terminology of J. B. S. Haldane (1949), Varley and Gradwell defined the "killing power" of a particular agent of mortality i as $k_i = -\log S_i$, and the total killing power exerted upon a population as the sum of each component, $K = k_1 + k_2 + k_3 + \ldots$. (The expression $\log \lambda = \log F + \log S_1 + \log S_2 + \ldots$ would have served equally well.) Their approach then followed directly from Morris's in using correlation and regression to determine which variables were most strongly correlated with variation in $[\log N(t + 1) - \log N(t)]$ or, equivalently, K. Because of their particular terminology, Varley and Gradwell's approach is often referred to as "*k*-factor analysis."

During the 1960s, entomologists disputed the relative merits of the Morris (1963) and Varley-Gradwell approaches to key-factor analysis, finally resolving the issue by the end of the decade (Varley and Gradwell 1970, Harcourt 1971, Luck 1971), by which time Morris's analysis was less persuasive. The dispute is unimportant now, except historically, because it arose from unknowledgeable application of statistics and not from any disagreement over biological issues. Regardless of such statistical subtleties, the application of Morris's and Varley-Gradwell's analysis to insect populations has led to identification of stages of the life history during which demographic processes contribute most to variation in population size. The analysis by Canadian economic entomologist D. G. Harcourt (1963) of the diamondback moth, a worldwide pest of cabbage and related crops, is typical of such studies.

In southern Ontario, the diamondback moth usually has four, but exceptionally up to six, generations per year. During each growing season between 1958 and 1961, Harcourt sampled populations at each stage of the life cycle (egg, several larval instars, pupa, and adult). He determined the rate of parasitism from extensive rearings of larvae and pupae, and estimated fecundity by raising adult moths in the laboratory; mortality of the adults was estimated indirectly by comparing the number of eggs present on the cabbage plants at the beginning of a generation with the number expected if all the adults from the previous generation had survived to lay eggs (Table 18-1). Mortality during each stage varied from as little as 1 per cent of eggs to 89 per cent of adults (that is, 89 per cent of the potential fecundity of adult females). The major causes of mortality also varied from one stage of the life history to the next. Small larvae were particularly vulnerable to heavy rainfall, whereas large larvae and pupae were heavily parasitized by wasps. Adults usually succumbed to weather factors.

Each of the causes of mortality itself varied considerably from generation to generation. Harcourt assessed the contribution of variation in survival rate during each stage to variation in numbers in the population at the beginning of each generation by correlating $[\log N(t+1) - \log N(t)]$ with each of the survival rates. Not surprisingly, 73 per cent of the variation in N derived from variation in adult survival; mortality rate was greatest during that stage and its variation more or less governed the population trend. But high mortality does not necessarily imply cause of variation in N between generations. Rainfall during the first larval period caused the death of 55 per cent of the individuals in the population, but variation in this factor was not significantly related to the trend in the population from generation to generation.

D. G. Harcourt and E. J. Leroux (1967) summarized key-factor analyses for twelve Canadian agricultural and forest insect pests. In these diverse insects, the critical stage at which key factors acted, as well as the nature of

Table 18-1
Life table for the second generation of the diamondback moth (*Plutella maculipennis*) on early cabbage in southern Ontario, 1961

		Mortality			
Stage	Number per 100 plants	Cause	Number per 100 plants	Percentage of stage	Killing factor (k_x)
Egg	1580	Infertility	25	1.6	0.016
Larva					
period 1	1555	Rainfall	1199	77.1	1.474
period 2	356	Rainfall	36		
		Parasitism by *M. plutellae*	52	24.7	0.284
period 3	268	Parsitism by *H. insularis*	69	25.7	0.297
Pupa	199	Parasitism by *D. plutellae*	92	46.2	0.620
Adult	107	Inclement weather	20	18.7	0.207
Reduction in fecundity		Photoperiod		73.6	1.332

Source: Harcourt and Leroux 1967.

the key factor itself, varied from species to species. In only one of the species, the Colorado potato beetle (Harcourt 1971), was food supply a key factor. Weather was the most important factor in three of the species, including the diamondback moth; and disease, parasitism, and emigration from local populations ranked highly in the rest. In 9 of the 12 studies, the key factor was determined to be density-dependent in the sense that its value was found to vary inversely with the initial size of the population. A more extensive summary of such studies may be found in H. Podoler and D. Rogers (1975) (see also Elliot 1985).

The characteristic return time of a population depends on its innate capacity for increase.

The fluctuations in populations of insects, and other organisms, may be thought of as resulting from changes in birth and death rates brought about by changes in the environment. When a population is regulated by a strong density-dependent factor during at least one stage of the life history, even a density-independent change in fecundity or survival will bring density-restoring mechanisms into play and reveal the density-dependent factor. Changes in density-dependent factors alter the value of K itself and thereby put the population out of equilibrium. In either case, the population is continually in the process of returning to K or catching up with a varying value of K. How well it does so depends on the capacity of the population to grow relative to the period of variation in density-independent factors or in K. Populations with high potential growth rates achieve their equilibria quickly, and thus track variation in K closely. Populations with low growth potential cannot take advantage of increases in K and are also likely to resist decreases in K. Hence such populations vary little.

Population biologists compare the response of populations to change by their characteristic return times (T). T is simply the inverse of the exponential rate of growth of a population free of the effects of crowding; hence $T = 1/r$. Theoretical analyses have shown that a population will track the environment closely when its characteristic return time is less than the period of the environmental fluctuation (time from peak to peak) divided by 2π (Nisbet and Gurney 1982). When T is much longer than the period of the pertinent environmental variation, the population varies little. Small insects have values of T much shorter than 1 year and their populations track seasonal variation in conditions; the same variation evokes little response from populations of long-lived mammals.

Population cycles may result from intrinsic demographic processes.

Except for factors associated with daily, lunar (tidal), and seasonal cycles, environmental fluctuations tend to be irregular rather than periodic. Histor-

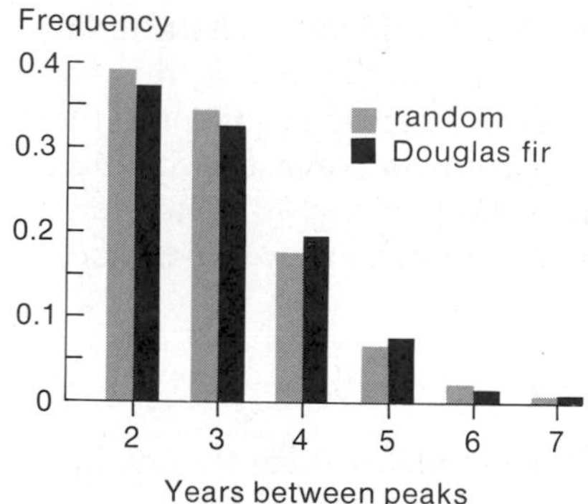

Figure 18-7
Frequency distributions of intervals between peaks in the widths of growth rings of Douglas fir and a series of random numbers. (After Cole 1951.)

ical records reveal that years of abundant rain or drought, extreme heat or cold, or such natural disasters as fires and hurricanes occur irregularly, perhaps even at random. Biological responses to these factors are similarly aperiodic. For example, widths of the growth rings of trees vary in direct relation to temperature and rainfall; patterns of successive ring widths cannot be distinguished from a random series (Figure 18-7).

Trends in the sizes of many populations do, however, change with periodic frequency (for example, see Figure 18-4). For many years, ecologists believed that explanations for such cycles must be sought among environmental factors exhibiting similar periodic variation. The regular 11-year cycle in sunspot numbers was frequently mentioned (see, for example, Criddle 1930, 1932; MacLulich 1936), but the sunspot cycle never matched population cycles well (MacLulich 1937) and no model could be devised to link the two. By 1940, the idea had been abandoned (Elton 1942).

With the development of population models in the 1920s and 1930s it became evident that because of their inherent dynamical properties, populations subjected to even minor, random environmental fluctuation could be caused to oscillate. Such cycling can result from time delays in the response of births and deaths to changes in the environment. Just as the momentum imparted to a pendulum by the acceleration of gravity carries it past the equilibrium point and causes it to swing back and forth periodically, the "momentum" imparted to a population by high birth rates at low density or high death rates at high density carries the population past its equilibrium when demographic responses are time-delayed.

Cyclic population behavior has been explored in a number of theoretical treatments, largely confirmed by laboratory experiments. Time delays that cause populations to oscillate when displaced from their equilibria are inherent to models based upon discrete generations. According to these models, populations respond by discrete increments from one time to the next and therefore cannot continuously readjust growth rate as population size approaches equilibrium. As we shall see below, this can cause the population to overshoot the equilibrium, first in one direction and then the other as N draws closer to K. In models of continuously growing populations, oscillations can be induced by introducing a time lag in the term for density dependence.

The discrete logistic equation can produce population cycles.

Populations with discrete generations may be described as growing according to the expression $N(t+1) = N(t)\lambda$ or, in logarithmic form $\log N(t+1) = \log N(t) + \log\lambda$. According to logistic growth, $\log\lambda$ decreases in direct relation to the size of the population; that is, $\log\lambda = r[1 - N(t)/K]$, where r is the intrinsic exponential growth rate of the population free of density effects. Now we may write an equation for discrete logistic growth as

$$\log N(t+1) - \log N(t) = r(1 - N(t)/K) \tag{18-1}$$

The behavior of a population growing accordingly is difficult to characterize because equation (18-1) has terms with both $\log N$ and N; that is, it is not linear in N. In such circumstances, one may approximate the equation in the region close to the equilibrium point by a linear form. In the case of equation (18-1), as $N(t)$ nears K, $[1 - N(t)/K]$ approaches $[\log K - \log N(t)]$. Therefore, close to the equilibrium point K, change in population size may be described reasonably well by

$$\log N(t+1) - \log N(t) = r[\log K - \log N(t)] \qquad (18\text{-}2)$$

This equation reveals that the behavior of the population near the equilibrium depends critically on the value of r. When r is less than 1, the increase in the population between t and $t + 1$ will be less than the difference between the population size and the equilibrium. Therefore, the population will approach the equilibrium directly, as shown in Figure 18-8. When r exceeds 1 but is less than 2, the population will overshoot the equilibrium, because $\Delta \log N$ exceeds the difference between $\log K$ and $\log N$, but it will end up closer to the equilibrium than before. Thus the population will oscillate back and forth across the equilibrium value, getting closer with each generation. This behavior is called damped oscillation.

When r exceeds 2, the population ends up farther from the equilibrium each generation and the oscillations increase. With increasing r, these oscillations take on very complex, eventually unpredictable forms referred to as "chaos." The full range of behavior of this model and a variety of related ones is described by May (1974, 1976, 1977), Man and Oster (1976), Schaffer (1984), Pool (1989), and, in a very thorough presentation, by Nisbet and Gurney (1982). Briefly, as deviations from the equilibrium become large, we can no longer deal with the linearized form of equation (18-2) and must encompass its full expression in our thinking. This step will lead us to the concept of the limit cycle.

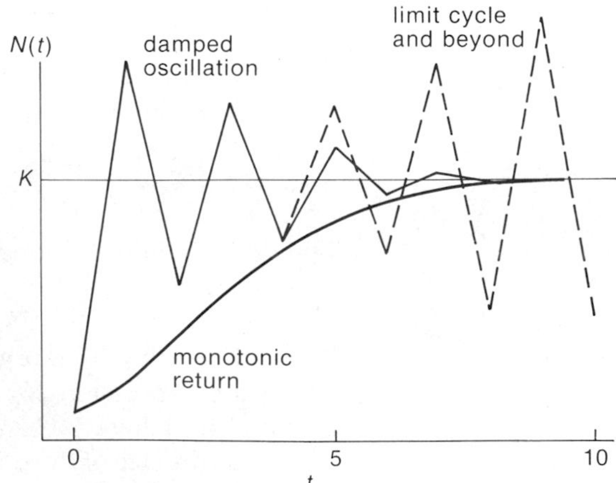

Figure 18-8
Approach to equilibrium according to a discrete logistic process when $r < 1$ (monotonic return), $1 < r < 2$ (damped oscillation), and $r > 2$ (limit cycle).

Limit cycles are stable oscillations.

The discrete logistic equation,

$$N(t+1) = N(t)e^{r[1-N(t)/K]} \tag{18-3}$$

is portrayed as plots of $N(t+1)$ versus $N(t)$, and $N(t)$ versus t, for values of r of 0.5, 1.5, and 2.5, in Figure 18-9. When r is less than 1 (top graphs), for values of $N(t)$ less than K, values of $N(t+1)$ are also less than K, and the population increases (or decreases) monotonically toward the equilibrium. When the value of r is between 1 and 2, for some values of $N(t)$ less than K,

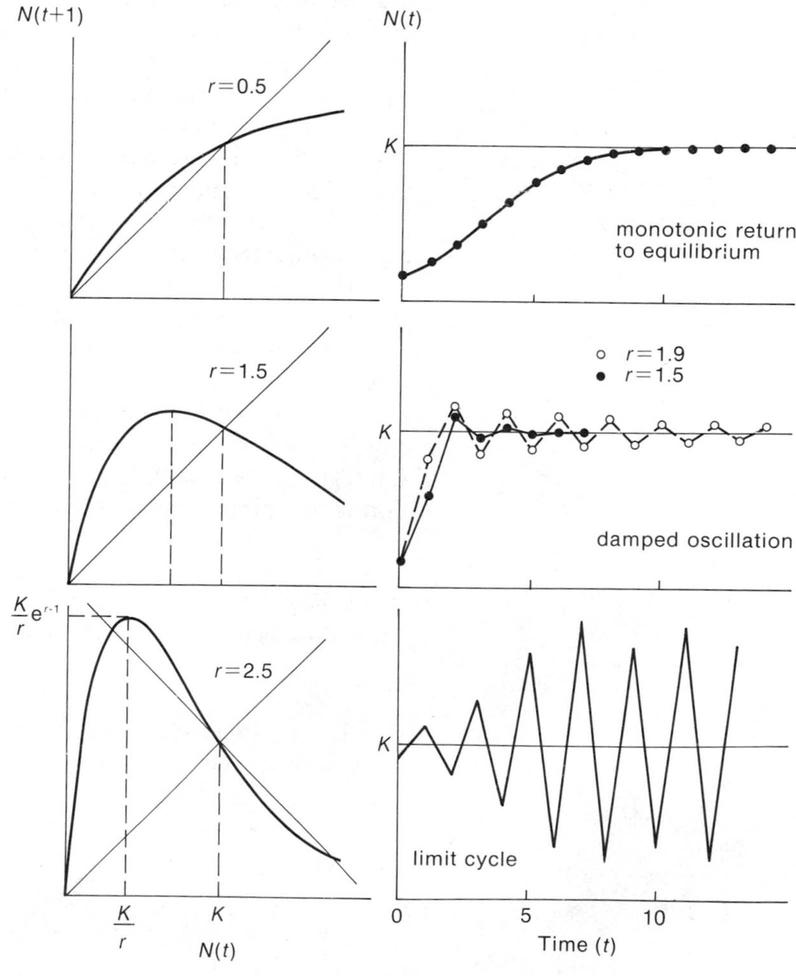

Figure 18-9
Left: Relationship between population size in one time period and that in the previous period for various values of r in a discrete logistic process. *Right:* Corresponding time courses of population growth in each case.

$N(t+1)$ exceeds K and the population overshoots. The oscillation is nonetheless damped and the population eventually reaches the equilibrium, as we determined from our analysis of the behavior of the model close to the equilibrium. The position of the maximum of $N(t+1)$ with respect to $N(t)$, determined by differentiating equation (18-3), is $N(t) = K/r$, at which point $N(t+1) = (K/r)e^{r-1}$.

Inspection of the graph shows that oscillations will be damped so long as the slope of the relationship between $N(t+1)$ and $N(t)$ is less negative than -1 at K. Under this condition, regardless of the overshoot, the population at $t+1$ will always be closer to K than it was at t. The slope $dN(t+1)/dN(t)$ at K is simply $(1-r)$; thus it is less negative than -1 when r is less than 2. When r exceeds 2, however, $dN(t+1)/dN(t)$ is more negative than -1 and the population moves farther from the equilibrium—up to a point. Below K, the curve relating $N(t+1)$ to $N(t)$ eventually bends back down toward the origin of the graph and, above K, the curve eventually flattens out as $N(t+1)$ approaches 0. As a result, a line with slope -1 passing through $N(t+1) = N(t) = K$ crosses the curve at two points corresponding, let us say, to N_{low} and N_{high}. Clockwise of this line, the population exhibits increasing oscillations, but counterclockwise of the line the oscillations are damped. This means that close to the equilibrium the oscillations increase, but at high amplitude the oscillations are damped. The result is a stable oscillation, with the population eventually alternating between N_{low} and N_{high} (Figure 18-9, lower right).

The models described above pertain to simple populations without age structure. In the more complicated case, fluctuations in a population are accompanied by changes in the age distribution of individuals and, as a result, the projection of population size becomes very difficult. P. H. Leslie (1959) examined the behavior of an age-structured population with density-dependence using a projection matrix. In his particular example with four age classes ($x = 0, 1, 2, 3$), the population exhibited a damped oscillation with a period of about 9 to 10 years.

Time delays cause oscillations in continuous-time models.

Oscillations are produced in continuous-time models when the response of population growth to density is time delayed; that is, when the effect of density dependence reflects the density of the population τ time units in the past (τ is the lower-case Greek letter tau). Modified thusly, the logistic equation becomes

$$dN/dt = rN(t)[1 - N(t-\tau)/K] \qquad (18\text{-}4)$$

G. E. Hutchinson (1948) pointed out that this model produces damped oscillations in N so long as the product $r\tau$ is less than $\pi/2$ (about 1.6). Below $r\tau = e^{-1}$ (0.37), the population increases or decreases monotonically—without oscillation—to the equilibrium point. For $r\tau$ greater than $\pi/2$, the

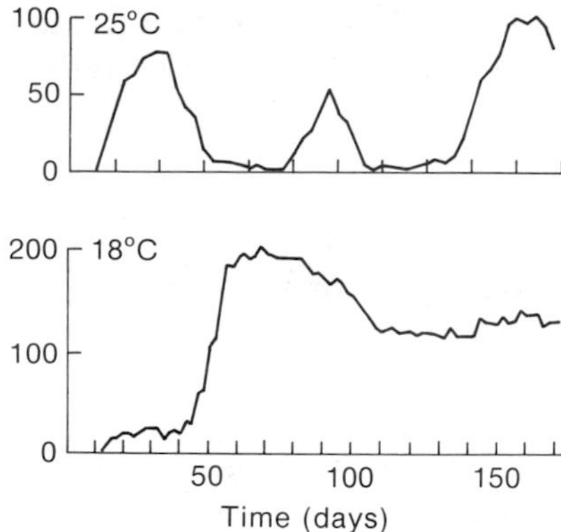

Figure 18-10
Growth of *Daphnia magna* populations at 18°C (*below*) and 25°C (*above*), showing the development of population cycles at the warmer temperature. (After Pratt 1944.)

oscillations increase until the maximum population size reaches $N/K = e^{r\tau}$. Thus for $r\tau = 2$, oscillations increase in amplitude until the maximum value of N is $e^2 = (7.4)$ times K. Population biologists refer to such stably maintained oscillations as limit cycles. Their periods, from peak to peak, increase from about 4 times τ to more than 5 times τ with increasing r (May 1976).

Population cycles have been observed in many laboratory cultures of single species. David Pratt's (1943) observations on the water flea *Daphnia magna* have been widely quoted, partly because the populations exhibited marked oscillations when cultured at 25°C but strong damping at 18°C (Figure 18-10). The period at 25°C appeared to be just over 40 days for two cycles, suggesting a time delay in the density-dependent response of about 10 days; this is on the order of the average age of water fleas giving birth at 25°C. The time lag arose in the following manner. As population density increased, reproduction decreased, to near zero when the population exceeded 50 individuals. In contrast to fecundity, survival was less sensitive to density and adults lived at least 10 days, even at the highest densities. As a result, the pulse of deaths in the population lagged about 10 days behind the pulse of births at the beginning of each upswing of the cycle. With births prevented by high density early in the cycle, when the population fell to densities low enough to permit reproduction it contained only senescent, nonreproducing individuals. Thus the beginning of a new cycle awaited the buildup of young, fecund individuals. The length of the time delay was approximately the average adult life span at high density.

At the lower temperature, reproductive rate fell quickly with increasing density, and life span increased greatly over that at 25°C at all densities. Populations at the colder temperature apparently lacked at a time delay because death was more evenly distributed over ages; and some individuals gave birth, even at high population densities; consequently generations overlapped more broadly. At the higher temperature, the *Daphnia* behaved according to a discrete-generation model with its built-in time delay of one generation. At lower temperature, they behaved according to a continuous generation model with little or no time delay.

Clyde Goulden and his colleagues at the Academy of Natural Sciences of Philadelphia have shown that the storage of lipid reserves by some species of water fleas reduces the sensitivity of mortality to density and therefore introduces a time delay into the population processes. *Daphnia galeata,* a large species, stores energy in the form of lipid droplets during periods of abundant food (that is, low density), which it can then live off of when food supplies are reduced by overgrazing at high population density (Figure 18-11). Females also pass lipid to each offspring through oil droplets in the eggs, thereby increasing the survival of young, prereproductive water fleas under poor feeding conditions (Tessier et al. 1983). The smaller *Bosmina longirostris* stores little lipid and therefore starvation increases directly in response to increases in population density. The consequences for population growth are predictable: *Daphnia* exhibits pronounced limit cycles with a period of 15 to 20 days under the conditions of the experiment; *Bosmina* populations grow quickly to an equilibrium with perhaps a single strongly damped overshoot (Figure 18-12). For *Daphnia,* Goulden et al. (1982) estimated that r was about 0.3 days^{-1}. With a cycle period of 15 to 20

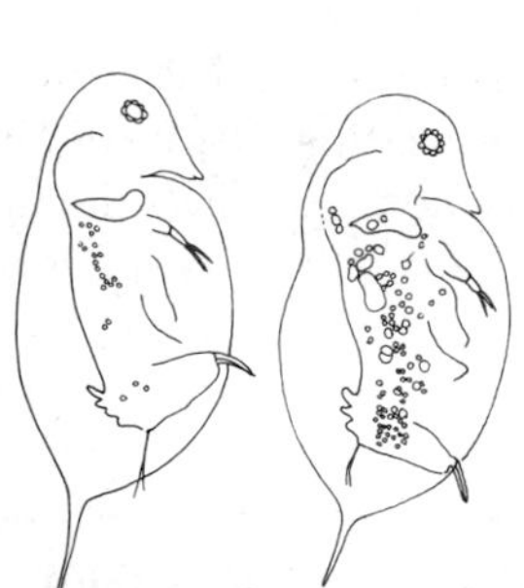

Figure 18-11
Distribution of energy stores in the form of oil droplets in individuals of *Daphnia galeata* grown under low (*left*) and high (*right*) food abundance. (From Goulden and Hornig 1980.)

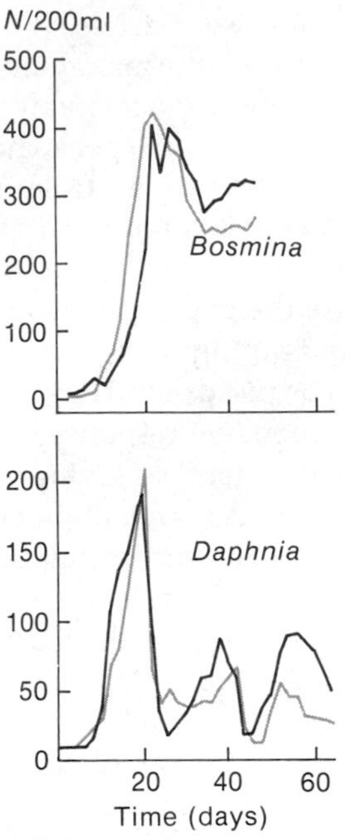

Figure 18-12
Densities of two populations of the cladoceran *Bosmina longirostris* (*above*) and two populations of its larger relative *Daphnia galeata* (*below*). (From Goulden et al. 1982.)

days, τ must have been about 4 to 5 days, and therefore $r\tau$ was about 1.2 to 1.5. Because the value of $r\tau$ was somewhat less than $\pi/2$, the cycles in the *Daphnia* population should have damped out eventually.

A. J. Nicholson's experiments with blowfly populations exhibit oscillations produced by time lags.

The behavior of a population with respect to its equilibrium is sensitive to many aspects of life history that govern time delays in responses to density. Slight differences in culture conditions or the intrinsic properties of species can tip the balance between a monotonic approach to equilibrium and a limit cycle (see, for example, Fujii's [1968] results for the stored products beetle *Callosobruchis*). A. J. Nicholson's (1958) experimental manipulation of time delay in laboratory cultures of the sheep blowfly *Lucilia cuprina* provide a dramatic demonstration of the relationship of time delays to population cycles. Under one set of culture conditions, Nicholson provided the larvae with 50 grams of liver per day while giving the adults unlimited food. The number of adults in the population cycled through a maximum of about 4000 to a minimum of 0 (at which point all the individuals were either eggs or larvae), with a period of between 30 and 40 days (Figure 18-13).

In this experiment, regular fluctuations of the blowfly populations were caused by a time delay in the response of fecundity and mortality to the density of adults in the cages. When adults were numerous, many eggs were laid, resulting in strong larval competition for the limited food supply. None of the larvae that hatched from eggs laid during adult population peaks survived, primarily because they did not grow large enough to pupate. Therefore, large adult populations gave rise to few adult progeny, and because adults lived less than 4 weeks the population soon began to decline. Eventually so few eggs were laid on any particular day that most of the larvae survived, and the size of the adult population began to increase again.

Nicholson's result may be interpreted as a time-delayed logistic process, which provides a good fit to the observed oscillations in one of Nicholson's experiments with a value of $r\tau = 2.1$. This model predicts the ratio of the maximum to the minimum population to be 84 and the cycle period to be 4.54τ (May 1975). Nisbet and Gurney (1982) devote an entire chapter to the analysis of Nicholson's data, ending up with a somewhat more complex model to account for some of the departures of the cycles from the time-delayed logistic model. R. W. Poole (1979) also presents a detailed treatment of the data based on forecasting from time-series data (Box and Jenkins 1970). Regardless of these details, however, the experiment clearly reveals that density-dependent factors did not immediately affect the mortality rates of adults as the population increased, but were felt a week or so later when the progeny were larvae. Larval mortality was not expressed in the size of the adult population until those larvae emerged as adults about 2 weeks after eggs were laid. The blowfly population resembled Pratt's *Daphnia* popula-

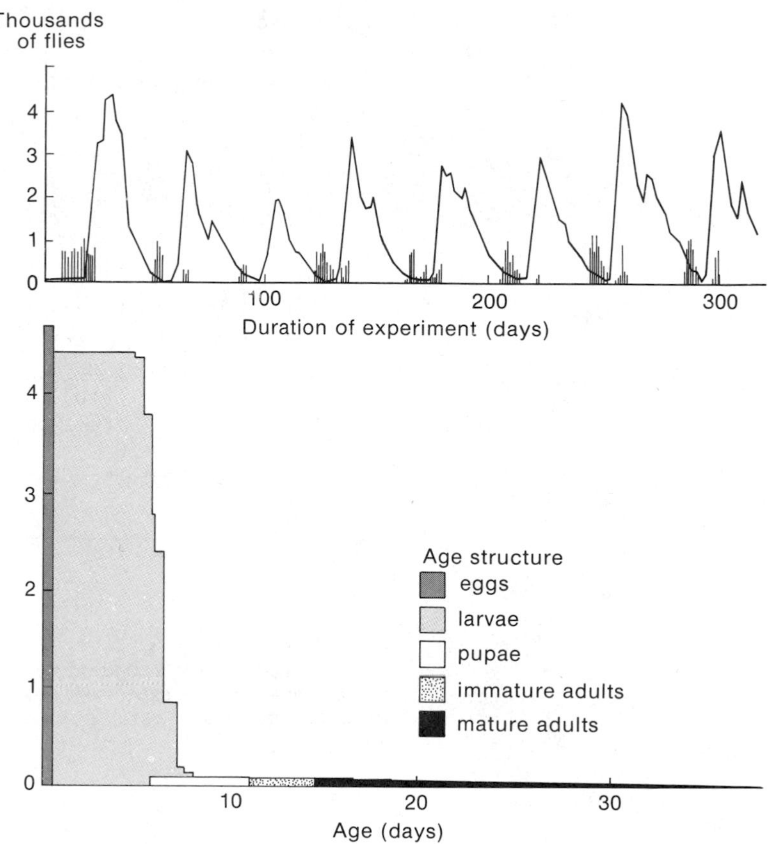

Figure 18-13
Above: Fluctuations in laboratory populations of the sheep blowfly *Lucilia cuprina.* Larvae were provided with 50 grams of liver per day; adults were given unlimited supplies of liver and water. The continuous line represents the number of adult blowflies in the population cage. The vertical lines are the number of adults that eventually emerged from eggs laid on the days indicated by the lines. *Below:* The average age structure of the population. (After Nicholson 1958.)

tion at high temperature in which crowding created discrete, nonoverlapping generations of individuals with an inherent time delay equal to the larval development period, about 10 days. Michael Hassell, John H. Lawton, and Robert M. May (1976) explored discrete generation models of the blowfly population; their paper should be consulted for further discussion.

The hypothesis that population cycles were caused by time delays could be tested directly by eliminating the time delay in the density-dependent response; that is, by making the deleterious effects of resource depletion at high density felt immediately. Nicholson did this by adjusting the amount of food so that food limited adults as severely as larvae. Because adults require protein to produce eggs, by restricting the liver available to adults to 1 gram per day, Nicholson cut egg production to a level determined by the availability of liver rather than the number of adults in the popula-

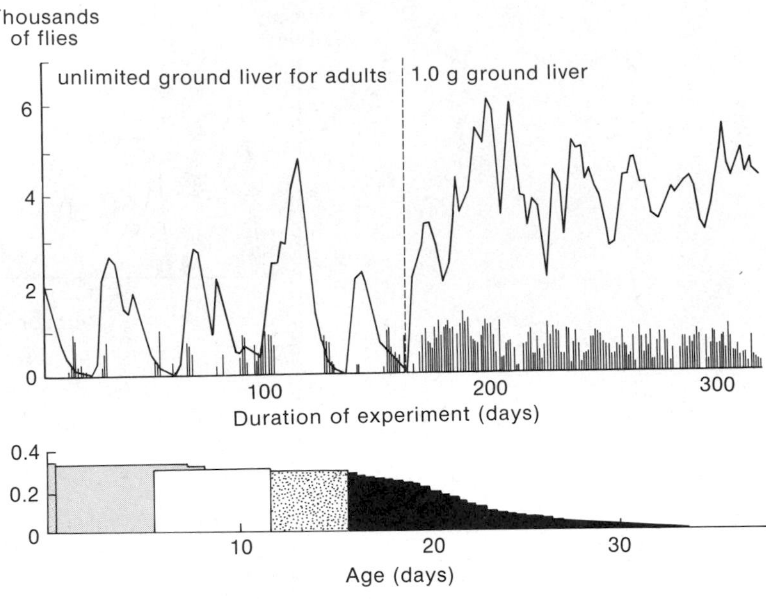

Figure 18-14
Effect on fluctuations in a population of the sheep blowfly of limiting the amount of liver available to adults. The experiment was similar to that depicted in Figure 18-13 in all other respects. Average structure is shown for the latter half of the experiment. (After Nicholson 1958.)

tion. Under these conditions, the recruitment of new individuals into the population was determined at the egg-laying stage by the influence of food supply on per capita fecundity, and most of the larvae survived. As a result, fluctuations in the population all but disappeared (Figure 18-14).

We have seen that responses of populations to density can be delayed by development time and by the storage of nutrients, both of which put off deaths to a later point in the life cycle or to a later time. Density-dependent effects on fecundity can act with little delay when eggs are produced quickly from resources accumulated over a short period. Populations controlled primarily by such factors would not be expected to exhibit marked oscillations.

Time delays longer than a single generation may be introduced through the transmission of maternal effects to progeny through the egg or by selection of different genotypes at different population densities. Studies on the western tent caterpillar by W. G. Wellington (1960) suggested a time delay through maternal effects. A 4-year survey on Vancouver Island, Canada, included a peak year for the population, 1956, followed by a rapid decline through 1959. The larvae of the tent caterpillar were classified as either "active" or "sluggish" depending upon their behavior, which, in turn, was determined by the level of nutrition of their mother. Broods of tent caterpillars composed primarily of active larvae differed from those with a higher proportion of sluggish larvae in that they constructed more tents with a more elongate structure, foraged over greater distances along the limbs of the tree, ate more leaves, and, as a result of their greater food intake, developed more rapidly. Active caterpillars survived better than sluggish ones.

The proportion of active larvae in a particular brood was influenced by the past history of the infestation. Broods in areas in which the population had been low in previous years were likely to contain more active caterpillars than broods from areas of recent high levels of infestation. The activity level of the larvae depended on the nutrition provided by the eggs from which they developed. Most moths laid similar numbers of eggs, regardless of their ability to provision each individual egg; how well each female provisioned her eggs with nutrients depended on how much she consumed as a caterpillar. When food was relatively scarce, few eggs were properly provisioned and broods contained few active caterpillars. Thus the quality of caterpillars in broods of a particular year depended on the availability of food to the female parent when she was a caterpillar during the previous year. Furthermore, because the activity level of each caterpillar partly determined how much food it consumed, the food supply of its parent during the previous year affected the success of its progeny during the following year. In this way, populations could exhibit a time delay of more than 1 year in their response to density or to changes in the availability of food.

Regardless of the time delay in the density-dependent response, a population at its equilibrium point will remain there until perturbed by some outside influence, whether a change in the equilibrium level (K) or a catastrophic change in population size (N). Once displaced from the equilibrium, some populations will move toward stable limit cycles, depending on the nature of the time delay and the response time. Others will return to the equilibrium directly or through damped oscillations. Cycles may be reinforced through interactions with other species—prey, predators, parasites, perhaps even competitors—with similar time constants.

SUMMARY

Although ecologists find it easier to think about populations in equilibrium, most populations fluctuate, either because their size reflects variation in the environment or because they express oscillatory properties intrinsic to their dynamics. In this chapter, we have characterized factors that govern changes in populations with the goal of understanding the response of populations to external perturbations.

1. Canadian and English entomologists developed the method of "key-factor" analysis to identify factors that caused fluctuations in population size. Their techniques were based on the regression of change in population size $\Delta \log N$ on the logarithms of the survival rates (S) of individuals at risk to various factors. Field studies revealed that population-governing factors were often those that caused high and variable mortality from gener-

ation to generation. Many of these factors acted in density-dependent fashion.

2. The characteristic return time of a population is defined as the inverse of the exponential growth rate in the absence of crowding; that is, $T = 1/r$. Populations can track environmental variation closely when the characteristic return time is less than the period of environmental variation divided by 2π.

3. Discrete-time models of populations with density dependence tend to oscillate when perturbed. For r between 0 and 1, population size (N) approaches equilibrium (K) monotonically. For r between 1 and 2, N undergoes damped oscillations, and eventually settles down to K. For r greater than 2, oscillations in N increase in amplitude until either a stable limit cycle is achieved or the population fluctuates irregularly (chaos).

4. Continuous-time models can produce cyclic population change when the density-dependent response is time delayed. Defining the time delay as τ, such models exhibit monotonic damping when the product $r\tau$ lies between 0 and e^{-1} (0.37), damped oscillations when $r\tau$ lies between e^{-1} and $\pi/2$ (1.6), and limit cycles with period 4τ or more when $r\tau$ exceeds $\pi/2$.

5. Many laboratory populations of animals exhibit oscillations that arise from time delays in the response of individuals to density. The time delays are related to the period of development from egg to adult and may be enhanced by the storage of nutrients. In laboratory populations of sheep blowflies, A. J. Nicholson experimentally removed the time delay and was able to eliminate cycles in numbers.

6. Time delays may be increased beyond the length of a generation by maternal influences on the quality of offspring, as suggested by W. G. Wellington for the western tent caterpillar, and by selection of different genotypes at different population densities.

19

Evolution, Social Behavior, and Population Regulation

Until a phenomenon can be understood thoroughly through observation, theory, and experiment, the development of ideas relating to that phenomenon often follows the path of opinion. Without understanding to light the way, the path of opinion is free to wander the landscape and, as often as not, may diverge in several directions. This has certainly been true with regard to thinking about the regulation of population size. Two directions were taken early on: one by the "climate" school of density independence (especially Uvarov and, later, Andrewartha and Birch), the second by the "balance" school of density dependence (especially Pearl, Lotka, and Nicholson). The different opinions of these schools of thought were based upon fragments of observation and theory, like the different ideas of an elephant entertained by blind men, each feeling a different appendage. For many years, neither school enjoyed the final judgment of evidence, and opinions swayed back and forth with each new revelation. Although the idea of the balance of nature seemed firmly rooted during the 1930s and 1940s, the fact that Andrewartha and Birch could, in 1954, breathe so much life back into the idea of density-independent regulation indicated the tenuous hold of the balance school. But controversy also stimulated resolution, and we understand population regulation much better for having been forced to rethink old ideas from time to time and to recognize what is opinion and what is knowledge.

During the 1950s and 1960s opinions as old as utopian ideals joined with fresh observations on populations and, especially, social behavior prompted a major reevaluation of ideas concerning population regulation. The utopian ideal was simply that nature could not be so wasteful as Darwin had supposed; instead, both man and animals used self-control to regulate their populations at a level that could be sustained without the ravages of starvation and disease. Darwin's main point about nature, and his major impact on contemporary thinking, was to destroy this idealistic view of a rational and harmonious world. In *On the Origin of Species* he had written:

> In looking at Nature, it is most necessary to keep the foregoing considerations always in mind—never to forget that every single organic being may be said to be striving to the utmost to increase in numbers; that each lives by a struggle at some period of its life; that heavy destruction inevitably falls either on the young or old, during each generation or at recurrent intervals. Lighten any check, mitigate the destruction ever so little, and the number of the species will almost instantaneously increase to any amount.

Resistance to this revelation, and a strong clinging to the honest and proud belief that somehow we can do better for ourselves, has periodically erupted in thinking about the regulation of population size.

The idea of self-regulation has expressed itself repeatedly in the context of population regulation.

The balance school of population regulation was based upon mechanistic connections between resources and population processes. The field and laboratory work stimulated by the balance concept revealed that changes in behavior and social relationships often mediated the response of population processes to density. These observations led some investigators back to ideas about intrinsic behavioral control of population size below the ultimate limit imposed by resources. The links in this chain of thought are simple. First, one supposes that the external factors which limit population density express themselves through social behavior (territoriality and emigration) and physiology (impairment and postponement of reproduction). Second, through experimental manipulation one observes that many of these mechanisms operate at low densities in the absence of food or other resource limitation. Third, one then concludes that populations possess mechanisms to limit their growth before individuals feel the direct effects of crowding exerted by external factors.

From the standpoint of today's ideas, the different viewpoints of nature were first contrasted in the late 1940s by two ornithologists, David Lack and Alexander Skutch, in their interpretation of the lower reproductive rates of tropical birds compared to those at higher latitudes. Songbirds typically lay a clutch of 4 to 6 eggs in temperate and boreal regions of North America and Europe, but only 2 or 3 in the tropics. The trend is general, affecting virtually all groups of birds in all regions of the world. Lack, as we have seen, was a strong proponent of the balance tradition of Nicholson. He argued further that the reproductive rate had evolved by natural selection to the greatest possible level within the limits of resources; genetic factors that caused one individual to leave more descendants than another would eventually predominate (Lack 1947). (No wonder that populations should have so great a potential for increase.) In Lack's view, populations are regulated by external factors that balance the intrinsic force of growth. Birds lay fewer eggs in the tropics because parents can gather fewer resources to provide their young. In the meantime, however, Skutch (1949) had interpreted the latitudinal gradient in clutch size quite differently. He supposed that clutch size had evolved, not to maximize reproductive rate, but rather to produce just enough offspring to offset adult mortality and maintain a constant population size. Offspring over and above the number needed to balance the population would be wasteful of individuals, he argued, and an inefficient use of the available resources. Skutch agreed with Lack that population size was ultimately tied to resource levels; they differed over the mechanism by which population growth was controlled. On one hand, Lack believed that regulation was imposed by external factors opposing the constant tendency of the population to overstep its resource limits. On the other hand, Skutch believed that populations were self-regulated—the prudential restraint from marriage that Darwin thought impossible—to prevent the detrimental effects of crowding.

Ideas about self-regulation appeared in many guises during the next two decades, but the controversy always paralleled the argument between Lack and Skutch, pitting profusion against prudence. The debate over self-regulation centered primarily upon four phenomena. First, territorial behavior and emigration from local populations result in some organisms being excluded from breeding. One may interpret such behavior as a spacing mechanism that regulates the density of the population. Second, as we have just seen, many people have regarded variation in reproductive rate as evidence of the adjustment of fecundity to balance mortality in the population. Third, the appearance of debilitating physiological pathology at high population densities, often resulting in a cessation of breeding, has suggested mechanisms for self-regulation. Fourth, research on cyclic species has indicated that population declines may result from changes in the frequency of certain genotypes selected in response to population density but independently of resources.

Each of these ideas sprang from studies that produced evidence suggestive of self-regulation and undoubtedly played upon previously held opinions about population control. It remained until 1962 for the British ecologist V. C. Wynne-Edwards to bring together the evidence for self-regulation and articulate the concept in the full context of ecology and population biology. This he did in a masterful book, *Animal Dispersion in Relation to Social Behavior,* which stands next to Lack's *Natural Regulation of Animal Numbers* and Andrewartha and Birch's *The Distribution and Abundance of Animals* in its influence on population biology. Thus exposed and given credibility, the idea of self-regulation came under the full force of criticism from the scientific community; we shall put off the resolution of the controversy until after we have had a look at some of the phenomena.

Territorial behavior may limit population density.

Individuals interact with others of the same species in many ways. Sometimes one individual senses another only through its effect on the resources that they share. In other cases, individuals may confront each other directly.

The larvae of the grain beetle *Rhizopertha*, which so captivated Crombie, fought to the death over a kernel of wheat. For *Rhizopertha*, a single kernel is life; without one, nothing. For a sparrow or a mouse, life is a succession of thousands of kernels and no single one is worth an all-out defense. The thousands can be safeguarded only by defending the area in which they may be found. Such defense of a resource against the intrusion of others is one basis for territorial behavior in animals.

The modern concept of territory is due to H. E. Howard, who, in 1920, published *Territory in Bird Life*. He recognized that male white-throated sparrows vehemently singing from perches on fine spring mornings were advertising their claims to areas or territories. These areas had well-defined boundaries and were defended against the intrusion by other males. Soon females would be attracted to the areas and the pairs would mate and rear their offspring within the territory.

A territory may be defined as any defended area, but because one rarely observes territorial defense, a more practical definition is "any exclusive area." Although territoriality is most conspicuous in birds, it is now well-known in insects, mollusks, crustacea, fish, reptiles, mammals, and others. In 1941, Margaret Nice classified territories based on the object defended, which varies from feeding and breeding areas to nests, mates, and roosts, and other resources. Territories may be permanent, providing a home area for life, or established for a few minutes on a mudflat where a migrating shorebird has located a rich concentration of food.

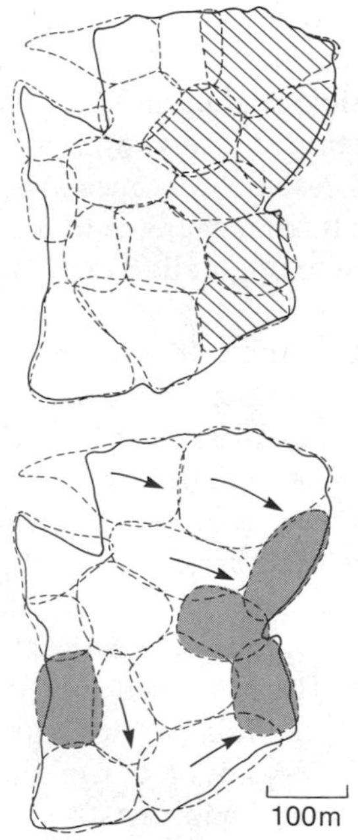

In 1956, the British behaviorist Robert Hinde discussed the purpose of territories. He recognized many advantages accruing to the holders of territories: exclusive use of a resource, reduced predation and disease owing to greater spacing from neighbors, predator escape facilitated by familiarity with an area. But he also recognized that advantages might accrue to a population of territorial animals if their behavior kept the density of individuals below a level that would overeat the food supply and otherwise damage the environment. Such a population would be less likely to fluctuate, owing to time-delayed response to density, or crash to extinction.

One result of territorial defense is the exclusion of individuals from breeding. Although the size of defended areas may vary in response to both resource level and population density, territorial behavior sets an upper limit to the number of individuals that occupy an area; others must go elsewhere. Great tits *(Parus major)*, European relatives of North American chickadees and titmice, defend territories during the breeding season. The tits nest in natural holes in trees, or in boxes provided by humans. Their preferred habitat is woodland, although individuals will nest in hedgerows or similarly broken patches of woods. Birds nesting in hedgerows are usually 1-year-olds and do not often breed successfully. The older individuals that secure territories in woodlands raise 90 per cent of their broods. When one removes territorial males from woodland habitat, newcomers from hedgerows—mostly first-year birds—rapidly replace them; established territories may also expand to fill the void (Figure 19-1). Removal experiments of this sort have revealed a "floating" population of individuals excluded from breeding because they cannot defend territories against others.

100m

Figure 19-1
Replacement of birds removed from a population of great tits. Six pairs were shot in late March 1969 (territories indicated by striped areas). Within 3 days, four new pairs had taken up residence in the wood (shaded areas) and the remaining area was filled by expansion of existing territories (arrows). (After Krebs 1971.)

One of the most thorough studies of the role of territorial behavior in population processes has been that of Adam Watson and his colleagues on the red grouse *(Lagopus lagopus)* in Scotland (see Watson and Moss [1980] for a review). The red grouse, which North Americans call the willow ptarmigan, is a bird of heath and tundra. Male grouse defend territories to which they may attract many hens; some, of course, attract none; and others do not even hold territories. The grouse breed in summer, and population density is greatest in the late summer after the chicks of the year have grown up. Territories are taken up anew each year in the fall and defended through the beginning of the next breeding season. Birds without territories usually die or disappear before then. Breeding success depends primarily upon chick survival (the key factor in this case), which depends on the quality of the egg, which in turn depends on the food supply in the territory for the hen. Hence territory provides a link between resources and population processes.

Factors affecting red grouse populations have been worked out by extensive observation and experimentation. Removal of territorial males resulted in an influx of previously excluded individuals, many of them young (Watson and Jenkins 1968). Further observations showed that winter mortality is socially induced—territory holders survive, others don't—and that territorial behavior sets an upper limit to the size of the spring population. Social dominance determines the ability of a male to hold a territory. When hormones were implanted to increase aggressiveness, experimental males had more frequent encounters with neighbors, displayed more often, and some enlarged their territories; nonterritorial males with implanted hormones gained small territories in spite of losing most encounters (Watson and Parr 1981). The relationship of territory size to resources was shown by fertilizing areas of heath to increase production of the heather shoots upon which grouse subsist. In the fertilized areas, males defended smaller territories but these evidently contained more resources because hens reared larger broods on fertilized areas than on adjacent control areas (Miller et at. 1970).

In populations of many species of small mammals, the effects of aggression increase with population density, and juveniles suffer the most. This has been shown in many studies (Bondrup-Nielsen and Imms 1986, Flowerdon 1974, Gilbert et al. 1986, Healey 1967, Krebs and Myers 1974, Sadlier 1965), but I shall illustrate the point with the results of experiments by R. Boonstra (1978) on Townsend's vole *Microtis townsendii*. Voles are mouselike rodents, actually more closely related to lemmings, which inhabit meadows and other habitats with abundant grasses and herbs, their principal food. Near Vancouver, Canada, Boonstra trapped large numbers of free-living juvenile voles, introduced them to areas in which the local populations had been manipulated, and followed their subsequent survival. In one area, all voles except the introduced juveniles were removed when trapped. In a second area, all adult males were removed, leaving behind only adult females and the introduced juveniles. In a third area, the control, adults were trapped but immediately released.

On the control plot fewer than 10 per cent of the juveniles survived the first month. Where males were removed, about 15 per cent survived, not significantly higher than on the control plot. Where both sexes were removed, however, about 60 per cent of the juveniles survived the first month. Furthermore, they came into sexual maturity at a lower weight (median, 42 grams) than on either the control or male removal plots (48 and 51 grams). Boonstra's results show that under the conditions of the study, adult voles reduced the local survival of juveniles and delayed sexual maturation. Adult females evidently were responsible for the effect because the removal of males did not change the situation. Juveniles could enter the breeding population only where adults were few.

Aggression forces many individuals to emigrate from dense populations.

A series of studies on populations of house mice, conducted during the late 1940s and early 1950s at the University of Wisconsin, emphasized the importance of emigration to population processes. R. L. Strecker and J. T. Emlen, Jr. (1953) established populations in large, escape-proof rooms by introducing equal numbers of adult male and female mice, trapped from populations in the city of Madison. Water and nesting sites were provided in abundance, but food was limited to a constant ration each day. In this experiment, the populations grew rapidly until the mice consumed all the food provided each day, at which time reproduction ceased abruptly. The authors concluded that an external factor—in this case, food—limited population growth.

In another experiment, Strecker (1954) established a population of house mice in a large room in the subbasement of a University of Wisconsin building. The room had several natural escape routes—ventilating shafts that ran throughout the building—so the mice could, and did, emigrate from the population. As in the previous experiment, this population also was provided with a constant ration of food. Within 8 months, the population had become limited by food supply, and the emigration of individuals from the population, determined by the appearance of mice in offices and laboratories throughout the building, rose to a high level. Both subadult and adult animals, and equal numbers of males and females, left the population. Because emigration resulted in a low resident population, reproduction at home remained high even after food had become a limiting factor. Social pressure maintained the population at a sufficiently low level with respect to the food supply that mice could continue to breed even when food was limited.

The house mouse experiment has been repeated with many species of small mammal under a variety of conditions, with a consistent outcome (for a review, see Lidicker 1975). For example, when Charles J. Krebs and his coworkers fenced large plots of grassland in Indiana, vole *(Microtus)* populations shot up to three times the average density on control plots, and the vegetation was heavily overgrazed (Krebs et al. 1969).

J. H. Myers and C. J. Krebs (1971) documented emigration from unfenced populations of the field vole *Microtus pennsylvanicus* and *M. ochrogaster*. They trapped and removed all the residents in several small plots of meadow and then assumed that voles subsequently recovered in these removal plots had emigrated from some other area. Vole populations varied through tremendous cycles of abundance, and emigration varied through the cycle. During periods of increase, emigration accounted for 60 to 70 per cent of the disappearance of voles from control trapping grids. During periods of population decrease, few individuals dispersed. Among females, young rather than old individuals were caught entering removal areas and these emigrants appeared to mature at a younger age than females that stayed behind. Male dispersers were less active than residents when placed in cages, suggesting that they might be less aggressive. Perhaps most surprising was the observation that dispersers and residents tended to have different genotypes with respect to certain blood proteins. Indeed, W. E. Howard had suggested, in 1960, that certain individuals might be genetically predisposed to leave the area of their birth and certain others, to stay behind. Here was some supporting evidence.

Dispersal in response to population density is not limited to birds and mammals. Probably it is common to most species with an active stage; some spectacular examples, including the migratory locusts of Africa, are well known among species of insects (Dempster 1963, 1968; Smith 1972).

Dispersal has always been the waif of population ecology. Its importance has been underscored by experimental studies, yet it is difficult to measure and its consequences for the individual are all but unknowable. In this regard, William Lidicker (1975) remarked:

> Commonly, dispersers have been either ignored or considered to be of little significance by population ecologists. In part this has been due to considerable difficulty in measuring dispersal, but it is also widely assumed that such movements are demographically unimportant. For example, dispersers are typically considered to be individuals that leave home only when conditions become intolerable, and then have a near-zero probability of surviving long enough to reproduce elsewhere.

Lidicker distinguished two types of dispersal, each having different demographic consequences. The first, what he called "saturation dispersal," occurs when a population becomes crowded with respect to resources and aggressive individuals force others to leave the area. The future for such animals is probably quite bleak, but perhaps better than that offered by the hopeless conditions back home. The second, "presaturation dispersal," involves individuals that leave their natal areas well before conditions deteriorate due to crowding. Lidicker and others have suggested that such animals have an innate predisposition to wander and that they are liable to enjoy high fecundity and long life. Myers and Krebs's (1971) observation that many voles disperse during the increase phase of the population cycle and that dispersers and residents exhibit genotypic differences support the idea of presaturation dispersal. But both the extent and ultimate purpose of such behavior is poorly understood and, therefore, open to a variety of interpretations.

Individuals may exhibit physiological impairment of reproduction in crowded populations.

Mouse populations provided limited resources cannot increase beyond a certain density, as we have seen from the results of experiments at the University of Wisconsin. When individuals can disperse from such populations, aggressive behavior keeps density low enough that the remaining individuals continue to breed, although at reduced levels. When dispersal is prevented, the crowded adults use all the resources and breeding ceases. One would think, then, that the density of a population would increase in direct relation to the level of resources provided, as Chapman (1928) showed for populations of flour beetles. To test this idea, Charles H. Southwick (1955) extended investigations on the Wisconsin mice to confined populations provided unlimited food. Under these conditions, the mice continued to breed regardless of the density of the population, but the survival of embryos and young was so reduced that the population leveled off in spite of the unlimited resource (Figure 19-2). These populations were regulated primarily by increased mortality of preweaned young (and also of some adults) as density increased, rather than by cessation of reproduction, as Strecker and Emlen had observed in the limited-food situation. Deaths of preweaned mice were directly related to the presence of mice other than the mother, especially to males, in the nest box (Brown 1953, Southwick 1955). With increasing population density, the frequency of fighting and the number of diseased and wounded mice, especially males, increased sharply.

Similar observations had been made by John B. Calhoun (1949), working with the Rodent Ecology Project at Johns Hopkins University, on a population of Norway rats. The rats were confined to a 100-foot-square pen, in the center of which was a 20-foot-square pen provided with abun-

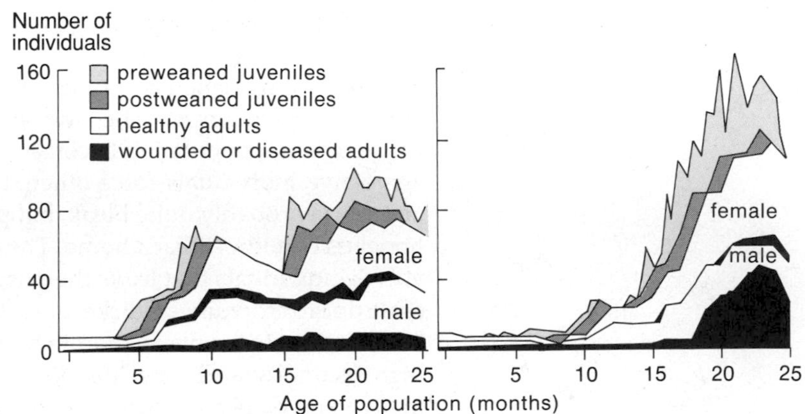

Figure 19-2
Growth of two confined populations of house mice *(Mus musculus)* supplied with unlimited food, showing the age structure and incidence of diseased or wounded adults. (After Southwick 1955.)

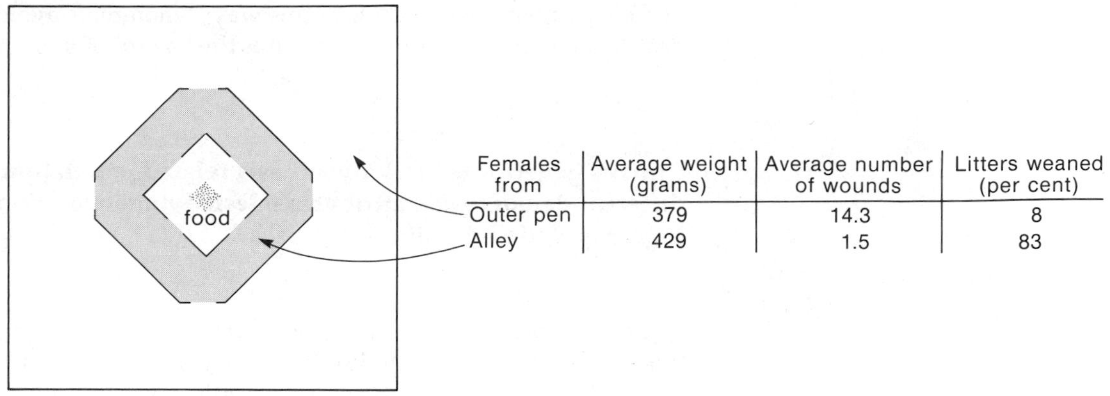

Females from	Average weight (grams)	Average number of wounds	Litters weaned (per cent)
Outer pen	379	14.3	8
Alley	429	1.5	83

Figure 19-3
Weight, fecundity, and evidence of fighting of females from the alley and outer pen in the experiment of Calhoun (1940).

dant food. The food area was surrounded by a 20-foot-wide alley. There were four passageways (3-inch clay drain pipes) between the food area and the alley, and four between the alley and the outer part of the pen. Animals living in the outer part had to pass through the alley to get to food and they frequently suffered the aggression of the alley rats. Calhoun determined that compared to rats born in the outer part of the pen, alley rats grew larger, had fewer wounds, and were more successful in weaning litters (Figure 19-3). Based on these observations, Calhoun concluded that "any environmental conditions which affect the maternal physiology to the extent that foetal nutrition is lowered will slow down the rate of population growth due to early post-parturition mortality . . ."; furthermore, "this process of upsetting maternal physiology may be one of the limiting factors in population growth" and "physiological and psychological disturbance in socially inhibited individuals . . . may be a potent factor in population control among mammals."

John J. Christian and David E. Davis (1964) gave the ideas of Calhoun and others a firm physiological foundation and further developed the notion that physiological response to social behavior could regulate population size. They pointed out that the social stress of crowded laboratory populations often led to a variety of abnormal physiological symptoms, collectively referred to as the "general adaptive syndrome": (1) the adrenal glands enlarge considerably; (2) growth and reproduction are curtailed in both sexes; (3) sexual maturation is delayed or inhibited; (4) spermatogenesis of the male is delayed; (5) the estrus cycle of the female is prolonged and rates of ovulation and implantation are diminished; (6) mortality of embryos within the uterus is greatly increased (see Christian 1961); (7) lactation is often inadequate, leading to stunting of the young at weaning; and (8) susceptibility to disease may increase. Christian and Davis suggested that the behavioral-endocrine feedback system that produces the general adaptive syndrome can restrict populations considerably below the ultimate limit of food availability and that, in this way, "mammals avoid the hazard of destroying their environment, and thus the hazard of their own extinction."

Chitty's genetic-feedback hypothesis related population cycles to different demographic attributes selected under conditions of low or high population density.

A last example of the frequent appearance of the idea of self-regulation is due to Dennis Chitty (1960, 1967), whose research focused upon fluctuating populations of small mammals, especially voles. Finding no evidence of starvation, disease, increased predation, or stress at high densities, Chitty suggested that population declines are brought about through a deterioration in the "quality" of individuals in the population, even in the absence of adverse conditions. By quality, Chitty meant attributes, such as growth rate, aggressiveness, and parental behavior, that contributed to an individual's fecundity and that might be determined by genetic or maternally inherited factors. Whereas Christian and Davis viewed the general adaptive syndrome as being induced by the environment, Chitty felt that some individuals were more sensitive to crowding than others. The proportion of the more sensitive individuals might increase during periods of low population density, thereby increasing the eventual adverse response of the population to crowding. During the ensuing population peak and crash, selection might favor individuals less sensitive to crowding, thus establishing a cycle of gene frequency in the population that could enhance the population's natural tendency to cycle. Chitty's student Charles J. Krebs produced evidence supporting this theory, as we have seen, although the data have been interpreted differently (see Gaines and Whittam 1980).

Chitty (1960) concluded that "voles probably exemplify a general law that all species are capable of limiting their own population densities without either destroying the food resources to which they are adapted, or depending upon enemies or climatic accidents to prevent them from doing so." Chitty was influenced by the early work of Christian (1950) on the general adaptive syndrome and referred to examples of changes in the behavior of insects (migratory locusts and tent caterpillars) with increasing density. He also recognized that

> self-regulatory mechanisms have presumably been evolved through natural selection, and arguments in support of this view can certainly be advanced. In the present paper, however, the only argument required is the purely methodological one that it is best to start with the fewest and simplest explanations possible, and to add to them only when it is clear that there are fundamental differences between similar phenomena in related species.

Where Chitty sidestepped the problem posed by the evolution of self-regulation, V. C. Wynne-Edwards not only articulated the self-regulation viewpoint but provided an evolutionary mechanism. This, as we shall see, led to the downfall of self-regulation, as a house of cards on a weak foundation eventually collapses in ruin.

V. C. Wynne-Edwards's theory of animal dispersion elaborated the idea of self-regulation.

Wynne-Edwards agreed with most ecologists that resources, particularly food, ultimately limit population density, but he denied that population processes were governed by external factors acting through the physiology of the organism. In his books *Animal Dispersion in Relation to Social Behavior* (1962), and more recently, *Evolution through Group Selection* (1986), he argued that such manifestations of poor environmental conditions as reduced fecundity, delayed maturity, and emigration are adaptations evolved in order to keep populations from overusing their resources. Wynne-Edwards joined Skutch, Christian and Davis, Chitty, and others of the self-regulation school in believing that evolution has minimized wastefulness in populations by limiting the recruitment of young individuals to a level that just balances adult mortality. He went one step further in suggesting that animals assess the density of their populations through various group displays, such as flocking and territorial singing. Such behavior, he presumed, has evolved to facilitate the estimation of population density, whereby animals adjust their fecundity. Of course organisms need not be consciously aware of population density nor must they willfully adjust their fecundity accordingly; they must only be adapted to respond appropriately to levels of interaction within the population. As Wynne-Edwards (1963) summarized his theory of population control:

> The mechanisms involved work homeostatically, adjusting the population density in relation to fluctuating levels of resources; where the limiting resource is food, as it most frequently is, the homeostatic system prevents the population from increasing to densities that would cause overexploitation and the depletion of future yields. The mechanisms depend in part on the substitution of conventional prizes, namely, the possession of territories, homes, living space and similar real property, or of social status as the proximate objects of competition among the members of the group concerned, in place of the actual food itself.

Adaptations of social behavior involved in territoriality, dominance hierarchies, and sensitivity to the stress of crowding would require a special mechanism of evolution if animals were to prudently control their populations. Self-regulation implies restraint. But among individuals with greater or lesser reproductive rate, the more fecund must always be selected. Wynne-Edwards overcame this problem by postulating selection of groups of organisms based upon the efficiency with which they utilized their resources.

The idea of intergroup selection is an old one, formally introduced by the population geneticist Sewall Wright (1931, 1945) to explain the evolution of behaviors by which the adapted individual accepts a cost in order to benefit some other member of the group. Such "altruistic" behavior cannot be selected by choosing among individuals because the donor of the behavior suffers where the recipient gains. But groups having altruists may fare better over the long run than those without because of the greater cooperation within the altruistic group. It was a small step for Wynne-Edwards to place restraint from breeding in the category of such cooperative, altruistic behavior:

> In the context of the social group a difficulty appears, with selection acting simultaneously at the two levels of the group and the individual. It is that the homeostatic control of population density frequently demands sacrifices of the individual; and while population control is essential to the long-term survival of the group, the sacrifices impair fertility and survivorship in the individual.

Wynne-Edwards also recognized the problem that altruism posed for evolutionary theory.

> One may legitimately ask how two kinds of selection can act simultaneously when on fundamental issues they are working at cross purposes. At first sight there seems to be no easy way of reconciling this clash of interests; and to some people consequently the whole idea of intergroup selection is unacceptable.

Then he elaborated on how group selection might operate:

> One of the most important premises of intergroup selection is that animal populations are typically self-perpetuating, tending to be strongly localized and persistent on the same ground. . . . Isolation is normally not quite complete, however. Provision is made for an element of pioneering, and infiltration into other areas; but the gene-flow that results is not commonly fast enough to prevent the population from accumulating heritable characteristics of its own. Partly genetic and partly traditional, these differentiate it from other similar groups.
>
> Local groups are the smallest racial units capable of continuous existence for long enough to undergo evolutionary differentiation. In the course of generations some die out; others survive, and have the opportunity to spread into new or vacated ground as it becomes available, themselves subdividing as they grow. In so far as the successful ones take over the habitat left vacant by the unsuccessful, the groups are in a relation of passive competition.

Thus Wynne-Edwards recognized that for intergroup selection to work, populations must be subdivided into local, discrete groups capable of replacing others that die out. The life span of a group is its time to extinction; the fecundity of a group is related to number of dispersers or ability to expand into unoccupied areas. So long as groups have heritable variation, including variation in culturally transmitted traits, those traits may then be selected. But within each group, selection also operates to increase the frequency of the most fecund individuals. And so the final outcome must represent the balance between selection at the group and individual levels. Wynne-Edwards argued that group selection is the stronger force:

> What would be the effect of selection, for example, on individuals the abnormal and socially undesirable fertility of which enabled them and their hereditary successors to contribute an ever-increasing share to future generations?
>
> Initially, groups containing individuals like this that reproduced too fast, so that the overall recruitment rate persistently tended to exceed the death-rate, must have repeatedly exterminated themselves . . . by over-taxing and progressively destroying their food resources.

In Wynne-Edwards's view, adaptations resulting from intergroup selection were properly assigned to the group and transcended the individuality of its members; the advantage of particular members is subordinated to the advantage of the group. He stressed this point by a simple analogy:

> A football team is made up of players individually selected for such qualities as skill, quickness and stamina, material to their success as members of the team. The survival of the team to win the championship, however, is determined by entirely distinct criteria, namely, the tactics and ability it displays in competition with other teams, under a particular code of conventions laid down for the game. There is no difficulty in distinguishing two levels of selection here, although the analogy is otherwise very imperfect.

The self-regulation issue has stimulated new ideas in population biology and evolution.

Wynne-Edwards crystallized the extreme self-regulation view into a logical structure that spanned the breadth of ecology, behavior, and evolution. Thus made so conspicuous, his theory became the most controversial topic in ecology during the 1960s and stimulated a barrage of criticism aimed at utterly demolishing the idea.

The difficulties with Wynne-Edwards's theory may be summed up as follows. First, the predictions of his hypothesis do not differ substantially from predictions based upon selection of maximum reproductive rate within groups. That is, what we observe in nature can be interpreted in many ways; natural history cannot be claimed to favor one theory or another without additional observation and experimentation. This is not to say that either Wynne-Edwards or his critics were wrong, only that the data were equivocal.

Second, the mechanism required to select traits that benefit the group at the expense of the individual must be based on competition between groups or populations rather than between individuals. Although such a mechanism is plausible, it depends on the existence of discrete, persistent groups, the development of heritable variation among the groups with respect to mechanisms of population control, and the expression of that variation in terms of group fitness. Furthermore, even if group selection could be shown to operate, critics pointed out that individual selection will normally be the more powerful force because individuals are replaced within groups more rapidly than discrete groups are replaced within populations (Wiens 1966). Where group fitness and individual fitness conflict, the latter is more likely to prevail (Williams 1966b, Lack 1966).

At present, few ecologists believe that populations are self-regulated. The theoretical difficulties are overwhelming. Moreover, one need not resort to group selection to explain adaptations that Wynne-Edwards envisioned as part of the social convention of self-regulation. Territoriality serves the purpose of the individual who secures his plot of ground. Individuals that are subordinate because of age, experience, or genetic makeup may well be better served by acquiescing in their position and hoping for a change in their situations tomorrow than by fighting losing battles for social status today. Dispersal from a crowded area may be more fruitful than trying to tough it out at home.

The general adaptive syndrome may express the inability of organisms to adapt to all conditions. If physiological responses to encounters with others have evolved under predominantly moderate densities, physiological dysfunction in the same individuals placed at abnormally high densities might be expected, just as any machine is likely to break down when operated under conditions for which it is not designed.

Behavior that seemingly reduces the individual's fitness could also arise at high densities in fluctuating populations. If reproductive physiology were adapted to the conditions of low densities, at high densities the same adaptations might function to restrict reproduction (Williams 1975). Although behavioral and physiological responses to high density might limit population growth, one cannot interpret such responses as adaptations to curtail population growth rate. Behavior appropriate to most situations may be detrimental to the individual in some others; the organism cannot adapt perfectly to all conditions.

Wynne-Edwards, and those whose views he articulated so well, greatly advanced the study of ecology and evolution. The ensuing controversy stimulated others to sharpen their thinking about the evolution of adaptations involved in the regulation of population size. Their ideas also led directly to a complete reevaluation of life history patterns as adaptations, beginning with Robert H. MacArthur (1960, 1961, 1962), George Williams (1966), and Martin Cody (1966). John Maynard Smith (1964) and W. D. Hamilton (1964) responded to Wynne-Edwards's group-selection model of the evolution of social behavior by their own idea of kin selection, from which sprang the new discipline of sociobiology. We should also remember that Wynne-Edwards, Skutch, and others raised many questions about the function of such behaviors as group displays and flocking, which remain unanswered. Finally, and somewhat ironically, the stigma of group selection has faded sufficiently that theoreticians and experimentalists are reconsidering group-selection models for natural populations (Boorman and Levitt 1973, Gilpin 1975, Wade 1977, 1978, D. S. Wilson 1980, 1983, E. O. Wilson 1973). Whatever else, the debate over self-regulation has provided a rich legacy of behavioral and ecological phenomena that will challenge behaviorists, ecologists, and evolutionary biologists for years to come.

SUMMARY

Just as the controversy over density dependence was being resolved, the wavering convictions of population ecologists were buffeted from an entirely different quarter. Observations of the response of social behavior and physiology to crowding persuaded several groups of workers that populations adjusted their sizes through the self-restraint of individuals from breeding. Such behavior was seen as directed toward the idealistic goal of preventing the population from overexploiting its resources and avoiding the crash that would inevitably follow upon imprudently persistent breeding.

1. Territorial behavior can place an upper limit to population size and exclude individuals from breeding. Such behavior clearly may benefit individuals holding territories, but it also may stabilize population size and prevent overexploitation of resources.

2. Aggression in crowded populations causes subordinate individuals to emigrate. Although long-distance movements are difficult to monitor, some evidence suggests that dispersers usually are younger and less aggressive than residents. Furthermore, genetic or maternally inherited factors may predispose some individuals to emigrate.

3. Many mammals exhibit physiological responses to crowding—collectively known as the general adaptive syndrome—that reduce fecundity or inhibit breeding altogether in dense experimental populations. Because the phenomenon appears even when food and other resources are abundant, some investigators, particularly John B. Calhoun, John J. Christian, and David E. Davis, regarded such responses as evidence of self-regulation.

4. Lacking evidence for other factors, Dennis Chitty suggested that periodic declines of cycling populations of small mammals are brought about by changes in the intrinsic quality of individuals as breeders. He postulated that selective pressures varying through the population cycle select individuals in sparse populations whose behavior is inappropriate to the conditions of dense populations.

5. V. C. Wynne-Edwards assembled ideas about self-regulation into a general theory of population control through social conventions. Social interaction enabled individuals to assess the density of their populations and adjust their fecundity accordingly. Hence the population was kept from

overexploiting its resources and thereby persisted longer, at the expense of individual advantage within the population.

6. Wynne-Edwards suggested that self-regulation could evolve by means of intergroup selection. The theory required the existence of discrete, persistent groups and sufficient heritable variance in "fitness" (that is, persistence and colonizing ability) among them to overcome selection among individuals within populations. Most ecologists now agree that these conditions are unlikely and that behavior interpreted as self-regulatory by Wynne-Edwards can be explained on the basis of individual selection.

V

Population Interactions

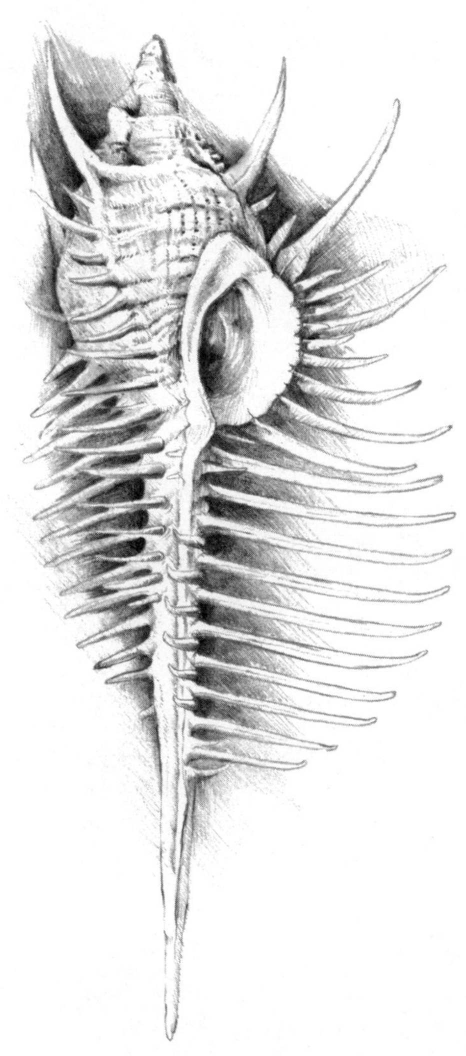

20

Resources and Consumers

The logistic equation, which has been so important to the development of ecological concepts, describes the behavior of a population abstracted from its environment. The variables r and K—the written symbols of population dynamics—express the relationship of the population to its environment: r is the per capita population growth rate in the absence of crowding in a particular environment; K is the number of individuals that can be supported by a particular environment.

The environment includes the set of conditions—such as temperature, humidity, salinity, and resources (including food and hiding places)—that determine birth and death rates of a population. The various conditions and some types of resources have values that are intrinsic properties of the environment and independent of consumer populations. But other resources, including populations of prey organisms, may be greatly influenced by their consumers. Such resource-consumer interactions make it impossible to characterize the dynamics of a single population in isolation.

Populations interact primarily through feeding relationships. These may be direct, as when predators feed upon prey, or indirect, as when two populations share a common resource or a common consumer. Taken together, these interactions influence the dynamics of both consumer and resource populations.

A biological community consists of many populations tied together in a web of feeding relationships. During the 1920s, Charles Elton and others emphasized the utility of describing biological communities by means of feeding relationships. But the eminent ecologist G. Evelyn Hutchinson, of Yale University, was the first to state explicitly the link between feeding relationships and the regulation of community structure. In a famous paper published in 1959, provocatively entitled "Homage to Santa Rosalia, or Why Are There So Many Kinds of Animals," Hutchinson sought bases for explaining the numbers of species and their relationships one to another within communities. He laid the groundwork for much of ecology during recent decades by a simple intuition:

> In any study of evolutionary ecology, food relations appear as one of the most important aspects of the system of animate nature. There is quite obviously much more to living communities than the raw dictum "eat or be eaten," but in order to understand the higher intricacies of any ecological system, it is most easy to start from this crudely simple point of view.

Hutchinson considered the relationship between predator and prey to be fundamental. In a sense, however, this is a special case of the more general relationship between consumers and their resources.

What are resources?

In his 1982 book *Resource Competition and Community Structure,* ecologist David Tilman of the University of Minnesota defines a resource as

> any substance or factor which can lead to increased [population] growth rates as its availability in the environment is increased, and which is consumed by an organism. For instance, the growth rate of a plant [population] . . . may be increased by the addition of nitrate, which is consumed by the plant. There may be concentrations of nitrate at which the addition of more nitrate will lead to decreased growth and death. Still, by the definition offered above, nitrate is a resource—a consumable factor which can potentially limit the growth rate of the population.

Two properties are key to Tilman's definition of resources. First, a resource is consumed and its amount is thus reduced. Second, a resource is utilized by the consumer for its own maintenance and growth. By this definition, food is always a resource, even though some components of the diet may not be. For example, we consume cellulose and other forms of plant fiber in our diet, but these pass through our digestive tracts largely unused. Thus plant fibers are not themselves resources of human consumers, but the food that contains them most certainly is. Water is a resource for terrestrial plants and animals. Water is consumed and it is critical to maintenance and growth; furthermore, when its availability is reduced, biological processes are so affected as to reduce population growth.

"Consumption" extends beyond the act of eating in this definition of resources. As Tilman points out, "For sessile animals, space (open sites) may be a resource. Increases in the amount of open space can cause increased reproductive rates, and the animals "consume" the open sites as they colonize and grow on them." Among barnacles growing on rocks within the intertidal zone, individuals require space to grow and larvae require space to settle and take up adult life (Figure 20–1). Hence crowding increases adult mortality and reduces fecundity by limiting growth of adults and recruitment of larvae. Hiding places and other safe sites comprise another kind of resource. Each area of habitat has a limited number of holes, crevices, or patches of dense cover in which an organism may escape predation or seek

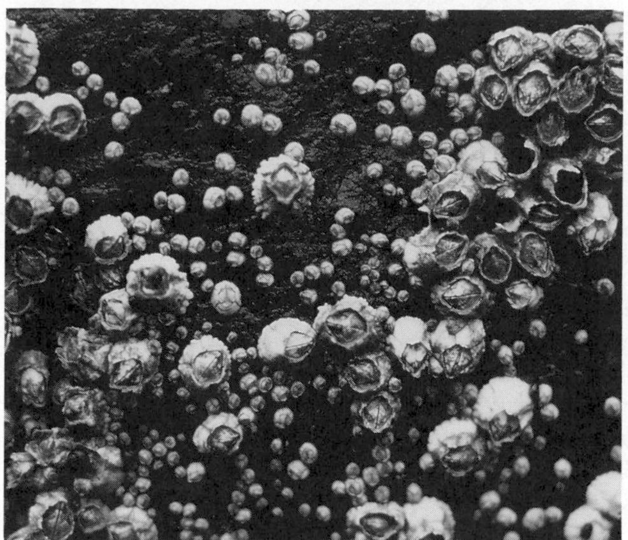

Figure 20–1
Competition for space among barnacles on the Maine coast. Above their optimum range in the intertidal zone the barnacles are sparse and young can settle in the bare patches *(left)*. Lower in the intertidal, dense crowding of barnacles precludes further population growth *(right)*; young barnacles can settle only on older individuals. (Courtesy of the American Museum of Natural History.)

refuge from adverse weather (Martin 1988). As individuals occupy, or "consume," the best sites, others must settle for less favorable places; they may suffer higher mortality as a consequence.

What factors are not resources? According to Tilman, "temperature is not a resource. The reproductive rate of a species may increase with increases in temperature, through some range, but the species does not consume temperature. This is not to imply that temperature and other nonconsumable physical and biological factors are not important, but that they must be considered in a different way from resources." Temperature, like humidity, salinity, hydrogen ion concentration (pH), bouyancy, and viscosity, are conditions that influence the rates of processes and, therefore, the individual's ability to consume resources, but they are not themselves used and thereby transformed by the activities of organisms.

A water-flea consumer and an algal resource reveal the characteristics of a consumer-resource system.

Resources are linked to population processes by many paths. I shall illustrate this point here by the findings of Karen Porter and her colleagues at the University of Georgia on the response of the water flea *Daphnia magna* to the density of their algal resource *Chlamydomonas reinhardi.* Water fleas feed by using a set of legs specialized as filtering appendages to draw a current of water past the feeding apparatus surrounding the mouth. This apparatus grabs algal cells from the water stream and stuffs them into the mouth opening. Because of the simple feeding method and the homogeneous distribution of the food resource in the environment, this consumer-resource system can be brought into the laboratory and controlled experimentally.

Porter et al. (1982, 1983) established laboratory cultures under conditions of water quality, temperature, and resource densities that matched the natural lake environment of *Daphnia.* Algal concentrations were maintained between 0 and 1 million (that is, 10^6) *Chlamydomonas* cells per cubic centimeter (cm^3) of water. The investigators measured rates of water filtering and ingestion, and several population variables, as a function of resource concentration. They estimated filtering rate by the uptake of cells (determined by accumulation of radioactive label from cultures of *Chlamydomonas* labeled with carbon-14) and the density of cells in the culture. For example, if each water flea ingested 10^4 cells in a 60-minute trial and the algal concentration was 10^5 cells cm^{-3}, or, inversely, 10^{-5} cm^3 $cell^{-1}$, the filtering rate would be 0.1 cubic centimeter of water per hour (10^4 cells $h^{-1} \times 10^{-5}$ cm^3 $cell^{-1}$). Estimated in this way, filtering rate includes both the movement of water by the filtering appendages and the ingestion of cells by the mandibular apparatus.

At algal concentrations below 10^3 cells cm^{-3}, *Daphnia* fed at a barely detectable rate and could not maintain themselves. At 10^3 cells cm^{-3}, they ingested fewer than 5000 cells h^{-1}. Rate of feeding increased at higher concentrations, but leveled off at about 25,000 cells h^{-1} (about 7 per second) at concentrations between 10^4 and 10^5 cells cm^{-3} (Figure 20–2). Over the same range of concentrations, the apparent filtering rate decreased from 4 cm^3 h^{-1} to less than 1 cm^3 h^{-1}. Although the beat of the filtering appendages slowed from about 6 to 3.5 per second, this could not have accounted for the apparent drop in filtering rate. The rate of movement of the mandibular appendages also did not vary appreciably over algal densities between 10^3 and 10^6 cells cm^{-3}. Apparently, therefore, concentrations exceeding $10^{3.5}$ cells cm^{-3} provide so much food that *Chlamydomonas* cells stream by the stuffed mouths of the *Daphnia* uneaten, and ability to process rather than capture food limits ingestion.

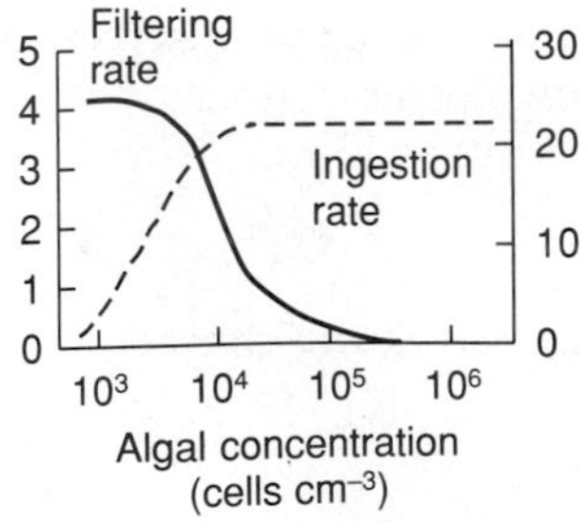

Figure 20–2
Filtering rate (cm^3 ind^{-1} h^{-1}) and ingestion rate (10^3 cells ind^{-1} h^{-1}) of *Daphnia magna* feeding on cultures of the alga *Chlamydomonas* at different food densities. (From Porter et al. 1982.)

Although the rate at which *Daphnia* ingests resources increases up to algal concentrations between 10^3 and 10^4 cells cm^{-3}, and then levels off, resource concentration exerts a more complex influence on population processes. As the feeding rate increases between food concentrations of 10^3 and 10^4 cells cm^{-3}, the intrinsic rate of exponential increase *(r)* also increases, as one would expect. But as the feeding rate levels off above concentrations of 10^4 cells cm^{-3}, *r* first rises to a peak at 10^5 cells cm^{-3} and then drops to 10^6 cells cm^{-3} (Table 20–1). Between 10^4 and 10^5 cells cm^{-3}, average life span decreased somewhat, but fecundity increased greatly, more individuals survived to breed, and reproduction started earlier in life, even though ingestion rates remained constant. At high food concentrations, the efficiency of food gathering might have increased, as suggested by the decreased beat rate of the filtering appendages. A small increase in resource availability (assimilation) can lead to a large increase in resources available

Table 20-1
Reproductive parameters for *Daphnia magna* cohorts fed a range of *Chlamydomonas* concentrations at 20°C

	Food concentration (cells cm^{-3})			
	10^3	10^4	10^5	10^6
Per cent reproducing	50	87	97	50
Eggs per brood	2.8	2.6	15.5	21.1
Broods per female	1.7	7.5	8.2	3.4
Days between broods	5.4	3.6	3.1	3.3
Age at first brood (days)	23.4	16.9	9.8	9.1
Net reproductive rate (R_0)	2.25	16.23	99.33	34.80
Exponential rate of increase (r)	0.03	0.10	0.28	0.20

Source: Porter et al. 1983.

for reproduction (assimilation − maintenance) when resources are close to threshold levels.

At the highest resource concentration, brood size continued to increase but life expectancy, especially of younger water fleas, decreased markedly and fewer individuals reproduced; thus the number of broods per female decreased markedly. Perhaps *Chlamydomonas* produces a toxic by-product at high concentration or the sheer numbers of prey clog the feeding apparatus of juvenile *Daphnia.* Regardless of the cause, the *Daphnia–Chlamydomonas* example highlights the complex relationship between consumers and their resources.

Resources may be distinguished as renewable and nonrenewable.

We may classify resources into two major types according to how they are affected by their consumers. *Nonrenewable resources,* such as space, are not altered by use. Once occupied, space becomes unavailable; it is "replenished" only when the consumer leaves. In contrast, *renewable resources* are constantly regenerated, or renewed. Births in a population of prey continually supply food items for predators. By continually decomposing the organic detritus in the soil, microorganisms provide a fresh supply of nitrate-nitrogen to plant roots.

Among renewable resources, we may recognize three types (Figure 20–3). The first of these have their source external to the system, beyond the influence of their consumers: sunlight strikes the surface of the earth regardless of whether plants "consume" it or not; local precipitation is largely independent of the consumption of water by plants, even though transpiration plays a major role in returning water to the atmosphere; for all practical purposes, detritus rains down from the sunlit surface of the sea to

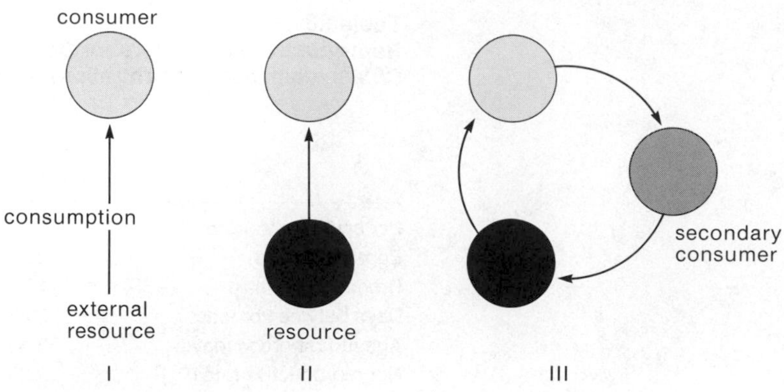

Figure 20–3
Three types of relationships between consumers and renewable resources.

the abyssal depths uninfluenced by the consumers groping there in everlasting darkness.

Renewable resources of the second type are generated within a system and are directly affected by the activities of consumers. Most predator-prey, plant-herbivore, and parasite-host interactions depend upon resources of this type. Predation strongly influences growth rates of prey populations, particularly when predators depress prey populations below the carrying capacity of the environment, activating density-dependent response mechanisms.

Renewable resources of the third kind issue from within a system, but resource and consumer are linked indirectly either through other resource-consumer steps or abiotic processes. For example, in the nitrogen cycle of a forest, plants assimilate nitrate from the soil. Herbivores and detritivores consume plant biomass, returning large quantities of organic nitrogen compounds to the soil. These are attacked by microorganisms, which release the nitrogen in a form the plants can use. Consumption thus follows the cycle: soil → plant → detritivore → microorganism → mineral soil. Uptake of nitrate by plants can have little direct effect on its release by detritivores. Similarly, consumption of detritus cannot immediately influence plant production. Clearly, however, detritivores and microorganisms do influence plant production indirectly through the rate at which they release nutrients into the soil.

Resources and consumers are dynamically coupled.

Most equations developed to describe single-population growth assume a constant environment; in particular, resources and conditions do not vary with respect to the size of the population. In the logistic equation, for example, values of r and K are determined by characteristics of the species and environment independently of population size. To be sure, the decrease in per-capita population growth rate with increasing density derives from the interaction of the population with its resources, but the interaction is not made explicit in the logistic equation, nor are resources formally endowed with any particular dynamics. Once one recognizes that resources also have renewal dynamics, it becomes necessary to both formalize these dynamics and to link the resulting equations for consumers and resources by terms expressing their mutual effects.

In the case of a renewable resource whose supply is unaffected by its consumer, the rate of supply reflects factors outside the consumer-resource system, and may thus be considered a constant value,

$$\frac{dR}{dt} = k_R$$

where R is the quantity of resource and k_R is the rate at which it is supplied.

The rate of increase of the consumer population depends both on the provisioning of its resource and its own density (C); that is, $dC/dt = f(k_R, C)$. Now suppose that each individual must consume resources at rate a just to maintain itself. A population of C individuals would then use resources at rate aC for maintenance, leaving resources supplied at rate $k_R - aC$ for reproduction. Further suppose that each individual converts resources available for reproduction into population growth (births) with efficiency b. With these assumptions, we may write an equation for population growth rate of the consumer,

$$\frac{dC}{dt} = bC(k_R - aC)$$

By rearranging this equation to

$$\frac{dC}{dt} = bk_R C\left(1 - \frac{aC}{k_R}\right) \qquad (20\text{–}1)$$

we see that it is in the form of the logistic with $r = bk_R$ and $K = k_R/a$. According to this equation, the intrinsic growth rate of the population increases with the rate of resource supply and the carrying capacity of the environment is simply the resource supply divided by the maintenance level of resource requirement per individual.

Now let us suppose that consumers gain resources in direct relation to their abundance rather than their rate of supply; hence $dC/dt = f(R, C)$. Again we shall assume that the rate of supply of the resource is independent of the consumer; however, consumers deplete resources at a rate governed by both the number of consumers and the abundance of the resource (that is, for a given number of consumers, resources are consumed more rapidly when abundant than scarce). So, $dR/dt = f(k_R, C, R)$. The following set of equations is one of many reasonable interpretations of these assumptions:

for the resources,

$$\frac{dR}{dt} = k_R - gCR, \tag{20-2}$$

where g is the efficiency of resource gathering; for the consumers,

$$\frac{dC}{dt} = bR\left(1 - a\frac{C}{R}\right) \tag{20-3}$$

Because both C and R vary, the dynamics of the system are more complex than those portrayed by equation (20–1).

Whereas for the logistic equation we could express the equilibrium value of C in terms of a constant (k_R above), the equilibrium of equation (20–3) depends on the variable (R) whose own equilibrium value depends on the magnitude of C. We may, however, solve this system of two equations by finding the joint equilibria of the two variables. To do this, we rearrange equation (20–2) to obtain $dR/dt = 0$ when $R = k_R/gC$ and rearrange equation (20–3) to obtain $dC/dt = 0$ when $C = R/a$. Now, by substituting $C = R/a$ into the equation for R, we obtain the equilibrium values $R = \sqrt{ak_R/g}$ and $C = \sqrt{k_R/ag}$. At the intersection of these two values, both the resources and the consumers coexist in a steady state.

Systems of equations, such as the one presented above, sometimes produce nonintuitive results. It would have been difficult to guess before going through the math that the levels of both resources and consumers varied in proportion to the square root of the rate of resource supply. Furthermore, while it makes sense that the equilibrium level of resources should decrease when resources are consumed more efficiently (that is, higher g), it may come as a surprise that more efficient consumers also assume lower equilibrium numbers. In fact, efficient consumers may eat their resources down to low levels and thereby limit their own populations.

The joint equilibrium values of R and C tell only a small part of the story of their dynamics. What happens when R and C are displaced from their joint equilibrium? Do one or both of the variables increase or decrease, and under what conditions? Do R and C return to their joint equilibrium and, if so, what path do they take? Such questions about resource-consumer systems have been dealt with in great detail with respect to the interaction between predators and their prey, which we shall consider in the next chapter. Before doing so, however, we shall investigate further the role of resources in limiting the size of consumer populations, particularly when consumers require more than a single resource.

Resources that constrain population size are called limiting resources.

Consumption reduces the level of a resource. What is used by one cannot be used by another. By diminishing their resources, consumers limit their own

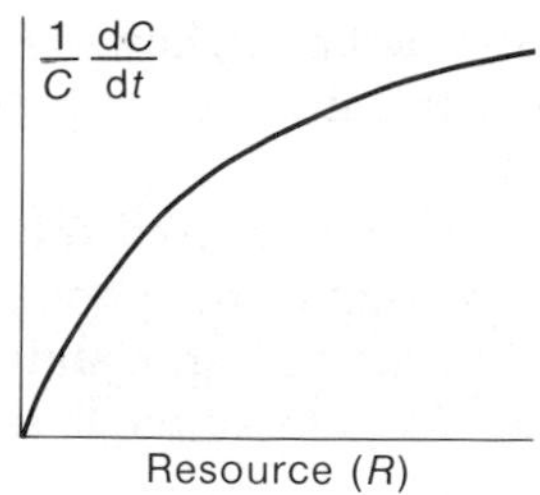

Figure 20–4
The per capita rate of growth of a population of consumers (*C*) as a function of resource availability (*R*). At high levels of a particular resource *R* growth rate levels off (saturation) as other resources or factors limit population growth.

population increase. As a population grows, its overall resource requirement grows as well, eventually to be balanced by a decreasing supply of resources available to fulfill the need. But whereas all resources, by definition, are reduced by their consumers, not all resources limit consumer populations. All animals require oxygen, for example, but they do not depress its level in the atmosphere even noticeably before some other resource, such as food supply, limits population growth.

The potential of a resource to limit population growth depends on its availability relative to demand. At one time, ecologists believed that populations were limited by the single resource having the greatest relative scarcity. This principle has been called Liebig's law of the minimum, after Justus Liebig, who expounded upon the idea in 1840. Accordingly, each population increases until the supply of some resource (the limiting resource) no longer satisfies the population's requirement for it. Although we now know that two or more resources can interact to limit population growth, Leibig's perspective first placed population regulation in the context of resource supply.

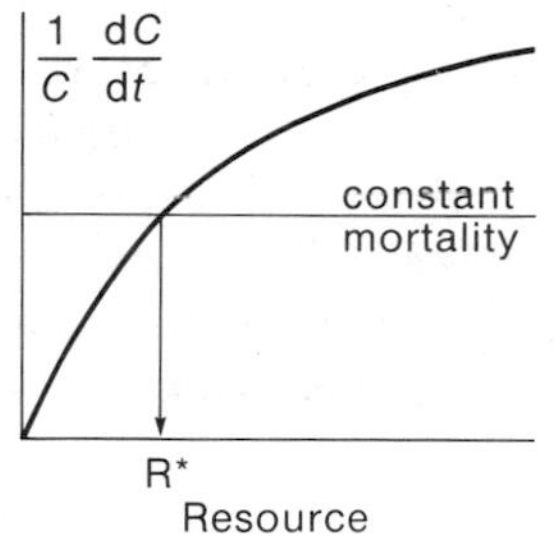

Figure 20–5
A critical level of resource (*R**) and its corresponding level of population growth potential is required to balance the removal of individuals from the population.

The Monod equation relates population growth rate to the abundance of a single resource.

The dynamics of the resource-consumer system can be summarized in a graph plotting the per capita rate of population growth as a function of the level of a limiting resource (Figure 20–4). Consumer dynamics are explicit in the vertical axis (dC/Cdt). Resource dynamics are implicit in the horizontal axis (R) in the sense that the value of R continually decreases as the population of consumers multiplies. As this relationship is portrayed in Figure 20–4, the consumer population exhibits positive growth at all levels of the resource. As a result, consumers continually increase in number, but at a slowing rate as the resources are depressed, until population growth stops at a resource level of zero.

As drawn, the curve implies that consumers don't require resources for maintenance. This is impossible, of course (unless one considers the special case of space as a resource). We can add a needed bit of reality to the model by introducing an intrinsic level of mortality (m) to the consumer population (Figure 20–5). Now, a certain positive contribution to population growth is required to balance mortality (in this sense, mortality is equivalent to a maintenance requirement). With mortality added to the graph, the consumer population is balanced ($dC/Cdt = 0$) at that level of resources (R^*) at which the positive component of population growth equals the intrinsic mortality.

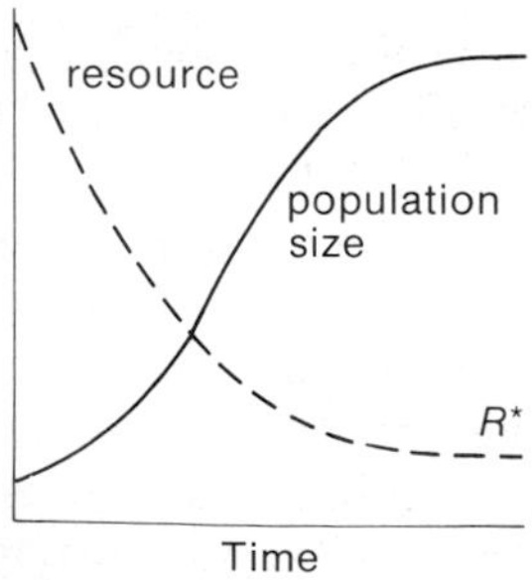

Figure 20–6
Over time, as population size increases resource availability decreases until, at the critical level R^*, the population ceases to grow further.

The change in both consumer and resource levels over time is implied by the relationship between consumer growth rate and resource level, as shown in Figure 20–6, even though the size of the consumer population and the value of $\hat{C}$ are not explicit. From the perspective of resource depletion, the growth rate of the consumer population determines whether the

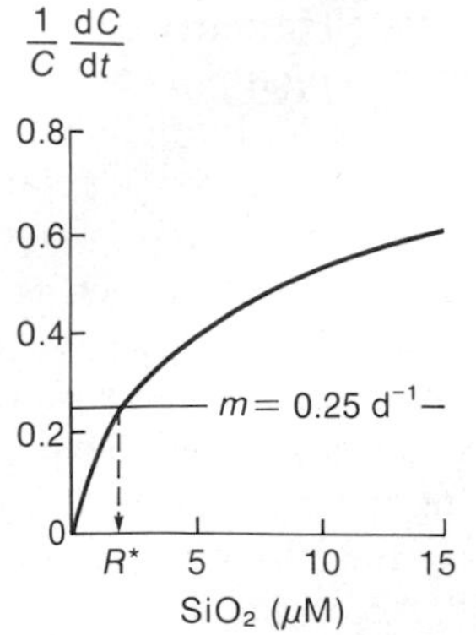

Figure 20–7
Relationship between per capita growth rate and concentration of silicon available to a population of the diatom *Asterionella*. (After Tilman et al. 1981.)

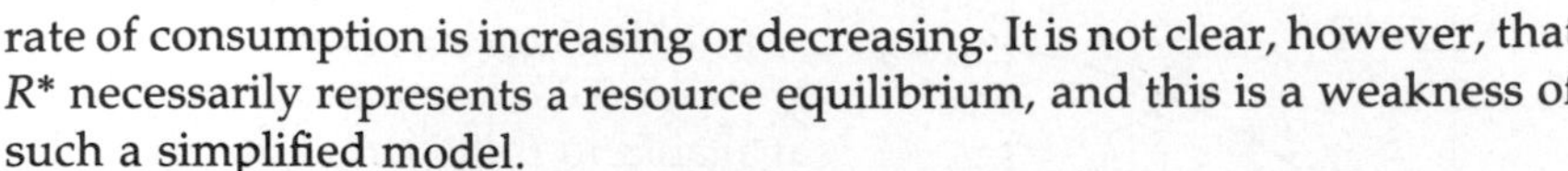

rate of consumption is increasing or decreasing. It is not clear, however, that R^* necessarily represents a resource equilibrium, and this is a weakness of such a simplified model.

The value of R^* defined in Figure 20–5 can be calculated from the growth rate response curve and the intrinsic mortality of the population. The French microbiologist Jacques Monod (1950) derived an equation to express the growth of a bacterial population as a function of the level of a limiting resource. His expression

$$\frac{1dC}{Cdt} = \frac{bR}{k+R}$$

provides an adequate fit to many such relationships and has been applied widely.

The Monod equation has biologically realistic properties. In particular, as resource levels decrease to zero, the population growth rate of the consumer decreases to zero. At high resource levels—that is, when R is very large compared to the constant k—the expression approaches bR/R, which is just the constant b. Thus the growth rate of the population has an upper limit at resource levels that satiate the consumers. Notice that b is equivalent to the intrinsic capacity of the population for increase (r) because it is the growth rate of the population in the absence of crowding (that is, when there are many resources per individual). (The constant k is the amount of resource at which the growth rate of the consumer population is exactly one-half of the maximum [b].)

When a term m is included in Monod's equation to represent the intrinsic mortality rate of the population, one may solve for R^* as

$$R^* = \frac{mk}{b-m} \qquad (20\text{–}4)$$

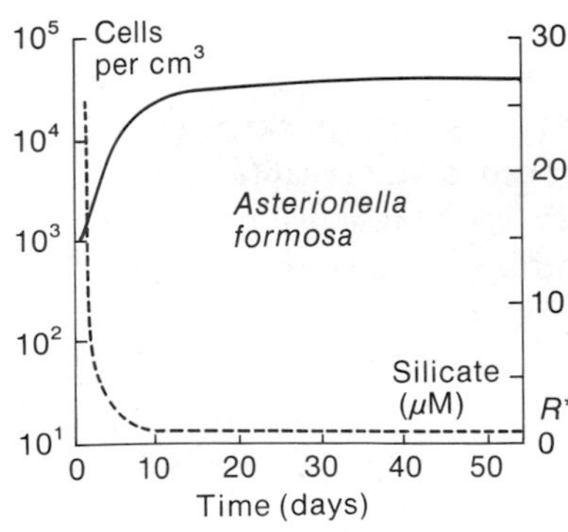

Figure 20–8
Growth of a diatom population and depletion of its silicon resource in a chemostat experiment. (From Tilman et al. 1981.)

David Tilman and his colleagues (1981) determined the relationship between population growth rate and the concentration of a lilmiting nutrient for the diatom *Asterionella formosa* cultured in media with limiting silicon (SiO_2), which diatoms require to secrete their silicate "shell." Diatom populations were established at silicate concentrations ranging from 0 to 40 micromoles (μM); population growth rates were then measured over brief periods, so that the silicate levels did not decrease substantially during the course of the experiment. Such short-term measurements provide a series of snapshots of the consumer-resource dynamics: single points on the curve relating dC/Cdt to R. Because the experiments were brief, mortality was not a factor and only the positive contribution of resources to population growth was revealed. In this particular experiment, *Asterionella* had a maximum growth rate (b) of about 0.6 d^{-1} and the concentration of silicate at which half the maximum growth rate occurred (k) was about 9 μM (Figure 20–7). Hence one could describe the relationship between population growth and resource level (R, μM SiO_2) by Monod's expression as $dC/Cdt = 0.6R/(9+R)$.

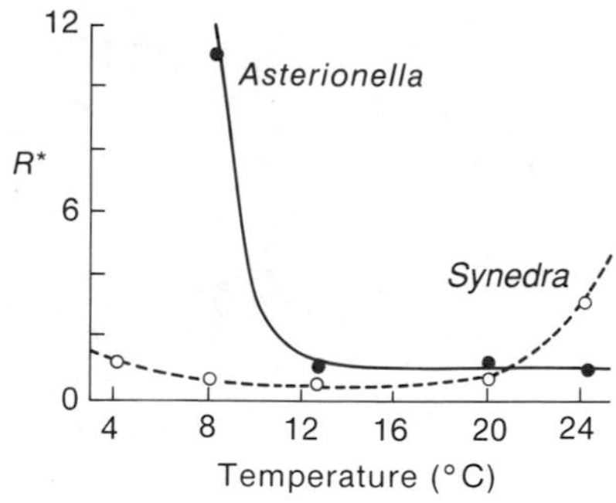

Figure 20–9
Relationship of critical levels of silicon for populations of the diatoms *Asterionella* and *Synedra* to temperature. *Synedra* is a superior competitor only at temperatures exceeding 20°C. (After Tilman et al. 1981.)

Tilman also established long-term cultures of *Asterionella* in a chemostat to study the course of the diatom-silicate relationship over time. A chemostat is a continuous-flow culture vessel to which fresh nutrient medium is added, and from which culture is removed, at a constant rate. The volume of the culture remains constant but its contents are replaced with a time constant defined by the ratio of the flux to the volume. For example, when a 100 cm^3 (0.1 liter) of culture medium is added each day to a flask containing 1 liter, the time constant of the chemostat is 0.1 day^{-1}. Growing a culture of diatoms in such a chemostat is equivalent to imposing a mortality rate of 0.1 day^{-1} on the population because, as one-tenth of the culture is removed each day, so are one-tenth of the diatoms growing in the culture.

In one experiment, Tilman et al. (1981) established a chemostat with a time constant of 0.11 day^{-1}. According to the relationship of diatom growth to resource level obtained under the same conditions ($b = 0.6\ d^{-1}$, $k = 9\ \mu M$) and a mortality rate of 0.11 day^{-1}, equation (20–4) predicts that the equilibrium resource level R^* should be $0.11 \times 9/(0.6 - 0.11)$, or approximately 2 μM, which was close to the level obtained in the chemostat after the diatom population had reached an equilibrium (Figure 20–8). In general, the value of R^* depends on the time constant of the chemostat, the conditions of the experiment, and the species of diatom. For example, when Tilman et al. (1981) conducted similar chemostat experiments using *Asterionella* and a second species of diatom, *Synedra ulna*, at temperatures between 4 and 24°C, they found that *Asterionella* could maintain positive population growth at lower resource levels than *Synedra* at temperatures below 21°C, but not at higher temperatures (Figure 20–9).

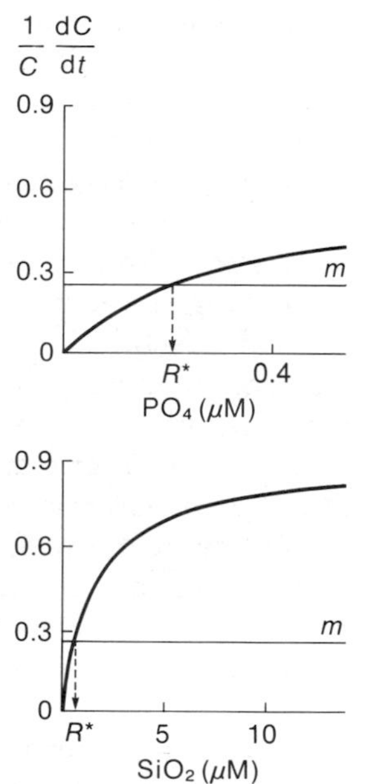

Figure 20–10
Experimentally determined requirements of the diatom *Cyclotella* for phosphorus and silicon in a chemostat with a turnover rate of 0.25 day^{-1}. The critical resource levels were 0.2 μM PO_4 and 0.6 μM SiO_2. (After Tilman 1982.)

Two resources can simultaneously limit a consumer population.

The growth of a population under a given set of conditions responds uniquely to the level of each of its resources. For the diatom *Cyclotella meneghiniana* grown under silicate and phosphate limitation in a chemostat with a time constant of 0.25 day^{-1}, R^* is 0.2 μM for phosphate and 0.6 μM for silicate (Figure 20–10). According to Leibig's law of the minimum, whichever of these resources is reduced to its value of R^* first will limit the growth of the *Cyclotella* population.

Because silicate and phosphate resources have different dynamics, one cannot compare them on a single resource axis. Instead, one can portray the response of population growth rate on a three-dimensional graph with the base defined as the level of silicate on one horizontal axis and the level of phosphate on the other (Figure 20–11). Accordingly, $R^*_{silicate}$ and $R^*_{phosphate}$ define a point in a two-dimensional resource space at which both of the resources are equally limiting.

Tilman (1982) devised a simple graphical analysis of this situation, delimiting a region of consumer population increase with respect to the levels of two resources (Figure 20–12). When both nutrients potentially limit population growth, and individuals consume each resource indepen-

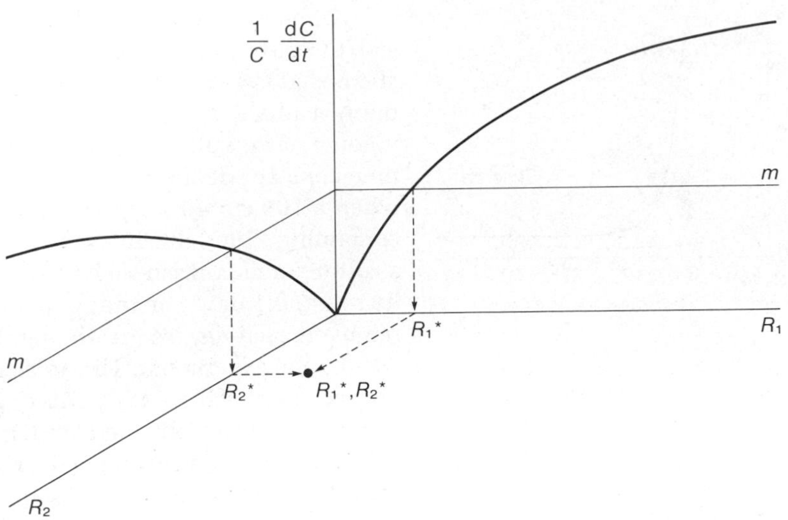

Figure 20–11
Determination of the joint critical levels of two resources R_1^* and R_2^*, at which point both equally limit the population.

dently of the other, the portion of the graph to the right of R_1^* defines a region of positive population growth with respect to resource 1; to the left, population growth is negative. With respect to resource 2, the portion of the graph above R_2^* defines a region of positive population growth; that below, negative. The lines separating these regions are the "zero-growth isoclines" for each of the resources. When both resources are considered together, the consumer population can increase only when levels of the resources fall in the upper right-hand quadrant of the graph. In the region below, resource 2 is limiting; in the region to the left, resource 1 is limiting; in the quadrant below and to the left, both resources limit population growth.

When two resources independently affect the growth of the consumer population, the consumer will reduce both of the resources until one of them reaches its value of R^*, at which point the population will stop growing. Hence the eventual outcome of a chemostat experiment may settle upon any point along the lines defining the boundary of the region of consumer increase. The particular endpoint in any situation will depend upon the initial concentrations and relative rates of supply of the resources, but one or the other resource will limit the population in accordance with Leibig's law of the minimum.

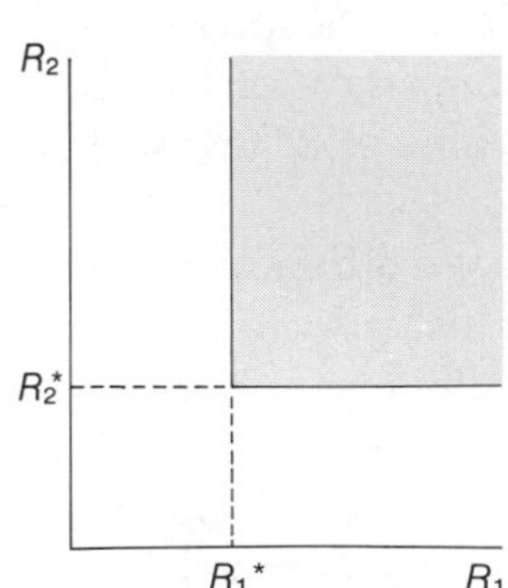

Figure 20–12
Two dimensional representation of the space determined by two resources within which a population can increase (shaded area).

Leibig's law applies, however, only to resources having independent influence on the consumer. In many cases, two or more resources interact to determine the growth rate of a consumer population. That is, the growth rate of a consumer at a particular level of one resource depends on the level of one or more other resources (Figure 20–13). Tilman referred to nutrients as essential when they limit consumer populations independently. For diatoms, silicon and phosphorus are essential resources; the consumer requires both. When two essential nutrients interact to determine the growth rate of

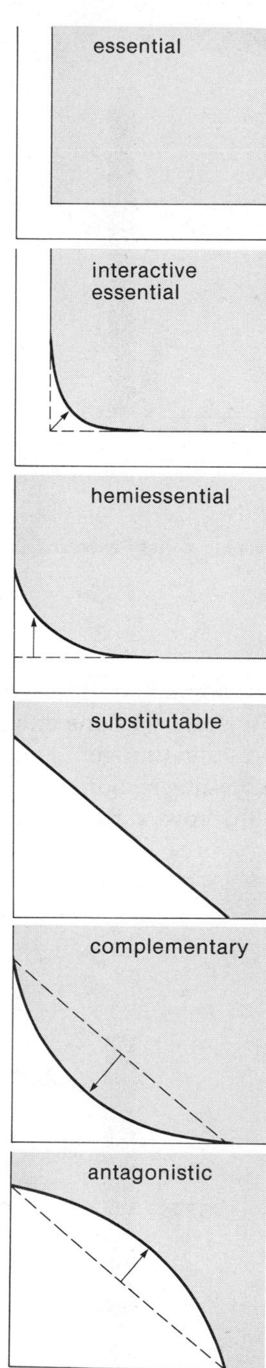

Figure 20–13
Graphical representation of the joint influence of two resources on population growth as related to the degree of interdependence of requirements for the two resources.

the consumer population, as one resource approaches its value of R^*, the consumer requires higher levels of the other limiting resource. Nonlimiting resources may also affect the value of R^* of an essential resource, a relationship that Tilman refers to as hemi-essential.

When one resource may be substituted for the other, they are defined as substitutable, complementary, or antagonistic, depending on how they interact to determine consumer growth. Substitutable resources usually are alternative forms of the same requirement, such as the sugars glucose and fructose as sources of carbon for bacteria. In all cases, the level of one may be depressed below its R^* provided that a second, substitutable resource is present (Figure 20–13). When resources are complementary, small amounts of one will replace relatively large amounts of the other; hence at equilibrium smaller amounts of two resources are consumed in combination than the amount of either resource consumed alone. When resources are antagonistic, substitution requires relatively large amounts of one in the place of the other.

When increases in two resources enhance the growth of a consumer population more than the sum of both individually, the resources are said to be synergistic (from the classical roots *syn* = together and *ergon* = work). Tilman's complementary resources are therefore synergistic in promoting consumer growth. Numerous studies illustrate synergism in natural systems; W. J. H. Peace and P. J. Grubb's (1982) study of *Impatiens parviflora* provides an excellent example in this tradition. *Impatiens parviflora* is a small herbaceous plant common in woodlands of England. Studies on other shade-tolerant forest herbs had shown that individual plants can grow in deeper shade on alkaline soils than on strongly acidic soils. For example, such plants growing at a pH of 7.5, just on the alkaline side of neutral, can tolerate light levels as low as 1 to 2 per cent of full sun. When grown on soils with an acid pH of 4.5, the same plants cannot grow at light levels below 5 to 6 per cent. One explanation for this difference in response is that slightly alkaline conditions solubilize mineral nutrients more readily; in acid soils some metals become highly reactive and form insoluble complexes with phosphorus.

Peace and Grubb investigated the relationships among light, phosphorus, and nitrogen by following the growth of *Impatiens* plants under controlled laboratory conditions. They used a woodland soil of moderate fertility and near-neutral pH (6.6) to which they added a nitrate and phosphate fertilizer under different intensities of light. In one experiment, fertilized and nonfertilized treatments were exposed to different levels of light from the time of seed germination until the end of the experiment at 37 days. Added light enhanced growth of the fertilized plants more than that of the controls (Figure 20–14); hence the ability of *Impatiens* to utilize light depends on the presence of other resources. Plant growth requires both the carbon reduced by photosynthesis, as a source of energy and for structural carbohydrates, and nitrogen and phosphorus for the synthesis of proteins and amino acids.

Peace and Grubb also examined the interaction of nitrogen and phosphorus at the highest light intensities used in the previous experiment. Plants were collected, dried, and weighed, after 7 days and after 35 days.

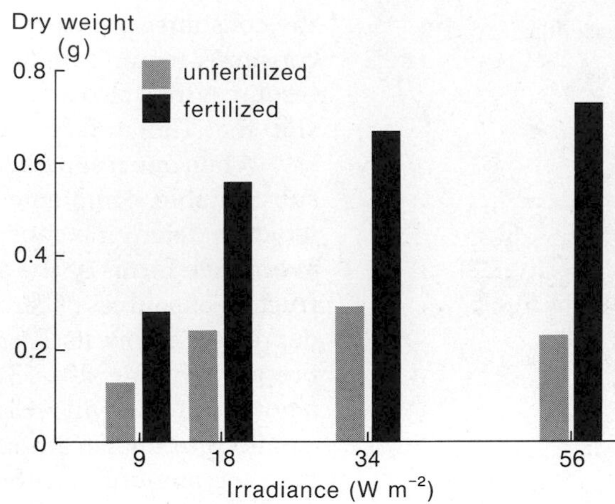

Figure 20–14
Joint influence of light levels and fertilizer on growth of *Impatiens.* (After Peace and Grubb 1982.)

The addition of nitrogen in the absence of phosphorus had little effect on growth during the first week (Figure 20–15). Phosphorus alone enhanced growth as much as the combination of phosphorus and nitrogen. Thus at its concentration in natural soil, nitrogen was not a limiting resource for young plants. During the subsequent month of growth, however, nitrogen and

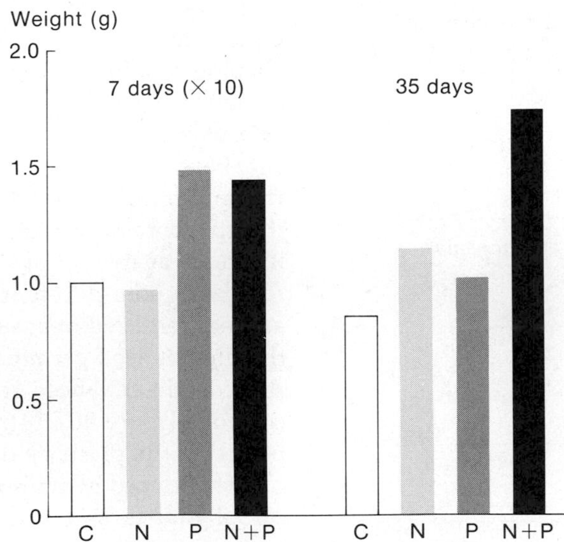

Figure 20–15
Joint influence of nitrogen and phosphorus fertilization on growth of *Impatiens;* C = control without fertilization. (After Peace and Grubb 1982.)

phosphorus were synergistic in promoting plant growth. The addition of either alone had no appreciable effect, whereas both together increased dry weight by about 50 per cent.

Peace and Grubb thought that phosphorus might limit growth early on because it is relatively immobile in the soil and plants must therefore obtain their supply from the soil immediately surrounding the roots. Being more mobile, nitrogen can enter the area of the roots from a greater distance. As a result, seedlings deplete local levels of phosphorus quickly and phosphorus thus tends to limit plant growth before nitrogen does. Eventually most of the more readily available nitrogen in the soil is taken up by the plant and it too becomes limiting.

If this were the correct explanation, Peace and Grubb reasoned that dense plantings would deplete soil nitrogen more quickly than plantings of one seedling per pot, and nitrogen limitation would be felt earlier in the growth period. Indeed, when *Impatiens* were planted 10 to the pot instead of 1, as in previous experiments, nitrogen, rather than phosphorus, proved to be the limiting resource during the first week.

Consumer-resource systems may involve more than two levels or more than one consumer.

Although one may conceive of systems dominated by a single consumer and a single resource, most species are both; furthermore, they generally require more than one resource and must share their resources with other consumers. The consequences of these interactions for population dynamics will be discussed fully in subsequent chapters, but they are introduced here by a general class of resource-consumer interaction involving more than two species.

Consider a predator food chain in which a second species (an herbivore) feeds upon the first (a plant) and a third (a carnivore) feeds upon the second. In this most simple food chain, the first species has no biological resource and the third is a consumer only. The consequences of this arrangement for the regulation of resource and consumer populations have intrigued ecologists for decades. In 1960, in a provocative and controversial paper in the *American Naturalist* entitled "Community structure, population control, and competition," ecologists N. G. Hairston, F. E. Smith, and L. B. Slobodkin — then at the University of Michigan — argued that populations of plants, herbivores, and predators are regulated by different factors. They observed that

> cases of obvious depletion of green plants by herbivores are exceptions to the general picture, in which the plants are abundant and largely intact. Moreover, cases of obvious mass destruction by meteorological catastrophes are exceptional in most areas. Taken together, these two observations mean that producers are neither herbivore-limited nor catastrophe-limited, and must therefore be limited by their own exhaustion of a

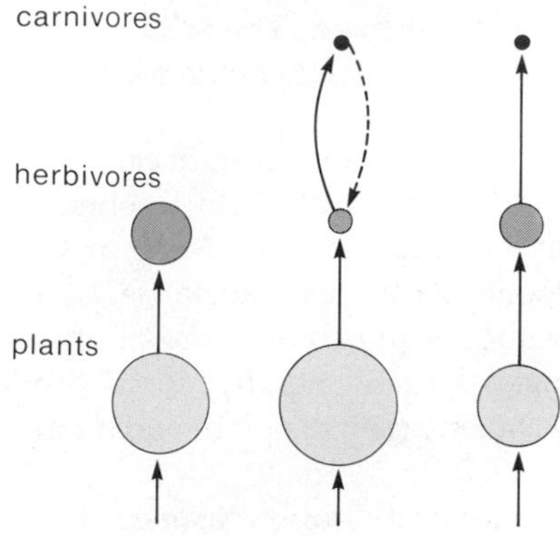

Figure 20–16
Food chains with and without a third (carnivore) level. In the center, carnivores considerably depress populations of herbivores, allowing the plant resource populations to increase; at right, they do not.

> resource. In many areas, the limiting resource is obviously light, but in arid regions water may be the critical factor, and there are spectacular cases of limitation through the exhaustion of a critical mineral.

Thus plant populations are limited by resources but are little affected by consumers. The authors then pointed out that

> there are temporary exceptions to the general lack of depletion of green plants by herbivores. This occurs when herbivores are protected either by man or natural events, and it indicates that the herbivores are able to deplete the vegetation whenever they become numerous enough, . . . It therefore follows that the usual condition is for populations of herbivores not to be limited by their food supply.

Hairston, Smith, and Slobodkin went on to suggest that herbivore populations are limited by their consumers, the predators. Furthermore, because predators themselves have no consumers, they must be limited by their resources, the herbivores.

The three-level food chain envisioned by Hairston and his colleagues is contrasted with the simpler two-species consumer-resource system in Figure 20–16. Accordingly, the addition of predators reduces the herbivore population, which, in turn, may result in an increase in the abundance of plants. The merits of these arguments have been debated heatedly (see, for example, Ehrlich and Birch 1967, Murdoch 1966, Pimentel 1988) and will be considered in the light of theory and evidence in the following chapters. For the present, we should be aware that the behavior of complex natural systems may surprise us.

Hairston, Smith, and Slobodkin also had something to say about the interaction of species sharing the same resource (Figure 20–17). When two species share the same limiting resource, the one that can persist on the lower resource level inevitably excludes the other. When predators are superimposed on such a system, Hairston et al. suggested that because the herbivore populations are predator-limited, they compete less intensely for common resources and therefore should more likely coexist. Here again, counterarguments can be, and have been, voiced. But the issue of the relationship between the diversity of species and consumer-resource relationships has continued to play a prominent role in ecology. As we shall see, this theme persistently recalls Hutchinson's question, "Why are there so many kinds of animals?"

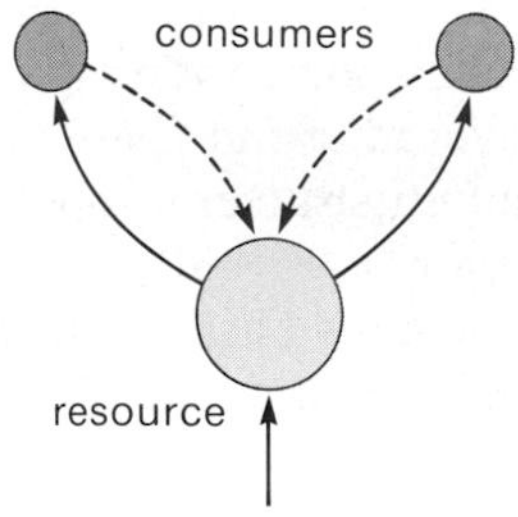

Figure 20–17
Food web diagram illustrating two consumers feeding upon, and depressing the level of, a single resource. Whichever consumer can continue to grow on the lower resource level will persist.

SUMMARY

The dynamics of population growth and regulation cannot be separated from the interactions of populations with their resources and consumers. In an effort to provide a framework for much of the detailed consideration of population interactions to follow, some general features of resource-consumer systems were described.

1. A resource may be defined as any factor whose increase promotes population growth and which is consumed. Thus light, food, water, mineral nutrients, and space are resources. Temperature, salinity, and other such conditions are not.

2. Resources may be distinguished as being nonrenewable (space) or renewable (light, food), and the latter may be further distinguished according to whether the consumer has no influence on the provisioning of the resource, a direct influence, or an indirect influence through other consumers.

3. The interaction between resource and consumer can be formalized by a set of equations describing the rate of increase of each.

4. Of all the resources consumed only one or a few will limit population growth of the consumer. These will normally be those whose supply relative to demand is least. This principle is known as Leibig's law of the minimum.

5. The relationshp between the rate of growth of a consumer population (C) and the level of its resources (R) may be described by Monod's equation, $dC/Cdt = bR/(k + R)$, where b is the intrinsic exponential rate of increase of the consumer in the absence of crowding and k is the level of resources at which the consumer growth rate is one-half b.

6. David Tilman used Monod's relationship to devise a graphical analysis of the use of several resources by one consumer and to categorize the interaction of resources in promoting consumer growth.

7. Laboratory and field experiments have revealed widespread synergism among resources in their effect on consumers.

8. Finally, considerations of resource-consumer relationships involving many species have stimulated new avenues of research on the regulation of community structure and the diversity of species.

21

Predation, Parasitism, and Herbivory

The central theme of population biology is understanding the regulation of population size, and considerations of predators and parasites have played a major role throughout its development. W. F. Fiske (1910) and W. R. Thompson (1924), both economic entomologists concerned with the control of insect pests, devised simple mathematical models relating parasitism of the larvae of pest species to the density of parasitic species of wasps and flies. The problem was of practical interest to entomologists, who were just learning the technique of biological control by parasites and who needed to know how many parasites to release to achieve control of pest populations within a reasonable period.

In the mid-1920s, A. J. Lotka (1925) and V. Volterra (1926) formalized the general relationship between a predator and its prey by a system of equations describing the rate of change in each with respect to the other. A. J. Nicholson (1933) and Nicholson and V. A. Bailey (1935) used a different approach to arrive at an alternative formalization, but together with the Lotka and Volterra models and the logistic equation for density-dependent population regulation, they provided the basis for most subsequent work on population interactions.

The basic question of population biology is this: What factors influence the size and stability of populations? In previous chapters, we saw the influences of regulatory factors through density dependence and time delays in the responses of birth and death rates to population density. As we expand our perspective on population regulation to include the interactions between species, the basic question of population biology is elaborated. When a species is both consumer and resource, it becomes necessary to ask, as did Hairston, Smith, and Slobodkin, Are populations limited primarily by what they eat or by what eats them? Furthermore, time delays in population responses derive from the resource-consumer interaction itself. Can predation therefore recruit a damped resource population into part of an oscillating consumer-resource system? Or would predation more likely stabilize a prey population that otherwise would oscillate with its own resources?

None of these questions has a simple answer, in part because ecologists don't yet fully know the answers, in part because the answers may be both yes and no. Our positions on these issues at present are the products of more than 60 years of theoretical and experimental investigation, which is as vital now as it was in the early years of Lotka's and Nicholson's work.

Population biologists distinguish predators, parasites, and parasitoids.

Consumers go by many names, the most familiar of which are predators, parasites, parasitoids, herbivores, and detritivores. From the standpoint of population interactions, some of these are useful distinctions, while others are confusing. Let's start with predators. The images of an owl eating a mouse and a spider eating a fly capture the essentials of predation. Predators catch individuals and consume them, thereby removing them from the prey population. In contrast, a parasite consumes a living host. While it may increase the probability of a host's dying from other causes or reduce its fecundity, a parasite does not by itself remove an individual from the resource population.

The word parasitoid is applied to species of wasps and flies whose larvae consume the tissues of living hosts, usually the eggs, larvae, and pupae of other insects, inevitably leading to the death of the host (Waage and Greathead 1986). Parasitoids are like parasites in consuming living tissue; in most other ways they are like predators, but with important differences: first, a single host may support more than one parasitoid; second, and this is the major distinction, the act of parasitism does not remove the host individual from the resource population. For predators, every prey individual encountered is a suitable resource. For parasitoids, only intact, nonparasitized hosts are suitable in most cases. When more than one parasitoid egg is laid on a host, usually only the first survives. Even when a single host can support more than one parasitoid, superparasitism may lead to reduced growth and survival.

Herbivores eat whole plants or parts of plants. But the term has no special status in the formalization of consumer-resource interactions because herbivores may function either as predators, consuming whole plants, or as parasites, consuming living tissues but not killing their victims. The sparrow that eats a seed is a predator by its act because it kills the entire living embryo of a plant contained in that seed. The deer that browses on shrubs and trees is no less a parasite than a mosquito (or a vampire bat) taking its blood meal.

Special properties of plants and parasites complicate consumer-resource interactions.

Population biologists have used predator-prey and parasitoid-host relationships as bases for general models of consumer-resource interactions.

Predation is a "clean" demographic event that readily lends itself to modeling. Predator-prey and, especially, parasitoid-host systems can be brought into the laboratory and subjected to experimentation.

Population regulation in most parasites and disease organisms is greatly complicated by their complex life cycles, with different sets of factors affecting each stage (Anderson 1982, Anderson and May 1982, Bacon 1985, Rollinson and Anderson 1985). Many parasites interact with both primary and intermediate hosts (the distinction being that parasites reproduce sexually in their primary hosts), and with factors in the external environment during free-living stages. Problems of locating hosts (contagion) figure importantly in models designed specifically for parasite and, particularly, disease populations. In the case of endoparasites, population growth may occur within a single host, resulting in a subdivision of the parasite population and fostering population interactions among the descendants of infecting individuals.

Plants have posed special problems for population biologists: first, because many consumers don't kill them; second, because plant growth varies greatly in response to physical factors and herbivory; and, third, they reproduce in number both sexually and by asexual reproduction, which is a form of growth (Crawley 1983). But regardless of characteristics particular to each kind of consumer and resource, their interactions also reflect more general considerations common to all (Kuno 1987). One group of these considerations includes the demographic properties of the consumer and resource populations: densities, temporal and spatial patterns of abundance, age structure, and time delays in demographic responses. To have any validity, models must include at a minimum the densities of both resource and consumer populations (Hassell and May 1979). As other considerations are incorporated, models begin to lose their generality, becoming special cases that can be tested only by the behavior of a particular system, and not by general patterns in consumer-resource relationships. Theoreticians must balance generality and reality (that is, particular application).

The consumer-resource interaction also is embedded in the matrix of other interactions throughout the community. The validity of two-species models may be tested by observations on isolated two-species systems. But how can we be confident of their application to more complicated natural systems? Suppose, for example, that oscillations in a two-species system can be eliminated by removing a time delay in some demographic response. Does this mean that noncycling populations have no such time delays? Or could many time-delayed interactions of different period cancel each other and damp the oscillations intrinsic to any one of them when isolated?

A last consideration is the mutual evolution of consumer and resource populations. Evolution is an adaptive, genetic response to environmental change. In the case of consumers and biological resources, each is part of each other's environment. As one changes in response to the other it also stimulates further change in the other. Does this evolutionary cat and mouse game continue indefinitely? Can the consumer-resource system achieve a stable evolutionary configuration, or do its dynamics change continuously?

Where to start with all this? As is often the case, the best points of departure are the natural phenomena themselves.

Mite-mite and other interactions demonstrate that predators may effectively limit prey populations.

The cyclamen mite is a pest of strawberry crops in California. Populations of the mites are usually kept under control by a species of predatory mite of the genus *Typhlodromus.* Cyclamen mites typically invade a strawberry crop shortly after it is planted, but their populations do not reach damaging levels until the second year. Predatory mites usually invade fields during the second year, rapidly subdue the cyclamen mite populations, and keep them from reaching damaging levels a second time.

Greenhouse experiments have demonstrated the role of predation in keeping the cyclamen mites in check (Huffaker and Kennett 1956). One group of strawberry plants was stocked with both predator and prey mites; a second group was kept predator-free by regular application of parathion, an insecticide that kills the predatory species but does not affect the cyclamen mite. Throughout the study, populations of cyclamen mites remained low in

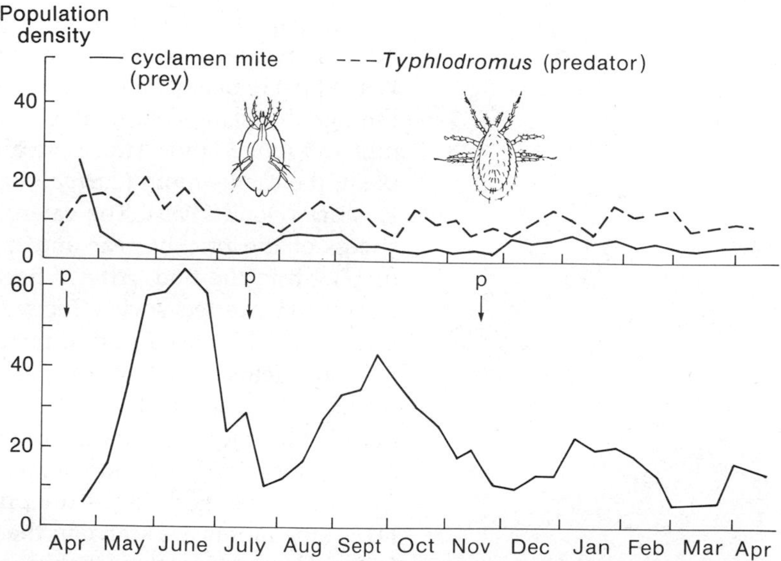

Figure 21-1
Infestation of strawberry plots by cyclamen mites *(Tarsonemus pallidus)* in the presence of the predatory mite *Typhlodromus (above)* and in its absence *(below).* Prey populations are expressed as numbers of mites per leaf; predator levels are the number of leaflets in 36 which had one or more *Typhlodromus.* Parathion treatments are indicated by p's. (After Huffaker and Kennett 1956.)

plots shared with *Typhlodromus*, but their infestation attained damaging proportions on predator-free plants (Figure 21–1). In field plantings of strawberries, the cyclamen mites also reached damaging levels where predators were eliminated by parathion, but they were effectively controlled in untreated plots (a good example of an insecticide having the wrong effect). When cyclamen mite populations began to increase in an untreated planting, the predator populations quickly responded to reduce the outbreak. On average, cyclamen mites were about twenty-five times more abundant in the absence of predators than in their presence.

Typhlodromus owes its effectiveness as a predator to several factors in addition to its voracious appetite (Huffaker and Kennett 1969). Its population can increase as rapidly as that of its prey. Both species reproduce parthenogenetically; female cyclamen mites lay 3 eggs per day over the 4 or 5 days of their reproductive life span; female *Typhlodromus* lay 2 or 3 eggs per day for 8 to 10 days. Seasonal synchrony of *Typhlodromus* reproduction with the growth of prey populations, ability to survive at low prey densities, and strong dispersal powers also contribute to the predatory efficiency of *Typhlodromus*. During winter, when cyclamen mite populations dwindle to a few individuals hidden in the crevices and folds of leaves in the crown of the strawberry plants, the predatory mites subsist on the honeydew produced by aphids and white flies. They do not reproduce except when they are feeding on other mites. Whenever predators appear to control prey populations, one usually finds a high reproductive capacity compared with that of the prey, combined with strong dispersal powers and the ability to switch to alternate food resources when primary prey are unavailable.

Inadequate dispersal is perhaps the only factor that keeps the cactus moth from exterminating its principal food source, the prickly pear cactus. When prickly pear *(Opuntia)* was introduced to Australia, it spread rapidly through the island continent, covering thousands of acres of valuable pasture and range land. After several unsuccessful attempts to eradicate the plant, the cactus moth *(Cactoblastis cactorum)* was introduced from South America (Dodd 1959). The caterpillar of the moth feeds on the growing shoots of the prickly pear and quickly destroys the plant—literally by nipping it in the bud. After it became established in Australia, the moth population exerted such effective control that, within a few years, the prickly pear became a pest of the past (Figure 21–2).

The cactus moth has not eradicated the prickly pear because the cactus manages to disperse to predator-free areas, thereby keeping one jump ahead of the moth and maintaining a low-level equilibrium in a continually shifting mosaic of isolated patches. Indeed, one would probably not guess that the cactus moth keeps the prickly pear at its present low population levels; the moths are scarce in the remaining stands of cactus in Australia today. The same moth probably controls prickly pear populations in some areas of Central and South America, but its decisive role might have gone unnoticed if the appropriate experiment had not been performed in Australia.

Experiments on the effect of sea urchins on populations of algae have demonstrated predator control in some biological communities of rocky

Figure 21 – 2
Photographs of a pasture in Queensland, Australia, 2 months before *(left)* and 3 years after *(right)* introduction of the cactus moth to control the prickly pear cactus. (From Dodd 1959; courtesy of W. H. Haseler, Dept. of Lands, Queensland, Australia.)

shores. The simplest experiments consist of removing sea urchins, which feed on attached algae, and following the subsequent growth of their algal prey (Paine and Vadas 1969). When urchins are kept out of tidepools and subtidal rock surfaces, the biomass of algae quickly increases, indicating that predation reduces algal populations below the level that the environment can support. Different kinds of algae also appear after predator removal. Large brown algae flourish and begin to replace both coralline algae (whose hard shell-like coverings deter grazers) and small green algae (whose short life cycles and high reproductive rates enable algal population growth and reestablishment to keep ahead of grazing pressure by sea urchins). In subtidal plots kept free of predators, brown kelps became established in thick stands that shaded out most small species.

Grazing by urchins plays an important role in the dynamics of kelp beds, which occur in shallow waters off many of our rocky coasts. Experimental studies of this interaction have also revealed complex influences of other consumers and resources (North 1970, Mann 1981). Kelps are large brown algae that form the underwater equivalent of terrestrial forests. The tallest fronds, buoyed by gas-filled floats, spread out along the surface of the water where they are exposed to the full intensity of the sun's light. Kelp beds are understandably one of the most productive ecosystems and they have been harvested extensively by man as a source of iodine and fertilizer.

Kelps are long-lived compared to other marine plants. Once established a bed may last between 1 and 10 years, depending on the locality and exposure to waves. Death of adult plants occurs mostly from storms, high water temperature, and grazing by fish and sea urchins. Once a bed has been devastated, by whatever cause, it normally becomes reestablished within a few years.

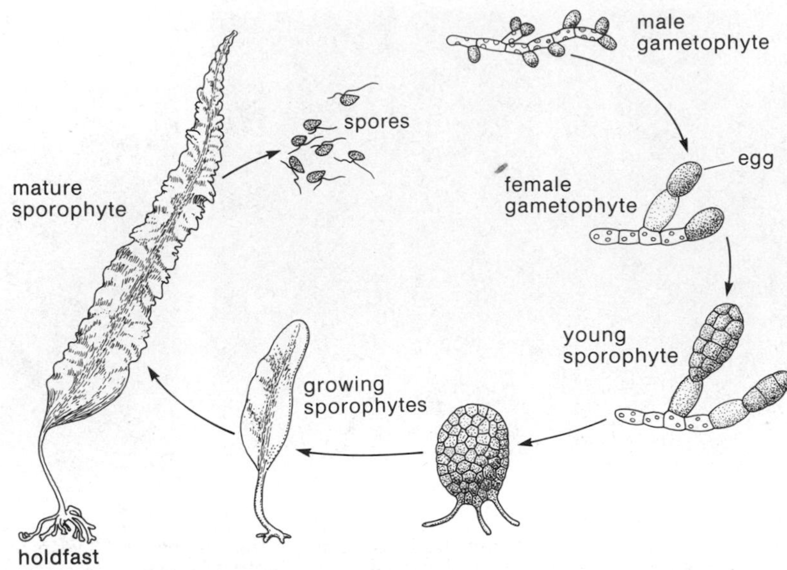

Figure 21–3
Life cycle of the prominent kelp *Laminaria.*

The life cycles of most large species of kelp begin when spores are released directly into the water by mature plants. These settle on hard substrates on the bottom at depths to 25 meters where they grow into minute haploid, gametophyte plants that immediately undergo sexual reproduction. The resulting embryo, which usually remains attached to the female gametophyte, secures itself to the bottom by a holdfast while its fronds enlarge (Figure 21–3). During this early growth phase, the young kelp plants are vulnerable to grazing (literally predation at this stage) by small urchins. Even mature kelps are sometimes attacked by large urchins when alternative, and usually preferred, foods are scarce.

Beginning in 1940, kelp beds in areas near Los Angeles and San Diego all but disappeared. Disposal of sewage was suggested as a factor because deterioration of the beds started near sewage outlets. Areas in which kelp beds had disappeared were practically devoid of all types of algae, and instead swarmed with dense populations of immature sea urchins. Normally, when a kelp bed disappears, the urchins disappear with it, giving newly settled plants a chance to regenerate. Pollutants did not affect mature kelp plants; a few beds persisted within polluted areas in sheltered spots free of wave damage. The kelp beds disappeared because no young kelp grew where normal causes of mortality had removed mature plants. The young kelp plants were eaten by the urchins.

Sea urchins apparently can obtain nourishment from suspended and dissolved organic matter in seawater. Sewage maintains an urchin population in the absence of adequate algal food in the same way that honeydew from aphids maintains predatory mites when their normal prey, the cyclamen mite, is unavailable. When young *Macrocystis* plants reinvaded a devastated kelp bed, urchins were there to meet them and quickly devoured the newly settled plants.

When urchin populations were controlled by dumping quicklime (calcium oxide) into devastated areas, and when these areas were reseeded with young *Macrocystis* plants, the kelp beds quickly returned. Moreover, when, in 1963, San Diego stopped dumping sewage directly into the ocean, kelp beds quickly grew back even where urchins had not been destroyed. These beds now contain fewer, but larger urchins, whose grazing does not seriously depress algal growth under most circumstances.

We must add yet two other components to the dynamics of the kelp community—sea otters and their human predators. Once abundant along the west coasts of the United States and Canada, sea otters were hunted to the verge of extinction during the nineteenth century by Russian and American fur traders. Now under close protection, the otters have staged a successful comeback. Wherever otters have reached their former densities, kelp beds also flourish. Needless to say, otters eat urchins (Estes et al. 1978, Van Blaricom and Estes 1988). The situation may be even more complex along our Atlantic coasts, where lobsters prey upon the crabs that normally keep urchins in check. In this case, overfishing of the lobsters has allowed populations of the predators of urchins to increase, thereby depressing grazing on young kelp plants.

The preceding examples demonstrate strong depressing effects of consumers on the abundance of their resources, but each example involves greatly simplified systems and conspicuous or vulnerable prey. Defensive mechanisms of prey and the searching behavior of predators do not figure in the dynamics of each of these systems until the prey have been reduced to very low levels. But in many resource-consumer interactions, searching by predators and active defense or escape on the part of prey play important roles. Perhaps, in such relationships, advantage goes to the prey, whose populations thereby escape the depressing effect of predation. Such systems usually involve prey that are highly mobile or widely dispersed, making them difficult to study. Only recently have investigators begun to experimentally exclude predators in such systems in order to evaluate their effect on prey populations (Pacala and Roughgarden 1984; Joern 1986; Schoener and Spiller 1987).

T. R. Torgerson and R. W. Campbell (1982) studied the problem of avian predation on the western spruce budworm *(Choristoneura occidentalis),* a moth that causes extensive damage to spruce trees during outbreak years. They constructed wire cages around branches infested by third to sixth instar caterpillars and compared the numbers of pupating moth larvae on the caged branches with those on uncaged control branches a month later. At two out of three experimental sites, densities of pupae on caged branches were about three times those on control branches (5.1 versus 1.6 per 100 current-season shoots). At the third site, caged and control branches did not differ significantly.

In a similar experiment on broad-leaved trees in the Hubbard Brook Forest, New Hampshire, R. T. Holmes et al. (1979) estimated that birds removed an average of 37 per cent of caterpillars per week, but that other potential prey, such as beetles, leafhoppers, and spiders, were not detectably affected. Predation clearly was important under the circumstances of these experiments, but not so devastating as in the mite-mite, cactus-moth, and kelp-urchin systems.

Predation may establish coupled oscillations of predator and prey populations.

Populations of some predators and prey vary in what appear to be closely linked cycles, as we have seen in the case of the snowshoe hare and its predator, the lynx. Because such cycles persist, they appear to represent a stable interaction between predator and prey. One of the earliest goals of population biologists was to establish such cycles in experimental populations, for which the dynamics of the relationship could be worked out.

When azuki bean weevils *(Callosobruchus chinensis)* are maintained in cultures with parasitoid braconid wasps, the populations of predator and prey fluctuate out of phase with each other in regular cycles of population change (Figure 21–4). Introduced to a population of weevils ultimately limited by a constant ration of seeds, the wasps rapidly increase in number. As their population grows, parasitism becomes a major source of mortality for the weevils. When parasitism exceeds the reproductive capacity of the weevil population, the number of weevils begins to decline but, because *Heterospilus* is an efficient parasitoid, it continues to prey heavily even as the population is reduced. Eventually, the weevils are nearly exterminated, and then the predator population, lacking adequate food, decreases rapidly. The wasp is not so efficient that all weevil larvae are attacked; hence a small but persistent reserve of weevils always remains to initiate a new cycle of prey population growth after the predators have become scarce.

Extremely efficient predators often eat their prey populations to extinction, and then follow suit. This hopeless situation can be stabilized, however, if some of the prey can find refuges in which they can escape predators. G. F. Gause (1934) demonstrated this principle in some of the earliest

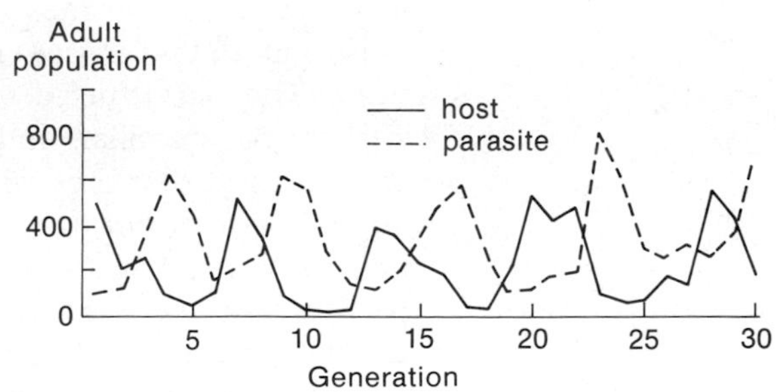

Figure 21–4
Population fluctuations of the azuki bean weevil (host) and its braconid wasp parasitoid, *Heterospilus*. (After Utida 1957.)

experimental studies on predator-prey systems. Gause employed *Paramecium* as prey and another ciliated protozoan, *Didinium,* as predator. In one experiment, predator and prey individuals were introduced to a nutritive medium in a plain test tube. By creating so simple an environment, Gause had stacked the deck against the prey; the predators readily found all of them, and when the last *Paramecium* had been consumed, the predators starved. In a second experiment, Gause added some structure to the environment by placing glass wool, in which the *Paramecium* could escape predation, at the bottom of the test tube. The tables having thus been turned, the *Didinium* population starved after consuming all readily available prey, but the *Paramecium* population was restored by individuals concealed in the glass wool.

Gause finally achieved recurring oscillations in the predator and prey populations by periodically adding small numbers of predators—restocking the pond, so to speak. The repeated addition of individuals to the culture corresponds, in natural predator-prey interactions, to repopulation by colonists from other areas of a locality in which extinction of either predator or prey has occurred. This is reminiscent of the interaction between the cactus moth and prickly pear, in which the cactus escapes complete annihilation by dispersing to predator-free areas.

C. B. Huffaker's experiments on mites demonstrated how environmental heterogeneity can stabilize predator-prey systems.

C. B. Huffaker, a University of California biologist who pioneered the biological control of crop pests, attempted to produce a mosaic environment in the laboratory that would allow predator and prey to persist without restocking either population (Huffaker 1958). The six-spotted mite *(Eotetranychus sexmaculatus)* was prey; another mite, *Typhlodromus occidentalis,* was predator; oranges provided the prey's food. Huffaker established experimental populations on trays within which he could vary the number, exposed surface area, and dispersion of the oranges (Figure 21–5).

Each tray had 40 positions arranged in 4 rows of 10 each; where oranges were not placed, rubber balls of about the same size were substituted. The exposed surface area of the oranges was varied by covering the oranges with different amounts of paper, the edges of which were sealed in wax to keep the mites from crawling underneath. In most experiments, Huffaker first established the prey population with 20 females per tray, then introduced 2 female predators 11 days later. Both species reproduce parthenogenetically; males are not required, thank you.

When six-spotted mites were introduced to the trays alone, their populations leveled off at between 5500 and 8000 mites per orange area. When predators were added, their numbers increased rapidly and soon they wiped out the prey population. Their own extinction followed shortly.

Although predators always eliminated the six-spotted mites, the position of the exposed areas of oranges influenced the course of extinction.

Figure 21-5
One of Huffaker's experimental trays with 4 oranges, half exposed, distributed at random among the 40 positions in the tray. Other positions are occupied by rubber balls. Each orange was wrapped with paper and its edges sealed with wax. The exposed area was divided into numbered sections to facilitate counting the mites. (From Huffaker 1958; courtesy of C. B. Huffaker.)

When the orange areas were in adjacent positions, minimizing dispersal distance between food sources, the prey reached maximum populations of only 113 to 650 individuals and were driven to extinction within 23 to 32 days after the beginning of the experiment. The same area of exposed oranges randomly dispersed throughout the 40-position tray supported prey populations that reached maxima of 2000 to 4000 individuals and persisted for 36 days. Thus survival of the prey population could be prolonged by providing remote areas of suitable habitat to which predators disperse slowly.

Huffaker reasoned that if predator dispersal could be further retarded, the two species might coexist. To accomplish this, he increased the complexity of the environment and introduced barriers to dispersal. The number of possible food positions was increased to 120 and the equivalent area of 6 oranges was dispersed over all 120 positions. A mazelike pattern of Vaseline barriers was placed among the food positions to slow the dispersal of the predators; *Typhlodromus* must walk to get where it is going, but the six-spotted mite spins a silk line that it can use like a parachute to float on wind currents. To take advantage of this behavior, Huffaker placed vertical wooden pegs throughout the trays, which the mites used as jumping-off points in their wanderings. This arrangement finally produced a series of three population cycles over 8 months (Figure 21-6). The distribution of the predators and prey throughout the trays continually shifted as the prey, exterminated in one feeding area, recolonized the next a jump ahead of their predators.

In spite of the tenuousness of the predator-prey cycle achieved, we see that a spatial mosaic of suitable habitats allows predator and prey populations to coexist stably. But, as we saw in Gause's experiments with protozoa, predator and prey also may coexist locally if some prey can take refuge in

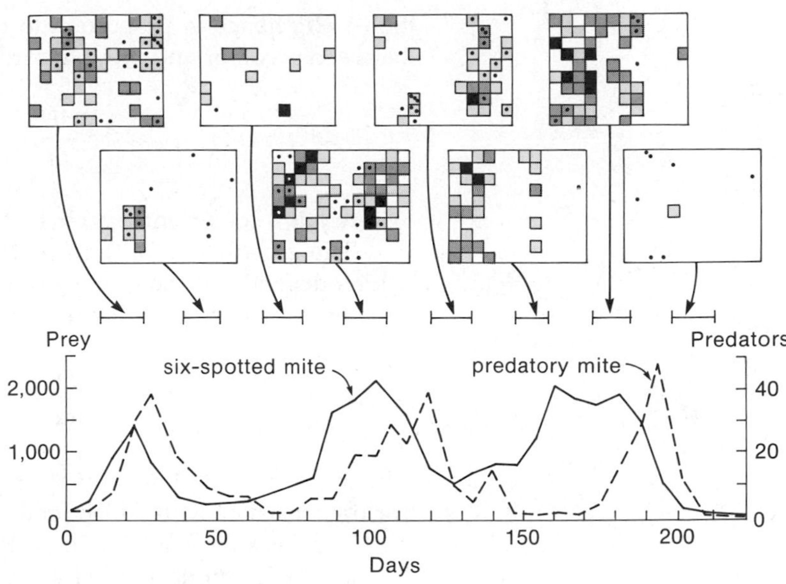

Figure 21-6
Below: Population cycles of the six-spotted mite and the predatory mite *Typhlodromus* in a laboratory situation. *Above:* The boxes show the relative density and positions of the mites in the trays: shading indicates the relative density of six-spotted mites; circles indicate the presence of predatory mites. (After Huffaker 1958.)

hiding places. And when the environment is so complex that predators cannot easily find scarce prey, stability again can be achieved, as we shall see.

Simple predator and prey models predict oscillations in population size.

A. J. Lotka (1925) and the Italian biologist V. Volterra (1926) independently provided the first mathematical descriptions of predator-prey interactions. In discussing these models, we shall follow the common practice of designating the number of predator individuals by P and the number of prey by H (think of H for herbivore). Lotka and Volterra expressed the rate of growth of both populations by differential equations having the form $dH/dt = f(H,P)$ and $dP/dt = g(P,H)$, where f and g designate arbitrary functions of the variables H and P.

The growth rate of the prey population was seen as having two components: the unrestricted exponential growth of the prey population in the absence of predators, rH, where r is the difference between the per capita birth and death rates; and the removal of prey by predators, over and above other causes of death. Lotka and Volterra both assumed that predation varied in direct proportion to the product of the prey and predator populations, HP, hence in proportion to the probability of a random encounter between predator and prey. Accordingly,

$$\frac{dH}{dt} = rH - pHP \tag{21-1}$$

where p is a coefficient expressing the efficiency of predation.

The growth rate of the predator population balances the birth rate, which depends on the number of prey captured, against a constant death rate imposed from outside the system (that is, by weather),

$$\frac{dP}{dt} = apHP - dP \tag{21-2}$$

The birth term is the number of prey captured (pHP) times a coefficient (a) for the efficiency with which food is converted to population growth. The death rate is a constant (d) times the number of predator individuals. The Lotka-Volterra model proved successful in at least one respect: it predicted that predator and prey populations would oscillate, as we shall see below.

When both predator and prey populations are in equilibrium ($dH/dt = 0$ and $dP/dt = 0$), $rH = pHP$ and $apHP = dP$. These equations can be rearranged to give

$$\hat{P} = \frac{r}{p} \quad \text{and} \quad \hat{H} = \frac{d}{ap}$$

where $\hat{P}$ and $\hat{H}$ are the equilibrium sizes of the predator and prey populations. Notice that both $\hat{P}$ and $\hat{H}$ are constant values, each being independent of the abundance of the other population.

According to the Lotka-Volterra model, when the populations are displaced from their joint equilibrium, rather than returning to the equilibrium point, the populations oscillate around it in a continuous cycle. Lotka (1925) showed that when changes in the predator and prey populations are expressed in terms of each other, that is,

$$\frac{dH}{dP} = \frac{H(r - pP)}{P(apH - d)}$$

and integrated, the solution in the region close to the joint equilibrium point of the populations is

$$\frac{d(P - b/p)^2}{b/p} + \frac{b(H - d/ap)^2}{d/ap} = \text{constant} \tag{21-3}$$

Expressing P and H as deviations from their equilibrium values, (for example, $P^* = P - \hat{P} = P - b/p$), and combining constants into single variables a and b, equation (21–3) can be seen to have the form $aP^{*2} + bH^{*2} = c$, which is the equation for an ellipse. The larger the value of c, the larger the ellipse

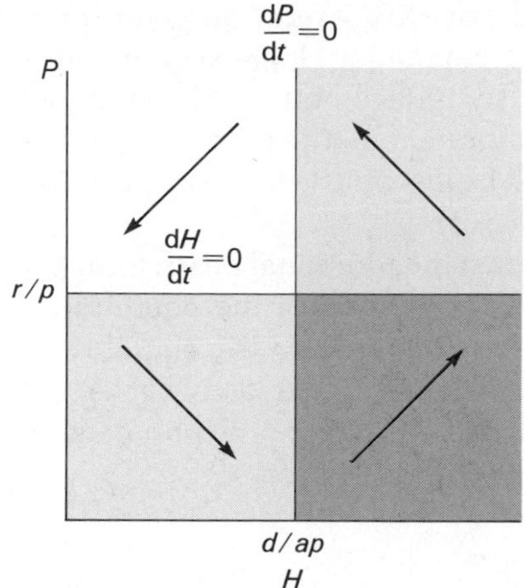

Figure 21-7
Representation of the Lotka-Volterra predator-prey model on a population graph. The trajectories of the populations show that the predator and prey will continually oscillate out of phase with each other.

and hence the greater the amplitude of the oscillation of each population about its equilibrium point. The period of the oscillation when the populations are close to their equilibria is $T = 2\pi/\sqrt{bd}$. Hence the higher the birth rate of the prey or the death rate of the predator—that is, the higher the rate of population turnover—the faster the system oscillates.

Lotka's solution of the predator-prey equations is an approximation accurate only for small displacements from equilibrium. The complete solution may be found by remembering that $dP/Pdt = d\log P/dt$ and recasting equations (21-1) and (21-2) in the form

$$\frac{d\log H}{d\log P} = \frac{r - pP}{apH - d}$$

which may be integrated to obtain

$$p(C\log C - C) + ap(P\log P - P) = \text{constant}$$

(Elseth and Baumgardner 1981).

The relationship between predator and prey is most conveniently portrayed on a graph of which the axes are the sizes of the populations. By convention, predator numbers increase along the vertical axis and prey numbers, along the horizontal axis (Figure 21-7). The equilibrium population values of the predator (P) and prey (H) partition the graph into four regions. The line $P = r/p$, representing the condition $dH/dt = 0$, is called the equilibrium isocline of the prey. For any combination of predator and prey numbers that lies in the region below the line, the prey increase because there are few predators to eat them. In the region above the prey isocline, prey populations decrease because of overwhelming predator pressure. For the predators, population increases are limited to the region to the right of the predator equilibrium isocline ($H = d/ap$), where prey are abundant enough to sustain population growth. To the left of the line, predator populations decrease owing to insufficient prey.

The change in both populations simultaneously follows an arrow, shown in each of the four sections of the graph, that adds the individual vectors of change of the predator and prey populations. In the lower right, for example, both predator and prey increase and the joint population trajectory moves up and to the right. The vectors in the four regions together define a counterclockwise cycling of the predator and prey population one-quarter cycle out of phase, with the prey population increasing and decreasing just ahead of that of its predator (Figure 21-8).

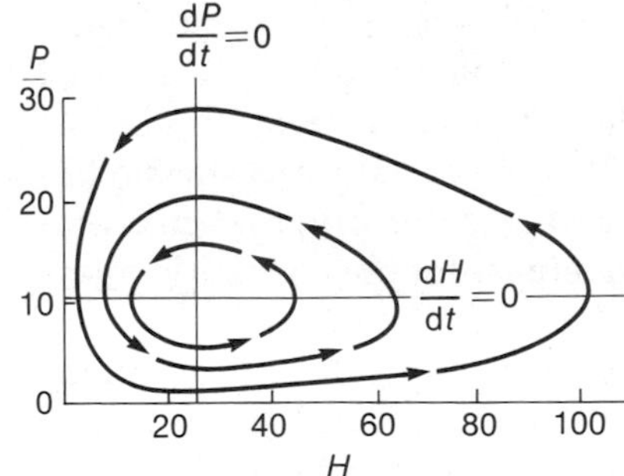

Figure 21-8
Simulated population trajectories of predator and prey according to a Lotka-Volterra model. The degree of oscillation about the equilibrium point reflects the initial population sizes in the simulation. (From Elseth and Baumgartner 1981.)

The equilibrium isocline for the predator ($dP/dt = 0$) defines the minimum level of prey ($H = d/ap$) that can sustain the growth of the predator population. That of the prey ($dH/dt = 0$) defines the greatest number of predators ($P = r/p$) that the prey population can sustain. If the reproductive rate of the prey (r) increased, or the hunting efficiency of the predators (p) decreased, or both, the prey isocline (r/p) would increase—that is to say, the prey population would be able to bear the burden of a larger predator population. If the death rate of the predators (d) increased and either the predation efficiency (p) or reproductive efficiency of the predators (a) decreased, the predator isocline (d/ap) would move to the right, and more prey would be required to support the predator population. Increased predator hunting efficiency (p) would simultaneously reduce both isoclines; fewer prey would be needed to sustain a given capture rate (the predator isocline decreases), and the prey population would be less able to support the more efficient predators (the prey isocline decreases).

Volterra (1926) pointed out that if a constant additional death term (D) were added to both the predator and the prey species, the equilibrium number of predators would decrease [$P = (r - D)/p$] and the equilibrium number of prey would increase [$H = (d + D)/ap$]. Hence a cold snap, or indiscriminate application of pesticide, will likely enhance prey populations (often injurious crop pests) at the expense of their predators.

A. J. Nicholson and V. A. Bailey proposed an alternative to the Lotka-Volterra model.

According to the Lotka-Volterra model, when either the predator or prey population is displaced from its equilibrium the system will oscillate in a closed cycle. Any further perturbation of the system will give the population fluctuations a new amplitude and duration until some other outside influence acts. This equilibrium is said to be neutral because no internal forces act to restore the populations to the intersection of the predator and prey isoclines. Therefore, random perturbations will eventually increase the fluctuations to the point that the trajectory strikes one of the axes of the predator-prey graph, and one or both populations die out. This property in itself suggests that the Lotka-Volterra equations greatly oversimplify nature.

Lotka (1925) pointed out that adding higher order terms to the differential equations for predators and prey would tend to create an inward spiraling of the population trajectory toward the joint equilibrium. For example, the expression $dH/dt = rH - pHP - cH^2$—simply a logistic equation ($dH/dt = rH - cH^2$) with a predation term (pHP) added—leads to a damped oscillation and a stable equilibrium. Density dependence in the predator population has the same effect. Other biologically realistic modifications include time lags, which tend to make a predator-prey system unstable and lead to ever greater oscillations, and refuges for prey, which increase the likelihood of persistence.

Nicholson and Bailey (1935) were concerned about difficulties with the predation term of the Lotka-Volterra model (that is, pHP), according to which the number of prey caught increased in direct proportion to the number of predators. Surely as their density increases, predators must exhibit some mutual interference that reduces their efficiency. To remedy this situation, Nicholson and Bailey developed models of their own. Because Nicholson worked with parasitoids and their hosts, the Nicholson-Bailey models were formulated with the attributes of these organisms explicitly in mind.

Because P and H may stand for "parasitoid" and "host" as well as "predator" and "herbivore," our notation will remain unchanged. The biology of the parasitoid-host system is fairly simple. A certain number of hosts (H_a) are attacked each generation. Each of these yields a certain number of parasitoid offspring (c). The number of hosts in the next generation is simply the number of larvae or pupae not attacked times the birth rate of individual hosts (b). Generations are discrete and nonoverlapping, and so Nicholson and Bailey used difference equations for their model:

$$H(t + 1) = b[H(t) - H_a] \qquad (21-5)$$

and

$$P(t + 1) = cH_a \qquad (21-6)$$

It now remains to write an expression for the attack rate (H_a) in terms of the densities of both host and parasitoid. Nicholson (1933: 141) summed up their approach to this problem:

> Generally speaking, animals appear to be limited in density, either directly or indirectly, by the difficulty they experience in finding the things they require for existence, or by the ease with which they are found by natural enemies. We are thus faced with the problem of the competition that exists among animals when searching, and the whole of the investigation that follows is concerned with this important problem.

As we have seen, a distinctive feature of parasitoid-host interactions is that a successful attack does not remove the host from the system. Therefore, hosts may be attacked many times. Because the number of parasites that can be reared on a host is relatively limited, the dynamics of both the parasitoid and host depend on the number of hosts attacked rather than the total number of hosts encountered by each parasitoid. If parasitoids encountered hosts at random, particularly with regard to previous history of attack as Nicholson and Bailey assumed, the probability distribution of encounters per host (X) would be described by the Poisson series, $p(X) = u^X e^{-u}/X!$, where u is the mean number of encounters per host. The proportion of hosts not encountered ($X = 0$) is $p(0) = e^{-u}$; the proportion of hosts attacked (that is, encountered at least once) is $1 - e^{-u}$. Now if the probability of encounter is directly proportional to the number of parasitoids, so will be the mean number of encounters per host; that is, $u = aP(t)$. Substituting $H_a = H(t)\{1 - \exp[-aP(t)]\}$ into equations (21–5) and (21–6) gives us the basic form of the Nicholson-Bailey model:

$$H(t + 1) = bH(t)[e^{-aP(t)}] \qquad (21-7)$$

and

$$P(t + 1) = cH(t)[1 - e^{-aP(t)}] \qquad (21-8)$$

As the number of parasitoids increases, the rate of attack on the hosts increases steeply at first and then less rapidly as parasitoids encounter

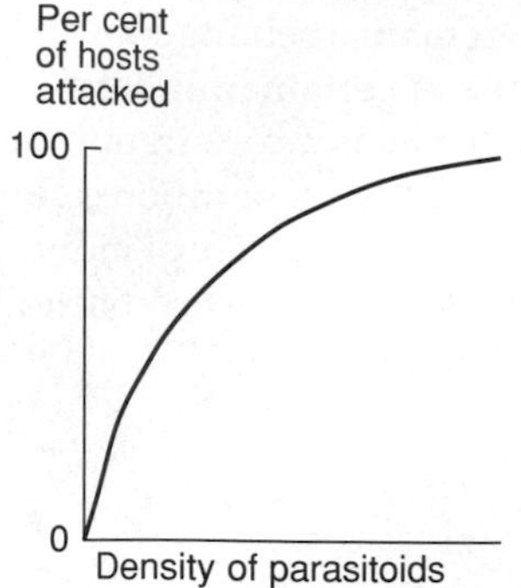

Figure 21–9
The relationship between density of parasitoids and the attack rate upon the host population according to the Nicholson-Bailey model.

previously attacked hosts more frequently (Figure 21–9). When searching is random, just by chance some of the hosts are never attacked regardless of how dense the population of parasitoids.

The Nicholson-Bailey model presented in equations (21–7) and (21–8) has three parameters (a, b, and c), all of which can be determined experimentally and applied to data on parasitoid-host populations. For example, T. Burnett (1958) modeled the interaction between the greenhouse whitefly *Trialeurodes vaporariorum* and its chalcid wasp parasitoid *Encarsia formosa* (Figure 21–10). The values $a = 0.068$, $b = 2$ (experimentally imposed), and $c = 1$ gave a reasonable fit to the data over several population cycles.

As formulated above, the Nicholson-Bailey model is unstable; population fluctuations continually increase in amplitude until either the host or parasitoid is exterminated. While at variance with the neutrally stable Lotka-Volterra model, the instability of the Nicholson-Bailey equations is due to their formulation as discrete time processes. Recast as a set of difference equations, the Lotka-Volterra model behaves in the same fashion. One can achieve some degree of stability in the Nicholson-Bailey models by adding a density-dependent term to the population equation for the host. This and other modifications are treated in detail by Hassell (1978).

The response of predators to prey density is not linear.

Nicholson and Bailey criticized the Lotka-Volterra model because its linear relationship between attack rate and predator (or parasitoid) density

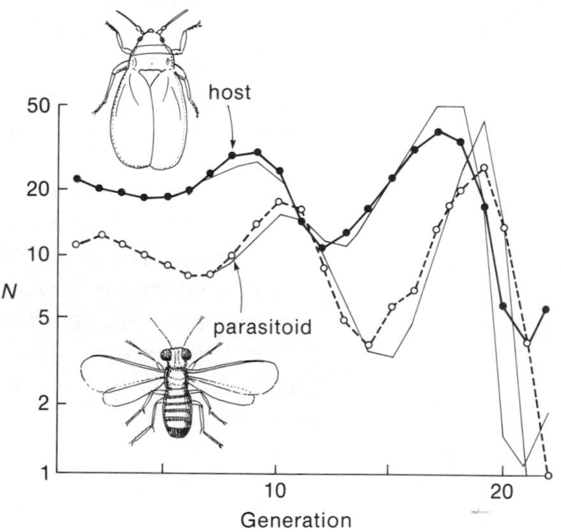

Figure 21–10
Population fluctuations of the whitefly *Trialeurodes vaporariorum* (solid symbols) and its chalcid parasitoid *Encarsia formosa* (open symbols). The thin lines are trajectories predicted by the Nicholson-Bailey model. (From Begon and Mortimer 1981; after Hassell 1978.)

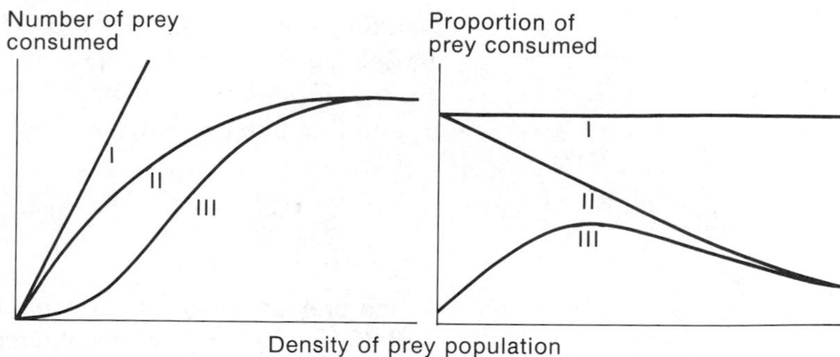

Figure 21–11
The functional response of predators to increasing prey density: (I) predator consumes a constant proportion of the prey population regardless of its density; (II) predation rate decreases as predator satiation sets an upper limit to food consumption; (III) predator response lags at low prey density owing to low hunting efficiency or absence of search image. *Left:* The functional response in terms of the *number* of prey consumed; *right:* the functional response in terms of the *proportion* of prey consumed.

seemed unrealistic. The Canadian entomologist C. S. Holling (1959) voiced a similar complaint about the linear relationship between the number of prey consumed per predator and the density of prey. In the Lotka-Volterra model, the rate at which individuals are removed from the prey population is described by the term (pHP). Thus for a given density of predators (P), the rate of exploitation increases in direct proportion to the density of prey (H). Many biological factors ought to alter the form of this relationship and perhaps thereby alter the dynamics of the predator and prey populations.

The relationship of an individual predator's rate of food consumption to prey density has been labeled the "functional response" by Holling (1959). Three general types of curves are illustrated in Figure 21–11. Type I is the linear relationship of the Lotka-Volterra model. Type II describes a situation in which the number of prey consumed per predator initially rises quickly as the density of prey increases but then levels off with further increase in prey density. Type III resembles type II in having an upper limit to prey consumption, but differs in that the response of predators to prey is depressed at low density.

Two factors dictate that the functional response should reach a plateau. First, predators may become satiated—continually stuffed—at which point their rate of feeding is limited by the rate at which they can digest and assimilate food. Second, as the predator captures more prey, the time spent handling and eating the prey cuts into searching time. Eventually, the two reach a balance and prey-capture rate levels off.

Holling (1959) described this second factor by a simple expression known as the "disc equation" because of its application to experiments in which blindfolded human subjects were required to discover and pick up small discs of paper on a flat surface. Any such task, including the subduction and eating of prey, requires a certain handling time, T_h. The total handling time is therefore the handling time per item times the number of encounters, E; the time left over for searching is the total time minus the total handling time; that is, $T_s = T - T_hE$. The number of encounters can, itself, be defined as the product of search time, prey density (P), and a constant (a) for the efficiency of searching; that is, $E = a(T - T_hE)P$. The expressions for T_s and E can be combined and solved for

$$E = \frac{aPT}{1 + aPT_h} \tag{21-9}$$

When prey are scarce, the denominator term aPT_h is small compared to 1 and the number of prey encountered approaches aPT. Hence encounters are directly proportional to prey density. When prey are dense, aPT_h is large compared to 1 and E approaches the ratio T/T_h, which is a constant value defining the maximum number of prey that can be captured in time T. Thus at high prey density, search time drops to near zero and the number of prey captured is limited only by how long the predator requires to handle each one; the shorter the handling time, the more prey can be captured (Figure 21–12).

Hassell (1978) fitted equation (21–9) to the relationship between attack rate and prey density observed in several laboratory systems (Table 21–1). As one would expect, both handling time and search efficiency varied widely, depending on the characteristics of the predator and prey and the structure of the laboratory environment. The very high handling time of the parasitoid *Nasonia* on housefly *(Musca domestica)* hosts reflects the fact that many pupae were rejected because they had already been parasitized (encountered). In that study, the number of encounters referred only to those not previously attacked, but handling time included time spent rejecting any pupa encountered, parasitized or not.

The encounter rate defined by Holling's disc equation can be incorporated in the Nicholson-Bailey model, with appropriate provision for differ-

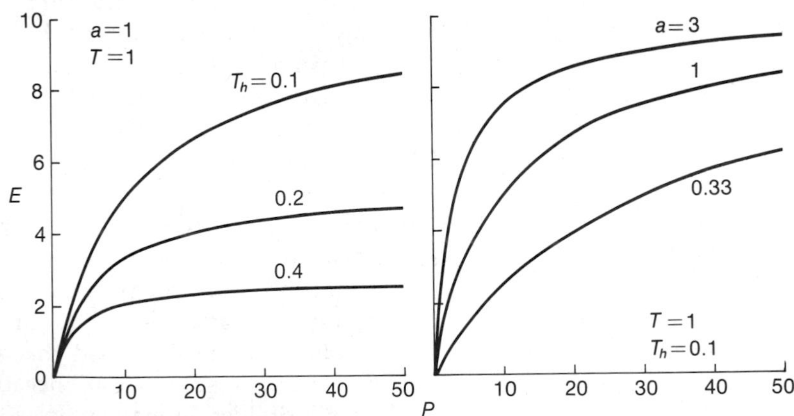

Figure 21–12
The relationship between encounter rate (E) and prey density (P) according to Holling's disc equation. *Left:* The curves represent different handling times (T_h) *per prey; right:* different levels of hunting efficiency (a).

Table 21-1
Estimated values of handling time in consumer-resource equations for a selection of parasitoids and arthropod predators

Parasitoid or predator	Host	Handling time (hrs) T_h	T_h/T
Parasitoids			
Nemeritis canescens	*Ephestia cautella*	0.007	< 0.0001
Chelonus texanus	*Ephestia kuhniella*	0.12	< 0.001
Dahlbominus fuscipennis	*Neodiprion lecontei*	0.24	< 0.003
Pleolophus basizonus	*Neodiprion sertifer*	0.72	< 0.02
Dahlbominus fuscipennis	*Neodiprion sertifer*	0.96	< 0.01
Cryptus inornatus	*Loxostege sticticalis*	1.44	> 0.02
Nasonia vitripennis	*Musca domestica*	12.00*	< 0.1
Predators			
Anthocoris confusus	*Aulacorthum circumflexus*	0.38	< 0.001
Notonecta clauca	*Daphnia magna*	0.76	< 0.005
Ischnura elegans	*Daphnia magna*	0.82	< 0.002
Harmonia axyridis	*Aphis craccivora*	1.61	< 0.002
Phytoseiulius persimilis	*Tetranychus urticae*	1.87	< 0.005

* This figure is the handling time for each host, in which a female lays many eggs. The handling time per egg is roughly 0.4 hours.
Source: From Hassell 1978.

ences between parasitoids and predators (Royama 1971; Rogers 1972). This addition of handling time to the model always makes the system of equations less stable because predators consume a smaller fraction of the prey or host population as prey or host density increases (Hassell and May 1973). This, of course, is inverse to the density-dependent expression of factors that regulate population size.

Predators having a type III functional response can stabilize prey populations.

At high prey densities, type II and type III response curves differ little; they both are inversely density-dependent. Over the lower range of prey density, however, type III responses differ from those of type II in that the proportion of prey consumed increases with density of prey.

Several factors may lead to decreased predator response at lower prey density, hence to a type III functional response: (1) a heterogeneous habitat may afford a limited number of safe hiding places, which protect a larger proportion of the prey at lower densities than at higher densities; (2) lack of reinforcement of learned searching behavior owing to low rate of prey encounter may reduce hunting efficiency at low prey density; and (3) switching to alternative sources of food when prey are scarce reduces hunting pressure. The relationship of prey consumption to prey density depends upon change in average prey vulnerability in the case of (1), search and capture efficiency in the case of (2), and motivation to hunt or searching time in the case of (3).

Learning influences the searching behavior of many predators, especially vertebrates with complex nervous systems. An object is always easier to find if one has a preconception of what it looks like and where it occurs. Such "search images" (Tinbergen 1960) are acquired by experience; that is, learning. The denser a prey population, the more frequently predators encounter prey; search efficiency increases and handling time decreases as a result of experience. At low prey density, neither are so well developed and the capture rate is therefore depressed. The search image concept stimulated a number of studies on predator behavior (for example, Gibb 1958, 1962; Mook, Mook, and Heikens 1960; Tinbergen et al. 1962; Royama 1970; Murton 1971; Pearson 1985), but its role in altering the functional response still is not well understood.

Predators often direct their hunting effort toward the most abundant prey.

Regardless of the mechanisms underlying the type III response, predators often switch to a second prey as the density of the first is reduced (Murdoch 1969). When John H. Lawton et al. (1974) presented the predatory water bug *Notonecta glauca* two types of prey—an isopod, *Ascellus aquaticus,* and

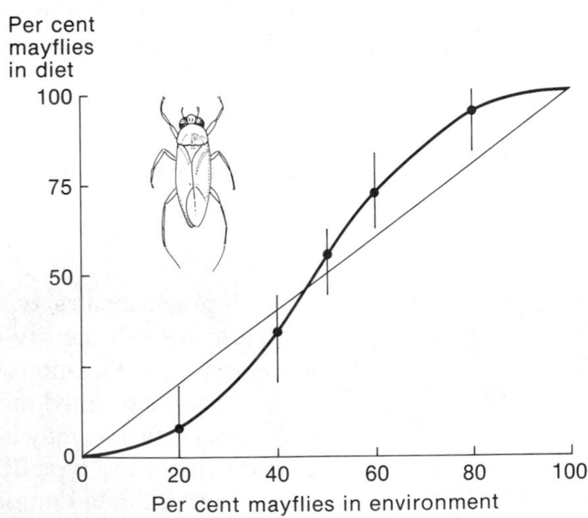

Figure 21–13
The percentage of mayfly larvae in the diet of the water bug *Notonecta* as a function of their relative abundance among available prey. The straight line indicates no preference; data are presented as means and ranges of separate trials. (From Begon and Mortimer 1981; after Lawton et al. 1974.)

larvae of the mayfly *Cloeon dipterum*—they found that the predators consumed the more abundant prey species in greater proportion than its percentage occurrence (Figure 21–13). This switching depended to some degree on variation in attack success on isopods as a function of their relative density. When water bugs encountered them infrequently, fewer than 10 per cent of attacks were successful. At higher densities, and therefore higher encounter rates, attack success rose to almost 30 per cent (also see Cornell and Pimentel 1978).

The consequences of switching for predator-prey stability have been explored in some detail by W. W. Murdoch and A. Oaten (1975) and Oaten and Murdoch (1975). Using continuous-time models they found that the type III response characteristic of switching could stabilize predator and prey populations. M. P. Hassell and H. N. Comins (1978) pointed out, however, that switching does not produce the same effect in discrete-time, difference-equation models because of the overwhelming time delay in the response of predators to prey population density.

Predator populations can respond to an increase in prey density by growth and immigration.

Individual predators can increase their consumption of prey only to the point of satiation. Predator response to increasing prey density above the level that results in satiation can be achieved only through increase in the number of predators, either by immigration or population growth, which together constitute the numerical response. Populations of most predators grow slowly, especially when the reproductive potential of the predator is much less than that of the prey, and the life span longer. Immigration from surrounding areas contributes importantly to the numerical response of mobile predators, which may opportunistically congregate where resources become abundant. The bay-breasted warbler, a small insectivorous bird of eastern North America, exhibits such behavior during periodic outbreaks of the spruce budworm. During years of outbreak in a particular area, the density of warblers may reach 120 pairs per 100 acres, compared with about 10 pairs per 100 acres during nonoutbreak years (Morris et al. 1958). This population behavior clearly shows how a predator may take advantage of a shifting mosaic of prey abundance.

In a study of the larch sawfly in tamarack swamps in Manitoba, C. H. Buckner and W. J. Turnock (1965) found that while avian predators were more abundant in areas of high sawfly density, they consumed a smaller proportion of the sawflies there than in areas with low sawfly populations. The sawflies varied in number over three orders of magnitude while the numerical response of the birds resulted in only a doubling of population size. So while 6 per cent of larvae and 65 per cent of adult sawflies were eaten in one area of low population density, only 0.5 per cent and 6 per cent were consumed where sawflies were 50 times more abundant (see also Crawford and Jennings 1989).

Table 21–2
Response of predatory birds to different densities of the brown lemming near Barrow, Alaska

	1951	1952	1953
Brown lemming (ind per acre)	1 to 5	15 to 20	70 to 80
Pomarine jaeger	Uncommon, no breeding	Breeding pairs 4 mi^{-2}	Breeding pairs 18 mi^{-2}
Snowy owl	Scarce, no breeding	Breeding pairs 0.2 to 0.5 mi^{-2} many nonbreeders	Breeding pairs 0.2 to 0.5 mi^{-2} few nonbreeders
Short-eared owl	Absent	One record	Breeding pairs 3 to 4 mi^{-2}

Source: Pitelka et al. 1955.

Three predatory birds, the pomarine jaeger, the snowy owl, and the short-eared owl, each respond in a different manner to varying densities of lemmings on the arctic tundra (Table 21–2). Lemming populations exhibit great fluctuations; high and low points in a population cycle may differ by a factor of 100. At Barrow, Alaska, during the summer of 1951, when lemmings were scarce, none of the predatory birds bred; short-eared owls did not even appear in the area. During the following summer, one of moderate lemming density, both the jaeger and snowy owl bred, but short-eared owls again were absent. In 1953, a peak year for lemmings, all three species of avian predators bred. Jaegers were four times more abundant in 1953 than in 1952, showing a strong numerical response. In contrast, the density of snowy owls did not increase, but each pair of birds instead reared more young. Whereas most snowy owl nests contained 2 to 4 eggs during the year of moderate lemming abundance, clutches of up to 12 eggs were laid during the peak year of 1953.

Graphical analyses demonstrate the conditions for stability in predator-prey systems.

Studies of the functional and numerical responses of predators, as well as the population dynamics of prey populations, have revealed the biological limitations of simple predator-prey models, such as the Lotka-Volterra equations. Many complications of biological reality have been incorporated into models and their contributions to the stability of predator-prey interactions studied (May 1975; Hassell 1978). But we can accomplish the same goal and avoid much of the difficult mathematics involved by using two graphical approaches to the problem of stability in predator-prey systems. The first of these is based on modifications of the predator and prey isoclines drawn on a graph of their joint abundances (for example, Figure 21–7). The second, which shall be taken up below, builds upon the functional response graph shown in Figure 21–11.

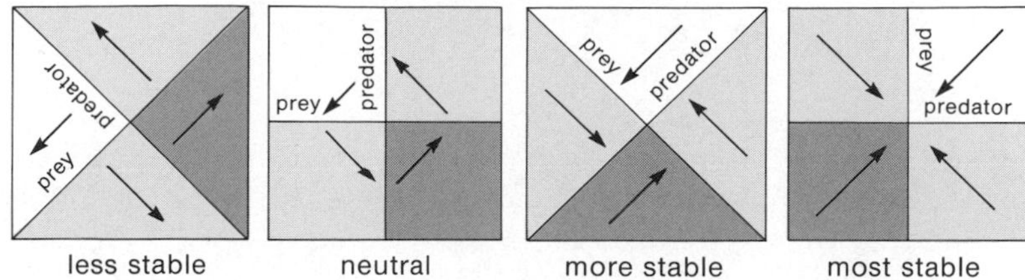

Figure 21–14
The effect of rotating the predator and prey isoclines in the stability of the interaction. Clockwise rotations are stabilizing; counterclockwise rotations are destabilizing.

The isoclines for the Lotka-Volterra model are reproduced in the second panel of Figure 21–14. When the predator isocline is vertical and the prey isocline is horizontal, as it is in the second panel, the system is in neutral equilibrium; in the absence of perturbation, predator and prey populations continue to cycle indefinitely. But rotate the isoclines counterclockwise (first panel) and the trajectory of the populations then spirals outward from the equilibrium in cycles of increasing amplitude. Hence the system becomes unstable. Rotate the isoclines clockwise, however, and the equilibrium becomes stable and the population trajectory spirals in toward the intersection of the isoclines (third and fourth panels). The farther the isoclines are rotated in a clockwise direction, the more rapidly populations move toward their equilibria.

Michael L. Rosenzweig and Robert H. MacArthur (1963) showed that certain biological considerations altered the predator-prey graph in such a way as to affect the rotation of the isoclines and thereby change the stability of the system. First, in the absence of predators, the prey population is limited by the carrying capacity of the environment (K), determined by the availability of food or other resources. As a result, the prey isocline bends down toward the prey axis as the number of prey increase, intersecting the axis at K (Figure 21–15). This is equivalent to a clockwise rotation of the prey axis and therefore has a stabilizing effect.

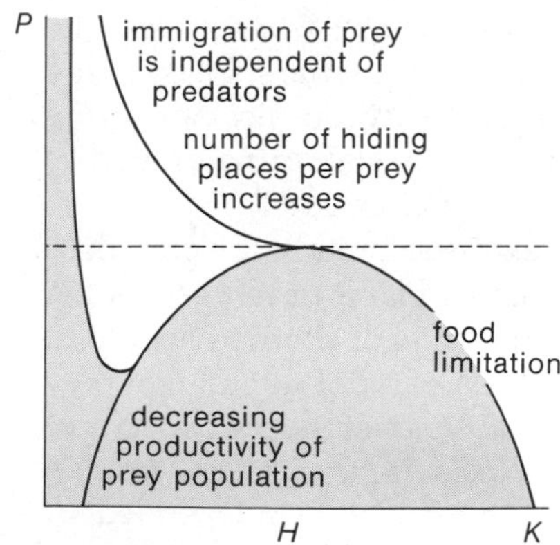

Figure 21–15
Graphical representation of the prey equilibrium isocline, incorporating several biological properties of natural systems. The isocline for the Lotka-Volterra equation is indicated by the dashed line.

Two opposing factors influence the shape of the prey isocline at low densities. On one hand, small populations can support fewer predators than large populations because the recruitment rate of smaller populations is lower. As pointed out by Rosenzweig (1969), this gives the prey isocline a characteristic humped shape, with the line bending down toward the prey axis at low densities (counterclockwise rotation), as well as at high densities. On the other hand, when scarce prey are more difficult to locate than abundant prey because, for example, a larger proportion can find good hiding places, predation is reduced and the prey population can persist in the presence of denser predator populations. Graphically, this corresponds to an upward swing (clockwise rotation) in the prey isocline at the left. Whether the isocline swings down or up at low prey densities depends largely on the heterogeneity of the habitat and the availability of alternative prey.

The predator isocline also bends under the weight of biological reality

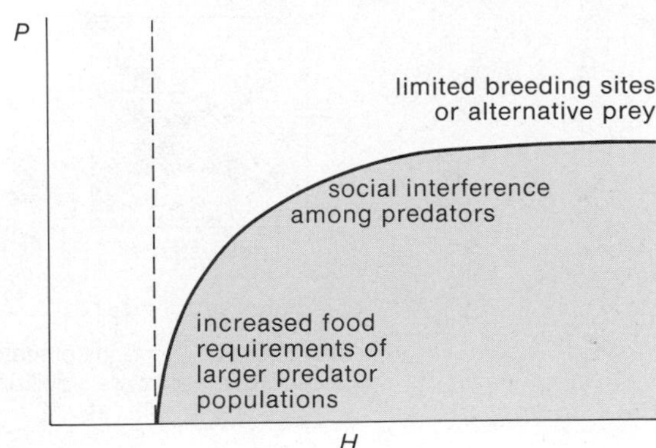

Figure 21–16
Graphical representation of the predator equilibrium isocline, incorporating several biological properties of natural systems. The isocline for the Lotka-Volterra equation is indicated by the dashed line.

(Figure 21–16). As a population of predators increases, its food requirement increases and predators must capture more prey to maintain their population at a constant level. The predator isocline, therefore, should be rotated clockwise. Furthermore, social interference among predators (territoriality, for example) would reduce the efficiency with which they utilize prey resources and would bend the predator isocline further to the right. The availability of suitable breeding sites or of alternative food resources also could limit the size of the predator population independently of the abundance of the prey, causing the predator isocline to become horizontal at high prey population densities.

Biological considerations alter the stability of predator-prey systems in many ways. For a humped prey isocline, the system is stable when the predator isocline lies to the right of the hump and unstable when it lies to the left. That is to say, increasing predator efficiency reduces the stability of the system, all else being equal. When a vertical predator isocline lies close to the center of the hump, the system may exhibit internal stabilizing properties that lead to stable limit cycles (May 1972, 1975; Gilpin 1974).

As the graphical approach also shows, extrinsic limits to the growth of the predator population, the availability of hiding places or refuges for the prey, and, under some circumstances, the availability of alternative prey all act to stabilize the predator-prey interaction. The use of alternative prey allows predator populations to increase in the absence of the prey species of concern; hence the predator isocline rotates clockwise toward a horizontal position, at which extreme its growth would be completely independent of the density of the one type of prey.

Graphical analyses suggest that under most biologically realistic conditions, we should expect predator and prey interactions to be highly stable. Population cycles might even be the exception. How then can we account for the widespread existence of population cycles in nature? One possibility is that these represent stable limit cycles. But the conditions required to

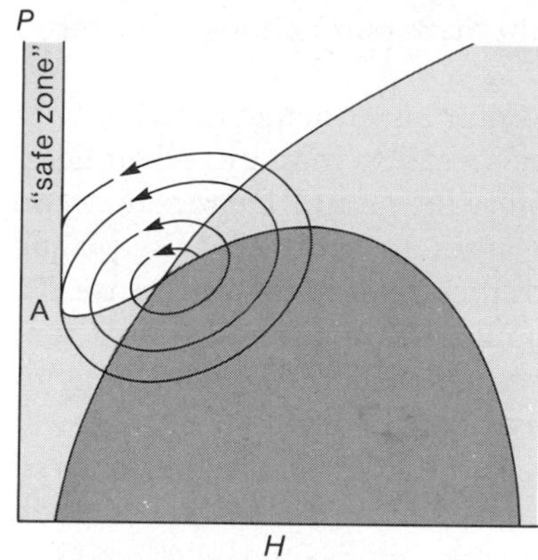

Figure 21–17
Possible stability conditions for an oscillating predator-prey system. The cycle is maintained by the balance between nonstabilizing influences at the intersection of the isoclines and the stabilizing influence of a "safe" zone.

maintain such stable oscillations are narrowly restricted and depend on a very special geometry of the predator and prey isoclines. It is more likely that robust conditions for stability will occur when two powerful forces oppose each other and bring a system into an equilibrium reflecting the balance between the two. This balance may come about in two ways.

When a predator is efficient but its prey has suitable refuges or hiding places that protect it at very low population densities, the geometry of the predator and prey isoclines can create a stable cycle (Figure 21–17). Whereas the intersection of the predator and prey isoclines may result in an outward-spiraling population trajectory, this spiral is finally arrested by the presence of refuges or, equivalently, of some constant immigration. Upon hitting the upturned prey isocline, the population trajectory would descend along it as if it were a "safe" corridor until predators became scarce enough that the prey could increase again. This would result in a stable oscillation because no matter where the population trajectory first strikes the safe zone, it would always continue to the same point (A) before embarking on a new cycle.

A second, and perhaps more likely, way in which predator-prey cycles are stabilized is by a balance between the natural stabilizing geometry of the predator and prey isoclines and the destabilizing effects of time lags, which seem to be properties of most biological interactions. In such a case, the farther a spiraling population trajectory extends from the joint predator-prey equilibrium, the stronger the stabilizing tendency of certain isocline geometries (Figure 21–18).

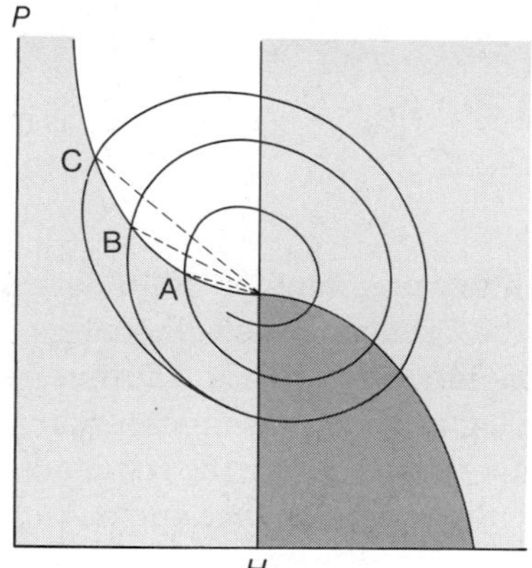

Figure 21–18
A predator-prey model demonstrating how a rise in the prey isocline at low densities can stabilize an outward spiraling population trajectory caused, for example, by a time lag in the predator response. The steepness—hence the stability of the apparent prey isocline (dashed lines)—increases as the oscillations increase from A to B to C.

The dynamics of parasite-host systems incorporate complex life cycles and the immune responses of hosts.

Parasitism resembles predation in that a living resource is consumed, but differs in that the resource is not killed, at least not immediately. Parasites nonetheless have profound physiological and behavioral consequences for their hosts (Schall et al. 1982, Smith Trail 1980, Burden 1987, Dobson 1988), adversely affect reproduction in many instances (Hudson 1986), and may limit geographical distribution (Cornell 1974, Van Riper et al. 1986). Parasitism is more difficult to model than the predator-prey interaction because the population effects of consumption on the resource are more subtle and complicated (for example, Anderson 1982, Bacon 1985, Rollinson and Anderson 1985). The life cycles of parasites and their interactions with hosts also may be complex (Mattingly 1969, Burnet and White 1972). Unlike prey, hosts can respond to parasites during their lifetimes through the mechanisms of immunity.

On a population level, a serious outbreak of a viral or bacterial disease is often followed by a period during which most of the individuals in a population have achieved some degree of immunity to reinfection. Until the immunity is lost or until susceptible individuals are recruited into the population, the disease organisms may be unable to spread. The history of epidemics of bubonic plague—the Black Death—in human populations provides a striking example of this.

The bubonic plague organism, a bacterium, is endemic in many wild populations of rodents. The disease can be spread to man by rodent fleas, particularly the rat flea. Several times in history the natural balance between the bacteria, rodents, and fleas has been upset so badly that the plague spread to the human population, in which it caused epidemic disease. To produce a major plague epidemic, rat populations must be so great that hordes of rats, searching for food in houses, come into close contact with humans. Furthermore, rat fleas must heavily infest rats before they will abandon their preferred hosts for humans. These conditions have occurred infrequently, even in the crowded, garbage-ridden conditions prevalent in the cities and towns of medieval Europe. But once the plague takes hold in a human population, its course runs swiftly and surely.

A typical epidemic of the Black Death initially spreads rapidly, infects a large portion of the population, and takes a high toll in human lives. But as susceptible individuals either die or become immune to the disease organism, the number of lethal cases and the mortality rate drop almost as rapidly as the disease first strikes. This is illustrated by data for a localized outbreak of the plague in India between 1953 and 1959 (May 1961; quoted by Watt 1968):

Year	Contracted cases	Per cent lethal
1953	20,539	70.5
1954	6670	84.5
1955	705	23.1
1956	331	20.5
1957	44	0
1958	26	0
1959	37	0

The great plague of the fourteenth century originated in 1346 during the siege of Caffa, a small military post on the Crimean Straits. From there, the epidemic spread to Italy and the south of France by 1347, and during the next year it had reached all of Europe. The plague did not disappear entirely until 1357. It reappeared in Europe three more times during the fourteenth century, at intervals of 10 to 13 years, but with lower incidence and mortality rate.

The plague also visited London three times during the seventeenth century, at intervals of 22 and 40 years. But unlike the epidemic waves that struck Europe during the fourteenth century, the effects were not attenuated during each successive outbreak, during which 13 to 15 per cent of the population of London died (Creighton 1891). The longer interval increased the severity of successive epidemics in two ways. First, many more individuals lost their immunity over a 22-year or 40-year period than over a decade. Second, the proportion of the population born since a plague epidemic, and therefore not immune, was much larger after intervals of 22 and 40 years than after the shorter intervals during the fourteenth century. Because more humans were susceptible, the plague organism spread more rapidly through the population.

Man is not a common host of the bubonic plague and the incidence of infections in human populations depends more on the ecological conditions surrounding the plague-rat system than the joint dynamics of the plague and human populations. Other diseases, such as measles, whooping cough, and smallpox, infect people, primarily, and their incidence often exhibits strong cycles with periods of 2 to 20 years. For example, although the smallpox virus brought to Mexico by Spanish colonists in the sixteenth century remained at endemic levels in the native population, epidemics occurred only at 11-year to 19-year intervals; for example, in 1520, 1531, 1545, 1564, and 1576 (Cartwright and Biddiss 1972).

The observation of cycles of disease prevalence stimulated the interest of epidemiologists (those who study population attributes of disease processes; Glass 1986) in general problems concerning the regulation of population size and the generation of cycles in abundance. Mathematical approaches to the problem have been summarized by Roy M. Anderson and Robert M. May (1979, 1980) and May and Anderson (1979). In analyzing models developed specifically for forest insect-pathogen (Stairs 1972; Tinsley 1979) and fox-rabies (Baer 1975; Macdonald 1980) interactions, Anderson (1981) noted that cyclic fluctuations are favored by "high disease pathogenicity in conjunction with low host reproductive potential, long lived free-living infective agents and long incubation periods during which hosts are infected but not infectious." The high pathogenicity and low recruitment rate favor strong immune responses and a high proportion of immune individuals for a long period following an epidemic. A disease refuge provides for reinfection when the host population is again susceptible and guarantees that the pathogen will not disappear. Long incubation periods, like immunity, contribute to the overall time delay in the responses of pathogen and host populations to each other. As in predator-prey interactions, it is these time delays that underlie the oscillatory behavior of the system.

Herbivores can strongly influence plant populations.

The varied chemical and structural defenses of plants suggest that herbivores can seriously damage vegetation. Why else would plants protect themselves so strongly? The impact of herbivory on plant production has been measured in many studies of the amount of plant biomass or proportion of seeds consumed by herbivores. The impact of herbivory on plant population dynamics is most readily revealed when herbivores are either added or removed (see, for example, Brubaker and Greene 1979, Louda 1984, Marquis 1984, McNaughton 1985, Whitham and Mopper 1985).

We have seen the role of the cactus moth in controlling the population of prickly pear cactus in Australia. Herbivorous insects have been employed in many other situations to control imported weeds (DeBach 1974). Consider the example of Klammath weed, a European species toxic to livestock (and the source of the drug hypericin, its toxic ingredient), which accidently became established in northern California in the early 1900s. By 1944, the weed had spread over 2 million acres of range land in 30 counties. Biological control specialists borrowed a herbivorous beetle of the genus *Chrysolina* from an Australian control program. In a success of similar proportions to that of the war against prickly pear, within 10 years after the first beetles were released, Klammath weed was all but obliterated as a range pest. Its abundance was estimated to have been reduced by more than 99 per cent.

The proportion of net primary production consumed by herbivores is least in forests, intermediate in grasslands, and greatest in aquatic environments (Slobodkin et al. 1967). In the littoral zones of aquatic communities, herbivores consume most of the plant production; in terrestrial habitats, the bulk of it moves up the food chain through the detritus pathway. The difference between the functional roles of herbivores in these two habitats is related to the greater digestibility of aquatic compared to terrestrial vegetation.

In grasslands, herbivores, mostly insects and grazing mammals, consume 30 to 60 per cent of the aboveground vegetation. Their influence on plant production is strikingly revealed by exclosure experiments. In one study in California, George Batzli and Frank Pitelka (1970) used wire fences to keep voles out of small areas of grassland. They then measured seed production and species composition of the standing crop of plants for 2 years. When they compared the experimental exclosures to unfenced, but otherwise similar, control plots, Batzli and Pitelka found that grazing by voles in the unfenced plots reduced the abundance and seed production of food plants (mostly annual grasses) but did not affect perennial grasses and herbs not included in the voles' diet (Figure 21–19).

Exclosure experiments demonstrate that grazing herbivores can affect seed production through reducing the vegetative growth required to support reproduction. Seed predators themselves exert a more direct effect on plant demography by consuming anywhere between 10 and 100 per cent of the seeds available (Janzen 1971). Although they consume relatively little of the total biomass of the plant, seed predators attack a vital stage in the life cycle and can influence plant populations greatly.

Although herbivores rarely consume more than 10 per cent of forest vegetation (Morrow 1977), occasional outbreaks of tent caterpillars, gypsy moths, and other insects can completely defoliate or otherwise eradicate entire forests. Long-term studies of growth and survival of trees after defoliation by insects demonstrate that there may be a considerable lag between an infestation and the expression of its effects (Belyea 1952; Duncan and Hobson 1958; Churchill et al. 1964; Kulman 1971).

Studies of *Eucalyptus* stands in Australia have shown that suppression of growth by insect attack can be the rule rather than the exception (Morrow and LaMarche 1978). In some areas, 30 to 50 per cent of the leaf area of young trees is consumed each year by herbivorous insects, primarily sawfly larvae, true bugs, and beetles. Following treatment by insecticide during 1 year, the widths of growth rings in the wood increased two- to threefold

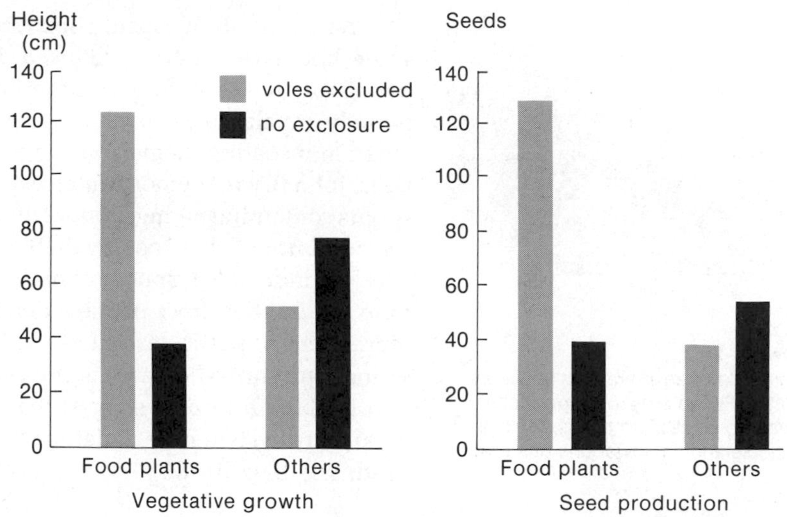

Figure 21–19
Relative biomass (summed height per 100 cm²) and seed production (seeds per 100 cm²) of food plants and nonfood plants in grassland plots fenced to exclude voles and in unfenced control plots. The bar graphs present results of the experiment after 2 years. Food plants are mostly annual grasses; nonfood plants include perennial grasses and herbs. (After Batzli and Pitelka 1970.)

over unsprayed controls, and the effect persisted for at least 3 years. Furthermore, comparison of ring-widths of the same trees for the 20 years before the beginning of the experiment indicated that the trees suffered chronic suppression of growth by herbivory. Why Australian forests suffer greater insect damage than temperate forests elsewhere remains a mystery.

Responses of plants to grazing may cause oscillations in herbivore-plant systems.

Grazing resembles infection by parasites and pathogens in that, like the host, the individual plant usually is not killed. Thus plants have time to respond to grazing and browsing by producing defensive chemical compounds in a manner recalling the immune mechanisms of animals. Depending on how long the plant takes to produce these defenses, time lags may be incorporated into the grazer-plant system, thereby promoting oscillations.

Plants have many inducible defenses against herbivory, most of them chemical. In response to wounding, toxic, noxious, or nutrition-reducing compounds may be produced—either in the area of the wound or systemically throughout the plant—that reduce subsequent herbivory. In some cases, these responses may take only minutes or hours; in others, they require a new season of growth.

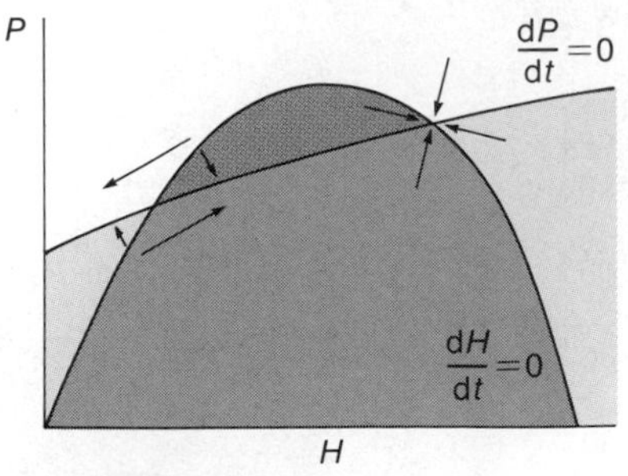

Figure 21–20
An example of an interaction between predator (*P*) and prey (*H*) populations having two equilibrium points, the higher one stable and the lower one unstable.

When shoots of aspen, poplar, birch, and alder are heavily browsed by snowshoe hares, shoots produced during the following year have exceptionally high concentrations of terpene and phenolic resins, which are extremely unpalatable to the hares (Bryant 1981). By extracting the resins with ether, and coating them in varying concentrations on shoots of unbrowsed trees, John Bryant demonstrated experimentally that hares would not touch shoots containing 80 mg or more of resin per gram dry weight, regardless of the amount of other food available.

The inducible response of browsed plants could result in a considerable time lag in the affect of hare density on subsequent hare survival and reproduction, particularly if second-year regrowth also was protected by resins. The snowshoe hare is the classic example of an oscillating population, and Bryant's data suggest that the hare–plant system rather than the lynx–hare system may underlie the predator-prey oscillation (but see Lindroth and Batzli 1986).

Predator-prey systems may have two stable equilibria.

Grazing resists analysis as a predator-prey system because population effects resulting from the consumption of vegetation are difficult to determine. I. Noy-Meir (1975) showed, however, that one may treat plant biomass increase in the same manner as population increase and determine the stable points of herbivore-plant interactions by analyzing graphical analogs of predator-prey equations, as we have done earlier. Hence the following consideration of multiple stable points in predator-prey interactions applies equally well to herbivore-plant systems.

When the equilibrium isocline of the prey is humped and the predator can switch to alternative prey, the predator and prey equilibrium lines may cross in two places (Figure 21–20). As we have seen before, the right-hand equilibrium point is stable. That on the left, however, is not. In the proximity of the left-hand equilibrium point, combinations of predator and prey populations within the region of prey increase will tend to move toward the higher, stable equilibrium; those combinations outside the region will tend to move toward the predator axis, hence to the elimination of the prey. Because populations tend to move directly away from the left-hand equilibrium point when displaced from it, it is referred to as unstable.

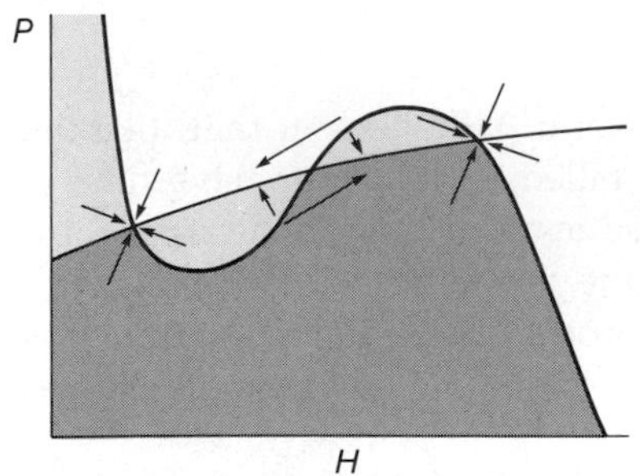

Figure 21–21
An example of an interaction between predator (*P*) and prey (*H*) populations in which there are upper and lower stable equilibria and an intermediate unstable point.

When prey have a refuge at low density, the isocline of a predator with alternative prey may cross that of the prey at three points, two of which are stable equilibria (Figure 21–21). In such cases, the predator-prey system may be regulated at one of two points and certain conditions of the environment may cause it to switch from one point to the other. Because of the potential role of predator switching in creating this isocline geometry, predator-prey interactions have been subjected to a second kind of graphical analysis, based upon the functional response curve, which superimposes per capita rates of recruitment to a prey population and predation upon that population (Figure 21–22). The recruitment curve is the net contribution of

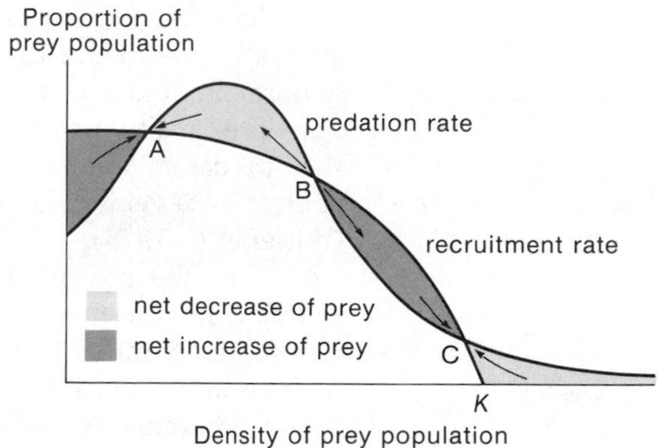

Figure 21-22
Predation and recruitment rates in a hypothetical predator-prey system. When predation exceeds recruitment, prey populations decrease, and vice versa (as shown by arrows). Points A and C are stable equilibria for the prey population; the lower point (A) represents population control by predators; the upper point (C) represents population control by food and other resources.

births and deaths in the absence of predators. Hence the per capita recruitment is high when the prey population is small and goes to zero as the population approaches its carrying capacity. The predation curve is the sum of the functional and numerical responses of predator populations. Predation rate may be low at low prey density owing to switching or difficulty of locating scarce prey; it also tails off at high prey densities because of predator satiation and extrinsic limits to predator populations. The predation curve in the graph thus represents a type III functional response.

The recruitment and predation curves in Figure 21-22 were drawn to produce three equilibrium points, which are homologous to the equilibria in Figure 21-21. The highest and lowest points represent stable equilibria around which populations are regulated; the middle equilibrium is unstable. The lower equilibrium point (A) corresponds to the situation in which predators regulate a prey population substantially below its carrying capacity (*K*). The upper equilibrium (C) corresponds to the situation in which a prey population is regulated primarily by availability of food and other resources; predation exerts a minor depressing influence on population size. The implications of Figure 21-22 for such practical concerns as the control of crop pests are clear.

Predators maintain a shaky hold on prey populations at point A. If a heavy frost or an introduced disease reduced the predator population long enough to allow the prey population to slip above point B, the prey would continue to increase to the higher stable equilibrium point (C), regardless of whether the predator population recovered. To the farmer, this means that a crop pest, normally controlled at harmless levels by predators and parasites, suddenly becomes a menacing epidemic. After such an outbreak, predators could exert little control over the pest population until some quirk of the environment brought its numbers below point B, back within the realm of predator control.

Outbreaks of tent caterpillars in the prairie provinces of Canada are generally preceded 2 to 4 years earlier by a year in which the winter is abnormally cold and the spring unusually warm (Ives 1973). These conditions presumably upset the normal balance between tent caterpillars and their predators and parasites. Infestations are subsequently brought under control by several cold winters that kill most of the tent caterpillar eggs (Witter et al. 1975).

Using the predation-recruitment diagram in Figure 21–22, we can examine the consequences of different levels of predation for prey population control (Figure 21–23). Inefficient predators cannot regulate prey populations at low densities; they depress prey numbers slightly, but the prey population remains near the equilibrium level set by resources (upper-left diagram, point C). Increased predation efficiency at low prey density can result in predator control at point A (upper-right diagram).

When functional and numerical responses are sufficient to maintain high densities of predators, predation may effectively limit prey growth under all circumstances, and equilibrium point C disappears (lower-left diagram). Finally, predation may be so intense at all prey densities that the prey are eaten to extinction (lower-right diagram, no equilibrium point). We might expect this situation only in simple laboratory systems or when predator populations are maintained at high densities by the availability of some alternative, but less preferred prey (hence no switching). Indeed, many ecologists have advocated providing parasites and predators of the pest with innocuous alternative prey to enhance biological control. At the very least, the curves suggest that the position of a predator-prey equilibrium, whether it is at very low levels of prey or close to their carrying capacity, may

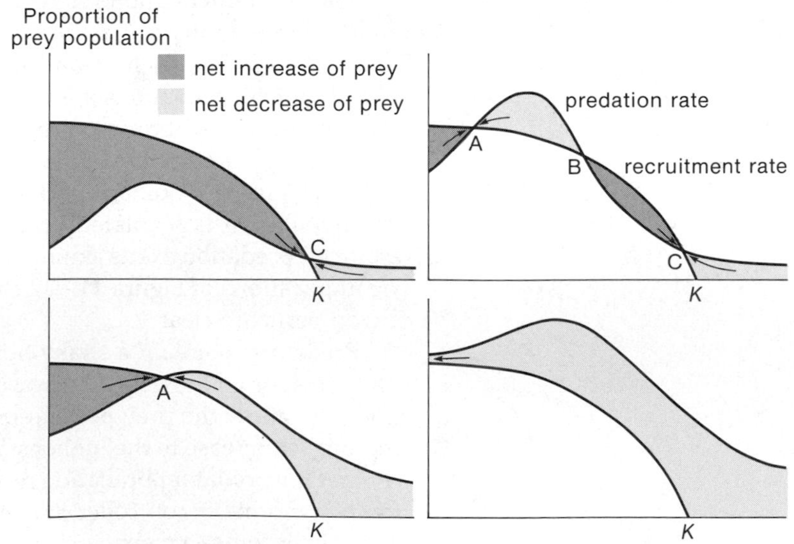

Figure 21–23
Predation and recruitment curves at different intensities of predation, showing the effect on the number of equilibria.

shift between extremes with small changes in closely matched predation and recruitment curves. Such considerations would appear to make equilibria at intermediate prey densities—that is, close to the hump of the prey equilibrium isocline—very unlikely.

Predator-prey systems achieve characteristic population ratios.

The ability of a prey population to support predators varies with its density. A small prey population can support few predators because, while each prey individual's reproductive potential may be high, the total recruitment rate of a small population is low. Prey populations near their carrying capacities also are unproductive because, although numerous, each individual's reproductive potential is severely limited by the effects of crowding.

At some intermediate density, the overall recruitment rate of the prey population reaches a maximum (Ricker 1954; Beverton and Holt 1957; Watt 1968). Because predators can remove a number of individual prey equivalent to the annual recruitment rate without reducing the size of the prey population, the prey population density that yields the maximum recruitment generally will support the greatest number of predators. This point corresponds to the peak of the hump of the prey isocline in a predator-prey graph. The rate of recruitment at this point is known as the maximum sustainable yield (see May et al. 1979).

Ranchers and game managers are clearly concerned with maintaining populations of beef cattle, deer, and geese at their most productive levels to maximize man's ability to harvest these species without reducing their populations. We may ask: Do predators also prudently manage their prey populations to maximize the productivity of their own populations? And, if so, how can such behavior evolve?

Territorial animals, which exclude competitors from their feeding areas, could indeed space themselves with respect to their prey to achieve maximum yields. When the feeding areas of predators overlap, however, intraspecific competition dictates that each predator maximizes its immediate harvest at the expense of long-term yields. Man behaves no differently. Intelligently managed ranches, with fences to exclude competing livestock, can achieve maximum sustainable yields. Alas, in highly competitive situations—fishing in international waters, to name one—man has proved to be pathetically shortsighted and imprudent, and stocks of many species of fish and other seafoods have declined dramatically (Beverton and Holt 1957; Cushing 1975). Overexploitation of whale populations has similarly led to the near extinction of some species, and has doomed the whaling industry (Laws 1962, 1977; McVay 1966).

In most cases, the level of exploitation of a prey population is determined by the ability of the predator to capture prey compared to the ability of the prey to avoid being captured. Both skills are evolved characteristics of the population. Regardless of whether predators act prudently to manage prey populations for maximum sustainable yield, they often achieve a char-

Table 21-3
The relationship between populations of predators and their prey in several localities

Locality	Predator	Principal prey	Density of predators (ind 100 mi^{-2})	Ratio between predator and prey populations	
				Numbers	Biomass
Jasper Nat'l Park[1]	Wolf	Elk, mule deer	1	1:100	1:250
Wisconsin[2]	Wolf	White-tailed deer	3	1:300	1:300
Isle Royale[3]	Wolf	Moose	10	1:30	1:175
Algonquin Park[4]	Wolf	White-tailed deer	10	1:150	1:150
Canadian arctic[5]	Wolf	Caribou	1.7	1:84	1:186
Utah[6]	Coyote	Jackrabbits	28	1:1000	1:100
Idaho primitive area[7]	Mountain lion	Elk, mule deer	7.5	1:116	1:524
Ngorongoro Crater, Tanzania[8]	Hyena	Ungulates	440	1:135	1:46
Nairobi Park, Kenya[9]	Felids	Ungulates	96	1:97	1:140
Alaska[10]	Pomarine jaeger	Lemmings		1:1263	1:90

References: (1) Cowan 1947; (2) Thompson 1952; (3) Mech 1966; (4) Pimlott et al. 1967; (5) Kelsall 1968; (6) Clark 1972; Wagner and Stoddart 1972; (7) Hornocker 1970; (8) Kruuk 1969; (9) Foster and Coe 1968; (10) Maher 1970.

acteristic equilibrium with their prey populations. The relationship between wolves and various prey populations in several areas demonstrates this equilibrium particularly well (Table 21-3). Population ratios and, particularly, biomass ratios (1 pound of wolf for each 150 to 300 pounds of prey) are relatively constant despite the fact that the species and density of the principal prey of the wolf vary considerably with locality.

Different predator-prey systems may achieve different equilibria. The population ratio of mountain lions to deer in California is 1:500-600, which is equivalent to a biomass ratio of about 1:900, and the exploitation rate is only 6 per cent, compared to values of 18 and 37 per cent for wolves in two other areas. A mountain lion-elk-mule deer system in Idaho has a biomass ratio of 1:524, and an exploitation rate of 5 per cent for the elk population and 3 per cent for the mule deer population. Evidently, wolves exploit their prey more efficiently than mountain lions, perhaps because of their social hunting habits.

Where predators feed on more abundant populations of prey, as in savanna, grassland, and tundra habitats, predators not only are more numerous, but they achieve higher biomass ratios (1:50-1:150, Table 21-3). Such conditions seem to enhance both prey productivity and predator efficiency. Large cats — lions and cheetahs — remove 16 per cent of prey biomass in Nairobi Park, Kenya, where their biomass ratio is 1 : 140.

SUMMARY

The study of predator and prey interactions attempts to answer two major questions. First, do predators reduce the size of their prey populations well below the carrying capacity? Second, do the dynamics of the predator-prey interaction cause the populations to oscillate? The first question is of great practical concern to those interested in the management of crop pests, game populations, and endangered species. It also has far-reaching implications for understanding the interactions among species that share resources and, therefore, for understanding the regulation of biological communities. The second question was motivated by observations of predator-prey cycles in nature and directly addresses the question of stability in natural systems. Ecologists have approached these problems with a combination of observation, theory, and experiment.

1. Ecologists distinguish three basic kinds of consumer: predators, which remove each prey individual from the prey population as they consume it; parasitoids, mostly small flies and wasps that kill their hosts, but only after the parasitoid larvae have pupated; and parasites and grazers, both of which consume portions of the living animal or plant organism. From the standpoint of population theory, each of these has been treated differently.
2. Experimental studies of a number of pest species and their natural predators demonstrated that, in many cases, consumers could reduce resource populations far below their carrying capacities.
3. Experimental studies demonstrated that predator and prey populations could be made to oscillate in the laboratory. Maintenance of population cycles usually requires a complex environment in which prey are able to establish themselves in refuges.
4. A. J. Lotka and V. Volterra, in the 1920s, devised simple models of predator and prey dynamics that predicted population cycles. The models used differential equations in which the rate of prey removal was directly proportional to the product of the predator and prey populations.
5. In 1933 and 1935, A. J. Nicholson and Bailey proposed alternative models in which the rate of prey removal was an asymptotic function of predator density. These models were based explicitly on the relationship between parasitoids and their hosts.
6. In 1959, C. S. Holling introduced the concept of the functional response, which described the asymptotic relationship between prey removal rate per predator and the density of the prey. Whereas the Lotka-Volterra and Nicholson-Bailey models were inherently unstable, one form of the functional response curve (type III) resulted in stable regulation of the prey population at low density.

7. The numerical response describes the response of a predator population to increasing prey density by population growth and immigration.

8. Graphical analyses of predator-prey interactions based on simple models but with such features as density dependence, prey refuges, and alternative prey added to give reality demonstrate the conditions for stability in predator-prey interactions. In general, stability is promoted by density-dependence of either the predator or prey, refuges or hiding places in which the prey can escape predation, reduced predator efficiency, and, under some circumstances, availability of alternative prey. Stable population cycles in nature apparently express the balance of these stabilizing factors and the destabilizing influence of time delays in population responses.

9. One source of time delay in parasite and grazing systems is the acquisition of immunity or other resistance. Plants may exhibit chemical responses to browsing and grazing that discourage further consumption. Time delays in these responses and their eventual loss may be responsible for cycles observed in disease outbreaks and in populations of herbivores (and the carnivores that prey on them!).

10. Models of consumers suggest that systems can have two stably regulated points between which populations may move, depending on environmental conditions. The lower equilibrium is determined by the strong depressing influence of predators on prey populations; the upper equilibrium is close to the carrying capacity of the prey in the absence of predation. Changes in climate may shift a system from one to the other of these points, resulting in successions of endemic and epidemic conditions.

11. Each prey population has a density at which harvesting that population without causing population decrease is greatest. This is called the point of maximum sustainable yield. When a single predator can completely control its prey, as man can do in the cases of many domestic and game species, maximum sustainable yields can be achieved. When predators compete for the same resources, maximization of short-term yields generally precludes the achievement of maximum sustainable yield.

22

Competition

When two or more consumers utilize the same resource, and when the levels of the resource are affected by consumption and also affect birth and death rates of the consumers, the consumers may be said to compete. One dictionary definition of competition is "a contest between rivals; a match." The connotation of direct interaction between rivals only partly applies to competition in ecological systems. As is so often the case, ecological usage more closely parallels economic usage: "The effort of two or more parties, acting independently, to secure the custom of a third party by offering most favorable terms." In the ecological context, however, the third party is rarely given a choice in the matter; "custom," or business, should read "life, energy, and nutrients"; and "most favorable terms" may be freely translated as "most efficient predatory or other acquisitive behavior." The heart of the analogy lies in the truism that the resource eaten by one consumer is not available in the same form to another, its competitor.

Competition between individuals of the same species (intraspecific competition) leads to density dependence of birth and death rates. When individuals of different species compete (interspecific competition), birth and death rates of one vary with the population density of the other and, generally, vice versa. An important distinction between intraspecific and interspecific competition is the difference in their population consequences. Whereas intraspecific competition leads to the stable regulation of population size within limits imposed by the environment, interspecific competition may cause the extinction of one of the competing populations. Short of this, competition from any one species may profoundly affect the population dynamics and carrying capacity of another, either through its effect on their mutual resources or by direct interference.

Predation has obvious effects on the prey individual and the prey population. When predators are either removed from or added to a system experimentally, prey populations usually respond so dramatically as to leave no doubt about the dynamical link between the two. But competition, especially when it results from the consumption of shared resources, is more subtle. No one individual dies because a competitor eats a potential prey item or excludes it from a foraging area. Because competition produces less obvious effects than does predation, the development of thinking about competition and its integration into our perceptions of ecological systems have been slow and tentative.

The concept of competition has nonetheless been a part of ecology almost from its beginning. Charles Darwin, who borrowed heavily from economics for insight and analogy, fully understood its implications; evolution, of course, is the expression of competition within populations between individuals having different genotypes. But Darwin also appreciated the implications of interspecific competition for populations. For example, in discussing the natural checks on populations in *On the Origin of Species,* he mentions "the prodigious number of plants which in our gardens can perfectly well endure our climate, but which never become naturalised, for they cannot compete with our native plants. . . ."

Whereas intraspecific competition was implicit in natural selection, interspecific competition left less of an imprint on Darwin, preoccupied as he was with the direct effects on populations of climate, predators, and food supplies. Charles Elton's book *Animal Ecology,* published in 1927 and the first modern, general account of ecology, similarly elevates these factors, especially the feeding (or trophic) relations among animals, to a preeminent position compared to competition. Elton discussed interspecific competition only in relation to ecological succession, where the replacement of one species by a second, having similar ecological requirements, suggested the possibility of interaction between the two. With respect to the process by which one species replaces another, Elton asked:

> Does it drive the other one out by competition? and if so, what precisely do we mean by competition? Or do changing conditions destroy or drive out the first arrival, making thereby an empty niche for another which quietly replaces it without ever becoming "red in tooth and claw" at all? Succession brings the ecologist face to face with the whole problem of competition among animals, a problem which does not puzzle most people because they seldom if ever think out its implications at all carefully. At the present time it is well known that the American grey squirrel is replacing the native red squirrel in various parts of England, but it is entirely unknown why this is occurring, and no good explanation seems to exist. And yet more is known about squirrels than about most other animals.

Elton ended this ambivalent paragraph on interspecific competition with a plea that holds true today: "There is plenty of work to do in ecology."

The mathematical treatment of population processes, including competition, by A. J. Lotka, V. Volterra, and others between 1925 and 1935 created a wave of laboratory experimental work in response (Crombie 1947). Predator-prey and host-parasitoid interactions dominated early efforts, but the Russian biologist G. F. Gause (1932, 1934), who was strongly influenced by the work of Raymond Pearl and Lotka, examined competition between species in laboratory populations of yeast and protozoa; his efforts were soon followed by work on flour beetles, fruit flies, and others amenable to experimentation. But in spite of the convincing laboratory demonstration of interspecific competition, the process played a relatively small role in ecological thinking for decades afterward. The benchmark 1949 text, *Principles of Animal Ecology,* by W. C. Allee and his colleagues in Chicago, recounted laboratory work on competition in detail, but assigned competition a minor role in the workings of natural communities.

The role of competition in determining the population sizes of species utilizing the same resources and in governing the partitioning of resources among competing species was fully recognized only in the mid-1940s, when its importance was persuasively argued by the English ornithologist David Lack (1944, 1947). Lack sought to reconcile Gause's experimental result, that two similar species could not coexist in a simple laboratory environment, with the observation that natural habitats often harbored several species with similar ecological requirements. He reasoned that the process of species formation may be accompanied by sufficient divergence in ecological requirements (hence reduced competition) between two species to permit their coexistence: "When two related bird species meet in the same region, they tend to compete, and both can persist there only if they are isolated ecologically either by habitat or food" (1947:162).

By the mid-1940s, Elton (1946) had edged toward Lack's position, but not wholeheartedly:

> We simply do not understand exactly why populations of, say, a Pentatomid bug, a grasshopper, a moth caterpillar, a vole, a rabbit and an ungulate should be able to draw upon the same common resource (grassland vegetation) and yet remain in equilibrium at any rate sufficiently to form a stable animal community over long periods of years. . . . I think it has usually been assumed . . . that the equilibrium is made possible by some specialized division of labour [Lack's point], and that the animals do not come into direct competition at all; or else that the amount of resources is generally sufficient to provide for all the populations present because they are limited by factors other than food in the increase of those populations. The second idea is on the whole supported by the general evidence that animals do not normally become limited in numbers by starvation, and that the biomass of phanerogamic vegetation [higher plants] is far beyond that of animals dependent on it [a point picked up later by Hairston, Smith, and Slobodkin (1960)]."

Elton's and others' ambivalence notwithstanding, the tide of opinion was definitely turning in favor of competition. Lack's perception of competition in natural communities became firmly established (ordained, one might say) in the doctrine of modern ecology by G. E. Hutchinson's famous "concluding remarks" paper of 1957:

> Volterra (1926) demonstrated by elementary analytic methods that under constant conditions two species utilizing, and limited by, a common resource cannot coexist in a limited system. . . . Gause (1934, 1935) confirmed this general conclusion experimentally in the sense that if the two species are forced to compete in an undiversified environment one inevita-

> bly becomes extinct. If there is a diversification in the system so that some parts favor one species, other parts the other, the two species can coexist. These findings have been extended and generalised to the conclusion that two species, when they co-occur, must in some sense be occupying different niches. The present writer believes that properly stated as an empirical generalization . . . the principal is of fundamental importance and may be properly called the Volterra-Gause principal.

After reviewing much of the literature on the coexistence of ecologically similar species, Hutchinson concluded, at least tentatively, that "animal communities appear qualitatively to be constructed as if competition were regulating their structure. . . ."

A. G. Tansley provided the first experimental demonstration of competition.

The British botanist A. G. Tansley (1917) was the first to devise an experiment to determine the existence of competition between closely related species. Tansley prefaced his report with the observation that closely related species occurring in the same region often grow in different habitats or on different types of soil. The observation was not new nor was the suggestion that the ecological segregation of such species might have resulted from competition, leading to exclusion of one species or the other, depending on local ecological conditions. But no one had experimentally investigated the correctness of that hypothesis, or of its alternative, that two species had such different ecological requirements that each could not grow where the other flourished.

Tansley selected a pair of species of bedstraw (genus *Galium,* in the madder family [Rubiaceae]), which are small, perennial, herbaceous plants. One species, *G. saxatile,* normally lives on acid, peaty soils whereas the other, *G. sylvestre,* inhabits limestone hills and pastures. These he planted as seeds, both singly and together, in soils taken from areas on which each species grows. Because the seeds were planted together in a "common garden," only soil type and presence or absence of the other species were experimental treatments.

Tansley's experiments were plagued by such technical problems as poor germination and lapses in watering. His results were nonetheless quite clear. When planted singly, each of the species grew and maintained itself on both types of soil, although germination and growth were most vigorous on the soil on which one normally encounters the species in nature. When grown together on calcareous (limestone) soils, *G. sylvestre* plants overgrew and shaded out those of *G. saxatile.* The reverse occurred on the more acid, peaty soil typical of *G. saxatile* habitat. Tansley concluded:

> In the case investigated, the calcifuge [absent from calcareous soils] species *(Galium saxatile)* is heavily handicapped, especially in the seedling stage, as

> a direct effect of growing on calcareous soil, and is thus unable to compete effectively with its calcicole [confined to calcareous soil] congener, *Galium sylvestre.* The calcicole species is handicapped as a result of growing on acid peat and is therefore reduced to a subordinate position in competition with its calcifuge rival, which is less handicapped. Both species can establish and maintain themselves — at least for some years — on either soil. If these results are of general application they would explain the observed distribution in the case of other similar pairs of species, viz. that they are "bodenstet" [German, meaning narrowly restricted with respect to soil type] where both members of the pair occur and "bodenvag" [broadly distributed with respect to soil type] where only one occurs. Where however the handicapping is very severe, as in the case of *G. saxatile* on calcareous soils, it is unlikely that seedlings germinating on such soils would survive the general competition of the other vegetation even in the absence of plants of the congeneric competitor, and this would explain the absence of *G. saxatile* on calcareous soils in this country outside the area of distribution of *G. sylvestre.*

Thus in this brief paper Tansley put on record (1) that the presence or absence of a species could be determined by interspecific competition; (2) that the conditions of the environment affected the outcome of competition; (3) that competition might be felt very broadly (that is, from "other vegetation") throughout the community; and (4) that the present ecological segregation of species might have resulted from competition in the past. But although Lack's and Hutchinson's concepts contain little that was not presaged by Tansley, further experimentation on competition did not come until after the publication of mathematical treatments of population interactions by Lotka, Volterra, and others.

Setting 1930, or thereabouts, as the dividing point between the earlier naturalistic tradition and the later population tradition in ecology, one must consider Gause's (1932, 1934) experiments on yeast and protozoa as the first modern research on competition. Gause's protocol, essentially identical to that of Tansley, was to grow two species, both singly and together, under controlled conditions. The difference in the population growth of one species in the presence and absence of the other was a measure of competition between them.

As shown in Figure 22–1, when the protozoans *Paramecium aurelia* and *P. caudatum* were established separately on the same type of nutritive medium, both populations grew rapidly to limits eventually imposed by resources. When grown together, however, only *P. aurelia* persisted. Similar experiments with fruit flies, mice, flour beetles, and annual plants have always produced the same result: one species persists and the other dies out, usually after 30 to 70 generations (Miller 1967).

Thomas Park's (1954, 1962) work on competition between two species of flour beetles demonstrated another of Tansley's points, that the outcome of competition depends on the conditions of the environment. When Park established *Tribolium castaneum* and *T. confusum* together in vials of wheat flour under cool, dry conditions, *T. castaneum* usually excluded *T. confusum.* Under warm, moist conditions, however, it was *T. confusum* that persisted

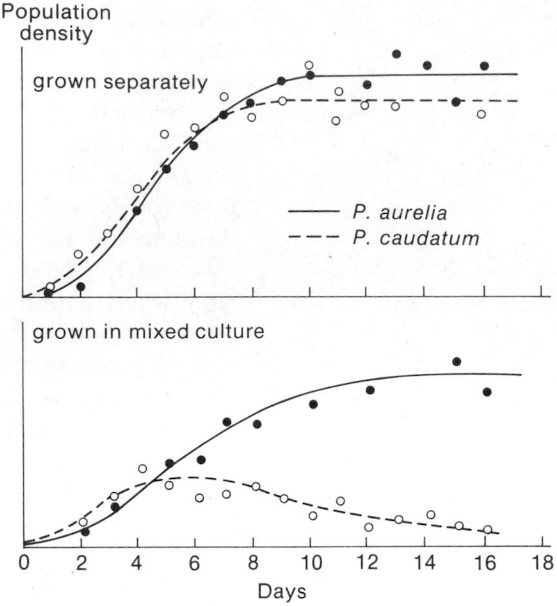

Figure 22–1
Increase in populations of two species of *Paramecium* when grown in separate cultures *(above)* and when grown together *(below)*. Although both species thrive when grown separately, *P. caudatum* cannot survive together with *P. aurelia.* (After Gause 1934.)

(Table 22–1). The experiments showed not only that physical conditions were critical to the outcome of competition, but also that only the relative performance of the two species mattered. *T. castaneum* consistently achieved higher population densities under moist than dry conditions when grown alone, but just as consistently was excluded under moist conditions by *T. confusum,* whose populations when grown alone were yet higher.

Table 22–1
Competition between two species of flour beetles of the genus *Tribolium* at different temperatures and relative humidity

		Equilibrium population size*		Per cent of contests† won by	
Temperature	Humidity	*T. confusum*	*T. castaneum*	*T. confusum*	*T. castaneum*
Cool	Dry	26.0	2.6	100	0
	Moist	28.2	45.2	71	29
Moderate	Dry	29.7	18.8	87	13
	Moist	32.9	50.1	14	86
Warm	Dry	23.7	9.6	90	10
	Moist	41.2	38.3	0	100

* Number of adults, larvae, and pupae per gram of flour.
† Based on 20–30 contests in each combination of temperature and humidity.
Source: Park 1954, 1962.

The competitive exclusion principle states that two species cannot coexist on a single limiting resource.

The accumulating results of laboratory experiments on competition eventually appeared so general as to warrant their elevation to the status of a principle. It has variously been named, after its principal authors, Gause's Principle (Lack 1944) and the Volterra-Gause Principle (Hutchinson 1957), but Garrett Hardin's (1960) Competitive Exclusion Principle seems to have stuck.

Hardin traced the birth of the principle to a meeting on March 21, 1944, of the British Ecological Society, the topic of which was the ecology of closely related species. A report of the meeting published in the *Journal of Animal Ecology* (13:176, 1944), noted that "a lively discussion . . . centred about Gause's contention (1934) that two species with similar ecology cannot live together in the same place. . . . Mr. Lack, Mr. Elton, and Dr. Varley supported the postulate. . . . Capt. Diver made a vigorous attack on Gause's concept, on the grounds that the mathematical and experimental approaches had been dangerously oversimplified."

The experimental results that led to the competitive exclusion principle can be summarized thusly: two species cannot coexist on the same limiting resource. "Limiting" is required in the definition of the principle because competition expresses itself only when consumption depresses resources and thereby limits population growth. Hardin expanded the concept but shortened the statement to: complete competitors cannot coexist.

Similar species do, however, coexist in nature. As we shall see in later chapters, detailed observations always reveal ecological differences between such species, often based on subtle differences in habitat or diet preference. The principle,which states that identical species cannot coexist, and the observation, that coexisting species are never identical, together prompt us to ask, How much ecological segregation is sufficient to allow coexistence? We can begin to answer that question in this chapter by considering the mathematical approaches which Capt. Diver thought so dangerously oversimplified. We shall find, however, that the captain was justified in his criticism to the extent that no compelling answer has emerged from the development of theory itself; hence much of the present controversy in ecology. But simplified as they are, mathematical approaches to the study of competition have provided the impetus for most of the observation and experimentation on the problem over the last half-century. Modern as Tansley's thinking was in 1917, no one paid attention to competition until Volterra's models had shown what was possible.

The logistic equation can be modified to incorporate interspecific competition.

Most competition theory is based upon the formulations of Lotka (1925, 1932), Volterra (1926), and Gause (1934), who used the logistic equation for population growth as their starting point. Remember that according to the logistic equation, the rate of increase of population i is expressed by

$$\frac{dN_i}{dt} = r_i N_i \left(1 - \frac{N_i}{K_i}\right) \quad (22\text{–}1)$$

where r is the exponential rate of increase in the absence of competition and K is the number of individuals that can be supported by the environment. Intraspecific competition appears as the term N/K; as N approaches K, N/K approaches 1 and the quantity $(1 - N/K)$ approaches 0. As we have seen before, a stable equilibrium is reached when $N = K$.

Volterra incorporated interspecific competition in the logistic equation by adding the term $a_{ij}N_j/K_i$ to the quantity within the parentheses. Hence

$$\frac{dN_i}{dt} = r_i N_i \left(1 - \frac{N_i}{K_i} - \frac{a_{ij}N_j}{K_i}\right) \quad (22\text{–}2)$$

where N_j is the number of individuals of a second species (j), and a_{ij} is the coefficient of competition; that is, the effect of an individual of species j on the exponential growth rate of the population of species i. The term a_{ij} is a dimensionless constant whose value is a fraction of the effect of an individual of species i on its own population growth rate. Strictly speaking, the term N_i/K_i should have coefficient a_{ii}, but this is assumed to be 1 and left out. Although it need not be, the value of a_{ij} usually is less than 1; individuals of the same species likely compete more intensely than individuals of different species.

Because each species of a pair exerts an effect on the other, the mutual relationship between them requires two equations: one, presented above, for the effect of species j on species i; and a second equation for the effect of species i on species j,

$$\frac{dN_j}{dt} = r_j N_j \left(1 - \frac{N_j}{K_j} - \frac{a_{ji}N_i}{K_j}\right) \quad (22\text{–}3)$$

One may think of the competition coefficient a_{ji} as the degree to which individuals of species i utilize the resources of individuals of species j. That is why $a_{ji}N_i$ is divided by K_j, the carrying capacity of species j. It is the degree to which individuals of species i usurp the resources of species j—expressed by the value K_j—that determines the effect of i on j's rate of population growth.

Equilibrium of competition equations reveal conditions for coexistence under logistic competition.

If two species are to coexist, the populations of both must reach a stable size greater than 0. That is, both dN_i/dt and dN_j/dt must both equal 0 at some combination of positive values of N_i and N_j. From equations (22–2) and (22–3), we see that $dN_i/dt = 0$ when

$$1 - N_i/K_i - a_{ij}N_j/K_i = 0$$

or

$$\hat{N}_i = K_i - a_{ij}N_j \qquad (22\text{–}4)$$

Similarly, $dN_j/dt = 0$ when

$$\hat{N}_j = K_j - a_{ji}N_i \qquad (22\text{–}5)$$

The little hats over the *N*s indicate that they are equilibrium values. In the absence of interspecific competition ($a_{ij} = 0$), the equilibrium population size N_i is equal to K_i, a measure of the resources available to species *i*. Equation (22–4) shows that interspecific competition reduces the effective carrying capacity of the environment for species *i* by amount $a_{ij}N_j$; that is, in proportion to the population size and coefficient of competition of the second species.

By substituting $K_j - a_{ji}\hat{N}_i$ for $\hat{N}_j$ in equation (22–4), and doing a little algebra, we find that

$$\hat{N}_i = \frac{(K_i - a_{ij}K_j)}{(1 - a_{ij}a_{ji})} \qquad (22\text{–}6)$$

and, similarly, that

$$\hat{N}_j = \frac{(K_j - a_{ji}K_i)}{(1 - a_{ij}a_{ji})} \qquad (22\text{–}7)$$

Because competition coefficients generally are less than 1, the denominators of these equations normally will assume positive values. Therefore, the equilibrium values $\hat{N}_i$ and $\hat{N}_j$ will both be positive only when the numerators of both equations are positive; hence when a_{ij} is less than the ratio K_i/K_j and a_{ji} is less than the ratio of K_j/K_i. In the most general terms, coexistence requires that

$$a_{ij}a_{ji} < 1 \qquad (22\text{–}8)$$

A graphical representation illustrates the basic features of logistic competition.

When two species compete, the growth rate of each population depends upon the number of individuals in the populations of both species. This is shown graphically in Figure 22–2, which portrays dN_i/dt as a function

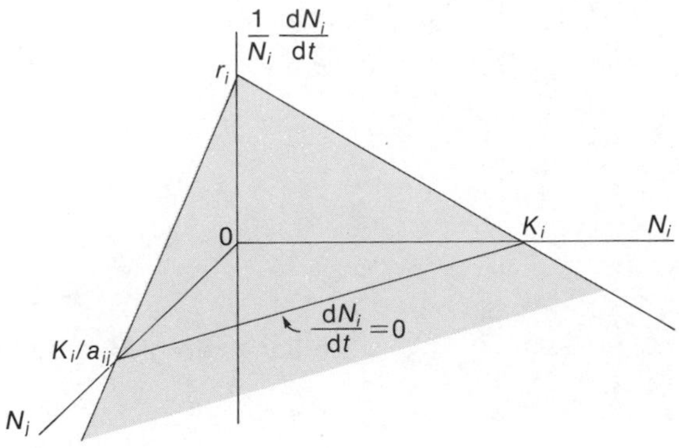

Figure 22–2
The growth rate of species i ($dN_i/dN_i dt$) as a function of its population and the size of a competing population (N_j).

of N_i and N_j. Values of dN_i/dt lie on a plane describing a single value of dN_i/dt for every combination of values of N_i and N_j. In the absence of species j ($N_j = 0$), the population of species i would, according to logistic growth, increase along the N_i axis to K_i. When the population of species i is very small (N_i close to 0, hence little intraspecific competition), dN_i/dt is positive so long as the population of the competitor j is less than K_i/a_{ij}. This intercept represents the level of population j at which its use of the resources of population i equals that of K_i individuals. This intercept is derived algebraically from equation (22–2) with the assumption that N_i is, for all intents and purposes, 0; hence $dN_i/dt = 0$ when $(1 - a_{ij}N_j/K_i) = 0$, or $N_j = K_i/a_{ij}$.

To comprehend how two species coexist, we must understand the conditions under which the populations either increase or decrease. As you can see in Figure 22–2, population growth rate dN_i/dt is positive for certain combinations of N_i and N_j, and negative for others. The dividing line between the two regions, which is called the equilibrium isocline ($dN_i/dt = 0$), is the intersection of two planes along the line defined by equation (22–4): $\hat{N}_i = K_i - a_{ij}N_j$. The equilibrium isocline for N_i can be portrayed on a two-dimensional graph the axes of which are the population sizes of species i and j (Figure 22–3). Within the region that includes the origin of the graph, population i increases; outside the region it decreases, as indicated by the arrows. Species j has an analogous equilibrium isocline, shown in the small graph in Figure 22–4.

The behavior of species i and j together depends on the relative positions of their equilibrium isoclines. When the isocline of one species lies outside that of the second along its entire length, the first is the superior competitor and it eliminates the second. In Figure 22–4, the isocline of species i lies outside that of species j. When both populations are small (point A), both increase and the trajectory of the populations jointly moves up and to the right. When both populations are large (point B), both decrease. But within the region between the isoclines of the two species j decreases (it is outside its equilibrium isocline) and species i increases (it is

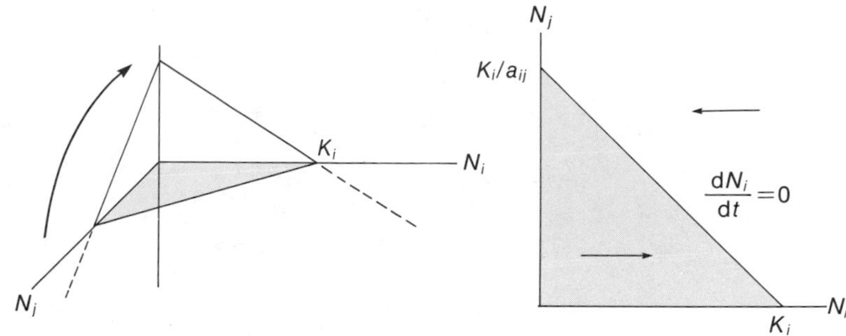

Figure 22–3
Simplification of the competition graph by considering only the plane in Figure 22–2 bound by the N_i and N_j axes. The growth rate of population *i* (dN_i/dt) exceeds 0 for combinations of populations of species *i* and *j* within the shaded area.

within its isocline). As a result, the joint trajectory of the populations moves down and to the right, eventually reaching K_i on the N_i axis ($\hat{N}_i = K_i, \hat{N}_j = 0$). Thus species *i* eliminates species *j* from the system. A simulation of the time course of such an interaction is shown in Figure 22–5, which closely resembles the outcome of Gause's experiment on *Paramecium* (Figure 22–1).

The graphical analysis of competition clearly indicates that coexistence can be achieved only when the equilibrium isoclines of species *i* and *j* cross (Figure 22–6). At the point of their joint intersection, corresponding to $\hat{N}_i$ and $\hat{N}_j$, the growth rates of both populations are 0. The joint population

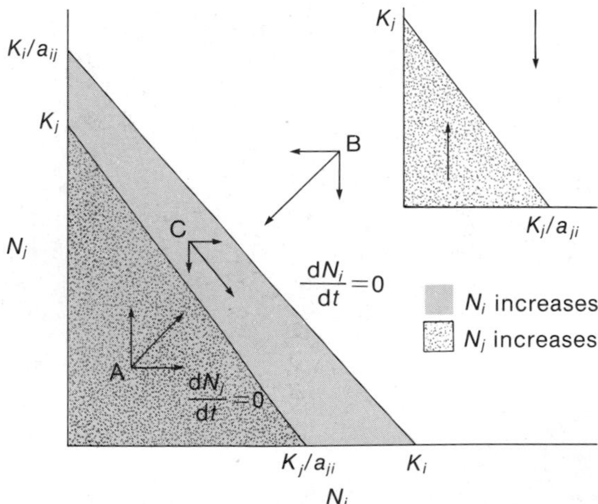

Figure 22–4
Graphical representation of the equilibrium conditions for two species where species *i* is the better competitor. The component of the population trajectory attributed to species *j* is shown in the small graph. Areas in which populations can increase are indicated by shading for species *i* and stippling for species *j*. Both species increase in region A, both decrease in region B, and *i* increases and *j* decreases in region C.

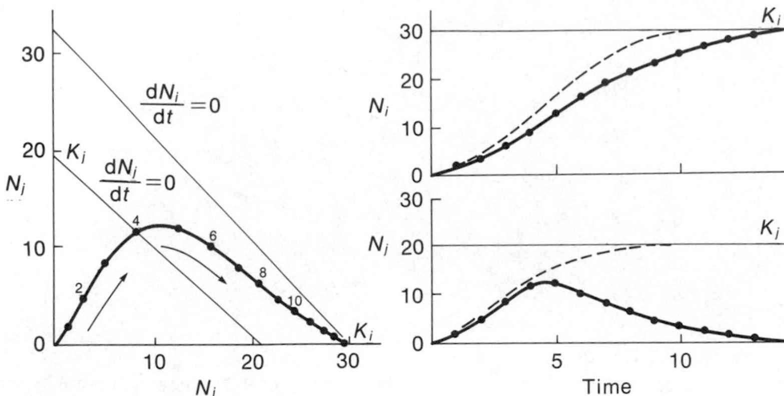

Figure 22–5
The course of competition between two populations portrayed on a competition graph *(left)* and as the change in population size with respect to time *(right)*. The time intervals are indicated on the competition graph by numerals next to the sample points. Dashed lines indicate population increase when species are grown separately.

trajectories in each of the four sections of the graph defined by the particular isoclines in Figure 22–6 show that the joint equilibrium point is stable. Accordingly, the conditions for coexistence are the inequalities $K_i < K_j / a_{ji}$ and $K_j < K_i / a_{ij}$. By rearranging these inequalities, we obtain $a_{ij} < K_i / K_j$ and $a_{ji} < K_j / K_i$, as we have seen before. Biologically, these inequalities correspond to the situation in which each species limits itself (K_i) more than it limits the other (K_j / a_{ji}). For coexistence, intraspecific competition must

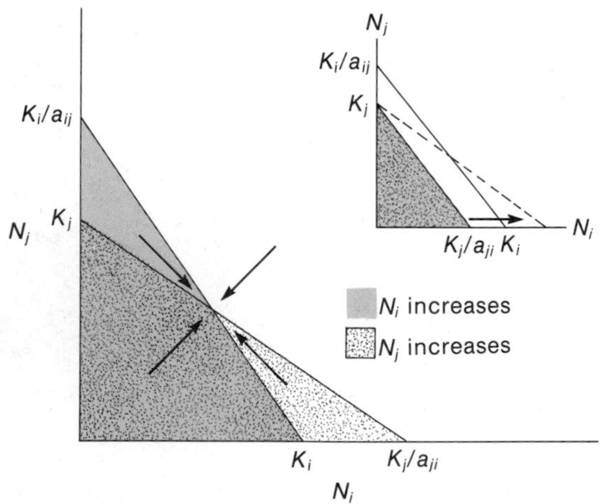

Figure 22–6
Conditions for the stable coexistence of two competing species. The graph can be obtained from Figure 22–4 by reducing the competition coefficient of species *i* (the better competitor) against species *j*, as shown in the upper right. When a_{ij} decreases, the intercept K_i/a_{ij} increases.

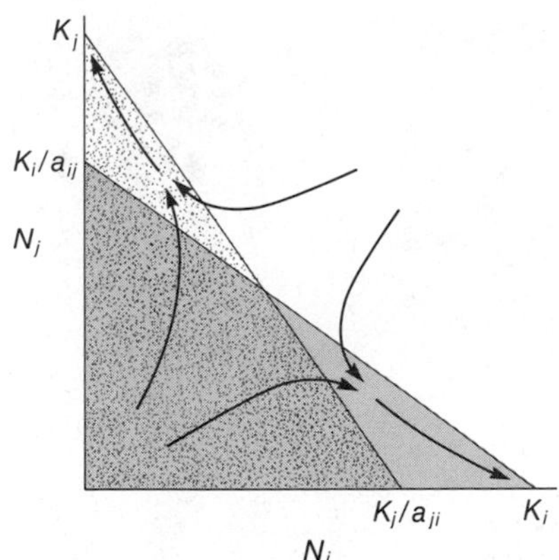

Figure 22–7
Outcome of competition between two species that are both more strongly limited by interspecific competition than by intraspecific competition (a_{ij} and a_{ji} are both large).

exceed interspecific competition (see, for example, Vance 1985). Notice also that r_i and r_j are immaterial to the outcome of competition; they affect only its time course.

What happens when both populations are more strongly limited by interspecific competition than by intraspecific competition? This situation, shown in Figure 22–7, has an unstable equilibrium point where the isoclines cross, away from which the joint population trajectories move. One species or the other prevails, depending on which initially holds the numerical advantage. This situation requires the unlikely circumstance of each species being superior to the other in a predominantly intraspecific environment (that is, when abundant) and inferior in a predominantly interspecific environment (that is, when rare). Conceivably, if each species inhibited the population growth of the other by the release of species-specific toxic chemicals, whichever species occurred initially in greater numbers would likely have a competitive edge. But such conditions seem unlikely in natural systems and ecologists have paid little attention to this unstable equilibrium.

We may estimate competition coefficients from the results of competition experiments.

In his laboratory studies of competition between pests of stored grain products, A. C. Crombie (1945) estimated values of a_{ij} and a_{ji} from changes in population size when species i and j were grown together. According to logistic interspecific competition,

$$\frac{dN_i}{dt} = r_i K_i \left(1 - \frac{N_i}{K_i} - \frac{a_{ij} N_j}{K_i}\right) \tag{22–9}$$

The term a_{ij} may be estimated when r_i and K_i are known from the rate of change in population i. Rearranging equation (22–9) gives

$$a_{ij} = \frac{(K_i - N_i)}{N_j} - \frac{dN_i}{dt} \frac{K_i}{r_i N_i N_j} \tag{22–10}$$

Crombie obtained values of r_i and K_i from populations of species i grown without j. For competition between the beetle *Rhizopertha dominica* (species R) and the moth *Sitotraga cerealella* (species S), Crombie estimated $r_R = 0.05$ and $r_S = 0.10\ d^{-1}$, $K_R = 338$ and $K_S = 200$, and $a_{RS} = 1.0$ and $a_{SR} = 1.3$. Because the product $a_{RS}a_{SR}$ exceeded 1, we would not have expected the species to coexist. Furthermore, because K_R exceeded K_S/a_{SR}, *Rhizopertha* should have outcompeted *Sitotraga,* as indeed it did (Figure 22–8).

Rhizopertha and *Sitotraga* resemble each other ecologically in that the larvae of both species burrow into grains of wheat and feed upon the germ. Because the larvae are aggressive toward one another, we should not be surprised that their interspecific competition coefficients are high. Crombie also investigated competition between these species and a second species of

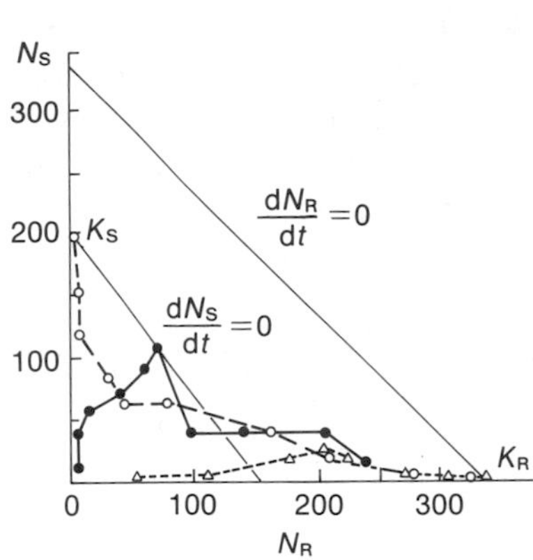

Figure 22-8
Population trajectories of competing grain pests, the beetle *Rhizopertha* (N_R) and the moth *Sitotraga* (N_S) in renewed wheat cultures with different initial densities of the two species. (From Crombie 1945.) The photograph shows a similar situation of rice weevils infesting grains of rice. (Courtesy U.S. Dept. Agriculture.)

beetle, *Oryzaephilus surinamensis.* Unlike the first two, however, larvae of *Oryzaephilus* feed on the outside of the wheat grain. Accordingly, competition coefficients between *Rhizopertha* and *Oryzaephilus* are low (about 0.2 and 0.1) and the two species coexist with little mutual inhibition (Figure 22-9).

When two species coexist, competition coefficients may be estimated for logistic competition by

$$a_{ij} = \frac{K_i - \hat{N}_i}{\hat{N}_j} \tag{22-11}$$

(equation [22-4] rearranged). Attempts to reconcile the outcomes of competition experiments with logistic competition equations have, however, pointed to some of the oversimplifications that worried Capt. Diver. For example, when Francisco Ayala (1970) grew two species of fruit flies *(Drosophila)* together, his estimates of competition coefficients seemed to conflict with Volterra's simple model. *D. pseudoobscura,* from the western United States, and *D. serrata,* from New Guinea, coexist indefinitely in laboratory cultures at 23°C. At higher temperatures, *serrata* replaces *pseudoobscura;* at lower temperatures, *pseudoobscura* persists at the expense of *serrata.* The sizes of populations of each species maintained separately and together are summarized in Table 22-2. Entering these values for *K*s and $\hat{N}$s into equa-

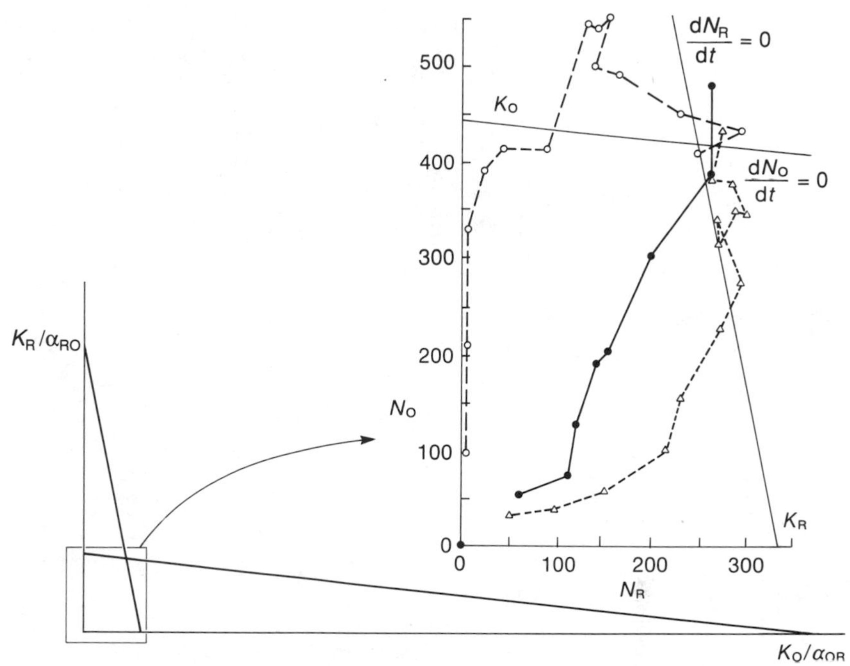

Figure 22–9
Population trajectories of competing beetles *Rhizopertha* (N_R) and *Oryzaephilus* (N_O) in renewed wheat cultures, with different initial densities of the two species. (From Crombie 1945.)

tion (22–11), we obtain estimates of $a_{ps} = (K_p - \hat{N}_p)/\hat{N}_s = (664 - 252)/278 = 1.49$, and similarly $a_{sp} = 3.86$. The product $a_{ps}a_{sp} = 5.75$ violates the general condition for coexistence, yet neither species outcompetes the other!

Rather than refuting a general principle of competition theory, Ayala's result shows that simple linear models, like the logistic equation, may not

Table 22–2
Equilibrium population sizes in competition experiments in which *Drosophila pseudoobscura* coexists with *D. serrata*

	Mathematical expression according to logistic equation	Adult population size
Species raised separately		
*D. pseudoobscura**	$N_p = K_p$	664
D. serrata	$N_s = K_s$	1251
Species raised together		
D. pseudoobscura	$\hat{N}_p = K_p - a_{ps}\hat{N}_s$	252
D. serrata	$\hat{N}_s = K_s - a_{sp}\hat{N}_p$	278
		Total = 530

* Arrowhead chromosome arrangement. Similar experiments with the Chiricahua chromosome arrangement had nearly identical outcomes.
Source: Ayala 1970.

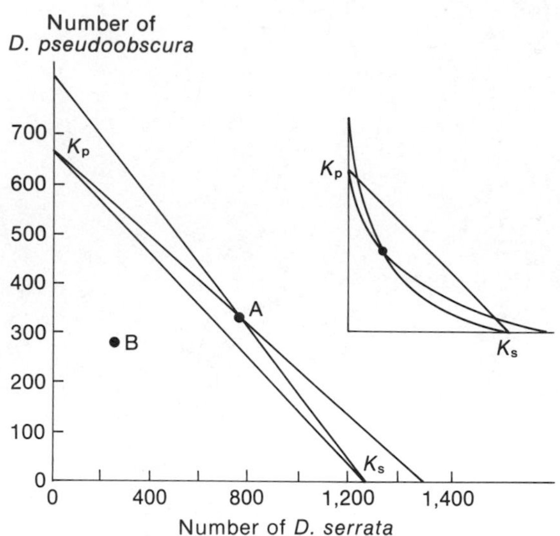

Figure 22–10
Comparison between a prediction of the logistic competition model for coexisting species (A) and the outcome of an experiment with coexisting species of *Drosophila* (B). When coexistence occurs, the model predicts that the equilibrium population (point A) must lie outside a line connecting the two carrying capacities (K_S and K_P). Ayala's result is indicated by point B. Modification of the graphical model to accommodate point B is shown in the small graph.

adequately describe population dynamics in some situations (Law and Watkinson 1987). Ayala's result can be reconciled with the conditions for coexistence ($K_i < K_j / a_{ji}$) by bending the equilibrium isoclines into the curves shown in Figure 22–10 (Gilpin and Justice 1972, 1973). Such bending could result when each species competes intensely with the other when rare, but less so when each is close to its own carrying capacity. Imagine the situation in which there are two resources, A and B. Each species feeds on both, but species i uses resource A much more efficiently than does species j, and specializes on it when rare; the reverse applies to species j and resource B. When i is rare and j is abundant, most of the competition between them results from their utilization of resource A, and i is the better competitor. When j is rare and i is abundant, competition centers upon resource B, and j is the better competitor (Ayala 1970, 1971; Gilpin and Ayala 1973).

David Tilman represented two-species competition on graphs relating population change to resource availability.

Logistic competition theory is based on the dynamics of the consumer populations involved; it does not explicitly consider changes in resources

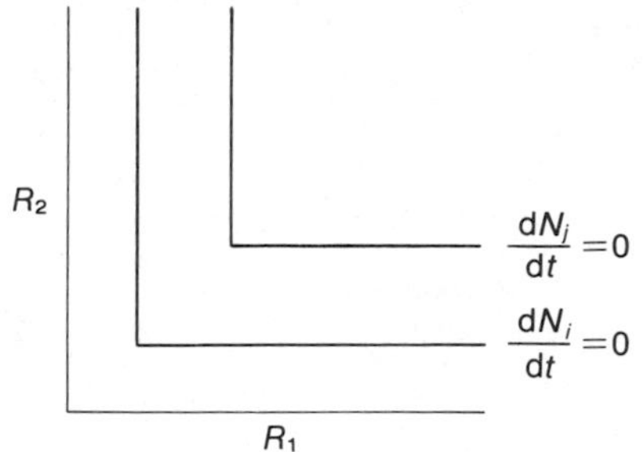

Figure 22–11
Tilman's portrayal of competition between two species in which one (i) can maintain population growth at lower levels of both limiting resources than can the second (j). In this case, species i wins the competition and j is excluded from the system. (From Tilman 1982.)

utilized by the competitors. As we have seen, Tilman (1982) treated the regulation of population size from the standpoint of resource dynamics. Two species of consumers competing for two essential resources are portrayed in Figure 22–11. Here, species i can increase at lower levels of both resources than species j; that is, species i requires less of both resources. In this situation, i outcompetes j.

Coexistence may occur when the equilibrium isoclines, what Tilman calls the Zero Net Growth Isoclines, or ZNGI for short, cross, as in Figure 22–12. But two additional factors influence the nature of the joint equilibrium. The first includes the rates of consumption of each of the resources by each of the species. These are indicated by the vectors C_i and C_j projecting down and to the left through the joint equilibrium. The second is the position of the supply point of the resource, which is the point in the graph representing the levels of each of the essential resources in the absence of the consumers.

Two conditions are required for stable coexistence. The first is that each species must consume relatively more of the resource that limits its growth at equilibrium. In Figure 22–12, species i is limited by resource 2 at the joint equilibrium point; because C_i indicates that species i uses resource 2 more rapidly than resource 1 at this point, this condition is met. The second is that the resource supply point be located in the region between the consumption vectors (shaded in Figure 22–12). Outside of this region, one species has an

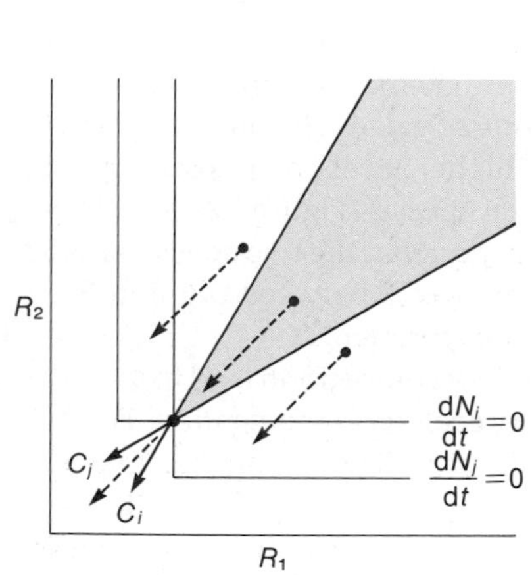

Figure 22–12
Tilman's portrayal of competition in which the Zero Net Growth Isoclines cross at a stable equilibrium point, shown with a dot. The point is stable because each species consumes relatively more of the resource that limits its growth at the equilibrium. (From Tilman 1982.)

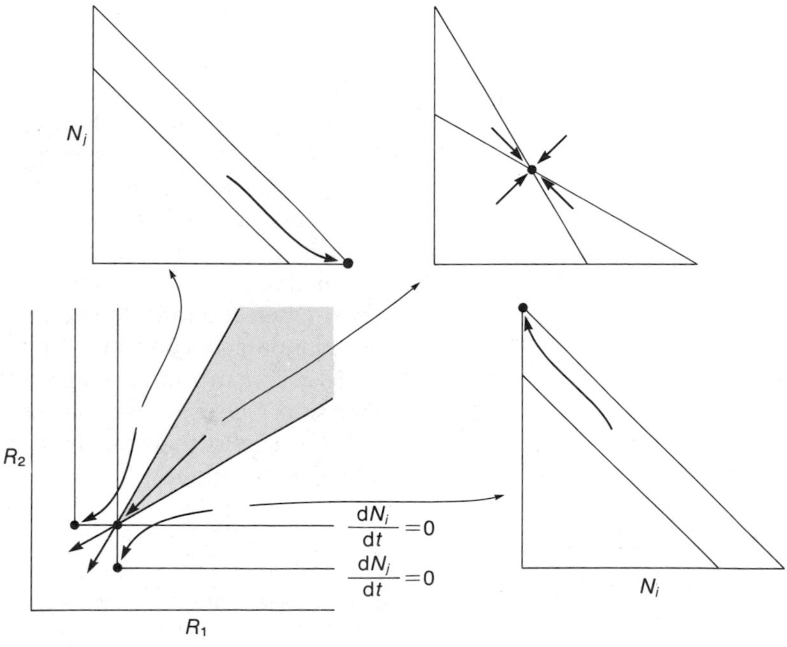

Figure 22–13
Comparison of Tilman's competition graphs with graphs showing the outcome of Lotka-Volterra competition. Whether one species excludes the other, or both coexist, depends on the initial conditions of availability of the two limiting resources.

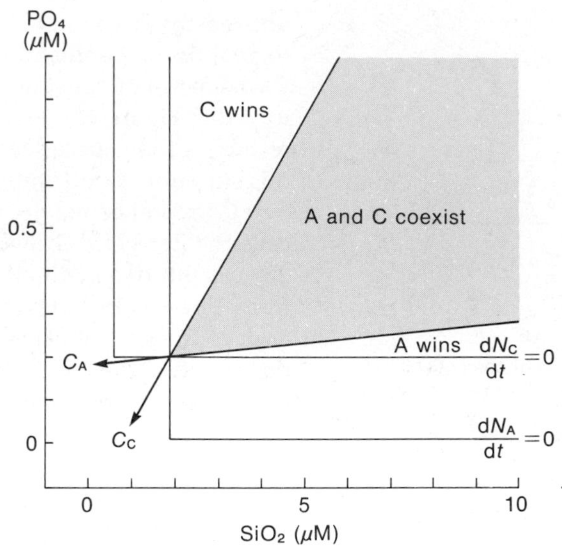

Figure 22–14
Observed Zero Net Growth Isoclines and consumption vectors for the diatoms *Asterionella* and *Cyclotella* predict stable coexistence between the two species. (After Tilman 1982.)

overwhelming advantage over the other because its limiting resource is supplied at a much greater rate than the limiting resource of the other species. (Figure 22–13 illustrates the correspondence between Tilman's graph and the ones for logistic competition.)

Tilman tested his model by growing two species of diatoms, *Asterionella formosa* and *Cyclotella meneghiniana,* in chemostats under controlled rates of nutrient supply. *Asterionella* requires relatively higher levels of silicon and *Cyclotella* requires relatively higher levels of phosphorus, as shown by the position of the ZNGI for each species (Figure 22–14). These combined with the consumption vectors suggest that the two species should coexist when the ratio of silicon to phosphorus is between 5.6 and 97, a prediction largely confirmed by chemostat experiments.

Both Tilman's and the logistic portrayals of competition lead to a similar conclusion. Coexistence is made possible by ecological segregation. For two species to persist together, each must excel at utilizing some requirement that limits the other. When two species are forced to compete for the same resource in laboratory cultures, one inevitably excludes the second. When species are limited by different resources and cultured under heterogeneous conditions or under ratios of resources that are limiting to both, they can co-occur without either one's gaining a competitive advantage.

No natural system is as simple as the two-species microcosm concocted in the laboratory. Natural environments are heterogeneous over space and variable over time. Competition extends beyond pairs of species. Consumers also are consumed. With the essentials of competition firmly established by simple theory and laboratory experiment, and with the growing sense in the 1950s and 1960s of the potential importance of competition in regulating the structure of biological communities, ecologists turned their attention to demonstrating the existence of competition in natural settings, which is the topic of the next chapter.

SUMMARY

1. Competition is the utilization or contesting of a resource by more than one individual consumer. When the individuals belong to the same species, their interaction is called intraspecific competition; when they belong to different species, it is called interspecific competition. Intraspecific competition is expressed demographically as density dependence and may lead to the regulation of population size. Interspecific competition is expressed as a reduction in the carrying capacities of competing populations and, in its extreme, may lead to the exclusion of a species.

2. Competition was recognized as a fundamental ecological process by Darwin, who based his theory of evolution by natural selection on competition between individuals having different genotypes. Modern studies on competition date from the theoretical investigations of V. Volterra and A. J. Lotka during the 1920s and early 1930s. During subsequent decades, interspecific competition was a fashionable subject for laboratory investigation, but not until the 1940s and 1950s did David Lack, G. E. Hutchinson, and others recognize the potential role of interspecific competition in regulating the structure of biological communities.

3. Competition may be demonstrated in the laboratory and field by change in the population size of one species following the addition or removal of another. When two species compete strongly, the population of the first is sensitive to changes in numbers of the second, and vice versa.

4. Thomas Park's experiments on flour beetles *(Tribolium)* showed that the outcome of competition depends on the conditions of the environment for any given pair of species.

5. Volterra's theoretical investigation and the laboratory studies that it stimulated led to the generalization that no two species of competitors could coexist on the same limiting resource. This has come to be known as the Competitive Exclusion Principle.

6. Volterra's and Lotka's mathematical treatments of competition were based on the logistic equation of population growth in which the term N/K represents the intensity of intraspecific competition. Competition between species i and j is incorporated into the equation for the population growth rate of species i ($\mathrm{d}N_i/\mathrm{d}t$) by the analogous term $a_{ij}N_j/K_i$. The term a_{ij} is called the coefficient of competition and expresses the effect of individuals of species j on the growth rate of population i. The dynamics of population i are thus described by

$$\frac{\mathrm{d}N_i}{\mathrm{d}t} = r_iN_i\left(1 - \frac{N_i}{K_i} - \frac{a_{ij}N_j}{K_i}\right)$$

7 The joint equilibrium of species i and j ($dN_i/dt = 0$, $dN_j/dt = 0$) is described by

$$\hat{N}_i = \frac{K_i - a_{ij}K_j}{1 - a_{ij}a_{ji}}$$

and the analogous equation for species j. In the most general terms, coexistence ($\hat{N}_i > 0$, $\hat{N}_j > 0$) requires that $a_{ij}a_{ji} < 1$.

8. The mathematical description of logistic competition may be portrayed on a graph whose axes are the sizes of populations i and j (that is, N_i and N_j). Each combination of these values describes a point at which dN_i/dt and dN_j/dt are either positive, zero, or negative. The relative positions of the equilibrium isoclines (lines describing combinations of N_i and N_j for which dN_i/dt or $dN_j/dt = 0$) determine whether two competitors will coexist, or one will exclude the other.

9. The competition coefficient, a_{ij}, can be estimated from the dynamics of competing populations or, if two populations coexist, from their equilibrium population sizes in the presence and absence of each other. These estimates depend on the assumption of linearity implicit in logistic competition.

10. Competition between species can be understood in terms of David Tilman's analysis of resource dynamics. The outcome of competition depends on the relative positions of the Zero Net Growth Isoclines of each species, the vectors of resource consumption, and the relative supplies of each resource in the absence of consumption (the resource supply point).

23

Competition in Nature

Theory usually is kept simple. Most models extract a single one of the many potential interactions between species. Laboratory experiments also are designed with simplicity in mind. Indeed, experiments are often intended to create a direct analog of a mathematical system, using organisms in the place of symbols. The degree to which the mathematical and biological constructions match is a measure of the adequacy of the theory. By and large, logistic competition theory and the behavior of mixed species populations in the laboratory match closely. Exceptions, such as the results of Ayala's experiments on fruit flies, indicate a higher degree of complexity than that built into the model.

But if exceptions can arise in the laboratory, we should question the value of simple theory as a representation of the real world. Competition certainly can be made to occur in the laboratory. But does it have similar apparency in nature? Or are the different ecological requirements of different species sufficient to reduce competition between them to insignificance? Hairston, Smith and Slobodkin (1960) echoed Elton's reservations about competition when they suggested that herbivores were limited by predators rather than resources, and therefore did not compete among each other strongly. Do predator-prey interactions, mutualisms, and the responses of populations to the physical environment reduce the expression of competition to the point that it is drowned out by the noise of the natural world? Is competition important among some groups of organisms, but not among others?

Elimination of species following introduction of competitors demonstrates the population effects of competition.

Competitive exclusion is a transient phenomenon. The evidence of exclusion having taken place is lost when the poorer competitor disappears. We can observe competitive exclusion in the laboratory because we can mix populations according to whim and follow the course of their interaction. The closest natural analogy to the laboratory experiment is the accidental or intended introduction of species by man.

When many species of parasites have been introduced simultaneously to control a weed or insect pest, the control species are brought together in the same locality to prey on, or parasitize, the same resource. We should not be surprised that competitive exclusion has occurred under these conditions.

Between 1947 and 1952, the Hawaii Agriculture Department released 32 potential parasites to combat several species of fruit pests, including the Oriental fruit fly (Bess et al. 1961). Thirteen of the species became established, but only 3 kinds of braconid wasps proved to be important parasites of fruit flies. Populations of these species, all closely related members of the genus *Opius*, successively replaced each other from early 1949 to 1951, after which only *Opius oophilus* was commonly found to parasitize fruit flies (Figure 23–1). As each parasite population was replaced by a more successful species, the level of parasitism of fruit flies by wasps also increased, suggesting superior competitive ability.

A similar pattern of replacement involving wasps that parasitize scale insects has been more thoroughly documented in southern California (DeBach and Sundby 1963; DeBach 1966). Scale insects are pests of citrus groves, capable of causing extensive damage to the trees. As the evolution of resistance by pests reduced the effectiveness of chemical pesticides, agricultural biologists turned to the importation of insect parasites and predators (DeBach 1974). Yellow scales have infested California citrus groves since oranges and lemons were first planted there. In the late 1800s, the red scale was accidentally introduced and has replaced yellow scale almost completely, perhaps itself a case of competitive exclusion (DeBach et al. 1978; Andrewartha and Birch 1984). Of the many species introduced in an effort

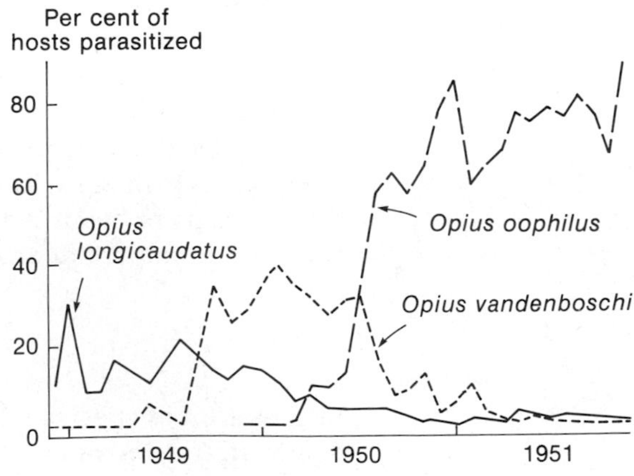

Figure 23–1
Successive change in predominance of three species of wasps of the genus *Opius* parasitic on the Oriental fruit fly. (After Bess et al. 1961.)

to control citrus scale, tiny parasitic wasps of the genus *Aphytis* (from the Greek *aphyo,* to suck) have been most successful. One species, *A. chrysomphali,* was accidentally introduced from the Mediterranean region and became established by 1900.

The life cycle of *Aphytis* begins when adults lay their eggs under the scaly covering of hosts. The newly hatched wasp larva uses its mandibles to pierce the body wall of the scale and proceeds to consume the body contents. After the wasp pupates and emerges as an adult, it continues to feed on scales while producing eggs. Each female can raise 25 to 30 progeny under laboratory conditions, and the development period is so short (egg to adult in 14 to 18 days at 27°C [80°F]) that populations may produce 8 to 9 generations per year in the long growing season of southern California.

In spite of its tremendous population growth potential, *A. chrysomphali* did not effectively control scale insects, particularly not in the dry interior valleys. In 1948, a close relative from southern China, *A. lingnanensis,* was introduced as a control agent. This species increased rapidly and widely replaced *A. chrysomphali* within a decade (Figure 23–2). When both species were grown in the laboratory, *A. lingnanensis* was found to have the higher net reproductive rate, whether the two species were placed separately or together in population cages.

Although *A. lingnanensis* had excluded *A. chrysomphali* throughout most of southern California, it still did not provide effective biological control of scale insects in the interior valleys because cold winter temperatures greatly reduced parasite populations. Wasp larval development slows to a standstill at temperatures below 16°C (60°F), and adults cannot tolerate temperatures below 10°C (50°F).

In 1957, a third species of wasp, *A. melinus,* was introduced from areas

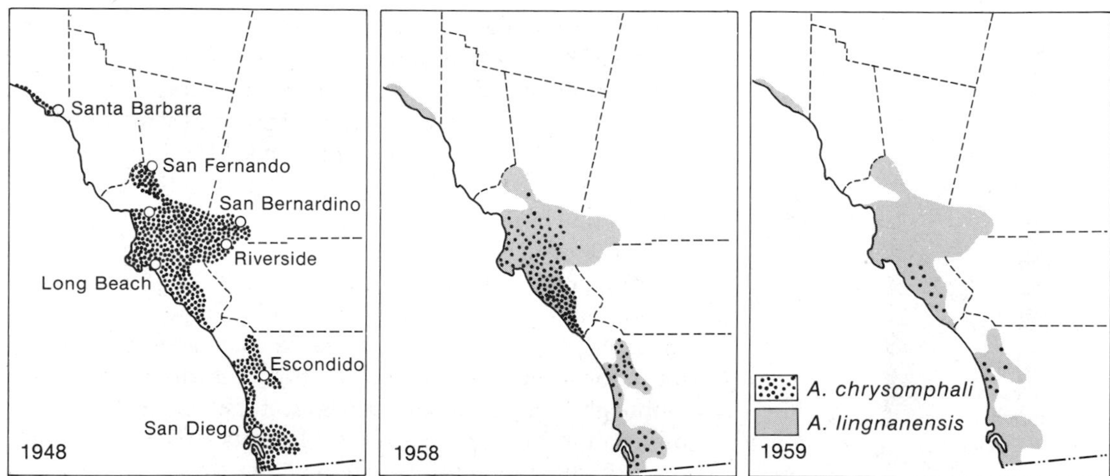

Figure 23–2
Successive changes in the distribution of *Aphytis chrysomphali* and *A. lignanensis*, wasp parasites of citrus scale, in southern California. *A. lignanensis* was first released in 1948 and rapidly replaced *A. chrysomphali* throughout the region. (After DeBach and Sundby 1963.)

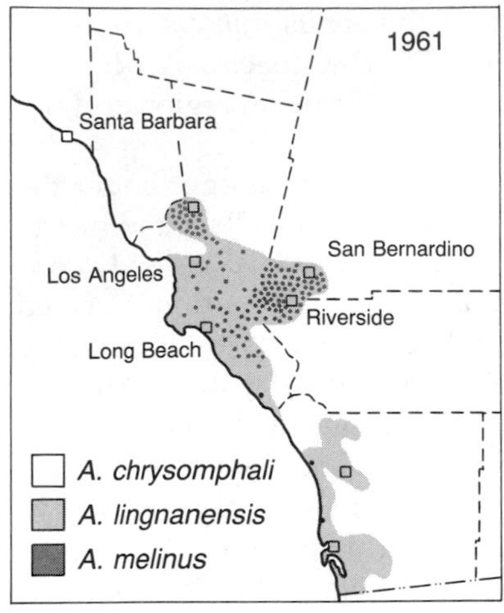

Figure 23–3
Distribution of three species of *Aphytis* in southern California in 1961. *A. melinus* predominates in the interior valley, while *A. lignanensis* is more abundant near the coast. (After DeBach and Sundby 1963.)

in northern India and Pakistan where temperatures range from below freezing in winter to above 40°C in summer. As hoped for, *A. melinus* spread rapidly throughout the interior valleys of southern California, where temperatures resemble the wasp's native habitat, but it did not become established in the more mild coastal areas (Figure 23–3). P. DeBach and R. A. Sundby (1963) demonstrated by laboratory experiment that at a temperature (27°C) and relative humidity (50 per cent) resembling the typical climate of coastal areas more closely than that of the interior valleys, *A. lingnanensis* was the superior competitor.

Removal or addition of populations have provided experimental demonstrations of competition in natural environments.

Rates of plant growth vary with respect to resource levels and therefore can provide a sensitive index to the intensity of competition (Harper 1977). The depressing effect of intraspecific competition on the growth of trees has been demonstrated in forest-thinning experiments. For example, the acceleration of growth of young longleaf pine trees in response to selective thinning of trees more than 15 inches in diameter is shown in Figure 23–4. Each core of wood was obtained by boring into a tree's trunk, from the bark to the center, with a long, tubular device called an increment borer. The core of wood removed by the borer tube provides a record of annual growth without cutting down the entire tree. These cores showed an increased growth rate, particularly in summer (light wood), during the 18 years between the time the forest was thinned and the time the cores were taken.

Similar competitive effects have been demonstrated between species of forest trees. One experiment in tropical forests of Surinam was initiated to determine whether the growth of commercially valuable trees could be improved by removing species of little economic importance (Schulz 1960). Foresters poisoned 70 per cent of undesirable trees having girths greater than 30 centimeters in one area, and greater than 15 centimeters in another, leaving the desirable species untouched. The increase in girth of the desirable trees was then measured over a year in the experimental plots and in control plots that had not been selectively thinned (Figure 23–5).

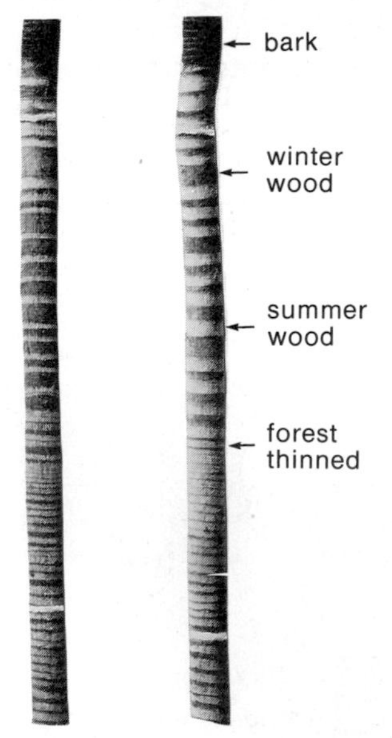

Figure 23–4
Cores of two longleaf pine trees obtained near Birmingham, Alabama, showing the effect of removing large trees on subsequent growth. (Courtesy of the U.S. Forest Service.)

Removing the larger trees increased the penetration of light to the forest floor by a factor of six. The additional light greatly stimulated the growth of trees remaining on the experimental plot. Improvement was greatest among small individuals, which had been most shaded by others before thinning; trees whose girth exceeded 100 centimeters did not grow appreciably faster. Although removal of small trees (15 to 30 cm girth) further increased light penetration by only one-third, it led to a striking response in growth rate, particularly among the remaining large trees. The improved growth could not have been caused by increased light because many of the trees that responded were much taller than the trees that were poisoned. The added growth probably resulted, therefore, from reduced competition for either water or mineral nutrients in the soil.

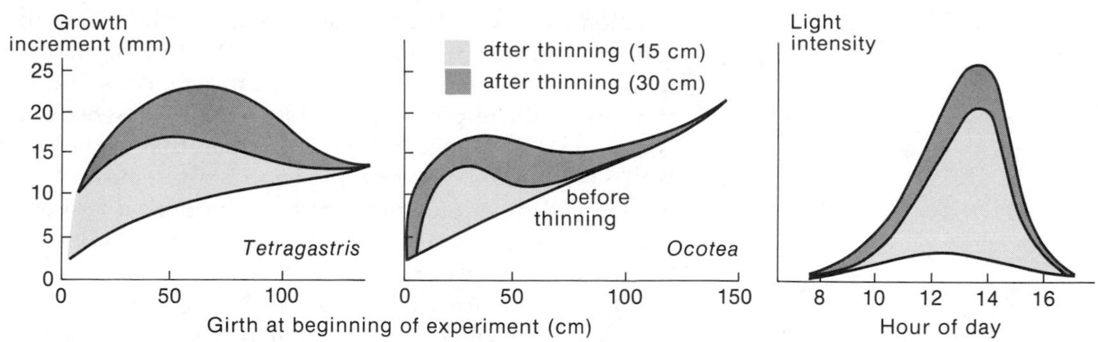

Figure 23–5
Effect on the increase in girth of two species of tropical forest trees, *Ocotea* and *Tetragastris*, achieved by removing competing trees greater than 15 centimeters or 30 centimeters in girth. The increase in light intensity that resulted from thinning is shown at right. (After Schulz 1960.)

With smaller species of plants, it is practical to study competition between species by transplanting individuals to natural sites, with or without close competitors, and following their subsequent growth and reproduction (Cavers and Harper 1967). W. G. Smith (1975) applied this technique to the study of competition between two species of *Desmodium*, which are small herbaceous legumes common in oak woodlands in the midwestern United States. Small individuals of each species, *D. glutinosum* and *D. nudiflorum*, were planted either 10 cm from a large individual of the same species (intraspecific test), 10 cm from a large individual of the other species (interspecific test), or at least 3 m from any *Desmodium* plant (control). As an index to subsequent growth, Smith measured total increase in length of all leaves, both old and new, added together.

The results of the experiment (Figure 23–6) showed that both species grew best in the absence of individuals of either species (although surrounded by unrelated plants that occur in the habitat). It was also clear, however, that the growth of *D. nudiflorum* was depressed more by interspecific competition than by intraspecific competition, and that interspecific competition was asymmetrical, with *D. glutinosum* exerting the stronger effect and *D. nudiflorum* the weaker effect.

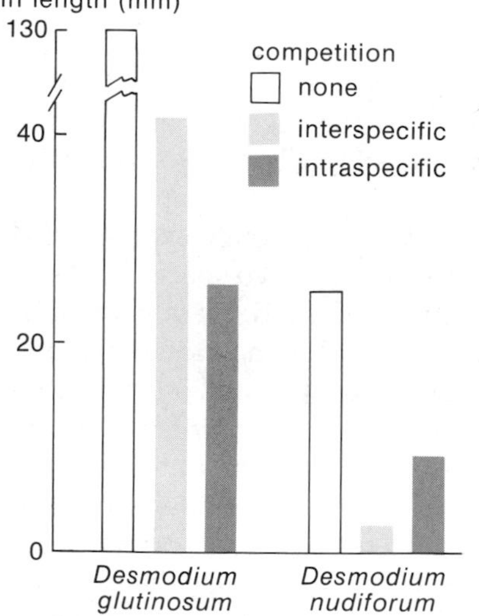

Figure 23–6
The growth responses of two species of *Desmodium* when planted near individuals of the same species, near individuals of the other species, and at a distance from individuals of either species. (After Smith 1975.)

Addition and substitution experiments have been an important tool in the study of plant competition.

In most studies of competition in nature, investigators assess the response of individuals following removal of others of the same or of different species. Addition experiments usually are more difficult because their success depends on the germination of seeds, survival of transplanted seedlings, or willingness of animals to stay where they are put. But both removal and addition experiments are limited in that, while they reveal interspecific competition, they do not allow a quantitative comparison of the relative magnitudes of interspecific and intraspecific competition. The problem is simply this: the response of individuals to the presence of competing species is measured at a higher total density than that of the intraspecific control. When individuals of competing species are removed, the performance of individuals under pure intraspecific conditions does not allow direct comparison with performance under interspecific conditions because the total density of animals or plants is reduced.

If responses to density were strictly linear, addition and removal experiments would allow one to estimate competition coefficients from the change in dN/dt with the addition or removal of individuals of competing species. According to logistic competition, one predicts a change of $-a_{ij}N_j/K_i$. But because one cannot assume that the expression of competition is linear with density, one must compare the change resulting from interspecific competition to that resulting from interspecific competition (that is, $-a_{ii}N_i/K_i$) over the same range in density. Furthermore, even when the population response (dN_i/dt) is linear with density, it is often impractical to measure population changes; instead, indices to survival and production, such as individual growth rate, seed production, or survival through a part of the life cycle, are used to determine response. These indices rarely are linear with respect to either dN/dt or density.

In order to circumvent these problems, C. T. de Wit and his colleagues in the Netherlands (de Wit 1960) developed the substitution or "replacement series" experiment, in which one keeps the total density of plants constant while varying the ratio of individuals of two species. Thus, for example, at a total density of 256 individuals per pot, a planting of 32 of species A (12.5 per cent) and 224 (87.5 percent) of species B creates conditions of strong interspecific competition for A and strong intraspecific competition for B. By varying the ratio of the species, the degree of intraspecific and interspecific competition can be varied for each, while maintaining the total intensity of competition overall.

D. R. Marshall and S. K. Jain (1969) applied de Wit's experimental design to a study of competition in wild oats *(Avena)*. Two species, *A. fatua* and *A. barbata*, which co-occur in California grasslands, were planted in pots at combined densities of 8, 16, 32, 64, 128, and 256 individuals per pot. Because one cannot follow populations in pots, Marshall and Jain recorded various indices to growth, survival, and reproduction in order to assess response to density. Under the conditions of the experiment, survival to reproduction was high and independent of density. The effects of both intraspecific and interspecific competition were expressed primarily in growth and seed production. In annual plants, such as *Avena*, that mature, set seed, and die within a single season, the number of seeds produced is most relevant to long-term population change.

When Marshall and Jain sowed their wild oats in pure populations, the number of spikelets per pot (that is, for the total planting) increased with the number of individuals initially and then leveled off for each species (Figure 23–7). No matter how many seeds were planted in each pot, the number harvested at the end of the experiment never exceeded about 500 to 600 per

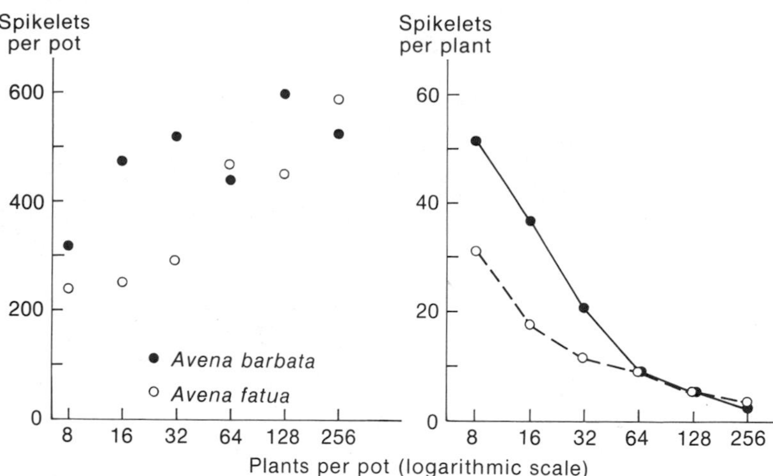

Figure 23–7
When pure cultures of two species of oats *(Avena)* were planted at different densities, the total number of spikelets per pot (a measure of reproductive output) increased *(left)*, but the number of spikelets per plant decreased *(right)*. (After Marshall and Jain 1969.)

pot. Thus each plant produced many spikelets (seed heads) per plant in low density plantings, but fewer as density and intraspecific competition increased (right-hand graph, Figure 23–7).

Marshall and Jain studied the effects of interspecific competition by planting the two species at ratios of 0 : 8, 1 : 7, 4 : 4, 7 : 1, and 8 : 0 at each of the 6 densities. The results of mixed species plantings are usually portrayed in a replacement series diagram in which the response variable (number of spikelets in this case) is plotted for each species separately as a function of the planting ratio at a particular density. Possible outcomes for different levels of intraspecific and interspecific competition are illustrated for a single species in Figure 23–8. When seed production by one species is affected equally by the density of individuals of both species ($a_{ij} = a_{ii} = 1$), the effective density for each individual of species *i* or *j* is independent of the planting ratio and the total output of *i* or *j* individuals increases in proportion to the number planted. Such a result appears as a straight diagonal line on the replacement series diagram.

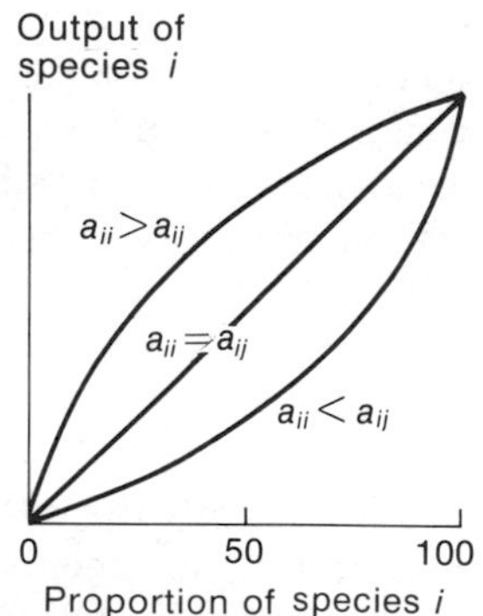

Figure 23–8
The relationship between output of one species *(i)* as a function of the proportion of that species in mixed plantings for different relationships between the coefficients for intraspecific competition (a_{ii}) and interspecific competition (a_{ij}).

When intraspecific competition exceeds interspecific competition ($a_{ii} > a_{ij}$), individuals of species *i* are more productive at lower planting ratios and the observed output of seeds per pot lies above the diagonal. When interspecific competition exceeds intraspecific competition ($a_{ij} > a_{ii}$), production rises disproportionately as the planting ratio increases (hence as interspecific competition decreases). As a result, the output of seeds lies below the line.

The results of mixed species plantings at the low and high densities of 32 and 256 plants per pot are compared to those for pure populations (no interspecific competition) in Figure 23–9. At the lower density, the numbers of spikelets produced by *A. fatua* lie above the diagonal and do not differ

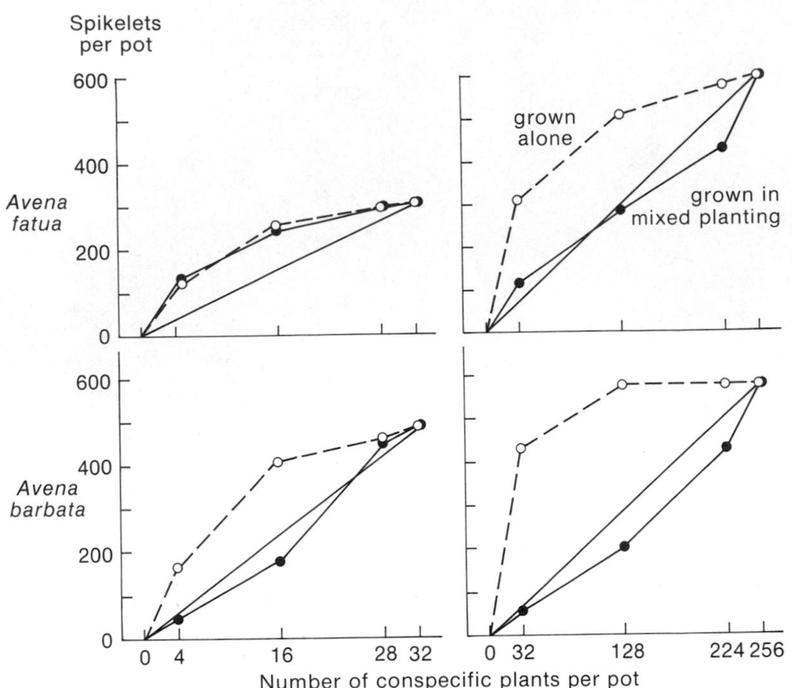

Figure 23–9
Production of spikelets by two species of oats grown in the absence (open circles) or presence (solid circles) of interspecific competition at different densities. (From Marshall and Jain 1969.)

from the numbers produced at the same density in pure populations. Thus at a density of 32 plants per pot, individuals of *A. fatua* are indifferent to the presence of individuals of *A. barbata* and behave as if subjected only to intraspecific competition. *A. barbata,* however, is influenced by competition from *A. fatua,* even at this low density. Because numbers of spikelets produced by this species lie close to the diagonal, we can infer that competition from individuals of *A. fatua* is of the same order of magnitude as intraspecific competition (that is, $a_{fb} = 0$ and $a_{bf} = 1$).

At the highest density, 256 plants per pot, *A. fatua* clearly is influenced by interspecific competition; for *A. barbata* the effects of interspecific competition even exceed those of intraspecific competition.

R. L. Hall (1974a) used a replacement series to investigate the interaction between a grass *(Chloris)* and a legume *(Stylosanthes)* in greenhouse experiments. Because the latter has symbiotic nitrogen-fixing bacteria in its roots, it should be relatively insensitive to the reduction of nitrogen availability in the soil by the grass. The results of the experiment showed that soil nitrogen was limiting overall and that *Stylosanthes* was the superior competitor owing to its unique source of nitrogen fixed from the atmosphere.

In a second set of experiments, Hall (1974b) tested two species, *Setaria anceps* and *Desmodium intortum,* on soils either deficient in potassium or with potassium added. On the potassium-deficient soil, *Setaria* was the superior competitor and potassium limited plant growth. By adding potassium, Hall reduced the competitive effect of *Setaria* on *Desmodium* to the level of the effect of intraspecific competition (a_{DS} close to 1). That potassium no longer exerted a common limit on yield of the two species was indicated by the fact that total production was greater in mixed plantings than with plantings of either species alone.

Replacement series experiments may not adequately reflect the conditions of plants in nature. Yet the range and control of densities and physical conditions that can be achieved in greenhouses points up the limitations of removal experiments in nature. Whereas one can infer interspecific competition from such field experiments, it is difficult to assess its strength relative to intraspecific competition (Connell 1983) or to investigate the nature of the limiting resource itself.

Experimental studies have revealed competition among species of animals.

Plants occupy space. In dense plantings, roots and leaves crowd together so closely that individuals constantly vie for sunlight, water, and soil nutrients. Among animals, the space-occupying invertebrates of rocky shores resemble plant systems most closely. Among the most prominent of these are the barnacles, which may form dense, continuous populations. Just as plants rely on light as a source of energy, barnacles gather food in the form of plankton in the water that washes over them. And just as plant production per unit area (or per pot) is limited by space rather than by number of plants, the productivity of barnacles is independent of the number of spats (immatures) that settle a bare surface, above a certain critical density.

One of the first experimental demonstrations of competition in the field resulted from the work of Joseph Connell (1961) on two species of barnacles within the intertidal zone of the rocky coast of Scotland. Adults of *Chthamalus stellatus* normally occur higher in the intertidal zone than those of *Balanus balanoides*, the more northerly of the two species. Although the vertical distributions of newly settled larvae of the two species overlap broadly within the tide zone, the line between the vertical distributions of adults is sharply drawn.

Connell demonstrated that adult *Chthamalus* are restricted to the portion of the intertidal zone above *Balanus* owing not to physiological tolerance limits but rather to interspecific competition. When Connell removed *Balanus* from rock surfaces, *Chthamalus* thrived in the lower portions of the intertidal zone where they were normally absent.

The two species compete directly for space. The heavier-shelled *Balanus* grow more rapidly than *Chthamalus*, and as individuals expand, the shells of *Balanus* edge underneath those of *Chthamalus* and literally pry them off the rock. *Chthamalus* can occur in the upper parts of the intertidal zone because they are more resistant to dessication than *Balanus*. So when surfaces in the upper levels are kept free of *Chthamalus, Balanus* do not invade.

Connell's work revealed most of the principles of competition outlined by Tansley and subsequently demonstrated in laboratory populations. As in Park's studies on *Tribolium,* the outcome of the interaction depended on the environment. As in Marshall and Jain's studies on *Avena,* competition between the species was asymmetrical, with *Balanus* exerting the stronger interspecific influence except where limited by physical factors.

Competition between barnacles results from physical interference rather than differential exploitation of food or other resources. But even more mobile animals may exhibit similar interference competition through occasional aggressive encounters. For example, two species of voles (small mouse-like rodents of the genus *Microtus*) co-occur in some areas of the Rocky Mountain states. In western Montana, the meadow vole inhabits both dry habitats and the wetter habitats surrounding ponds and water courses, whereas the montane vole is restricted to dry habitats. When meadow voles were trapped and removed from an area of wet habitat, montane voles began to move in from surrounding dry habitats (Koplin and Hoffman 1968). R. E. Stoecker (1972) obtained the complementary result at another site in Montana. After trapping for 9 days an area of dry habitat, which initially was occupied solely by the montane vole, meadow voles began to be caught, presumably after having moved in from more moist surrounding habitats. Each of the species excludes the other from one habitat by aggressive behavior or else avoids habitats where the other species is common. Both types of behavior, which occur widely among rodents, tend to sharpen the boundaries between the ecological distributions of closely related species (Grant 1972; Heller 1971; Merideth 1977; Sheppard 1971).

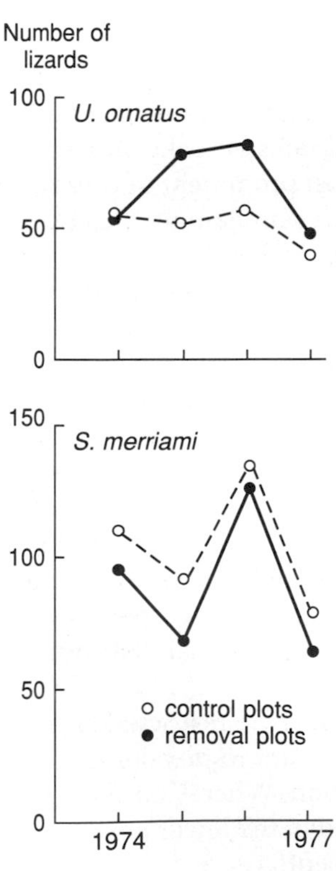

Figure 23–10
Populations of two species of lizards per hectare on plots from which one or the other of the species was removed, and on control plots where both populations remained. (From Dunham 1980.)

Although territorial defense and social aggression occur frequently within species, they are more the exception between species, where competition more usually occurs through exploitation of resources. Because exploitative competition indirectly expresses its effects through differential survival and reproduction of individuals of different species, its detection may be difficult (Belovsky 1984, Kleeberger 1984, Lenski 1984).

Arthur Dunham (1980) removed populations to investigate the interaction between two species of lizards in Big Bend National Park, Texas. In the study area, the canyon lizard *Sceloporus merriami* and the tree lizard *Urosaurus ornatus* search for insect prey on exposed surfaces of large rocks. Dunham established experimental areas from which one or the other species was removed and then compared population densities, survival, feeding rates, and body growth between these areas and undisturbed control areas over 4 years, 1974–1977.

Population densities on the control and experimental plots are shown in Figure 23–10. Where *Sceloporus* were removed, numbers of *Urosaurus* increased over controls during 1975 and 1976. There were no differences during 1977. In contrast, where *Urosaurus* were removed, *Sceloporus* populations did not respond. Once again, we find a marked asymmetry in the effect of one species on another. Fluctuations in control and experimental populations of *Sceloporus* were closely related to rainfall (1975 and 1977 were dry years), suggesting that physical factors may more severely limit

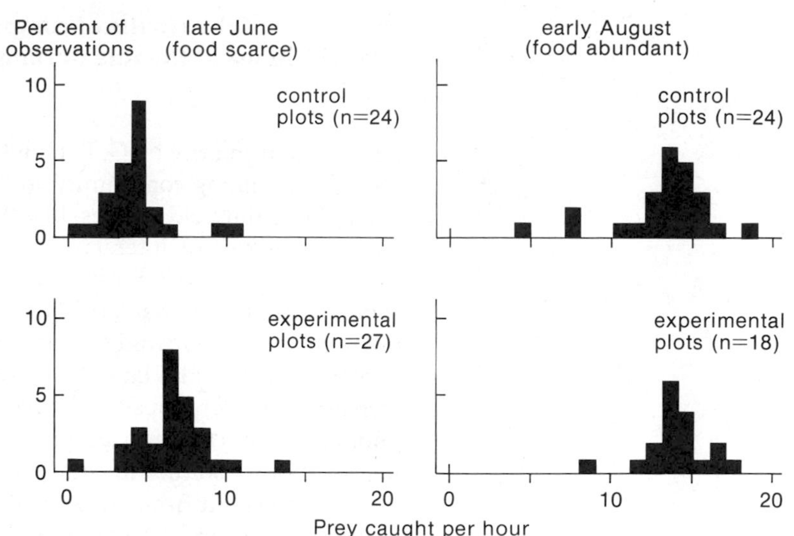

Figure 23-11
Feeding rates of adult male *Urosaurus ornatus* on experimental plots from which *Sceloporous merriami* had been removed and on unmanipulated control plots. (From Dunham 1980.)

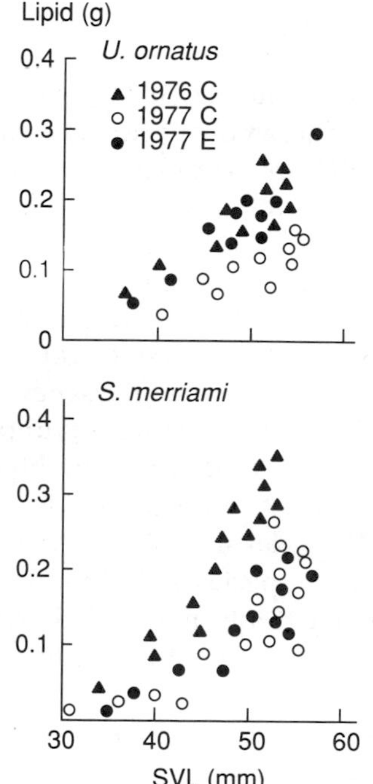

Figure 23-12
Total prehibernation lipid contents of female *Sceloporus merriami* and *Urosaurus undulatus* on experimental (E) plots from which the second species had been removed and on control (C) plots. SVL denotes snout-vent length, a measure of body size. (From Dunham 1980.)

Sceloporus while competition may more strongly influence *Urosaurus*, a situation recalling the relative ecological positions of *Balanus* and *Chthamalus* in the intertidal zone.

Differences between experimental and control populations of *Urosaurus* additionally suggest that the intensity of competition varies, perhaps depending on food resources at a particular time. For example, feeding rates of adult male *Urosaurus* were significantly higher on experimental than on control plots during a period of low food abundance, but not during a period of high food abundance (Figure 23-11). Hence the competitive effect of a second species may be expressed only during periods of low resource availability (Wiens 1977).

The ultimate measure of competition between species is its effect on survival and fecundity. Recaptures of marked individuals indicated that *Sceloporus* adults lived longer on the experimental plots, as one might expect as a result of reduced competition, but that the adult life span of *Urosaurus* decreased, perhaps owing to increased intraspecific interaction associated with higher density.

Estimating fecundity poses a problem because investigators have difficulty locating clutches of eggs. Fecundity does, however, depend strongly on the amount of fat that females store prior to hibernation for the winter. Measurements of total lipids (Figure 23-12) show that for both species of lizards, more fat was stored during 1976—a wet year—than during 1977—a dry, unproductive year. Comparisons also were made between experimental and control plots at the end of the 1977 growing season. For *Urosaurus*, but not for *Sceloporus*, removal of individuals of the other species resulted in a significant increase in prehibernation lipid, to levels matching those in 1976.

Overviews of the results of competition studies have led to different interpretations of the role of competition in biological communities.

Stimulated primarily by G. E. Hutchinson's insights into the role of competition in regulating community structure, ecologists have conducted scores of experimental field studies, like those of Connell and Dunham, to determine the degree of interaction between species in natural communities. These have been summarized and reviewed several times, with different conclusions. To Connell (1975), the results of these studies suggested that predation overshadowed competition as the more important ecological interaction. During the late 1970s, the primacy of competition came under increasing attack because of (1) the failure of some studies to demonstrate its population effects (for example, Schroder and Rosenzweig 1975); (2) the failure of other studies to show the ecological segregation between species expected to result from strong competition (for example, Wiens 1977; Rotenberry 1980); and (3) a growing lack of confidence in the statistical procedures used to demonstrate ecological segregation in natural communities (Simberloff 1978; Strong et al. 1979; Simberloff and Boecklin 1981; Connor and Simberloff 1979, 1984).

Two recent summaries of the literature have addressed the existence of competition in natural communities in comprehensive detail. These studies, one by Thomas Schoener (1983) and the other by Connell (1983), addressed only the effects of experimental removal and addition of species on potential competitors. But because of slightly different samples of studies, methods of analysis, and predispositions of the authors, somewhat different conclusions were reached. Schoener found that authors claimed evidence for competition in 90 per cent of all the studies examined, and for 76 per cent of all the species included in the studies, "indicating its pervasive importance in ecological systems." Exploitative and interference competition were about equally represented overall, but varied in frequency depending on the group considered. Schoener also found evidence to support the contention of Hairston, Smith, and Slobodkin (1960) that herbivore numbers are controlled by predation rather than by competition.

Connell examined studies involving 215 species in 527 experiments, finding evidence for competition in about half the species and two-fifths of the experiments. But he emphasized that negative results may be underrepresented in the literature. He also pointed out that investigators rarely bothered to determine the relative strengths of intraspecific and interspecific competition. Most studies are designed to examine the competitive influence of species j on species i by removing or adding individuals of species j. Rarely are individuals of species i added or removed in the same study. Where they were, Connell reported that intraspecific competition was as strong or stronger than interspecific competition in three-quarters of the experiments (for example, Underwood 1978).

Unlike Schoener, Connell found no evidence supporting the thesis of Hairston, Smith, and Slobodkin (1960). That is, plants, herbivores, and carnivores exhibited similar frequencies of competition in all habitats considered (marine, freshwater, and terrestrial). Connell further emphasized that evidence for competition appears unevenly among groups, implying that competitive interactions probably do not organize some communities, and that competitive effects vary over time, as emphasized by John Wiens (1977). He also cited several positive interactions: either functional mutualisms or interactions expressed indirectly through third species.

In this chapter we sought to determine whether interspecific competition could be identified in the field as well as under simplified laboratory conditions. To this question, we must answer definitely yes, but probably not for all groups of species, and probably to a greater or lesser degree depending on the habitat type, and probably not all the time. The results are sufficiently variable and ambiguous overall that two investigators could come to strikingly different conclusions from the same literature; clearly preconception has not been crushed under the present weight of evidence.

How does competition occur?

Besides showing that one can find evidence for competition in some of the species some of the time, and that one can fail to find evidence for competition in some of the species some of the time, field studies have also revealed the different mechanisms by which competition occurs. Park (1962) distinguished "exploitative" competition, in which individuals, by using resources, deprive others of the benefits of those resources, and "interference" competition, in which individuals cause direct harm to others by physical means (fighting, for example) and chemical means (toxins). Schoener (1983) further subdivided competition according to its mechanisms into six categories:

1. consumptive competition, based on the utilization of some renewable resource;
2. preemptive competition, based on the occupation of open space;
3. overgrowth competition, which occurs when one individual grows upon or over another, thereby depriving the second of light, nutrient-laden water, or some other resource;
4. chemical competition, by production of a toxin that acts at a distance after diffusing through the environment;
5. territorial competition; that is, the defense of space; and
6. encounter competition, involving transient interaction over a resource which may result in physical harm, loss of time or energy, and theft of food.

These mechanisms of competition are defined in terms of the capabilities of organisms and the habitats in which they occur; hence their distribution among organisms and habitats is predictably heterogeneous (Table 23–1). Preemptive and overgrowth competition appear among sessile space-users, primarily terrestrial plants and marine macrophytes and animals living on hard substrates; territorial and encounter competition,

Table 23-1
A survey of proposed mechanisms of interspecific competition in experimental field studies

	Mechanism						
Group	Consumptive	Preemptive	Overgrowth	Chemical	Territorial	Encounter	Unknown
Freshwater							
Plants	0	0	1	1	0	0	0
Animals	13	1	0	1	1	5	2
Marine							
Plants	0	6	4	1	0	0	0
Animals	9	10	6	0	7	6	0
Terrestrial							
Plants	28	3	11	7	0	1	9
Animals	21	1	0	1	11	15	6
Total	71	21	22	11	19	27	17

Source: Schoener 1985.

among actively moving animals; chemical competition, among terrestrial plants (toxins are diluted too readily in aquatic systems, but see Jackson and Buss 1975). By Schoener's count, consumptive competition is the commonest, especially in terrestrial environments; preemptive and overgrowth competition predominate in marine habitats, in part because most studies have involved sessile organisms living on hard substrates.

Of Schoener's categories, one that we have not talked about, and which merits some discussion, is chemical competition, or allelopathy (Whittaker and Feeny 1971; Rice 1984; Harborne 1982; Putnam and Tang 1986). Although the causing of injury (-pathy) to other individuals (allelo-) by chemical means has been reported most frequently in terrestrial plants, such interactions may take on a variety of forms. Thus some parasites appear to exclude other, potentially competing parasites from a host by stimulating the host's immune system against them (Schad 1966, Cohen 1973). It has also been suggested that the abundant oils in the eucalyptus trees of Australia promote frequent fires in the leaf litter, which kill the seedlings of competitors (Mutch 1970). More frequently, it is the direct effect of a toxic substance that does the damage. In an early study, A. B. Massey (1925) showed that toxic substances released into the soil by walnut trees inhibit seedling growth in other species of plants.

In shrub habitats in southern California, several species of sage of the genus *Salvia* apparently use chemicals to inhibit the growth of other vegetation (C. H. Muller 1966, 1970; Muller et al. 1968). Clumps of *Salvia* usually are surrounded by bare areas separating the sage from neighboring grassy areas (Figure 23–13). Observed over long periods, *Salvia* may be seen to expand into the grassy areas. But because sage roots extend only to the edge of the bare strip and not beyond, it is unlikely that a toxin is extruded into the soil directly by the roots. The leaves of *Salvia* produce volatile terpenes (a class of organic compounds that includes camphor and gives foods spiced with sage part of their distinctive taste) that apparently affect nearby plants directly through the atmosphere (W. H. Muller 1965; Muller et al. 1964).

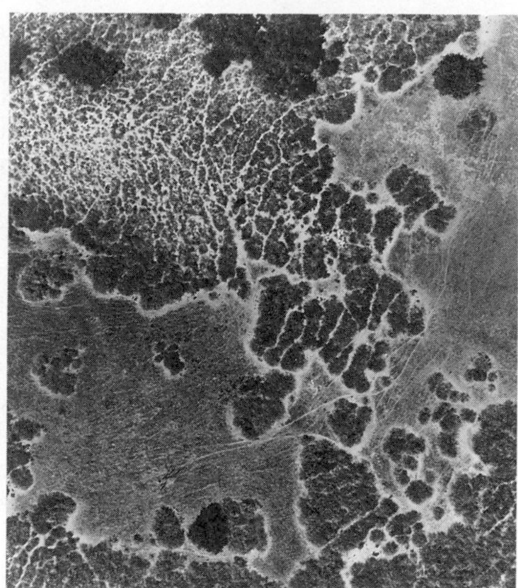

Figure 23–13
Bare patch at edge of a clump of sage includes a 2-meter wide strip with no plants (A-B) and a wider area of inhibited grassland (B-C) lacking wild oat and bromegrass, which are found with other species to the right of (C) in unaffected grassland. Aerial view shows sage and California sagebrush invading annual grassland in the Santa Inez Valley of California. (Courtesy of C. H. Muller; from Muller 1966.)

B. Bartholomew (1970) suggested that the halo zone around *Salvia* could be caused by grazing and by seed-eating birds and mammals, and that one does not require plant toxins, regardless of their efficacy, to explain the phenomenon. In order to test this hypothesis, he placed cages in the halo zone to keep out small birds and mammals; controls were constructed in the same manner, but with one of the sides of the cage left open. After 1 year, the exclosure cages contained about 20 times as much plant biomass as the control, and about the same amount as found in the grass area beyond the halo. John Harper (1977: 378) interpreted these results as suggesting that "the toxin hypothesis is unnecessary to account for the observed pattern of vegetation." J. B. Harborne (1982: 215) drew a different conclusion from the same data: "The possible role of animals, especially birds and rodents, in producing the bare zones was experimentally investigated by Muller (1970) and Bartholomew (1970), but no convincing evidence that they play a causal role could be established in numerous experiments." Both are correct in that although both toxins and herbivores have demonstrable effects on grasses, decisive experiments to show the role of one or the other (or both!) in the formation of bare zones have not been conducted.

Asymmetry is the general rule in competition.

Both Connell (1983) and Schoener (1983) referred to many cases in which one member of a species pair responded to the addition or removal of individuals of the other, but not vice versa. Among 98 reciprocal tests of competition between species, Connell tabulated no interaction for 44 pairs, reciprocal negative effects for 21 pairs, and response by only one species in 33 pairs. Schoener concluded that asymmetry was the rule in competitive interactions, rather than the exception (see Morin and Johnson 1988).

Asymmetry in competition must derive from asymmetry in ecology. Moreover, it will almost always be the case that the superior competitor is more strongly limited by some factor—environmental tolerances or predators—external to the competitive interaction. In the case of the barnacles studied by Connell (1961a, 1961b), *Balanus* could effectively exclude *Chthamalus* from the lower levels of the intertidal zone, but it lacked the physiological tolerance that allowed *Chthamalus* to exist higher up. Therefore, when *Balanus* were removed from the lower portion of the tidal zone, where they dominated, *Chthamalus* could become established, showing a strong effect of *Balanus* on *Chthamalus* (a_{CB} high). However, when *Chthamalus* were removed from rocks at high tide levels, where they dominated, *Balanus* failed to colonize (a_{BC} negligible).

Predators frequently balance strong asymmetry among competitors by their preference for the dominant species. On marine hard substrates (rocky shores, reef rubble), rapidly growing, leafy species of algae often outcompete slowly growing, encrusting forms, which they overgrow and shade out of existence. But such predators as sea urchins and herbivorous fish readily graze the leafy forms of algae while leaving the less accessible, encrusting ones untouched (Hay 1981; Hay et al. 1983; Lubchenco 1978). In this situation, competition appears asymmetric: removal of leafy algae results in the establishment of encrusting forms (an experiment possible only in the absence of herbivores); removal of encrusting forms (whose presence depends on those herbivores) does not lead to the establishment of leafy species, because these are eaten as soon as they colonize the rock surface.

Along the shores of the Caribbean, patch reefs present three ecological zones: reef flat, the topmost part of the reef covered by shallow water and frequently exposed at low tide; reef slope, the outer edge of the reef as it descends to deep water; and sandy plain, the area of sandy bottom between patches of coral reef. Because few fish can inhabit the shallow water on the reef flat, herbivory there is slight. On the sand plain, predatory fish drive away herbivorous species, which cannot find hiding places in the exposed habitat. Therefore, only the structurally complex reef slope experiences strong herbivory by fish. And, as a result, only species of algae typical of the reef slope effectively resist herbivory.

Mark Hay et al. (1983) discovered this by transplanting species of algae from each of the three habitats to locations within each habitat. As one might have expected, algae from the reef slope were virtually untouched in all the habitats, but species from the reef flat and sand plain were heavily grazed (40 to 90 per cent of their biomass consumed in 48 hours) when transplanted to the reef slope (Figure 23–14). In another experiment, Hay transplanted algae to the reef slope, where some individuals were completely enclosed in cages and others were exposed to herbivory in partial cages. The sand-plain species were grazed heavily in the partially exposed cages, but grew vigorously when protected from herbivores; reef-flat spe-

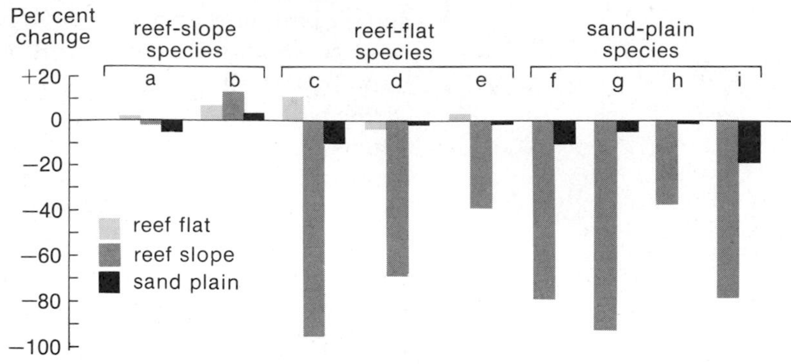

Figure 23–14
Per cent change in biomass of species of algae growing in three coral-reef habitats after transplanting to each of the other habitats. Reef-slope species resist grazing by fish; other species are heavily grazed when transplanted to the reef slope. (After Hay et al. 1983.)

cies transplanted to the same conditions did not grow well when protected, possibly because of low levels of light.

Sand-plain species of algae grow more rapidly than those from the slope and flat; they would likely exclude species from other habitats through overgrowth competition were it not for their vulnerability to herbivory on the reef slope and their inability to tolerate the physically stressful conditions of desiccation and temperature fluctuation on the reef flat (Hay 1981). Reef flat species form a dense turf of algae, resistant to both herbivory and dessication: the price of this adaptation is slow growth (Hay 1981). These examples illustrate that asymmetric competition results when species that share one resource (space or food) are differently specialized with respect to other aspects of the environment (physical conditions or consumers).

Competition may occur between unrelated species.

Darwin emphasized that competition should be most intense between closely related species or organisms. In *On the Origin of Species,* he remarked: "As species of the same genus have usually, though by no means invariably, some similarity in habits and constitution, and always in structure, the struggle will generally be more severe between species of the same genus, when they come into competition with each other, than between species of distinct genera." Darwin reasoned that similar structure indicated similar ecology, especially resource requirements. Most experimental studies of competition have adopted this viewpoint in the sense that pairs of species tested often are close relatives and frequently are congeners. But many resources are jointly utilized by distantly related organisms. Barnacles and mussels, as well as algae, occupy space in the intertidal zone. Both fish and aquatic birds prey on aquatic invertebrates (Eriksson 1979). Krill *(Euphausia superba)*, shrimp-like crustacea that abound in subantarctic waters, are fed upon by virtually every other larger type of animal, including fish, squid, diving birds, seals, and whales. Recent increases in seal and penguin populations in the southern oceans have been related by several authors to decreased competition from whales, whose populations have been decimated by commercial exploitation (Croxall and Prince 1979; Laws 1977).

In terrestrial habitats, invertebrates in forest litter are consumed by spiders, ground beetles, salamanders, and birds. Birds also compete with lizards for many of the same prey in other habitats. One experimental study of competition for seeds between ants and rodents has attracted considerable attention (Brown and Davidson 1977; Brown et al. 1979). Seeds are a major resource for ants, rodents, and birds in desert ecosystems (Brown et al. 1979). Furthermore, species consume many of the same kinds of seeds although rodents tend to take more large seeds, and ants consume more of those less than a millimeter in length. Because many of the ants and rodents in deserts eat little else than seeds, the opportunity for interspecific (interphylum!) competition is certainly present.

James H. Brown and Diane Davidson established study plots from which all rodents were removed, all ants were removed, and neither were removed (controls). The number of ant colonies and the number of rodents were monitored on two plots of each type over 2 years (Table 23–2). Ants and rodents together clearly depress the abundance of seeds. Where both were removed, seeds were two to four times more abundant than on control plots from which either ants or rodents had been removed.

The experiments also suggest that ants and rodents compete for the same seeds. Where ants were removed, rodent numbers increased 20 per cent compared to controls. In the absence of rodents, ant colonies were 70 per cent more numerous. The results present some ambiguity, however, because most of the colonies of ants, and most of the observed response to rodent removal, belong to the genus *Pheidole,* whose tiny workers harvest small seeds, mostly below the range gathered by rodents (Davidson 1977).

Although Brown and Davidson's study leaves some doubt as to how ants and rodents interact to produce the observed population effects, it does emphasize that one cannot apply taxonomic restrictions to resource overlap. Ants and rodents clearly do eat many of the same seeds. Whether their

Table 23–2
Results of experiments in which ants or rodents were eliminated from experimental plots in Arizona and the remaining taxon was counted repeatedly over the study period

	Rodents removed	Ants removed	Control	Per cent increase relative to control
Number of ant colonies	543	—	318	70.8
Rodent individuals	—	151	126	19.8

Source: Brown, Davidson, and Reichman 1979.

populations respond to their mutual competitive interaction or to other factors remains to be seen. I would guess, however, that as one considers less closely related species, factors other than their shared resources are more likely to limit each population.

SUMMARY

Laboratory experiments present clear evidence of competition among species. But natural populations are limited by physical conditions and consumers as well as by shared resources. Since the early 1960s, ecologists have conducted numerous field studies designed to reveal the influence of competition on the sizes of natural populations.

1. In pest control programs, the disappearance of one parasite following the introduction of a second similar species provided indirect evidence of competition.

2. Increased growth rates of some species of trees following removal of other species provided direct evidence of competition for light, water, and nutrients in plants. Transplant experiments with small species of plants, creating varying conditions of intraspecific and interspecific competition, often showed striking effects of interspecific competition.

3. A common method of studying plant competition is the substitution experiment, developed by C. T. de Wit, in which the ratios of two species of plants are varied but their total density is held constant. The results are portrayed on replacement series diagrams in which one can visualize the relative strengths of intraspecific and interspecific competition. Experiments with oats *(Avena)* and other plants have demonstrated strong asymmetry in interspecific competition.

4. Removal experiments involving intertidal invertebrates have demonstrated strong competition among such space-filling animals as barnacles, mussels, and encrusting sponges. Competitive exclusion is accomplished by direct physical interaction.

5. Rapid invasion of a habitat by a species of small mammal following removal of another, related species expresses direct interference through aggressive behavior.

6. Exploitation competition is most convincingly demonstrated in studies that show appropriate changes in resource levels accompanying demographic response of one species after removal of a competitor.

7. Recent reviews of studies in natural systems demonstrate the pervasiveness of competition. But different authors have reached different conclusions concerning its importance, relative to herbivory or predation and physical factors, in regulating population size and the presence and absence of species.

8. Field studies have revealed many mechanisms of competition. Thomas Schoener classified these as consumptive, preemptive, overgrowth, chemical, territorial, and encounter. The first is usually refered to as

exploitative competition and the last three as interference competition. Preemptive and overgrowth competition are based upon utilization of space and renewable resources, respectively, but involve close contact of competing individuals.

9. Experiments have demonstrated the prevalence of asymmetry in competition, in which the effect of one species is clearly greater than the other. Coexistence is possible when the asymmetry is balanced by physical conditions or by consumers, whose actions favor the poorer competitor.

10. Finally, although competition is likely strongest between similar species, distantly related organisms also may share resources. Ants, rodents, and birds consume the seeds of desert annuals; barnacles and algae occupy the same rock surfaces in intertidal zones.

24

Interactions in Complex Systems

In nature, many species interact within complex biological communities. In contrast, theoretical models usually include few species, primarily because the equations required to describe multi-species interactions frequently cannot be solved or do not have unique solutions. Experiments also are limited to few species because the number of treatments (combinations of species) required to sort out all interactions increases disproportionately as species are added. A two-species system (A and B) can be sorted out with three treatments: A and B separately and the combination AB. Seven treatments are needed to fully characterize a three-species system: A, B, C, AB, AC, BC, ABC; a four-species system requires 15 treatments, and so on.

As a result, the terrain between model or experimental systems and natural communities is mostly uncharted. How well do simple Lotka-Volterra models elucidate stability and coexistence in more complex biological systems? Ecologists are uncertain! But they are beginning to explore this problem from two directions. One has been to enlarge theoretical and experimental systems. If additional species do not change the predictions of system behavior based on simpler models, then one might extrapolate results to complex natural communities more confidently. A second direction has been to forsake the population biologist's need to understand the dynamics of each species in favor of casting both theory and its experimental tests in terms of community properties, such as numbers of species and the numbers and types of interactions among them.

The behavior of a system derives from events that affect each of its components. As one climbs to higher levels of complexity, more and more components are added. For example, while an isolated individual responds to the many factors in its environment, a population grows according to the sum of these individual behaviors plus the interactions of individuals among themselves. That is, whereas population density may affect the environment of the individual, it is only at the population level that density dependence has meaning as a regulating force.

As we consider progressively higher levels of interaction, we tend to summarize and simplify lower-level interactions. Thus while ecological studies of individuals record responses in terms of morphological, physiological, and behavioral change, studies of populations provide brief summaries of these responses in the form of life table variables. When we characterize interactions between two populations, we further distill these responses to the essences of *r*, *K*, and *a*. Extrapolated to the level of the biological community, a useful synopsis of all underlying interactions would include number of species (static) and rates of addition and deletion of species (dynamic).

The validity of successive extrapolation to higher levels of organization is debatable. Disciplines such as cellular, molecular, and developmental biology are devoted to the philosophy that one can understand a system only by dissecting it in minute detail and describing the dynamics of each part, and each part of a part. An opposing philosophy has surfaced since the 1950s largely owing to G. E. Hutchinson and R. H. MacArthur, whose method embraces properties that are accorded a conceptual independence from underlying processes, and are often called "emergent" properties (Harre 1972). George Salt (1979) suggested the following definition: "An emergent property of an ecological unit is one which is wholly unpredictable from observation of the components of that unit," adding as a corollary, "An emergent property of an ecological unit is only discernable by observation of that unit itself." Salt was careful to distinguish emergent properties from "collective" properties, such as life tables, which merely summarize the components of a system.

A scientific philosophy based on emergent properties asserts that underlying processes are mutually adjusted within a larger organization so as to maintain the integrity of the next higher level. One way to test this notion is to search for regularity and predictability at each level of organization. If the properties of each resulted directly from independent underlying processes, one would expect increasing variability among systems at each higher level of organization. Remember by analogy that the variance of a sum is equal to the sum of the component variances in the absence of

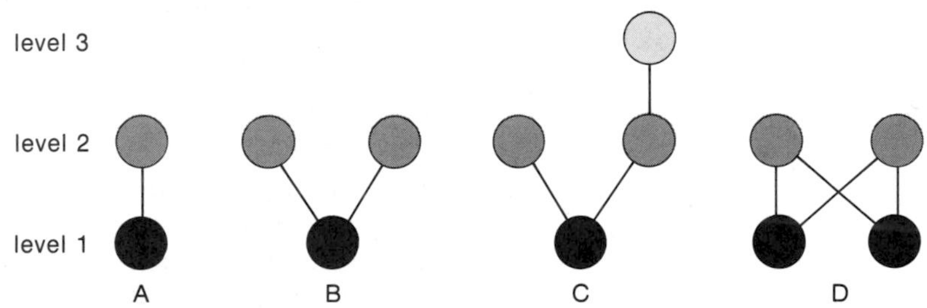

Figure 24–1
Examples of simple food webs. Level 1 includes resource populations; levels 2 and 3, consumer populations. Web B is unstable and reduces to web A through competitive exclusion. Two consumers can coexist either when one is subject to predation (C) or when there are two resources (D).

internal mutual adjustment (covariances). Hence uniformity of community structure and function would argue for determination and regulation at the community level, just as the uniformity among individuals in a population argues for such determination at the organism level (achieved by means of covariances — that is, regulatory feedbacks — among component processes).

The relationship between levels of organization, particularly between the population and the community, is likely to preoccupy ecologists for decades. It will be covered in more detail in Part VII on community ecology. In the meantime it is appropriate for us to consider the validity of extending the simple theory of population interactions to more complex systems.

Owing to their interactions, some combinations of species are more probable than others.

Interactions between species can be diagrammed as food webs (Figure 24–1). Each species is represented by a circle and the interactions between them by solid lines. Each level in the diagram is a trophic level; primary producers occupy the lowest level and top predators the highest. Some of the simplest food webs — the subjects of models considered in the previous chapters — are pictured in Figure 24–1. As we have seen, Lotka's predator-prey model, represented by a single species on one trophic level and a single species on the next higher trophic level, is generally stable, although its dynamics may lead to the extinction of one or both species when population fluctuations are large.

G. W. Harrison (1980) has shown that food chains of any length are qualitatively stable in the sense that all species' equilibria are positive, but as Stuart Pimm and John Lawton (1977) pointed out, the longer the food chain the longer the time for return to equilibrium following perturbation — that is, the greater the dynamic fragility of the system.

The second diagram in Figure 24–1, representing two predators feeding on the same prey population, is not stable owing to the competitive exclusion principle; such a system will reduce itself to a two-species predator-prey system with the elimination of one of the competing predators. In contrast, the addition of a single predator of the better competitor permits their coexistence: one is limited by its food resource, the other by its predator. The principle that predators can increase the probability of coexistence on the prey trophic level has given predators an important role in the regulation of community diversity in the eyes of many ecologists. Clearly, a more complex system can sometimes be more stable than a simpler configuration of species.

The two competitors depicted at the right of Figure 24–1 coexist stably when each shares the resources (prey) of the other but is specialized to utilize one or the other resource more effectively (hence self-regulation is stronger than interspecific interaction).

The mathematical analysis of coexistence and competitive exclusion can be extended to multispecies systems.

Lotka-Volterra competition models can be formulated for an arbitrarily large number of species *(m)*, the dynamics of each of which are described by

$$\frac{dN_i}{dt} = r_iN_i\left(1 - \frac{N_i}{K_i} - \sum_{j \neq i} \frac{a_{ij}N_j}{K_i}\right) \tag{24-1}$$

and for which the equilibrium population sizes are described by

$$\hat{N}_i = K_i - \sum_{j \neq i} a_{ij}N_j \tag{24-2}$$

If m species are to coexist, then all values of N_i ($i = 1, \ldots, m$) must be positive. If any one is negative, it will be excluded from the system and the number of species will decrease.

The solution to these equations is most readily obtained by matrix algebra. Assigning an intraspecific competition coefficient (a_{ii}) to each species, we may rearrange equation (24–2) to

$$K_i = \sum_{j=1}^{m} a_{ij}\hat{N}_j \tag{24-3}$$

All the analogous equations for species 1 through m can be combined in the single matrix equation

$$\mathbf{K} = \mathbf{AN} \tag{24-4}$$

where **K** is a column vector of carrying capacities, **A** is the $m \times m$ matrix of competition coefficients, and **N** is the column vector of equilibrium population sizes. The equation can be rearranged to solve for

$$\mathbf{N} = \mathbf{A}^{-1}\mathbf{K} \tag{24-5}$$

The matrix **A** is referred to as the community matrix. Some of its properties will be described later.

A graphical model illustrates the conditions for coexistence of three species.

Multispecies interaction in a strictly Lotka-Volterra setting may be examined graphically by adding an axis for a third species. Equilibrium isoclines no longer are straight lines on a two-dimensional graph; rather they become planes in a three-dimensional graph (Figure 24–2). Any three nonparallel

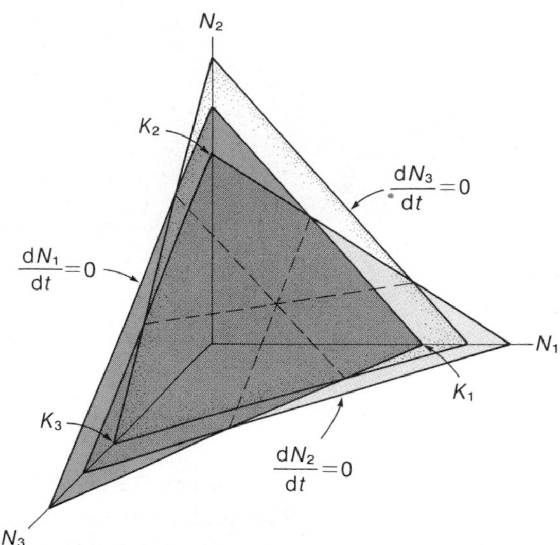

Figure 24–2
Competition graph for the three-species case. Equilibrium conditions for each species ($dN/dt = 0$) are represented by planes whose intersection designates a stable equilibrium for the community. The dashed lines represent the intersection of each pair of planes.

planes intersect at one, and only one, point. If that point occurs within the boundaries of the graph—that is, if N_1, N_2, and N_3 are all positive—the three species will coexist. When the point of intersection is negative for one or more of the species, they do not coexist.

In Figure 24–2, the intersection of each pair of planes is indicated by a dashed line, all of which cross at a single point within the boundaries of the graph. Thus, Figure 24–2 represents a stable equilibrium. Notice that all *K*s are less than K/as. Thus the condition for coexistence in the two-species case, that intraspecific competition more strongly limits population size than interspecific competition, also applies in multispecies systems.

The key to coexistence is weak interaction, hence low values of a_{ij}. So long as each species can specialize upon a different resource, it will limit itself more than it will be limited by other species. As a general rule, then, as many species can coexist as there are different resources.

MacArthur and Levins (1967) explored theoretically the situation in which species compete for resources that are distinguished by continuous variation in some attribute, such as prey size. Accordingly, each species of consumer is specialized to utilize a different portion of the resource continuum, as shown in Figure 24–3. MacArthur and Levins assumed that the coefficient of competition between two species is proportional to the overlap between the utilization curves of each. Hence a_{12}, a_{32}, and their reciprocals are large compared to a_{13} and a_{31}. According to logistic competition theory, the middle species (2) can increase when its carrying capacity exceeds the combined effects of both species upon its population; that is, when $K_2 > a_{21}N_1 + a_{23}N_3$. MacArthur and Levin's analysis showed that if the *K*s for all three species were equal, species 2 could invade between species 1

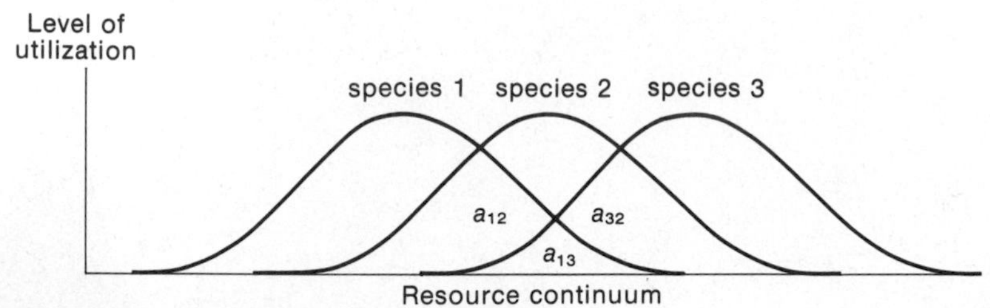

Figure 24–3
Utilization of a resource continuum by three species. The regions of overlap represent the coefficients of competition.

and 3 when the competition coefficients were less than about 0.5. Invasion was made more likely as K_2 increased, but above a certain value of K_2 species 2 would outcompete both 1 and 3 and eliminate them from the system.

When species are ordered along a single resource continuum, their competition coefficients are placed in a fixed relationship to each other and the conditions for coexistence are more stringent than when species can specialize on discrete resources. Increasing the number of dimensions of the resource continuum removes some of the ordering between species and coexistence becomes easier to achieve. For example, on a two-dimensional resource continuum, three species can have equal competition coefficients among themselves, permitting coexistence so long as, in general, $a_{ij}a_{jk}a_{ik} < 1$. The theory of competition along a resource continuum has been explored more fully by Abrams (1975), Goh (1976, 1977), May and MacArthur (1972), Roughgarden (1974, 1976), and Turelli (1978).

Higher order interactions between competitors may alter conditions for coexistence.

Coexistence under Lotka-Volterra competition requires the assumption that all interactions between species are strictly additive and are independent of the presence or absence of other species. That is, only the two-species interactions (*ij, ik, jk*) can be important. The community matrix comprises all these coefficients but does not include higher-order interactions (*ijk*.) Several ecologists have attempted to determine whether the interaction between a pair of species is independent of the presence of a third species, as it must be for the Lotka-Volterra representation to be valid. E. A. Bender et al. (1984) have shown how experiments can be designed to tease apart community interactions.

John H. Vandermeer (1969) estimated the competition coefficients between the six possible pairs of four species of protozoa raised together in the laboratory. He estimated K_i by growing each of the four species in isolation. The vector **K** and the community matrix **A** based on two-species competi-

Table 24–1
Matrix of two-way competition coefficients, values of K and estimated equilibrium population sizes in three-way and four-way competition among four species of protozoa

A	PA	PC	PB	BL	K	Four-way $\hat{N}$	Three-way $\hat{N}$			
PA	1.00	1.75	−2.00	−0.65	671	−267	—	−562	−57	−326
PC	0.30	1.00	0.50	0.60	366	488	571	—	403	496
PB	0.50	0.85	1.00	0.50	230	−28	−149	−62	—	−65
BL	0.25	0.60	−0.50	1.00	194	−46	−221	23	−34	—

Source: Vandermeer 1969.

tion are presented in Table 24–1. Note the negative values of a_{ij} in the community matrix, indicating that species j enhances the population size of i ($N_i > K_i$), possibly because the by-products of j's metabolism provide a resource to species i.

The outcome of simultaneous interaction between all four species can be predicted from the two-way interactions between species, provided there are no higher order interactions, using the matrix equation (24–5). Accordingly, should the four species be grown together, the equilibrium populations of all but *P. caudatum* would be negative, and we would not expect the species to coexist. As also shown in Table 24–1, the same is true of all four of the three-species combinations.

Vandermeer used two-species combinations to determine the values of a_{ij}; of the six possible pairs, three coexisted and the predicted equilibrium sizes for these were all positive (Table 24–2). Which of these three two-species will persist when all four species compete with each other? The answer to this question lies in determining which of the pairs (i,j) can be invaded by a third species (k); because no trios are stable, no invasible pair can persist and the community eventually will consist of those two species that can both coexist and resist invasion by the other two. Successful invaders are those for which $dN_k/dt = K_k - a_{ki}N_i - a_{kj}N_j$ exceeds 0; that is,

Table 24–2
Ability of a third species of protozoa to invade stable two-species systems as judged by rate of population growth of the third species

	Stable two-species systems (i,j)					
	PA	PC	PA	BL	PB	BL
Equilibrium N	64	346	686	23	106	247
Can species k invade (dN_k/dt)?						
PA	—		—		1044 (yes)	
PC	—		146 (yes)		165 (yes)	
PB	−96 (no)		−125 (no)		—	
BL	−27 (no)		—			

Source: Vandermeer 1969.

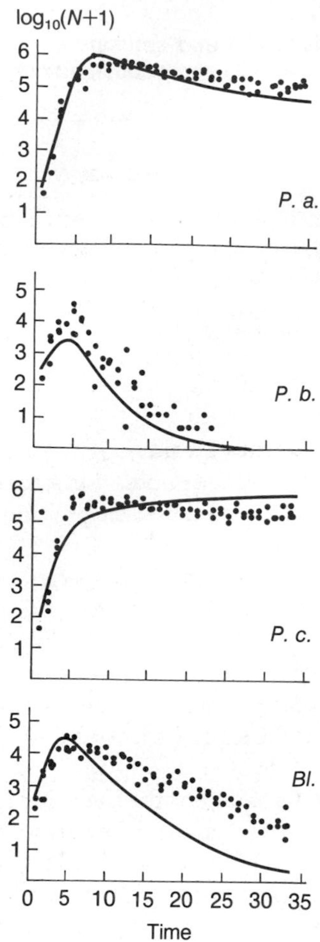

Figure 24–4
Population trajectories of four species of protozoa engaged in a four-species competition experiment; solid lines are the estimated trends based on pair-wise competition experiments. (From Vandermeer 1969.)

those which can increase when rare in the face of equilibrium populations of the other two species. As shown in Table 24–2, neither *P. bursaria* nor *Blepharisma* can invade a community consisting of *P. caudatum* and *P. aurelia,* whereas the other combinations are invasible. Therefore, we would expect the four-species system to become simplified through competitive exclusion to coexisting populations of *P. caudatum* and *P. aurelia.* Because the results of the four-way competition experiment conformed closely to prediction (Figure 24–4), Vandermeer concluded that higher-order and multispecies interactions were absent.

Other studies, notably those of Nelson G. Hairston, Sr. et al. (1968) on protozoa and, especially, Henry Wilbur (1972) on salamanders, reported evidence for multispecies interactions between competitors. Hairston et al. showed that the outcome of competition (that is, $\hat{N}_i$) between any two species of protozoa could be altered by the presence of a third species. But we have already seen that such a result might be predicted from two-species interaction coefficients. To demonstrate Salt's so-called emergent properties, one must show that the coefficient of interaction between two species is altered by the presence of a third.

Wilbur (1972) searched for multispecies interactions directly by raising immature stages of three species of salamanders *(Ambystoma)* at all combinations of three densities (0, 32, and 64 individuals per cage) in enclosed portions of natural habitat at the edge of a pond. Hence all species were raised alone and in combination with each and both of the other two. Wilbur recorded survival, length of larval period, and mass at metamorphosis. He analyzed these response variables in such a way as to reveal statistical effects due to the density of the population, competition from the other two species, and any interactions involving all three species uniquely. Such multispecies interactions were found to be quite important (Figure 24–5).

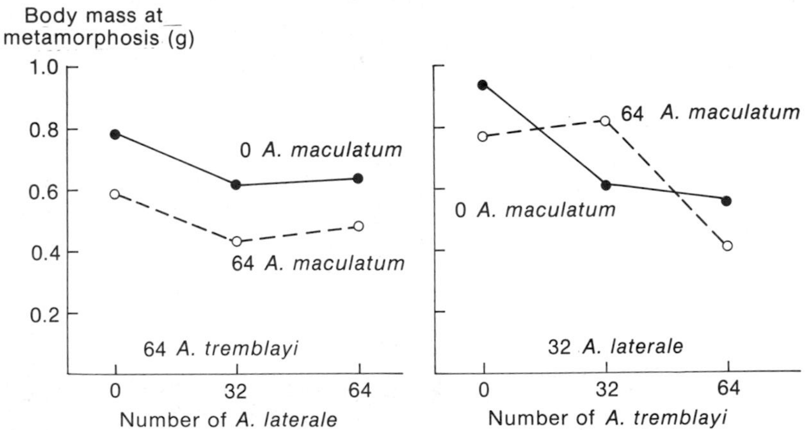

Figure 24–5
Body mass at metamorphosis of the salamanders *Ambystoma tremblayi* and *A. laterale* when placed in competition with each other and in the presence or absence of a third species, *A. maculatum.* (From Wilbur 1972.)

Table 24–3
Equilibrium population sizes in laboratory microcosms composed of different combinations of competing aquatic crustaceans

	Equilibrium populations of			
Experiment	*Alonella*	*Ceriodaphnia*	*Simocephalus*	*Hyalella*
Hyalella present	417	294	48	312
Remove C	614	—	83	327
Remove S	431	331	—	325
Remove C and S	739	—	—	330
Hyalella absent	2287	911	1407	—
Remove C	3624	—	1512	—
Remove S	2186	1411	—	—

*Results are modified somewhat to simplify table. As a result, calculated values of a_{ij} do not correspond exactly to Neill's values.
Source: Neill 1974.

William Neill's (1974) study of competition between freshwater crustacea showed directly that one species can influence the interaction between populations of two others. Neill maintained the crustacea on diets of algae in 1500 ml laboratory microcosms. He determined the equilibrium densities of each species after 2 to 4 months in the presence of each one of the other three species (two-species interactions), combinations of two of the other species (three-way interactions), and with all three other species (the four-way interaction). The experiments revealed asymmetrical coefficients of competition in many cases (Table 24–3). For example, *Hyalella* was influenced little by competition from any species; its numbers did not vary (range 312–331), regardless of which other species it was raised with. In contrast, competition from *Hyalella* greatly reduced equilibrium populations of all three of the other species. So while the coefficients a_{Hi} were all close to zero, the values of a_{iH} were quite large.

Apropos of the question of multispecies interactions, the outcome of competition between some pairs of species was strikingly influenced by a third. When *Ceriodaphnia* (C; $\hat{N} = 294$) was removed from the four-species system, the population of *Alonella* (A) increased by 197 individuals ($614 - 417 = 197$). According to the equation, $a_{ij} = (K_i - N_i)/N_j$, $a_{AC} = 197/294 = 0.67$ in the presence of *Simocephalus* (S) and *Hyallela* (H); similarly, $a_{AS} = (431 - 417)/48 = 0.29$ in the presence of species C and H. If competition between *Ceriodaphnia* and *Alonella* were independent of competition between *Simocephalus* and *Alonella*, the increase in *Alonella* when both other species were absent would equal the sum of the increases when *Ceriodaphnia* or *Simocephalus* were removed individually; that is $197 + 14 = 211$. Neill found, however, that removal of both resulted in an increase in *Alonella* of 322 individuals ($739 - 417$). So a_{AC} and a_{AS} were not strictly additive, or linear. When a relationship $Y = f(X)$ is linear, it may be described by $Y = a + bX$; when it is not linear, its description requires higher order and multispecies terms (X^2, X^3, XZ, . . .). The outcome of such interactions cannot be predicted by the strictly linear Lotka-Volterra models.

By comparing a_{AC} and a_{AS} in the presence of *Hyallela* (+H) and its absence (−H), we may judge whether a third species can affect the interaction between two others. Without *Hyalella* $a_{AC(-H)} = (3624 - 2287)/911 = 1.47$, and $a_{AS(-H)} = (2186 - 2287)/1407 = -0.072$, which probably does not differ significantly from 0. These values do, however, differ from $a_{AC(+H)} = 0.67$ and $a_{AS(+H)} = 0.29$. Hence three-way interactions appear to be important in this system and it would have been unsound to extrapolate simple competition coefficients obtained in two-species competition to a prediction of the behavior of the more complicated four-species system.

M. J. Pomerantz (1980) and M. D. Thomas and Pomerantz (1981) have suggested that apparent multispecies and higher-order interactions in complex systems could result simply from nonlinearities in the response of population growth to density. While such nonlinearities have been observed, and while they may explain nonadditivity in interactions, they further emphasize that Lotka-Volterra models, based on linear responses to density both within and between species, do not adequately represent the natural world. Biological considerations would seem to make nonadditivity and higher order interactions inevitable. As John M. Emlen (1984) points out:

> Suppose that species A, B, and C compete, and that species C, via interference, keeps B from microhabitats where it most frequently encounters A. It seems likely, then, that removal of C will result in closer contact between A and B and thus higher a_{AB} and a_{BA} values.

This recalls the interaction between bluegill, pumpkinseed, and green sunfish described by E. E. Werner and D. J. Hall (1976, 1977), in which the three species utilized different pond microhabitats when reared together but converged in weedbeds when reared alone. Emlen continued with a second possibility:

> Suppose that species C is highly attractive as a food to predators that also eat A and B. If C is locally removed, the predators are apt to move elsewhere, relieving predation pressure on A and B. With alleviation of danger from predators, these species may forage more widely, or differently, or at different times than before, all changes that have potential for altering their competitive relationship. As these simple scenarios suggest, it is probably optimistic folly to hope for additivity in competition intensity.

Facilitation may be an indirect consequence of multispecies competition.

Predation benefits the consumer at the expense of the resource population; competitors interact to their mutual detriment. But many organisms also interact to their mutual benefit. When such relationships result from the direct involvement of two species in each others' lives—as in the case of nitrogen-fixing bacteria or mycorrhyzae associated with the roots of plants, the gut flora of ruminants, and ants defending plants against herbivores—they are called mutualisms.

Competitors may also benefit each other indirectly as well. Consider three competing species, each of which depresses the population growth of the other two (all a_{ij}s positive). If A and B compete intensely, as do B and C, then as A becomes more numerous, the population of C may also increase because B—C's strong competitor—is depressed by A.

The change in a species population in a complex system of competitors is determined by the matrix equation $\mathbf{N} = \mathbf{A}^{-1}\mathbf{K}$. Hence it is the inverse of the community matrix ($\mathbf{A}^{-1}$) that determines the response of the populations to a change, for example, in environmental conditions that alter the values of K of one or more species.

Positive elements in the matrix $\mathbf{A}^{-1}$ indicate facilitation between two species indirectly through their relationships with others (Levine 1976; Lawlor 1979). Consider the competition matrix

$$\mathbf{A} = \begin{bmatrix} 1.00 & 0.23 & 0.10 \\ 0.28 & 1.00 & 0.88 \\ 0.11 & 0.75 & 1.00 \end{bmatrix}$$

which is part of a larger matrix compiled by Diane Davidson (1980) for species of ants in a desert habitat in the western United States. Species B and C compete intensely; A and B, and A and C, less so. The inverse of this simplified community matrix is

$$\mathbf{A}^{-1} = \begin{bmatrix} 1.10 & -0.50 & 0.33 \\ -0.60 & 3.21 & -2.77 \\ 0.32 & -2.35 & 3.04 \end{bmatrix}$$

(don't worry about the algebra of the matrix inversion). Suppose we arbitrarily set carrying capacities to $K_A = 100$, $K_B = 120$, and $K_C = 100$; this results in a stable three-species system with equilibrium populations $\hat{N}_A = 83$, $\hat{N}_B = 48$, and $\hat{N}_C = 54$. Now consider the effect of increasing the carrying capacity of species C from 100 to 110 individuals. The new equilibrium values are $\hat{N}_A = 90$ (an increase of 7), $\hat{N}_B = 21$ (a decrease of 27), and $\hat{N}_C = 84$ (an increase of 30 even though K_C was increased by only 10). The depressing effect of C on B relieved some of the competition from B on population A and indirectly allowed it to increase (facilitation), even though A and C also compete directly.

Predators can influence the outcome of competition between species.

The principle that the coexistence of species can be altered by the presence or absence of consumers has been appreciated for more than a century.

Darwin probably was the first to report an experimental investigation of the phenomenon (1859):

> If turf which has long been mown, and the case would be the same with turf closely browsed by quadrupeds, be let to grow, the more vigorous plants gradually kill the less vigorous, though fully grown plants; thus out of twenty species growing on a little plot of mown turf (three feet by four) nine species perished, from the other species being allowed to grow up freely.

Following Darwin's early lead, A. G. Tansley and R. S. Adamson (1925) conducted more systematic and extensive experiments on the effect of rabbit grazing on the composition of British chalk grasslands. Over a 6-year period, the vegetation within rabbit-free exclosures became dominated by the grass *Zerna erecta* (now called *Bromus erectus*), a dominant competitor normally held in check by grazing. Similar studies on the effects of grazing animals on the composition of grasslands are summarized in detail by John Harper (1977: 435).

It should be obvious that a selective grazer can alter the outcome of competition, even to the point of ensuring coexistence where it is not possible in a pure competition process (Noy-Meir 1981; Gleeson and Wilson 1986). To show this mathematically (just to check up on our intuition), suppose that in the absence of predation, species 1 outcompetes species 2 (that is, $K_1 > K_2/a_{21}$ and $K_2 < K_1/a_{12}$). Now suppose that a predator removes individuals of species 1 at rate m per capita. The dynamics of species 1 may now be described by

$$\frac{dN_1}{dt} = r_1 N_1 \left(1 - \frac{N_1}{K_1} - \frac{a_{12}N_2}{K_1}\right) - m_1 N_1 \qquad (24-6)$$

and equilibria ($dN/dt = 0$) are achieved when

$$N_1 = \frac{K_1(1 - m_1/r_1) - a_{12}K_2}{1 - a_{12}a_{21}} \qquad (24-7)$$

and

$$N_2 = \frac{K_2 - a_{21}K_1(1 - m_1/r_1)}{1 - a_{12}a_{21}} \qquad (24-8)$$

These conditions resemble the equilibria under pure competition except that the value of K_1 is discounted by the term $(1 - m_1/r_1)$. Two points are worth noting here. First, the maximum exponential growth rate of species 1 (r_1) influences the equilibrium populations of both species. In the absence of predation, equilibria are determined only by as and Ks. Second, N_2 can be positive even when $a_{21}K_1$ exceeds K_2 (that is, $K_1 > K_2/a_{21}$). The term $(1 - m_1/r_1)$ varies between 0 and 1 depending on the intensity of predation (when m exceeds r, the population is eaten to extinction). Thus the higher the value of m_1, and therefore the lower $(1 - m_1/r_1)$, the more probable is coexistence of species 1 and 2. By rearranging equation (24–8), we find that species 2 coexists with 1 when m_1 lies between $r_1(1 - K_2/a_{21}K_1)$ and r_1.

G. F. Gause (1935; recounted by Slobodkin 1961) explored theoretically the situation in which both species of competitors suffered predation at the same level (hence nonselective predation). This is equivalent to both species having the same externally imposed mortality rate m. In the solution to these equations, K_2 is now adjusted by factor $(1 - m/r_2)$, as K_1 is by factor $(1 - m/r_1)$. The joint equilibrium finds both N_1 and N_2 positive when

$$a_{12} < \frac{K_1(1 - m/r_1)}{K_2(1 - m/r_2)} \text{ and } a_{21} < \frac{K_2(1 - m/r_2)}{K_1(1 - m/r_1)} \qquad (24\text{–}9)$$

Whereas in the absence of predation or other density-independent mortality ($m = 0$) the growth potential of the population (r) did not affect the joint equilibrium, with the introduction of a mortality term it does. In fact, nonselective mortality can promote coexistence where, in its absence, one species would exclude the other; it also can destroy a stable coexistence.

These properties are shown graphically in Figure 24–6, in which the graph of the positions of equilibrium isoclines ($dN_i/dt = 0$) on a plane defined by N_i (individuals of species i) and N_j is given a third axis perpendicular to that plane for the level of mortality m. On this three-dimensional graph, isoclines become isoplanes that intercept the axes at points K_i on the N_i axis, K_i/a_{ij} on the N_j axis, and r_i on the m axis. Now the joint equilibrium of the two species is the line defined by the intersection of the two isoplanes.

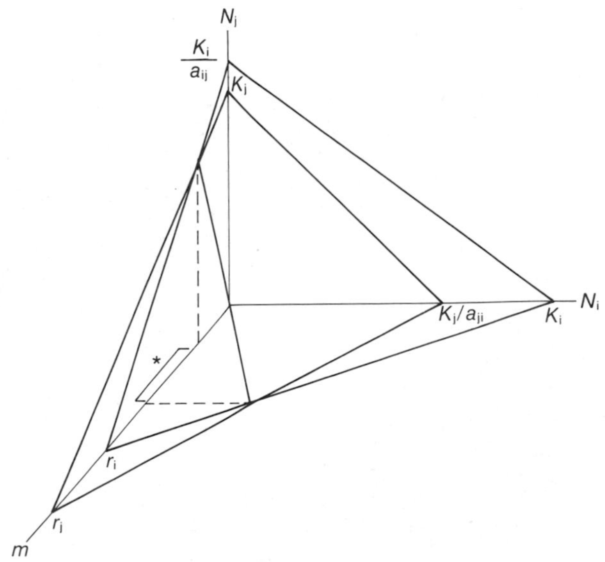

Figure 24–6
The influence of indiscriminant mortality (m) on the outcome of competition between two species (i and j). The asterisk denotes the range of mortality rates over which the two species coexist.

As you can see in the figure, a species excluded at low m will achieve an equilibrium at intermediate m if its potential rate of increase (r) is higher than that of the other species.

The range of values of m over which two species coexist depends on the geometrical comformation of their isoplanes. When the isoclines on the N_i–N_j plane are parallel ($a_{ij} = 1/a_{ji}$), the junction of the isoplanes will be parallel to the N_i–N_j plane and coexistence is achieved at a single value of m. Under any other circumstance, coexistence occurs when m lies between the values

$$\frac{r_i r_j(K_i - a_{ij}K_j)}{(r_jK_i - r_i a_{ij}K_j)} \text{ and } \frac{r_i r_j(K_j - a_{ji}K_i)}{(r_iK_j - r_j a_{ji}K_i)} \qquad (24\text{–}10)$$

A numerical example will illustrate this point. Suppose that $K_i = K_j$ and $a_{ij} = 1.5$ and $a_{ji} = 0.8$. In the absence of predation ($m = 0$), these values violate a required condition for coexistence because $a_{ij} > K_i/K_j$, and species j excludes species i. Now suppose that $r_i = 2$ and $r_j = 1$. The two species will coexist over a range of values of m because r_i exceeds r_j. This range is calculated from the above expressions to lie between $m = 0.50$ and $m = 0.33$. When r_i is less than r_j, the two species cannot coexist at any level of m. Suppose, for example, that $r_i = 1$ and $r_j = 2$. The range of values of m for "coexistence" (equation [24–10]) is now between -2.0 and -0.2, which is not possible biologically.

As a final numerical example, suppose that species i and j coexist at $m = 0$; say $a_{ij} = a_{ji} = 0.8$ and $K_i = K_j$. Further suppose that $r_i = 2$ and $r_j = 1$.

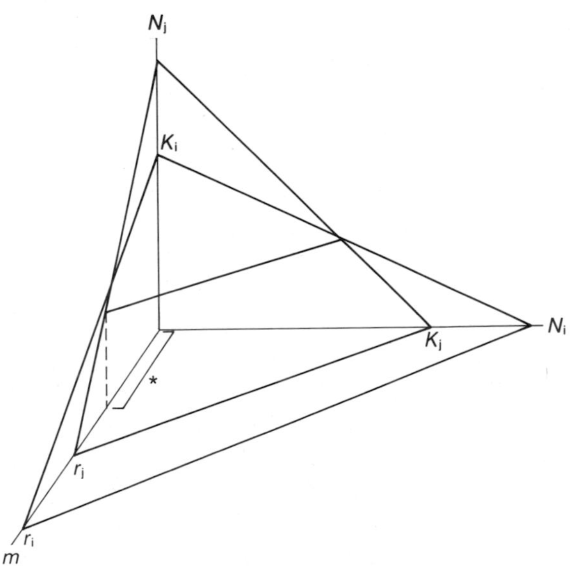

Figure 24–7
An example in which indiscriminant mortality can lead to the exclusion of one of a pair of species that would otherwise coexist.

Now the range of values of m over which the species can coexist lies between -0.67 and 0.33, which includes the biologically realistic range $m = 0$ to $m = 0.33$. When m exceeds 0.33, species i excludes species j. This example is graphed in Figure 24–7. One conclusion to be taken from this analysis is that predation or other nonselective mortality can either promote or reduce coexistence of competing species depending on the circumstances. For additional discussion of such models, see Comins and Hassell (1976), Cramer and May (1972), Parrish and Saila (1970), and Roughgarden and Feldman (1975).

Experimental analyses demonstrate the effects of consumers on the coexistence of resource populations.

We have already seen that grazing can maintain a high diversity of plants in grasslands (Harper 1969, 1977). In the absence of grazers, dominant competitors grow rapidly and exclude others. Similar results have been obtained from experiments on marine algal communities under the pressure of grazing by limpets, snails, and urchins (Lubchenco 1978, Paine and Vadas 1969, Witman 1987). These studies indicate that predation has a strong hand in shaping the structure of communities.

In fact, ecologists have become sharply divided between a predation school inspired by J. H. Connell and R. T. Paine, and a competition school inspired by David Lack, G. E. Hutchinson, and R. H. MacArthur. Studies that swelled enrollment in the predation school during the 1970s grew out of the experimental tradition fostered by Connell's and Paine's work in rocky shore habitats, and the results are striking (Connell 1972; Paine 1977).

The classic is Paine's (1966, 1974) work on the exposed rocky coast of the state of Washington. Within the intertidal zone, several species of barnacles, gooseneck barnacles, mussels, limpets, and chitons (a kind of grazing mollusc) dominate the habitat; these are preyed upon by the starfish *Pisaster* (Figure 24–8). One study area, 8 meters in length and 2 meters in vertical extent, was kept free of starfish by physically removing them; an adjacent control area was left undisturbed. Following the removal of the starfish, the number of prey species in the experimental plot decreased rapidly, from 15 at the beginning of the study to 8 at the end.

Diversity declined in the experimental areas when populations of barnacles and mussels increased and crowded out many of the other species. Paine concluded that starfish were a major factor in maintaining the diversity of the area. The crown-of-thorns starfish *(Acanthaster)* similarly enhances the diversity of coral reefs near the Pacific coast of Central America by voraciously consuming a species of coral, *Pocillopora,* that would otherwise crowd out many other species (Figure 24–9; Porter 1972, 1974).

Sources of mortality other than predation may also enhance coexistence (Witman 1987). Particularly on rough coasts, physical disturbance by wave action and battering by floating ice and logs haphazardly knocks individuals or whole patches of individuals off rocks. The newly opened

Figure 24–8
Congregation of starfish *(Pisaster)* at low tide on the coast of the Olympic Peninsula, Washington. The starfish, shown at lower left, is an important predator on mussels (lower right).

Figure 24–9
Crown-of-thorns starfish consuming a coral head in Panama. (Courtesy of J. W. Porter; from Porter 1972b.)

space becomes available for colonization, often by species that do not compete well. Although these are eventually excluded from a particular patch as it becomes overgrown by superior competitors, populations of the poorer competitors may be maintained within a larger area when new patches are made available for colonization. On a regional basis, ability to colonize new places is a suitable measure of r and rate of patch clearing is a measure of m. Therefore, even though a species may be excluded locally if a spot remains undisturbed ($m = 0$), it may coexist regionally by more rapidly colonizing open space when it does appear ($m > 0$). The dynamics of this process have been explored by Levin and Paine (1974).

The theory of competition in the face of extrinsic mortality also predicts that coexistence at low m may shift to exclusion at high m. This point was made by a series of experiments on communities of protozoa and other small organisms living in pitcher plants, in which John Addicott (1974) showed that predation by mosquito larvae reduced the number of prey species. In the simple, test-tube-like environment of the pitcher plant, mosquito larvae are such efficient predators that they drive all resource populations to extinction save those with the highest potential growth rates (*r*). It would appear therefore that the influence of predation on the coexistence of resource populations will vary with the complexity of the habitat and the efficiency and abundance of the predators.

In experiments with communities established in artificial ponds, Peter Morin (1981) showed that predatory salamanders can reverse the outcome of competition among frog and toad tadpoles. Morin seeded each of the ponds with 200 hatchlings of the spadefoot toad *(Scaphiopus holbrooki)*, 300 of the spring peeper *(Hyla crucifer)*, and 300 of the southern toad *(Bufo terrestris)*. To replicates of the ponds he added either 0, 2, 4, or 8 of the predatory broken-striped newt *(Notophthalmus viridescens)*. In the absence of newt predation, *Scaphiopus* tadpoles grew rapidly, survived well, and dominated the ponds along with smaller numbers of *Bufo*. *Hyla* tadpoles were all but eliminated (Figure 24–10). *Notophthalmus* apparently prefer toad tadpoles and as Morin increased the number of predators in the tanks, the survival of both *Scaphiopus* and *Bufo* decreased markedly. With fewer toads per pond, levels of food increased and the survival and growth of *Hyla* tadpoles improved immensely, as did the growth of surviving *Scaphiopus* and *Bufo* tadpoles. (More extensive experiments with this system are reported by Wilbur et al. [1983], Morin [1983], Morin et al. [1983], Wilbur and Alford [1985], and Alford and Wilbur [1985]. Additional experimental studies of complex interactions in a variety of systems may be found in Horvitz and Schemske [1984]; Connor and Quinn [1984], Sih et al. [1985], Harrold and Reed [1985], Fowler and MacGarvin [1985], Witman [1985, 1987], Koptur [1985], Hay and Taylor [1985], Hay [1986], Paul and Hay [1986], and Karban et al. [1987].)

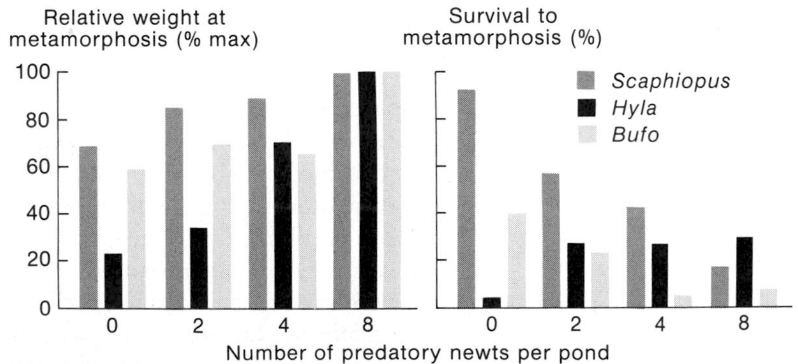

Figure 24–10
Effect of predators on per cent survival and weight of metamorphosis in three species of anurans (frogs and toads) raised in large tanks. (From Morin 1981.)

Species interactions determine the properties of large food webs.

Simple graphical and algebraic techniques are sometimes adequate for exploring the interactions between a few species, but they are generally useless for more complex systems in the natural world. To bridge the gap between population interactions and community structure, new applications of matrix algebra were developed during the 1970s (MacArthur 1972, Vandermeer 1972, May 1973, Cohen 1978, Pimm 1982, Cohen and Newman 1985, Cohen et al. 1985). The simplest of these applications were based on Lotka-Volterra interactions between species analogous to the models we have seen.

In the most general case, the population growth rate of species i in a community of m species can be expressed as

$$\frac{dN_i}{dt} = N_i\left((b_i + \sum_{i=1}^{m} a_{ij}N_j\right)$$

where b_i is the intrinsic exponential growth rate of population i in the absence of both intraspecific and interspecific competition (all N_js including N_i close to zero), and a_{ij} is the coefficient of interaction between species i and species j, including the intraspecific interaction term a_{ii}. Thus b_i is equivalent to r_i in the equation for logistic competition; a_{ij} is equivalent to $-a_{ij}r_i/K_i$, and is the influence of species j on the growth rate of population i, normalized by r_i and K_i. The signs of the interaction coefficients depend on the kind of interaction between two species:

Species i	Species j	a_{ij}	a_{ji}
predator	prey	+	−
mutualist	mutualist	+	+
competitor	competitor	−	−

Interactions of the form (0,+) and (0,−), or no interaction between two species (0,0), are also possible.

Equilibrium populations in such a community are calculated as we have done before, by evaluating the system of equations, one for each species, at $dN_i/dt = 0$. Hence $b_i = -\sum a_{ij}N_j$. In matrix terms, $\mathbf{b} = -\mathbf{AN}$ and the column vector of equilibrium population sizes is $\mathbf{N} = -\mathbf{bA}^{-1}$. Such a community of m species persists if all Ns exceed 0; if some do not, then the system is unstable and will simplify itself by eliminating of one or more species.

When all species in a system described by $\mathbf{A}$ can coexist, ecologists sometimes estimate its relative stability by how rapidly the species return to their equilibria when perturbed. Both the qualitative stability (Do species coexist or not?) and relative stability (How fast does a qualitatively stable

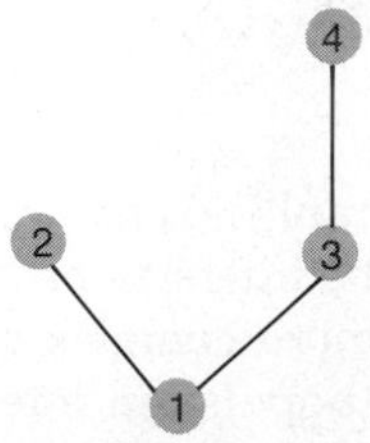

species i \ species j	1	2	3	4
1	+	−	−	0
2	+	+	0	0
3	+	0	+	−
4	0	0	+	+

Figure 24–11
The Jacobian matrix for a simple food web consisting of one resource population (1), two consumer populations (2 and 3), and a single selective predator (4).

system approach equilibrium?) of complex systems may be explored through the properties of a matrix having elements $a_{ij}N_i$, thus

$$\mathbf{C} = \begin{bmatrix} a_{11}N_1 & a_{12}N_2 & a_{13}N_1 & \cdots \\ a_{21}N_2 & a_{22}N_2 & a_{23}N_2 & \cdots \\ a_{31}N_3 & a_{32}N_3 & a_{33}N_3 & \cdots \\ \cdots & \cdots & \cdots & \cdots \end{bmatrix}$$

Such a matrix is so important to studying the behavior of large systems that it is given the special name Jacobian matrix. It can be constructed for any imaginable stable food web, as shown by example in Figure 24–11.

The qualitative stability of many systems has been investigated by C. Jeffries (1974). Quantitative stability can be estimated from the largest eigenvalue of **C**. The mathematical details are not necessary here. Most ecologists feel that the quantitative stability of a system is important because species that do not return quickly to their equilibria are vulnerable to extinction in the wake of external perturbations.

May (1972) attempted to utilize these mathematical techniques to explore the stability of complex systems. He constructed arbitrary communities of M species by choosing interaction coefficients among them at random. In a community of M species, May defined its connectance (C) as the proportion of interspecific interactions not equal to 0. The intensity of the interaction (a_{ij}) was a random variable with mean 0 and variance I^2. In constructing communities, pairs of species were chosen to interact and then assigned a value of a_{ij} at random. Hence by chance, some species had coefficients (a_{ij}, a_{ji}) that were qualitatively (0,0), (0,+), (0,−), (+,+), (+,−), and (−,−).

May determined that as a rule, a community is qualitatively stable when

$$I(CM)^{1/2} < 1$$

Thus increases in the number of species (M), connectance (C), and intensity of interaction (I) all tend to reduce the stability of the system. To illustrate, with an intensity $I = 0.5$, communities of 10 species would be stable so long as the connectance was less than 0.4, but C must be less than 0.2 for stability in communities of 20 species, and less than 0.08 for 50 species. In these three cases, the number of interacting species pairs $CM(M-1)$ are fewer than 36, 76, and 196, respectively, and the average numbers of interactions per species are less than 3.6, 3.8, and 3.9. For large numbers of species (M), $CM(M-1)$ approaches CM^2 and the averages of the number of interactions per species approaches CM^2/M or CM. Because a criterion for a qualitatively stable system is $CM < I^{-2}$, when $I = 0.5$, this is equivalent to $CM < 4$.

May's result indicates that no matter how large the system, the numbers of interactions per species has an upper limit determined by the average intensity of interaction. Hence it is unlikely that more diverse systems are more complex because more complex systems are not stable.

May's matrices were arbitrary and biologically unrealistic (Lawlor 1978). For example, by random choice of interaction coefficients it would be possible to have predators with nothing to eat, or circular predator-prey systems (A eats B, B eats C, C eats A) that are improbable in the natural world. Extensive exploration of biologically probable food webs by S. L. Pimm, J. H. Lawton, and P. Yodzis (summarized by Pimm 1982) permit a number of general conclusions about the characteristics of food webs that contribute to their stability. First, consistent with May's finding, food webs should not be too complex; in particular, an increase in number of species must be accompanied by a reduction in connectance, all other things being equal.

Second, food chains generally should be short, perhaps not more than two or three predators in length. Although efficiency of energy transfer limits the number of trophic levels that can support predator populations, dynamical considerations also are pertinent. Long food chains may be qualitatively stable but, as species are added, the time required for return to equilibrium increases. Hence longer food chains are dynamically more fragile—susceptible to outside perturbations—than shorter food chains.

Third, omnivores—that is, species that feed on more than one trophic level—should be scarce. Adding the extra link in the food web tends to destabilize the system (Jeffries 1974). Where omnivores do occur, they should feed on adjacent trophic levels rather than on separated trophic levels (Figure 24–12). That is, a top carnivore (level 4) would be more likely to feed on levels 2 and 3, if omnivorous, than on levels 1 and 3; the latter is intrinsically less stable.

Other characteristics of food webs will be discussed in Part VII on community ecology. In this chapter, I wished to point out a possible bridge between the dynamics of simple interactions between few species and the behavior of large, more complex systems of species that we call communities. The models that provide the basis of this approach are demonstrably simpler than nature; responses to species densities are linear and without higher-order interactions, time lags, and other biological details. Models are not spatially subdivided, as nature is, and have little room for temporal diversification.

Ecologists also disagree on whether the structure of food webs is determined by the dynamical considerations embodied in simple models (Pimm 1982) or, alternatively, by the constraints of energy flow through the community and the feeding adaptations of species. Clearly we are far from

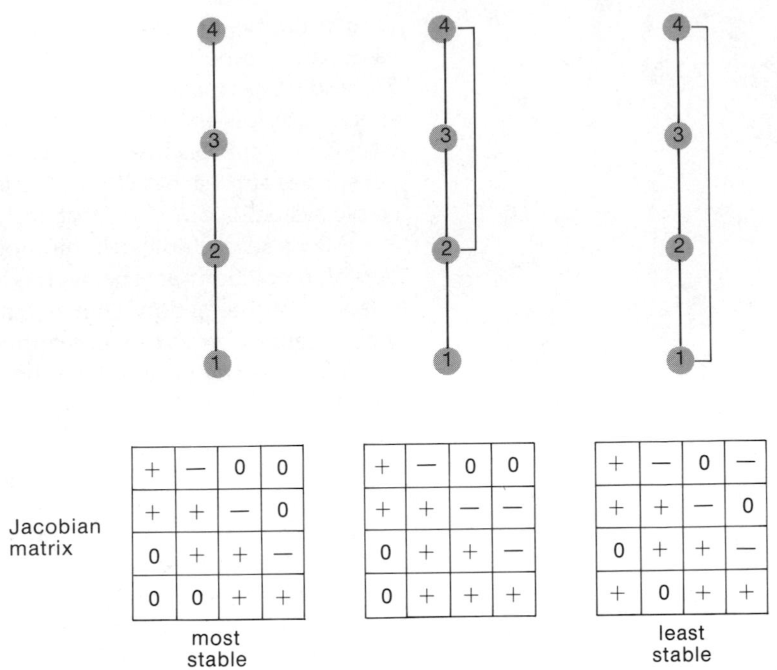

Figure 24–12
Simple food webs and their associated Jacobian matrices illustrating the general rule that omnivory (feeding on more than one trophic level) reduces the stability of the system. (From Pimm 1982.)

resolving some of these questions, and the ultimate resolution, if it is possible, will likely come from reconciling a combination of viewpoints that, at present, focus separately on dynamical control, energetics, and adaptations of individual species.

SUMMARY

The dynamics of complex communities are determined by the interactions between all possible pairs of species, and by unique interactions among combinations of three or more species. Because Lotka-Volterra models are defined only in terms of interactions between pairs of species, it has been crucially important to determine the extent to which multispecies interactions occur in natural communities, as these would affect the validity of extrapolating from simple models to the behavior of complex systems.

1. When communities display characteristics that cannot be inferred from the properties of their component parts, these are referred to as emergent properties. Depending on the nature of interactions within assemblages of species, community structure and dynamics may be emergent properties.

2. Assemblages of species are qualitatively stable if all component population equilibria are positive. Quantitative stability refers to the rate of return to equilibrium after a perturbation.

3. The outcome of competition within a multispecies system that obeys Lotka-Volterra interactions can be predicted from the matrix algebra expression $\mathbf{N} = \mathbf{A}^{-1}\mathbf{K}$, where $\mathbf{N}$ is the column vector of population equilibria ($\hat{N}$), $\mathbf{K}$ is the column vector of carrying capacities, and $\mathbf{A}^{-1}$ is the inverse of the community matrix $\mathbf{A}$. The community matrix includes all the pairwise coefficients of interaction between species, a_{ij}.

4. Three-species systems can be shown graphically to be qualitatively stable when each species is more strongly limited by intraspecific competition than by interspecific competition. For consumers of discrete resources, this requires that there be at least as many resources as species that consume them.

5. When resources are continuously distributed with respect to qualities such as size, there is a limit to the amount of overlap between the joint use of resources that permits coexistence in a multispecies system. In general, a third species cannot invade the space on the resource continuum between two others when the coefficient of competition between the latter two exceeds about 0.5.

6. Experimental studies of competition have demonstrated higher-order interactions among competitors and among competitors and their predators. Some such interactions might arise from nonlinearities in the intraspecific response of population growth to density. Others undoubtedly are caused by shifts in feeding behavior resulting from the effects of certain competitors on food resources or in order to reduce predation.

7. Even in a multispecies Lotka-Volterra system, an increase in one competitor can result in an increase in the population of another through effects on a third mutual competitor. This type of dynamic is referred to as facilitation.

8. Predation or another source of mortality can either promote the coexistence of incompatible competitors or destroy the compatibility of others, depending on its intensity. In models incorporating the demographic effects of predation, the potential exponential growth rates (r) of competing populations, in addition to values of a and K, influence population equilibria.

9. Experiments in which predators are either removed from or added to communities frequently show that the action of predators promotes coexistence by selectively depressing dominant competitors.

10. Investigations of the qualitative and quantitative stability of large food webs have led to several general conclusions: (1) The average number of other species with which one interacts should be independent of species diversity. That is to say, the connectance of food webs (the percentage of potential interactions between species pairs that are nonzero) should decrease as the number of species in the community increases. (2) Food chains should be short, generally not more than three or four species, because the quantitative stability of longer food chains is much reduced. (3) Omnivores (species that feed on more than one trophic level) should be scarce because such food chain links decrease the qualitative stability of systems.

25

Evolutionary Responses and Extinction

When discussing competition and predation, we assumed that carrying capacities, competition coefficients, and predation efficiencies were uniform characteristics of each population. In fact, each of these traits varies according to genetically determined adaptations of individuals, and each is therefore subject to evolutionary change. Populations of predators, prey, and competitors form part of the environment of every species. Each selects traits in the others that tend to alter their interactions. For example, by capturing easily found prey, a predator leaves behind—hence selects—more cryptic prey. These reproduce and pass on to future generations their protective coloring or other effective defenses against predators. As a result, the efficiency with which the population of predators exploits that prey population as a whole tends to diminish over evolutionary time.

When two populations interact, both respond by evolutionary change. When their relationship is antagonistic, as it is between predator and prey, the species can become locked into an evolutionary battle to increase their own fitness, each at the other's expense. Such a struggle can lead to an evolutionary stalemate in which both antagonists continually evolve in response to each other or run out of the genetic variation needed to fuel further evolutionary change. In either event, the net outcome of their interaction may reach an equilibrium.

Adaptations determine how species interact. What we symbolize in models as predator efficiencies and coefficients of competition represent the population consequences of adaptations of organisms to their environments. Even without knowing the details of the adaptations, we may inquire how evolutionary changes in interaction coefficients might alter the equilibrium sizes and temporal stability of interacting populations.

In this chapter, we shall approach this problem primarily by examining the behavior of experimental systems both in terms of the theory of population interactions and with an eye to generalizing results to natural communities. Because evolved responses require many generations, they are difficult to appreciate except in theoretical models and experiments with short-lived organisms. Furthermore, if most natural systems achieve approximate evolutionary equilibria, we cannot perceive their evolutionary dynamics except following disturbance. But changes in the environment and the appearance of new species in the community occasionally stir the bubbling evolutionary pot of population interactions, as we shall see.

The lethal myxoma virus became benign to rabbits through the rapid evolution of rabbit and virus populations.

Shortly after the release of a few pairs on a ranch in Victoria in 1859, the European rabbit became a major pest in Australia. The hundreds of millions of rabbits distributed throughout most of the continent destroyed range and pasture lands and threatened wool production. The Australian government tried poisons, predators, and other potential controls, all without success. After much investigation, the answer to the rabbit problem seemed to be a myxoma virus (a relative of smallpox) discovered in populations of a related South American rabbit. Myxoma produced a small, localized fibroma (a fibrous cancer of the skin). Its effect on South American rabbits was not severe, but a European rabbit infected by the virus died quickly of myomatosis.

In 1950, myxoma virus was introduced locally in Victoria and an epidemic of myxomatosis broke out and spread rapidly (Fenner and Ratcliffe 1965; Fenner 1971). The virus was transmitted primarily by mosquitoes, which bite infected areas of the skin and carry the virus on their snouts. The first epidemic killed 99.8 per cent of the individuals in infected rabbit populations. But during the following myxomatosis season (coinciding with the mosquito season), only 90 per cent of the remaining population was killed, and during the third outbreak only 40 to 60 per cent of infected rabbits succumbed.

The decline in the lethality of myxomatosis in the Australian rabbits resulted from evolutionary responses in both the rabbit and the virus populations. Before the introduction of myxoma, some rabbits had genetic factors that conferred resistance to the disease. These were strongly selected by the myxoma infection until most of the surviving rabbit population consisted of their resistant progeny. At the same time, virus strains with less virulence predominated because reduced virulence lengthened the survival time of infected rabbits and thus increased the mosquito-borne dispersal of the virus (mosquitoes bite only living rabbits). A virus organism that kills its host quickly has little chance of being carried by mosquitoes to other hosts.

Left to its own, the Australian rabbit-virus system would probably evolve to an equilibrial state of benign, endemic disease, as it is in the population of South American rabbits from which the myxoma was isolated. Pest management specialists keep the system out of equilibrium and maintain the effectiveness of myxoma as a control agent by finding new strains of the virus to which the rabbits have yet to evolve immunity.

Population interactions of predator and prey are affected by their mutual evolutionary responses.

Selection of decreased virulence in myxoma virus was accomplished by the mosquito vector. Most predators do not rely on a third party to find prey, and rather than evolve toward a benign equilibrium of restraint and tolerance, predator and prey more likely become locked in an evolutionary battle of persistent intenseness. The outcome of the battle depends on which population gets the evolutionary upper hand at the moment.

David Pimentel and R. Al-Hafidh (1963) explored the evolution of host-parasitoid relationships with the housefly and a wasp parasitoid of the fly pupae, *Nasonia vitripennis* (Figure 25–1). In one population cage, *Nasonia* was allowed to parasitize a fly population that was kept at a constant level by replenishment from a stock that had not been exposed to the wasp. None of the flies that escaped wasp parasitism were returned to the population cage; hence, the wasps were provided only with evolutionarily "naive" hosts.

In a second population cage, the fly hosts were kept at the same constant number, but because emerging flies were allowed to remain the population could evolve resistance to the wasps. The population cages were maintained for about 3 years, long enough for evolutionary change to occur. Over the course of this experiment, the reproductive rate of the wasps dropped from 135 to 39 progeny per female, and their longevity decreased from 7 to 4 days. The average level of the parasite population also decreased (1900 adult wasps versus 3700 in the nonevolving system), and the population size was more constant. These results suggested that when subjected to intense parasitism, flies evolved additional defenses.

Experiments were then established in 30-compartment population cages in which the numbers of flies were allowed to vary freely. One such experiment was started with flies and wasps that had no previous contact with each other, and a second was established with animals from the evolving population discussed above. In the first cage, wasps were efficient parasites, and the system underwent severe oscillations. In the second cage, however, the wasp population remained low and the flies attained a high and relatively constant population (Figure 25–2). This result strongly reinforced the conclusion drawn from earlier experiments that the flies had evolved resistance to the wasp parasites.

Figure 25–1
The wasp *Nasonia* parasitizing a pupa of the housefly. (Courtesy of D. Pimentel; from Pimentel 1968.)

The genetics of parasites and diseases of economic crops and the genetic basis of crop resistance have been scrutinized closely by agronomists (for example, Plank 1968; Watson 1970). Control of such pathogens as wheat rust is accomplished primarily by breeding strains of wheat with genes that confer resistance to the pathogen. But defenses provided by resistance genes have proven to be easily circumvented by rapid evolutionary change in the pathogen, probably a single gene replacement in many cases. Agricultural geneticists keep track of such changes in plant disease organisms by routinely exposing different genetic strains of the crop plant to a variety of races of the pathogen and recording the virulence (Green 1971). A surprising result of a Canadian survey of wheat rust—a fungus—was

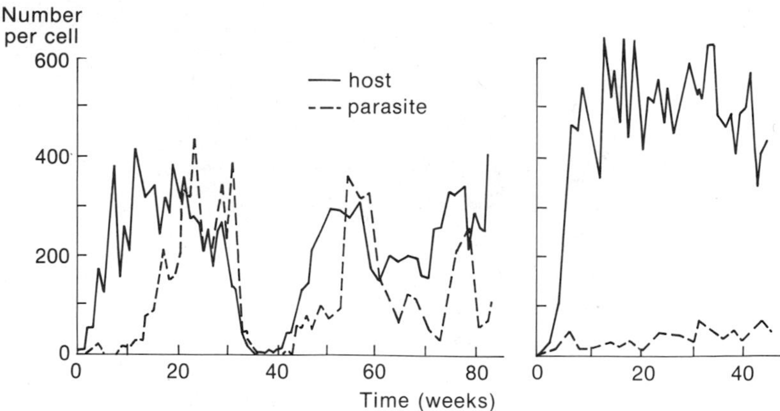

Figure 25–2
Populations of houseflies and a wasp parasitoid, *Nasonia vitripennis*, in 30-celled laboratory cages. Control *(left):* flies had no previous experience with the wasp; experimental *(right):* flies had been exposed to wasp parasitism for more than 1000 days. (After Pimentel 1968.)

that, for a given race, virulence on strains of wheat with different resistance genes appeared and disappeared sporadically (Green 1975).

Altered virulence in this system apparently derives from changes in single genes. For example, in 1969, race 15 B-1L of the wheat rust was virulent on strains of wheat with resistance genes (Sr) 8, 10, and 11, and it was avirulent on strains with genes 15 and 17. In 1970, a subrace of the rust was isolated that had become avirulent on Sr 11. In 1971, a new subrace appeared that was virulent on Sr 15. That same subrace lost its virulence on Sr 8 the following year; another lost its virulence on Sr 11. In 1973, a new subrace virulent on Sr 17 appeared.

The significance of virulence changes for the rust is not clear. The virulence of different resistance genes was determined only in experimental plantings. Hence the changes in virulence did not appear in the population of rust as a whole. What the rust story does reveal, however, is that natural populations of predators and parasites continually generate genetic variants that challenge the resistance of their prey and hosts, which presumably respond in kind. Of course, changes in many aspects of predatory behavior and prey defense do not have such simple genetic bases as virulence and resistance in the wheat-rust system. Nevertheless, it is clear that some consumers and their resource populations are locked into an endless evolutionary struggle.

Predator and prey adaptations may achieve an evolutionary equilibrium.

The evolutionary responses of predator and prey populations can be depicted by a simple graphical model that relates the rates of evolution of the

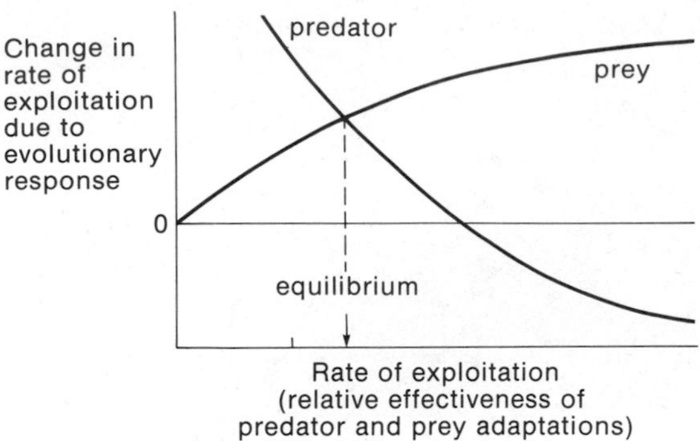

Figure 25–3
A graphical model of the evolutionary equilibrium between predator and prey adaptations that determine the level of exploitation in the system. When exploitation rates are low, there is little selection on the prey but predators are strongly selected to increase rate of prey consumption. The reverse is true at high exploitation rates.

two to the efficiency of predation (Rosenzweig 1973; Figure 25–3). For the prey, the rate at which new adaptations to escape or avoid predators are selected should vary in direct proportion to the predation rate. In the absence of predation, there can be no selection of adaptations for predator avoidance. But as predation increases, so does selection and evolutionary response, at least up to limits set by availability of genetic variation.

Selection of new adaptations of the predator to exploit the prey should vary in opposite fashion. When a particular prey species is not heavily exploited, adaptations of predators to utilize this resource will be selected and predation upon that prey population will increase. As exploitation of the prey increases, however, intraspecific competition among predators reduces the selective value of further increase. Very high rates of predation conceivably could select individuals that shifted their diets toward other prey species. Hence evolution by a predator population could result in decreased efficiency of utilizing a particular prey species, as indicated in Figure 25–3.

In this simple model, the countervailing influences of predator and prey adaptation achieve a stable evolutionary steady state where the two curves cross. When predator adaptations are relatively effective and the prey are exploited at a high rate, selection on the prey population will tend to improve its escape mechanisms relatively faster than selection on the predator population will improve its ability to exploit the prey. Conversely, when the exploitation rate is low, the prey evolve relatively more slowly than the predators. This steady state between the adaptations of predators and prey should result in a relatively constant rate of exploitation regardless of the specific predator and prey adaptions.

Pimentel's experiments on host-parasite interactions, described above, illustrate the dynamics of the predator-prey equilibrium. The housefly (host) and the parasitic wasp *Nasonia* undoubtedly have achieved an evolutionary equilibrium in their natural habitats. When brought into a simple laboratory habitat, *Nasonia* wasps are able to exploit housefly populations at a greatly increased rate because they require little time to search out hosts. This is equivalent to shifting the exploitation rate of *Nasonia* on houseflies far above the equilibrium level in Figure 25–3. This shift increases selective pressure on the housefly to escape parasitism relative to selective pressure on the predator to further increase its exploitation rate. As a result, the ability of the housefly to escape parasitism increased and the level of exploitation by *Nasonia* decreased toward a new steady state.

Evolution of higher predation efficiency and higher prey production tend to destabilize predator-prey models.

The simple graphical model of predator-prey interaction developed in Chapter 21 can be used to examine the effects of changes in predator and prey adaptations on their joint equilibrium and its relative stability. On a graph whose horizontal axis is the number of prey and whose vertical axis is the number of predators, the prey isocline ($dH/dt = 0$) can be represented as a hump-shaped curve touching the horizontal axis at 0 and K_H; the predator isocline ($dP/dt = 0$) can be represented as a vertical line (Figure 25–4). The intersection of the predator isocline and the prey axis represents the number of prey required to maintain the population of predators at a constant level. As you will remember, when the predator isocline lies to the left of the peak of the hump in the prey isocline the system is qualitatively unstable near the equilibrium point. To the right of the hump it is qualitatively stable.

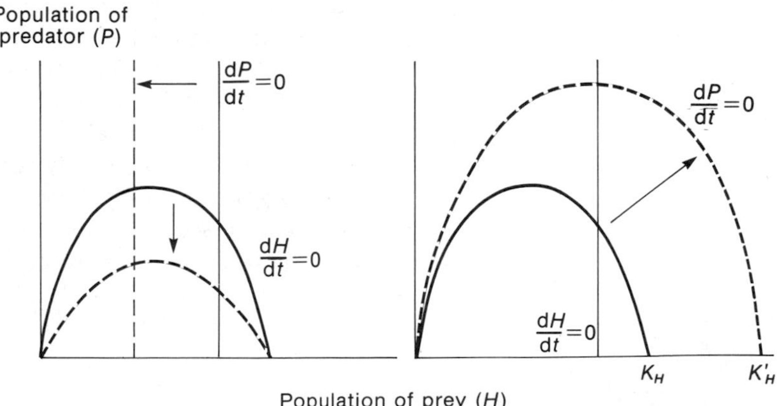

Figure 25–4
Evolution of predator and prey isoclines. *Left:* An increase in predator efficiency shifts the predator isocline to the left and reduces the prey isocline, tending to reduce the stability of the system. *Right:* Increased efficiency of the prey at exploiting its own food resource greatly enlarges the region of prey population increase but can also reduce the stability of the predator-prey interaction.

Evolution of increased predator efficiency reduces the prey isocline and shifts the predator isocline to the left. As a result, both equilibria decrease. Perhaps more importantly, the predator isocline may shift to the left of the hump and lead to increasing oscillations of predator and prey populations. Decreased predation efficiency has the opposite result.

When prey exploit their own resources more efficiently, the prey population may achieve a higher carrying capacity (K_H) and sustain a larger predator population. The resulting shift in the prey isocline could result in the predator's isocline lying to the left of the hump in the prey isocline and could, therefore, reduce the stability of their interaction.

In general, increased efficiency of resource consumption by either the predator or prey tends to destabilize a predator-prey system. This tendency might normally be balanced by the evolution of adaptations of prey to escape predation. But although an evolutionary equilibrium is likely, population stability and coexistence of predator and prey populations are not guaranteed. When adaptations favored by selection of traits among individuals tend to destabilize the predator-prey system, the system itself can be selected against in the sense that such overly efficient predators drive their prey and, subsequently, themselves to extinction. Such selection is analogous to group selection of traits that reduce intraspecific competition and promote population stability. The efficacy of such selection in nature is unknown (Gilpin 1975).

Many have argued that the evolution of reduced virulence in myxoma virus is just such as case. Within an individual rabbit, the most virulent genotypes of virus predominate because they proliferate most rapidly; but such viruses do not persist in the population at large because their virulence reduces their ability to colonize new hosts. Paul Ewald (1983) took exception to this idea. He argued that when a disease organism is transmitted by biting vectors high virulence should be selected because the probability of being transferred from host to host increases with the density of the parasite in the blood. In support of his argument he pointed out that most diseases transmitted by animal vectors (for example, malaria and sleeping sickness) are associated with high mortality rates in humans, whereas diseases transmitted by contact or in aerosols (influenza and most venereal diseases) typically have low mortality rates. Other interpretations of these data are possible and the question is by no means resolved. But the rapid evolution of disease organisms in response to such drugs as penicillin (a type of host defense) underscores the fact that resource-consumer systems have evolutionary as well as demographic dynamics.

The outcome of competition experiments may depend on the genotypes of the competitors.

Evolution proceeds by the replacement of genotypes within a population—an expression of intraspecific competition between individuals having different genes. When two species are closely matched in competition, the different competitive abilities of genotypes within a population may result in a different outcome when a population is predominated by one genotype than when it is predominated by another. Sometimes the phenotypic expressions of such genes express themselves in the phenotype so subtly as to avoid detection by examining individuals directly. Instead, they must be inferred from the outcome of competition. In this sense, competitive ability summarizes the interaction of a phenotype with its environment. The experiments discussed below demonstrate genetic variation for competitive ability. With such genetic variation it is also clear that competitive ability can evolve.

In competition experiments involving the flour beetles *Tribolium confusum* and *T. castaneum,* Thomas Park (1954, 1962) found that within certain ranges of temperature and humidity, within which the competitive abilities of the two species were closely matched, the outcome was uncertain. At 29 °C and 70 per cent relative humidity, for example, *T. castaneum* "won" in about five-sixths of the experiments, and *T. confusum* in one-sixth.

Park began all his experiments with 2 pairs of each species. When I. M. Lerner and F. K. Ho (1961) began experiments under identical conditions with 10 pairs of each species, *T. castaneum* won 20 out of 20 times. These results suggested that in Park's experiments, genetic variation in the parent individuals might have altered the outcome of competition. The larger parental populations used by Lerner and Ho probably resembled the average genetic makeup of the species more closely, thereby producing more consistent results.

To test this hypothesis, Lerner and E. M. Dempster (1962) developed several strains of each species that were inbred to minimize their genetic variation, and then tested the strains against each other in competition experiments. They found that if the populations were inbred for more than 12 generations, certain strains of *T. confusum* would consistently outcompete certain strains of *T. castaneum*—a reversal of their normal competitive relationship. These experiments indicated that changes, even minor ones, in the genetic constitution of a population can greatly influence its competitive ability (Park et al. 1964). Similar results have been found in competition experiments involving rice (Sakai 1961), *Erodium* (geranium family; Martin and Harding 1981), clover (*Trifolium;* Turkington and Harper 1979), and fruit flies (*Drosophila;* Ayala 1969, 1972), among others (see, especially, Turkington and Aarssen 1984).

In Ayala's (1969) experiments on competition between *Drosophila serrata* and *D. nebulosa,* the competitive ability of *D. serrata* appeared to increase during the course of the experiment. When the two species were established in population cages at 19 °C, they quickly achieved a pattern of stable coexistence with 20 to 30 per cent *D. serrata* and 70 to 80 per cent *D. nebulosa.* In one experiment, but not in a replicate, the frequency of *D. serrata* began to increase after the twentieth week and attained about 80 per cent by the thirtieth week, a reversal of the initial predominance of *D. nebulosa.* When individuals of both species were removed from the experimental populations after the thirtieth week and tested against stocks maintained in single-species cultures, the competitive ability of each species was found to have increased after exposure to the other in the competition experiment. In one of the replicates, the competitive ability of *D. serrata* evidently had evolved much more rapidly than that of *D. nebulosa,* and their equilibrium frequencies were greatly altered. The difference between these replicates also appeared in the competitive ability of *D. serrata* tested against the unselected stocks of *D. nebulosa.*

Hairston (1983) reported a parallel situation in natural populations of salamanders of the genus *Plethodon* living in the Appalachian Mountains. In one section of the mountains, the vertical ranges of *P. jordani,* a high-altitude species, and *P. glutinosum,* a low-altitude relative, overlapped by only about 100 meters. In another section, their altitudinal overlap was about 1200 meters. Hairston (1973) predicted that competition between the two species is more intense where their altitudinal overlap is narrower, and he later demonstrated this by competition experiments in the field (Hairston 1980). He also transplanted individual *P. jordani* from the region of narrow overlap to plots within the region of broad overlap and found that the populations of local *P. glutinosum* declined as a result (Hairston 1983). The simplest interpretation of this result is that the competition coefficient a_{gj} of individuals from the area of narrow overlap exceeds that of individuals from the area of broad overlap; *P. jordani* evidently is a better competitor against *P. glutinosum* in the first area than in the second.

Interspecific competitive ability may be selected more strongly in sparse populations.

In experiments similar to those of Ayala, Pimentel et al. (1965) have shown that a poor competitor can evolve a competitive advantage (judged by relative population density) over a formerly superior adversary. J. B. S. Haldane (1932) suggested that when populations were rare, intraspecific competition would be greatly reduced, permitting the evolution of greater efficiency in interspecific competition.

> [T]hat natural selection will always make an organism fitter in its struggle with the environment . . . is clearly true when we consider the members of a rare and scattered species. It is only engaged in competing with itself against inorganic nature. But as soon as a species becomes fairly dense matters are entirely different. Its members inevitably begin to compete with one another. I am not thinking only of the active and often conscious competition between higher animals, but also of the struggle for mere space which goes on between neighboring plants of closely packed associations. And the results may be biologically advantageous for the individual, but ultimately disastrous for the species.

To test this idea, Pimentel et al. (1965) conducted laboratory experiments with flies to determine whether two species could coexist on one food resource by frequency-dependent evolutionary changes in their competitive ability. That is, could one species, as it is being excluded by the second and becoming rare, evolved increased interspecific competitive ability rapidly enough to gain back the upper hand? J. A. Moore (1952) and Lerner and Dempster (1962) had earlier shown that competitive ability could be selected.

The housefly *(Musca domestica)* and the blowfly *(Phaenicia sericata)*, which have similar ecological requirements and comparable life cycles (about 2 weeks), were chosen for the experiments. Both species feed on dung and carrion in nature, and they are often found together on the same food resources. The flies were raised in small population cages at 27°C, with a mixture of agar and liver provided as food for the larvae and sugar for the adults. The outcomes of an initial series of four competition experiments between individuals from wild populations of the housefly and the blowfly were split two each. The mean extinction time for the blowfly, when the

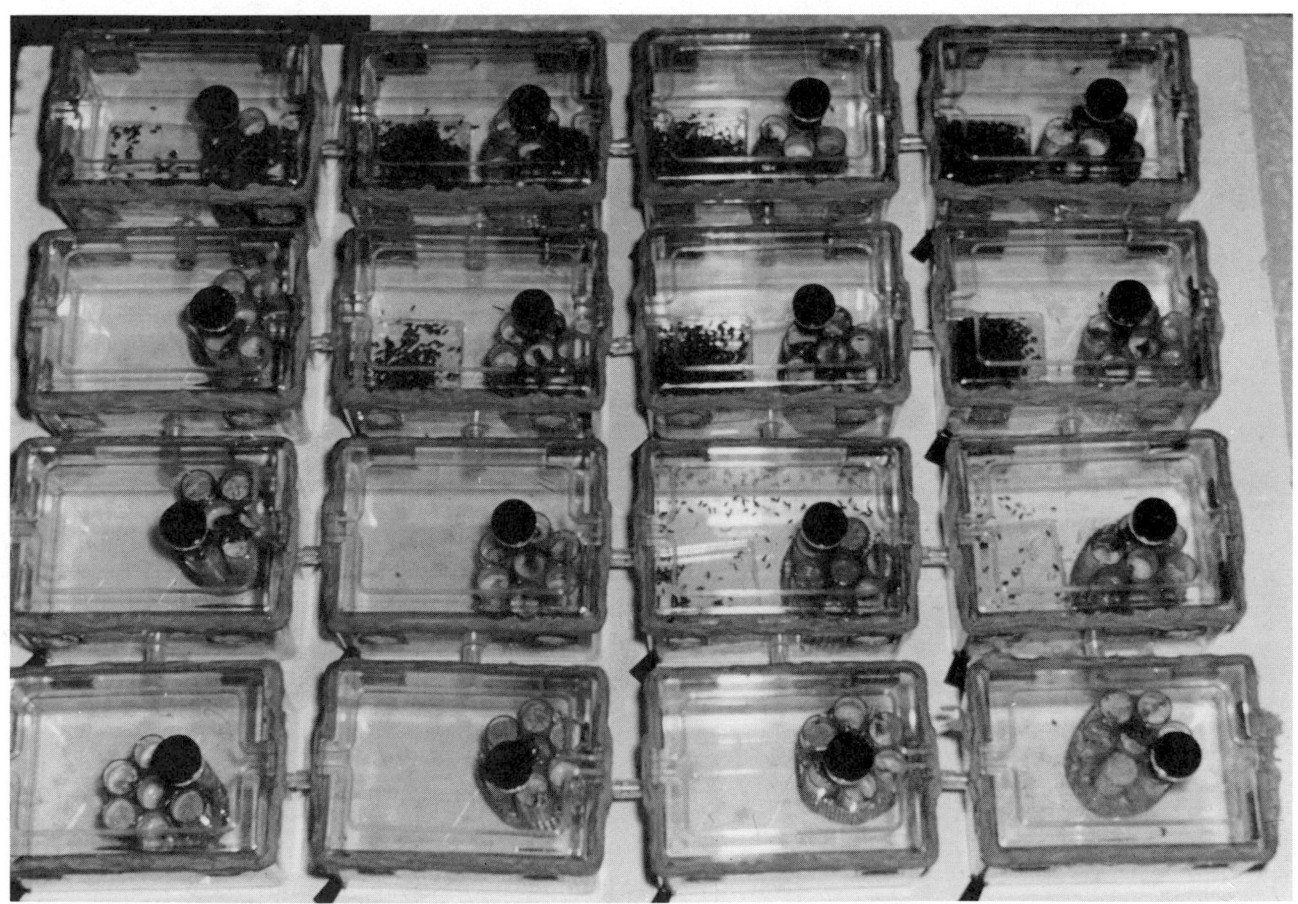

Figure 25–5
The 16-cell cage used by Pimentel to study competition between populations of flies. Note the vials with larval food in each cage and the passageways connecting the cells. The dark objects concentrated in the upper-right-hand cells are fly pupae. (Courtesy of D. Pimentel; from Pimentel et al. 1963.)

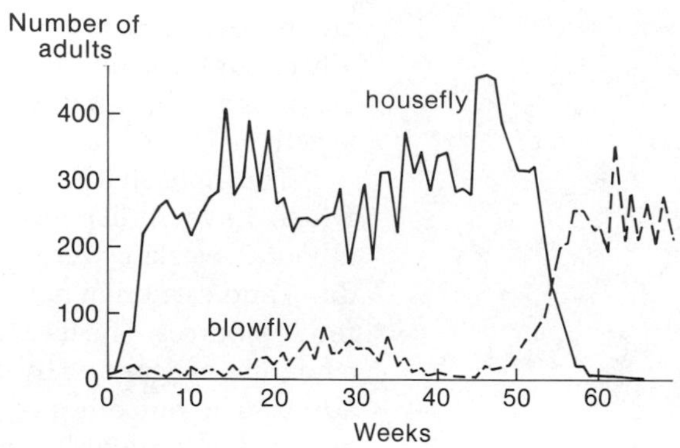

Figure 25–6
Changes in competing populations of houseflies and blowflies in a 16-celled cage. (After Pimentel et al. 1965.)

housefly won, was 92 days; it was 86 days for the housefly when the blowfly won. Thus the two species were close competitors, but the small cages used did not allow enough time for evolutionary change before one of the populations was excluded.

To prolong the housefly-blowfly interaction, Pimentel and his colleagues started a population in a 16-cell cage, which consisted of single cages in 4 rows of 4 cages with connections between them (Figure 25–5). Under these conditions, populations of houseflies and blowflies coexisted for almost 70 weeks, and showed a striking reversal of number between the two species (Figure 25–6). After 38 weeks, when the blowfly population was still low and just a few weeks prior to its sudden increase, individuals of both species were removed from the population cage and tested in competition with each other and with wild strains of the housefly and blowfly. Captured wild blowflies turned out to be inferior competitors to wild and experimental strains of the housefly. But blowflies that had been removed from the population cage at 38 weeks consistently outcompeted both wild and experimental populations of the housefly. Apparently, the experimental blowfly population had evolved superior competitive ability while it was rare and on the verge of extermination. For further discussion of the evolutionary reversal of competitive dominance, see C. M. Pease (1984).

Selection can alter the coefficients of Lotka-Volterra competition.

Competition between genotypes within a population leads to a replacement of individuals by others with higher intrinsic capacity for increase *(r)* at a given population size. The effect of raising the population growth rate is to increase the equilibrium population size *(K)*, as shown in Figure 25–7. When a population is limited by availability of resources, selection will tend to

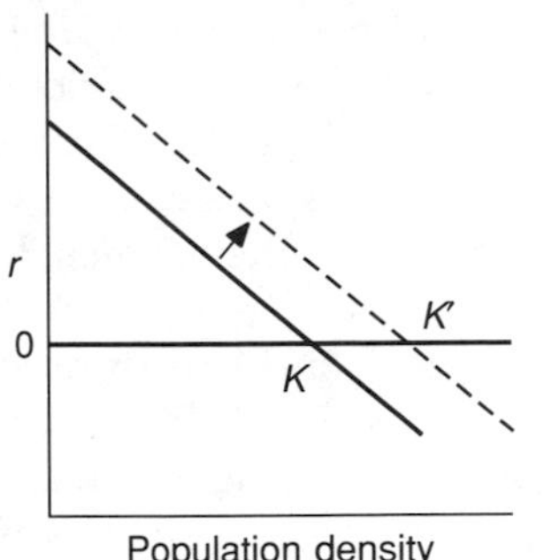

Figure 25–7
The effect of intraspecific competition on the evolution of the curve relating population growth rate to density. Selection favors individuals with higher intrinsic capacity for increase and results in an increase in the carrying capacity of the population from K to K' in the absence of balancing forces.

increase its carrying capacity by favoring adaptations that increase the efficiency with which individuals utilize resources. This tendency is balanced by correlated responses to selection having opposite effect and by the adaptations of competitors, predators, and prey. At length, selection must exhaust the genetic variation and evolutionary opportunities available for increasing K.

In the case of interspecific competition, both species respond by adaptation to each other (Levin 1971; Leon 1974; Lawlor and Maynard Smith 1976; Roughgarden 1974, 1976; Arthur 1980). Selection of increased carrying capacity (K) shifts the equilibrium isocline $(dN_i/dt = 0)$ outward from the origin of the population graph and thus tends to increase competitive ability (Figure 25–8). But with selection applied equally to both species, the eventual outcome of the evolutionary relationship of two competitors is difficult to predict.

Selection may also be applied to the coefficients of competition (a_{ij}) between the two species (Gill 1974, Hairston 1983). Ecological divergence of two species tends to reduce their competition coefficients. The resulting shift in the equilibrium isoclines tends to stabilize the interaction and allows the species to coexist indefinitely (Figure 25–8).

Because a decrease in a_{ij} without a change in K_i invariably results in an increase in the equilibrium population N_i, selection will always favor reduced a_{ij}, all other things being equal. Why, then, do we observe competition between species in nature at all? From the standpoint of Lotka-Volterra competition, the answer is that decreases in a_{ij} inevitably lead to decreases in K_i, which more than offset the advantage of reduced a_{ij}. From the standpoint of the species' ecology, this means that a shift in resource utilization neces-

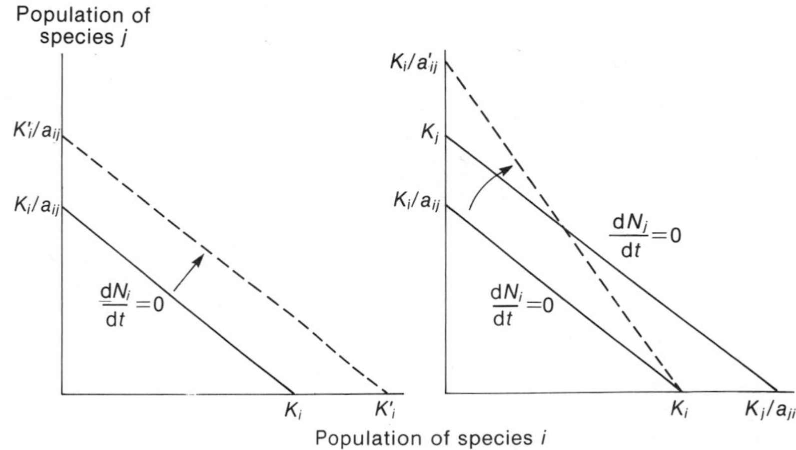

Figure 25–8
Evolution of the isoclines of competing populations. *Left:* Selection increases the carrying capacity (K_i) and shifts the isocline away from the origin of the graph. Both intercepts (K_i and K_i/a_{ij}) increase by the same factor and i's ability to exclude competitors increases. *Right:* The angle of the isocline shifts following a decrease in a_{ij} caused by ecological divergence; accordingly, the probability of coexistence increases.

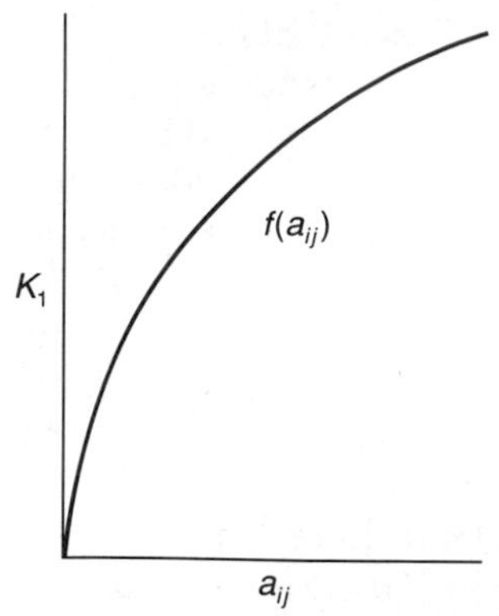

Figure 25–9
An arbitrary function relating carrying capacity (K_i) to degree of ecological overlap with other species (a_{ij}) based on the assumption that competitors utilize resources yielding the greatest return.

sary to reduce a_{ij} may also reduce the amount of resource available to a population. In the absence of competition, evolutionary adaptations will result in species converging on the best possible resources. When other consumers are present, resource utilization by each species balances the intrinsic qualities of resources against the depression of their availability by competitors.

What is the optimum level of a_{ij}? Suppose that K_i and a_{ij} are related by a function $[K_i = f(a_{ij})]$ like the one graphed in Figure 25–9. In this case, K_i increases as a_{ij} increases. Presumably, species j utilizes the best resources, and so the more similar i is to j, the higher its carrying capacity. Now we wish to determine the combination of a_{ij} and K_i lying on this curve that results in the highest value of N_i. We may rearrange the equation for equilibrium under logistic competition to

$$K_i = a_{ij}(K_j - a_{ji}N_i) + N_i \qquad (25-1)$$

For given values of a_{ji} and K_j, this equation describes a family of straight lines with slope $(K_j - a_{ji}N_i)$ and intercept N_i, a different line for each value of N_i. The answer to the question of which a_{ij} results in the highest N_i is the value of a_{ij} at which point the curve $K_i = f(a_{ij})$ touches the line representing the highest value of N_i. This is shown graphically in Figure 25–10, which also illustrates some basic relationships. First, the higher the value of a_{ji}, the flatter will be the lines described by equation (25–1), and the higher will be

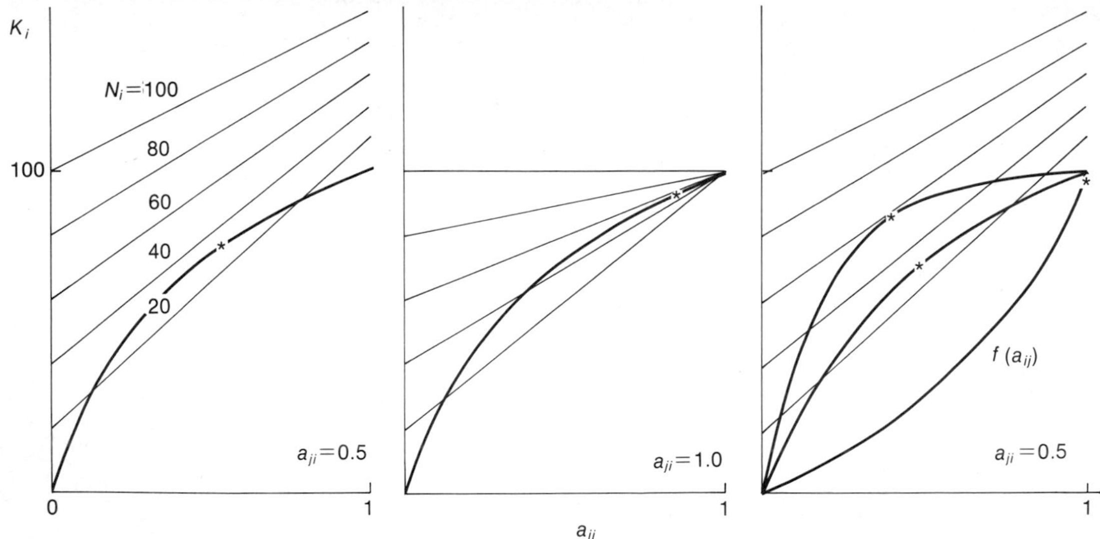

Figure 25–10
Left: Family of lines of equal population size of species 1 (N_i) in the space defined by a_{ij} and K_i for $a_{ji} = 0.5$. For the function relating K_i and a_{ij}, indicated by the heavy line, the optimum level of a_{ij} is that point defining the highest value of N_i (indicated by *). *Center:* When species i is a stronger competitor ($a_{ji} = 1$), greater ecological overlap is favored. *Right:* When a_{ij} can be reduced without affecting K_i substantially, lower values of a_{ij} are selected. When only a few extremely profitable resources are available, competitors may converge on them.

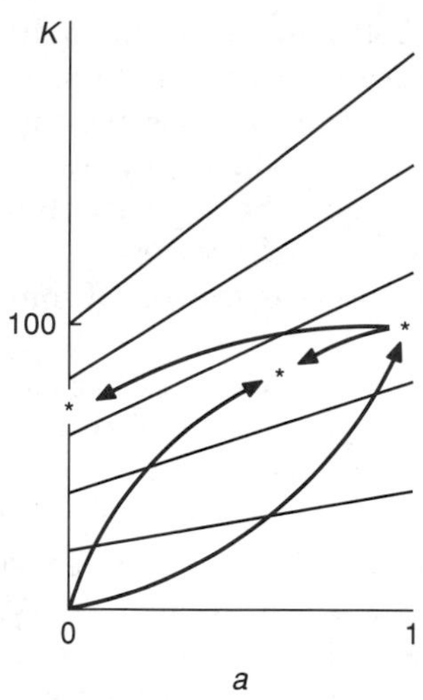

Figure 25–11
Graphical analysis of selection for optimum ecological overlap *(a)* when competition is symmetrical and both species evolve simultaneously.

the selected value of a_{ij}. Hence the better species *i* is at reducing the population of *j* (higher a_{ji}), the more strongly individuals similar to *j* ecologically will be selected. Second, the more the function $f(a_{ij})$ bulges outward—that is, the less K_i increases with a_{ij} as a_{ij} approaches 1—the lower will be the selected value of a_{ij}.

When values of *a* are based upon the exploitation of mutual resources, a change in a_{ji} will also likely result in a similar change in a_{ji}. For the special case in which competition is symmetrical and the two values are the same ($a_{ij} = a_{ji} = a$), equation (25–1) becomes $K_i = a(K_j - aN_i) + N_i$. When one further assumes for the sake of simplification that species *j* evolves in response to species *i* in identical fashion to the response of *i* to *j* ($K_i = K_j = K$), then equation (25–1) reduces to $K = N + aN$, the lines for which are shown in Figure 25–11. The tangent of one of these lines to the curve $f(a)$ is the joint equilibrium for both species.

If the environment held few kinds of resources, so that slight divergence between the species would lead to a more limited resource supply, the curve of $f(a)$ would bulge downward and the optimum value of *a* would be 1.0. Hence both species would converge on the same resources and, in all probability, one would outcompete the other. Conversely, in the presence of many kinds of resources, specialization does not reduce *K* substantially, except at the extreme, and a lower value of *a*, conceivably even $a = 0$, would be selected. These simple models suggest that the optimum degree of divergence and specialization depend upon the spectrum of resources available, which, in turn, depend upon the characteristics of the environment and the presence of other consumers.

Is competition a potent evolutionary force in nature?

Theory suggests that if resources are sufficiently varied, competitors should diverge and specialize. Laboratory experiments have shown that competitive ability may have a genetic component and therefore be under the influence of selection, although particular adaptations have not been identified. But if competition exerts a potent evolutionary force in nature, one should be able to find evidence that competitors have partly molded each other's adaptations.

The most straightforward responses to such selection are ecological divergence and specialization. The observation that species differ in their ecological requirements is commonplace (Lack 1971). But even so, differences between species have not necessarily evolved as a result of their interaction (Connell 1980). An alternative explanation for such differences is that the species became adapted to different resources in different places, and when their populations subsequently became overlapping due to range extensions these ecological differences remained.

One way to get around this objection is to compare the ecology of a species in an area in which it co-occurs with a competitor to its ecology in another area where the competitor is absent. If the areas are otherwise similar, then any differences in the species between the areas could be uniquely explained by the presence or absence of a competing species. Specifically, we may infer that divergence has occurred because of competition when two ecologically similar species differ more where they are found together than in nonoverlapping parts of their ranges. The demonstration of this phenomenon, referred to as character displacement (Brown and Wilson 1956), has become an important issue in the debate over the role of competition in structuring natural communities.

William Brown and E. O. Wilson considered a number of examples that seemed to fit the pattern of character displacement, one of which was an example from David Lack's work on the ground finches *(Geospiza)* of the Galapagos Islands (also see Schluter et al. 1985; Grant 1986). On islands with more than one species, the finches usually have beaks of different size, indicating different ranges of preferred food sizes. For example, on Marchena and Pinta islands, the ranges in beak size of the three resident species of ground finches do not overlap (Figure 25–12). On Floreana and San Cristobal islands, the two species *G. fuliginosa* and G. fortis have different-sized beaks. On Daphne Island, however, where *G. fortis* occurs in the absence of *G. fuliginosa,* its beak is intermediate in size between those of the two species on Floreana and San Cristobal. On Los Hermanos Island, G.

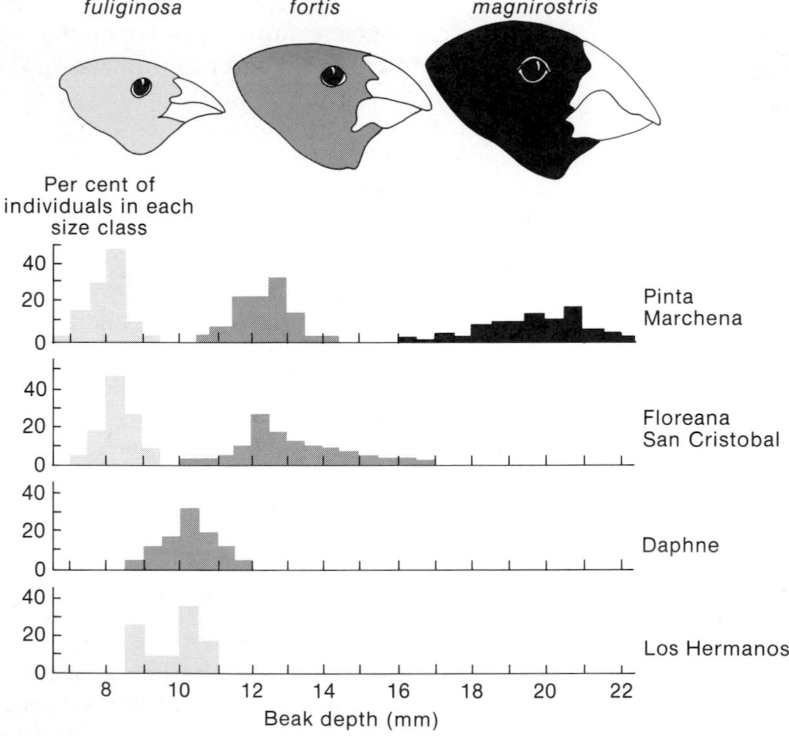

Figure 25–12
Proportions of individuals with beaks of different sizes in populations of ground finches *(Geospiza)* on several of the Galapagos Islands. (After Lack 1947.)

fuliginosa occurs in the absence of *G. fortis,* and its beak is intermediate in size.

Another example of character displacement featured by Brown and Wilson was that of two species of rock nuthatches *(Sitta)* that co-occur over part of their ranges in the Middle East. Peter Grant (1975) undertook a detailed examination of this case, including extensive fieldwork on the ecology and behavior of both species; the original description had been based primarily on examination of museum specimens (Vaurie 1950). Grant found a confusing situation with only weak and inconsistent suggestions of character displacement. In many areas of geographical overlap, the two species occupied different habitats, which by itself might have accounted for divergence. In addition, both species exhibited clines of variation in some characters over their entire ranges, which resulted in differences within the area of overlap. Summarizing his findings, Grant reiterated his earlier (1972), generally negative results:

> Ecological character displacement has remained difficult to establish despite some good recent efforts with modern (Huey and Pianka, 1974; Huey et al., 1974) and paleontological (Eldredge, 1974; Gingerich, 1974) material. Both in its extent and frequency, the ecological aspect of character displacement may have been overemphasized as an evolutionary process.

Joseph Connell (1980) similarly dismissed evidence for character displacement as being weak, including that from the oft-cited studies of Raymond Huey et al. (1974) on two species of African burrowing lizards and Thomas Fenchel (1975) on mud snails. Connell rightly pointed out that the ecological conditions in areas of overlap and nonoverlap differed so much in both studies that one could not attribute character differences exclusively to the presence of competitors. After all, whether a competitor is present or absent must depend to some extent on local ecological conditions. Thus although competition can be demonstrated in nature, has a genetic basis, and can certainly undergo evolutionary change in the laboratory, it has yet to be clearly demonstrated as playing a role in the ecological segregation of species.

Extinction expresses the failure of species to adapt.

In 1810, the American ornithologist Alexander Wilson observed an immense flock of passenger pigeons in the Ohio River Valley. For days the column of birds, perhaps a mile wide, passed overhead in numbers to darken the sky. Wilson estimated that there were more than 2 billion birds. The last passenger pigeon died in the Cincinnati Zoological Garden just over a century later. With its extinction on September 1, 1914, the passenger pigeon joined a growing list of species that have vanished from the earth.

Perhaps many of these, including the passenger pigeon, would have persisted had it not been for man's activities. But the fossil record reveals that virtually all lineages have become extinct without leaving descendants. The several million living species of plants and animals are derived from a small fraction of those alive at any time in the distant past.

Extinction is a fact of life. It is also an important expression of population processes, analogous to the death of an individual. Why do species go extinct? Clearly the answer must lie in change in the environment and the failure of the species to adapt quickly enough to counter the demographic consequences of such change. For this reason, the study of extinction ought to elucidate many aspects of ecology and evolution, but it has largely failed to do so. This failure can be traced directly to the fact that extinction is so infrequent (on a human time scale) and difficult to predict that direct observation is not feasible; extinction is merely the final event in a long sequence of subtle evolutionary and ecological processes leading to the demise of a population. In spite of the difficulties, however, it is important for us to try to understand extinction because it culminates the relationship between a population and its environment.

The stochastic probability of extinction increases as population size decreases.

Birth and death rates of populations depend on a variety of ecological factors, but whether a particular individual dies or successfully rears one or more progeny during a particular period is largely a matter of chance. When the annual probability of death is one-half, for example, some individuals live and, on average, an equal number die. There exists a finite probability, however, that all the individuals in such a population will die, just as 10 coin tosses could all come up tails with a small but finite probability (we expect this to happen once in 1024 trials on average).

Changes in populations owing to chance events are called stochastic fluctuations and their force is more strongly felt in small populations (Pimm et al. 1988). This becomes clear when we consider that the probability of obtaining 5 tails in a row with successive tosses of a coin is 1 in 32, compared with the smaller chance of 1 in more than 1000 for obtaining twice as many tails in a row. When we visualize each individual in the population as a coin, and turning up tails is equivalent to death, we see clearly that a population of 5 individuals has a higher probability of extinction than one of 10.

The probability of extinction of populations has received considerable attention from theorists, who have derived a mathematical expression relating probability of extinction at time t [$p_0(t)$] to birth rate (b), death rate (d), and population size (N),

$$p_0(t) = \left[\frac{e^{(b-d)t} - 1}{(b/d)e^{(b-d)t} - 1}\right]^N \tag{25-2}$$

(Pielou 1969: 17ff). For understanding extinction of species occurring in communities at population equilibrium, the most pertinent application of

Table 25-1
Probability of extinction when birth rate = death rate = 0.5 per year, for populations of initial size *i* within period *t*

Population size (*i*)	Time (*t*)			
	1	10	100	1000
1	0.33	0.83	0.98	0.998
10	$<10^{-4}$	0.16	0.82	0.980
100	$<10^{-48}$	$<10^{-7}$	0.14	0.819
1000	$<10^{-99}$	$<10^{-79}$	$<10^{-8}$	0.135

equation (25–2) is for the case in which b and d are equal—that is, when births balance deaths and the average change in population size is zero—for which

$$p_0(t) = \left[\frac{bt}{1+bt}\right]^N \tag{25-3}$$

Accordingly, the probability of extinction decreases with increasing population size and it increases with larger b and d, indicating more rapid population turnover. The relationship of extinction probability to population size (N) within time period (t) is shown in Table 25–1 for a population in which $b = d = 0.5$. These are reasonable values for adult death and recruitment in a population of terrestrial vertebrates. We see, for example, that for a population with 10 individuals the probability of extinction is 0.16 within a 10-year period, 0.82 within a 100-year period, and virtually certain (0.98) within 1000 years. Even for an initial population size of 1000, the probability of extinction is more than 10 per cent within a millennium and becomes virtually certain (0.999) within a million years.

The calculations in Table 25–1 assume that b and d are not density dependent but remain constant as the size of the population fluctuates stochastically from its initial value. Although this assumption is not reasonable for many populations, it probably does apply to relatively rare species for which interspecific competition predominates over intraspecific competition.

When such populations dwindle to small size, they become more and more susceptible to extinction, particularly on small islands where populations are restricted geographically and are not frequently augmented by immigration. In fact, extinction occurs frequently enough on small islands that its probability can be determined from historical records. For example, Jared Diamond (1969) compared species lists compiled in 1917 and 1968 for birds on the Channel Islands, off the coast of southern California. In the 51-year interval between censuses, there were numerous cases of species disappearances, varying from 70 per cent of the avifauna (7 out of 10 species) on Santa Barbara Island (3 km^2 in area) to 17 per cent (6 out of 36 species) on Santa Cruz Island (249 km^2). On the annual basis, these figures can be expressed as between 0.10 and 1.7 per cent of the avifauna per year, with extinction rate and island size inversely related. Comparable rates have been determined for two tropical islands: 0.2 per cent per year on Karkar, an island of 368 km^2 located 16 km off the coast of New Guinea (Diamond 1971); and 0.23 per cent per year on Mona, 26 km^2, located between Puerto Rico and Hispaniola in the Greater Antilles (Terborg and Faaborg 1973).

Populations of birds on an island of 20 km^2 probably number in the hundreds for species of small body size, and so, based on Table 25–1, these extinction probabilities seem too high. J. F. Lynch and N. K. Johnson (1974) presented evidence to demonstrate that reported extinctions may reflect inadequate census procedures, extirpations by man or introduced pests, or habitat changes on islands, more than stochastic disappearances. But populations on small islands appear more vulnerable to extinction than those on large islands.

Patterns of distribution among and within islands suggest that extinction may result from a decrease in competitive ability.

In tossing a coin some trials come up heads, others come up tails, but the probability of each outcome is the same for each trial. Similarly, some species persist and others disappear, but accumulating evidence suggests that the probability of extinction differs greatly among species. Some teeter perilously close to their inevitable fate and falter with the least ecological setback; others are resilient and productive, able to withstand the perturbations of their environments. Such differences in probability of extinction can be inferred from patterns of geographical distribution and taxonomic differentiation of populations inhabiting groups of islands, such as the West Indies.

Immigrants to islands appear to be excellent competitors initially. Colonizing species are usually abundant and widespread on the mainland; these qualities make good immigrants. Many invaders of an island exhibit increased population size and expand into habitats not occupied by the parent population on the mainland (Crowell 1962; Grant 1966; MacArthur et al. 1972; Cox and Ricklefs 1977). After an immigrant population becomes established, however, its competitive ability appears to wane; its distribution among habitats becomes restricted, and local population density decreases (Ricklefs 1970; Ricklefs and Cox 1978). These trends eventually can lead to extinction.

We can judge the relative ages of populations on islands by their patterns of geographical distribution and by differences in their appearance from the appearance of mainland forms from which they were derived (Ricklefs and Cox 1972). Range maps of representative species of birds in the Lesser Antilles (Figure 25–13) demonstrate the progressive changes in distribution and differentiation of species with time, called the "taxon cycle" by Wilson (1961). On the basis of such distribution patterns, R. E. Ricklefs and George W. Cox (1972) assigned populations to one of four arbitrary stages (Table 25–2): expanding (I), differentiating (II), fragmenting (III), and endemic (IV). Similar patterns have been described for ants on

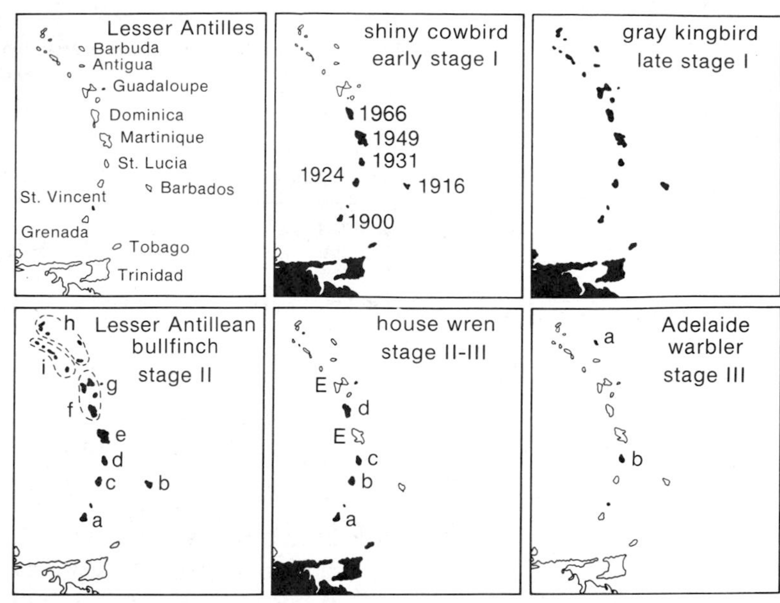

Figure 25–13
Distribution patterns and taxonomic differentiation of several birds in the Lesser Antilles, illustrating progressive stages of the taxon cycle. The shiny cowbird has expanded its range in the islands (dates of arrival are indicated); the house wren has become extinct (E) on several islands during this century. Lower-case letters designate subspecies. (After Ricklefs and Cox 1972.)

islands in the southwestern Pacific Ocean (Wilson 1961), and for birds and some insects in the Solomon Islands (Greenslade 1968, 1969). Among birds of the West Indies, species in late stages of the taxon cycle exhibit reduced population densities and restriction to a narrower range of habitats, often including montane forests (Figure 25–14).

Populations become more vulnerable to extinction as the taxon cycle progresses (Table 25–3). More endemic species (stage IV) of West Indian

habitats per species
occurrences per habitat
8
6
4
2
0
I II III IV
Stage of taxon cycle

Figure 25–14
Relative indices of population density and ecological distribution of songbirds in the West Indies as a function of stage of taxon cycle. Figures are based on censuses in 9 habitats on Jamaica, St. Lucia, and St. Kitts. (After Ricklefs and Cox 1978.)

Table 25–2
Characteristics of distribution and taxonomic differentiation of species in the stages of the taxon cycle

Stage of cycle	Distribution among islands	Differentiation between island populations
I	Expanding or widespread	Island populations similar to each other
II	Widespread over many neighboring islands	Widespread differentiation of populations on different islands
III	Range fragmented due to extinction	Widespread differentiation
IV	Endemic to one island	—

Source: Ricklefs and Cox 1972.

Table 25-3
Rate of extinction of island populations of birds in the West Indies and the Hawaiian Islands as a function of stage of the taxon cycle

	Stage of cycle			
	I	II	III	IV
West Indies				
Number of recently extinct or endangered populations	0	8	12	13
Total number of island populations	428	289	229	57
Per cent extinct or endangered	0	2.8	5.2	22.8
Hawaiian Island drepanids				
Number recently extinct	2	2	9	7
Total number of island populations	23	12	12	10
Per cent extinct	8	16	75	70

Source: Ricklefs and Cox 1972; Hawaiian data from Amadon 1950.

birds have become extinct since 1850, or are currently in grave danger of extinction, than island populations of widespread species (stage I to III). The same is true of songbirds in stages III and IV in the Hawaiian Islands.

From patterns of distribution and taxonomic differentiation we may infer that the time scales of ecological and evolutionary processes increase in the order: immigration, evolutionary differentiation, extinction. Current or recent immigrants to the West Indies show little or no taxonomic differentiation, and some species have expanded their ranges in the archipelago at the rate of one island every 10 to 50 years. There are no cases of species with gaps in their distributions, indicating extinctions of island populations, that are not highly differentiated among islands. The fact that extinction follows the inception of visible evolutionary change on islands suggests that vulnerability to extinction may have an evolutionary component.

Ricklefs and Cox (1972) proposed that the decrease in competitive ability of island populations with age may be caused by evolutionary responses of an island's biota to new species. According to their scenario, immigrants are relatively free of parasites, predators, and efficiently specialized competitors when they colonize an island, so that their populations increase rapidly and become widespread. Having reached this stage, new immigrants constitute a larger part of the environment of many other species, which then evolve to exploit, avoid exploitation by, or outcompete the newcomer. Conceivably, a large number of species, when adapting to a single abundant new population, can evolve faster than the new species can adapt to meet their evolutionary challenge. Competitive ability of the immigrants is progressively reduced by counteradaptations of island residents until the once-abundant new species becomes rare. Species are eventually forced to extinction by subsequent arrivals from the mainland that are more efficient competitors. When a species becomes rare, other species no longer gain evolutionary advantage by adapting to it, and evolutionary pressure upon the rare species is released, similar to the effect Pimentel observed in his experiments on competition between houseflies and blowflies. If this occurred before a species' decline had proceeded too far, the species might again increase and begin a new cycle of expansion throughout the island.

This apparently has occurred many times; species distributions provide ample evidence of secondary expansions within the West Indies (for example, Figure 25–13, lower left).

Although the scenario of the taxon cycle is speculative and has not found acceptance, even as a reasonable hypothesis, by some biologists (Pregill and Olson 1981), it is clear that islands offer promise for the study of extinction. The fact that one sees relationships between ecological distribution and geographical distribution suggests that sister populations of extinct island populations (species in stage III and IV) are, themselves, ecologically closer to their final demise than are populations of stage I and II species. Thus the course that leads to extinction is at least somewhat deterministic and predictable.

The relatively uncharted area of evolutionary response and extinction should crown our thinking about population interactions. Populations are genetically heterogeneous; they respond to one another by genetic change, thereby altering their interrelationships. The importance of such evolutionary accommodation to the development and regulation of community organization remains unclear, but nonetheless has stimulated much debate in ecology. Even though exclusion in laboratory experiments can occur rapidly, the mutual adjustment of populations in natural communities probably allows time for evolution to occur. Except when caused by introduced disease or predators, extinction of species almost certainly follows upon long-term evolutionary changes.

Salt (1979) argued, however, that emergent properties of systems, which he believes are derived from mutual evolution (coevolution) of their parts, only occur when the parts have had a long history of association. Hence emergent properties are less likely in communities as a whole than in smaller biological systems, particularly the organism. Connell (1980) similarly suggested that long-term association is needed for coevolution to occur. It is possible between predator and prey, he argued, because the two exist in continual association—the first cannot live without the second and actively seeks it out. Competitors are less likely to coevolve because they interact at an ecological distance through their mutual resources.

We shall examine the role of evolution in shaping biological communities in greater detail in Part VII. However, in order to provide a better background for integrating evolution and ecology at the community level, we shall consider their general relationship as expressed in the adaptations of organisms in Part VI.

SUMMARY

Populations of predators, prey, and competitors respond to each other by evolutionary change in characteristics that determine predation efficiency and coefficients of competition. Hence their interactions have evolutionary as well as population dynamics.

1. Evidence of such evolutionary changes in consumer-resource systems has been obtained in laboratory studies on host-parasite interactions. After periods of co-occurrence, rates of parasitism decreased and host populations increased, apparently following the selection of improved defenses against parasites. Further evidence of this change was obtained when stocks of hosts from these experiments were tested versus fresh stocks of parasites.

2. Studies on pathogens of plant crops—wheat rust, for example—have revealed a simple genetic basis to the outcome of the interaction.

3. Because selection for defenses of prey increases in proportion to predation rate, and selection for predatory efficiency decreases as predation rate increases, predator and prey reach an evolutionary steady state at some intermediate level of predation.

4. In simple predator-prey models, increases in consumer efficiency, either of the predator or of the prey upon its own resources, tend to reduce the stability of the predator-prey interaction.

5. The outcome of competition in many experimental systems has been shown to depend on the genotypes of the competitors, indicating a genetic basis to the coefficient of competition (a_{ij}), the carrying capacity (K), or both.

6. Experiments with competition between species of flies have revealed reversals of competitive ability over the course of tens of generations. By testing populations against unselected controls, genetic changes in competing populations were confirmed.

7. Competition selects reduced competition coefficients and thus tends to stabilize interactions between competitors and promote coexistence. The optimum level of a_{ij} varies inversely with the breadth of resources over which competing populations can diversify and specialize.

8. A test of whether competition can result in evolutionary divergence in nature is to compare ecological (or related morphological) traits of a population in the presence and absence of a competitor. When the two differ, the pattern is referred to as character displacement, but ecologists have found little solid evidence for such patterns.

9. Extinction expresses the failure of species to adapt, hence it is a key to understanding the evolutionary consequences of predation and competition. Extinction has been difficult to study because it cannot be predicted and it occurs over long periods.

10. Probability of extinction by chance events is greater in small populations than in large ones. The biologically important problem is how species become so reduced (to fewer, say, than 1000 individuals) as to be susceptible to the coup de grace of stochastic extinction.

11. Patterns of distribution of island taxa indicate that species pass through a sequence of stages (the taxon cycle) leading from initial ecological breadth and abundance to eventual extinction. Species in later stages of the cycle have reduced habitat breadths and abundances within habitats, and increased probabilities of extinction.

12. One hypothesis to account for this process is that the demographic productivity of immigrants, though high initially, decreases due to adaptations of their predators, prey, and competitors, until the population is driven to extinction by new immigrants that are superior competitors.

VI

Evolutionary Ecology

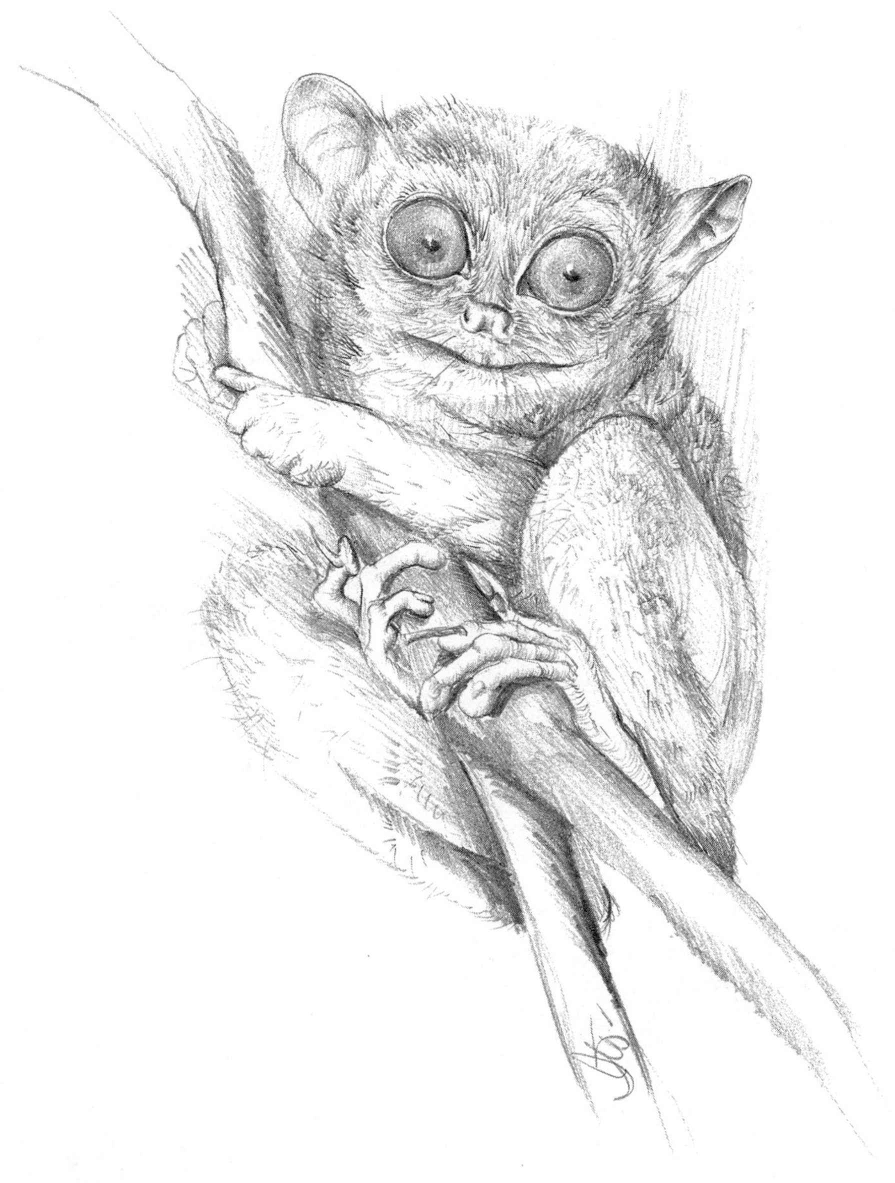

26

Evolution and Adaptation

The study of ecology encompasses many points of view. One of these holds that ecological relationships manifest interactions between the gene pools of populations and selective factors in the environment. The form and function of organisms are adaptations with genetic bases; understanding their evolution requires a knowledge of the environment, the organism, and the genetic potential of the gene pool. The environment embodies the selective pressures that establish fitness differences between individuals with different genotypes. Evolutionary responses of the phenotype to these pressures depend on functional interrelationships that limit form and function to combinations of traits that work, in the sense that they obey physical laws. Responses also depend upon the availability of genetic variation in the population, upon which selection acts. Moreover, genetic variation itself makes advantageous certain types of adaptations, principally of the breeding structure of the population, that optimally manage the expression of genetic variation in the progeny of each individual.

Mutual evolutionary responses of interacting populations, discussed in Chapter 25, affirm that a consideration of evolution is central to the study of ecological systems. Indeed, from its beginning ecology has formed links to three related biological disciplines: genetics, which studies the mechanism of heredity; evolution, which includes the study of change in the genetic makeup of populations; and development, which represents the realization of the genetic blueprint in the form of the individual. A distinct subdiscipline, called evolutionary ecology, concentrates especially on the interpretation of form and function in animals and plants as adaptations to their environments.

Evolutionary ecology has five major programs of inquiry. The first is to understand the mechanisms of evolutionary change from a genetical standpoint, including limits imposed by the availability of genetic variation. This is discussed in detail in texts on population genetics and evolution (for example, Futuyma 1986, Hartl 1980, Milkman 1982) and will be touched upon only briefly here. Second, genetic variation in populations confronts individuals with the problem of choosing the genotypes of their mates so as to optimize the genotypes of their offspring. This problem may be solved through modification of the breeding system and pattern of mate choice, as we shall see below.

The third program, to which we shall devote most of our attention, is the interpretation of form and function in the context of adaptation. This program, often called the adaptationist paradigm (Gould and Lewontin 1979) and what Eric L. Charnov (1982) calls "selection thinking," presumes that evolution can achieve that combination of traits best-suited to any particular environment and permissible within the bounds set by physical laws. The major goals of this program are to learn how physical limits on form and function prescribe the possible phenotypes among which the environment selects; to determine appropriate measures of fitness, especially with respect to traits that govern the interactions of individuals within a population; and to understand where the adaptationist paradigm can be applied validly and where it cannot.

The fourth program elaborates the last point: it is to determine the degree and mechanisms of matching of phenotypes to environments. Matching can occur both by selection of the phenotype by the environment and selection of the environment by the phenotype. The issue is the degree to which response to the environment is a property of the individual or of the gene pool of the population. Furthermore, organisms retain tangible evidence of their evolutionary history in the form of traits that are shared with related species regardless of the environment. For example, plants in the rose family have flowers with five-part symmetry; those in the lily family are endowed with a three-part or six-part symmetry. Whatever the adaptive significance of flower symmetry, it is lost in the dim evolutionary history of these groups, and does not stand at issue in their present-day evolutionary response to the environment, as do the size, color, and form of individual flower parts.

Finally, the fifth program of evolutionary ecology is to determine the extent to which properties of larger ecological systems — communities and ecosystems — depend upon evolutionary relationships among their parts.

Evolutionary response to selection proceeds by the substitution of genes within populations.

Much of the variation among organisms within a population has a genetic basis. In some cases, the differences between individuals express variation at a single genetic locus. The principal color groups of eyes (blue or brown) in humans and typical (salt-and-pepper) and melanic (black) individuals of the peppered moth result from different forms of the same genes (alleles). In other cases, traits come under the influence of many loci whose effects may be additive, complementary, or modifying. Evolutionary change reflects the substitution of new alleles for old ones at individual genetic loci. When selection is applied to some traits, such as the amount of black pigment in the wing scales of moths, a single gene locus or a few gene loci are affected. Selection upon other traits, such as body shape or patterns of social behavior, may affect numerous genes responsible for producing the trait, many of which in turn influence other characteristics of the phenotype.

The evolutionary mechanics of selection and genetic responses are the subject of population genetics (Crow and Kimura 1970, Hartl 1980). A primary task of population geneticists since the late 1920s has been to develop quantitative predictions of changes in gene frequencies in response to selection. In such simple cases as selection upon single genetic loci with simple genetic dominance between alleles, these models describe mathematically the intuitive result that the allele whose bearers leave the most descendants eventually predominates in the gene pool. The equations of population genetics also allow one to predict rates of change in gene frequency, hence how rapidly a population can respond genetically to a change in the environment.

The time required for a dominant gene to replace its recessive allele depends on its frequency at the beginning and end of the substitution process and the strength of selection. In the case of the replacement of the typical form of the peppered moth by the *carbonaria* form in polluted woods of England, this substitution is known to have taken about a century. From the results of H. B. D. Kettlewell's experiments on the peppered moth, one can estimate that the fitness of the allele for typical coloration was only 47 per cent that of the *carbonaria* allele; hence the fitness differential, or strength of selection, was 0.53. Plugging this value into equations predicting the time required for allele substitution, a change in frequency of the *carbonaria* allele from 0.05 to 0.95 would have required 47 generations, while a change from 0.01 to 0.99 would have required 204 generations (Ricklefs 1979). The peppered moth has 1 generation each year, and so the population genetics equations appear to be consistent with the observed time course of the gene substitution, assuming an initial frequency prior to the industrial revolution of about 1 per cent. Had selection been only a tenth as strong, the same change in frequency would have required 10 times as long.

The same equations may be turned around to estimate fitness differentials from changes in gene frequency. For example, since the introduction of pollution control programs in England, the frequency of the *carbonaria* allele has decreased at a rate consistent with a 12 per cent selective disadvantage (Cook et al. 1986). Records of trapped red and silver foxes kept by the Moravian mission posts in Labrador for a hundred years, 1834–1933, show that the proportion of silver foxes in the catch declined from about 0.15 to 0.05 (Elton 1942). Because silver coat is a recessive phenotype, the frequency of the gene must have decreased from 0.39 to 0.22 over a period of 100 years, a change that would have required a fitness differential of 0.035, or 3.5 per cent. It does not seem unreasonable that a greater demand for silver fox furs might have led trappers to cause an annual mortality of silver foxes 3 per cent in excess of that for red foxes.

Many traits of ecological interest have a polygenic basis.

The relationships between organisms and their environments often depend upon modifications of continually varying traits, such as lengths of appendages, body size and shape, thickness of hair or cuticle, and continuous gradations of behavior. Because variation in these characters depends on the contributions of many genetic loci, their response to selection cannot be analyzed by the simple population genetics models that have been applied to single-gene traits such as melanism in moths. Animal and plant breeders interested in such traits as milk production, oil content of seeds, and rate of egg production have developed a mathematical treatment of continuously varying traits and their responses to selection, known as quantitative genetics (Bulmer 1980, Falconer 1981). The theory of quantitative genetics rests on the assumption that variation within a population results from the additive contributions of many genes with similar effect. Thus the length of an appendage may come under the influence of a dozen genetic loci each of which may cause a small increase or decrease in length relative to the population average, depending on the allele. Individuals with a net excess of length-incrementing alleles at these 12 loci would have appendages longer than the average.

Although quantitative genetics greatly simplifies our concept of gene action (Thompson and Thoday 1979), its models have proven successful in predicting the results of selective breeding programs. To the extent that artificial selection mimics selection in nature, quantitative genetics can help us to understand the evolutionary responses of natural populations.

The heart of quantitative genetics is the variance of a trait within a population. Variance is a statistical measure of variation; specifically, it is the average of the squared deviations of individuals from the mean of the population. Hence the variance (V; also Var and s^2) in a trait X (a measurement) in a population of n individuals is

$$V = \frac{1}{n} \sum_{i=1}^{n} (X_i - \overline{X})^2 \qquad (26-1)$$

where X_i is the value for each individual i ($i = 1$ to n), and $\overline{X}$ is the mean value of X in the population. The calculation of a variance is shown by example in Table 26–1.

Continuously variable traits frequently exhibit a bell-shaped distribution of values in natural populations, with most individuals clustered near the mean and with the frequency diminishing at extreme values (Fig. 26–1). Each individual's value of a particular trait (the phenotypic value) is determined by deviations from the population mean caused by genetic and environmental influences. Because both sources of deviation enter into the calculation of variance for all values in the population, we may speak of phenotypic variance (V_P) as having two components, one attributable to genetic constitution (V_G) and one resulting from environmental factors (V_E). The two components added together equal the total phenotypic variance, or $V_P = V_G + V_E$.

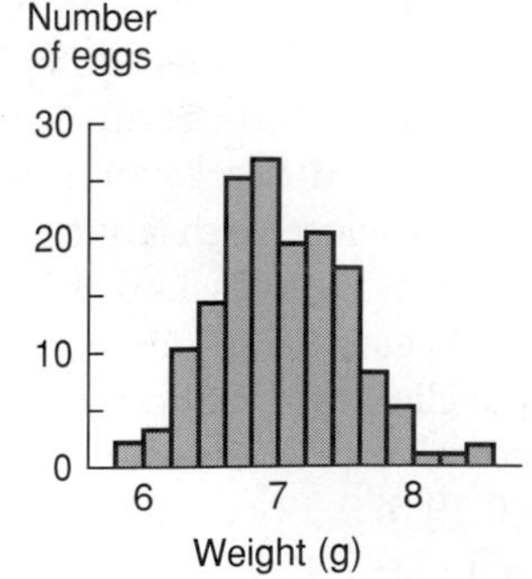

Figure 26–1
Frequency distribution of the weights of eggs of the European starling *Sturnus vulgaris* near Philadelphia, Pennsylvania.

Table 26-1
Calculation of variance in height in a small population ($n = 10$)

Individual (i)	Height (inches) (x_i)	Deviation from mean ($x_i - \bar{x}$)	Squared deviation ($(x_i - \bar{x})^2$)
1	75	5	25
2	73	3	9
3	72	2	4
4	72	2	4
5	71	1	1
6	70	0	0
7	68	−2	4
8	67	−3	9
9	67	−3	9
10	65	−5	25
	Total = 700	Sum of squared deviations =	90
	Mean ($\bar{x}$) = 70	Variance =	9.5

The genotypic variance can be subdivided further: V_A is the additive variance determined by the expression of alleles in homozygous form; V_D is the dominance variance determined by the interaction of alleles in heterozygous form; and V_I is the interaction variance, comprising the influences of different genes on the expression of alleles at a particular locus. In general, $V_G = V_A + V_D + V_I$, although variance may be further incremented by the correlation between particular genotypes and particular environments and other, usually minor factors.

The principal task of quantitative genetics has been to estimate the magnitude of the several components of phenotypic variance. This is made necessary by the fact that response to selection derives only from the additive genetic component of variance, because only V_A reflects the genetic diversity of the population—that is, the different alleles that replace, and are replaced by, others during evolutionary change. Phenotypic variance can be partitioned into its several components by statistical analyses of the results of breeding programs designed for the purpose (Falconer 1981). Let it suffice to say that these analyses utilize correlations of phenotypic values between close relatives, usually between parents and their offspring or between siblings.

The proportion of the phenotypic variance due to additive genetic factors is often expressed as their ratio, called the heritability (h^2) of a trait: $h^2 = V_A/V_P$. Most studies of heritability involve traits of commercial value in livestock, poultry, and crops, for which representative values of h^2 appear in Table 26-2. The data indicate that sizes have higher heritabilities (0.50–0.70), hence less environmental influence, than weights (0.20–0.35). Among traits related to production and fecundity, those creating the greater drain on energy and nutrients have the lower heritabilities. Thus the percentage of butterfat in milk is under strong genetic control ($h^2 = 0.60$), while total milk production has a low heritability (0.30); variation in egg size in chickens has a large additive genetic component ($h^2 = 0.60$), while rate of egg production has a lower heritability (0.30). Any character that requires a

Table 26–2
Heritabilities of several traits in domesticated and laboratory animals and plants

Character	Organism	Heritability
Size or length		
plant height	corn	0.70
root length	radish	0.65
tail length	mice	0.60
length of wool	sheep	0.55
body length	pigs	0.50
Weight		
body weight	sheep	0.35
body weight	pigs	0.30
body weight	chickens	0.20
Production		
butterfat content of milk	cattle	0.60
egg weight	chickens	0.60
thickness of back fat	pigs	0.55
weight of fleece	sheep	0.40
milk yield	cattle	0.30
Fecundity		
egg production	chickens	0.30
yield	corn	0.25
litter size	pigs	0.15
litter size	mice	0.15
conception rate	cattle	0.05
Life history		
age at onset of laying	chickens	0.50
age at puberty	rats	0.15
viability	chickens	0.10

Source: Falconer 1960, Brewbaker 1964.

large commitment of resources must be sensitive to environmental variation in those resources. The heritabilities of fecundity and life history characteristics are generally low (0.05 to 0.50). Heritability estimates accumulating for traits of individuals in wild populations resemble those of domestic animals and crops (Boag 1983, Travis 1983, Price et al. 1984, Noordwijk 1984).

Artificial selection of quantitative traits illustrates some characteristics of evolution in natural populations.

Artificial selection usually is accomplished with large fitness differentials applied over short periods; in this way, it caricatures the less intense selection that occurs in natural populations in that selection is exaggerated and some results would have no chance of surviving in the wild. Despite these differences, selective breeding programs suggest the kinds of evolutionary response that one might observe under natural conditions.

The change in a quantitative trait resulting from a single generation of selection (R, for response) depends on the deviation of selected individuals from the mean value of the population (S, for selection differential) and the

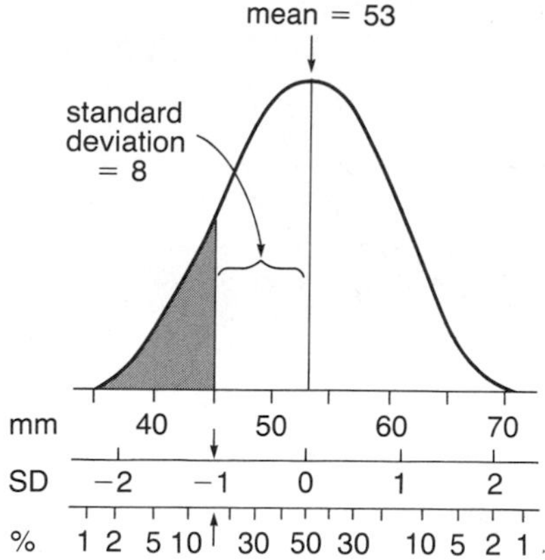

Figure 26–2
Schematic diagram of variation in a hypothetical trait showing the relationship between the measurement scale, the standard deviation, and the proportion of the population with phenotypic values more extreme than a particular value. For example, 16 per cent of the population have phenotypic values greater than one standard deviation above or below the mean (shaded portion).

heritability of the trait, according to the relationship

$$R = h^2S \tag{26-2}$$

For example, if h^2 were 0.5 and males and females 10 size units larger than the population average were bred together, their progeny would be 5 size units overall above the average of the unselected population. The greater the heritability of the trait, the more rapidly it can respond to selection.

Values of R and S are conveniently expressed as multiples of the standard deviation of measurements within the population, the standard deviation (SD or s) being the square root of the variance. Each value expressed in standard deviation units corresponds to a particular percentile rank in the population. Zero SD units is the population mean; when values are symmetrically distributed about the mean, half the individuals lie above that value and half below. When values have a normal distribution, as in Figure 26–2, 31 per cent of individuals lie above +0.5 SD and 31 per cent lie below −0.5 SD from the mean; 16 per cent have values more extreme than +1.0 SD or − 1 SD; 7 per cent exceed 1.5 SD; and only 2.3 per cent exceed 2.0 SD in each direction from the mean. As a result, the more intense is selection (the larger the value of S), the smaller the number of individuals selected and the smaller the number of resulting progeny for the next generation of selection. When selection is too strong the population dwindles, eventually to extinction. Even in artificial selection programs, the strength of selection is limited by the size of the stock population and the reproductive rate of selected individuals, which must be at least as large as the number of individuals eliminated by selection each generation.

The relationship between selection intensity (per cent of individuals selected), phenotypic selection differential (S), and response (R) is illustrated by a program of selection for rate of egg laying in the flour beetle *Tribolium castaneum* (Ruano et al. 1975). In the stock population, the number of eggs laid from 7 to 11 days after adult emergence (the phenotypic trait investigated) had a mean of 19.0, a standard deviation of 11.8, and a heritability (h^2) of 0.30. One control line and five lines with different levels of selection (ranging from 50 to 5 per cent) were established (Table 26–3).

Table 26–3
Selection procedure in six lines of *Tribolium castaneum* under selection for fecundity

Line	Number of families scored per generation	Number of females scored per family	Total scored	Total selected	Selection intensity (per cent removed)	Selection differential (S)*
A	10	20	200	10	95	2.0
B	20	10	200	20	90	1.8
C	40	5	200	40	80	1.4
D	66	3	198	66	67	1.1
E	100	2	200	100	50	0.8
F	200	1	200	200	0	0.0

* Standard deviation units.
Source: Ruano et al. 1975.

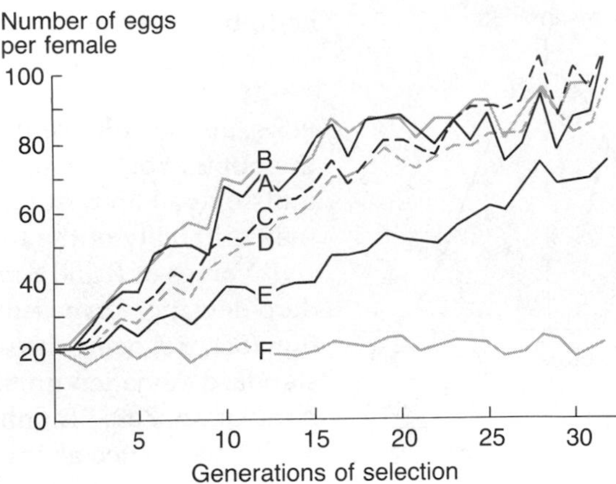

Figure 26–3
Change in rate of egg laying in *Tribolium castaneum* lines exposed to different levels of selection, increasing from F to A. (From Ruano et al. 1975.)

Knowing the variability, heritability, and selection differential, one can estimate the initial response of the population to selection. For example, in the C line a selection intensity of 20 per cent corresponds to a selection differential of 1.4 SD (Falconer 1981), or 16.5 eggs (1.4×11.8). With a heritability of 0.30, the response to selection should be about 5.0 eggs per generation ($R = h^2S = 0.30 \times 16.5$). The observed response fell somewhat short of this prediction (about 3.0 eggs per generation), probably because the estimate of heritability included maternal and dominance effects as well as additive genetic variation. But the beetle population did behave as predicted in that the rate of response varied in direct proportion to the intensity of selection (Figure 26–3).

With continued selection pressure on experimental populations, the response to selection eventually stops, as seen in Figure 26–3. The slowed response is the result of two factors: erosion of genetic variation by selection and the application of opposing selection by correlated changes in other traits. Selection works only when individuals vary genetically within a population. When all unfit alleles are removed by selection, evolution pauses until new mutations or gene combinations appear.

The maximum long-term response to selection in the absence of new genetic variation can be estimated from properties of the probability distributions governing quantitative traits (Robertson 1970; Dudley 1977). Suppose that a measurement is influenced equally by n genetic loci each having two alleles: (−) and (+). An individual homozygous for (−) alleles at each of the n loci exhibits the minimum measurement. Each (+) allele in the genotype adds a single unit increment to the measurement, so that the maximum possible is $2n$ units greater than the minimum (there are two copies of each gene in diploid organisms). Suppose the frequency in the population of the (+) allele is the same at each locus and is p; then the mean value of the measurement in the population is $2np$ greater than the minimum possible. If the alleles segregated randomly, probability theory tells us that the variance

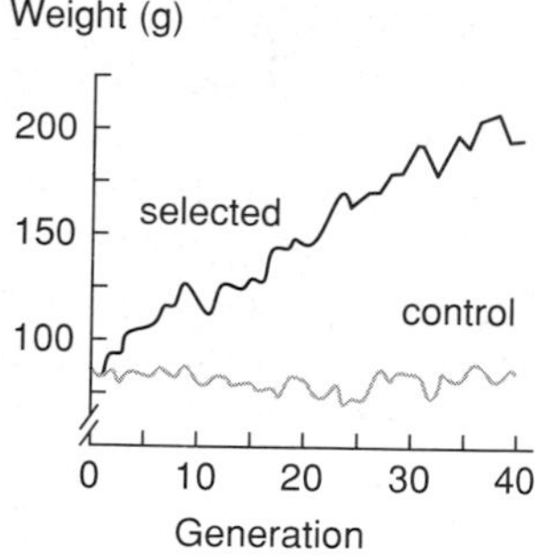

Figure 26–4
Four-week body weights of selected and control lines of Japanese quail. (After Marks 1978.)

in number of (+) alleles among individuals would be $2npq$, hence the standard deviation would be $\sqrt{2npq}$. How much of a selection response can be expected when the (+) alleles are strongly selected? Without the addition of new genetic variation, the maximum possible measurement is $2n$ above the minimum. Thus the maximum response (maximum phenotype − average phenotype) is $2n - 2np$, which may be expressed as $2n(1-p)$ or $2nq$. In terms of numbers of standard deviations, the magnitude of this response is $2nq/\sqrt{2nqp}$, which may be rearranged to give $\sqrt{2nq/p}$. For example, when a trait is controlled by 8 loci at which the frequency of (+) alleles is 0.5, the maximum phenotypic response to selection is $(2 \times 8 \times 0.5/0.5) = 4$ standard deviations above the mean. Somewhat surprisingly, therefore, selection can produce phenotypes well beyond the extreme value observed in an unselected population. Probability theory reminds us that the proportion of individuals receiving 16 (+) alleles at 8 loci just by chance, when the frequency of (+) is 0.5, is less than 1 in 66,000 (that is, 0.5^{16}); 12 or more out of 16 (+) alleles (1.4 SD units above the mean) occur in less than 4 per cent of individuals. So we see that the short-term evolutionary potential of quantitatively varying traits may extend far beyond the distribution of phenotypes in a population, providing that the multilocus, additive model of quantitative variation is reasonable.

Laboratory studies show that the large gains predicted by quantitative genetics models can, in fact, be realized. Mature body weights of unselected Japanese quail average about 91 g with a standard deviation of 8 g. After 40 generations of selection for high body weight, the population average in one study increased to 200 g, or almost 14 SD units above the mean of the unselected population (Figure 26–4; Marks 1978). Realized heritabilities ($h^2 = R/S$) decreased from 0.30–0.45 in generations 1 to 10, to 0.15–0.20 in generations 11 to 30, and 0.05–0.10 in generations 31 to 40.

What is often more difficult to explain than the response of a trait to selection is the leveling off the response, often after only modest progress. This usually is not due to exhaustion of genetic variation for the trait, because reverse selection (back toward the mean of the unselected population) typically produces an immediate response, which can result only from genetic variation remaining in the population. Furthermore, when selection is merely relaxed, so that selection coefficients are zero, a selected trait will sometimes return toward the preselection measurement, apparently by itself. The most reasonable explanation for these results is that selection applied to one trait causes changes in other traits that affect the fitness of the organism. Increase in rate of egg laying, for example, may cause physiological or morphological changes that reduce viability and thus oppose the artificial selection regime.

Correlated responses to selection limit evolutionary response.

Evolutionary responses often include characters other than the one selected. Both development and function integrate the parts of organisms, bringing about an interdependence of phenotypic traits, particularly among those involving size or rate of growth and production. Animal and plant breeders have estimated genetic correlations between traits in many domestic species. For example, in poultry the genetic correlation between body weight and egg weight is 0.50; between body weight and egg production it is −0.16. Therefore, one cannot apply selection to body weight without also obtaining a relatively rapid increase in egg size and a slower, but steady decrease in rate of laying. Simultaneous selection of large egg size and small body size goes against the grain of genetic correlation, and usually is unsuccessful (Nordskog 1977).

Many characters, such as body and egg size, are linked in developmentally, genetically, or functionally related groups that tend to respond to selection in concert. For example, Larry Leamy (1977) determined genetic correlations among skeletal measurements of mice and identified four clusters of traits highly integrated genetically among themselves, but relatively independent of each other: (1) skull length; (2) skull width, body weight, and tail length; (3) skull width (providing a link to group 2), scapula length, and other measurements associated with the pectoral girdle; and (4) limb bones and total body length. In mice, therefore, selection for body weight produces responses in tail length and the proportions of the skull, as well as in body weight itself.

When Peter Dawson (1966) selected for fast and slow larval development in *Tribolium castaneum* and *T. confusum,* he observed a number of correlated responses that decreased fitness regardless of the direction of selection. Selection for rapid development resulted in decreased size and, in *T. castaneum,* decreased larval survival and an increase in incidence of adult abnormalities. Selection for slow development resulted in increased adult weight, increased frequency of adult abnormalities, and decreased fertility of females.

Dawson assessed the fitness of each of the selected strains of flour beetles by placing them in competition with unselected stocks of the other species (1967). Both the "fast" and "slow" lines suffered decreased fitness during early stages of selection, suggesting that development rate is optimized with respect to its effect on fitness. (As Dawson continued his selection experiments, the fitness of one of the selected lines began to improve, apparently because of increased cannibalism by the selected larvae. Evolution frequently springs such surprises on us.)

Results similar to Dawson's were obtained by M. W. Verghese and A. W. Nordskog (1968), who examined correlated responses in reproductive fitness in selected lines of chickens. Selection for both increase and decrease in body weight and egg weight caused a decline in reproductive fitness, as indicated by rate of egg production, hatch rate, and survival of chicks to 9 months. Regardless of the character or direction of selection, fitness in the selected lines varied from 54 to 85 per cent of the nonselected line. These and other experiments emphasize the fact that natural populations are balanced genetically, and their adaptations are both well-tuned to the environment and finely adjusted to each other.

Russell Lande (1979) used the concept of genetic correlation to examine the short-term response of many characters to simultaneous selection. Genetic correlation between two traits [$r_G(XY)$] measures the degree to which genetic variation in one is related to genetic variation in the other. One also may calculate the environmental correlation (r_E) and the total phenotypic correlation (r_P) between two traits. Separation of these components of covariation requires statistical analysis of particular breeding programs, as does the separation of genetic components of variance.

Correlation is calculated from the covariance between two traits, estimated by an equation analogous to that for variance,

$$Cov(XY) = \frac{1}{n} \sum_{i=1}^{n} (X_i - \overline{X})(Y_i - \overline{Y}) \qquad (26-3)$$

The square of the correlation (r^2) is equal to $Cov(XY)/[Var(X)Var(Y)]$. Now, whereas the response of trait X to selection directly upon itself is $R(X) = h^2(X)S(X)$, the correlated response of trait Y to selection on trait X is

$$R(Y) = h(X)h(Y)r_G(XY)S(X)\,\frac{V_P(X)}{V_P(Y)} \qquad (26-4)$$

where V_P is the phenotypic variance and h is the square root of heritability. From equations (26–3) and (26–4), one can see that the ratio of the response in Y to that in X, resulting from selection on X, is

$$\frac{R(Y)}{R(X)} = r_G(XY)\,\frac{h(Y)V_P(Y)}{h(X)V_P(X)} \qquad (26-5)$$

Lande illustrated the implications of correlated response by substituting the following values for brain and body weight of mice: $r_G(\text{brain-body}) = 0.68$, $h(\text{brain}) = 0.8$, $h(\text{body}) = 0.6$, $V_P[\log(\text{brain})] = 0.058$, and $V_P[\log(\text{body})] = 0.145$. For these values, $R(Y)/R(X) = 0.36$. Because measurements X and Y were log-transformed values, the ratio $R(Y)/R(X)$ expresses the allometric relationship between the selected and unselected populations. Lande used matrix algebra to generalize his equations for many gene loci, and emphasized that some patterns of phenotypic correlations between two traits could result fortuitously from selection on only one of the traits, or even on a third trait genetically correlated to the first two (also see Lande and Arnold 1983, Arnold 1983, Arnold and Wade 1984).

While genetic correlation may steer short-term responses to selection, both the matrix of genetic correlations relating different traits and the vector of selection coefficients change as the phenotype responds. Two situations illustrated in Figure 26–5 show how the constraining influence of genetic correlations may eventually be overcome. In the first case, selection pushes both X and Y toward some optimum value, but genetic correlation dictates that the initial response deviates from the shortest phenotypic route to the goal. But as the position of the population changes in phenotypic space, the direction of the selection vector also shifts, pushing the population in a curved, perhaps even spiral course toward the goal. Achievement of the optimum is delayed, but not prevented, so long as the genetic correlation is not perfect. In the second case, selection is applied on one trait only (X), but

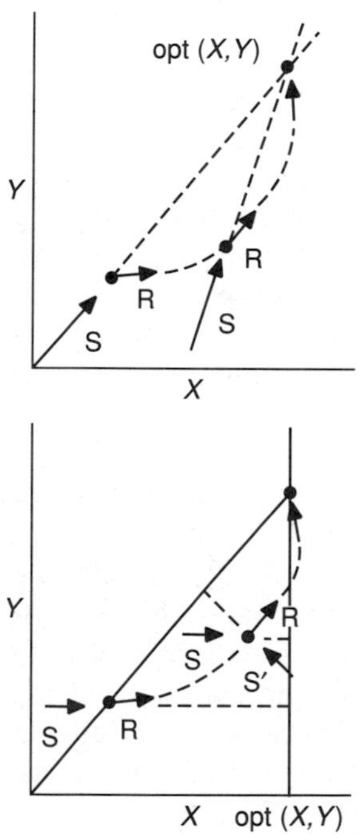

Figure 26–5
Above: Response (*R*) to selection (*S*) for optimum phenotypic values of a pair of traits when genetic correlations deflect the response trajectory. *Below:* Response to selection on a single trait *X* when there is an optimum ratio between *X* and *Y*. Deviation from that ratio established secondary selection (*S′*) tending to restore the ratio.

there is an optimum ratio between X and Y (perhaps an optimum shape). The selection response causes the population to deviate from the optimum ratio, which establishes secondary selection on Y, leading in a curved path to both the optimum value of X and the optimum ratio of X to Y.

Individuals manage genetic variation by adaptations of their breeding systems.

The gene pool of a population represents a balance between selection, which tends to reduce genetic variation, and several processes that increase genetic variation. Most genetic variation in a local population arises from mutation, immigration of alleles from other areas having different selective environments, temporally varying selective factors, and changes in gene frequency from generation to generation arising purely by chance in small populations (genetic drift). Given that such variation exists in all populations, we may ask: How much variation is good for individuals? And what strategies can individuals adopt to obtain the best genetic constitution for their offspring?

The answer to the first question depends upon the situation. When the individual's environment is highly predictable from generation to generation, and the individual is well-adapted, imposition of genetic variation reduces fitness. In such a case, individuals should reproduce asexually, as by parthenogenesis in animals and vegetative reproduction in plants, producing offspring genetically identical to the parent. Although most organisms or their close relatives can reproduce sexually, many abandon the practice seasonally, some mix both sexual and asexual reproduction, while others commit their futures wholly to asexual modes of procreation (Smith 1978, Bell 1982).

Much has been written about the evolutionary consequences of sexual reproduction both for the fitness of the individual within populations and for the evolutionary future of the population as a whole (Michod and Levin 1988, Stevens and Bellig 1988). Most evolutionists agree that sex is somehow necessary as a source of the new gene combinations that build the genetic foundation of long-term evolutionary change (see J. M. Smith [1978] for a highly readable account). We can infer this because most higher taxa of organisms (genera, families) contain sexual forms, and few strictly asexual groups are thought to have had long evolutionary histories. Yet there is a tremendous fitness incentive for the individual to give up sex. Each offspring of a female bears one set of her genes and one set of those of her mate. When females reproduce parthogenetically, each offspring contains two sets of the genes of the parent. So, unless a mate is required to help rear offspring—certainly not the case in most organisms—the fitness of a genotype that destines an individual to be parthenogenetic has twice the fitness of the genotype of a sexual individual, all other things being equal.

Sex remains a mystery to biologists. Most attempts to explain how sexual reproduction can be maintained in the face of an overwhelming short-term advantage to asexual reproduction have proposed that the compensating advantage to sex resides either with increase in the genetic variation among progeny or reduction of deleterious genetic variation. Variable offspring may confer a fitness advantage in a highly unpredictable environment where variation assures that at least some progeny will be well-suited to whatever environmental conditions should come their way. In strongly competitive environments, variation among progeny increases the chance that at least some will have the extremely high fitness necessary to persist (Williams and Mitton 1973, Williams 1975).

Alternative explanations for sex argue that genetic recombination made possible by sexual reproduction is necessary to eliminate deleterious mutations from the germ line (Muller 1964, Felsenstein 1974). In an asexual clone, deleterious mutations can be eliminated only by selection among progeny. When the entire mutation rate per genome exceeds the number of selective deaths per generation, mutations will accumulate and reduce the vitality of the clone. In contrast, individuals in sexual populations continually exchange genetic material between family lines. Because the entire population is joined into a common gene pool, selection picks and chooses among continually reshuffled genotypes. The deleterious mutation that crops up in one line can be exchanged for a beneficial allele through recombination.

Whatever the raison d'etre of sex, it provides three potential benefits: long-term evolutionary flexibility of the population, medium-term elimination of deleterious mutations from the family line, and short-term production of variation among the progeny of the individual.

Breeding systems manage genetic variation in sexual populations.

Accepting as given that most populations reproduce sexually and that most gene pools contain variation, some of it deleterious, ecological geneticists are now beginning to address the effects of different sexual strategies on fitness (Willson and Burley 1983, Wyatt 1983, Stephenson and Bertin 1983, Schoen 1982, 1983). A major problem is the optimal degree of outcrossing for a population. Everyone knows that close inbreeding is bad (Ralls et al. 1988). Brother-sister matings and, where possible (especially in plants), selfing may result in the expression of deleterious recessive genes. The mechanism for this is simple. Suppose an individual is heterozygous for a rare, deleterious, recessive gene (the gene pool is full of them and most individuals have some). If it were to mate with an individual from the general population, which probably does not have the same rare allele, half of their progeny would be heterozygous, like the one parent, and half would be homozygous for the common form of the gene, like the other parent. None of the progeny would be disadvantaged by the union. If, however, the individual selfed, one-quarter of its offspring would be homozygous for the deleterious allele and would suffer loss of fitness as a result. Matings between close relatives produce the same result, only less frequently.

Most species employ mechanisms, including dispersal of progeny, recognition of close relatives, and negative assortative mating, to reduce the occurrence of inbreeding. Hermaphroditic species of plants, in which individuals bear both male and female sexual organs, have additional mechanisms to prevent selfing, including self-incompatibility, temporal separation of male and female function, and elaborate flower structures designed to make self-fertilization difficult (Willson 1983).

While close inbreeding generally creates problems, it may confer benefits as well (Lloyd 1979, 1980, Wells 1979). In particular, selfers can guarantee fertilization of their flowers in habitats lacking suitable pollinators or where individuals are widely spaced (Baker 1965, Ghiselin 1969). Many weedy species that colonize isolated patches of disturbed habitat (for example, dandelions) are selfers (Jain 1976). One assumes that most deleterious variation was weeded out of such populations as it was exposed in homozygous individuals during the transition between outcrossing and selfing (Ritland and Ganders 1987).

Outcrossing at great distance may also reduce fitness when populations of plants include spatially defined ecotypic variation over small scales of distance, particularly in complex, heterogeneous environments. In such cases, local adaptation to particular habitat patches enhances fitness and receiving pollen from individuals adapted to different habitat conditions may reduce the fitness of progeny that become established near the female parent.

Several studies have reported an optimal outcrossing distance in populations of plants (Levin 1984). Nearby individuals are likely to be close relatives, which raises the specter of inbreeding. Distant individuals are likely to be adapted to different conditions. M. V. Price and N. M. Waser (1979) fertilized flowers of the larkspur *Delphinium nelsoni* in central Colorado with pollen obtained from the same individual and from individuals located at distances of 1, 10, 100, and 1000 meters. Their results showed that the number of seeds set per flower was greatest when the pollen source came from a distance of 10 meters, and was least for selfed pollen and that obtained 1000 meters distant. Furthermore when these seeds were planted, survivorship to 1 and 2 years greatly favored matings across the intermediate distance of 10 meters.

Although Price and Waser's study indicated an optimum outcrossing distance, larkspurs cannot exploit this advantage. The plants are pollinated primarily by bees and hummingbirds, which tend to visit nearby flowers in succession. As a result, colored dye particles placed along with natural pollen on the male anthers were recovered primarily on flowers within 2 meters from its source. Perhaps no adaptation of flower structure could modify the behavior of pollinators (which have little interest in the plant's fitness) to achieve the optimal outcrossing distance.

Another possibility for managing genetic variation is the selective abortion of developing ovules on the basis of the genotype of the embryo (Willson and Burley 1983). Most plants produce many more flowers than they can mature as fruits; flowers are relatively cheap, fruits expensive. Excess fertilized ovules are reduced in part by predation or other extrinsic damage and in part by programmed abortion (Casper 1984). Abscission of flowers and fruits can be highly selective with respect to number of ovules fertilized per flower or amount of herbivore damage. These phenomena have been reviewed by Andrew Stephenson (1981), who cited evidence that some species more likely abort fruit when flowers are self-fertilized than when they are outcrossed.

Noting that plants appear able to "recognize" the father of particular offspring, Daniel H. Janzen (1977) suggested that the abortion of developing ovules within fruits might be the result of a mechanism by which the female parent can select among its offspring according to their genotypes. So while plants may be unable to control the behavior of their pollinators, they may through overproduction of flowers and ovules within flowers exercise control over the genotypes of their progeny. At present, this is an untested hypothesis with exciting potential. Animals can choose mates more actively than plants and there is abundant evidence of mate selection based upon kinship and genotype, often with a rare mating advantage (Greenberg 1979, Bateson 1983, Holmes and Sherman 1983).

Evolutionary ecologists interpret form and function as adaptations to the environment.

Organisms are well adapted to their environments. This assumption has provided a powerful tool for conceptualizing the organism-environment interaction and understanding design limitations placed upon the responses of organisms to environmental change (Wake 1982). This research program also has led to the discovery of new components of the fitness of organisms and has emphasized the important roles of interactions between individuals—predators, competitors, mutualists, society members, mates, parents, and offspring—in directing the course of adaptation. Therefore, even if the assumption of adaptation is not fully correct, it has expanded our ecological concept tremendously.

The primary question addressed by evolutionary ecology is: How do the adaptations of organisms reflect their environment? To answer this question requires an understanding of selective factors in the environment and the evolutionary responsiveness of the phenotype. To have scientific validity, an idea concerning the phenotype-environment relationship must include a suitable criterion for fitness of phenotypes, the genetic basis of phenotypic variation, and a model to link aspects of form and function that determine fitness to each other and to conditions of the environment (Arnold 1983).

One class of models of adaptation, that of phenotypic optimization, does not incorporate genetic variation explicitly, but rather assumes that phenotypic variation has a parallel genetic basis and that selection of optimum phenotypes brings about appropriate genetic change. For example, particular conditions may dictate an optimum allocation of resources between producing reproductive structures and continuing to grow; a certain genotype presumably produces that optimum phenotype and it will be

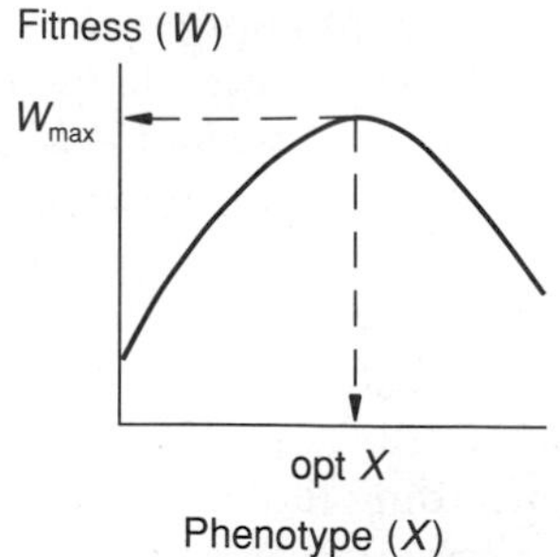

Figure 26–6
Relationship between fitness and phenotypic value when there is a single optimum phenotype.

selected irrespective of the particular genetic basis of the trait. The core of an optimization model is the relationship between phenotype and fitness (Figure 26–6). Each such relationship is unique for a particular environment. To visualize the relationship, one must either measure the fitnesses of a range of phenotypes directly or devise a realistic model that predicts the fitnesses of nonexistent phenotypes. The degree to which observed phenotypes match the predictions of the model measures the validity of the mechanics embodied in that model, although we must keep in mind that alternative models may make similar predictions.

An Evolutionarily Stable Strategy (ESS) resists invasion by all other phenotypes.

A second approach to understanding adaptation, particularly useful in the case of discretely varying phenotypes and when phenotypes interact with one another, is that of determining the Evolutionarily Stable Strategy (Smith and Price 1973; Smith 1974, 1982; Parker 1984). An Evolutionarily Stable Strategy, or ESS, is that phenotype or combination of phenotypes, which when constituting a population make it impossible for individuals with alternative phenotypes to invade the population. As John Maynard Smith (1982) puts it, an ESS is "a strategy such that, if all the members of a population adopt it, no mutant strategy can invade." Referring back to the relationship between phenotype and fitness in Figure 26–6, let us suppose that only two phenotypes, A and B, occur (Figure 26–7). A population consisting only of B individuals can be invaded by A individuals—that is, the phenotype will increase when rare—owing to the superior fitness of A. A population consisting only of A individuals resists invasion by the B phenotype; hence A is the ESS. Considering the range of all possible phenotypes, it is clear that only C, which confers the greatest fitness, will resist invasion by all other phenotypes. Therefore, a phenotype identified by the maximum-fitness criterion is also an Evolutionarily Stable Strategy.

The utility of ESS thinking becomes more apparent when phenotypes interact with one another and the fitness of each depends on the proportions of other phenotypes in the population (Smith 1982, Parker 1984). Hence the concept of the Evolutionarily Stable Strategy has found broad application in the study of social behavior and mating systems, as we shall see a few chapters hence.

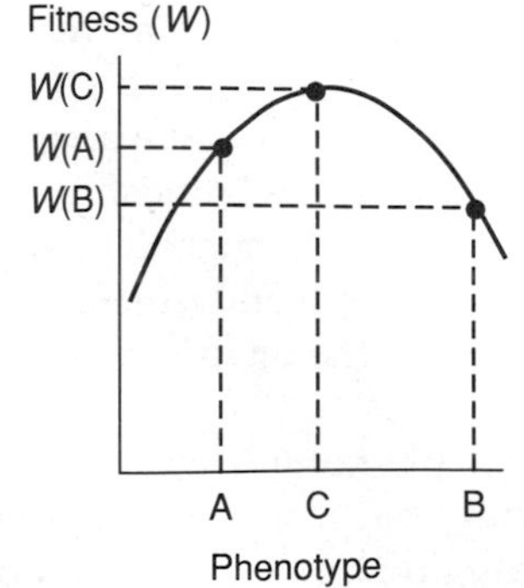

Figure 26–7
Phenotype C is an Evolutionarily Stable Strategy because no other phenotype can invade a population of C individuals.

The influence of the phenotype-environment interaction on fitness is the key to understanding adaptation.

Regardless of the fitness criterion adopted, and irrespective of whether genotypic or phenotypic models are employed, the principal challenge to the evolutionary ecologist is to understand how small changes in phenotype affect fitness. To accomplish this goal, ecologists must understand the interrelationship of the many characteristics of the organism and the fullness of its interaction with all aspects of its environment. At any given time, our maturing concept of the organism-environment complex can be summarized in the form of models, whose consistency with observations on natural systems is continually scrutinized by further research, followed by reformulation of ideas when inconsistencies arise.

I shall illustrate this process with an example from my own work on the life histories of birds. In 1968 David Lack, the prominent English ecologist, suggested that growth rates of birds optimally balance two environmental factors: predation and other sources of mortality, which select rapidly growing individuals that pass through vulnerable developmental stages quickly; and food supply, which selects more slowly growing individuals that demand less food and thereby allow their parents to provide for more offspring. From this simple idea, Lack suggested that growth rate should vary in direct proportion to vulnerability of nestlings to mortality factors. But observations on birds revealed, to the contrary, that growth rate is relatively insensitive to variation in mortality rate, suggesting that Lack's hypothesis was not sufficient (Ricklefs 1969).

In an attempt to resolve this problem, I developed a model to describe the influence of growth rate on fitness through the consequences of growth rate for number of chicks produced per brood and number of surviving broods per season (Ricklefs 1984). The details of such a model require knowledge of the food requirements of growing chicks and the initiation of

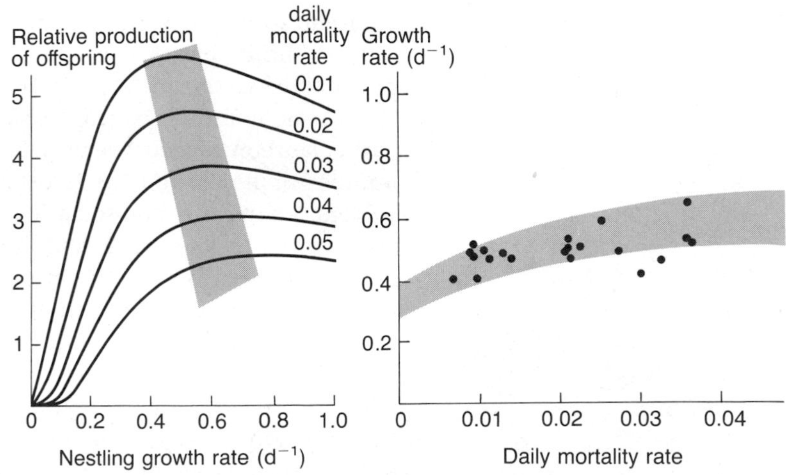

Figure 26–8
Left: Relative fitness (offspring produced) as a function of nestling growth rate in a model relating the two through the influence of growth rate on brood size, number of broods per season, and brood survival for brood mortality rates from 0.01 to 0.05 day^{-1}. The shaded area indicates the optimum growth rate for each level of mortality. *Right:* Observed growth rates of various species as a function of daily mortality rate, the shaded area corresponding to the optimum determined by the model. (After Ricklefs 1984.)

new nesting attempts following both success and failure of the previous attempt. A set of expressions for each of these effects resulted in equations relating an index of fitness (number of young reared per season) to growth rate. The curves defined by these equations have maxima at intermediate growth rates close to rates measured in nature (Figure 26–8). Furthermore, when nest mortality rate is varied over a wide range in the model, the position of the optimum growth rate changes little, contrary to Lack's expectation but consistent with observation. This is related to the fact that birds rapidly replace destroyed clutches and so the impact of mortality on number of young reared per season is correspondingly diminished. The consistency of the model's predictions with observation suggests that Lack's concept of how growth rate influences fitness is essentially correct and that factors other than the ones included in the model are not important to the evolutionary optimization of growth rate.

The adaptationist program has many difficulties.

My confidence in understanding growth rates in birds rests firmly on the assumption that the organism-environment relationship reflects solely the fitness relationships of phenotypes with selective factors in their surroundings. This general view has been challenged on a number of grounds. One of these is that many processes oppose the perfection of adaptation: lack of suitable genetic variation for response to the environment; continuous production of less fit phenotypes by mutation and immigration; changes in environment that leave previously well-adapted phenotypes behind. In addition, some phenotypes of potentially great fitness are not possible because of limits imposed by physical laws. Some things, such as rates of physical diffusion, are beyond the realm of biological influence. Evolution also drags along the baggage of past adaptation, which may contain no particular relevance for the present. Body plans of major taxa have different number of limbs: four in terrestrial vertebrates, six in insects, eight in spiders, ten in crabs, and so on. The number can be changed by reduction of appendages (snakes have zero limbs), or the modification of other parts, such as antennae or mouth parts, to function partly as limbs, but the basic number of limbs characteristic of each major group has no particular meaning, other than historical, and may restrict future evolutionary potential. Finally, some adaptations have fortuitous consequences for developmentally or genetically linked features of the phenotype. Lande's observation that selection of body size alone automatically results in a particular increase in brain size, owing to their genetic intercorrelation, makes this point.

Some difficulties with adaptationist thinking were eloquently argued by Harvard evolutionary biologists Stephen J. Gould and Richard Lewontin (1979) in a paper entitled "The spandrels of San Marco and the Panglossian paradigm: A critique of the adaptationist programme." Spandrel refers to the space, generally triangular, between an arch supporting a ceiling or other horizontal structure and the ceiling itself. In many cathedrals, these were decorated with painted scenes. Gould and Lewontin's point was that the spandrel arises as a consequence of architecture, but is not a key feature of building design, as are arches and the horizontal members they support. The analogy to organism architecture cautions us that some structures may be fortuitous or secondary consequences of other, strongly selected adaptations. As Gould and Lewontin put it, "One must not confuse the fact that a structure is used in some way . . . with the primary evolutionary reason for its existence." They summarized their reservations about the adaptationist program by questioning the ubiquity of its basic assumption:

> This [adaptationist] programme regards natural selection as so powerful and the constraints upon it so few that direct production of adaptation through its operation becomes the primary cause of nearly all organic form, function, and behaviour. . . . We would not object so strenuously to the adaptationist programme if its invocation, in any particular case, could lead in principle to its rejection for want of evidence.

In this passage Gould and Lewontin suggest that evolutionary ecologists should test the basic assumption of adaptation as well as determine the validity of adaptationist explanations for patterns in nature (for one approach, see Clutton-Brock and Harvey 1979, 1984).

Taxonomically useful characters illustrate that, once established, some adaptations resist further change.

Evolutionists divide the characteristics of organisms into one set that reveals the phylogenetic history of a group and another set that responds easily to selection and reflects the contemporary environment. Characters do not, of course, fall discretely into taxonomically revealing and ecologically revealing sets. Rather they are arranged along a continuum from evolutionarily stable to labile. Different characters are useful in distinguishing different levels of taxonomic groups (Mayr et al. 1953, Crowson 1970). Among birds, for example, orders are distinguished primarily by skeletal features, such as the structure of the palate and arrangement of bones in the skull, whereas families are distinguished by variations in the beak, pattern of scales on the tarsus (lower leg), and number and relative length of the primary (flight) feathers. The lower taxonomic categories — genera and species — are often based upon small differences in measurements and ratios of measurements, as well as plumage coloration and song.

The differences that distinguish mammals from reptiles, truly a drastic reorganization of certain parts of the body plan, physiological processes, and patterns of reproduction, evolved over tens of millions of years through intermediate stages that enjoyed varying levels of success, judging from their abundance in the fossil record. Eventually, however, evolution arrived upon a particularly providential set of characteristics, those shared by all modern mammals, and the group underwent an explosion of evolutionary diversification between 70 and 60 million years ago. Regardless of subsequent modifications that now distinguish bears, rabbits, mice, seals, and wildebeests, all mammals retain the fundamental class characteristics, including warm-bloodedness and nursing of the young. At each lower taxonomic level, other traits have been set aside by evolution.

How do characters at each level in this hierarchy match up with the ecological distributions of organisms? Which ones may be thought of as ecological characters in the sense of the adaptationist program, and which are too remote to infer the ecological mold into which they were cast? The matching of organisms to their environments comes about in two ways: by the organism's choosing its environment to match its evolved features, and by selective modification of the gene pool of a population. Larger taxonomic groups appear to be widespread with respect to climate and other aspects of the physical environment, but often specialized with respect to diet. Thus the insect order Homoptera (leafhoppers, cicadas) may be found wherever vascular plants occur but, because their mouth parts are specialized for sucking plant juices, their local ecological roles are narrowly prescribed. All the evolutionary modifications of this group have taken place within an ecological context established by the ordinal characters. As Gould (1982) has put it, "current utility permits no necessary conclusion about historical origin. Structures now indispensible for survival may have arisen for other reasons and been 'coopted' by functional shift for their new role."

Smaller taxonomic groups are often distinguished by differences in body size, habitat, microhabitat, and selection of diet according to prey or host species, rather than manner of feeding. We may presume that the modifications responsible for such diversification are evolutionarily more malleable and, therefore, more amenable to study by evolutionary ecologists. As a general rule of thumb, adaptationist thinking may provide a useful tool for understanding aspects of the organism-environment interaction based on characters that distinguish close relatives.

Do large systems have uniquely evolved properties?

The functioning of biological communities and ecosystems is determined by the collective adaptations of all their constituent species. To a large degree these species constitute important aspects of each other's selective environments and so each has evolved with respect to the evolution of others in the system. This mutual accommodation of species to each other is broadly referred to as coevolution, a term that has also been given some specialized meanings, as we shall see in Chapter 31.

Ecologists have often wondered about the importance of coevolution to the maintenance of community function, and whether there is a criterion for fitness at the level of the system as well as that of the individual genotype. One approach to this problem would be to perform an experiment—impossible to pull off other than with laboratory microcosms—in which ecosystems are constituted with species appropriate to a natural system, that is, representing all the ecological roles found in a natural system, but obtained from different places and thereby having independent evolutionary histories. If such artificially concocted systems functioned less well than corresponding natural systems, one could conclude that "coevolved" properties conferred unique qualities to large ecological systems.

The degree to which communities and ecosystems have an evolved genetic integrity has not been resolved by ecologists, and may not be for decades, if ever. Smaller parts of this larger question, concerning the role of adaptation in suiting the organism to its environment and the mutual coevolution of populations, have, however, contributed importantly to our concepts and understanding of ecology, as we shall see in the chapters that follow.

SUMMARY

The study of evolutionary ecology is based upon the assumption that differences between the adaptations of organisms can be interpreted as evolutionary responses to different selective pressures in the environment. An important consideration for ecologists is the degree to which adaptation to local environments is determined by availability of suitable genetic variation. Another concern is the extent to which organisms are evolved to manage genetic variation within the population.

1. Population genetics models demonstrate how evolution can proceed by the substitution of alleles according to their relative fitnesses.
2. Many adaptations of ecological interest involve modifications of continuously varying traits. Variation in a trait within a population is described by its variance, which has environmental and genetic components. The science of quantitative genetics has developed statistical analyses to tease apart these components from the results of certain breeding programs.
3. Heritability (h^2) is the ratio of additive genetic variance to phenotypic variance; its value ranges between 0 and 1. Heritabilities of traits in natural populations are on the order of 0.5–0.7 for many size traits, but often lower for production-related traits.
4. The response of a trait (R) to selection is equal to the heritability times the selection differential (S). In animal and plant breeding, the stronger the selection, the faster the response, so long as enough offspring are produced to replace individuals selectively removed.
5. Response to selection levels off when genetic variation is exhausted or, more frequently, when correlated responses in other traits reduce the fitness of selected individuals.
6. The response in one trait to selection applied on a second depends on the genetic correlation between them. The phenotype often consists of groups of genetically intercorrelated characters that tend to respond to selection in concert. Such correlations may inhibit the independent evolutionary response of two traits to opposing selective pressures.

7. Genetic variation maintains the long-term evolutionary potential of a population, and it has been argued that a primary function of sexual reproduction is to increase genetic variability through recombination of genotypes.

8. Genetic variation may also may impair the ability of individuals to adapt to local variation in the environment. To some degree, individuals can minimize the effects of genetic variation on the genotypes of their offspring by selective mating to control level of inbreeding and outcrossing distance.

9. Assuming that evolutionary response mechanisms are permissive, evolutionary ecologists interpret modifications of form and function as adaptations to the environment. To study such adaptations requires a suitable definition of evolutionary fitness and a model linking the organism-environment interaction to fitness.

10. Given these two elements, the outcome of evolution may be interpreted as a phenotypic optimum or an Evolutionarily Stable Strategy (ESS). Phenotypic optimization is a mathematical technique of finding the point on a phenotypic continuum having the greatest fitness. An ESS is a trait that, if adopted by all members of a population, is superior to all other possible alternative traits. ESS analysis is particularly useful for adaptations involving interactions between members of a population.

11. Adaptationist thinking should be applied with caution to the extent that evolutionary response cannot keep up with environmental change, or that adaptation cannot be perfected owing to mutation and immigration of locally less fit alleles. Furthermore, some imaginable phenotypes are not possible owing to limits imposed by physical laws beyond the reach of evolution.

12. Stephen J. Gould and Richard Lewontin have emphasized that many attributes of form and function are fortuitous consequences of selection on other traits, and that adaptations may have evolved in response to environmental conditions of the past that are no longer relevant to their present-day utility.

13. Characteristics that distinguish higher taxonomic groups (orders, families) usually are thought of as being less responsive to selection than those that vary between lower taxonomic categories (genera, species). Whether the characteristic traits of higher orders lack appropriate genetic variation, are too highly integrated through genetic correlation to have evolutionary latitude, are protected from selection pressures in the environment, or place restrictions on habitat use by their bearers is poorly understood.

27

Adaptation to Heterogeneous Environments

The world is anything but constant and uniform. A single genotype cannot be best suited to all the conditions encountered by each individual. These facts pose a fundamental and difficult challenge to understanding adaptation, raising several related issues to prominence in the ecological literature of the past three decades. One is the definition of fitness under variable conditions. That is, how does one best integrate the varied success of a genotype under different conditions into a single measure of fitness? A second issue concerns whether environmental heterogeneity facilitates coexistence of more than one genotype in a population. Interest in this problem was greatly stimulated by discovery in the 1960s of high levels of genetic variation in most populations. Third, heterogeneous environments present individuals with choices concerning habitat use, prey selection, mate selection, and so on. The behaviors that govern these choices are, to a large extent, under genetic control and therefore are evolved adaptations. To comprehend such adaptations, one must understand the consequences of rules adopted for making choices.

Richard Levins's fitness set analysis addressed the problem of evolution in a heterogeneous environment.

Arguing over the maintenance of genetic variation within populations has been a traditional pasttime of population geneticists for decades (Lewontin 1974). Until the mid-1960s, when biochemical techniques uncovered vast amounts of previously unknown genetic variation, obvious polymorphisms of structure and color were thought to arise from the superior fitness of heterozygotes, as in the maintenance of the sickle-cell gene in human populations by the resistance of heterozygotes to malaria (Allison 1956, Cavalli-Sforza and Bodmer 1971). But experiments with flies and other organisms

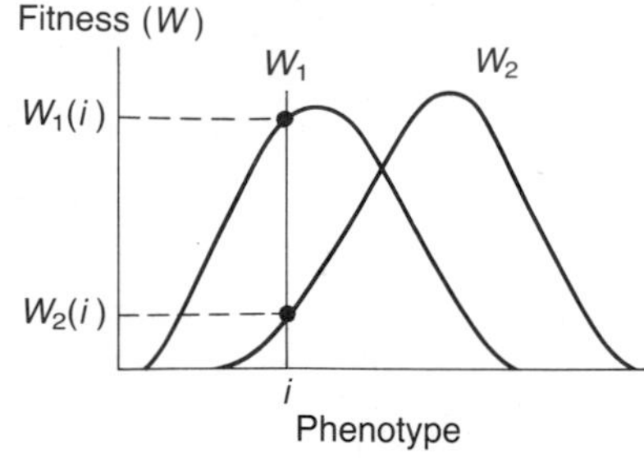

Figure 27–1
Relationship between phenotype and fitness in two different environments (1 and 2). The fitnesses of a single phenotype (i) in the two environments (W_1 and W_2) are indicated.

had also shown that constant laboratory conditions revealed no such heterosis; rarely did two or more morphs or genotypes coexist in population cages. This led A. B. da Cunha and Theodosus Dobzhansky (1954), Richard Lewontin (1958), and others to suggest that polymorphism persisted only when different genotypes were selectively favored in different parts of the environment or at different times.

Starting with these general concepts, Richard Levins (1962) set out to "attempt to explore systematically the relationship between environmental heterogeneity and fitness in populations." His theoretical approach took a giant step beyond contemporary bounds of empirical data and experimental result.

The heart of Levins's theory was fitness set analysis. A fitness set portrays phenotypes on a graph whose axes are components of fitness. One then superimposes a fitness criterion upon this graph to identify that phenotype having the greatest fitness. For simplicity, Levins restricted his analysis to the case of a population exposed to two different environments (1 and 2) for which the components of fitness are the respective fitnesses of each phenotype in each of the two environments (Figure 27–1). When all possible phenotypes are positioned on a graph according to their fitnesses in environment 1 (W_1) and environment 2 (W_2), the solid figure outlined is the fitness set (Figure 27–2). Only those phenotypes lying on the periphery of the figure are of interest because at least one of these will have a fitness superior to any phenotype within. But which point on the periphery represents the greatest fitness? When conditions are constant, and individuals experience only environment 1 or only environment 2, the selected phenotype will be the one having the greatest fitness in that environment.

To determine the most fit phenotype in a heterogeneous environment Levins devised the "adaptive function," a mathematical statement combining fitnesses under each of the conditions (for example, 1 and 2) into a measure of the overall fitness of a phenotype in a heterogeneous environment. When the environment varies spatially and organisms encounter habitat patches of type i in direct proportion to their frequency (p_i), the overall fitness of the phenotype (W) is the average of the fitnesses in each habitat patch weighted by their frequency. Thus, $W = p_1W_1 + p_2W_2$. By rearranging this expression, we obtain $W_2 = (W/p_2) - (p_1/p_2)W_1$, which is the equation of a straight line on a graph whose axes are W_1 and W_2 (Figure 27–3), the same axes used to define the fitness set. This equation defines a family of lines having slope $-p_1/p_2$ and intercept, or distance from the origin of the graph, determined by the value W—the overall fitness. Each line represents the combinations of W_1 and W_2 resulting in the same overall fitness. These lines of equal fitness—or adaptive functions, as Levins called them— superimposed upon the fitness set identify the phenotype with the highest overall fitness as the point on the periphery of the fitness set touched by the adaptive function lying farthest from the origin of the graph, hence representing the highest value of W among all phenotypes (Figure 27–4). This adaptive function is a tangent to the fitness set.

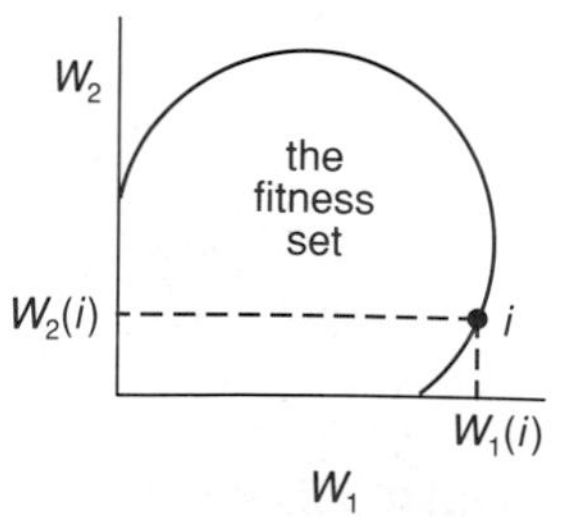

Figure 27–2
The fitness set portrays the relationship between the fitnesses of all possible phenotypes in environments 1 and 2. Each point on the perimeter of the set represents a single phenotype (i) whose position on the graph is determined by its fitnesses W_1 (i) and W_2 (i).

As the ratio of habitat patches varies from place to place, the slope of the adaptive function ($-p_1/p_2$) varies and the tangent adaptive function touches the fitness set at different points (Figure 27–5). From this result,

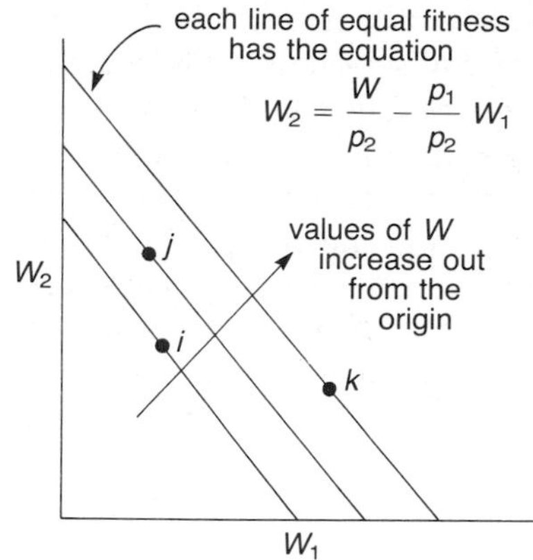

Figure 27–3
Adaptive functions are lines of equal fitness (W) whose positions on the graph are determined by the relative proportions of environments 1 and 2. The fitness values of points i, j, and k are ordered according to $W_i < W_j < W_k$.

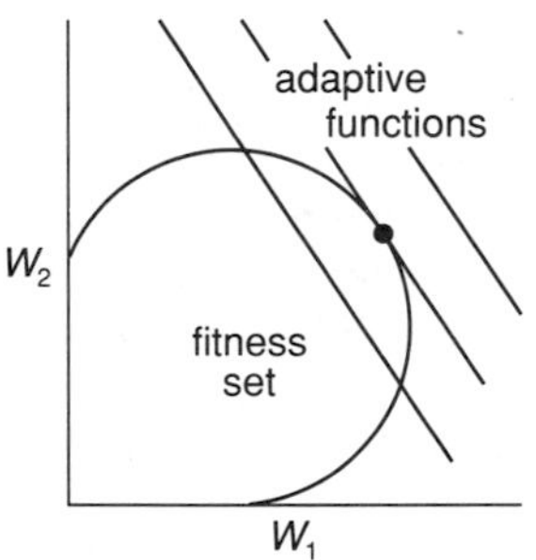

Figure 27–4
The superimposition of adaptive functions on the fitness set reveals the most fit phenotype as that touched by the adaptive function of highest value (the tangent to the fitness set).

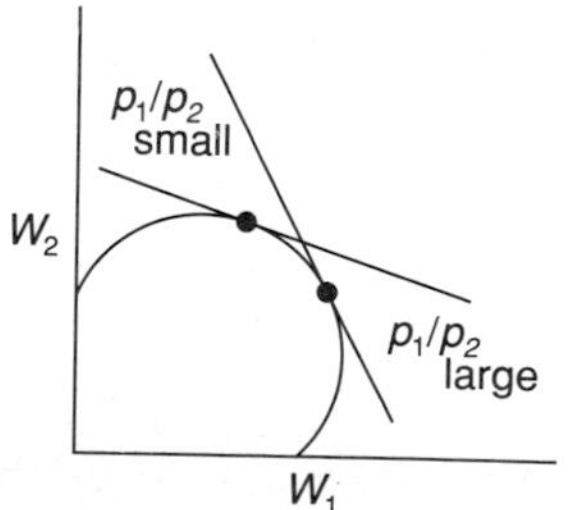

Figure 27–5
As the slope of the adaptive function changes with changes in the proportion of environments 1 and 2, the most fit phenotype also changes.

Levins suggested that along a cline of frequency in habitat patches one should expect to see a cline in selected phenotype.

The fitness set portrayed in Figures 27–2, 27–4, and 27–5 was based on habitat patches sufficiently similar that the fitnesses of each phenotype in the two habitats did not differ greatly. Levins also explored the situation in which the habitat patches present markedly different selective conditions. In such cases, the fitness set has a depression in its center, which results in a different outcome of selection (Figure 27–6). Linear adaptive functions superimposed on this fitness set show that phenotypes lying along the inward bulging part of the fitness set's periphery cannot be selected. As the ratio of habitat patches changes, the phenotype with maximum fitness shifts abruptly from one lobe of the fitness set to the other, skipping over intermediates. Intuitively, when habitats differ greatly, phenotypes intermediate between those superior in each type of habitat are not well suited to either. This result gave the adage "jack of all trades, master of none" considerable currency in ecological thinking. It also led to an interesting prediction: along a gradual cline of frequency of habitat patches, one could expect to see an abrupt change in phenotype (see Endler 1977 for more recent work on this problem).

When the environment varies temporally, so that habitat patches are encountered in a time sequence, the adaptive function may be more appropriately calculated as a geometric mean, rather than an arithmetic mean. This is because fitnesses represent factorial increases in the numbers of phenotypes from generation to generation. When population growth rate varies over generations, the long-term expectation of population size is directly proportional to the product of the population growth rates experienced during each time segment. Therefore, in a temporally varying environment, the average fitness of a phenotype is $W = W_1 p_1 W_2 p_2$, which can be rearranged to give the adaptive function $W_2 = (W/W_1 p_1)^{1/p_2}$. This equation is portrayed graphically in Figure 27–7.

At this point in the development of his theory, Levins abandoned logic and asserted the existence of an extended fitness set (F'), which he defined as "the set of all points in the fitness space that represent possible populations, that is, all possible mixtures of the phenotypes represented in the set F'' (Figure 27–8). According to Levins, selection favored a mixture of genotypes represented by the tangent of the hyperbolic adaptive function on the extended fitness set. Thus he believed that he had achieved an explanation for polymorphism in populations, and further predicted that along a cline of frequency of temporally heterogeneous conditions, one should observe a cline in the frequency of two phenotypes (a morph-ratio cline).

Levins erred in thinking that in a temporally varying environment mixtures of two morphs assume the arithmetic mean of their component fitnesses. In fact, the "fitness" (that is, population growth rate) of a mixture of phenotypes is the average of the geometric mean of the component fitnesses of each phenotype. Hence Levine's extended fitness set does not exist. The absurdity of the concept can be appreciated most readily when the fitness set bows inward from one axis to the other. A mixture of two phenotypes, each well suited to a different type of condition but having zero fitness in the alternative condition, has zero fitness overall: neither of the

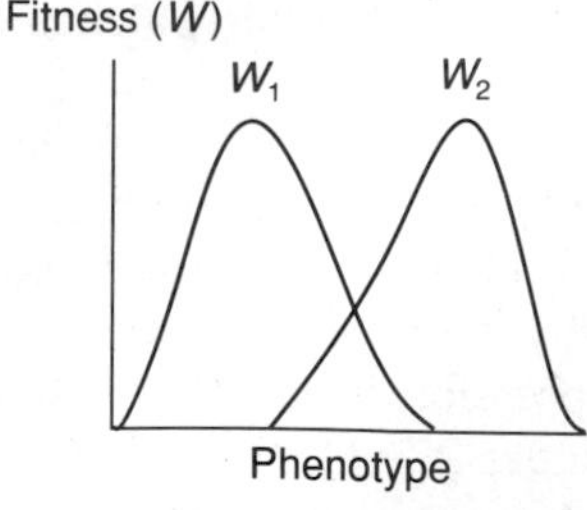

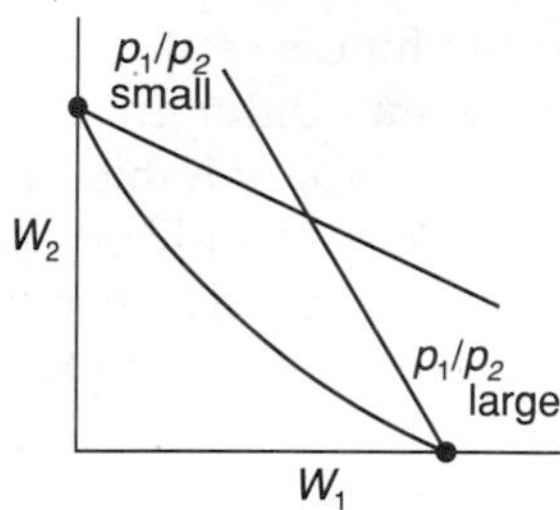

Figure 27–6
When environments 1 and 2 differ greatly, the fitness set is concave and changes in the proportion of environments cause a stepwise shift in optimum phenotype.

phenotypes can persist in a temporally varying environment. Levins made this mistake because he ignored the mechanics of selection in his model.

Although he claimed to have done so, Levins failed to solve the problem of genetic polymorphism in the absence of heterozygote advantage (Templeton and Rothman 1974). Other hypotheses (for example, Gillespie 1978, 1982) similarly rely on temporal variation in the environment, but have proper genetic bases and explicitly incorporate the mechanics of selection. But in spite of the shortcomings of Levins's analysis it left a legacy of concepts, explored initially by Levins's collaborator Robert MacArthur. Among these is the concept of environmental grain, which we touched upon earlier in this book.

MacArthur and Levins (1964) were interested in the outcome of competition between populations, or of competition between genotypes within populations, in heterogeneous environments. From the beginning, it was clear that the outcome would depend on the degree to which organisms could choose among the different habitat patches:

> To make these ideas more precise, we first consider an imaginary habitat in which there is a scattering of uniform units or grains of resource 1 and another scattering of uniform grains of resource 2. In such an environment we can distinguish as "fine-grained" an individual or species which utilizes both resources in the proportion in which they can occur. (If the actual grain size of the resources were so fine the species could not discriminate and select, then the species would have to be "fine-grained," hence the terminology.) An individual or a species will be called "coarse-grained" if it discriminates and selected only grains of one of the resources.

Levins's earlier treatment of fitness in a heterogeneous environment assumed that patches of habitat or resource were fine-grained; that is, individuals confronted habitat patches in direct proportion to their frequency. When patches are coarse-grained, the individual may use both types of patches or specialize on one or the other. But when an individual can choose, what is the best strategy to follow? MacArthur and Levins argued that "normally, coarse-grained utilization [i.e., specialization] will be expected only where the time and energy lost due to neglecting the other possible resource is slight compared with the benefits of specialization."

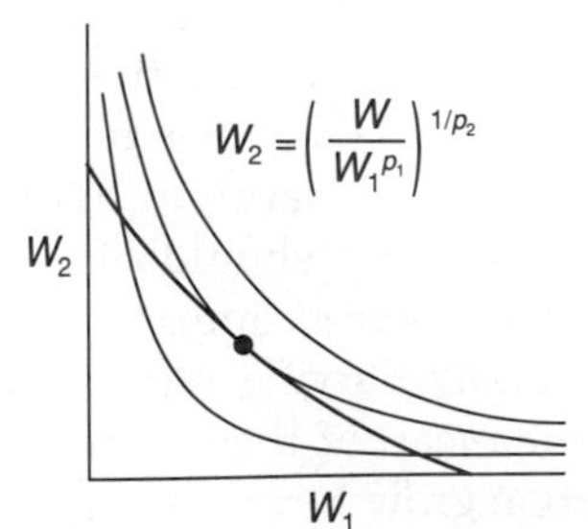

Figure 27–7
When environments are encountered in sequence the adaptive functions are hyperbolas and may favor an intermediate phenotype even for a concave fitness set.

The theory of optimal foraging addresses the problem of choice among resources or habitats.

When an animal moves through a habitat in search of food, it sequentially encounters potential prey. With each encounter, the organism is confronted with the choice of pursuing and eating the prey, which requires time and the expenditure of energy, and results in either the acquisition of food energy or the passing up of potential prey in favor of continuing to search. Intuitively, when a predator meets potential prey infrequently compared to the time it

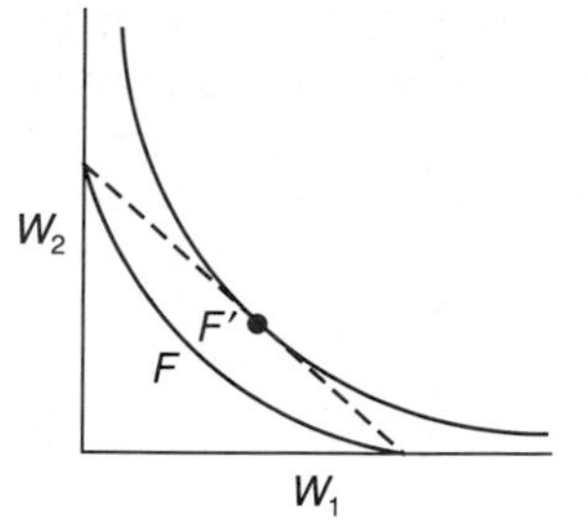

Figure 27–8
Levins's concept of the extended fitness set (F') as a mixture of different proportions of extreme phenotypes, which he thought could explain the maintenance of polymorphisms in populations.

requires to subdue and consume them, it should eat all prey types. But when a predator encounters prey often, it should pass by less desirable types because it will soon find better ones.

MacArthur and Eric Pianka (1966) formalized these arguments in a graphical model, which employed a cost-benefit analysis to determine the number of kinds of prey that a predator should include in its diet. They assumed that the best diet selection minimized the average time required to find and consume an individual prey. To simplify the model, we shall consider only the case in which prey species have identical nutritive value but vary in abundance and predator escape tactics. Increasing diet breadth by adding a new kind of prey affects the rate of consumption of prey in two ways: first, a broader diet avails the predator of more potential prey individuals; opposing this benefit, however, the average ease of pursuit, capture, and consumption decreases, provided that the predator always includes prey species in its diet in descending order of suitability. A predator should broaden its diet until the decrease in average quality of its prey more than offsets any decrease in searching time due to greater abundance of prey included within the diet.

In Figure 27–9, the average search time (T_s) and pursuit and handling time (T_p) per prey are graphed as a function of diet breadth, with prey ranked according to their suitability. The optimum diet breadth is that which results in the lowest sum of search and pursuit time ($T_s + T_p$). The slope of the search-time curve becomes less steep with increasing diet breadth because each new prey species adds proportionately less to the total diet. The slope of the pursuit-time curve increases because as the suitability of the prey decreases the capture and handling time increase more rapidly. Because of the shapes of the T_s and T_p curves, the sum of the two usually assumes a U-shaped curve, the lowest point of which defines the optimum diet breadth.

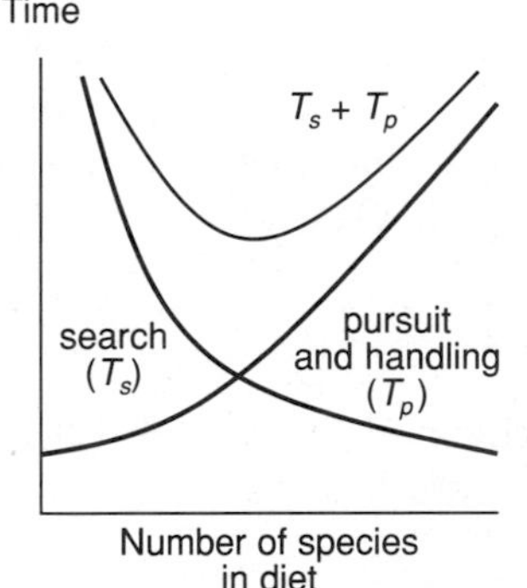

Figure 27–9
A graphical model of the optimization of diet breadth by a predator. Searching time (T_s) and pursuit and handling time (T_p) are drawn as a function of diet breadth. The sum of the two ($T_s + T_p$) is inversely related to rate of prey consumption. Hence optimum diet breadth corresponds to the minimum of the ($T_s + T_p$) curve.

A change in the overall abundance of prey, resulting from a change in the productivity of the habitat or the number of competing species, will change search time, but not pursuit and handling time; only the quantity of prey vary, not their quality. When productivity increases, the T_s curve decreases and the optimum diet breadth shifts to the left. Thus increased productivity favors specialization. When competition increases, prey become more scarce and the T_s curve increases, shifting the optimum to the right and favoring increased generalization (Figure 27–10).

How does prey diversity affect diet breadth? When the number of potential prey species increases without increasing their overall variety or total production, the T_s curve increases because each species becomes proportionately less abundant, the T_p curve decreases somewhat because the number of easily caught prey species increases, and optimum diet breadth shifts to the right. But although the diet includes more species of prey, the proportion of potential prey eaten and the variety of prey, assessed by their predator-escape characteristics, do not change. When the number of potential prey species increases by increasing the variety of species (adding species more difficult to capture), only the search time curve increases (each prey becomes less abundant), and although diet breadth increases slightly, proportion of total prey eaten as well as rate of prey capture decreases. In

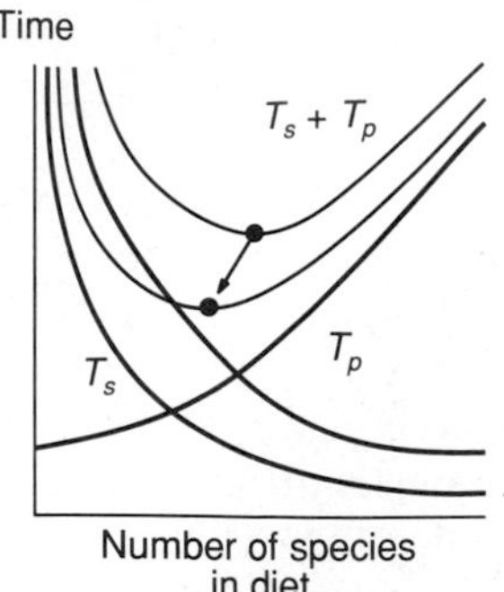

Figure 27–10
When availability of prey increases, due either to increased habitat production or reduced competition, searching time (T_s) decreases and the optimum diet breadth narrows.

general, then, increased production leads to diet specialization and increased rate of consumption of prey. Increased competition is equivalent to reduced production and has an opposite effect. Increased variety of prey species may lead to a somewhat broader food niche, but reduced rate of prey consumption, all other things being equal.

The "classical" model of optimal foraging incorporates the concepts of encounter rate and profitability of individual prey.

Optimization of choice by predators has been explored theoretically by a number of authors (Schoener 1969a, 1969b, 1971; Rapport 1971; Pulliam 1973; Cody 1974; Pyke et al. 1977; Orians and Pearson 1979; Townsend and Hughes 1981; Sih 1984, Charnov and Stephens 1988; for a general treatment see Stephens and Krebs 1986). These treatments agree with MacArthur and Pianka's (1966) prediction that increased production of resources favors increased diet specialization. Mathematically, the problem of optimal diet has been characterized by the "classical" model of prey choice. The exposition here follows that of John Krebs and N. B. Davies (1981), who, for the sake of simplicity, considered a predator hunting two types of prey (1 and 2, again). The pertinent characteristics of the prey are encounter rate (λ), reward (E), handling time (h), and profitability (E/H). A predator searches until it encounters the first prey (i). It then pursues, subdues, and consumes that prey with handling time h_i. During this handling period, the predator cannot search and therefore passes up any other prey that it might have encountered had it continued searching. When prey i is consumed, the predator once again starts searching.

When two prey are sought, the total encounter rate of prey is the sum of the encounter rates of each individually ($\lambda_1) + \lambda_2$). The total number of prey encountered in a series of search intervals of total time T_s is equal to $T_s(\lambda_1 + \lambda_2)$, and the expected reward obtained from these prey is

$$E = T_s(\lambda_1 E_1 + \lambda_2 E_2)$$

The total handling time associated with search time T_s is $T_s(\lambda_1 h_1 + \lambda_2 h_2)$, and therefore the total time required to obtain reward E is

$$T = T_s + T_s(\lambda_1 h_1 + \lambda_2 h_2)$$

the sum of the search and the handling times. Now, the amount of reward received per unit of time when the diet includes prey items 1 and 2 is therefore

$$\frac{E(1,2)}{T(1,2)} = \frac{\lambda_1 E_1 + \lambda_2 E_2}{1 + \lambda_1 h_1 + \lambda_2 h_2}$$

The reward per unit time for selecting only one of the two prey types is, similarly,

$$\frac{E(1)}{T(1)} = \frac{\lambda_1 E_1}{1 + \lambda_1 h_1}$$

Supposing that prey type 1 provides more reward per unit time than type 2 [that is, $E(1)/T(1) > E(2)/T(2)$], the predator should specialize on prey type 1 rather than include both 1 and 2 in its diet when $E(1)/T(1) > E(1,2)/T(1,2)$, or

$$\frac{1}{\lambda_1} < \frac{E_1}{E_2}(h_2 - h_1)$$

In words, the more specialized diet results in a higher feeding rate when the time to encounter the next item of type 1 ($1/\lambda_1$) is less than the ratio of the food value (E_1/E_2) times the difference in the handling times ($h_2 - h_1$). Specialization is therefore favored by high encounter rate and food value, and low handling time. Notice that regardless of the abundance or food value of prey type 1, if it takes longer to handle than prey type 2 the predator should always include both types in its diet.

Krebs et al. (1977) attempted to test the "classical" foraging model by putting great tits *(Parus major)* through an ingenious prey choice situation. Pieces of meal worms of two sizes were placed on a conveyor belt that passed under an opening, allowing the caged subject access to the "prey." Reward (E) was related to the size of the prey, which was under the experimenters' control, handling time (h) was measured, and encounter rate (λ_1) could be varied. According to the theory, when the benefit of selecting the single, most profitable prey type [$E(1)/T(1) - E(1,2)/T(1,2)$] exceeded zero, the subjects should have switched from picking prey items off the belt at random to selecting only the larger pieces. The results of the experiment conformed generally to prediction (Figure 27–11) but, contrary to expectation, prey type 1 was never selected to the exclusion of prey type 2.

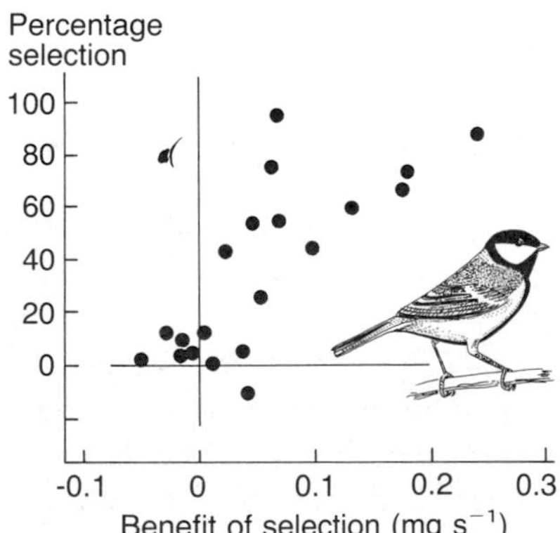

Figure 27–11
A test of the "classical" optimal diet model with great tits showing the relationship between preference for a more profitable prey and the benefit of specializing on that prey. The model predicts an abrupt switch when the benefit of specialization exceeds zero, a prediction to which the subjects only partially conformed. (After Krebs and Davies 1984.)

Krebs and R. H. McCleery (1984) discussed many possible explanations for deviations of the experimental results from theory, but their bottom line was that the theory tremendously oversimplifies both the food resource and the behavioral response of an astute predator to that resource. Certainly errors in discriminating prey types, long-term learning of preferences for one type or another, short-term learning during runs of one or the other prey type, simultaneous encounter of more than one prey, and non-substitutability of prey because each provides a different essential requirement add complications that are difficult to incorporate into simple models. Further considerations include the reduction of risks associated with variable prey availability (Barnard and Brown 1983, Caraco 1983, Caraco et al. 1980, Stephens 1981, Weissburg 1986, Werner et al. 1983, Real and Caraco 1986, Wunderle et al. 1988) and spatial distributions of resources with respect to nesting, perching, or roosting sites (Orians and Pearson 1979, Kacelnik 1984; Krebs and Avery 1985, Tamm 1989).

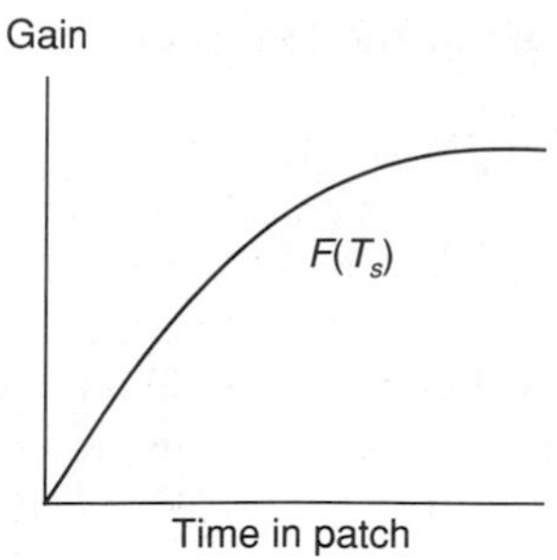

Figure 27–12
Cumulative amount of resource consumed as a function of amount of time in a patch of resource. The curve levels off as the consumer depletes the resource and rate of aquisition declines.

Optimal patch use depends on patch quality and traveling time between patches.

Most prey occur in patches of high abundance separated by unsuitable habitat. Predators must travel between patches, and so they are confronted with choices concerning prey use within patches and must decide when to leave a patch to search elsewhere. As in optimal prey choice, discussed above, this problem has received considerable attention from theoreticians (Charnov 1976, McNair 1982) and has been investigated experimentally in some laboratory systems. Most models of the patch-choice problem derive from the idea that the quality of a patch decreases as the predator captures prey and reduces the level of resource within. Eventually, the predator receives greater reward by moving to a new patch than by continuing to search in a depleted one. The amount of time in a patch before leaving is called the giving-up time.

As a predator initially searches within a patch, it encounters and consumes food at a high rate. As it depletes the resource, its return per unit of time decreases and the total gain from the patch begins to level off (Figure 27–12). The rate of gain decreases to zero as the resource is exhausted. During a period of search (T_s), the average rate of gain from a patch [$F(T_s)$ is the amount of food consumed [$P(T_s)$] divided by the sum of the search time and the time required to travel between patches (T_t). Thus $F(T_s) = P(T_s)/(T_s + T_t)$. Assuming that T_t is fixed according to the spatial distribution of patches, the optimum giving up time (T_s^*) can be found by taking the derivative of this expression and solving to find the maximum (that is, $dF/dT_s = 0$). This is shown graphically in Figure 27–13. A major prediction of this model is that optimal giving-up time decreases as travel time between patches decreases. Intuition tells us that when a new patch with high resource level can be reached quickly, one cannot gain by staying long in a patch of diminishing quality.

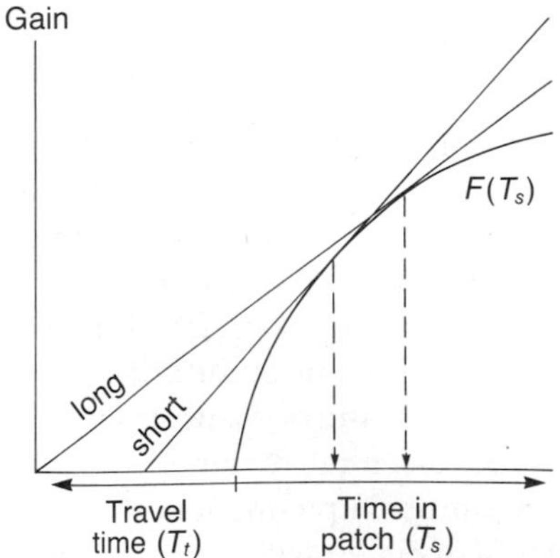

Figure 27–13
Graphical portrayal of the influence of travel time between patches on optimal giving-up time for a consumer within a patch. The greater the distance between patches, the longer the consumer should stay in each.

The graphical result can be given some quantitative substance by specifying the predation function $P(T_s)$. Let N be the total quantity of resources in a patch; a is a constant representing predation efficiency. Now, when the predation function is, for example, $P(T_s) = aNT_s/(1 + aT_s)$, the shape of which is illustrated in Figure 27–13, one can solve for $T_s^* = \sqrt{(T_t/a)}$. This result leads to three conclusions: optimal giving-up time (T_s^*) is (1) independent of the total amount of resource per patch (N); (2) decreases as the rate of prey depletion (predation efficiency, a) increases; and (3) increases as T_t increases.

Tests of the theory of optimal giving-up time have been hard to come by. They rely on matching the behavior of a predator to quantitative predictions obtained from specific formulations of the model. In order to make the comparison, one must be able to measure the predation function $P(T_s)$ and the travel time between patches (T_t) directly. This is possible only in contrived laboratory situations, among which parasitoid-host systems and birds feeding on mealworms have proved especially suitable. Although results often conform to the predictions of models (for example, Cook and Hubbard 1977, Cowie 1977), predators in such situations may be performing more like small biological analogues of simplistic models than revealing the richness of their behavior in the natural world. Model systems do, however, provide insights into the behavioral capacities of organisms (perception, learning, memory, flexibility) relevant to making choices among the variety of opportunities presented by the environment (see Shettleworth 1984). Perhaps it is naive to think that behavior patterns, evolved in richly complex natural settings, will function "optimally" in simplified, unnatural laboratory experiments. But it is too early to disavow this approach as misleading, particularly inasmuch as it has already greatly enlarged the ecologist's concept of the organism-environment relationship.

The ideal free distribution equalizes gains among individuals in a population.

Models of optimal choice discussed to this point address the behavior of a single individual confronted with environmental heterogeneity. Frequently, however, many individuals face the same options and optimal behavior may be influenced by the decisions of other members of the population. For example, a patch becomes less attractive to newcomers as the number of predators exploiting it increases. Such a patch likely has fewer remaining resources relative to undiscovered patches; furthermore, competing individuals may precipitate costly behavioral conflicts. Presented with many patches of varying intrinsic quality, consumers having complete knowledge and freedom of choice should occupy or exploit patches in direct proportion to their quality. Those with the highest levels of resources should attract the most consumers, those with the lowest levels the fewest.

Each consumer bases its choice of a patch on criteria that maximize its own rate of gain of resources. Imagine two patches, one having more resources than the second. At first, consumers will choose the intrinsically better patch. But as the population of consumers builds up in that patch, its apparent quality decreases, owing to depletion of resources and antagonistic interactions, until the second patch becomes the better choice. At this point, additional consumers choose the second and first patch alternately as the quality of both continues to decrease. As a result, each individual in the population will exploit a patch of equal realized quality, regardless of the variation in intrinsic patch quality in the absence of consumers. This is called the "ideal free distribution," a term used by Stephen Fretwell and H. L. Lucas (1970) to describe the filling of different habitat patches by territorial species of birds.

The existence of ideal free distributions of individuals in nature is suggested by several lines of evidence, that for habitat selection by birds being among the most compelling. In migratory species, individuals arrive on the breeding grounds over a period of several weeks. Early arrivals generally fill certain habitats that confer high breeding success before newcomers begin to establish territories in poorer habitats. When individuals are removed from "good" habitats, their places are quickly taken by others moving in from "poor" habitats (Krebs 1971). When population densities are low, perhaps following a particularly severe winter, occupancy of poor habitats decreases more than that of good habitats (for example, O'Connor 1980). Usually, however, breeding success in poor habitats does not match that in good habitats, thus contravening one expectation of an ideal free distribution. This discrepancy may reflect unequal competitive ability within a population based on age or social dominance; dominant individuals may be able to monopolize resources by defending large territories and set an upper limit to population density below the density attained under an ideal free distribution.

The concept of the ideal free distribution has been applied to several phenomena, perhaps most successfully by G. A. Parker (1974, 1978), who examined the number of male dungflies *(Scatophaga)* that compete for mates on cowpats (also known as meadow muffins and cow pies) of different size, hence quality and attractiveness to females. The distribution of males among these patches of "habitat" approximates an ideal free distribution, in which males gain similar numbers of matings regardless of patch quality.

The achievement of an ideal free distribution has also been investigated in the laboratory, where the quality of patches can be controlled. M. Milinski (1979) conducted experiments in which six stickleback fish were provided with food (water fleas) at different rates at opposite ends of an aquarium. Each end could be regarded as a patch, and the system had the following conditions conducive to establishing an ideal free distribution: (1) the two patches differed in profitability; (2) profitability decreased as the number of fish using a patch increased; (3) the fish were free to move between patches, hence they had free choice.

Hungry fish were placed in the aquarium about 3 hours before the start of the experiment. During trials, the numbers of fish in each half of the tank were recorded at the end of each 20-second interval. Prior to adding any food to the tank the fish were distributed equally between the two halves. In one experiment, water fleas were added at a rate of 30 per minute to one end of the tank and 6 per minute to the other, a ratio of 5 to 1. Within 5 minutes, the fish had distributed themselves between the two halves in the same ratio as predicted for an ideal free distribution. In a second experiment, water fleas were provided at rates of 30 and 15 per minute, a 2 to 1 ratio. Again the distribution of fish followed suit, and when the better and poorer patches were reversed in the tank, the fish reversed their distribution within about 5 minutes. Behavioral mechanisms used to achieve an ideal free distribution were not determined, but cues for behavioral choices must incorporate both the rate of food provisioning and the number of competitors within the patch. Such experiments demonstrate considerable sensitivity of organisms to conditions of their environment as well as behavioral flexibility in making choices.

SUMMARY

The environment of the individual varies both temporally and spatially, and includes a diverse array of resources. Analyses based on the performance of organisms in different environments and upon different resources provide insights into the rules governing decisions about diet and habitat selection.

1. Fitness set analysis, devised by Richard Levins, portrays phenotypes in a fitness space whose axes are component fitnesses in different habitat or resource patches. The adaptive function provides a criterion by which one can identify phenotypes in the fitness set having the highest fitness.
2. Specific predictions of Levins's analysis pertained to geographic clines in the ratios of different patch types. When patches are of similar quality, so that fitnesses in each differ only slightly, the optimum phenotype should change gradually over the cline. Beyond a threshold dissimilarity between patches, the optimum phenotype should shift abruptly from one well-suited to the first type of patch to one well-suited to the second.
3. Richard Levins and Robert MacArthur introduced the concept of environmental grain to distinguish situations in which organisms utilized resources or habitat patches in proportion to their occurrence (fine-grained) or were able to choose among them (coarse-grained).
4. Simple models of prey choice, based upon considerations of search time, handling time, and reward, predict that predators should specialize more when prey are abundant than when prey are scarce.
5. The optimum period that a consumer should remain within a patch of resources before moving on to another patch is the optimal giving-up time. According to simple models, this increases as the initial level of resources within patches and the traveling time between patches increase.
6. When many individuals choose among patches of habitats, assessment of habitat quality is influenced by the presence of other, competing individuals. In theory, individuals should distribute themselves among habitat patches so as to equalize their fitnesses, resulting in what is called an ideal free distribution.

28

Evolution of Life Histories

Adaptation is a form of response to the environment. It also resolves conflicting requirements of organisms. An individual cannot be everywhere at once; it cannot do everything at once. It must divide limited time, energy, and nutrients among competing demands. A mammal's fur conserves heat when at rest in the cold, but impedes the dissipation of heat when active, especially in a warm climate. As an adaptation, the thickness of fur optimally resolves this conflict between heat conservation and dissipation for a particular type of organism in a particular climate.

In Chapter 27, we discussed decisions concerning prey, patch, and habitat choices. Each choice an organism makes influences its allocation of time between activities having different costs and benefits. Thus a decision to pursue an item of food yields a potential increment of energy and nutrients consumed but exacts a price in the currency of potential prey passed by. Each decision resolves a choice; the rules according to which each individual makes decisions express evolved properties of its nervous system.

Body insulation and prey choice seem remote to the adaptationist's measure of evolutionary fitness, partly because each constitutes a small part of the overall phenotype exposed to selection and only indirectly influences survival and number of offspring. Other characteristics, such as number of eggs, age at first reproduction, and parental care, seem more directly tied to fitness components and so they have received special attention from evolutionary ecologists. These so-called life history adaptations resolve conflicts over the allocation of limited time and resources between activities and processes that more or less directly enhance fecundity and those that enhance survival. In this chapter, we shall examine the contributions of particular life history adaptations to fitness, the linking of such adaptations through competing demands on limited time and resources, and criteria by which such conflicts are resolved in the best evolutionary interests of the organism.

The study of life history adaptation focuses upon traits closely associated with the life table.

R. J. Lincoln et al. (1982), in their *Dictionary of Ecology, Evolution and Systematics,* define a life history as "the significant features of the life cycle through which an organism passes, with particular reference to strategies influencing survival and reproduction." One of my skeptical colleagues unmasked the meaning of this definition when he quipped, "A life history is everything an organism is and does." While such a broad definition deprives the term of any meaning whatsoever, there has nonetheless developed a substantial and important literature dealing with the evolution of life histories. A perusal of this literature conveys the impression that by life history ecologists mean traits whose variation directly influences life table schedules of fecundity and survival, hence the population growth rate of individuals having particular values of those characters. Age at first reproduction is directly visualized in the life table itself; number of eggs directly influences fecundity, all other things being equal. The challenge to understanding such adaptations derives, of course, from the fact that all other things are not equal. In a world of limited resources, one must pay for more eggs by less growth, storage, and maintenance, leading to increased risk of death and fewer eggs produced in the future by survivors.

The study of life histories is largely a legacy of David Lack, whose influence is ubiquitous in population biology and evolutionary ecology. During the early 1940s, Reginald Moreau, a British colleague of Lack who had worked in Africa for many years, called attention to the fact that songbirds in the tropics laid fewer eggs—2 or 3, on average—than their counterparts at higher latitudes—generally 4 to 10, depending on the species (Moreau 1944). Lack, who had been thinking much about the role played by adaptations for feeding in the process of speciation and the coexistence of ecologically similar organisms in communities (Lack 1944), turned his attention to this latitudinal gradient in clutch size and to other variations in clutch size both within and among species. In his first paper on the subject, published in 1947, Lack clearly recognized that any genetically determined increase in clutch size would be strongly selected owing to the greater fecundity of the bearer, unless counteracting forces reduced survival of the offspring. He presumed that the ability of adults to gather food for their young was limited, and so broods with more than a certain critical number of offspring, determined by availability of food, would be undernourished and the survival of the chicks thereby reduced. In Lack's (1947) words:

> The average clutch-size is ultimately determined by the average maximum number of young which the parents can successfully raise in the region and at the season in question, i.e. . . . natural selection eliminates a disproportionately large number of young in those clutches which are higher than the average, through the inability of the parents to get enough food for their young, so that some or all of the brood die before or soon after fledging, with the result that few or no descendants are left with their parent's propensity to lay a larger clutch.

Lack further suggested that because of the longer day length at high latitudes, temperate and arctic-zone birds could gather more food and therefore rear more offspring than birds breeding in the tropics, where day length remains close to 12 hours year round.

Further developments in the evolutionary study of life histories will be reviewed a bit later. But before going on, we must clearly distinguish between life history adaptations and life table variables. As we have seen in earlier chapters on populations, the population growth rate (fitness) of individuals bearing particular life history traits can be determined from their age-specific schedules of survival and fecundity. These life table entries express the interaction between the adaptations of the organism and its environment; they are not themselves adaptations, even though their values vary among genotypes. The distinction between adaptation and life table becomes apparent when one recognizes that identical individuals would have different life tables in different environments.

If life history adaptations are not equivalent to life table variables, then the question arises once more, What is a life history? Fitness is a rate—that of population growth. Rates are inversely related to time. In the life table, time's place is taken by the age of the individual. Thus adaptations that influence the age of births and deaths also influence the rate of increase, or fitness. They thus comprise one set of life history adaptations. Among these are rates of development, including the timing of such discrete events in the development program as metamorphosis (tadpole to frog, veliger to snail, caterpillar to moth, and so on) and the onset of sexual maturity (Werner 1986). The life table entries themselves are age-specific expectations of fecundity and survival; these are influenced by sets of partially overlapping adaptations. For example, parental care increases realized fecundity but also increases the risk of parental death. Fecundity at each age reflects the allocation of resources between growth and reproduction at an earlier age. Adaptations that influence more than one element in the life table pose the greatest challenge because their modification alters fitness in a complicated fashion. Finally, there is the problem of termination of life. In some species, such as the salmon, chambered nautilus, and century plant (agave), individuals reproduce once and die. In others, reproductive vigor and survival probability fall off with age as the individual undergoes physiological senescence. But because fitness is enhanced by long life and procreation, genetically determined decline and death are enigmatic.

Interest in life history adaptations has been stimulated by their variation between species.

Life history traits and life table parameters vary with respect to environmental conditions and to each other. Consistent patterns of variation, such as the general decrease in reproductive rate of animals from the poles to the tropics, have aroused curiosity and at the same time have suggested possible relationships between adaptation and environment that govern the pattern.

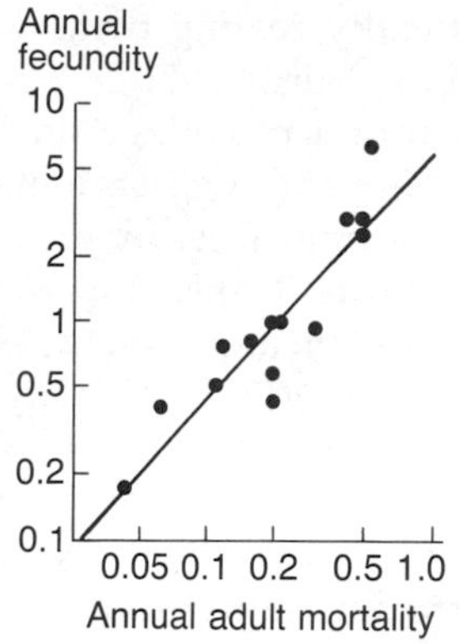

Figure 28–1
Relationship between annual fecundity and adult mortality in several populations of birds ranging from albatross (low) to sparrow (high). (From data in Ricklefs 1977).

Lack noted that clutch size and day length vary together, although plausible explanations for their correlation have been proposed based on many other covarying environmental conditions. The manner in which traits vary with respect to each other has also suggested mechanisms of functional integration of the phenotype. The observation that fecundity and adult mortality are strongly correlated (Figure 28–1) has suggested to some (perhaps reinforced by their own experience) that parental care burdens the individual with risks.

Some sets of life history traits are generally associated. At one extreme, elephants, albatrosses, giant tortoises, and oak trees exhibit long life, slow development, delayed maturity, high parental investment, and low reproductive rate; at the other extreme, one finds mice, fruit flies, and weedy plants (Pianka 1970). In broad comparisons within the plant and animal kingdoms, such associations of traits vary in close relation to body size and undoubtedly reflect the relative slowness of all life processes in large organisms (Peters 1983, Calder 1984, Schmidt-Nielsen 1984). But even among organisms of similar size and body plan, different environments produce widely divergent life histories. Storm petrels, which are seabirds the size of thrushes, rear at most a single chick each year, do not begin to reproduce until 4 or 5 years of age, and may live to 30 or 40 years. Thrushes, themselves, may produce several broods of 3 or 4 young each year beginning with their first birthday, but rarely live beyond 3 or 4 years. Similarly varied life histories may be found even among different populations of the same species (Schaffer and Elson 1975, Leggett and Carscadden 1978, Tinkle and Ballinger 1972, Jerling 1988, Fleming and Gross 1989).

Beyond such strong associations of life history traits with environmental conditions, many taxonomic groups also exhibit characteristic values of life history adaptations. Thus ducks (Anseriformes) usually lay 8 to 10 eggs per clutch, shorebirds (Charadriiformes) 4, hummingbirds (Apodiformes) 2, and petrels (Procellariiformes) 1. These differences probably reflect ways in which taxonomically conservative traits affect the selection of habitats by organisms and their particular interactions with the environment. Ducklings feed themselves and so their number is not limited by the ability of parents to gather food for them (what does limit their number remains a mystery; Rohwer 1985, Lessells 1986). Shorebirds typically lay very large eggs whose number may be limited by the ability of parents to incubate them successfully. Taxonomic affinity of a trait does not, however, reveal the significance of large egg size to shorebirds (why don't other birds lay large eggs?) or why shorebirds don't modify nest structure (usually a shallow depression in the ground) to accommodate more eggs (ducks incubate much larger egg masses than shorebirds).

No one knows why all species of hummingbirds, from the tropics to the arctic, lay 2 eggs per clutch. Apparently clutch size is constrained by some other adaptation or set of adaptations of these birds to their peculiar way of life. The single-egg clutch of all petrels, representatives of which are found in all oceans of the world and which vary in size from 30-gram storm petrels to 10-kilogram albatrosses, is thought to reflect a sparse and unpredictable food supply, which greatly limits the ability of adults to provide food to their chicks. But other seabirds that lay only 1 egg per year, such as gannets and swallow-tailed gulls, appear to have little difficulty rearing twins when provided an additional egg or chick (Nelson 1964, Harris 1970).

Life history variation in every group of organisms provides a similarly rich phenomenology that both raises one's curiosity and suggests tentative explanations for the adaptive basis of life history traits (see, for example, Eisenberg [1981] for mammals; Lack [1968] for birds; Tinkle et al. [1970], Shine and Bull [1979] for reptiles; Balon [1975], Thibault and Schultz [1978], Potts and Wooton [1984] for fish; Thorson [1950], Calow [1978], Maltby and Calow [1986] for invertebrates; Solbrig [1980], Silvertown [1982], Wiens [1984], Primack [1987] for plants). This natural history provides the setting within which theoretical and experimental studies have attempted to resolve how adaptation and environment interact to determine the life table of a population.

Life history theory began to develop rapidly during the 1960s.

The history of evolutionary ecology begins with Charles Darwin, who both identified and provided tentative explanations for some of the phenomena treated in this part of the book. But with regard to the types of life history traits discussed in this chapter, Darwin was uncharacteristically silent. He recognized the great reproductive capacity of all living beings but regarded the environment as an agent of selective mortality, undiscriminating with regard to birth rate.

Darwin did not have the benefit of life table analysis to guide his thinking. But even its development by A. J. Lotka and others, and its application to evolutionary phenomena by the population geneticist Ronald Fisher in his classic book *The Genetical Theory of Natural Selection* (1930), did not generate sufficient interest to inaugurate a new field of inquiry. In fact, the next major step toward the establishment of life history as an important focus for ecologists was the publication of Moreau's and Lack's papers on clutch size in birds during the mid-1940s. Although these attracted considerable attention, they still did not spark the flame that was to begin burning two decades later. In part, Lack's theory was too simplistic; it isolated clutch size (or reproductive rate more generally) as a single adaptation unrelated to other aspects of the phenotype. The ability of adults to deliver food to their offspring was accepted as determined by food availability in the environment. Neither the effort devoted to gathering food nor the time and effort devoted to caring for the young entered into the equation for fitness. Indeed, Lack's approach was decidedly nonquantitative. In addition, population biology during the late 1940s and most of the 1950s was embroiled in a controversy over the role of density dependence in the regulation of population size. Absorbed as ecologists were over the "balance of nature" (Egerton 1973), their attention was diverted from inquiry into life histories.

The early 1960s marked a turning point in population studies and saw the birth of modern evolutionary ecology. It is sometimes difficult to know what forces urge a discipline in one direction or another. The year 1959 was the centennial of the publication of *On the Origin of Species.* Dover Publications reprinted Ronald A. Fisher's *Genetical Theory of Natural Selection* in 1958. At that time George Williams, at the State University of New York at Stonybrook, was pondering the adaptive bases for the evolution of senescence (Williams 1957) and insect societies (Williams and Williams 1957). In 1960, papers were published by A. W. F. Edwards, W. A. Kolman, and H. Kalmus and C. A. B. Smith on the adaptive significance of the 1 : 1 ratio of males to females in most populations, a topic not touched since Fisher's treatment in 1930. This was a period of reunification of ecology and evolution.

Life history study burst upon this arena in 1966 with the publication of papers by Martin Cody and George Williams. Cody made two points of lasting significance. First, he applied Levin's ideas about fitness sets to life history evolution, calling attention to the fact that different components of fitness may be under conflicting selective pressures. Adaptation, he said, is largely the resolution of compromises in the allocation of time and energy to competing demands. Second, Cody introduced the concept that different life history adaptations are favored under conditions of high and low population density, relative to the carrying capacity of the environment. At high density, selection favors adaptations that enable individuals to survive and reproduce with few resources; hence efficiency carries a premium. At low density, adaptations promoting rapid population increase are selected; hence high rates of productivity, regardless of efficiency, increase fitness. These contrasting strategies were referred to as K-selected and r-selected traits, respectively, after the variables of the logistic equation for population growth (Boyce 1985).

George C. Williams's (1966) paper explored the demographic coupling between life history traits. He pointed out that each increment of reproductive effort influenced both contemporary fecundity and survival to reproduce in the future. Moreau (1944) had recognized 20 years earlier that fecundity and adult survival could be linked:

> It is not unreasonable to suggest that [the number of broods the parents can produce during their reproductive lives] may be affected by both the number of broods in the season and the number of young in the brood. It is possible, for example, that B/5 [a brood size of 5 offspring], at least in some circumstances, might put a significantly bigger strain on the parents than B/4, so that they were prevented from raising a larger total number as the product of the smaller broods in the same season; or that a succession of B/5 would so shorten the reproductive lives of the parents that their total of off-spring, produced in smaller, less exacting broods, would be greater.

But Williams quantified present and future components of fitness in the uniform currency of reproductive value, based on life table calculations, and indicated how the conflict between the effect of a life history modification on present and future reproduction was resolved according to the relative values of present offspring and the expectation of future offspring. A simple illustration of this principle compares evolution in two populations, one in which individuals have a high probability of survival between breeding seasons and the other, a low probability. In the first population, the expectation of future reproduction has a high value and selection tends to diminish reproductive investment in favor of protecting an otherwise high survival rate. In the second, because most individuals die before having further opportunity to breed, selection favors a high investment in the current crop of offspring.

W. D. Hamilton completed the 1966 triptych of life history papers with a theoretical contribution showing explicitly how variation in each of the life table entries — that is, fecundity and survival at each age — results in variation in fitness. This result and other related derivations formed the mathematical basis for subsequent development of most life history theory. Papers by Madhav Gadgil and William H. Bossert (1970), showing that the outcome of selection depended on the form of the relationship between life table values determined by particular life history adaptations, and by Garth Murphy (1968) and William M. Schaffer (1974), raising the issue of life history evolution in a variable environment, pretty much completed the groundwork.

Natural selection adjusts the allocation of limited time and resources among competing demands.

Because life history evolution is now seen as an optimal resolution of conflicting demands on the organism, a critical part of the study of life histories has been to understand the allocation of limited time and resources to competing functions. Time spent searching for food cannot be used to care for offspring directly nor to watch for predators. Energy and nutrients allocated to reproduction cannot be used to support growth. Besides time, energy, and nutrients, organisms also must partition their body structure — even cells within tissues — between competing functions (Ricklefs 1979). In many species, eggs are produced in direct proportion to the size of gonads, seeds in direct proportion to flower number. Photosynthetic rate depends in part on how much of its production a plant has allocated to photosynthetic tissue at the expense of root and support tissue. Plants also are built upon a modular body plan. In some species, nodes at points of leaf attachment may produce either lateral branches or flowers, but not both, thus trading off growth form against reproduction.

Selection of an increase in one or another function in an organism places demands on the individual to increase delivery of energy and nutrients or to increase allocation of production toward that function. We think of energy, nutrients, and production as being in limited supply because selection drives these functions to the physiological limit. Hence any increase in one component of demand, as by increased reproductive effort, results in a decrease in delivery to another component of demand.

Demonstrating the validity of this assumption of trade-offs that is built into most life history theory has been difficult. Adding and subtracting eggs in nests of several species of birds has, in some cases, revealed an inverse

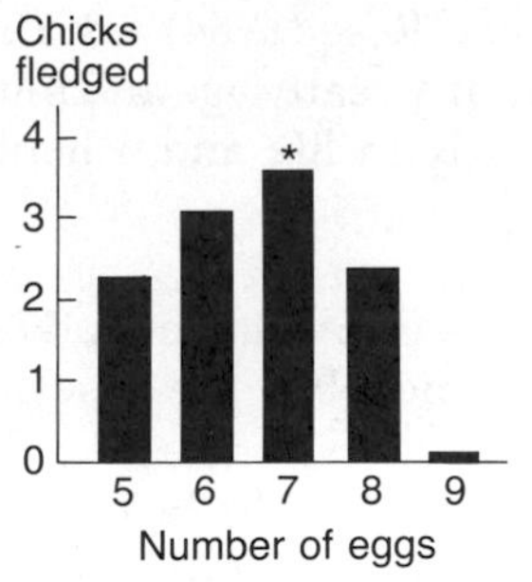

Figure 28–2
Number of chicks fledged from nests of European magpies in which 7 eggs were laid but the experimenter added or removed eggs to make up manipulated clutches between 5 and 9. The most productive clutch size was 7. (After Hogstedt 1980.)

relationship between the number of chicks in the nest and their survival. This often results in maximum productivity from intermediate brood sizes, as predicted by David Lack. For example, Goren Hogsted (1980) showed that the clutch laid by individual female magpies produced the maximum number of chicks that she could nourish. Either adding or subtracting eggs resulted in fewer offspring fledged (Figure 28–2).

When Diane DeSteven (1980) performed a similar clutch addition experiment with tree swallows, she found that adults were able to feed two extra chicks. The same proportion of young left the nest, and at about the same weight, in enlarged broods as in normal broods. Furthermore, DeSteven was unable to detect any difference in year-to-year survival of adults depending on the number of chicks they reared, although her samples were too small to detect subtle changes. However, Nadav Nur (1988), in similar experiments with the blue tit, found that among parents rearing enlarged broods, fewer survived to the following breeding season and these produced fewer offspring.

David Reznik (1983) altered allocation of resources to reproduction in guppies by preventing females from mating with males. If growth and reproduction competed for allocation of assimilated resources, the experimental fish should have attained larger size by the end of the study period. But, in fact, little of the difference in reproductive tissue accumulated between mated and unmated females was converted to growth (Figure 28–3).

Reznick (1985) summarized the literature purporting to address the question of trade-offs between life history traits and found few studies whose results could be interpreted unambiguously. The evidence presented in published studies falls into four categories:

1. Phenotypic correlations between traits within populations. These are difficult to interpret because each trait may respond independently to some third character not measured in the study. For example, low reproductive rate and long life span might be consequences of low metabolism, rather than having any direct interrelationship.

2. Experimental manipulations of the sort performed by Hogsted, DeSteven, Nur, and Reznick. These may indicate direct functional relationships between traits, because other variables can be controlled in an experiment. But because manipulated values are not genetically determined, the outcomes of the experiments do not directly address the evolution of life history patterns. They instead assess phenotypic plasticity, which may or may not reflect evolutionary tradeoffs.

3. Genetic correlations between traits. These indicate the degree to which two traits will respond in concert to selection on either or both. But, as we have seen, selection can produce an evolutionary trajectory perpendicular to genetic correlations, given enough time. Thus while genetic correlations can predict the short-term response, they do not necessarily predict the long-term course of life history evolution.

4. Correlated responses to selection. Selection experiments certainly are the most direct approach to studying life history evolution. The difficulty of performing such experiments has greatly limited their application, as one might expect. Revealing examples, however, are the experiments of

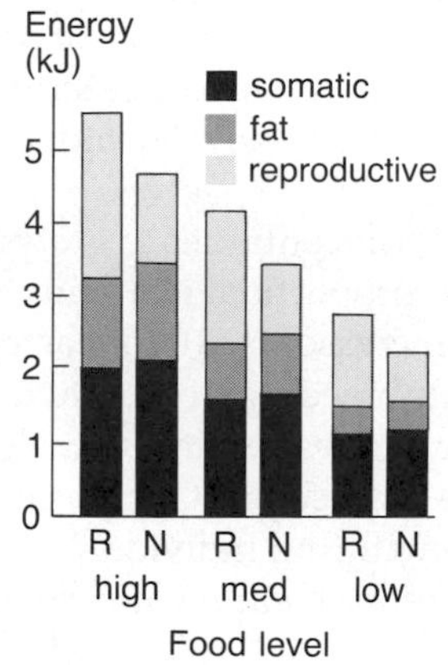

Figure 28–3
Contents of energy in somatic tissues, fat deposits, and reproductive tissues (including eggs) in female guppies raised at three food levels and which were either permitted to be (R), or prevented from being (N), courted and inseminated by males. (From Reznick 1983.)

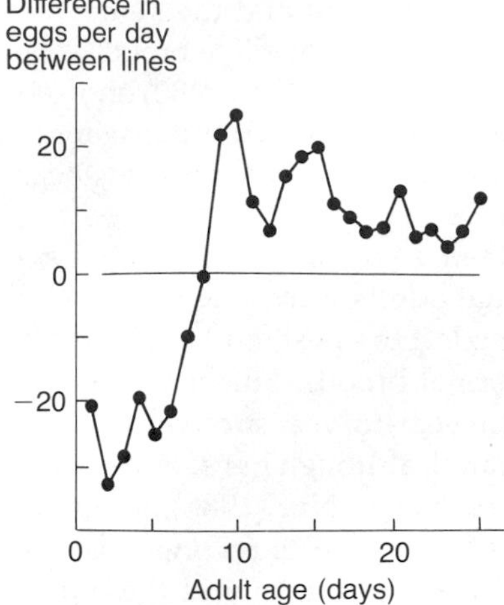

Figure 28–4
Difference in daily egg production as a function of age between a control line of fruit flies and a line in which life span was terminated at an early age. Negative values indicate increased egg production rate among short-lived flies. (From Rose and Charlesworth 1981.)

Michael Rose and Brian Charlesworth (1981) and Rose (1984), who artificially terminated the life span of *Drosophila* flies at an early age and found a heritable increase in rate of egg production early in life and a heritable decrease in natural life span (Figure 28–4).

According to Reznick's tabulation, the evidence of available studies revealed costs of reproduction more often than not, but the case needs strengthening by more direct experimentation.

Most issues concerning the evolution of life histories can be phrased in terms of three questions: When should I begin to produce offspring? How often should I breed? How many offspring should I attempt to produce in each breeding episode? Each of these questions expresses in a different way the fundamental trade-off between fecundity and adult survival.

Age at first reproduction generally increases in direct relation to adult life span.

When should an animal or plant begin to breed? Long-lived organisms typically begin to reproduce at an older age than short-lived ones. What selective forces could produce this result? We shall assume that age at first reproduction has genetic variation and can be selected independently of other life history characteristics, although it also may reflect rate of development and physiological processes selected for other reasons. At every age, an individual must choose between attempting to reproduce and abstaining from breeding. When young individuals resolve this choice in favor of abstention, they may delay the onset of sexual maturity. Thus age at first reproduction can be understood in terms of the benefits and costs of breeding at a particular age. The benefit appears in the life table as an increase in fecundity at that age. The cost may appear as reduced survival to older ages, reduced fecundity at older ages, or both.

Consider the following example. A type of fish continues to grow only until sexual maturity. Its fecundity is directly proportional to body size. Suppose that the number of eggs laid per year increases by 10 for each year that reproduction is delayed, so that individuals breeding in their first year produce 10 eggs and the same number each year thereafter, individuals first breeding in their second year produce 20 eggs, and so on. Comparing the cumulative egg production of early- and late-maturing individuals (Table 28–1), one can see that the optimal age at first reproduction varies in direct proportion to expected life span.

For organisms that do not grow after their first year (most birds, for example), the choice between breeding or not depends on balancing current reproduction against survival. Nonbreeding individuals avoid the risks of preparing for reproduction—courtship, nest building, migration to breeding areas. Presumably, life experience gained with age also reduces the risks of breeding, increases the realized fecundity of a certain level of parental investment, or both, favoring delayed reproduction. Balancing this are

Table 28-1
Total eggs produced by individuals in a hypothetical population as a function of life span and age at first reproduction (*a*)

	Life span							
a	1	2	3	4	5	6	7	8
1	**10***	**20**	30	40	50	60	70	80
2	0	**20**	**40**	**60**	80	100	120	140
3	0	0	30	**60**	**90**	**120**	150	180
4	0	0	0	40	80	**120**	**160**	**200**
5	0	0	0	0	50	100	150	**200**
6	0	0	0	0	0	60	120	180

* Boldface indicates most productive age at first reproduction for a given life span.

many factors that reduce expectation of future reproduction, including high predation rates, encroaching senescence at old age, and, for organisms that live a single year or less (annuals) in seasonal environments, the end of the productive season.

Perennial life histories are favored by high and relatively constant adult survival.

Another issue in the evolution of life history adaptations has been the number of breeding episodes that the individual undertakes. Plants and animals either reproduce during a single season and die (annuals) or have the potential to reproduce over a span of many seasons (perennials). Population biologists have pondered the relative advantages of each habit in terms of the trade-off between survival probability and fecundity. To survive the nonreproductive "winter" period, a perennial plant must allocate resources to storage of materials in roots and formation of freeze-resistant or drought-resistant buds, presumably at the expense of production. But at some point the advantages of the perennial habit must outweigh the costs in reduced fecundity relative to annuals.

Following the earlier lead of Lamont Cole (1954), Charnov and Schaffer (1973) compared annuals and perennials in an algebraic model. Suppose a population of plants contains some individuals that produce a large number of seeds at the end of the first growing season and then die (annual), and others that produce fewer seeds but survive through the winter to reproduce in subsequent growing seasons (perennial). Which has the greater fitness? For the purposes of their model, Charnov and Schaffer assumed that annual and perennial plants have the same probability of survival during their first (in the annual's case, only) growing season (S_0) and that perennials have a constant probability of survival thereafter (S_p).

The factor by which a population of an annual plant grows (λ) equals the number of seeds each individual produces (B_a) times their survival to reproductive age (S_0), or $\lambda_a = B_a S_0$. The increase of a population of a perennial plant equals the number of seeds (B_p) times their survival (S_0), plus the probability of survival of the parent (S_p), hence $\lambda_p = B_p S_0 + S_p$. The population growth rate of the annual exceeds that of the perennial ($\lambda_a > \lambda_p$) when $B_a S_0 > B_p S_0 + S_p$. Dividing both sides of the inequality by S_0, we obtain $B_a > B_p + S_p/S_0$ [or $B_a - B_p > S_p/S_0$]. Accordingly, an annual life history is favored when the number of seeds produced by the annual exceeds the fecundity of the perennial by the ratio S_p/S_0. When few perennials survive from one breeding season to the next or when perennials produce relatively few seeds, the annual habit is favored. Where individuals survive well once established but seedlings survive poorly (high S_p/S_0), the annual habit must result in extremely high fecundity to be favored. It is no wonder that annuals predominate among the floras of deserts, where few adult plants can survive drought periods, and perennials predominate among the floras of the tropics, where competition and predator pressure make difficult the establishment of seedlings. Adding complexity to the model, by incorporating growth from year-to-year and annual variation in survival probabilities, does not destroy the basic qualitative conclusion that the life history strategy is determined primarily by the ratio of adult survival to juvenile (Stearns 1976, Bulmer 1985).

Optimal reproductive effort varies inversely with adult survival.

For annual plants, expectation of life beyond the first breeding season is so small that all resources are devoted to current reproduction. Perennials, however, must allocate resources between current reproduction and adaptations that prolong life. When particular adaptations affect both fecundity and survival, the trade-off between the two must be optimized. Intuitively, when life span is short regardless of the consequences of reproduction, the balance of allocation should tip in favor of current fecundity. When potential life span is great, current fecundity should not unduly jeopardize future reproduction. This can be shown algebraically quite simply. First, we partition adult survival into two components, one directly related to reproduction (S_R), and the other independent of reproduction (S). Now, fitness may be expressed as

$$\lambda = SS_R + S_0 B$$

Certain reproductive adaptations that cause small changes in the values of survival (ΔS_R) and fecundity (ΔB) will influence fitness according to

$$\Delta\lambda = S\Delta S_R + S_0 \Delta B$$

When changes that enhance fecundity (ΔB positive) also reduce survival (ΔS_R negative), their effects on $\Delta\lambda$ depend on the relative values of S and S_0. In general, when S is large compared to S_0, selection favors adaptations that increase adult survival at the expense of fecundity, and vice versa. Thus one expects parental investment in offspring to decrease with increasing adult life span.

When survival and fecundity vary with age, models of life-history evolution must be based on the life table.

In the preceding models, we assumed that fecundity and adult survival were constant values, unvarying over age. In reality, however, among animals and plants that reproduce repeatedly, rates of survival and fecundity vary with age within the reproductive period. To the degree that differences in these variables represent the outcome of genetically determined modifications of the life history, it is important to understand the relative strengths of selection acting on changes in life table variables at different ages.

The characteristic equation of a population relates the rate of geometric or exponential increase in population size to the life table variables by

$$1 = \sum \lambda^{-x} l_x b_x \tag{28-1}$$

where survivorship to age x (l_x) is the product of the individual survival rates up to that age ($l_x = s_0 s_1 \ . \ . \ . \ s_{x-1} = \Pi_{i=0}^{x-1} s_i$). This equation allows us to determine how small changes in s_x and b_x change fitness, and therefore indicates the strength of selection on adaptations that affect these life table variables. As William D. Hamilton (1966) and John Merrit Emlen (1970) have shown, a small change in fecundity at age x influences λ according to

$$\Delta\lambda = \frac{\lambda^{-x} l_x \Delta b_x}{\lambda^{-1} \sum x \lambda^{-x} l_x b_x} \tag{28-2}$$

This equation shows that the strength of selection on b_x diminishes with age in direct proportion to the decrease in survivorship. For example, if 50 per cent of individuals survived to age 1 and only 25 per cent at age 2, then selection on change in fecundity at age 1 would be twice as strong as that on the same change in fecundity at age 2. If, under these conditions, an increase in fecundity at the first age caused that at the second to decrease, so long as the gain in b_1 was more than twice the loss in b_2, the modification would be selected.

The relative strength of selection at different ages also is influenced by the rate of growth of the population (λ). When population size is approximately constant ($\lambda = 1$), the term λ^{-x} is 1 at all ages. But in an increasing population ($\lambda > 1$), the term λ^{-x} falls off with age and modifications affecting fecundity at young ages have relatively greater effect on fitness. Symmetrically, the amount by which adaptive changes in fecundity at younger ages are favored is reduced in declining populations (see Hoogendyk and Estabrook 1984).

In any population, fewer individuals live to older ages. Hence a smaller proportion of genes that affect life table variables at older ages are expressed—exposed to selection—and therefore strength of selection declines with l_x. In a growing population, each individual born today is a larger fraction of the total population than is each individual born in the future; as the population expands the value of each individual diminishes. Therefore, from the standpoint of the life table, offspring born late in an individual's life have relatively less value than offspring born to the same individual early in its life, when they constitute a larger proportion of the total population.

The effect of a change in survival rate at age x (Δs_x) on fitness is given by

$$\Delta\lambda = \frac{\lambda^{-x} l_x s_x^{-1} \Delta s_x \sum_{i=x+1} \lambda^{x-i} b_i \prod_{j=x}^{i-1} s_j}{\lambda^{-1} \sum x\lambda^{-x} l_x b_x} \qquad (28-3)$$

Not surprisingly, the strength of selection on a fractional change in survival rate ($\Delta s_x/s_x$) varies in direct proportion to the survivorship to age x, but also in proportion to the expectation of reproduction at older ages. As with changes in fecundity, the relative strength of selection is diminished with age in increasing populations and augmented in decreasing populations.

Optimization of the life history is a matter of resolving conflicts in the expression of adaptations in different components of fitness, the life table variables. Equations (28–2) and (28–3) give us a quantitative basis for evaluating conflicting selective forces and predicting general patterns of adaptation expressed in the life table. For example, in conflicts between early and late reproduction, which might come about through use of resources for current reproduction that might otherwise be stored and used for future reproduction, conflict will be resolved in favor of reproduction at younger ages and fecundity might be expected to decline with age. This pattern should be more pronounced in populations with low adult survival rates and following selection during phases of rapid population growth. When conflict arises between fecundity at age x and survival to the following breeding period, the relative strengths of selection on b_x and s_x balance the value of current offspring against the expected value of future offspring according to

$$\Delta\lambda = \frac{\lambda^{-x} l_x \left(\Delta b_x + \Delta s_x \sum_{i=x+1} \lambda^{x-i} b_i \prod_{j=x}^{i-1} s_j \right)}{\lambda^{-1} \sum x\lambda^{-x} l_x b_x} \qquad (28-4)$$

(The term $\Sigma\lambda^{x-i} b_i \Pi\, s_j$ is simply the expectation of future reproduction of an individual of age $x + 1$ weighted by population growth.) Therefore, adaptations that increase current fecundity prevail over those that increase survival when adult survival is low and when the population is rapidly growing owing to high fecundity. By this reasoning, species with characteristically long life spans should put less effort into producing offspring, and more into avoiding predation and other sources of mortality, than species that, by their nature and habitat, are characteristically short-lived.

Many plants and invertebrates, and some fish, reptiles, and amphib-

Table 28–2
Numerical comparisons of the strategies of slow growth–high fecundity (A) and rapid growth–low fecundity (B) in two hypothetical fish

	Year					
	1	2	3	4	5	6
Strategy A						
Body weight*	10	12	14.4	17.3	20.8	25.0
Growth increment†	2	2.4	2.9	3.5	4.2	5.0
Weight of eggs	8	9.6	11.5	13.8	16.6	20.0
Cumulative weight of eggs*	8	17.6	29.1	42.9	59.5	79.5
Strategy B						
Body weight*	10	15	22.5	33.8	50.7	76.1
Growth increment†	5	7.5	11.3	16.9	25.4	38.1
Weight of eggs	5	7.5	11.3	16.9	25.4	38.1
Cumulative weight of eggs*	5	12.5	23.8	40.7	66.1	104.2

* Body weight + growth increment = next year's body weight; cumulative weight of eggs to last year + weight of eggs = cumulative weight of eggs to this year.
† Growth increment and weight of eggs in each year are equal to the body weight.

ians do not have a characteristic adult size. They grow, at a continually decreasing rate, throughout their adult lives (indeterminate growth). Fecundity is directly related to body size in most species with indeterminate growth. Because egg production and growth draw upon the same resources of assimilated energy and nutrients, increased fecundity during one year must be weighted against reduced expectation of fecundity in subsequent years. For organisms having longer life expectancies, growth should be favored over fecundity during each year. For organisms with less chance of living to reproduce in future years, resources allocated to growth instead of to eggs are largely wasted.

Consider two hypothetical fish, each weighing 10 grams at sexual maturity, but which allocate resources to growth and reproduction differently. Both gather enough food each year to reproduce their weight in new tissue or eggs. Fish A allocates two-tenths of its production to growth and eight-tenths to eggs, whereas fish B allocates one-half each to growth and eggs. Calculated growth, fecundity, and accumulated fecundity (Table 28–2) show that for fish living 4 or fewer years, on average, high fecundity and slow growth give the greater overall productivity, whereas for fish living longer than 4 years, more rapid growth and lower fecundity are superior. Adult mortality, therefore, determines the optimum allocation of resources between growth and reproduction.

Extensive preparation for breeding and uncertain or ephemeral environmental conditions may favor a single, all-consuming reproductive episode.

Some species of salmon have adopted a course of rapid growth for several years, culminating in a single immense reproductive effort, in which a large portion of the body tissues is converted to eggs, followed shortly after spawning by death. Gadgil and Bossert (1970) reasoned that because salmon make so great an effort to migrate upriver just to reach their spawning grounds, it may be to their advantage to make the trip just once, at which time they should produce as many eggs as possible, even if this supreme reproductive effort requires the conversion of muscle and digestive tissue to eggs and ensures their death.

The salmon live history pattern is sometimes referred to as "big-bang" reproduction but more properly as semelparity. This term comes from the Latin *semel* (once) and *pario* (to beget); it is contrasted with iteroparous, from *itero* (to repeat). Semelparity is rarely encountered among animals and plants that live for more than 1 or 2 years. Usually, the effort and allocation of resources required to survive between growing seasons are so much

Figure 28–5
Stages in the life cycle of the Kaibab agave *(Agave kaibabensis)* in the Grand Canyon of Arizona. The plant grows as a rosette of thick, fleshy leaves for up to 15 years. Then it rapidly sends up its flowering stalk and sets fruit, after which the entire plant dies (left-hand rosette).

greater than those used to prepare for breeding that, once a perennial life form has been adopted, reproduction every year seems the most productive pattern.

The best-known cases of semelparous reproduction in plants occur in the agaves (century plants; Schaffer and Schaffer 1977) and the bamboos (Janzen 1976), two distinctly different groups. Most bamboos are tropical or warm-temperate-zone plants that form dense stands in disturbed habitats. Reproduction does not appear to require substantial preparation or resources, as needed to grow a heavy flowering stalk. But opportunities for successful seed germination probably are rare. Once established, a bamboo plant increases by asexual reproduction, continually sending up new stalks, until the habitat in which it germinated is fairly packed with bamboo. Only at this point, when vegetative growth becomes severely limited, do plants benefit from producing seeds, which may colonize disturbed sites.

The environment and habits of agaves are at opposite ends of the spectrum from those of bamboos. Most species of agave inhabit arid climates with sparse and erratic rainfall. Plants grow vegetatively for several years, the number varying from species to species, then send up a gigantic flowering stalk. After producing its seeds, the agave dies (Figure 28–5). One curious fact about agaves is that they frequently live side by side with yuccas, a group of plants with a similar growth form, but which flower year after year. The root systems of agaves do, however, differ from those of yuccas; yucca roots descend deeply to tap persistent sources of groundwater. Agaves have shallow, fibrous roots that catch water percolating through the surface layers of desert soils after rain showers, but that are left high and dry during drought periods. The erratic water supply of the agave may prevent successful seed production or seedling establishment every year, and the period between suitable years may be very long. Under these conditions, it may be most advantageous for the agave to grow and store nutrients until an unusually wet year comes—perhaps 1 in 10 or even 1 in 100—and then to put all resources into reproduction.

Senescence evolves because of the reduced strength of selection at old age.

While few organisms exhibit programmed death associated with reproduction, most do experience a gradual increase in mortality and decline in fecundity resulting from deterioration of physiological function, known as senescence (Kohn 1971, Rockstein 1974, Lamb 1977, Calow 1978). For example, rates of most physiological functions in humans decrease in a roughly linear fashion between the ages of 30 and 85 years, to 80 to 85 per cent of the value in 30-year-old individuals in nerve conduction and basal metabolism, 40 to 45 per cent in the volume of blood circulated through the kidneys, and 37 per cent for maximum breathing capacity (Mildvan and

Strehler 1960). Birth defects and infertility generally occur with increasing prevalence in women progressively older than 30 years (Milham and Gittelsohn 1965, Menken et al. 1986). Senescence, reproductive decline, and death in old age do not result from abrupt physiological change. Rather, the demographic consequences of senescence result from a gradual decrease in physiological function with age. Such changes are found throughout the animal kingdom (Comfort 1956, Strehler 1960).

How can senescence evolve? Why is senescence not eliminated by selection when survival presumably is advantageous to an individual at any age? The answer to these questions is generally thought to originate in the declining strength of selection on genes expressed at progressively greater age, owing to the fact that fewer individuals bearing those genes survive to express them (Medawar 1957, Williams 1957, Hamilton 1966). Given this age-dependence of selection, senescence can be thought of arising in two ways. First, deleterious genes are constantly being added to the population by mutation, whose rate probably varies little with respect to age of expression of the gene. The ability of selection to remove these alleles declines with age, and so deleterious alleles with later age of expression will build up to higher levels in the population. Individuals that do survive to old age will be more likely to express bad genes as reduced physiological function.

Second, some alleles may act pleiotropically to enhance fitness at early ages but reduce fitness in later life (Rose and Charlesworth 1981, Rose 1982). Such alleles will tend to be incorporated into the gene pool because effects expressed at young age contribute more to fitness than do those expressed at old age. Although there is relatively little evidence for such pleiotropic alleles (indeed, for some genes, alleles beneficial at one age ought to be so at all ages), fruit flies selected for increased early survival have reduced survival rates later in life (Rose 1984).

When should we expect senescence to begin? How rapidly should it encroach upon old age? The answers to these questions depend upon differences in the strength of selection between ages. Changes in fitness caused by changes in survival rate at a given age (x) are equal to

$$\Delta\lambda = \frac{s_x^{-1}\Delta s_x \sum_{i=x+1} \lambda^{-i} l_i b_i}{\lambda^{-1} \sum x\lambda^{-x} l_x b_x} \qquad (28-5)$$

which is another way of expressing equation (28–3). Until the age at first reproduction, the sum of the $\lambda^{-i}l_ib_i$ terms is constant because all reproduction (terms in b_i) lies in the future. Hence the strength of selection on changes in survival rate remains constant until the onset of reproduction, and senescence should not manifest itself until after that point, which is certainly the case in humans (Figure 28–6). Furthermore, senescence should increase faster in populations with higher characteristic mortality rates because the sum of the $\lambda^{-i}l_ib_i$ terms drops off faster with age after reproduction begins. Thus, in the Dall sheep, whose minimum annual adult mortality rate is about ten times that of humans, senescence encroaches much more rapidly as well (Figure 28–6).

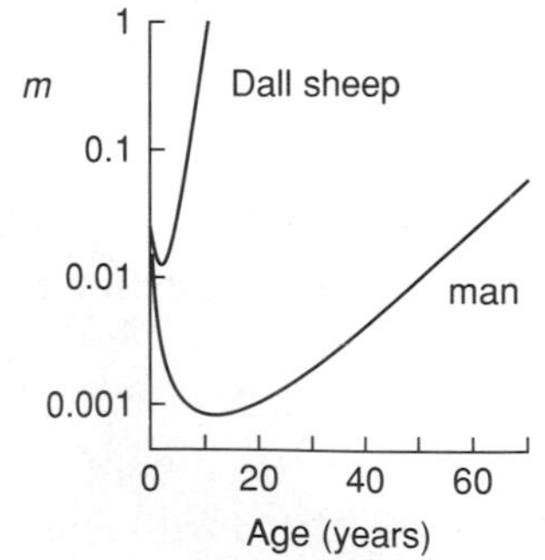

Figure 28–6
Relationship between annual mortality rate and age in the Dall mountain sheep (from Deevey 1947) and in a human population (Costa Rican females, 1963; from Keyfitz and Flieger 1968), showing that senescence encroaches more rapidly in a population having higher minimum adult mortality.

Life history patterns vary according to the growth rate of the population.

The relative strength of selection on life history traits expressed at different ages depends on the growth rate of the population, as we have seen. This has had important consequences for thinking about life history evolution. For example, reproductive rate has been linked to the growth rate of populations to explain latitudinal variation in fecundity (Cody 1966, Skutch 1949, MacArthur and Wilson 1967). The argument runs as follows. In temperate and arctic regions, populations are periodically reduced by catastrophic weather and individuals die with little regard to their genotypes. Population crashes are followed by longer periods of population increase during which adaptations that increase intrinsic population growth rate (r)—including increased fecundity and earlier maturity—are selected. In "constant" tropical environments, where populations fluctuate little, populations remain near the limit imposed by resources (K), and adaptations that improve competitive ability and efficiency of resource utilization are selected.

The distinction between temperate and tropical patterns has been described as the r-and K-selection spectrum (Pianka 1970). The term r refers to the growth capacity (exponential growth rate) of the population, and K denotes the carrying capacity of the environment for population—the upper resource limit to population size. Although the naming of the concept set off a minor semantic battle among population ecologists (Hairston et al. 1970, Pianka 1972, Wilbur et al. 1974, Boyce 1985), r- and K-selection occupy an important place in current thinking about life history patterns (Stearns 1976, Boyce 1985).

Eric Pianka (1970) listed a variety of traits that could be considered as either r-selected or K-selected (Table 28–3). Selection favoring r-selected traits under conditions of population growth could arise in two ways. First,

Table 28–3
Some attributes of *r*- and *K*-selected species

	r-selection	K-selection
Mortality	Variable and unpredictable	More constant and predictable
Population size	Variable, below carrying capacity	Constant, close to carrying capacity
Intra- and inter-specific competition	Variable, often weak	Usually strong
Selection favors:	Rapid development	Slow development
	High r_m	Low resource thresholds
	Early reproduction	Delayed reproduction
	Small body size	Large body size
	Semelparity	Iteroparity
Length of Life	Usually shorter	Usually longer
Leads to	High productivity	High efficiency

Source: Planka 1970.

individuals in populations reduced below their carrying capacities, and therefore presented with abundant resources, should be able to grow more rapidly, reproduce at an earlier age, and produce more progeny than individuals in populations at the carrying capacity. This provides a resource-based explanation for *r*- and *K*-selected traits in different populations. In populations regulated by density-dependent processes, all modifications of the phenotype influence the relationship between population growth rate and density. Modifications that enhance growth rate at low population density but reduce growth rate at high density are favored only when population density is low (and presumably growing); hence these are distinctively *r*-selected traits. Conversely, modifications that enhance population growth rate at high density, even at the expense of growth rate at low density, are *K*-selected traits.

A second mechanism for generating divergent *r*- and *K*-selected traits derives from the dependence of strength of selection at different ages on the rate of population growth (λ). As we have seen, in a growing population modifications of traits expressed at later ages are relatively more weakly selected than those expressed at earlier ages. As a result, in a growing population selection favors early reproduction at the expense of longevity and continued fecundity. Early reproduction and high reproductive rates are traits listed by Pianka (1970) as *r*-selected.

Although the theory is plausible, a direct relationship between population growth rate or population fluctuations and life history characteristics has not been established. Pianka (1970) placed insects at the *r*-selected end of the spectrum and mammals at the *K*-selected end, reasoning that insect populations fluctuate more than mammal populations. But the differences in life history traits between the two groups could be attributable to differences in body size, over which differences arise in the time and power scale of all physiological processes. Small organisms move more rapidly relative to body length, use more energy relative to body weight, and have more rapid development and shorter generations than large animals. These traits may be inherently correlated to size through physical and physiological relationships—just as a pendulum swings at a rate inversely related to its length—and thus largely insensitive to environmental influence. The importance of *r*-and-*K*-selection theory, relative to other sources of variation in life histories, depends on demonstrating a direct link between differences in population fluctuations and life history traits in pairs of otherwise similar organisms. A further difficulty with the interpretation of adaptations according to *r*-and-*K*-selection theory is that the traits attributed to different levels of population fluctuation are similar to those predicted for different levels of adult mortality and population turnover, even among populations with constant size.

Several investigators have attempted to contrast genetic responses to *r*-selected and *K*-selected regimes in laboratory populations. Francisco Ayala (1965) found that when populations of *Drosophila* were maintained for long periods under crowded conditions the numbers of adults per cage gradually increased, presumably owing to selection of traits that improved fecundity and survival at high density. Further experiments in which *Drosophila* populations were kept considerably below carrying capacity by removing adults (Mueller and Ayala 1981, Taylor and Condra 1980) confounded the selective effects of low density with those of high mortality (Reznick 1985). Similar experiments on laboratory populations of bacteria (Luckinbill 1978, 1984) and protozoa (Luckinbill 1979) have also produced ambiguous or negative results.

Bet hedging minimizes reproductive failure in an unpredictable environment.

When the environment varies unpredictably over the life span of the individual, selection may favor the spreading of reproduction over many seasons or concentrating it early in life, depending on the circumstances (Goodman 1979, Hastings and Caswell 1979, Murphy 1968, Schaffer 1974). When recruitment of offspring is unpredictable from year to year, selection favors adult survival at the expense of present fecundity, a strategy referred to as "bet-hedging" by Stephen Stearns (1976). The logic of the strategy is best appreciated by considering the extreme case, breeding only once. If conditions fluctuated such that in some years breeding success were zero, semelparous breeders would occasionally fail to reproduce and their lines would die out. Spreading reproduction over several years, even at the expense of annual fecundity, would be favored under such conditions.

The strength of selection for bet-hedging strategies is difficult to calculate analytically; it depends on the amount of variation in life table variables and the distribution of that variation. Because populations increase geometrically, some authors have proposed that the geometric mean of population growth rate over years provides a better measure of fitness than the arithmetic mean. For example, a population that alternated between growth rates (λ) of 1 and 3 would grow more slowly than one that had a consistent growth rate of 2 (the same arithmetic average) every year. In 4 years, the first would grow to 9 times its present size ($1 \times 3 \times 1 \times 3$; a geometric mean of 1.73), the second, 16 ($2 \times 2 \times 2 \times 2$; geometric mean of 2). But the importance of bet-hedging in natural populations will not be fully appreciated until more data are collected on variation in life table parameters and the biological constraints involved in bet-hedging adaptations are understood. At present, it is not possible to draw any conclusions about the role of environmental variation in molding the life histories of plants and animals.

SUMMARY

1. The life history traits of organisms are those clearly associated with the life table, including reproductive rate, age at first reproduction, and life span. Their evolution can be interpreted as resolving conflicts over the allocation of limited time and resources between various activities.

2. Life table values express the interaction between adaptations and the environment. One can quantify the fitness consequences of changes in life history adaptations by means of their effects on life table values.

3. Variation and correlation of life history traits among species constitute the phenomenological basis for the study of life history evolution. Delayed reproduction, long life, and low reproductive rate—or their opposites—frequently are associated.

4. Theories of life history variation among species, including correlations among life history traits, are based on the evolutionary optimization of allocation between competing activities, or density-dependent relationships between traits through their effects on the environment.

5. Life history evolution involves the balance between selection on fecundity and adult survival, and between selection on traits expressed early and late in life. Low nonreproductive mortality favors delayed sexual maturity and low reproductive rate.

6. When survival between breeding seasons is low or requires large sacrifices in fecundity, annual life histories are favored relative to perennial life histories.

7. When reproduction requires costly preparation, selection may favor a single all-consuming reproductive event followed by death, as in salmon, agaves, and bamboos.

8. Senescence arises owing to the declining strength of selection upon traits expressed at progressively older ages. Selection wanes primarily because fewer individuals survive to older ages, and so to prosper or suffer because of genes expressed at older ages.

9. In rapidly growing populations, traits expressed at young ages are selected more strongly compared to those expressed at older ages, and vice versa.

10. In variable environments, selection favors traits least affected by change and, particularly, may favor bet-hedging strategies that reduce the risk of poor reproductive performance by the individual.

29
Sex and Adaptation

The study of natural selection and adaptation is made all the more interesting by sex—by the fact that most animals and plants reproduce sexually, that is. We measure fitness in terms of number of descendants, and because each sexually produced descendant arises from the union of one female and one male gamete, individuals can gain fitness through either male or female function. Many populations, including that of humans, consist of a mixture of individuals that are either male or female. Others consist of a single type of individual that develops both male and female function. In both cases, individuals must divide their resources between male and female functions. With regard to unisexual individuals we may ask, What is the optimum ratio of males to females in their progeny? With regard to bisexual individuals, What is the optimum allocation of resources between male and female functions?

The presence of male and female functions within a single population gives rise to three additional considerations. The first is choosing among potential mates having different genotypes, made necessary by genetic variation within the population. The second is the mating system; that is, the number of other individuals that one mates with, and how one organizes these arrangements and manages them behaviorally. The third is the selection of traits in one sex by choices exercised by the second, referred to as sexual selection. In this chapter, we shall examine these consequences of sexuality for selection and adaptation beginning with the conditions under which sexual function should be combined in each organism or separated between males and females.

At the outset, we shall accept sex and the distinct qualities of maleness and femaleness as general properties of present-day life. The origin of sexual reproduction, whose initial purpose presumably was associated with advantages conferred by genetic recombination, has been lost in the dim evolutionary past. Sex came along early in the evolution of life, judging from its near ubiquity among eukaryotic organisms. The maintenance of sexual reproduction in many groups of organisms, as opposed to adopting parthenogenetic reproduction as some animals and plants have, is also something of a mystery, with no single hypothesis being particularly compelling (Smith 1978).

Sex is the union of two haploid gametes to form a diploid cell. No law of nature states that the gametes must be different (that is, male and female), and indeed in such single-celled organisms as yeast, protozoa, and some green algae the gametes are identical (a condition known as isogamy) and "sexes" are distinguished only as genetically defined mating types. The evolution of maleness and femaleness—basically the divergence of big (female) and little (male) gametes—followed upon the initial establishment of sexual reproduction. Conditions for the evolution of anisogamy are discussed by Jeffrey A. Parker et al. (1972) and Parker (1978). Judging by the absence of isogamy in multicellular organisms, and the extreme specialization of the female gamete, which provides nourishment for the developing embryo, and the male gamete—a set of chromosomes with a way of getting around—anisogamy must be powerfully selected. But it is at this point, with the assumptions of sex and male/female function behind us, that we shall begin our inquiry.

Separation of sexes is favored when fixed costs of sexual function are high and the sexes compete strongly for resources.

As with many issues, botanists and zoologists have divergent views about sex. This is not to say that one or the other has more fun, but just that their perspectives differ. In most species of plants, individuals have both male and female function, leading plant ecologists to ask, What conditions favor separation of the sexes? In most animal species, male and female functions are separated between individuals, and animal ecologists ask, What conditions favor the joining of both sexual functions in a single individual? These questions are, of course, opposite sides of a common coin. Before attempting to answer them, however, some definitions.

The condition of separate male and female individuals is referred to as gonochoristic when applied to animals, from the Greek *gonos* pertaining to procreation (for example, gonad) , and *choris,* apart or separate. Botanists refer to the same condition as dioecious, from the Greek *di-* (two) and *oikos* (dwelling).

When both sexes are together, the individual is referred to as an hermaphrodite, after Hermaphroditus, son of Hermes and Aphrodite, who while bathing became joined in one body with a nymph. (The circumstances of this happy union are not important here.) Hermaphrodites may be simultaneous, as in the case of many snails and most worms, or they may be sequential (that is, first one sex, then the other). When male function is followed by female function, as in some molluscs and echinoderms, the individual is said to be protandrous (Greek *andros,* male). When female function comes first, as in some fishes, the individual is protogynous (Greek *gyne,* woman). Botanists apply an additional term, monoecious, to the con-

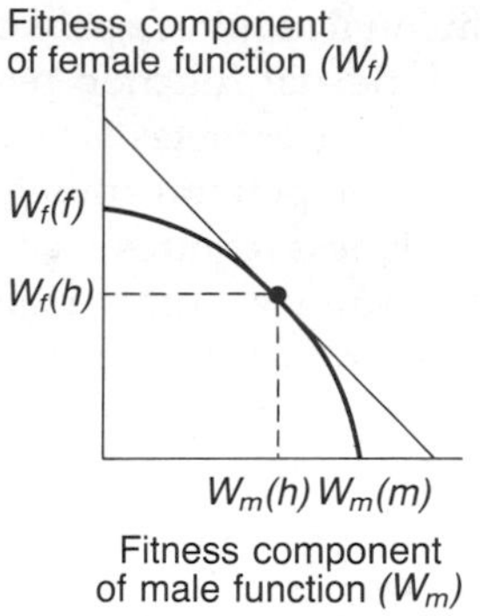

Figure 29–1
Convex fitness set for allocation to male function between 0 and 1. The adaptive function indicates that the Evolutionarily Stable Strategy (ESS) is an hermaphrodite with intermediate allocation between male and female sexual functions. (After Charnov 1982.)

dition of an individual plant bearing separate male and female flowers, rather than so-called perfect flowers with both male and female parts. While perfect-flowered hermaphrodites are the rule among plants (72 per cent of species, by one estimate), almost all imaginable combinations of sexual patterns are known, including populations with hermaphrodites and either male or female plants, populations with male, female, and monoecious plants, and hermaphrodite individuals with perfect flowers and which also bear either male or female flowers (Yampolsky and Yampolsky 1922, Charnov 1982, Willson 1983).

Situations favoring hermaphroditism have been investigated theoretically by specifying the conditions under which an hermaphrodite strategy can invade a population consisting of separate males and females, and, conversely, when a population consisting of hermaphrodites can resist invasion by either males, females, or both. The approach, that of determining the Evolutionarily Stable Strategy (ESS), was outlined by Eric A. Charnov et al. (1976) and elaborated by Charnov (1982). We shall consider only the simplest situation here, that of simultaneous hermaphroditism. Suppose that males, females, and self-incompatible hermaphrodites occur in a population with frequencies m, f, and h, respectively ($m + f + h = 1$). Each male contributes to future generations through N matings, and each female through n seeds. Hermaphrodites contribute through both male and female functions, at a level aN and bn, respectively. The proportion of male contribution to progeny from each male is $W(m) = 1/(m + ah)$, and the proportion of female contribution from each female is $W(f) = 1/(f + bh)$. For each hermaphrodite, the combined contribution through male and female function is $W(h) = a/(m + ah) + b/(f + bh)$.

An hermaphrodite can invade a population of males and females only when hermaphrodites are more fit than either males or females; because males and females have equal fitness in an outcrossing population $W(h)$ must exceed $W(f)$. A little algebra gives the necessary inequality $a(f + bh) + b(f + ah) > m + ah$. When hermaphrodites are rare and h is therefore close to 0, and when males and females are equally abundant in the population ($m = f = 0.5$), the inequality reduces to $a + b > 1$. This inequality can be shown to be a sufficient condition to prevent the invasion of a population of hermaphrodites by either males or females. Therefore, hermaphroditism is an ESS when the sum of the combined male and female contributions exceeds the sum of the contributions from each sex separately (Figure 29–1).

Charnov's analysis showed that when a female can achieve a certain amount of male function by giving up a smaller amount of her female function, she (subsequently it or they) should do so. Similarly, males should add female function when it does not cut deeply into their male productivity. For the flowers of female plants to produce a little bit of pollen, or increase the rate of pollen transfer, seemingly few resources would have to be transferred from female to male function (Bell 1985, Stanton et al. 1986). And so the strategy should be adopted frequently, as it is among plants. For hermaphrodites to be excluded from a population, the fitness set must bulge inward so that gains in adding function of one sex are more than offset by losses in function of the other (Figure 29–2). This may occur when the

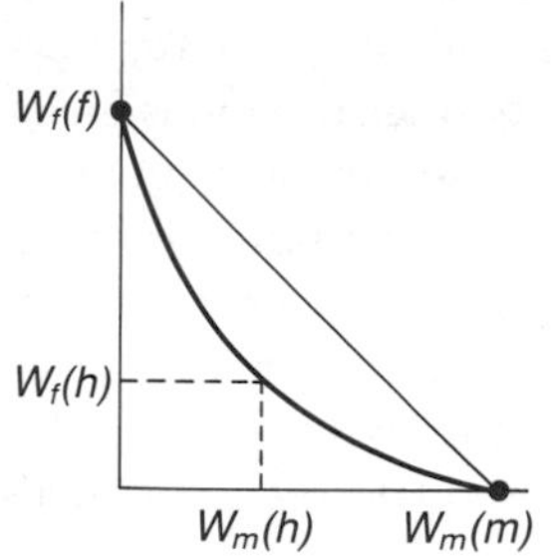

Figure 29–2
When the fitness set for allocation between male and female function is concave, extreme phenotypes (males and females) are selected and hermaphrodites are excluded from the population. (After Charnov 1982.)

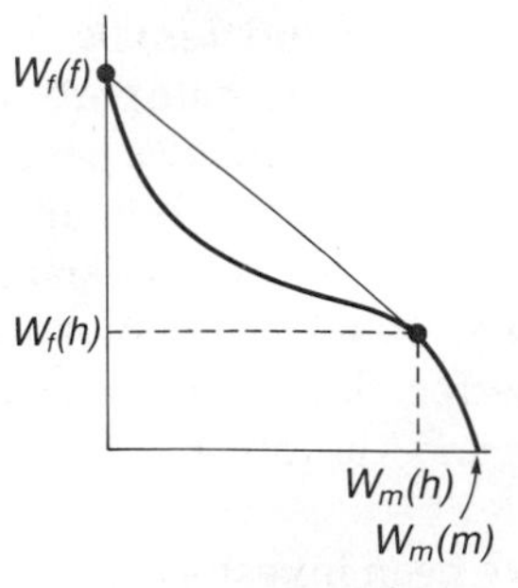

Figure 29–3
Fitness set representation of the conditions required to maintain a mixed population of females and hermaphrodites. (After Charnov 1982.)

establishment of new sexual function in an individual requires a substantial fixed cost before *any* gametes can be produced. Sexual function requires gonads, ducts, and other structures for transmitting gametes, secondary sexual characteristics for attracting mates and for competition among individuals within sexes. In many animals, where maleness requires specializations for mate attraction and antagonistic interaction with other males, or where femaleness requires specializations for egg production or brood care, fixed costs may be high and hermaphroditism disadvantaged compared to sexual specialization. In fact, hermaphroditism is rare among active animal species and those with brood care; it is particularly common among sedentary aquatic species that shed gametes into the water.

Some populations include hermaphrodites and individuals of one sex, usually females. Fitness set analysis shows that this situation may arise when female function is gained at large cost in male function (Figure 29–3). For females to invade a population of hermaphrodites, $W(f)$ must exceed $W(h)$ when $m = f = 0$ and $h = 1$. The condition for this is $b < 0.5$. It can be shown that the fitness of both females and hermaphrodites is frequency dependent and that an ESS is achieved when the proportion of females is $f = h(1 - b)$ (Charnov 1982).

ESS and fitness set analysis elucidate general guidelines for thinking about sex allocation, but the method of inquiry is so new that little has yet been done to measure the shape of fitness sets relating male and female function, establish values of a and b, and verify basic assumptions of the theory, such as degree of self-compatibility and level of inbreeding, in nature (Charlesworth and Charlesworth 1981, Willson 1983). So, why are so many species of plants hermaphroditic? Charnov (1982) outlined eight possibilities that are associated either with self-fertilization or the interaction of male and female function in the same plants:

> (1) Hermaphroditism allows self-fertilization which is adjusted to an intermediate optimum. (2) It allows facultative selfing. (3) Pollen is a primary attractant for pollinating agents; a female would simply not be visited. (4) Pollen limits seed set and the visit of a pollinating agent achieves gains for both genders. Both genders share the flower resource, or cost of attraction, which makes the fitness set convex [bulge outward]. (5) Temporal displacement of availability of resource for the two sex functions means that female resource is simply not available for male function. (6) Seed set is resource limited; pollen (the male gain curve) fitness saturates because of limitations in mate availability . . . or saturation of pollen vectors. (7) Seed set is resource limited, but limited fruit dispersal results in sib competition and thus a law of diminishing returns through seeds. (8) Male and female function may depend upon (be limited by) different resources (e.g., protein versus carbohydrate).

In sum, low overlap in resource requirements of male and female function, and "cost-sharing" in which sexual organs, such as flowers, can serve both male and female function, tend to produce an outward-bulging fitness set and favor hermaphroditism.

Sequential hermaphroditism reflects changing costs and benefits of male and female sexual function as an organism grows. In some marine gastropods having internal fertilization, such as the slipper shell *Crepidula,* insemination requires the production of only small amounts of sperm. Hence male function requires few resources and has little effect on somatic growth. As a consequence, many such species are protandrous hermaphrodites, being male when small and female when they are large and can produce correspondingly large clutches of eggs (Hoagland 1978). When male competition for mates is important, large size can be an advantage. This apparently has led to protogynous hermaphroditism in some species of fish that inhabit reefs, where males compete among themselves for breeding territories (Warner 1975, Charnov 1979).

Optimal sex ratio balances contributions to fitness through male and female function.

Two facts are remarkable about sex ratio. First, so many populations have very nearly equal numbers of males and females. Second, there are so many exceptions to this rule. Because of its prevalence, symmetry, and application to the human population, the even, or 1:1, sex ratio is considered the standard condition, deviations being special cases.

R. A. Fisher (1930) summed up the theoretical basis of sex-ratio theory when he observed that every product of a sexual union has exactly one mother and one father. One consequence of this truism is that individuals of the rarer sex in the population will enjoy greater fitness because they compete with fewer others of the same sex for matings. For example, when a population of 5 males and 10 females produces 100 offspring, each male contributes 20 sets of genes but each female contributes only 10 sets of genes. With such a skewed sex ratio, any parental genotype giving rise to a larger proportion of male offspring would be favored and the frequency of males in the population would increase. Similarly, if females were the rarer sex, genotypes that increased the proportion of female progeny would be favored, and the frequency of females would increase in the population. Fitnesses are balanced, and there is no selective pressure to alter the sex ratio, when males and females occur with equal frequency, in which case individuals of both sexes contribute equally to future generations and the frequencies of males and females among the progeny of an individual are of no consequence to its fitness.

With respect to selection upon the sex ratio, the relevant measure of fitness is the number of grandprogeny, not progeny. This is because a gene that influences the sex ratio expresses itself in the proportion of males or females among the progeny, not the total number of progeny, of the individual bearing that gene; its consequences are not felt until the progeny have passed their genes onto the following generation.

Mathematical development of this theory of sex ratio also provides a quantitative basis for understanding selection of deviations from the even sex ratio. Consider a population in which the frequencies of males and females among zygotes (the primary sex ratio) are M and F, respectively ($M + F = 1$). Because all individuals have one father and one mother, the fitness contribution of an individual through its male offspring is proportional to $1/M$ and its fitness contribution through female offspring is proportional to $1/F$. Now suppose that a mutant appears whose offspring are male and female in the frequencies m and f ($m + f = 1$). Provided that mutant and typical individuals produce the same number of offspring, the fitness of the mutant is proportional to $(m/M) + (f/F)$; scaled in the same way, the fitness of the typical genotype is proportional to $(M/M) + (F/F) = 2$ (Shaw and Mohler 1953). Now, the difference in fitness (D) between the mutant and typical individual is $D(m,f) = (m/M) + (f/F) - 2$, which, with a little rearranging, is

$$D = \frac{(m - M)(F - M)}{MF} \tag{29-1}$$

(Crow and Kimura 1970). We see that D is equal to 0 (evolutionary equilibrium) either when $m = M$, which is uninteresting because there is no genetic variation in fitness, or when $M = F$ (an even sex ratio). The condition $M = F$ represents a stable equilibrium; when the frequency of males exceeds that of females in the population ($F - M < 0$), a mutant that produces a lower frequency of male offspring than the population average ($m - M < 0$) will increase [$D(m,f) > 0$]. The survival of the zygote to reproductive maturity does not bear on the evolution of sex ratio in this case because the fitness value of the zygote is unaffected. That is, if fewer males survive than females, each has an enhanced fitness value as an adult, but fewer make it to that stage. The total contribution of each sex to future populations is identical.

Whichever sex requires less investment should be produced in greater proportion.

Under some circumstances an individual male or female offspring requires more resources to produce than an individual of the other sex. This may pertain especially when sexual dimorphism in size results in different parental investment in individual offspring of each sex. Suppose that males cost some fraction (a) of females to produce such that $F + aM = 1$ and $F = 1 - aM$. Now equation (29-1) becomes $D(m,f) = (m - M)(F - aM)/MF$ and $D(m,f) = 0$ when $F = aM$, or $F/M = a$. Thus if males are the more expensive sex ($a > 1$), the equilibrium sex ratio will favor females. Furthermore, the total expenditure on the sexes will be equal, as indicated by the expression $F = aM$.

Differential mortality of the sexes can affect equilibrium sex ratio if deaths are compensated by production.

Suppose that males survive less well than females, by fraction s, to the termination of parental care at time T. If resources not delivered to deceased males could instead be invested in females, then equilibrium sex ratio should be adjusted in favor of the more poorly surviving sex. When such transfer of resources, referred to as compensation, applies, $F + sM = 1$ and $D(m,f) = 0$ when $F = sM$, or $F/M = s$. Therefore, when males survive less well than females ($s < 1$), the equilibrium sex ratio favors males. Furthermore, at the termination of parental care (time T), the sex ratio should be even, owing to the greater mortality of the more prevalently produced sex.

Differential costs and differential early survival of male and female offspring should lead to deviations in sex ratios of populations from 1 : 1. Yet surprisingly few data for vertebrates suggest either substantial deviations in the sex ratio or genetic variation among individuals in the sex ratio of their offspring (Williams 1979). Indeed, the situation is so uniform that George Williams (1979) considered the 1 : 1 sex ratio a trivial consequence of the sex-determining mechanism. In most vertebrates, one pair of chromosomes (the sex chromosomes) for which the population is polymorphic (for example, X and Y) controls an individual's sex. When an individual is homozygous for one of the chromosomes it is either male or female, depending on the type of organism; when heterozygous, it is the other sex. Thus, in humans, XX homozygotes are female and XY heterozygotes are male. Because matings occur between XX and XY individuals, the genotypes of the offspring are half XX and half XY—half male and half female. The heterogametic sex may be either male (mammals) or female (birds and butterflies), but the sex-determining mechanism will produce equal numbers of males and females in either case.

Charnov (1982) objected to Williams's "nonselection" thinking in this case on two grounds. First, the occurrence of sex chromosomes may signal the predominant selective advantage of the even sex ratio. That is, rather than sex ratio being the inevitable consequence of a particular sex-determining mechanism, the mechanism itself may have evolved to its prevalance because it produced the most fit ratio of the sexes among progeny. After all, animals employ many other sex-determining mechanisms (Bull 1983), including temperature-sensitive sex ratios in many reptiles (Bull 1980). Moreover, even in the case of XY sex determination, meiotic drive and sperm competition could produce unbalanced sex ratios, should they be selected (Hamilton 1967).

Charnov's second objection to Williams's thinking is that the existence of sex chromosomes may make it difficult to evolve toward some alternative system of sex determination. Thus even when selection favors an unbalanced sex ratio, there may be no genetic variation in sex ratio to work on. This does not invalidate selection thinking about sex ratio, but rather illustrates how the genetic system might limit evolution.

In certain situations, mothers should vary the sex of their offspring in relation to their own breeding condition.

In many species, competition among individuals of one sex (usually males) for matings leads to tremendous variance in fitness contributions among individuals (Arnold and Wade 1984, Trail 1985, Clutton-Brock 1988). Presumably, where competition is keen, some males achieve many matings, others none. In harem-forming species, for example, a few males control most of the females and others lower in social dominance have little access to mates. Generally contests among males are won by the largest individuals (see, for example, Howard 1979). Robert Trivers and D. E. Willard (1973) suggest that if the condition of male progeny is directly influenced by the condition of the mother, then females in good condition should produce male offspring, which will grow to large size and fare well in male-male competition for mates; females in poor condition should place their investment in female offspring, which will mate successfully regardless of the parental care they are given. Certainly in mammals, in which the female cares directly for her offspring through the periods of gestation and lactation, the condition of the mother will likely influence that of her offspring.

Experimental confirmation of Trivers and Willard's idea has emerged from an experimental laboratory study on reproducing female wood rats *(Neotoma floridana)* (McClure 1981). In this species, as in most mammals, females normally invest equally in male and female offspring. But when Polly Ann McClure restricted food drastically during the first 3 weeks of the lactation period to below the maintenance level of a nonreproductive female, male offspring were selectively starved (mothers actively rejected their attempts to nurse) and the sex ratio at 3 weeks was altered to about 0.5 males for each female (also see Gosling 1986).

In hymenoptera and other haplodiploid invertebrates, the sex of offspring is controlled facultatively in response to local mate competition.

The hymenoptera (bees, ants, and wasps) have an unusual sex-determining mechanism by which fertilized eggs produce females and unfertilized eggs produce males. As a result, females are diploid and males are haploid, giving rise to the term "haplodiploid" sex-determining mechanism. Reproductive females can control the sex of their offspring simply by fertilizing eggs or not. The haplodiploid system raises questions about its evolutionary origin and adaptive significance, but also offers the hope of testing certain aspects of sex-ratio theory because sex ratio is under the direct control of the female, presumably through mechanisms that have genetic variation.

Haplodiploidy has evolved many times from more usual diploid sex-determining systems. Two steps are required. First, genes must arise in the maternal genome that enable the female to suppress the fertilization of eggs;

these eggs also must spontaneously initiate development without being fertilized and develop into males. Second, genes must arise that suppress the development of maleness in diploid individuals. The first step confers tremendous advantage to the genes responsible for it. The haploid males contain only genes from the mother, including the genes responsible for haploidy. These are transmitted to the next generation at twice the frequency of maternal genes transmitted through sexually produced offspring, in which half of the genome comes from the father. Therefore, if uniparental males make at least half the fitness contribution as biparental males, then genes favoring haplodiploidy can invade a population (Bull 1983). Why isn't the uniparental sex female? Presumably this would lead to strict parthenogenesis with haploid clones of asexually reproducing females. In spite of initial advantages in the transmission of genes to future generations, such clones might die out owing to the lack of sexual reproduction and the reduced genetic potential of haploids compared to diploids (heterozygotes are not possible).

Many wasps are parasitoids on other insects or complete their larval development within the fruits of certain plants. For some of these species, hosts are so scarce and mates so difficult to find that females mate on the host on which they grew up before dispersing to find new hosts on which to lay their own eggs. When a host is parasitized by a single female wasp, females within each brood are limited to mating with their brothers. Under this circumstance, because the number of grandprogeny of a female is directly proportional to the number of female offspring that disperse to search out new hosts, male offspring have reduced fitness value. One might therefore expect a reduced proportion of males in each brood. This is, when males compete only with their brothers for the opportunity to mate, one male offspring is as good as several from the standpoint of their mother. Thus it is not surprising that most species of parasitoid wasps have sex ratios skewed greatly in favor of females. Several produce only one male per brood. Such males are often wingless and, in extreme cases, males fertilize females as larvae within the host, or, as in viviparous pyemotid mites, within the mother (Hamilton 1967, 1979). In the last case, males become sexually functional as larvae and never develop into adult forms. These observations confirm the idea that when local mate competition occurs only among sibs, the fitness value of males is greatly reduced and, where possible, mothers limit the number of males among their progeny (Herre 1986). In normally outcrossing haplodiploid species, a sex ratio closer to 1 : 1 is the rule.

In many parasitoid species, hosts are superparasitized, meaning that more than one female may lay her eggs in the same host. Also, in most cases, ovipositing wasps can use chemical cues to determine whether eggs have been previously laid by another female or not. When a female attacks a "virgin" host, and if few hosts are superparasitized, her daughters will likely engage in sib-mating because their brothers frequently will be the only males around. When a female attacks a host that has been parasitized recently, her sons can compete to mate with the daughters of the first parasitoid female, and their value will be considerably enhanced. One might expect superparasite females to produce more male offspring per brood than the first female to lay on a particular host. And this is, in fact, frequently the case. Moreover, the proportion of sons in the progeny of the second female ought to increase as the ratio of the number of offspring of the first female to the number of her own offspring increases, because opportunities for outcrossing increase in direct proportion to this ratio (Hamilton 1967).

Jack Werren (1980) established conditions for testing this idea in laboratory experiments with the wasp *Nasonia,* which parasitizes larvae of the fly *Sarcophaga.* The results conformed to the predictions of William D. Hamilton's quantitative model of male and female contributions to fitness (Figure 29–4). The progeny of primary parasites were 9 per cent male. The proportion of males produced by secondarily parasitic females varied from a similarly low level when their own offspring predominated in the host to 100 per cent males when their offspring were few compared to those of the primary parasites.

Haplodiploid wasps also provide an opportunity to test Trivers and Willard's (1973) idea about controlling the sex of offspring in relation to their expected fitness contributions in a variable environment (King 1988). In many parasitoids, larger hosts result in the production of larger female offspring with greater fecundity than the smaller females that emerge from smaller hosts. Males presumably are less disadvantaged by developing to small size within a small host because male sexual function is less dependent on body size than is female sexual function. In his 1979 paper on sexual selection, Charnov quoted the results of an experiment by the Russian Ivan Chewyreuv, published in 1913, on the ichneumonid wasp *Pimpla instigator,* which lays a single parasitic egg on the pupae of moths and butterflies.

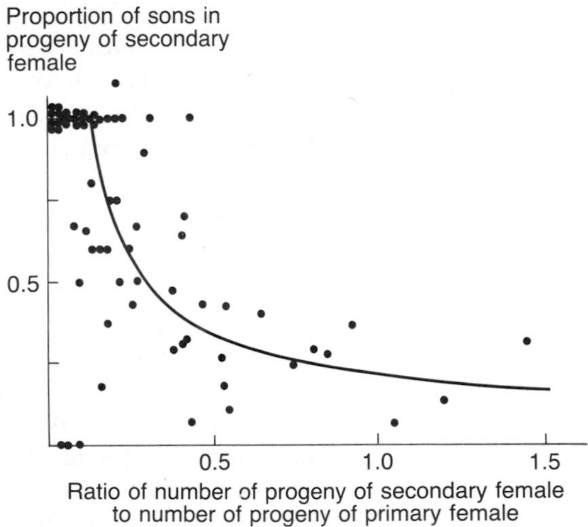

Figure 29–4
Relationship between the sex ratio in progeny of secondary females of the parasitoid *Nasonia* and the probability that their offspring will mate with the offspring of different females. (From Werren 1980.)

When Chewyreuv presented females with large hosts, all of 23 offspring produced were female; when he presented them with small hosts, 38 out of 47 offspring were male. Large and small hosts could be alternated, in which case the female would either fertilize the egg or not as appropriate. In these laboratory experiments, pupae of different species of moth were used to present hosts of different sizes. But Chewyreuv also showed that when the wasp *Exenterus* parasitized cocoons of the moth *Lophyrus* in nature, the larger host pupae tended to produce a preponderance of females. In *Lophyrus,* males are only half the size of females. Seventy-nine per cent of the wasps emerging from the larger, female pupae were females, compared to only 47 per cent of those emerging from the smaller, male pupae.

Werren's and Chewyreuv's results show that when females *can* control the sex of their progeny, they do produce more of the sex with the greater fitness contribution. This contribution depends on the expected fecundity of progeny of each sex, which depends in turn on resources for egg production in female wasps and availability of potential outcrossed matings for males.

Mating systems in populations depend on the degree to which individuals of one sex can monopolize resources.

The theory of sex allocation discussed above deals with the distribution of resources by a parent to male and female offspring. Adults also are selected to obtain as many matings of as high quality as possible. It is a basic asymmetry of life that the fitness contribution of a female is limited by her ability to make eggs and otherwise provide for her offspring, while the fitness contribution of a male is usually determined by the number of matings that he can procure. Because the female gamete is larger than the male gamete, it requires more resource to produce and a female's fecundity is likely to be limited by her ability to gather resources to make eggs. When males contribute no resources to their mate, as in some species they can do by defending territories for the females to feed in or by feeding the female directly, or when males do not care for their young, a male can increase its own fecundity only by mating with additional females.

When a male mates with as many females as he can locate and properly persuade, and provides his offspring nothing more than a set of genes, he is said to be promiscuous. Among animal taxa as a whole, promiscuous mating is by far the commonest system. Promiscuity generally precludes a lasting pair bond. When adopted by a population, it also tends to increase variance in mating success among males, some individuals obtaining perhaps dozens of matings while others get none. Over the population as a whole, the number of matings must average out to one per period of female receptivity.

When males can contribute to the fecundity of their mate by procuring resources for the female or caring for the offspring directly, pair bonds may outlast the copulatory rapture of promiscuous species. As males increasingly contribute to the realized fecundity of a single mating, mating systems progress from promiscuity (least care), through polygamy and serial polygamy, to strict monogamy (most care, relative to the female). Polygamy describes the situation in which a single individual of one sex forms long-term pair bonds with more than one individual of the opposite sex. Normally males are mated to more than one female, in which case the system is referred to as polygyny (literally, many females). Polygyny may be expressed through the defense of several females against the mating attempts of other males (a harem), or the defense of territory or nesting sites to which more than one female are attracted to breed as well as mate. Thus polygyny may arise through the ability of a male to control matings by defending females, in which the contribution of the male to its progeny may be primarily genetic, or through the ability of a male to control or provide resources necessary to the female to reproduce.

In serial polygamy, an individual of one sex forms a durable bond with one individual of the opposite sex, but eventually abandons its mate, leaving it to care for their offspring, while it goes off to seek a new mate. There is never more than one pair bond formed at a time in such systems, but one sex always takes the primary responsibility for caring for the offspring of a mating. Most commonly it is the male that abandons the female to seek greener pastures (serial polygyny), but the reverse situation (serial polyandry) is also known.

Monogamy refers to the formation of a pair bond between one male and one female that often persists through the period required to rear the offspring of a mating, and which may last until one of the pair dies. Monogamy arises primarily in situations in which males can make a large contribution to the number and survival of offspring. Hence it is most common in species with prolonged dependence of offspring, in which both sexes can provide for the young. Monogamy is not common in mammals (Eisenberg 1981) because providing milk is a specialized task of the female. But it is common among birds, especially those in which parents feed their offspring, a task of which both sexes are equally capable.

Mating systems are associated with habitat and diet.

Because the habits of birds have been so keenly scrutinized by biologists, and because birds encompass the full range of mating systems and express these diverse interactions between the sexes in a spectacular variety of structures and behaviors, it is not surprising that studies of birds have provided the seed from which theory concerning mating systems has grown. An early and crucial step toward interpreting mating systems as evolved adaptations was recognizing associations between mating systems of populations and their habitats, habits, and diets.

The English behaviorist John Crook (1964, 1965) was among the first to recognize that mating system and ecology went hand in hand. Surveying species of African weaverbirds, he noted that monogamy predominated among insectivorous species inhabiting forest and savanna, whereas polygyny was the rule among seed-eating species of open savannas and grass-

Table 29-1
Relationship of habitat and food to mating system in African ploceid finches

Habitat			Food		Pair bond		Sociality		
Forest	Savanna	Grassland	Insects	Seeds	Monogamous	Polygynous	Solitary	Grouped territories	Colonial
+			+		17	0	15	0	1
	+		+		5	0	4	0	2
+			+	+	3	0	2	0	0
	+		+	+	1	4	1	0	4
		+	+	+	1	1	1	0	0
	+			+	2	10	0	1	16
		+		+	0	15	0	13	3

Source: Crook 1964, 1965, as summarized by Lack 1968 and presented by Clutton-Brock and Harvey 1984.

lands (Table 29-1). Furthermore, most of the monogamous species bred as solitary pairs on widely dispersed territories while polygynous species were invariably colonial. In a broader study of North American songbirds, Jered Verner and Mary Willson (1966, 1969) found a similarly strong association of polygyny with grassland, prairie, and marsh habitats. But simple correlations between variables cannot explain how these associations arose. Empirical patterns, such as those noted by Crook and Verner and Willson, may, however, suggest hypotheses or models that embody a plausible mechanism relating selective factors in the environment to evolved behaviors.

Verner (1964) was the first to propose a model for the origin of polygynous mating systems. His reasoning was summarized by Verner and Willson (1966):

> It seems clear that polygyny would be advantageous to a male whenever the total number of successful offspring (those that survive to reproduce) from all his females exceeds the number that would be reared from one nest if he mated monogamously. Among females also, selection will favor those leaving the greatest number of successful offspring, whether these are reared in monogamous or polygamous associations. Consequently, in those species in which females select mates from among available males, there will be selection against polygyny unless this results in greater reproductive success for the females as well as for the males. Verner (1964) contends that polygyny can be advantageous for females if, within the limited area from which a female is likely to select a mate, the difference between two males' territories is sufficient that a female is able to rear more offspring on the better territory, by herself, than she could rear on the poorer one even with full assistance from the male. This difference between territories can be regarded as a "polygyny threshold," since it is likely that polygyny will be favored by natural selection whenever the difference exceeds a certain level. The parameters of a male's territory that might operate in this regard include all requisites of successful breeding (food, cover, space, nest sites, etc.).

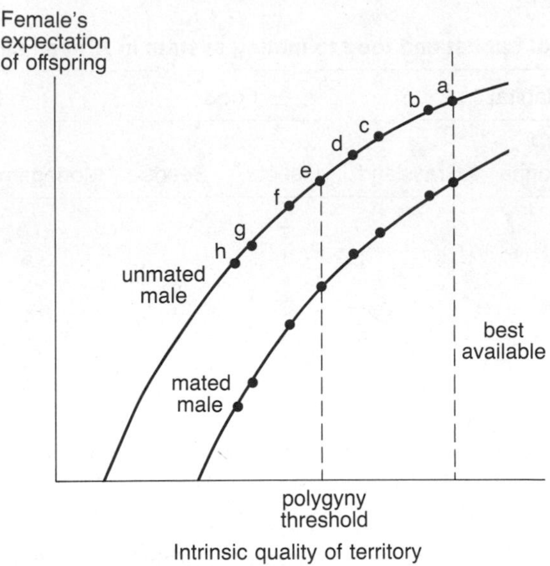

Figure 29–5
The polygyny threshold model of Verner and Willson. Female fitness varies in relation to the intrinsic quality of a male's territory; presence of a mated female reduces the quality of territory to subsequent females. Females should select unmated males (a–e) until the territory of unmated males (f) drops below that of the best mated male (a). At this point, the polygyny threshold, female choice will alternate between mated males (a–e, which then become bigamists) and unmated males (f–h). (After Orians 1969.)

Gordon Orians, a behavioral ecologist at the University of Washington, with whom both Verner and Willson worked as students, devised a simple graphical representation of the process of mate selection according to the Verner-Willson polygyny threshold model (Figure 29–5). The model contrasts the fitness of a female mated to a monogamous male with that of a female mated to a polygynous male as a function of the quality of the male's territory. Several predictions can be made from this model. First, polygyny will occur only when the quality of territories varies among males so much that some females will have higher fitness mated to a polygynous male on a territory of high quality than they would mated monogamously to a male (and receiving his undivided attention) on a territory of poor quality. This is Verner's polygyny threshold of territory quality. Second, if females can assess the quality of territories, the first individuals to pair should mate monogamously, latecomers choosing mates with territories of progressively lower quality until the polygyny threshold is reached. At this point, females should be ambivalent about pairing up with unmated or previously mated males (Alatalo et al. 1982).

According to the polygyny threshold hypothesis, polygynous males should occupy territories, or otherwise defend resources, of intrinsically higher quality than those of monogamous or unmated males. Several studies have attempted to measure territory quality and evaluate the polygyny threshold model, the most direct being those of Sarah Lenington (1980) on red-winged blackbirds and W. K. Pleszczynska (1978) on lark buntings.

Figure 29-6
Male red-winged blackbird. (Photograph by A. Morris, courtesy of VIREO, Academy of Natural Sciences, Philadelphia.)

Red-winged blackbirds (Figure 29-6) frequently breed in cattail marshes where, in the early spring, males establish territories, among which females choose to breed (Orians 1980, Nero 1984). Males accept as many females into their territories as will mate with them; 2 or 3 are common, but up to 12 have been observed. Because the sex ratio of blackbirds is even, many males obtain no mates. Females of polygynous males defend small territories of their own against other females within the male's space.

To a female, the quality of a territory depends on characteristics that contribute to the number of offspring she can rear: those that influence the amount of food she can gather for her brood (males do not feed the young) and that determine the safety of her nest from storms and predators. In her study, Lenington was able to relate both components of a female's productivity to physical characteristics of the territory: number of young fledged per successful nest (an index to feeding conditions) was correlated with territory size and the amount of edge of cattail habitat; safety of the nest was related to the density of the cattails, thus the effectiveness of the vegetation as cover (Table 29-2).

Using these attributes of territories, Lenington ranked males according to the number of offspring that a single female could anticipate. Although Lenington expected the first females to arrive on the breeding grounds to choose males with territories of the highest "quality," observations at a marsh near Princeton, New Jersey, failed to show any particular order with respect to her criteria. But when only food-related aspects of territory quality were considered, and the quality of potential nest sites was ignored, the order of choice corresponded much more closely to territory quality (Figure 29-7). A second prediction, that secondary females should choose only the highest quality territories, still was not supported. A third prediction, that the reproductive success of females on a territory of a given quality should be inversely related to number of females, was, however, supported.

Pleszczynska (1978) studied a small songbird, the lark bunting, nesting in a 4-hectare alfalfa field in South Dakota. The buntings nest somewhat colonially in high densities. Males are polygynous. Broods of primary females receive care from their fathers; those of secondary females do not. Because most food for the offspring is gathered from areas outside the

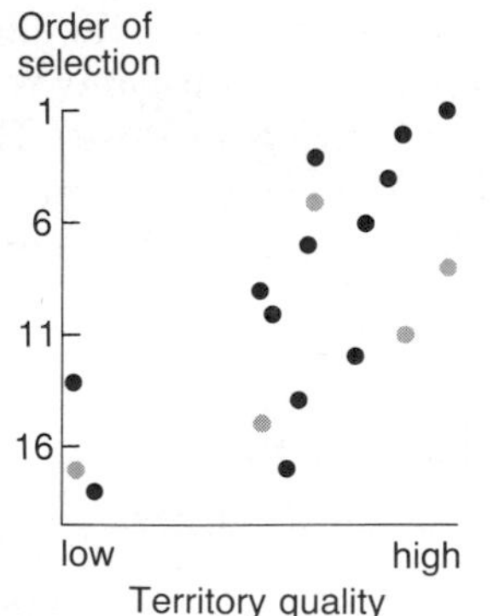

Figure 29-7
Order of selection of males as a function of territory quality by primary females (black symbols) and secondary females (gray symbols) in a marsh-nesting population of red-winged blackbirds in New Jersey. (After Lenington 1980.)

Table 29-2
Expected reproductive success of red-winged blackbirds as a function of the safety of the nest (cattail density) and food supply within the territory (territory size and length of cattail edge)

	Food availability	
Cattail density	High ($F = 3.7$)	Low ($F = 2.0$)
High ($S = 0.58$)	2.16	1.62
Low ($S = 0.26$)	0.97	0.73

Note: S is proportion of nests that fledge young; F is number of young fledged per nest. Values in the body of the table are the products of the values of S and F.
Source: Lenington 1980.

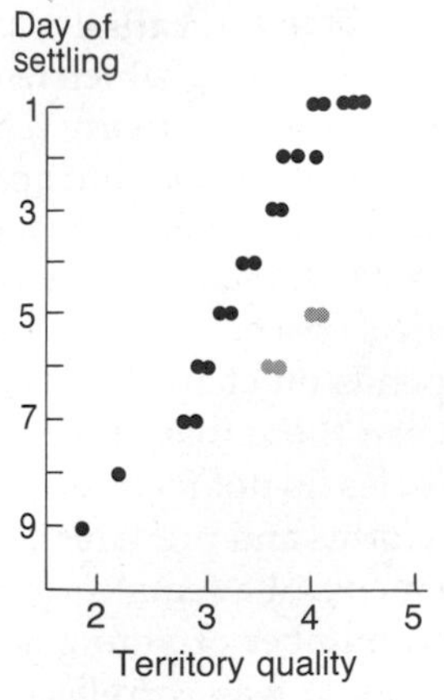

Figure 29–8
In lark buntings, early-arriving females mate with males holding territories of the highest quality. After males on the best territories were mated, some of these were able to secure secondary mates (gray symbols) before males on poor territories acquired their first mate. (After Pleszczynska and Hansell 1980.)

territory, the quality of a territory is related to characteristics of potential nest sites, primarily the density of vegetation. Indeed, vegetation cover (assessed by light levels at nest sites) was strongly correlated with nest success.

As in red-winged blackbirds, male lark buntings establish territories before females arrive from the wintering grounds. In Pleszczynska's study area, females settled into the alfalfa field over a 9-day period in the spring. The order in which they settled territories of different "quality" corresponded almost precisely to Orians's model. Females arriving over the first 4 days mated monogamously with males on high-quality territories; those settling on days 5 and 6 tended to become secondary females; those arriving on days 7 through 9 mated usually monogamously with males on territories of lower quality (Figure 29–8).

According to Orians's model of mate selection, once the polygyny threshold has been reached the choice between becoming a secondary female on a territory of higher intrinsic quality or a primary female on a territory of lower intrinsic quality is ambivalent. And, indeed, Pleszczynska found that primary and secondary females that choose mates on the same days had nearly identical nesting success, about 0.6 young fledged per egg laid.

Pleszczynska investigated the role of vegetation cover in nest success experimentally by placing plastic leaves over nest sites to simulate denser cover. Over 3 years of such manipulations, nest success increased to between 65 and 76 per cent from control levels of between 42 and 52 per cent. Confident that female lark buntings could judge nesting success, therefore territory quality, by vegetation cover, Pleszczynska predicted the eventual mating status of males in locations in Colorado and in North and South Dakota by this criterion, proving to be correct in 52 out of 58 cases. Finally, Pleszczynska and R. Hansell (1980) experimentally altered territory quality in the alfalfa patch by either stripping alfalfa plants of leaves in areas of high cover or adding plastic leaves to alfalfa plants where cover was sparse. The distribution of females settling into the patch was consistent with the altered territory values, even to the point of creating some cases of trigamy where plastic leaves were added to enhance territory quality (Figure 29–9). These experiments lend considerable force to the idea that polygyny can arise where females perceive large differences in quality between males, which is most likely to occur in heterogeneous habitats.

Promiscuous mating systems can occur only when the male is emancipated from parental care.

Although promiscuous mating is the rule in the animal and plant kingdoms, it surfaces rarely and sporadically among birds, being predominant only among grouse and a few groups of tropical, frugivorous species — the manakins, cotingas, and birds-of-paradise. Representatives of all these groups have communal mating areas, called leks, in which congregated males

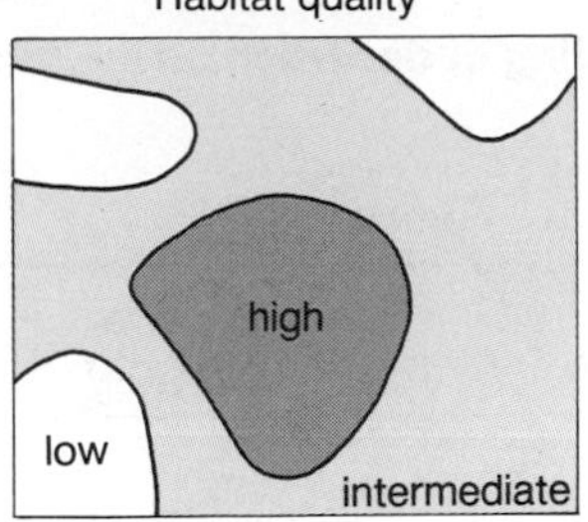

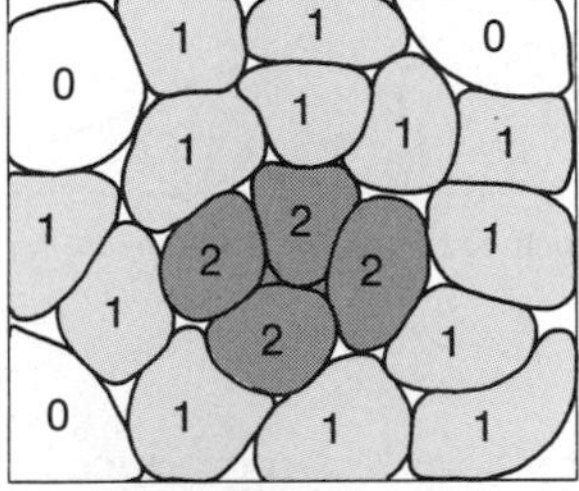

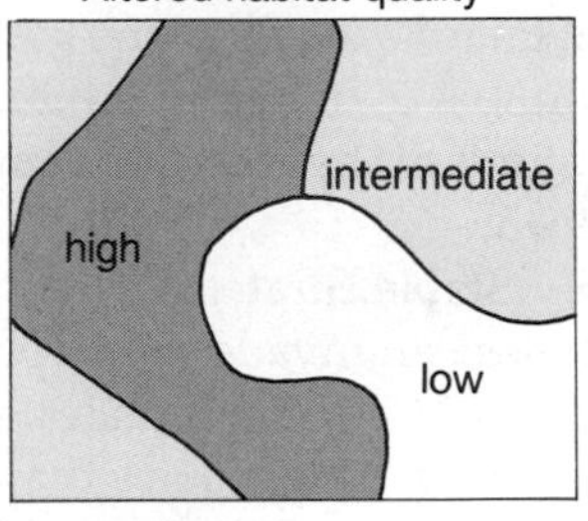

Status of males

Figure 29–9
The effect of habitat quality on number of mates per male lark bunting in natural and subsequently manipulated habitat. (After Pleszczynska 1978 and Pleszczynska and Hansell 1980.)

perform elaborate displays to attract females to mate with them (Cooper and Forshaw 1977, Johnsgard 1983, Lill 1974, Payne 1984, Snow 1982). Because such mating systems are so unusual among birds, evolutionary ecologists have paid considerable attention to the surfacing of promiscuity on the calm sea of avian monogamy (Bradbury 1981, Emlen and Oring 1977, Wiley 1974, Wittenberger 1978).

The males of most species of birds contribute substantially to the fecundity of their mates (and therefore to their own), either through defending territories and their included resources or through directly providing for the young. The key to promiscuity has, therefore, been perceived as the emancipation of the male from parental investment without substantially reducing the fecundity of his mates. This may be possible in one of several situations: first, the young of precocial birds feed themselves and so fecundity may be limited by factors other than the provisioning of food by the parents. Females of such species, including the promiscuously mating grouse and their relatives, may be able to rear nearly as many young by themselves as they could with the help of males. Second, in mammals, the female is specialized to nurse the young, and potential male contributions are limited to defending resources, protecting the litter, and providing food directly to the female, none of which elevates the potential contribution of the male to near that of the female. Of course, this explanation begs the question of biparental nursing in mammals. Third, resources may be conspicuous and of fixed quantity, in which case a female alone may be able to gather as much as a male and female working together. This explanation may apply to fruit-eating species (Ricklefs 1980).

The emancipation of parents from parental care can be analyzed as an Evolutionarily Stable Strategy.

The evolution of promiscuous mating from a monogamous system depends on the emancipation of the male from caring for the offspring directly or indirectly. In a few birds, and more commonly among other organisms, such as fish, it is the female that abandons the eggs or young to the care of the male parent. The choice between caring for the young or deserting them resides with each individual parent, but the consequences depend in large part on the choice made by its mate. Frequently one parent may abstain from caring for the offspring with little reduction in fitness so long as the other parent remains faithful to the clutch or brood; but when both parents abandon their progeny few or none survive. It is generally assumed that by reducing care for one set of offspring, a parent can go on to lay more eggs or obtain more matings. The evolutionarily stable strategy for the two parents can be investigated by game theory analysis. This discussion follows closely upon the treatment of John Maynard Smith (1982).

Each parent may either care for the offspring of a mating, which Smith refers to as guarding, or may desert them. The male may obtain an additional mating with probability p if he guards the brood and P if he deserts. It

Table 29-3
Payoffs to the male and female parent for guarding of desertion on the part of either one*

		Female	
		Guards (v eggs)	Deserts (V eggs)
Male	guards (1 + p matings)	vS_2 / $vS_2(1 + p)$	VS_1 / $VS_1(1 + p)$
	deserts (1 + P matings)	vS_1 / $vS_1(1 + P)$	$VS_{\emptyset}$ / $VS_{\emptyset}(1 + P)$

* Payoffs for females are indicated in the upper right half of each box; payoffs for males in the lower left.

is assumed that $P > p$. The number of eggs that a female can produce is V if she deserts her clutch and v if she remains to guard ($V > v$). The probability of survival of the eggs is S_0 when neither parent guards, S_1 when one parent guards, and S_2 when both parents guard. The number of offspring produced by each parent under the four possible combinations of the male and female guarding and deserting are shown in Table 29–3.

Both parents guarding is an Evolutionarily Stable Strategy when the strategy of desertion by neither males nor females can invade the population. That is, biparental care is an ESS when $vS_2 > VS_1$ for females and $S_2(1 + p) > S_1(1 + P)$ for males. In words, these conditions require that remaining to care for the offspring increases fitness more than abandoning them increases the number of matings for both parents; that is $S_2/S_1 > V/v$ and $(1 + P)/(1 + p)$. The ESS conditions for all four combinations of behavior (Table 29–4) reveal that when biparental care is an ESS, uniparental care cannot be; also, when biparental abandonment is an ESS, uniparental abandonment (=uniparental care) cannot be. Both biparental care and biparental abandonment may be ESSs under the special conditions $S_2/S_1 > V/v > S_1/S_0$ and $S_2/S_1 > (1 + P)/(1 + p) > S_1/S_0$. Because neither type of uniparental care is an ESS under these conditions, the evolutionary step between biparental and uniparental care may be impossible to achieve, even if the alternative strategy conferred greater fitness on both males and females. When these inequalities are reversed, both female and male uniparental care can be Evolutionarily Stable Strategies and the outcome of evolution is ambivalent.

Unambiguous female uniparental care occurs when the ESS for the female is to guard no matter what the male does—that is, $vS_2 > VS_1$ and $vS_1 > VS_0$—and for the male to desert so long as the female cares, $S_1(1 + P) > S_2(1 + p)$. These inequalities can be rearranged to give the conditions S_2/S_1 and $S_1/S_0 > V/v$ for females and $S_2/S_1 < (1 + P)/(1 + p)$ for males. Quite simply, the female's gain must be greater through caring for offspring than with leaving eggs either unattended or in the care of a male; the male

Table 29-4
ESS conditions for different combinations of male and female parental care

1. Both parents care for offspring
   female $vS_2 > VS_1$
   male $S_2(1 + p) > S_1(1 + P)$
2. Female cares, male deserts
   female $vS_1 > VS_0$
   male $S_1(1 + P) > S_2(1 + p)$
3. Male cares, female deserts
   female $VS_1 > vS_2$
   male $S_1(1 + p) > S_0(1 + P)$
4. Neither care for offspring
   female $VS_0 > vS_1$
   male $S_0(1 + P) > S_1(1 + p)$

Table 29–5
Relationship of location of fertilization to parental care in fish (above) and amphibia (below)

	Care delivered by			
Fertilization	Both parents	Male	Female	Neither
External	8	28	6 }	191
Internal	0	2	10 }	
External	0	14	8	10
Internal	0	2	11	0

Source: Smith 1978, after Breder and Rosen 1966; Gross and Shine 1981).

must be better off deserting when his mate remains with the brood. The conditions for male uniparental care are the reverse.

In birds, the ratio S_1/S_0 is almost always very large; in no species do both parents abandon the eggs, although the newly hatched chicks of megapodes are free of all parental care. Female uniparental care is quite common, being the situation in promiscuous mating systems, but male uniparental care is known only in rheas, the mallee fowl (a megapode), and a few polyandrous shorebirds, in which the female lays a set of eggs that are incubated by her first mate while she lays a second set of eggs fertilized by a second male (Pitelka et al. 1974). The predominance of female uniparental care is undoubtedly due to the fact that $(1 + P)/(1 + p)$ is usually much larger than V/v; when either sex abandons the clutch, males can gain additional matings more rapidly than females can produce more eggs.

Under the ambivalent conditions where either parent's abandoning the clutch is an ESS, which one stays and which leaves depends on the relationship between fertilization and egg-laying (Dawkins and Carlisle 1976, Ridley 1978). When fertilization is internal, the male's mating function is completed before the eggs are laid and the male is free to leave the female—holding her bag of eggs, so to speak. When fertilization is external, the eggs are laid first and the male must subsequently fertilize them. Smith (1978) summarized data of C. N. Breder and D. E. Rosen (1966) showing a strong relationship between external fertilization and male uniparental care in fish (Table 29–5).

A final consideration pertinent to male uniparental care is the certainty of paternity (Trivers 1972). When fertilization is external and eggs are laid in a nest prepared and guarded by the male, the male can be certain that the offspring are his own progeny and that subsequent care is likely to enhance the survival of his own genes. When fertilization is internal and females mate at random within the population, males cannot know for certain that they have fertilized eggs laid by a particular female, unless they guard the female throughout her receptive period (see, for example, Burton 1985).

Selection of mating behavior can lead to a mixed ESS for one sex.

There is more than one way to skin a cat—or for a male to obtain matings. If the fitness of two alternative tactics increases as their frequency in the population decreases, such frequency-dependent selection may lead to stable polymorphism, or mixed Evolutionarily Stable Strategy. At the evolutionary equilibrium, selective pressures favoring one tactic and the other exactly balance and individuals expressing either one have the same fitness (Gadgil 1972). Several cases of alternative mating tactics within the same population seem to exist. Although the degree to which individuals (usually males) differ genetically often is not known, K. D. Kallman et al. (1973) have described a sex-linked genetic locus in platyfish *Xiphophorus maculatus* that controls the age (hence size) of maturation of males. This undoubtedly influences mating tactics, as shown in a similar case of the bluegill sunfish *Lepomis macrochirus* by Gross and Charnov (1980). Two situations with more extensive behavioral observations involve dispersal to find mates in fig wasps and an unusual lek-forming bird, the ruff.

Fig wasps lay their eggs in the flowers of figs. This is a mutualism in which the adult female effects fertilization of the flowers, but the larvae consume some of the fig's developing seeds (Janzen 1979, Wiebes 1979, Beck and Lord 1988). William Hamilton (1979) has described the mating behavior of 18 species of fig wasps associated with 2 species of figs in tropical American forests. In the most abundant species, several females frequently lay their eggs in a single fig flower. Males developing from these broods are both flightless and have enormous mandibles and heads (for a wasp) with which they fight other males for matings within the same fruit. Evidently the usual presence of unrelated females in the same fruit selects strongly for males to compete among themselves to mate locally. Females, of course, must fly to find flowers in which to lay their eggs.

In the rarest species, males tend to be winged and to disperse to find mates, perhaps to avoid sister mating. It may be less difficult for such species to find mates than the males of parasitoid species because females are attracted to rather obvious resources, the fig flowers. Some of the less common species are, however, polymorphic in that some of the males developing within broods have wings and disperse to find mates while others are wingless and remain within the fruit of their birth to mate.

The ruff (*Philomachus pugnax,* literally a quarrelsome lover of combat) derives its English name from the distinctive collar of elongated feathers ringing the necks of the males; it derives its scientific name from the intense contests among males on their leks. This system of mating behavior, described by A. I. Hogan-Warburg (1966) and I. G. Van Rhijn (1973), is unusual among the group of wading birds to which the ruff belongs and is further distinguished by polymorphism in appearance and behavior among the males. Males are of two types: independents, which establish territories on the mating arena, and satellites, which do not attempt to establish their own courting spot but are tolerated by independent males. In appearance, independent males have a predominantly dark plumage, with much variation among individuals; satellite males are predominantly white. Males having intermediate plumage may be either satellites or independents. The polymorphism almost certainly is genetic, although there is no direct evidence; the behavior and plumage may be pleiotropically linked through the effects of a few genes on the levels of certain hormones that affect both traits. Observations show that both independent and satellite males mate with females attracted to the lek. Independent males may tolerate satellites on their territories to enhance the attractiveness of their mating court to females.

In each case of dimorphism in appearance and mating behavior of males, the two forms would appear to balance conflicting selective pressures. Independent ruffs may obtain more matings than satellite males, but they assume greater risk of injury through combat and of being taken by predators. W. Cade (1979) described a situation in field crickets *(Gryllus)* in which calling males are more likely to attract females than are silent males, but they are also more likely to be parasitized by a fly that is attracted to the mating call.

Sexual selection has led to the elaboration of courtship behavior.

Sexual selection is the situation in which one sex determines the fitness of traits expressed in the other by exercising choice in mating (Harvey and Arnold 1982, Partridge and Halliday 1984, Bradbury and Anderson 1987). The usual result of sexual selection is strong sexual dimorphism, especially of ornamentation, coloration, and courtship behavior. Darwin, in his book *Descent of Man and Selection in Relation to Sex* (1871), was the first to propose that sexual dimorphism could be explained by selection applied differentially to one sex. Evolutionary ecologists pretty much agree that sexual dimorphism can arise in three different ways. First, the different sexual roles of males and females may place each in a different relationship to the environment, causing differential selection and response. For example, because females produce large gametes fecundity often is directly related to body size; this may provide a basis for the larger size of females in many species (Figure 29 – 10). In addition, the special nutritional requirements for egg production, and for protecting the eggs and young, which usually falls upon the female, may lead to females using the environment differently than males, thereby bringing different selective factors upon themselves. Simply having to find suitable nest sites may require females to utilize different habitats than males during the nesting season.

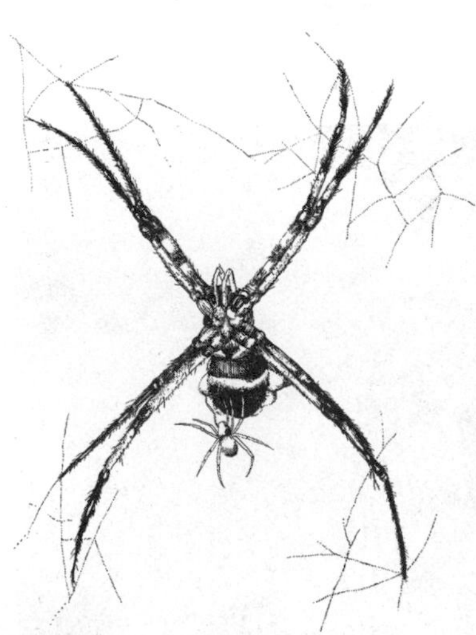

Figure 29 – 10
Extreme sexual dimorphism in size in the garden spider *Argiope argentata.* The male is much smaller than the female, which is portrayed in the normal resting position at the hub of her web.

Second, sexual dimorphism may also arise through contests between males for opportunities to mate with females. Such contests may select elaborate weapons for combat, such as the antlers of deer and the horns of the mountain sheep (Figure 29 – 11). Although some authors include male-male competition in their definitions of sexual selection, it is more properly distinguished as intrasexual selection. Because evolution following upon intrasexual competition presents no conceptual difficulties, in spite of its prevalence it will be given little attention here.

Figure 29–11
Elk have immense antlers that are used during contests between males to establish control over harems of females. (Courtesy of U. S. Department of Interior.)

Third, sexual dimorphism may arise through intersexual selection; that is, the direct exercise of choice among individuals of the opposite sex based upon their appearance and behavior. With few exceptions, females do the choosing, and males respond with magnificent displays of vainglorious courtship ritual (Figure 29–12). Why females choose, and males compete among themselves for the opportunity to mate, derives from the general

Figure 29–12
Two male sharp-tailed grouse displaying on a communal courting area (lek) in southern Michigan (left). The female (right), has a dull plumage. (Courtesy of U. S. Soil Conservation Service.)

asymmetry of parental investment that defines the male and female conditions (Bateman 1948, Trivers 1972). Males enhance their fecundity in direct proportion to the number of matings they obtain; females are limited in number of offspring by the number of eggs they can produce, but they stand to gain in quality of offspring by choosing to mate with males bearing superior genotypes.

The goal of female choice presumably is to recognize and favor males with particular genetic constitutions. The dilemma presented by sexual dimorphism is that many of the traits selected, such as bright coloration, long feathers, and elaborate courtship behavior, would seem to put males at great risk. How can females select traits that appear to reduce the fitness of males among their progeny? Darwin was aware of this problem, but not until the mid-1970s did the beginnings of a resolution appear.

R. A. Fisher (1930) was the first to provide a detailed explanation for the sexual selection of male adornment, by a mechanism he termed "runaway sexual selection." The elements of Fisher's model progressed logically as follows: (1) variation among males in a fitness-related trait made some individuals more desirable as mates than others; (2) females that perceived this difference among males and selected mates accordingly had higher fitness than nonselective females; (3) persistent female choice for extreme values of the male trait under selection leads to continued male response and eventually to the bizarre courtship antics of, say, the birds-of-paradise, to pick a conspicuous example. Fisher's model includes the origin of both female choice and male sexual-selected traits as adaptive modifications. Also, the "runaway" behavior of the model can be achieved only if female choice is based on comparison of a trait among males, rather than upon some absolute ideal. In the latter case, selection would stop when male adaptations coincided with the ideal. In the former case, sources of new variation would always provide a superior male by comparison with others in the population, just as, in artificial selection programs, breeders continually up the ante even as the population pays through selective deaths to stay in the game.

The difficulty with Fisher's model has persistently been that female choice must initially be based on traits that intrinsically confer advantage to males in the absence of female choice. It is difficult to imagine how some sexually selected traits could have originated in this way, considering the ridiculous extremes to which they have been carried. Models designed to explore the process of sexual selection in more detail have been presented by Trivers (1972), Halliday (1978), and, incorporating specific genetic bases for selected traits, by O'Donald (1980), Lande (1980, 1981), Kirkpatrick (1982, 1987), and Ten Cate and Bateson (1988). These theoretical investigations generally confirm that runaway sexual selection is feasible under a nonrestrictive range of conditions, but they fail to address the origin of female choice. Once the system gets going, male advantage can be sustained when female choice more than compensates any increased encumbrance endured by favored males.

Female choice is certainly a fact of life, experienced at some level by most males (Searcy and Andersson 1986, Hendrick 1988). Richard D. Howard (1978) demonstrated that female bullfrogs strongly preferred to mate with larger males. Male bullfrogs are territorial, and females may perceive that the larger males are able to secure larger territories in male-male competition, or perhaps the overall genetic quality of an individual is reflected in its body size.

A particularly compelling demonstration of female choice was the experimental study of Malte Andersson (1982) on tail length of male long-tailed widowbirds *(Euplectes progne)*. This polygynous species inhabits open grasslands of central Africa. Females, about the size of a sparrow, are mottled brown, short-tailed, altogether ordinary in appearance. The males are black, except for a red shoulder patch, and sport a half-meter-long tail conspicuously displayed in courtship flights. Males may attract a half-dozen females to nest in their territories, but they provide no care for their offspring. Tremendous variance in male reproductive success provides the classic conditions for sexual selection. Andersson's experiment was simple and straightforward. He cut the tail feathers of some males to shorten them, and glued the clipped feathers onto the ends of other males' tails to lengthen them, and observed the subsequent success of males in attracting females to their territories. Controls were unclipped males and males whose tails were cut and glued back into place.

Andersson found that length of tail had no effect on a male's ability to maintain a territory, and fitness apparently is not influenced by the effect of tail length on contests between males. But, strikingly, males with experimentally elongated tails attracted significantly more mates than those with shortened or unaltered tails (Figure 29–13). This result strongly suggests choice of mate on the basis of tail length and that preference is relative; that is, selection persists regardless of response. Even though male widowbirds have gone a long way to accommodate female preference, longer tails still look better to females, whether they are in the best interests of the male or not. Andersson's study was too brief to determine whether an extra-long tail reduced a male's survival.

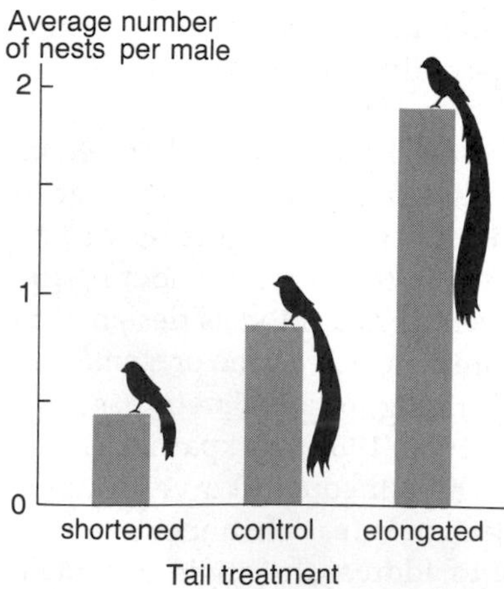

Figure 29–13
Relative reproductive success of male long-tailed widowbirds with artificially shortened and lengthened tails. (After Andersson 1982.)

Even though the basis for runaway selection in female choice has been demonstrated, its origin is still shrouded in mystery. One possibility is that outlandish traits had their origin in more conservative, intrinsically adaptive beginnings, but that establishment of strong female choice took hold of the processes and predominated environmental selective factors. Another, intriguing possibility is the "handicap principle" suggested by the Israeli behavioral ecologist Amotz Zahavi (1975). He viewed male secondary sexual characteristics as handicaps: that a male can survive while bearing such a handicap signals a female that he has an otherwise superior genotype. It may sound crazy, but if you wanted to demonstrate your strength to someone, you might carry around a large set of weights. A weaker individual couldn't do it, so there is little chance of a lesser genotype falsely advertising strength. According to Zahavi, the greater the handicap borne, the greater the ability of the individual to offset the handicap by other virtues.

You may wonder why a female should choose males that cancel out their superior constitutions by taking on burdens. The strong man may be superior intrinsically, but can he dance while holding a set of weights? This difficulty has stimulated a flood of comment and countercomment on the idea of the handicap principle (Smith 1976, 1978, Davis and O'Donald 1976, Zahavi 1977, Bell 1978, Eshel 1978; Dominey 1983, Kodric-Brown and Brown 1984, Kirkpatrick 1986) without much agreement. But there are several possible bases for resolution. One is that the superior traits of males are passed on to both male and female offspring, but their handicaps are expressed only in males. So by choosing the handicapped male, a female may produce superior daughters. If true, this would reinforce the restriction of male purpose to providing genes, and make proper choice of genotype by females all the more important. It also would establish a conflict between satisfying female choice and being able to succeed in other phases of life, including male-male contest.

Another plausible scenario for the origin of female choice is that male handicaps may arise from normal, environmentally promoted variation within the population. Vagaries of upbringing may cause some males to have, for example, longer tails than others. If long tails are a handicap in some way, only males with otherwise superior genotypes will be able to survive and prosper with them. A female would be wise to choose such a male, because tail length has already done the work of weeding out weaker genotypes for her, but since the long tail of the male is not genetically determined, the handicap would not be passed on to her male offspring. Given this situation, females that choose longer-tailed members of the male population would be strongly selected. Once that choice had become genetically established in the population, there would be strong selection of males having long tails arising from genetic factors, because their male offspring would then benefit from the previously established female choice. At this point, the process can take care of itself and runaway sexual selection is at hand.

Clearly, we have much to learn about sexual selection. Females may simply prefer odd males, as seems to be the case in *Drosophila* flies (Petit and Ehrman 1969, Ehrman 1972), perhaps because they are most likely to be genetically distinct and not cause inbreeding depression in the progeny. Also, comparisons of related taxa suggest that female preference changes over time and diverges between isolated populations. The many species of highly distinctive birds-of-paradise must have arisen from a common ancestor with highly sexually selected plumage and behavior. Yet, in each isolated line of bird-of-paradise, female preference has wandered to focus upon different parts of the plumage, colors, feather shapes, and displays, as if saying to the male, "Do whatever you wish, but it better be good."

SUMMARY

In most species reproductive function is divided between two sexes, creating conflicts over the allocation of resources between male and female sexual function, and affecting interactions between individuals of the same and different sexes.

1. Separation of sexual function between individuals is relatively rare among plants but common among animals. It is favored when the fixed costs of sexual function are high and the two sexes compete strongly.

2. The optimal sex ratio in a population balances contributions of gene complements to progeny through male and female function. In general, because the rarer sex is favored, frequency-dependent selection produces an even sex ratio in most populations. Sex ratio may, however, be influenced by differential costs of producing males and females, differential mortality, and variation in the condition of the female parent.

3. In some parasitic wasps, males compete with siblings for matings, and the sex ratio is shifted in favor of females.

4. Mating systems may be monogamous (a lasting bond formed between one male and one female), polygynous (with more than one mate, usually female, per individual), or promiscuous (mating at large within the population, without lasting pair bonds).

5. Polygyny arises when individuals of one sex can monopolize either resources or mates through intrasexual competition. In birds, polygyny is associated with heterogeneous habitats, such as grasslands and marshes, in which the quality of breeding sites varies greatly. Accordingly, some females gain greater fitness by joining a mated male that holds a superior territory than by joining an unmated male on an inferior territory.

6. Promiscuity may arise when males contribute little, other than genes, to the number or survival of their offspring, which is the common condition in all plants and most animals.

7. When either parent may care for the offspring, but two are little better than one, the sex of the caring parent may depend on whether fertilization is internal, in which case the female remains with the eggs she lays, or external, in which case the male remains with the eggs he fertilizes.

8. In a few species, individuals of one sex, usually males, may adopt different strategies of gaining matings. These usually contrast dispersing versus nondispersing forms, or territory-holding versus nonterritorial, opportunistic forms.

9. When males compete among each other for mates, females are able to choose among them. This leads to sexual selection of traits in males that are believed to indicate male fitness. These may be in the form of "handicaps" that only the more fit males can bear without encumbrance. Strong female preferences established within a population are believed to result in "runaway" selection of outlandish traits that serve no purpose of the male other than attracting mates and may be otherwise detrimental.

30

Evolution of Family and Social Behavior

Humans are the most social of all animals. Societies are sustained by role specialization among members, the interdependence attendant upon specialization, and the cooperation that satisfies the needs of this interdependence. Yet humans also are competitive, to the point of violence, within this mutually supportive structure. Social life balances contrasting tendencies toward mutual help and conflict.

Theories of population biology discussed earlier in this book incorporate as a basic premise competition between individuals of the same species. In population biology, progeny are the measure of success, the prize of the future going to those individuals that outsurvive and outproduce others—those that gather more resources than the average and leave fewer for competitors.

In the previous chapter we discussed conditions under which adults might cooperate to rear their offspring. But in many species sociable behavior extends far beyond mates: to progeny, extended families, and large groups of unrelated individuals. Some animal populations exhibit much of the complexity of human societies. The social insects—ants, bees, termites—are remarkable for their division of labor and behavioral integration of the hive or nest. Similar subtlety of social interaction, including role specialization and altruistic behavior, is being discovered increasingly among mammals and birds.

Mated pairs of animals seem to get on quite well together. But most behavioral ecologists believe that the decision to stick together, or not, is subject to selfish motives. Is this also true of higher levels of social behavior, especially where unrelated individuals cooperate to their mutual benefit? Is all social behavior personally or genetically self-serving? We humans like to think not. In the past, such beliefs have provided pragmatic justification for conflict, from petty squabbles to abhorrent acts of war. If there were no intrinsic value to societal cooperation (which then surfaces only as an iceburg's tip buoyed up by the societal conflict beneath), then lofty goals of peaceful coexistence, whether between cousins or countries, would diminish beside the practical concern of keeping the peace.

Social behavior includes all types of interactions between individuals, from cooperation to antagonism. Outright conflict can assume the ritualized appearance of males posturing for social rank or access to mates. Sometimes social behavior provides a means of organizing and making orderly the expression of basic conflicts within groups. Defense of territories and the establishment of dominance hierarchies serve this purpose with a minimum of social strife. How is it that contestants can agree upon "gentlemanly" conduct to resolve their disputes? One also sees instances that appear as true helping of others at personal cost or risk. How are such behaviors to be reconciled with the exigencies of evolutionary fitness?

Genetic relationship divides social interaction into three categories, each with different evolutionary contexts. First, mates generally are unrelated individuals that join together in the production of offspring. Because individuals of each sex have veto power over the mating act, their common interest in progeny requires cooperation at least to the point of fertilization. Second, unrelated individuals have little common genetic interest in the future; conflict is the usual basis of their interaction. Third, behavior directed toward closely related individuals is modified by the fact of common genetic descent. When individuals share parents, grandparents, or some more distant set of ancestors, they also share portions of their unique genotypes, thereby giving them a common genetic interest and a basis for cooperative involvement, as we shall see shortly.

Because we are the supremely social species, the study of social behavior stimulates a particular interest in us. But we must be wary of drawing facile analogies between animal societies and our own. Our lives may sometimes seem like the toil of ants in an ant hill, but human and ant societies differ in fundamental aspects of organization, as we shall see. Probably no other animal society has the cultural tradition of behavior as ours does, and cultural and genetic evolution may offer substantially different opportunities for achieving social integration. Nevertheless, animal societies will be understood for their own sake, and this knowledge will be applied, rightfully or not, to our own circumstance.

Our goal in this chapter is to understand the ecological conditions and attributes of organisms that combine to nurture various types of social behavior between individuals. As with any aspect of science, understanding comes from the application of observation, theory, and experiment to a common problem. In the study of social behavior, in particular, one can see the important roles that instances of each of these has played in the subsequent development of the discipline.

Dominance hierarchies and territoriality organize social interactions within populations.

Rarely do individuals of the same species encounter one another closely without arousing some type of behavior. Each individual has a personal space within which other individuals from the general population are not tolerated. The distance to which this space is defended—called individual distance (Conder 1949)—varies from species to species and with the circumstances of the individual, as one might expect. In many cases, individual distance extends no farther than one's reach at a particular moment. At one extreme, chickadees and other birds sometimes roost closely huddled together on cold winter nights; at the other, animals sometimes go out of their way to chase others. Individual distance satisfies the practical need of space for unobstructed movement. Birds cannot take off without risk of injury when others are within a wing's length. The path of motion of any animal is restricted by others close at hand.

Defense of territory is the extension of individual distance to space and objects (Hinde 1956, Brown 1964, Brown and Orians 1970, Morse 1980). When such critical resources as food or breeding sites are defendable, organisms may enlarge their individual distance to include them. Any areas or the spaces around objects defended against the intrusion of others may be regarded as territories. These may be transient or more or less permanent depending on the lability of the resource and the individual's need of it. Shorebirds may defend a particularly good feeding area on a beach until the rising tide covers it over; male hummingbirds may defend a bush against others so long as it is in flower; kittiwake gulls defend their nesting ledge through the breeding season until the young are able to fly. The object of a territorial claim may vary from food to nesting site or mate, but defense typically takes the form of highly ritualized displays, usually without physical contact, at perceived boundaries of exclusive areas.

At times, dispersion of individuals on territories may not be practical because of the social pressure of high population density, transience of the critical resource, or overriding benefits of living in groups (Morse 1980). In such circumstances, conflict among individuals also is resolved by contest, with social rank rather than space going to the winner. Once individuals order themselves into a hierarchy of social status, subsequent contests between them are resolved quickly in favor of the higher-ranking individual. When a social hierarchy is linearly ordered, the first ranked member of the group dominates all others, the second ranked dominates all but the first, and so on down the line to the last ranked individual, who dominates none.

Territory and dominance hierarchy are alternative expressions of the same social tendencies. We see this most clearly when a population switches from one to the other as circumstances change. For example, the dragonfly *Leucorrhinia rubicunda* switches from a territorial system with strong site fidelity at low density to one of broadly overlapping feeding areas at high density (V. I. Pajunen 1966). In one area of Finland, dragonflies in a sparse population were spaced 3 to 7 meters apart along the edge of a pool; in a dense population individuals were a half meter apart on average. Pajunen scored the intensity of interactions between males (only males are territorial) from low values of 1 for no interaction, through intermediate values corresponding to varying degrees of threat and pursuit, to the highest score (5) for threats followed by fighting. Where the dragonflies were sparse, the

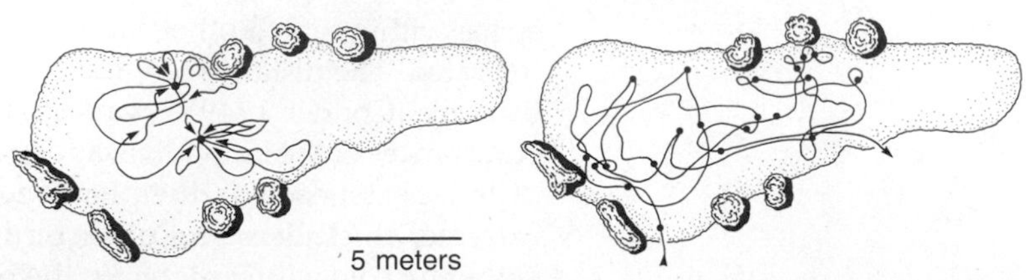

Figure 30–1
Examples of male flight activity in the dragonfly *Leucorrhinia rubicunda* showing well-developed site attachment *(left)* and weak site attachment *(right)*. The outline of a pool and several bushes are indicated; periods of observation were 1 and 13 minutes, respectively. (After Pajunen 1966.)

frequency and intensity of aggressive interactions were high (average score 4.2) compared to densely inhabited areas (3.1) because of frequent territorial defense in the first area. In dense populations, individual dragonflies rarely returned to resting sites, which are usually stems of emergent vegetation. Instead, they flew over larger areas of the ponds and alighted to rest less frequently than did dragonflies in less dense populations (Figure 30–1). Both the level of territorial defense and the level of site tenacity decreased as the density of dragonflies in an area increased. Territoriality also is more likely to be manifested as resources become more defendable and valuable. In an experimental study, Anna hummingbirds defended artificial feeders more aggressively as the provision of sugar water was increased to a substantial proportion of the daily energy requirement (Ewald and Carpenter 1978, Ewald and Orians 1983).

The position of an individual in a dominance hierarchy is sometimes reflected by its spatial position within a group. In large foraging flocks of wood pigeons, individuals low in the dominance hierarchy tend to be at the periphery, where they are more exposed to predators than the dominant individuals occupying the center of the flock (Murton 1967). Peripheral birds appeared to be nervous, and because they spend much of their time looking up from feeding, they are often undernourished. Birds in the center of the flock are generally calmer and feed more because they are protected from the surprise attack of a predator by the vigilance of individuals at the periphery (Murton 1970).

Territory and dominance hierarchy raise a number of questions concerning the evolutionary origin of such behaviors and their consequences for the population. Behaviorists perceive the decision to defend a territory or not as balancing the costs of defense against the benefits of a secured resource. J. Peterson Myers et al. (1979) showed that the size of territories defended by sanderlings (small shorebirds) feeding on California beaches varied inversely with the density of conspecifics. At high densities, intrusions were frequent and the cost of territorial defense therefore increased. Territory size was smallest in areas of high prey density, primarily because of the larger number of sanderlings that congregated upon such rich sites.

Figure 30–2
Examples of variation in throat plumage of Harris's sparrows. (After Morse 1980.)

The communication of social dominance usually is ritualized.

Most behaviors involved in determining rank or defending territorial boundaries involve ritualized behaviors that rarely lead to risky physical struggle. Certain appearances or behaviors appear to signal higher status than others. Why, then, don't less dominant individuals assume the behavior of their betters? That is, why don't they cheat the system and deceive other members of the population into thinking that they are more aggressive than they really are? Part of the answer is that some ritualized behaviors allow contestants to judge each other's size, which is difficult to cheat about and which often determines the outcome of earnest combat. Another part is that signaling status and being able to back it up must go together. This has been demonstrated by a remarkable set of experiments by Sievert Rohwer (Rohwer 1977, 1982, Rohwer and Rohwer 1978, Rohwer and Ewald 1981) on Harris's sparrows.

Harris's sparrows, which are related to the more familiar white-throated and white-crowned sparrows, breed in the Canadian arctic and winter in small flocks in the central United States. Social status within flocks is highly correlated with the amount of dark coloration in the plumage of the throat and upper breast (Figure 30–2). Upon first encounters, lighter birds generally avoid darker individuals, and so dark coloration meaningfully signals status. When Rohwer dyed the plumage of light individuals dark, they were mercilessly attacked by others who easily saw through the ruse. But dark individuals bleached to a lighter color found themselves constantly having to attack naturally light birds to regain their status. The system works because plumage and aggression go together — there are no cheaters. When Rohwer implanted testosterone into lighter birds to raise their level of aggression, such individuals that also were dyed darker rose in the dominance hierarchy while those left with their light plumage, but given dark emotions, were persecuted by naturally dark birds and could not rise in status. Behavior and plumage probably are affected by common physiological conditions associated with hormone levels. When the two do not match, birds get into trouble either because they do not get their due respect or are treated as imposters.

Rohwer's experiments raise a difficult question. If social status depends on the level of a hormone whose production imposes little physiological cost, why don't all members of the population have the highest level of aggression? If a sparrow's social rank can be elevated with a nickel's worth of hormone and dye, what's to stop any individual from assuming the trappings of high rank? Similar hormone implants have turned monogamous song sparrows into bigamists (Wingfield 1984) and nonterritorial red grouse into landowners (Watson and Parr 1981). At this point, we can only presume that status either is linked to age and experience, according to which hormone-mediated levels of behavior are adjusted to serve other purposes, or that variation in rank is a mixed Evolutionarily Stable Strategy in which dominance has its costs and all members of the hierarchy have roughly equivalent fitness.

Given that status truly represents an individual's capabilities in behavioral encounters, one might also ask: Why do low-ranking individuals, which are excluded from food and mates, and which are exposed to higher risks, remain with the flock? The answer must simply be that it is better to be a low-ranking individual in a flock, perhaps rising in rank with age, than to be a loner. Both the establishment of flocks and the size of flocks must balance costs and benefits to the members. One presumes that individuals won't associate with others unless it is to their personal advantage.

Group living confers advantages and disadvantages.

Animals get together for a variety of reasons. Sometimes they are independently attracted to suitable habitat or resources and fortuitously form aggregations, like those of vultures around a carcass or dungflies on a cowpat. Within such groups, individuals may interact, usually to contest space, resources, or mates. In other cases, progeny remain with their parents to form family groups. In this case, aggregation results from a failure to disperse. True social groups arise through the attraction of unrelated individuals to each other—that is, through joining together. The evolutionary motivation for such behavior presumably resides in increased individual fitness either by facilitating feeding, protecting group members from predators, or providing increased access to mates.

Feeding in groups may be facilitated by reducing the time required to look out for predators, as we have seen in the wood pigeon, providing information concerning the location of food, and cooperation in obtaining food. Peter Ward and Amotz Zahavi (1973) suggested that breeding colonies, roosting congregations, and other group formations may serve as "information centers" for food-finding, where the food supply is unpredictable in time and space. Ward and Zahavi envisioned individuals spreading out over vast areas in search of food, conveying information about feeding conditions when they return to the assemblage and, as they leave to feed, following others that have located good foraging areas. Certainly animals are capable of learning from the behavior of others (Klopfer 1959, Alcock 1969, Mackintosh 1974, Mason and Reidinger 1981). But although field observations (Krebs 1974, Brown 1986) and laboratory experiments (de Groot 1980) suggest that individuals can utilize the behavior of others to locate resources, the fitness components of such behavior deriving from group living are not understood.

Cooperation involves the coordination of individual behavior toward a common goal (Packer and Ruttan 1988): defense of the herd by male musk oxen, pack-hunting by wolves and killer whales, and other mammals (Meck 1970, Kruuk 1972, Schaller 1972), and by some birds (Bednarz 1988); the coordinated behavior of aquatic birds that corral fish into a small area where they can be fed upon easily (Bartholomew 1942, Emlen and Ambrose 1970). Such instances of cooperation are not common, but where they do occur social behavior serves more to foster mutualism between individuals than to organize competition between them: individual behavior becomes subservient to the group, and groups may assume a characteristic behavior of their own.

The social organization of the pinyon jay exhibits strong integration of the group.

During the last two decades, studies of social behavior, particularly of birds and mammals, have revealed near-human levels of individual recognition, cooperation, and social subservience. I could pick any of dozens of well-known species, each with its own peculiarities, to illustrate the subtleties of social behavior in animals. One of my favorites is the pinyon jay *(Gymnorhinus cyanocephalus)*, a native of the southwestern United States (Figure 30–3). As a taxonomic group, the jays (family Corvidae) exhibit a wide range of social organization, from simple territoriality in the Steller's jay (Brown 1963) to single family groups on territories in the Florida scrub jay (Woolfenden and Fitzpatrick 1984), larger social groups of several families with group territories in the Mexican jay (Brown 1970), communal breeding,

Figure 30–3
The pinon jay *(Gymnorhinus cyanocephalus)*. (After Balda and Bateman 1973.)

with several females laying in the same nest, in the Dickey jay (Crossin 1967), and large, socially integrated flocks in the pinyon jay (Balda and Bateman 1971, 1973, Balda and Balda 1978).

Flock size in the pinyon jay varies up to a maximum of about 250 individuals in the area surrounding Flagstaff, Arizona. The flocks have a complicated organization: year-old birds are excluded from breeding, some cooperative feeding of young exists, sentinels watch for predators while the rest of the flock feeds, individuals show strong behavioral cohesion and tolerance of others, and pinyon nuts are stored in large communal caches.

So long as the nut crop of the pinyon pine is good, the flock remains within a small area, typically about 20 square kilometers, which is not intruded upon by other flocks. During years of nut crop failure, the flock may join others in wandering over hundreds of kilometers in search of food, but always returning eventually to its home base. The flock habitat includes ponderosa pine forest for nesting and pinyon-juniper woodland for gathering pine nuts. Foraging for insect larvae occurs widely during certain times of the year.

In autumn, roughly October to January, loosely organized flocks of 200 to 250 individuals of all age groups are spread over about 1 hectare while feeding and 2 to 3 hectares during inactive periods. There is virtually no aggression within the flock, even though individuals feeding on snow-free patches of ground or at a salt lick may be crowded shoulder to shoulder. The flock usually moves 2 to 3 kilometers per hour while feeding, but long-distance group movements also occur after considerable mutual stimulation within the group. At first, a few birds fly off a short distance and return to the flock, apparently encouraging other birds to do the same. Eventually this leaving-and-returning behavior overwhelms the flock as a whole, and all the birds fly off together.

In autumn, foraging individuals concentrate on insect grubs in the soil and on the few pinyon nuts left over in cones from the summer crop. Seeds not eaten are buried on the south sides of tree trunks, where the winter sun first melts the snow and thaws the ground. The jays also choose their mates during the autumn, sealing the bond with a ceremony in which the male feeds seeds to his prospective mate. These ceremonies become more intense as the year wears on, and by winter, mated pairs stay together within the flock.

An unusual feature of the jay flock are the sentries, numbering 4 to 12, which remain at the periphery of the flock perched on high vantage points. Individuals take turns being sentries. At the appearance of hawks, foxes, and other predators, the sentries give warning calls that send the flock into trees and to protective cover; occasionally the sentries mob the intruder while the rest of the flock moves off to a new area.

Breeding begins in earnest in February, when courtship becomes more frequent, often involving chases during which the pair flies rapidly through and over trees, performing sharp turns and steep dives, and mated pairs begin to leave the flock to feed. By this time, the flock consists mostly of first-year birds during the day, but is joined by mated pairs in the early evening after a calling ceremony. Young birds begin the ceremony by uttering soft calls, which then become louder and culminate in a long flight, during which the courting birds rejoin the flock.

Nest-building commences in March over an area of 50 hectares, within which nests are spaced 15 to 150 meters apart. First-year birds stay in a small flock that remains within 1 kilometer of the nesting area. While females incubate their eggs, males form a separate feeding flock, which individuals occasionally leave in order to feed their mates on the nest.

Nesting is synchronized, with the first eggs of the entire flock being laid within a 3- or 4-day period. As a result, most of the young are of the same age and leave their nests at the same time. During most of the nest period, young are fed by their own parents, but 4 to 5 days before fledging, adults begin feeding chicks of other pairs, as well as their own, with as many as 7 adults providing food at any one nest.

Foster feeding continues for 3 weeks after fledging, during which time the adults and young from nearby nests form an aggregation. At any one time, a few of the adults act as sentinels for the aggregation, others remain hidden with the young in bushes, and the rest forage in a group. Within the whole flock, the individual aggregations remain distinct, even after the young begin to forage for themselves, and do not mix with others.

During August, when the seed crop of the pinyon trees has ripened, all the aggregations converge in the pinyon woodland to gather the nuts and store them in the breeding area for future use. The harvest lasts 2 to 3 weeks and marks the reformation of the large group flock and the completion of the annual cycle.

The particular circumstances of habitat, food, and evolutionary history that promoted the remarkably complex social organization of the pinyon jay are a mystery. Such extreme sociality—by no means unique—illustrates the extent to which the individual can become integrated into, and even subservient to, a larger social unit. Absence of aggression suggests that resources and mates are not strongly contested. When quietly crowded around a bird feeder, the jays seem almost self-domesticated. Sentinel duty, feeding the offspring of other pairs, and storing nuts in communal caches altruistically promote the fitness of other individuals.

The evolutionary modification of social interaction balances the costs and benefits of social behaviors.

Any social interaction other than mutual display can be dissected into a series of behavioral acts by one individual (the donor of the behavior) directed toward the other (the recipient). One individual delivers food, the other receives it; one threatens, the other is threatened. When one individual attacks another, it may be thought of as the donor of a behavior. The attacked individual usually responds by standing its ground or fleeing; in either case, it becomes the donor of a behavior. The donor-recipient distinction is useful when one considers that each behavioral act results in a change in the fitness of both the donor and receiver of the behavior. These increments may be positive or negative, depending on the interaction. Four combinations of cost and benefit for donor and recipient organize the outcomes of social interactions into four categories:

Change in fitness of		
Donor	**Recipient**	**Category of behavior**
+	+	Cooperation, mutualism
+	−	Selfishness
−	+	Altruism
−	−	Disruption, spite

These categories are displayed graphically in Figure 30–4, which will prove useful for later discussion of behavioral interactions among relatives.

Isolated acts of behavior probably rarely benefit both donor and recipient. By their nature, most social behaviors are asymmetric, instant by instant, although a donor may be a receiver of the same behavior from the recipient individual at some other time. If the consequences of behavior for fitness could be judged from the outcome of isolated acts, cooperation would be strongly selected; it probably occurs rarely, however. Spiteful behavior—reducing another's fitness despite the cost to your own—cannot be favored under any circumstance. Selfish and altruistic behaviors remain the most likely possibilities; selfishness should normally be selected to the exclusion of altruism because it increases the fitness of the donor. As we shall see below, however, altruism may be selected, in theory, when selfless behavior is reciprocated, when interactions occur between close relatives, or when altruism enhances the fitness of discrete social groups.

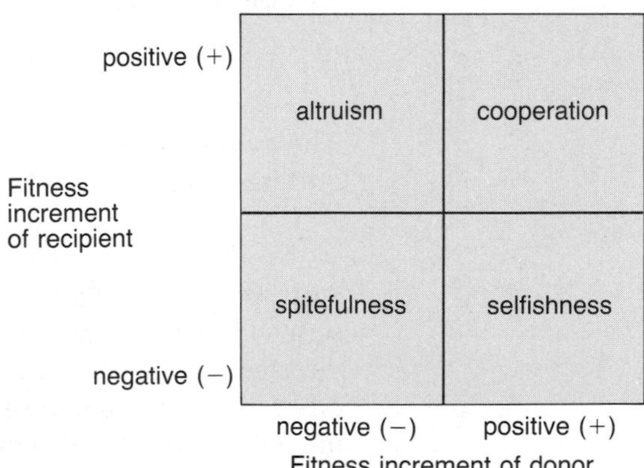

Figure 30–4
Four types of behavior classified according to the fitness increments of the donors and recipients of the actions.

An important caveat in our discussion of the evolution of behavior: selection results from the fitness contributions of all aspects of the phenotype influenced by genetic variation; isolated acts of behavior rarely will have independent genetic determination. Consistent patterns of behavior are likely to be the simplest expression of genetic variation within a population. But because each act of behavior results from a long chain of events extending through the nervous and endocrine systems from stimulus receptors to motor effectors, modulated by physiological consequences of experience and by internal rhythms and spontaneous activities, no genetic change is likely to have a simple expression in behavior, with a simple attendant fitness contribution.

Game theory indicates how individuals should interact socially within a large population.

A remarkable aspect of the behavior of pinyon jays is that individuals appear to share resources without conflict. The usual situation in populations lacking such social integration is for individuals to contest disputes over resources. Contests rarely result in bloodshed because conflict is organized by territorial boundaries or by dominance hierarchies. Yet neither of these systems can become established without some reliable method of evaluating relative rank. Normally this is accomplished by learning to associate certain attributes of other individuals with the probability that they will prevail in a dispute. Individuals may continually change their social status as they age, grow, or acquire particular experiences. In such a system, variation among individuals would be determined environmentally and developmentally rather than be under direct genetic control. The results of evolution must, therefore, be systems of behavioral exploration of social status and agreement over status signals within the population as a whole.

An approach to understanding the evolution of social interaction is to analyze the outcomes of interactions between individuals with different behavior programs. This can be done through the theory of games (Smith 1982), a simple example of which will be considered here by way of illustration. Suppose that individuals are either aggressive or submissive in contests. These opposing behavior programs, or strategies, may be referred to as hawk and dove. Furthermore, the object of each contest potentially provides a benefit (B) to the winner, measured in terms of increments of fitness. The aggressive behavior of hawks has a cost (C) measured in the same units, but doves avoid the cost by avoiding confrontation. Now, when two hawks contest a unit of resource, they split the cost and benefit equally; when two doves meet over a resource, they agreeably split the benefit. When a hawk and a dove meet, the dove always demurs, leaving the hawk the benefit without the cost of confrontation. Hence the hawk-dove interaction is an asymmetric one of selfishness and altruism.

The payoffs in the hawk-dove game are shown in Figure 30–5. In a pure population of hawks, the fitness contribution (W) to each individual

		Strategy of opponent	
		Hawk	Dove
Strategy of player	Hawk	1/2 ($V-C$)	V
	Dove	0	1/2 V

Figure 30-5
Payoffs in the hawk-dove game.

resulting from playing the hawk-dove game, $W(H)$ is $0.5(B-C)$. In a pure population of doves, $W(D)$ is equal to $0.5\ B$. Clearly the doves have the better arrangement because resources are shared without the costs of confrontation. Nevertheless, to determine the evolutionary potential of each behavior, their interaction must be compared within the same population.

When hawks and doves encounter each other at random in a mixed population, the payoff of the hawk-dove game to each depends on their relative frequencies. Suppose the frequency of hawks in the population is p (that of doves is $1-p$); of encounters over potential resources by both types of individuals, proportion p will be with hawks and proportion $(1-p)$ with doves. Therefore, payoffs of the hawk-dove game in a mixed population are

$$W(H) = \frac{p}{2}(B-C) + (1-p)B$$

and

$$W(D) = \frac{(1-p)}{2}B$$

A little algebra shows that, all other things being equal, the proportion of hawks increases—that is, $W(H) > W(D)$—when $B > C$, regardless of the frequency p. It is a simple matter to show that H is an Evolutionarily Stable Strategy, and that D is not, whenever the benefits of contesting a resource outweigh the costs. A dove can invade a pure population of hawks ($p = 1$) only when $0 > 0.5(B-C)$; hence H is an ESS when $B > C$. Moreover, a hawk can invade a pure population of doves ($1-p=1$) whenever $B > 0.5B$, which is always. A mixed strategy can result only when costs of conflict outweigh the benefits ($C > B$); as the proportion of hawks increases, they encounter each other more frequently, and therefore more often suffer the costs of confrontation. A mixed ESS occurs when $p = B/C$, at which point the payoffs to both hawks and doves are the same.

The lesson of the hawk-dove game is that altruism is difficult to maintain as an evolutionarily viable strategy whenever altruists can be taken advantage of by selfish genotypes. Does this mean that altruism does not exist? Certainly a direct answer to this question depends upon whether altruism has been observed in nature, or not. Although many behaviors have been interpreted as being selfless, critics have argued each case, finding selfish motivation behind the most saintly of acts. Individuals that utter alarm calls, for example, which many think of as accepting personal risk by making themselves conspicuous in order to warn others of danger, have also been thought of as provoking panic among others in the population so as to divert the attention of a potential predator (Sherman 1985). Such differences of opinion have not been resolved because the fitness consequences of individual acts of behavior are difficult to isolate.

A second issue raised by the altruism question concerns the mechanisms by which altruistic genotypes can be maintained in populations, provided they exist.

Group selection, kin selection, and reciprocal altruism have been proposed to explain the occurrence of altruistic behavior.

The evolutionary problem posed by altruism has been appreciated for a long time. In *On the Origin of Species* Darwin recognized

> one special difficulty, which at first appeared to me insuperable, and actually fatal to the whole theory of evolution by natural selection. I allude to the neuters or sterile females in insect-communities; for these neuters often differ widely in instinct and in structure from both the males and fertile females, and yet, from being sterile, they cannot propagate their kind.

The distinctive instinct of sterile castes of insects is that they forgo personal reproduction and devote themselves to the fertility of the colony as a whole. Although Darwin was more interested in the evolution of divergent morphology between sterile and fertile individuals, and between different sterile castes, his explanation applies equally well to the evolution of sterility itself:

> This difficulty, though appearing insuperable, is lessened, or, as I believe, disappears, when it is remembered that selection may be applied to the family, as well as to the individual, and may thus gain the desired end. Breeders of cattle wish the flesh and fat to be well marbled together: an animal thus characterised has been slaughtered, but the breeder has gone with confidence to the same stock and has succeeded.

With the concept of selection among families as his point of view, Darwin explained the evolutionary potential of sterile castes of social insects accordingly:

> by the survival of the communities with females which produced most neuters having the advantageous modifications, all the neuters ultimately came to be thus characterized.

Darwin thus considered insect colonies as family groups whose fitness was measured in competition with other colonies. Adaptations of the sterile castes—by extension the existence of sterile castes at all—were selected according to their contribution to the productivity of the colony as a whole.

This concept of intergroup selection, originally applied to insect societies, eventually was adopted more widely to explain the evolution of social behavior in all kinds of animal groups (Wright 1931, 1945). The idea was accepted with little interest and more or less uncritically for many years. Then, in the early 1960s, the Scottish zoologist V. C. Wynne-Edwards (1962, 1963) advocated group selection so forcefully, and saw its magnificent consequences for social behavior so universally, that evolutionary ecologists were forced to consider the argument more carefully.

The hornet's nest of controversy aroused by Wynne-Edwards, and those who differed with him, was discussed earlier in connection with the regulation of population size. Wynne-Edwards believed that individuals restrained themselves from reproduction in order for the population to avoid overtaxing its resources. Restraint from reproduction is altruistic toward selfish individuals, which attempt to monopolize resources and enhance their personal fitnesses within the population. Wynne-Edwards argued that selection of groups that altruistically nurtured their resources predominated over selection of selfish individuals within groups, which led to overexploitation. Opponents of Wynne-Edwards's views offered two counterarguments: first, few populations have the kind of group structure required of intergroup selection and, second, even where such selection could exist, it was bound to be weaker than individual selection because groups displace each other within populations more slowly than individuals displace each other within groups.

By the mid-1960s, individual-selectionists had clearly won the day (Lack 1966, Williams 1966) and group-selection theory faded from view. But without group selection, how was one to explain cases of altruistic behavior? One solution was to argue that such behavior did not exist; another was an idea proposed by John Maynard Smith (1964) and William D. Hamilton (1964), which they called kin selection.

Kin selection may favor altruistic acts between related individuals.

Kin selection results from differences in the fitnesses of genes based upon interactions between related individuals, which have a finite probability of inheriting copies of the same gene from an ancestor. For example, when an individual directs an altruistic act toward a sibling it enhances the personal fitness of an individual with which it shares a substantial part of its genotype. If the altruistic act occurred because of a single-gene mutation inherited from one parent, a brother or sister would have a copy of the same gene with a probability of one-half. Therefore, the fitness of the altruistic gene would exceed that of its selfish alternative so long as the cost to the altruist was less than one-half the benefit to the recipient. In general, a single altruistic act between individuals with genetic relationship r increases the fitness of the altruist gene so long as the cost (C) is less than the benefit (B) times the coefficient of relationship; that is, $C < Br$.

Hamilton (1964) generalized this cost-benefit relationship into a concept of inclusive fitness of a gene. Suppose that the fitness contributions of a single act of behavior were w_s to the donor (self) and w_i to the recipient. The net fitness contribution of the act would be $w_s + r_i w_i$; this is inclusive fitness because it includes components (which may be gains or losses) realized through the personal fitness of both the donor and the recipient. Summed over all the interactions of the individual, the total fitness contribution is

$$W = w_s + \sum f_i r_i w_i$$

where f_i is the proportion of interactions with individuals having coefficient of genetic relationship r_i. (This equation applies strictly only when genes for the behavior in question are rare. Otherwise, distantly related individuals may bear the same gene with high probability. It is therefore most useful in evaluating behaviors as Evolutionarily Stable Strategies.)

For behavior directed toward individuals having coefficient of relationship i, $W > 0$ when $w_s > -r_i w_i$, or $w_i > -w_s/r_i$. This relationship between w_i and w_s that increases inclusive fitness is shown graphically in Figure 30–6. Notice that the concept of inclusive fitness applies to both altruistic and selfish behavior; there are limits to both. When siblings interact, the evolutionarily sound limits of generosity are $-w_s < 0.5w_i$ and the evolutionarily tolerable limits of selfishness are $w_s < -2w_i$. As the coefficient of relationship decreases below 0.5, altruistic behavior becomes less likely to increase inclusive fitness ($-w_s < r_i w_i$) and selfish behavior more likely ($w_s < -w_i/r_i$), as shown in Figure 30–7.

The concepts of inclusive fitness and kin selection are based on the

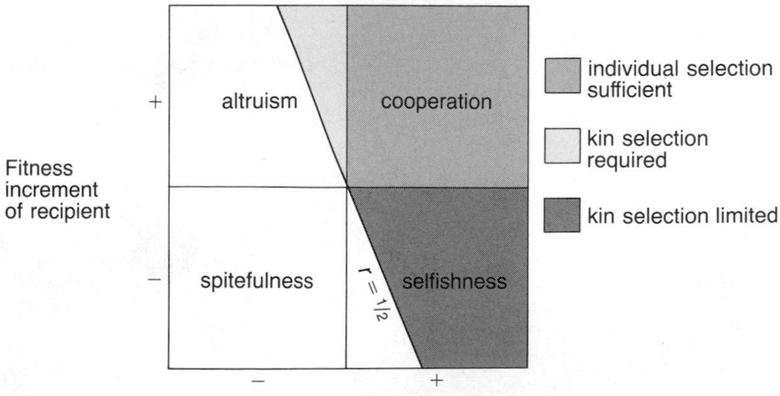

Figure 30–6
Shaded areas indicate combinations of fitness increments to donors and recipients of behaviors having positive inclusive fitnesses when the genetic relationship between donor and recipient is $\frac{1}{2}$ (full sib).

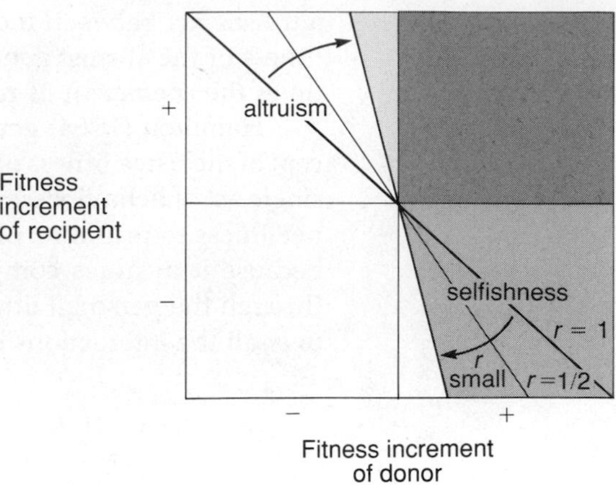

Figure 30–7
Reducing the genetic relationship between individuals increases the inclusive fitness of selfish behaviors and reduces that of altruistic behaviors.

coefficient of relationship between the donors and recipients of behavior. This coefficient may be defined as the probability that one individual has an exact copy (identical by descent from a common ancestor) of a gene carried by another. Consider the geneological diagram shown in Figure 30–8. When one individual has inherited a particular gene, what is the probability

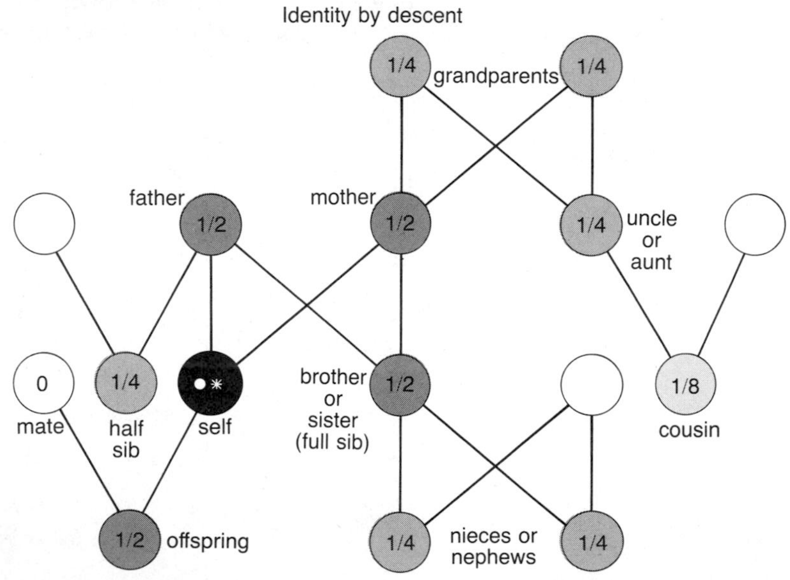

Figure 30–8
Degrees of genetic relationship (identity by descent: probability of occurrence of a copy of a gene carried by self) among relatives.

Table 30–1
Probabilities of identity by descent for individual alleles at single gene loci in individuals with various degrees of relationship

Parent	0.50
Offspring	0.50
Full sibling	0.50
Half sibling	0.25
Grandparent	0.25
Grandchild	0.25
Uncle or aunt	0.25
Nephew or niece	0.25
First cousin	0.125

that a relative has inherited a copy of the same gene? By this criterion, the relationship between an individual and one of its parents is 0.5, since it must have received the gene in question from one parent or the other. Reciprocally, because each parent contributes half of its genes to each of its offspring the probability that a son or daughter will inherit a particular gene is also 0.5. The genetic relationship between full siblings is 0.5 because the gene possessed by one individual was inherited from either its mother or its father which, in either case, would pass the gene onto a sibling with a probability of 0.5. Other coefficients of genetic relationship (probabilities of identity by descent) are summarized in Table 30–1.

The maintenance of altruistic behavior by kin selection requires that such behavior be restricted to close relatives within larger groups. Certainly individuals of many species tend to associate in family groups. When dispersal is limited, nearby individuals with which interactions are most frequent are likely to be close relatives. Moreover, individuals seem to be sensitive to their degree of relationship to others even when they have had no family experience with which to learn who are related directly (Hoogland 1982; Bateson 1983; Carlin and Holldobler 1983; Holmes and Sherman 1985, 1985; Lacy and Sherman 1983; Klahn and Gamboa 1983; Fletcher and Michener 1987; Waldman 1988). The cues that indicate relationship must involve some subtle matching of chemical, acoustic, or visual signals that are under genotypic control and are extremely variable within populations.

Altruistic behavior might evolve by reciprocity among nonrelated individuals.

Because group selection had been discredited and kin selection could produce altruistic behavior directly only toward close relatives, an opening remained for a mechanism that could account for cooperation distributed widely throughout a population. One candidate for this position—reciprocal altruism—was proposed by Robert Trivers in 1971. Trivers (1985) defines reciprocal altruism as the exchange of altruistic acts between individuals. As long as the cost to the donor of each act is less than the benefit to the recipient, and two individuals give and take more or less equally, the arrangement of reciprocal altruism benefits the participants. To avoid bestowing altruistic behavior on individuals who do not respond in kind, organisms keep track of who reciprocates and who doesn't, and confine their altruism accordingly. Under this system, nonreciprocators are at a disadvantage and may be selected against, leaving a population of highly cooperative individuals. As Trivers (1971, 1985) points out, however, reciprocal altruism can be expected to arise only when individuals have a long association with each other and can discriminate between those that reciprocate and those that don't.

Several studies, such as G. S. Wilkinson's (1984) on food sharing in vampire bats, claim to present evidence for reciprocal altruism, but support

		Strategy of recipient		
		Accept	Offer	Reciprocate
Strategy of donor	Accept	0	2	0
	Offer	−1	1	1
	Reciprocate	0	1	1

Figure 30-9
Payoffs in a reciprocal altruism game. (After Smith 1982.)

is still relatively weak. And although organisms certainly distinguish individuals within a population (Trivers 1985), it remains to be shown that altruistic behavior is distributed among individuals according to their past history of interaction.

A further difficulty with reciprocal altruism is that because it requires a reciprocating partner in order to work, it can confer little fitness when reciprocators are rare. In fact, an altruistic gene cannot easily invade a nonreciprocating population. Reciprocal altruism *can* be an Evolutionarily Stable Strategy. R. Axelrod and Hamilton (1981), Axelrod and Dion (1988), and Smith (1982) used game theory analyses to show that reciprocation can be a successful strategy for a player. For example, suppose that a population consists of three genotypes: selfish individuals, which always take what is offered, with a gain of two units of fitness (the accept strategy, A); altruistic individuals, which always donate help at a cost of one unit (offer, O); and reciprocators (R), which always do what an individual last did to them, except that, being ever hopeful, their first act is to offer. The payoff matrix for repeated interactions between two individuals is shown in Figure 30-9. In a population consisting of all three types, fitnesses are ranked in order $W_A > W_R > W_O$. As the altruistic offerers are eliminated from the population, the reduced payoff matrix in the absence of offerers favors reciprocators over acceptors, and reciprocal altruism prevails. But one can also see that in a pure population of acceptors, the reciprocator has no net fitness advantage. Hence the evolution of altruism through reciprocity requires a mechanism to increase the proportion of reciprocators initially to the point that they can do each other some good. This may be particularly difficult because reciprocation is a complicated strategy involving responses conditioned on prior experience.

Several behavior systems suggest the operation of kin selection.

As ecologists turned their attention to the evolution of social behavior in the 1960s, several phenomena provided foci for discussion. One of these was the evolution of alarm calling. Although many dismissed this behavior as

Figure 30–10
Belding's ground squirrel at Tioga Pass, California. (Courtesy of P. W. Sherman; from Sherman 1977.)

being selfish, Paul Sherman's (1977) observations on the behavior of Belding's ground squirrels in the Sierra Nevada of California offered strong support for the hypothesis that alarm trills given in the presence of terrestrial predators are maintained by kin selection. (Alarm whistles to aerial predators appear to be selfish [Sherman 1985].)

Because ground squirrels are social creatures and exposed to view in their montane meadow habitat (Figure 30–10), the threat of predation is great. Reducing this threat with alarm calls warning of danger is an important factor in their lives. Sherman found that alarm callers to terrestrial predators were adult and first-year females; males and yearling females did not exhibit the behavior. He also found that males disperse widely from their place of birth, whereas females are sedentary. Moreover, females called more frequently when they had occupied the same territory for many years, hence female relatives were likely to be nearby, or when relatives were known to be present in the population (Figure 30–11).

Sherman also determined that alarm calling to terrestrial predators has a cost. When predators approached groups of ground squirrels, at least one

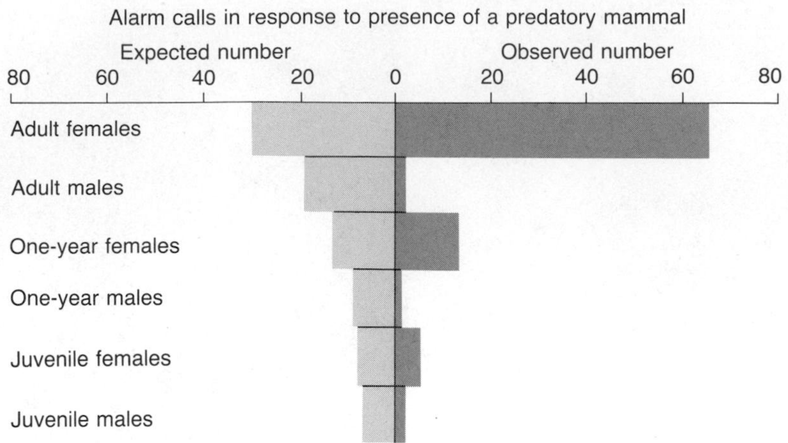

Figure 30–11
Expected and observed frequencies of alarm calling by sex and age classes of Belding's ground squirrels. Expected values were computed by assuming that individuals call in direct proportion to the number of times they are present when a predatory mammal appears. (From Sherman 1977.)

of whom gave an alarm call, 14 of 107 calling individuals were attacked (13 per cent) compared with 8 of 168 silent individuals (5 per cent). These observations are consistent with kin selection playing a role in the evolution of alarm calling.

Kin selection arguments also have been applied to explain the phenomenon of helping at the nest, observed in many species of birds. Initial reports of the behavior, by Alexander Skutch (1935, 1961) and C. Hilary Fry (1972), indicated that in several species, particularly in the tropics, the young in a single nest were fed by more than a single pair of adults. Subsequent observations on populations in which birds were colorbanded for individual recognition revealed that the helpers were usually the pair's offspring from previous nestings. Hence "helping" usually involves individuals assisting their parents to raise younger siblings, or assisting their siblings to raise their nieces and nephews. Frequently, helpers are birds raised the previous year, hence they are at least a year old and physiologically capable of breeding on their own.

Helping is viewed as altruistic in the sense that helpers forgo producing their own offspring for the sake of enhancing the personal fitness of their parents. But helping may either benefit or hinder the parents, and it may benefit or cost the helping individual. Because helpers are related to both their parents and their siblings by coefficient 0.5, the combinations of cost and benefit to donor and recipient favored by kin selection are fairly straightforward (Figure 30–12). Most studies indicate that helpers increase the productivity of their parents substantially, but rarely to the extent of individuals breeding on their own (for example, Figure 30–13). As Glen Woolfenden and John Fitzpatrick (1984) point out, individuals may also enhance their own personal fitness by remaining with their parents: they are more likely to survive on familiar territory, and they may inherit the

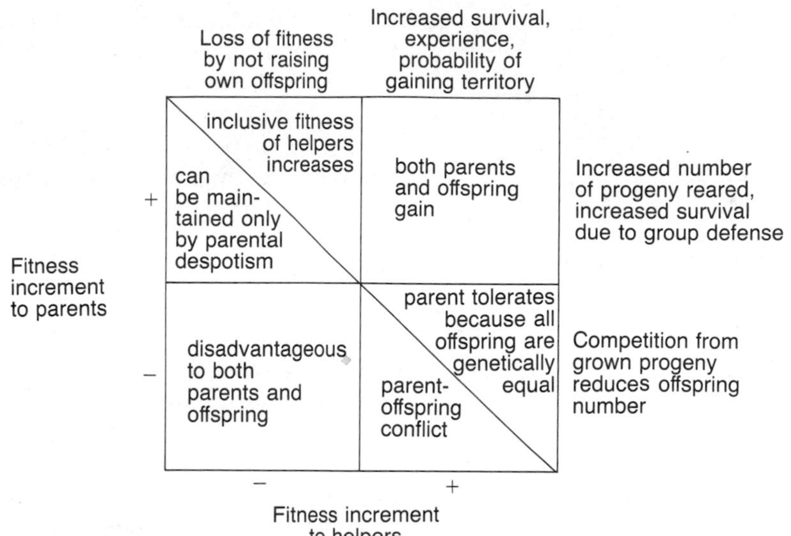

Figure 30–12
Explanation of fitness increments to parents and offspring arising through helping at the nest (offspring helping their parents to rear siblings). Because the coefficient of relationships between siblings is 0.5, the diagonal line has a slope of −2.0.

family territory when their father dies. Helping also may substitute for investment of the true parents, thereby increasing their survival.

Why is helping more frequent in the tropics? High adult survival rates in the tropics allow family groups to remain intact from year to year with high probability; low frequency of death and replacement of one mate also means that helpers assist in rearing full siblings more frequently than half siblings (Brown 1974, Ricklefs 1975). In addition, when populations are regulated by density-dependent factors, and when recruitment of one-year-old birds exceeds death of adults, young birds would have to compete with more experienced older birds for breeding territories. Under such conditions, which seem common in the tropics, the inclusive fitness of genes which dictate that a young bird remain with its parents may exceed the inclusive fitness of genetic tendencies to set out on one's own. Although helping could evolve under conditions in which such cooperation enhanced the personal fitness of both parent and offspring, inclusive fitness considerably widens the range of costs and benefits to each under which the behavior is favored.

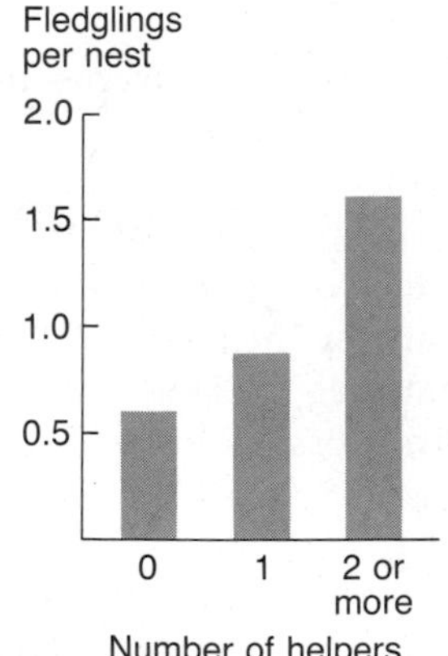

Figure 30–13
Relationship between nesting productivity and number of helpers in the white-fronted bee-eater. (After Emlen 1981.)

Kin selection has been implicated in the evolution of warning coloration.

Warning coloration signals predators that potential prey items are distasteful, poisonous, or both. One can see straight off that conspicuous advertis-

Appearance	Dispersion of larvae: Aggregated	Solitary
Aposematic	9 species	11
Cryptic	0	44

Figure 30–14
Relationship between appearance of larvae of various species of lepidoptera and their degree of aggregation. (From Jarvi et al. 1981.)

ing of this quality benefits the vulnerable prey if it dissuades would-be captors; recognizing the meaning of the display also benefits a predator, who avoids the regret of unwise eating. Adopted by all members of both prey and predator populations, these conventions of advertisement and appropriate response are unquestionably Evolutionarily Stable Strategies. But how does the system get started? Can a genetic mutation leading to warning coloration invade a population of distasteful, cryptic prey hunted mercilessly by visually oriented predators which cannot distinguish them from other palatable species? A single individual with warning coloration would be rendered conspicuous and vulnerable in a population of otherwise cryptic organisms, and would likely be nailed by an unsuspecting predator. Noxiousness and warning coloration can confer fitness only if potential predators have opportunities to learn their meaning—by eating someone else.

In the case of a novel mutation, the most likely other individual in a population to have the same rare gene (identical by descent) is a parent or sibling. Half of the individuals in a single brood are likely to have the same unique gene inherited from one parent. If the conditioning stimulus is strong enough, one attempt to eat a noxious, aposematic individual may save the lives of several of its brothers and sisters. With its inclusive fitness thus enhanced, the mutation could spread rapidly through the population.

This scenario requires that members of a brood remain in close proximity so that the experience of a single predator is translated into the rejection of a sibling. Among butterflies and moths, warning coloration and aggregation of broods of larvae are strongly associated (Sillen-Tullberg and Leimar 1988), which is consistent with the idea that kin selection is important to the evolution of warning coloration (Figure 30–14). Among species that disperse their eggs widely over many food plants, warning coloration is less frequent.

The optimum level of parental investment can differ for parents and their offspring.

Parental investment represents an arrangement between parents and their offspring. Rather than passively accepting whatever their parents offer, most offspring actively solicit care. Young animals beg for food and solicit brooding; eggs actively take up yolk from the ovarian tissues or bloodstream of the mother. For the most part, the fitness interests of parent and offspring are compatible; when progeny thrive, so do their parents' genes. But when the selfish accumulation of resources by one offspring reduces the overall fecundity of the parents, parent and offspring can come into conflict. Trivers (1974) was the first to give parent-offspring conflict a theoretical basis. Each act of parental care benefits the offspring by enhancing its survival, but costs the parent by decreasing the number of other contemporary or future offspring. Resources allocated to one child cannot be delivered to others; prolonged care delays the birth of subsequent children; risks of caring for

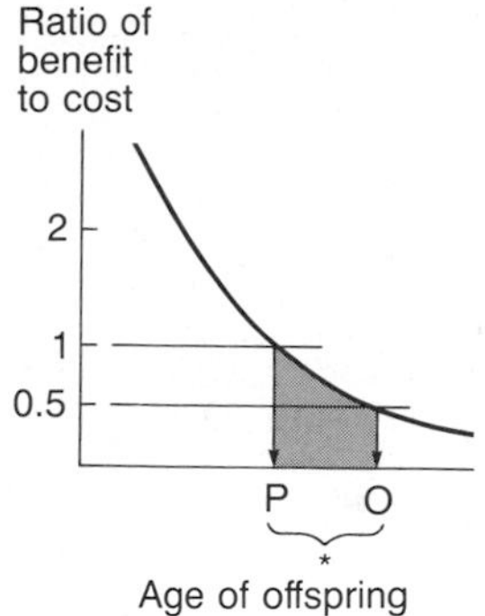

Figure 30–15
Ratio of benefit to cost of an act of parental care toward an offspring decreases with age as the offspring grows and becomes self-sufficient. Because offspring are of equal genetic value to parents, parents should shift care to succeeding offspring when the ratio falls below 1. Because siblings have a genetic relationship of one-half, however, inclusive fitness arguments dictate that they should solicit parental care to a cost-benefit ratio of one-half, thereby establishing a region of parent-offspring conflict. (After Trivers 1972.)

today's children decrease the probability that parents will survive to rear tomorrow's.

From the standpoint of the individual offspring and relative to the benefit bestowed upon its own genotype, the cost of parental care—measured by a reduction in the number of its siblings—must be discounted by one-half; a sibling shares only one-half of an individual's genes identically by descent. Therefore, when an individual possesses a genetically determined trait that increases the delivery of parental care to it, that trait is favored so long as the cost, in terms of number of siblings, is less than twice the benefit to the individual.

As offspring develop and become more self-sufficient, the ratio of benefit to cost of continuing to care for them decreases (Figure 30–15). The cost of a particular parental act may change little over the age of the offspring but, as the young mature and are better able to care for themselves, the benefits of parental care dwindle. When the benefit/cost ratio drops below 1, the parent should cease to provide care to that offspring in favor of producing additional ones. Suppose, however, that a child has a gene that increases its solicitation of parental care. Because the inclusive fitness of the child discounts the cost of care to the parent by one-half, from the standpoint of the child, parental acts should be continued until the benefit/cost ratio is one-half. Thus the period of time between the ages at which B/C is 1 and 0.5 is one of conflict between parent and offspring. With respect to the timing of weaning, Trivers stated:

> Weaning conflict is usually assumed to occur either because transitions in nature are assumed always to be imperfect or because such conflict is assumed to serve the interests of both parent and offspring by informing each of the needs of the other. In either case, the marked inefficiency of weaning conflict seems the clearest argument in favor of the view that such conflict results from an underlying conflict in the way in which the inclusive fitness of mother and offspring are maximized. Weaning conflict in baboons, for example, may last for weeks or months, involving daily competitive interactions and loud cries from the infant in a species otherwise strongly selected for silence (DeVore, 1963). Interactions that inefficient *within* a multicellular organism would be cause for some surprise, since, unlike mother and offspring, the somatic cells within an organism are identically related.

ESS reasoning has been applied to the problem of parent-offspring conflict in a series of difficult papers by G. A. Parker and M. R. MacNair (1978, 1979) and MacNair and Parker (1978, 1979). The major perplexity of the evolutionary resolution of parent-offspring conflict is the fact that conflicting offspring themselves grow up to be parents of conflicting offspring, presumably themselves suffering the same way they made their own parents suffer (Alexander 1974, West Eberhard 1975). This can be shown not to be a difficulty by a simple model of the fate of a conflictor gene passed down through progeny to grandprogeny (Figure 30–16). The examples clearly show that such a gene will invade a population of nonconflictors as long as the cost/benefit ratio is less than 2. When conflict affects the number of

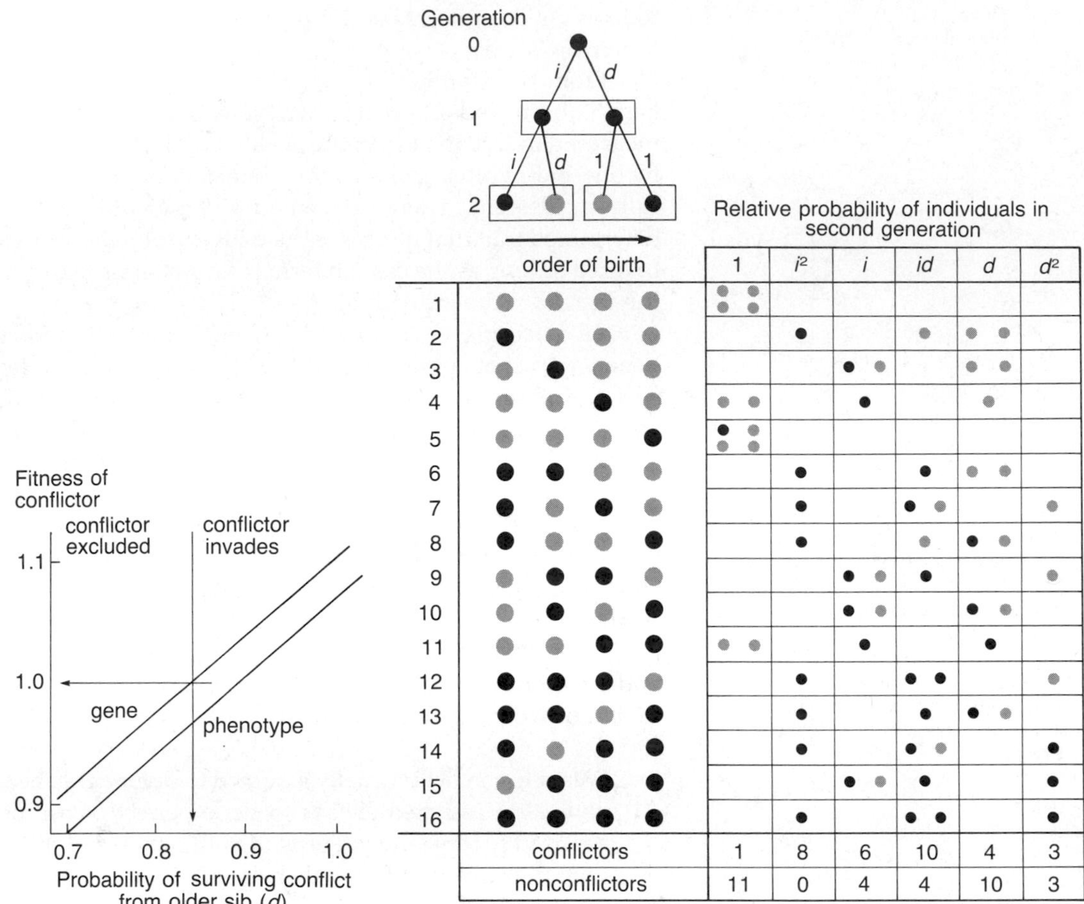

Figure 30–16
Fitness consequences of sibling interaction in a simple family history. Each parent has 2 offspring; if the firstborn is a conflictor its fitness is incremented to *i* and the fitness of the second-born is decremented to *d* whether it is a conflictor or not. This simple family tree has 16 possible inheritance patterns of the conflictor gene each with equal probability; the fitness increments are shown at right. At left is shown the total fitness of the conflictor gene and phenotype when $i = 1.1$ for various values of *d*. In this particular example, the conflictor gene can invade the population when *d* is above about 0.83.

half-siblings ($r = \frac{1}{4}$), rather than full siblings ($r = \frac{1}{2}$), the cost/benefit ratio must be less than 4, and both the intensity and duration of parent-offspring conflict will be greater.

Given that conflict exists, to what extent can parents control how much care they deliver to their offspring? Parents certainly are physically in control in the sense of behavioral dominance. But parents also must rely on solicitation signals from their offspring to determine how much care should be delivered. Young can sense their state of nutrition better than their parents can, and so parents have to rely on communication from the offspring to adjust feeding rates. Offspring might increase their rate of solicitation of care to support increased growth rate, greater activity and perhaps behavioral dominance over siblings, or storage of nutrients for use after weaning. If they changed any of these things, increased demand would accurately reflect their greater requirements, which could be provided only at the expense of other, more reticent young. This could be countered from the parent's standpoint only through reduced responsiveness to solicitation, which would in turn favor offspring with lower requirements.

Regardless of whether parents are in control or not, Amotz Zahavi (1977) has pointed out that, by punishing or withholding resources from selfish offspring, parents reduce their own fitness. Ousting an unrelated competitor from one's territory or killing a microbe that has invaded the body is one thing, but kicking a child out of the house or destroying a growing ovum would be self-defeating. Marcus Feldman and I. Eshel (1982) took a population genetics approach to this intriguing problem. They showed, first, that an altruist cannot invade a brood of selfish offspring when the decrease in personal fitness of the altruist compared to its selfish siblings is more than twice the benefit of increased fecundity of the parent. Hence conflicting up to a B/C ratio of 2 is an ESS. Then they asked whether a gene for parental interference could invade if it depressed the fitness of the selfish individual and redistributed that fitness to the rest of the brood. They found that when the overall fitness of a brood is lowered by parental interference, and when there is free recombination between the conflictor and interferor gene loci, fixation of selfish offspring and noninterfering parents is stable. When the loci are tightly linked genetically, that combination may be unstable and parental interference may be able to prevent the invasion of the conflictor gene. These models greatly oversimplify the genetic basis of parental care, and their ambivalence in spite of this simplicity suggests that the issue will be difficult to resolve.

Evidence bearing on conflict between parent and offspring is summarized by Trivers (1984) and Shaanker et al. (1988). There is little doubt that such conflict occurs and that both the level of care and its duration are worked out with some difficulty (for example, Davies 1976). The issue of conflict over weaning time is clouded by the development of offspring, whose behavioral capabilities (hence ability to solicit actively) continually increase with age. Perhaps the apparently rising "conflict" over weaning observed in many species merely reflects growth and maturation of the young. Furthermore, the costs and benefits of "conflicting" behavior have yet to be measured to determine whether the cutoff line is $B/C = 1$ or $\frac{1}{2}$. At present, therefore, the concept of parent-offspring conflict should provide a point of reference for discussing behavior and physiology governing parental investment in offspring.

Insect societies are based upon sibling altruism and parental despotism.

The complex societies of the termites, ants, bees, and wasps have presented a formidable challenge to evolutionary ecologists, primarily because of the existence of nonreproductive castes (Brockman 1984, West Eberhard 1981). Darwin recognized that the evolutionary future of such castes was assured by their contribution to the fitness of the whole colony, and suggested that evolution proceeds in such species by a process similar to family selection in artificial breeding programs. George C. Williams and Doris C. Williams (1957) placed the evolution of sterile castes of insects in the context of competition between families. Subsequent discussion of the problem has become more complicated as arguments based upon kin selection, cooperation, and parental manipulation have been raised. Thus insect societies have caught the attention of all camps concerned with the evolution of social and family behavior more generally.

Before considering the evolutionary issues raised by insect societies, we shall discuss a bit of their natural history. The best reference remains E. O. Wilson's (1971) book *The Insect Societies.* Wilson defines several grades of sociality, the highest of which is eusociality. This grade is characterized by several adults living together in groups, overlapping generations — that is, parents and offspring together in the same colony — cooperation in nest building and brood care, and reproductive dominance, including the presence of sterile castes. Defined thusly, eusociality is limited among insects to the termites (Isoptera) and the ants, bees, and wasps (Hymenoptera), although the elements of eusociality are present in one mammal, the African naked mole rat.

The complex organization of the insect society is dominated by one or a few egg-laying queens. Nonreproductive progeny of the queen gather food and care for developing brothers and sisters, some of which become sexually mature, leave the colony to mate, and establish new colonies. Most insect societies are huge extended families.

Bee societies are simply organized; females are divided among a sterile worker caste and a reproductive caste that is produced seasonally. Whether an individual will become a sterile worker or a fertile reproductive is controlled by the quality of nutrition given the developing larvae (Light 1942–1943). In general, differentiation of sterile castes is stimulated by environmental, usually nutritional, factors (Brian 1979, 1980). The development of sexual forms can be inhibited by substances produced by the queen and fed to the larvae. In bees, the worker caste represents an arrested stage in the development of the reproductive female, stopped short of sexual maturity.

Ant and termite colonies often have a continuous gradation of worker castes, ranging from very small individuals that are primarily responsible for the nutrition of the colony to larger individuals that are specialized morphologically to defend against intruders (Oster and Wilson 1978; Figure 30–17). The so-called soldiers are often equipped with formidable mandibles and stings; those of some species of termites can produce noxious gases *(Rhinotermes)* or direct a stream of noxious fluid at an intruder *(Nasutotermes).* In termite colonies, workers are both male and female; in hymenopteran societies, workers are all females produced from fertilized eggs. Males, which develop from unfertilized eggs, appear in colonies only as reproductives (drones) which leave to seek mates.

The queen in many insect societies is highly specialized as an egg-laying machine. The queens of some termites may lay as many as 6000 to 7000

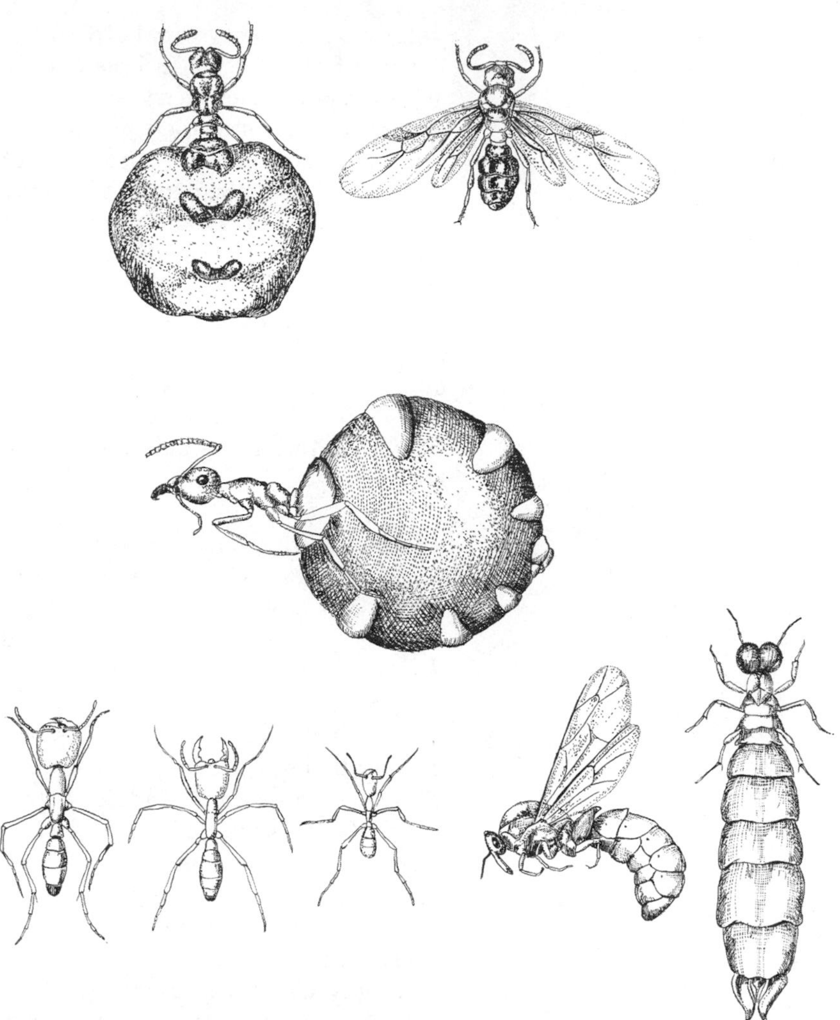

Figure 30–17
Castes of several species of ants: a virgin queen *(top right)* and an old egg-laying queen *(top left)* of the workerless social parasite *Anergates atralulus* of Europe; a replete of the honey ant *Myrmecosystus melliger* from Mexico *(center);* three sizes of blind workers, a winged male, and a queen, blind and wingless, of the African visiting ant *Dorylus nigricans* *(from left at bottom).* (After Wheeler 1923 and Grasse 1951.)

eggs per day for many years. Termite queens do not store sperm, so sexually active males must accompany them continuously. By contrast, most bee and wasp queens can store millions of sperm for many years and produce an entire colony from a single mating. Some ant queens have been known to be sexually active for as long as 15 years, laying more than a million eggs.

The extreme specialization of the queen is a problem at colony-founding time because there are no progeny to take care of the queen's first brood. The female must initially provide food for her young. Honeybee and army ant queens have overcome this problem by taking part of the parental colony in which they were born to help found their own colonies and to help care for the first broods (Schnierla 1956). In most other species of ants and in termites, new queens feed their first larvae with secretions from their body, and then wait until their brood are large enough to gather food before continuing to reproduce. In some of the more primitive ponerine ants of Australia, the queen herself, when founding a new colony, builds the first shelter and gathers food for her brood (Wheeler 1933). Colony growth in these ants is slow at first because the queen must first perform most of the tasks eventually provided by workers before she can develop into an efficient egg-laying machine.

The evolution of eusociality continues to challenge biologists.

From its distribution across taxonomic groups, it is clear that eusociality has evolved independently many times in the bees, wasps, and ants (Snelling 1981). It is less clear what route was followed. The most widely accepted sequence of evolutionary steps in the development of insect societies is that envisioned by William M. Wheeler (1928). The steps must have included a lengthened period of parental care of the developing brood, either guarding the nest or continual provisioning of the larve in a manner similar to birds feeding their young. If a parent lived and continued to produce eggs after its first progeny emerged as adults, then the offspring would be in a position to help raise the subsequent brood. This overlapping of generations, which is not common among insects, and extensive parental care are necessary ingredients in the recipe for eusociality. Once progeny remain with their mother after they attain adulthood, the way is open to relinquishing their own reproductive function solely to support hers.

Only Charles D. Michener (1958, 1969) and N. Linn and Michener (1972) have proposed a serious alternative to Wheeler's scenario. They suggested that insect societies originated in the aggregation of unrelated adult individuals, most of which eventually became subordinated to a single reproductive female. Some possible early stages of this sequence are known from bees and wasps, and while most cases of several females occupying the same nest are parasitic or mutually aggressive, benefits may accrue to each participant through joint defense of the nest against predators and parasites (Michener and Brothers 1974, Abrams and Eickwort 1981). Kenneth M. Noonan (1981) has shown, however, that foundresses may be close relatives, actively chosen through kin recognition and contesting nest sites with unrelated individuals (Craig and Crozier 1979, Crozier 1979, Lester and Selander 1981; but see Quellar et al. 1988).

Regardless of their origins, insect societies are not necessarily peacefully maintained. There is considerable conflict between colonies, reinforced by acute senses of kin recognition, and between queens within colonies (see, for example, Ratnicks 1988). Workers and queens also dispute egg-laying; one encounters a spectrum from species in which workers regularly produce eggs, to those in which workers become fecund only after a queen dies, and those in which the workers are permanently sterile. Indeed, the workers of slave-making ants establish dominance hierarchies among themselves (Franks and Scovell 1983).

Most ideas about the evolution of eusociality have centered on the sterile caste as an alternative strategy of reproduction that maximizes the sterile individual's contribution to the inclusive fitness of its genes (West Eberhard 1981). This could occur where there is strong local resource competition among close relatives for some resource essential to reproduction, and this competition can be minimized by specialization of function, as between egg-laying and food gathering/brood tending.

Inclusive fitness arguments have dominated thinking about the evolution of eusociality since William D. Hamilton (1964, 1972) pointed out the curious asymmetry of genetic relationship within colonies of social hymenoptera resulting from their haplodiploid breeding system. Remember that males are haploid and themselves have no father; females are diploid offspring of two parents. Therefore, of the genotype of a female wasp, a male sibling contains only one-quarter of the genes identically by descent (none of the male parent, and one-half of the female parent), whereas a female sibling contains three-quarters of the genes identically by descent (the entire set of the father and half the set of the mother) (Figure 30–18). Because the genetic relationship between females and either their sons or daughters is one-half, Hamilton reasoned that a female wasp should prefer to raise sisters than to raise its own offspring. This should greatly favor the evolution of eusociality and worker sterility.

Richard D. Alexander (1974) argued against Hamilton's idea on the grounds that female workers in bee and ant colonies produced roughly equal numbers of both male and female reproductives, whose average degree of genetic relatedness is one-half, and that termites have highly developed eusociality but the more usual sex-determining mechanism in

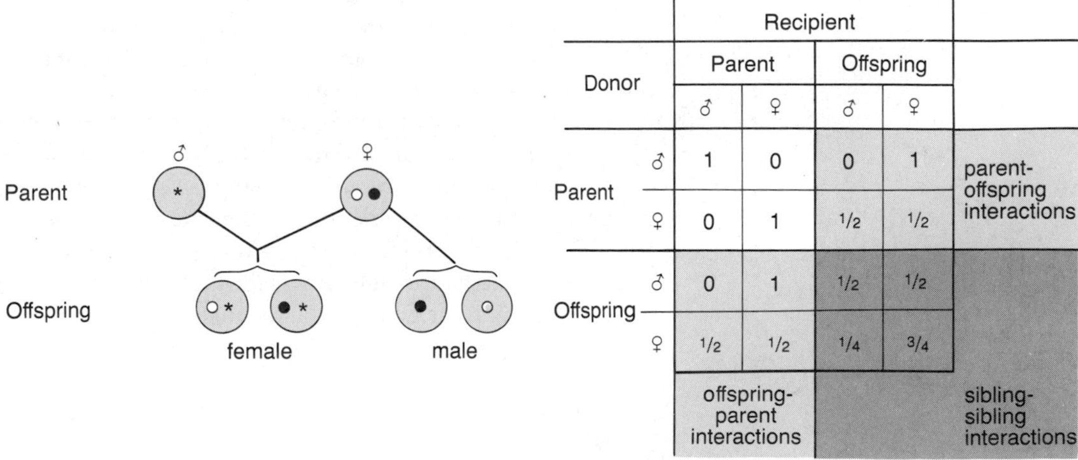

Figure 30–18
Coefficients of genetic relationship in a haplodiploid mating system.

which both males and females develop from fertilized eggs and are interrelated with genetic coefficients of correlation of one-half. Alexander preferred the idea that insect societies evolved through parental manipulation of broods to raise additional offspring rather than to lay their own eggs or go off on their own to found new colonies. Quite simply, any worker that attempted to mature sexually is killed by the queen. Only when a queen dies can workers become reproductives, and then there may be vicious fighting to determine which ascends the throne. As colonies increase in size, and the odds that an individual worker can become a queen following the death of its parent diminish, the best strategy of offspring may be to increase inclusive fitness by working, so long as the queen lives. In colonies of millions of individuals, the queen signals her presence by producing a distinctive and reassuring odor, or pheromone.

SUMMARY

Selection imposed by behavioral interactions with members of one's family and with unrelated individuals within one's population provide the evolutionary basis for social behavior.

1. Territoriality is the defense of an object or area from intrusion by other individuals. Territories are maintained when the resources gained are rewarding and defensible.

2. Dominance hierarchies order social groups by rank, which is established by direct confrontation. Because rank is generally respected, dominance relationships reduce conflict within the group.

3. Depending on resource dispersion and population density, populations may switch between territorial and dominance organization.

4. Living in large social groups may benefit the individual through better detection of, and defense against, predators, and through more efficient location of food. Groups may form to the extent that these advantages outweigh competition between group members.

5. Pinyon jays provide an example of a highly socialized species. They live in large, socially integrated groups, within which individuals altruistically benefit others by posing as sentinels, communally feeding offspring, and storing pine nuts in communal caches.

6. Isolated acts of social behavior involve a donor and recipient. When both benefit, the behavior is termed cooperation or mutualism; when the donor benefits at a cost to the recipient, the behavior is selfish; when the recipient benefits at a cost to the donor, the behavior is altruistic.

7. Game theory analysis shows that altruistic behavior is difficult to maintain in a population when altruists can be taken advantage of by selfish individuals.

8. The presence of apparently altruistic acts in populations has variously been explained by group selection, kin selection, and reciprocation.

Group selection has been largely dismissed because it is weak and because the requisite group structure has been difficult to identify in nature.

9. Kin selection arises because, when an individual interacts with a relative, it affects the fitness of a portion of its own genotype inherited directly from a common ancestor. Inclusive fitness expresses the benefit (or cost) of a behavior to the donor plus the benefit or cost to the recipient, the latter adjusted by degree of relationship. For interactions between siblings, which are related by one-half, selection will favor any altruistic behavior whose cost to the donor is less than one-half the benefit to the recipient.

10. Reciprocal altruism is the exchange of altruistic acts between individuals. Such behaviors are selected when the cost to the donor is less than the benefit to the recipient. Because reciprocation is required, behavior must be conditioned upon previous experience with an individual. Moreover, reciprocation has little selective value when rare, and so its origination is difficult to envision.

11. The ability of kin selection to produce altruistic behavior has been argued in the case of warning calls in mammals, warning coloration in insects, and cooperative breeding in birds.

12. Parent and offspring may conflict over the optimum level of parental investment. All siblings are genetically equal in the eyes of their parents, but siblings are genetically related by only one-half. Therefore, individual offspring should prefer unequal parental investment in themselves even when parental fitness is reduced as a result.

13. The social insects (termites, ants, wasps, and bees) represent extended family groups in which most offspring are retained in the colony as sterile workers, increasing their mother's fitness by rearing reproductive siblings.

14. Insect societies most likely arose through direct parental repression of sexual maturation of the offspring. The inclusive fitness of sterile workers may, however, be as high as that of reproductives which leave the colony to breed on their own.

31
Coevolution

The organism evolves as a set of integrated, cooperative organs and activities because interactions between components of the organism-system determine the individual's fitness. Interactions between individuals within populations also influence the evolution of social behavior and other adaptations, and may lead to highly structured societies based on mutual cooperation. Extrapolating these phenomena to higher levels of ecological organization, one is led to ask whether biological communities have evolved properties of structure and function arising from interactions among their component species.

We have seen that one species may apply selection upon another, as a predator does upon its prey, and that this interaction may result in an adaptive response—perhaps protective coloration or the production of a defensive chemical. Such responses in the second species alter the environment of the first; adaptations selected among prey make the prey population more difficult to exploit on the whole, thereby applying pressure on the predator to modify its tactics or alter its diet. Evolutionary responses would also seem to link species involved in mutualistic interactions, which depend on specialized adaptations of each participant to respond appropriately to the other.

Reciprocal evolutionary responses between populations are refered to as coevolution, although, as we shall see below, there is considerable disagreement over what coevolution is. The importance of the topic in the minds of ecologists is emphasized by recent reviews and edited volumes (Gilbert and Raven 1975, Nitecki 1983, Futuyma and Slatkin 1984, Boucher 1985). Several issues are involved. First, do pairs of populations undergo reciprocal evolution, or do "coevolved" traits arise from the response of populations to general selective pressures exerted by a variety of other species, followed by ecological sorting out of species with compatible features? Second, are species organized into interacting sets based on their evolved adaptations, whether "coevolved" or not? And third, do such adaptations enhance such system properties as productivity of the biological community and its resistance to perturbation?

Coevolution is the interdependent evolution of two or more species having an obvious ecological interaction.

This definition (Lincoln et al. 1982) refers to a process of mutual selection and adaptive response within a restricted set of species leading to properties that are unique to the system. Coevolution is more frequently ascribed to apparent matching of adaptations between pairs or small groups of species, without evidence bearing upon the evolutionary history of the relationship itself. The logical difficulty of inferring coevolution from matching adaptations is revealed by considering adaptations of organisms to physical characteristics of the environment. Adaptation and environment are clearly matched, yet the physical environment does not respond adaptatively and the match between organism and environment cannot be called coevolution. Much of the controversy over what coevolution is has resulted from inferring a process from patterns that such a process, as well as others, might have produced.

Modern discussion of coevolution stems from a paper by Paul Ehrlich and Peter Raven (1964), entitled "Butterflies and Plants: A Study in Coevolution." The paper opens with an admonition that reciprocal evolution, generally ignored before then, holds a key to interpreting patterns of organic diversity:

> One of the least understood aspects of population biology is community evolution—the evolutionary interactions found among different kinds of organisms where exchange of genetic information among the kinds is assumed to be minimal or absent. Studies of community evolution have, in general, tended to be narrow in scope and to ignore the reciprocal aspects of these interactions. Indeed, one group of organisms is all too often viewed as a kind of physical constant. In an extreme example a parasitologist might not consider the evolutionary history and responses of hosts, while a specialist in vertebrates might assume species of vertebrate parasites to be invariate entities. This viewpoint is one factor in the general lack of progress toward the understanding of organic diversification.

Ehrlich and Raven devote most of the paper to the relationships of groups of butterflies specialized to feed on particular groups of plants, and speculate on how feeding preferences are based upon chemical characteristics, particularly defensive compounds, of the leaves. The patterns suggest coevolution; plants evolving new chemicals to improve their defenses, insects evolving detoxification mechanisms. This interpretation was accepted by most ecologists. Coevolutionary interpretations of predator-prey, host-pathogen, and competitive relationships, and, of course, mutualisms, became commonplace.

It is interesting to note that data similar to those set forth by Ehrlich and Raven had suggested a similar interpretation at an earlier time, but the idea did not take hold. C. T. Brues (1920), in an important paper on "The Selection of Food Plants by Insects, With Special Reference to Lepidopterous Larvae," recognized the same patterns of food plant specialization detailed by Ehrlich and Raven:

> If we examine the food-plants of the genera or higher groups of butterflies, we find that most of them exhibit well-marked preference for certain, usually related plants. . . . This must not be understood to mean that the individual species of insects affect indiscriminantly many or all members of the plant group, but that their normal food-plant or plants do not fall outside the group. . . . [T]he fixity of the instinct to feed on only certain kinds of plants is all the more extraordinary, for we cannot readily dismiss it as a physiological or nutritional necessity.

Brues then specifically addressed the possibility of coevolution:

> On account of the very close biological association between insects and plants in many ways it is true that the two have been mutually specialized until they have become highly modified in reference to one another, but this is not the case with food-plants, as no benefit ordinarily accrues to the plants and any idea of parallel evolution must be restricted to a development of undesirable attributes on the part of the plants and adaptations on the part of the insects to overcome such barriers to feeding.

The idea of undesirable attributes evolved as anti-herbivore defenses had been recognized many decades earlier by the German naturalist E. Stahl (1888; translated by G. S. Fraenkel 1959):

> We have long been accustomed to comprehend many manifestations of the morphology [of plants], of vegetative as well as reproductive organs, as being due to the relations between plants and animals, and nobody, in our special case here, will doubt that the external mechanical means of protection of plants were acquired in their struggle [for existence] with the animal world. The great diversity of mechanical protection does not appear to us incomprehensible, but is fully as understandable as the diversity in the formation of flowers. In the same sense, the great differences in the nature of chemical products, and consequently of metabolic processes, are brought nearer to our understanding, if we regard these compounds as means of protection, acquired in the struggle with the animal world. Thus, the animal world which surrounds the plants deeply influenced not only their morphology, but also their chemistry.

Brues (1922) added the possibility that animals might change in response to plant defensive adaptations, Fraenkel (1959) drove home the defensive nature of exotic plant chemicals, which were being discovered and characterized in great number at the time, and Ehrlich and Raven (1964) placed the

Figure 31 – 1
The thorns of *Acacia hindsii*, like those of *A. cornigera*, are greatly enlarged and have a soft pith that the ants excavate for nests. (Courtesy of D. H. Janzen; from Janzen 1966.)

system in the then new context provided by the merging of ecology and evolutionary biology during the late 1950s and early 1960s.

Mutualisms provide convincing examples of coevolution.

The best illustrations of what appear to be coevolved traits come from obligate mutualisms in which two species are inextricably bound through mutual dependence (Boucher et al. 1982, Janzen 1985). Daniel Janzen's (1966, 1967) study of interdependence between certain kinds of ants and swollen-thorn acacias in Central America is exemplary. The acacia plant provides food and nesting sites for ants in return for protection that the ants provide from insect pests. The bull's-horn acacia *(Acacia cornigera)* has large horn-like thorns with a tough woody covering and a soft pithy interior (Figure 31 – 1). To start a colony in the acacia, a queen ant of the species *Pseudomyrmex ferruginea* bores a hole in the base of one of the enlarged thorns and clears out some of the soft material inside to make room for her brood. In addition to housing the ants, the acacias provide food for the ants in nectaries at the bases of their leaves, and in the form of nodules, called Beltian bodies, at the tips of some leaves (Figure 31 – 2). As the colony grows, more and more of the thorns on the plant are filled; in return, the ants protect the plant from insect pests. A colony may grow to more than a thousand workers within a year, and eventually may have tens of thousands of workers. At any one time, about a quarter of the ants are outside the nest actively gathering food and defending the plant against herbivorous insects. The relationship between *Pseudomyrmex* and *Acacia* is obligatory: neither the ant nor the acacia can survive without the other. Other ant-acacia associations are facultative. That is, the ant and the acacia can co-occur to mutual benefit, but they can both exist independently as well. Species of acacia that lack the protection of ants altogether frequently produce toxic compounds that defend their leaves against herbivores (Rehr et al. 1973).

The mutualism between ants and acacias has been accompanied by adaptations of both parties to increase the effectiveness of the association. For example, *Pseudomyrmex* is active both night and day, an unusual trait for ants, and thereby provides continuous protection for the acacia. In a similar adaptive gesture, the acacia retains its leaves throughout the year, and thereby provides a continuous source of food for the ants. Most related species lose their leaves during the dry season.

To test the influence of ants on the growth and survival of acacia plants, Janzen kept ants off new acacia shoots and compared their growth to shoots that had ants. After 10 months, the shoots lacking ants weighed less than one-tenth those with intact ant colonies, and produced fewer than half the number of leaves and a third the number of swollen thorns. The mutual benefits of the ant-acacia relationship and the highly specialized adaptations of both the plants and the ants provide a strong case for coevolution based upon a long evolutionary association between the two species.

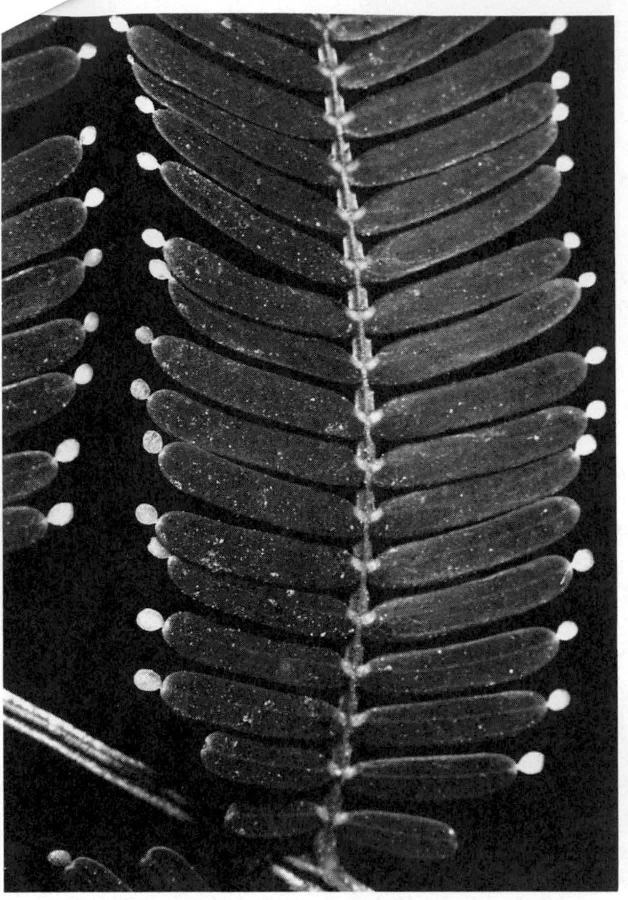

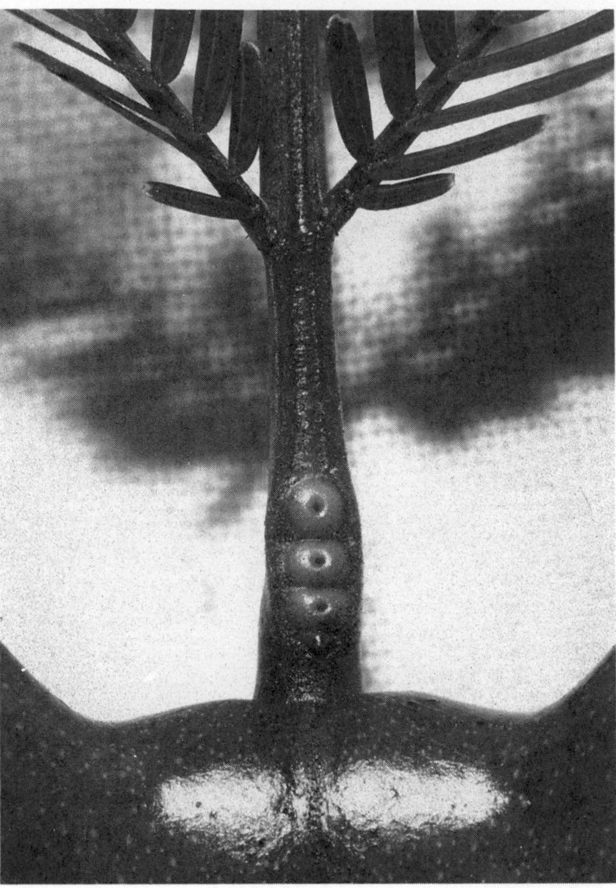

Figure 31 – 2
The leaves of *Acacia collinsii*, like those of *A. cornigera*, provide ants with food in the form of Beltian bodies at the tips of leaflets *(left)* and nectaries at the leaf base *(right)*. (Courtesy of D. H. Janzen; from Janzen 1966.)

A similar situation in which ants protect aphids and leafhoppers from predators and harvest the nutritious "honeydew" that they excrete is more difficult to interpret (Way 1963, Buckley 1987). One such system involving aphids and leafhoppers on ironweed *(Veronia noveboracensis)* in New York State has been studied experimentally by Catherine M. Bristow (1984). Aphids *(Aphis)* are small, sedentary, and form dense colonies on the inflorescences. Also occurring on ironweed is the larger membracid (leafhopper) *Publilia,* which sucks plant juices from the leaves. These insects are tended by three species of ants. One, in the genus *Tapinoma,* is tiny (2 – 3 mm) but abundant. The other two *(Myrmica)* are larger (4 – 6 mm) and more aggressive, but less common. The two genera of ants rarely co-occur on the same plant.

The presence of *Tapinoma* greatly enhances survival of aphid colonies but has less effect on the survival of leafhoppers. The larger *Myrmica* offers substantial protection to leafhoppers but is less effective in warding off predators of aphids. Where Bristow excluded both species of ants, predators were more numerous; where she added the predatory larvae of ladybird beetles, *Myrmica* and, to a lesser extent, *Tapinoma* effectively reduced their predation on leafhoppers.

The system has all the elements expected of coevolution, but it is not clear that the adaptations of the ant and homopteran participants evolved in response to each other. Most insects that suck plant juices produce large volumes of excreta from which they either do not or cannot extract all the nutrients. Thus honeydew production may reflect diet rather than being an adaptation to encourage protection by ants. Ants are voracious generalists, likely to attack any insect they encounter. Hence no special adaptation may be required to confer benefit on the aphids and leafhoppers upon whose excreta they also feed. The fact that the different genera of ants more effectively protect different honeydew sources may simply reflect their different sizes and levels of aggression, likely evolved in response to unrelated environmental factors.

Why don't ants eat the aphids and leafhoppers they tend? Perhaps this restraint is an evolved character of ants which facilitates the ant-homopteran mutualism. It may even have arisen as an extension of the common ant behavior of defending plant structures that produce nectar—flowers or specialized nectaries (for example, Figure 31–2). The point has not been addressed experimentally in this system, but a similar situation in the Cape region of South Africa highlights the importance of particular adaptations of ants, whether they are coevolved or not, to maintaining an ant-plant mutualism. There, many species of plants in the family Proteaceae have seeds with fleshy, edible, attached structures, called elaiosomes as illustrated for some Australian plants in Figure 31–3. Foraging ants pick up seeds and transport them to their underground nests, where the elaiosomes are eaten. The seeds themselves, which the ants cannot eat, are then discarded either in underground chambers or in refuse heaps on the surface. In many regions, these disposal sites are suitable for seed germination and seedling establishment (see Berg 1975, Culver and Beattie 1978). But in the fynbos (brushy, chaparral-like habitat), germination of many species of plants occurs only after fires have swept the habitat, and the only seeds that germinate are those stored by ants in underground nest chambers.

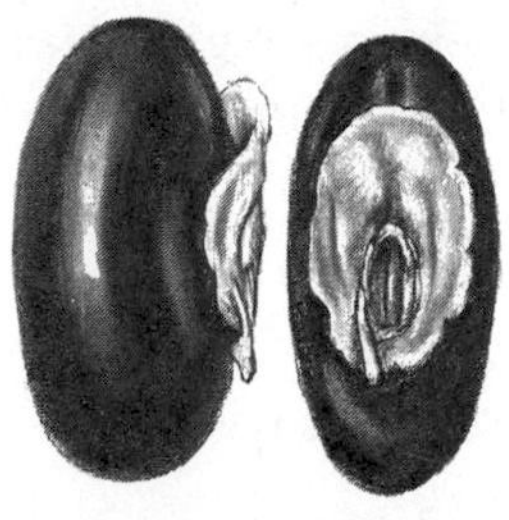

Figure 31–3
Ant-dispersed seeds of two Australian plants, *Kennedia rubicunda (top)* and *Beyeria viscosa (bottom)*, showing the edible, light-colored appendage (elaiosome) that attracts ants. (After Berg 1975.)

Recently, the Argentine ant *Iridomyrmex humilis* has invaded areas of fynbos shrublands and displaced many of the less aggressive native ants. *Iridomyrmex* differs from the native ants in not storing seeds within its nests; the elaiosomes are removed on the surface where the seeds are dropped. W. Bond and P. Slingsby (1984) found that germination of one species of *Mimetes* following fire was drastically reduced in areas invaded by *Iridomyrmex* (Figure 31–4). With the continued persistence of the Argentine ant, it is likely that much of the native Cape flora will disappear as underground seed reserves are depleted. In a similar case, Stanley Temple (1977) has suggested that the virtual extinction of the tree *Calvaria major* on the island of Mauritius followed upon the extinction more than 300 years ago of the dodo bird, the only native species capable of effectively dispersing *Calvaria*.

Clearly, particular adaptations of ants influence their effectiveness as seed dispersers. But these adaptations may be evolutionarily independent of

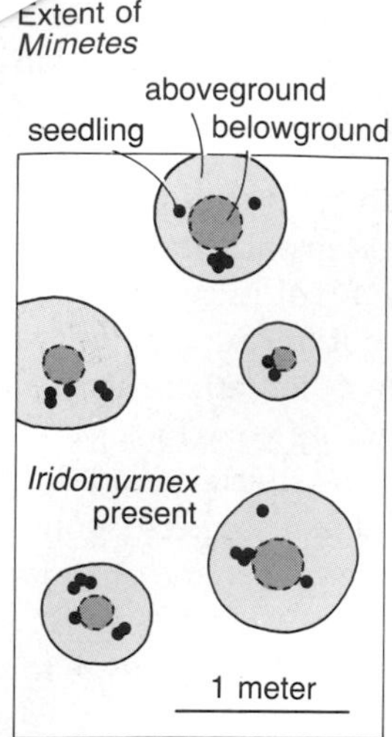

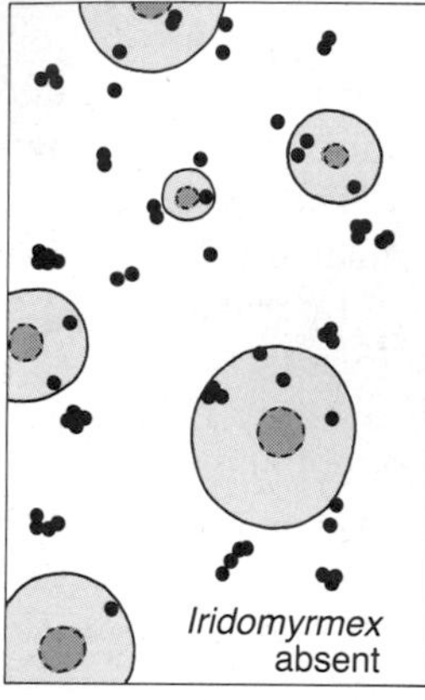

Figure 31–4
Seedling dispersion in *Mimetes cucullatus* populations after a burn in the presence of *Iridomyrmex (top)* and in its absence *(bottom)*. The extent of aboveground and below ground parts of mature plants is indicated. (From Bond and Slingsby 1984.)

the plants themselves. The ant-dispersers are diet generalists and their food-caching behavior likely evolved for reasons unrelated to their role as seed-dispersers of *Mimetes* and its relatives. The plants may have merely evolved to take advantage of this fortuitous element in their environment without any reciprocal evolution on the part of the ant. Whereas some mutualisms, such as that between ant and acacia described above, clearly involve specialized adaptations of both parties and seem to represent cases of coevolution, many mutualisms may involve more serendipitous arrangements.

Returning to the problem of feeding relationships of herbivores on plants, and of parasites on hosts, Janzen (1980) questioned the usual coevolutionary interpretation of defensive behaviors, structures, and chemical, and their circumvention by consumers. He suggested that the defenses of plants and hosts were not usually selected by the herbivores and parasites that can circumvent them. Suppose, for example, that plant P is eaten by herbivore H. Mutation P* arises in the plant population, which results in the production of a substance toxic to H. P* quickly increases relative to P, and the population of H declines. Coevolutionists would suppose that H might respond by mutation H*, whose bearers rendered the toxin ineffective by metabolizing or sequestering it. The result would be a system of P* and H* whose populations were reciprocally adapted to one another.

Janzen suggested alternatively that mutation P* might be circumvented by some other species, say F, already capable of handling the new defense, perhaps even recognizing a potential food plant by its presence. Therefore, rather than leading to highly coevolved groups of small numbers of species, evolutionary responses to herbivores and plant defenses might instead lead to a reshuffling of the feeding relationships within a community or even the invasion of the community by new species.

Plant-pathogen systems provide genetic models for coevolution.

Plant geneticists have developed strains of domestic crops, such as flax and wheat, that are resistant to particular genetic strains of various pathogens, such as rusts (teliomycetid fungi). The crop strains differ from one another by a few, perhaps single, genetic changes that make them either susceptible or resistant to infection by particular strains of rust (Williams 1975; Bennetzen et al. 1988). Over the course of crop improvement programs, when new strains of rust have appeared, either by mutation or by immigration from other areas, crop geneticists select new resistant strains of the crop by exposing experimental populations to the pathogen.

Mode (1958) used the crop-rust system as the basis for the first explicit genetic model of coevolution. He prefaced his work with the basic premise of coevolution:

> it seems reasonable to assume . . . that such obligate parasites as the rust, smut, and mildew fungi have evolved in association with their hosts. The

> genetic system of host and parasite, therefore, [has] very likely been established in response to two types of opposing selection pressures, namely, the selection pressure exerted on the host by the parasite, and the selection pressure exerted on the parasite by the host.

Mode's model assumed that virulence factors in the rust and resistance factors in flax were alleles of single genetic loci whose fitness relationships exhibited strong interactions between the genotypes of the rust and flax (Table 31–1). Mode found that, providing certain conditions of fitness relationships were met,

> a host-pathogen system operating under complementary genetic systems of the host and parasite will eventually reach a state of stable equilibrium [which is] advantageous to both the host and parasite. In the first place, a stable pathogen population solves the host's problem of maintenance of resistance to disease, and secondly the pathogen is able to survive without eliminating its host.

The stable equilibrium achieved in Mode's model could easily be upset by the appearance of a new virulence gene in the pathogen or resistance gene in the host. Hence stability in the coevolved system depends upon constancy of genotypes within each population.

Perhaps a more usual situation is the appearance of new genotypes either by migration or mutation, creating continual evolutionary flux in such a system. This seems to have been the case in the rust-wheat system described by G. J. Green (1975) in which new virulence genes of the rust *Puccinia graminis* appeared from time to time and swept through the population (Figure 31–5). Genetic races of wheat rust are characterized both by physiological characteristics and by virulence when tested on lines of wheat containing different resistance alleles. As shown in Table 31–2, most of the virulence strains within a single physiological race of the rust differ by only one gene and it is possible to outline a plausible evolutionary relationship

Table 31–1
A simple model for coevolution between resistance genes of a host (R) and virulence genes of a pathogen (A)

		Host		
		R_1	R_2	frequency
Pathogen	A_1	R	S	x
	A_2	S	R	$1-x$
	frequency	p	$1-p$	

Each gene locus has two alleles. The particular combination of host and pathogen alleles determines whether the host is resistant (R) or susceptible (S).
If the phenotype is R, the host fitness is 1 and the pathogen fitness is 0, and vice versa for phenotype S. The fitnesses of each allele depend on the allele frequencies in the other population: $W(A_1) = 1 - p$; $W(A_2) = p$; $W(R_1) = x$; $W(R_2) = 1 - x$.
Source: Mode 1958.

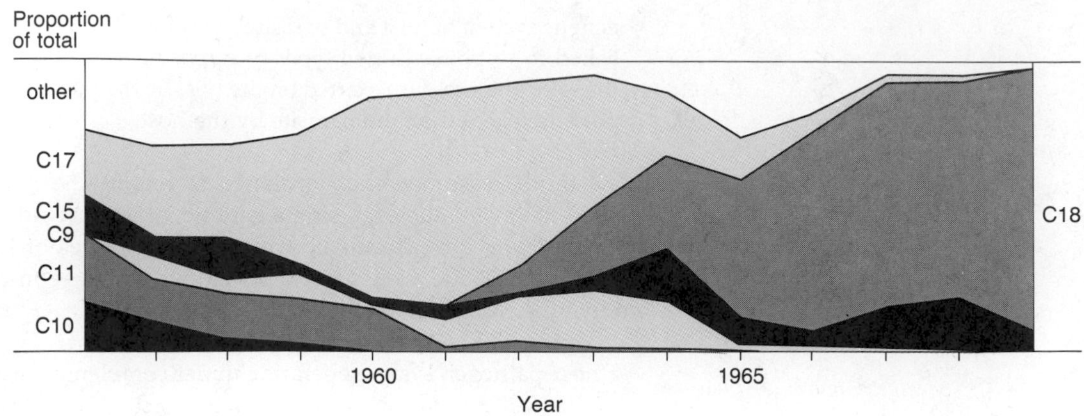

Figure 31–5
Relative proportions of virulence genes in the rust *Puccinia graminis* infecting Canadian wheat. (From Green 1975.)

between the races. For example, strains 1 and 24 differ only by virulence on wheat with resistance gene 9b; strains 43 and 20, although in different physiological races, differ only in their virulence with respect to resistance gene 6.

The rust-wheat system contains the essential element of Mode's concept of coevolution, namely a strong interaction between the fitnesses of

Table 31–2
Physiological races of wheat rust *Puccinea graminis* and virulence formulas based on the ability of strains to infect wheat with certain resistance genes

Physiological race	Virulence formula	Resistance gene 5	6	7a	8	9a	9b	10	11
11	C12	−	+	+	−	+	+	+	+
	C56	−	+	−	+	+	+	−	+
	C20	−	−	+	+	−	−	−	+
32	C21	−	−	−	−	+	−	−	+
	C22	−	−	−	−	+	−	−	−
	C32	−	−	−	−	+	+	−	+
	C34	−	+	+	−	+	+	−	+
	C43	−	+	+	+	−	−	−	+
17	C1	+	+	+	−	+	+	+	+
	C24	+	−	+	−	+	+	+	−
	C29	+	+	+	−	+	−	+	+
29	C3	+	+	−	−	+	−	−	+
	C5	+	−	−	−	+	+	−	+
	C6	+	−	−	−	+	+	−	+
	C30	−	−	−	−	+	+	−	−

Note: + indicates lack of virulence, hence effective host genes against a particular strain of rust.
Source: Green 1975.

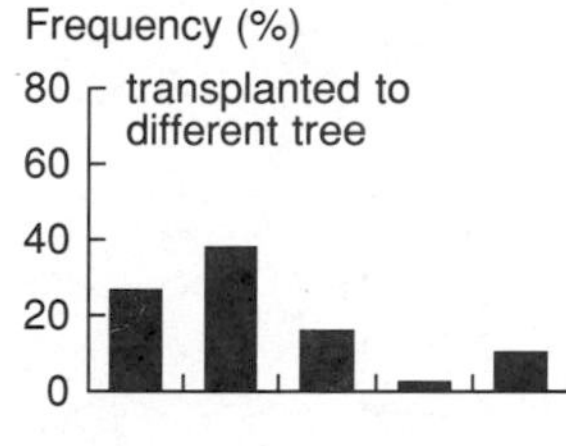

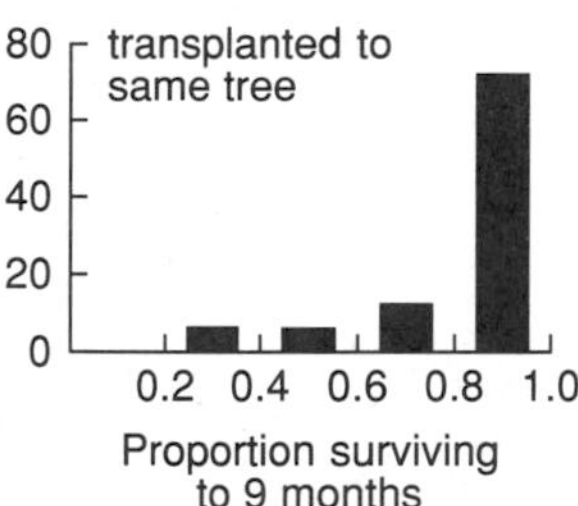

Figure 31 – 6
Black pineleaf scale on needles of ponderosa pine illustrating the damage caused by feeding. The survival of individual scale insects, which depends on adaptation of localized populations to the genotypes of particular trees, decreases markedly when scales are transplanted to different trees. (Courtesy of D. N. Alstad and G. F. Edmunds, Jr.; from Edmunds and Alstad 1978).

genotypes of the host and those of the pathogen. Rather than going to equilibrium, the system is kept in flux by the introduction of new virulence genes in the rust and, perhaps, by new resistance genes in the wheat, although the latter are pretty much controlled by plant geneticists.

Genotype-genotype interactions have been found in several natural systems and may turn out to be the rule in populations of plants and herbivores, or hosts and pathogens. George F. Edmunds and Donald N. Alstad (1978, 1981, 1983) demonstrated that variation between trees in the defenses of ponderosa pines were matched by variation in genotypes of scale insects that infest them (Figure 31 – 6). The scales are extremely sedentary, with so little migration from tree to tree that local populations (demes) on individual trees have evolved independently of those on other trees. This local adaptation is revealed when scales are transferred both between trees and between branches within the same tree. The survival of scales after inter-tree transplants is greatly reduced compared to control transfers within the same tree. It is reasonable to assume that the differences between trees and demes of scales are genetic, hence this represents a case of genotype-genotype interaction. C. Wiklund, Jr. (1981) and Michael Singer (1983) demonstrated intraspecific variation in preferences of lepidoptera among different species of host plants and Sara Via (1984) has shown genotype-host plant interactions in growth performance of larvae of the agromyzid fly *Liriomyza sativae* on cow pea and tomato. Hence the genetic background for coevolution seems to be in place.

Variation in plant defensive chemistry has a genetic basis.

Differences in the defensive chemicals of plants can be related to genetic changes, particularly when the pathways of biochemical synthesis and responsible enzymes are known. May Berenbaum (1978, 1981, 1983) has placed elements of the relationship between certain butterflies and their umbelliferous host plants in the context of coevolution. Umbellifers produce many noxious chemicals, among the most prominent of which are the furanocoumarins. The biosynthetic pathway (Figure 31 – 7) leads from para-coumaric acid, which, being a precursor of lignin, is found in virtually all plants, to hydroxycoumarins, such as umbelliferone, and then to furanocoumarins. The last include linear and angular forms, which are produced directly from hydroxycoumarins by different enzyme reactions. As one proceeds down the biosynthetic pathway from *p*-coumaric acid to hydroxycoumarins, and linear or angular furanocoumarins, toxicity increases and occurrence among plant families decreases. Hydroxycoumarins possess some biocidal properties; linear furanocoumarins bind with pyrimidine bases and interfere with DNA replication in the presence of ultraviolet light; angular furanocoumarins interfere with growth and reproduction quite generally, although the mechanisms of action have not been detailed.

Para-coumaric acid is widespread among plants, occurring in at least 100 families; Berenbaum (1983) lists only 31 families in which hydroxycou-

lignin

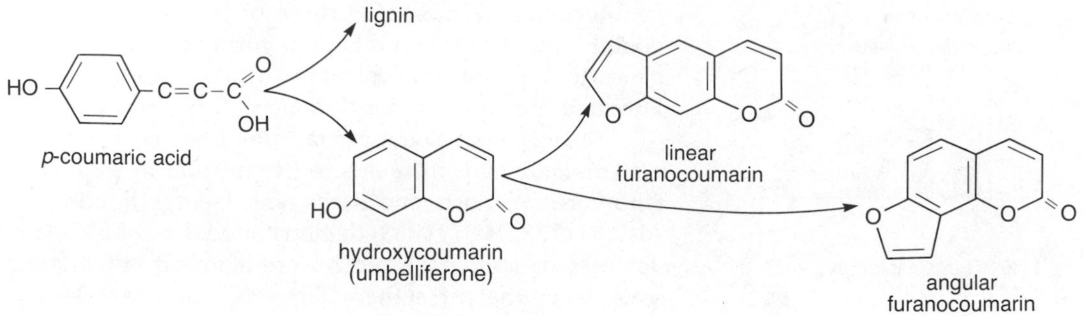

HO–C₆H₄–C=C–C(=O)OH

p-coumaric acid

linear furanocoumarin

HO

hydroxycoumarin (umbelliferone)

angular furanocoumarin

Figure 31–7
Biosynthetic pathway of furanocoumarins.

marins have been found. Linear furanocoumarins (LFCs) are restricted to 8 plant families and are widely distributed only in 2—Umbelliferae (parsley family) and Rutaceae (citrus family). Angular furanocoumarins (AFCs) are known only from 2 genera of Leguminosae (pea family), and 10 genera of Umbelliferae.

Among species of herbaceous umbellifers in New York, some (especially those growing in woodland sites with low levels of UV light) lack furanocoumarins, others have linear furanocoumarins only, and some have both linear and angular furanocoumarins. From a survey of the herbivorous insects collected from these species, Berenbaum (1981) concluded that host plants containing angular and linear furanocoumarins were attacked by more species of insects than found on plants with only linear furanocoumarins, or none; that the herbivores on AFC/LFC plants tended to be extreme diet specialists, most having been found on no more than 3 genera of plants; and that these specialists tended to be abundant compared to the numbers of the few generalists found on AFC/LFC plants and compared to levels of any herbivores on either LFC plants or on umbellifers lacking furanocoumarins.

Although linear and, especially, angular furanocoumarins are extremely effective deterrents to most species of herbivorous insects, some genera that have evolved to tolerate these chemicals have become successful specialists. Berenbaum (1983) makes a strong case for coevolution here in Janzen's restricted sense. The taxonomic distribution of hydrocoumarins, linear furanocoumarins, and angular furanocoumarins across host plants suggests that plants containing LFCs are a subset of those containing hydroxycoumarins, and those containing AFCs are an even smaller subset of those containing LFCs. This is consistent with an evolutionary sequence of plant defenses progressing from hydroxycoumarins to LFCs and AFCs. Furthermore, insects specialized on plants containing LFCs belong to groups that characteristically feed on plants containing hydroxycoumarins, and those specialized on AFCs have close relatives that feed on plants containing LFCs. Although phylogenetic relationships and the history of host-plant utilization have not been matched, the taxonomic distributions of insects across host plants produce patterns that would be expected of a coevolved system.

Diffuse coevolution may occur broadly within ecological communities.

The controversy over whether coevolution takes place within small groups of organisms locked into either evolutionary struggles or cooperative ventures is giving way to a more generalized view of coevolution that recognizes far-reaching and overlapping evolutionary relationships between the species within a biological community (Fox 1981, Howe 1984). Although both members of mutualist and antagonist species pairs inevitably exert selection on each other, the general overlapping of species relationships into webs of interaction undoubtedly results in corresponding webs of adaptation. Speaking about the evolution of mutualisms, including the interactions of plants with their pollinators and seed dispersers, Henry F. Howe (1984) states that "For coevolution to occur, organisms must have distinctive selective effects on each other. For obligate mutualism to evolve, effects must be both unique and of long duration." Howe suggests that most mutualisms are "diffuse," by which he means that adaptation occurs in the context of a matrix of species interactions.

This expanded view of coevolution leads naturally into the final part of this book, which deals with the ecology of entire communities (assemblages of species). Before moving on, however, I would like to recount briefly one case of an obligate mutualism, clearly involving coevolution, which emphasizes the strong interdependencies that can arise in nature and the problems of arriving at a satisfactory mutual "agreement" between coevolved parties.

The yucca moth is a coevolved pollinator of the yucca.

The curious pollination relationship that occurs between species of yucca plants (*Yucca,* in the lily family) and moths of the genus *Tegeticula* (Figure 31–8) was first described by C. V. Riley nearly a century ago (1892) and has been considerably elaborated since (Powell and Mackie 1966). The moth enters the yucca flower and deposits 1 to 5 eggs on the ovary. Later, when the eggs hatch, the larvae burrow into the ovary, where they feed on the developing seeds. But after the moth has laid her eggs, she scrapes pollen off the anthers in the flower and rolls it into a small ball, which she grasps with specially modified mouthparts. She then flies to another plant, enters a flower, and proceeds to place the pollen ball onto the stigma of the flower before laying another batch of eggs.

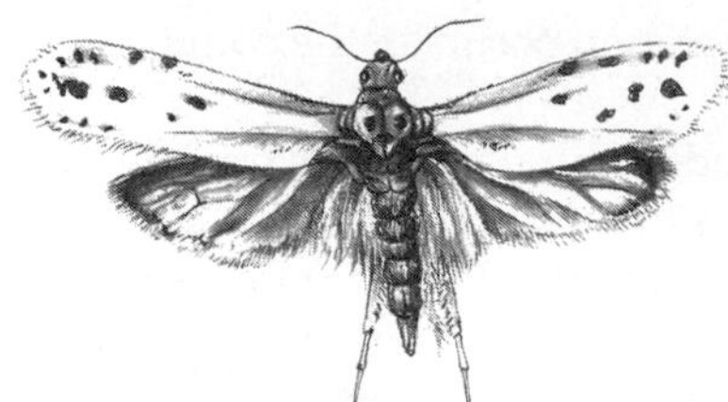

Figure 31–8
The mohave yucca *(Yucca shidigera)* and the yucca moth *(Tegeticula maculata)* that pollinates it. (After Powell and Mackie 1966.)

The relationship between the moth and the yucca is obligatory. *Tegeticula* can grow nowhere else, *Yucca* has no other pollinator. In return for pollinating its flowers, the yucca seemingly tolerates the moth larvae feeding on its seeds, but the extent of this loss of potential reproduction is small, rarely exceeding 30 per cent, and more nearly half that value on average, in *Yucca whipplei* (Powell and Mackie 1966). *Yucca* and *Tegeticula* are specialized with respect to each other. The moth has a highly idiosyncratic pollination behavior. The yucca is specially adapted to make the moth's behavior effective: the pollen is sticky and can easily be formed into a ball and the stigma is specially modified as a receptable.

A puzzling aspect of the relationship is the restraint the moth exercises in laying a small number of eggs in each flower. Over the short term, it would seem that moths laying larger numbers of eggs per flower might have higher individual fitnesses, even though such behavior over the long term might lead to extinction of the yucca. Interpretation of this system poses the same difficulty as altruistic behaviors within populations, without th tion of falling back upon kin selection. One possibility is that yuccas effectively regulate the relative fitness of moths laying different numbers eggs per flower by selective abortion of developing fruits that are too highl infested by moth larvae to produce seed. Selective abortion of insect-damaged fruits is well known (Janzen 1971, Stephenson 1981), and yuccas are known to possess mechanisms for fruit abortion (Udovic and Aker 1981, Aker 1982). The obvious experiment of transferring large numbers of moth eggs to single flowers of *Yucca,* followed by measuring both fruit maturation and emergence of fully developed larvae, has not been tried. The particular evolutionary steps of moth behavior leading to the present system are lost in time, but the advantage to the moth of ensuring that the flower within which it lays its eggs is properly pollinated should be obvious. Most pollination relationships are neither so intricate nor so specialized as that of the yucca and moth. What "preadaptations" of *Yucca* and *Tegeticula* ancestors, or what accidents of history, started them along their coevolutionary pathway are considerations for the next generation of ecologists.

SUMMARY

1. Coevolution is the interdependent evolution of species that interact ecologically. The interactions may be antagonistic (consumer-resource) or cooperative (mutualism). Because each species in the coevolved pair is an important component of the environment of the other, changes in one select adaptive responses in the other, and vice versa.

2. The concept of coevolution originally developed around the evolution of defenses against herbivores by plants, and responses of herbivores allowing them to circumvent those defenses.

3. The most convincing demonstrations of coevolution involve cases of mutualism, such as the obligate interdependence of *Pseudomyrmex* ants and *Acacia.* The ant keeps the plant free of herbivores while the plant provides the ant with food and housing; both have adaptations of structure and behavior or phenology that promote the relationship.

4. Plant-pathogen systems (such as fungal rusts on crop plants) have provided genetic models of the coevolutionary process. Genes that increase the virulence of the rust select resistance genes in the plants, leading to a continual cycle of coevolutionary change.

5. Analysis of metabolic pathways shows the genetic (enzyme) steps in the development of toxic chemicals in plants. When variations in these pathways (and in the abilities of insects to detoxify the chemicals) are overlaid upon taxonomic relationships within each group, one can infer the evolutionary history of the plant-insect interaction.

6. The yucca moth–yucca interaction is an obligate mutualism in which the moth pollinates the plant but its larvae consume developing seeds. How the level of seed predation is controlled in this coevolved system presents a challenging problem.

op-
can
of

VII

Community Ecology

32

Concept of the Community

Every place on earth—each meadow, each pond, each rock at the edge of the sea—is shared by many coexisting organisms. These plants, animals, and microorganisms are tied one to another by their feeding relationships and other interactions, forming a complex whole often referred to as the biological community. Interrelationships within the community determine such functional attributes as the flow of energy and the cycling of elements within the ecosystem. These interrelationships also exercise feedback and control upon population processes, whereby they determine the relative abundances of organisms. Finally, interrelationships within the community select among genotypes and therefore influence the evolution of coexisting species.

These three types of interactions—which we may think of as thermodynamic, cybernetic, and evolutionary—have been discussed at length in earlier parts of this book. The mechanisms underlying these interactions have been dissected by observation and experiment, and they are reasonably understood in broad outline. But there is a set of related issues that have perplexed and polarized ecologists for decades, and for which no clear resolution is in sight. The common element underlying these issues is the century-old debate over the unity of the community. At one extreme is the frequently surfacing view that the community is a superorganism whose function and organization can be appreciated only by considering its place in nature as a whole entity. Common sense tells us that it would be absurd to ponder a kidney's function apart from that of the organism to which it belongs. Many have argued that it is equally pointless to consider the soil bacterium without reference to the detritus it feeds upon, its own predators, or the plants nourished by its wastes. The presence of each species can be understood only in terms of its contribution to the dynamics of the whole system. At the other extreme is the individualistic concept, which argues that community structure and function simply express interactions of the individual species comprising the local association, and do not reflect any organization, purposeful or otherwise, above the species level. The composition of the local community itself is determined in part by factors external to the contemporary, local scene.

Regardless of the subtle nature of the community, ecologists have described a number of community-level properties. The simplest is the number of species, or diversity. Early naturalists knew that more species lived in tropical localities than in temperate and boreal zones. Barro Colorado Island, a 16-square-kilometer island in Gatun Lake, Panama, supports 211 species of trees that grow to be taller than 10 meters (Croat 1978), more species than are found in all of Canada. High tropical diversity characterizes most groups of organisms, except those especially adapted to conditions unique to higher latitudes (Fischer 1960, MacArthur 1972, Brown and Gibson 1983). Further comparisons of communities have been based on numbers of species at each trophic level (that is, primary producers, herbivores, carnivores) and, within trophic levels, among different "guilds" (Root 1967) distinguished by method or location or foraging (for example, herbivores comprise leaf-eaters, stem-borers, root-chewers, nectar-sippers, bud-nippers).

Another community-level property is relative abundance. In most associations, a few species are abundant and many more are rare. That the distribution of species over abundance classes follows regular patterns was first appreciated by the Danish botanist Christen Raunkaier (1918; see Kenoyer 1927), who laid the empirical groundwork for continuing debate over the significance of these patterns.

That one may describe community properties and relate them to physical conditions of the environment does not argue for or against a holistic interpretation of community organization. By analogy to the organism, purposefully integrated communities should exhibit a strong community-environment correlation. But one would also expect such correlations among individualistic communities, because organisms and populations independently respond to the environment and to each other, and the properties of associations reflect the properties of their members.

We shall return to this issue of the regulation of community structure, particularly with respect to the diversity of species, over the next several chapters. First, however, we shall examine the properties of communities and the nature of change in communities in greater detail.

The community is an association of interacting populations.

The term community has been given a variety of meanings by ecologists. It usually is applied to a group of populations that occur together, but there ends any similarity among definitions. Throughout the development of ecology as a science, the term has often been tacked on to associations of plants and animals that are spatially delimited and that are dominated by one or more prominent species or by a physical characteristic (Shimwell 1971, Daubenmire 1968, Slobodkin 1961). One speaks of an oak community, a sagebrush community, and a pond community, meaning all the plants and animals found in the particular place dominated by its namesake. Used in this way, community is unambiguous; it is spatially defined and includes all the populations within its boundaries.

Ecologists also define communities on the basis of interactions among associated populations. This is a functional rather than descriptive use of the term. Association is sometimes used for groups of populations that occur in the same area without regard to their interactions; community will be used to denote an association of interacting populations.

Communities are difficult to delimit when interactions among populations extend beyond arbitrary spatial boundaries. The migration of birds between temperate and tropical regions link the communities in each area; within some tropical localities, as many as half the birds present during the northern winter are migrants. Salamanders, which complete their larval development in streams and ponds but pursue their adult existence in the surrounding woods, tie together aquatic and terrestrial communities, just as trees do when they shed their leaves into streams, thereby supporting aquatic, detritus-based food chains.

Community structure and function are manifestations of a complex array of interactions, directly or indirectly tying all members of a community together into an intricate web. The influence of a population extends to ecologically distant parts of the community through its competitors, predators, and prey. Insectivorous birds do not eat trees, but they do prey on many of the insects that feed on foliage or pollinate flowers. By preying upon pollinators, birds indirectly may affect the number of fruits produced, the amount of food available to animals that feed upon fruits and seedlings, and the predators and parasites of those animals. The ecological and evolutionary impact of a population extends in all directions throughout the trophic structure of the community by way of its influence on predators, competitors, and prey, but this influence dissipates as it passes through each successive link in the chain of interaction. Its impact similarly spreads through space and forward through time by way of the movements of individuals and the inertia of population processes. Because of this inertia, the present-day community also is imprinted with a record of the past.

Is there a natural unit at the community level of ecological organization?

Ecologists describe functional relationships among an association of species just as physiologists relate the various parts of the body. The analogy between community and organism is obvious, and well noted in the ecological literature, as we have seen. For example, V. E. Shelford (1931), in a paper entitled "Some Concepts of Bioecology," wrote: "It is an old practice to liken organisms to cosmic systems, and cosmic systems to organisms. Again in this case, it is convenient to liken the biome (plant-animal formation) to an amoeboid organism, a unit of parts, growing, moving, and manifesting internal processes which may be likened to metabolism, locomotion, etc., in an organism." He even drew parallels between certain processes of community change and wound healing.

Certainly the most influential person to espouse the organismic viewpoint was the American plant ecologist F. E. Clements (1916, 1936), who perceived communities as discrete units with sharp boundaries, to which he attributed a unique organization. Clements's view was reinforced by the conspicuousness of many dominant vegetation types. A forest of ponderosa pines, for example, appears distinct from the fir forests that grow in moister habitats and from the shrubs and grasses typical of drier sites. The boundaries between these community types are often so sharp as to be crossed within a few meters along a gradient of climate conditions. Some community boundaries, such as that between deciduous forest and prairie in the midwestern United States, or between broad-leaved and needle-leaved forest in southern Canada, are respected by most species of plants and animals.

An opposite view of community organization was held by H. A. Gleason (1926, 1939), who suggested that the community, far from being a distinct unit like an organism, was merely a fortuitous association of organisms whose adaptations enabled them to live together under the particular physical and biological conditions that characterize a particular place. A plant association, he said, was "not an organism, scarcely even a vegetational unit, but merely a coincidence."

Clements's and Gleason's concepts of community organization predict different patterns in the distribution of species over ecological an geographical gradients. On one hand, Clements believed that the species belonging to a community were closely associated with each other; the ecological limits of distribution of each species coincided with the distribution of the community as a whole. This type of community organization is commonly called a closed community. On the other hand, Gleason believed that each species was distributed independently of others that co-occurred in a particular association, an organization referred to as an open community. Such open communities have no natural boundaries; therefore, their limits are arbitrary with respect to the geographical and ecological distributions of their component species, which may extend their ranges independently into other associations.

Ecotones occur at sharp physical boundaries or where habitat-dominating growth forms change.

The structure of closed and open communities is depicted schematically in Figure 32–1. In the left-hand diagram, the distributions of species in each community are closely associated along a gradient of environmental conditions — for example, from dry to moist. Closed communities are natural ecological units with distinct boundaries. The edges of such communities, called ecotones, are regions of rapid replacement of species along the

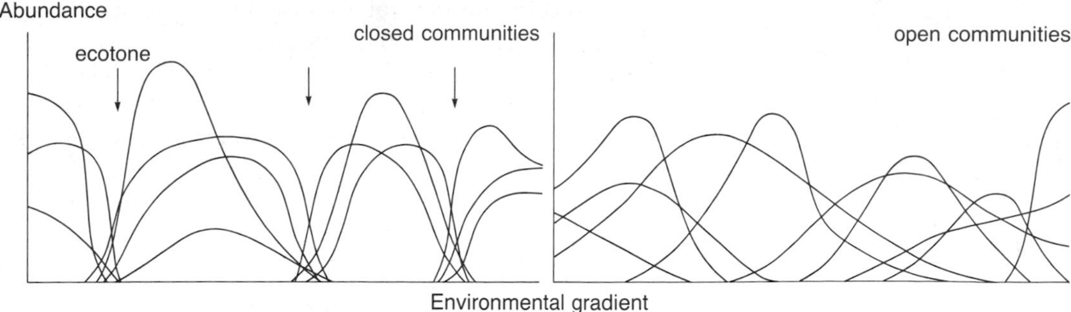

Figure 32–1
Hypothetical distributions of species organized into distinct assemblages (closed communities, *left*) or distributed at random along a gradient of environmental conditions (open communities, *right*). Ecotones between closed communities are indicated by arrows.

gradient. In the right-hand diagram, species are distributed at random with respect to each other, giving an open structure. We may arbitrarily delimit a "community" at some point, perhaps a dry forest community near the left-hand end of the moisture gradient, while recognizing that some of the species included are more characteristic of drier portions of the gradient, while others reach their greatest abundance in wetter sites.

The separate concepts of both open and closed communities have some validity in nature. We observe distinct ecotones between associations under two circumstances: first, when the physical environment changes abruptly —for example, at the transition between aquatic and terrestrial communities, between distinct soil types, or between north-facing and south-facing slopes of mountains—and second, when one species or life form so dominates the environment that the edge of its range signals the distributional limits of many other species.

A change in soil acidity often accompanies the transition between broad-leaved and coniferous forest. The decomposition of needles produces more organic acids than does the breakdown of leaves; furthermore, because needles tend to decompose slowly, a thick layer of organic duff accumulates at the soil surface. At the boundary between grassland and shrubland, or between grassland and forest, sharp changes in surface temperature, soil moisture, light intensity, and burning frequency result in many species replacements. The boundaries between grasslands and shrublands often are sharp because when one or the other vegetation type holds a slight competitive edge, it dominates the community (Schultz et al. 1955). Grasses prevent growth of shrub seedlings by reducing the moisture content of the surface layers of the soil; shrubs depress growth of grass seedlings by shading them. Fire evidently maintained a sharp boundary between prairies and forest in the midwestern United States (Borchert 1950). Perennial grasses resist fire damage that kills tree seedlings outright, but fires do not penetrate deeply into the moister forest habitats.

Sharp physical boundaries create well-defined ecotones. Such boundaries occur at the interface between most terrestrial and aquatic (especially marine) communities (Johannesson 1989, Figure 32–2) and where underlying geological formations cause the mineral content of soil to change abruptly. The ecotone between plant and associations on serpentine-derived soils and nonserpentine soils in southwestern Oregon is shown in more detail by the diagrams of soil minerals and occurrence of plant species

Figure 32–2
A sharp community boundary (ecotone) in the Bay of Fundy, New Brunswick, associated with an abrupt change in the physical properties of adjacent habitats. Seaweeds extend only to the high tide mark. Between the high tide mark and the spruce forest, waves wash soil from the rocks and salt spray kills pioneering land plants, leaving the area devoid of vegetation.

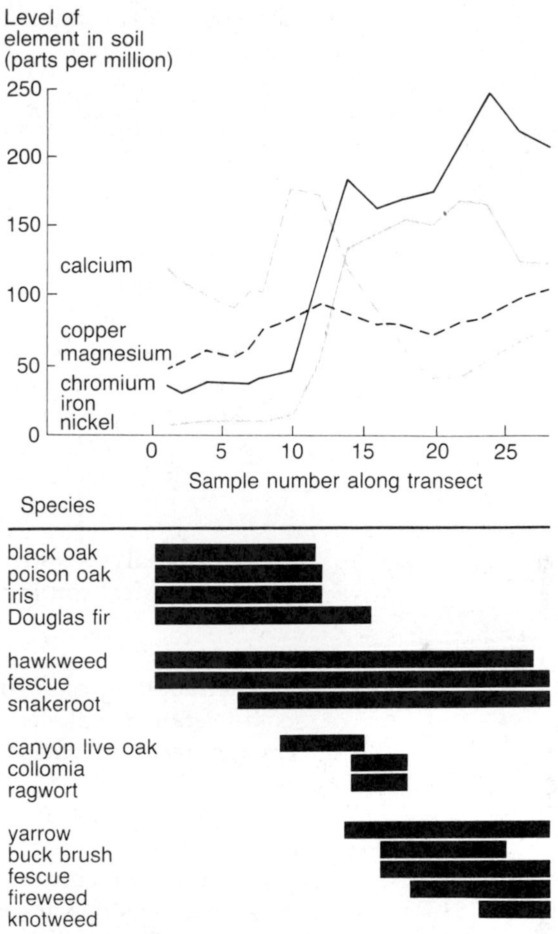

Figure 32-3
Changes in the concentration of elements in soil *(above)* and replacement of plant species *(below)* across the boundary between nonserpentine (samples numbers 1 to 10) and serpentine soils (sample numbers 18 to 28) in southwestern Oregon. The transect diagrammed here is somewhat atypical in that magnesium does not increase as abruptly as usual across the serpentine ecotone. (After White 1971.)

in Figure 32-3. Levels of nickel, chromium, iron, and magnesium increase across the boundary into serpentine soils; copper and calcium contents of the soil drop off. The edge of the serpentine soil marks the boundaries of many species that are either excluded from, or restricted to, serpentine outcrops. A few species are found only within the narrow zone of transition; others, seemingly unresponsive to variation in soil minerals, extend across the ecotone.

Plant ecologists have long recognized the influence of climate on plant associations. For example, Forrest Shreve (1936) described the chaparral-desert transition in Baja California in relation to moisture and freezing temperatures. Desert species, particularly the cacti, do not tolerate prolonged frost and quickly drop out north of the frost line; chaparral species drop out to the south within the transition zone owing to water stress. As Shreve concluded:

> plants of the desert are more sharply confined to their own formation than are the species and genera of the chaparral and other northern types of vegetation. This appears to be due to the fact that the only requirement for the long southward extension of a chaparral plant is the occurrence in the desert region of relatively moist habitats, however restricted in area, while the northward extension of a desert plant requires a well-drained soil, a high percentage of sunshine and freedom from freezing temperatures of more than a few hours' duration. These more exacting requirements are met only in close proximity to the edge of the desert or else in light soils or on steep south slopes near the sea.

Gradient analysis portrays the structure of natural communities along ecological continua.

The deciduous forests of eastern North America are bounded to the north by cold-tolerant needle-leaf forests, to the west by drought and fire- and drought-resistant grasslands, and to the southeast by fire-resistant pine forests. Within the region of their distribution, deciduous forests present a physiognomically uniform appearance. As a result of early botanical explorations, ecologists were aware that different species of trees and other plants occurred in different areas within the forest biome. According to Clements's closed-community viewpoint, the distinctive vegetation of each area represented a distinct community separated by sharp vegetational transitions from other communities. But as plant distributions were described in more detail, their associations were found to fit less and less well the closed-community concept; classifications of plant communities became more and more finely split until absurd levels of distinction were reached.

Out of this mounting chaos, there arose a new concept of community organization, referred to as the continuum. Within broadly defined habitats, such as forest, grassland, and estuary, populations of plants and animals gradually replace each other along gradients of physical conditions. The environments of the eastern United States form a continuum, with a north-south temperature gradient and an east-west rainfall gradient. Species of trees found in any one region—for example, those native to eastern Kentucky—have different geographical ranges, suggesting a variety of evolutionary backgrounds (Figure 32–4). Some species reach their northern limits in Kentucky, some their southern limits. Because few species have broadly overlapping geographical ranges, associations of plant species found in eastern Kentucky do not represent closed communities. Each species has a unique evolutionary history and present-day ecological position, with a variable degree of association with other species in the local community.

Figure 32–4
Geographical range limits of 12 species of trees found in plant associations in eastern Kentucky. (After Fowells 1965.)

A more detailed view of Kentucky forests would reveal that many of the tree species are segregated along local gradients of conditions. Some are found along ridge tops; others along moist river bottoms; some on poorly developed, rocky soils; others on rich organic soils. The species represented in each of these more narrowly defined associations might exhibit correspondingly closer ecological distributions, but the open community concept would still dominate our thinking about these associations.

Populations of species can be ordinated with respect to physical or derived gradients.

A first step toward resolving the problem of open versus closed communities was to devise methods of portraying the distributions of species along ecological gradients. One way in which this has been done is to plot the abundances of species along some continuous gradient of ecological conditions (Loucks 1962). This is often called gradient analysis (Whittaker 1967). In gradient analysis, closed community organization can be identified by the presence of sharp ecotones in species distributions, as indicated earlier in Figure 32–1. The gradient itself might be based on any number of physical variables, such as moisture, temperature, salinity, exposure, or light level. It is usually constructed by measuring both the abundances of species and the physical conditions at a number of localities and then plotting the abundances of each species as a function of the value of the physical condition. Study localities may be situated at regular intervals along a known physical gradient, such as that of temperature as it decreases up an elevation gradient.

A second method is to place communities along one or more artificial axes constructed from the degree of floristic or faunistic similarity between local associations. The mathematical procedures to accomplish this are together referred to as ordination (Gauch 1982). Because the positions of associations along the ordination axes depend upon their similarities and differences, closed communities would appear as distinct clusters of local associations, well-separated from others; local samples of an open community structure would be positioned more or less evenly throughout the ordination space.

An early, widely used ordination technique was referred to as polar ordination because its first step was to determine those communities representing opposite ends of the primary ordination axis. Polar ordination is really only a step away from gradient analysis because associations are selected from some previously perceived continuum of physical conditions, the end points of which—whether wettest versus driest or hottest versus coldest—were fairly obvious. Intermediate associations were placed along the gradient in accordance with their similarity to the associations at each of the poles of the gradient. Alternatively, the poles of the ordination axis were chosen as the two most dissimilar associations. In either case, the computations were relatively simple (Curtis and McIntosh 1951, Bray and Curtis 1957, Loucks 1962).

With the advent of large computers, ordination techniques grew in complexity and abstractness. The most advanced methods—principal components analysis, reciprocal averaging (correspondence analysis), and detrended correspondence analysis (Gauch 1982)—are similar in their ability to derive ordination axes without any preconception about gradients of physical conditions to which the abundances of plants and animals respond. For example, reciprocal averaging works by maximizing the correlation between species and samples. Table 32–1 is a species-by-sample incidence matrix, in which the abundance of each species of plant in each sample locality is recorded. The ordering of the samples across columns and of the species down rows are taxonomic, geographic, chronological, or arbitrary, but not ecological.

Reciprocal averaging simultaneously reorganizes the ordering of the samples and species along continuous scales so as to maximize the correlation (or correspondence) between the two, as shown in Figure 32–5. In effect, a continuous gradient has been extracted from the information in an incidence matrix.

Table 32–1
Composition of forests in the beech-maple forest region of the midwestern United States

	Percentage of trees in stand									
Species	A	B	C	D	E	F	G	H	I	J
Acer rubrum						8	19		9	
Acer saccharum	17	13		14	7	28	4	6		49
Carya ovata	6	6	7	5			3		6	
Fagus grandifolia	33	21	5	17	72	40	7			
Fraxinus americanus	3	2		7	5	1	8	7	5	4
Juglans nigra		1		10				4		
Liriodendron tulipifera	21	15	2	5	10	1	1			
Nyssa sylvatica	4				2	6	1			
Quercus alba	8	1	63	7	15	46	3	13	8	
Quercus borealis	5	2	18	2			8	7	21	19
Quercus macrocarpa								4	1	
Tilia americana		13		2				31	19	16
Ulmus americana		1		9			3	36	25	1

Not all columns sum to 100 per cent owing to minor species excluded from the table. Locations are: A–D, Turkey Run State Park, Indiana; E, Hueston's Woods, Oxford, Ohio; F, Canfield, Ohio; G, Graber Woods, Wayne County, Ohio; H–J, Harms Woods, Evanston, Illinois.
Source: Braun 1950.

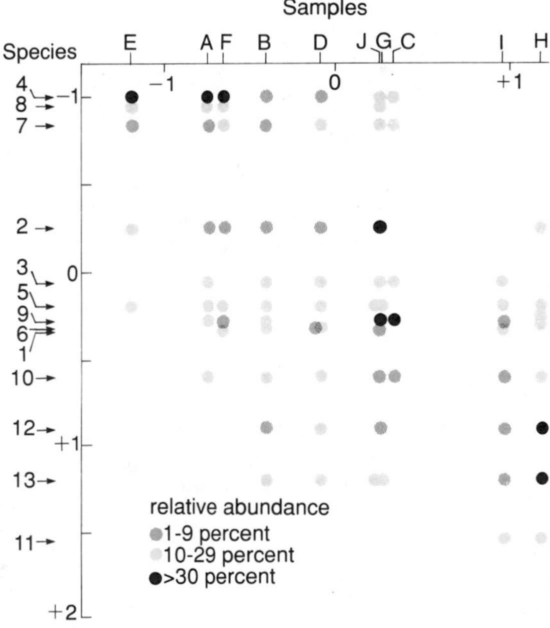

Figure 32–5
Reciprocal averaging ordination of species of trees and sample locations from the beech-maple forest region of eastern North America (data in Table 32–1). The species and samples are placed on derived axes so as to maximize the correlation between them.

Derived gradients reflect similarities and differences in species composition between sample localities. If the distributions of species are sensitive to ecological conditions, then the gradient may also reflect the underlying continuum of these conditions over which the sample localities are distributed. Reciprocal averaging, as well as other ordination techniques, may generate several axes, indicating a multidimensional continuum. It also scales both rows and columns of an incidence matrix; hence it ordinates the trees with respect to each other as well as the sample localities with respect to each other. As an ecological tool, ordination allows the investigator to relate the positions of sample localities in ordinated space to ecological measurements taken at the sample sites, whereby one may identify the physical conditions underlying the derived continuum. One may also relate the positions of trees in their ordinated space to various physiological, morphological, and life history characteristics, thereby revealing the adaptations and phenotypic responses of plants to physical gradients.

Communities appear to have predominantly open structures.

Because ordination is a relatively new technique, most studies of distributions along ecological continua have involved direct gradient analysis in which species abundance is portrayed along one or more axes of physical conditions. Cornell University ecologist Robert Whittaker (1967) was the principal proponent of this technique, and his work was influential in putting to rest the extreme Clementsian view of the closed community.

Most of Whittaker's work was conducted in mountainous areas where

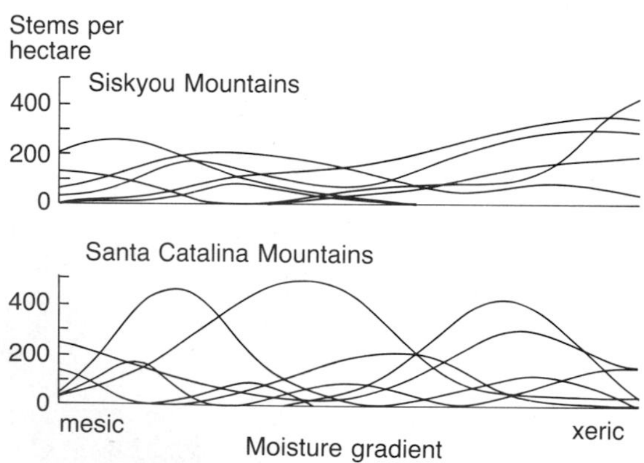

Figure 32–6
Distribution of species along moisture gradients at 460–470 meters elevation in the Siskyou Mountains of Oregon and at 1830–2140 meters elevation in the Santa Catalina Mountains of southeastern Arizona. Species in the more diverse Arizona flora occupy narrower ecological ranges; thus, in spite of the greater total number of species in the flora of the Santa Catalina Mountains, they and the Siskyou Mountains have a similar number of species at each sampling locality. (After Whittaker 1960, Whittaker and Niering 1965.)

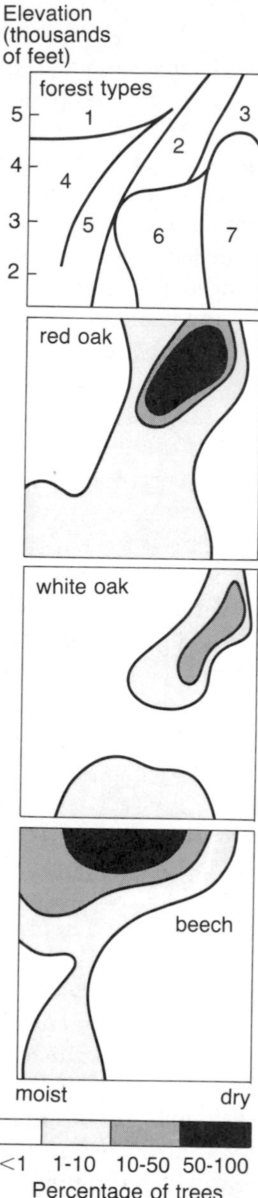

Figure 32–7
Distribution of red oak, white oak, and beech with respect to altitude and soil moisture in the Great Smoky Mountains of Tennessee. The approximate boundaries of the major forest associations are shown in the top diagram. Forest types are: 1, beech; 2, red oak-chestnut; 3, white oak-chestnut; 4, cove; 5, hemlock; 6, chestnut oak-chestnut; 7, pine. Relative abundance, represented by degree of shading, corresponds to percentage of tree stems more than 1 centimeter in diameter in samples of approximately 1000 stems. (After Whittaker 1956.)

moisture and temperature vary over short distances according to elevation, slope, and exposure, which in turn determine light, temperature, and moisture levels at a particular site. When Whittaker plotted the abundances of each species at sites at the same elevation distributed along a continuum of soil moisture, he found that the species occupied unique ranges with peaks of abundance scattered along the environmental gradient (Figure 32–6). Compared to communities in the mountains of southeastern Arizona, fewer species of plants occur in the mountains of Oregon, but each species has a wider ecological distribution, on the average.

In the Great Smoky Mountains of Tennessee, dominant species of trees are widely distributed outside the plant associations that bear their names (Figure 32–7). For example, red oak is most abundant in relatively dry sites at high elevations, but its distribution extends into forests dominated by beech, white oak, chestnut, and even hemlock (an evergreen, coniferous species), and extends throughout the entire range of elevation in the Smoky Mountains. Beech prefers moister situations than red oak, and white oak reaches its greatest abundance in drier situations, but all three species occur together in many areas. The distributions of insect species in the same area were also independent of each other (Whittaker 1952). The distributions of species of birds along an elevation gradient in Peru similarly failed to reveal evidence of distinct ecotones between associations of species (Terborgh 1971, 1985). Even Shreve (1936), who emphasized the distinct boundary between desert and chaparral plant associations, recognized that

> as is true of the meeting ground between any two great plant formations, the dominant plants of each formation are found to vary in the distance to which they extend into the other. This indicates that their habitat requirements are not so nearly identical as their close association in the midst of their respective formations would suggest.

Can the community be a unit of adaptation?

The community is an association of interacting individuals. Community function is the sum of what individuals do and thus reflects the adaptations of individuals. We may, however, ask: Do the attributes of a community express more than the evolved properties of individuals, selected to suit the individual's purpose? Is the community itself a unit of adaptation having properties that can be interpreted only in terms of community function? For example, in a paper on natural selection at the level of the ecosystem, M. J. Dunbar (1960) wrote:

> As to the mechanisms by which selection might take effect at this [ecosystem] level, they are of the ordinary Darwinian sort except that the criterion for selection is survival of the system rather than of the individual or even the species. For instance, suppose an ecosystem, locally defined, begins to develop oscillations to a lethal degree, a degree such that one or more vital

> parts are not able to survive; the resulting empty environmental space, as in Cuvierian cataclysms, is available for occupation by communities from the adjacent regions; and these adjacent systems, as their survival suggests, are not of precisely the same constitution as the extinguished system. . . . One or more of the specific elements will have growth rates, breeding potential and/or metabolic adjustment to temperature different from the former system; and if the difference is favorable to the continued survival of the system, its chances of survival are enhanced. In this way the system dominant in any geographic region changes, and changes (if present assumptions are correct) in the direction of greater stability.

Such holistic viewpoints as Dunbar's parallel organismic conceptions of the community, and are fraught with the same difficulties. The evolution of unique system-level properties requires heritable variation between systems in properties affecting their "fitness." At the very least, such evolution requires the existence of discrete, competing systems—for which there is little evidence—that differ by heritable (that is, stable) traits determining their function. But if such system-level traits were not advantageous to organisms comprising the system, these traits would quickly be altered by natural selection within the system. Hence such traits could not be stable in the system over time—not heritable. If traits that enhance system function also were advantageous to individuals, then they would be maintained by individual selection and the doubtful and inevitably weaker force of system selection would have no unique consequence.

Although the evolution of system properties is unlikely, there is certainly much coevolution with systems. Predator-prey accommodations, mutualistic interactions, diversification of competitors, all have evolved components. Analyses of population dynamics have shown that many such coevolved adaptations enhance the stability of interacting populations and thus may promote local community function. We may set down, at least tentatively, a basic ecological principle: community efficiency and stability increase in direct proportion to the degree of evolutionary adjustment between associated populations. The action of this principle is shown quite clearly when foreign species are introduced to a community. In most cases they cannot successfully invade the community and so die out. But occasionally, exotic species gain a foothold and rapidly come to dominate the local association. Such species can upset the delicate balances achieved between members of the community and, in so doing, disrupt communication function. Outbreaks of introduced pests like the European pine sawfly and gypsy moth can, in fact, almost destroy a community by defoliating its major primary producers. The effects of such introduced diseases as the chestnut blight and Dutch elm disease have been equally pervasive. Avian pox and malaria had a strong hand in the extinction of much of the native Hawaiian avifauna. Introduced arboreal snakes have nearly eliminated the land-bird fauna of Guam. The list goes on and on.

The evolutionary adjustment of species to one another depends upon their degree of association (Howe 1984). When two species always occur together, their interaction exerts an important influence on the evolution of each. When, however, the ecological and geographical ranges of two species are mostly nonoverlapping, each exerts only a small portion of all the selective influences on the population of the other. A community cannot exhibit strong coevolutionary adjustment among its members if the adaptations of its species are molded primarily by relationships in other communities.

The actual degree of association between species lies somewhere between two extremes: the one of obligate association, as we find among many pairs of mutually interdependent organisms, and the other of independence, each species being distributed randomly with respect to others. Spatial contact between populations is a necessary component of association. Temporal association is equally important.

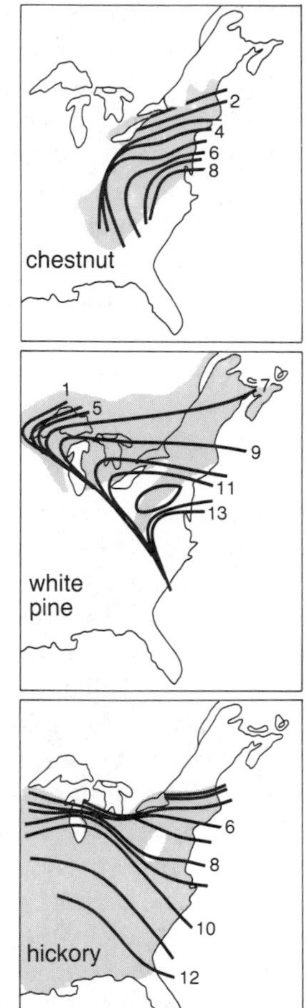

Figure 32–8
Migration of four species of trees in eastern North America from Pleistocene refuges to their present distributions following the retreat of the glaciers. Numerals associated with distribution lines indicate thousands of years before the present. (From Davis 1976.)

The historical record reveals both change and continuity in communities.

The historical record preserved in fossils, fragmentary as it is concerning the organization of communities, equally supports the opposing perspectives of change and of stable association. New species arise and others disappear from the fossil record, to be sure, but their individual spans usually are long enough for much evolutionary adjustment to their cohabitants. And whereas associations of major codominants in systems appear ephemeral in the long history of the fossil record, associations between some of these dominant species and the smaller things that live on or within them may have arisen many millions of years ago.

Pollen grains deposited in lakes and bogs left by retreating glaciers in the northeastern United States record the coming and going of plant species. The composition of plant associations in the past has changed by extension and integration of new species. For example, H. E. Wright, Jr. (1964, 1968) showed that the sequence of reforestation following the last major glaciation contained intermediate forest associations that are not found in the area today. The general pattern of reforestation began with spruce forest, which dominated the area until about 10,000 years ago, followed by extensive associations of pine and birch, which were later replaced by more temperate elm and oak forests.

The migration of tree species from their southern refugia since the height of the last glaciation has been summarized by Margaret Davis (1965, 1976) and is illustrated by maps for representative species in Figure 32–8. For such species as beech and hickory, postglacial migration involved northerly range extension from southern regions across most of the eastern United States. In contrast, white pine and chestnut appear to have emerged from refugia in the Carolinas and expanded their ranges to the west as much as to the north.

In contrast with the seeming flux in species composition of trees in broad-leaved temperate zone forests is the stability of many associations involving plants and animals. Historical information concerning these associations is often indirect. Consistencies between taxonomic relationships among hosts and taxonomic relationships among herbivores or disease organisms suggest a long evolutionary history of association (Linsley 1961, Brooks 1985). In some cases, fossils reveal evidence of consumer activities indistinguishable from traces left by present-day consumers. For example, Paul Opler (1973) reported the characteristic tracks of present-day leaf-mining moths from oak leaves fossilized more than 20 million years ago. S. A. Smith et al. (1985) found drill holes of gastropod predators in Devonian-age fossils of marine brachiopods indistinguishable from those occurring more than 100 million years later in other fossils and also made by living gastropods. Mycorrhizal associations appear in fossils more than 100 million years old (Stubblefield et al. 1987). While the species in these interactions undoubtedly were not the same over that span of time, the evidence reveals long associations between groups of interacting organisms over their mutual evolutionary history.

Biogeographic evidence can also reveal prolonged associations, particularly when groups of species share relictual distributions. During the early part of the Cenozoic period, about 60 million years ago, vast areas of North America, Asia, and Europe were covered by temperate forest more or less continuously distributed across the land areas of high latitudes. As the climate of the earth cooled, and temperate vegetation retreated southward, the Asian and American remnants of this forest were separated until, at present, relict groups of species occur in isolated areas (Graham 1972, Thorne 1972). A part of this vast and flora remains as sets of closely related species, including magnolias, rhododendrons, tulip trees, and gums, in the southeastern United States and in southeastern Asia. The affinities of the floras of these two areas have been recognized for more than a century (Gray 1860, Fernald 1929, Li 1952; Wolfe 1975, 1981). At present, the components of this flora constitute a sizable fraction of the perennial plants in woodlands and wetlands (Table 32–2), suggesting that these plant communities and the animals associated with them might have had a long history of coevolution. Other components of the woodland flora have had different biogeographic histories, and the floras of fields and disturbed habitats clearly have different origins, yet the mixing of floras that makes up

Table 32–2
Proportions of genera classified by life form and habitat having relict, disjunct distributions in eastern North America and eastern Asia

	Trees	Shrubs	Vines	Perennial herbs	Annual herbs
Aquatic				6.7	
Wet		4.3		6.7	0.0
Woodlands	16.3	13.2	20.7	10.7	0.0
Montane	10.0	15.4		0.0	0.0
Fields				0.0	0.0
Roadsides, wastes			13.3	0.0	0.0

Values are reported only when a category contains more than 10 genera.

the distinctive composition of present-day communities may not have occurred so rapidly as to prevent evolutionary accommodations among their species.

Evolutionary history may leave a distinctive imprint on community ecology.

The distribution of life forms over the surface of the earth is by no means uniform. Some regions lack groups that are abundantly represented elsewhere. Many irregularities in distribution patterns are linked to major climatic patterns; for example, snakes and lizards cannot tolerate the cold of arctic environments. Historical accidents of distribution, caused by geographical barriers to dispersal, have also played important roles in the distribution of major groups of animals and plants. Anomalies of distribution are most obvious on islands. Australia lacks most groups of mammals except for marsupials and bats, which are highly diversified there. Few species of any kind reach small remote islands.

Distributional heterogeneity is not limited to islands. The major continental land masses have been sufficiently isolated to reduce the exchange of forms evolved in each area. Major barriers to dispersal are reflected in the major biogeographic regions of the earth, first described in detail by Alfred Russell Wallace (1876), codiscoverer with Darwin of the theory of evolution by natural selection (Figure 32–9).

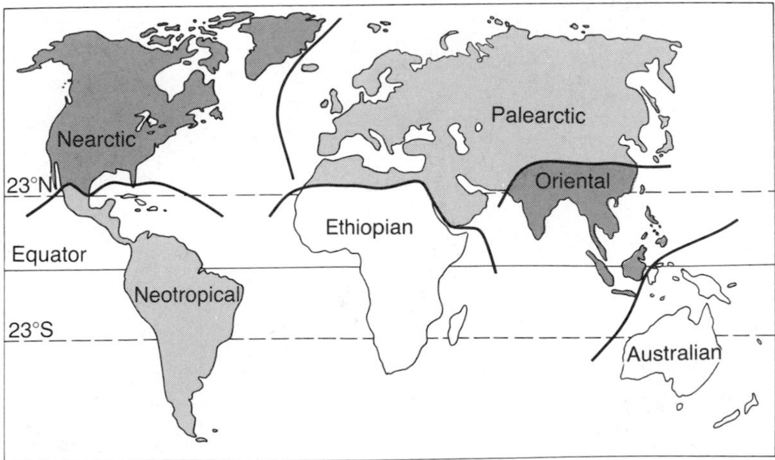

Figure 32–9
Major biogeographic regions of the earth. This scheme, which is widely accepted today, was originally described by A. R. Wallace in 1876. (From Brown and Gibson 1982.)

The different evolutionary histories and taxonomic affinities of plants and animals of the earth's regions are in part obliterated by convergence in form and function. Where woodpeckers are missing from a fauna, other species may adapt to fill their role (Figure 32–10). Rainforests in Africa and South America are inhabited by plants and animals with different evolutionary origins but having remarkably similar appearances (Keast 1972; Bourliere 1973; Figure 32–11). Plants and animals of North and South America deserts are more similar in morphological characteristics than one would expect from their different phylogenetic origins (Mares 1976, Orians and Solbrig 1977). Similarities also have been noted in the behavior and ecology of Australian and North American lizards, despite the fact that they belong to different families and have been separated for, perhaps, 100 million years (Pianka 1971). Plants in the Mediterranean climate zones of southern Europe, South Africa, California, Chile, and Australia are remarkably similar in morphological and physiological adaptations to their winter-rainfall, summer-drought environments (Mooney and Dunn 1970, Mooney 1977).

Wherever one looks one finds convergence, and this reinforces our belief that community organization depends on local conditions of the environment more than it does on the evolutionary origins of the species that comprise the community. In many instances, species-for-species matching have been made (Cody 1974, Fuentes 1976, Mares 1976), suggesting that environments may closely specify the particular characteristics of species that inhabit them and that these specifications depend only on climate and other physical factors; plants and animals have little additional modifying influence on each other by reason of their different phylogentic backgrounds.

This view is much too simple, however. Cases of species-for species matching have not stood up under close scrutiny (Ricklefs and Travis 1980). In fact, detailed studies of convergence are as likely to turn up remarkable differences between the plants and animals in superficially similar environments. In spite of striking convergences, Michael Mares (1976) noted that

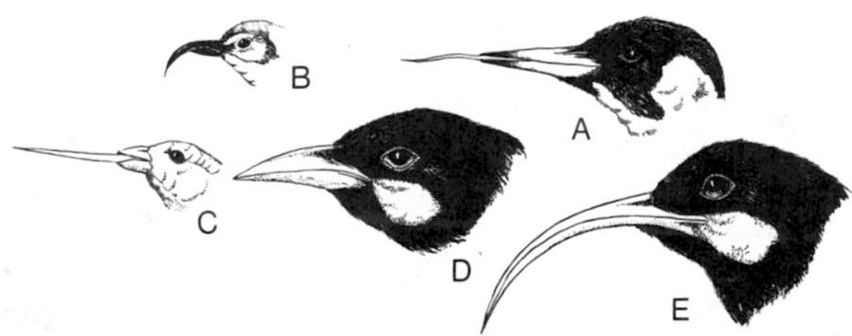

Figure 32–10
Unrelated birds that have become adapted to extract insects from wood: (A) the European green woodpecker excavates with its beak and probes with its long tongue; (B) the Hawaiian honeycreeper *(Heterorhynchus)* taps with its short lower mandible and probes with its long upper mandible; (C) the Galapagos woodpecker-finch trenches with its beak and probes with a cactus spine; and the New Zealand huia (now extinct) divided foraging roles on the basis of sex—the male (D) excavated with his short beak and the female (E) probed with her long beak. (After Lack 1947.)

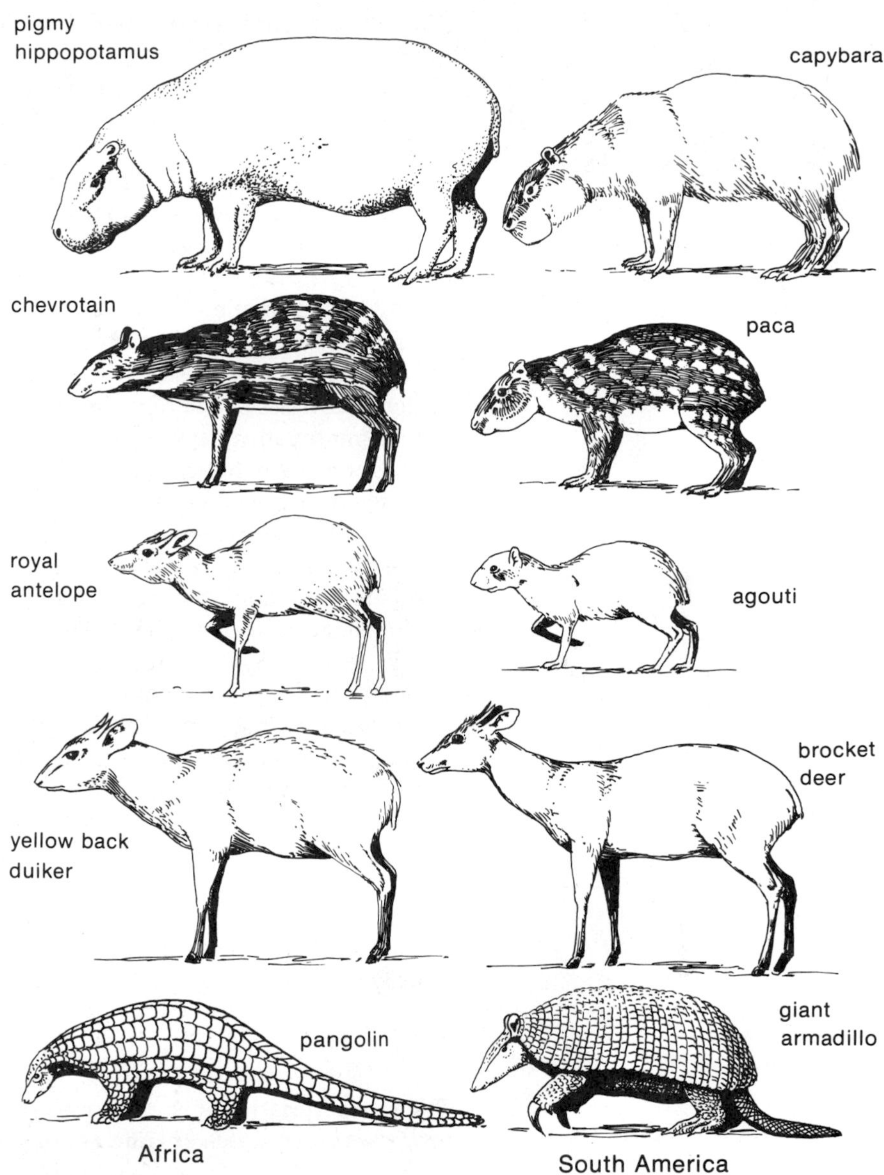

Figure 32–11
Morphological convergence among unrelated African *(left)* and neotropical *(right)* rainforest mammals. Each pair is drawn to the same scale. (After Bourliere 1973.)

the ancient Monte Desert of South America is the only desert region of the world lacking bipedal, seed-eating, water-independent rodents, like the kangaroo rats of North America and gerbils of Asia. Among frogs and toads, however, several South American forms have carried adaptation to desert environments a step further than their North American counterparts: they construct unique nests of foam in which their eggs are kept from drying

(Blair 1975). Differences between the Australian agamid lizard *Amphibolurus inermis* and its North American iguanid analogue, *Dipsosuarus dorsalis,* include diet, optimum temperature for activity, burrowing behavior, and annual cycle, even though at first glance the species are dead ringers for each other (Pianka 1971).

Coevolved relationships between species also may reveal the unique biogeographic position of each region. Unfortunately, little information has been gathered to compare such features of community organization, but one example drawn from the dispersal of seeds by ants illustrates the complexity of the problem. This interrelationship—a type of mutualism—is encouraged by the presence of edible appendages, called elaiosomes, on the seeds. Ants gather these, with the seeds attached, and carry them into underground nests, whereby they effectively plant the seeds. This seed trait is uncommon in most of the world, being restricted primarily to a few species of trees in mesic environments. In Australia and South Africa, however, the trait is well represented among xerophytic shrubs, and it is associated with ecological and morphological features that are lacking in ant-dispersed plants elsewhere (Berg 1975). Whether this difference between plants in Australia and those elsewhere is a consequence of the unique evolutionary history of the Australian flora is not resolved. Indeed, Milewski (1982, 1983) has suggested that the poor soils of Australia and South Africa make it costly for plants to produce the nutritionally expensive fleshy fruits that are dispersed by birds and mammals in most parts of the world. He suggests, therefore, that the dispersal of seeds by ants in Australia and South Africa is an accident of unique local geology rather than indicating the unique historical origins of the flora and fauna. In another paper, A. V. Milewski (1981) suggested that many of the distinctive attributes of the reptile fauna and Australia resulted not from their unique evolutionary history, but from the absence of avian predators, whose scarcity could be attributed ultimately to the poor nutrient status of the vegetation and the paucity of insects (also see Morton and James 1988, Westoby 1988).

The characteristics of the community emerge from a hierarchy of processes over scales of time and space.

There is little agreement among ecologists about what a community is and how its structure is regulated. It is clear, however, that the concept of community cannot usefully be applied to a contemporary, physically circumscribed place, because its characteristics are molded to some extent by historical, broadly regional, and even global processes. Rather than limit our perception to a delineated area, we might think of community as a single point of reference in time and space from which population and evolutionary influence emanates, with a force that diminishes over time and distance. Just as a point emanates such influence, it also receives similar influence from other points, the strength of which decrease with distance. The characteristics of a community at a given point express all these influences; a community therefore has no fixed boundary but is more like a sphere of influence with a density that is great at the center and diminishes with increasing radius. Thus the area of interest to the community ecologist must encompass the scale of process important in shaping community structure at a given point.

Several kinds of process are important, each with a different characteristic scale of time and space. Scale in space varies between the activity range of the individual, geographical and ecological dispersal of individuals within populations, and expansion and contraction of the geographical ranges of populations. Scale in time varies according to the rates of individual and population movements, the dynamics of population interactions, and the selective replacement of genotypes within populations. The relative predominance of local, contemporary processes versus regional, historical processes in shaping community attributes depends on the relative spatial and temporal scales of these processes. The fate of a population at a particular point depends on the balance between the tendency of intolerable physical conditions, interspecific competition, and predation to exclude a population locally and the rate of dispersal of individuals to the point from surrounding areas of population surplus (Shmida and Wilson 1982, Pulliam 1988). Diversity of species at a point depends on the balance between local rates of extinction—resulting from predators, disease, competitive exclusion, change in the physical environment—and regional rates of species production and immigration (MacArthur 1969). Every point on earth has limited access via dispersal to sources of colonizing species. Local diversification depends not only on the capacity of the environment to support a variety of species, but also on the access of a region to colonists, the capacity of a region to generate new forms through speciation, and its ability to sustain varieties in the face of varying environments. Although ecology has traditionally focused upon local, contemporary systems, it clearly must expand its concept to embrace global and historical processes.

In the chapters that follow, we shall pursue this theme in discussing, first, the dynamics of communities, particularly in response to disturbance, then the structure and organization of biological communities, the concept of ecological niches and, finally, the regulation of diversity in biological communities.

SUMMARY

1. The concept of community encompasses associations of interacting populations. Questions about communities address the regulation of species diversity, the relationship between community organization and stability, and the evolutionary origin of community properties.

2. Generally speaking, communities are not discrete units separated by abrupt transitions in species composition. Species tend to be distributed over ecological gradients of conditions independently of the distributions of other species. This pattern is referred to as open community structure.

3. Discontinuities between associations of plants and animals, called ecotones, sometimes occur at sharp physical boundaries or accompany change in habitat-dominating growth forms. The aquatic-terrestrial transition is an example of the first kind of ecotone; the prairie-forest transition is an example of the second.

4. To analyze the distributions of species with respect to environmental conditions and each other, ecologists have devised various types of gradient analysis. Sample localities may be ordinated with respect to gradients of physical conditions or along derived gradients calculated from faunal and floral similarities.

5. The distributions of species along these gradients, or environmental continua, emphasize the open structure of communities.

6. Because communities are not discrete units, and because their traits cannot have high heritabilities, communities almost certainly are not units of adaptation. The properties of local communities are the sum of the evolved properties of their component species.

7. Fossil records and biogeographic evidence indicate that communities may change through time with the differential migration and dispersal of their component populations.

8. Studies of communities developing under similar ecological conditions in widely separated parts of the world suggest that, in spite of the convergence of many adaptations, history may play an important role in shaping the attributes of local communities.

9. The characteristics of the community emerge from the interaction of a hierarchy of processes acting over different scales of time and space. Ecological investigation of communities must encompass the study of regional, historical processes as well as local, contemporary processes.

33

Community Development

Communities exist in a continual state of flux. Organisms die and others are born to take their places. Energy and nutrients pass through the community. Yet the appearance and composition of most communities do not change over time. Oaks replace oaks, squirrels replace squirrels, and so on, in continual self-perpetuation. But when a habitat is disturbed—a forest cleared, a prairie burned, a coral reef obliterated by a hurricane—the community slowly rebuilds. Pioneering species adapted to the disturbed habitat are successively replaced by others until the community attains its former structure and composition.

The sequence of changes initiated by disturbance is called succession and the ultimate association of species achieved is called a climax. These terms describe natural processes that caught the attention of early ecologists, whose work has been reviewed by William Drury and Ian Nisbet (1973) and Robert McIntosh (1974, 1985). By 1916, University of Minnesota ecologist Frederic Clements had outlined the basic features of succession, supporting his conclusions by detailed studies of change in plant communities in a variety of environments. Since then, the study of community development has grown to include the processes that underlie successional change, adaptations of organisms to the different conditions of early and late succession, and interactions between colonists and the species that replace them. Ecologists have come to realize that succession and community perpetuation are different expressions of the same processes and, further, that so-called climax communities are patchwork quilts of successional stages following upon localized disturbances. In this chapter, we shall examine the course and causes of succession both from the traditional viewpoint of community development and in the light of recent studies on community succession.

Succession follows an orderly pattern of species replacements.

The creation of any new habitat—a plowed field, a sand dune at the edge of a lake, an elephant's dung, a temporary pond left by a heavy rain—invites a host of species particularly adapted to be good invaders. These first colonists are followed by others slower to take advantage of the new habitat but eventually more successful than the pioneering species. In this way, the character of the community changes with time. Successional species themselves change the environment. For example, plants shade the earth's surface, contribute detritus to the soil, and alter soil moisture. These changes often inhibit the continued success of the species that cause them, and make the environment more suitable for other species which then exclude those responsible for the change.

The opportunity to observe succession is almost always at hand in abandoned fields of various ages (Figure 33–1). On the piedmont of North Carolina, bare fields are quickly covered by a variety of annual plants. Within a few years, most of the annuals are replaced by herbaceous perennials and shrubs. The shrubs are followed by pines, which eventually crowd out the earlier successional species; pine forests are in turn invaded and then replaced by a variety of hardwood species that constitute the last stage of the successional sequence (Oosting 1942). Change is rapid at first. Crabgrass quickly enters an abandoned field, hardly allowing time for the plow's furrows to smooth over. Horseweed and ragweed dominate the field in the first summer after abandonment, aster in the second, and broomsedge in the third. The pace of succession falls off as slower growing plants appear. The transition to pine forest requires 25 years. Another century must pass before the developing hardwood forest begins to resemble the natural climax vegetation of the area.

The transition from abandoned field to mature forest is only one of several successional sequences leading to the same climax. In the eastern United States and Canada, forests are the end point of several different successional series, or seres, each having a different beginning (Christensen and Peet 1984). The sequence of species on newly formed sand dunes at the southern end of Lake Michigan differs from the sere that develops on abandoned fields a few miles away (Cowles 1899, Olson 1958). The sand dunes are first invaded by marram and bluestem grasses. Plants of these species established in soils at the edge of a dune send out rhizomes (runners) under the surface of the sand, from which new shoots sprout. These grasses stabilize the dune surface and add organic detritus to the sand. Numerous annuals follow the perennial grasses onto the dunes, further enriching and stabilizing them and gradually creating conditions suitable for the establishment of shrub species. Sand cherry, dune willow, bearberry, and juniper form shrub layers before pines become established. As in the abandoned fields in North Carolina, pines persist for only one or two generations, with little reseeding after initial establishment, giving way in the end to the beech-oak-maple-hemlock forest characteristic of the region.

Succession follows a similar course on Atlantic coastal dunes, where

Figure 33–1
Stages of secondary succession in the oak-hornbeam forest in southern Poland. From A to F, the time since clear-cutting progresses from 0 to 7, 15, 30, 95, and 150 years. (Photographs by Z. Glowacinski, courtesy of O. Jarvinen; from Glowacinski and Jarvinen 1975.)

Figure 33–2
Initial stages of plant succession on sand dunes along the coast of Maryland. *Top:* Beach grass on the frontal side of a dune. This grass is used widely to stabilize dune surfaces. *Bottom:* Invasion of back dune areas by bayberry and beach plum. (Courtesy of the U.S. Soil Conservation Service.)

beach grass initially stabilizes the dune surface, followed by bayberry, beach plum, and other shrubs (Oosting 1954). Shrubs act like the snow fencing often used to keep dunes from blowing out; they are called dune-builders because they intercept blowing sand and cause it to pile up around their bases (Figure 33–2). Succession in estuaries leading to the establishment of terrestrial communities begins with salt-tolerating plants and progresses as sediments and detritus build the soil surface above the water line (Redfield 1972).

Primary succession develops in habitats newly exposed to colonization by plants and animals.

Beginning with Clements, ecologists have classified seres into two groups according to their origin. The establishment and development of plant communities in newly formed habitats previously without plants — sand dunes, lava flows, rock bared by erosion or exposed by a receding glacier — is called primary succession. The return of an area to its natural vegetation following a major disturbance is called secondary succession. The distinction between the two is blurred because disturbances vary in the degree to which they destroy the fabric of the community and its physical support systems. A tornado that levels a large area of forest usually leaves the soil's bank of nutrients, seeds, and sproutable roots intact. In contrast, a severe fire may burn through the organic layers of the soil, destroying hundreds or thousands of years of biologically mediated development.

Species colonizing the thin deposits of clay left by receding glaciers in the Glacier Bay region of southern Alaska (Figure 33–3) must cope with deficiencies of nutrients, particularly nitrogen, and with stressful wind and cold. Here the sere begins with mat-forming mosses and sedges, and then progresses through prostrate willows, shrubby willows, alder thicket, sitka spruce, and, finally, to spruce-hemlock forest. Succession is rapid, reaching the alder thicket stage within 10 to 20 years, and tall spruce forest within 100 years (Crocker and Major 1955, Lawrence et al. 1967).

Succession is a means by which dry land is reclaimed from certain aquatic habitats, such as the bogs that form in kettleholes or beaver ponds in cool north temperate and subarctic regions. Bog succession begins when rooted aquatic plants become established at the edge of the pond (Figure 33–4). Some species of sedges (rush-like plants) form mats on the water surface extending out from the shoreline. Occasionally these mats grow completely over the pond before it is filled in by sediments, producing a more or less firm layer of vegetation over the water surface, a so-called quaking bog. The detritus produced by the sedge mat accumulates in layers of organic sediments on the bottom of the pond, where the stagnant water contains little or no oxygen to sustain microbial decomposition. Eventually these sediments become peat, used by man as a soil conditioner and, sometimes, as a fuel for heating (Figure 33–5).

Figure 33–3
A valley exposed by a receding glacier, visible at top center, in North Tongass National Forest, Alaska. The recently bared rock surfaces at the bottom of the valley just below the glacier have not yet been recolonized by shrubby thickets. (Courtesy of the U.S. Forest Service.)

As the bog is filled in by sediments and detritus, sphagnum moss and shrubs, such as Labrador tea and cranberry, become established along the edges, themselves adding to the development of a soil with progressively more terrestrial qualities. The shrubs are followed by black spruce and larch, which eventually are replaced by climax species of forest trees, including birch, maple, and fir, depending on the locality.

Figure 33-4
Stages of bog succession illustrated by a bog formed behind a beaver dam in Algonquin Park, Ontario. The open water in the center is stagnant, poor in minerals, and low in oxygen. These conditions result in the accumulation of detritus from the vegetation at the edge and lead to a gradual filling-in of the bog, passing through stages dominated by shrubs and, later, black spruce.

The intensity and extent of disturbance influence the character of the sere.

Breaks in the canopy of a forest tend to be closed by individuals taking advantage of new opportunities to garner space in the sun. A small gap—perhaps produced by a fallen limb—is quickly filled by growth of branches from surrounding trees. A big gap left by a fallen tree may provide saplings in the understory with a chance of reaching the canopy and a permanent place in the sun. A large area cleared by fire may have to be colonized anew by seed blown or carried in from the surrounding intact forest. Even when reseeding initiates the successional sequence, the size and type of disturbance will influence which species become established first (Denslow 1980, Orians 1982, Runkle 1985). Some require abundant sunlight for germination and establishment, and their seedlings are intolerant of competition from other species (Grubb 1977, Parrish and Bazzaz 1982). These usually have strong powers of dispersal, often owing to small seeds that are easily blown about and can reach the centers of large disturbances inaccessible to members of the climax community.

Figure 33–5
A 1-meter vertical section through a peat bed in a filled-in bog in Quebec, Canada. The layers represent the accumulation of organic detritus from plants that successively colonized the bog as it was filled in. The peat beds are probably several meters thick. Vegetation on the surface of the bog consists mostly of sphagnum, blueberry, and Labrador tea.

The influence of gap size on succession has been investigated in several marine habitats, where disturbance and recovery frequently follow upon each other. Working in southern Australia, Michael Keough (1984) investigated the colonization of artificially created patches, ranging in size from 25 to 2500 cm^2 (5 to 50 cm on a side), by various encrusting invertebrates. The major epifaunal taxa vary considerably in colonizing ability and competitive ability, which are generally inversely related (Table 33–1). When patches of

Table 33–1
Summary of life history attributes of the major epifaunal taxa at Edithburgh, southern Australia

Taxon	Form	Colonizing ability	Competitive ability	Capacity for vegetative growth
Tunicates	Colonial	Poor	Very good	Very extensive, up to 1 m^2
Sponges	Colonial	Very poor	Good	Very extensive, up to 1 m^2
Bryozoans	Colonial	Good	Poor	Poor, up to 50 cm^2
Serpulid polychaetes	Solitary	Very good	Very poor	Very poor, up to 0.1 cm^2

Source: Keough 1984.

different sizes were created within larger areas of uniform habitat, the exposed areas were quickly filled by growth of tunicates and sponges from the surrounding, intact areas. In this case, patch size had little influence on community development because the distance from the edge to the center of the patches (<25 cm) was easily spanned by growth. The many bryozoan and polychaete larvae that settled the patches were quickly overgrown.

Among isolated patches, which were hard substrates placed in sand mimicking the shells of *Pinna* clams, size was very important. Just by chance, few of the small patches were colonized by tunicates and sponges, which produce relatively few propagules, thereby allowing bryozoans and polychaetes to obtain a foothold. Because they were bigger targets, many of the large patches were settled by a few larvae of tunicates and sponges, which then spread rapidly and eliminated other types of species that had colonized along with them. As a result, tunicates and sponges predominated the larger isolated patches, but bryozoans and polychaetes, which, once established, can deter the colonization of tunicate and sponge larvae, were able to dominate many of the smaller patches. In this system, bryozoans and polychaetes are disturbance-adapted species—what botanists call weeds. They get into open patches quickly, mature and produce offspring at an early age, and then often are eliminated by more slowly colonizing but superior competitors. Such weedy species require frequent disturbances to stay in the system.

Predators and herbivores may interact with patch size to influence the sere, either because the behavior of consumers is sensitive to patch size or because consumers require the cover of intact habitat, from whose edge they venture to feed in a newly exposed area. Rabbits rarely feed far from the cover of brush or trees to avoid being seen by predators far from safety. Limpets (grazing molluscs) similarly do not venture far from the safety of mussel beds to feed on algae. Wayne Souza (1984) demonstrated this point forcefully in an intertidal, rocky shore habitat in central California, where he cleared patches of either 625 or 2500 cm^2 in mussel beds and excluded limpets from half the patches in each of these sets by applying a barrier of copper paint along their edges. He then monitored the colonization of the cleared patches by several species of algae during the following 3 years. Limpets live in the crevices between mussels when they are not feeding, so as to avoid predation. Because this behavior limits their foraging range (Figure 33–6), it is not surprising that the densities of limpets in small patches (surrounded by much edge compared to area) were greater than those in larger patches. Also not surprisingly, the density of algae through the course of the experiment was greater in the larger patches.

Where limpet grazing was prevented, total cover by all species of algae was high and did not differ between patches of different size. As one would expect, limpet grazing depressed the establishment and growth of most species of algae, but favored three species of rare, presumably inferior competitors: the brown alga *Analipus,* the green *Cladophora,* and the red *Endocladia.* All of these have a low-lying, crustose growth form that makes them vulnerable to shading and overgrowth by other species. In addition, establishment of *Endocladia* was sensitive to patch size, generally being more common in larger patches regardless of limpet grazing. Abundance of

Figure 33–6
A natural cleared patch in a bed of mussels *(Mytilus californianus)* on the central coast of California. The patch is about 1 meter across and has been colonized by a heavy growth of the green alga *Ulva*. Note the distinct browse zone around the perimeter of the patch created by limpets, which seek refuge in the mussel bed. (Courtesy of W. Sousa; from Sousa 1984.)

colonizing mussels, which eventually crowd out all other space-occupying species, was greater in large patches when limpets were allowed to graze. This is due to the interaction of patch size and limpet grazing, rather than patch size per se, because mussels colonized areas protected by grazing independently of patch size.

The climax is the local end point of the sere.

Succession traditionally is viewed as leading inexorably toward an ultimate expression of community development, the climax community (Clements 1936, Shimwell 1971). Early studies of succession demonstrated that the many seres found within a region, each developing under a particular set of local environmental circumstances, progress toward the same climax (Cooper 1913, Oosting 1956). These observations led to the concept of the mature community as a natural unit, even as a closed system, clearly stated by Frederic Clements in 1916:

> The developmental study of vegetation necessarily rests upon the assumption that the unit or climax formation is an organic entity. As an organism

the formation arises, grows, matures, and dies. Its response to the habitat is shown in processes or functions and in structures which are the record as well as the result of these functions. Furthermore, each climax formation is able to reproduce itself, repeating with essential fidelity the stages of its development. The life history of a formation is a complex but definite process, comparable in its chief features with the life history of an individual plant.

Clements recognized 14 climaxes in the terrestrial vegetation of North America, including 2 types of grassland (prairie and tundra), 3 types of scrub (sagebrush, desert scrub, and chaparral), and nine types of forest, ranging from pine-juniper woodland to beech-oak forest. The nature of the local climax was thought to be determined solely by climate. Aberrations in community composition caused by soils, topography, fire, or animals (especially grazing) were thought to represent interrupted stages in the transition toward the local climax—immature communities.

In recent years, the concept of the climax as an organism or unit has been greatly modified, to the point of outright rejection by many ecologists, with the recognition of communities as open systems whose composition varies continuously over environmental gradients. Whereas in 1930, plant ecologists described the climax vegetation of much of Wisconsin as a sugar maple-basswood forest, by 1950 ecologists placed this forest type on an open continuum of climax communities extending both over broad, climatically defined regions and local, topographically defined areas (Whittaker 1953, McIntosh 1967, Peet and Loucks 1977). To the south, beech increased

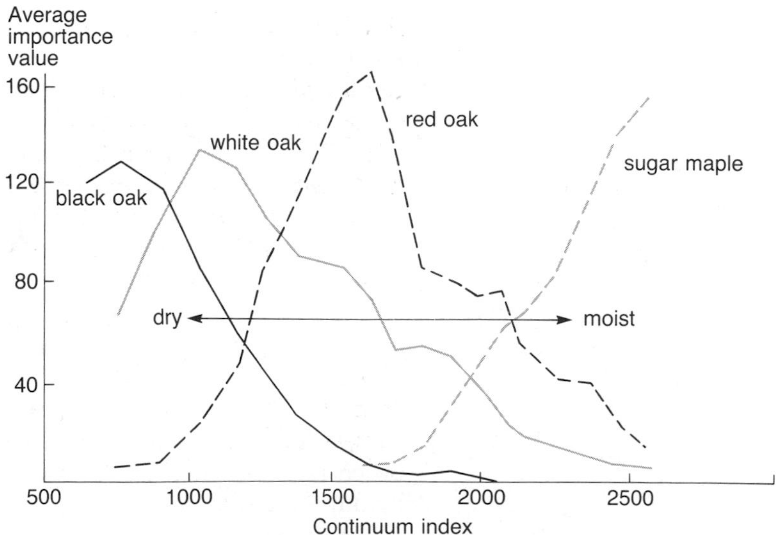

Figure 33–7
Relative importance of several species of trees in forest communities of southwestern Wisconsin arranged along a continuum index. Soil moisture, exchangeable calcium, and pH increase to the right on the continuum index. (After Curtis and McIntosh 1951.)

in prominence; to the north, birch, spruce, and hemlock were added to the climax community; in drier regions bordering prairies to the west, oaks became prominent. Locally, quaking aspen, black oak, and shagbark hickory, long recognized as successional species on moist, well-drained soils, came to be accepted as climax species on drier upland sites.

Mature stands of forest in Wisconsin, representing the end points of local seres, have been ordered along a continuum index ranging from dry sites dominated by oak and aspen to moist sites dominated by sugar maple, ironwood, and basswood (Curtis and McIntosh 1951). A continuum index for Wisconsin forests was calculated from the species composition of each forest type, and its value varied between arbitrarily set extremes of 300 for a pure stand of bur oak to 3000 for a pure stand of sugar maple. Although increasing values of the index correspond to seral stages leading to the sugar maple climax, they may also represent local climax communities determined by topographic or soil conditions. Thus the so-called climax vegetation of southern Wisconsin is actually a continuum of forest (and, in some areas, prairie) types (Figure 33–7).

Succession results from variation in the ability of organisms to colonize disturbed habitats and from changes in the environment following the establishment of new species.

Two factors determine the position of a species in a sere: the rate at which it invades a newly formed or disturbed habitat, and changes in the environment over the course of succession. Some species disperse slowly, or grow slowly once established, and therefore become dominant late in the sequence of associations in a sere. Rapidly growing plants that produce many small seeds, carried long distances by the wind or by animals, have an initial advantage over species that are slow to disperse, and they dominate early stages of the sere. Where fire is a regular feature of a habitat, many species have fire-resistant seeds or root crowns that germinate or sprout soon after a fire and quickly reestablish their populations (Hanes 1971, Vogl and Schorr 1972, Vogl 1973, Kozlowski and Algren 1974, Mooney et al. 1981, Christensen 1985, Riggan et al. 1988).

Early successional species sometimes modify the environment so as to allow later-stage species to become established. Growth of herbs on a cleared field shades the soil surface and helps retain moisture, providing ameliorated conditions for the establishment of less tolerant plants. Conversely, the establishment of some species may inhibit the entrance of others into the sere either by superior competition for limiting resources or by direct interference.

This diverse array of processes governing the course of succession was summarized by Joseph Connell and R. O. Slatyer (1977) under three classes of mechanisms—facilitation, inhibition, and tolerance—relating the role of early stages of succession to the development of later stages. Facilitation, inhibition, and tolerance describe the effect of the presence of one species on the probability of establishment of a second, whether it is positive, negative, or neutral.

Facilitation embodies Clements's view of succession as a developmental sequence in which each stage paves the way for the next just as structure follows structure during an organism's development and during the building of a house. Colonizing plants enable climax species to invade, just as wooden forms are essential to the pouring of a concrete wall but have no place in the finished building. As mentioned above, early stages facilitate the development of later stages by contributing to the nutrient and water levels of soil and by modifying the microenvironment of the soil surface. Alder trees *(Alnus)*, which harbor nitrogen-fixing bacteria in their roots, provide an important source of nitrogen to soils developing on sand bars in rivers and areas exposed by retreating glaciers (Van Cleve et al. 1971). Black locust plays the same role in early succession in the southern Appalachian region of the United States (Boring and Swank 1984).

Soils do not develop in marine systems, but facilitation is often encountered when one species enhances the quality of settling and establishment sites for another. Working with experimental panels placed subtidally in Delaware Bay, T. A. Dean and L. E. Hurd (1980) found that for some species combinations, the presence of one inhibited the establishment of a second, but that hydroids enhanced the settlement of tunicates, and both facilitated the settlement of mussels. In southern California, early-arriving, fast-growing algal stands provide dense, protective cover for the reestablishment of kelp plants following their removal by winter storms. When Larry G. Harris et al. (1984) kept areas clear of early successional species of algae, kelp sporophytes that settled were quickly removed by grazing fish. Teresa Turner (1983) found that the establishment of the surfgrass *Phyllospadix scouleri* in rocky intertidal communities depended on the presence of certain successional algae to which its seeds cling and then germinate (Figure 33–8). In the absence of these algae, the seagrass cannot invade the community.

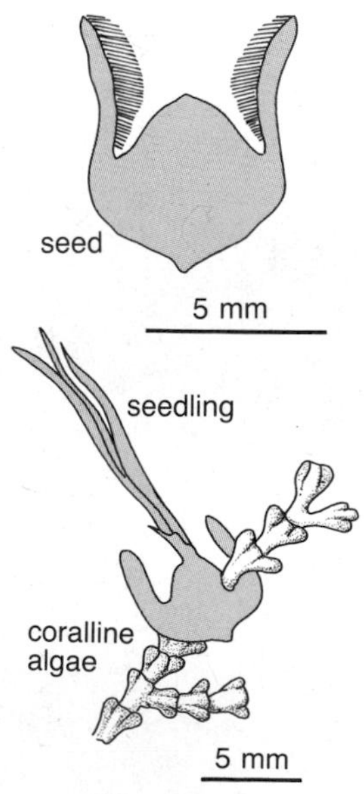

Figure 33–8
Seeds of the surfgrass *Phyllospadix* have barbs that allow them to become attached to certain types of erect algae. (After Turner 1983.)

Inhibition of one species by the presence of another is a common phenomenon, which we have discussed in detail in the parts of this book dealing with competition and predation. One species may inhibit another by eating it, by reducing resources below a level the second species can barely subsist upon, or by direct conflict with chemicals and antisocial behavior (see, for example, Brown and Gauge 1989). With respect to succession, climax species by definition inhibit species characteristic of earlier stages: the latter cannot invade the climax community except following disturbance.

Because inhibition is so intimately connected with species replacement, it is an integral part of the orderly succession from early stages of the sere through the climax. Inhibition can give rise to an interesting situation when the outcome of an interaction between two species depends upon which becomes established first. Colonizing propagules often are the most sensitive stage of the life history, and sometimes neither species of a pair can become established in the presence of the competitively superior adults of the other. In this case, the course of succession depends upon precedence. Precedence, in turn, may be strictly random depending on which species reaches a disturbed site first, or it may follow upon certain properties of the disturbed site—its size, location, season, and so on. We have seen such a case in the subtidal of southern Australia, where bryozoans can prevent the establishment of tunicates and sponges when they become established first. Because of their stronger powers of dispersal, this is more likely to happen on small, isolated substrates.

According to Connell and Slatyer's "inhibition model," succession follows upon the establishment of one species or another only through the death and replacement of established individuals. Thus, just by chance, successional change moves toward predominance of the longer-lived species.

The tolerance model of Connell and Slatyer (1977) holds that "succession leads to a community composed of those species most efficient in exploiting resources, presumably each specialized on different kinds or proportions of resources." According to this model, species are equally capable of invading newly exposed habitat and becoming established. The ensuing sere is then determined by the life spans and competitive abilities of the colonists. Early stages will be dominated by poor competitors with short

Figure 33–9
An old field on the piedmont of North Carolina. Such habitats developed after the abandonment of agricultural land.

life cycles but which become established quickly; climax species will be superior competitors, but may grow more slowly and not express their dominance in the sere until others have grown up and reproduced.

Old-field succession on the piedmont of North Carolina illustrates the early development of the sere.

Clearly all three of Connell and Slatyer's mechanisms — facilitation of establishment, inhibition of establishment, and competitive exclusion (replacement of established populations) — together with the life history characteristics of successional species, are important factors in every sere; none operates exclusively of the others. Early stages of plant succession on old fields in the piedmont region of North Carolina (Figure 33–9) demonstrate how these factors combine in a particular sere (Oosting 1950, Keever 1950, Monk and Gabrielson 1985). The first 3 to 4 years of old-field succession are dominated by a small number of species that replace each other in rapid sequence: crabgrass, horseweed, ragweed, aster, and broomsedge. The life history cycle of each species partly determines its place in the succession (Figure 33–10). Crabgrass, a rapidly growing annual, is usually the most conspicuous plant in a cleared field during the year in which the field is abandoned. Horseweed is a winter annual, whose seeds germinate in the fall. Through the winter, the plant exists as a small rosette of leaves; it blooms by the following midsummer. Because horseweed disperses well and develops rapidly, it usually dominates 1-year-old fields. But because seedings require full sunlight, horseweed is quickly replaced by shade-tolerant species.

Ragweed is a summer annual; seeds germinate early in the spring and the plants flower by late summer. Ragweed dominates the first summer of succession in fields that are plowed under in the late fall, after horseweed normally germinates. Aster and broomsedge are biennials that germinate in

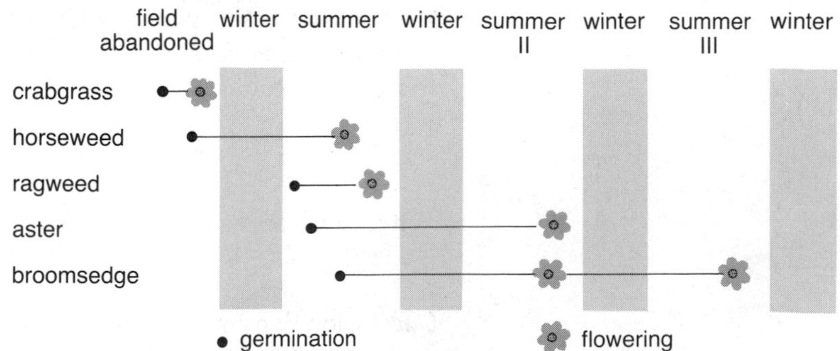

Figure 33–10
Schematic summary of the life histories of five early successional species of plants that colonize abandoned fields in North Carolina.

the spring and early summer, exist through the winter as small plants, and bloom for the first time in their second autumn. Broomsedge persists and flowers during the following autumn as well.

Horseweed and ragweed both disperse their seeds efficiently and, as young plants, tolerate desiccation. These abilities allow them to invade cleared fields rapidly and produce seed before competitors become established. Decaying horseweed roots stunt the growth of horseweed seedlings; this self-inhibiting effect, whose function and origin are not understood, cuts short the life of horseweed in the sere. Such growth inhibitors presumably are the by-products of other adaptations that increase the fitness of horseweed during the first year of succession. One might postulate that if horseweed plants had little chance of persisting during the second year, owing to invasion of the sere by superior competitors, self-inhibition would have little negative selection value. At any rate, the phenomenon is fairly common in early stages of succession (Rice 1984).

Aster successfully colonizes recently cleared fields, but it grows slowly and does not dominate the habitat until the second year. The first aster plants to colonize a field thrive in the full sunlight; the seedlings, however, are not shade-tolerant and adult plants shade their progeny out of existence. Furthermore, asters do not compete effectively with broomsedge for soil moisture. Catherine Keever (1950) observed this when she cleared a circular area, 1 meter in radius, around several broomsedge plants and planted aster seedlings at various distances. After 2 months, the dry weight of asters planted 13, 38, and 63 cm from the bases of the broomsedge plants averaged 0.06, 0.20, and 0.46 gram; available soil water at these distances was 1.7, 3.5, and 6.4 grams per 100 grams of soil.

Early succession on the piedmont suggests the importance of inhibition of seedling establishment and replacement through competitive exclusion (the tolerance model). To demonstrate facilitation, one would have to show that late successional species cannot become established unless they are preceded by earlier colonists. On experimental old-field plots in southeastern Pennsylvania, Jack McCormick (unpublished) painstakingly picked seedlings of early successional species and discovered that the later successional species still invaded the plots and became established. Probably, agricultural clearing does not disturb the soil so much as to reduce its nutrient status and water-holding ability below levels sufficient to support the establishment of most species in the sere. Facilitation is undoubtedly more conspicuous in primary succession.

When does succession stop?

Succession continues until the addition of new species to the sere and the exclusion of established species no longer change the environment of the developing community. The progression of different growth forms modifies conditions of light, temperature, moisture, and, for primary seres, soil nutrients. The replacement of grasses by shrubs and then by trees on aban-

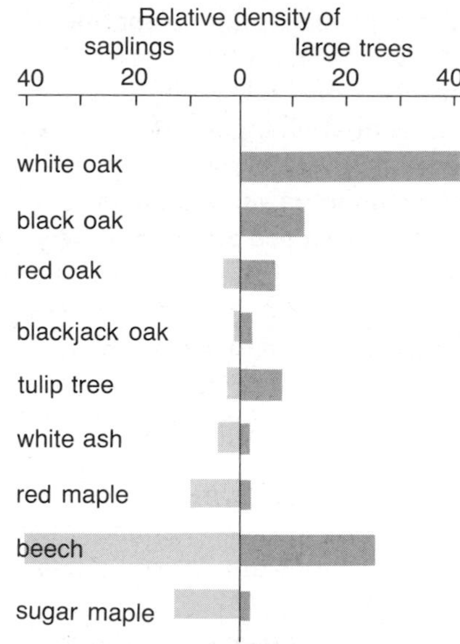

Figure 33–11
Composition of a forest undisturbed for 67 years near Washington, D.C. The relative predominance of beech and maple saplings in the understory foretells a gradual successional change in the community beyond the present oak-beech stage. (After Dix 1957.)

doned fields brings a corresponding modification of the physical environment. Conditions change more slowly, however, when the vegetation reaches the tallest growth form that the environment can support. The final biomass dimensions of the climax community are limited by climate independently of events during succession.

Once forest vegetation establishes itself, patterns of light intensity and soil moisture do not change, except in the smallest details, with the introduction of new species of trees. For example, beech and maple replace oak and hickory in northern hardwood forests because their seedlings are better competitors in the shade of the forest-floor environment, but beech and maple seedlings probably develop as well under their own parents as they do under the oak and hickory trees they replace. At this point, succession reaches a climax; the community has come into equilibrium with its physical environment (Leak 1970; Waggoner and Stephens 1970). To be sure, subtle changes in species composition usually follow the attainment of the climax growth form of a sere. For example, a site near Washington, D.C., left undisturbed for nearly 70 years developed a tall forest community dominated by oak and beech. The community had not reached an equilibrium at the time it was studied because the youngest individuals—the saplings in the forest understory, which eventually replace the existing trees—included neither white nor black oak (Dix 1957). In another century, the forest will likely be dominated by species with the most vigorous reproduction, namely red maple, sugar maple, and beech (Figure 33–11).

The composition and age structure of a forest in northwestern Wisconsin having had minimal human disturbance over 200 years indicated a transitory state, perhaps toward the end of a sere, between oak dominance and a basswood-maple climax (Eggler 1938). At the time of that study, red oak was the commonest large tree in the forest, but basswood and, especially, maple were reproducing much more vigorously (Table 33–2). The ratios of seedlings and saplings (less than 1 inch diameter) to large trees (greater than 10 inches diameter) were maple 186, basswood 155, red oak 18, and white oak 37. (White oak and bitternut are close to the northern edge

Table 33–2
Number of trees of different species in a 2000 m² forest area in northwestern Wisconsin (individuals are separated into size classes)

	Diameter of trunk (inches)			
Species	Less than 1	1 to 3	4 to 9	Greater than 10
Sugar maple	3913	2	16	21
Basswood	931	22	21	6
Red oak	781	1	34	44
White oak	75	3	9	2
Bitternut	88	4	0	0
White pine	0	0	1	2
Ironwood	1606	40	3*	0

* Maximum size class of ironwood, an understory species.
Source: Eggler 1938.

of their ranges in northern Wisconsin and did not form a major component of the forest.)

The preponderance of red oak in the canopy of the Wisconsin forest and the evidence in the understory of successional changes yet to come indicate that the forest had been disturbed in some way, allowing seral species to enter and setting into motion the machinery of succession, or that the climate had changed in recent time to favor basswood and maple over oak, thereby pointing the community toward a new equilibrium. The age structure of the tree populations, determined by increment borings, suggested that fire destroyed much of the forest sometime between 1840 and 1850 (Figure 33–12). Most of the sugar maples were more than 150 years old, indicating that they withstood fire damage. In fact, two-thirds of the sugar maple cores were so badly scarred by fire that their growth rings could not be counted accurately. Red oak and basswood both exhibited periods of rapid proliferation starting about 1850. Red oak gained its predominant position in the forest at that time and will not be excluded until existing trees die and are replaced by basswood or maple seedlings.

The time required for succession to proceed from a cleared habitat to a climax community varies with the nature of the climax and the initial quality of the soil. Clearly, succession is slower to gain momentum when starting on bare rock than on a recently cleared field. A mature oak-hickory forest climax will develop within 150 years on cleared fields in North Carolina (Oosting 1942). Climax stages of western grasslands are reached in 20 to 40 years of secondary succession (Schantz 1917). On the basis of radiocarbon dating methods, John Olson (1958) suggested that complete primary succession to a beach-maple climax forest on Michigan sand dunes requires up to 1000 years. In the humid tropics, forest communities regain most of their

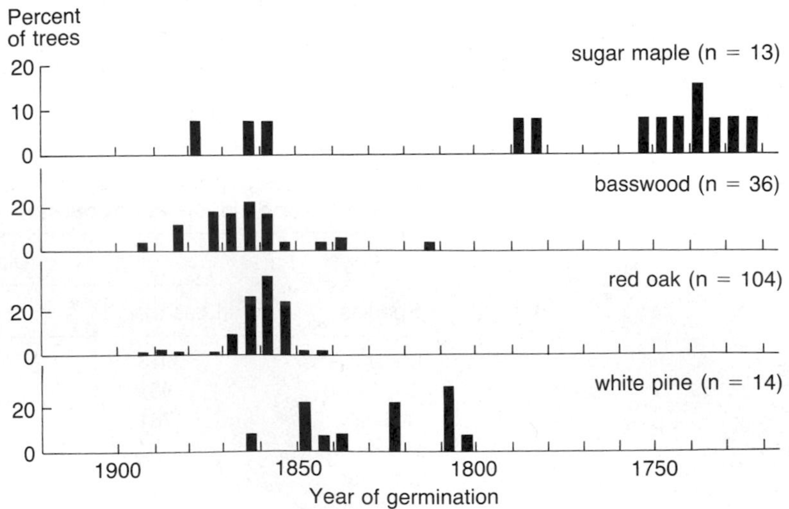

Figure 33–12
Age groups of sugar maple, basswood, red oak, and white pine in a northern Wisconsin forest. (After Eggler 1938.)

climax elements within 100 years after clear-cutting, provided that the soil is not abused by farming or prolonged exposure to sun and rain (Budowski 1965). But the development of a truly mature tropical forest devoid of any remnants of successional species requires many centuries.

Succession can be modeled as a Markov process.

During the course of succession, individuals of various species become established, grow, reproduce, and die. Those having died are replaced by growth of neighbors or the establishment of new individuals of the same or different species. Each change in the system over time may be represented as a transition from one state to another over a time interval; each transition occurs with a certain probability. Suppose we designate open space (or a recent disturbance) state O, and individuals of the several species that may be present in the system as states A, B, . . . and so on. The death of an individual of species A would be the transition A → O; the replacement of an individual of species B by one of A would be B → A. Each of these transitions would occur with a certain probability (P_{AO}, P_{BA}); a system could be represented by a matrix of such transition probabilities (Figure 33–13). This is known as a Markov process. Starting with any mix of species and open space, whose proportions are designated X_O, X_A, . . . the transition matrix allows one to simulate the course of change in the system (succession) over time intervals (for example, t to $t + 1$) and to determine the equilibrium configuration, turnover, and stability of the system.

The transition matrix simplifies nature in ways that make specific application difficult. Systems are divided into discrete subunits that correspond only loosely to individuals or areas of habitat; in nature, transition matrices undoubtedly vary with season, age of stand, and other factors (Lippe et al. 1985); transition probabilities may be influenced by prior transitions. Of course, a probability matrix can be elaborated to match one's ability to gather validating data for a particular system; in fact, very complex models have been constructed to predict the course of succession on particular forest stands (Botkin et al. 1972, Shugart 1984). More importantly, the transition matrix incorporates hypotheses about biological interactions whose consequences may be assayed by the behavior of the model.

	State at time t		
State at time $t+1$	O	A	B
O	P_{OO}	P_{AO}	P_{BO}
A	P_{OA}	P_{AA}	P_{BA}
B	P_{OB}	P_{AB}	P_{BB}

Figure 33–13
Transition probabilities in a system in which two species (A and B) colonize open space (O) or replace each other. Open space is generated by the death of individuals without replacement.

The transition matrix representations of Connell and Slatyer's facilitation, inhibition, and tolerance models are presented in Figure 33–14, along with another configuration that results in cyclic succession. In each case, so long as continuing disturbance creates open space, all species will remain in the system. The equilibrium levels will be determined by the particular transition probabilities and may be calculated analytically (Horn 1975).

When succession is initiated by a unique event (such as land clearing) following which the community remains undisturbed, the transition matrix representation would shift from a transient state in which P_{AO}, $P_{BO} = 1$ (clear-cutting) to the usual state in which P_{AO} and P_{BO} were very low. When all of the system is in state O, it is out of equilibrium and changes begin to

Facilitation model

Future state	Present state O	A	B
O	+	+	+
A	+	+	0
B	0	+	+

Inhibition model

Future state	Present state O	A	B
O	+	+	+
A	+	+	0
B	+	0	+

Tolerance model

Future state	Present state O	A	B
O	+	+	+
A	+	+	0
B	+	+	+

Cyclic model

Future state	Present state O	A	B
O	+	+	+
A	+	0	+
B	+	+	0

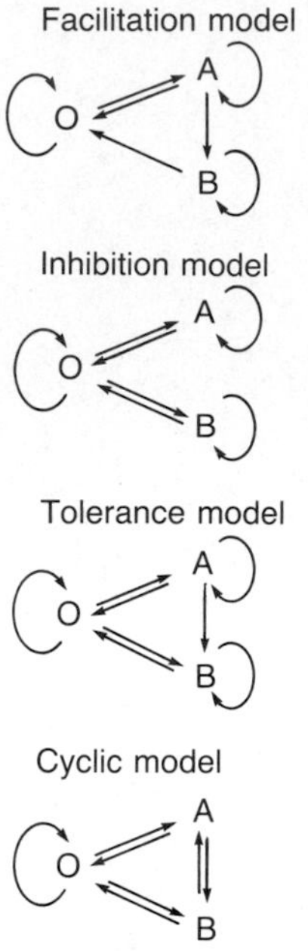

Figure 33–14
Transition probabilities for successional systems behaving according to facilitation, inhibition, tolerance, and cyclic models of species interaction.

restore the equilibrium condition. Initially, the species with the better colonizing potential (P_{OA} and P_{OB}) predominates. Eventually, however, the species with the greater competitive ability (P_{AB} or P_{BA}) increases in abundance, and it may exclude the other when there is no space-forming disturbance.

As an illustration of transitions in a particular system, consider the matrix in Table 33–3, which is abstracted from data compiled by Henry Horn (1975) for a forest near Princeton, New Jersey. More species were

Table 33–3
Proportions of saplings of five species of trees under mature individuals of each species

	Canopy				
Saplings	G	O	H	M	B
G	39	10	1	13	1
O	9	11	16	11	1
H	9	11	5	26	1
M	34	45	58	18	7
B	9	23	20	32	90
Total	100	100	100	100	100

Species are: G = sweet gum, O = red oak, H = hickory, M = red maple, B = beech.
Source: Horn 1975.

present in the forest, but the five listed were among the most common and illustrate the principle. The transition probabilities are the proportion of seedlings and saplings of the five species under the canopies of mature individuals of each species. Strictly speaking, these proportions do not represent transition probabilities because a lot can happen between the opening of a gap in the canopy and its being filled by another individual. But let's assume they do. Another feature of Horn's matrix is that transition probabilities do not take life span and growth rates into account. Moreover, they are dependent on the proportion of adults of each species in the canopy, from which the seeds for replacement must come. Finally, you will note that open space is not a state in the model; this is equivalent to saying that the time interval is such that deceased individuals are replaced from one sampling period to the next by one of the seedlings growing up underneath.

The abundances of each species in the canopy are represented by X_i, for which the subscripts (i) for the different species are G (sweet gum), O (red oak), H (hickory), M (red maple), and B (beech). From inspecting the transition probabilities, it seems clear that beech should come to predominate the forest as it is the only species that replaces itself with high probability (Figure 33–15). Changes in abundance may be projected through time (t) by a set of equations whose terms are the transitions among the five species. That is, for example,

$$X_M(t+1) = 0.34\ X_G(t) + 0.34\ X_O(t) + 0.58\ X_H(t) + 0.18\ X_M(t) + 0.07\ X_B(t)$$

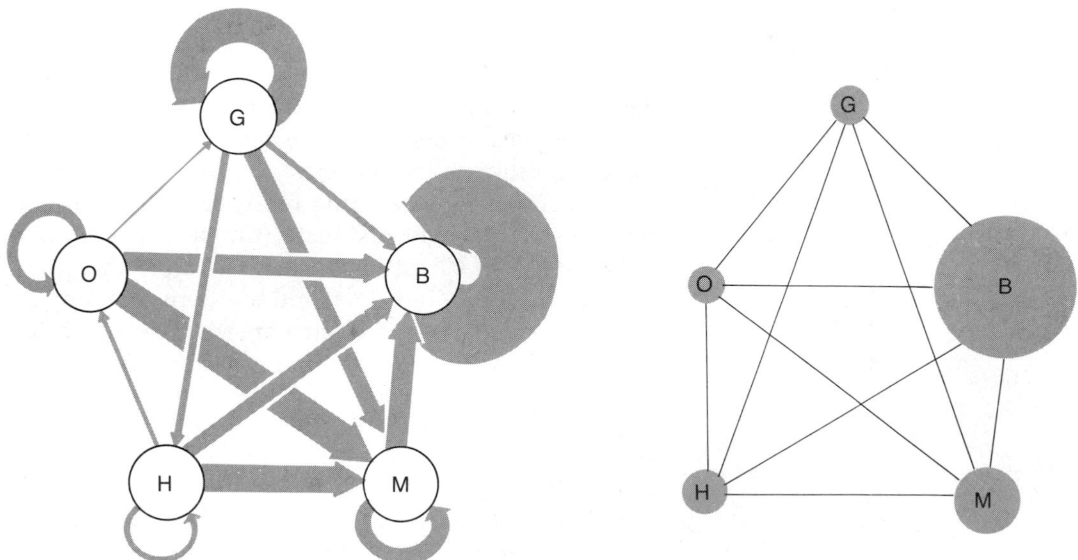

Figure 33–15
Left: The thicknesses of arrows are proportional to transition probabilities among a subset of trees in a deciduous forest in New Jersey (see Table 33–3). *Right:* The relative abundances of each species of tree after the system has come into equilibrium according to the transition probabilities at left.

In matrix notation, $\mathbf{PX}(t) = \mathbf{X}(t+1)$, where $\mathbf{P}$ is the matrix in Table 33–3 and $\mathbf{X}$ is a column vector of abundances of each species. At equilibrium, for each species, $X_i(t+1) = X_i(t)$; the gain of new individuals when they replace individuals of other species must equal the loss of individuals replaced by different species. Hence for maple, $dX_M = X_M(t+1) - X_M(t) = 0$ when $0.34\ X_G + 0.34\ X_O + 0.58\ X_H - 0.82\ X_M + 0.07\ X_B = 0$ [because only 18 per cent of red maple are replaced by individuals of the same species, 82 per cent (-0.82) are lost to the populations of other species]. Solving the set of equations for the system reveals that a stationary distribution (all changes in abundance equal to 10) will occur when $X_G = 0.05$, $X_O = 0.04$, $X_H = 0.06$, $X_M = 0.14$, and $X_B = 0.71$. In this case, all the species remain in the system with abundances skewed toward species with high probabilities of self-replacement (beech) or interspecific replacement (red maple, which is a prolific colonizer).

The character of the climax is determined by local conditions.

Clements's idea that a region had only one true climax (the monoclimax theory) forced botanists to recognize a hierarchy of interrupted or modified seres by attaching names like subclimax, preclimax, and postclimax. This terminology naturally gave way before the polyclimax viewpoint, which recognized the validity of many different vegetation types as climaxes, depending on the habitat. More recently, the development of the continuum index and gradient analysis fostered the broader pattern-climax theory of Robert Whittaker (1953), which recognizes a regional pattern of open climax communities whose composition at any one locality depends on the particular environmental conditions at that point.

Many factors determine the climax community, among them soil nutrients, moisture, slope, and exposure. Fire is an important feature of many climax communities, favoring fire-resistant species and excluding others that otherwise would dominate (Cooper 1961, Kozlowski and Ahlgren 1974, Gill 1975, Christensen 1985, Riggan et al. 1988). The vast southern pine forests in the Gulf Coast and southern Atlantic Coast states are maintained by periodic fires. The pines are adapted to withstand scorching under conditions that destroy oaks and other broad-leaved species (Figure 33–16). Some species of pines do not even shed their seeds unless triggered by the heat of a fire passing through the understory below. After a fire, pine seedlings grow rapidly in the absence of competition from other understory species.

Any habitat that is occasionally dry enough to create a fire hazard but normally wet enough to produce and accumulate a thick layer of plant detritus is likely to be influenced by fire. Chaparral vegetation in seasonally dry habitats in California is a fire-maintained climax that is replaced by oak woodland in many areas when fire is prevented. The forest-prairie edge in the midwestern United States separates climatic climax and fire climax communities (Borchert 1950). Frequent burning eliminates seedlings of

Figure 33–16
Top: A stand of longleaf pine in North Carolina shortly after a fire. Although the seedlings are badly burned *(lower left),* the growing shoot is protected by the dense, long needles (shown on an unburned individual, *lower right*) and often survives. In addition, the slow-growing seedlings have extensive roots which store nutrients to support the plant following fire damage.

hardwood trees but the perennial grasses sprout from their roots after a fire (Daubenmire 1968). The forest-prairie edge occasionally shifts back and forth across the countryside, depending on the intensity of recent drought and the extent of recent fires. After prolonged wet periods the forest edge advances out onto the prairie as tree seedlings grow up and begin to shade out the grasses. Prolonged drought followed by intense fire can destroy tall forest and allow rapidly spreading prairie grasses to gain a foothold. Once prairie vegetation is established, fires become more frequent owing to the rapid buildup of flammable litter. Reinvasion by forest species then becomes more difficult. By the same token, mature forests resist fire and rarely become damaged enough to allow the encroachment of prairie grasses. Hence the stability of the forest-prairie boundary.

Grazing pressure also can modify the climax (Harper 1969). Grassland can be turned into shrubland by intense grazing. Herbivores kill or severely damage perennial grasses and allow shrubs and cacti unsuitable for forage to establish themselves. Most herbivores graze selectively, suppressing favored species of plants and bolstering competitors that are less desirable as food. On the African plains, grazing ungulates follow a regular succession of species through an area, each using different types of forage (Vesey-Fitzgerald 1960, Gwynne and Bell 1968, Jarman and Sinclair 1979, McNaughton 1979, Walker 1981). By excluding wildebeest, the first of the

Figure 33–17
Zebras and Thompson's gazelles feed side by side in the Serengeti ecosystem of East Africa but utilize different plants.

successional species, from large fenced-off areas, Sam McNaughton (1976) was able to show that the subsequent wave of Thompson's gazelles preferred to feed in areas previously used by wildebeest or other large herbivores (Figure 33–17). Apparently, heavy grazing by wildebeest stimulates growth of the preferred food plants of gazelles and reduces cover within which predators of the smaller gazelles could conceal themselves.

Transient and cyclic climaxes develop where climax conditions are unstable.

We view succession as a series of changes leading to a climax, determined by, and in equilibrium with, the local environment. Once established, the beech-maple forest is self-perpetuating and its general appearance does not change in spite of the constant replacement of individuals within the community. Yet not all climaxes are persistent. A simple case of a transient climax would be the development of animal and plant communities in seasonal ponds—small bodies of water that either dry up in the summer or freeze solid in the winter and thereby regularly destroy the communities that become established each year during the growing season. Each spring the ponds are restocked either from larger, permanent bodies of water, or from spores and resting stages left by plants, animals, and microorganisms before the habitat disappeared the previous year.

Succession recurs whenever a new environmental opportunity appears. For example, excreta (Mohr 1943) and dead organisms are a resource for a wide variety of scavengers and detritus feeders. On African savannas, carcasses of large mammals are fed upon by a succession of vultures (Figure 33–18), beginning with large, aggressive species that devour the largest masses of flesh, followed by smaller species that glean smaller bits of meat from the bones, and finally by a kind of vulture that cracks open bones to feed on the marrow (Kruuk 1967, Houston 1979). Scavenging mammals, maggots, and microorganisms enter the sere at different points and assure that nothing edible remains. This succession has no climax because all the scavengers disperse when the feast is concluded. We may, however, consider all the scavengers a part of a climax, which is the entire savanna community.

In simple communities, particular life history characteristics in a few dominant species can create a cyclic climax. Suppose, for example, that species A can germinate only under species B, B can germinate only under species C, and C only under A. This situation would create a regular cycle of species dominance in the order A, C, B, A, C, B, A, . . . with the length of each stage determined by the life span of the dominant species.

Stable cyclic climaxes, which are known from a variety of localities, usually follow the scheme presented above, often with one of the stages being bare substrate (Watt 1947, Forcier 1975, Sprugel 1976). Wind or frost heaving sometimes drives the cycle. When heaths and other vegetation forms suffer extreme wind damage, shredded foliage and broken twigs

Figure 33–18
Vultures feeding on a wildebeest carcass in Masai Mara Park, Kenya.

create an opening for further damage and the process becomes self-accelerating. Soon a wide swath is opened in the vegetation; regeneration occurs on the protected side of the damaged area while wind damage further encroaches upon the exposed vegetation. As a result, waves of damage and regeneration move through the community in the direction of the wind (Figure 33–19). If we watched the sequence of events at any one point, we would witness a healthy heath being reduced to bare earth by wind damage and then regenerating in repeated cycles (Figure 33–20). Similar cycles occur where hummocks of earth form in windy regions around the bases of clumps of grasses. As the hummocks grow, the soil becomes more exposed and better drained. With these changes in soil quality, shrubby lichens take over the hummock and exclude the grasses around which the hummock formed. Shrubby lichens are worn down by wind erosion and eventually are replaced by prostrate lichens, which resist wind erosion but, lacking roots, cannot hold the soil. Eventually the hummocks are completely worn down and grasses once more become established and renew the cycle.

Figure 33–19
Waves of regeneration in balsam fir forests on the slopes of Mt. Katahdin, Maine. (Courtesy of D. G. Sprugel; from Sprugel and Bormann 1981.)

Mosaic patterns of vegetation types are common to any climax community where the death of individuals alters the environment. Treefalls open the forest canopy and create patches of habitat that are dry, hot, and sunlit compared to the forest floor under unbroken canopy. These openings are often invaded by early seral forms, which persist until the canopy closes (Aubreville 1938, Forcier 1975, Williamson 1975, Brokaw 1985). Treefalls thus create a mosaic of successional stages within an otherwise uniform community. Indeed, adaptation by some species to grow in particular conditions created by different-sized openings in the canopy could enhance the overall diversity of the climax community (Denslow 1980, 1985, Orians 1982, Pickett 1983). Similar models have been developed for intertidal regions of rocky coasts, where wave damage and intense predation continually open new patches of habitat (Dayton 1971, Levin and Paine 1974, Connell 1978, Sousa 1985, Connell and Keough 1985).

Cyclic patterns of changes and mosaic patterns of distribution must be incorporated into the concept of the community climax. The climax is a dynamic state, self-perpetuating in composition, even if by regular cycles of change. Persistence is the key to the climax. If a cycle persists, it is inherently as much a climax as an unchanging steady state.

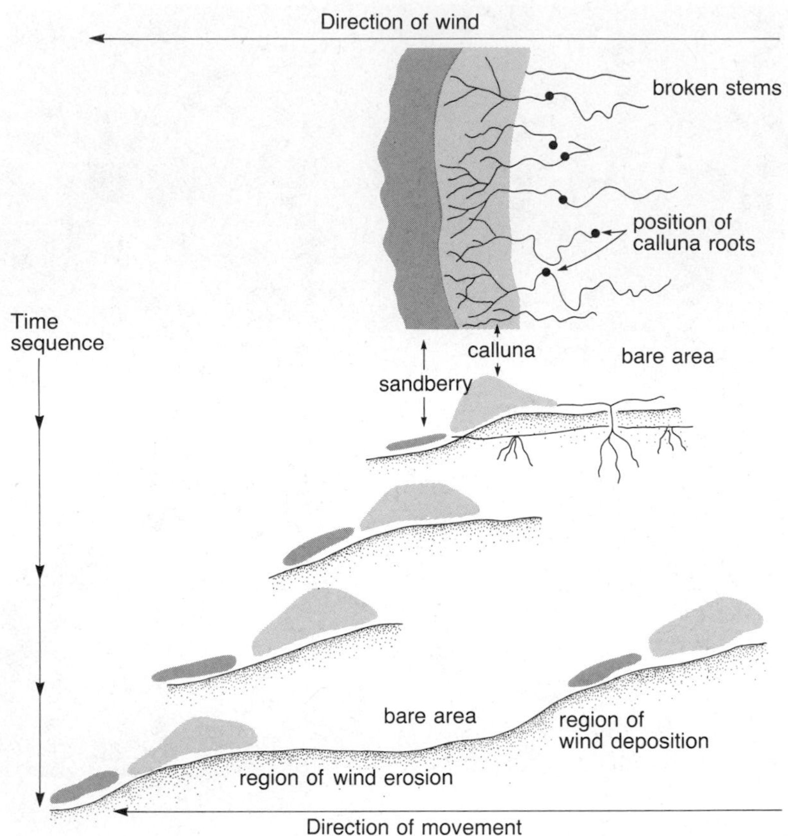

Figure 33–20
Sequence of wind damage and regeneration in the dwarf heaths of northern Scotland. (After Watt 1947.)

The characteristics of dominant species change during succession.

Succession in terrestrial habitats entails a regular progression of plant forms. Plants characteristic of early stages and late stages of succession employ different strategies of growth and reproduction (Grubb 1977). Early-stage species are opportunistic and capitalize on high dispersal ability to colonize newly created or disturbed habitats rapidly. Climax species disperse and grow more slowly, but shade tolerance as seedlings and large size as mature plants give them a competitive edge over early successional species. Plants of climax communities are adapted to grow and prosper in the environment they create, whereas early successional species are adapted to colonize unexploited environments.

Some characteristics of early and late successional stage plants are compared in Table 33–4. To enhance their colonizing ability, early seral species produce many small seeds that usually are wind-dispersed (dandelion and milkweed, for example). Their seeds are long-lived and can remain

Table 33-4
General characteristics of plants during early and late stages of succession

Character	Early stage	Late stage
Seeds	Many	Few
Seed size	Small	Large
Dispersal	Wind, stuck to animals	Gravity, eaten by animals
Seed viability	Long, latent in soil	Short
Root/shoot ratio	Low	High
Growth rate	Rapid	Slow
Mature size	Small	Large
Shade tolerance	Low	High

dormant in soils of forests and shrub habitats for years until fires or treefalls create the bare-soil conditions required for germination and growth (Harper 1977). The seeds of most climax species, being relatively large, provide their seedlings with ample nutrients to get started in the highly competitive environment of the forest floor (Salisbury 1942).

The survival of seedlings in shade is directly related to seed weight (Figure 33-21). The ability of seedlings to survive the shade conditions of climax habitats is inversely related to their growth rate in the direct sunlight of early successional habitats (Grime and Jeffrey 1965). When placed in full sunlight, early successional herbaceous species grew ten times more rapidly than shade-tolerant trees. Shade-intolerant trees, like birch and red maple, had intermediate growth rates. Shade tolerance and growth rate must be balanced against each other; each species must reach a compromise between those adaptations best suited for its place in the sere.

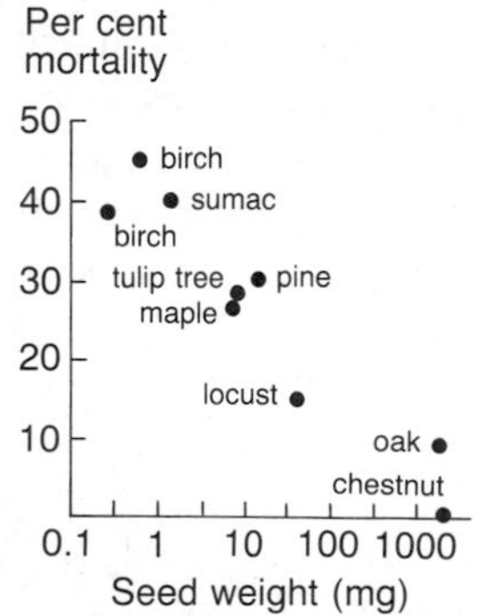

Figure 33-21
Relationship between seed weight and the survival of seedlings after 3 months under shaded conditions. (After Grime and Jeffrey 1965.)

The rapid growth of early successional species is due partly to the relatively large proportion of seedling biomass allocated to leaves (Abrahamson and Gadgil 1973). Leaves carry on photosynthesis, and their productivity determines the net accumulation of plant tissue during growth. Hence the growth rate of a plant is influenced by the allocation of tissue to the root and the aboveground parts (shoot). In the seedlings of annual herbaceous plants, the shoot typically comprises 80 to 90 percent of the entire plant; in biennials, 70 to 80 per cent; in herbaceous perennials, 60 to 70 per cent; and in woody perennials, 20 to 60 per cent (Monk 1966).

The allocation of a large proportion of production to shoot biomass in early successional plants leads to rapid growth and production of large crops of seeds. Because annual plants must produce seeds quickly and copiously they never attain large size. Climax species allocate a larger proportion of their production to root and stem tissue to increase their competitive ability; hence they grow more slowly. The progression of successional species is therefore accompanied by a shift in the balance between adaptations promoting dispersal and adaptations enhancing competitive ability.

The biological properties of a developing community change as species enter and leave the sere. As a community matures, the ratio of biomass to productivity increases; the maintenance requirements of the community also increase until production no longer can meet the demand, at which point the net accumulation of biomass in the community stops (Odum 1969, Whittaker 1975, Peet 1981). The end of biomass accumulation does not necessarily signal the attainment of climax; species may continue to invade the community and replace others whether the biomass of the community increases or not. The attainment of a steady-state biomass does mark the end of major structural change in the community; further changes are limited to the adjustment of details.

As plant size increases with succession, a greater proportion of the nutrients available to the community are tied up in organic materials. Furthermore, because the vegetation of mature communities has more supportive tissue, which is less digestible than photosynthetic tissue, a larger proportion of their productivity enters the detritus food chain rather than the consumer food chain. Other aspects of the community change as well (see, for example, Vitousek and White 1981). Soil nutrients are held more tightly in the ecosystem because they are not exposed to erosion; minerals are taken up more rapidly and stored to a greater degree by the well-developed root systems of forests; the environment near the ground is protected by the canopy of the forest; conditions in the litter are more favorable to detritus-feeding organisms.

Ecologists generally agree that communities become more diverse and complex as succession progresses (Margalef 1968, Odum 1969), although Whittaker (1975) suggested that, in some seres, intermediate stages of succession may be more diverse because they contain elements of early seral stages as well as elements of the climax community. It is not known whether the increase in the diversity of a community during its early stages of succession is related to increased production, greater constancy of physical characteristics of the environment, or greater structural heterogeneity of the habitat. Furthermore, there is no reason to suspect that gradients of diversity along a successional continuum are related to the same factors that determine diversity along a structurally analogous gradient of mature communities.

Succession emphasizes the dynamical nature of the biological community. By upsetting their natural balance, disturbance reveals to us forces that determine the presence or absence of species within a community and processes responsible for the regulation of community structure. Succession also emphasizes that the structure of the community comprises a patchwork mosaic of successional stages and that community studies must consider disturbance cycles on many scales of time and space. The next three chapters address the structure of the community and its regulation. The dynamics and structural consequences of disturbance and succession will never be far from the surface of our discussions.

SUMMARY

1. Succession describes change in a community following either disturbance or the colonization of newly exposed substrate. The particular

sequence of communities at a given point is referred to as a sere, and the ultimate stable association of plants and animals is the climax.

2. Succession on newly formed substrates, such as sand dunes, landslides, and lava flows—referred to as primary succession—involves substantial modification of the environment by early colonists. Moderate disturbances, which leave much of the physical structure of the ecosystem intact, are followed by secondary succession.

3. The initial stages of the sere depend upon the intensity and extent of the disturbance, but its endpoint is determined primarily by climate and topography. That is, within a region, seres tend to converge upon a single climax.

4. The entrance and persistence of a species in a sere, especially for secondary succession, depend upon its colonizing and competitive abilities. Members of early stages tend to disperse well and grow rapidly; those of later stages tend to tolerate low resource levels or dominate direct interactions with other species.

5. Joseph Connell and R. O. Slatyer categorized processes governing succession as facilitation, inhibition, and tolerance, referring to the effect of one established species on the probability of colonization of a second, potential invader. Facilitation is most prominent in early stages of primary succession. Inhibition is a more common feature of secondary succession, and may be expressed in priority effects, conferring competitive dominance on the first arrival.

6. Succession continues until the community is dominated by species that are capable of becoming established in their own and each other's presence. At this point the community becomes self-perpetuating.

7. The interactions that drive succession may be described by a transition matrix of probabilities of replacement of species i by species j. The dynamics of succession and the composition of the local climax may be determined from the properties of this matrix.

8. The character of the climax may be influenced profoundly by local conditions, such as fire and grazing, that alter interactions between seral species.

9. Transient climaxes develop on ephemeral resources and habitats, such as vernal pools and the carcasses of individual animals. In such cases, the regional climax may be thought of as including transient seres.

10. Cyclic local climaxes may develop where each species can become established only in association with some other one. Cyclic climaxes are often driven by harsh physical conditions, such as frost and strong winds. The regional climax would include the local cyclic sere.

11. Characteristics of species vary according to their place in the sere, and so the overall structure and function of the community change accordingly. In general, biomass increases with time whereas net production and diversity tend to be greatest in middle stages.

12. Succession emphasizes the dynamical nature of the community and reveals processes involved in the regulation of community structure.

34

Structure of the Community

Biological communities have poorly defined boundaries, and it is therefore difficult to characterize their structure precisely and unambiguously. It has proven convenient, however, to approach the problem of structure from several perspectives. One addresses patterns within small areas of uniform habitat. Such local areas may be thought of as ecologically homogeneous patches encompassing the daily activities of individuals. Thus local descriptions of community structure represent the ecological interrelationships of individuals co-occuring at a single point: their adaptations and responses to the physical environment, and their interactions, including direct conflict over resources, predation, disease, chemical interference, and so on.

Such local descriptions of community organization do not, however, match the scales of processes that determine community structure. Adaptations of individuals reflect properties of a much larger gene pool, which extends over space through a broad variety of ecological circumstances and which reflects a long history of selection, perhaps in situations that have been altered over time or no longer exist. Furthermore, the presence of individuals at particular places may reflect population processes over wide areas linked together by dispersal. The local presence of an individual or a population does not reveal whether the species is maintained locally or whether its population is continually replenished by immigration of individuals from elsewhere (Shmida and Wilson 1985, Pulliam 1988).

Within local areas, investigators have characterized the number of species present, their relative abundances, their feeding relationships, and the way in which they partition the spectrum of ecological resources.

A second perspective on community structure addresses patterns of distribution of species over larger, ecologically varied areas within which individuals restrict their activities to particular habitats (Cody 1985). Each habitat has a distinctive assemblage of plants, animals, and microorganisms. Each population is seen to be specialized with respect to habitat. The presence of species at particular places depends, in part, on the adaptations of the individuals; that is, whether they can tolerate local conditions, but also upon demographic interactions between populations (Rosenzweig 1981, 1987). Thus as we broaden our perspective from the local, uniform habitat to the larger region encompassing many habitats, we witness first the realization of the adapted attributes of, and direct interactions between, individuals that co-occur at a given point, and second that of the dynamical interactions between populations.

Within the local community, structure is defined by the abundances and activities of individuals of each species: where they live, what they eat, and so on. Within the region, structure is defined by patterns of habitat selection. There are two additional levels of pattern, which we shall discuss in Chapter 36. The first of these arises from processes responsible for the production of new species—the geographical, ecological, and behavioral segregation of species into independently evolving lineages. Speciation arises from the genetical structure of populations, which depends in part upon mating system and dispersal characteristics, and in part upon the ecological circumstances of the population. The final level of pattern derives primarily from historical happenstance. At this level, the presence or absence of a species is not the predictable result of population characteristics and interactions but rather the product of biogeographical routes of dispersal and barriers to dispersal between ecologically suitable regions; of unlikely biological and physical circumstances, such as an extended drought, local volcanism, or a new, virulent disease; and of the uncertainties of stochastic fluctuations. Together, such factors result in unique assemblages of species at each place depending on its history of climate and geography.

In this chapter and the next, we shall examine ways in which biological communities have been described, beginning with the attributes of local communities and progressing to distributions of species among habitats within larger regions. As always, one must remember that ecological scales differ between organisms. What constitutes a region of varied habitat for a population of soil arthropods may lie within the activity space of an individual mouse.

Lists of species provided the first descriptions of biological communities.

During the latter part of the nineteenth century, European naturalists turned their attention from describing new species to characterizing local floras according to their species composition (Shimwell 1971, Mueller-Dombois and Ellenberg 1974). This avenue of study, sometimes called floristic analysis or phytosociology (Braun-Blanquet 1932, 1965), led directly to those functional concepts of the community explored during the early part of this century by Frederic Clements and others. But the initial concern of floristic analysis was one of classification, in which species composition provided a basis for showing the relationships among plant associations, just as morphological characters indicate relationships among species. But whereas species have, on the whole, proven to be discrete types more or less distinguishable from one another, associations, as a rule, merge without sharp boundary. As floristic data became more complete, this difficulty became more insistent and the original premise of the endeavor began to unravel.

As the rationale for floristic analysis weakened (and also became the subject of great debate), community studies were yielding new insights. The first of these had to do with the description of functional relationships among species within the community—the food web. The second addressed quantitative patterns of relative abundance among species.

Food webs describe functional relationships among species in a community.

Because floristic analysis encompassed only plants, feeding relationships were not an issue. And even though competition between plants had been discussed by Charles Darwin and shown experimentally by Alfred Tansley and others early in this century, it had little impact on traditional botany. System-oriented viewpoints came from different branches of ecology. In his essay "The Lake as a Microcosm," the American limnologist S. A. Forbes (1887) argued convincingly that local assemblages of species should be considered as functional wholes along with their physical environment:

> A lake is an old and relatively primitive system, isolated from its surroundings. Within it matter circulates, and controls operate to produce an equilibrium comparable with that in a similar area of land. In this microcosm, nothing can be fully understood until its relationship to the whole can be clearly seen. . . . The lake appears as an organic system, a balance between building up and breaking down in which the struggle for existence and natural selection have produced an equilibrium, a community of interest between predator and prey.

Forbes was more interested in function than structure, and his insights pointed aquatic ecologists toward the emerging concept of the ecosystem and the importance of trophic relationships in organizing aquatic systems.

Ecosystem analysis places species in functional groups with similar trophic position. Thus plants are lumped together as producers, all herbivores—from ant to zebra—share the herbivore label, and so on. Because trophic structure described in this way conforms to certain thermodynamic requirements, functional descriptions of communities tend to emphasize similarities. Placing species together in functional categories obscures the distinctiveness of communities arising from differences in numbers of species or their evolutionary history.

Food web analysis, although based on functional relationships, emphasizes the connections between populations and recognizes, for example, that not all producers are consumed by all herbivores. Because food web analysis includes species-level information about the community, it has greater power than ecosystem analysis to differentiate structure (Cohen 1989).

Food web analysis has gone through two phases: descriptive and analytical. The descriptive phase began early in this century. Its principal expression was the food web diagram in which arrows connect the species in the community according to their feeding relationships. Such diagrams usually were so complex that ecologists were unable to devise suitable statistics to describe food webs or to compare them between communities. To some, food web diagrams merely emphasized the overwhelming complexity of natural systems and the need to simplify structure into trophic groups.

The analytical phase of food web analysis, which began in the mid-1950s, followed upon a theoretical insight and an experimental manipulation of a natural food web. In 1955, Robert MacArthur, then a graduate student at Yale University, published a brief paper entitled "Fluctuations of Animal Populations and a Measure of Community Stability." In it, he suggested that the more complex the community, the greater its stability. His reasoning was simple. When predators have alternative prey, their own numbers depend less on fluctuations in the numbers of a particular prey species. Where energy can take many routes through a system, a disruption of one pathway merely shunts more energy through another and the overall flow is not interrupted. MacArthur's idea linked community stability directly to species diversity and food web complexity, and his important insight stimulated a flurry of theoretical, comparative, and experimental work to explore this insight (see May 1973, Goodman 1975, Goh 1975, 1977, McNaughton 1977, Van Emden and Williams 1974). These studies have yet to produce a consensus, partly because both structure and stability elude definition in ways that would suggest unambiguous measurements in natural communities, and partly because different theories make different predictions (Saunders and Bazin 1975). For example, an alternative to the idea that diversity generates stability goes like this: as communities become more diverse, the species exert greater influence on each other through various interactions; these biological links in turn may create pervasive time lags in population processes, and therefore destabilize diverse systems.

Experimental studies of community structure gained impetus from Robert Paine's investigation of food web complexity.

Joseph Connell's studies during the late 1950s, on the role of competition and predation in determining the vertical distribution of species of barnacles in the intertidal zone, impressed ecologists very deeply with the power of experimentation. Of course, Connell's were by no means the first experiments performed by field ecologists, but they came at a time when population and community ecology were undergoing a great renaissance under the leadership of David Lack, G. Evelyn Hutchinson, Robert MacArthur, and

Richard Levins. As theory developed, the results of experiments took on more meaning. It was upon this stage that Robert T. Paine, of the University of Washington, began his influential research on the role of consumers in determining the structure of rocky-shore, intertidal communities.

It is no accident that the rocky shore environment has hosted such studies. The substrate is firm, which allows one to attach cages and other devices easily, and ensures that one can return to the same spot time after time to follow the progress of experimental and control treatments. Most of the important organisms are sessile or slow-moving. And because space is the most limited resource, one can manipulate competition between species and follow their interaction by observing their occupation of space.

In an important paper summarizing his early work and impressions arising from it, Paine (1966) emphasized the role of predators in determining the overall structure of the community. He compared food webs in the Gulf of California and on the coast of Washington, both of which were dominated by imposing predators, the sea stars *Pisaster* and *Heliaster* (Figure 34–1). The crucial role of the predators in maintaining the structure of the community was demonstrated when Paine removed sea stars from experimental areas on the coast of Washington. Released from predation, mussels *(Mytilus)* spread very rapidly, crowding other organisms out of the study areas and reducing the diversity and complexity of the local food web. Removal of the urchin *Strongylocentrotus,* an herbivore, similarly allowed a small number of competitively superior algae to dominate the system, crowding out many ephemeral or grazing-resistant species (Paine and Vadas 1969). Paine had shown that predators and herbivores can manipulate competitive relationships among species at lower trophic levels, and thereby control the structure of the community. Such species as *Pisaster* are called keystone predators because, when removed, the entire edifice of the community tumbles down.

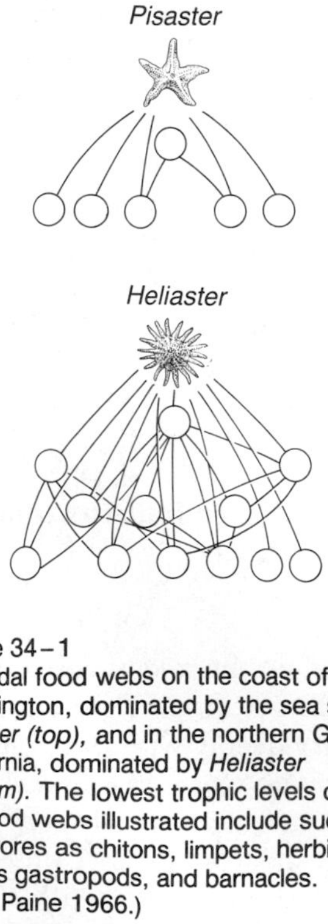

Paine's influence can be measured by the volume of similar studies carried out over the last two decades. Characteristically, however, what began as a promising inquiry into the relationship between structure and function in communities, has ended up raising doubts about most of the initial premises. Paine himself (1980) admitted that experimental community ecology

> has not been characterized by stunning breakthroughs, ecological stability remains a frustrating issue, and to a field ecologist, the ties between model and reality at times appear remote. All but ignored in these recent developments is an insightful recognition that trophic pathways might contribute little to ecosystem stability, and that the answers lie in the spatial patterning of the environment.

Pisaster

Heliaster

Figure 34–1
Intertidal food webs on the coast of Washington, dominated by the sea star *Pisaster (top)*, and in the northern Gulf of California, dominated by *Heliaster (bottom)*. The lowest trophic levels of the food webs illustrated include such herbivores as chitons, limpets, herbivorous gastropods, and barnacles. (After Paine 1966.)

Whereas Hutchinson, MacArthur, and Levins emphasized interactions among competitors on the same trophic level, Paine and others who have followed his example stressed the additional importance of consumer-resource relationships as a key to understanding community organization. Paine (1980) has also pointed out the different ways of conceptualizing food webs. His connectedness, energy flow, and functional webs (Figure 34–2)

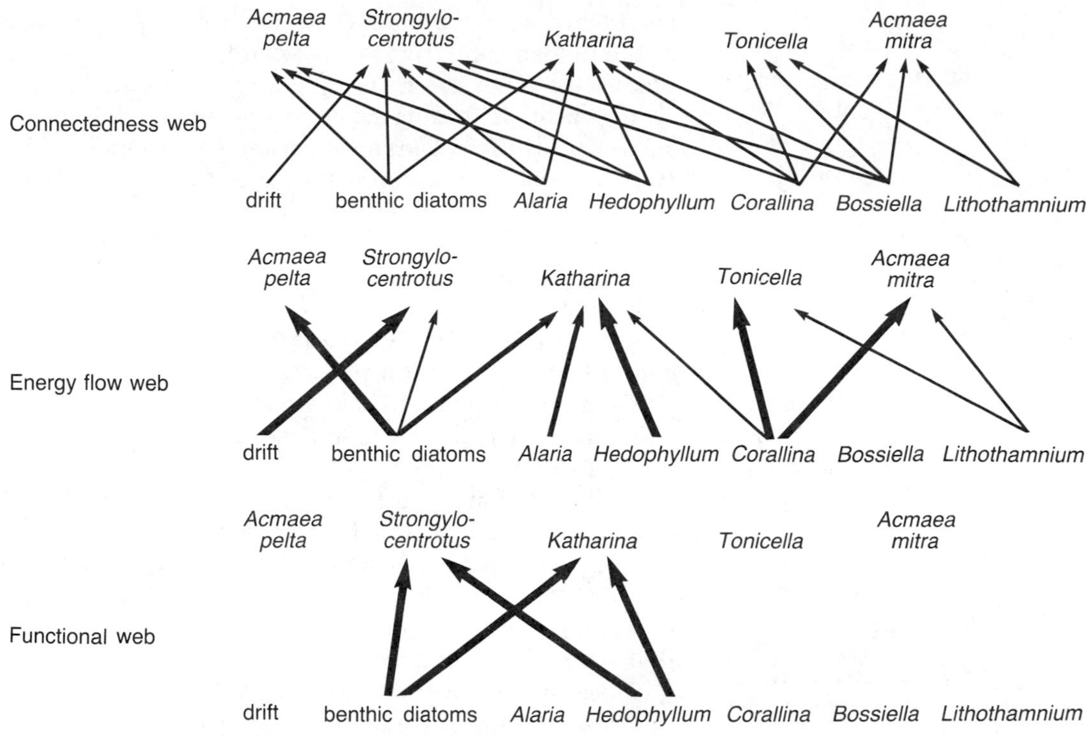

Figure 34–2
Three approaches to depicting trophic relationships, illustrated for the same set of species in a rocky intertidal habitat. (From Paine 1980.)

correspond to the distinctions made earlier between the structural, functional, and cybernetic organization of the community. The first, made popular by Charles Elton (1927), emphasized the feeding relationships among organisms, which he portrayed as links in a food web. Later Raymond Lindeman (1942) suggested that the connections between species could be quantified by the flux of energy between a resource and its consumer. Lastly, the importance of each link in maintaining the integrity of the community is a measure of its cybernetic influence. This controlling role, which is revealed only by experiment, need not correspond to the functional importance of a feeding link in an intact community, as shown dramatically for an intertidal-zone food web in Figure 34–2.

The theoretical influence of food web structure on stability emphasizes the need to characterize community structure in terms of feeding relationships.

As Paine and others were using food web diagrams to portray the structure of biological communities, a few ecologists were beginning to wonder how differences in the structure of webs would affect the dynamics, stability, and

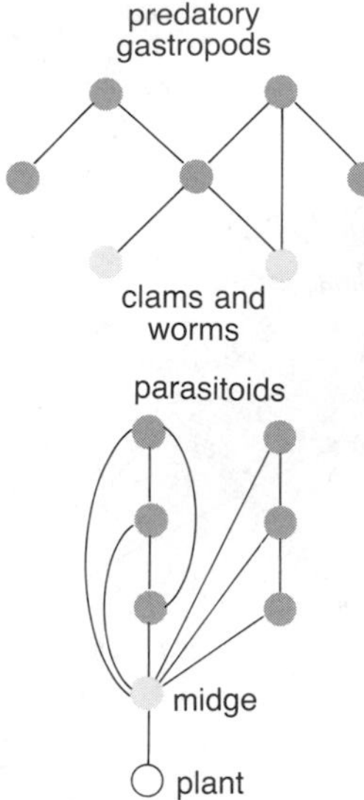

Figure 34–3
Examples of food webs with little *(top)* and frequent *(bottom)* omnivory (feeding on more than one trophic level). The top web is based on intertidal gastropods, bivalves, and their prey; the bottom on a plant *Bacharis*, its insect herbivores, and their parasitoids. (After Pimm 1982.)

persistence of communities. The issue is a crucial one in ecology, being a part of the fundamental contrast between holistic and individualistic philosophies. One raises the issue by posing such questions as, Is a particular arrangement of feeding relationships among species intrinsically more stable than a different arrangement among the same number of species? and How important is the consideration of food web stability in shaping the structure of natural communities?

The two food webs illustrated in Figure 34–3 have similar number of species organized into completely different structures. The mudflat community described by Paine (1963) is relatively simple, having 8 links among 7 species, and with only 1 species preying on more than one trophic level. By contrast, the plant-insect-parasitoid system described by Donald Force (1974) is complex, with 13 links among 8 species, several cases of feeding on more than one trophic level, and one case in which 2 species feed upon each other. Theoretical work by Stuart Pimm and John Lawton (1978) indicates that, all other things being equal, increasing the degree of feeding on more than one trophic level (omnivory) reduces a food web's stability, that is, its resistance to perturbation and ability to return to equilibrium.

With respect to the food web structure of the community, theory and observation clearly are divided by a great chasm. Pimm (1980) has argued that attributes of food web design are consistent with and depend upon qualities that enhance the intrinsic dynamic stability of the food web. Yet we observe radically different web designs in nature, such as those shown in Figure 34–3. Does this variation mean that the rules of food web stability vary depending upon the organisms and ecological circumstances involved, or that feeding relationships are overwhelmingly a consequence of the attributes of species and little influenced by the effects of design on system stability?

These are tough questions. An important first step toward answering them must be to characterize community organization in ways that match theory. Pimm (1982) has emphasized five attributes of communities: diversity, connectance (basically, complexity), food chain length, omnivory (feeding on different trophic levels), and compartmentalization (or subdivision of the food web). The last of these once again raises the issue of the discreteness of the community. A highly compartmentalized community is one in which subsets of species interact among themselves, but only infrequently with species in other subsets. At the extreme, each compartment could be considered a separate community. Compartments might be established in association with patches of habitats or distinctive species within habitats. For example, in southern Canada, broad-leaved trees, pines, firs, and hemlocks each have distinctive associations of lepidopteran herbivores, with few species or genera of moths and sawflies feeding on plants in more than one of the groups. In mixed forests, therefore, one might consider the faunas of broad-leaved and needle-leaved trees as separate communities. By way of contrast, moths feed widely among species of broad-leaved trees and, although most are specialized to greater or lesser degree, they seem not to distinguish discrete subsets. Indeed, after analyzing a number of food webs, Pimm and Lawton (1980) found little evidence for compartmentalization in spite of considerable feeding specialization within associations of

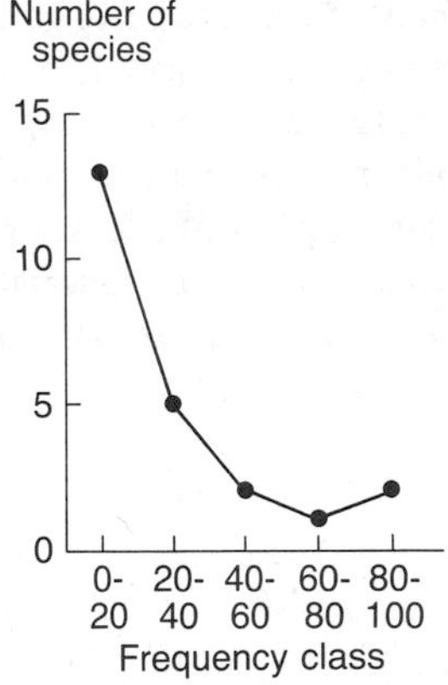

Figure 34-4
Number of species of plants in a peat bog near Kalamazoo, Michigan, in each of five frequency classes based on percentages of 25 0.1 m^2 sampling areas occupied. (From data in Kenoyer 1927.)

species occupying more or less uniform habitats. Other attributes of food-web structure have been described by Briand (1983), Briand and Cohen (1984, 1987), and Cohen (1989).

The single greatest stumbling block to pursuing food web analysis is the empirical description of inherently complex systems. It is difficult to measure the feeding relationships of every species in an association. It is more difficult to measure the functional and cybernetic strengths of each interaction. Simple summary statistics, such as predator-prey ratios, obscure much detail. For the near future, investigations of food web complexity will focus on the feeding relationships of individual species and the adaptations that are associated with specialization and omnivory.

Up to the present, most considerations of community structure have been restricted to summary measures, two of which—diversity and relative abundance—will be discussed in the remainder of this chapter. In Chapter 35, we shall turn to the species attributes of resource specialization and habitat selection that determine the overall food web structure of the community.

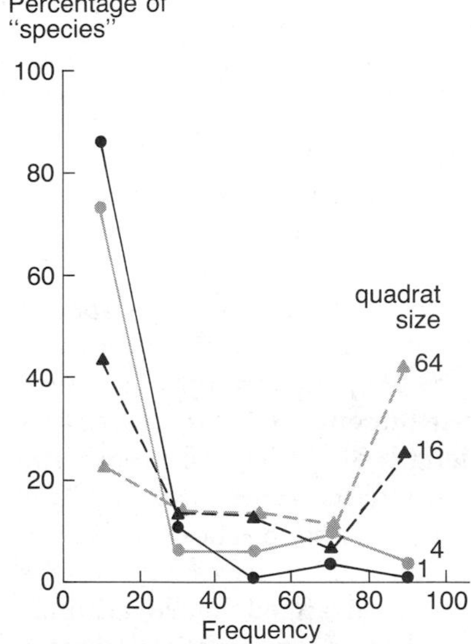

Figure 34-5
The shape of the species-frequency distribution depends on the size of the sampling area, as demonstrated by Curtis and McIntosh (1950) for an artificial community composed of seeds, screws, paper clips, nails, and so on, scattered at random over a surface and sampled as one would a natural plant community using different-sized quadrats.

The abundances of species within associations have been described by a variety of statistical distributions.

C. Raunkiaer (1918) noted early in this century that the abundances of populations within local assemblages assumed regular distributions. When he plotted the numbers of species in each of several abundance classes, the points followed a reversed J-shape, as shown in Figure 34-4. This pattern suggested that within a particular community a few species attained high abundance—they were the dominants in the community—while most of the others were represented by relatively few individuals. [To put Raunkaier's work in perspective, J. T. Curtis and Robert McIntosh (1950) demonstrated that the J-shaped distribution can be generated as an artifact of the sampling methods and definition of abundance classes (Figure 34-5).]

Raunkiaer did not use a mathematical expression to describe his "law" of frequency, but its regularity and pervasiveness among different assemblages have tempted others to mathematical description ever since. Mathematics can serve two purposes. One is to describe many data (species abundances, in this case) with a simple equation whose variables may be used to make comparisons among different samples of species. A second is to use the logic of the mathematical model to infer processes that produce the distributions observed. Each of the models described below has a corresponding logic for relative abundances, and a few of the models were proposed specifically to test predictions following upon certain community processes.

Historically, an important attempt to characterize patterns of relative abundance mathematically was made by R. A. Fisher, A. S. Corbet, and C. B. Williams (1943) for a sample of moths caught at a light trap in England. They suggested that the number of species in each abundance class

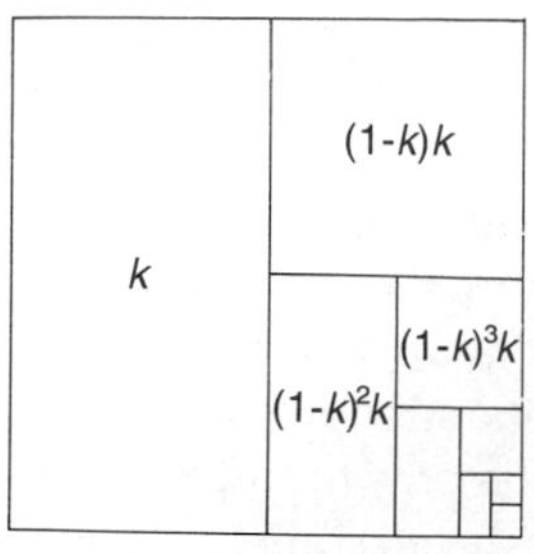

Figure 34–6
Geometric representation of the niche preemption model of relative abundance in which each successive species utilizes a constant fraction (k) of the remaining resources.

followed a logarithmic series. Such series consist of terms of the form ax, $ax^2/2$, $ax^3/3$, . . . ax^i/i, where x is a number between 0 and 1, and a is proportional to the number of species in the sample. Each term in the series is equal to the number of species in the sample that are represented by i individuals. The term ($ax^3/3$), for example, is the number of species represented by three individuals. The logarithmic series provides a reasonable fit to some kinds of samples, but although Williams (1964) used it extensively, the series has not been applied widely by others to describe relative abundance.

Related distributions are the geometric series and the broken stick distribution, both of which were devised to represent underlying processes molding community organization. The geometric series is produced by a model, referred to as niche preemption, in which species colonize an area in succession and each manages to preempt a constant fraction (k) of the remaining resource. Assuming abundance to be directly proportional to resource, the abundance of the first species would be proportional to k, that of the second to $k(1-k)$, that of the third to $k(1-k)^2$, and the ith species to $k(1-k)^{i-1}$, as shown schematically in Figure 34–6 for $k = 0.5$. When the logarithm of species abundance is plotted against the rank of species abundance, assemblages with a geometric distribution fall upon a straight line. When the fraction k varies from species to species in the sequence, the distribution of abundances approaches the logarithmic series (Clark et al. 1964).

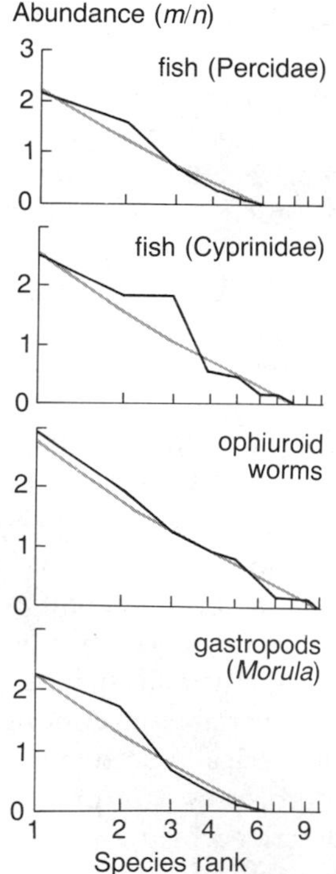

Figure 34–7
Abundances in units of m individuals divided by n species, of species ordered according to abundance rank; the ranks are portrayed on a logarithmic scale. The gray line is the prediction of the broken stick model. (From King 1964.)

Robert MacArthur's broken stick model assumes that species apportion resources among themselves at random and with equal access.

In 1957, Robert MacArthur proposed that the abundance of each species within an assemblage was determined by processes that resembled the random partitioning of resources distributed along a continuum of resource types. For the purposes of his model, MacArthur envisioned resources as if they were distributed evenly along the length of a stick. To predict the relative abundances of N species according to this caricature, $N-1$ points are picked at random along the length of the stick, and the stick is broken at those points. The length of each segment of the stick corresponds to the abundance of a species. When the segments are arranged on a logarithmic scale of decreasing rank, the expected distribution of stick lengths decreases approximately linearly (Figure 34–7). Birds, fish, ophiuroid worms, and predatory gastropods were found to fit the broken stick distribution well (MacArthur 1957, 1960, King 1964), whereas the abundances of many small-bodied, short-lived organisms, such as soil arthropods, nematodes, protozoa, and phytoplankton, were found to have less equitably distributed abundances, with fewer very common species and many more rare ones (Hairston 1959, King 1964, Batzli 1969, Whittaker 1965, 1972).

The geometric, logarithmic, and broken stick distributions represent models of resource apportionment within communities, but the fit of data to any one of these distributions can reveal little about processes that determine relative abundance. Models that generate the broken stick and logarithmic series are probabilistic. That is, each expected value of the abundance of the *i*th species is associated with a probability distribution. As one can see from Figure 34–8, the same model can produce quite different patterns of relative abundance. So, in order to evaluate a particular model one must sample many independent communities of the same size and evaluate both the average of the ranked abundances and their variation among samples. Moreover, as Joel Cohen (1968) pointed out, many similar distributions of relative abundance can be produced by models having different properties. To illustrate his point, Cohen proposed a "balls-and-boxes" model and an "exponential" model, both of which produce the same distribution of relative abundances as the broken stick. In the balls-and-boxes model, the environment is subdivided and the abundance of each species is determined by throwing units of abundance into the subdivisions until each species has at least one subdivision to itself; clearly the number of species cannot exceed the number of subdivisions, or boxes, in the system. In the exponential model, species abundances fluctuate as random variables with time, independently of each other, and with a frequency distribution $P(x_i \geq x) = e^{-kx}$. That is, the probability that the abundance of species i (x_i) is equal to or greater than x is an exponentially decreasing function of x, where k is a constant.

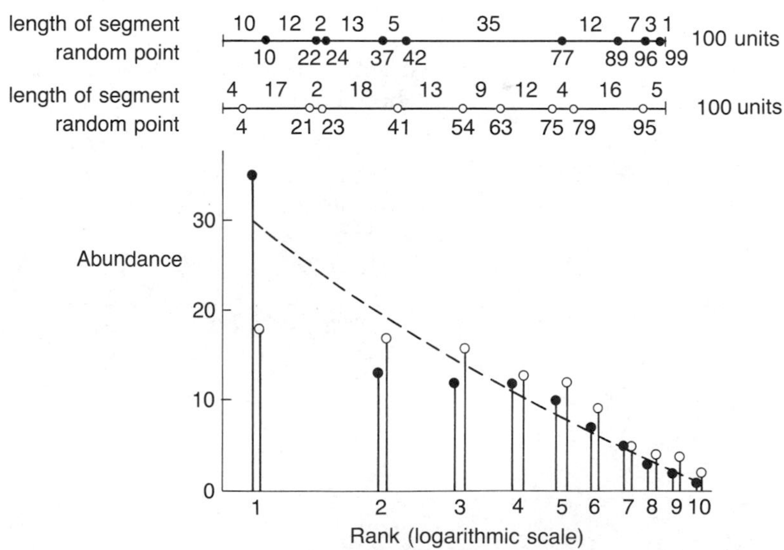

Figure 34–8
Two simulations of relative abundances of species in a community according to a broken stick process. The abundances of n species is determined by $n - 1$ random points (numerals below each line), which divides the line into n segments whose lengths are indicated by numerals above the line. The segments are then portrayed along a scale of the logarithm of rank abundances. The expected (average) distribution of abundances according to the broken stick model for a sample of 10 species and 100 individuals is indicated by the dashed line.

The lognormal distribution of species abundances provides a useful description of large assemblages.

As Robert May (1975) pointed out, broken stick, geometric, and logarithmic distributions are the expected outcomes of various modes of resource partitioning in very simple communities. More realistically, the abundance of a particular species reflects the balance between a large number of factors and processes, variations in each of which result in small increments or decrements in abundance. Statisticians have shown that the sums of many independent factors with small effects tend to assume a normal distribution, the familiar bell-shaped curve. Because factors affecting population size tend to exert a multiplicative influence, one might expect a sample of population sizes to be lognormally distributed, with many species having intermediate levels of abundance and relatively fewer rare or common species.

In 1948, Frank Preston published a seminal paper entitled "The Commonness, and Rarity, of Species," in which he characterized the distribution of species abundances by a lognormal curve. Preston assigned species to classes of abundance based on a logarithmic scale of numbers of individuals per species: 1 – 2 individuals, 2 – 4 individuals, 4 – 8 individuals, 8 – 16 individuals, and so on. Preston called these classes "octaves" because each is twice as large as the preceding class. (In the musical scale, the vibration frequency of each note is twice that of the note one octave lower.) Species whose numbers in the sample fall on the class boundaries (2,4,8,16, . . .) are placed half in the class above the boundary and half in the class below the boundary. In a large sample of individuals, frequencies of species are often distributed normally over the logarithmic abundance categories, as shown in Figure 34 – 9.

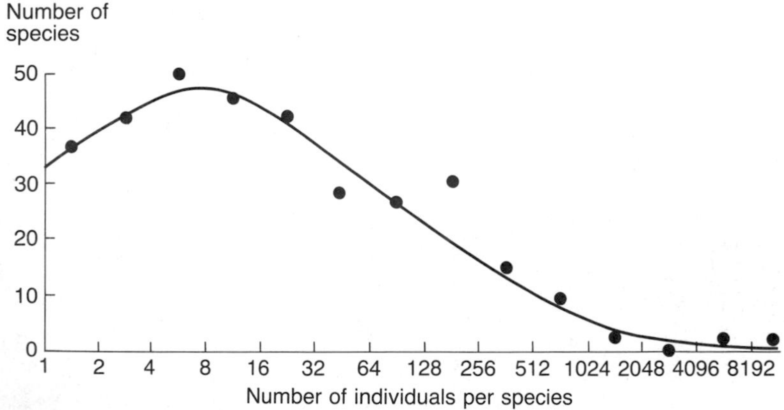

Figure 34 – 9
Relative abundances of species of moths attracted to light traps near Orono, Maine. Size classes (octaves), which increase by a factor of two from one class to the next, are on the horizontal axis; the number of species in each size class is plotted on the vertical axis. The distribution of abundances is hump-shaped with a mode of 48 species in the 4 – 8 individuals size class. (After Preston 1948.)

The normal distribution is described by the equation

$$n_R = n_0 e^{-(1/2)(R/s)^2}$$

in which n_R is the number of species whose abundance is R octaves greater or less than the modal abundance of species within the community, n_0 is the modal number of species (that is, the number in the most abundant class), and s (the standard deviation) is a measure of dispersion (the breadth of the normal curve). These relationships can be seen in a graph of the lognormal distribution (Figure 34–10). The dispersion of the curve—whether it is narrow or broad—is proportional to the constant s.

In theory, at least, the entire lognormal distribution of species abundances in a community can never be fully sampled. Some species are too rare to be represented by one or more individuals in a sample of any size. These species fall below the "veil line" of the distribution and their presence can be revealed only by increasing the total number of individuals examined. When the size of a sample is doubled, the modal abundance of species is moved one octave to the right (the abundance of all species is doubled, on the average) and additional species, each represented by one individual, appear in the distribution at the veil line.

A useful feature of Preston's lognormal curve is that it takes sample sizes into account. One can predict the total number of species (N) in a community, including those not represented in the sample, knowing only the number of species in the modal abundance class (n_0) and the dispersion of the lognormal distribution (s), by the equation $N = n_0\sqrt{(2\pi s^2)} = 2.5\ sn_0$. For the sample of moths graphed in Figure 34–9, n_0 is 48, and s is 3.4 octaves; therefore $N = 2.5 \times 3.4 \times 48 = 408$ species. The actual sample of over 50,000 specimens contained only 349 species, 86 per cent of the num-

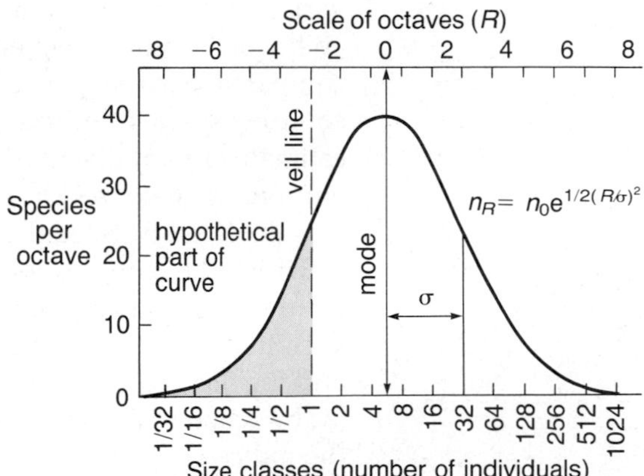

Figure 34–10
Lognormal distribution of species abundances. The part of the curve to the left of the "veil line" (shaded), which corresponds to species with less than one individual in the sample, and thus not represented, is hypothetical. The scale of octaves (R) begins at the modal octave (0). One standard deviation (σ) on both sides of the mode includes about two-thirds of all the species in the sample (After Preston 1948.)

ber theoretically present in the sample area if abundances were distributed lognormally. In practice, confidence limits on estimates of N are so broad that the estimate is of limited value. When the sample is small enough, the veil line may fall close to, or even to the right of the mode, and the revealed lognormal distribution then resembles geometric and logarithmic series, as well as the broken stick distribution. In this sense, the lognormal can be considered a more general statistical description of species abundances.

The dispersion (s) of lognormal curves fitted to large samples of associations is remarkably similar for various groups of organisms. Preston (1948) obtained values of 2.3 for birds and 3.1–4.7 for moths; Ruth Patrick et al. (1954) obtained values of 2.8–4.7 for diatoms. May (1975) suggested that this range of values is to be expected from statistical sampling properties and is independent of the influence of the environment on the community. But, as MacArthur (1969) pointed out, dispersion values do vary with environment. Working with censuses of forest birds, MacArthur obtained values of $s = 0.98$ for lowland tropical localities, 1.36 for temperate localities, and 1.97 for islands. MacArthur's data indicate that there are greater discrepancies between abundances of species on islands compared to temperate and, especially, to tropical mainland areas. Patrick (1963) found similar variations in the dispersion of the lognormal curve fitted to diatom samples from streams with different water conditions in the eastern United States.

Diversity indices take into account variation in abundance among species.

Differences in the abundances of species in communities confront ecologists with two practical problems. First, the total number of species included varies with sample size, because as more individuals are sampled, the probability of encountering very rare species increases. Thus diversity cannot be compared between areas sampled with different intensities merely by counting species. Second, not all species should count equally toward one's estimate of total diversity because their functional roles in the community vary, to some degree, in proportion to their overall abundance.

Ecologists tackled the second problem by formulating diversity indices to which the contribution of each species is in some way weighted by its relative abundance (Pielou 1966, 1977, Whittaker 1972, May 1975). Two such indices are widely used in ecology: Simpson's index (Simpson 1949) and the Shannon-Weaver index (Shannon and Weaver 1949), made popular by Ramon Margalef (1958) and MacArthur (1955, 1957). In both cases, the indices are calculated from the proportions (p_is) of the species (i) in the total sample of individuals. Simpson's index is

$$D = 1/\Sigma p_i^2$$

For any particular number of species in a sample (S), the value of D can vary from 1 to S depending on the evenness of species abundances. When five

Table 34–1
Comparison of diversity indices D, H, and e^H, for artificial "communities" of five species having different relative abundances

Proportion of sample represented by species							
A	B	C	D	E	D	H	e^H
0.25	0.25	0.25	0.25	0.00	4.00	1.386	4.00
0.20	0.20	0.20	0.20	0.20	5.00	1.609	5.00
0.24	0.24	0.24	0.24	0.04	4.31	1.499	4.48
0.25	0.25	0.25	0.25	0.001	4.02	1.393	4.03
0.50	0.30	0.10	0.07	0.03	2.81	1.229	3.42

species are equally abundant, each p_i is 0.20. Therefore, each $p_i^2 = 0.04$, the sum of the p_i^2s is 0.20, and the reciprocal of the sum is 5, the number of species in the sample. Similar calculations for some pencil and paper communities are presented in Table 34–1, where you can see that rarer species contribute less to the value of the diversity index than do common species.

The Shannon-Weaver index, developed from information theory, is calculated by the equation

$$H = -\Sigma p_i \log_e p_i$$

and, like Simpson's index, gives less weight to rare species than to common ones. Because H is roughly proportional to the logarithm of the number of species, it is sometimes preferable to express the index as e^H, which is proportional to the number of species. The values in Table 34–1 are presented in this way so that they may be compared to Simpson's index.

The relative merits of various diversity indexes and their applications have been argued back and forth (Sheldon 1969, Hurlbert 1971, Fager 1972, Peet 1975, Pielou 1977). In fact, the results of most studies are relatively insensitive to which index of diversity is applied, or to whether an index of any kind is used in place of a simple count of the species present. This indifference may be attributed to the fact that communities have characteristic patterns of relative abundance among their member species. As a result, each diversity index tends to bear a consistent relationship to all other indexes as well as to the number of species in the community.

The number of species increases with the size of the sample.

A more serious problem in estimating the number of species in an association is the fact, readily apparent from the lognormal distribution, that the number of species increases in direct relation to the number of individuals

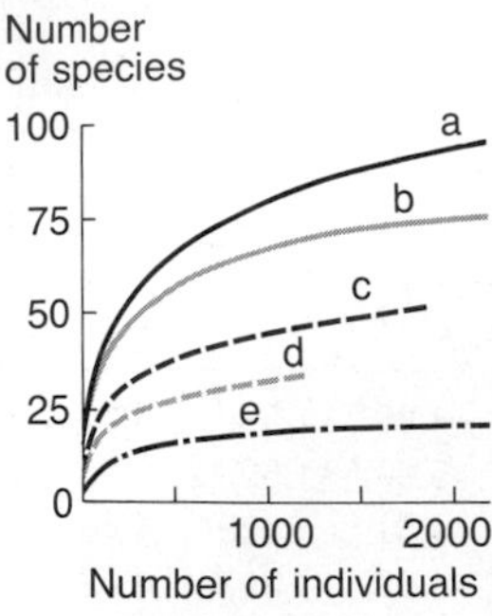

Figure 34–11
Number of species of bivalves and polychaete worms as a function of sample size for different marine environments. Tropical and deep sea faunas tend to be more diverse than faunas in less constant environments. Habitats are (a) tropical shallow water, (b) slope (deep sea), (c) outer continental shelf, (d) tropical shallow water, and (e) boreal shallow water. (After Sanders 1969.)

sampled. If measurements of diversity are to be standardized for comparison, they must be based on comparable samples. When samples include different numbers of individuals, comparability may be achieved by a statistical procedure known as rarefaction (Simberloff 1979), in which subsamples of individuals of equal size are drawn at random from the total. The effect of rarefaction on number of species, which can be thought of as the relationship between number of species and sample size, is portrayed for samples of benthic marine organisms dredged from soft sediments in various localities by Howard Sanders (1969) of the Woods Hole Institute of Oceanography. The total number of specimens per sample varied owing to the densities of organisms in the substrate and unavoidable variation in sampling procedures, although there was a general relationship between sample size and diversity in the original samples. Sanders could not have known whether that relationship was an artifact of sampling or reflected consistent differences between localities without rarifying the samples to make them comparable (Figure 34–11). As specimens were randomly removed from the samples, diversity in all the samples decreased, ultimately being constrained by the truism that there can be no more species than individuals. But the rarefaction curves themselves clearly distinguished localities, showing that for comparable samples, diversity varied considerably.

In many situations, it is difficult to define what we mean by "comparable" samples. For example, suppose that we wished to compare the diversity of herbivores feeding on rare and common species of tree. When individual trees are sampled uniformly, the total sample for a common species usually is larger than that for a rarer species and differences in diversity therefore may express a sampling artifact. When two species of trees are sampled with the same intensity, each individual of the commoner one will be sampled less intensely, which could lead to an underestimate of diversity if there were differences in the herbivore faunas of individual trees associated with, say, habitat, genotype, surrounding species of trees, and accidents of history. This hypothetical problem illustrates that what we mean by comparability is influenced by the processes generating species distribution patterns.

A practical illustration of this difficulty arose in a comparison of the number of species of lepidopteran (moth) larvae on species of broad-leaved trees in southern Ontario, where the Canadian Forest Insect Survey has collected forest insect pests for many decades. At face value, the samples suggested that commoner species of trees in the area harbored more species of herbivores (Figure 34–12). But because the Survey organized its efforts by areas rather than by species of trees, commoner species of trees were sampled more frequently; not surprisingly, the number of moth species also varied in direct proportion to the total sample of individuals collected from each species of tree. Using a statistical analysis that accounted for the correlation between local abundance of a tree and the effort expended in collecting specimens from it, Richard Karban and R. E. Ricklefs (1983) showed that sample size, not local abundance, was responsible for variation in species diversity. Appropriate samples for comparison were, therefore, equal numbers of specimens per species of tree, rather than per individual tree or area of habitat occupied.

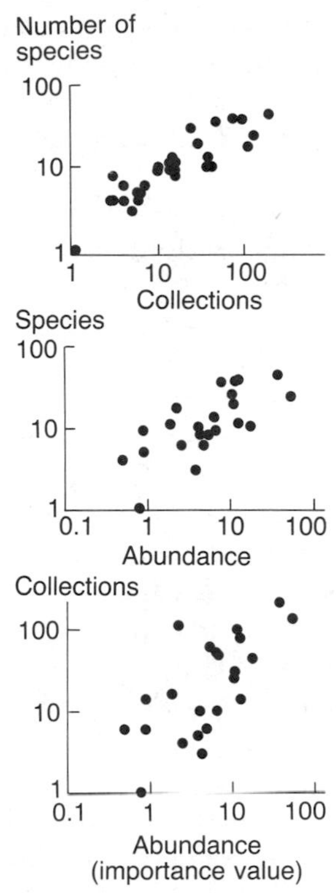

Figure 34–12
Relationships between number of species of moths recorded from deciduous trees in a portion of southern Ontario, Canada. Each species of tree is characterized for comparison by the number of sample collections obtained from it *(top)* and its relative abundance in the local flora *(middle)*. Number of collections and relative abundance *(bottom)* are correlated because individual trees were sampled more or less at random. (From Karban and Ricklefs 1983.)

The number of species encountered increases in direct relationship to the area searched.

Within large areas, one can find more species than within small areas. This species-area relationship was first formalized by Olaf Arrhenius (1921), who unwittingly began a decades-long debate over the meaning of the mathematical constants by which the species-area relationship could be described. This debate began almost immediately; H. A. Gleason (1922) wrote:

> Although [the relationship between species and area] attracted the interest of some of the early phytogeographers, Arrhenius is, so far as I know, the first to develop an equation for the expression of his results. Since this equation appears to be wholly erroneous, a brief discussion of the question may not be out of place.

Common practice for several decades has been to portray the relationship between species (S) and area (A) with a power function of the form

$$S = cA^z$$

where c and z are constants fitted to data (Preston 1960, 1962, MacArthur and Wilson 1967). Species-area relationships usually are portrayed graphically by plotting the logarithm of species number versus the logarithm of area, as in Figure 34–13. After log-transformation, the species-area relationship becomes

$$\log S = \log c + z \log A$$

which is the equation for a straight line. [It should be noted that the practice in some studies, particularly on plants, is to relate the number of species to the logarithm of area; that is, $S = k\log A$ (for example, Gleason 1922, Hopkins 1955); empirical support can be found for both relationships (Schoener 1974)].

Analysis of species-area relationships among many groups of organisms during the 1960s and 1970s revealed that most values of z fell within the range 0.20–0.35 (May 1975, Connor and McCoy 1979). The persistent consistency of z suggested two possibilities. First, observed z values might be a simple statistical consequence of the lognormal distribution of species abundances. As one increases the area of a sample, one usually increases the number of individuals included within it, and as sample size increases the veil line is displaced to the left, exposing more and more species. May (1975) argued that the approximation of the slope of the species-area relationship to $\frac{1}{4}$ was a mathematical consequence of the lognormal sampling distribution. George Sugihara (1980) has pointed out, however, that a wide range of values of z is possible within the context of a lognormal sampling scheme, and suggested that the $z = \frac{1}{4}$ approximation provides information about the processes responsible for the generation of community structure (see Wright

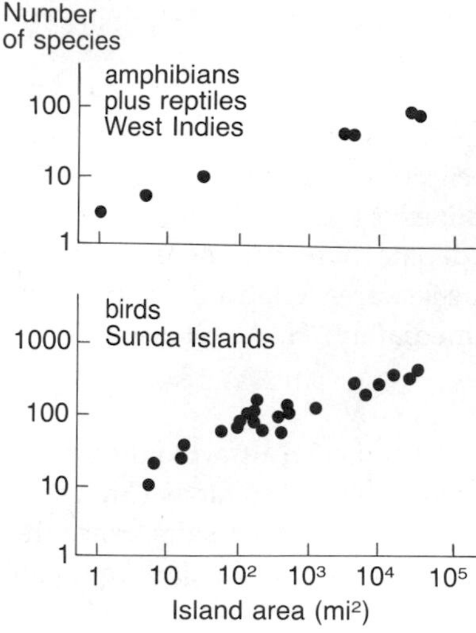

Figure 34–13
Species-area curves for amphibians and reptiles in the West Indies and for birds in the Sunda Islands, Malaysia. (From MacArthur and Wilson 1963, 1967.)

1988). Empirical studies have demonstrated z-values differing greatly from 0.25 in association with differences in biological attributes of samples (for example, Ricklefs and Cox 1972). Indeed, z values obtained for continental areas of different size tend to be lower than for series of islands within a comparable size range.

Where a flora or fauna is perfectly known (that is, all species have been sampled), the sampling properties of the lognormal distribution cannot be held accountable for any relationship between species and area; no species are hiding behind the veil line. In spite of the fact that a new species was discovered on the island of Puerto Rico during the 1960s, we possess near-perfect knowledge of the land bird fauna of the West Indies, among which there is a pronounced species-area relationship with a slope of about $z = 0.24$ (Figure 34–14). Here, differences in diversity between large and small islands must express differences in their intrinsic qualities. Likely candidates are habitat heterogeneity, which undoubtedly increases with the size and consequent topographic heterogeneity of the island, and size per se. Larger islands are better "targets" for potential immigrants from mainland sources of colonization. In addition, the larger populations on larger islands probably persist longer, owing to greater genetic diversity, broader distributions over area and habitat, and numbers large enough to avoid chance extinction.

Larger islands usually are more heterogeneous than small ones. High mountains create elevational gradients of habitats associated with temperature and moisture. In the West Indies, such large islands as Puerto Rico and Guadaloupe have rainforest and desert, and the whole range of habitats in between. But there is also sufficient difference among islands in height, habitat diversity, and distance from the mainland and nearby islands to statistically separate the contributions of each of these factors to the species-area relationship. William D. Hamilton and his colleagues (1963) were the first to attempt such an analysis, using data on the numbers of plant species on islands in the Galapagos Archipelago of Ecuador. They found that elevation and degree of isolation predicted species diversity better than did island area.

Islands in the Lesser Antilles host from 11 to 42 species of resident land birds (Figure 34–14 and Table 34–2). A statistical analysis comparing number of species to area, altitude, distance from mainland, and distance to nearest island shows that each of the independent variables makes a unique contribution to variation in species diversity. As one would expect, diversity increases with area and elevation, and decreases with degree of isolation. Barbados is larger than St. Vincent but, being low and isolated, harbors only half as many species of birds. Such analyses suggest that area per se may be an important determinant of diversity, but it is still possible that island size is associated with variation in habitat diversity over and above that attributable to elevation.

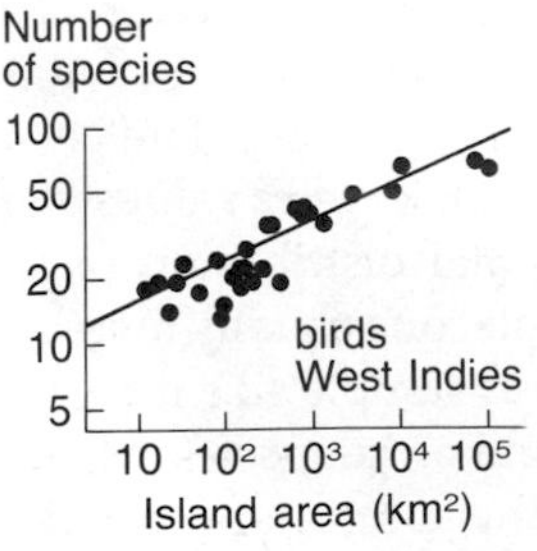

Figure 34–14
Species-area curve for land birds of the West Indies, including both the Greater and Lesser Antilles. (After Ricklefs and Cox 1972.)

After all that has been written about the relationship between the species-area curve and the lognormal distribution, empirical characterizations of species-area curves on islands arguably are not relevant because the areas sampled are heterogeneous with respect to access to colonists and, perhaps, to general aspects of their ecology. What we are interested in

Table 34–2
Size, altitude, geographical position, and number of species of land birds on islands of the Lesser Antilles

Island	Area (km^2)	Altitude (m)	Distance from mainland (km)	Nearest island (km)	Number of resident species of land birds
Anguilla	90	300	850	7	11
St. Martin	85	410	800	7	13
St. Bartholomew	25	300	800	20	12
Saba	12	860	750	25	18
St. Eustatius	21	600	750	15	18
St. Kitts	180	1140	750	3	21
Nevis	130	1100	700	3	19
Barbuda	160	300	800	45	20
Antigua	280	400	700	60	20
Monserrat	100	910	650	35	22
Guadeloupe	1500	1500	600	40	34
Desirade	27	280	600	5	19
Marie Galante	24	300	600	35	14
Dominica	800	1450	550	40	39
Martinique	1100	1340	450	30	38
St. Lucia	600	960	350	30	42
St. Vincent	350	1240	300	40	35
Barbados	430	340	400	250	16
Bequia	19	300	300	10	19
Carriacou	34	300	200	25	21
Grenada	310	840	150	100	35

Islands are listed more or less from north to south.
Source: Lack 1976.

knowing is how species are accumulated with increasing area. The answer depends upon how populations are distributed geographically. Clearly, species are restricted with respect to habitat (Cody 1985), and as larger areas include more habitats, more species will be included in the total sample. But it is also known that larger areas of uniform habitat usually have more species than smaller areas of different, but equally uniform habitat. A classic example of this relationship is the number of insects collected from species of trees with different geographical ranges and abundance within those ranges. Although Karban and Ricklefs (1983) showed that the species diversity of herbivore faunas was independent of local abundance of the host, T. R. E. Southwood and his colleagues (Southwood 1961, Southwood et al. 1982) have demonstrated convincingly that faunas increase with increasing geographic range of the host. This relationship could result from three causes:

1. More complete sampling from a larger area may expose more of the lognormal curve of species abundances (for example, Kuris et al. 1980; see Lawton et al. 1981 and Rey et al. 1981).

2. The mode (n_0) and breadth (s), or both, of the lognormal curve may be larger for hosts with greater geographical extent. Perhaps hosts with larger ranges accumulate more herbivore species because they are better targets for ecological and evolutionary colonization (Janzen 1968, 1973), or perhaps the larger populations on more widespread hosts are demographically more stable and have lower probabilities of extinction.

3. Larger areas contain a larger number of habitat subdivisions each with its own, partially independent lognormal distribution of species abundance.

T. R. E. Southwood et al. (1982) and Donald Strong et al. (1984) believe that the British insect fauna of trees has been so thoroughly sampled that it is essentially perfectly known. If this were true, we could rule out the first factor as a cause of the species-area relationship. Distinguishing between possibilities (2) and (3) requires systematic local sampling from within widely distributed and narrowly distributed species. George Stevens (1983) has accomplished this with wood-boring beetles (Scolytidae) by setting out measured lengths of cut wood in five localities within the eastern deciduous forest of the United States. By his method, each species of tree was sampled to exactly the same degree in each of the forest areas in which it was present. He found that although the number of wood-boring beetle species attracted to baits in all five localities increased with the geographical distribution of the host, hence the number of localities in which it was sampled, local diversity was independent of the host area. That is, the number of species of beetles collected from baits of a particular species in a single locality, say in North Carolina, was unrelated to the geographical range of the host. This particular result favors alternative (3) and indicates that there is geographical turnover of herbivore species within the population of a single host, perhaps associated with variation in climate, parasites, or resource quality of the host.

Local diversity, relative abundance, and the species-area relationship are statistical manifestations of biological properties of organisms. If taking a community-pattern viewpoint is looking at the problem from the top down, then the view from the bottom up would bring such individual and population traits as ecological specialization within habitats and selection of living places among habitats into close focus. This perspective is the subject of the next chapter.

SUMMARY

1. Because biological communities usually do not have natural boundaries, descriptions of community structure are arbitrarily divided between patterns observed within small areas of uniform habitat and patterns observed within larger regions containing a variety of habitats.

2. Within local areas, ecologists have characterized communities by

the number of species present, their relative abundances, and their feeding and other ecological relationships.

3. Historically, much of community ecology grew out of attempts to classify plant associations on the basis of their species composition. Indeed, floristic analysis provided data that forced ecologists to abandon a concept of communities as discrete entities.

4. Functional viewpoints in community ecology developed from system considerations and the description of feeding relationships, which ultimately led to analytical studies of food webs and experimental studies of the roles of individual species in the maintenance of community structure.

5. The complexity of the community has forced ecologists to seek summarizing statistics of structure from which comparative and deductive studies may follow. Two of these statistics are the number of species in a community and the frequency distribution of their relative abundances.

6. Several models of abundance, including the logarithmic series, geometric series, and broken stick distributions, have expressed developing concepts of relative abundance and species interactions. But the premises of these models cannot be tested by comparison with observed frequency distributions of abundances owing to similarities among the predictions of different models and the weakness of statistical inference based on sampled community properties.

7. The lognormal distribution of species abundances has proved to be a useful empirical device, characterizing frequency distributions by a modal abundance class (n_0) and dispersion of abundances about the mode (s). The term n_0 depends on the size of the sample, but the value of s is an intrinsic property of the community. Species with predicted abundances less than 1 individual—those beyond the veil line—are not detected unless the size of the sample is increased. The lognormal concept stresses the dependence of diversity estimates upon sample size.

8. Various indices of diversity, most notably Simpson's index and the Shannon-Weaver (information) index, have been devised to take into account variation in abundance when making comparisons between samples. Number of species also increases as sample size increases, as predicted by the lognormal distribution of abundances. As a result, ecologists have devised rarefaction procedures and other statistical techniques to make samples of species ("communities") comparable.

9. Number of species increases in direct relation to the area sampled. In part, this results from larger areas giving rise to larger total samples. But studies of well-known faunas and floras also indicate that larger areas are more heterogeneous ecologically, providing opportunities to sample more kinds of habitats, and additionally that larger islands have more species because they are better targets for colonization and because larger populations better resist extinction.

35

The Niche Concept in Community Ecology

Frequency distributions of species abundances and plots relating species diversity to area suggest pattern in the organization of species within communities (Brown and Maurer 1989) but they reveal neither the details of community structure nor the processes that regulate it. The local abundance of a species reflects the way in which individuals utilize local resources. Common species obtain a bigger share than do rare ones, because they either utilize a broader variety of resources, are specialized on resources that are more abundant, or compete more effectively for shared resources. The distribution of species across habitats also reflects the manner in which individuals utilize resources. Compared to narrowly distributed species, more widespread ones either specialize on resources that are widespread or are ecologically flexible enough to utilize different resources in different areas. Finally, local abundance and distribution across habitats may be interrelated to the extent that they reflect a common set of specific and environmental attributes.

Relative abundance and distribution ultimately manifest specialization of individuals within habitats and of populations among habitats. These arise from (1) the pattern of resource availability (an attribute of the environment), (2) the particular adaptations of species to exploit environmental resources, and (3) interactions among species that partly determine the success of the species' strategy for resource use. This perspective, particularly the idea that interactions may determine the presence or absence of species at a particular place, has prompted ecologists to describe community structure by the patterns of resource utilization and resource overlap among species. The central concept in these studies has been that of the niche.

The niche concept expresses the relationship of the individual to all aspects of its environment.

Recent approaches to the study of community ecology have focused upon the concept of the niche, particularly as it has been envisioned by G. Evelyn

Hutchinson (1957) and others following in his path. Before ecology and population biology had become thoroughly fused, the niche had been given a variety of meanings. The California naturalist Joseph Grinnell (1917) used the niche to describe the habitats and habits of birds; to Charles Elton (1927), the niche was the species' place in the biological environment—its relationship to food and enemies. G. F. Gause (1934) suggested that the intensity of competition between two species reflected the degree to which their niches overlapped. David Lack (1947) realized that niche relationships could provide a basis for evolutionary diversification of species.

But it remained for Hutchinson, of Yale University, to define the niche concept formally. Ideally, he said, one could describe the activity range of each species along every dimension of the environment, including such physical and chemical factors as temperature, humidity, salinity, and oxygen concentration, and such biological factors as prey species and resting backgrounds against which an individual may escape detection by predators. Each of these dimensions may be thought of as a dimension in space. If there were n dimensions, then the niche would be described in n-dimensional space. Of course, we cannot visualize a space with more than three dimensions; the concept of the n-dimensional niche is an abstraction. But we may deal with multidimensional concepts mathematically and statistically, and depict their essence by physical or graphical representations in three or fewer dimensions. For example, a graph relating biological activity to a single environmental gradient represents the distribution of a species' activity along one niche dimension. The level of activity, whether oxygen metabolism as a function of temperature or consumption rate as a function of prey size, conveys the ability of an individual to exploit resources in a particular part of the niche space and, conversely, the degree to which the environment can support the population of that species. In two dimensions, the individual's niche may be depicted as a hill, with contours representing the various levels of biological activity. In three dimensions, we must think of a cloud in space whose density conveys niche utilization. Beyond three, the mind boggles.

The niche of each species occupies a part of the n-dimensional volume that represents the total resource space available to the community. We may think of the total niche space of a community as a volume into which the niches of all the species fit, as do balls of various sizes packed into a box. The number of species in the community therefore depends upon the total amount of niche space and the average size of each species' niche. Community ecologists are interested in the factors that determine both these quantities. But one must first find ways to measure the niche.

Niche dimensions describe physical conditions and resource qualities.

We may conveniently think of the niche as a volume in a multidimensional space having coordinates defined by the values of continuously varying resource attributes. While conceivable, this abstraction has proven difficult to transfer to natural communities. The problem is in part practical and in

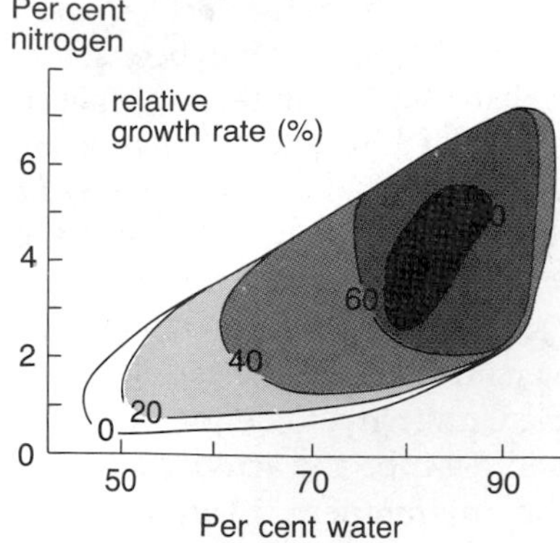

Figure 35–1
Relative daily growth rates of lepidoptera larvae reared on species of plants with different combinations of nitrogen and water concentrations in the leaves. (After Scriber and Slansky 1981.)

part conceptual. On the practical side, niches have many axes—virtually everything that one can measure about the ecology of a species—any subset of which may provide the basis for resource partitioning. So, how do we determine the important niche dimensions? Another practical difficulty is that resources and conditions of the environment often vary qualitatively or discretely and so cannot be described by continuous axes. Host plants are identifiable resources for herbivores, yet Latin names cannot be scaled along continuous axes. Measurement of host plant qualities, such as the chemical composition of the foliage, enables one to ordinate discrete resources along continuous axes, as shown in Figure 35–1. But this raises the conceptual problem of how to define the volume of a niche (that is, degree of specialization) when part of the niche space is devoid of resources. Does an herbivore feeding on five species of plants occupying a small part of the space defined by measurements of foliage quality have a larger niche than another herbivore feeding on only three species of plants more widely separated in niche space? Further complicating the definition of niche space are unique, qualitative attributes of resources, such as the presence or absence of a toxic chemical, a specialized predator, or a disease organism.

Another challenge to ecologists presented by the niche concept is how to characterize the occupancy of niche space. The presence or absence of a species within a habitat depends on the local growth of the population plus the net migration of individuals into and out of the area. It seems reasonable therefore that an appropriate measure of niche occupancy takes into account the contribution of each point in niche space to population growth.

Each organism has an activity space, a range of environmental conditions and resource qualities within which it can live. Each condition (temperature, insolation, acidity, humidity, salinity, soil particle size) and each attribute of resource (prey size, defenses, nutritional value, soil moisture and nutrient content, light intensity) is a single axis of the community niche space. At any particular instant, the individual resides at a single point representing the values of each niche dimension encountered at that time. Within this instant the individual may be gathering resources, escaping a predator, caring for its offspring, or interacting with other adults. Each of these activities makes a small increment to longevity or fecundity which, when summed over the individual's lifetime, define its contribution to population growth.

Clearly, a vast distance separates concept from practical application. What one can measure in nature includes the relative numbers or amounts of various prey items consumed, proportions of time spent in various microhabitats, and distributions of individuals over resource types. Some investigators have measured the relative qualities of different types of resources; for example, in terms of growth rates of lepidoptera larvae on foliage of different species of trees (Scriber and Feeny 1979, Hough and Pimentel 1978; Figure 35–1), and relative energetic efficiencies and handling times of rodents utilizing seeds of different species of plants (Rosenzweig and Sterner 1970) or seabirds utilizing coastal shellfish (Irons et al. 1986; Table 35–1). But the relationship of resource attributes to increments of population growth and the rate of resource depression continues to elude ecologists.

Table 35–1
Foraging parameters for glaucous-winged gulls feeding on intertidal organisms at Chichagof Harbor, Attu Island, Alaska

Prey item	Search time (s)	Handling time (s)	Energy per prey (kJ)	Net rate of energy gain (kJ h^{-1})
Urchins	35.8	8.3	7.45	606.7
Chitons	37.9	3.1	24.52	2153.9
Limpets	9.9	1.5	2.93	1020.5
Mussels	18.9	2.9	1.42	243.3
Barnacles	14.1	2.1	0.16	27.6

Source: Irons et al. 1986.

Assuming that one could define niche axes and quantify niche occupancy, a last set of practical problems concerns the boundaries of the niche itself. The considerations are as follows: individuals and populations may include within their niches parts of the resource spaces from a variety of habitats; a particular population may be differently specialized—for example, to different host plants (Fox and Morrow 1981)—in different habitats; different individuals within a habitat may exhibit different specializations associated with age, sex, size, genetic polymorphism, or individual preference (for example, Norton-Griffiths 1969, Roughgarden 1974).

Keeping in mind the practical limitations of field ecology, we may nonetheless begin to characterize niche relationships within biological communities by patterns of resource utilization and microhabitat occurrence of community members. Imagine for a moment that the niche occupancy of species can be portrayed as distributions along a single niche axis or dimension, such as those for the two species portrayed in Figure 35–2. Each distribution can be characterized by its size and position: the location of its peak along the niche axis, its breadth, and its height. The relationship between two areas of niche occupancy can be described by the distance between niche peaks and the degree of niche overlap. Many studies of communities have sought to describe structure in terms of the niche breadths and overlaps among species. In most of these studies investigators consider only three or four different niche "axes" (Schoener 1974).

Eric Pianka's investigation of lizard communities in desert habitats constitutes a particularly thorough study of niche relationships. Because he conducted similar studies in three widely separated regions having different species diversity (North America, Africa, and Australia), Pianka could determine which aspects of niche relationships varied in association with change in the number of species present in local communities. To characterize the niche dimensions of each species, Pianka (1973) used time of activity, microhabitat preference (open sun, shade; ground, grass, bush, tree), and type of prey eaten (termites, beetles, locusts, ants, and so on). He calculated the niche breadth (B_j) of species j by an equation analogous to Simpson's measure of diversity; that is, $B_j = 1/\Sigma(p_{ij}^2)$, where p_{ij} represents the proportion of records for species j in each category (i) of a particular niche dimension. Note that niche variables consist of sets of discrete entities without a

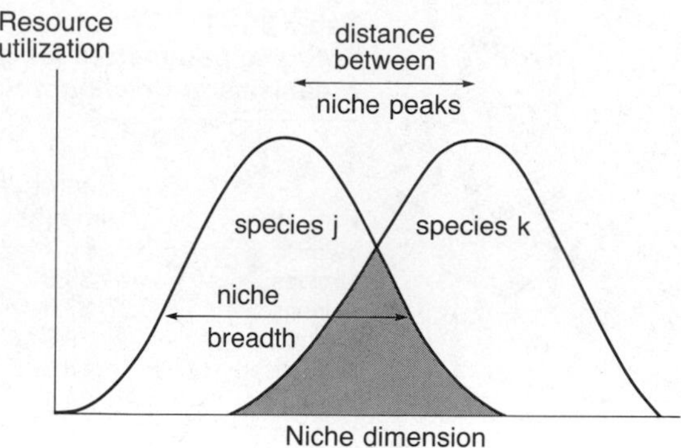

Figure 35–2
Activity curves of two species along a single resource dimension showing niche breadth, the distance between the centers of the niches, and niche overlap.

defined relationship among them; in calculating niche breadths, the difference between, for example, ants and beetles, is assumed to equal the difference between locusts and termites. B_j takes on values between 1 (low niche breadth; specialization on a single category of niche entities) and the number of categories (n) recognized for a given niche axis. For example, if the proportions of beetles, termites, and grasshoppers in the diet of lizard j were 0.10, 0.20, and 0.70, then its niche breadth for the diet axis would be $B_j = 1/(0.102 + 0.202 + 0.702) = 1.85$ (out of a total possible of three categories).

Because Pianka's niche axes are not continuous variables, it is not possible to define the position of each species' niche or the distance between the niches of two species. One can, however, calculate coefficients of overlap (O_{jk}) between the niches of two species (j and k) by expressions such as

$$O_{jk} = \frac{\Sigma(p_{ij}p_{ik})}{[\Sigma(p_{ij}^2)\Sigma(p_{ik}^2)]}$$

where p_{ij} is the proportion of species j's activity or diet recorded from category i of a particular resource dimension.

Robert MacArthur (1972) proposed an expression relating number of species in a community (S), total niche space available to those species (R), average niche space occupied per species (U), average number of other species co-occupying each species niche space (C), and average overlap in niche space between two such species (A), as follows:

$$S = \frac{R}{U}(1 + CA)$$

The ratio of R to U is the number of niches of size U that can be fit into a space of size R without overlap. The product CA is the number of species-equivalents sharing the niche space of each species. Thus if each species overlaps with five others an average of 0.20, then $CA = 1$ and 2 species ($1 + CA$), on average, occupy each region of the total community niche space.

Pianka's measurements allow us to evaluate the quantities in MacArthur's equation for each of the three regions, as shown in Table 35–2. In each of the three regions, lizards occupy niches of similar size, but the total niche space occupied by all lizards in Australia exceeds that in Africa and, especially, in North America. In order to make MacArthur's equation balance, average values of CA in the habitats within each region would have to be 1.6 in North America, 1.3 in Africa, and 1.2 in Australia. Differences in species diversity are more closely associated with total variety of resources utilized by the entire community than with specialization of individual species or degree of niche overlap.

Approaches similar to that of Pianka have been employed in a number of studies of niche organization among organisms other than desert lizards (Schoener 1974), including ants (Culver 1974, Davidson 1977), stem-boring insects (Rathke 1976), lizards (Schoener 1968), fish (Roughgarden 1974, Werner 1977), birds (Cody 1974), rodents (Brown and Lieberman 1973, Brown 1975), and plants (Parrish and Bazzaz 1976, Werner and Platt 1976, Yeaton and Cody 1979, Yeaton et al. 1985). Such studies were ardently pursued during the 1970s but failed to reveal general patterns of community organization. For example, Pianka's results indicated that the average size of a species' niche is conserved from region to region and that variation in species diversity paralleled variation in community niche space; in contrast, Martin Cody's work (1974) on birds suggested that more diverse communities exhibited denser packing of species in ecological space.

Discrepancies between studies may arise in part from the fact that niche relationships in different communities and different kinds of species are governed by different processes (Schoener 1983). But at least part of the problem is methodological. Each investigator uses different measurements of niche dimension, and the measurements themselves often depend on characteristics of the environment. Categories of microhabitat use in deserts and forests cannot be matched or compared. The foods eaten by birds, fish, and beetles are similarly incomparable.

Table 35–2
Relationship of local species diversity to resource diversity, average niche breadth, and degree of interspecific competition for lizards in desert habitats

Region	D_s	D_r	D_u	D_r/D_u	$(1 + ca)$
North America	4–10	25.9	7.7	3.34	1.2–3.0
Kalahari (Africa)	10–16	68.9	10.9	6.34	1.6–2.5
Australia	18–36	107.5	8.5	12.62	1.4–2.9

Values are presented for variables in the equation $D_s = D_r(1 + ca)/D_u$, where D_s is local species diversity, D_r is resource diversity, D_u is average niche breadth, c is the average number of niche neighbors, and a is the average niche overlap of neighbors.
Source: Pianka 1973.

Morphological similarity provides a habitat-free measure of ecological similarity.

Habitat bias in the measurement of ecological overlap may be avoided by considering an organism's structure as an expression of its niche relationships. Morphological adaptations are molded by the environment and therefore must reflect the position of the species in ecological niche space. In fact, the basic premise of ecomorphological analysis is that the positions of species in morphological space correspond to their positions in ecological niche space. The idea that one may use morphology to assess ecological relationship arose largely out of G. E. Hutchinson's (1959) observation that closely related species occurring in the same habitats and eating the same types of food often differed in body length by a factor of about 1.3. Accordingly, differences in body size suggested differences in the size of prey consumed; such ecological segregation enabled otherwise similar species to avoid strong competition and coexist. Both the existence and meaning of Hutchinson's ratio of 1.3 : 1 have become the subjects of intense debate in the literature of community ecology, as we shall see in Chapter 36. But his insight nonetheless generated an important approach to the description of community pattern.

Early attempts to analyze morphological differentiation within communities were those of Thomas Schoener (1965) and Henry Hespenheide (1971, 1973), who focused upon ratios of beak size among co-occurring birds. Since then the scope of morphological studies and the analytical methods employed have been greatly extended (Lederer 1984, Leisler and Winkler 1985). The validity of the ecomorphological analogy has been investigated and largely validated (Karr and James 1975, Cody and Mooney 1978, Miles and Ricklefs 1984, Gilbert 1985, Niemi 1985). Morphology offers several advantages over direct ecological analysis: measurements are independent of habitat context and readily comparable among studies. But there are also disadvantages: morphology is relatively fixed and therefore insensitive to subtle ecological shifts. Furthermore, species may be interrelated only by distance; morphology allows no direct analogy to ecological overlap.

In ecomorphological analyses of community structure, the degree of packing of species into niche space is estimated by morphological distances between nearest neighbors or by the average morphological volume occupied per species. The volume of morphological space occupied by the entire community is derived from the variances of the species distributions along each of the morphological axes (Travis and Ricklefs 1983).

Morphological analyses of community structure have consistently revealed relatively constant density of species packing in morphological space; diversity among communities varies in direct relation to the total morphological volume occupied. This finding suggests that added species increase the variety of ecological roles played by members of the community. In a study of bat communities in temperate and tropical localities, M. V. Fenton (1972) used two dimensions to define morphological space: the ratio of ear length to forearm length and the ratio of the lengths of the third and fifth digits of the hand bones in the wing. The first—a measure of ear length relative to body size—is related to the bat's sonar system and, thus, to the type and location of its prey. The second dimension is related to the shape of the wing—whether it is long and thin or short and broad—therefore providing an index to flight characteristics of the bat and, in turn, to types of prey it can pursue and habitats within which it can capture prey efficiently.

When one plots each species of bat in a community on a graph whose axes are the two morphological dimensions, one can visualize the niche relationships among species (Figure 35–3). In the less diverse community in Ontario, the species have similar morphology; all of the species are small insectivores. The more diverse community in Cameroun occupies a much larger volume of morphological space, corresponding to the greater variety of ecological roles played by bats there. To small insectivorous species are added fruit-eaters, necter-eaters, fish-eaters, and large, predatory bat-eaters. More detailed analyses of bat communities (Findley 1973, 1976; Findley and Black 1983; and Schum 1984) have verified Fenton's finding that morphological space increases in direct proportion to number of species. The same is true of comparisons of birds between tropical and temperate localities (Karr and James 1975).

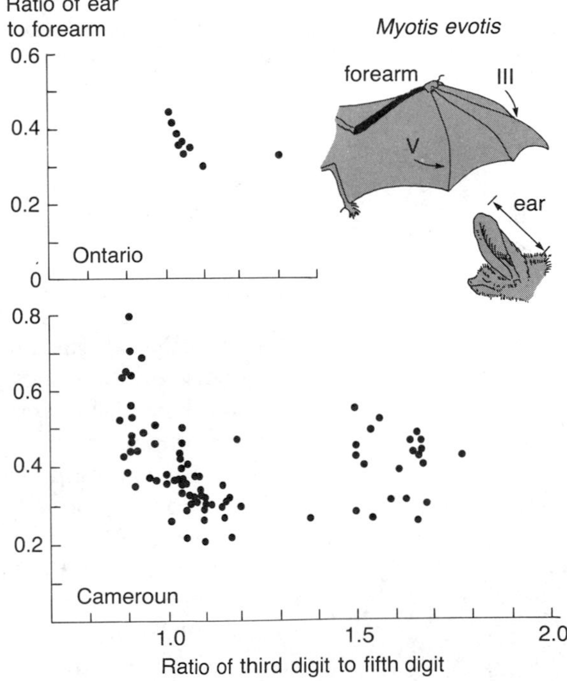

Figure 35–3
Distribution in morphological space of species in the aerial-feeding bat faunas of southeastern Ontario *(top)* and Cameroun *(bottom)*. The horizontal axis is the ratio of the lengths of the third and fifth digits of the hand and the vertical axis is the ratio of ear length to forearm length. (After Fenton 1972.)

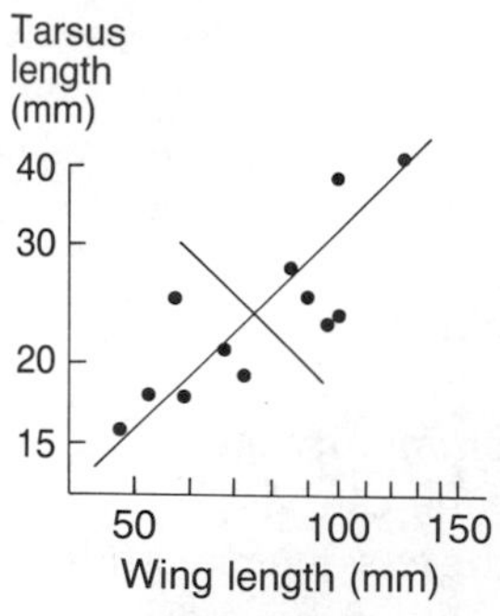

Figure 35-4
Relationship between the logarithm of wing length and the logarithm of tarsus length among passerine birds in a chaparral habitat in California.

In some cases, ecological and morphological analyses of the same communities have led to different conclusions. Ricklefs and Joseph Travis (1980) analyzed morphological relationships among birds living in shrub habitats in western North America and Chile, whose ecological relationships were described by Cody (1974). Morphological space was defined by eight dimensions, which were the logarithms of eight measurements that can be obtained from museum skins (lengths of the bill, wing, tarsus [lower leg], and so forth). Figure 35-4 shows how the morphological space looks when portrayed in two dimensions, in this case tarsus length and wing length. Ricklefs and Travis used logarithms of measurements for two reasons: when variables are log-transformed, nearest neighbor distances tend to be more uniform throughout the morphological space than they are when portrayed as linear dimensions, and straight lines in morphological space represent ratios and products of measurements and thus correspond to size and shape. For example, in Figure 35-4, values represented by points on a line that runs diagonally from lower left to upper right are positively correlated with wing and tarsus length. These values are therefore a linear combination of log(wing) and log(tarsus) and represent the product of the two measurements [log(wing) + log(tarsus) = log(wing × tarsus). Similarly, values on a diagonal line from the upper left to the lower right represent the ratio of two measurements [log(tarsus) − log(wing) = log(tarsus/wing)]. At one end of this line, one would find such ground-feeding birds as thrushes, having relatively long legs and short wings; on the other end, one would find such aerial species as swallows, having relatively long wings and short legs.

The relationships between two species in morphological space can be assessed by their Euclidean distance, the length of the straight line connecting their positions. The equation for this distance (D_{ij}) between species *i* and *j* is that for the hypotenuse of a right triangle (Figure 35-5). In a space defined by axes *X* and *Y*,

$$D_{ij} = [(X_j - X_i)^2 + (Y_j - Y_i)^2]^{1/2}$$

We may calculate distance in a space of higher dimension by adding terms for each additional axis [that is, $(Z_j - Z_i)^2$] within the square root. The average distance between each species and its nearest neighbor provides a measure of packing in morphological space.

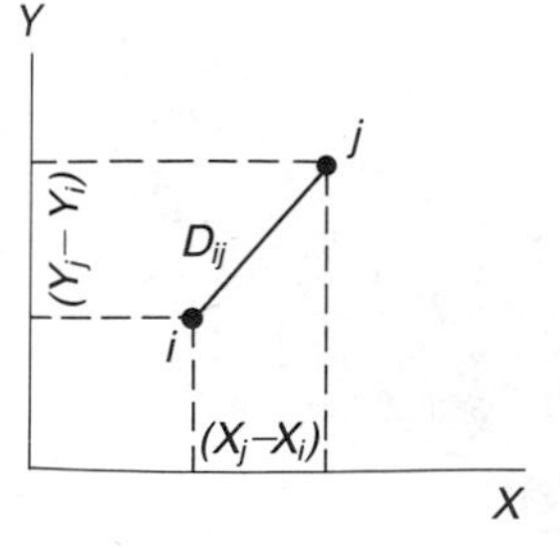

Figure 35-5
Euclidian distance (D_{ij}) between points *i* and *j* in two-dimensional space.

For the communities studied by Cody (1974), which consisted of between 6 and 17 species of passerine birds, average nearest neighbor distances were independent of diversity and averaged 0.21 $\log_{10}$ units. Thus as diversity increases, the total morphological (and, presumably, ecological) space occupied expands. This result is consistent with analyses of Gordon Orians (1969) and John Terborgh (1980) who found lesser ecological variety in the bird communities of temperate forests compared to tropical forests. The latter abound with fruit-eaters and nectar-feeders and with specialized insect-eaters without counterparts at higher latitudes. The conservation of species packing apparent in the morphological analysis differs, however, from Cody's result. According to Cody, degree of ecological overlap between species increased as more species were added. Among the communi-

Table 35–3
Average nearest neighbor distances (NND) and farthest neighbor distances (FND) based on morphological characteristics of lizards in desert habitats

Region	Number of localities	Species per locality	NND Median	NND Range	FND Median	FND Range
Australia	8	18–36	0.21	0.19–0.30	2.16	2.12–2.22
Africa	12	10–15	0.15	0.13–0.19	0.87	0.70–0.99
North America	13	4–10	0.32	0.21–0.47	1.06	0.93–1.46

Source: Data in Ricklefs et al. 1981.

ties of lizards studied by Pianka, a morphological analysis revealed similar nearest neighbor distances in the Australian (most diverse) and North American (least diverse) faunas. But, by contrast, the African species were packed more densely in morphological space, perhaps because many of the African species are similarly specialized to feed on termites (Table 35–3).

Escape space adds dimensions to the niche.

Niche dimensions are not limited to feeding and to tolerance of physical conditions. Avoidance of predation is equally important to population processes, and avenues of predator escape constitute dimensions of the niche along which species may be diversified (Lawton and Strong 1981). One would think that predators would be most efficient when they focused their attention on portions of the niche space most densely occupied by prey species (Martin 1988). Where many prey species use the same mechanisms to escape predation, predators having adaptations or learned behaviors that enable them to exploit these prey will be favored. Thus these prey populations will suffer increased mortality. Conversely, prey having unusual adaptations for predator escape should be strongly selected—a variant of the odd-man-out idea. As a result, predation pressure should diversify prey with respect to escape mechanisms; such species should therefore tend to become uniformly distributed within available escape space (Blest 1963, Rand 1967). The quality of a particular place within niche space depends upon predator attributes (hunting methods, body size, visual acuity and color perception, and so on) and escape space (color and pattern of resting background for cryptic prey, availability of hiding places and other refuges, structure of the vegetation for those relying on fleeing to escape, and so on). But compared to the analysis of trophic niche relationships, the positions of species with respect to escape space has been paid little attention outside of papers on cryptic insects (Sargent 1966, Ricklefs and O'Rourke 1975, Otte and Joern 1977, Endler 1978, 1984, Pearson 1985).

Ricklefs and O'Rourke used morphology to estimate the packing of species of moths in escape space, in this case the variety of backgrounds against which day-resting moths conceal themselves to avoid detection by diurnal visual predators. Among cryptic species, appearance has evolved to match the background against which the species rests. Hence we assume that the morphological appearance, or aspect, reflects characteristics of the resting place and the searching techniques of predators to be avoided (Figure 35–6). The variety of cryptic patterns has been referred to by A. Stanley Rand (1967) as aspect diversity.

Ricklefs and O'Rourke described the appearance of moths by 12 characters, including morphology and position of the legs, coloration, and reaction to disturbance (Figure 35–7). Samples of moths were analyzed from three areas: a spruce-aspen forest in Colorado (43 species), a Sonoran desert

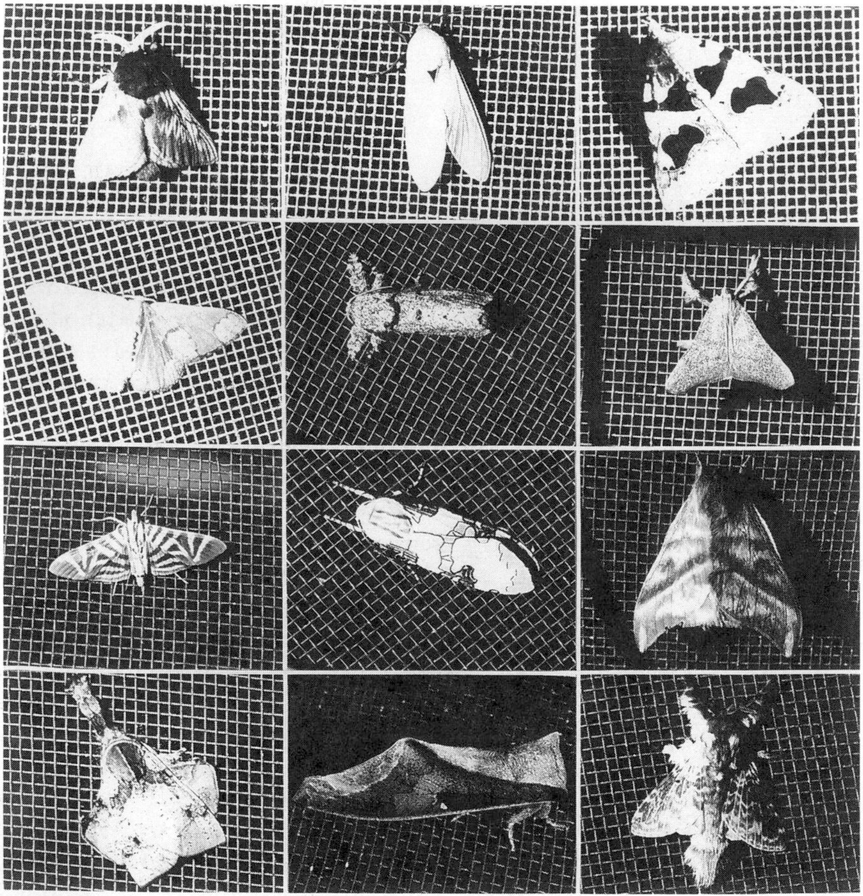

Figure 35–6
Representative species of moths from Panama photographed against the window screens to which they were attracted by ultraviolet lights. The moths show the variety of appearances in the community. (From Ricklefs and O'Rourke 1975.)

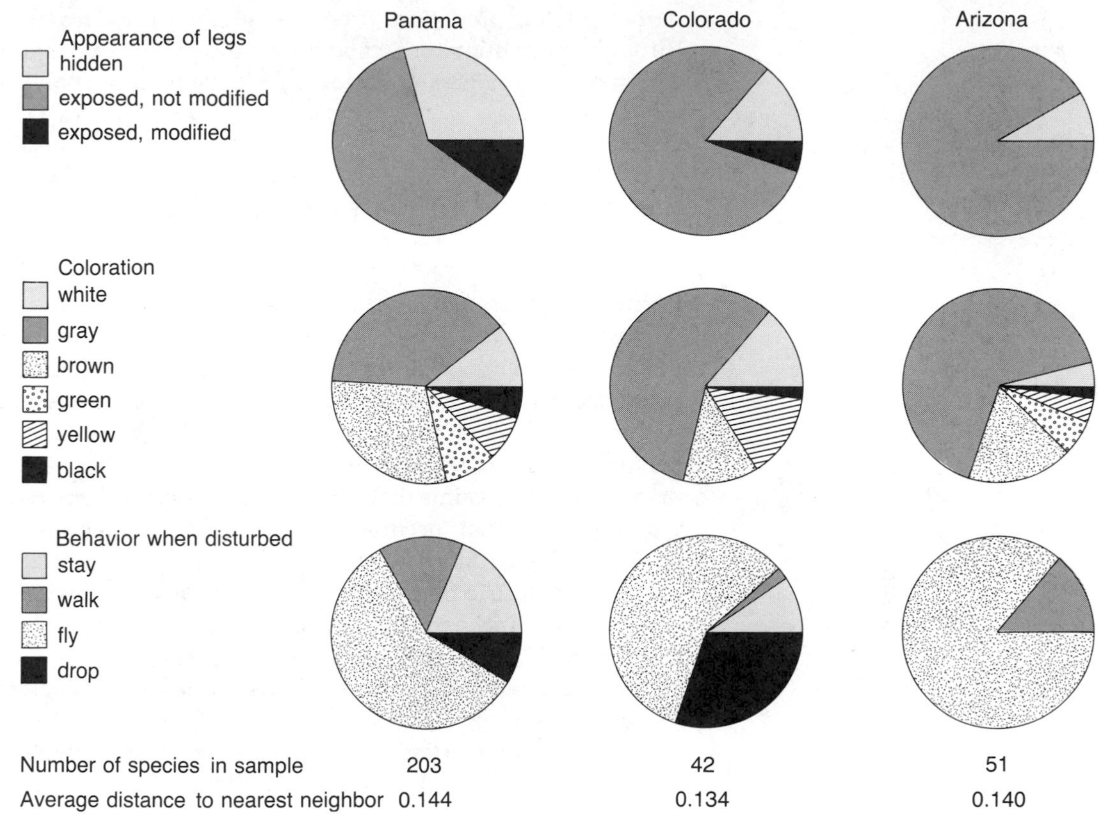

Figure 35–7
Diversity in the appearance and behavior of moths from three localities exemplified by 3 of the 12 characters utilized by Ricklefs and O'Rourke (1975). Variety is greatest in the most diverse sample, that from Panama.

habitat in Arizona (51 species), and a lowland rainforest habitat in Panama (203 species). The number of moth species sampled probably represents the differences in diversity between the areas.

The total volume of escape space utilized, as revealed by the variety of morphological characters represented in each sample, was greater in Panama than in both Colorado and Arizona. A Euclidian distance measure of the differences between species further revealed that the density of species packing in the morphologically defined escape space was nearly identical in the three samples, varying between 0.134 and 0.144 units. These distances were calculated differently than the morphological distances in the bird communities described above and are not comparable. But the result is the same: species are added to a community by expansion of the niche space utilized rather than by denser packing of the same space. Variation in the amount of escape space utilized could arise from a number of factors: variation in the diversity of escape possibilities presented by the environment, greater predation pressure, greater pressure for diversification through competition for escape space, and greater opportunity for diversification owing to a greater variety of independently evolving populations.

Commonness and rarity are related to occupation of niche space.

Within any given habitat or study area, some species are encountered frequently, others less often, and a few are so rare as to be represented by unique specimens or observations; an unknown number of species are present but unrecorded. As we have seen, distributions of abundances within communities appear to follow regular patterns. Abundant species certainly utilize more of the resources available within a habitat than do rare ones. One must assume that the common species either are very generalized with respect to their occupancy of niche space, specialized on a predominant resource within that space, or extremely efficient at utilizing the resources available to them. This leads us to consider how relative abundance depends upon niche relationships. Do generalists tend to be more abundant than specialists? Do some portions of the niche space support more consumers than others? To what extent is species packing related to density of resource within niche space?

None of these questions can be answered definitively, except in a few special or trivial cases. For example, Deborah Rabinowitz (1981) found that most rare species of plants in midwestern prairies are specialized to occupy distinctive, infrequently occurring microhabitats, such as the mounds of dirt thrown up by burrowing animals. On these mounds themselves, such specialists are abundant and predictable, but over the prairie as a whole they are rare. Other rare species include those that only incidentally enter a study area, but are common in other habitats.

In a study of populations of birds in lowland tropical forests in Panama and Costa Rica, Karr (1977) determined that many species which are locally rare or irregular are more abundant in other habitats. Of 331 species of birds recorded from a 100-hectare plot of forest at La Selva, Costa Rica, 82 were considered rare. Of these, 51 were species primarily associated with other habitats; they occasionally happened to wander into the forest (Table 35–4). An additional 12 species were difficult to detect, because they either had large home ranges and did not often cross the study area (large, raptorial birds), were secretive, or were active only at night. Another 7 of the rare species occupied the study area on a seasonal basis, migrating to other regions or habitats for the rest of the year. Only 6 of the 82 were considered rare because of specialized feeding habits. Two of these were ant-followers, feeding on the insects flushed by moving swarms of army ants. Four had unusual feeding behaviors; for example, falcons of the genus *Micraster* provoke small birds to mob them and then pick off individuals whose attention lapses (Smith 1969). In the last 6 of the 82 rare species, low population density could not be associated with any particular factor. Karr's analysis leads us to wonder whether any species are truly rare, or whether

Table 35-4
Ecological correlates of rarity among species of birds in lowland rainforest habitat in central Panama and eastern Costa Rica

	Panama	Costa Rica
Species associated with other habitats		
Aquatic	5	12
Second growth, forest edge	16	13
Foothill species	6	20
Dry forest	4	6
Sampling problems		
Large species with large home range	15	6
Species difficult to observe	4	4
Nocturnal species	4	2
Seasonal movements		
Out in the dry season	6	4
In during the dry season	7	3
Specialized species		
Ant-following species	4	2
Specialized hunting technique	3	4
Unknown	3	6
Total number of "rare" species	77	82

Source: Karr 1977.

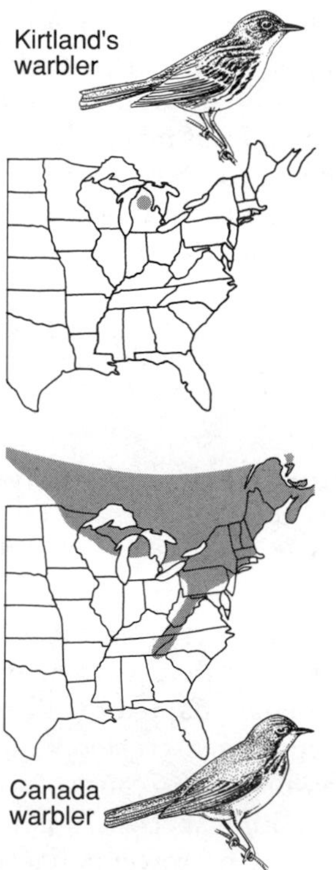

Figure 35-8
Geographical ranges of two closely related species of bird illustrating extremes of current ecological success. (From Peterson 1980.)

rarities are merely individuals out of place, such as the European birds infrequently carried to the shores of North America by storms or errors in navigation.

Species pushed to the brink of extinction by man aside, many populations of animals and plants must nonetheless be considered infrequent wherever they are found; indeed, their total populations may be very small. One example of such a species is the Kirtland's Warbler *(Dendroica kirtlandii),* which breeds solely within a small area of Michigan, but which superficially resembles closely related species whose populations number in the millions and spread across most of North America (Figure 35-8). Even within its small breeding range, which is restricted to regenerating stands of jack pine, the 500 or so breeding pairs of Kirtland's warbler are sparsely scattered over fewer than 100 surveyor's sections (square miles) at densities of 1 pair per 10 to 100 acres of suitable habitat; that is, a tenth to a hundredth of the densities achieved by other species of *Dendroica* (Mayfield 1960). Why Kirtland's warbler is so rare is not known. It lives in a restricted habitat, but does not use all the habitat available. It does not appear to be peculiarly specialized within its habitat compared to other species of warblers.

Rather than focus on particular species, it may be more instructive to examine the distribution of species and their abundances over the available niche space. Practical problems of defining niche space have made this goal difficult to achieve. Here we shall try to catch a glimpse of niche relationships by examining the distribution of herbivores over plant species and the distribution of parasites over host species. In both cases, resources are discrete and may be easily descibed taxonomically, even though the qualities that distinguish food plants and hosts in the eyes of herbivores and parasites have not been characterized. We shall use data collected by the Canadian

Forest Insect Survey. The survey employs an extensive, albeit haphazard sampling scheme in which field personnel collect all the lepidoptera larvae and other species of herbivorous insects from selected individual trees. Each of these samples is called a collection, for which the individuals and species of herbivore are carefully enumerated. In addition, hymenopteran and dipteran parasites were reared from more than 50,000 caterpillars collected by survey teams.

These data provide two matrices of information on niche relationships: the incidences of each species of moth on each host tree, and the incidences of each species of parasite on each species of moth. Of the first matrix, one may ask whether parts of the niche space (in this case, species of trees) are more suitable to herbivores than others. Richard Karban and Robert Ricklefs (1984) analyzed CFIS data to answer this question. Because species number varies directly in relation to sampling intensity (Karban and Ricklefs 1983), the species richness of herbivores was adjusted according to number of collections. Using this criterion, it was found that some species of trees harbored unusually large numbers of species of lepidoptera (red and bur oaks, white birch, basswood, and bitternut hickory) whereas others were comparatively depauperate of herbivore species (cherries, hophornbeam). But differences in the diversity of herbivores were not related to any of 14 different physical and chemical measures of leaf quality. The number of caterpillars obtained per collection also varied among the common, well-sampled species, from more than 40 in trembling aspen, black and white ashes, Manitoba maple, and pin cherry, to fewer than 20 in white birch, basswood, bitternut hickory, white oak, sugar and red maples, and hophornbeam. But the number of individuals per collection was associated only with available carbohydrates among leaf quality traits, and the diversity of caterpillars was not related to their total abundance. Thus to herbivores, host plants appear to differ in "quality," but the determinants of quality are not obvious and they seem to affect abundance and diversity differently.

Do specialists utilize their restricted niche space more efficiently than generalists?

According to the old adage "Jack of all trades, master of none," specialization is synonymous with efficiency. Species of moths that lay their eggs on only a single species of host plant ought to be better adapted to utilize that plant than generalists confronted with overcoming a greater variety of antiherbivore tactics. In fact, many ecologists assume that a cause of specialization is strong competition for resources from other species, forcing each to increase its competitive ability by mastering a narrower range of resources. Thus specialists should utilize their preferred resource more efficiently than do those generalists that include the resource within their niche. Returning again to moth larvae, experiments designed to test the specialization/efficiency hypothesis have produced ambiguous results. John Smiley (1978) failed to find a relationship between specialization and larval growth

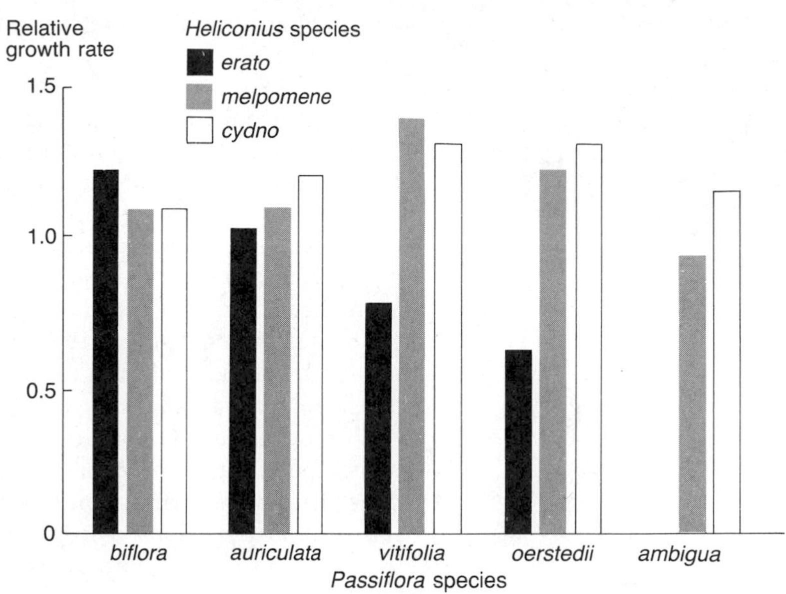

Figure 35–9
Growth rates of the larvae of three species of butterflies of the genus *Heliconius* on various species of *Passiflora*, the normal host genus for *Heliconius*. The usual host of *H. erato* is *P. biflora*, while that of *H. melpomene* is *P. oerstedii; H. cydno* is a generalist. (From Smiley 1978.)

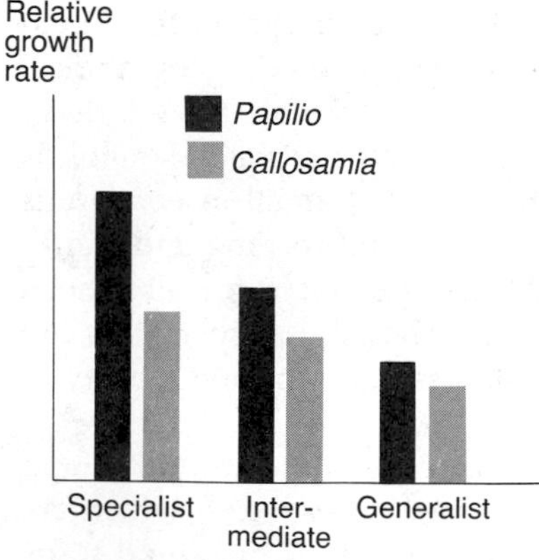

Figure 35–10
Relative growth rates of larvae of the lepidopteran genera *Papilio* (swallowtail butterflies) and *Callosamia* (silk-moths) in relation to degree of larval food-plant specialization. (After Scriber and Feeny 1979.)

performance among butterflies of the genus *Heliconius*, which feed on various species of passion flowers *(Passiflora)* (Figure 35–9). In contrast, however, J. M. Scriber and Paul Feeny (1979) found that specialized members of the swallow-tailed butterfly genus *Papilio* and the silk-moth genus *Callosamia* grew better on their normal host plants than did closely related generalists (Figure 35–10).

Whether differences in larval growth rates and feeding efficiencies are translated into differences in population size and occupancy of niche space raises a further question. Returning to the Canadian Forest Insect Survey data, Dean and Ricklefs (1980) found that the abundance of each species of moth in collections was independent of the number of genera of trees included among its host plants. Generalists were just as abundant as specialists. In addition, the number of individuals of each species of parasite reared per host (parasite efficiency) was positively correlated with the diversity of moth species attacked by parasitic wasps and unrelated to diversity among parasitic flies (Dean and Ricklefs 1979). Again, specialization does not appear to bring about expected increases in populations or resource use efficiency. One reason for this lack of relationship may be that specialization is forced upon consumers by reduced productivity and competitive ability resulting from increased predation or detrimental change in physical conditions.

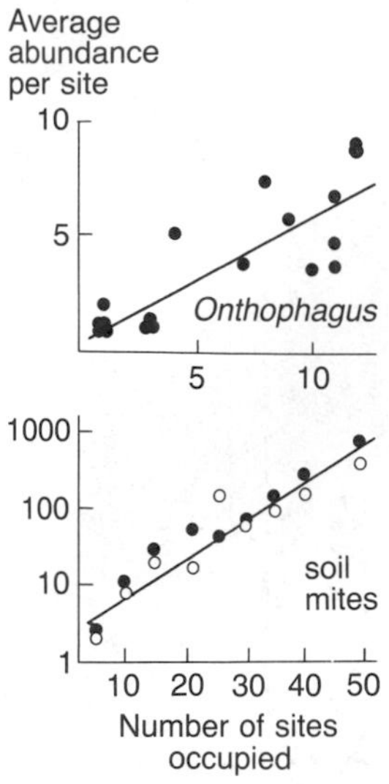

Figure 35–11
Relationship between local abundance and distribution for scarabid beetles of the genus *Onthophagus* and oribatid mites. (From Hanski 1982.)

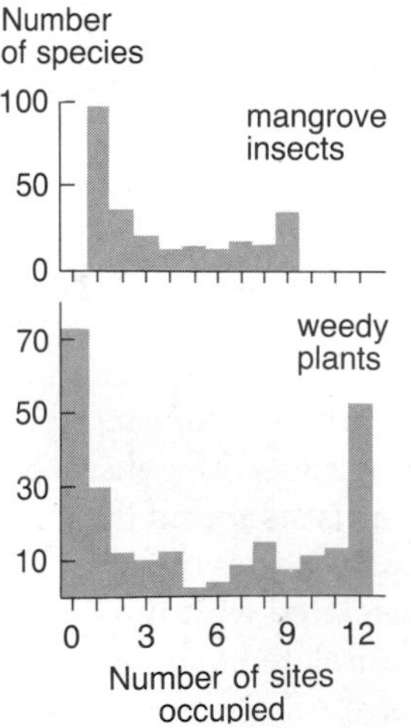

Figure 35–12
Frequency distributions of species in a sample of insects on Florida mangrove trees and a sample of weedy plants in Russian villages. (From Hanski 1982.)

Are local ecological specialization and habitat specialization related?

The complementarity of local ecology and geographical distribution has been recognized from the earliest days of the discipline (Willis 1922, Elton 1927, Pitelka 1941, Andrewartha and Birch 1954). Broad surveys of habitat distributions of animals and plants have consistently revealed that the more widespread species tend to be the more abundant locally (McNaughton and Wolf 1970, Price 1971, Able and Noon 1976, Parrish and Bazzaz 1976, Muhlenberg et al. 1977, Ricklefs and Cox 1977, Hanski and Koskela 1977, 1978, Dueser and Shugart 1979, Bock and Ricklefs 1983; Figure 35–11). Indeed, I. Hanski (1982) has suggested that the relationship is so prevalent that it amounts to an ecological "rule." S. J. McNaughton and L. L. Wolf (1970) suggested that both local abundance and geographical distribution reflect specialization on abundant resources or prevalent ecological conditions. They and, more recently, J. H. Brown (1984) have implied that resources and conditions which can support large numbers of individuals locally are also likely to be most extensive geographically and across habitats.

The McNaughton-Wolf hypothesis can be tested only when one can determine the occupation of niche space by a species over its geographic range. Several lines of evidence suggest that their idea of specialization on abundant resources or pervasive conditions does not provide a universal explanation for the positive correlation between local abundance and geographical/habitat distribution. First, many widespread but locally specialized herbivores shift host plants from one part of their range to the next (Fox and Morrow 1981). This phenomenon has not been studied systematically to assess its prevalence, and one could argue that differences between species of host plants, many of which are closely related, are inconsequential to the particular specializations that determine population size. Second, R. T. Paine (1980) has documented a parallel situation in the starfish *Pisaster* in which its principal prey vary so much over its broad geographical range that the diets of northern and southern individuals have little in common (Table 35–5). Third, patterns of habitat distribution of songbirds in deciduous forests of North America indicate that small insectivorous species are greater habitat specialists than are ground-feeding and trunk-feeding species, although vegetation structure is suitable for all groups throughout the habitat range (Ricklefs 1972). The habitat specialists are merely replaced by other species with similar habits from one locality to another.

According to Hanski's (1982) view, with respect to distribution and local abundance, species fall into two categories: core species, which are abundant and widely distributed, and satellite species, which tend to be restricted and scarce. This pattern is reminiscent of Raunkiaer's J-shaped distribution of species abundances, examples of which are illustrated in Figure 35–12. Guided by this premise, Hanski suggested that population processes possess a saddle point, to one side of which the population increases toward ubiquity, and to the other side of which the population decreases toward extinction. In thinking about this problem, Hanski forsook

Table 35-5
Geographical variation in the diet of the predatory sea star

	Proportion of prey in diet at				
	Punta Baja, Mexico	Monterey Bay, California	Outer Coast, Washington	Friday Harbor, Washington	Torch Bay, Alaska
Major prey taxa					
Mytilus mussels	0.30	0.17	0.18	0.12	0.80
Balanus barnacles		0.30	0.54	0.51	0.18
Tetraclita barnacles	0.45	0.26			
Chthamalus barnacles			0.10		
Pollicipes goose-neck barnacles	0.10	0.04	0.03		
Carnivorous gastropods		0.02	<0.01	0.04	0.02
Herbivorous gastropods	0.10	0.11	0.12	0.25	
Chitons	0.03	0.04	0.01	0.07	

Source: Paine 1980, after various sources.

the specific niche relationships of species to concentrate on a more general proposition, namely that resistance to local extinction is directly related to spatial distribution, or proportion of possible sites occupied. He further proposed that probability of reinvasion of a local area is also directly proportional to the number of other areas occupied. When Hanski combined these relationships in a model with a moderate level of environmental variation (producing variation in probability of extinction), the steady-state distribution of species among the possible number of sites becomes bimodal, with concentrations of species toward the extremes of ubiquity and global extinction.

The inverse relationship between extinction probability and distribution seems empirically justified in studies of island faunas, where local extinction (disappearance from an island) and distribution (number of islands occupied) can be measured unambiguously (Simberloff 1976, Ricklefs and Cox 1972). Whether other aspects of Hanski's mathematical model are realistic remains to be determined. It is doubtful to me that bimodality of abundance is a general phenomenon. We have seen in Chapter 34 that Raunkaier's J-shaped distribution of species abundances is sensitive to sampling design. Bimodality in number of sites occupied may suffer from the same type of sampling artifact.

The pertinence of Hanski's work, and earlier models of populations in spatially segregated environments (Cohen 1970, Levins and Culver 1971, Slatkin 1974, Levin 1976), derives from the fact that local and regional persistence are linked by migration between localities (or habitats). To the degree that migration couples spatially segregated populations, the niche of a species and its niche relationships should be considered over the entire extent of the population (Pulliam 1988). Thus geographical distribution need not be thought of in terms of specialization to particular factors coupled with the distribution of those factors. Rather, each species should be thought of as being specialized within the entire geographical and ecological range accessible to it. Habitat specialization extends individual ecologi-

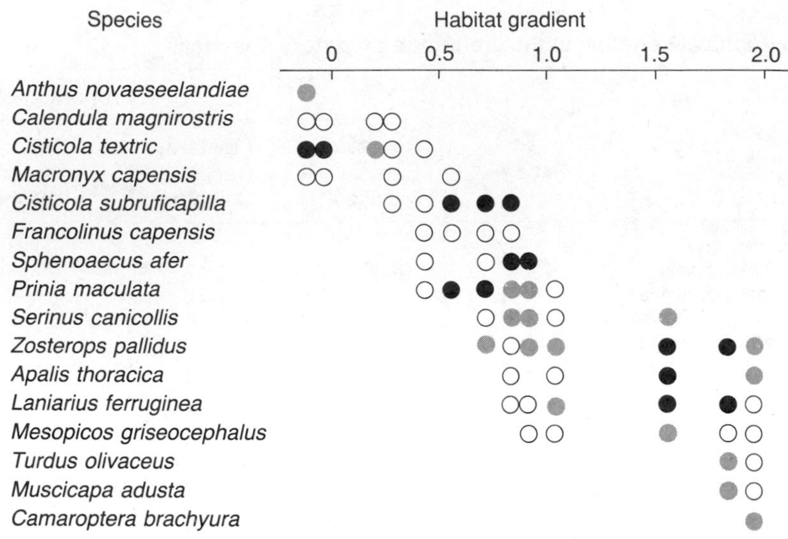

Figure 35–13
Occurrences of species of songbirds along a habitat gradient in the Mediterranean climate region of South Africa. The habitat gradient is roughly proportional to the height of the vegetation. Denser populations of birds are indicated by dark symbols. (Data from Cody 1975.)

cal specialization to the level of the population. To the degree that individuals migrate between local populations, linking the populations both dynamically and evolutionarily, specialization becomes a general property of the species rather than of individuals in a local population.

Consider the distribution of species of songbirds along an ecological gradient of habitat structure within the Mediterranean climate region of South Africa. Species are distributed along a habitat axis over the geographic region (Figure 35–13) much as they would be along an ecological dimension of the niche space within a locality. The entire region is accessible to all the species. Behavioral flexibility, phenotype-dependent habitat selection (Herrera 1978), and local adaptation (ecotypic differentiation) would allow individuals to respond to changes in niche space along the ecological gradient. Now, geographical distribution becomes in part a function of specialization on widespread ecological opportunities, as envisioned by McNaughton and Wolf (1970), phenotypic flexibility in response to environmental change, and the ability of population surpluses in favorable habitats to maintain populations in less suitable areas through emigration. Particular resources utilized may not be so important as, or no more important than, other factors, such as predation, disease, and climate, which influence population growth.

This idea of the regional niche and the interaction of species within that enlarged concept of ecological space is little more than a concept at present. It shall, however, be useful when we consider factors responsible for the regulation of community structure in Chapter 36.

SUMMARY

1. The niche concept expresses the relationship of the individual to physical and biological aspects of its environment. Each factor may be thought of as an axis of a multidimensional niche space, within which organisms and populations occupy characteristic spaces. The niche of a species is defined by the range of values on each niche axis within which the individual or population can persist.

2. The ecological relationship between two species can be described by the degree to which their niches overlap, which is a function of niche breadth and niche separation.

3. Field studies have attempted to measure niche relationships by quantifying overlaps among species with respect to their utilization of microhabitats and food items. Comparisons between habitats are difficult because niche measures are habitat dependent.

4. Morphological similarity provides a habitat-free estimate of ecological similarity (niche separation), although not of ecological (niche) overlap. Comparisons of the packing of species in morphological space have, with few exceptions, indicated that differences in diversity are associated with differences in total space occupied rather than distances between neighboring species.

5. Opportunities for escaping predators are an important component of the niche and can be considered a resource for which prey species compete.

6. Ecologists have yet to determine the relationship between local abundance and the occupation of niche space. Most species that are rare locally are common elsewhere. Globally rare species do not appear to share any particular ecological traits and may be superficially indistinguishable from more abundant relatives.

7. Although one assumes that specialists, which occupy a small niche space, must be more efficient at utilizing that space than generalists, few data support this proposition. The ecological and evolutionary significance of specialization remains a challenging problem.

8. Species that are widespread geographically also tend to be abundant locally. Such species may be specialists on abundant and widespread resources, or they may be ecologically so flexible as to exploit a wide variety of resources locally and different resources in different places.

9. Different local populations are coupled demographically by migration of individuals between them. Thus a comprehensive concept of niche relationships must include the distribution of species across habitats and within regions.

36

Regulation of Community Structure

Comparisons of communities of plants and animals have revealed certain patterns which suggest that community properties may be regulated. One example of such a pattern is the trophic organization of the community, for which the laws of thermodynamics dictate that energy flux decreases at each higher level in the food chain. This organizing principle produces certain regularities in the distribution of numbers of individuals and biomass among trophic levels within the community (Brown and Maurer 1987, 1989).

Ecologists have also noted patterns in communities that are seemingly indifferent to energetic constraints. The most important of these are certain regularities in the numbers of species within communities, or species diversity. As we have noted earlier, larger islands tend to support more species than smaller islands, suggesting that diversity is somehow regulated with respect to area or to some ecological factor correlated with area. To cite another example, biologists have found more kinds of organisms in the tropics than at higher latitudes. This pervasive pattern, which is equally apparent in the Northern and Southern Hemispheres, and in the New and Old Worlds, provides a focus for the present chapter.

The great naturalist explorers of the last century — Charles Darwin, H. W. Bates, Alfred Russel Wallace, and others — clearly recognized that the tropics held a great store of undescribed species, many having bizarre forms and habits. This remains true to the present. Taxonomists have so far described fewer than 2 million species. But by extrapolating the rate of discovery of new insects and other life forms, some biologists have estimated that as many as 30 million kinds of animals and plants may inhabit the earth (Wilson 1988).

Why there are so many species of organisms in the tropics (why are there so few toward the poles)? This question and, more generally, the factors that regulate the diversity of natural communities provide the theme of this chapter. Biologists hold two kinds of views. One is that diversity increases without limit over time; tropical habitats, being much older than temperate and arctic habitats, have had time to accumulate more species. The second is that diversity reaches an equilibrium at which point factors removing species from a system balance those that add species. Accordingly, factors that add species weigh more heavily in the balance as one moves toward the tropics.

Throughout the first half of this century, the nonequilibrial viewpoint found the broadest favor. Tropical habitats, one believed, had persisted since the beginning of time; vicissitudes of climate, particularly during the last ice age, occasionally destroyed most temperate and arctic habitats, resetting the diversity clock, so to speak. More recently, however, with the integration of population ecology into community theory, ecologists have considered diversity as an equilibrium between opposing diversity-dependent processes, just as equilibrium population size represents the balance between opposing density-dependent birth and death processes. This viewpoint challenges ecologists to identify the processes responsible for increasing and decreasing the number of species in communities and to understand why the balance between these processes differs systematically from place to place (Roughgarden 1989).

During the past decade, ecologists have intensely debated the regulation of diversity. Arguments over the relative importance of competition, predation, and disturbance in "structuring" communities have been particularly vehement. Indeed, the equilibrium view itself has come under increasing pressure from contrary evidence. In the pages that follow, I shall attempt to introduce the principal views on the regulation of community diversity and outline the evidence bearing upon them. The diversity problem is crucial to ecology because it measures how well we understand the integration of ecological processes in natural systems. We most likely have identified all the factors that impinge on community diversity. Species production and diversification, migration of species between areas, competitive interactions between populations, overexploitation of prey resources by predators and disease organisms, evolutionary specialization of populations both within and among habitats, and the consequences of climate changes for populations have all piqued the interests of ecologists and evolutionary biologists over the last 100 years. But how these processes act in a diversity-dependent manner to determine rates of species production and extinction remains largely beyond our grasp. And so the diversity issue acts as a barometer of some of the most fundamental areas of understanding in the natural sciences. While we watch with justified pride as ecology matures as a science, a quick glance at the diversity barometer tells us that much of the best lies ahead.

More species occur in tropical regions than in temperate and arctic regions.

Within most large taxonomic groups of organisms—plant, animal, and perhaps microbial—the number of species increases markedly, with a few exceptions, toward the equator (Fischer 1960, Stehli 1968, Stevens 1989).

Figure 36–1
Desert and marsh vegetation illustrate extremes of production with an inverse relation between production and diversity. (Sonoran Desert of Baja California and Malheur Refuge, Oregon; courtesy of U. S. Department of Interior.)

For example, within a small region at 60° north latitude one might find 10 species of ants; at 40° there may be between 50 and 100 species; and in a similar sampling area within 20° of the equator, between 100 and 200 species. Greenland is home to 56 species of breeding birds, New York boasts 105, Guatemala 469, and Colombia 1395 (Dobzhansky 1950). Diversity in marine environments follows a similar trend: arctic waters harbor 100 species of tunicates, but more than 400 species are known from temperate regions, and more than 600 from tropical seas (Fischer 1960).

Within a given belt of latitude around the globe, the number of species may vary widely among habitats according to productivity, degree of structural heterogeneity, and suitability of physical conditions (see, for example, Kotler and Brown 1988, Gentry 1988). For example, censuses of birds in small areas (usually 5–20 hectares) of relatively uniform habitat reveal about 6 species of breeding birds in grasslands, 14 in shrublands, and 24 in floodplain deciduous forests (Table 36–1). In determining species diversity, habitat structure apparently overrides productivity. Marshes are productive but structurally uniform, and they have relatively few species. By contrast, desert vegetation is less productive but its greater variety of structure apparently makes room for more kinds of inhabitants (Figure 36–1). The relationship between structure and diversity is apparent to bird watchers and other naturalists, but Robert MacArthur and John MacArthur (1961) were the first to set it forth in a plain, graphical form for ecologists by plotting the diversity of birds observed in different habitats according to foliage-height diversity—a measure of the structural complexity of vegetation (Figure 36–2). M. H. Greenstone (1984) has shown a similar relationship for web-

Table 36–1
Plant productivity and the number of species of birds in representative temperate zone habitats

Habitat	Approximate productivity (g m^{-2} yr^{-1})	Average number of bird species
Marsh	2000	6
Grassland	500	6
Shrubland	600	14
Desert	70	14
Coniferous forest	800	17
Upland deciduous forest	1000	21
Floodplain deciduous forest	2000	24

Source: Tramer 1969; productivity data from Whittaker 1970.

building spiders, noting that the diversity of species varies in direct relationship to heterogeneity in the heights of the tips of vegetation to which such spiders attached their webs. Pianka (1967) found a close relationship between the number of species of lizards and the total volume of vegetation per unit area in desert habitats of the southwestern United States.

On a regional basis, the number of species is sensitive to the suitability of physical conditions, the heterogeneity of habitats, and isolation from centers of dispersal. In North America, the number of species in most groups of animals and plants increases from north to south, but the influence of geographical heterogeneity and the isolation of peninsulas is apparent. George Gaylord Simpson (1964) tabulated the number of species of mammals in 150-mile-square blocks distributed over the entire area of North America to the Isthmus of Panama. The number of species per block increased from 15 in northern Canada to more than 150 in Central America (Figure 36–3). Across the same latitude in the middle of the United States, more species of mammals live in the topographical heterogeneous western mountains (90 to 120 species per block) than in the more uniform environments of the east (50 to 75 species per block). The number of species of breeding land birds follows a similar pattern (MacArthur and Wilson 1967, Cook 1969), but reptile and amphibian faunas do not (Kiester 1971). Reptiles are more diverse in the eastern half of the United States than in the mountainous western regions; amphibians are strikingly underrepresented in the deserts of the southwest, owing to the requirement of most species for abundant water.

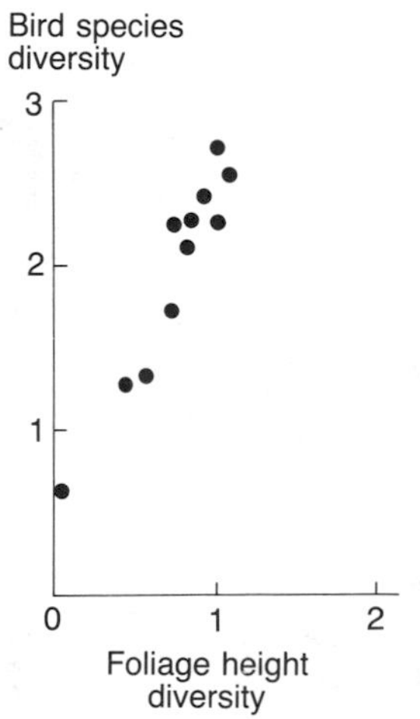

Figure 36–2
Relationship between bird species diversity and foliage height diversity determined for areas of deciduous forest in eastern North America. (From MacArthur and MacArthur 1961.)

Niche relationships change as diversity increases.

The community may be thought of as a group of species occupying niches within a space defined by axes of resource quality and ecological condition.

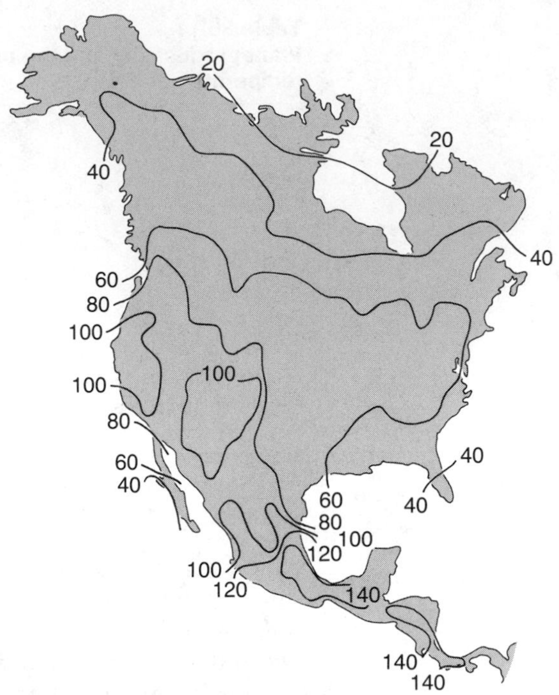

Figure 36–3
Species density contours for mammals in 150-mile-square quadrats in continental North America. (From Simpson 1964.)

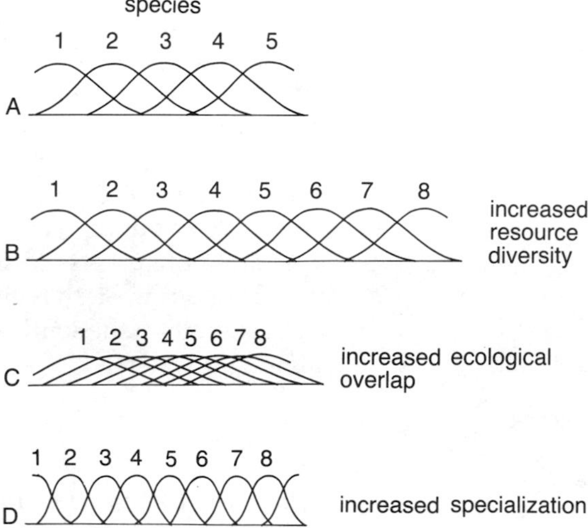

Figure 36–4
Schematic diagram showing how resource utilization along a continuum can be altered to accommodate more species.

Within this concept, adding or removing species has certain geometrical consequences. The possibilities are illustrated in Figure 36–4 by distributions of species' niches along a single continuous dimension, perhaps size of prey item or height distribution within the intertidal zone. Furthermore, let us characterize niche relationships by measures of niche breadth and overlap, and assume that the availability of resources is uniform throughout niche space. Now, the addition of species to the community can be accommodated by one, or any combination, of three types of adjustments. First, without changing niche relationships, the total size of the community niche may expand in direct proportion to the number of species. Note that niche size refers to variety of resources, not their amounts. Second, without changing niche breadth, increased diversity may be accommodated by increased niche overlap. In this case, the average productivity of each species is reduced owing to increased sharing of resources. Third, without increasing niche overlap, increased specialization may accommodate additional species in the community niche space. Here, too, average productivity decreases, because each species has access to a narrower range of resources.

Most ecologists agree that the high diversity of the tropics results at least in part from there being a greater variety of ecological roles. That is, the total niche space occupied is greater near the equator than it is toward the poles. For example, Gordon Orians (1969) suggested that part of the increase in the number of species of birds toward the tropics is related to an increase in frugivorous and nectarivorous species, and in insectivorous species that hunt by searching for their prey while quietly sitting on perches — a type of behavior uncommon among species in temperate regions. Among mammals, the increase in number of species between temperate and tropical areas results primarily from the addition of bats to tropical communities (Wilson 1974). Terrestrial mammals are no more diverse at the equator than they are in the United States and other temperate regions at a similar latitude, although their variety does decrease as one goes farther to the north.

In streams and rivers, the number of species in most taxonomic groups increases from the headwaters to the mouth of the river. One presumes that as a river increases in size, it presents a greater variety of ecological opportunities and its physical conditions become more stable and therefore reliable. Local communities reflect these changes. For example, a headwater spring in the Rio Tamesi drainage of east-central Mexico supported only one species of fish, a detritus-feeding platyfish (Figure 36–5). Further downstream, three species occurred: the platyfish, plus a detritus-feeding molly *(Poecilia)* that prefers slightly deeper water than the platyfish, and a mosquito fish *(Gambusia)* that eats mostly insect larvae and small crustacea. Species that appear in the community farther downstream include additional carnivores — among them, fish-eaters — and other fish that feed primarily on filamentous algae and vascular plants. None of the species drops out of the community downstream from any of the sampling localities. Thus diversity increases as the stream becomes larger and presents more kinds of habitats and a greater variety of food items.

Although these and other examples suggest an increase in total niche space with increasing diversity, two considerations must be borne in mind. First, changes in niche breadth (specialization) and overlap may also accom-

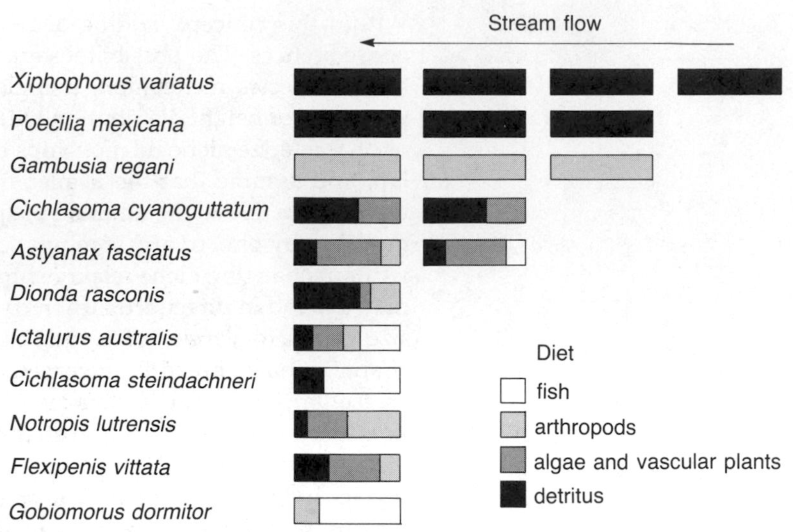

Figure 36–5
Food habits of fish species in four communities (vertical columns) from a headwater spring with 1 species *(right)* to downstream communities with up to 11 species *(left)*. The communities sampled were in the Rio Tamesi drainage of east-central Mexico. (From Darnell 1970.)

pany variation in diversity, although these are more difficult to measure. Second, increased community niche space may be in part a consequence of diversity rather than an underlying cause or permissive factor. The mere presence of species makes for more possible roles—a greater variety of interactions. The larger proportion of specialized, parasitic species in collections of insects from sites with high diversity (Janzen and Schoener 1968) suggests enhancement of diversity through biotic interaction.

Habitat specialization increases in direct relation to diversity.

Robert Whittaker (1972) distinguished two kinds of diversity, which he called alpha (or local) diversity and gamma (or regional) diversity. Local diversity is the number of species in small areas of more-or-less uniform habitat. Clearly, local diversity is sensitive to definition of habitat and to area and intensity of sampling effort. Regional diversity is the total number of species observed in all the habitats within a region. When the same species are present in all the habitats of a region, then local and regional diversities are the same. When each habitat has a unique flora and fauna, regional diversity is equal to the average local diversity times the number of habitats in the region. Whittaker called the turnover of species from one habitat to the next, beta diversity. Therefore, gamma diversity = alpha diversity × beta diversity. It is not practical to measure beta diversity directly because the habitat distributions of species overlap one another. But we can calculate the equivalent number of unique habitats recognized by species within a region from the relationship, beta diversity = gamma diversity/alpha diversity.

Where many species coexist within a region, each occurs in relatively few kinds of habitats (MacArthur et al. 1966). Changes in gamma diversity generally result from parallel changes in both alpha and beta diversity. This relationship has been most carefully noted in comparisons of islands and mainland regions, where one may examine a range of species diversity (resulting from different degrees of geographic isolation) within similar ranges of physical conditions. Islands usually have fewer species than comparable mainland areas; island species often attain greater densities than their mainland counterparts and expand into habitats that would normally be filled by other species on the mainland (Crowell 1962). These phenomena, called density compensation and habitat expansion or compression (MacArthur et al. 1972, Wright 1980), are collectively referred to as ecological release. On the island of Puerto Rico, MacArthur and his co-workers found that many species of birds occupied most of the habitats on the island. In Panama, which has a similar variety of tropical habitats, species occupied fewer habitats, often a single type.

Surveys of bird communities in five tropical localities with different numbers of species — Panama, Trinidad, Jamaica, St. Lucia, and St. Kitts — show that where fewer species occur, each is likely to be more abundant and live in more habitats (Table 36–2). Similar numbers of individuals of all species added together were seen in each of the five localities although the total number of species (regional diversity) differed by a factor of almost seven between Panama and St. Kitts. In each habitat in Panama (mainland),

Table 36–2
Relative abundance and habitat distribution of resident land birds in five tropical localities within the Caribbean Basin*

Locality	Number of species observed (regional diversity)	Average number of species per habitat (local diversity)	Habitats per species	Relative abundance per species per habitat (density)	Relative abundance per species	Relative abundance of all species
Panama	135	30.2	2.01	2.95	5.93	800
Trinidad	108	28.2	2.35	3.31	7.78	840
Jamaica	56	21.4	3.43	4.97	17.05	955
St. Lucia	33	15.2	4.15	5.77	23.95	790
St. Kitts	20	11.9	5.35	5.88	31.45	629

* Based on 10 counting periods in each of 9 habitats in each locality. The relative abundance of each species in each habitat is the number of counting periods in which the species was seen (maximum 10); this times the number of habitats gives relative abundance per species; this times the number of species gives relative abundance of all species together.
Source: Cox and Ricklefs 1977.

about three times as many species (alpha diversity) were recorded and populations of each species were about half as dense as in the corresponding habitat on St. Kitts (small island). The beta diversity (species turnover between habitats, or equivalent number of habitats recognized) increased by a factor of almost three between St. Kitts and Panama.

The time hypothesis suggests that more diverse areas are older.

One explanation for the diversity of the tropics is that tropical conditions appeared earlier on the earth's surface than more polar environments, allowing time for the evolution of a greater variety of plants and animals. Although this idea has been voiced recently (Fischer 1960, Margalef 1963, 1968), it is hardly new. It was fully stated in 1878 by the English naturalist Alfred Russel Wallace:

> The equatorial zone, in short, exhibits to us the result of a comparatively continuous and unchecked development of organic forms; while in the temperate regions there have been a series of periodical checks and extinctions of a more or less disastrous nature, necessitating the commencement of the work of development in certain lines over and over again. In the one, evolution has had a fair chance; in the other, it has had countless difficulties thrown in its way. The equatorial regions are then, as regards their past and present life history, a more ancient world than that represented by the temperate zones, a world in which the laws which have governed the progressive development of life have operated with comparatively little check for countless ages, and have resulted in those wonderful eccentricities of structure, of function, and of instinct—that rich variety of colour, and that nicely balanced harmony of relations which delight and astonish us in the animal productions of all tropical countries.

Because the tropical zone girdles the earth about its equator—the earth's widest point—the tropics include more area than temperate and arctic regions. The earth's climate has undergone several cycles of warming and cooling, which have been discovered by records, in sediments and fossils, of their influence on vegetation and ocean temperature. As the climate of the earth warmed, as it last did during the Oligocene Epoch, perhaps 30 million years ago, the area of the tropics and subtropics expanded, reaching what is now the United States and southern Canada, and the temperate and arctic zones were squeezed into smaller areas closer to the poles. During the last 25 million years, the climate of the earth has become cooler and drier, and the tropics have contracted.

Both high and low latitudes experienced drastic fluctuations in climate during the ice ages of the last 2 million years. Temperate and arctic areas witnessed the expansion and retreat of glaciers, causing major habitat zones to be displaced geographically and, possibly, disappear. Periods of glacial expansion were coupled with high rainfall in the tropics (pluvial periods). The Amazonian rainforest, which today covers vast regions of the Amazon River's drainage basin, was repeatedly restricted to small, isolated refuges during periods of drought (Haffer 1969, Prance 1982). Restriction and fragmentation of the rainforest habitat could have caused the extinction of many species; conversely, the isolation of populations in patches of rainforest could have facilitated the formation of new species. Allen Keast (1961) proposed a scheme of alternate expansion and contraction of humid habitats to account, in part, for the diversity of the present Australian avifauna. Other animals and plants were probably similarly affected; for example, there are about 500 species in the genus *Eucalyptus,* a type of tree or shrub native only to Australia, but now widely planted around the world (Pryor and Johnson 1971).

Two additional types of information could shed light on the time hypothesis. First, if tropical climates antedate others, the fauna and flora of cooler regions should be clearly derived from those of the tropics. In particular, one could reject the time hypothesis if elements of more polar faunas and floras were clearly as old as those in the tropics. Considering botanical evidence, we note that coniferous trees, which predominate in the forests of cold climates, generally predate the flowering trees that have replaced them in the tropics. Conifers did not necessarily evolve in temperate zones, but cold-hardy spruces and firs almost certainly occurred in habitats similar to those in which they are found today for many tens of millions of years. Other groups of distinctive temperate-zone plants, such as the oaks and maples, have fossil histories that carry their origins back into the Mesozoic, over 60 million years ago. It is difficult to argue, therefore, that the relatively depauperate temperate and arctic communities to which these species belong do not have continuity in time. To be sure, the areas of temperate and arctic habitats may have been so restricted in the past that few representatives survived, but the habitats themselves and the lineages of plants and animals that live there appear to be ancient.

A second test of the time hypothesis could come from the direct observation of species diversity in the fossil record. In particular, it would be reasonable to reject the time hypothesis if diversity did not increase over time. The fossil record is so fragmentary that this test can be applied to few taxa and is restricted to certain types of habitats, particularly marine habitats (for example, see Stehli et al. 1969, Hallam 1977). Clearly, worldwide diversity has increased greatly from the beginning of the Paleozoic Era, when fossilized remains of organisms first become abundant, to the present, particularly as plants and animals invade new adaptive zones (Simpson 1969, Sepkoski 1978, Sepkoski et al. 1981). Less information is available for particular habitats or communities, or for established groups of organisms. M. A. Buzas (1972) and T. G. Gibson and Buzas (1973) have reported that the diversity of temperate zone communities of foraminifera (marine, shelled protozoa) has not changed during the last 15 million years. But unlike other groups of organisms, the species themselves also have changed little, and so the taxon may not be representative of others in which evolution continues to proceed more rapidly.

Equilibrium theories of diversity have predominated the last three decades of thinking about community organization.

The integration of population thinking into community studies during the 1950s and 1960s led ecologists to postulate that diversity might be regulated by local interactions among species. If competitive exclusion set limits to the ecological similarity of species, then communities could become saturated with species, additions being balanced by local extinctions (MacArthur 1972, Cody 1975). Thus the number of species would reach an equilibrium: new species added to the local community by regional diversification and migration of populations would be compensated by the local exclusion of close competitors. The equilibrium point itself would be affected by physical conditions, variety of resources, predators, environmental variability, and perhaps other factors. Thus conditions in the tropics might allow greater numbers of species to coexist locally by reducing the intensity or consequences of competition (Connell and Orias 1964, Pianka 1966).

Ecologists were attracted to this view because it placed at least part of the problem of species diversity within their conceptual domain of present-day processes taking place within small areas. The earlier idea that diversity was a reflection of history left ecologists with little to say about the matter. The alternative, what we might call local determinism, became so attractive, however, that ecologists embraced it nearly to the exclusion of regional/historical factors having an influence on local diversity (Ricklefs 1987). Instead of viewing diversity as a balance between regional production of species and local extinction, ecologists began to feel that levels of diversity were determined largely by local conditions. Hence their investigations concentrated almost exclusively on local interactions and niche relationships among populations.

One can trace a logical progression of ideas leading to the hypothesis of local determinism. The transition from the regional/historical viewpoint to the local/contemporary one of the 1960s and 1970s began with A. J. Lotka's and V. Volterra's mathematical representations of species interactions. G. F. Gause then demonstrated experimentally what had been predicted by theory, that species could not coexist on a single limiting resource, the so-called "competitive exclusion principle." In the 1940s, David Lack applied the concept of competitive interaction to the problems of ecological diversification, culminating in his landmark treatise on Darwin's finches in the Galapagos Islands (Lack 1947). Then, in the 1950s, G. E. Hutchinson developed the concept of species packing in multidimensional niche space. Remarking upon the uniformity of size ratios between pairs of ecological counterparts, Hutchinson (1957, 1959) planted the seed of the idea of limiting similarity—that there was an upper limit to the ecological similarity of species set by their competitive interactions. During the 1960s, the consequences of community interactions were formalized by Robert MacArthur, Richard Levins, and Robert May, who explored the deterministic, equilibrium properties of systems characterized by matrices of interactions among their component species (MacArthur and Levins 1967, Vandermeer 1972, May 1973).

The concepts of limiting similarity, saturation, and beta diversity supported the local/deterministic view of community regulation.

While emphasizing the importance of local population interactions in determining local diversity, ecologists were still faced with the reality of regional processes. Almost subconsciously, a large part of the discipline adapted its thinking in such a way that one could divorce the problems of local and regional diversity and attribute each to different causes having their own scales in space and time. The line of reasoning goes like this. Local interactions, taking place within a milieu of local conditions, determine the number of species that can coexist in the local community. This is the saturation point beyond which no new species can be added to the community. Regional processes, such as species production and migration, and historical accidents of geographical location, determine regional diversity. The difference between the two is accommodated by differences in the degree of habitat specialization, or beta diversity, which is adjusted to maintain the number of species locally in accordance with local conditions while the number of species in the region may vary.

This portrayal somewhat caricatures a train of thought to which no individual fully subscribed. But regional/historical and local/deterministic viewpoints nonetheless presented a clear dichotomy with which ecologists have struggled. MacArthur (1965) expressed the collective conscience of many ecologists in the mid-1960s: "if the areas being compared are not saturated with species, an historical answer involving rates of speciation and length of time available will be appropriate; if the areas are saturated with species, then the answer must be expressed in terms of the size of the niche space and the limiting similarity of coexisting species." In 1972, MacArthur furthered the distinction and placed ecologists clearly in the local/deterministic camp: "The ecologist and the physical scientist tend to be machinery oriented, whereas the paleontologist and most biogeographers tend to be history oriented. They tend to notice different things about nature." His remark that history may "leave an indelible mark even upon the equilibrium so dear to the ecologist" emphasized the philosophical gulf that had developed between regional and local viewpoints. This adoption of local determinism by many ecologists is what Sharon Kingsland (1985), in her treatise on the history of population biology, has called the "eclipse of history."

The number of species on islands depends on immigration and extinction rates.

Even while MacArthur was leading community ecology away from regional/historical perspectives, he and E. O. Wilson developed their famous equilibrium theory of island biogeography, which states that the number of species on islands balances regional processes governing immigration

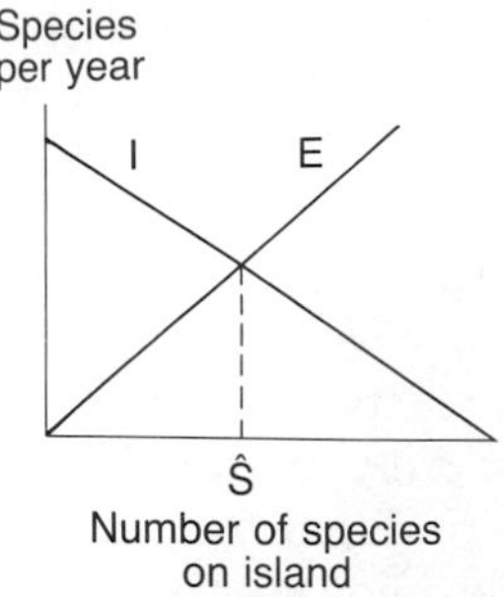

Figure 36–6
Equilibrium model of the number of species on islands. The equilibrium number of species *(Ŝ)* is determined by the intersection of the immigration (I) and extinction (E) curves. (After MacArthur and Wilson 1963, 1967.)

against local processes governing extinction. On islands too small to support speciation through geographical isolation of populations, the number of species increases solely by immigration from other islands or from the mainland. Whereas we know little about rates of speciation within continents, we may reasonably assume that fewer species immigrate to islands the greater the distance that separates them from the mainland. Hence the advantage of the island model.

Consider an offshore island. The flora and fauna of the adjacent mainland comprise the species pool of potential colonists to the island. The rate of immigration of new species to the island decreases as the number of species on the island increases; that is, as more and more of the potential mainland colonists are found on the island, few immigrants belong to new species (Figure 36–6). When all the mainland species occur on the island, the immigration rate of new species must be zero. If species disappeared at random, the number of extinctions per unit of time would increase with the number of species present on the island. Where the immigration and extinction curves cross, the corresponding number of species on the island is an equilibrium (*S*).

Immigration and extinction rates probably do not vary in strict proportion to the number of potential colonists and the number of species established on the island. Some species undoubtedly are better colonizers than others and they reach the island first. Thus the rate of immigration to the island initially decreases more rapidly than it would if all mainland species had equal potential for dispersal; and, as a result, the relationship between immigration rate and island diversity follows a curved line. Competition between species on islands probably abets extinction, so the extinction curve rises progressively more rapidly as species diversity increases (Figure 36–7).

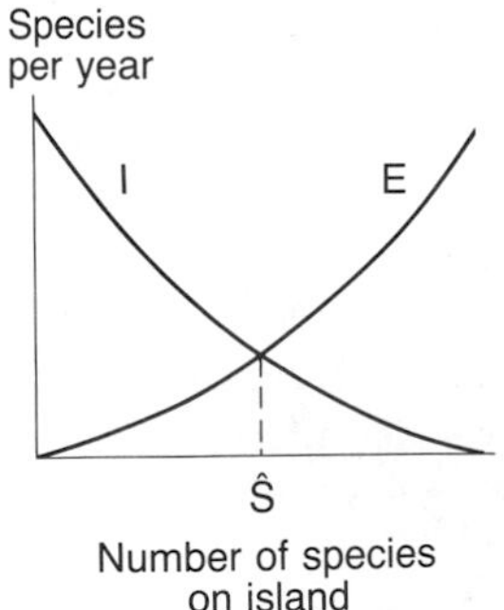

Figure 36–7
Several biological considerations influence the shape of the immigration and extinction curves. The immigration rate initially drops off rapidly as the best colonists become established on the island. The extinction rate increases more rapidly at high species number because of increased competition between species. (After MacArthur and Wilson 1963, 1967.)

If the probability of extinction increased as absolute population size decreased, extinction curves for species on small islands would be higher than for those on larger islands. Therefore, small islands would support fewer species than large islands (Figure 36–8). If the rate of immigration to islands decreased with distance from mainland sources of colonists, the immigration curve would be lower for far islands than for near islands, and the equilibrium number of species for distant islands should lie to the left of the equilibrium for islands close to the mainland (Figure 36–9). These predictions have been verified for islands throughout the world (MacArthur and Wilson 1967).

If the number of species on an island behaved according to the equilibrium model, a change in the number of species would lead to a response tending to restore the equilibrium diversity. A natural test of this prediction was begun quite spectacularly in 1883 when the island of Krakatoa, located between Sumatra and Java in the East Indies, blew up after a long period of repeated volcanic eruptions. At least half the island disappeared beneath the sea and hot pumice and ash covered its remaining area. The entire flora and fauna of the island were certainly obliterated. During the years that followed the explosion, plants and animals recolonized Krakatoa at a surprisingly high rate: within 25 years, more than 100 species of plants and 13 species of land and freshwater birds were found there. During the ensuing

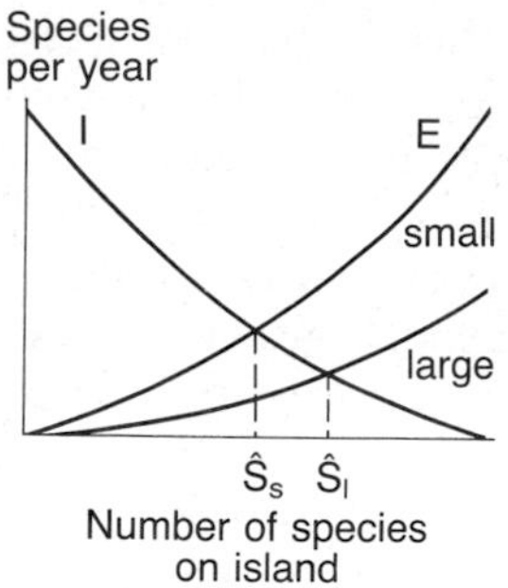

Figure 36–8
According to the MacArthur-Wilson equilibrium model, small islands are thought to support fewer species owing to higher extinction rates.

13 years, 2 species of birds disappeared and 16 were gained, bringing the total to 27. During the next 14-year period between exploring expeditions, the number of species on the island did not change, but 5 species disappeared and 5 new ones arrived, suggesting that the number of species had reached an equilibrium, at a level that would be expected for an island the size of Krakatoa in the East Indies. The turnover rate during this period was 1.3 per cent of species per year [5 species/(27 species × 14 years)]. Experimental studies, involving the colonization of glass slides by diatoms (Patrick 1967), sponges by protozoans (Cairns et al. 1967), water-filled vials by microorganisms (Maguire 1963), and mangrove islands by arthropods (Simberloff 1969, Simberloff and Wilson 1969), reinforce the pattern of colonization and attainment of an equilibrium seen on Krakatoa (Figure 36–10).

The equilibrium theory can be applied to the number of species in mainland communities.

An equilibrium view of diversity can be applied to mainland assemblages of species as well as to those on islands. The major difference is that regional production of species augments the addition of species from outside the system by immigration (MacArthur 1969, Rosenzweig 1975, Brown and

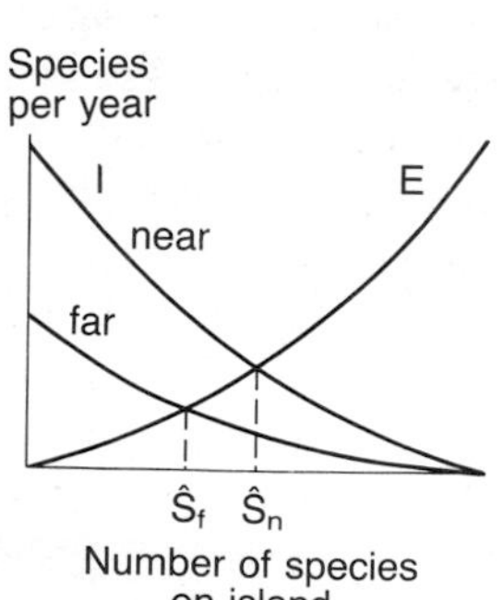

Figure 36–9
According to the MacArthur-Wilson equilibrium model, islands close to the mainland are thought to support more species owing to higher immigration rates.

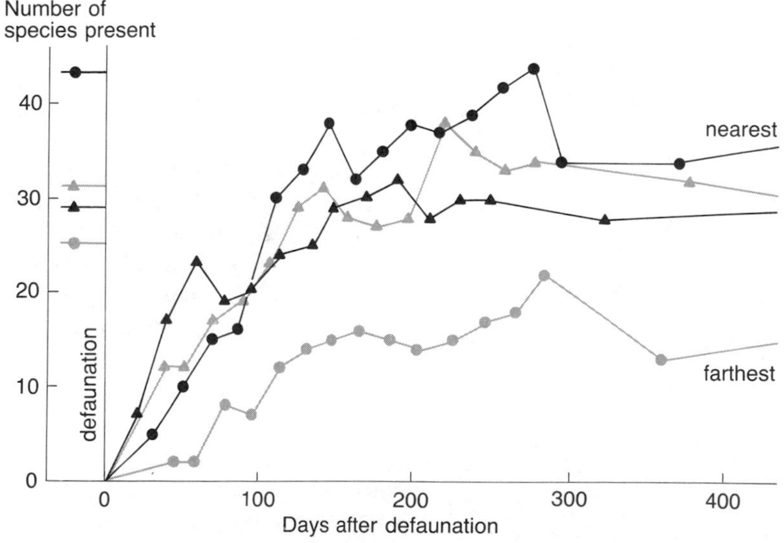

Figure 36–10
Recolonization curves of four small mangrove islands in the lower Florida Keys whose entire faunas, consisting almost solely of arthropods, were exterminated by methyl bromide fumigation. Estimated numbers of species present before defaunation are indicated at left. Species accumulated more slowly and achieved lower equilibrium numbers on more distant islands. (From Simberloff and Wilson 1970.)

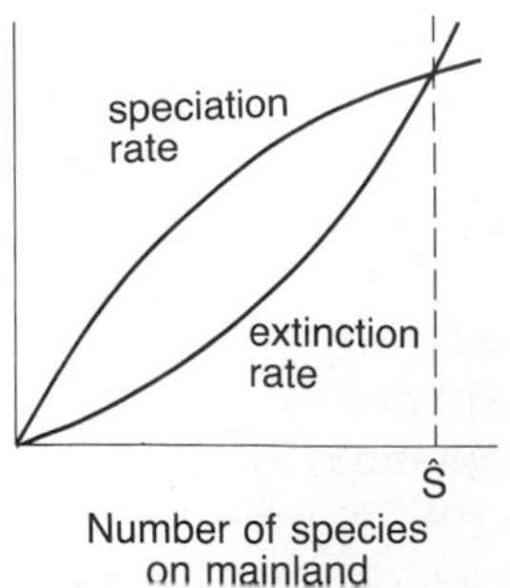

Figure 36–11
Equilibrium model of the number of species in a mainland region with a large area; new species are generated by the evolutionary process of speciation rather than immigration from elsewhere. (After MacArthur 1969.)

Gibson 1983). For a large region isolated from others by barriers to dispersal (an island continent, for example), all new species must be generated by speciation events within the region. The curves relating rates of species production and extinction to regional diversity might look like those drawn in Figure 36–11. The curvature of the lines would vary depending on the processes that produce them: probability of extinction per species could increase if competitive exclusion increased with diversity, whereas it could decrease if mutualisms and alternative paths to energy flow buffered diverse communities from external perturbations; speciation rate per species could level off if opportunities for further diversification were restricted by increasing diversity, but the rate could increase if diversity led to greater specialization and higher probability of reproductive isolation of subpopulations.

Regardless of the particular shape of the immigration, speciation, and extinction curves, most biologically reasonable models will define an equilibrium level of diversity. And so while such models provide a perspective, they do not explain variation in diversity: most causes can be incorporated into an equilibrium model. The time hypothesis would be appropriate if systems were far from equilibrium. Or, as MacArthur (1969) pointed out, local communities may be saturated with species while diversity continues to increase regionally, the growing difference being made up by increasing beta diversity. If diversity were in equilibrium, the positions of species production and extinction curves would be affected by a variety of factors (Connell and Orias 1964, Pianka 1966, Buzas 1972), each of which could shift the equilibrium.

Can reduced competition explain high diversity?

Intense competition promotes exclusion of species from a community. Many ecologists have argued that less severe competition for shared resources would allow more species to coexist. How may interspecific competition be reduced? Greater ecological specialization, greater resource availability, reduced resource demand, and intensified predation have all been cited as possibilities. In every case, however, adjustment of populations to these factors is driven by competition among individuals of the same species, as well as between individuals of different species. I find it difficult to imagine reduced competition as an intrinsic property of any population. Density-dependent selection tends to increase population size. Therefore, if variation in competition were to explain variation in diversity, then the relative strengths of intraspecific competition and interspecific competition must vary in their influence on the gene pool of the population.

Perhaps differences in these relative strengths may be envisioned in terms of the geometry of niche relationships. In a niche space having few dimensions, each population has relatively few neighbors. As the number of dimensions increases, the same number of species are packed differently into the niche space, and each has a greater number of neighbors (Figure

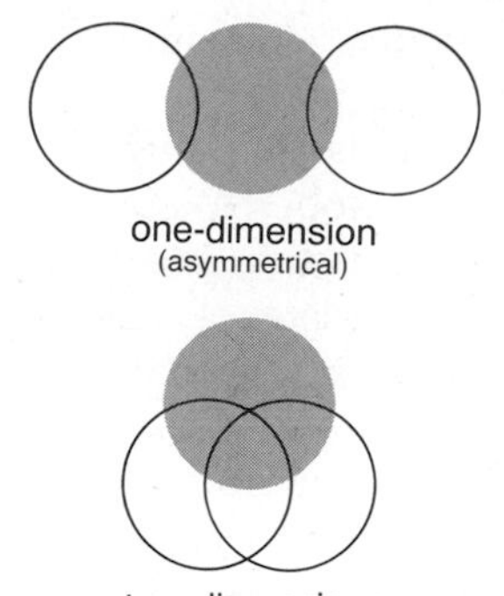

Figure 36–12
Schematic diagram of ecological relationships between three species organized along one dimension *(top)* and two dimensions *(bottom)*. Relationships are symmetrical on two dimensions, making it difficult for two species to exclude the third, or vice versa.

36–12). This variety may prevent each individual population from adapting to compete effectively with its neighbors in communities of high niche dimension, preventing the emergence of competitive dominants. High dimensionality may occur in physically ameliorated environments, such as the wet tropics or the abyssal depths of the oceans; in harsher environments, a few physical factors may dominate the character of the niche space by establishing a small number of critically important dimensions. Furthermore, because many niche dimensions are biologically generated, the dimensionality of the niche may increase as biological communities build over time, further enhancing their capacity to support many species against the evolutionary tendency of a small number of species to dominate.

Variation in plant species diversity has many plausible explanations.

Because the diversity of plants as resources rather straightforwardly determines the diversity of animals, the most rigorous tests of general explanations for diversity lie in their application to plant communities. Egbert Leigh (1982) summarized the various answers of ecologists to the question, Why are there so many different kinds of trees in the tropics? Explanations may be grouped into five categories based upon: (1) environmental heterogeneity in space, (2) environmental variability in time, (3) environmental heterogeneity produced by disturbances to the uniform structure of the forest, (4) rates of production (by speciation or immigration) of competitively equivalent species, and (5) favoring of rarer species by consumer pressure.

Peter Ashton (1969, 1977) argued that trees are diverse in proportion to the heterogeneity of the environment. Abundant evidence suggests that tropical forest trees may be specialized to certain soil and climate conditions. But could variability in the physical environment in the tropics account for a ten-fold or even greater diversity of plants compared to temperate zone forests? David Tilman (1982) proposed a potential solution to this problem based on his model of competition among consumers. Consider a graph whose axes are the levels of two resources (Figure 36–13). The population of each species can increase when each of the two resources are more abundant than some specified level; hence each can persist in the absence of competition within a specified area of the graph. When competitors are present, the area of the graph within which each species can persist is reduced to a narrow segment (Figure 36–14). Within each geographical region, the natural heterogeneity of conditions (resource levels) may be depicted by an enclosed area superimposed on the resource-level graph. Species whose boundaries of persistence fall within the area may coexist (Figure 36–15). Now, notice that at low resource levels, a given range of environmental heterogeneity intersects the persistence areas of more species than a similar range of heterogeneity at high resource levels. An inverse relationship between resource level and diversity has some empirical and experimental support (Tilman 1982). Possibly, therefore, the low resource levels of soils under many tropical forests may promote their high diversity.

Resource 2

area of persistence

Resource 1

Figure 36–13
A model for the situation in which persistence of a species requires minimum critical levels of two resources that independently limit the population.

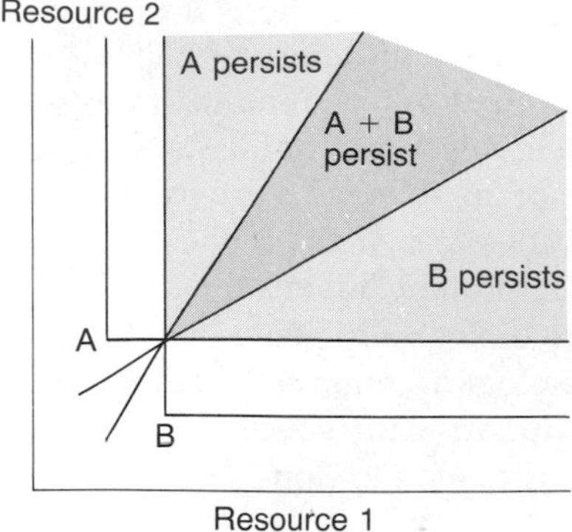

Figure 36–14
Conditions for the coexistence of two species according to Tilman's resource model.

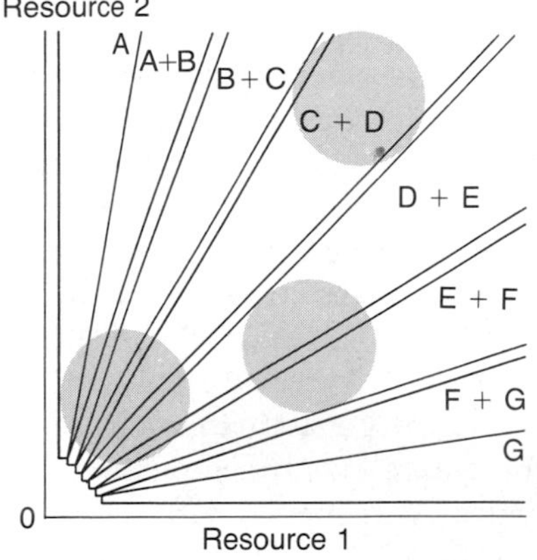

Figure 36–15
Seven-species competition for two resources, showing the regions of coexistence (never more than two species). When local environmental variability is considered (shaded areas), habitats with lower overall resource abundance are likely to support more species over the range of variability, provided the absolute range is comparable between habitats. (From Tilman 1982.)

Basically, Tilman suggests that heterogeneity is relative, depending on supply and demand for limiting resources.

Unpredictable temporal heterogeneity generally is not thought to promote diversity (May 1974). But Peter Chesson and Robert Warner (1981) have suggested that year-to-year variation in reproductive rate, such that each species is favored in some years, may lead to coexistence. The mechanism relies on a kind of frequency dependence. Suppose that a certain fraction of individual trees dies each year and each one is replaced by an individual of the same or different species in direct proportion to the production of seeds by each of the species. When a species is very rare, a bad year for seed production has little effect on probability of seedling establishment, but a good year can bring a comparative bonanza—a little less of almost nothing is almost nothing, but a little more represents a huge relative increase! Although this mechanism may work in theory, it acts strongly only for very rare species. Thus any competitive inferiority will push a species into the very rare category. Moreover, because each species must experience some years during which it produces more seed than all other species, relatively few species can coexist solely by this mechanism.

The Chesson-Warner model was inspired by coral reef fish, which exhibit a diversity comparable to tropical forest trees and exhibit little niche diversification (Ehrlich 1975, Sale 1980). For that system, considerable controversy has attended the issue of competition among species (Anderson et al. 1981, Sale and Williams 1982, Clant 1988). Peter Sale (1977, 1978) suggested that juvenile fish colonize coral heads at random, with opportunity spread evenly among individuals of all species to take the place of adults that die or otherwise leave their territories in the reef. This mechanism became known as the lottery, or random-access, hypothesis. Colonization by lottery reduces competitive exclusion to chance extinction in the coral reef community, and thereby allows greater diversity. But this model cannot explain the diversity of larval fish in the plankton from which coral residents come.

Consideration of tropical rainforests and coral reefs led Joseph Connell (1978) to relate high diversity to intermediate levels of disturbance to communities. We have already touched upon the Intermediate Disturbance Hypothesis. Briefly, disturbances, caused by physical conditions, predators, or other factors, open space for colonization and initiate a cycle of succession by species adapted to colonizing disturbed sites. With a moderate level of disturbance, the community becomes a mosaic of patches of habitat at different stages of regeneration; together these patches contain the full variety of species characteristic of the successional sere. In order to account for differences in diversity between regions, especially on the magnitude of the latitudinal gradient of tree species diversity, one requires comparable differences in level of disturbance. Rates of turnover of individual forest trees (that is, the inverse of average life span) do not differ systematically between temperate and tropical areas (Table 36–3). Thus, while disturbance may promote diversity (Watt 1947, Grime 1977), it seems unlikely to account for much of the observed variation in diversity among forests or, indeed, other types of communities.

A somewhat different model by which gap formation may generate

Table 36–3
Turnover of canopy trees in primary forests in tropical and temperate localities

Locality	Turnover time (years)	Turnover rate (% yr^{-1})
Tropical		
Panama	62–114	0.9–1.6
Costa Rica	80–135	0.7–1.3
Venezuela	104	1.0
Gabon	60	1.7
Malaysia	32–101	1.0–3.1
Temperate		
Great Smoky Mountains	49–211	0.5–2.0
Tionesta, Pennsylvania	107	0.9
Hueston Woods, Ohio	78	1.3

Turnover time does not include the years required to grow into the canopy, estimated to be 54–185 years for various temperate zone species.
Source: Data in Runkle 1982, Putz and Milton 1982, Browkaw 1985.

diversity was proposed by R. E. Ricklefs (1977), based upon the idea that disturbances create a range of conditions for seed germination and seedling establishment within which different species of trees may specialize (Denslow 1980, 1987; Orians 1983; Pickett 1983). Disturbance creates transient environmental heterogeneity during a critical life stage of plants and other organisms. Furthermore, the interaction of the structural gap in the forest canopy with physical conditions of the environment may create greater heterogeneity in the tropics than in temperate regions (Figure 36–16). Gaps in forest canopies created by treefalls admit light to the forest floor, thereby changing the physical conditions for seedling establishment and the decomposition of organic detritus. Furthermore, with fewer living tree roots in the forest gap, nutrients are more readily leached out of the soil. These

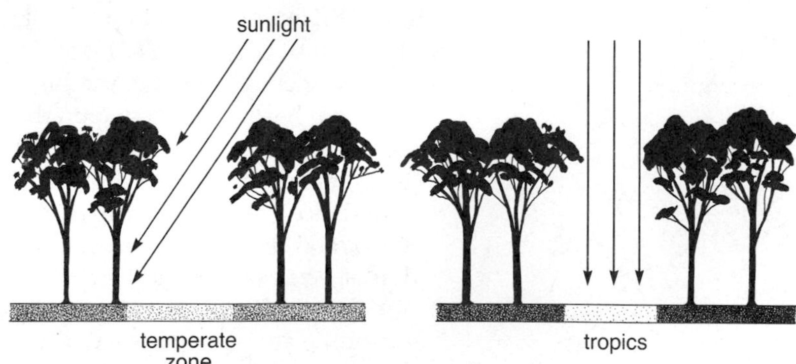

Figure 36–16
Schematic diagram of the interaction of physical factors with gaps in the canopies of forests to produce a wider range of environmental conditions in tropical forests than in temperate forests.

processes occur more intensely in the tropics: sunlight enters gaps from directly overhead rather than at the shallower angles of higher latitudes; more rain falls in the tropics, so decomposition and leaching proceed more rapidly. Furthermore, conditions under the intact canopy are more protected in tropical forests than in temperate zones owing to denser canopy and year-round persistence of leaves. As a result, the forest-environment interaction creates a greater variety of physical conditions available for seedling germination in the tropics than at higher latitudes. If trees specialize on this gradient of conditions, then perhaps their coexistence becomes easier in the tropics. It may be relevant that the latitudinal gradient in diversity that characterizes trees does not appear among shrubs, herbs, and grasses, which lack the distinctive physical structure of the forest canopy.

By reducing populations of competitors, predators may enhance community diversity.

When predators reduce populations of their prey below the carrying capacity of their resources, they may reduce competition among competitors and promote coexistence (Parrish and Saila 1970). Moreover, selective predation on superior competitors may allow competitively inferior species to persist in a system (for example, Paine 1966, Harper 1967). Both accidental and intentional experiments have revealed the immense capacity of predators to reduce populations of their prey under some situations. Community effects have been particularly well documented in aquatic systems, where the introduction of a predatory sea star, salamander, or fish can utterly change the community of primary consumers and producers (Paine 1966, 1974, Zaret and Paine 1973; Porter 1972; Zaret 1980; Wilbur et al. 1983; Carpenter et al. 1987, 1988).

From Darwin's time, at the least, naturalists have opined that both selective and nonselective herbivory may influence the diversity of plant species (Harper 1967). In particular, several authors, including R. L. Doutt (1960), J. B. Gillette (1962), and Daniel Janzen (1970) have suggested that herbivory could promote the high diversity of tropical forests. Janzen argued that herbivores feed upon the buds, seeds, and seedlings of abundant species so efficiently as to reduce their densities. In turn, this allows other, less common species to grow in their place. The key to this idea is that abundance per se, rather than the intrinsic quality of individuals as food items, makes a species vulnerable to consumers. Consumers locate abundant species easily and their own populations grow to high levels.

Several lines of evidence support the pest pressure hypothesis. For example, attempts to establish plants in monoculture frequently are doomed by infestations of herbivores. Dense plantations of rubber trees in their native habitats of the Amazon Basin, where many species of herbivores have evolved to exploit them, have met with singular lack of success. But rubber tree plantations thrive in Malaya, where specialist herbivores are not (yet) present. Attempts to grow many other commercially valuable crops in single-species stands in the tropics have frequently met the same disastrous end that befell the rubber plantations. Cacao, the South American plant that is the source of chocolate, provides a conspicuous exception. Insect pests in cacao plantations are no more numerous or diverse within the native range than in plantations in Africa and Asia (Strong 1974).

The difficulties of monoculture extend beyond the tropics. Temperate zone trees have their herbivores; few acorns escape predation by squirrels and weevils, and seedlings are attacked by herbivores and pathogens just as they are in the tropics. If pest pressure does promote greater diversity in the tropics, it must operate differently in different latitude belts. In particular, tropical herbivores and plant pathogens must be either more specialized with respect to species of host plant or their populations must be more sensitive to the density and dispersion of host populations.

One prediction of the pest pressure hypothesis is that the establishment of seedlings should be more difficult close to adults of the same species than at a distance (Janzen 1970). Adult individuals may harbor populations of specialized herbivores and pathogens that readily infest nearby progeny; furthermore, because most seeds germinate close to their parent, herbivores may be attracted to the abundance of seedlings there while overlooking the few that germinate at some distance. The prediction of distance-dependent germination and establishment success has been tested in a number of studies, with varied outcomes but generally supportive results (Augspurger 1983, Clark and Clark 1984, Howe et al. 1985).

If seedlings establish themselves more readily at greater distance from the parent or other conspecifics, individuals of the same species should be widely dispersed within a forest rather than randomly associated or clumped. This prediction may be tested by mapping individuals of each species and applying one of the several statistics to test degree of dispersion. Most studies have shown that many species of trees, especially the more common ones, have significantly clumped distributions (for example, Hubbell 1979, Hubbell and Foster 1983). This result weighs against a major role of pest pressure in forest community structure, but degree of clumping is relative, and herbivores and pathogens may cause greater dispersion of mature individuals than would occur otherwise.

Are species produced more rapidly in the tropics than at higher latitudes?

Environmental heterogeneity and consumer activities may influence local population processes so as to affect species diversity. Regional processes are more difficult to contemplate, much less observe directly. Correspondingly little has been written about their role in determining local and regional diversity. In nonequilibrium systems, rates of species production and extinction as well as the age of the region or habitat type directly determine diversity. Where diversity is maintained at an equilibrium level, rate of species production will influence regional diversity and it will influence

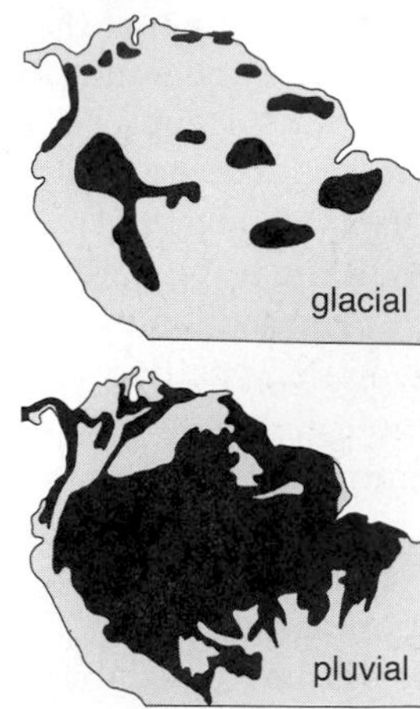

Figure 36–17
Approximate distribution of lowland rainforest in South America during the height of glacial periods in the Northern Hemisphere *(top)* and at present *(bottom)*.

local diversity unless local communities become saturated with species at levels determined solely by local ecological conditions.

Many systematists and biogeographers have suggested that the rate of species production or the temporal history of an environment contributes to tropical diversity (Willis 1922, Haffer 1969, Simpson and Haffer 1978, Prance 1982). But because thousands of generations may be needed for isolated populations of a species to develop barriers to reproduction and themselves become distinct species, ecologists can neither observe the process nor experiment with it.

A few theoretical papers have addressed the problem of tree species diversity in the context of species production. For example, Stephen Hubbell (1979) proposed a lottery-type model of replacement of forest trees. In his model, the probability that the individual that fills a gap in the forest belongs to a particular species varies in direct proportion to the frequency of that species in the forest. Hence gaps are more likely to be filled by more abundant species than by rare ones. Without the appearance of new species, such a system will eventually become dominated by a single species, just by chance. Thus, in such a system, diversity must be maintained, and its level determined, by the rate of species production (that is, the random appearance of unique individuals). Hubbell's model predicts that the relative abundance of species approach a lognormal distribution, which indeed reasonably characterizes many forests. But the model provides no means of independently verifying whether differences in diversity are caused by differences in rates of species production.

Jurgen Haffer (1969, 1974) and Gillian Prance (1982) have suggested that fragmentation of tropical forests during the periodic dry periods of the recent ice age (Figure 36–17) provided opportunities for allopatric speciation in the tropics at a time when harsh conditions and restriction of habitats in temperate and arctic zones may have caused an increase in extinction

Table 36–4
Taxonomic levels of diversity, genus/family, and species/genus ratios among forest trees in several regions

	Families	Genera	Species	Genera per family	Species per genus
Deciduous trees of eastern North America	33	59	205	1.8	3.5
Rainforests of Golfo Dulce, Costa Rica	64	260	417	4.1	1.6
Indonesia					
Malaya		113	210		1.9
Sabah		99	198		2.0
Central America					
Barro Colorado Island, Panama		80	112		1.4

Source: Ricklefs 1989, after various sources.

rate. If differences in diversity between temperate and tropical forests express recent high rates of species production in the tropics, one would expect to find more species per genus in tropical forests than in temperate zone counterparts. But, in fact, tropical forests present their tremendous diversity to us as much at the family and genus levels as at the species level. Samples of a similar number of forest trees from small areas in the Republic of Panama (46 species) and Ontario, Canada (33 species), for example, reveal many more species per genus (including 4 species of *Populus*) in the temperate forest but similar numbers of genera per family in both localities (Table 36–4). The difference in diversity between the species-rich tropical site (with a total of more than 300 species of trees: Knight 1975, Croat 1978, Gentry 1988) and the temperate zone site, having fewer than 50 species (for example, Braun 1950, Beschel et al. 1962), resides primarily at the family level. In fact, the tropical forests are decidedly depauperate in closely related species. Their great number of higher taxa belies the ancient roots of diversity there. I conclude that if differences in speciation rate are responsible for differences in diversities of forests, they have been persistent differences acting over very long periods.

Do communities reveal evidence of interactions among species?

Ecologists have yet to agree on the processes responsible for variation in diversity over the face of the globe. Competition has received the most attention from vertebrate ecologists while predation seems to be favored by many aquatic ecologists. Disturbance has become a primary concern of many interested in marine communities and forests. Ecologists have concentrated much of their effort during the past two decades on showing that particular processes occur in natural communities (Sih et al. 1985). In earlier chapters, we reviewed evidence showing that competition and predation can have strong demographic consequences; the importance of disturbance on a wide range of scales is obvious (Delcourt et al. 1983). The existence of a process or class of interaction does not, however, demonstrate its responsibility for pattern in communities. Predation is a feature of all assemblages, but to conclude that variation in predation is responsible for variation in diversity would require additional evidence.

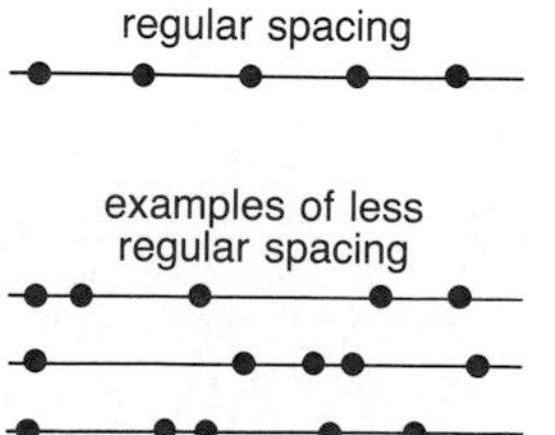

Figure 36–18
Comparison of regular and more or less random spacing along a single dimension.

Ecologists do, however, agree that if competition wields a strong hand in molding the structure of communities, its effect should be evident in niche relationships among coexisting species. In particular, competition should lead to regular spacing between the positions of species within niche space. For example, concerning the average point of occupancy of species along a one-dimensional niche space, if the position of each species were sensitive to the position of others — that is, if the coexistence of species depended on somehow minimizing competition between them — then they should exhibit regular spacing (Figure 36–18). Conversely, if the positions of species were not influenced by interactions among each other, then their positions in niche space would be independently, or randomly determined.

Randomness of distribution can be tested statistically, and ecologists began to apply such tests in the late 1970s. Most of these studies have failed to reveal pervasive evidence of interaction (nonrandomness), but an equal number of authors have raised objection to the results, questioning either the analytical procedures or the interpretations. Most of the tests have involved patterns of geographical distributions on islands or the positions of species in either morphological or ecological (niche) space. In the first case, significant nonoverlap of geographical distribution among close relatives would be consistent with competitive interaction between them. Jared Diamond (1975) illustrated this point with several examples of distribution of closely related (conspecific) birds in the region of New Guinea (Figure 36–19). For example, two species of cuckoo-doves occupy 6 and 14 islands, respectively; neither occur on 13 islands in the area, but, most significantly, no island is inhabited by both species. The probability of such a distribution occurring by chance, when different species inhabit islands independently of each other (that is, are randomly distributed), is less than one in 40 (Table 36–5). Hence Diamond concluded that interaction played an important role in distribution and local community diversity.

E. F. Connor and D. S. Simberloff (1979) questioned the validity of Diamond's analysis. They reasoned that if the distributions of *Macropygia* species might have occurred by chance 1 time in 40, then for every 40 pairs of species combinations examined one would likely observe one as nonrandom in appearance as the example of the cuckoo-doves. To test interaction adequately, Connor and Simberloff reasoned, one had to examine all pairs of species; they then did this for birds in the West Indies. In their analysis

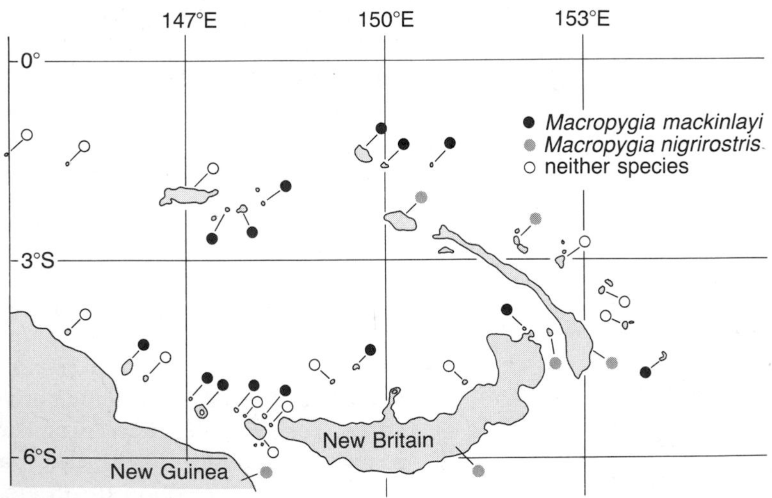

Figure 36–19
Distribution of cuckoo-doves of the genus *Macropygia* in the Bismarck Archipelago. Most islands have one of the two species, no island has both, and some have neither. (From Diamond 1975.)

Table 36-5
The occurrence of fruit pigeons of the genus *Macropygia* on islands in the Bismark Sea

		Macropygia mackinlayi		
		Present	Absent	Total
Macropygia nigrirostris	Present	0	6	6
	Absent	14	13	27
	Total	14	19	33

Islands with each species		Number of islands		
M. mackinlayi	*M. nigrirostris*	Observed	Expected	χ^2
Present	Present	0	2.5	2.1
Present	Absent	14	11.5	0.5
Absent	Present	6	3.5	1.5
Absent	Absent	13	15.5	0.3
	Total	33	33.0	4.4 ($P < 0.05$)

Source: Diamond 1975.

they accepted the number of species per island and the number of islands occupied per species as fixed properties of the system. Within these constraints, they then randomized the distributions of the 211 species in the sample several times and examined the results. Of the 22,155 possible pairs of species (all combinations of n species equals $n(n-1)/2$), 12,448 had exclusive distributions (no co-occurrence on any island), on average, in the randomized sets of species. The exclusive distributions among pairs of species in the actual avifauna of the West Indies number 12,757, so close that one must accept general agreement. Connor and Simberloff found similar agreement between observed and randomized distributions in a number of other tests, and concluded that interaction between species was not an important determinant of their geographical distributions.

Jared Diamond and Michael Gilpin (1982) responded to the Connor-Simberloff analysis, protesting that by looking at all species pairs one reduced the chances of recognizing interaction, which one would expect only among ecologically similar species. They also objected that the constraints on Connor and Simberloff's randomization algorithm were so great as to nearly guarantee a resemblance between randomized communities and the observed communities from which they were derived. The problem here is the procedure for producing the random community, or null model, as it has been called (Harvey et al. 1983, Quinn and Dunham 1983, Simberloff 1983). What is the appropriate species pool and how should species be allocated to the random communities? Does the process of producing random communities incorporate any biological assumptions that are unrealistic?

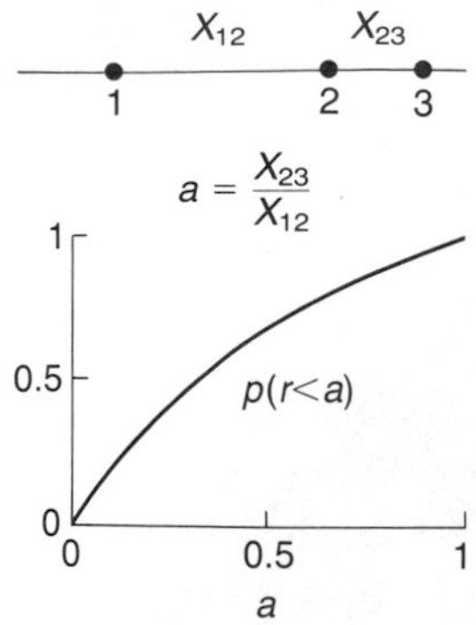

Figure 36–20
The probability that the ratio between the smaller and larger of two randomly determined line segments will be less than the value *a*.

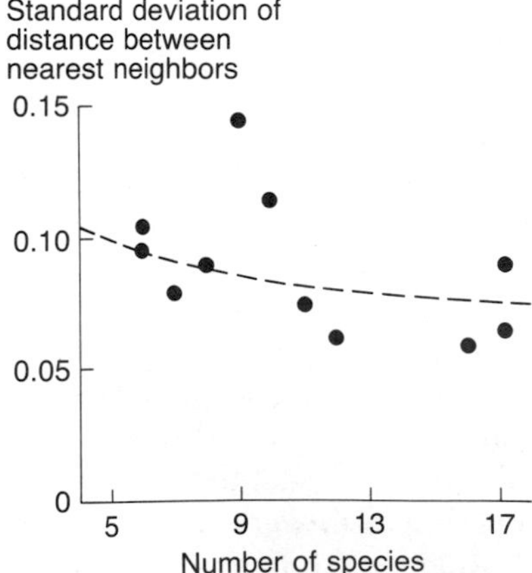

Figure 36–21
Standard deviation of average nearest neighbor distance in morphological space among passerine birds inhabiting scrub communities. The dashed line is the predicted value for randomly generated communities. (After Ricklefs and Travis 1980.)

The problem of constructing the null model is illustrated by another of the early attempts to identify pattern in communities. Donald Strong et al. (1979) investigated size ratios of confamilial species of birds inhabiting the Tres Marias Islands off the western coast of Mexico. In an earlier study of these communities, Peter Grant (1966) had suggested that size ratios exceeded those found in mainland communities, indicating that competition and ecological divergence had exerted important influence on the island communities. Strong and his colleagues tested this hypothesis by drawing sets of species at random from the "pool" of species inhabiting the nearby mainland of Mexico. They found that size ratios of related birds on the Tres Marias Islands were no greater than among those drawn at random from the mainland, and thus rejected Grant's claim. But while they tested the hypothesis of interaction among species, Strong et al. (1979) also tested the assumption that all mainland species have equal probability of colonizing offshore islands. This assumption can be rejected easily; ecologists well know that colonizing ability varies with habitat, abundance, and behavior. Some species are better colonists than others. One can say with certainty, therefore, that the null model is unrealistic, but devising a better procedure for constructing random communities poses a great challenge.

Daniel Simberloff and William Boeklin (1981) partly avoided the problem of a source pool of species by assuming a certain statistical distribution of, for example, sizes of species from which to draw. For the purposes of their analysis, they assumed that species occupied a uniform probability distribution of sizes. Three points (species) drawn at random from such a distribution define two intervals, that between the small and medium-sized species and that between the medium and the largest (Figure 36–20). The ratio *(a)* between the smallest and largest intervals in randomly drawn trios of species itself has an expected probability distribution between 0 and 1, against which one may test the size intervals separating trios of actual species. The probability of observing a ratio *(r)* less than some value a in a randomly drawn community is $1/[(1/2) + (1/2a)]$. Thus the probability of $r < 1$ is 1.0 while the probability of $r < 0.5$ is 0.67 and that of $r = 0.33$ is 0.5; one would expect half the ratios to exceed 0.33 and half to be less than that value. Simberloff and Boeklin worked out a similar distribution for the expected size of the smallest of $n - 1$ segments produced by drawing n points at random. If similarity between species limits coexistence, one should find fewer small ratios than expected at random. These distributions provide a "null" hypothesis against which one may compare observed distributions of species in natural communities. If species were spaced more evenly than expected of random distributions, the distribution of r would be shifted from that of a toward 1. After making such comparisons, Simberloff and Boeklin found little support for nonrandomness in the data for which claims for even spacing had been made.

Ricklefs and Joseph Travis (1980) took a similar null-model approach to assess regularity of spacing of bird species in the scrub habitats described by Cody (1974). They constructed random communities of 5, 9, 13, and 17 species either by drawing species at random from the total species pool or by generating points (synthetic species) at random from the morphological space occupied by the species. If interaction influenced community struc-

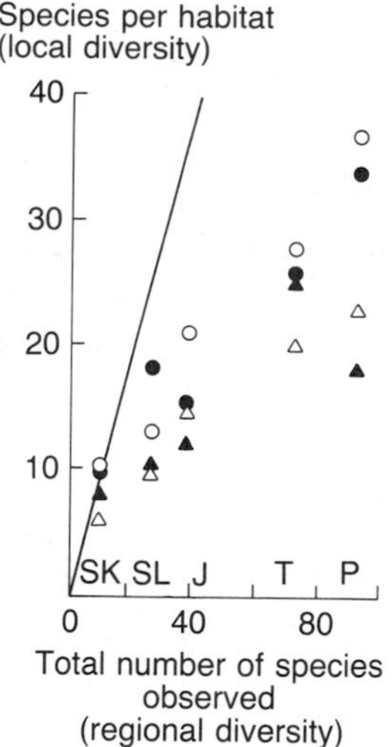

Figure 36–22
Local species richness of birds in the Caribbean region is sensitive to regional diversity, which is determined primarily by biogeographic considerations. Areas are St. Kitts (SK), St. Lucia (SL), Jamaica (J), Trinidad (T), and central Panama (P). Within each area, standardized counts were made of songbird species within small, homogeneous areas of habitat, each type indicated by a different symbol. (From Ricklefs 1987.)

ture, one would expect species to space themselves more evenly in natural communities than in the null communities. The criterion for spacing was the standard deviation of nearest neighbor distances, which decreases to zero as spacing becomes perfectly even. As one can see in Figure 36–21, natural communities do not differ from randomly generated ones; thus, morphological spacing does not reveal species interaction.

Absence of nonrandom spacing need not imply that competition is absent (experiments demonstrate that it is present in many cases!) or unimportant in regulating diversity. We may conclude only that evolutionary adjustment of species or their selective establishment and extinction locally do not produce nonrandom patterns. Because species extend geographically and ecologically over large areas, we might not expect their evolved properties to reflect local ecological conditions. Furthermore, because each species may compete with many others in the system over a large number of ecological dimensions, small variation in proximity between nearest neighbors along a single dimension of niche space may have a negligible influence on the demography of any particular population. That is, selection for even spacing may be very weak.

Are local communities saturated with species?

While we may identify the particular factors responsible for patterns of species diversity with difficulty, several general classes of questions can be answered in a more straightforward fashion: Do species fill ecological space in communities up to some fixed capacity determined by their local interactions? Alternatively, can species cram themselves into niche space, and the sizes of their niches be compressed, without reaching a threshold of limiting similarity or minimum critical niche size for coexistence or persistence? If species packing reaches saturation locally, one would expect the local diversity of similar habitats to be similar irrespective of regional diversity, the discrepancy being made up by beta diversity (turnover of species among habitats). This idea can be tested by plotting the relationship between local and regional diversity for many areas with similar environments. Using this technique, John Terborg and John Faaborg (1980) found evidence for saturation in bird communities of the West Indies; George W. Cox and Ricklefs (1977), working with slightly different samples of the same system, did not (Figure 36–22). In two other tests of this hypothesis to date, H. V. Cornell (1985) failed to find evidence for saturation in the local diversity of cynipine gall-forming wasps on oak trees in California (Figure 36–23), whereas the results of George Stevens (1986) supported the hypothesis of saturation for wood-boring scolytid beetles in hardwood stands of eastern North America. Clearly, we cannot reach definitive conclusions about ecological saturation at this point. To the extent that supporting evidence surfaces, it will weigh in favor of local processes and against regional/historical processes in regulating local species diversity.

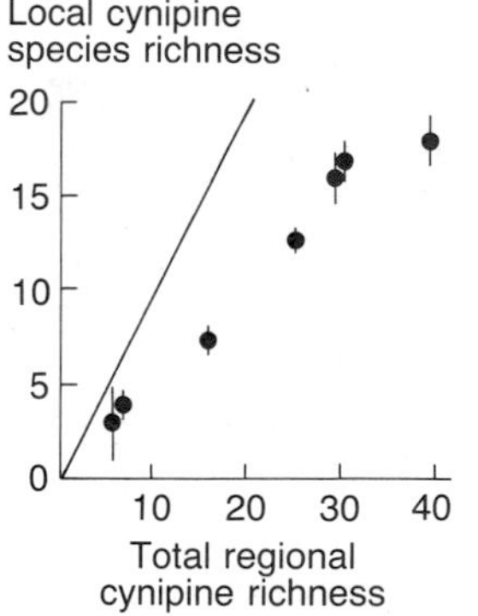

Figure 36–23
Local richness of cynipine wasps on oaks is directly related to the total number of species of cynipines recorded from throughout the range of each oak species (represented by a single data point). The solid line indicates local diversity equal to regional diversity. (From Cornell 1985.)

Is community structure convergent?

Local-process theories for species diversity also predict that communities occurring under similar physical conditions should have similar numbers of species, regardless of geographical location or history. Although plants and animals pervasively reveal convergence of form and function (for example, Orians and Solbrig 1977, Mooney 1977), counterexamples to convergence of community structure—particularly, species diversity—are accumulating about as rapidly as the idea is tested (Orians and Paine 1983, Lawton 1984). Botanists have long recognized that tropical rainforests of Malaysia harbor greater plant diversity than similar forests in South America and, especially, in Africa (Richards 1952; but see Gentry 1988). I shall emphasize the case against convergence by two further plant examples. In the tropics of the New World and western Africa, mangrove communities consist of the same 3 or 4 species of specialized trees distributed throughout the region (Davis 1940). In physically similar environments in Malaysia, mangrove vegetation comprises 17 "principal" and 23 "subsidiary" species (Watson 1928), hence an order of magnitude greater diversity than in the New World. Second, comparisons of chaparral and coastal sage vegetation in Mediterranean climate regions has revealed four times the number of species of plants in Israel compared to southern California, and almost twice the number of species in local (0.1 hectare) samples (Shmida 1981). (In some cases, apparent lack of convergence has been explained by peculiar local conditions; for example, see Morton and James [1988], Westoby [1988].)

These examples of nonconvergence and the lack of consistent evidence for local saturation of ecological communities suggest that regional/historical considerations may play a large role in the generation and maintenance of biological diversity and that the region, as well as the local patch of uniform habitat, is a suitable scale for ecological study.

SUMMARY

1. A conspicuous pattern revealed by studies of biological communities is the tendency of species diversity in tropical regions to greatly exceed that at higher latitudes.

2. Diverse tropical communities contain a greater ecological variety, as well as number, of species, compared to temperate communities.

3. In regions of high species diversity, individual species tend to be habitat specialists compared to their counterparts in regions of low diversity. Hence while more species occupy each habitat (alpha diversity), species also distinguish differences between habitats more finely (beta diversity).

4. One explanation for tropical diversity—the time hypothesis—suggests that tropical habitats are older than temperate habitats, and the former have had more time to accumulate species. According to this hy-

pothesis, diversity is nonequilibrial, and the number of species within a region continually increases.

5. During the last three decades, thinking about diversity has been dominated by equilibrium theories, which state that habitats become saturated by species, reaching numbers determined by local ecological conditions. Accordingly, the greater species diversity in the tropics compared to higher latitudes is due to differences in resources and in interactions between species such that more species can coexist.

6. Differences in the number of species on islands emphasize the importance of regional processes—immigration from the mainland or other islands, in this case—to the maintenance of species diversity. On continents, immigration of species to local areas reflects, in part, the rate of production of new species, which is also a regional process.

7. Several explanations for high plant diversity in the tropics focus on the role of disturbance in creating mosaics of successional stages of different age and in establishing heterogeneous conditions for seedling establishment within gaps in the forest canopy. These ideas have not been evaluated adequately.

8. Predators are thought to enhance diversity among their prey by reducing populations (hence competition for resources), thereby easing conditions for coexistence. Evidence that predators and diseases may act in a density-dependent manner also favors this hypothesis. Density-dependent and frequency-dependent predation favor the persistence of rare species and enhance diversity.

9. Several issues in community ecology have not been adequately resolved. One concerns the rate of species production, particularly whether environmental and species-specific characteristics promote speciation more strongly in tropical regions, where diversity is high. Resolving this problem is outside the traditional discipline of ecology and will require a new synthesis of ecological and evolutionary work.

10. A second issue is whether communities show evidence of being structured by species interactions. If interactions were important, then species should exhibit regular distribution in ecological space. This prediction has been scrutinized by statistical tests with mixed result. More importantly, ecologists must resolve the more philosophical issue concerning whether community pattern is capable of revealing the processes responsible for community structure.

11. Finally, two related predictions of community theory based upon local interaction are being put to the test: (1) local communities should be saturated with species at levels determined by local environmental conditions, and (2) diversity in ecologically similar but geographically distant localities should converge toward the same level. These ideas are just beginning to be tested. But already, accumulating evidence for a pervasive role of regional and historical processes in shaping local biological communities requires that ecologists expand the spatial and temporal scale of their concepts in order to address community phenomena.

A Survey of Biological Communities

The photographs on the pages that follow illustrate most of the important plant formations in temperate North America and tropical Central and South America, with a few examples from other continents. The sequence of terrestrial habitats illustrates the influence of temperature and moisture on vegetation structure. In addition, there are photographs of several freshwater and marine communities.

Environments without life

Life can gain a foothold in regions with almost any combination of temperature and moisture found on earth, providing the moisture is available and other nutrients are present. But life also is excluded from a few environments. The extreme cold on the slopes of Mount Denali, Alaska *(below)*, freezes life to a standstill. Water occurs only as ice and is therefore unavail-

able to plants. Water is a problem as well on the shifting sand dunes of Death Valley, California *(above)*. The little rain that falls either evaporates or percolates through the coarse sand. Temperatures at White Sands, New Mexico *(below)*, are favorable for life, and the region's rainfall supports desert shrubs in the surrounding valley, but the pure gypsum sand (calcium sulfate) does not contain the nutrients needed to support life. (Courtesy of W. J. Smith and the U.S. National Park Service.)

The humid tropics

Year-round warm temperatures and plentiful moisture in the humid tropics create conditions for the most luxuriant and diversified communities in the world. One hectare may include 200 or more species of trees, the tallest of which soar more than 60 meters above the forest floor. Vegetation forms include vines that drape the trees in a lowland forest in Panama *(facing page)* and air plants that clothe trees in a mist-enshrouded cloud forest in Guatemala *(above)*. Because soils are impoverished of nutrients except near the surface, root systems of tropical trees tend to be shallow and the trunks of many trees are buttressed for support *(left)*. (Courtesy of W. J. Smith.)

Tropical mountains

Temperature decreases about 6 °C for each 1000-meter increase in elevation. Plant productivity parallels the lower temperatures of montane habitats, creating cold and almost barren deserts in the tropics. The mean annual temperature and rainfall would support forest or woodland in seasonal temperate climates with warm summers, but the year-round cold of tropical mountains does not permit such luxurious growth. On the paramo of the high Andes in Colombia at about 3700 meters *(above)*, the temperature hovers around 5 °C throughout the year. One is struck by the silence, broken only by the relentless wind. Many plants have dwarfed life forms, with small, thick leaves clustered tightly around the plant stem providing protection from the cold wind *(below)*.

Subtropical deserts

Belts of seasonally hot, dry climate girdle the earth at about 30° north and south of the equator. These are harsh environments where only drought-adapted species of plants and animals thrive. Whereas light and nutrients are critical in the humid tropics, the bare ground exposed in deserts testifies that these resources go wanting where rainfall limits plant growth. Cacti have greatly reduced leaves to decrease water loss. Their thick, succulent stems have taken over the function of photosynthesis. Numerous thorns hinder desert animals from getting at their stored water. Desert shrub habitats of the Sonoran Desert of Arizona and northern Mexico *(below)* are among the most diverse vegetation types of arid regions. Giant saguaro cacti and paloverde trees dominate the landscape. The Joshua tree, a treelike yucca, occurs primarily in the Mohave Desert of southern California *(following page, above)*. (Courtesy of the U.S. Fish and Wildlife Service and the National Park Service.)

Temperate woodland and shrubland

In temperate habitats with more abundant water and lower summer temperatures than deserts, succulent cacti are replaced by bushes, shrubs, and small trees. The wide spacing and low growth form of plants in the Great Basin region of the western United States, exemplified in Zion National Park *(below)*, indicate that water remains a critical factor. At higher elevations, in Coconino National Forest of Arizona, an open woodland dominates the landscape *(facing pages, above)*. Juniper woodland develops at about 2000

meters elevation in this area, where snow covers the ground for much of the winter and summers are cool. The milder Mediterranean climate of the southern California coast, characterized by warm, dry summers and cool, moist winters, supports a characteristic dense shrubland called chaparral *(below)*. In moist canyons and valleys, oak woodland tends to replace chaparral species, but frequent fires often prevent this natural succession and maintain the fire-adapted chaparral vegetation. (Courtesy of the U.S. Forest Service and the U.S. Soil Conservation Service.)

Temperate forests

Tall forests of broad-leaved, deciduous trees occur throughout the temperate zone where rainfall is plentiful and winters are cold. Oak, beech, maple, hickory, and other hardwoods dominate temperate forests. Seasonal patterns of summer activity and winter dormancy are characteristic. The stand of Indiana hardwoods dominated by white oak *(above)* has a well-developed understory of small shrubs. Abundant sugar maple saplings suggest that the composition of the forest is changing. In the Appalachian Moun-

tains of West Virginia, red spruce and hemlock occur with broad-leaved trees to form mixed forests *(facing page, below)*. In the southeastern United States, sandy soils are too poor for broad-leaved trees. Pines are widely distributed in vast forests that are managed and harvested for paper pulp. In Florida, the palmetto frequently forms a dense understory *(above)*. In the northern United States and Canada, and in mountainous regions of the West, birch and aspen frequently invade disturbed sites, often representing the farthest incursion of broad-leaved forests into cold regions *(below)*. (Courtesy of the U.S. Forest Service and J. Lane, Archbold Biological Station.)

Grasslands

Grasslands develop under a variety of temperate climates with cold winters and summer drought. True prairie, remnants of which can be found in Kansas *(above)*, Texas *(below)*, and other midwestern states, is characterized by grasses and forbs with extensive root systems. Tall grass prairies grow on fertile soil and are maintained by periodic fires that keep trees from becoming established. Farther to the west, lower rainfall supports sparser vegeta-

tion, the shortgrass prairies to the east of the Rocky Mountains and in western interior valleys. In many regions of the tropics, particularly at high elevations, grasslands develop under climates with seasonal drought, as in the Masai Mara Reserve of southern Kenya *(above)*. These habitats are, however, extremely productive during the rainy seasons and support large populations of migratory herbivores. Savanna vegetation, typified by drought-adapted trees interspersed by grassland, may develop under similar but generally hotter climatic conditions, as in the Samburu region of northern Kenya *(below)*. (Courtesy of the U.S. Department of Agriculture and the U.S. Soil Conservation Service.)

Temperate needle-leaf forests

Forests of pine, spruce, fir, hemlock, redwood, and others grow under a variety of temperature, soil, moisture, and fire conditions that favor drought-resistant needle-leaf species over less tolerant broad-leaved trees. Poor soils and frequent fires favor pines throughout much of the southeastern United States. Dry summers and cold winters characterize the environments of coniferous forests at high elevations in the western mountains. Piñon pine-juniper-cedar woodland is found near Flagstaff, Arizona, where the climate is too dry to support a closed forest *(facing page, above)*. A moister site in Inyo National Forest, California, is dominated by tall Jeffrey pines *(below)*. Undergrowth is sparse in the dry, acid soil, and trees are widely spaced. In contrast, abundant winter rainfall and cool, foggy summers create ideal conditions for redwoods in the temperate rainforests of northern California *(facing page, below)*. They bear little resemblance to the humid forests of the tropics, however, because they lack diversity in species and plant forms and are relatively unproductive. (Courtesy of the U.S. Forest Service.)

Temperate montane environments

Montane habitats are much colder and are often drier than the surrounding lowlands. Trees reach their upper limits of elevation at about 3000 meters in the Cascade Mountains of Oregon (*above*). Above timberline, snow persists well into summer in habitat that can support only the low grassy vegetation characteristic of the alpine tundra, as in the Rocky Mountains of Colorado at

3700 meters *(facing page, below)*. The differential melting of the snow cover creates habitat heterogeneity, increasing the variety of plants that coexist in the tundra. Lichens are the first plants to colonize bare rock surfaces in these habitats *(above)* and start the slow process of soil formation. Wind-driven ice strips bark and branches from trees near the timberline in Colorado *(below)*. (Courtesy of the U.S. Forest Service.)

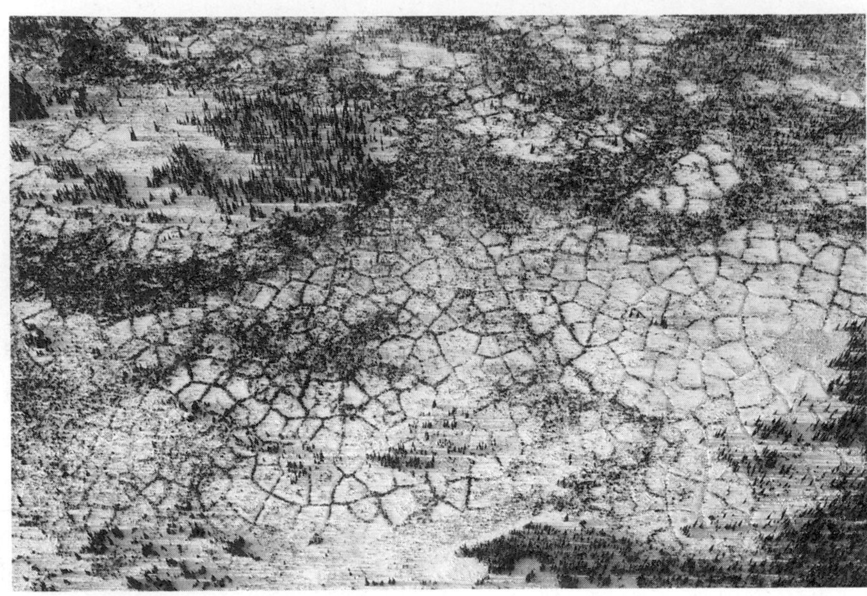

The arctic tundra

Permanently frozen soils underlie the arctic tundra habitat. Warm summer temperatures thaw the ground to a depth of a few centimeters or perhaps a meter, briefly creating a shallow, often waterlogged layer of soil on which arctic vegetation develops. Repeated freezing and thawing creates characteristic polygonal patterns in the ground surface of some areas, as shown in

the aerial photograph *(facing page, above)*. At Cold Bay, Alaska, lichens, mosses, and grasses are found on the hummocky, frost-heaved soil *(facing page, below)*. Kettle lakes formed by the melting of large blocks of ice, left by retreating glaciers, are a prominent feature of the Kuskokwim River Delta, Alaska *(above)*. Montane tundra in the arctic is better drained than lowland habitats and spruce trees occasionally get a foothold in protected valleys, such as those near Mount Denali, Alaska *(below)*. (Courtesy of the U.S. Soil Conservation Service.)

Freshwater habitats

Freshwater covers a small fraction of the earth's surface, yet freshwater habitats display remarkable diversity. Variation in water movement, mineral and oxygen content of the water, and size and shape of the stream or lake basin all contribute to this variety. Communities in deep lakes and fast-moving streams consist mostly of phytoplankton and thin layers of diatoms on the surfaces of rocks. Vegetation shows above the surface only where water is shallow and still as in an artificially flooded marsh in Maine *(facing page)* or a cattail marsh in New York *(above)*. Floating water hyacinths choke a deeper channel in Louisiana, buoyed up by gas trapped in their stems *(below)*. (Courtesy of the U.S. Department of Agriculture.)

The land meets the sea

The topography of coastal areas often determines the character of plant communities at the edge between the land and the sea. Shallow, sloping, sandy shores, like those of Cape Cod, Massachusetts *(above)*, create shifting dune habitats colonized by a few species of plants that stabilize the dune and allow other species to establish a foothold. Salt marshes develop in

more protected bays and in river estuaries, as at Barnstable Harbor on Cape Cod *(facing page, below)*. At the rockbound coast of Maine, an abrupt meeting of land and sea promotes little intermingling of the two environments *(above)*. The intertidal zone, which becomes exposed to air twice each day, nonetheless may support prolific growth of seaweeds specialized to tolerate the drying conditions *(below, right)*. On sandy beaches of the eastern United States, horseshoe crabs come ashore in spring to mate and lay their eggs, presumably seeking a refuge from predation for the developing embryos *(below, left)*. (Courtesy of F. B. Bowles, W. J. Smith, and the U.S. National Marine Fisheries Service.)

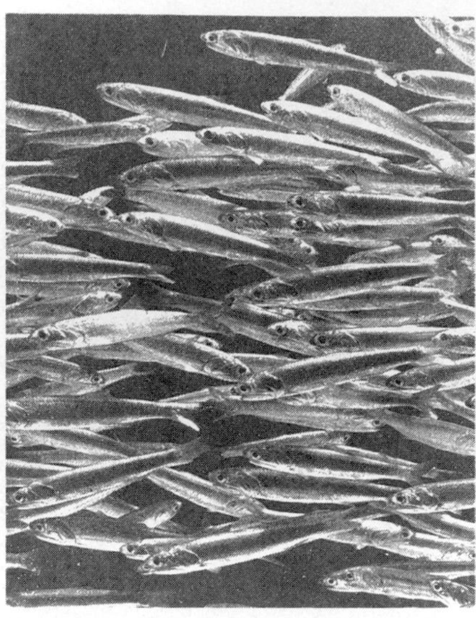

The marine environment

The little-explored mantle of water covering most of the surface of the earth contains a wide variety of habitats and life forms. Open waters create a vast realm for the tiny phytoplankton and zooplankton, the fish that exploit them *(above, left)*, and the seabirds and other predators that eat the herbivorous and planktivorous fish *(below)*. Other fish more closely resemble terrestrial grazers and predators, feeding on algae and small animals near

the bottom and among reefs *(facing page, above, right)*. Tropical habitats are dominated by corals, as on the Caribbean coast of Panama *(above)*, but reef-building species are restricted to sunlit depths because they rely on symbiotic green algae in their tissues for much of their nutrition. In rich, cold waters, one frequently encounters kelp forests, whose structure provides feeding and refuge for many species of fish and invertebrates *(below)*. (Courtesy of J. Porter, W. J. Smith, and P. Dayton.)

Glossary

Abscission. The process by which a part of an organism separates from the rest.

Acclimation. A reversible change in the morphology or physiology of an organism in response to environmental change; also called acclimatization.

Active transport. Movement of ions or other substances across a membrane against a concentration gradient, requiring the expenditure of energy.

Activity space. The range of environmental conditions suitable for the activity of an organism.

Adaptation. A genetically determined characteristic that enhances the ability of an individual to cope with its environment; an evolutionary process by which organisms become better suited to their environments.

Adaptive function. In fitness set analysis, a mathematical expression combining the fitnesses of a phenotype in each of several environments into a measure of overall fitness in a heterogeneous environment.

Additive genetic variance (V_A). Variation in a phenotypic value within a population due to the difference in expression of alleles in the homozygous state.

Age class. Individuals in a population of a particular age.

Alkaloids. Nitrogen-containing compounds, such as morphine and nicotine, that are produced by plants and are toxic to many herbivores.

Allele. One of several alternative forms of a gene.

Allelopathy. Direct inhibition of one species by another using noxious or toxic chemicals.

Allochthonous. Referring to materials transported into a system, particularly minerals and organic matter transported into streams and lakes. *Compare with* Autochthonous.

Allometric constant. Slope of the relationship between the logarithm of one measurement of an organism to the logarithm of another, usually its overall size.

Allometry. Relative increase in a part of an organism or a measure of its physiology or behavior in relation to some other measure, usually its overall size.

Allopatric. Occurring in different places, usually referring to geographical separation of populations.

Alluvial. Referring to sediment deposited by running water.

Alpha diversity. The variety of organisms occurring in a particular place or habitat; often called local diversity.

Altruism. In an evolutionary sense, enhancing the fitness of an unrelated individual by acts that reduce the evolutionary fitness of the altruistic individual.

Ambient. Referring to conditions of the environment surrounding the organism.

Amino acid. One of about thirty organic acids containing the group NH_2, which are the building blocks of proteins.

Ammonification. Metabolic breakdown of proteins and amino acids with ammonia as an excreted by-product.

Anaerobic. Without oxygen.

Anion. A part of a dissociated molecule carrying a negative electrical charge.

Anisogamous. Having gametes unequal in size and behavior; usually a large, sedentary (female) and a small, motile (male) gamete. *See* Isogamous.

Annual. Referring to an organism that completes its life cycle from birth or germination to death within a year.

Anoxic. Lacking oxygen; anaerobic.

Antagonistic resource. A resource that interacts with another such that a consumer requires more of a mixture of resources than of either one alone.

Aposematism. *See* Warning coloration.

Arena. *See* Lek.

Arithmetic mean. The sum of a series of values divided by the number of values; a measure of the central tendency of a sample. *Compare with* Geometric mean.

Arrhenotoky. A sex determination system in which males develop only from unfertilized haploid eggs, and thus have only one chromosome complement.

Artificial selection. Manipulation by man of the fitnesses of individuals in a population to produce a desired evolutionary response.

Aspect diversity. Variety of outward appearances of species that live in the same habitat and are eaten by visually hunting predators.

Assimilation. Incorporation of any material into the tissues, cells, and fluids of an organism.

Assimilation efficiency. A percentage expressing the proportion of ingested energy that is absorbed into the bloodstream.

Assimilatory. Referring to a biochemical transformation that results in the reduction of an element to an organic form, hence its gain by the biological compartment of the ecosystem.

Association. A group of species living in the same place.

Asymmetrical competition. Interaction between species in which one exploits a particular resource more efficiently than the other; the second may persist by better avoiding predation or by subsisting on an exclusive resource.

Autecology. The study of organisms in relation to their physical environment. *Compare with* Synecology.

Autochthonous. Referring to materials produced within a system, particularly organic matter produced, and minerals cycled, within streams and lakes. *Compare with* Allochthonous.

Autotroph. An organism that assimilates energy from either sunlight (green plants) or inorganic compounds (sulfur bacteria). *Compare with* Heterotroph.

Balance school. In reference to population regulation, individuals who advocate the idea that biotic factors, such as competition, predation, and parasitism, acting in density-dependent fashion, control population size.

Balanced polymorphism. Maintenance of more than one allele in a population by the selective superiority of the heterozygote over both homozygotes. *See* Heterosis.

Barren. An area with sparse vegetation owing to some physical or chemical property of the soil.

Basal metabolic rate (BMR). The energy expenditure of an organism that is at rest, fasting, and in a thermally neutral environment.

Batesian mimicry. Resemblance of an edible species (mimic) to an unpalatable species (model) to deceive predators.

Bedrock. Unweathered solid rock underlying soil.

Benthic. Bottom dwelling in rivers, lakes, and oceans.

Bet hedging. Spreading of risk; reducing the risk of catastrophic failure or death. In life-history evolution, bet hedging leads to more frequent or prolonged, but less intense, reproduction.

Beta diversity. The variety of organisms within a region arising from turnover of species among habitats.

Biennial. Requiring two years to complete the life cycle.

Biological community. *See* Community.

Biological oxygen demand (BOD). The amount of oxygen required to oxidize organic material in water samples; high values in aquatic habitats often indicate pollution by sewage and other sources of organic wastes, or the overproduction of plant material resulting from overenrichment by mineral nutrients.

Biomass. Weight of living material, usually expressed as a dry weight, in all or part of an organism, population, or community. Commonly presented as weight per unit area, a biomass density.

Biomass accumulation ratio. The ratio of weight to annual production.

Biota. Fauna and flora together.

Birth rate (*b*). The average number of offspring produced per individual per unit of time, often expressed as a function of age (x).

Bohr effect. A shifting of the oxygen dissociation curve of hemoglobin to the left under acid conditions, which facilitates the unloading of oxygen from the bloodstream to tissues.

Boreal. Northern; often refers to the coniferous forest regions that stretch across Canada, northern Europe, and Asia.

Boundary layer. A layer of still or slow-moving water or air close to the surface of an object.

Breeding system. Degree of polygyny, outcrossing, and selective mating within a population; the adaptations by which organisms adjust these attributes.

Broken-stick model. A model of relative abundance of species obtained by random division into segments of a line representing the resources of the environment.

Brood parasite. An organism that lays its eggs in the nest of another species or that of another individual of the same species.

C_3 photosynthesis. Photosynthetic pathway in which carbon dioxide is initially assimilated into a three-carbon compound, phosophoglyceraldehyde (PGA), in the Calvin cycle.

C_4 photosynthesis. Photosynthetic pathway involving the initial assimilation of carbon dioxide into a four-carbon compound, such as oxaloacetic acid (OAA) or malate.

Calcification. Deposition of calcium and other soluble salts in soils where evaporation greatly exceeds precipitation.

Caliche. An alkaline salt deposit on the soil surface, usually occurring in arid regions with groundwater close to the surface.

Calvin cycle. The basic assimilatory sequence of photosynthesis during which an atom of carbon is added to the 5-carbon ribulose bisphosphate (RuBP) molecule to produce phosophoglyceraldehyde (PGA) and then glucose.

CAM photosynthesis. Photosynthetic pathway in which the initial assimilation of carbon dioxide into a four-carbon compound occurs at night; found in some succulent plants in arid habitats.

Carbonic acid. A weak acid (H_2CO_3) formed when carbon dioxide dissolves in water.

Carbonate ion. An anion (CO_3^{2-}) formed by the dissociation of carbonic acid or one of its salts.

Carnivore. An organism that consumes mostly flesh.

Carrying capacity (*K*). Number of individuals in a population that the resources of a habitat can support; the asymptote, or plateau, of the logistic and other sigmoid equations for population growth.

Caste. Individuals within a social group sharing a specialized form or behavior.

Cation. A part of a dissociated molecule carrying a positive electrical charge.

Cation exchange capacity. The ability of soil particles to absorb positively charged ions, such as hydrogen (H^+) and calcium (Ca^{2+}).

Cellulose. A long-chain molecule made up of glucose subunits found in the cell walls and fibrous structures in plants.

Chaos. Erratic change in the size of populations governed by difference equations and having high intrinsic rates of growth.

Character displacement. Divergence in the characteristics of two otherwise similar species where their ranges overlap, caused by the selective effects of competition between the species in the area of overlap.

Characteristic equation. An equation relating the exponential growth rate of a population to age-specific survivorship and fecundity under stable-age conditions; Euler's equation.

Characteristic return time. In reference to populations, the inverse of the exponential growth rate; an index to the rate at which a population regains its equilibrium value after perturbation.

Chemoautotroph. An organism that oxidizes inorganic compounds (often hydrogen sulfide) to obtain energy for synthesis of organic compounds; for example, sulfur bacteria.

Chemostat. A culture vessel with a constant flux of nutrient medium.

Clay. A fine-grained component of soil, formed by the weathering of granitic rock, composed primarily of hydrous aluminum silicates.

Climate school. In reference to population regulation, individuals who adhere to the idea that climate factors are preeminent in controlling population size.

Climax. The end point of a successional sequence, or sere; a community that has reached a steady state under a particular set of environmental conditions.

Climograph. A diagram on which localities are represented by the annual cycle of their temperature and rainfall.

Cline. Gradual change in population characteristics or adaptations over a geographic area.

Closed-community concept. The idea, popularized by F. C. Clements, that communities are distinctive associations of highly interdependent species.

Clutch size. Number of eggs per set; usually with reference to the nests of birds.

Coadaptation. Evolution of characteristics of two or more species in response to changes in each other, often to mutual advantage.

Coarse-grained. Referring to qualities of the environment that occur in large patches with respect to the activity patterns of an organism and, therefore, among which the organism can select.

Coefficient of competition (a). *See* Competition coefficient.

Coefficient of determination (R^2). The proportion of variation in one (dependent) variable that is related to variation in one or more other (independent) variables.

Coevolution. Occurrence of genetically determined traits (adaptations) in two or more species selected by the mutual interactions controlled by these traits.

Coexistence. Occurrence of two or more species in the same habitat; usually applied to potentially competing species.

Cohort. A group of individuals of the same age recruited into a population at the same time.

Cohort life table. *See* Dynamic life table.

Community. An association of interacting populations, usually defined by the nature of their interaction or the place in which they live.

Community matrix. The symmetrical matrix of coefficients of interaction between species in a community.

Compartmentalization. Subdivision of a food web into groups of strongly interacting species somewhat isolated from other such groups.

Compartment model. Representation of a system in which the various parts are portrayed as units (compartments) that receive inputs from, and provide outputs to, other such units.

Compensation point. Depth of water or level of light at which respiration and photosynthesis balance each other; the lower limit of the euphotic zone.

Competition. Use or defense of a resource by one individual that reduces the availability of that resource to other individuals, whether of the same species (intraspecific competition) or other species (interspecific competition).

Competition coefficient (a). A measure of the degree to which one consumer utilizes the resources of another, expressed in terms of the population consequences of the interaction.

Competitive Exclusion Principle. The hypothesis that two or more species cannot coexist on a single resource that is scarce relative to demand for it.

Complementary resource. A resource that interacts with another synergistically; that is, fewer mixed resources are required than either one alone.

Condition. Physical or chemical attributes of the environment that, while not being consumed, influence biological processes and population growth; for example, temperature, salinity, acidity. *Compare with* Resource.

Conductance. Capacity of heat, electricity, or a substance to pass through a particular material.

Conduction. The ability of heat to pass through a substance.

Conformer. An organism that allows its internal environment to vary with external conditions.

Congeneric. Belonging to the same genus.

Connectance (*C*). Proportion of interspecific interactions in a community matrix not equal to zero.

Conspecific. Belonging to the same species.

Consumer. Individual or population that utilizes a particular resource.

Continental shelves. Areas of relatively shallow seas surrounding continents and lying on the edges of the continental plates.

Continuum. A gradient of environmental characteristics or of change in the composition of communities.

Continuum index. A scale of an environmental gradient based upon changes in physical characteristics of community composition along that gradient.

Convection. Transfer of heat by the movement of a fluid (for example, air or water).

Correlated response. Response of the phenotypic value of one trait to selection upon another trait.

Counteradaptation. Evolution of characteristics of two or more species based upon their mutual antagonism, as in competitive and consumer-resource systems.

Countercurrent circulation. Movement of fluids in opposite directions on either side of a separating barrier through which heat or dissolved substances may pass.

Covariance. The average product of deviations of two variables from their respective means.

Crassulacean acid metabolism. *See* CAM photosynthesis.

Cross-resistance. Resistance or immunity to one disease organism resulting from infection by another, usually closely related organism.

Crypsis. An aspect of the appearance of organisms whereby they avoid detection by others.

Cybernetic. Pertaining to feedback controls and communication within systems.

Cycle. Recurrent variation in a system periodically returning to its starting point.

Cyclic climax. A steady-state, cyclic sequence of communities, none of which by itself is stable.

Cyclic succession. Continual community change through a repeated sequence of stages.

Damped oscillation. Cycling with progressively smaller amplitude, as in some populations approaching their equilibria.

Death rate (d_x). The percentage of newborn dying during a specified interval. *Compare with* Mortality.

Deme. A local population within which mating occurs among individuals more or less at random.

Demographic. Pertaining to populations, particularly their growth rate and age structure.

Dentrification. Biochemical reduction, primarily by microorganisms, of nitrogen from nitrate (NO_3^-) eventually to molecular nitrogen (N_2).

Density. Referring to a population, the number of individuals per unit area or volume; referring to a substance, the weight per unit volume.

Density compensation. Increase in population size in response to reduction in the number of competing populations; often observed on islands.

Density dependent. Having influence on individuals in a population that varies with the degree of crowding within the population.

Density independent. Having influence on individuals in a population that does not vary with the degree of crowding.

Deterministic. Referring to the outcome of a process that is not subject to stochastic (random) variation.

Detritivore. An organism that feeds on freshly dead or partially decomposed organic matter.

Detritus. Freshly dead or partially decomposed organic matter.

Developmental response. Acquisition of one of several alternative forms by an organism depending on the environmental conditions under which it grows.

Diapause. Temporary interruption in the development of insect eggs or larvae, usually associated with a dormant period.

Difference equation. Equation describing the change of a quantity over an interval, as, for example, ΔN describes the change in number of individuals in a population between time t and $t + 1$.

Differential equation. Equation describing the instantaneous rate of change in a quantity, as, for example, dN/dt describes the change in number of individuals in a population (N) over time (t).

Diffuse coevolution. The evolution of traits influencing species interactions, subject to selection from a wide variety of species interacting with different intensities.

Diffuse competition. The sum of weak competitive interactions with species that are ecologically distantly allied.

Diffusion. Movement of particles of gas or liquid from regions of high to low concentration by means of their own spontaneous motion.

Dimorphism. Occurrence of two forms of individuals within a population.

Dioecy. In plants, the occurrence of reproductive organs of the male and female sex on different individuals. *See* Monoecy.

Diploid. Pertaining to cells or organisms having two sets of chromosomes. *See* Haploid; Meiosis.

Direct competition. Exclusion of individuals from resources by aggressive behavior or use of toxins. *See also* Indirect competition.

Disc equation. An equation relating the rate of encounter of prey by a predator to searching efficiency, prey abundance, and handling time.

Dispersal. Movement of organisms away from the place of birth or from centers of population density.

Dispersion. The spatial pattern of distribution of individuals within populations.

Dissimilatory. Referring to a biochemical transformation that results in the oxidation of the organic form of an element, hence its loss from the biological compartment of the ecosystem.

Dissociation. The breaking up of a compound into its component parts; especially the formation of ions in an aqueous solution.

Dissolution. The entry of a substance in solution with water.

Distribution. The geographical extent of a population or other ecological unit.

Diversity. The number of species in a local area (alpha diversity) or region (gamma diversity). Also, a measure of the variety of species in a community that takes into account the relative abundance of each species.

Dominance hierarchy. Orderly ranking of individuals in a group, based on the outcome of aggressive encounters.

Dominance variance (V_D). Variation in a phenotypic value within a population due to the unequal expression of alleles in the heterozygous state.

Dormancy. An inactive state, including hibernation, diapause, and seed dormancy, usually assumed during an inhospitable period.

Dynamic life table. The age-specific survival and fecundity of a cohort of individuals in a population followed between birth and death of the last individual; cohort life table.

Dynamic steady state. Condition in which fluxes of energy or materials into and out of a system are balanced.

Ecocline. A geographical gradient of vegetation structure associated with one or more environmental variables.

Ecological efficiency. Percentage of energy in the biomass produced by one trophic level that is incorporated into the biomass produced by the next higher trophic level.

Ecological isolation. Avoidance of competition between two species by differences in food, habitat, activity period, or geographical range.

Ecological release. Expansion of habitat and resource utilization by populations in regions of low species diversity, resulting from reduced interspecific competition.

Ecology. The study of the natural environment and of the relations of organisms to each other and to their surroundings.

Ecomorphology. The study of the relationship between the ecological relations of an individual and its morphology.

Ecosystem. All the interacting parts of the physical and biological worlds.

Ecotone. A habitat created by the juxtaposition of distinctly different habitats; an edge habitat; a zone of transition between habitat types.

Ecotype. A genetically differentiated subpopulation that is restricted to a specific habitat.

Ectomycorrhyzae. Mutualistic fungal association with the roots of plants in which the fungus forms a sheath around the outside of the root.

Ectoparasite. A parasite, for example, a tick, that lives on, or attached to, the host's surface.

Ectothermy. Capacity to maintain body temperature by gaining heat from the environment, either by conduction or by absorbing radiation.

Edaphic. Pertaining to, or influenced by, the soil.

Effective population size (N_e). The average size of a population expressed in terms of individuals assumed to contribute genes equally to the next generation; generally smaller than the actual size of the population, depending on the variation in reproductive success among individuals.

Egestion. Elimination of undigested food material.

Eigenvalue. One of n roots of n simultaneous linear equations. The first eigenvalue of the population matrix is the exponential rate of population growth.

Eigenvector. Column of mathematical elements that provides a solution to a matrix multiplication; in population biology, the first eigenvector of the population matrix is the stable age distribution.

Electrical potential (E_h). The relative capacity, measured in volts, of one substance to oxidize another.

Electron acceptor. A substance that readily accepts electrons, hence that is capable of oxidizing another substance.

El Niño. A warm current from the tropics that intrudes each winter along the west coast of northern South America.

Eluviation. The downward movement of dissolved soil materials from the topmost (A) horizon, carried by percolating water.

Emergent property. A feature of a system not deducible from lower order processes.

Emigration. Movement of individuals out of a population.

Endemic. Confined to a certain region.

Endomycorrhyzae. Mutualistic fungal association with the roots of plants in which part of the fungus resides within the root tissues.

Endoparasite. A parasite that lives within the tissues or bloodstream of its host.

Endothermy. Capacity to maintain body temperature by the metabolic generation of heat.

Energetic efficiency. The ratio of useful work or energy storage to energy intake.

Energy. Capacity for doing work.

ENSO. El Niño/Southern Oscillation; an occasional shift in winds and ocean currents, centered in the South Pacific region, with worldwide consequences for climate and biological systems.

Environment. Surroundings of an organism, including the plants, animals, and microbes with which it interacts.

Environmental grain. A concept of the spatial or temporal heterogeneity of the environment relative to the activities of an organism.

Environmental variance (V_E). Variation in a mensural trait within a population due to the influence of environmental factors.

Enzyme. Organic compound in a living cell or secreted by it that accelerates a specific biochemical transformation without itself being affected.

Epidemiology. The study of factors influencing the spread of disease through a population.

Epifaunal. Pertaining to animals living on the surface of a substrate.

Epilimnion. The warm, oxygen-rich surface layers of a lake or other body of water. *Compare with* Hypolimnion.

Equilibrium. A state of balance between opposing forces.

Equilibrium isocline. A line on a population graph designating combinations of competing populations, or predator and prey populations, for which the growth rate of one of the populations is zero.

Equitability. Uniformity of abundance in an assemblage of species. Equitability is greatest when species are equally abundant.

Escape space. Refuge from predators and parasites; often reflected in the adaptations of prey organisms to fight, flee, or escape detection.

Essential resource. A resource that limits a consumer population independently of other resources; that is, without synergism or the possibility of substitution.

Estuary. A semienclosed coastal water, often at the mouth of a river, having a high input of freshwater and great fluctuation in salinity.

Euclidean distance. The straightline distance between two points in multidimensional space.

Euler's equation. *See* Characteristic equation.

Euphotic zone. Surface layer of water to the depth of light penetration at which photosynthesis balances respiration. *See* Compensation point.

Eusocial. Referring to the complex social organization of termites, ants, and many wasps and bees, dominated by an egg-laying queen that is tended by nonreproductive offspring.

Eutrophic. Rich in the mineral nutrients required by green plants; pertaining to an aquatic habitat with high productivity.

Eutrophication. Enrichment of water by nutrients required for plant growth; often overenrichment caused by sewage and runoff from fertilized agricultural lands and resulting in excessive bacterial growth and oxygen depletion.

Evapotranspiration. The sum of transpiration by plants and evaporation from the soil. Potential evapotranspiration is the amount of evapotranspiration that would occur, given the local temperature and humidity, if water were superabundant.

Evolutionarily Stable Strategy (ESS). A strategy such that, if all members of a population adopt it, no alternative strategy can invade.

Evolutionary ecology. The integrated science of evolution, genetics, adaptation, and ecology; interpretation of the structure and function of organisms, communities, and ecosystems in the context of evolutionary theory.

Excretion. Elimination from the body, by way of the kidneys, gills, and dermal glands, of excess salts, nitrogenous waste products, and other substances.

Expectation of further life (e_x). The average remaining lifetime of an individual of age x.

Exploitation. Removal of individuals or biomass from a population by consumers.

Exploitative competition. Interaction between individuals by way of their reduction of shared resources.

Exponential mortality rate (k_x). Negative logarithm of the probability of survival between ages of x and $x + 1$. *See* Killing power.

Exponential rate of increase (r). Rate at which a population is growing at a particular time, expressed as a proportional increase per unit of time. *See* Geometric rate of increase.

External forcing function. In systems modeling, a material input from outside the system or a condition of the environment of the system that influences its structure and function.

Extinction. Disappearance of a species or other taxon from a region or biota.

Facilitation. Enhancement of a population of one species by the activities of another, particularly during early succession.

Facultative. Being able to adjust to a variety of conditions or circumstances; optional for the organism. *Compare with* Obligate.

Fall bloom. The rapid growth of algae in temperate lakes following the autumnal breakdown of thermal stratification and mixing of water layers.

Fall overturn. Vertical mixing of water layers in temperate lakes in autumn following breakdown of thermal stratification.

Fecundity. Rate at which an individual produces offspring.

Fermentation. Anaerobic, energy-releasing transformation of organic substances; often involving the transformation of pyruvate to ethanol or lactic acid.

Field capacity. The amount of water that soil can hold against the pull of gravity.

Filter feeder. An organism that strains tiny food particles from its aqueous environment by means of sieve-like structures; for example, clams and baleen whales.

Fine-grained. Referring to qualities of the environment that occur in small patches with respect to the activity patterns of an organism, and among which the organism cannot usefully distinguish.

Fitness. Genetic contribution by an individual's descendants to future generations of a population.

Fitness set. A graphical portrayal of phenotypes within a population according to their values for various components of fitness.

Floristic. Referring to the species composition of plant communities.

Flux. Movement of energy or material into or out of a system.

Food chain. A representation of the passage of energy through populations in the community.

Food chain efficiency. *See* Ecological efficiency.

Food web. A representation of the various paths of energy flow through populations in the community.

Frequency dependence. Referring to the condition in which the expression of a process varies with the relative proportions of phenotypes in a population.

Front. A meeting of two water masses having different characteristics.

Functional response. Change in the rate of exploitation of prey by an individual predator as a result of a change in prey density. *See also* Numerical response.

Gamete. A haploid cell that fuses with another haploid cell of opposite sex during fertilization to form the zygote. In animals the male gamete is called the sperm and the female gamete, the egg or ovum.

Game theory. Analysis of the outcomes of social interactions between individuals with different behavior programs.

Gamma diversity. The inclusive diversity of all the habitat types within an area; regional diversity.

Gene. Generally, a unit of genetic inheritance. In biochemistry, gene refers to the part of the DNA molecule that encodes a single enzyme or structural protein.

Gene flow. Exchange of genetic traits between populations by movement of individuals, gametes, or spores.

Gene frequency. The proportion of a particular allele of a gene in the gene pool of a population.

Gene locus. Segment of a chromosome on which a gene resides.

General adaptive syndrome. The set of abnormal physiological responses to the stress of social interaction in dense populations.

Generalist. A species with broad food or habitat preferences.

Generation time. Average age at which a female gives birth to her offspring, or the average time for a population to increase by a factor equal to the net reproductive rate.

Genetic drift. Change in allele frequency due to random variations in fecundity and mortality in a population.

Genetic feedback. Evolutionary response of a population to the adaptations of competitors, predators, or prey.

Genetic feedback hypothesis. The idea that selection at high population density favors individuals genetically predisposed to low reproductive rates, and vice versa, leading to population cycling.

Genetic variance (V_G). Variation in a phenotypic value within a population due to the expression of genetic factors.

Genotype. All the genetic characteristics that determine the structure and functioning of an organism; often applied to a single gene locus to distinguish one allele, or combination of alleles, from another.

Geometric mean. The *n*th root of the product of *n* values; alternatively, the antilogarithm of the mean of the logarithms of a series of values. *Compare with* Arithmetic mean.

Geometric rate of increase (λ). Factor by which the size of a population changes over a specified period. *See* Exponential rate of increase.

Geometric series. A series in which each element is formed by multiplying the previous one by some factor. A characterization of the relative abundances of species in which the logarithm of abundance decreases linearly with the abundance rank of the species.

Gonochoristic. In animals, the condition of separate male and female individuals; in plants, called dioecious.

Gradient. Difference in condition or concentration of a substance between two points or across a boundary.

Gradient analysis. Portrayal and interpretation of the abundances of species along gradients of physical conditions.

Grain. The scale of heterogeneity of habitats in relation to the activities of organisms.

Greenhouse effect. Warming of the earth's climate owing to the increased concentration of carbon dioxide and certain other pollutants in the atmosphere.

Gross production. The total energy or nutrients assimilated by an organism, a population, or an entire community. *See also* Net production.

Gross production efficiency. The percentage of ingested food utilized for growth and reproduction by an organism.

Groundwater. Water that percolates through the soil and through cracks and interstices in bedrock.

Group selection. Elimination of groups of individuals with a detrimental genetic trait, caused by competition with other groups lacking the trait; often called intergroup selection.

Habit. In plants, the general life-form of an individual (for example, erect, prostrate, woody).

Habitat. Place where an animal or plant normally lives, often characterized by a dominant plant form or physical characteristic (that is, the stream habitat, the forest habitat).

Habitat compression. Restriction of habitat distribution in response to increase in number of competing species.

Habitat expansion. Increase in average breadth of habitat distribution of species in depauperate biotas, especially on islands, compared with species in more diverse biotas.

Habitat patch. An area of distinct habitat type.

Habitat selection. Preference for certain habitats.

Handicap principle. The idea that elaborate, sexually selected displays and adornments act as handicaps that demonstrate the generally high fitness of the bearer.

Haplodiploid. A sex-determining mechanism by which females develop from fertilized eggs and males from unfertilized eggs.

Haploid. Referring to a cell or organism that contains one set of chromosomes.

Heat. A measure of the kinetic energy of the atoms or molecules in a substance.

Heat of melting. Amount of heat energy added to a substance to make it melt.

Heat of vaporization. Amount of energy added to a substance to make it vaporize.

Herbivore. An organism that consumes living plants or their parts.

Heritability (h^2). The proportion of variance in a phenotypic trait due to the effects of additive genetic factors.

Hermaphrodite. An organism that has the reproductive organs of both sexes.

Heterogeneity. The variety of qualities found in an environment (habitat patches) or a population (genotypic variation).

Heterosis. Situation in which the heterozygous genotype is more fit than either homozygote; also called overdominance. *See* Balanced polymorphism.

Heterotroph. An organism that utilizes organic materials as a source of energy and nutrients. *Compare with* Autotroph.

Heterozygous. Containing two forms (alleles) of a gene, one derived from each parent.

Hibernation. State of winter dormancy associated with lowered body temperature and metabolism.

Higher-order interaction. The influence of one species on the interaction between two others.

Homeostasis. Maintenance of constant internal conditions in the face of a varying external environment.

Homeothermy. Ability to maintain constant body temperature in the face of fluctuating environmental temperature; warm-blooded.

Homology. The condition of having similar evolutionary origin.

Homozygous. Containing two identical alleles at a gene locus.

Horizon. A layer of soil distinguished by its physical and chemical properties.

Humus. Fine particles of organic detritus in soil.

Hydrarch succession. Progression of terrestrial plant communities developing in an aquatic habitat such as a bog or swamp. *Compare with* Xerarch succession.

Hydrological cycle. Movement of water throughout the ecosystem.

Hydrolysis. A biochemical process by which a molecule is split into parts by the addition of the parts of water molecules.

Hygroscopic water. Water held tightly by surface adhesion to particles in the soil, generally unavailable to plants.

Hyperdispersion. Pattern of distribution in which distances between individuals are more even than expected from random placement; overdispersion.

Hyperosmotic. Having an osmotic potential (generally, salt concentration) greater than that of the surrounding medium.

Hyperparasitism. Occurrence of parasitoids upon other parasitoids.

Hypertonic. Having a salt concentration greater than that of the surrounding medium.

Hypolimnion. The cold, oxygen-depleted part of a lake or other body of water that lies below the zone of rapid change in water temperature (thermocline). *Compare with* Epilimnion.

Hypo-osmotic. Having an osmotic potential (generally, salt concentration) less than that of the surrounding medium.

Hypotonic. Having a salt concentration less than that of the surrounding medium.

Ideal free distribution. The distribution of individuals across resource patches of different intrinsic quality that equalizes the net rate of gain of each when competition is taken into account.

Identical by descent. Genes in different individuals that are direct copies of a gene in a common ancestor.

Illuviation. The accumulation of dissolved substances within a soil layer, usually the middle (B) horizon.

Imago. The adult stage of an insect.

Inbreeding. Mating among related individuals.

Incidence matrix. Table of occurrences of species in samples.

Inclusive fitness. The total fitness of an individual and the fitnesses of its relatives, the latter weighted according to degree of relationship; usually applied to the consequences of social interaction between relatives.

Indirect competition. Exploitation of a resource by one individual that reduces the availability of that resource to others. *See also* Direct competition.

Individual distance. The distance within which one individual does not tolerate the presence of another.

Inducible response. Any change in the state of the organism caused by an external factor; usually reserved for the response of organisms to parasitism and herbivory.

Industrial melanism. The evolution of dark coloration by cryptic organisms in response to industrial pollution, especially by soot, of their environments.

Information center. An aggregation of individuals within which information about food or other resources is obtained, generally by observing the return and departure of foraging individuals.

Infrared (IR) radiation. Electromagnetic radiation having a wavelength longer than about 700 nm.

Inhibition. The suppression of a colonizing population by another that is already established, especially during successional sequences.

Innate capacity for increase (r_0). The intrinsic growth rate of a population under ideal conditions without the restraining effects of competition.

Intergroup selection. *See* Group selection.

Intermediate disturbance hypothesis. The idea that species diversity is greatest in habitats with moderate amounts of physical disturbance, owing to the coexistence of early and late successional species.

Intermediate host. A host that harbors an asexual stage of the life cycle of a parasite or disease organism.

Internal control feedback. In systems modeling, the influence of one component on other components within the system.

Interspecific competition. Competition between individuals of different species.

Intraspecific competition. Competition between individuals of the same species.

Intrinsic rate of increase (r_m). Exponential growth rate of a population with a stable age distribution, that is, under constant conditions.

Ion. The dissociated parts of a molecule, each of which carries an electrical charge, either positive (cation) or negative (anion).

Isocline. A line on a population graph designating combinations of competing populations, or predator and prey populations, for which the growth rate of one of the populations is zero.

Isogamous. Having gametes similar in size and behavior; not differentiated into unequal (male and female) gametes. *See* Anisogamous.

Iteroparity. The condition of reproducing repeatedly during the lifetime.

Jacobian matrix. A symmetrical matrix (**C**) of rows (i) and columns (j) having elements $a_{ij}N_i$ where a is the coefficient of interaction and N is population size.

Key factor analysis. A statistical treatment of population data designed to identify factors most responsible for change in population size.

***k*-factor analysis.** Analysis of the relationship between the killing power of various environmental factors and change in population size. *Compare with* Key factor analysis.

Killing power (k_x). Logarithm of the probability of survival from one age class (x) to the next; k-value. *See* exponential mortality rate.

Kinetic energy. Energy associated with motion.

Kin selection. Differential reproduction among lineages of closely related individuals based upon genetic variation in social behavior.

Kranz anatomy. Arrangement of tissues in the leaves of C_4 plants in which photosynthetic cells having chloroplasts are grouped in sheaths around vascular bundles.

Larva. Immature stage of an invertebrate that is fundamentally unlike the adult and must undergo some kind of metamorphosis.

Laterite. A hard substance rich in oxides of iron and aluminum, frequently formed when tropical soils weather under alkaline conditions.

Laterization. Leaching of silica from soil, usually in warm, moist regions with an alkaline soil reaction.

Leaching. Removal of soluble compounds from leaf litter or soil by water.

Legume. Any of a number of species of plant belonging to the pea family (Leguminosae).

Lek. A communal courtship area on which several males hold courtship territories to attract and mate with females; sometimes called an arena.

Leslie matrix. A matrix of values of age-specific fecundity and survivorship used to project the size and age structure of a population through time; population matrix.

Liebig's law of the minimum. The idea that the growth of an individual or population is limited by the essential nutrient present in the lowest amount relative to requirement.

Life history. The set of adaptations of an organism that more or less directly influence life table values of age-specific survival and fecundity; hence, reproductive rate, age at first reproduction, reproductive risk, and so on.

Life table. A summary by age of the survivorship and fecundity of individuals in a population.

Life zone. A more or less distinct belt of vegetation occurring within, and characteristic of, a particular latitude or range of elevation.

Lignin. A long-chain, nitrogen-containing molecule made up of phenolic subunits, occurring in woody structures of plants and being highly resistant to digestion by herbivores.

Limestone. A rock formed chiefly by the sedimentation of shells and precipitation and the sedimentation of calcium carbonate ($CaCO_3$) in marine systems.

Limit cycle. Oscillation of predator and prey populations occurring when stabilizing and destabilizing tendencies of their interaction balance.

Limiting resource. A resource that is scarce relative to demand for it.

Limiting similarity. Minimum degree of ecological similarity compatible with the coexistence of two or more populations.

Limnology. The study of freshwater habitats and communities, particularly lakes, ponds, and other standing waters.

Littoral. Pertaining to the shore of the sea.

Loam. Soil that is a mixture of coarse sand particles, fine silt, clay particles, and organic matter.

Local mate competition. The situation in which males compete for mating with females at or near their place of birth, hence frequently with close relatives.

Logarithmic series. A characterization of the relative abundances of species in which the number of species in each progressively larger abundance class (i) has the form ax^i/i, where x lies between 0 and 1 and a is proportional to the number of species in the sample.

Logistic equation. Mathematical expression for a particular sigmoid growth curve in which the percentage rate of increase decreases in linear fashion as population size increases.

Lognormal distribution. A characterization of the number of species in logarithmically scaled abundance classes, according to which most species have moderate abundance and fewer have either extremely high or low abundance.

Lower critical temperature (T_c). Surrounding temperature below which warm-blooded animals must generate heat to maintain their body temperature.

Markov process. A stochastic process where the future state of a system depends only upon its present state.

Mark-recapture method. A way of estimating the size of a population by the recapture of marked individuals.

Mating system. Pattern of matings between individuals in a population, including number of simultaneous mates, permanence of pair bond, and degree of inbreeding.

Matrix. A rectangular array (rows and columns) of mathematical elements (such as the coefficients of simultaneous equations) that is subject to special mathematical manipulations.

Maximum sustainable yield (MSY). The greatest rate at which individuals may be harvested from a population without reducing the size of the population; that is, at which recruitment equals or exceeds harvesting.

Mediterranean climate. A pattern of climate found in middle latitudes characterized by cool, wet winters and warm, dry summers.

Meiofaunal. Pertaining to animals that live within a substrate, such as soil or aquatic sediment.

Meiosis. A series of two divisions by cells destined to produce gametes, involving pairing and segregation of homologous chromosomes, and reducing chromosome number from diploid to haploid.

Melanism. Occurrence of black pigment, usually melanin.

Mesic. Referring to habitats with plentiful rainfall and well-drained soils.

Mesophyte. A plant that requires moderate amounts of moisture.

Metabolism. Biochemical transformations responsible for the building up and breaking down of tissues and the release of energy by the organism.

Metamorphosis. An abrupt change in form during development that fundamentally alters the function of the organism.

Micelle. A complex soil particle resulting from the association of humus and clay particles, with negative electric charges at its surface.

Michaelis-Menton constant (K_m). The concentration of a substrate at which the velocity of an enzyme-catalyzed reaction proceeds at half its maximum rate.

Microhabitat. The particular parts of the habitat that an individual encounters in the course of its activities.

Mimic. An organism adapted to resemble another organism or an object.

Mimicry. Resemblance of an organism to some other organism or object in the environment, evolved to deceive predators or prey into confusing the organism and that which it mimics.

Mineralization. Transformation of elements from organic to inorganic forms, often by dissimilatory oxidations.

Mixed Evolutionarily Stable Strategy. An ESS comprising more than one phenotype within a population; generally the outcome of frequency-dependent fitnesses of the phenotypes.

Model. An organism, usually distasteful or otherwise noxious, upon which a mimic is patterned.

Monod equation. An equation describing the growth rate of a population as a monotonic, asymptotic function of the concentration of resources.

Monoecy. In plants, the occurrence of reproductive organs of both sexes on the same individual, either in different flowers (hermaphrodite) or in the same flowers (perfect flowers). *See* Dioecy.

Monogamy. The situation in which each individual mates with only one of the opposite sex, generally involving a strong and lasting pair bond. *See* Polygamy.

Morph. A specific form, shape, or structure.

Morph ratio cline. A gradual geographic change in the frequency of morphs in a population, usually associated with a gradual change in ecological conditions.

Mortality (m_x). Ratio of the number of deaths to individuals at risk, often described as a function of age (x). *Compare with* Death rate.

Müllerian mimicry. Mutual resemblance of two or more conspicuously marked, distasteful species to enhance predator avoidance.

Mutation. Any change in the genotype of an organism occurring at the gene, chromosome, or genome level; usually applied to changes in genes to new allelic forms.

Mutualism. Relationship between two species that benefits both parties.

Mycorrhizae. Close association of fungi and tree roots in the soil that facilitates the uptake of minerals by trees.

Myxomatosis. A viral disease of some mammals, transmitted by mosquitoes, that causes a fibrous cancer of the skin in susceptible individuals.

Natural selection. Change in the frequency of genetic traits in a population through differential survival and reproduction of individuals bearing those traits.

Negative feedback. Tendency of a system to counteract externally imposed change and return to a stable state.

Neighborhood size. The number of individuals in a population included within the dispersal distance of a single individual.

Net aboveground productivity (NAP). Accumulation of biomass in aboveground parts of plants (trunks, branches, leaves, flowers, and fruits), over a specified period; usually expressed on an annual basis (NAAP).

Net production. The total energy or nutrients accumulated by the organism by growth and reproduction; gross production minus respiration.

Net production efficiency. The percentage of assimilated food utilized for growth and reproduction by an organism.

Net reproductive rate (R_0). The expected number of offspring of a female during her lifetime.

Neutral equilibrium. The particular state of a system that has no forces acting upon it.

Niche. The ecological role of a species in the community; the many ranges of conditions and resource qualities within which the organism or species persists, often conceived as a multidimensional space.

Niche breadth. The variety of resources utilized and range of conditions tolerated by an individual, population, or species.

Niche overlap. The sharing of niche space by two or more species; similarity of resource requirement and tolerance of ecological conditions.

Niche preemption. A model in which species successively procure a proportion of the available resources, leaving less for the next.

Nitrification. Breakdown of nitrogen-containing organic compounds by microorganisms, yielding nitrates and nitrites.

Nitrogen fixation. Biological assimilation of atmospheric nitrogen to form organic nitrogen-containing compounds.

Nodules. Swellings on the roots of some legumes and other plants within which nitrogen fixation is carried out by symbiotic bacteria.

Nonlinearity. Dependence of interaction coefficients on population density.

Nonrenewable resource. A resource present in fixed quantity, such as space, that can be utilized completely by consumers.

Normal distribution. A bell-shaped statistical distribution in which the probability density varies in proportion to exp $(-x^2/2)$, where x is a distance (the standard deviation) from the mean.

Null model. A set of rules for generating community patterns, presupposing no interaction between species, against which observed community patterns can be compared statistically.

Numerical response. Change in the population size of a predatory species as a result of a change in the density of its prey. *See also* Functional response.

Nutrient. Any substance required by organisms for normal growth and maintenance.

Nutrient cycle. The path of an element through the ecosystem, including its assimilation by organisms and its regeneration in a reusable inorganic form.

Nutrient use efficiency (NUE). Ratio of production to uptake of a required nutrient.

Nymph. Adult-like immature stage of an arthopod, especially an insect, with direct development.

Obligate. Referring to a way of life or response to particular conditions without alternatives. *Compare with* Facultative.

Oligotrophic. Poor in the mineral nutrients required by green plants; pertaining to an aquatic habitat with low productivity.

Omnivore. An organism whose diet is broad, including both plant and animal foods; specifically, an organism that feeds on more than one trophic level.

Omnivory. In the sense of food-web analysis, feeding on more than on trophic level.

Open community. A local association of species having independent and only partially overlapping ecological distributions.

Open-community concept. The idea, advocated by H. A. Gleason and R. H. Whittaker, that communities are the local expression of the independent geographic distributions of species.

Optimal foraging. A set of rules, including breadth of diet, by which organisms maximize food intake per unit of time or minimize the time required to meet their food requirements; risk of predation may also enter the equation for optimal foraging.

Optimum giving up time (GUT). The time that an organism should remain within a patch of resources before moving onto the next in order to maximize its rate of food intake.

Order of magnitude. A factor of 10; for example, three orders of magnitude is a factor of 1000.

Ordination. A set of mathematical methods by which communities are ordered along physical gradients or along derived axes over which distance is related to dissimilarity in species composition.

Organism concept of the community. The idea that species are functionally integrated into discrete associations in which each species has evolved to serve the well-being of the whole.

Organismic viewpoint. The idea that a community is a discrete, highly integrated association of species within which the function of each species is subservient to the whole.

Oscillation. Regular fluctuation through a fixed cycle above and below some mean value.

Osmoregulation. Regulation of the salt concentration in cells and body fluids.

Osmosis. Diffusion of substances in aqueous solution across the membrane of a cell.

Osmotic potential. The attraction of water to an aqueous solution owing to the concentration of ions and other small molecules; usually expressed as a pressure.

Outcrossing. Mating with unrelated individuals within a population.

Overdispersion. *See* Hyperdispersion.

Overlapping generations. The co-occurrence of parents and offspring in the same population as reproducing adults.

Oxic. Having oxygen.

Oxidation. Removal of one or more electrons from an atom, ion, or molecule.

Oxygen tension. The partial pressure of oxygen dissolved in an aqueous solution, such as blood.

Parasite. An organism that consumes part of the blood or tissues of its host, usually without killing the host.

Parasitoid. Any of a number of so-called parasitic insects whose larvae live within and consume their host, usually another insect.

Parental investment. An act of parental care that enhances the survival of individual offspring or increases their number.

Parent material. Unweathered rock from which soil is derived.

Parent-offspring conflict. The situation arising when the optimum level of parental investment in a particular offspring differs between the parent and that offspring. This conflict derives from the fact that offspring are genetically equivalent to their parents but siblings carry identical copies of only half the genes of each individual.

Parthenogenesis. Reproduction without fertilization by male gametes, usually involving the formation of diploid eggs whose development is initiated spontaneously.

Partial pressure. The proportional contribution of a particular gas to the total pressure of a mixture.

Pathogenicity. The capacity of an organism to produce disease in its host.

Pelagic. Pertaining to the open sea.

Per capita. Expressed on a per-individual basis.

Perennial. Referring to an organism that lives for more than one year; lasting throughout the year.

Perfect flower. A flower having both male and female sexual organs (anthers and pistils).

Permeability. Capacity of a material to pass through something, such as a biological membrane.

pH. A scale of acidity or alkalinity; the logarithm of the concentration of hydrogen ions.

Phase plane. A plane whose axes are the abundances of two interacting populations.

Phenolic. Pertaining to compounds, such as lignin, based on the phenol chemical structure, a hydroxylated six-carbon ring (C_6H_5OH).

Phenolics. Aromatic hydrocarbons produced by plants, many of which exhibit antimicrobial properties.

Phenotype. Physical expression in the organism of the interaction between the genotype and the environment; outward appearance and behavior of the organism.

Phenotypic value. The measure of a particular phenotypic trait in a particular organism.

Phenotypic variance (V_p). A statistical measure of the variation in a measure of structure or function (phenotypic value) among individuals in a population.

Pheromones. Chemical substances used for communication between individuals.

Photic. Pertaining to surface waters to the depth of light penetration.

Photoautotroph. An organism that utilizes sunlight as its primary energy source for the synthesis of organic compounds.

Photoperiod. Length of the daylight period each day.

Photosynthates. Organic products of photosynthesis; that is, simple carbohydrates.

Photosynthesis. Utilization of the energy of light to combine carbon dioxide and water into simple sugars.

Photosynthetic efficiency. Percentage of light energy assimilated by plants based either on net production (net photosynthetic efficiency) or on gross production (gross photosynthetic efficiency).

Phytoplankton. Microscopic floating aquatic plants.

Pit organ. A recessed organ in the heads of pit vipers for detecting infrared radiation (heat) given off by prey organisms.

Plankton. Microscopic floating aquatic plants (phytoplankton) and animals (zooplankton).

Pleiotropy. Influence of one gene on the expression of more than one trait in the phenotype.

Plotless sampling. A method of estimating the density and dispersion of populations from the distribution of distances between individuals.

Podsolization. Breakdown and removal of clay particles from the acidic soils of cold, moist regions.

Poikilothermy. Inability to regulate body temperature; cold-bloodedness.

Poisson distribution. A statistical description of the random distribution of items among categories, often applied to the distribution of individuals among sampling plots.

Polar ordination. A method of ordering communities according to similarity in species composition in which the most dissimilar communities define the opposite ends (poles) of the ordination axis.

Polyandry. The situation in which a female mates with more than one male at the same time or in quick succession.

Polygamy. A mating system in which a male pairs with more than one female at one time (polygyny) or a female pairs with more than one male (polyandry). *See* Monogamy.

Polygenic. Determined by the expression of more than one gene locus.

Polygyny. The situation in which a male mates with more than one female at the same time or in quick succession.

Polygyny threshold. The difference between the intrinsic values of territories or males such that the realized values of an unmated male on a poorer territory and a mated male on a better territory are equal in the eyes of an unmated female.

Polymorphism. Occurrence of more than one distinct form of individual or genotype in a population.

Polysaccharide. A long-chain carbohydrate, such as cellulose, made up of many monosaccharides (simple sugars).

Population cycle. Recurrent variation in population size where numbers fluctuate between extremes over a regular period.

Population matrix. *See* Leslie matrix.

Population trajectory. The sum of the vectors of two populations on a graph portraying the abundance of one as a function of the abundance of the other; a representation of the simultaneous change in the sizes of two, usually interacting, populations.

Potential energy. The energy of a substance or object deriving from its position, particularly with respect to gravity.

Potential evapotranspiration (PE). The amount of transpiration by plants and evaporation from the soil that would occur, given the local temperature and humidity, if water were not limited.

Predator. An animal (rarely a plant) that kills and eats animals.

Primary consumer. An herbivore, the lowermost eater on the food chain.

Primary host. A host that harbors the sexual stage of the life cycle of a parasite or disease organism.

Primary producer. A green plant that assimilates the energy of light to synthesize organic compounds.

Primary production. Assimilation (gross primary production) or accumulation (net primary production) of energy and nutrients by green plants and other autotrophs.

Primary succession. Sequence of communities developing in a newly exposed habitat devoid of life.

Production. Accumulation of energy or biomass.

Promiscuous. Mating with many individuals within a population, generally; without the formation of strong or lasting pair bonds.

Protandry. Course of development of an individual during which its sex changes from male to female.

Protogyny. The sequence of sexual change by an individual from female to male.

Proximate factor. An aspect of the environment that the organism uses as a cue for behavior; for example, daylength. (Proximate factors often are not directly important to the organism's well-being.) *Compare with* Ultimate factor.

Pupa. In insects with indirect development (complete metamorphosis), the transition stage between larva and adult.

Pyramid of energy. The concept that the energy flux through a given link in the food chain decreases at progressively higher trophic levels.

Pyramid of numbers. Charles Elton's concept that the sizes of populations decrease at progressively higher trophic levels; that is, as one progresses along the food chain.

Quantitative genetics. Study of the inheritance and response to selection of continuously varying traits having polygenic inheritance.

Quantitative trait. A trait having continuous variability within a population and revealing the expression of many gene loci.

Radiation. Energy emitted in the form of waves.

Rain shadow. Dry area on the leeward side of a mountain range.

r- and K-selection. Alternative expressions of selection on traits that determine fecundity and survival to favor rapid population growth at low population density (r) or competitive ability at densities near the carrying capacity (K).

Rarefaction. A method of determining the relationship between species diversity and sample size by randomly deleting individuals from a sample.

Reciprocal altruism. The exchange of altruistic acts between individuals.

Recruitment. Addition of new individuals to a population by reproduction; often restricted to the addition of breeding individuals.

Redox potential (Eh). The relative capacity of an atom or compound to donate or accept electrons, expressed in volts (electrical potential). Higher values indicate more powerful oxidizers.

Reduction. Addition of one or more electrons to an atom, ion, or molecule.

Regulator. Organism that maintains an internal environment different from external conditions.

Regulatory response. A rapid, reversible physiological or behavioral response by an organism to change in its environment.

Relative abundance. Proportional representation of a species in a sample or a community.

Renewable resource. A resource that is continually supplied to the system so that it cannot be fully depleted by consumers.

Replacement series. An experimental investigation of competition between two species in which the ratio of the species is varied while the density is held constant.

Replacement series diagram. Portrayal of vegetative growth or reproduction of two species of plants in a competition experiment as a function of the proportion of one species when the initial density of the two is held constant.

Reproductive effort. Allocation of time or resources, or the assumption of risk, in order to increase fecundity.

Reproductive value (v_a). The expected reproductive output of an individual of age a relative to that of a newborn individual at the same time.

Residence time. Ratio of the size of a compartment to the flux through it, expressed in units of time; thus, the average time spent by energy or a substance in the compartment.

Residual reproductive value (RRV). The reproductive value of an individual discounted by its current reproduction.

Resource. A substance or object required by an organism for normal maintenance, growth, and reproduction. If the resource is scarce relative to demand, it is referred to as a limiting resource. Nonrenewable resources (such as space) occur in fixed amounts and can be fully utilized; renewable resources (such as food) are produced at a rate that may be partially determined by their utilization.

Respiration. Use of oxygen to break down organic compounds metabolically for the purpose of releasing chemical energy.

Resting metabolic rate (RMR). The metabolic rate of an organism in an inactive, fasting, thermoneutral state.

Riparian. Along the bank of a river or lake.

RuBP. Ribulose bisphosphate, a five-carbon carbohydrate to which a carbon atom is attached during the assimilatory step of the Calvin cycle in photosynthesis.

RuBP carboxylase. Enzyme in the Calvin cycle of photosynthesis responsible for the reaction of ribulose bisphosphate and carbon dioxide to form two molecules of phosphoglyceraldehyde.

Ruderal. Pertaining to or inhabiting highly disturbed sites. *See* Weed.

Rumen. An elaboration of the forepart of the stomach of certain ungulate mammals within which cellulose is broken down by symbiotic bacteria.

Runaway sexual selection. The situation in which females persistently choose the most extreme male phenotypes in a population, leading to continuous elaboration of secondary sexual characteristics.

Salt gland. Organ in birds and reptiles used to excrete sodium chloride in high concentration.

Saturation. In reference to biological communities, an upper limit to the number of species that can coexist locally, set by their competitive interactions.

Saturation point. With respect to primary production, the amount of light that causes photosynthesis to attain its maximum rate.

Search image. A behavioral selection mechanism that enables predators to increase searching efficiency for prey that are abundant and worth capturing.

Secondary host. A host that harbors an asexual stage of the life cycle of a parasite or disease organism.

Secondary plant compounds. Chemical products of plant metabolism specifically for the purpose of defense against herbivores and disease organisms.

Secondary succession. Progression of communities in habitats where the climax community has been disturbed or removed entirely.

Selection. Differential survival or reproduction of individuals in a population owing to phenotypic differences among them.

Selection differential (*S*). Difference in the mean phenotypic value of selected individuals and that of the population from which they were drawn.

Selection response (*R*). Difference in the mean phenotypic value of the offspring of selected individuals and that of the population from which the parents were drawn.

Selfing. Mating with oneself; applicable, of course, only to individuals (usually plants) having both male and female sexual organs.

Self-incompatible. Unable to mate with oneself; in the case of an hermaphrodite, owing to structural or biochemical factors that prevent fertilization.

Self-regulation. The idea that population size is regulated with respect to its resources by individual restraint from overexploitation.

Self-thinning curve. In populations of plants limited by space or other resource, the characteristic relationship between logarithm and biomass.

Semelparity. The condition of having only one reproductive episode during the lifetime.

Senescence. Gradual deterioration of function in an organism with age, leading to increased probability of death; aging.

Sere. A series of stages of community change in a particular area leading toward a stable state.

Serial polygamy. The situation in which an individual mates with several individuals of the opposite sex in succession; the broods of each mating are generally cared for, in part, simultaneously.

Serpentine. An igneous rock rich in magnesium that forms soils toxic to many plants.

Sex ratio. Ratio of the number of individuals of one sex to that of the other sex in a population.

Sexual selection. Selection by one sex for specific characteristics in individuals of the opposite sex, usually exercised through courtship behavior.

Shannon-Weaver index (*H*). A logarithmic measure of the diversity of species weighted by the relative abundance of each.

Simpson's index (*D*). A measure of the diversity of species weighted by the relative abundance of each.

Social behavior. Any direct interaction among distantly related individuals of the same species; usually does not include courtship, mating, parent-offspring, and sibling interactions.

Social dominance. Physical domination of one individual over another, initiated and sustained by aggression within a population.

Sociobiology. Study of the biological basis of social behavior.

Soil. The solid substrate of terrestrial communities resulting from the interaction of weather and biological activities with the underlying geological formation.

Soil horizon. A layer of soil formed at a characteristic depth and distinguished by its physical and chemical properties.

Soil profile. Characterization of the structure of soil vertically through its various horizons, or layers.

Soil skeleton. Physical structure of mineral soil, referring principally to sand grains and silt particles.

Solar equator. The parallel of latitude that lies directly under the sun at any given season.

Specialist. An organism with restricted use of habitats or resources.

Species. A group of actually or potentially interbreeding populations that are reproductively isolated from all other kinds of organisms.

Specific heat. Amount of energy that must be added or removed to change the temperature of a substance by a specific amount. By definition, one calorie of energy is required to raise the temperature of one gram of water by one degree Celsius.

Spite. A social interaction in which the donor of a behavior incurs a cost in order to reduce the fitness of a recipient.

Spring bloom. An increase in phytoplankton growth during early spring in temperate lakes associated with vertical mixing of the water column.

Spring overturn. Vertical mixing of water layers in temperate lakes in spring as surface ice disappears.

Stable age distribution. Proportion of individuals in various age classes in a population that has been growing at a constant rate.

Stable equilibrium. The particular state to which a system returns if displaced by an outside force.

Standard deviation (*s*). A measure of the variability among items in a sample, such as individuals in a population; the square root of the variance, hence the square root of the average squared deviation from the mean.

Static life table. The age-specific survival and fecundity of individuals of different ages within a population at a given time; time-specific life table.

Steady state. Condition of a system in which opposing forces or fluxes are balanced.

Stochastic. Referring to patterns resulting from random effects.

Stratification. The establishment of distinct layers of temperature or salinity in bodies of water based upon the different densities of warm and cold water or saline and fresh water.

Substitutable resource. A resource that can satisfy a particular requirement in the place of some other resource; that is, interchangeably.

Succession. Replacement of populations in a habitat through a regular progression to a stable state.

Superorganism. An association of individuals in which the function of each promotes the well-being of the entire system.

Superparasitism. Occurrence of more than one individual of a particular parasitoid species per host.

Survival (l_x). Proportion of newborn individuals alive at age x; also called survivorship.

Survivorship. *See* Survival.

Switching. Change in diet to favor items of increasing suitability or abundance.

Symbiosis. Intimate, and often obligatory, association of two species, usually involving coevolution. Symbiotic relationships can be parasitic or mutualistic.

Sympatric. Occurring in the same place, usually referring to areas of overlap in species distributions.

Synecology. The relationship of organisms and populations to biotic factors in the environment. *Compare with* Autecology.

Synergism. The interaction of two causes such that the total effect is greater than the sum of two acting independently.

Systematics. The classification of organisms into a hierarchical set of categories (taxa) emphasizing their evolutionary interrelationships.

Systems ecology. The study of an ecological structure as a set of components linked by fluxes of energy and nutrients or by population interactions; frequently applied to ecosystems.

Systems modeling. Portrayal of a system's function by mathematical functions describing the interactions among its components.

Tannins. Polyphenolic compounds, produced by most plants, that bind proteins, thereby impairing digestion by herbivores and inhibiting microbes.

Taxon cycle. Cycle of expansion and contraction of the geographical range and population density of a species or higher taxonomic category.

Taxonomy. The description, naming, and classification of organisms.

Temperature profile. The relationship of temperature to depth below the surface of water or the soil, or the height above the ground.

Territoriality. Situation in which individuals defend exclusive spaces, or territories.

Territory. Any area defended by one or more individuals against intrusion by others of the same or different species.

Thermal conductance. Rate at which heat passes through a substance.

Thermal stratification. Sharp delineation of layers of water by temperature, the warmer layer generally laying over the top of the colder layer.

Thermocline. The zone of water depth within which temperature changes rapidly between the upper warm water layer (epilimnion) and lower cold water layer (hypolimnion).

Thermodynamic. Relating to heat and motion.

Three-halves power law. A generalization proposing that the relationship between the logarithms of biomass and density of a population of plants has a slope of $-3/2$.

Time delay. Delay in the response of a population or other system to conditions of the environment; time lag.

Time lag. Delay in response to a change.

Time-specific life table. *See* Static life table.

Tolerance. In reference to succession, the indifference of establishment of one species to the presence of others.

Torpor. Loss of the power of motion and feeling, usually accompanied by greatly reduced rate of respiration.

Transfer function. In systems modeling, the equation describing the flux of energy or a substance between one component and another.

Transition matrix. A matrix of probabilities of changing from one state to another.

Transit time. Average time that a substance or energy remains in the biolgical realm or any compartment of a system; ratio of biomass to productivity.

Transpiration. Evaporation of water from leaves and other parts of plants.

Transpiration efficiency. The ratio of net primary production to transpiration of water by a plant, usually expressed as grams per kilogram of water; water use efficiency.

Trophic. Pertaining to food or nutrition.

Trophic level. Position in the food chain, determined by the number of energy-transfer steps to that level.

Trophic structure. Organization of the community based on feeding relationships of populations.

Turbulent mixing. Mixing in fluids resulting from irregular motion, referable to wind-induced vertical mixing in shallow water.

Ultimate factor. An aspect of the environment that is directly important to the well-being of an organism (for example, food). *Compare with* Proximate factor.

Ultraviolet (UV) radiation. Electromagnetic radiation having a wavelength shorter than about 400 nm.

Unstable equilibrium. The particular state of a system upon which forces are precisely balanced, but away from which the system moves when displaced.

Upwelling. Vertical movement of water, usually near coasts and driven by offshore winds, that brings nutrients from the depths of the ocean to surface layers.

Urea. Nitrogenous waste product excreted by mammals and some other organisms; also employed in the bloodstream of sharks for osmoregulation.

Vapor pressure. The pressure exerted by a vapor that is in equilibrium with its liquid form.

Variance (*V*). A statistical measure of the dispersion of a set of values about its mean.

Vector. In matrix algebra, a row or column of mathematical elements that is subject to special mathematical manipulations.

Veil line. The point on a lognormal distribution of species abundances below which one expects fewer than one individual per species; hence the veil line separates observed from potentially observed species.

Vertical mixing. Exchange of water between deep and surface layers.

Vesicular-arbuscular mycorrhyzae. A type of endomycorrhyzal association of a fungus with the roots of plants distinguished by the branching pattern of growth of the fungus within the root tissues.

Viscosity. The quality of a fluid that resists internal flow.

Warning coloration. Conspicuous patterns or colors adopted by noxious organisms to advertise their distastefulness or dangerousness to potential predators; aposematism.

Water potential. Force by which water is held in the soil by capillary and hygroscopic attraction.

Watershed. Drainage area of a stream or river.

Water use efficiency. *See* Transpiration efficiency.

Weathering. Physical and chemical breakdown of rock and its component minerals at the base of the soil.

Weed. A plant or animal, generally having high powers of dispersal, capable of living in highly disturbed habitats.

Wilting coefficient. The minimum water content of the soil at which plants can obtain water.

Xerarch succession. Progression of terrestrial plant communities developing in habitats with well-drained soil. *Compare with* Hydrarch succession.

Xeric. Referring to habitats in which plant production is limited by availability of water.

Xerophyte. A plant that tolerates dry (xeric) conditions.

Zero net growth isocline (ZNGI). An equilibrium isocline representing the lowest level of a resource, or combination of resources, for which population growth is positive.

Zooplankton. Tiny floating aquatic animals.

Zygote. Diploid cell formed by the union of male and female gametes during fertilization.

Bibliography

ABEGGLEN, J. J. 1984. On Social Organization in Hamadryas Baboons. Associated Univ. Presses, Cranbury, New Jersey.

ABLE, K. P., & B. R. NOON. 1976. Avian community structure along elevational gradients in the northeastern United States. Oecologia 26: 275–294.

ABRAHAMSON, W. G., & M. GADGIL. 1973. Growth form and reproductive effort in goldenrods (*Solidago*, Compositae). Amer. Nat. 107: 651–661.

ABRAMS, J., & G. C. EICKWORT. 1981. Nest switching and guarding by the communal sweat bee *Agapostemon virescens* (Hymenoptera, Halictidae). Insectes Sociaux 28: 105–116.

ABRAMS, P. 1975. Limiting similarity and the form of the competition coefficient. Theoret. Pop. Biol. 8: 356–375.

ACKERMAN, R. A., G. C. WHITTOW, C. V. PAGANELLI, & T. N. PETTIT. 1980. Oxygen consumption, gas exchange, and growth of embryonic wedge-tailed shearwaters (*Puffinus pacificus chlororhynchus*). Physiol. Zool. 53: 210–221.

ADDICOTT, J. F. 1974. Predation and prey community structure: an experimental study of the effect of mosquito larvae on the protozoan communities of pitcher plants. Ecology 55: 475–492.

AKER, C. L. 1982. Regulation of flower, fruit and seed production by a monocarpic perennial, *Yucca whipplei*. J. Ecol. 70: 357–372.

ALATALO, R. V., A. LUNDBERG, & K. STAHLBRANDT. 1982. Why do pied flycatcher females mate with already-mated males? Anim. Behav. 30: 585–593.

ALBREKTSON, A., H. A. ARONSSON, & C. O. TAMM. 1977. The effect of forest fertilization on primary production and nutrient cycling in the forest ecosystem. Silva Fenn. 11: 233–239.

ALCOCK, J. 1969. Observational learning in three species of birds. Ibis 111: 308–321.

ALEXANDER, R. D. 1974. The evolution of social behavior. Ann. Rev. Ecol. Syst. 5: 325–383.

ALFORD, R. A., & H. M. WILBUR. 1985. Priority effects in experimental pond communities: competition between *Bufo* and *Rana*. Ecology 66: 1097–1105.

ALLEE, W. C. 1934. Concerning the organization of marine coastal communities. Ecol. Monogr. 4: 541–554.

ALLEE, W. C., O. PARK, A. E. EMERSON, T. PARK, & K. P. SCHMIDT. 1949. Principles of Animal Ecology. Saunders, Philadelphia.

ALLEN, T. F. H., & T. B. STARR. 1982. Hierarchy. Perspectives for Ecological Complexity. Univ. Chicago Press, Chicago.

ALLISON, A. C. 1956. Sickle cells and evolution. Sci. Amer. 195: 87–94.

ALSTAD, D. N., & G. F. EDMUNDS, JR. 1983. Adaptation, host specificity, and gene flow in the black pineleaf scale. Pp. 413–426 in R. F. Denno and M. S. McClure (Eds.), Variable Plants and Herbivores in Natural and Managed Systems. Academic Press, New York.

ALTMAN, P. L., & D. S. DITTMER. (Eds.). 1964. Biology Data Book. Fed. Amer. Soc. Exp. Biol., Washington, D.C.

AMADON, D. 1950. The Hawaiian honeycreepers (Aves, Drepanidae). Bull. Amer. Mus. Nat. Hist. 95: 151–262.

ANDERSON, G. R. V., A. H. EHRLICH, P. R. EHRLICH, J. D. ROUGHGARDEN, B. C. RUSSELL, & F. H. TALBOT. 1981. The community structure of coral reef fishes. Amer. Nat. 117: 476–495.

ANDERSON, N. H., & J. R. SEDELL. 1979. Detritus processing by macroinvertebrates in stream ecosystems. Ann. Rev. Ent. 24: 351–377.

ANDERSON, R. M. 1981. Infectious disease agents and cyclic fluctuations in host abundance. Pp. 47–80 in R. W. Hiorns and D. Cooke (Eds.), The Mathematical Theory of the Dynamics of Biological Populations II. Academic Press, London.

ANDERSON, R. M. (Ed.). 1982. Population Dynamics of Infectious Diseases. Chapman and Hall, London.

ANDERSON, R. M., & R. M. MAY. 1979. Population biology of infectious diseases, Part I. Nature 280: 361–367.

ANDERSON, R. M., & R. M. MAY. 1980. Infectious diseases and population cycles of forest insects. Science 210: 658–661.

ANDERSON, R. M., & R. M. MAY. (Eds.). 1982. Population Biology of Infectious Diseases. Springer-Verlag, New York.

ANDERSSON, M. 1982. Female choice selects for extreme tail length in a widowbird. Nature 299: 818–820.

ANDERSSON, M., & J. KREBS. 1978. On the evolution of hoarding behavior. Anim. Behav. 26: 707–711.

ANDREWARTHA, H. G., & L. C. BIRCH. 1954. The Distribution and Abundance of Animals. Univ. Chicago Press, Chicago.

ANTONOVICS, J. 1971. The effects of a heterogeneous environment on the genetics of natural populations. Amer. Sci. 59: 593–599.

APPLEBAUM, S. W. 1964. Physiological aspects of host specificity in the Bruchidae—I: General considerations of developmental compatibility. J. Insect Physiol. 10: 783–788.

APPLEBAUM, S. W., B. GESTETNER, & Y. BIRK. 1965. Physiological aspects of host specificity in the Bruchidae—IV: Developmental incompatibility of soybeans for *Callosobruchus.* J. Insect Physiol. 11: 611–616.

AR, A., & H. RAHN. 1980. Water in the avian egg: overall budget of incubation. Amer. Zool. 20: 373–384.

AR, A., H. RAHN, & C. V. PAGANELLI. 1979. The avian egg: mass and strength. Condor 81: 331–337.

ARNOLD, S. J. 1983. Morphology, performance and fitness. Amer. Zool. 23: 347–361.

ARNOLD, S. J., & M. J. WADE. 1984. On the measurement of natural and sexual selection: theory. Evolution 38: 709–719.

ARNOLD, S. J., & M. J. WADE. 1984. On the measurement of natural and sexual selection: applications. Evolution 38: 720–734.

ARRHENIUS, O. 1921. Species and area. J. Ecol. 9: 95–99.

ARTHUR, W. 1982. The evolutionary consequences of interspecific competition. Adv. Ecol. Res. 12: 127–187.

ASHMOLE, N. P. 1963. The regulation of numbers of tropical oceanic birds. Ibis 103b: 458–473.

ASHTON, P. S. 1969. Speciation among tropical forest trees: some deductions in the light of recent research. Biol. J. Linn. Soc. 1: 155–196.

ASHTON, P. S. 1977. A contribution of rain forest research to evolutionary theory. Ann. Mo. Bot. Gard. 64: 694–705.

ATSATT, P. R., & D. J. O'DOWD. 1976. Plant defense guilds. Science 193: 24–29.

AUBRÉVILLE, A. 1938. La fôret coloniale: les fôrets de l'Afrique occidentale française. Ann. Acad. Sci. Colon. Paris 9: 1–245.

AUGSPURGER, C. K. 1983. Seed dispersal of the tropical tree, *Platypodium elegans,* and the escape of its seedlings from fungal infection. J. Ecol. 71: 759–771.

AUGSPURGER, C. K. 1983. Offspring recruitment around tropical trees: changes in cohort distance with time. Oikos 20: 189–196.

AUSTIN, G. T. 1974. Nesting success of the cactus wren in relation to nest orientation. Condor 76: 216–217.

AXELROD, R., & D. DION. 1988. The further evolution of cooperation. Science 242: 1385–1390.

AXELROD, R., & W. D. HAMILTON. 1981. The evolution of cooperation. Science 211: 1390–1396.

AXLER, R. P., G. W. REDFIELD, & C. R. GOLDMAN. 1981. The importance of regenerated nitrogen to phytoplankton productivity in a subalpine lake. Ecology 62: 345–354.

AYALA, F. J. 1965. Evolution of fitness in experimental populations of *Drosophila serrata.* Science 150: 903–905.

AYALA, F. J. 1969. Evolution of fitness. IV. Genetic evolution of interspecific competitive ability in *Drosophila.* Genetics 61: 737–747.

AYALA, F. J. 1970. Competition, coexistence, and evolution. Pp. 121–158 in M. K. Hecht and W. C. Steere (Eds.), Essays in Evolution and Genetics. Appleton-Century-Crofts, New York.

AYALA, F. J. 1971. Competition between species: frequency dependence. Science 171: 820–824.

AYALA, F. J. 1972. Darwinian versus non-Darwinian evolution in natural populations of *Drosophila.* Proc. 6th Berkeley Symp. Math. Stat. Prob. 5: 211–236.

BACON, P. J. (Ed.). 1985. Population Dynamics of Rabies in Wildlife. Academic Press, London.

BAER, G. M. (Ed.). 1975. The Natural History of Rabies, Vols. I and II. Academic Press, New York.

BAKER, H. G. 1965. Characteristics and modes of origin of weeds. Pp. 147–172 in H. G. Baker and G. L. Stebbins (Eds.), The Genetics of Colonizing Species. Academic Press, New York.

BAKER, J. R. 1938. The evolution of breeding seasons. Pp. 161–177 in G. R. de Beer (Ed.), Evolution: Essays on Aspects of Evolutionary Biology. Oxford Univ. Press, London and New York.

BAKER, R. R. 1978. The Evolutionary Ecology of Animal Migration. Holmes and Meier, New York.

BAKUS, G. J., N. M. TARGETT, & B. SCHULTE. 1986. Chemical ecology of marine organisms: an overview. J. Chem. Ecol. 12: 951–987.

BALANDRIN, M. F., J. A. KLOCKE, E. S. WURTELE, & W. H. BOLLINGER. 1985. Natural plant chemicals: sources of industrial and medicinal materials. Science 228: 1154–1160.

BALDA, R. P., & J. H. BALDA. 1978. The care of young piñon jays *(Gymnorhinus cyanocephalus)* and their integration into the flock. J. f. Ornithol. 119: 146–171.

BALDA, R. P., & G. C. BATEMAN. 1971. Flocking and annual cycle of the piñon jay, *Gymnorhinus cyanocephalus.* Condor 73: 287–302.

BALDA, R. P., & G. C. BATEMAN. 1973. The breeding biology of the piñon jay. Living Bird 11: 5–42.

BALL, R. C., & F. F. HOOPER. 1963. Translocation of phosphorus in a trout stream ecosystem. Pp. 217–228 in V. Schultz and A. W. Klement (Eds.), Radioecology. Reinhold, New York.

BALON, E. K. 1975. Reproductive guilds of fishes: a proposal and definition. J. Fish. Res. Bd. Canad. 32: 821–864.

BARBER, D. A. 1972. "Dual isotherms" for the absorption of ions by plant tissues. New Phytol. 71: 255–262.

BARBER, R. T., & F. P. CHAVEZ. 1983. Biological consequences of El Niño. Science 222: 1203–1210.

BARBOUR, A. D. 1982. Schistosomiasis. Pp. 180–208 in R. M. Anderson (Ed.), The Dynamics of Infectious Diseases: Theory and Applications. Chapman and Hall, London and New York.

BAREL, C. D. N., R. DORIT, P. H. GREENWOOD, G. FRYER, N. HUGHES, P. B. N. JACKSON, H. KAWANABE, R. H. LOWE-MCCONNELL, M. NAGOSHI, A. J. RIBBINK, E. TREWAVAS, F. WITTE, & K. YAMAOKA. 1985. Destruction of fisheries in Africa's lakes. Nature 315: 19–20.

BARNARD, C. J., & C. A. J. BROWN. 1983. Risk-sensitive foraging in common shrews (*Sorex aramacus* L.). Behav. Biol. Sociobiol. 16: 161–164.

BARNES, R. K., & K. H. MANN. (Eds.). 1980. Fundamentals of Aquatic Ecosystems. Blackwell, Oxford.

BARR, T. C. 1968. Cave ecology and the evolution of troglobytes. Evol. Biol. 2: 35–102.

BARRON, D. H., & G. MESCHIA. 1954. A comparative study of the exchange of the respiratory gasses across the placenta. Cold Spring Harbor Symp. Quant. Biol. 19: 93–101.

BARROWCLOUGH, G. F. 1980. Gene flow, effective population sizes, and genetic variance components in birds. Evolution 34: 789–798.

BARRY, R. G., & R. J. CHORLEY. 1970. Atmosphere, Weather, and Climate. Holt, Rinehart & Winston, New York.

BARTHOLOMEW, B. 1970. Bare zone between California shrub and grassland communities: the role of animals. Science 170: 1210–1212.

BARTHOLOMEW, G. A. 1942. The fishing activities of double-crested cormorants on San Francisco Bay. Condor 44: 13–21.

BATEMAN, A. J. 1948. Intra-sexual selection in *Drosophila.* Heredity 2: 349–368.

BATESON, P. (Ed.). 1983. Mate Choice. Cambridge Univ. Press, Cambridge.

BATZLI, G. O. 1969. Distribution of biomass in rocky intertidal communities on the Pacific coast of the United States. J. Anim. Ecol 38: 531–546.

BATZLI, G. O., & F. A. PITELKA. 1970. Influence of meadow mouse populations on California grassland. Ecology 51: 1027–1039.

BAUST, J. G. 1973. Mechanisms of cryoprotection in freezing tolerant animal systems. Cryobiology 10: 197–205.

BAUST, J. G., & R. E. MORRISSEY. 1975. Supercooling phenomenon and water content independence in the overwintering beetle, *Coleomegilla maculata.* J. Insect Physiol. 21: 1751–1754.

BEALS, E. W., & J. B. COPE. 1964. Vegetation and soils in eastern Indiana woods. Ecology 45: 777–792.

BECK, N. G., & E. M. LORD. 1988. Breeding system in *Ficus carica,* the common fig. I. Floral diversity. Amer. J. Bot. 75: 1913–1922.

BECK, S. D. 1965. Resistance of plants to insects. Ann. Rev. Entomol. 10: 207–232.

BECK, S. D. 1980. Insect Photoperiodism (2nd ed.). Academic Press, New York.

BECKMAN, W. A., J. W. MITCHELL, & W. P. PORTER. 1973. Thermal model for prediction of a desert iguana's daily and seasonal behavior. J. Heat Transfer, May 1973: 257–262.

BEDNARZ, J. C. 1988. Cooperative hunting in Harris' hawks *(Parabuteo unicinctus).* Science 239: 1525–1527.

BEETON, A. M. 1965. Eutrophication of the St. Lawrence Great Lakes. Limnol. Oceanogr. 10: 240–254.

BEGON, M., & M. MORTIMER. 1986. Population Ecology (2nd ed.). Sinauer, Sunderland, Massachusetts.

BELL, G. 1978. The handicap principle in sexual selection. Evolution 32: 872–885.

BELL, G. 1982. The Masterpiece of Nature: The Evolution and Genetics of Sexuality. Univ. California Press, Berkeley.

BELL, G. 1985. On the function of flowers. Proc. Roy. Soc. Lond. B 224: 223–265.

BELOVSKY, G. E. 1984. Moose and snowshoe hare competition and a mechanistic explanation from foraging theory. Oecologia 61: 150–159.

BELYEA, R. M. 1952. Death and deterioration of balsam fir weakened by spruce budworm defoliation in Ontario. J. Forestry 50: 729–738.

BENDER, E. A., T. J. CASE, & M. E. GILPIN. 1984. Perturbation experiments in community ecology: theory and practice. Ecology 65: 1–13.

BENNETZEN, J. L., W. E. BLEVINS, & A. H. ELLINGBOE. 1988. Cell-autonomous recognition of the rust pathogen determines *Rpl*-specified resistance in maize. Science 241: 200–210.

BERENBAUM, M. R. 1978. Toxicity of a furanocoumarin to armyworms: a case of biosynthetic escape from insect herbivores. Science 201: 532–534.

BERENBAUM, M. R. 1981. Patterns of furanocoumarin distribution and insect herbivory in the Umbelliferae: plant chemistry and community structure. Ecology 62: 1254–1266.

BERENBAUM, M. R. 1983. Coumarins and caterpillars: a case for coevolution. Evolution 37: 163–179.

BERG, R. Y. 1975. Myrmecochorous plants in Australia and their dispersal by ants. Aust. J. Bot. 23: 475–508.

BERKNER, L. V., & L. C. MARSHALL. 1965. History of major atmospheric components. Proc. Natl. Acad. Sci. 53: 1215–1225.

BERRY, J. A. 1975. Adaptation of photosynthetic processes to stress. Science 188: 644–650.

BERRY, J. A., & O. BJORKMAN. 1980. Photosynthetic response and adaptation to temperature in higher plants. Ann. Rev. Plant Physiol. 31: 491–543.

BESCHEL, R. E., P. J. WEBBER, & R. TIPPETT. 1962. Woodland transects of the Frontenac Axis Region, Ontario. Ecology 43: 386–396.

BESS, H. A., R. VAN DEN BOSCH, & F. A. HARAMOTO. 1961. Fruit fly parasites and their activities in Hawaii. Proc. Haw. Ent. Soc. 17: 367–378.

BEVERTON, R. J. H., & S. J. HOLT. 1957. On the dynamics of exploited fish populations. Fish. Invest. 19: 1–533.

BILLINGS, W. D., & H. A. MOONEY. 1968. The ecology of arctic and alpine plants. Biol. Rev. 43: 481–529.

BINKLEY, D., & D. RICHTER. 1987. Nutrient cycles and H^+ budgets of forested ecosystems. Adv. Ecol. Res. 16: 1–51.

BIRCH, L. C. 1953. Experimental background to the study of the distribution and abundance of insects, I: The influence of temperature, moisture and food on the innate capacity for the increase of three grain beetles. Ecology 34: 698–711.

BIRCH, L. C. 1953. Experimental background to the study of the distribution and abundance of insects, II: The relation between innate capacity for increase in numbers and the abundance of three grain beetles in experimental populations. Ecology 34: 712–726.

BIRCH, L. C., & D. P. CLARK. 1953. Forest soil as an ecological community with special reference to the fauna. Quart. Rev. Biol. 28: 13–36.

BISHOP, J. A., & L. M. COOK. 1980. Industrial melanism and the urban environment. Adv. Ecol. Res. 11: 373–404.

BJORKMAN, O. 1968. Further studies on differentiation of photosynthetic properties of sun and shade ecotypes of *Solidago virgaurea.* Physiol. Plant. 21: 84–99.

BJORKMAN, O., M. R. BADGER, & P. A. ARNOLD. 1980. Adaptation to high temperature stress. Pp. 231–249 in N. C. Turner and P. J. Kramer (eds.), Adaptation of Plants to Water and High Temperature Stress. Wiley, New York.

BJORKMAN, O., & J. BERRY. 1973. High efficiency photosynthesis. Sci. Amer. 229: 80–93.

BLAIR, W. F. 1960. The Rusty Lizard, a Population Study. Univ. Texas Press, Austin.

BLAIR, W. F. 1975. Adaptation of anurans to equivalent desert scrub of North and South America. Pp. 197–222 in D. W. Goodall (Ed.), Evolution of Desert Biota. Univ. Texas Press, Austin.

BLEST, A. D. 1957. The function of eye-spot patterns in Lepidoptera. Behaviour 11: 209–256.

BLEST, A. D. 1963. Relations between moths and predators. Nature 197: 1046–1047.

BLEST, A. D. 1964. Protective display and sound production in some New World arctiid and ctenuchid moths. Zoologica 49: 161–181.

BLISS, C. L., & R. A. FISHER. 1953. Fitting the negative binomial distribution to biological data and a note on the efficient fitting of the negative binomial. Biometrics 9: 176–200.

BLOCK, B. A. 1987. Strategies for regulating brain and eye temperatures: a thermogenic tissue in fish. Pp. 401–420 in P. Dejours, L. Bolis, C. R. Taylor, and E. R. Weibel (Eds.), Comparative Physiology: Life in Water and on Land. Fidia Research Series, IX. Liviana Press, Padova, Italy.

BLOOM, B. R. 1979. Games parasites play: how parasites evade immune surveillance. Nature 279: 21–26.

BOAG, P. T. 1983. The heritability of external morphology in Darwin's finches *(Geospiza)* on Isla Daphne Major, Galapagos. Evolution 37: 877–894.

BOCK, C. E., & R. E. RICKLEFS. 1983. Range size and local abundance of some North American songbirds: a positive correlation. Amer. Nat. 122: 295–299.

BOGERT, C. M. 1949. Thermoregulation in reptiles, a factor in evolution. Evolution 3: 195–211.

BOHN, H., B. MCNEAL, & G. O'CONNOR. 1979. Soil Chemistry. Wiley, New York.

BOJE, R., & M. TOMCZAK. (Eds.). 1978. Upwelling Ecosystems. Springer-Verlag, New York.

BOND, W., & P. SLINGSBY. 1984. Collapse of an ant-plant mutualism: the Argentine ant *(Iridomyrmex humilis)* and myrmecochorous Proteaceae. Ecology 65: 1031–1037.

BONDRUP-NIELSEN, S., & R. A. IMS. 1986. Reproduction and spacing behaviour of females in a peak density population of *Clethryionomys glareolus.* Holarctic Ecol. 9: 109–112.

BONNELL, M. L., & R. K. SELANDER. 1974. Elephant seals: genetic variation and near extinction. Science 184: 908–909.

BOONSTRA, R. 1978. Effect of adult Townsend voles *(Microtus townsendi)* on survival of young. Ecology 59: 242–248.

BOORMAN, S. A., & P. R. LEVITT. 1973. Group selection on the boundary of a stable population. Theoret. Pop. Biol. 4: 85–128.

BOOTH, W. 1988. Reintroducing a political animal. Science 241: 156–158.

BORCHERT, J. R. 1950. The climate of the central North American grassland. Ann. Assoc. Amer. Geogr. 40: 1–39.

BORING, L. R., & W. T. SWANK. 1984. The role of black locust *(Robinia pseudoacacia)* in forest succession. J. Ecol. 72: 749–766.

BORMANN, F. H. 1958. The relationships of ontogenetic development and environmental modification of photosynthesis in *Pinus taeda* seedlings. Pp. 197–215 in K. V. Thimann (Ed.), The Physiology of Forest Trees. Ronald Press, New York.

BOTKIN, D. B., J. F. JANAK, & J. R. WALLIS. 1972. Some ecological consequences of a computer model of forest growth. J. Ecol. 60: 849–872.

BOTKIN, D. B., P. A. JORDAN, A. S. DOMINSKI, H. S. LOWENDORF, & G. E. HUTCHINSON. 1973. Sodium dynamics in a northern ecosystem. Proc. Natl. Acad. Sci. 70: 2745–2748.

BOTKIN, D. B., G. M. WOODWELL, & N. TEMPEL. 1970. Forest productivity estimated from carbon dioxide uptake. Ecology 51: 1057–1060.

BOUCHER, D. H. (Ed.). 1985. The Biology of Mutualism. Croom Helm, London.

BOUCHER, D. H., S. JAMES, & K. H. KEELER. 1981. The ecology of mutualism. Ann. Rev. Ecol. Syst. 13: 15–47.

BOURLIÈRE, F. 1973. The comparative ecology of rainforest mammals in Africa and tropical America: some introductory remarks. Pp. 279–292 in B. J. Meggars, E. S. Ayensu, and W. D. Duckworth (Eds.), Tropical Forest Ecosystems in Africa and South America: A Comparative Review. Smithson. Inst. Press, Washington, D.C.

BOWEN, G. D. 1973. Mineral nutrition of mycorrhizas. Pp. 151–201 in G. C. Marks and T. T. Kozlowsky (Eds.), Ectomycorrhizas. Academic Press, New York and London.

BOX, G. E. P., & G. M. JENKINS. 1970. Time Series Analysis: Forecasting and Control. Holden-Day, San Francisco.

BOYCE, M. S. 1984. Restitution of r- and K-selection as a model of density-dependent natural selection. Ann Rev. Ecol. Syst. 15: 427–447.

BRADBURY, J. W. 1981. The evolution of leks. Pp. 138–169 in R. D. Alexander and D. W. Tinkle (Eds.), Natural Selection and Social Behavior. Chiron Press, New York.

BRADBURY, J. W., & M. B. ANDERSSON. (Eds.). 1987. Sexual Selection. Testing the Alternatives. Wiley-Interscience, New York.

BRADY, N. C. 1974. Nature and Property of Soils (8th ed.). Macmillan, New York.

BRAUN, E. L. 1950. Deciduous Forests of Eastern North America. Free Press, New York. Reprinted 1974, Hafner, New York.

BRAUN-BLANQUET, J. 1932. Plant Sociology: The Study of Plant Communities. Transl. by G. D. Fuller and H. S. Conard, McGraw-Hill, New York.

BRAUN-BLANQUET, J. 1965. Plant Sociology: The Study of Plant Communities. Revised transl., Hafner, New York.

BRAY, J. R., & J. T. CURTIS. 1957. An ordination of the upland forest communities of southern Wisconsin. Ecol. Monogr. 27: 325–349.

BRAY, J. R., & E. GORHAM. 1964. Litter production in forests of the world. Adv. Ecol. Res. 2: 101–157.

BREDER, C. N., & D. E. ROSEN. 1966. Modes of Reproduction in Fishes. Natural History Press, New York.

BRETT, J. R., & T. D. D. GROVES. 1979. Physiological energetics. Pp. 279–352 in W. A. Hoar, D. J. Randall, and J. R. Brett (Eds.), Fish Physiology, Vol. 7. Academic Press, New York.

BRIAN, M. V. 1979. Caste differentiation and division of labor. Pp. 121–222 in H. R. Hermann (Ed.), Social Insects, Vol. I. Academic Press, New York.

BRIAN, M. V. 1980. Social control over sex and caste in bees, wasps and ants. Biol. Rev. 55: 379–415.

BRIAND, F. 1983. Environmental control of food web structure. Ecology 64: 253–263.

BRIAND, F., & J. E. COHEN. 1984. Community food webs have scale-invariant structure. Nature 307: 264–267.

BRIAND, F., & J. E. COHEN. 1987. Environmental correlates of food chain length. Science 238: 956–960.

BRINKHURST, R. O. 1959. Alary polymorphism in the Gerroidae (Hemiptera-Heteroptera). J. Anim. Ecol. 28: 211–230.

BRISTOW, C. M. 1984. Differential benefits from ant attendance to two species of Homoptera on New York ironweed. J. Anim. Ecol. 53: 715–726.

BROCK, T. D. 1970. High temperature systems. Ann. Rev. Ecol. Syst. 1: 191–220.

BROCK, T. D. 1985. Life at high temperatures. Science 230: 132–138.

BROCK, T. D., & G. K. DARLAND. 1970. Limits of microbial existence: temperature and pH. Science 169: 1316–1318.

BROCKMANN, H. J. 1984. The evolution of social behaviour in insects. Pp. 340–361 in J. R. Krebs and N. B. Davies (Eds.), Behavioural Ecology. An Evolutionary Approach (2nd ed.). Sinauer, Sunderland, Massachusetts.

BRODA, E. 1975. The history of inorganic nitrogen in the biosphere. J. Mol. Evol. 7: 87–100.

BROKAW, N. V. L. 1985. Treefalls, regrowth, and community structure in tropical forests. Pp. 53–69 in S. T. A. Pickett and P. S. White (Eds.), The Ecology of Natural Disturbance and Patch Dynamics. Academic Press, Orlando, Florida.

BROOKS, D. R. 1985. Historical ecology: a new approach to studying the evolution of ecological associations. Ann. Mo. Bot. Gard. 72: 660–680.

BROUGHTON, W. J. (Ed.). 1983. Nitrogen Fixation, Vols. 1, 2, 3. Oxford Univ. Press, New York.

BROWER, J. V. Z. 1958. Experimental studies of mimicry in some North American butterflies: Part I, The monarch, *Danaus plexippus* and viceroy, *Limenitis archippus;* Part II, *Battus philenor* and *Papilio troilus, P. polyxenes,* and *P. glaucus;* Part III, *Danaus gilippus berenice and Limenitis archippus floridensis.* Evolution 12: 32–47; 123–136; 273–285.

BROWER, J. V. Z., & L. P. BROWER. 1962. Experimental studies of mimicry: Part 6, The reaction of toads *(Bufo terrestris)* to honeybees *(Apis mellifera)* and their dronefly mimics *(Eristalis vinetorum).* Amer. Nat. 96: 297–308.

BROWER, L. P. 1969. Ecological chemistry. Sci. Amer. 220: 22–29.

BROWN, C. M., D. S. MCDONALD-BROWN, & J. L. MEERS. 1974. Physiological aspects of inorganic nitrogen metabolism. Adv. Microbial Physiol. 11: 1–52.

BROWN, C. R. 1986. Cliff swallow colonies as information centers. Science 234: 83–85.

BROWN, J. H. 1975. Geographical ecology of desert rodents. Pp. 315–341 in M. L. Cody and J. M. Diamond (Eds.), Ecology and Evolution of Communities. Harvard Univ. Press, Cambridge, Massachusetts.

BROWN, J. H. 1984. On the relationship between abundance and distribution of species. Amer. Nat. 124: 255–279.

BROWN, J. H., & D. W. DAVIDSON. 1977. Competition between seed-eating rodents and ants in desert ecosystems. Science 196: 880–882.

BROWN, J. H., D. W. DAVIDSON, & O. J. REICHMAN. 1979. An experimental study of competition between seed-eating desert rodents and ants. Amer. Zool. 19: 1129–1143.

BROWN, J. H., & A. C. GIBSON. 1983. Biogeography. C. V. Mosby, St. Louis, Missouri.

BROWN, J. H., & G. A. LIEBERMAN. 1973. Resource utilization and coexistence of seed-eating desert rodents in sand dune habitats. Ecology 54: 788–797.

BROWN, J. H., & B. A. MAURER. 1987. Evolution of species assemblages: effects of energetic constraints and species dynamics on the diversification of the North American avifauna. Amer. Nat. 130: 1–17.

BROWN, J. H., & B. A. MAURER. 1989. Macroecology: the division of food and space among species on continents. Science 243: 1145–1150.

BROWN, J. L. 1963. Aggressiveness, dominance and social organization in the Stellar jay. Condor 65: 460–484.

BROWN, J. L. 1964. The evolution of diversity in avian territorial systems. Wilson Bull. 76: 160–169.

BROWN, J. L. 1970. Cooperative breeding and altruistic behavior in the Mexican jay, *Aphelocoma ultramarina.* Anim. Behav. 18: 366–378.

BROWN, J. L. 1974. Alternate routes to sociality in jays—with a theory for the evolution of altruism and communal breeding. Amer. Zool. 14: 63–80.

BROWN, J. L., & E. R. BROWN. 1981. Kin selection and individual fitness in babblers. Pp. 244–256 in R. D. Alexander and D. W. Tinkle (Eds.), Natural Selection and Social Behavior. Chiron Press, New York.

BROWN, J. L., & G. H. ORIANS. 1970. Spacing patterns in mobile animals. Ann. Rev. Ecol. Syst. 1: 239–262.

BROWN, R. Z. 1953. Social behavior, reproduction and population changes in the house mouse. Ecol. Monogr. 23: 217–240.

BROWN, V. K., & A. C. GANGE. 1989. Differential effects of above- and below-ground insect herbivory during early plant succession. Oikos 54: 67–76.

BROWN, W. L., JR., & E. O. WILSON. 1956. Character displacement. Syst. Zool. 5: 49–64.

BROWNIE, C., D. R. ANDERSON, K. P. BURNHAM, & D. S. ROBSON. 1978. Statistical inference from band recovery data—a handbook. U.S. Fish and Wildl. Serv., Resource Publ. No. 131: 1–212.

BRUBAKER, L. B., & S. K. GREENE. 1979. Differential effects of insects on radial growth in fir. Canad. J. For. 9: 95–105.

BRUES, C. T. 1920. The selection of food plants by insects, with special reference to lepidopterous larvae. Amer. Nat. 54: 313–332.

BRYANT, J. P. 1981. Phytochemical deterrence of snowshoe hare browsing by adventitious shoots of four Alaskan trees. Science 213: 889–890.

BRYANT, J. P., & P. J. KUROPAT. 1980. Selection of winter forage by subarctic browsing vertebrates: the role of plant chemistry. Ann. Rev. Ecol. Syst. 11: 261–285.

BUCHSBAUM, R. 1948. Animals Without Backbones (2nd ed.). Univ. Chicago Press, Chicago.

BUCKLEY, R. C. 1987. Ant-plant-homopteran interactions. Adv. Ecol. Res. 16: 53–85.

BUCKLEY, R. C. 1987. Interactions involving plants, homoptera, and ants. Ann. Rev. Ecol. Syst. 18: 111–135.

BUCKNER, C. H., & W. J. TURNOCK. 1965. Avian predation on the larch sawfly, *Pristiphora erichsonii* (Htg.) (Hymenoptera: Tenthredinidae). Ecology 46: 223–236.

BUDOWSKI, G. 1965. Distribution of tropical American rain forest species in the light of successional processes. Turrialba 15: 40–42.

BULL, J. J. 1980. Sex determination in reptiles. Quart. Rev. Biol. 55: 3–21.

BULL, J. J. 1983. Evolution of Sex Determining Mechanisms. Benjamin/Cummings, Menlo Park, California.

BULLOCK, T. H. 1982. Electroreception. Ann. Rev. Neurosci. 5: 121–170.

BULMER, M. G. 1980. The Mathematical Theory of Quantitative Genetics. Oxford Univ. Press, Oxford.

BULMER, M. G. 1985. Selection for iteroparity in a variable environment. Amer. Nat. 126: 63–71.

BÜNNING, E. 1967. The Physiological Clock (rev. 2nd ed.). Springer-Verlag, New York.

BUNT, J. S. 1973. Primary production: marine ecosystems. Human Ecol. 1: 333–345.

BUNTING, B. T. 1967. The Geography of Soil (rev. ed.). Aldine, Chicago.

BUOL, S. W., F. D. HOLE, & R. J. MCCRACKEN. 1973. Soil Genesis and Classification. Iowa State Univ. Press, Ames.

BURDON, J. J. 1987. Diseases and Plant Population Biology. Cambridge Univ. Press, New York.

BURKHARDT, D., W. SCHLEIDT, & H. ALTNER. 1967. Signals in the Animal World. Transl. by K. Morgan, McGraw-Hill, New York.

BURNET, M., & D. O. WHITE. 1974. Natural History of Infectious Disease (4th ed.). Cambridge Univ. Press, London.

BURNETT, T. 1958. Dispersal of an insect parasite over a small plot. Canad. Entomol. 90: 279–283.

BURTON, R. F. 1973. The significance of ionic concentrations in the internal media of animals. Biol. Rev. 48: 195–231.

BURTON, R. S. 1985. Mating system of the intertidal copepod *Tigriopus californicus.* Mar. Biol. 86: 247–252

BURTON, R. S., & M. W. FELDMAN. 1982. Changes in free amino acid concentrations during osmotic response in the intertidal copepod *Tigriopus californicus.* Comp. Biochem. Physiol. 73A: 441–445.

BUZAS, M. A. 1972. Patterns of species diversity and their explanation. Taxon 21: 275–286.

CABLE, D. R. 1975. Influence of precipitation on perennial grass production in the semidesert southwest. Ecology 56: 981–986.

CADE, W. 1979. The evolution of alternative male reproductive strategies in field crickets. Pp. 343–379 in M.

S. Blum and N. A. Blum (Eds.), Sexual Selection and Reproductive Competition in Insects. Academic Press, New York.

CAIRNS, J., M. L. DAHLBERG, K. L. DICKSON, N. SMITH, & W. T. WALLER. 1969. The relationship of fresh-water protozoan communities to the MacArthur-Wilson equilibrium model. Amer. Nat. 103: 439–454.

CALDER, W. A. 1968. Nest sanitation: A possible factor in the water economy of the roadrunner. Condor. 70: 279.

CALDER, W. A., & J. R. KING. 1974. Thermal and caloric relations of birds. Pp. 259–413 in D. S. Farner and J. R. King (Eds.), Avian Biology, Vol. 4. Academic Press, New York.

CALDER, WILLIAM A., III. 1984. Size, Function, and Life History. Harvard Univ. Press, Cambridge, Massachusetts.

CALHOUN, J. B. 1949. A method for self-control of population growth among mammals living in the wild. Science 109: 333–335.

CALOW, P. 1978. Life Cycles. An Evolutionary Approach to the Physiology of Growth, Reproduction, and Ageing. Chapman and Hall, London.

CAMPBELL, B., & E. LACK. (Eds.). 1984. A Dictionary of Birds. Buteo Books, Vermillion, South Dakota.

CAMPBELL, J. W. (Ed.). 1970. Comparative Biochemistry of Nitrogen Metabolism. Vols. I and II. Academic Press, New York.

CANE, M. A. 1983. Oceanographic events during El Niño. Science 222: 1189–1194.

CARACO, T. 1983. White-crowned sparrows *(Zonotrichia leucophrys):* foraging preferences in a risky environment. Behav. Ecol. Sociobiol. 12: 63–69.

CARACO, T. S., S. MARTINDALE, & T. W. WHITHAM. 1980. An empirical demonstration of risk-sensitive foraging preferences. Anim. Behav. 28: 820–830.

CAREY, F. G. 1982. A brain heater in the swordfish. Science 216: 1327–1329.

CAREY, F. G., J. M. TEAL, J. W. KANWISHER, & K. D. LAWSON. 1971. Warm-bodied fish. Amer. Zool. 11: 137–145.

CARLIN, N. F., & B. HOLLDOBLER. 1983. Nestmate and kin recognition in interspecific mixed colonies of ants. Science 222: 1027–1029.

CARLISLE, A., A. H. F. BROWN, & E. J. WHITE. 1966. Organic matter and nutrient elements in the precipitation beneath a sessile oak *(Quercus petraea)* canopy. J. Ecol. 54: 87–98.

CARLSON, T. 1913. Über Geschwindigkeit und Grösse der Hefevermehrung in Würze. Biochem. Z. 57: 313–334.

CARPENTER, E. J., & J. L. CULLINEY. 1975. Nitrogen fixation in marine shipworms. Science 187: 551–552.

CARPENTER, S. R. 1983. Resource limitation of larval treehole mosquitoes subsisting on beech detritus. Ecology 64: 219–223.

CARPENTER, S. R., & J. F. KITCHELL. 1987. The temporal scale of variance in limnetic primary production. Amer. Nat. 129: 417–433.

CARPENTER, S. R., J. F. KITCHELL, J. R. HODGSON, P. A. COCHRAN, J. J. ELSER, M. M. ELSER, D. M. LODGE, D. KRETCHMER, X. HE, & C. N. VON ENDE. 1987. Regulation of lake primary productivity by food web structure. Ecology 68: 1863–1876.

CARPENTER, S. R., P. R. LEAVITT, J. J. ELSER, & M. M. ELSER. 1988. Chlorophyll budgets: response to food web manipulation. Biogeochem. 6: 79–90.

CARTWRIGHT, F. F., & M. D. BIDDISS. 1972. Disease and History. Crowell, New York.

CASPER, B. B. 1984. On the evolution of embryo abortion in the herbaceous perenniel *Cryptantha flava.* Evolution 38: 1337–1349.

CAUGHLEY, G. 1977. Analysis of Vertebrate Populations. Wiley, New York and London.

CAVALLI-SFORZA, L. L., & W. F. BODMER. 1971. The Genetics of Human Populations. W. H. Freeman, San Francisco.

CAVERS, P. B., & J. L. HARPER. 1967. Studies in the dynamics of plant populations. I. The fate of seed and transplants introduced into various habitats. J. Ecol. 55: 59–71.

CENTER, T. D., & C. D. JOHNSON. 1974. Coevolution of some seed beetles (Coleoptera: Bruchidae) and their hosts. Ecology 55: 1096–1103.

CHAETUM, E. L., & C. W. SEVERINGHAUS. 1950. Variations in fertility of white-tailed deer related to range conditions. Trans. North Amer. Wildl. Conf. 15: 170–189.

CHAPIN, F. S., III. 1980. The mineral nutrition of wild plants. Ann. Rev. Ecol. Syst. 11: 233–260.

CHAPMAN, R. 1928. The quantitative analysis of environmental factors. Ecology 9: 111–122.

CHARLESWORTH, E., & B. CHARLESWORTH. 1981. Allocation of resources to male and female functions in hermaphrodites. Biol. J. Linn. Soc. 14: 57–74.

CHARNOV, E. L. 1976. Optimal foraging, the marginal value theorem. Theoret. Pop. Biol. 9: 129–136.

CHARNOV, E. L. 1979. Natural selection and sex change in pandalid shrimp: test of a life history theory. Amer. Nat. 113: 715–734.

CHARNOV, E. L. 1982. The Theory of Sex Allocation. Princeton Univ. Press, Princeton, New Jersey.

CHARNOV, E. L., & W. M. SCHAFFER. 1973. Life history consequences of natural selection: Cole's result revisited. Amer. Nat. 107: 791–793.

CHARNOV, E. L., J. M. SMITH, & J. BULL. 1976. Why be an hermaphrodite? Nature 263: 125–126.

CHARNOV, E. L., & D. W. STEPHENS. 1988. On the evolution of host selection in solitary parasitoids. Amer. Nat. 132: 707–722.

CHESSON, P. L., & R. R. WARNER. 1981. Environmental variability promotes coexistence in lottery competitive systems. Amer. Nat. 117: 923–943.

CHEW, R. M., & A. E. CHEW. 1970. Energy relationships of the mammals of a desert shrub. Ecol. Monogr. 40: 1–21.

CHEWYREUV, I. 1913. Le rôle des femelles dans la determination du sexe de leur descendance dans le groupe des Ichneumonides. C. R. Soc. Biol. Paris 74: 695–699.

CHITTY, D. 1960. Population processes in the vole and their relevance to general theory. Canad. J. Zool. 38: 99–113.

CHITTY, D. 1967. The natural selection of self-regulatory behavior in animal populations. Proc. Ecol. Soc. Australia 2: 51–78.

CHRISTENSEN, N. L. 1985. Shrubland fire regimes and their evolutionary consequences. Pp. 85–100 in S. T. A. Pickett and P. S. White (Eds.), The Ecology of Natural Disturbance and Patch Dynamics. Academic Press, Orlando, Florida.

CHRISTENSEN, N. L., & R. K. PEET. 1984. Convergence during secondary forest succession. J. Ecol. 72: 25–36.

CHRISTIAN, J. J. 1950. The adreno-pituitary system and population cycles in mammals. J. Mamm. 31: 247–259.

CHRISTIAN, J. J., & D. E. DAVIS. 1964. Endocrines, behavior and population. Science 146: 1550–1560.

CHURCHILL, G. B., H. H. JOHN, D. P. DUNCAN, & A. C. HOBSON. 1964. Long-term effects of defoliation of aspen by the forest tent caterpillar. Ecology 45: 630–633.

CLARK, D. A., & D. B. CLARK. 1984. Spacing dynamics of a tropical rain forest tree: evaluation of the Janzen-Connell model. Amer. Nat. 124: 769–788.

CLARK, F. W. 1972. Influence of jackrabbit density on coyote population change. J. Wildl. Mgmt. 36: 343–356.

CLARK, H., & D. FISCHER. 1957. A reconsideration of nitrogen excretion by the chick embryo. J. Exp. Zool. 136: 1–15.

CLARK, P. J., P. T. ECKSTROM, & L. C. LINDEN. 1964. On the number of individuals per occupation in a human society. Ecology 45: 367–372.

CLARK, P. J., & F. C. EVANS. 1954. Distance to nearest neighbor as a measure of spatial relationships in populations. Ecology 35: 445–453.

CLARKE, R. D. 1988. Chance and order in determining fish-species composition on small coral patches. J. Exp. Mar. Biol. Ecol. 115: 197–212.

CLARKSON, D. T., & J. B. HANSON. 1980. The mineral nutrition of higher plants. Ann. Rev. Plant. Physiol. 31: 239–298.

CLAUSEN, J., D. D. KECK, & W. M. HIESEY. 1948. Experimental studies on the nature of species, III: Environmental responses of climatic races of *Achillea.* Carnegie Inst. Wash. Publ. 581: 1–129.

CLAYTON, H. H. 1944. World weather records. Smithson. Misc. Coll. 79: 1–1199.

CLAYTON, H. H., & F. L. CLAYTON. 1947. World weather records 1931–1940. Smithson. Misc. Coll. 105: 1–646.

CLEMENTS, F. E. 1916. Plant succession: Analysis of the development of vegetation. Carnegie Inst. Wash. Publ. 242: 1–512.

CLEMENTS, F. E. 1936. Nature and structure of the climax. J. Ecol. 24: 252–284.

CLOUD, P. 1968. Pre-metazoan evolution and the origins of the metazoa. Pp. 1–72 in E. T. Drake (Ed.), Evolution and Environment. Yale Univ. Press, New Haven, Connecticut.

CLOUD, P. 1974. Evolution of ecosystems. Amer. Sci. 62: 54–66.

CLUTTER, M. E. (Ed.). 1978. Dormancy and Developmental Arrest: Experimental Analysis in Plants and Animals. Academic Press, New York.

CLUTTON-BROCK, T. H. (Ed.). 1988. Reproductive Success. Studies of Individual Variation in Contrasting Breeding Systems. Univ. Chicago Press, Chicago.

CLUTTON-BROCK, T. H., F. E. GUINNESS, & S. D. ALBON. 1982. Red Deer. Behavior and Ecology of Two Sexes. Univ. Chicago Press, Chicago.

CLUTTON-BROCK, T. H., & P. H. HARVEY. 1979. Comparison and adaptation. Proc. Roy. Soc. Lond. B 205: 547–565.

CLUTTON-BROCK, T. H., & P. H. HARVEY. 1984. Comparative approaches to investigating adaptation. Pp. 7–29 in J. R. Krebs and N. B. Davies (Eds.), Behavioural Ecology: An Evolutionary Approach. Sinauer, Sunderland, Massachusetts.

CLUTTON-BROCK, T. H., M. MAJOR, & F. E. GUINNESS. 1985. Population regulation in male and female red deer. J. Anim. Ecol. 54: 831–846.

CODY, M. L. 1966. A general theory of clutch size. Evolution 20: 174–184.

CODY, M. L. 1974. Optimization in ecology. Science 183: 1156–1164.

CODY, M. L. 1974. Competition and the Structure of Bird Communities. Princeton Univ. Press, Princeton, New Jersey.

CODY, M. L. 1975. Towards a theory of continental species diversities: bird distributions over Mediterranean habitat gradients. Pp. 214–257 in M. L. Cody and J. M. Diamond (Eds.), Ecology and Evolution of Communities. Harvard Univ. Press, Cambridge, Massachusetts.

CODY, M. L. (Ed.). 1985. Habitat Selection in Birds. Academic Press, New York.

CODY, M. L., & H. A. MOONEY. 1978. Convergence versus nonconvergence in Mediterranean-climate ecosystems. Ann. Rev. Ecol. Syst. 9: 265–321.

COGGER, H. G. 1974. Thermal relations of the mallee dragon *Amphibolurus fordi* (Lacertilia: Agamidae). Aust. J. Zool. 22: 319–339.

COHEN, J. E. 1968. Alternate derivations of a species-abundance relation. Amer. Nat. 102: 165–172.

COHEN, J. E. 1970. A Markov contingency-table model for replicated Lotka-Volterra systems near equilibrium. Amer. Nat. 104: 547–560.

COHEN, J. E. 1973. Heterologous immunity in human malaria. Quart. Rev. Biol. 48: 467–489.

COHEN, J. E. 1978. Food Webs and Niche Space. Princeton Univ. Press, Princeton, New Jersey.

COHEN, J. E. 1989. Food webs and community structure. Pp. 181–202 in J. Roughgarden, R. M. May, and S. A. Levin (Eds.), Perspectives in Ecological Theory. Princeton Univ. Press, Princeton, New Jersey.

COHEN, J. E., & C. M. NEWMAN. 1985. A stochastic theory of community food webs. 1. Models and aggregated data. Proc. Roy. Soc. Lond. B 224: 421–448.

COHEN, J. E., C. M. NEWMAN, & F. BRIAND. 1985. A stochastic theory of community food webs. 2. Individual webs. Proc. Roy. Soc. Lond. B 224: 449–461.

COLE, L. C. 1951. Population cycles and random oscillations. J. Wildl. Mgmt. 15: 233–252.

COLE, L. C. 1954. The population consequences of life history phenomena. Quart. Rev. Biol. 29: 103–137.

COLEMAN, D. C., C. P. P. REID, & C. V. COLE. 1983. Biological strategies of nutrient cycling in soil systems. Adv. Ecol. Res. 13: 1–55.

COLEY, P. D. 1983. Herbivory and defensive characteristics of tree species in a lowland tropical forest. Ecol. Monogr. 53: 209–233.

COMFORT, A. 1956. The Biology of Senescence. Rinehart, New York.

COMINS, H. N., & M. P. HASSELL. 1976. Predation in multiprey communities. J. Theoret. Biol. 62: 93–114.

COMSTOCK, J., & J. EHLERINGER. 1984. Photosynthetic responses to slowly decreasing leaf water potentials in *Encelia frutescens.* Oecologia 61: 241–248.

CONDER, P. J. 1949. Individual distance. Ibis 91: 649–655.

CONNELL, J. H. 1961. The influence of interspecific competition and other factors on the distribution of the barnacle *Chthamalus stellatus.* Ecology 42: 710–723.

CONNELL, J. H. 1961. The effects of competition, predation by *Thais lapillus,* and other factors on natural populations of the barnacle, *Balanus balanoides.* Ecol. Monogr. 31: 61–104.

CONNELL, J. H. 1972. Community interactions on marine rocky intertidal shores. Ann. Rev. Ecol. Syst. 31: 169–192.

CONNELL, J. H. 1975. Some mechanisms producing structure in natural communities: a model and evidence from field experiments. Pp. 460–490 in M. L. Cody and J. Diamond (Eds.), Ecology and Evolution of Communities. Harvard Univ. Press, Cambridge, Massachusetts.

CONNELL, J. H. 1978. Diversity in tropical rain forests and coral reefs. Science 199: 1302–1310.

CONNELL, J. H. 1980. Diversity and the coevolution of competitors, or the ghost of competition past. Oikos 35: 131–138.

CONNELL, J. H. 1983. On the prevalence and relative importance of interspecific competition: evidence from field experiments. Amer. Nat. 122: 661–696.

CONNELL, J. H., & M. J. KEOUGH. 1985. Disturbance and patch dynamics of subtidal marine animals on hard substrata. Pp. 125–151 in S. T. A. Pickett and P. S. White (Eds.), The Ecology of Natural Disturbance and Patch Dynamics. Academic Press, Orlando, Florida.

CONNELL, J. H., & E. ORIAS. 1964. The ecological regulation of species diversity. Amer. Nat. 98: 399–414.

CONNELL, J. H., & R. O. SLATYER. 1977. Mechanisms of succession in natural communities and their role in community stability and organization. Amer. Nat. 111: 1119–1144.

CONNOR, E. F., & E. D. MCCOY. 1979. The statistics and biology of the species-area relationship. Amer. Nat. 113: 791–833.

CONNOR, E. F., & D. S. SIMBERLOFF. 1979. The assembly of species communities: Chance or competition? Amer. Nat. 113: 791–833.

CONNOR, E. F., & D. SIMBERLOFF. 1984. Neutral models of species co-occurrence patterns. Pp. 316–331 in D. R. Strong, Jr., D. Simberloff, L. G. Abele, and A. B. Thistle (Eds.), Ecological Communities: Conceptual Issues and the Evidence. Princeton Univ. Press, Princeton, New Jersey.

CONNOR, V. M., & J. F. QUINN. 1984. Stimulation of food species growth by limpet mucus. Science 225: 843–844.

CONOVER, R. J. 1966. Assimilation of organic matter by zooplankton. Limnol. Oceanogr. 11: 338–345.

COOK, L. M., C. S. MANI, & M. E. VARLEY. 1986. Postindustrial melanism in the peppered moth. Science 231: 611–613.

COOK, R. E. 1969. Variation in species density in North American birds. Syst. Zool. 18: 63–84.

COOK, R. E. 1977. Raymond Lindeman and the trophic-dynamic concept in ecology. Science 198: 22–26.

COOK, R. M., & S. F. HUBBARD. 1977. Adaptive strategies in insect parasites. J. Anim. Ecol. 46: 115–125.

COOPER, C. F. 1961. The ecology of fire. Sci. Amer. 204: 150–160.

COOPER, W. S. 1913. The climax forest of Isle Royale, Lake Superior and its development. Bot. Gaz. 55: 1–44, 115–140, 189–235.

CORMACK, R. M. 1979. Models for capture-recapture. Pp. 217–255 in R. M. Cormack, G. P. Patil, and D. S. Robson (Eds.), Sampling Biological Populations. Intern. Co-op. Publ. House, Fairland, Maryland.

CORMACK, R. M. 1981. Loglinear models for capture-recapture experiments on open populations. Pp. 197–215 in R. W. Hiorns and D. Cooke (Eds.), The Mathematical Theory of the Dynamics of Biological Populations II. Academic Press, London and New York.

CORMACK, R. M., G. P. PATIL, & D. S. ROBSON. (Eds.). 1979. Sampling Biological Populations. Intern. Co-op. Publ. House, Fairland, Maryland.

CORNELL, H. 1974. Parasitism and distributional gaps between allopatric species. Amer. Nat. 108: 880–883.

CORNELL, H. & D. PIMENTAL. 1978. Switching in the parasitoid *Nasonia vitripennis* and its effect on host competition. Ecology 59: 297–308.

CORNELL, H. V. 1985. Species assemblages of cynipid gall wasps are not saturated. Amer. Nat. 126: 565–569.

COSGRAVE, D. J. 1977. Microbial transformations in the phosphorus cycle. Adv. Microbial. Ecol. 1: 95–134.

COSSINS, A. R., M. J. FRIEDLANDER, & C. L. PROSSER. 1977. Correlations between behavioral temperature adaptations of goldfish and the viscosity and fatty acid composition of their synaptic membranes. J. Comp. Physiol. 120: 109–121.

COTT, H. B. 1940. Adaptive Coloration in Animals. Methuen, London.

COWAN, I. M. 1947. The timber wolf in the Rocky Mountain National Parks of Canada. Canad. J. Res. 25: 139–174.

COWIE, R. J. 1977. Optimal foraging in great tits *Parus major.* Nature 268: 137–139.

COWLES, H. C. 1899. The ecological relations of the vegetation on the sand dunes of Lake Michigan. Bot. Gaz. 27: 95–117, 167–202, 281–308, 361–391.

COWLES, R. B., & C. M. BOGERT. 1944. A preliminary study of the thermal requirements of desert reptiles. Bull. Amer. Mus. Nat. Hist. 83: 265–296.

COX, G. W. 1984. The distribution and origin of Mima mound grasslands in San Diego County, California. Ecology 65: 1397–1405.

COX, G. W., & D. W. ALLEN. 1987. Soil translocation by pocket gophers in a Mima moundfield. Oecologia 72: 207–210.

COX, G. W., & R. E. RICKLEFS. 1977. Species diversity, ecological release, and community structuring in Caribbean land bird faunas. Oikos 29: 60–66.

COYNE, J. A., J. BUNDGAARD, & T. PROUT. 1983. Geographic variation of tolerance to environmental stress in *Drosophila pseudoobscura.* Amer. Nat. 122: 474–488.

CRAIG, R., & R. H. CROZIER. 1979. Relatedness in the polygynous ant *Myrmecia pilosula.* Evolution 33: 335–341.

CRAMER, N. F., & R. M. MAY. 1972. Interspecific competition, predation, and species diversity: a comment. J. Theoret. Biol. 34: 289–293.

CRAWFORD, D. L., & R. L. CRAWFORD. 1980. Microbial degradation of lignin. Enzyme and Microbial Technol. 2: 11–22.

CRAWFORD, H. S., & D. T. JENNINGS. 1989. Predation by birds on spruce budworm *Choristoneura fumiferana:* functional, numerical, and total responses. Ecology 70: 152–163.

CRAWLEY, M. J. 1983. Herbivory. The Dynamics of Animal-Plant Interactions. Univ. California Press, Berkeley.

CREIGHTON, C. 1891. A History of Epidemics in Britain from A.D. 664 to the Extinction of Plaque. Cambridge Univ. Press, Cambridge.

CROAT, T. B. 1978. Flora of Barro Colorado Island. Stanford Univ. Press, Stanford, California.

CROCKER, R. L. 1952. Soil genesis and the pedogenic factors. Quart. Rev. Biol. 27: 139–168.

CROCKER, R. L., & J. MAJOR. 1955. Soil development in relation to vegetation and surface age at Glacier Bay, Alaska. J. Ecol. 43: 427–448.

CROGHAN, P. C. 1958. The osmotic and ionic regulation of *Artemia salina* (L). J. Exp. Biol. 35: 219–233.

CROMBIE, A. C. 1944. On intraspecific and interspecific competition in larvae of graminivorous insects. J. Exp. Biol. 20: 135–151.

CROMBIE, A. C. 1945. On competition between different species of graminivorous insects. Proc. Roy. Soc. Lond. B 132: 362–395.

CROMBIE, A. C. 1947. Interspecific competition. J. Anim. Ecol. 16: 44–73.

CROOK, J. H. 1964. The evolution of social organization and visual communication in the weaver birds (Ploceinae). Behavior, Suppl. 10: 1–178.

CROOK, J. H. 1965. The adaptive significance of avian social organisations. Symp. Zool. Soc. Lond. 14: 181–218.

CROOK, J. H. 1966. Gelada baboon herd structure and movement. Symp. Zool. Soc. Lond. 18: 237–258.

CROOK, J. H. 1970. The Socio-ecology of Primates. Pp. 103–166 in J. H. Crook (Ed.), Social Behavior in Birds and Mammals. Academic Press, New York.

CROSSIN, R. S. 1967. The breeding biology of the tufted jay. Proc. Western Found. Vert. Zool. 1: 265–299.

CROW, J. F., & M. KIMURA. 1970. An Introduction to Population Genetics Theory. Harper & Row, New York.

CROWELL, K. L. 1962. Reduced interspecific competition among the birds of Bermuda. Ecology 43: 75–88.

CROWSON, R. A. 1970. Classification and Biology. Aldine, Chicago.

CROXALL, J. P., & P. A. PRINCE. 1979. Antarctic seabird and seal monitoring studies. Polar Record 19: 573–595.

CROZIER, R. H. 1979. Genetics of sociality. Pp. 223–286 in H. R. Hermann (Ed.), Social Insects, Vol. I. Academic Press, New York.

CRUMPACKER, D. W., AND J. S. WILLIAMS. 1973. Density, dispersion, and population structure in *Drosophila pseudoobscura*. Ecol. Monogr. 43: 499–538.

CULVER, D. C. 1974. Species packing in Caribbean and north temperate ant communities. Ecology 55: 974–988.

CULVER, D. C., & A. J. BEATTIE. 1978. Myrmecochory in *Viola:* dynamics of seed-ant interactions in some West Virginia species. J. Ecol. 66: 53–72.

CUMMINS, K. W. 1974. Structure and function of stream ecosystems. BioScience 24: 631–641.

CURTIS, J. T., & R. P. MCINTOSH. 1950. The interrelations of certain analytic and synthetic phytosociological characters. Ecology 31: 434–455.

CURTIS, J. T., & R. P. MCINTOSH. 1951. An upland forest continuum in the prairie-forest border region of Wisconsin. Ecology 32: 476–496.

CUSHING, D. H. 1975. Marine Ecology and Fisheries. Cambridge Univ. Press, Cambridge.

DA CUNHA, A. B., & T. DOBZHANSKY. 1954. A further study of chromosomal polymorphism in *Drosophila willistoni* in relation to its environment. Evolution 8: 119–134.

DAFT, M. J., & T. H. NICHOLSON. 1972. Effect of *Endogone* mycorrhiza on plant growth. IV. Quantitative relationship between the growth of the host and the development of the endophyte in tomato and maize. New Phytol. 71: 287–295.

DAMIAN, R. T. 1964. Molecular mimicry: Antigen sharing by parasite and host and its consequences. Amer. Nat. 98: 129–147.

D'ANCONA, U. 1954. The Struggle for Existence. E. J. Brill, Leiden.

DANSEREAU, P. 1957. Biogeography—An Ecological Perspective. Ronald Press, New York.

DARWIN, C. 1859. On the Origin of Species. Murray, London.

DARWIN, C. 1871. The Descent of Man and Selection in Relation to Sex. Murray, London.

DARWIN, C. 1872. On the Origin of Species (6th ed.). Murray, London.

DAUBENMIRE, R. 1968. Ecology of fire in grasslands. Adv. Ecol. Res. 5: 209–266.

DAUBENMIRE, R. 1968. Plant Communities. A Textbook of Plant Synecology. Harper & Row, New York.

DAUBENMIRE, R., & D. C. PRUSSO. 1963. Studies of the decomposition rates of tree litter. Ecology 44: 589–592.

DAVIDSON, D. W. 1977. Foraging ecology and community organization in desert seed-eating ants. Ecology 58: 725–737.

DAVIDSON, D. W. 1980. Some consequences of diffuse competition in a desert ant community. Amer. Nat. 116: 92–105.

DAVIDSON, J. 1938. On the growth of the sheep population in Tasmania. Trans. Roy. Soc. South Australia 62: 342–346.

DAVIDSON, J., & H. G. ANDREWARTHA. 1948. Annual trends in a natural population of *Thrips imaginis* (Thysanoptera). J. Anim. Ecol. 17: 193–199.

DAVIDSON, J., & H. G. ANDREWARTHA. 1948. The influence of rainfall, evaporation and atmospheric temperature on fluctuations in the size of a natural population of *Thrips imaginis* (Thysanoptera). J. Anim. Ecol. 17: 200–222.

DAVIES, N. B. 1976. Parental care and the transition to independent feeding in the young spotted flycatcher *(Muscicapa striata)*. Behaviour 59: 280–295.

DAVIS, C. C. 1964. Evidence for the eutrophication of Lake Erie from phytoplankton records. Limnol. Oceanogr. 9: 275–283.

DAVIS, J. H. 1940. The ecology and geologic role of mangroves in Florida. Publ. Carnegie Inst., No. 517: 303–412.

DAVIS, J. W. F., & P. O'DONALD. 1976. Sexual selection for a handicap: a critical analysis of Zahavi's model. J. Theoret. Biol. 57: 345–354.

DAVIS, M. B. 1965. Phytogeography and palynology of northeastern United States. Pp. 377–401 in H. E. Wright, Jr., and D. G. Frey (Eds.), Quaternary of the United States. Princeton Univ. Press, Princeton, New Jersey.

DAVIS, M. B. 1976. Pleistocene biogeography of temperate deciduous forests. Geoscience and Man 13: 13–26.

DAWKINS, R., & T. R. CARLISLE. 1976. Parental investment, mate desertion and a fallacy. Nature 262: 131–133.

DAWSON, P. S. 1966. Correlated responses to selection for developmental rate in *Tribolium*. Genetica 37: 63–77.

DAWSON, P. S. 1967. Developmental rate and competitive ability in *Tribolium*. II: Changes in competitive ability following further selection for developmental rate. Evolution 21: 292–298.

DAYTON, P. K. 1971. Competition, disturbance, and community organization: the provision and subsequent utilization of space in a rocky intertidal community. Ecol. Monogr. 41: 351–389.

DAYTON, P. K., & M. T. TEGNER. 1984. Catastrophic storms, El Niño, and patch stability in a southern California kelp community. Science 224: 283–285.

DEAN, J. M., & R. E. RICKLEFS. 1979. Do parasites of Lepidoptera larvae compete for hosts? No! Amer. Nat. 113: 302–306.

DEAN, J. M., & R. E. RICKLEFS. 1980. Population size, variability, and aggregation among forest lepidoptera in southern Ontario. Canad. J. Zool. 58: 394–399.

DEAN, T. A., & L. E. HURD. 1980. Development in an estuarine fouling community: the influence of early colonists on later arrivals. Oecologia 46: 295–301.

DEBACH, P. 1966. The competitive displacement and coexistence principles. Ann. Rev. Entomol. 11: 183–212.

DEBACH, P. 1974. Biological Control by Natural Enemies. Cambridge University Press, London.

DEBACH, P., R. M. HENDRICKSON, JR., & M. ROSE. 1978. Competitive displacement: extinction of the yellow scale, *Aonidiella* (Coq.) (Homoptera: Diaspididae), by its ecological homologue, the California red scale *Aonidiella aurantii* (Mask.) in southern California. Hilgardia 46: 1–35.

DEBACH, P., & R. A. SUNDBY. 1963. Competitive displacement between ecological homologues. Hilgardia 34: 105–166.

DEEVEY, E. S., JR. 1947. Life tables for natural populations of animals. Quart. Rev. Biol. 22: 283–314.

DEGROOT, P. 1980. Information transfer in a socially roosting weaver bird (*Quelea quelea:* Ploceinae): an experimental study. Anim. Behav. 28: 1249–1254.

DELCOURT, H. R., P. A. DELCOURT, & T. WEBB III. 1983. Dynamic plant ecology: the spectrum of vegetational change in space and time. Quatern. Sci. Rev. 1: 153–175.

DELCOURT, P. A., & H. R. DELCOURT. 1987. Long-term Forest Dynamics of the Temperate Zone: A Case Study of Quaternary Forests in Eastern North America. Springer-Verlag, New York.

DELWICHE, C. C., & B. A. BRYAN. 1976. Denitrification. Ann. Rev. Microbiol. 30: 241–262.

DEMPSTER, J. P. 1963. The population dynamics of grasshoppers and locusts. Biol. Rev. 38: 490–529.

DEMPSTER, J. P. 1968. Intra-specific competition and dispersal: as exemplified by a psyllid and its anthocorid predator. Pp. 8–17 in T. R. E. Southwood (Ed.), Insect Abundance. Symp. Roy. Entomol. Soc. No. 4.

DENSLOW, J. S. 1980. Gap partitioning among tropical rainforest trees. Biotropica 12 (Suppl.): 47–55.

DENSLOW, J. S. 1985. Disturbance-mediated coexistence of species. Pp. 307–323 in S. T. A. Pickett and P. S. White (Eds.), The Ecology of Natural Disturbance and Patch Dynamics. Academic Press, Orlando, Florida.

DENSLOW, J. S. 1987. Tropical rainforest gaps and tree species diversity. Ann. Rev. Ecol. Syst. 18: 431–451.

DENTON, E. J. 1960. The buoyancy of marine animals. Sci. Amer. 203: 119–128.

DE STEVEN, D. 1980. Clutch size, breeding success, and parental survival in the tree swallow *(Iridoprocne bicolor)*. Evolution 34: 278–291.

DEVORE, I., & S. L. WASHBURN. 1963. Baboon ecology and human evolution. In F. C. Howell and F. Bourliere (Eds.), African Ecology and Human Evolution. Aldine, Chicago.

DEVRIES, A. L. 1980. Biological antifreezes and survival in freezing environments. Pp. 583–607 in R. Gilles (Ed.), Animals and Environmental Fitness. Pergamon Press, New York.

DEVRIES, A. L. 1982. Biological antifreeze agents in cold-water fishes. Comp. Biochem. Physiol. 73A: 627–640.

DEWITT, C. B. 1967. Precision of therm-regulation and its relation to environmental factors in the desert iguana, *Dipsosaurus dorsalis.* Physiol. Zool. 40: 49–66.

DIAMOND, J., & T. J. CASE. 1986. Overview: introductions, extinctions, exterminations, and invasions. Pp. 65–79 in J. Diamond and T. J. Case (Eds.), Community Ecology. Harper & Row, New York.

DIAMOND, J. M. 1969. Avifauna equilibria and species turnover rates on the Channel Islands of California. Proc. Natl. Acad. Sci. 67: 1715–1721.

DIAMOND, J. M. 1971. Comparison of faunal equilibrium turnover rates on a tropical island and a temperate island. Proc. Natl. Acad. Sci. 68: 2742–2745.

DIAMOND, J. M. 1975. Assembly of species communities. Pp. 342–444 in M. L. Cody and J. M. Diamond (Eds.), Ecology and Evolution of Communities. Harvard Univ. Press, Cambridge, Massachusetts.

DIAMOND, J. M., & M. E. GILPIN. 1982. Examination of the "null" model of Connor and Simberloff for species co-occurrences on islands. Oecologia 52: 64–74.

DICKINSON, C. H., & G. J. F. PUGH. (Eds.). 1974. Biology of Plant Litter Decomposition. Vols. I and II. Academic Press, London and New York.

DICKSON, R. C. 1949. Factors governing the induction of diapause in the Oriental fruit moth. Ann. Ent. Soc. Amer. 42: 511–537.

DIGGLE, P. J. 1979. Statistical methods for spatial point patterns in ecology. Pp. 95–150 in R. M. Cormack and J. K. Ord (Eds.), Spatial and Temporal Analysis in Ecology. Intern. Co-op. Publ. House, Fairland, Maryland.

DILLON, P. J., & F. H. RIGLER. 1974. The phosphorus-chlorophyll relationship in lakes. Limnol. Oceanogr. 19: 767–773.

DINGLE, H. (Ed.). 1978. Evolution of Insect Migration and Diapause. Springer-Verlag, New York.

DIX, R. L. 1957. Sugar maple in forest succession at Washington, D.C. Ecology 30: 663–665.

DIXON, A. F. G. 1959. An experimental study of the searching behaviour of the predatory coccinellid beetle *Adalia decempunctata* (L.). J. Anim. Ecol. 28: 259–281.

DOBSON, A. P. 1988. The population biology of parasitoid-induced changes in host behavior. Quart. Rev. Biol. 63: 139–165.

DOBZHANSKY, T. 1950. Evolution in the tropics. Amer. Sci. 38: 209–221.

DOBZHANSKY, T., & S. WRIGHT. 1943. Genetics of natural populations. X. Dispersion rates in *Drosophila pseudoobscura.* Genetics 28: 304–340.

DOBZHANSKY, T., & S. WRIGHT. 1947. Genetics of natural populations. XV. Rate of diffusion of a mutant gene through a population of *Drosophila pseudoobscura.* Genetics 32: 303–324.

DODD, A. P. 1959. The biological control of prickly pear in Australia. In A. Keast, R. L. Crocker, and C. S. Christian (Eds.), Biogeography and Ecology in Australia. Monogr. Biol., Vol. VIII.

DOLINGER, P. M., P. R. EHRLICH, W. L. FITCH, & D. E. BREEDLOVE. 1973. Alkaloids and predation patterns in Colorado lupine populations. Oecologia 13: 191–204.

DOMINEY, W. J. 1983. Sexual selection, additive genetic variance and the "phenotypic handicap." J. Theoret. Biol. 101: 495–502.

DOUTT, R. L. 1960. Natural enemies and insect speciation. Pan-Pacific Entomol. 36: 1–13.

DOWNES, R. W., & J. D. HESKETH. 1968. Enhanced photosynthesis at low O_2 concentrations: differential response of temperate and tropical grasses. Planta 78: 79–84.

DRENT, R. H., & S. DAAN. 1980. The prudent parent: energetic adjustments in avian breeding. Ardea 68: 225–252.

DRURY, W. H., & I. C. T. NISBET. 1973. Succession. J. Arnold Arboretum 54: 331–368.

DUDLEY, J. W. 1977. 76 generations of selection for oil and protein percentage in maize. Pp. 459–473 in E. Pollack, O. Kempthorne, and T. Bailey (Eds.), Proc. Intern. Conf. Quant. Genet., 1976. Iowa State Univ. Press, Ames.

DUESER, R. D., & H. H. SHUGART. 1979. Niche pattern in a forest-floor small mammal fauna. Ecology 60: 108–118.

DUNBAR, C. O. 1960. Historical Geology, 2nd. ed. Wiley, New York.

DUNBAR, M. J. 1960. The evolution of stability in marine environments: natural selection at the level of the ecosystem. Amer. Nat. 94: 129–136.

DUNBAR, R. I. M. 1983. Relationships and social structure in gelada and hamadryas baboons. In R. A. Hinde (Ed.), Primate Social Relationships: An Integrated Approach. Blackwell, Oxford.

DUNCAN, D. P., & A. C. HOBSON. 1958. Influence of the forest tent caterpillar upon the aspen forests of Minnesota. For. Sci. 4: 71–93.

DUNHAM, A. E. 1980. An experimental study of interspecific competition between the iguanid lizards *Sceloporus merriami* and *Urosaurus ornatus.* Ecol. Monogr. 50: 309–330.

DUNSON, W. A. 1976. Salt glands in reptiles. Pp. 413–445 in C. Gans and W. R. Dawson (Eds.), Biology of the Reptilia, Vol. 5. Academic Press, New York.

DUVIGNEAUD, P., & S. DENAYER-DE-SMET. 1970. Biological cycling of minerals in temperate deciduous forests. Pp. 199–225 in D. E. Reichle (Ed.), Analysis of Temperate Forest Ecosystems. Springer-Verlag, New York.

DUXBURY, A. C. 1971. The Earth and Its Oceans. Addison-Wesley, Reading, Massachusetts.

DYER, K. R. 1973. Estuaries: A Physical Introduction. Wiley, London.

DYMOND, J. R. 1947. Fluctuations in animal populations with special reference to those of Canada. Trans. Roy. Soc. Canad. 41: 1–34.

EBERHARDT, L. L. 1969. Population analysis. Pp. 457–495 in R. H. Giles, Jr. (Ed.). 1971. Wildlife Management Techniques (3rd ed.). Wildlife Society, Washington, D.C.

ECKHOLM, E., F. FOLEY, G. BARNARD, & L. TIMBERLAKE. 1984. Fuelwood: The Energy Crisis That Won't Go Away. Earthscan, Washington, D.C.

EDMONDSON, W. T. 1970. Phosphorus, nitrogen, and algae in Lake Washington after diversion of sewage. Science 169: 690–691.

EDMUNDS, G. F., JR., & D. N. ALSTAD. 1978. Coevolution in insect herbivores and conifers. Science 199: 941–945.

EDMUNDS, G. F., JR., & D. N. ALSTAD. 1981. Responses of black pineleaf scales to host plant variability. Pp. 29–38 in R. F. Denno and H. Dingle (Eds.), Insect Life History Patterns. Springer-Verlag, New York.

EDWARDS, A. W. F. 1960. Natural selection and the sex ratio. Nature 188: 960–961.

EDWARDS, C. A., & G. W. HEATH. 1963. The role of soil animals in breakdown of leaf material. Pp. 76–84 in J. Doeksen and J. Van Der Drift (Eds.), Soil Organisms. North-Holland, Amsterdam.

EDWARDS, G., & D. WALKER. 1983. C_3, C_4. Mechanisms, and Cellular and Environmental Regulation, of Photosynthesis. Univ. California Press, Berkeley.

EGERTON, F. N. 1973. Changing concepts in the balance of nature. Quart. Rev. Biol. 48: 322–350.

EGGLER, W. A. 1938. The maple-basswood forest type in Washburn County, Wis. Ecology 19: 243–263.

EHLERINGER, J. R. 1984. Ecology and ecophysiology of leaf pubescence in North American Desert plants. Pp. 113–132 in E. Rodriguez, P. Healey, and I. Mehta (Eds.), Biology and Chemistry of Plant Trichomes. Plenum Press, New York.

EHLERINGER, J., & I. FORSETH. 1980. Solar tracking by plants. Science 210: 1094–1098.

EHRLICH, P. R. 1975. The population biology of coral reef fishes. Ann. Rev. Ecol. Syst. 6: 211–247.

EHRLICH, P. R., & L. C. BIRCH. 1967. The "balance of nature" and "population control." Amer. Nat. 101: 97–107.

EHRLICH, P. R., J. HARTE, M. A. HARWELL, P. H. RAVEN, C. SAGAN, G. M. WOODWELL, J. BERRY, E. S. AYENSU, A. H. EHRLICH, T. EISNER, S. J. GOULD, H. D. GROVER, R. HERRERA, R. M. MAY, E. MAYR, C. P. MCKAY, H. A. MOONEY, N. MYERS, D. PIMENTEL, & J. M. TEAL. 1983. Long-term biological consequences of nuclear war. Science 222: 1293–1300.

EHRLICH, P. R., & P. H. RAVEN. 1965. Butterflies and plants: a study in coevolution. Evolution 18: 586–608.

EHRMAN, L. 1972. Genetics and sexual selection. Pp. 105–135 in B. Campbell (Ed.), Sexual Selection and the Descent of Man. Aldine, Chicago.

EICKMEIER, W., M. ADAMS, & D. LESTER. 1975. Two physiological races of *Tsuga canadensis.* Canad. J. Bot. 53: 940–951.

EINARSEN, A. S. 1942. Specific results from ring-necked pheasant studies in the Pacific northwest. Trans. North Amer. Wildl. Conf. 7: 130–145.

EINARSEN, A. S. 1945. Some factors affecting ring-necked pheasant population density. Murrelet 26: 39–44.

EISENBERG, J. F. 1981. The Mammalian Radiations. Univ. Chicago Press, Chicago.

EISNER, T., & J. MEINWALD. 1966. Defensive secretions of arthropods. Science 153: 1341–1350.

ELDREDGE, N. 1974. Character displacement in evolutionary time. Amer. Zool. 14: 1083–1097.

ELLIOTT, J. M. 1985. Population regulation for different life-stages of migratory trout *Salmo trutta* in a lake district stream, 1966–1983. J. Anim. Ecol. 54: 617–638.

ELSETH, G. D., & K. D. BAUMGARDNER. 1981. Population Biology. Van Nostrand, New York.

ELTON, C. 1924. Periodic fluctuations in the numbers of animals: their causes and effects. Brit. J. Exptl. Biol. 2: 119–163.

ELTON, C. 1927. Animal Ecology. Macmillan, New York.

ELTON, C. 1942. Voles, Mice and Lemmings. Problems in Population Dynamics. Clarendon Press, Oxford.

ELTON, C. 1946. Competition and the structure of ecological communities. J. Anim. Ecol. 15: 54–68.

EMERSON, R., & C. M. LEWIS. 1942. The photosynthetic efficiency of phycocyanin in *Chroococcus,* and the problem of carotenoid participation in photosynthesis. J. Gen. Physiol. 25: 579–595.

EMLEN, J. M. 1970. Age specificity and ecological theory. Ecology 51: 588–601.

EMLEN, J. M. 1984. Population Biology. The Coevolution of Population Dynamics and Behavior. Macmillan, New York, and Collier Macmillan, London.

EMLEN, J. T., JR. 1940. Sex and age ratios in the survival of California quail. J. Wildl. Mgmt. 4: 2–99.

EMLEN, S. T. 1981. Altruism, kinship, and reciprocity in the white-fronted bee-eater. Pp. 245–281 in R. D. Alexander and D. W. Tinkle (Eds.), Natural Selection and Social Behavior. Chiron Press, New York.

EMLEN, S. T., & H. W. AMBROSE III. 1970. Feeding interactions of snowy egrets and red-breasted mergansers. Auk 87: 164–165.

EMLEN, S. T., & L. W. ORING. 1977. Ecology, sexual selection, and the evolution of mating systems. Science 197: 215–223.

ENDLER, J. A. 1977. Geographic Variation, Speciation, and Clines. Princeton Univ. Press, Princeton, New Jersey.

ENDLER, J. A. 1978. A predator's view of animal color patterns. Evol. Biol. 11: 319–364.

ENDLER, J. A. 1984. Progressive background matching in moths, and a quantitative measure of crypsis. Biol. J. Linn. Soc. 22: 187–231.

EPLING, C. H., & T. DOBZHANSKY. 1942. Genetics of natural populations. VI. Microgeographic races in *Linanthus parryae.* Genetics 27: 317–332.

ERICKSON, R. O. 1945. The *Clematis fremontii* var. *riehlii* population in the Ozarks. Ann. Mo. Bot. Gard. 32: 413–460.

ERIKSSON, M. O. G. 1979. Competition between freshwater fish and goldeneyes *Bucephala clangula* (L.) for common prey. Oecologia 41: 99–107.

ESAU, K. 1960. Anatomy of Seed Plants. Wiley, New York.

ESHEL, I. 1978. On the handicap principle—a critical defense. J. Theoret. Biol. 70: 245–250.

ESPENSHADE, E. B., JR. (Ed.). 1971. Goode's World Atlas (13th ed.). Rand-McNally, Chicago.

ESTES, J. A., N. S. SMITH, & J. L. PALMISANO. 1978. Sea-otter predation and community organization in the western Aleutian Islands, Alaska. Ecology 59: 822–833.

EVANS, F. C. 1956. Ecosystem as the basic unit in ecology. Science 123: 1127–1128.

EWALD, P. W. 1983. Host-parasite relations, vectors, and the evolution of disease severity. Ann. Rev. Ecol. Syst. 14: 465–485.

EWALD, P. W., & F. L. CARPENTER. 1978. Territorial responses to energy manipulations in the Anna hummingbird. Oecologia 31: 277–292.

EWALD, P. W., & G. H. ORIANS. 1983. Effects of resource depression on use of inexpensive and escalated aggressive behavior: experimental tests using Anna hummingbirds. Behav. Ecol. Sociobiol. 12: 95–101.

EYRE, S. R. 1968. Vegetation and Soils: A World Picture (2nd ed.). Aldine, Chicago.

FAGER, E. W. 1972. Diversity: a sampling study. Amer. Nat. 106: 293–310.

FALCONER, D. S. 1981. Introduction to Quantitative Genetics (2nd ed.). Ronald Press, New York.

FEENY, P. P. 1968. Effect of oak leaf tannins on larval growth of the winter moth *Operophtera brumata.* J. Insect Physiol. 14: 805–817.

FEENY, P. P. 1969. Inhibitory effect of oak leaf tannins on the hydrolysis of proteins by trypsin. Phytochem. 8: 2119–2126.

FEENY, P. P. 1970. Seasonal changes in oak leaf tannins and nutrients as cause of spring feeding by winter moth caterpillars. Ecology 51: 565–581.

FELDMAN, G. C. 1984. Satellites, seabirds, and seals. Trop. Ocean-Atmos. Newsl. 28: 4–5.

FELDMAN, M. W., & I. ESHEL. 1982. On the theory of parent-offspring conflict: a two-locus genetic model. Amer. Nat. 119: 285–292.

FELSENSTEIN, J. 1974. The evolutionary advantage of recombination. Genetics 78: 737–756.

FENCHEL, T. 1975. Factors determining the distribution patterns of mud snails (Hydrobiidae). Oecologia 20: 1–17.

FENCHEL, T. 1975. Character displacement and coexistence in mud snails (Hydrobiidae). Oecologia 20: 19–32.

FENCHEL, T. 1988. Marine plankton food chains. Ann. Rev. Ecol. Syst. 19: 19–38.

FENCHEL, T., & T. H. BLACKBURN. 1979. Bacteria and Mineral Cycling. Academic Press, New York.

FENNER, F. 1971. Evolution in action: myxomatosis in the Australian wild rabbit. In A. Kramer (Ed.), Topics in the Study of Life. The Bio Source Book. Harper & Row, New York.

FENNER, F., & F. N. RATCLIFFE. 1965. Myxomatosis. Cambridge Univ. Press, London and New York.

FENTON, M. B. 1972. The structure of aerial-feeding bat faunas as indicated by ears and wing elements. Canad. J. Zool. 50: 287–296.

FERNALD, M. L. 1929. Some relationships of the floras of the northern hemisphere. Proc. Intern. Congr. Plant Sci., Ithaca 2: 1487–1507.

FINDLEY, J. S. 1973. Phenetic packing as a measure of faunal diversity. Amer. Nat. 107: 580–584.

FINDLEY, J. S. 1976. The structure of bat communities. Amer. Nat. 110: 129–139.

FINDLEY, J. S., & H. BLACK. 1983. Morphological and dietary structuring of a Zambian insectivorous bat community. Ecology 64: 625–630.

FINERTY, J. P. 1980. The Population Ecology of Cycles in Small Mammals. Yale Univ. Press, New Haven, Connecticut.

FISCHER, A. G. 1960. Latitudinal variation in organic diversity. Evolution 14: 64–81.

FISH, D., & S. R. CARPENTER. 1982. Leaf litter and larval mosquito dynamics in tree-hole ecosystems. Ecology 63: 283–288.

FISHER, R. A. 1930. The Genetical Theory of Natural Selection. Clarendon Press, Oxford.

FISHER, R. A., A. S. CORBET, & C. B. WILLIAMS. 1943. The relation between the number of species and the number of individuals in a random sample of an animal population. J. Anim. Ecol. 12: 42–58.

FISHER, R. A., & N. C. TURNER. 1978. Plant productivity in the arid and semiarid zones. Ann. Rev. Plant Physiol. 29: 277–317.

FISHER, S. G., & G. E. LIKENS. 1973. Energy flow in Bear Brook, New Hampshire: an integrative approach to stream ecosystem metabolism. Ecol. Monogr. 421–439.

FISKE, W. F. 1910. Superparasitism: an important factor in the natural control of insects. J. Econ. Entomol. 3: 88–97.

FLEMING, I. A., & M. R. GROSS. 1989. Evolution of adult female life history and morphology in a pacific salmon (Coho: *Oncorhynchus kisutch*). Evolution 43: 141–157.

FLETCHER, D. J. C., & C. D. MICHENER. (Eds.). 1987. Kin Recognition in Animals. Wiley-Interscience, New York.

FLOHN, H. 1968. Climate and Weather. McGraw-Hill, World Univ. Press, New York.

FLOWERDEW, J. R. 1974. Field and laboratory experiments on the social behaviour and population dynamics of the wood mouse *(Apodemus sylvaticus)*. J. Anim. Ecol. 43: 499–511.

FOCHTE, D. D., & W. VERSTRAETE. 1977. Biochemical ecology of nitrification and denitrification. Adv. Microbial Ecol. 1: 135–214.

FORBES, S. A. 1887. The lake as a microcosm. Bull. Sci. Assoc. Peoria, Ill.: 77–87.

FORCE, D. C. 1974. Ecology of insect host-parasitoid communities. Science 184: 624–632.

FORCIER, L. K. 1975. Reproductive strategies and the co-occurrence of climax tree species. Science 189: 808–809.

FORMAN, R. T. T. (Ed.). 1979. Pine Barrens: Ecosystem and Landscape. Academic Press, New York.

FORRESTER, J. W. 1961. Industrial Dynamics. M.I.T. Press, Cambridge, Massachusetts.

FOSTER, J. B., & M. J. COE. 1968. The biomass of game animals in Nairobi National Park, 1960–1966. J. Zool. Lond. 155: 413–425.

FOWELLS, H. A. 1965. Silvics of Forest Trees of the United States. Agric. Handbook No. 271. U.S. Dept. Agric., Washington, D.C.

FOWLER, S. V., & M. MACGARVIN. 1985. The impact of hairy wood ants, *Formica lugubris*, on the guild structure of herbivorous insects on birch, *Betula pubescens*. J. Anim. Ecol. 54: 847–855.

FOWLER, S. V., & M. MACGARVIN. 1986. The effects of leaf damage on the performance of insect herbivores on birch, *Betula pubescens*. J. Anim. Ecol. 55: 565–573.

FOX, L. R., & P. A. MORROW. 1981. Specialization: species property or local phenomenon? Science 211: 887–893.

FOX, R., S. W. LEHMKUHLE, & D. H. WESTENDORF. 1976. Falcon visual acuity. Science 192: 263–266.

FRAENKEL, G. S. 1959. The *raison d'etre* of secondary plant substances. Science 129: 1466–1470.

FRAENKEL, G. S. 1969. Evaluation of our thoughts on secondary plant substances. Ent. Exptl. Appl. 12: 474–486.

FRANK, P. W., C. D. BOLL, & R. W. KELLY. 1957. Vital statistics of laboratory cultures of *Daphnia pulex* De Geer as related to density. Physiol. Zool. 30: 287–305.

FRANKS, N. R., & E. SCOVELL. 1983. Dominance and reproductive success among slave-making worker ants. Nature 304: 724–725.

FRASER ROWELL, C. H. 1970. Environmental control of coloration in an acridid, *Gastrimargus africanus* (Saussure). Anti-Locust Bull. 47: 1–48.

FRETWELL, S. D., & H. L. LUCAS. 1970. On territorial behaviour and other factors influencing habitat distribution in birds. Acta Biotheoret. 19: 16–36.

FRISANCHO, A. R.. 1975. Functional adaptation to high altitude hypoxia. Science 187: 313–319.

FRY, C. H. 1972. The social organisation of bee-eaters (Meropidae) and cooperative breeding in hot-climate birds. Ibis 114: 1–14.

FRY, F. E. J., & J. S. HART. 1948. Cruising speed of goldfish in relation to water temperature. J. Fish. Res. Bd. Canad. 7: 169–175.

FUENTES, E. R. 1976. Ecological convergence of lizard communities in Chile and California. Ecology 57: 3–18.

FUJII, K. 1968. Studies on interspecies competition between the azuki bean weevil and the southern cowpea weevil: III, some characteristics of strains of two species. Res. Popul. Ecol. 10: 87–98.

FUTUYMA, D. J. 1979. Evolutionary Biology. Sinauer, Sunderland, Massachusetts.

FUTUYMA, D. J. 1986. Evolutionary Biology (2nd ed.). Sinauer, Sunderland, Massachusetts.

FUTUYMA, D. J., & M. SLATKIN. (Eds.). 1983. Coevolution. Sinauer, Sunderland, Massachusetts.

GADGIL, M. 1972. Male dimorphism as a consequence of sexual selection. Amer. Nat. 106: 574–580.

GADGIL, M., & W. H. ROSSERT. 1970. Life historical consequences of natural selection. Amer. Nat. 104: 1–24.

GAINES, M. S., & T. S. WHITTAM. 1980. Genetic changes in fluctuating vole populations: selective vs. nonselective forces. Genetics 96: 767–778.

GALLOWAY, J. N., G. E. LIKENS, & M. E. HAWLEY. 1984. Acid precipitation: natural versus anthropogenic components. Science 226: 829–831.

GANS, C. 1974. Biomechanics. Lippincott, Philadelphia.

GANS, C., & F. H. POUGH. (Eds.). 1982. Biology of the Reptilia, Vol. 12. Physiology C. Part I. Temperature Regulation and Thermal Relations. Academic Press, New York.

GATES, D. M. 1980. Biophysical Ecology. Springer-Verlag, New York.

GATHREAUX, S. A., JR. (Ed.). 1981. Animal Migration, Orientation and Navigation. Academic Press, New York.

GAUCH, H. G., JR. 1982. Multivariate Analysis in Community Ecology. Cambridge Univ. Press, Cambridge.

GAUSE, G. F. 1931. The influence of ecological factors on the size of population. Amer. Nat. 65: 70–76.

GAUSE, G. F. 1932. Experimental studies on the struggle for existence. I. Mixed population of two species of yeast. J. Exper. Biol. 9: 389–402.

GAUSE, G. F. 1934. The Struggle for Existence. Williams and Wilkins, Baltimore.

GAUSE, G. F. 1935. La théorie mathématique de la lutte pour la vie. Hermann, Paris.

GENTRY, A. H. 1988. Changes in plant community diversity and floristic composition on environmental and geographical gradients. Ann. Mo. Bot. Gard. 75: 1–34.

GHISELIN, M. T. 1969. The evolution of hermaphroditism among animals. Quart. Rev. Biol. 44: 189–208.

GIBB, J. A. 1958. Predation by tits and squirrels on the eucosmid *Ernarmonia conicolana* (Heyl.). J. Anim. Ecol. 27: 375–396.

GIBB, J. A. 1962. L. Tinbergen's hypothesis of the role of specific search images. Ibis 104: 106–111.

GIBSON, T. G., & M. A. BUZAS. 1973. Species diversity: patterns in modern and Miocene foraminifera of the eastern margin of North America. Geol. Soc. Amer. Bull. 84: 217–238.

GILBERT, B. S., C. J. KREBS, D. TALARICO, & D. B. CICHOWSKI. 1986. Do *Clethryionomys rutilus* females suppress maturation of juvenile females? J. Anim. Ecol. 55: 543–552.

GILBERT, F. S. 1985. Ecomorphological relationships in hoverflies. Proc. Roy. Soc. Lond. B 224L: 91–105.

GILBERT, L. E., & P. H. RAVEN. (Eds.). 1975. Coevolution of Animals and Plants. Univ. Texas Press, Austin.

GILL, A. M. 1975. Fire and the Australian flora. Aust. For. 38: 4–25.

GILL, D. E. 1974. Intrinsic rate of increase, saturation density and competitive ability. II. The evolution of competitive ability. Amer. Nat. 108: 103–116.

GILLES, R. 1975. Mechanisms of ion and osmoregulation. Pp. 257–347 in O. Kinne (Ed.), Marine Ecology, Vol. 2, Part 1. Wiley, London and New York.

GILLESPIE, J. H. 1978. A general model to account for enzyme variation in natural populations. V. The SAS-CFF model. Theoret. Pop. Biol. 14: 1–45.

GILLESPIE, J. H. 1982. A randomized SAS-CFF model of natural selection in a random environment. Theoret. Pop. Biol. 21: 219–237.

GILLETTE, J. B. 1962. Pest pressure, an underestimated factor in evolution. Syst. Assoc. Publ. 4: 37–46.

GILPIN, M. E. 1974. A model of the predator-prey relationship. Theoret. Pop. Biol. 5: 333–344.

GILPIN, M. E. 1975. Group Selection in Predator-Prey Communities. Princeton Univ. Press, Princeton, New Jersey.

GILPIN, M. E., & F. J. AYALA. 1973. Global models of growth and competition. Proc. Natl. Acad. Sci. 70: 3590–3593.

GILPIN, M. E., & K. E. JUSTICE. 1972. Reinterpretation of the invalidation of the principle of competitive exclusion. Nature 236: 273–274, 299–301.

GILPIN, M. E., & K. E. JUSTICE. 1973. A note on nonlineal competition models. Math. Biosci. 17: 57–63.

GINGERICH, P. D. 1974. Stratigraphic record of early Eocene *Hyopsodus* and the geometry of mammalian phylogeny. Nature 248: 107–109.

GLASS, R. I. 1986. New prospects for epidemiologic investigations. Science 234: 951–955.

GLEASON, H. A. 1922. On the relation between species and area. Ecology 3: 158–162.

GLEASON, H. A. 1926. The individualistic concept of the plant association. Torrey Bot. Club Bull. 53: 7–26.

GLEASON, H. A. 1939. The individualistic concept of the plant association. Amer. Midl. Nat. 21: 92–110.

GLEESON, S. K., & D. S. WILSON. 1986. Equilibrium diet: optimal foraging and prey coexistence. Oikos 46: 139–144.

GLOWACINSKI, Z., & O. JARVINEN. 1975. Rate of secondary succession in forest bird communities. Ornis Scand. 6: 33–40.

GLYNN, P. W. 1984. Widespread coral mortality and the 1982–83 El Niño warming event. Environ. Conserv. 11: 133–146.

GLYNN, P. W. 1988. El Niño–Southern Oscillation 1982–1983: nearshore population, community, and ecosystem responses. Ann. Rev. Ecol. Syst. 309–345.

GOH, B. S. 1975. Stability, vulnerability and persistence of complex ecosystems. Ecol. Model. 1: 105–116.

GOH, B. S. 1976. Nonvulnerability of ecosystems in unpredictable environments. Theoret. Pop. Biol. 10: 83–95.

GOH, B. S. 1977. Global stability in many-species systems. Amer. Nat. 111: 135–143.

GOLLEY, F. B. 1960. Energy dynamics of a food chain of an old-field community. Ecol. Monogr. 30: 187–206.

GOLLEY, F. B., & R. MISRA. 1972. Organic production in tropical ecosystems. BioScience 22: 735–736.

GOODMAN, D. 1975. The theory of diversity-stability relationships in ecology. Quart. Rev. Biol. 50: 237–266.

GOODMAN, D. 1979. Regulating reproductive effort in a changing environment. Amer. Nat. 113: 735–748.

GORDON, M. S. 1968. Animal Function: Principles and Adaptations. Macmillan, New York.

GOSLING, L. M. 1986. Selective abortion of entire litters in the coypu: adaptive control of offspring production in relation to quality and sex. Amer. Nat. 127: 772–795.

GOSLING, L. M. 1986. Biased sex ratios in stressed animals. Amer. Nat. 127: 893–896.

GOULD, S. J. 1982. Darwinism and the expansion of evolutionary theory. Science 216: 380–387.

GOULD, S. J., & N. ELDREDGE. 1977. Punctuated equilibria: the tempo and mode of evolution reconsidered. Paleobiology 3: 115–151.

GOULD, S. J., & R. C. LEWONTIN. 1979. The spandrels of San Marco and the Panglossian paradigm: a critique of the adaptationist programme. Proc. Roy. Soc. Lond. B 205: 581–598.

GOULDEN, C. E., L. L. HENRY, & A. J. TESSIER. 1982. Body size, energy reserves, and competitive ability in three species of Cladocera. Ecology 63: 1780–1789.

GOULDEN, C. E., & L. L. HORNIG. 1980. Population oscillations and energy reserves in planktonic cladocera and their consequences to competition. Proc. Natl. Acad. Sci. 77: 1716–1720.

GRAHAM, A. (Ed.). 1972. Floristics and Paleofloristics of Asia and Eastern North America. Elsevier, New York.

GRAHAM, N. E., & W. B. WHITE. 1988. The El Niño cycle: a natural oscillator of the Pacific Ocean–atmosphere system. Science 240: 1293–1302.

GRANT, P. R. 1966. Ecological imcompatibility of bird species on islands. Amer. Nat. 100: 451–462.

GRANT, P. R. 1966. The density of land birds on the Tres Marias Islands in Mexico, II: distribution of abundance in the community. Canad. J. Zool. 44: 1023–1030.

GRANT, P. R. 1972. Convergent and divergent character displacement. Biol. J. Linn. Soc. 4: 39–68.

GRANT, P. R. 1972. Interspecific competition among rodents. Ann. Rev. Ecol. Syst. 3: 79–106.

GRANT, P. R. 1975. The classical case of character displacement. Pp. 237–337 in T. Dobzhansky, M. K. Hecht, and W. C. Steere (Eds.), Evolutionary Biology, Vol. 8. Plenum, New York.

GRANT, P. R. 1986. Ecology and Evolution of Darwin's Finches. Princeton Univ. Press, Princeton, New Jersey.

GRANT, W. D., & P. E. LONG. 1981. Environmental Microbiology. Wiley (Halsted Press), New York.

GRASSÉ, P.-P. 1951. Traité de Zoologie. Vol. X. Insects Superieurs et Hemipteroides. Part II. Masson, Paris.

GRAVES, J. E., R. H. ROSENBLATT, & G. N. SOMERO. 1983. Kinetic and electrophoretic differentiation of lactate dehydrogenases of teleost species-pairs from the Atlantic and Pacific Coasts of Panama. Evolution 37: 30–37.

GRAVES, J. E., & G. N. SOMERO. 1982. Electrophoretic and functional enzymic evolution in four species of eastern Pacific barracudas from different thermal environments. Evolution 36: 97–106.

GRAY, A. 1860. Illustrations of botany of Japan and its relation to that of central and northern Asia, Europe, and North America. Proc. Amer. Acad. Arts Sci. 4: 131–135.

GRAY, I. E. 1954. Comparative study of the gill area of marine fishes. Biol. Bull 107: 219–225.

GREEN, G. J. 1971. Physiologic races of wheat stem rust in Canada from 1919 to 1969. Canad. J. Bot. 49: 1575–1588.

GREEN, G. J. 1975. Virulence changes in *Puccinia graminis* f. sp. *tritici* in Canada. Canad. J. Bot. 53: 1377–1386.

GREENBERG, L. 1979. Genetic component of bee odor in kin recognition. Science 206: 1095–1097.

GREENE, E., L. J. ORSAK, & D. W. WHITMAN. 1987. A tephritid fly mimics the territorial displays of its jumping spider predators. Science 236: 310–312.

GREENLAND, D. J., & J. M. L. KOWAL. 1960. Nutrient content of a moist tropical forest of Ghana. Plant Soil 12: 154–174.

GREENSLADE, P. J. M. 1968. Island patterns in the Solomon Islands bird fauna. Evolution 22: 751–761.

GREENSLADE, P. J. M. 1969. Land fauna: insect distribution patterns in the Solomon Islands. Phil. Trans. Roy. Soc. B 255: 271–284.

GREENSTONE, M. H. 1984. Determinants of web spider species diversity: vegetation structural diversity vs. prey availability. Oecologia 62: 299–304.

GREENWOOD, J. J. D. 1974. Effective population numbers in the snail *Cepaea nemoralis.* Evolution 28: 513–526.

GREGORY, P. T. 1982. Reptilian hibernation. Pp. 53–154 in C. Gans and F. H. Pough (Eds.), Biology of the Reptilia, Vol. 13. Academic Press, New York.

GRIEG-SMITH, P. 1964. Quantitative Plant Ecology (2nd ed.). Butterworths, London.

GRIFFIN, D. H. 1981. Fungal Physiology. Wiley, New York.

GRIFFIN, D. M. 1972. Ecology of Soil Fungi. Chapman and Hall, London.

GRIME, J. P. 1977. Evidence for the existence of three primary strategies in plants and its relevance to ecological and evolutionary theory. Amer. Nat. 111: 1169–1194.

GRIME, J. P. 1979. Plant Strategies and Vegetation Processes. Wiley, New York.

GRIME, J. P., & R. HUNT. 1975. Relative growth rate: its range and adaptive significance in a local flora. J. Ecol. 63: 393–422.

GRIME, J. P., & D. W. JEFFREY. 1965. Seedling establishment in vertical gradients of sunlight. J. Ecol. 53: 621–642.

GRINNELL, A. D. 1968. Sensory physiology. Pp. 396–460 in M. S. Gordon (Ed.), Animal Function: Principles and Adaptations. Macmillan, New York.

GRINNELL, J. 1917. The niche-relationships of the California thrasher. Auk 34: 427–433.

GRODZINSKI, W., & B. A. WUNDER. 1975. Ecological energetics of small mammals. Pp. 173–204 in F. B. Golley, K. Petrusewicz, and L. Ryszkowski (Eds.), Small Mammals: Their Productivity and Population Dynamics. Cambridge Univ. Press, Cambridge.

GROSS, A. D. 1947. Cyclic invasions of the snowy owl and the migration of 1945–1946. Auk 64: 584–601.

GROSS, F., & E. ZEUTHEN. 1948. The buoyancy of plankton diatoms: a problem of cell physiology. Proc. Roy. Soc. Lond. B 135: 382–389.

GROSS, M. R., & E. L. CHARNOV. 1980. Alternative male life histories in bluegill sunfish. Proc. Natl. Acad. Sci. 77: 6937–6940.

GROSS, M. R., R. M. COLEMAN, & R. M. MCDOWALL. 1988. Aquatic productivity and the evolution of diadromous fish migration. Science 239: 1291–1293.

GROSS, M. R., & R. SHINE. 1981. Parental care and mode of fertilization in ectothermic vertebrates. Evolution 35: 775–793.

GRUBB, P. J. 1977. The maintenance of species diversity in plant communities: the importance of the regeneration niche. Biol. Rev. 52: 107–145.

GUNN, D. L. 1960. The biological background of locust control. Ann. Rev. Entomol. 5: 279–300.

GWYNNE, M. O., & R. H. V. BELL. 1968. Selection of vegetation components by grazing ungulates in the Serengeti National Park. Nature 220: 390–393.

HADLEY, N. F. 1970. Desert species and adaptation. Amer. Sci. 60: 338–347.

HAFFER, J. 1969. Speciation in Amazonian forest birds. Science 165: 131–137.

HAFFER, J. 1974. Avian speciation in tropical South America. Publ. Nuttall Ornithol. Club, No. 14.

HAHN, W. E., & D. W. TINKLE. 1965. Fat body cycling and experimental evidence for its adaptive significance to ovarian follicle development in the lizard *Uta stansburiana.* J. Exptl. Zool. 158: 79–86.

HAINSWORTH, F. R., B. G. COLLINS, & L. L. WOLF. 1977. The function of torpor in hummingbirds. Physiol. Zool. 50: 214–222.

HAINSWORTH, F. R., & L. L. WOLF. 1970. Regulation of oxygen consumption and body temperature during torpor in a hummingbird, *Eulampis jugularis.* Science 168: 368–369.

HAIRSTON, N. G. 1959. Species abundance and community organization. Ecology 40: 404–416.

HAIRSTON, N. G. 1965. On the mathematical analysis of schistosome populations. Bull. World Health Org. 33: 45–62.

HAIRSTON, N. G. 1973. Ecology, selection, and systematics. Breviora 414: 1–21.

HAIRSTON, N. G. 1980. The experimental test of an analysis of field distributions: competition in terrestrial salamanders. Ecology 61: 817–826.

HAIRSTON, N. G. 1983. Alpha selection in competing salamanders: experimental verification of an a priori hypothesis. Amer. Nat. 122: 105–113.

HAIRSTON, N. G., J. D. ALLAN, R. K. COLWELL, D. J. FUTUYMA, J. HOWELL, M. D. LUBIN, J. MATHIAS, & J. H. VANDERMEER. 1968. The relationship between species diversity and stability: an experimental approach with protozoa and bacteria. Ecology 49: 1091–1101.

HAIRSTON, N. G., F. E. SMITH, & L. B. SLOBODKIN. 1960. Community structure, population control, and competition. Amer. Nat. 94: 421–425.

HAIRSTON, N. G., D. W. TINKLE, & H. M. WILBUR. 1970. Natural selection and the parameters of population growth. J. Wildl. Mgmt. 34: 681–690.

HALDANE, J. B. S. 1932. The Causes of Evolution. Longmans, Green, New York, London, and Toronto.

HALDANE, J. B. S. 1949. Suggestions as to quantitative measurement of rates of evolution. Evolution 3: 51–56.

HALDANE, J. B. S. 1949. Disease and evolution. Symposium sui fattori ecologici e genetici della speciazone negli animali. Ric. Sci. 19 (suppl.): 3–11.

HALL, K. R. L. 1965. Behaviour and ecology of the wild patas monkey, *Erythrocebus patas,* in Uganda. J. Zool. 148: 15–87.

HALL, R. L. 1974. Analysis of the nature of interference between plants of different species. I. Concepts and extension of the De Wit analysis to examine effects. Aust. J. Agric. Res. 25: 739–747.

HALL, R. L. 1974. Analysis of the nature of interference between plants of different species. II. Nutrient relations in a nandi *Setaria* and greenleaf *Desmodium* association with particular reference to potassium. Aust. J. Agric. Res. 25: 749–756.

HALLAM, A. (Ed.). 1977. Patterns of Evolution as Illustrated by the Fossil Record. Elsevier, Amsterdam and New York.

HALLIDAY, T. R. 1978. Sexual selection and mate choice. Pp. 180–213 in J. R. Krebs and N. B. Davies (Eds.), Behavioural Ecology: An Evolutionary Approach. Sinauer, Sunderland, Massachusetts.

HAMILTON, T. H., I. RUBINOFF, R. H. BARTH, JR., & G. L. BUSH. 1963. Species abundance: natural regulation of insular variation. Science 142: 1575–1577.

HAMILTON, W. D. 1964. The genetical evolution of social behaviour. J. Theoret. Biol. 7: 1–52.

HAMILTON, W. D. 1966. The moulding of senescence by natural selection. J. Thoeret. Biol. 12: 12–45.

HAMILTON, W. D. 1967. Extraordinary sex ratios. Science 156: 477–488.

HAMILTON, W. D. 1979. Wingless and fighting males in fig wasps and other insects. Pp. 167–220 in M. S. Blum and N. A. Blum (Eds.), Sexual Selection and Reproductive Competition in Insects. Academic Press, New York.

HAMMEL, H. T. 1968. Regulation of internal body temperature. Ann. Rev. Physiol 30: 641–710.

HAMMEL, H. T., F. T. CALDWELL, & R. M. ABRAMS. 1967. Regulation of body temperature in the blue-tongued lizard. Science 156: 1260–1262.

HAMMOND, E. C. 1938. Biological effects of population density in lower organisms. Part I. Quart. Rev. Biol. 13: 421–438.

HAMMOND, E. C. 1939. Biological effects of population density in lower organisms. Part II. Quart. Rev. Biol. 14: 35–59.

HANES, T. L. 1971. Succession after fire in the chaparral of southern California. Ecol. Monogr. 41: 27–52.

HANSKI, I. 1982. Dynamics of regional distribution: the core and satellite species hypothesis. Oikos 38: 210–221.

HANSKI, I., & H. KOSKELA. 1977. Niche relations among dung-inhabiting beetles. Oecologia 28: 203–231.

HANSKI, I., & H. KOSKELA. 1978. Stability, abundance, and niche width in the beetle community inhabiting cow dung. Oikos 31: 290–298.

HARBORNE, J. B. 1982. Introduction to Ecological Biochemistry (2nd ed.). Academic Press, London and New York.

HARCOURT, D. G. 1963. Major mortality factors in the population dynamics of the diamondback moth *Plutella maculipennis* (Curt.). Mem. Entomol. Soc. Canad. 32: 55–66.

HARCOURT, D. G. 1971. Population dynamics of *Leptinotarsa decemlineata* (Say) in eastern Ontario. III. Major population processes. Canad. Entomol. 103: 1049–1061.

HARCOURT, D. G., & E. J. LEROUX. 1967. Population regulation in insects and man. Amer. Sci. 55: 400–415.

HARDIN, G. 1960. The competitive exclusion principle. Science 131: 1292–1297.

HARDY, R. W., & U. D. HAVELKA. 1975. Nitrogen fixation research: a key to world food? Science 188: 633–642.

HARLEY, J. L. 1972. Fungi in ecosystems. J. Anim. Ecol. 41: 1–16.

HARLEY, J. L. (Ed.). 1979. The Soil-Root Interface. Academic Press, New York.

HARLEY, J. L., & S. E. SMITH. 1983. Mycorrhizal Symbiosis. Academic Press, London.

HARMON, M. E., J. F. FRANKLIN, F. J. SWANSON, P. SOLLINS, S. V. GREGORY, J. D. LATTIN, N. H. ANDERSON, S. P. CLINE, N. G. AUMEN, J. R. SEDELL, G. W. LIENKAEMPER, K. CROMACK, JR., & K. W. CUMMINS. 1986. Ecology of coarse woody debris in temperate ecosystems. Adv. Ecol. Res. 15: 133–302.

HARPER, J. L. 1967. A Darwinian approach to plant ecology. J. Ecol. 55: 247–270.

HARPER, J. L. 1969. The role of predation is vegetational diversity. Brookhaven Symp Biol. 22: 48–62.

HARPER, J. L. 1977. Population Biology of Plants. Academic Press, New York and London.

HARPER, J. L., J. T. WILLIAMS, & G. R. SAGAR. 1965. The behaviour of seeds in soil. J. Ecol. 51: 273–286.

HARRÉ, R. 1972. The Philosophies of Science. Oxford Univ. Press, London.

HARRIS, L. G., A. W. EBELING, D. R. LAUR, & R. J. ROWLEY. 1984. Community recovery after storm damage: a case of facilitation in primary succession. Science 224: 1336–1338.

HARRIS, M. P. 1970. Breeding ecology of the swallow-tailed gull, *Creagrus furcatus.* Auk 87: 215–243.

HARRISON, A. T., E. SMALL, & H. A. MOONEY. 1971. Drought relationships and distribution of two Mediterranean-climate California plant communities. Ecology 52: 869–875.

HARRISON, G. W. 1980. Global stability of food chains. Amer. Nat. 114: 455–457.

HARRISON, W. G. 1978. Experimental measurements of nitrogen remineralization in coastal waters. Limnol. Oceanogr. 23: 684–694.

HARROLD, C., & D. C. REED. 1985. Food availability, sea urchin grazing, and kelp forest community structure. Ecology 66: 1160–1169.

HART, J. S. 1957. Climatic and temperature induced changes in the energetics of homeotherms. Rev. Canad. Biol 16: 133–171.

HARTENSTEIN, R. 1986. Earthworm biotechnology and global biogeochemistry. Adv. Ecol. Res. 15: 379–409.

HARTL, D. 1980. Principles of Population Genetics. Sinauer, Sunderland, Massachusetts.

HARVEY, E. N. 1928. The oxygen consumption of luminous bacteria. J. Gen. Physiol. 11: 469–475.

HARVEY, P. H., & S. J. ARNOLD. 1982. Female mate choice and runaway sexual selection. Nature 297: 533–534.

HASLER, A. D. 1947. Eutrophication of lakes by domestic drainage. Ecology 28: 383–395.

HASSELL, M. P. 1978. The Dynamics of Arthropod Predator-Prey Systems. Princeton Univ. Press, Princeton, New Jersey.

HASSELL, M. P., & H. N. COMINS. 1978. Sigmoid functional responses and population stability. Theoret. Pop. Biol. 14: 62–67.

HASSELL, M. P., J. H. LAWTON, & R. M. MAY. 1976. Patterns of dynamical behaviour in single-species populations. J. Anim. Ecol. 45: 471–486.

HASSELL, M. P., & R. M. MAY. 1973. Stability in insect host-parasite models. J. Anim. Ecol. 42: 693–736.

HASSELL, M. P., & R. M. MAY. 1989. The population biology of host-parasite and host-parasitoid associations. Pp. 319–347 in J. Roughgarden, R. M. May, and S. A. Levin (Eds.), Perspectives in Ecological Theory. Princeton Univ. Press, Princeton, New Jersey.

HASTINGS, A., & H. CASWELL. 1979. Role of environmental variability in the evolution of life history strategies. Proc. Natl. Acad. Sci. 76: 4700–4703.

HATCH, M. D., & C. R. SLACK. 1966. Photosynthesis by sugar-cane leaves: a new carboxylation reaction and the pathway of sugar formation. Biochem. J. 101: 103–111.

HAUKIOJA, E. 1980. On the role of plant defenses in the fluctuation of herbivore populations. Oikos 35: 202–213.

HAXO, F. T., & L. R. BLINKS. 1950. Photosynthetic action spectra of marine algae. J. Gen. Physiol. 33: 389–422.

HAY, M. E. 1981. Herbivory, algal distribution, and the maintenance of between-habitat diversity on a tropical fringing reef. Amer. Nat. 118: 520–540.

HAY, M. E. 1981. The functional morphology of turf-forming seaweeds: persistence in stressful marine habitats. Ecology 62: 739–750.

HAY, M. E. 1986. Associational plant defenses and the maintenance of species diversity: turning competitors into accomplices. Amer. Nat. 128: 617–641.

HAY, M. E., T. COLBURN, & D. DOWNING. 1983. Spatial and temporal patterns in herbivory on a Caribbean fringing reef: the effects on plant distribution. Oecologia 58: 299–308.

HAY, M. E., & P. R. TAYLOR. 1985. Competition between herbivorous fishes and urchins on Caribbean reefs. Oecologia 65: 591–598.

HAYES, A. J. 1979. The microbiology of plant litter decomposition. Sci. Progr. 66: 25–42.

HAYNES, R. J., & K. M. GOH. 1978. Ammonium and nitrate nutrition of plants. Biol. Rev. 53: 465–510.

HEALEY, M. C. 1967. Aggression and self-regulation of population size in deermice. Ecology 48: 377–392.

HEATH, J. E. 1965. Temperature regulation and diurnal activity in horned lizards. Univ. Calif. Publ. Zool. 64: 97–136.

HEATWOLE, H. 1970. Thermal ecology of the desert dragon *Amphibolurus inermis*. Ecol. Monogr. 40: 425–457.

HEATWOLE, H. 1976. Reptile Ecology. Univ. Queensland Press, St. Lucia, Queensland.

HEBERT, D., & I. M. COWAN. 1971. Natural salt licks as a part of the ecology of the mountain goat. Canad. J. Zool. 49: 605–610.

HEDRICK, A. V. 1988. Female choice and the heritability of attractive male traits: an empirical study. Amer. Nat. 132: 267–276.

HEINRICH, B. 1979. Bumblebee Economics. Harvard Univ. Press, Cambridge, Massachusetts.

HEINRICH, B., & G. A. BARTHOLOMEW. 1971. An analysis of pre-flight warm-up in the sphinx moth, *Manduca sexta.* J. Exp. Biol. 55: 223–239.

HELLER, H. C. 1971. Altitudinal zonation of chipmunks *(Eutamias):* interspecific aggression. Ecology 52: 312–319.

HELLMERS, H. 1964. An evaluation of the photosynthetic efficiency of forests. Quart. Rev. Biol 39: 249–257.

HELLMERS, H., J. S. HORTON, G. JUHREN, & J. O'KEEFE. 1955. Root systems of some chaparral plants in southern California. Ecology 36: 667–678.

HENDERSON, L. J. 1913. The Fitness of the Environment. Macmillan, New York.

HERRE, E. A. 1985. Sex ratio adjustment in fig wasps. Science 228: 896–898.

HERRERA, C. M. 1978. Individual dietary differences associated with morphological variation in robins *Erithacus rubecula.* Ibis 120: 542–545.

HESPENHEIDE, H. A. 1971. Food preference and the extent of overlap in some insectivorous birds, with special reference to the Tyrannidae. Ibis 113: 59–72.

HESPENHEIDE, H. A. 1973. Ecological inferences from morphological data. Ann. Rev. Ecol. Syst. 4: 213–229.

HIESEY, W. M., & H. W. MILNER. 1965. Physiology of ecological races and species. Ann. Rev. Plant Physiol. 16: 203–216.

HILDEN, O. 1965. Habitat selection in birds: a review. Ann. Zool. Fenn. 2: 53–75.

HINDE, R. A. 1956. The biological significance of the territories of birds. Ibis 98: 340–369.

HINDE, R. A. (Ed.). 1983. Primate Social Relationships: An Integrated Approach. Blackwell, Oxford.

HOAGLAND, K. E. 1978. Protandry and the evolution of environmentally-mediated sex change: a study of the Mollusca. Malacologia 17: 365–391.

HOCHACHKA, P. W., & G. N. SOMERO. 1973. Strategies of Biochemical Adaptation. Saunders, Philadelphia.

HOCHACHKA, P. W., & G. N. SOMERO. 1984. Biochemical Adaptation. Princeton Univ. Press, Princeton, New Jersey.

HOGAN-WARBURG, A. J. 1966. Social behavior of the ruff, *Philomachus pugnax* (L.). Ardea 54: 109–225.

HOGSTED, G. 1980. Evolution of clutch size in birds: adaptive variation in relation to territory quality. Science 210: 1148–1150.

HOLDREN, G. C., & D. E. ARMSTRONG. 1980. Factors affecting phosphorus release from intact lake sediment cores. Environ. Sci. Technol. 14: 79–87.

HOLDRIDGE, L. 1967. Life Zone Ecology. Tropical Science Center, San Jose, Costa Rica.

HOLLING, C. S. 1959. The components of predation as revealed by a study of small mammal predation of the European pine sawfly. Canad. Entomol. 91: 293–320.

HOLMES, R. T., J. C. SCHULTZ, & P. NOTHNAGLE. 1979. Bird predation on forest insects: an exclosure experiment. Science 206: 462–463.

HOLMES, W. G., & P. W. SHERMAN. 1983. Kin recognition in animals. Amer. Sci. 71: 46–55.

HOOGENDYK, C. G., & G. F. ESTABROOK. 1984. The consequences of earlier reproduction in declining populations. Math. Biosci. 71: 217–235.

HOOGLAND, J. L. 1982. Prairie dogs avoid extreme inbreeding. Science 215: 1639–1641.

HOPKINS, B. 1955. The species-area relations of plant communities. J. Ecol. 43: 409–426.

HORN, H. S. 1975. Markovian properties of forest succession. Pp. 196–211 in M. L. Cody and J. M. Diamond (Eds.), Ecology and Evolution of Communities. Harvard Univ. Press, Cambridge, Massachusetts.

HORNOCKER, M. G. 1970. An analysis of mountain lion predation upon mule deer and elk in the Idaho Primitive Area. Wildl. Monogr. 21: 3–39.

HORVITZ, C. C., & D. W. SCHEMSKE. 1984. Effects of ants and an ant-tended herbivore on seed production of a neotropical herb. Ecology 65: 1369–1378.

HOUGH, A. F., & R. D. FORBES. 1943. The ecology and silvics of forests in the high plateaus of Pennsylvania. Ecol. Monogr. 13: 299–320.

HOUGH, J. A., & D. PIMENTEL. 1978. Influence of host foliage on development, survival, and fecundity of gypsy moth. Env. Entomol. 7: 97–102.

HOUSTON, D. C. 1979. The adaptations of scavengers. Pp. 263–286 in A. R. E. Sinclair and M. Norton-Griffiths (Eds.), Serengeti. Dynamics of an Ecosystem. Univ. Chicago Press, Chicago.

HOWARD, H. E. 1920. Territory in Bird Life. Murray, London.

HOWARD, L. O., & W. F. FISKE. 1911. The importation into the United States of the parasites of the gypsy moth and the brown-tailed moth. U.S. Dept. Agric. Bur. Entomol. Bull. 91: 1–312.

HOWARD, R. D. 1978. The evolution of mating strategies in bullfrogs, *Rana catesbiana.* Evolution 32: 850–871.

HOWARD, R. D. 1979. Estimating reproductive success in natural populations. Amer. Nat. 114: 221–231.

HOWARD, W. E. 1960. Innate and environmental dispersal of individual vertebrates. Amer. Midl. Nat. 63: 152–161.

HOWARTH, R. W. 1984. The ecological significance of sulfur in the energy dynamics of salt marsh and coastal marine sediments. Biogeochemistry 1: 5–27.

HOWARTH, R. W. 1988. Nutrient limitation of net primary production in marine ecosystems. Ann. Rev. Ecol. Syst. 19: 89–110.

HOWE, H. F. 1984. Constraints on the evolution of mutualisms. Amer. Nat. 123: 764–777.

HOWE, H. F., E. W. SCHUPP, & L. C. WESTLEY. 1985. Early consequences of seed dispersal for a neotropical tree *(Virola surinamensis).* Ecology 66: 781–791.

HOYT, D. F., & H. RAHN. 1980. Respiration of avian embryos —a comparative analysis. Respir. Physiol. 39: 255–264.

HSAIO, T. C. 1973. Plant responses to water stress. Ann. Rev. Plant Physiol. 24: 519–570.

HUBBELL, S. P. 1979. Tree dispersion, abundance, and diversity in a tropical dry forest. Science 203: 1299–1039.

HUBBELL, S. P., & R. B. FOSTER. 1983. Diversity of canopy trees in a neotropical forest and implications for conservation. Pp. 25–41 in S. Sutton, T. C. Whitmore, and A. Chadwick (Eds.), Tropical Rain Forest: Ecology and Management. Blackwell, Oxford.

HUBBELL, S. P., & R. B. FOSTER. 1986. Biology, chance, and history and the structure of tropical rain forest tree communities. Pp. 314–329 in J. Diamond and T. J. Case (Eds.), Community Ecology. Harper & Row, New York.

HUDSON, J. W. 1962. The role of water in the biology of the antelope ground squirrel, *Citellus leucurus.* Univ. Calif. Publ. Zool. 64: 1–56.

HUDSON, P. J. 1986. The effect of a parasitic nematode on the breeding production of red grouse. J. Anim. Ecol. 55: 85–92.

HUEY, R. B. 1974. Behavioral thermoregulation in lizards: importance of associated costs. Science 184: 1001–1003.

HUEY, R. B., & E. R. PIANKA. 1974. Ecological character displacement in a lizard. Amer. Zool. 14: 1127–1136.

HUEY, R. B., E. R. PIANKA, M. E. EGAN, & L. W. COONS. 1974. Ecological shifts in sympatry: Kalahari fossorial lizards *(Typhlosaurus).* Ecology 55: 304–316.

HUFFAKER, C. B. 1958. Experimental studies on predation: dispersion factors and predator-prey oscillations. Hilgardia 27: 343–383.

HUFFAKER, C. B., & C. E. KENNETT. 1956. Experimental studies on predation: predation and cyclamen-mite populations on strawberries in California. Hilgardia 26: 191–222.

HUFFAKER, C. B., & C. E. KENNETT. 1969. Some aspects of assessing efficiency of natural enemies. Canad. Entomol. 101: 425–447.

HUNTLEY, M., & C. BOYD. 1984. Food-limited growth of marine zooplankton. Amer. Nat. 124: 455–478.

HURLBERT, S. H. 1971. The nonconcept of species diversity: a critique and alternative parameters. Ecology 52: 577–586.

HUTCHINSON, G. E. 1948. Circular causal systems in ecology. Ann. N.Y. Acad. Sci. 50: 221–246.

HUTCHINSON, G. E. 1957. A Treatise on Limnology, Vol. 1: Geography, Physics, and Chemistry. Wiley, New York.

HUTCHINSON, G. E. 1957. Concluding remarks. Cold Spring Harbor Symp. Quant. Biol. 22: 415–427.

HUTCHINSON, G. E. 1959. Homage to Santa Rosalia, or Why are there so many kinds of animals? Amer. Nat. 93: 145–159.

HUTCHINSON, G. E. 1978. An Introduction to Population Ecology. Yale Univ. Press, New Haven and London.

HUTCHISON, V. H., H. G. DOWLING, & A. VINEGAR. 1966. Thermoregulation in a brooding female Indian python, *Python molurus bivittatus.* Science 151: 694–696.

IMMELMANN, K. 1971. Ecological aspects of periodic reproduction. Pp. 341–389 in D. S. Farner and J. R. King (Eds.), Avian Biology, Vol. 1. Academic Press, New York.

INGHAM, G. 1950. The mineral content of air and rain and its importance to agriculture. J. Agric. Sci. 4: 55–61.

INNIS, G. S. 1975. Role of total systems models in the grassland biome study. Pp. 14–47 in B. C. Patten (Ed.), Systems Analysis and Simulation in Ecology, Vol. III. Academic Press, New York.

INNIS, G. S. 1978. Grassland Simulation Model. Springer-Verlag, New York.

INNIS, G. S., & R. V. O'NEILL. (Eds.). 1979. Systems Analysis of Ecosystems. Statistical Ecology, Vol. 9. Intern. Co-op. Publ. House, Fairland, Maryland.

IRONS, D. B., R. G. ANTHONY, & J. A. ESTES. 1986. Foraging strategies of glaucous-winged gulls in a rocky intertidal community. Ecology 67: 1460–1474.

IRVING, L. 1966. Adaptations to cold. Sci. Amer. 214: 94–101.

ISLEY, F. B. 1944. Correlation between mandibular morphology and food specificity in grasshoppers. Ann. Entomol. Soc. Amer. 37: 47–67.

IVES, W. G. H. 1973. Heat units and outbreaks of the forest tent caterpillar, *Malacosoma disstria* (Lepidoptera: Lasiocampidae). Canad. Entomol. 105: 529–543.

JACKSON, G. J., R. HERMAN, & I. SINGER. (Eds.). 1969–1970. Immunity to Parasitic Animals, Vol. I and II. Appleton-Century-Crofts, New York.

JACKSON, J. B. C., & L. BUSS. 1975. Allelopathy and spatial competition among coral reef invertebrates. Proc. Natl. Acad. Sci. 72: 5160–5163.

JAIN, S. K. 1976. The evolution of inbreeding in plants. Ann. Rev. Ecol. Syst. 7: 469–495.

JANZEN, D. H. 1966. Coevolution of mutualism between ants and acacias in Central America. Evolution 20: 249–275.

JANZEN, D. H. 1967. Interaction of the bull's-horn acacia (*Acacia cornigera* L.) with an ant inhabitant (*Pseudomyrmex ferruginea* F. Smith) in eastern Mexico. Univ. Kansas Sci. Bull. 47: 315–558.

JANZEN, D. H. 1968. Host plants as islands in evolutionary and contemporary time. Amer. Nat. 102: 592–595.

JANZEN, D. H. 1969. Seed-eaters versus seed size, number, toxicity and dispersal. Evolution 23: 1–27.

JANZEN, D. H. 1970. Herbivores and the number of tree species in tropical forests. Amer. Nat. 104: 501–528.

JANZEN, D. H. 1971. Seed predation by animals. Ann. Rev. Ecol. Syst. 2: 465–492.

JANZEN, D. H. 1973. Host plants as islands. II. Competition in evolutionary and contemporary time. Amer. Nat. 107: 786–790.

JANZEN, D. H. 1976. Why bamboos wait so long to flower. Ann. Rev. Ecol. Syst. 7: 347–391.

JANZEN, D. H. 1977. A note on optimal mate selection by plants. Amer. Nat. 111: 365–371.

JANZEN, D. H. 1979. How to be a fig. Ann. Rev. Ecol. Syst. 10: 13–51.

JANZEN, D. H. 1985. The natural history of mutualisms. Pp. 40–99 in D. H. Boucher (Ed.), The Biology of Mutualism. Croom Helm, London.

JANZEN, D. H. 1988. Tropical ecological and biocultural restoration. Science 239: 243–244.

JANZEN, D. H., & T. W. SCHOENER. 1968. Differences in insect abundance and diversity between wetter and drier sites during a tropical dry season. Ecology 49: 96–110.

JARMAN, P. J., & A. R. E. SINCLAIR. 1979. Feeding strategy and the pattern of resource partitioning in ungulates. Pp. 130–163 in A. R. E. Sinclair and M. Norton-Griffiths (Eds.), Serengeti. Dynamics of an Ecosystem. Univ. Chicago Press, Chicago.

JARVI, T., B. SILLEN-TULBERG, & C. WIKLUND. 1981. The cost of being aposematic. An experimental study of predation of larvae of *Papilio machaon* by the great tit, *Parus major.* Oikos 36: 267–272.

JARVIS, P. G., & K. G. MCNAUGHTON. 1986. Stomatal control of transpiration: scaling up from leaf to region. Adv. Ecol. Res. 15: 1–49.

JEFFERS, J. N. R. 1978. An Introduction to Systems Analysis: With Ecological Applications. Edward Arnold, London, and Univ. Park Press, Baltimore.

JEFFRIES, C. 1974. Qualitative stability and digraphs in model ecosystems. Ecology 55: 1415–1419.

JENNY, H. 1941. Factors in Soil Formation. McGraw-Hill, New York.

JENNY, H. 1980. The Soil Resource. Origin and Behavior. Springer-Verlag, New York.

JERLING, L. 1988. Genetic differentiation in fitness related characters in *Plantago maritima* along a distributional gradient. Oikos 53: 341–350.

JOERN, A. 1986. Experimental study of avian predation on coexisting grasshopper populations (Orthoptera: Acrididae) in a sandhills grassland. Oikos 46: 243–249.

JOHANNESSON, K. 1989. The bare zone of Swedish rocky shores: why is it there? Oikos 54: 77–86.

JOHNSGARD, P. A. 1983. The Grouse of the World. Univ. Nebraska Press, Lincoln, Nebraska.

JOHNSON, D., D. W. COLE, & S. P. GESSEL. 1975. Processes of nutrient transfer in a tropical rain forest. Biotropica 7: 208–215.

JOHNSON, W. S., A. GIGON, S. L. GULMON, & H. A. MOONEY. 1974. Comparative photosynthetic capacities of intertidal algae under exposed and submerged conditions. Ecology 55: 450–453.

JOLLY, G. M. 1965. Explicit estimates from capture-recapture data with low death and immigration—stochastic model. Biometrika 52: 315–337.

JOLLY, G. M. 1979. Sampling of large objects. Pp. 193–201 in R. M. Cormack, G. P. Patil, and D. S. Robson (eds.), Sampling Biological Populations. Intern. Co-op. Publ. House, Fairland, Maryland.

JONES, F. R. H. 1968. Fish Migration. Edward Arnold, London.

JONES, M. L., S. L. SWARTZ, & S. LEATHERWOOD. (Eds.). 1984. The Gray Whale *Eschrichtius robustus.* Academic Press, Orlando, Florida.

JORDAN, C. F. 1985. Nutrient Cycling in Tropical Forest Ecosystems. Wiley, New York.

JORDAN, C. F., & R. HERRERA. 1981. Tropical rain forests: are nutrients really critical? Amer. Nat. 117: 167–180.

JORDAN, P., & G. WEBBE. 1969. Human Schistosomiasis. Heinemann Medical Books, London.

KACELNIK, A. 1984. Central place foraging in starlings *(Sturnus vulgaris).* I. Patch residence time. J. Anim. Ecol. 53: 283–299.

KALLMAN, K. D., M. P. SCHREIBMAN, & V. BORKOSKI. 1973. Genetic control of gonadotrop differentiation in the platyfish, *Xiphophorus maculatus* (Poecilidae). Science 181: 678–680.

KALMUS, H., & C. A. B. SMITH. 1960. Evolutionary origin of sexual differentiation and the sex-ratio. Nature 186: 1004–1006.

KANWISHER, J. 1959. Histology and metabolism of frozen intertidal animals. Biol. Bull. 116: 258–264.

KARBAN, R., R. ADAMCHAK, & W. C. SCHNATHORST. 1987. Induced resistance and interspecific competition between spider mites and a vascular wilt fungus. Science 235: 678–680.

KARBAN, R., & J. R. CAREY. 1984. Induced resistance of cotton seedlings to mites. Science 225: 53–54.

KARBAN, R., & R. E. RICKLEFS. 1983. Host characteristics, sampling intensity, and species richness of Lepidoptera larvae on broad-leaved trees in southern Ontario. Ecology 64: 636–641.

KARBAN, R., & R. E. RICKLEFS. 1984. Leaf traits and the species richness and abundance of Lepidopteran larvae on deciduous trees in southern Ontario. Oikos 43: 165–170.

KARR, J. R. 1977. Ecological correlates of rarity in a tropical forest bird community. Auk 94: 240–247.

KARR, J. R., & F. C. JAMES. 1975. Ecomorphological configurations and convergent evolution in species and communities. Pp. 258–291 in M. L. Cody and J. M. Diamond (Eds.), Ecology and Evolution of Communities. Harvard Univ. Press, Cambridge, Massachusetts.

KAZACOS, K. R., & R. E. THORSON. 1975. Cross-resistance between *Nippostrongylus brasiliensis* and *strongyloides ratti* in rats. J. Parasitol. 61: 525–529.

KEAST, A. 1961. Bird speciation on the Australian continent. Bull. Mus. Comp. Zool. 123: 305–495.

KEAST, A. 1972. Ecological opportunities and dominant families, as illustrated by the Neotropical Tyrannidae (Aves). Evol. Biol. 5: 229–277.

KEAST, A., & E. S. MORTON. 1980. Migrant Birds in the Neotropics: Ecology, Behavior, Distribution, and Conservation. Smithson. Inst. Press, Washington, D.C.

KEEVER, C. 1950. Causes of succession on old fields of the Piedmont, North Carolina. Ecol. Monogr. 20: 230–250.

KEISTER, A. R. 1971. Species density of North American amphibians and reptiles. Syst. Zool 20: 127–137.

KELSALL, J. P. 1968. The Migratory Barren-ground Caribou of Canada. Canadian Wildlife Service, Ottawa.

KENOYER, L. A. 1927. A study of Raunkaier's law of frequence. Ecology 8: 341–349.

KEOUGH, M. J. 1984. Effects of patch size on the abundance of sessile marine invertebrates. Ecology 65: 423–437.

KERSTER, H. W. 1964. Neighborhood size in the rusty lizard, *Sceloporus olivaceus.* Evolution 18: 445–457.

KESSEL, B. 1953. Distribution and migration of the European starling in North America. Condor 55: 49–67.

KETTLEWELL, H. B. D. 1955. Selection experiments on industrial melanism in the lepidoptera. Heredity 10: 287–301.

KETTLEWELL, H. B. D. 1956. Further selection experiments on industrial melanism in the lepidoptera. Heredity 10: 287–301.

KETTLEWELL, H. B. D. 1959. Darwin's missing evidence. Sci. Amer. 200: 48–53.

KEYFLITZ, N., & W. FLIEGER. 1968. World Population. An Analysis of Vital Data. Univ. Chicago Press, Chicago and London.

KEYNES, R. D., & H. MARTINS-FERREIRA. 1953. Membrane potentials in the electroplates of the electric eel. J. Physiol. 119: 315–351.

KIELANOWSKI, J. 1964. Estimates of the energy cost of protein deposition in growing animals. Pp. 13–20 in K. L.

Baxter (Ed.), Energy Metabolism. Academic Press, New York.

KIMURA, M., & G. H. WEISS. 1964. The stepping stone model of population structure and the decrease of genetic correlation with distance. Genetics 49: 561–576.

KING, B. H. 1988. Sex-ratio manipulation in response to host size by the parasitoid wasp *Spalangia cameroni:* a laboratory study. Evolution 42: 1190–1198.

KING, C. E. 1964. Relative abundance of species and MacArthur's model. Ecology 45: 716–727.

KING, D. L., & R. C. BALL. 1967. Comparative energetics of a polluted stream. Limnol. Oceanogr. 12: 27–33.

KING, J. R. 1974. Seasonal allocation of time and energy resources in birds. Pp. 4–70 in R. A. Paynter, Jr. (Ed.), Avian Energetics. Nuttall Ornithol. Club, Cambridge, Massachusetts.

KING, J. R., & D. S. FARNER. 1961. Energy metabolism, thermoregulation and body temperature. Pp. 215–288 in A. J. Marshall (Ed.), Biology and Comparative Physiology of Birds, Vol. II. Academic Press, New York.

KINGSLAND, S. E. 1985. Modeling Nature. Episodes in the History of Population Ecology. Univ. Chicago Press, Chicago.

KIRA, T., & T. SHIDEI. 1967. Primary production and turnover of organic matter in different forest ecosystems of the western Pacific. Jap. J. Ecol. 17: 70–87.

KIRK, T. K, W. J. CONNORS, & J. G. ZEIKUS. 1977. Advances in understanding the microbiological degradation of lignin. Rec. Adv. Phytochem. 11: 369–394.

KIRKPATRICK, M. 1982. Sexual selection and the evolution of female choice. Evolution 36: 1–12.

KIRKPATRICK, M. 1986. The handicap mechanism of sexual selection does not work. Amer. Nat. 127: 222–240.

KIRKPATRICK, M. 1987. Sexual selection by female choice in polygynous animals. Ann. Rev. Ecol. Syst. 18: 43–70.

KITCHING, R. L. 1983. Systems Ecology. An Introduction to Ecological Modelling. Univ. Queensland Press, St. Lucia, London, New York.

KLAHN, J. E., & G. J. GAMBOA. 1983. Social wasps: discrimination between kin and nonkin brood. Science 221: 482–484.

KLEEBERGER, S. R. 1984. A test of competition in two sympatric populations of desmognathine salamanders. Ecology 65: 1846–1856.

KLEIBER, M. 1961. The Fire of Life. Wiley, New York.

KLEIN, D. R. 1968. The introduction, increase and crash of reindeer on St. Matthew Island. J. Wildl. Mgmt. 32: 350–367.

KLOPFER, P. 1959. Social interactions in discrimination learning with special reference to feeding behavior in birds. Behaviour 14: 282–299.

KLUGE, M. & I. P. TING. 1978. Crassulacean Acid Metabolism. Springer-Verlag, Berlin.

KNIGHT, D. H. 1975. A phytosociological analysis of species-rich tropical forest: Barro Colorado Island, Panama. Ecol. Monogr. 45: 259–284.

KNUTSON, R. M. 1974. Heat production and temperature regulation in eastern skunk cabbage. Science 186: 746–748.

KODRIC-BROWN, A., & J. H. BROWN. 1984. Truth in advertising: the kinds of traits favored by sexual selection. Amer. Nat. 124: 309–323.

KOESTLER, A. 1967. The Ghost in the Machine. Macmillan, New York.

KOHN, R. R. 1971. Principles of Mammalian Aging. Prentice-Hall, Englewood Cliffs, New Jersey.

KOLMAN, W. A. 1960. The mechanism of natural selection for the sex ratio. Amer. Nat. 94: 373–377.

KOLLER, D. 1969. The physiology of dormancy and survival of plants in desert environments. Symp. Soc. Exptl. Biol. 23: 449–469.

KOPLIN, J. R., & R. S. HOFFMANN. 1968. Habitat overlap and competitive exclusion in voles *(Microtus).* Amer. Midl. Nat. 80: 494–507.

KOPTUR, S. 1985. Alternative defenses against hervibores in *Inga* (Fabaceae: Mimosoideae) over an elevational gradient. Ecology 66: 1639–1650.

KOTLER, B. P., & J. S. BROWN. 1988. Environmental heterogeneity and the coexistence of desert rodents. Ann. Rev. Ecol. Syst. 19: 281–307.

KOZLOVSKY, D. G. 1968. A critical evaluation of the trophic level concept. I. Ecological efficiencies. Ecology 49: 48–59.

KOZLOWSKI, T. T., & C. E. AHLGREN. (Eds.). 1974. Fire and Ecosystems. Academic Press, New York.

KRAMER, P. J. 1958. Photosynthesis of trees as affected by their environment. Pp. 157–186 in K. V. Thimann (Ed.), The Physiology of Forest Trees. Ronald, New York.

KRAMER, P. J. 1969. Plant and Water Relationships: A Modern Synthesis. McGraw-Hill, New York.

KRAMER, P. J. 1983. Water Relations of Plants. Academic Press, New York.

KREBS, C. J, B. L. KELLER, & R. H. TAMARIN. 1969. *Microtus* population biology: demographic changes in fluctuating populations of *M. ochrogaster* and *M. pennsylvanicus* in southern Indiana. Ecology 50: 587–607.

KREBS, C. J., & J. MYERS. 1974. Population cycles in small mammals. Adv. Ecol. Res. 8: 267–399.

KREBS, J. R. 1971. Territory and breeding density in the Great Tit, *Parus major* L. Ecology 52: 2–22.

KREBS, J. R. 1974. Colonial nesting and social feeding as strategies for exploiting food resources in the great blue heron *(Ardea herodias)*. Behaviour 51: 99–134.

KREBS, J. R., & M. I. AVERY. 1985. Central place foraging in the European bee-eater, *Merops apiaster*. J. Anim. Ecol. 54: 459–472.

KREBS, J. R., & N. B. DAVIES. 1981. An Introduction to Behavioural Ecology. Sinauer, Sunderland, Massachusetts.

KREBS, J. R., J. T. ERICHSEN, M. I. WEBBER, & E. L. CHARNOV. 1977. Optimal prey selection in the great tit *(Parus major)*. Anim. Behav. 25: 30–38.

KREBS, J. R., & R. H. MCCLEERY. 1984. Optimization in behavioral ecology. Pp. 91–121 in J. R. Krebs and N. B. Davies (Eds.), Behavioural Ecology. An Evolutionary Approach (2nd ed.). Sinauer, Sunderland, Massachusetts.

KRUCKENBERG, A. R. 1951. Intraspecific variability in the response of certain native plants to serpentine soil. Amer. J. Bot. 38: 408–419.

KRUCKENBERG, A. R. 1954. The ecology of serpentine soils: a symposium, III: Plant species in relation to serpentine soils. Ecology 35: 267–274.

KRUUK, H. 1967. Competition for food between vultures in East Africa. Ardea 55: 171–193.

KRUUK, H. 1969. Interactions between populations of spotted hyenas *Crocuta crocuta* (Erxleben) and their prey species. Pp. 359–374 in A. Watson (Ed.). Animal Populations in Relation to Their Food Resources. Blackwell, Oxford.

KRUUK, H. 1972. The Spotted Hyena. A Study of Predation and Social Behavior. Univ. Chicago Press, Chicago.

KUCHLER, A. W. 1949. A physiognomic classification of vegetation. Ann. Assoc. Amer. Geogr. 39: 201–210.

KUCHLER, A. W. 1964. Potential natural vegetation of the conterminus United States. Amer. Geogr. Soc., Spec. Publ. No. 36.

KUENZLER, E. J. 1961. Structure and energy flow of a mussel population. Limnol. Oceanogr. 6: 191–204.

KUIJT, J. 1969. The Biology of Flowering Parasitic Plants. Univ. California Press, Berkeley.

KULMAN, H. M. 1971. The effect of defoliation on tree growth. Ann. Rev. Entomol. 16: 289–324.

KUMMER, H. 1968. Social Organization of Hamadyras Baboons. Univ. Chicago Press, Chicago.

KUMMER, H. 1971. Primate Societies. Aldine Atherton, Chicago.

KUNO, E. 1987. Principles of predator-prey interaction in theoretical, experimental, and natural population systems. Adv. Ecol. Res. 16: 249–337.

KURIS, A. M., A. R. BLAUSTEIN, & J. J. ALIO. 1980. Hosts as islands. Amer. Nat. 116: 570–586.

KURTÉN, B. 1959. Rates of evolution in fossil mammals. Cold Spring Harbor Symp. Quant. Biol. 24: 205–215.

LACK, D. 1944. Ecological aspects of species formation in passerine birds. Ibis 86: 260–286.

LACK D. 1947. Darwin's Finches. Cambridge Univ. Press, Cambridge.

LACK, D. 1947. The significance of clutch-size, Parts 1 and 2. Ibis 89: 302–352.

LACK, D. 1954. The Natural Regulation of Animal Numbers. Oxford Univ. Press, London.

LACK, D. 1966. Population Studies of Birds. Clarendon Press, Oxford.

LACK, D. 1968. Ecological Adaptations for Breeding in Birds. Methuen, London.

LACK, D. 1971. Ecological Isolation in Birds. Harvard Univ. Press, Cambridge, Massachusetts.

LACY, R. C., & P. W. SHERMAN. 1983. Kin recognition by phenotype matching. Amer. Nat. 121: 489–512.

LAGLER, K. F., J. E. BARDACH, & R. R. MILLER. 1962. Ichthyology. Wiley, New York.

LAMB, M. J. 1977. Biology of Ageing. Blackie, Glasgow and London.

LAMB, R. J., & P. J. POINTING. 1972. Sexual morph determination in the aphid, *Acyrthosiphon pisum*. J. Insect Physiol. 18: 2029–2042.

LAMONT, B. 1983. The Biology of Mistletoes. Academic Press, Sydney, Australia.

LAMPERT, W., W. FLECKNER, H. RAI, & B. E. TAYLOR. 1986. Phytoplankton control by grazing zooplankton: a study on the spring clear-water phase. Limnol. Oceanogr. 31: 478–490.

LANDE, R. 1979. Quantitative genetic analysis of multivariate evolution, applied to brain : body size allometry. Evolution 33: 402–416.

LANDE, R. 1980. Sexual dimorphism, sexual selection, and adaptation in polygenic characters. Evolution 34: 292–305.

LANDE, R. 1981. Models of speciation by sexual selection of polygenic traits. Proc. Natl. Acad. Sci. 78: 3721–3725.

LANDE, R., & S. J. ARNOLD. 1983. Measuring selection on correlated characters. Evolution 37: 1210–1226.

LARCHER, W. 1980. Physiological Plant Ecology (2nd ed.). Springer-Verlag, New York.

LARCHER, W., A. CERNUSCA, L. SCHMIDT, G. GRABHERR, E. NÖTZEL, & M. SMEETS. 1975. Mt. Patscherkofel, Austria. Pp. 125–139 in T. Rosswall and O. W. Heal (Eds.), Structure and Function of Tundra Ecosystems. Swedish Natural Science Research Council, Stockholm, Ecol. Bull. 20.

LARSEN, D. P., D. W. SCHULTS, & K. W. MALUEG. 1981. Summer internal phosphorus supplies in Shagawa Lake, Minnesota. Limnol. Oceanogr. 26: 740–753.

LATHAM, M. C. 1975. Nutrition and infection in national development. Science 188: 561–565.

LAUGHLIN, R. 1965. Capacity for increase; a useful population statistic. J. Anim. Ecol. 34: 77–91.

LAW, R. 1975. Colonisation and the evolution of life histories in *Poa annua.* Ph.D. thesis, Univ. Liverpool (summarized in Begon and Mortimer, 1981).

LAW, R., & A. R. WATKINSON. 1987. Response-surface analysis of two-species competition. An experiment on *Phleum arenarium* and *Vulpia fasciculata.* J. Ecol. 75: 871–886.

LAWLER, G. H. 1965. Fluctuations in the success of year-classes of whitefish populations with special reference to Lake Erie. J. Fish. Res. Bd. Canad. 22: 1197–1227.

LAWLOR, L. R. 1978. A comment on randomly constructed ecosystem models. Amer. Nat. 112: 445–447.

LAWLOR, L. R. 1979. Direct and indirect effects of *n*-species competition. Oecologia 43: 355–364.

LAWLOR, L. R., & J. MAYNARD SMITH. 1976. The coevolution and stability of competing species. Amer. Nat. 110: 79–99.

LAWS, R. M. 1962. Some effects of whaling on the southern stocks of baleen whales. Pp. 137–158 in E. D. LeCren and M. W. Holdgate (Eds.), The Exploitation of Natural Animal Populations. Wiley, New York.

LAWS, R. M. 1977. The significance of vertebrates in the Antarctic marine ecosystem. Pp. 411–438 in G. A. Llano (Ed.), Adaptations within Antarctic Ecosystems. Smithson. Inst., Washington, D.C.

LAWTON, J. H., J. R. BEDDINGTON, & R. BONSER. 1974. Switching in invertebrate predators. Pp. 141–158 in M. B. Usher and M. H. Williamson (Eds.), Ecological Stability. Chapman and Hall, London.

LAWTON, J. H., H. CORNELL, W. DRITSCHILO, & S. D. HENDRIX. 1981. Species as islands: comments on a paper by Kuris *et al.* Amer. Nat. 117: 623–627.

LAWTON, J. H., & D. R. STRONG, JR. 1981. Community patterns and competition in folivorous insects. Amer. Nat. 118: 317–338.

LEAK, W. B. 1970. Successional change in northern hardwoods predicted by birth and death simulation. Ecology 51: 794–801.

LEAMY, L. 1977. Genetic and environmental correlations of morphometric traits in randombred house mice. Evolution 31: 357–369.

LEDERER, R. 1984. A view of avian ecomorphological hypotheses. Okol. Vogel 6: 119–126.

LEE, R. E., JR., C. CHEN, & D. L. DENLINGER. 1987. A rapid cold-hardening process in insects. Science 238: 1415–1417.

LEES, A. D. 1966. The control of polymorphism in aphids. Adv. Insect Physiol. 3: 207–277.

LE GALL, J., & J. R. POSTGATE. 1973. The physiology of sulphate-reducing bacteria. Adv. Microbial Physiol. 10: 81–128.

LEGGETT, W. C., & J. E. CARSCADDEN. 1978. Latitudinal variation in reproductive characteristics of American shad *(Alosa sapidissima).* Evidence for population specific life history strategies in fish. J. Fish. Res. Bd. Canad. 35: 1469–1478.

LEHMAN, J. T. 1980. Release and cycling of nutrients between planktonic algae and herbivores. Limnol. Oceanogr. 25: 620–632.

LEHMAN, J. T., AND D. SCAVIA. 1982. Microscale patchiness of nutrients in plankton communities. Science 216: 729–730.

LEIGH, E. G., JR. 1982. Introduction: why are there so many kinds of tropical trees? Pp. 63–66 in E. G. Leigh, Jr., A. S. Rand, and D. M. Windsor (Eds.), The Ecology of a Tropical Forest. Seasonal Rhythms and Long-term Changes. Smithson. Inst. Press, Washington, D.C.

LEISLER, B., & H. WINKLER. 1983. Ecomorphology. Current Ornithol. 2: 133–186.

LENINGTON, S. 1980. Female choice and polygyny in red-winged blackbirds. Anim. Behav. 28: 347–361.

LENSKI, R. E. 1984. Food limitation and competition: a field experiment with two *Carabus* species. J. Anim. Ecol. 53: 203–216.

LEON, J. A. 1974. Selection in contexts of interspecific competition. Amer. Nat. 108: 739–757.

LERNER, I. M., & E.M. DEMPSTER. 1962. Indeterminism in interspecific competition. Proc. Natl. Acad. Sci. 48: 821–826.

LERNER, I. M., & F. K. HO. 1961. Genotype and competitive ability of *Tribolium* species. Amer. Nat. 95: 329–343.

LESLIE, P. H. 1945. On the use of matrices in certain population mathematics. Biometrika 33: 183–212.

LESLIE, P. H. 1948. Some further notes on the use of matrices in population analysis. Biometrika 35: 213–245.

LESLIE, P. H. 1959. The properties of a certain lag type of population growth and the influence of an external random factor on a number of such populations. Physiol. Zool. 32: 151–159.

LESLIE, P. H., & T. PARK. 1949. The intrinsic rate of natural increase of *Tribolium castaneum* Herbst. Ecology 30: 469–477.

LESLIE, P. H., & R. M. RANSON. 1940. The mortality, fertility and rate of natural increase of the vole *(Microtus agrestis)* as observed in the laboratory. J. Anim. Ecol. 9: 27–52.

LESSELLS, C. M. 1986. Brood size in Canada Geese: a manipulation experiment. J. Anim. Ecol. 55: 669–689.

LESTER, L. J., & R. K. SELANDER. 1981. Genetic relatedness and the social organization of *Polistes* colonies. Amer. Nat. 117: 147–166.

LEVIN, B. R. 1971. A model for selection in situations of interspecific competition. Evolution 25: 249–264.

LEVIN, D. A. 1976. The chemical defenses of plants to pathogens and herbivores. Ann. Rev. Ecol. Syst. 7: 121–159.

LEVIN, D. A. 1984. Inbreeding depression and proximity-related crossing success in *Phlox drummondi.* Evolution 38: 116–127.

LEVINS, S. A. 1976. Population dynamic models in heterogeneous environments. Ann. Rev. Ecol. Syst. 7: 287–310.

LEVIN, S. A., & R. T. PAINE. 1974. Disturbance, patch formation, and community structure. Proc. Natl. Acad. Sci. 71: 2744–2747.

LEVINE, S. H. 1976. Competitive interactions in ecosystems. Amer. Nat. 110: 903–910.

LEVINS, R. 1962. Theory of fitness in a heterogeneous environment. I. The fitness set and adaptive function. Amer. Nat. 96: 361–373.

LEVINS, R. 1968. Evolution in Changing Environments. Some Theoretical Explorations. Princeton Univ. Press, Princeton, New Jersey.

LEVINS, R., & D. CULVER. 1971. Regional coexistence of species and competition between rare species. Proc. Natl. Acad. Sci. 68: 1246–1248.

LEWERT, R. M. 1970. Schistosomes. Pp. 981–1008 in G. J. Jackson, R. Herman, and I. Singer (Eds.), Immunity to Parasitic Animals, Vol. 2. Appleton-Century-Crofts, New York.

LEWIN, R. 1986. Damage to tropical forests, or Why were there so many kinds of animals? Science 234: 149–150.

LEWIS, E. G. 1942. On the generation and growth of a population. Sankhya 6: 93–96.

LEWONTIN, R. C. 1958. Studies on heterozygosity and homeostasis, II. Evolution 12: 494–503.

LEWONTIN, R. C. 1974. The Genetic Basis of Evolutionary Change. Columbia Univ. Press, New York.

LI, H.-L. 1952. Floristic relationships between eastern Asia and eastern North America. Trans. Amer. Phil. Soc., New Ser. 42: 371–429.

LIAO, C. F.-H., & D. R. S. LEAN. 1978. Nitrogen transformations within the trophogenic zone of lakes. J. Fish. Res. Board Canad. 35: 1102–1108.

LIDICKER, W. Z. 1975. The role of dispersal in the demography of small mammals. Pp. 103–128 in F. B. Golley, K. Petrusewicz, and L. Ryszkowski (Eds.), Small Mammals: Their Productivity and Population Dynamics. Cambridge Univ. Press, Cambridge.

LIEBIG, J. 1840. Chemistry in Its Application to Agriculture and Physiology. Taylor and Walton, London.

LIETH, H. 1973. Primary production: terrestrial ecosystems. Human Ecol. 1: 303–332.

LIGHT, S. F. 1942–43. The determination of castes in social insects. Quart. Rev. Biol. 17: 312–326; 18: 46–63.

LIKENS, G. E. 1972. Eutrophication and aquatic ecosystems. Amer. Soc. Limnol. Oceanogr., Spec. Symp. 1: 3–13.

LIKENS, G. E., F. H. BORMANN, N. M. JOHNSON, & R. S. PIERCE. 1967. The calcium, magnesium, potassium, and sodium budgets for a small forested ecosystem. Ecology 48: 772–785.

LIKENS, G. E., F. H. BORMANN, R. S. PIERCE, J. S. EATON, & N. M. JOHNSON. 1977. Biogeochemistry of a Forested Ecosystem. Springer-Verlag, New York.

LILL, A. 1974. Social organization and space utilization in the lek-forming white-bearded manakin, *M. manacus trinitatis* Hartert. Z. Tierpsych. 36: 513–530.

LIN, N., & C. D. MICHENER. 1972. Evolution of sociality in insects. Quart. Rev. Biol. 47: 131–159.

LINCOLN, R. J., G. A. BOXSHALL, & P. F. CLARK. 1982. A Dictionary of Ecology, Evolution and Systematics. Cambridge Univ. Press, Cambridge.

LINDEMAN, R. 1942. The trophic-dynamic aspect of ecology. Ecology 23: 399–418.

LINDROTH, R. L., & G. O. BATZIL. 1986. Inducible plant chemical defences: a cause of vole population cycles? J. Anim. Ecol. 55: 431–449.

LINDSAY, W. L., & E. C. MORENO. 1960. Phosphate equilibria in soils. Soil Sci. Soc. Amer. Proc. 24: 177–182.

LINSLEY, E. G. 1961. Bering Arc relationships of Cerambycidae and their host plants. Pp. 159–178 in J. L. Gressitt (Ed.), Pacific Basin Biogeography. Bishop Mus. Press, Honolulu.

LIPPE, E., J. T. DE SMIDT, & D. C. GLENN-LEWIN. 1985. Markov models and succession: a test from a heathland in the Netherlands. J. Ecol. 73: 775–791.

LIST, R. J. 1966. Smithsonian Meteorological Tables (6th rev. ed.). Smithson. Misc. Coll. 114: 1–527.

LLOYD, D. G. 1979. Some reproductive factors affecting the selection of self-fertilization in plants. Amer. Nat. 113: 67–69.

LLOYD, D. G. 1980. Demographic factors and mating patterns in Angiosperms. Pp. 67–88 in O. T. Solbrig (Ed.), Demography and Evolution of Plant Populations. Blackwell, Oxford.

LLOYD, P. J. 1967. American, German, and British antecedents to Pearl and Reed's logistic curve. Popul. Stud. 21: 99–108.

LOACH, K. 1967. Shade tolerance in tree seedlings. I. Leaf photosynthesis and respiration in plants raised under artificial shade. New Phytologist 66: 607–621.

LOEHLE, C., & J. H. K. PECHMANN. 1988. Evolution: the missing ingredient in systems ecology. Amer. Nat. 132: 884–899.

LOFTS, B. 1970. Animal Photoperiodism. Arnold, London.

LONG, J. L. 1981. Introduced Birds of the World. Universe Books, New York.

LOTKA, A. J. 1907. Relation between birth rates and death rates. Science 26: 21–22.

LOTKA, A. J. 1922. The stability of the normal age distribution. Proc. Natl. Acad. Sci. 8: 339–345.

LOTKA, A. J. 1925. Elements of Physical Biology. Williams and Wilkins, Baltimore.

LOTKA, A. J. 1932. The growth of mixed populations: two species competing for a common food supply. J. Wash. Acad. Sci. 22: 461–469.

LOUCKS, O. L. 1962. Ordinating forest communities by means of environmental scalars and phytosociological indices. Ecol. Monogr. 32: 137–166.

LOUCKS, O. L. 1977. Emergence of research on agro-ecosystems. Ann. Rev. Ecol. Syst. 8: 173–192.

LOUDA, S. M. 1984. Herbivore effect on stature, fruiting, and leaf dynamics of a native crucifer. Ecology 65: 1379–1386.

LOUW, G. N., & M. K. SEELY. 1982. Ecology of Desert Organisms. Wiley, New York.

LOWE, C. H., P. J. LARDNER, & E. A. HALPERN. 1971. Supercooling in reptiles and other vertebrates. Comp. Biochem. Physiol. 39A: 125–135.

LOWRY, W. P. 1969. Weather and Life. Academic Press, New York.

LUBCHENCO, J. 1978. Plant species diversity in a marine intertidal community: importance of herbivore food preference and algal competitive abilities. Amer. Nat. 112: 23–39.

LUCK, R. F. 1971. An appraisal of two methods of analyzing insect life tables. Canad. Entomol. 103: 1261–1271.

LUCKINBILL, L. S. 1978. r and K selection in experimental populations of *Escherichia coli*. Science 202: 1201–1203.

LUCKINBILL, L. S. 1979. Selection and the r/K continuum in experimental populations of protozoa. Amer. Nat. 113: 427–437.

LUCKINBILL, L. S. 1984. An experimental analysis of a life history theory. Ecology 65: 1170–1184.

LUSE, R. A. 1970. The phosphorus cycle in a tropical rain forest. Pp. H161–H166 in H. T. Odum and R. F. Pigeon (Eds.), A Tropical Rain Forest. U.S. Atomic Energy Commission, Washington, D.C.

LYMAN, C., ET AL. 1982. Hibernation and Torpor in Mammals and Birds. Academic Press, New York.

LYNCH, J. F., & N. K. JOHNSON. 1974. Turnover and equilibria in insular avifaunas, with special reference to the California Channel Islands. Condor 76: 370–384.

LYTHGOE, J. N. 1979. The Ecology of Vision. Clarendon Press, New York.

MACARTHUR, R. H. 1955. Fluctuations of animal populations and a measure of community stability. Ecology 36: 533–536.

MACARTHUR, R. H. 1957. On the relative abundance of bird species. Proc. Natl. Acad. Sci. 43: 293–295.

MACARTHUR, R. H. 1960. On the relative abundance of species. Amer. Nat. 94: 25–36.

MACARTHUR, R. H. 1960. On the relation between reproductive value and optimal predation. Proc. Natl. Acad. Sci. 46: 144–145.

MACARTHUR, R. H. 1961. Population effects of natural selection. Amer. Nat. 95: 195–199.

MACARTHUR, R. H. 1962. Some generalized theorems of natural selection. Proc. Natl. Acad. Sci. 48: 1893–1897.

MACARTHUR, R. H. 1965. Patterns of species diversity. Biol. Rev. 40: 510–533.

MACARTHUR, R. H. 1969. Patterns of communities in the tropics. Biol. J. Linn. Soc. 1: 19–30.

MACARTHUR, R. H. 1972. Geographical Ecology: Patterns in the Distribution of Species. Harper & Row, New York.

MACARTHUR, R. H., J. M. DIAMOND, & J. R. KARR. 1972. Density compensation in island faunas. Ecology 53: 330–342.

MACARTHUR, R., & R. LEVINS. 1964. Competition, habitat selection, and character displacement in a patchy environment. Proc. Natl. Acad. Sci. 51: 1207–1210.

MACARTHUR, R. H., & R. LEVINS. 1967. The limiting similarity, convergence, and divergence of coexisting species. Amer. Nat. 101: 377–385.

MACARTHUR, R. H., & J. MACARTHUR. 1961. On bird species diversity. Ecology 42: 594–598.

MACARTHUR, R. H., & E. R. PIANKA. 1966. On optimal use of a patchy environment. Amer. Nat. 100: 603–609.

MACARTHUR, R. H., H. RECHER, & M. CODY. 1966. On the relation between habitat selection and species diversity. Amer. Nat. 100: 319–332.

MACARTHUR, R. H., & E. O. WILSON. 1963. An equilibrium theory of insular zoogeography. Evolution 17: 373–387.

MACARTHUR, R. H., & E. O. WILSON. 1967. The Theory of Island Biogeography. Princeton Univ. Press, Princeton, New Jersey.

MACDONALD, D. W. 1980. Rabies and Wildlife: A Biologist's Perspective. Oxford Univ. Press, Oxford.

MACHIN, K. E., & H. W. LISSMANN. 1960. The mode of operation of the electric receptors in *Gymnarchus niloticus.* J. Exp. Biol. 37: 801–811.

MACINTOSH, N. J. 1974. The Psychology of Animal Learning. Academic Press, New York.

MACLULICH, D. A. 1937. Fluctuations in the numbers of the varying hare *(Lepus americanus).* Univ. Toronto Studies, Biol. Ser. No. 43.

MACMILLEN, R. E., & A. K. LEE. 1967. Australian desert mice: independence of exogenous water. Science 158: 383–385.

MACNAIR, M. R., & G. A. PARKER. 1978. Models of parent-offspring conflict. II. Promiscuity. Anim. Behav. 26: 111–122.

MACNAIR, M. R., & G. A. PARKER. 1979. Models of parent-offspring conflict. III. Intra-brood conflict. Anim. Behav. 27: 1202–1209.

MADDOCK, L. 1979. The "migration" and grazing succession. Pp. 104–129 in A. R. E. Sinclair and M. Norton-Griffiths (Eds.), Serengeti. Dynamics of an Ecosystem. Univ. Chicago Press, Chicago.

MAGUIRE, B. 1963. The passive dispersal of small aquatic organisms and their colonization of isolated bodies of water. Ecol. Monogr. 33: 161–185.

MAHER, W. J. 1970. The pomarine jaeger as a brown lemming predator in northern Alaska. Wilson Bull. 82: 130–157.

MALDONADO, J. F. 1967. Schistosomiasis in America. Editorial Cientifico-Medica, Barcelona.

MALTBY, L., & P. CALOW. 1986. Intraspecific life-history variation in *Erpobdella octoculata* (Hirundinea: Erpobdellidae). II. Testing theory on the evolution of semelparity and iteroparity. J. Anim. Ecol. 55: 739–750.

MALTHUS, R. T. 1798. An Essay on the Principle of Population as it Affects the Future Improvement of Society. Johnson, London.

MALY, E. J. 1969. A laboratory study of the interaction between the predatory rotifer *Asplanchna* and *Paramecium.* Ecology 50: 59–73.

MANN, K. H. 1973. Seaweeds: their productivity and strategy for growth. Science 182: 975–981.

MANN, K. H. 1981. Ecology of Coastal Waters: A Systems Approach. Univ. California Press, Berkeley.

MARES. M. A. 1976. Convergent evolution of desert rodents: multivariate analysis and zoogeographic implications. Paleobiol. 2: 39–63.

MARGALEF, D. R. 1958. Information theory in ecology. General Systems 3: 36–71.

MARGALEF, R. 1963. On certain unifying principles in ecology. Amer. Nat. 92: 357–374.

MARGALEF, R. 1968. Perspectives in Ecological Theory. Univ. Chicago Press, Chicago.

MARKS, G. C., & T. T. KOZLOWSKI. 1973. Ectomycorrhizae. Academic Press, New York.

MARKS, H. L. 1978. Long term selection for four-week body weight in Japanese quail under different nutritional environments. Theor. Appl. Genet. 52: 105–111.

MARQUIS, R. J. 1984. Leaf herbivores decrease fitness of a tropical plant. Science 226: 537–539.

MARSHALL, A. J., & H. S. DE S. DISNEY. 1957. Experimental induction of the breeding seasons in a xerophilous bird. Nature 180: 647.

MARSHALL, D. R., & S. K. JAIN. 1969. Interference in pure and mixed populations of *Avena fatua* and *A. barbata.* J. Ecol. 57: 251–270.

MARSHALL, J. S. 1962. The effects of continuous gamma radiation on the intrinsic rate of natural increase of *Daphnia pulex.* Ecology 43: 598–607.

MARTIN, M. M., & J. HARDING. 1981. Evidence for the evolution of competition between two species of annual plants. Evolution 35: 975–987.

MARTIN, M. M., & J. S. MARTIN.. 1984. Surfactants: their role in preventing the precipitation of proteins by tannins in insect guts. Oecologia 61: 342–345.

MARTIN, M. M., D. C. ROCKHOLM, & J. S. MARTIN. 1985. Effects of surfactants, pH, and certain cations on the precipitation of proteins by tannins. J. Chem. Ecol. 11: 485–494.

MARTIN, T. E. 1988. On the advantage of being different: nest predation and the coexistence of bird species. Proc. Natl. Acad. Sci. 85: 2196–2199.

MASON, J. R., & R. F. REIDINGER. 1981. Effects of social facilitation and observational learning on feeding behavior of the red-winged blackbird *(Agelaius phoeniceus)*. Auk 98: 778–784.

MASSEY, A. B. 1925. Antagonism of the walnuts (*Juglans nigra* L. and *J. cinerea* L.) in certain plant associations. Phytopathology 15: 773–784.

MATHER, M., & B. D. ROITBERG. 1987. A sheep in wolf's clothing: tephritid flies mimic spider predators. Science 236: 308–310.

MATTINGLY, P. F. 1969. The Biology of Mosquito-Borne Disease. American Elsevier, New York.

MAY, J. M. 1961. Studies in Disease Ecology. Hafner, New York.

MAY, R. M. 1972. Will a large complex system be stable? Nature 238: 413–414.

MAY, R. M. 1972. Limit cycles in predator-prey communities. Science 177: 900–902.

MAY, R. M. 1973. Qualitative stability in model ecosystems. Ecology 54: 638–641.

MAY, R. M. 1973. On relationships among various types of population models. Amer. Nat. 107: 46–57.

MAY, R. M. 1973. Stability and Complexity in Model Ecosystems. Princeton Univ. Press, Princeton, New Jersey.

MAY, R. M. 1974. Biological populations with nonoverlapping generations: stable points, stable cycles, and chaos. Science 186: 645–647.

MAY, R. M. 1975. Stability and Complexity in Model Ecosystems (2nd ed.). Princeton Univ. Press, Princeton, New Jersey.

MAY, R. M. 1975. Patterns of species abundance and diversity. Pp. 81–120 in M. L. Cody and J. M. Diamond (Eds.), Ecology and Evolution of Communities. Harvard Univ. Press, Cambridge, Massachusetts.

MAY, R. M. 1976. Simple mathematical models with very complicated dynamics. Nature 261: 459–467.

MAY, R. M., & R. M. ANDERSON. 1979. Population biology of infectious diseases: Part II. Nature 280: 455–461.

MAY, R. M., J. R. BEDDINGTON, C. W. CLARK, S. J. HOLT, & R. M. LAWS. 1979. Management of multispecies fisheries. Science 205: 267–277.

MAY, R. M., & R. H. MACARTHUR. 1972. Niche overlap as a function of environmental variability. Proc. Natl. Acad. Sci. 69: 1109–1113.

MAY, R. M., & G. F. OSTER. 1976. Bifurcations and dynamic complexity in simple ecological models. Amer. Nat. 110: 573–599.

MAYFIELD, H. 1960. The Kirtland's Warbler. Cranbrook Inst. Science, Bloomfield Hills, Michigan.

MAYR, E., E. G. LINSLEY, & R. L. USINGER. 1953. Methods and Principles of Systematic Zoology. McGraw-Hill, New York.

MCCARTHY, J. J., & J. C. GOLDMAN. 1979. Nitrogenous nutrition of marine phytoplankton in nutrient-depleted waters. Science 203: 670–672.

MCCLEAVE, J. D., ET AL. (Eds.). 1984. Mechanisms of Migration in Fishes. Plenum, New York.

MCCLELLAND, W. J. 1965. The production of cercariae by *S. mansoni* and *S. haematobium* and methods for estimating the numbers of cercariae in suspension. Bull. World Health Org. 33: 270–275.

MCCLURE, P. A. 1981. Sex-biased litter reduction in food-restricted wood rats *(Neotoma floridana)*. Science 211: 1058–1060.

MCCORMICK, J. 1970. The Pine Barrens: A Preliminary Ecological Inventory. New Jersey State Museum, Trenton.

MCGINNIS, S. M., & L. L. DICKSON. 1967. Thermoregulation in the desert iguana *Dipsosaurus dorsalis*. Science 156: 1757–1759.

MCGOWAN, J. A. 1984. The California El Niño, 1983. Oceanus 27: 48–51.

MCHARGUE, J. S., & W. R. ROY. 1932. Mineral and nitrogen content of the leaves of some forest trees at different times in the growing season. Bot. Gaz. 94: 381–393.

MCINTOSH, R. P. 1967. The continuum concept of vegetation. Bot. Rev. 33: 130–187.

MCINTOSH, R. P. 1974. Plant ecology 1947–1972. Ann. Missouri Bot. Gard. 61: 132–165.

MCINTOSH, R. P. 1985. The Background of Ecology. Concept and Theory. Cambridge Univ. Press, Cambridge and New York.

MCLACHLAN, A. 1986. Chironomid wing length: a measure of habitat duration and predictability? A reply to Vepsalainen. Oikos 46: 271–273.

MCLUSKY, D. S. 1981. The Estuarine Ecosystem. Wiley, New York.

MCMAHON, T. A. 1984. Muscles, Reflexes, and Locomotion. Princeton Univ. Press, Princeton, New Jersey.

MCMASTER, G. S., W. M. JOW, & J. KUMMEROW. 1982. Response of *Adenostoma fasciculatum* and *Ceanothus greggii* chaparral to nutrient additions. J. Ecology 70: 745–756.

MCMILLAN, C. 1956. Edaphic restriction of *Cupressus* and *Pinus* in the coast ranges of central California. Ecol. Monogr. 26: 177–212.

MCMILLAN, C. 1959. The role of ecotypic variation in the distribution of the central grassland of North America. Ecol. Monogr. 29: 285–308.

MCNAB, B. K. 1966. An analysis of the body temperatures of birds. Condor 68: 47–55.

MCNAIR, J. N. 1981. A stochastic foraging model with predator training effects. II. Optimal diets. Theoret. Pop. Biol. 19: 147–162.

MCNAIR, J. N. 1982. Optimal giving up times and the marginal value theorem. Amer. Nat. 119: 511–529.

MCNAUGHTON, S. J. 1973. Comparative photosynthesis of Quebec and California ecotypes of *Typha latifolia.* Ecology 54: 1260–1270.

MCNAUGHTON, S. J. 1976. Serengeti migratory wildebeest: facilitation of energy flow by grazing. Science 191: 92–94.

MCNAUGHTON, S. J. 1977. Diversity and stability of ecological communities: a comment on the role of empiricisms in ecology. Amer. Nat. 111: 515–525.

MCNAUGHTON, S. J. 1979. Grassland-herbivore dynamics. Pp. 46–81 in A. R. E. Sinclair and M. Norton-Griffiths (Eds.), Serengeti. Dynamics of an Ecosystem. Univ. Chicago Press, Chicago.

MCNAUGHTON, S. J. 1985. Ecology of a grazing ecosystem: the Serengeti. Ecol. Monogr. 55: 259–294.

MCNAUGHTON, S. J., & L. L. WOLF. 1970. Dominance and the niche in ecological systems. Science 167: 131–139.

MCPHEE, J. 1968. The Pine Barrens. Farrar, Straus & Giroux, New York.

MCVAY, S. 1966. The last of the great whales. Sci. Amer. 215: 13–21.

MECH, L. D. 1966. The Wolves of Isle Royale. U.S. Natl. Park Serv., Fauna Ser. No. 7.

MECH, L. D. 1970. The Wolf: The Ecology and Behavior of an Endangered Species. Natural History Press, New York.

MEDAWAR, P. B. 1957. The Uniqueness of the Individual. Methuen, London.

MEENTEMEYER, V., E. O. BOX, & R. THOMPSON. 1982. World patterns and amounts of terrestrial plant litter production. BioScience 32: 125–129.

MEIDNER, H., & D. W. SHERIFF. 1976. Water and Plants. Wiley, New York.

MENKEN, J., J. TRUSSELL, & U. LARSEN. 1986. Age and infertility. Science 233: 1389–1394.

MEREDITH, D. H. 1977. Interspecific agonism in two parapatric species of chipmunks *(Eutamias).* Ecology 58: 423–430.

MERRIAM, C. H. 1894. Laws of temperature control of the geographic distribution of terrestrial animals and plants. Nat. Geogr. Mag. 6: 229–238.

MICHENER, C. D. 1958. The evolution of social behavior in bees. Proc. 10th Intern. Congr. Entomol. (1956) 2: 441–448.

MICHENER, C. D. 1969. Comparative social behavior of bees. Ann. Rev. Entomol. 14: 299–342.

MICHENER, C. D., & D. J. BROTHERS. 1974. Were workers of eusocial Hymenoptera initially altruistic or oppressed? Proc. Natl. Acad. Sci. 71: 671–674.

MICHOD, R. E., & B. R. LEVIN. (Eds.). 1988. The Evolution of Sex. Sinauer, Sunderland, Massachusetts.

MILBANK, J. W., & K. A. KERSHAW. 1969. Nitrogen metabolism in lichens, I: Nitrogen fixation in cephalodia of *Peltigera aphthosa.* New Phytol. 68: 721–729.

MILDVAN, A. S., & B. L. STREHLER. 1960. A critique of theories of mortality. Pp. 216–235 in B. L. Strehler (Ed.), The Biology of Aging. Amer. Inst. Biol. Sci., Washington, D.C.

MILES, D. B., & R. E. RICKLEFS. 1984. The correlation between ecology and morphology in deciduous forest passerine birds. Ecology 65: 1629–1640.

MILEWSKI, A. V. 1981. A comparison of reptile communities in relation to soil fertility in the Mediterranean and adjacent arid parts of Australia and southern Africa. J. Biogeogr. 8: 493–503.

MILEWSKI, A. V. 1982. The occurrence of seeds and fruits taken by ants versus birds in Mediterranean Australia and southern Africa, in relation to the availability of soil potassium. J. Biogeogr. 9: 505–516.

MILEWSKI, A. V. 1983. A comparison of ecosystems in Mediterranean Australia and southern Africa: nutrient-poor sites at the Barrens and the Caledon Coast. Ann. Rev. Ecol. Syst. 14: 57–76.

MILHAM, S., JR., & A. M. GITTELSOHN. 1965. Parental age and malformations. Human Biol. 3: 13–22.

MILINSKI, M. 1979. An evolutionarily stable feeding strategy in sticklebacks. Z. Tierpsychol. 51: 36–40.

MILKMAN, R. 1982. Perspectives on Evolution. Sinauer, Sunderland, Massachusetts.

MILLER, G. R., A. WATSON, & D. JENKINS. 1970. Responses of red grouse populations to experimental improvement of their food. Symp. Brit. Ecol. Soc. 10: 323–334.

MILLER, P. C., W. A. STONER, & L. L. TIESZEN. 1976. A model

of stand photosynthesis for the wet meadow tundra at Barrow, Alaska. Ecology 57: 411–430.

MILLER, R. S. 1967. Pattern and process in competition. Adv. Ecol. Res. 4: 1–74.

MILNER, C., & R. E. HUGHES. 1968. Methods for the Measurement of the Primary Production of Grassland. Blackwell, Oxford.

MINDERMAN, G. 1968. Addition, decomposition and accumulation of organic matter in forests. J. Ecol. 56: 355–362.

MINSHALL, G. W. 1967. Role of allochthonous detritus in the trophic structure of a woodland springbrook community. Ecology 48: 139–149.

MINSHALL, G. W. 1978. Autotrophy in stream ecosystems. BioScience 28: 767–771.

M'KENDRICK, A. G., & M. K. PAI. 1911. The rate of multiplication of microorganisms: A mathematical study. Proc. Roy. Soc. Edinb. 31: 649–655.

MOAT, A. G. 1979. Microbial Physiology. Wiley, New York.

MODE, C. J. 1958. A mathematical model for the co-evolution of obligate parasites and their hosts. Evolution 12: 158–165.

MOHR, C. O. 1943. Cattle droppings as ecological units. Ecol. Monogr. 13: 275–298.

MOLLER, C. M., D. MULLER, & J. NIELSON. 1954. Graphic presentation of dry matter production of European beech. Forst. Fors Vaes. Danm. 21: 327–335.

MOMMSEN, T. P., & P. J. WALSH. 1989. Evolution of urea synthesis in vertebrates: the piscine connection. Science 243: 72–75.

MONK, C. 1966. Ecological importance of root/shoot ratios. Bull. Torrey Bot. Club 93: 402–406.

MONK, C. D., & F. C. GABRIELSON, JR. 1985. Effects of shade, litter and root competition on old-field vegetation in South Carolina. Bull. Torr. Bot. Club 112: 383–392.

MONOD, J. 1950. La technique de culture continue; théorie et applications. Ann. Inst. Pasteur 79: 390–410.

MOOK, J. H., L. J. MOOK, & H. S. HEIKENS. 1960. Further evidence for the role of "searching images" in the hunting behavior of titmice. Arch. Neerl. Zool. 13: 448–465.

MOONEY, H. A. (Ed.). 1977. Convergent Evolution in Chile and California. Dowden, Hutchinson & Ross, Stroudsburg, Pennsylvania.

MOONEY, H. A., & W. D. BILLINGS. 1961. Comparative physiological ecology of arctic and alpine populations of *Oxyria digyna.* Ecol. Monogr. 31: 1–29.

MOONEY, H. A., T. M. BONNICKSEN, N. L. CHRISTENSEN, J. E. LOTAN, & W. A. REINERS. (Eds.). 1981. Fire Regimes and Ecosystem Properties. Gen. Tech. Rep., U.S. Forest Ser., Washington, D.C.

MOONEY, H. A., & E. L. DUNN. 1970. Convergent evolution of Mediterranean-climate evergreen schlerophyll shrubs. Evolution 24: 292–303.

MOONEY, H. A., & E. L. DUNN. 1970. Photosynthetic systems of Mediterranean-climate shrubs and trees of California and Chile. Amer. Nat. 104: 447–453.

MOORE, J. A. 1952. Competition between *Drosophila melanogaster* and *Drosophila simulans,* II: The improvement of competitive ability through selection. Proc. Natl. Acad. Sci. 38: 813–817.

MORAN, N., & W. D. HAMILTON. 1980. Low nutritive quality as defense against herbivores. J. Theoret. Biol. 86: 247–254.

MOREAU, R. E. 1944. Clutch-size: a comparative study, with special reference to African birds. Ibis 86: 286–347.

MORIN, P. J. 1981. Predatory salamanders reverse the outcome of competition among three species of anuran tadpoles. Science 212: 1284–1286.

MORIN, P. J. 1983. Predation, competition, and the composition of larval anuran guilds. Ecol. Monogr. 53: 119–138.

MORIN, P. J., & E. A. JOHNSON. 1988. Experimental studies of asymmetric competition among anurans. Oikos 53: 398–407.

MORIN, P. J., H. M. WILBUR, & R. N. HARRIS. 1983. Salamander predation and the structure of experimental communities: responses of *Notophthalmus* and microcrustacea. Ecology 64: 1430–1436.

MORRIS, R. F. 1959. Single factor analysis in population dynamics. Ecology 40: 580–588.

MORRIS, R. F. 1963. The development of predictive equations for the spruce budworm based on key factor analysis. Mem. Entomol. Soc. Canad. 32: 16–22.

MORRIS, R. F., W. F. CHESIRE, C. A. MILLER, & D. G. MOTT. 1958. Numerical responses of avian and mammalian predators during a gradation of the spruce budworm. Ecology 39: 487–494.

MORROW, P. A. 1977. The significance of phytophagous insects in the *Eucalyptus* forests in Australia. Pp. 19–29 in W. Mattson (Ed.), The Role of Arthropods in Forest Ecosystems. Springer-Verlag, Berlin.

MORROW, P. A., & V. C. LAMARCHE, JR. 1978. Tree ring evidence for chronic insect suppression of productivity in subalpine *Eucalyptus.* Science 201: 1244–1246.

MORSE, D. H. 1980. Behavioral Mechanisms in Ecology. Harvard Univ. Press, Cambridge, Massachusetts.

MORTIMER, C. H. 1941–42. The exchange of dissolved substances between mud and water in lakes. J. Ecol. 29: 280–329; 30: 147–201.

MORTON, S. R., & C. D. JAMES. 1988. The diversity and abundance of lizards in arid Australia: a new hypothesis. Amer. Nat. 132: 237–256.

MOULTON, M. P., & S. L. PIMM. 1986. The extent of competition in shaping an introduced avifauna. Pp. 80–97 in J. Diamond and T. J. Case (Eds.), Community Ecology. Harper & Row, New York.

MROSOVSKY, N. 1976. Lipid programmes and life strategies in hibernators. Amer. Zool. 16: 685–697.

MUELLER, L. D., & F. J. AYALA. 1981. Tradeoff between r selection and K selection in *Drosophila* populations. Proc. Natl. Acad. Sci. 78: 1303–1305.

MUELLER-DOMBOIS, D., & H. ELLENBERG. 1974. Aims and Methods of Vegetation Ecology. Wiley, New York.

MUHLENBERG, M. D., D. LEIPOLD, H. J. MADER, & B. STEINHAUER. 1977. Island ecology of arthropods. II. Niches and relative abundances of Seychelles ants (Formicidae) in different habitats. Oecologia 29: 135–144.

MULLER, C. H. 1966. The role of chemical inhibition (allelopathy) in vegetational composition. Bull. Torrey Bot. Club 93: 332–351.

MULLER, C. H. 1970. Phytotoxins as plant habitat variables. Rec. Adv. Phytochem. 3: 106–121.

MULLER, C. H., R. B. HANAWALT, & J. K. MCPHERSON. 1968. Allelopathic control of herb growth in the fire cycle of California chaparral. Bull. Torrey Bot. Club 95: 225–231.

MULLER, C. H., W. H. MULLER, & B. L. HAINES. 1964. Volatile growth inhibitors produced by aromatic shrubs. Science 143: 471–473.

MULLER, H. J. 1964. The relation of recombination to mutational advance. Mutat. Res. 1: 2–9.

MULLER, W. H. 1965. Volatile materials produced by *Salvia leucophylla:* effects on seedling growth and soil bacteria. Bot. Gaz. 126: 195–200.

MURDOCH, C. L., J. A. JAKOBS, & J. W. GERDEMANN. 1967. Utilization of phosphorus sources of different availability by mycorrhizal and non-mycorrhizal maize. Plant Soil 27: 329–334.

MURDOCH, W. W. 1966. Population stability and life history phenomena. Amer. Nat. 100: 5–11.

MURDOCH, W. W. 1969. Switching in general predators: experiments on predator specificity and stability of prey populations. Ecol. Monogr. 39: 335–354.

MURDOCH, W. W., & A. OATEN. 1975. Predation and population stability. Adv. Ecol. Res. 9: 2–131.

MURIE, O. 1944. The Wolves of Mt. McKinley. U.S. Dept. Int., Natl. Park Ser., Fauna Ser. No. 5, Washington, D.C.

MURPHY, G. I. 1968. Patterns in life history and environment. Amer. Nat. 102: 390–404.

MURPHY, R. C. 1936. Oceanic Birds of South America. Amer. Mus. Nat. Hist., New York.

MURTON, R. K. 1967. The significance of endocrine stress in population control. Ibis 109: 622–623.

MURTON, R. K. 1970. Why do some bird species feed in flocks? Ibis 113: 534–535.

MURTON, R. K. 1971. The significance of a specific search image in the feeding behaviour of the wood-pigeon. Behaviour 40: 10–42.

MURTON, R. K., & N. J. WESTWOOD. 1977. Avian Breeding Cycles. Clarendon Press, Oxford.

MUTCH, W. R. 1970. Wildland fires and ecosystems—a hypothesis. Ecology 51: 1046–1051.

MYERS, J. H., & C. J. KREBS. 1971. Genetic, behavioral, and reproductive attributes of dispersing field voles *Microtus pennsylvanicus* and *Microtus ochrogaster.* Ecol. Monogr. 41: 53–78.

MYERS, J. P., P. G. CONNORS, & F. A. PITELKA. 1979. Territory size in wintering sanderlings: the effects of prey abundance and intruder density. Auk 96: 551–561.

MYERS, K. 1970. The rabbit in Australia. Pp. 478–506 in P. J. den Boer and G. R. Gradwell (Eds.), Dynamics of Populations. Centre Agric. Publ. Documentation, Wageningen, The Netherlands.

NAGY, K. A., D. K. ODELL, & R. S. SEYMOUR. 1972. Temperature regulation by the inflorescence of *Philodendron.* Science 178: 1196–1197.

NATIONAL ACADEMY OF SCIENCES. 1969. Eutrophication: Causes, Consequences, Correctives. National Academy of Sciences, Washington, D.C.

NEEDHAM, J. 1931. Chemical Embryology, Vols. I–III. Cambridge Univ. Press, Cambridge.

NEILL, W. E. 1974. The community matrix and the interdependence of the competition coefficients. Amer. Nat. 108: 399–408.

NELSON, J. B. 1964. Factors influencing clutch-size and chick growth in the North Atlantic gannet *Sula bassana.* Ibis 106: 63–77.

NERO, R. W. 1984. Redwings. Smithson. Inst. Press, Washington, D.C.

NEWBOULD, P. J. 1967. Methods for Estimating the Primary Production of Forests. Blackwell, Oxford.

NICHOLS, F. H., J. E. CLOERN, S. N. LUOMA, & D. H. PETERSON. 1986. The modification of an estuary. Science 231: 567–573.

NICHOLSON, A. J. 1933. The balance of animal populations. J. Anim. Ecol. 2: 132–178.

NICHOLSON, A. J. 1958. The self-adjustment of populations to change. Cold Spring Harbor Symp. Quant. Biol. 22: 153–173.

NICHOLSON, A. J. 1958. Dynamics of insect populations. Ann. Rev. Entomol. 3: 107–136.

NICHOLSON, A. J., & V. A. BAILEY. 1935. The balance of animal populations. Proc. Zool. Zoc. Lond. 3: 551–598.

NICOL, J. A. C. 1967. The Biology of Marine Animals (2nd ed.). Wiley, New York.

NIEMI, G. R. 1985. Patterns of morphological evolution in bird genera of New World and Old World peatlands. Ecology 66: 1215–1228.

NISBET, R. M., & W. S. C. GURNEY. 1982. Modelling Fluctuating Populations. Wiley, New York.

NITECKI, M. H. (Ed.). 1983. Coevolution. Univ. Chicago Press, Chicago.

NIXON, C. M. 1965. White-tailed deer growth and productivity in eastern Ohio. Game Res. Ohio 3: 123–136.

NOONAN, K. M. 1981. Individual strategies of inclusive-fitness-maximizing in *Polistes fuscatus* foundresses. Pp. 18–44 in R. D. Alexander and D. W. Tinkle (Eds.), Natural Selection and Social Behavior. Chiron Press, New York.

NOORDWIJK, A. J. VAN. 1984. Quantitative genetics in natural populations of birds illustrated with examples from the great tit, *Parus major.* Pp. 67–79 in K. Wohrmann and V. Loeschcke (Eds.), Population Biology and Evolution. Springer-Verlag, New York.

NORDSKOG, A. W. 1977. Success and failure of quantitative genetic theory in poultry. Pp. 569–586 in E. Pollack, O. Kempthorne, and T. B. Bailey (Eds.), Proc. Intern. Conf. Quant. Genet., 1976. Iowa State Univ. Press, Ames.

NORRIS, K. S. 1953. The ecology of the desert iguana, *Dipsosuarus dorsalis.* Ecology 34: 265–287.

NORTH, W. J. 1970. Kelp habitat improvement project. W. M. Keck Lab. Environ. Health Eng., Calif. Inst. Tech., Ann. Rept.

NORTON-GRIFFITHS, M. 1969. The organisation, control and development of parental feeding in the oystercatcher *(Haematopus ostralegus).* Behaviour 34: 55–114.

NOY-MEIR, I. 1975. Stability of grazing systems: an application of predator-prey graphs. J. Ecol. 63: 459–483.

NOY-MEIR, I. 1981. Theoretical dynamics of competitors under predation. Oecologia 50: 277–284.

NUR, N. 1988. The consequences of brood size for breeding blue tits. III. Measuring the cost of reproduction: survival, future fecundity, and differential dispersal. Evolution 42: 351–362.

NYE, P. H. 1961. Organic matter and nutrient cycles under moist tropical forest. Plant and Soil 13: 333–346.

NYE, P. H. 1977. The rate-limiting step in plant nutrient absorption from soil. Soil Sci. 123: 292–297.

OATEN, A., & W. W. MURDOCH. 1975. Functional response and stability in predator-prey systems. Amer. Nat. 109: 289–298.

OATEN, A., & W. W. MURDOCH. 1975. Switching, functional response, and stability in predator-prey systems. Amer. Nat. 109: 299–318.

O'CONNOR, R. J. 1980. Pattern and process in great tit *(Parus major)* populations in Britain. Ardea 68: 165–183.

O'DONALD, P. 1980. Genetic Models of Sexual Selection. Cambridge Univ. Press, Cambridge.

ODUM, E. P. 1953. Fundamentals of Ecology. Saunders, Philadelphia.

ODUM, E. P. 1959. Fundamentals of Ecology (2nd ed.). Saunders, Philadelphia.

ODUM, E. P. 1960. Organic production and turnover in old field succession. Ecology 41: 34–49.

ODUM, E. P. 1960. Factors which regulate primary productivity and heterotrophic utilization in the ecosystem. Trans. Sem. Algae Metrop. Wastes, U.S. Public Health Ser., R. A. Taft San. Eng. Center, Cincinnati, Ohio.

ODUM, E. P. 1962. Relationships between structure and function in the ecosystem. Jap. J. Ecol. 12: 108–118.

ODUM, E. P. 1968. Energy flow in ecosystems: a historical review. Amer. Zool. 8: 11–18.

ODUM, E. P. 1969. The strategy of ecosystem development. Science 164: 262–270.

ODUM, E. P. 1971. Fundamentals of Ecology (3rd ed.). Saunders, Philadelphia.

ODUM, E. P., & E. J. KUENZLER. 1963. Experimental isolation of food chains in an old-field ecosystem with use of phosphorus-32. Pp. 113–120 in V. Schultz and A. W. Klement (Eds.), Radioecology. Reinhold, New York.

ODUM, H. T. 1956. Primary production in flowing waters. Limnol. Oceanogr. 1: 102–117.

ODUM, H. T. 1957. Trophic structure and productivity of Silver Springs, Florida. Ecol. Monogr. 27: 55–112.

ODUM, H. T. 1970. Rain forest structure and mineral-cycling homeostasis. Pp. H-3–H-52 in H. T. Odum and R. F. Pigeon (Eds.), A Tropical Rain Forest. U.S. Atomic Energy Commission, Washington, D.C.

ODUM, H. T. 1983. Systems Ecology: An Introduction. Wiley, New York.

ODUM, H. T. 1988. Self-organization, transformity, and information. Science 242: 1132–1139.

ODUM, H. T., & R. C. PINKERTON. 1955. Time's speed regulator: the optimum efficiency for maximum power output in physical and biological systems. Amer. Sci. 43: 331–343.

ODUM, W. E. 1988. Comparative ecology of tidal freshwater and salt marshes. Ann. Rev. Ecol. Syst. 19: 147–176.

OGREN, W. L. 1984. Photorespiration: pathways, regulation, and modification. Ann. Rev. Plant Physiol. 415–442.

OLMSTED, C. E. 1944. Growth and development in range grasses. IV. Photoperiodic responses in twelve geographic strains of side oats grama. Bot. Gaz. 106: 46–74.

OLSON, J. S. 1958. Rates of succession and soil changes on southern Lake Michigan sand dunes. Bot. Gaz. 119: 125–170.

OLSON, J. S. 1963. Energy storage and the balance of producers and decomposers in ecological systems. Ecology 44: 322–331.

O'NEILL, R. V. 1968. Population energetics of the millipede, *Narceus americanus* (Beauvois). Ecology 49: 803–809.

OOSTING, H. J. 1942. An ecological analysis of the plant communities of Piedmont, North Carolina. Amer. Midl. Nat. 28: 1–126.

OOSTING, H. J. 1954. Ecological processes and vegetation of the maritime strand in the southeastern United States. Bot. Rev. 20: 226–262.

OOSTING, H. J. 1956. The Study of Plant Communities (2nd ed.). Dover, New York.

OPLER, P. A. 1973. Fossil lepidopterous leaf mines demonstrate the age of some insect-plant relationships. Science 179: 1321–1323.

ORIANS, G. H. 1969. The number of bird species in some tropical forests. Ecology 50: 783–801.

ORIANS, G. H. 1969. On the evolution of mating systems in birds and mammals. Amer. Nat. 103: 589–603.

ORIANS, G. H. 1980. Some Adaptations of Marsh-nesting Blackbirds. Princeton Univ. Press, Princeton, New Jersey.

ORIANS, G. H. 1982. The influence of tree falls in tropical forests on tree species richness. Trop. Ecol. 23: 255–279.

ORIANS, G. H., & R. T. PAINE. 1983. Convergent evolution at the community level. Pp. 431–458 in D. J. Futuyma and M. Slatkin (Eds.), Coevolution. Sinauer, Sunderland, Massachusetts.

ORIANS, G. H., & N. E. PEARSON. 1979. On the theory of central place foraging. Pp. 155–177 in D. J. Horn, R. Mitchell, and G. R. Stair (Eds.), Analysis of Ecological Systems. Ohio State Univ. Press, Columbus.

ORIANS, G. H., & O. T. SOLBRIG. (Eds.). 1977. Convergent Evolution in Warm Deserts. Dowden, Hutchinson & Ross, Stroudsburg, Pennsylvania.

OSAWA, A., & S. SUGITA. 1989. The self-thinning rule: another interpretation of Weller's results. Ecology 70: 279–283.

OSMOND, C. B. 1978. Crassulacean acid metabolism: a curiosity in context. Ann. Rev. Plant Physiol. 29: 379–414.

OSONUBI, O., & W. J. DAVIES. 1978. Solute accumulation in leaves and roots of woody plants subjected to water stress. Oecologia 32: 323–332.

OSTER, G. F., & E. O. WILSON. 1978. Caste and Ecology in the Social Insects. Princeton Univ. Press, Princeton, New Jersey.

OTTE, D. & A. JOERN. 1977. On feeding patterns in desert grasshoppers and the evolution of specialized diets. Proc. Acad. Natl. Sci. Phila. 128: 89–126.

OTTERMAN, J. 1974. Baring high-albedo soils by overgrazing: a hypothesized desertification mechanism. Science 186: 531–533.

OTTERMAN, J., Y. WAISEL, & E. ROSENBERG. 1975. Western Negev and Sanai ecosystems: comparative study of vegetation, albedo, and temperatures. Agro-ecosystems 2: 47–59.

OVERTON, W. S. 1969. Estimating the numbers of animals in wildlife populations. Pp. 403–455 in R. H. Giles, Jr. (Ed.). 1969. Wildlife Management Techniques (3rd ed.). Wildlife Society, Washington, D.C.

OVINGTON, J. D. 1957. Dry-matter production by *Pinus sylvestris* L. Ann. Bot. N. S. 21: 287–314.

OVINGTON, J. D. 1962. Quantitative ecology and the woodland ecosystem concept. Adv. Ecol. Res. 1: 103–192.

OVINGTON, J. D. 1965. Organic production, turnover, and mineral cycling in woodlands. Biol. Rev. 40: 295–336.

OVINGTON, J. D., & H. A. I. MADGWICK. 1959. The growth and composition of natural stands of birch, I: Dry-matter production. Plant Soil 10: 271–283.

PACKER, C., & L. RUTTAN. 1988. The evolution of cooperative hunting. Amer. Nat. 132: 159–198.

PAGANELLI, C. V. 1980. The physics of gas exchange across the avian eggshell. Amer. Zool. 20: 329–338.

PAINE, R. T. 1963. Trophic relationships of eight sympatric gastropods. Ecology 44: 63–67.

PAINE, R. T. 1966. Food web complexity and species diversity. Amer. Nat. 100: 65–75.

PAINE, R. T. 1971. The measurement and application of the calorie to ecological problems. Ann. Rev. Ecol. Syst. 2: 145–164.

PAINE, R. T. 1974. Intertidal community structure: experimental studies on the relationship between a dominant competitor and its principal predator. Oecologia 15: 93–120.

PAINE, R. T. 1977. Controlled manipulations in the marine intertidal zone, and their contributions to ecological theory. The Changing Scenes in Natural Sciences, 1776–1976, Acad. Natl. Sci. Phila., Spec. Publ. 12: 245–270.

PAINE, R. T. 1980. Food webs: linkage, interaction strength and community infrastructure. J. Anim. Ecol. 49: 667–685.

PAINE, R. T., & R. VADAS. 1969. The effects of grazing by sea urchins, *Strongylocentrotus* spp., on benthic algal populations. Limnol. Oceanogr. 14: 710–719.

PAJUNEN, V. I. 1966. The influence of population density on the territorial behavior of *Leucorrhinia rubicunda* L. (Odon., Libellulidae). Ann. Zool. Fenn. 3: 40–52.

PALMBLAD, I. G. 1968. Competition in experimental populations of weeds with emphasis on the regulation of population size. Ecology 49: 26–34.

PARK, T. 1954. Experimental studies of interspecific competition, II: Temperature, humidity and competition in two species of *Tribolium.* Physiol. Zool. 27: 177–238.

PARK, T. 1962. Beetles, competition, and populations. Science 138: 1369–1375.

PARK, T., P. H. LESLIE, AND D. B. MERTZ. 1964. Genetic strains and competition in populations of *Tribolium.* Physiol. Zool. 38: 289–321.

PARKER, G. A. 1974. The reproductive behavior and the nature of sexual selection in *Scatophaga stercoraria* L. (Diptera: Scatophagidae). IX. Spatial distribution of fertilization rates and evolution of male search strategy within the reproductive area. Evolution 28: 93–108.

PARKER, G. A. 1978. Evolution of competitive mate searching. Ann. Rev. Entomol. 23: 173–196.

PARKER, G. A. 1978. Selection on non-random fusion of gametes during the evolution of anisogamy. J. Theoret. Biol. 73: 1–28.

PARKER, G. A. 1984. Evolutionarily stable strategies. Pp. 30–61 in J. R. Krebs and N. B. Davies (Eds.), Behavioral Ecology: An Evolutionary Approach (2nd ed.). Sinauer, Sunderland, Massachusetts.

PARKER, G. A., R. R. BAKER, & V. G. F. SMITH. 1972. The origin and evolution of gamete dimorphism and the male-female phenomenon. J. Theoret. Biol. 36: 529–553.

PARKER, G. A., & M. R. MACNAIR. 1978. Models of parent-offspring conflict. I. Monogamy. Anim. Behav. 26: 97–110.

PARKER, G. A., & M. R. MACNAIR. 1979. Models of parent-offspring conflict. IV. Suppression: evolutionary retaliation by the parent. Anim. Behav. 27: 1210–1235.

PARR, J. C., & R. THURSTON. 1968. Toxicity of *Nicotiana* and *Petunia* species to larvae of the tobacco hornworm. J. Econ. Entomol. 61: 1525–1531.

PARRISH, J. A. D., & F. A. BAZZAZ. 1976. Underground niche separation in successional plants. Ecology 57: 1281–1288.

PARRISH, J. A. D., & F. A. BAZZAZ. 1982. Niche responses of early and late successional tree seedlings on three resource gradients. Bull. Torrey Bot. Club 109: 451–456.

PARRISH, J. D., & S. B. SAILA. 1970. Interspecific competition, predation, and species diversity. J. Theoret. Biol. 27: 207–220.

PARTRIDGE, L., & T. HALLIDAY. 1984. Mating patterns and mate choice. Pp. 222–250 in J. R. Krebs and N. B. Davies (Eds.), Behavioural Ecology. An Evolutionary Approach. Sinauer, Sunderland, Massachusetts.

PATRICK, R. 1963. The structure of diatom communities under varying ecological conditions. Ann. N.Y. Acad. Sci. 108: 353–358.

PATRICK, R. 1967. The effect of invasion rate, species pool, and size of area on the structure of the diatom community. Proc. Natl. Acad. Sci. 58: 1335–1342.

PATRICK, R., M. H. HOHN, & J. H. WALLACE. 1954. A new method for determining the pattern of the diatom flora. Natulae Naturae, No. 259.

PATTEN, B. C. 1981. Environs: the superniches of ecosystems. Amer. Zool. 21: 845–852.

PATTEN, B. C. 1982. Environs: relativistic elementary particles for ecology. Amer. Nat. 119: 179–219.

PAUL, V. J., & M. E. HAY. 1988. Seaweed susceptibility to herbivory: chemical and morphological correlates. Mar. Ecol. Prog. Ser. 33: 255–264.

PAYNE, R. 1977. The ecology of brood parasitism in birds. Ann. Rev. Ecol. Syst. 8: 1–28.

PAYNE, R. B. 1984. Sexual selection, lek and arena behavior, and sexual size dimorphism in birds. Ornithol. Monogr. No. 33.

PEACE, W. J. H., & P. J. GRUBB. 1982. Interaction of light and mineral nutrient supply in the growth of *Impatiens parviflora*. New Phytol. 90: 127–150.

PEAKER, M., & J. L. LINZELL. 1975. Salt Glands in Birds and Reptiles. Cambridge Univ. Press, Cambridge.

PEARCY, R. E., & J. EHLERINGER. 1984. Comparative ecophysiology of C_3 and C_4 plants. Plant, Cell Environ. 7: 1–13.

PEARL, R. 1921. The biology of death: V. Natural death, public health and the population problem. Sci. Monthly 13: 193–213.

PEARL, R. 1925. The Biology of Population Growth. Knopf, New York.

PEARL, R. 1927. The growth of populations. Quart. Rev. Biol. 2: 532–548.

PEARL, R., & S. L. PARKER. 1921. Experimental studies on the duration of life. I. Introductory discussion of the duration of life in *Drosophila*. Amer. Nat. 55: 481–509.

PEARL, R., & L. J. REED. 1920. On the rate of growth of the population of the United States since 1790 and its mathematical representation. Proc. Natl. Acad. Sci. 6: 275–288.

PEARSON, D. L. 1985. The function of multiple anti-predator mechanisms in adult tiger beetles (Coleoptera: Cicindelidae). Ecol. Entomol. 10: 65–72.

PEARSON, O. P. 1985. Predation. Pp. 535–556 in R. H. Tamarin (Ed.), Biology of New World *Microtus*. Amer. Soc. Mammalogists, Spec. Publ. No. 8.

PEASE, C. M. 1984. On the evolutionary reversal of competitive dominance. Evolution 38: 1099–1115.

PEET, R. K. 1975. Relative diversity indices. Ecology 56: 496–498.

PEET, R. K. 1981. Changes in biomass and production during secondary forest succession. Pp. 324–338 in D. C. West, H. H. Shugart, and D. B. Botkin (Eds.), Forest Succession. Concepts and Application. Springer-Verlag, New York.

PEET, R. K., & O. L. LOUCKS. 1977. A gradient analysis of southern Wisconsin forests. Ecology 58: 485–499.

PEHRSON, A. 1983. Digestibility and retension of food components in caged mountain hares *Lepus timidus* during the winter. Holarctic Ecol. 6: 395–403.

PENFOUND, W. T. 1956. Primary production of vascular aquatic plants. Limnol. Oceanogr. 1: 92–101.

PENNYCUICK, C. J. 1975. Mechanics of flight. Pp. 1–75 in D. S. Farner, J. R. King, and K. C. Parkes (Eds.), Avian Biology, Vol. 5. Academic Press, New York.

PETERS, R. H. 1983. The Ecological Implications of Body Size. Cambridge Univ. Press, Cambridge.

PETERS, T. M., & P. BARBOSA. 1977. Influence of population density on size, fecundity, and developmental rates of insects in culture. Ann. Rev. Entomol. 22: 431–450.

PETERSEN, R. C., & K. W. CUMMINS. 1974. Leaf processing in a woodland stream. Freshwater Biol. 4: 343–368.

PETERSON, B. J. 1980. Aquatic primary productivity and the ^{14}C-CO_2 method: a history of the productivity problem. Ann. Rev. Ecol. Syst. 11: 359–385.

PETERSON, R. T. 1980. A Field Guide to the Birds East of the Rockies. Houghton Mifflin, Boston.

PETIT, C., & L. EHRMAN. 1969. Sexual selection in *Drosophila*. Evol. Biol. 3: 177–223.

PHILLIPS, J. 1934. Succession, development, the climax and the complex organism: an analysis of concepts. I. J. Ecol. 22: 554–571.

PHILLIPS, J. 1935. Succession, development, the climax and the complex organism: an analysis of concept. II and III. J. Ecol. 23: 210–246, 488–508.

PHILLIPSON, J. 1966. Ecological energetics. Edward Arnold, London.

PIANKA, E. R. 1966. Latitudinal gradients in species diversity: a review of concepts. Amer. Nat. 100: 33–46.

PIANKA, E. R. 1967. On lizard species diversity: North American flatland deserts. Ecology 48: 333–351.

PIANKA, E. R. 1970. On *r* and *K* selection. Amer. Nat. 104: 592–597.

PIANKA, E. R. 1971. Comparative ecology of two lizards. Copeia 1971: 129–138.

PIANKA, E. R. 1972. *r* and *K* selection or *b* and *d* selection? Amer. Nat. 106: 581–588.

PIANKA, E. R. 1973. The structure of lizard communities. Ann. Rev. Ecol. Syst. 4: 53–74.

PIANKA, E. R. 1988. Evolutionary Ecology (4th ed.). Harper & Row, New York.

PICKETT, S. T. A. 1983. Differential adaptation of tropical species to canopy gaps and its role in community dynamics. Trop. Ecol. 24: 68–84.

PIELOU, E. C. 1966. Comment on a report by J. H. Vandermeer and R. H. MacArthur concerning the broken stick model of species abundance. Ecology 47: 1073–1074.

PIELOU, E. C. 1969. An Introduction to Mathematical Ecology. Wiley, New York and London.

PIELOU, E. C. 1977. Mathematical Ecology. Wiley, New York and London.

PIMENTEL, D. 1988. Herbivore population feeding pressure on plant hosts: feedback evolution and host conservation. Oikos 53: 289–302.

PIMENTEL, D., & R. AL-HAFIDH. 1963. The coexistence of

insect parasites and hosts in laboratory populations. Ann. Entomol. Soc. Amer. 56: 676–678.

PIMENTEL, D., E. H. FEINBERG, P. W. WOOD, & J. T. HAYES. 1965. Selection, spatial distribution, and the coexistence of competing fly species. Amer. Nat. 99: 97–109.

PIMLOTT, D. H. 1967. Wolf predation and ungulate populations. Amer. Zool. 7: 267–278.

PIMM, S. L. 1980. Properties of food webs. Ecology 61: 219–225.

PIMM, S. L. 1982. Food Webs. Chapman and Hall, London and New York.

PIMM, S. L., H. L. JONES, & J. DIAMOND. 1988. On the risk of extinction. Amer. Nat. 132: 757–785.

PIMM, S. L., & J. H. LAWTON. 1977. The number of trophic levels in ecological communities. Nature 268: 329–331.

PIMM, S. L., & J. H. LAWTON. 1978. On feeding on more than one trophic level. Nature 275: 542–544.

PIMM, S. L., & J. H. LAWTON. 1980. Are food webs divided into compartments? J. Anim. Ecol. 49: 879–898.

PINGREE, R. D., P. R. PUGH, P. M. HOLLIGAN, & G. R. FORSTER. 1975. Summer phytoplankton blooms and red tides along tidal fronts on the approaches to the English Channel. Nature 258: 672–677.

PITELKA, F. A. 1941. Distribution of birds in relation to major biotic communities. Am. Midl. Nat. 25: 113–137.

PITELKA, F. A., P. O. TOMICH, & G. W. TREICHEL. 1955. Ecological relations of jaegers and owls as lemming predators near Barrow, Alaska. Ecol. Monogr. 25: 85–117.

PLANK, J. E. VAN DER. 1968. Disease Resistance in Plants. Academic Press, New York.

PLESCZCZYNSKA, W. K. 1978. Microgeographic prediction of polygyny in the lark bunting. Science 201: 935–937.

PLESZCZYNSKA, W. K., & R. HANSELL. 1980. Polygyny and decision theory: testing of a model in lark buntings *(Calamospiza melanocorys).* Amer. Nat. 116: 821–830.

PODOLER, H., & D. ROGERS. 1975. A new method for the identification of key factors from life-table data. J. Anim. Ecol. 44: 85–114.

POMERANTZ, M. J. 1981. Do "higher order interactions" in competition systems really exist? Amer. Nat. 117: 583–591.

POOL, R. 1989. Is it chaos, or is it just noise? Science 243: 25–28.

POOL, R. 1989. Ecologists flirt with chaos. Science 243: 310–313.

POOLE, R. W. 1974. An Introduction to Quantitative Ecology. McGraw-Hill, New York.

POOLE, R. W. 1979. The statistical prediction of the fluctuations in abundance in Nicholson's sheep blowfly experiments. Pp. 213–246 in R. M. Cormack and J. K. Ord (Eds.), Spatial and Temporal Analysis in Ecology. Intern. Co-op. Publ. House, Fairland, Maryland.

PORTER, J. W. 1972. Predation by *Acanthaster* and its effect on coral species diversity. Amer. Nat. 106: 487–492.

PORTER, J. W. 1972. Ecology and species diversity of coral reefs on opposite sides of the isthmus of Panama. Bull. Biol. Soc. Wash. 2: 89–116.

PORTER, J. W. 1974. Community structure of coral reefs on opposite sides of the Isthumus of Panama. Science 186: 343–345.

PORTER, K. G., J. GERRITSEN, & J. D. ORCUTT, JR. 1982. The effect of food concentration on swimming patterns, feeding behavior, ingestion, assimilation, and respiration by *Daphnia.* Limnol. Oceanogr. 27: 935–949.

PORTER, K. G., J. D. ORCUTT, JR., & J. GERRITSEN. 1983. Functional response and fitness in a generalist filter feeder, *Daphnia magna* (Cladocera: Crustacea). Ecology 64: 735–742.

PORTER, W. P., & D. M. GATES. 1969. Thermodynamic equilibria of animals with the environment. Ecol. Monogr. 39: 245–270.

PORTER, W. P., J. W. MITCHELL, W. A. BECKMAN, & C. B. DEWITT. 1973. Behavioral implications of mechanistic ecology. Thermal and behavioral modeling of desert ectotherms and their microenvironment. Oecologia 13: 1–54.

POSTGATE, J. R., & S. HILL. 1979. Nitrogen fixation. Pp. 191–213 in J. M. Lynch and N. J. Poole (Eds.), Microbial Ecology: A Conceptual Approach. Blackwell, Oxford.

POTTS, G. W., & R. J. WOOTTON. (Eds.). 1984. Fish Reproduction. Strategies and Tactics. Academic Press, New York.

POTTS, W. T. W., & G. PARRY. 1964. Osmotic and Ionic Regulation in Animals. Pergamon, Oxford.

POUGH, F. H. 1980. The advantages of ectothermy for tetrapods. Amer. Nat. 115: 92–112.

POWELL, J. A., & R. A. MACKIE. 1966. Biological interrelationships of moths and *Yucca whipplei.* Univ. Calif. Publ. Entomol. 42: 1–59.

PRANCE, G. T. (Ed.). 1982. The Biological Model of Diversification in the Tropics. Columbia Univ. Press, New York.

PRATT, D. M. 1943. Analysis of population development in *Daphnia* at different temperatures. Biol. Bull. 85: 116–140.

PREGILL, G. K., & S. L. OLSON. 1981. Zoogeography of West Indian vertebrates in relation to Pleistocene climatic cycles. Ann. Rev. Ecol. Syst. 12: 75–98.

PREPAS, E. E., & D. O. TREW. 1983. Evaluation of the phosphorus-chlorophyll relationship for lakes off the Precambrian Shield in western Canada. Canad. J. Fish. Aquat. Sci. 40: 27–35.

PRESTON, F. W. 1948. The commonness, and rarity, of species. Ecology 29: 254–283.

PRICE, M. V., & N. M. WASSER. 1979. Pollen dispersal and optimal outcrossing in *Delphinium nelsoni.* Nature 277: 294–297.

PRICE, P. W. 1971. Niche breadth and dominance of parasitic insects sharing the same host species. Ecology 52: 587–596.

PRICE, T. D., P. R. GRANT, & P. T. BOAG. 1984. Genetic changes in the morphological differentiation of Darwin's ground finches. Pp. 49–66 in K. Wohrmann and V. Loeschcke (Eds.), Population Biology and Evolution. Springer-Verlag, New York.

PRIMACK, R. B. 1987. Relationships among flowers, fruits, and seeds. Ann. Rev. Ecol. Syst. 18: 409–430.

PROCTOR, J., & S. R. J. WOODELL. 1975. The ecology of serpentine soils. Adv. Ecol. Res. 9: 255–366.

PROSSER, C. L. 1973. Comparative Animal Physiology (3rd ed.). Saunders, Philadelphia.

PROSSER, C. L. 1986. Adaptational biology. Wiley, New York.

PROSSER, C. L., & F. A. BROWN. 1961. Comparative Animal Physiology (2nd ed.). Saunders, Philadelphia.

PRYOR, L. D., & L. A. S. JOHNSON. 1971. A Classification of the Eucalyptus. Australian National Univ., Canberra.

PULLIAM, H. R. 1973. On the advantages of flocking. J. Theoret. Biol. 38: 419–422.

PULLIAM, H. R. 1988. Sources, sinks, and population regulation. Amer. Nat. 132: 652–661.

PUTNAM, A. R., & C.-S. TANG. (Eds.). 1986. The Science of Allelopathy. Wiley, New York.

PUTZ, F. E., & K. MILTON. 1982. Tree mortality rates on Barro Colorado Island. Pp. 95–100 in E. G. Leigh, Jr., A. S. Rand, and D. M. Windsor (Eds.), The Ecology of a Tropical Forest. Smithson. Inst. Press, Washington, D.C.

PYKE, G. H., H. R. PULLIAM, & E. L. CHARNOV. 1977. Optimal foraging: a selective review of theory and tests. Quart. Rev. Biol. 52: 137–154.

QUELLER, D. C., J. E. STRASSMANN, & C. R. HUGHES. 1988. Genetic relatedness in colonies of tropical wasps with multiple queens. Science 242: 1155–1157.

QUINN, J. F., & A. E. DUNHAM. 1983. On hypothesis testing in ecology and evolution. Amer. Nat. 122: 602–617.

QUISPEL, A. (Ed.). 1974. The Biology of Nitrogen Fixation. North-Holland Publ. Col., Amsterdam.

RABINOWITCH, E., & GOVINDJEE. 1969. Photosynthesis. Wiley, New York.

RABINOWITZ, D. 1981. Seven forms of rarity. Pp. 205–217 in H. Synge (Ed.), The Biological Aspects of Rare Plant Conservation. Wiley, Chichester, New York.

RAHN, H., R. A. ACKERMAN, & C. V. PAGANELLI. 1977. Humidity in the avian nest and egg water loss during incubation. Physiol. Zool. 50: 269–283.

RAHN, H., & A. AR. 1980. Gas exchange of the avian egg: time, structure, and function. Amer. Zool. 20: 477–484.

RAHN, H., A. AR, & C. V. PAGANELLI. 1979. How bird eggs breathe. Sci. Amer. 240: 46–55.

RAHN, H., C. CAREY, K. BALMAS, B. BHATIA, & C. V. PAGANELLI. 1977. Reduction of pore area of the avian eggshell as an adaptation to altitude. Proc. Natl. Acad. Sci. 74: 3095–3098.

RAHN, H., & C. V. PAGANELLI. (Eds.). 1981. Gas Exchange in Avian Eggs. Department of Physiology, SUNY Buffalo, Buffalo, New York.

RAHN, H., C. V. PAGANELLI, & A. AR. 1974. The avian egg: air-cell gas tension, metabolism and incubation time. Respir. Physiol. 22: 297–309.

RAISON, J. K., J. A. BERY, P. A. ARMOND, & C. S. PIKE. 1980. Membrane properties in relation to the adaptation of plants to temperature stress. Pp. 261–273 in N. C. Turner and P. J. Kramer (Eds.), Adaptation of Plants to Water and High Temperature Stress. Wiley, New York.

RALLS, K., J. D. BALLOU, & A. TEMPLETON. 1988. Estimates of the cost of inbreeding in mammals. Conserv. Biol. 2: 185–193.

RALPH, C. J., & J. M. SCOTT. (Eds.). 1981. Estimating Numbers of Terrestrial Birds. Studies in Avian Biol. No. 6, Cooper Ornithol. Soc.

RAND, A. D. 1967. Predator-prey interactions and the evolution of aspect diversity. Atas do Simposio sobre a Biota Amazonica 5: 73–83.

RANDALL, D. J. 1968. Fish physiology. Amer. Zool. 8: 179–189.

RAPPORT, D. J. 1971. An optimization model of food selection. Amer. Nat. 105: 575–587.

RASMUSSEN, E. M. 1985. El Niño and variations in climate. Amer. Sci. 73: 168–177.

RASMUSSON, E. M., & J. M. WALLACE. 1983. Meterological aspects of the El Niño/Southern Oscillation. Science 222: 1195–1202.

RATHCKE, B. J. 1976. Competition and coexistence within a guild of herbivorous insects. Ecology 57: 76–87.

RATNIEKS, F. L. W. 1988. Reproductive harmony via mutural policing by workers in eusocial Hymenoptera. Amer. Nat. 132: 217–236.

RAUNKIAER, C. 1918. Recherches statistiques sur les formations végètales. K. Danske Vidensk Selsk. Biol. Meddel. 1: 1–47.

RAUNKIAER, C. 1934. The Life Forms of Plants and Statistical Plant Geography. Clarendon Press, Oxford.

RAUNKIAER, C. 1937. Plant Life Forms. Transl. by H. Gilbert-Carter, Clarendon Press, Oxford.

REAL, L., & T. CARACO. 1986. Risk foraging in stochastic environments. Ann. Rev. Ecol. Syst. 17: 371–390.

REDFIELD, A. C. 1972. Development of a New England salt marsh. Ecol. Monogr. 42: 201–237.

REHR, S. S., P. P. FEENY, & D. H. JANZEN. 1973. Chemical defense in Central American non-ant-acacias. J. Anim. Ecol. 42: 405–416.

REICH, P. B., & R. G. AMUNDSON. 1985. Ambient levels of ozone reduce net photosynthesis in tree and crop species. Science 230: 566–570.

REID, G. K. 1961. Ecology of Inland Waters and Estuaries. Reinhold, New York.

REIMOLD, R. J., & W. H. QUEEN. 1974. Ecology of Halophytes. Academic Press, New York.

RETTENMEYER, C. W. 1963. Behavioral studies of army ants. Univ. Kans. Sci. Bull. 44: 281–465.

RETTENMEYER, C. W. 1970. Insect mimicry. Ann. Rev. Entomol. 15: 43–74.

REUSS, J. O., & G. S. INNIS. 1977. A grassland nitrogen flow simulation model. Ecology 58: 379–388.

REY, J. R., E. D. MCCOY, & D. R. STRONG, JR. 1981. Herbivore pests, habitat islands, and the species-area relationship. Amer. Nat. 117: 611–622.

REYNOLDSON, T. B., & H. R. HAMILTON. 1982. Spatial heterogeneity in whole lake sediments—towards a loading estimate. Hydrobiologia 91: 235–240.

REZNICK, D. 1983. The structure of guppy life histories: the tradeoff between growth and reproduction. Ecology 64: 862–873.

REZNICK, D. 1985. Costs of reproduction: an evaluation of the empirical evidence. Oikos 44: 257–267.

RHEINHEIMER, G. 1980. Aquatic Microbiology (2nd ed.). Wiley, New York.

RHIJN, J. G. VAN. 1969. Behavioural dimorphism in male ruffs, *Philomachus pugnax* (L.). Behaviour 47: 153–229.

RHOADES, D. F. 1979. Evolution of chemical defense against herbivores. Pp. 3–54 in G. A. Rosenthal and D. H. Janzen (Eds.), Herbivores: Their Interactions with Secondary Plant Metabolites. Academic Press, New York.

RHODES, L. H., & J. W. GERDEMANN. 1975. Phosphate uptake zones of mycorrhizal and non-mycorrhizal onions. New Phytol. 75: 555–561.

RHODES, L. H., & J. W. GERDEMANN. 1980. Nutrient translocation in vesicular-arbuscular mycorrhizae. Pp. 173–195 in C. B. Cooks, P. W. Pappas, and E. D. Rudolph (Eds.), Cellular Interactions in Symbiosis and Parasitism. Ohio State Univ. Press, Columbus.

RICE, E. L. 1974. Allelopathy. Academic Press, New York and London.

RICE, E. L. 1984. Allelopathy (2nd ed.). Academic Press, Orlando, Florida.

RICHARDS, P. W. 1952. The Tropical Rainforest. Cambridge Univ. Press, London and New York.

RICKER, W. E. 1954. Stock and recruitment. J. Fish. Res. Bd. Canad. 11: 559–623.

RICKER, W. E. 1958. Handbook of computations for biological statistics of fish populations. Fish. Res. Bd. Canad. Bull. No. 119: 1–300.

RICKLEFS, R. E. 1969. Preliminary models for growth rates of altricial birds. Ecology 50: 1031–1039.

RICKLEFS, R. E. 1970. Stage of taxon cycle and distribution of birds on Jamaica, Greater Antilles. Evolution 24: 475–477.

RICKLEFS, R. E. 1972. Dominance and the niche in bird communities. Amer. Nat. 106: 538–545.

RICKLEFS, R. E. 1974. Energetics of reproduction in birds. Pp. 152–292 in R. A. Paynter, Jr. (Ed.), Avian Energetics. Publ. Nuttall Ornithol. Club, No. 15.

RICKLEFS, R. E. 1975. The evolution of co-operative breeding in birds. Ibis 117: 531–534.

RICKLEFS, R. E. 1977. On the evolution of reproductive strategies in birds: reproductive effort. Amer. Nat. 111: 453–478.

RICKLEFS, R. E. 1977. Environmental heterogeneity and plant species diversity: an hypothesis. Amer. Nat. 111: 376–381.

RICKLEFS, R. E. 1979. Ecology (2nd ed.). Chiron, New York.

RICKLEFS, R. E. 1979. Patterns of growth in birds. V. A comparative study of development in the starling, common tern, and Japanese quail. Auk 96: 10–30.

RICKLEFS, R. E. 1979. Adaptation, constraint, and compromise in avian postnatal development. Biol. Rev. 54: 269–290.

RICKLEFS, R. E. 1980. Geographical variation in clutch size among passerine birds: Ashmole's hypothesis. Auk 97: 38–49.

RICKLEFS, R. E. 1980. Commentary on the evolution of lek behavior in some tropical frugivorous passerines. Condor 82: 476–477.

RICKLEFS, R. E. 1983. Comparative avian demography. Pp. 1–32 in R. F. Johnston (Ed.), Current Ornithology, Vol. 1. Plenum, New York.

RICKLEFS, R. E. 1984. The optimization of growth rate in altricial birds. Ecology 65: 1602–1616.

RICKLEFS, R. E. 1984. Prolonged incubation in pelagic seabirds: a comment on Boersma's paper. Amer. Nat. 123: 710–720.

RICKLEFS, R. E. 1985. Modification of growth and development of muscles of poultry. Poultry science 64: 1563–1576.

RICKLEFS, R. E. 1987. Community diversity: relative roles of local and regional processes. Science 235: 167–171.

RICKLEFS, R. E., & G. C. BLOOM. 1977. Components of avian breeding productivity. Auk 94: 86–96.

RICKLEFS, R. E., D. COCHRAN, & E. R. PIANKA. 1981. A morphological analysis of the structure of communities of lizards in desert habitats. Ecology 62: 1474–1483.

RICKLEFS, R. E, & G. W. COX. 1972. Taxon cycles in the West Indian avifauna. Amer. Nat. 106: 195–219.

RICKLEFS, R. E., & G. W. COX. 1978. Stage of taxon cycle, habitat distribution, and population density in the avifauna of the West Indies. Amer. Nat. 112: 875–895.

RICKLEFS, R. E., & F. R. HAINSWORTH. 1968. Temperature dependent behavior of the cactus wren. Ecology 49: 227–233.

RICKLEFS, R. E., & F. R. HAINSWORTH. 1969. Temperature regulation in nestling cactus wrens. The nest environment. Condor 71: 32–37.

RICKLEFS, R. E., & K. K. MATTHEW. 1982. Chemical characteristics of the foliage of some deciduous trees in southeastern Ontario. Canad. J. Bot. 60: 2037–2045.

RICKLEFS, R. E., Z. NAVEH, & R. E. TURNER. 1984. Conservation of ecological processes. Environmentalist 4, Suppl. 8: 1–16.

RICKLEFS, R. E., & K. O'ROURKE. 1975. Aspect diversity in moths: a temperate-tropical comparison. Evolution 29: 313–324.

RICKLEFS, R. E., & J. TRAVIS. 1980. A morphological approach to the study of avian community organization. Auk 97: 321–338.

RICKLEFS, R. E., & T. WEBB. 1985. Water content, thermogenesis, and growth rate of skeletal muscles in the European starling. Auk 102: 369–376.

RIDLEY, M. 1978. Paternal care. Anim. Behav. 26: 904–932.

RIGGAN, P. J., S. GOODE, P. M. JACKS, & R. N. LOCKWOOD. 1988. Interaction of fire and community development in chaparral of southern California. Ecol. Monogr. 58: 155–176.

RIGGS, A. 1960. The nature and significance of the Bohr effect in mammalian hemoglobins. J. Gen. Physiol. 43: 737–752.

RILEY, C. V. 1892. The yucca moth and yucca pollination. Third Ann. Rept. Mo. Bot. Garden: 181–226.

RILEY, E. T., & E. E. PREPAS. 1984. Role of internal phosphorus loading in shallow, productive lakes in Alberta, Canada. Canad. J. Fish. Aquat. Sci. 41: 845–855.

RITLAND, K., & F. R. GANDERS. 1987. Covariation of selfing rates with parental gene fixation indices within populations of *Mimulus guttatus.* Evolution 41: 760–771.

RITLAND, K., & F. R. GANDERS. 1987. Crossability of *Mimulus guttatus* in relation to components of gene fixation. Evolution 41: 772–786.

RITTER, W. E. 1938. The California Woodpecker and I. A Study in Comparative Zoology. Univ. California Press, Berkeley.

ROBBINS, C. T., & A. N. MOEN. 1975. Composition and digestibility of several deciduous browses in the Northeast. J. Wildl. Mgmt. 39: 337–341.

ROBERTS, L. 1989. How fast can trees migrate? Science 243: 735–737.

ROBERTSON, A. 1970. A theory of limits in artificial selection with many linked loci. Pp. 246–288 in K. Kojima (Ed.), Mathematical Topics in Population Genetics. Springer-Verlag, New York.

ROBERTSON, B. 1908. On the normal rate of growth of an individual, and its biochemical significance. Arch. Entwicklungsmech. 25: 581–614.

ROBINSON, M. H. 1969. Defenses against visually hunting predators. Evol. Biol. 3: 225–259.

ROCKSTEIN, M. (Ed.). 1974. Theoretical Aspects of Aging. Academic Press, New York.

ROGERS, D. J. 1972. Random search and insect population models. J. Anim. Ecol. 41: 369–383.

ROHWER, F. C. 1985. The adaptive significance of clutch size of prairie ducks. Auk 102: 354–360.

ROHWER, S. 1977. Status signalling in Harris' sparrows: some experiments in deception. Behaviour 61: 107–129.

ROHWER, S. 1982. The evolution of reliable and unreliable badges of fighting ability. Amer. Zool. 22: 531–546.

ROHWER, S., & P. W. EWALD. 1981. The cost of dominance and advantage of subordination in a badge signalling system. Evolution 35: 441–454.

ROHWER, S., & F. C. ROHWER. 1978. Status signalling in Harris' sparrows: experimental deceptions achieved. Anim. Behav. 26: 1012–1022.

ROLLINSON, D., & R. M. ANDERSON. (Eds.). 1985. Ecology and Genetics of Host-Parasite Interactions. Academic Press, London.

ROOT, R. B. 1967. The niche exploitation pattern of the blue-gray gnatcatcher. Ecol. Monogr. 37: 317–350.

ROSE, M. R. 1982. Antagonistic pleiotropy, dominance, and genetic variation. Heredity 48: 63–78.

ROSE, M. R. 1984. Laboratory evolution of postponed senescence in *Drosophila melanogaster.* Evolution 38: 1004–1010.

ROSE, M. R., & B. CHARLESWORTH. 1981. Genetics of life history in *Drosophila melanogaster.* II. Exploratory selection experiments. Genetics 97: 187–196.

ROSENTHAL, G. A., D. L. DAHLMAN, & D. H. JANZEN. 1976. A novel means for dealing with L-canavanine, a toxic metabolite. Science 192: 256–258.

ROSENTHAL, G. A., & D. H. JANZEN. (Eds.). 1979. Herbivores: Their Interaction with Secondary Plant Metabolites. Academic Press, New York.

ROSENZWEIG, M. L. 1969. Why the prey curve has a hump. Amer. Nat. 103: 81–87.

ROSENZWEIG, M. L. 1973. Evolution of the predator isocline. Evolution 27: 84–94.

ROSENZWEIG, M. L. 1975. On continental steady states of species diversity. Pp. 121–140 in M. L. Cody and J. M. Diamond (Eds.), Ecology and Evolution of Communities. Harvard Univ. Press, Cambridge, Massachusetts.

ROSENZWEIG, M. L. 1981. A theory of habitat selection. Ecology 62: 327–335.

ROSENZWEIG, M. L. 1987. Habitat selection as a source of biological diversity. Evol. Ecol. 1: 315–330.

ROSENZWEIG, M. L., & R. H. MACARTHUR. 1963. Graphical representation and stability conditions of predator-prey interactions. Amer. Nat. 97: 209–223.

ROSENZWEIG, M. L., & P. W. STERNER. 1970. Population ecology of desert rodent communities: body size and seed-husking as bases for heteromyid coexistence. Ecology 51: 218–224.

ROTENBERRY, J. T. 1980. Dietary relationships among shrub steppe passerine birds: competition or opportunism in a variable environment? Ecol. Monogr. 50: 93–110.

ROTHSTEIN, S. I. 1971. Observation and experiment in the analysis of interactions between brood parasites and their hosts. Amer. Nat. 105: 71–74.

ROTHSTEIN, S. I. 1975. Evolutionary rates and host defenses against avian brood parasitism. Amer. Nat. 109: 161–176.

ROUGHGARDEN, J. 1974. Niche width: biogeographic patterns among *Anolis* lizard populations. Amer. Nat. 108: 429–442.

ROUGHGARDEN, J. 1974. Species packing and the competition function with illustrations from coral reef fish. Theoret. Pop. Biol. 5: 163–186.

ROUGHGARDEN, J. 1976. Resource partitioning among competing species—a coevolutionary approach. Theoret. Pop. Biol. 9: 388–424.

ROUGHGARDEN, J. 1989. The structure and assembly of communities. Pp. 203–226 in J. Roughgarden, R. M. May, & S. A. Levin (Eds.), Perspectives in Ecological Theory. Princeton Univ. Press, Princeton, New Jersey.

ROUGHGARDEN, J., & M. FELDMAN. 1975. Species packing and predation pressure. Ecology 56: 489–492.

ROVIRA, A. D. 1965. Interactions between plant roots and soil microorganisms. Ann. Rev. Microbiol. 19: 241–266.

ROWELL, T. E. 1967. Variability in the social organization of primates. Pp. 219–235 in D. Morris (Ed.), Primate Ethology. Aldine, Chicago.

ROYAMA, T. 1970. Factors governing the hunting behaviour and selection of food by the great tit *(Parus major* L.). J. Anim. Ecol. 39: 619–659.

ROYAMA, T. 1971. A comparative study of models for predation and parasitism. Res. Pop. Ecol., Suppl. 1: 1–91.

RUANO, R. G., F. OROZCO, & C. LOPEZ-FANJUL. 1975. The effect of different selection intensities on selection response in egg-laying of *Tribolium castaneum.* Genet. Res. 25: 17–27.

RUBINOFF, I. 1968. Central America sea-level canal: possible biological effects. Science 161: 857–861.

RUNKLE, J. R. 1982. Patterns of disturbance in some old-growth mesic forests of eastern North America. Ecology 63: 1533–1546.

RUNKLE, J. R. 1985. Disturbance regimes in temperate forests. Pp. 17–33 in S. T. A. Pickett and P. S. White (Eds.), The Ecology of Natural Disturbance and Patch Dynamics. Academic Press, Orlando, Florida.

RUSSELL, E. W. 1961. Soil Conditions and Plant Growth (9th ed.). Wiley, New York.

RYTHER, J. H. 1956. Photosynthesis in the ocean as a function of light intensity. Limnol. Oceanogr. 1: 61–70.

RYTHER, J. H., & W.M. DUNSTAN. 1971. Nitrogen, phosphorus, and eutrophication in the coastal marine environment. Science 171: 1008–1013.

RYTHER, J. H., & C. S. YENTSCH. 1957. The estimation of phytoplankton production in the ocean from chlorophyll and light data. Limnol. Oceanogr. 2: 281–286.

SADLIER, R. M. F. S. 1965. The relationship between agonistic behaviour and population changes in the deermouse, *Peromyscus maniculatus* (Wagner). J. Anim. Ecol. 34: 331–352.

SAI, F. T. 1984. The population factor in Africa's development dilemma. Science 226: 801–805.

SAKAI, K. 1961. Competitive ability in plants: its inheritance and some related problems. Symp. Soc. Exptl. Biol. 15: 245–263.

SALE, P. F. 1977. Maintenance of high diversity in coral reef fish communities. Amer. Nat. 111: 337–359.

SALE, P. F. 1978. Coexistence of coral reef fishes—a lottery for living space. Environ. Biol. Fishes 3: 85–102.

SALE, P. F. 1980. The ecology of fishes on coral reefs. Ann. Rev. Oceanogr. Mar. Biol. 18: 367–421.

SALE, P. F., & D. MCB. WILLIAMS. 1982. Community structure of coral reef fishes: are the patterns more than those expected by chance? Amer. Nat. 120: 121–127.

SALISBURY, E. J. 1942. The Reproductive Capacity of Plants. Studies in Quantitative Biology. G. Bell and Sons, London.

SALT, G. W. 1979. A comment on the use of the term *emergent properties.* Amer. Nat. 113: 145–161.

SANDERS, H. L. 1969. Benthic marine diversity and the stability-time hypothesis. Brookhaven Symp. Biol. 22: 71–81.

SANFORD, R. L., JR. 1987. Apogeotropic roots in an Amazon rain forest. Science 235: 1062–1064.

SARGENT, T. D. 1966. Background selections of geometrid and noctuid moths. Science 154: 1674–1675.

SARGENT, T. D. 1969. Background selections of the pale and melanic forms of the cryptic moth, *Phigalia titea* (Cramer). Nature 222: 585–586.

SAUNDERS, P. T., & M. J. BAZIN. 1975. Stability of complex ecosystems. Nature 256: 120.

SAUNDERS, P. T., & M. J. BAZIN. 1975. On the stability of food chains. J. Theoret. Biol. 52: 121–142.

SCHAD, G. A. 1966. Immunity, competition, and natural regulation of helminth populations. Amer. Nat. 100: 359–364.

SCHAFFER, W. M. 1974. Selection for optimal life histories: the effects of age structure. Ecology 53: 291–303.

SCHAFFER, W. M. 1974. Optimal reproductive effort in fluctuating environments. Amer. Nat. 108: 783–790.

SCHAFFER, W. M. 1984. Stretching and folding in lynx fur returns: evidence for a strange attractor in nature? Amer. Nat. 124: 798–820.

SCHAFFER, W. M., & P. F. ELSON. 1975. The adaptive significance of variations in life history among local populations of Atlantic salmon in North America. Ecology 56: 577–590.

SCHAFFER, W. M., & M. V. SCHAFFER. 1977. The adaptive significance of variations in reproductive habit in the Agavaceae. Pp. 261–276 in B. Stonehouse and C. Perrins (Eds.), Evolutionary Ecology. Macmillan, London.

SCHALL, J. J., A. F. BENNETT, & R. W. PUTNAM. 1982. Lizards infected with malaria: physiological and behavioral consequences. Science 217: 1057–1059.

SCHALLER, G. B. 1972. The Serengeti Lion: A Study of Predator-Prey Relations. Univ. Chicago Press, Chicago.

SCHEMSKE, D. W. 1984. Population structure and local selection in *Impatiens pallida* (Balsaminaceae), a selfing annual. Evolution 38: 817–832.

SCHINDLER, D. W. 1974. Eutrophication and recovery in experimental lakes: implications for lake management. Science 184: 897–899.

SCHINDLER, D. W. 1988. Effects of acid rain on freshwater ecosystems. Science 239: 149–157.

SCHLUTER, D., T. D. PRICE, & P. R. GRANT. 1985. Ecological character displacement in Darwin's finches. Science 227: 1056–1059.

SCHMIDT-NIELSON, K. 1964. Desert Animals: Physiological Problems of Heat and Water. Clarendon Press, Oxford.

SCHMIDT-NIELSEN, K. 1972. How Animals Work. Cambridge Univ. Press, Cambridge.

SCHMIDT-NIELSEN, K. 1975. Animal Physiology. Adaptation and Environment. Cambridge Univ. Press, London and New York.

SCHMIDT-NIELSEN, K. 1983. Animal Physiology. Adaptations and Environment (3rd ed.). Cambridge Univ. Press, London and New York.

SCHMIDT-NIELSEN, K. 1984. Scaling. Why Is Animal Size So Important? Cambridge Univ. Press, Cambridge.

SCHMIDT-NIELSEN, K., F. R. HAINSWORTH, & D. E. MURRISH. 1970. Counter-current heat exchange in the respiratory passages: effect on water and heat balance. Resp. Physiol. 9: 263–276.

SCHMIDT-NIELSEN, K., & B. SCHMIDT-NIELSEN. 1952. Water metabolism of desert mammals. Physiol. Rev. 32: 135–166.

SCHMIDT-NIELSEN, K., & B. SCHMIDT-NIELSEN. 1953. The desert rat. Sci. Amer. 189: 73–78.

SCHMIDT-NIELSEN, K., & C. R. TAYLOR. 1968. Red blood cells: why or why not? Science 162: 274–275.

SCHNEIDER, S. H. 1989. The greenhouse effect: science and policy. Science 243: 771–782.

SCHNEIRLA, T. C. 1956. A preliminary survey of colony division and related processes in two species of terrestrial army ants. Insects Sociaux 3: 49–69.

SCHOEN, D. J. 1982. Genetic variation in the breeding system of *Gilia achilleifolia.* Evolution 36: 361–370.

SCHOEN, D. J. 1983. Relative fitnesses of selfed and outcrossed progeny in *Gilia achilleifolia* (Polemoniaceae). Evolution 37: 292–301.

SCHOENER, T. W. 1965. The evolution of bill size differences among sympatric congeneric species of birds. Evolution 19: 189–213.

SCHOENER, T. W. 1968. The *Anolis* lizards of Bimini: resource partitioning in a complex fauna. Ecology 49: 704–726.

SCHOENER, T. W. 1969. Optimal size and specialization in constant and fluctuating environments: an energy time approach. Brookhaven Symp Biol. 22: 103–114.

SCHOENER, T. W. 1969. Models of optimal size for solitary predators. Amer. Nat. 103: 277–313.

SCHOENER, T. W. 1971. Theory of feeding strategies. Ann. Rev. Ecol. Syst. 2: 369–404.

SCHOENER, T. W. 1974. Resource partitioning in ecological communities. Science 185: 27–39.

SCHOENER, T. W. 1974. The species-area relation within archipelagos: models and evidence from island land birds. Proc. 16th Intern. Ornithol. Congr., Canberra: 629–642.

SCHOENER, T. W. 1983. Field experiments on interspecific competition. Amer. Nat. 122: 240–285.

SCHOENER, T. W., & D. A. SPILLER. 1987. Effect of lizards on spider populations: manipulative reconstruction of a natural experiment. Science 236: 949–952.

SCHOLANDER, P. F. 1955. Evolution of climatic adaptation in homeotherms. Evolution 9: 15–26.

SCHOLANDER, P. F. 1968. How mangroves desalinate seawater. Physiol. Plant. 21: 251–261.

SCHOLANDER, P. F., & W. E. SCHEVILL. 1955. Countercurrent vascular heat exchange in the fins of whales. J. Appl. Physiol. 8: 279–282.

SCHOPF, J. W. (Ed.). 1983. Earth's Earliest Biosphere. Its Origin and Evolution. Princeton Univ. Press, Princeton, New Jersey.

SCHREIBER, R. W., & E. A. SCHREIBER. 1984. Central Pacific seabirds and the El Niño/Southern Oscillation: 1982 to 1983 perspectives. Science 225: 713–716.

SCHRODER, G. D., & M. L. ROSENZWEIG. 1975. Perturbation analysis of competition and overlap in habitat utilization between *Dipodomys ordii* and *Dipodomys merriami.* Oecologia 19: 9–28.

SCHULTZ, A. M., J. L. LAUNCHBAUGH, & H. H. BISWELL. 1955. Relationship between grass diversity and brush seedling survival. Ecology 36: 226–238.

SCHULZ, J. P. 1960. Ecological Studies on Rain Forest in Northern Suriname. North-Holland Publ. Amsterdam.

SCHUM, M. 1984. Phenetic structure and species richness in North and Central American bat faunas. Ecology 65: 1315–1324.

SCHWAB, J. H. 1975. Suppression of the immune response by microorganisms. Bact. Rev. 39: 121–143.

SCRIBER, J. M. 1984. Host-plant suitability. Pp. 159–202 in W. J. Bell and R. T. Cardé (Eds.), Chemical Ecology of Insects. Chapman and Hall, London.

SCRIBER, J. M., & P. FEENY. 1979. Growth of herbivorous caterpillars in relation to feeding specialization and to the growth form of their food plants. Ecology 60: 829–850.

SCRIBER, J. M., & F. SLANSKY. 1981. The nutritional ecology of immature insects. Ann. Rev. Entomol. 26: 183–211.

SEARCY, W. A., & M. ANDERSSON. 1986. Sexual selection and the evolution of song. Ann. Rev. Ecol. Syst. 17: 507–533.

SEBER, G. A. F. 1965. A note on the multiple recapture census. Biometrika 52: 249–259.

SEBER, G. A. F. 1973. Estimation of Animal Abundance. Hafner, New York.

SEIGLER, D., & P. W. PRICE. 1976. Secondary compounds in plants: primary functions. Amer. Nat. 110: 101–105.

SEPKOSKI, J. J. 1978. A kinetic model of Phanerozoic taxonomic diversity. I. Analysis of marine orders. Paleobiology 4: 223–251.

SEPKOSKI, J. J., R. K. BAMBACH, D. M. RAUP, & J. W. VALENTINE. 1981. Phanerozoic marine diversity and the fossil record. Nature 293: 435–437.

SEYMOUR, R. S., & H. RAHN. 1978. Gas conductance in the eggshell of the mound-building brush turkey. Pp. 243–246 in J. Piiper (Ed.), Respiratory Function in Birds, Adult and Embryonic. Springer-Verlag, New York.

SHAANKER, R. U., K. N. GANESHAIAH, & K. S. BAWA. 1988. Parent-offspring conflict, sibling rivalry, and brood size patterns in plants. Ann. Rev. Ecol. Syst. 19: 177–205.

SHANKS, R. E., & J. S. OLSON. 1961. First-year breakdown of leaf litter in southern Appalachian forests. Science 134: 194–195.

SHANMUGAM, K. T., F. O'GARA, K. ANDERSEN, & R. C. VALENTINE. 1978. Biological nitrogen fixation. Ann. Rev. Plant Physiol. 29: 263–276.

SHANNON, C. E., & W. WEAVER. 1949. The Mathematical Theory of Communication. Univ. Illinois Press, Urbana.

SHAW, R. F., & J. D. MOHLER. 1953. The selective advantage of the sex ratio. Amer. Nat. 87: 337–342.

SHELDON, A. L. 1969. Equitability indices: dependence on the species count. Ecology 50: 466–467.

SHELFORD, V. E. 1931. Some concepts of bioecology. Ecology 12: 455–467.

SHELFORD, V. E., & W. P. FLINT. 1943. Populations of the chinch bug in the Upper Mississippi Valley from 1823 to 1940. Ecology 24: 435–455.

SHEPPARD, D. H. 1971. Competition between two chipmunk species *(Eutamias)*. Ecology 52: 320–329.

SHERMAN, P. W. 1977. Nepotism and the evolution of alarm calls. Science 197: 1246–1253.

SHERMAN, P. W. 1985. Alarm calls of Belding's ground squirrels to aerial predators: nepotism or self-preservation? Behav. Ecol. Sociobiol. 17: 313–323.

SHERMAN, P. W., & W. G. HOLMES. 1985. Kin recognition: issues and evidence. Fortsch. Zool., Exper. Behav. Ecol. 31: 437–460.

SHERRY, D. F. 1984. Food storage by black-capped chickadees: memory for the location and contents of caches. Anim. Behav. 32: 451–464.

SHETTLEWORTH, S. J. 1984. Learning and behavioural ecology. Pp. 170–194 in J. R. Krebs and N. B. Davies (Eds.), Behavioural Ecology. An Evolutionary Approach (2nd ed.). Sinauer, Sunderland, Massachusetts.

SHIMWELL, D. W. 1971. Description and Classification of Vegetation. Univ. Washington Press, Seattle.

SHINE, R., & J. J. BULL. 1979. The evolution of live-bearing in snakes and lizards. Amer. Nat. 113: 905–923.

SHMIDA, A., & M. V. WILSON. 1985. Biological determinants of species diversity. J. Biogeogr. 12: 1–20.

SHMIDA, H. 1981. Mediterranean vegetation in California and Israel: similarities and differences. Israel J. Bot. 30: 105–123.

SHORT, H. L. 1971. Forage digestibility and diet of deer on southern upland range. J. Wildl. Mgmt. 35: 698–706.

SHREVE, F. 1936. The transition from desert to chaparral in Baja California. Madroño 3: 257–264.

SHUGART, H. H. 1984. A Theory of Forest Dynamics. The Ecological Implications of Forest Succession Models. Springer-Verlag, New York.

SIGG, H., & A. STOLBA. 1981. Home range and daily march in a hamadryas baboon troop. Folia Primatol. 36: 40–75.

SIH, A. 1984. Optimal behavior and density dependent predation. Amer. Nat. 123: 314–326.

SIH, A., P. CROWLEY, M. MCPEEK, J. PETRANKA, & K. STROHMEIER. 1985. Predation, competition, and prey communities: a review of field experiments. Ann. Rev. Ecol. Syst. 16: 269–311.

SILANDER, J. A., JR. 1985. The genetic basis of the ecological amplitude of *Spartina patens*. II. Variance and correlation analysis. Evolution 39: 1034–1052.

SILBERGLIED, R. E., & T. EISNER. 1969. Mimicry of hymenoptera by beetles with unconventional flight. Science 163: 486–488.

SILLEN-TULLBERG, B., & O. LEIMAR. 1988. The evolution of gregariousness in distasteful insects as a defense against predators. Amer. Nat. 132: 723–734.

SILVERTOWN, J. W. 1982. Introduction to Plant Population Ecology. Longman, London.

SIMBERLOFF, D. S. 1969. Experimental zoogeography of islands: a model for insular colonization. Ecology 50: 296–314.

SIMBERLOFF, D. S. 1976. Species turnover and equilibrium island biogeography. Science 194: 572–578.

SIMBERLOFF, D. S. 1978. Using biogeographic distributions to determine if colonization is stochastic. Amer. Nat. 112: 723–726.

SIMBERLOFF, D. S. 1979. Rarefaction as a distribution-free method of expressing and estimating diversity. Pp. 159–176 in J. F. Grassle, G. P. Patil, W. K. Smith, and C. Taillie (Eds.), Ecological Diversity in Theory and Practice. Intern. Co-op. Publ. House, Fairland, Maryland.

SIMBERLOFF, D. S. 1983. Competition theory, hypothesis testing, and other community ecological buzzwords. Amer. Nat. 122: 626–635.

SIMBERLOFF, D. S., & W. BOECKLEN. 1981. Santa Rosalia reconsidered: size ratios and competition. Evolution 35: 1206–1228.

SIMBERLOFF, D. S., & E. O. WILSON. 1969. Experimental zoogeography of islands: the colonization of empty islands. Ecology 50: 278–296.

SIMPSON, B. B., AND J. HAFFER. 1978. Speciation patterns in the Amazonian forest biota. Ann. Rev. Ecol. Syst. 9: 497–518.

SIMPSON, E. H. 1949. Measurement of diversity. Nature 163: 688.

SIMPSON, G. G. 1964. Species density of North American recent mammals. Syst. Zool. 13: 57–73.

SIMPSON, G. G. 1969. The first three billion years of community evolution. Brookhaven Symp. Biol. 22: 162–177.

SINCLAIR, A. R. E., & J. M. FRYWELL. 1985. The Sahel of Africa: ecology of a disaster. Canad. J. Zool. 63: 987–994.

SINENSKY, M. 1974. Homeoviscous adaptation—a homeostatic process that regulates the viscosity of membrane lipids in *Escherichia coli.* Proc. Natl. Acad. Sci. 71: 522–525.

SINGER, M. 1983. Determinants of multiple host use by a phytophagous insect population. Evolution 37: 389–403.

SINGER, S. J., & G. L. NICHOLSON. 1972. The fluid mosaic model of the structure of cell membranes. Science 175: 720–731.

SINGLETON, R., JR., & R. E. AMELUNXEN. 1973. Proteins from thermophilic microorganisms. Bacteriol. Rev. 37: 320–342.

SKOGLAND, T. 1985. The effects of density-dependent resource limitations on the demography of wild reindeer. J. Anim. Ecol. 54: 359–374.

SKUTCH, A. F. 1935. Helpers at the nest. Auk 52: 257–273.

SKUTCH, A. F. 1949. Do tropical birds rear as many young as they can nourish? Ibis 91: 430–455.

SKUTCH, A. F. 1961. Helpers among birds. Condor 63: 198–226.

SLATKIN, M. 1974. Competition and regional coexistence. Ecology 55: 128–134.

SLATYER, R. O. 1967. Plant-Water Relationships. Academic Press, New York.

SLEIGH, M. 1973. The Biology of Protozoa. American Elsevier, New York.

SLOBODKIN, L. B. 1961. Growth and Regulation of Animal Populations. Holt, Rinehart & Winston, New York.

SLOBODKIN, L. B., F. E. SMITH, & N. G. HAIRSTON. 1967. Regulation in terrestrial systems, and the implied balance of nature. Amer. Nat. 101: 109–124.

SMALLEY, A. E. 1960. Energy flow of a salt marsh grasshopper population. Ecology 41: 672–677.

SMILEY, J. 1978. Plant chemistry and the evolution of host specificity: new evidence from *Heliconius* and *Passiflora.* Science 201: 745–747.

SMITH, F. E. 1961. Density dependence in the Australian thrips. Ecology 42: 403–407.

SMITH, H. S. 1935. The role of biotic factors in the determination of population densities. J. Econ. Entomol. 28: 873–898.

SMITH, J. M. 1964. Group selection and kin selection. Nature 201: 1145–1147.

SMITH, J. M. 1974. The theory of games and the evolution of animal conflicts. J. Theoret. Biol. 47: 209–221.

SMITH, J. M. 1976. Sexual selection and the handicap principle. J. Theoret. Biol. 57: 239–242.

SMITH, J. M. 1978. The handicap principle—a comment. J. Theoret. Biol. 70: 251–252.

SMITH, J. M. 1978. The Evolution of Sex. Cambridge Univ. Press, Cambridge.

SMITH, J. M. 1982. Evolution and the Theory of Games. Cambridge Univ. Press, Cambridge.

SMITH, J. M., & G. R. PRICE. 1973. The logic of animal conflict. Nature 246: 15–18.

SMITH, N. G. 1968. The advantage of being parasitized. Nature 219: 690–694.

SMITH, N. G. 1969. Provoked release of mobbing—a hunting technique of *Micraster* falcons. Ibis 111: 241–243.

SMITH, N. G. 1972. Migrations of the day-flying moth *Urania* in Central and South America. Carib. J. Sci. 12: 45–58.

SMITH, S. A., C. W. THAYER, & C. E. BRETT. 1985. Predation in the Paleozoic: gastropod-like drillholes in Devonian brachiopods. Science 230: 1033–1035.

SMITH, S. E. 1980. Mycorrhizas of autotrophic higher plants. Biol. Rev. 55: 474–510.

SMITH, S. M. 1975. Innate recognition of coral snake pattern by a possible avian predator. Science 187: 759–760.

SMITH, S. M. 1977. Coral-snake pattern recognition and stimulus generalisation by naive great kiskadees (Aves: Tyrannidae). Nature 265: 535–536.

SMITH, W. G. 1975. Dynamics of pure and mixed populations of *Desmodium glutinosum* and *D. nudiflorum* in natural oak forest communities. Amer. Midl. Nat. 94: 99–107.

SMITH TRAIL, D. R. 1980. Behavioral interactions between parasites and hosts: host suicide and the evolution of complex life cycles. Amer. Nat. 116: 77–91.

SMITHERS, S. R., R. J. TERRY, & D. J. HOCKLEY. 1969. Host antigens in schistosmiasis. Proc. Roy. Soc. London B 171: 483–494.

SMUTS, B. B., D. L. CHENEY, R. M. SEYFARTH, R. W. WRANGHAM, & T. T. STRUHSAKER. (Eds.). 1987. Primate Societies. Univ. Chicago Press, Chicago.

SNELLING, R. R. 1981. Systematics of social Hymenoptera. Pp. 369–453 in H. R. Hermann (Ed.), Social Insects, Vol. II. Academic Press, New York.

SNOW, D. 1982. The Cotingas. Cornell Univ. Press, Ithaca, New York.

SNYDER, L. L. 1947. The snowy owl migration of 1945–1946: second report of the Snowy Owl Committee. Wilson Bull. 59: 74–78.

SOLBRIG, O. T. 1980. Demography and Evolution in Plant Populations. Blackwell, Oxford.

SOLOMON, M. E. 1957. Dynamics of insect populations. Ann. Rev. Entomol. 2: 121–142.

SOMERO, G. N. 1978. Temperature adaptation of enzymes: biological optimization through structure-function compromises. Ann. Rev. Ecol. Syst. 9: 1–29.

SOMME, L. 1964. Effects of glycerol in cold hardiness in insects. Canad. J. Zool. 42: 87–101.

SOULSBY, E. J. L. 1968. Helminths, Arthropods, and Protozoa of Domestic Animals. Williams and Wilkins, Baltimore.

SOUSA, W. P. 1984. Intertidal mosaics: patch size, propagule availability, and spatially variable patterns of succession. Ecology 65: 1918–1935.

SOUSA, W. P. 1985. Disturbance and patch dynamics on rocky intertidal shores. Pp. 101–124 in S. T. A. Pickett and P. S. White (Eds.), The Ecology of Natural Disturbance and Patch Dynamics. Academic Press, Orlando, Florida.

SOUTHWICK, C. H. 1955. The population dynamics of confined house mice supplied with unlimited food. Ecology 36: 212–225.

SOUTHWICK, C. H. 1955. Regulatory mechanisms of house mouse populations: social behavior affecting litter survival. Ecology 36: 627–634.

SOUTHWOOD, T. R. E. 1961. The number of species on insect associated with various trees. J. Anim. Ecol. 30: 1–8.

SOUTHWOOD, T. R. E. 1978. Ecological Methods (2nd ed.). Chapman and Hall, London.

SOUTHWOOD, T. R. E., V. C. MORAN, & C. E. J. KENNEDY. 1982. The richness, abundance and biomass of the arthropod communities on trees. J. Anim. Ecol. 51: 635–649.

SPAIN, A. V. 1984. Litterfall and the standing crop of litter in three tropical Australian rainforests. J. Ecol. 72: 947–961.

SPRUGEL, D. G. 1976. Dynamic structure of wave-regenerated *Abies balsamea* forests in the north-eastern United States. J. Ecol. 64: 889–911.

SPRUGEL, D. G., & F. H. BORMANN. 1981. Natural disturbance and the steady-state in high-altitude balsam fir forests. Science 211: 390–393.

STAIRS, G. 1972. Pathogenic micro-organisms in the regulation of forest insect populations. Ann. Rev. Entomol. 17: 355–372.

STAMMBACH, E. 1987. Desert, forest and montane baboons: multilevel-societies. Pp. 112–120 in B. B. Smuts, D. L. Cheney, R. M. Seyfarth, R. W. Wrangham, and T. T. Struhsaker (Eds.), Primate Societies. Univ. Chicago Press, Chicago.

STANHILL, G. 1970. The water flux in temperate forests: precipitation and evapotranspiration. Pp. 242–256 in D. E. Reichle (Ed.), Analysis of Temperate Forest Ecosystems. Springer-Verlag, New York.

STANTON, M. L., A. A. SNOW, & S. N. HANDEL. 1986. Floral evolution, attractiveness to pollinators increases male fitness. Science 232: 1625–1627.

STARK, N., & C. F. JORDAN. 1978. Nutrient retention by the root mat of the Amazonian rain forest. Ecology 59: 434–437.

STEARNS, S. C. 1976. Life-history tactics: a review of the ideas. Quart. Rev. Biol. 51: 3–47.

STEELE, J. H. 1962. Environmental control of photosynthesis in the sea. Limnol. Oceanogr. 6: 137–150.

STEEMAN NIELSEN, E. 1951. Measurement of the production of organic matter in the sea by means of carbon-14. Nature. 167: 684–685.

STEFFAN, W. A. 1973. Polymorphism in *Plastosciara perniciosa.* Science 182: 1265–1266.

STEHLI, F. G. 1968. Taxonomic gradients in pole location: the recent model. Pp. 163–227 in E. T. Drake (Ed.), Evolution and Environment. Yale Univ. Press, New Haven, Connecticut.

STEHLI, F. G., R. G. DOUGLAS, & N. D. NEWELL. 1969. Generation and maintenance of gradients in taxonomic diversity. Science 164: 947–949.

STEPHENS, D. W. 1981. The logic of risk-sensitive foraging preferences. Anim. Behav. 29: 628–629.

STEPHENS, D. W., & J. R. KREBS. 1986. Foraging Theory. Princeton Univ. Press, Princeton, New Jersey.

STEPHENSON, A. G. 1981. Flower and fruit abortion: proximate causes and ultimate functions. Ann. Rev. Ecol. Syst. 12: 253–279.

STEPHENSON, A. G., & R. I. BERTIN. 1963. Male competition, female choice, and sexual selection in plants. Pp. 110–149 in L. Real (Ed.), Pollination Biology. Academic Press, New York.

STERNER, R. W. 1986. Herbivores' direct and indirect effects on algal populations. Science 231: 605–607.

STEVENS, G. C. 1983. Patterns of plant use by wood-boring insects. Ph.D. thesis, Univ. of Pennsylvania, Philadelphia.

STEVENS, G. C. 1986. Dissection of the species-area relationship among wood-boring insects and their host plants. Amer. Nat. 128: 35–46.

STEVENS, G. C. 1989. The latitudinal gradient in geographical range: how so many species coexist in the tropics. Amer. Nat. 133: 240–256.

STEVENS, G., & R. BELLIG. (Eds.). 1988. The Evolution of Sex. Harper & Row, San Francisco.

STEVENSON, R. D. 1985. The relative importance of behavioral and physiological adjustments controlling body temperature in terrestrial ecotherms. Amer. Nat. 126: 362–386.

STEWART, W. D. P. 1975. Nitrogen Fixation by Free-living Microorganisms. Cambridge Univ. Press, Cambridge.

STOECKER, R. E. 1972. Competitive relations between sympatric populations of voles (*Microtus montanus* and *M. pennsylvanicus*). J. Anim. Ecol. 41: 311–329.

STRECKER, R. L. 1954. Regulatory mechanisms in house mouse populations: the effect of limited food supply on an unconfined population. Ecology 35: 249–253.

STRECKER, R. L., & J. T. EMLEN, JR. 1953. Regulatory mechanisms in house mouse populations: the effect of limited food supply on a confined population. Ecology 34: 375–385.

STREHLER, B. L. (Ed.). 1960. The Biology of Aging. Amer. Inst. Biol. Sci., Washington, D.C.

STRICKLAND, J. D. H. 1960. Measuring the production of marine phytoplankton. Bull. Fish. Res. Bd. Canad. 12: 1–172.

STRICKLAND, J. D. H., & T. R. PARSONS. 1968. A manual of sea water analysis. Bull. Fish. Res. Bd. Canad. 125: 1–311.

STRONG, D. R. 1974. Rapid asymptotic species accumulation in phytophagous insect communities: the pests of cacao. Science 185: 1064–1066.

STRONG, D. R., J. H. LAWTON, & T. R. E. SOUTHWOOD. 1984. Insects on Plants. Community Patterns and Mechanisms. Harvard Univ. Press, Cambridge, Massachusetts.

STRONG, D. R., JR., L. A. SZYSKA, & D. S. SIMBERLOFF. 1979. Tests of community-wide character displacement against null hypotheses. Evolution 33: 897–913.

STROSS, R. G. 1969. Photoperiod control of diapause in *Daphnia*, II: Induction of winter diapause in the arctic. Biol. Bull. 136: 264–273.

STROSS, R. G., & V. C. HILL. 1965. Diapause induction in *Daphnia* requires two stimuli. Science 150: 1462–1464.

STUBBLEFIELD, S. P., T. N. TAYLOR, & J. M. TRAPPE. 1987. Fossil mycorrhizae: a case for symbiosis. Science 237: 59–60.

STUMM, W., & J. J. MORGAN. 1981. Aquatic Chemistry (2nd ed.). Wiley, New York.

SUBBA ROA, N. S. (Ed.). 1980. Recent Advances in Biological Nitrogen Fixation. Edward Arnold, London.

SUGIHARA, G. 1980. Minimal community structure: an explanation of species abundance patterns. Amer. Nat. 116: 770–787.

SUMMERHAYES, V. S., & C. S. ELTON. 1923. Contributions to the ecology of Spitsbergen and Bear Island. J. Ecol. 11: 214–286.

SVARDSON, G. 1957. The "invasion" type of bird migration. Brit. Birds 50: 314–343.

SVERDRUP, H. U. 1945. Oceanography for Meterologists. Allen and Unwin, London.

SVERDRUP, H. U. 1953. On conditions for the vernal blooming of phytoplankton. J. Cons. 18: 287–295.

SWAIN, T. 1979. Tannins and lignins. Pp. 657–682 in G. A. Rosenthal and D. H. Janzen (Eds.), Herbivores: Their Interactions with Secondary Plant Metabolities. Academic Press, New York.

SWANBERG, O. 1951. Food storage, territory and song in the thick-billed nutcracker. Proc. Tenth Inter. Congr. Ornithol. 545–554.

SWENSON, M. J. (Eds.). 1977. Dukes' Physiology of Domestic Animals (9th ed.). Cornell Univ. Press, Ithaca, New York.

TAMM, S. 1989. Importance of energy costs in central place foraging by hummingbirds. Ecology 70: 195–205.

TANSLEY, A. G. 1917. On competition between *Galium saxatile* L. (*G. hercynicum* Weig.) and *Galium sylvestre* poll. (*G. asperum* Schreb.) on different types of soil. J. Ecol. 5: 173–179.

TANSLEY, A. G. 1935. The use and abuse of vegetational concepts and terms. Ecology 16: 204–307.

TANSLEY, A. G., & R. S. ADAMSON. 1925. Studies of the vegetation of the English chalk. III. The chalk grasslands of the Hampshire-Sussex border. J. Ecol. 13: 177–223.

TAYLOR, B. R., D. PARKINSON, & W. F. J. PARSONS. 1989. Nitrogen and lignin content as predictors of litter decay rates: a microcosm test. Ecology 70: 97–104.

TAYLOR, C. R., & C. CONDRA. 1980. *r* and *K* selection in *Drosophila pseudoobscura.* Evolution 34: 1183–1193.

TEAL, J. M. 1957. Community metabolism in a temperate cold spring. Ecol. Monogr. 27: 283–302.

TEAL, J. M. 1962. Energy flow in the salt marsh ecosystem of Georgia. Ecology 43: 614–624.

TEERI, J., & L. STOWE. 1976. Climatic patterns and the distribution of C_4 grasses in North America. Oecologia 23: 1–12.

TEMPLE, S. A. 1977. Plant-animal mutualism: coevolution with dodo leads to near extinction of plant. Science 197: 885–886.

TEMPLETON, A. R., & E. D. ROTHMAN. 1974. Evolution in heterogenous environments. Amer. Nat. 108: 409–428.

TEN CATE, C., & P. BATESON. 1988. Sexual selection: the evolution of conspicuous characteristics in birds by means of imprinting. Evolution 42: 1355–1358.

TERAO, A., & T. TANAKA. 1928. Influence of temperature upon the rate of reproduction in the water-flea *Moina macrocopa* Strauss. Proc. Imper. Acad. Japan 4: 553–555.

TERBORGH, J. 1971. Distribution on environmental gradients: theory and a preliminary interpretation of distributional patterns in the avifauna of the Cordillera Vilcabamba, Peru. Ecology 52: 23–40.

TERBORGH, J. 1980. Causes of tropical species diversity. Acta XVII Congr. Intern. Ornithol. Berlin: 955–961.

TERBORGH, J. 1985. The role of ecotones in the distribution of Andean birds. Ecology 66: 1237–1246.

TERBORGH, J., & J. FAABORG. 1973. Turnover and ecological release in the avifauna of Mona Island, Puerto Rico. Auk 90: 759–779.

TERBORGH, J. W., & J. FAABORG. 1980. Saturation of bird communities in the West Indies. Amer. Nat. 116: 178–195.

TESSIER, A. J., L. L. HENRY, & C. E. GOULDEN. 1983. Starvation in *Daphnia:* energy reserves and reproductive allocation. Limnol. Oceanogr. 28: 667–676.

TEVIS, L., JR., & I. M. NEWELL. 1962. Studies on the biology and seasonal cycle of the giant red velvet mite, *Dinothrombium pandorae* (Acari, Thrombidiidae). Ecology 43: 497–505.

THIBAULT, R. E., & R. J. SCHULTZ. 1978. Reproductive adaptations among viviparous fishes (Cyprinodontiformes: Poeciliidae). Evolution 32: 320–333.

THOMAS, M. D. 1955. Effect of ecological factors on photosynthesis. Ann. Rev. Plant Physiol. 6: 135–156.

THOMPSON, D. Q. 1952. Travel, range and food habits of timber wolves in Wisconsin. J. Mammal. 33: 429–442.

THOMPSON, J. N., JR., & J. M. THODAY. (Eds.). 1979. Quantitative Genetic Variation. Academic Press, New York.

THOMPSON, W. R. 1924. La théorie mathématique de l'action des parasites entomophages et le facteur du hasard. Annls. Fac. Sci. Marseille 2: 69–89.

THORNE, R. F. 1972. Major disjunctions in the geographic ranges of seed plants. Quart. Rev. Biol. 47: 365–411.

THORNTHWAITE, C. W. 1948. An approach to a rational classification of climate. Geogr. Rev. 38: 55–94.

THORSON, G. 1950. Reproduction and larval ecology of marine bottom invertebrates. Biol. Rev. 25: 1–45.

THORSON, T. B., C. M. COWAN, & D. E. WATSON. 1967. *Potamotrygon* spp.: elasmobranchs with low urea content. Science 158: 373–377.

TILMAN, D. 1982. Resource Competition and Community Structure. Princeton Univ. Press, Princeton, New Jersey.

TILMAN, G. D. 1984. Plant dominance along and experimental nutrient gradient. Ecology 65: 1445–1453.

TILMAN, D., M. MATTSON, & S. LANGER. 1981. Competition and nutrient kinetics along a temperature gradient: an experimental test of a mechanistic approach to niche theory. Limnol. Oceanogr. 26: 1020–1033.

TINBERGEN, L. 1969. The natural control of insects in pinewoods, 1: factors influencing the intensity of predation by songbirds. Arch. Neerl. Zool. 13: 266–336.

TINBERGEN, N., G. J. BROEKHYSEN, F. FEEKES, J. C. W. HOUGHTON, H. KRUUK, & E. SZUK. 1962. Egg shell removal by the black-headed gull, *Larus ridibundus* L.: a behaviour component of camouflage. Behaviour 19: 74–117.

TINKLE, D. W., & R. E. BALLINGER. 1972. *Sceloporus undulatus:* a study of the intraspecific comparative demography of a lizard. Ecology 53: 570–584.

TINKLE, D. W., H. M. WILBUR, & S. G. TILLEY. 1970. Evolutionary strategies in lizard reproduction. Evolution 24: 55–74.

TINSLEY, T. W. 1979. The potential of insect pathogenic viruses as pesticidal agents. Ann. Rev. Entomol. 24: 63–87.

TOMBACK, D. F. 1980. How nutcrackers find their seed stores. Condor 82: 10–19.

TOMLINSON, P. B. 1986. The Botany of Mangroves. Cambridge Univ. Press, Cambridge.

TORGERSON, T. R., & R. W. CAMPBELL. 1982. Some effects of avian predators on the western spruce budworm in north central Washington. Environ. Entomol. 11: 429–431.

TOWNSEND, C. R., & R. N. HUGHES. 1981. Maximizing net energy returns from foraging. Pp. 86–108 in C. R. Townsend and P. Calow (Eds.), Physiological Ecology. An Evolutionary Approach to Resource Use. Blackwell, Oxford.

TRACY, C. R. 1982. Biophysical modeling in reptilian physiology and ecology. Pp. 275–321 in C. Gans and F. H. Pough (Eds.), Biology of the Reptilia, Vol. 12. Academic Press, New York.

TRAIL, P. W. 1985. The intensity of selection: intersexual and interspecific comparisons require consistent measures. Amer. Nat. 126: 434–439.

TRAMER, E. J. 1969. Bird species diversity: components of Shannon's formula. Ecology 50: 927–929.

TRANSEAU, E. N. 1926. The accumulation of energy by plants. Ohio J. Sci. 26: 1–10.

TRAVIS, J., & R. E. RICKLEFS. 1983. A morphological comparison of island and mainland assemblages of neotropical birds. Oikos 41: 434–441.

TRESHOW, M. 1970. Environment and Plant Response. McGraw-Hill, New York.

TREWARTHA, G. T. 1954. An Introduction to Climate. McGraw-Hill, New York.

TRIVERS, R. L. 1971. The evolution of reciprocal altruism. Quart. Rev. Biol. 46: 35–57.

TRIVERS, R. L. 1972. Parental investment and sexual selection. Pp. 136–179 in B. Campbell (Ed.), Sexual Selection and the Descent of Man 1871–1971. Aldine, Chicago.

TRIVERS, R. L. 1974. Parent-offspring conflict. Amer. Zool. 14: 249–264.

TRIVERS, R. L. 1985. Social Evolution. Benjamin/Cummings, Menlo Park, California.

TRIVERS, R. L., & D. E. WILLARD. 1973. Natural selection of parental ability to vary the sex ratio of offspring. Science 179: 90–92.

TRUDINGER, P. A. 1969. Assimilatory and dissimilatory metabolism of inorganic sulphur compounds. Adv. Microbial Physiol. 3: 111–158.

TURCO, R. P., O. B. TOON, T. P. ACKERMAN, J. B. POLLACK, & C. SAGAN. 1983. Nuclear winter: global consequences of multiple nuclear explosions. Science 222: 1283–1292.

TURESSON, G. 1922. The genotypic response of the plant species to the habitat. Hereditas 3: 211–350.

TURELLI, M. 1978. A reexamination of stability in randomly varying versus deterministic environments with comments on the stochastic theory of limiting similarity. Theoret. Pop. Biol. 13: 244–267.

TURKINGTON, R., & L. W. AARSSEN. 1984. Local-scale differentiation as a result of competitive interactions. Pp. 107–127 in R. Dirzo and J. Sarukhan (Eds.), Perspectives on Plant Population Ecology. Sinauer, Sunderland, Mass.

TURKINGTON, R., & J. L. HARPER. 1979. The growth, distribution and neighborhood relationships of *Trifolium repens* in permanent pasture. IV. Fine-scale biotic differentiation. J. Ecol. 67: 245–254.

TURNER, F. B. 1970. The ecological efficiency of consumer populations. Ecology 51: 741–742.

TURNER, T. 1983. Facilitation as a successional mechanism in a rocky intertidal community. Amer. Nat. 121: 729–738.

UDOVIC, D., & C. AKER. 1981. Fruit abortion and the regulation of fruit number in *Yucca whipplei.* Oecologia 49: 245–248.

UTIDA, S. 1957. Population fluctuation, an experimental and theoretical approach. Cold Spring Harbor Symp. Quant. Biol. 22: 139–151.

UVAROV, B. P. 1961. Quantity and quality in insect populations. Proc. Roy. Entomol. Soc. Lond. C 25: 52–59.

VALIELA, I., & J. M. TEAL. 1979. The nitrogen budget of a salt marsh ecosystem. Nature 280: 652–656.

VANBLARICOM, G. R., & J. A. ESTES. (Eds.). 1988. The Community Ecology of Sea Otters. Springer-Verlag, Berlin and New York.

VANCE, R. R. 1985. The stable coexistence of two competitors for one resource. Amer. Nat. 136: 72–86.

VAN CLEVE, K., L. A. VIERECK, & R. L. SCHLENTNER. 1971. Accumulation of nitrogen in alder *(Alnus)* ecosystems near Fairbanks, Alaska. Arctic Alpine Res. 3: 101–114.

VAN DYNE, G. M. 1966. Ecosystems, systems ecology, and systems ecologists. U.S. Atom. Energy Comm., Oak Ridge Natl. Lab. Rept. ORNL 3957: 1–31. Reprinted 1969, in G. W. Cox (Ed.), Readings in Conservation Ecology, Appleton-Century-Crofts, New York.

VAN DYNE, G. M. (Ed.). 1969. The Ecosystem Concept in Natural Resource Management. Academic Press, New York and London.

VAN DYNE, G. M., & J. C. ANWAY. 1976. A research program for and the process of building and testing grassland ecosystems models. J. Range Mgmt. 29: 114–122.

VANDERMEER, J. H. 1969. The competitive structure of communities: an experimental approach with protozoa. Ecology 50: 362–371.

VANDERMEER, J. H. 1972. Niche theory. Ann. Rev. Ecol. Syst. 3: 107–132.

VAN EMDEN, H. F., & G. F. WILLIAMS. 1974. Insect stability and diversity in agroecosystems. Ann. Rev. Entomol. 19: 455–475.

VAN RIPER III, C., S. G. VAN RIPER, M. L. GOFF, & M. LAIRD. 1986. The epizootiology and ecological significance of malaria in Hawaiian land birds. Ecol. Mongr. 56: 327–344.

VAN VALEN, L. 1971. The history and stability of atmospheric oxygen. Science 171: 439–443.

VARLEY, G. C. 1949. Population changes in German forest pests. J. Anim. Ecol. 18: 117–122.

VARLEY, G. C. 1963. The interpretation of change and stability in insect populations. Proc. Roy. Entomol. Soc. Lond. C 27: 52–57.

VARLEY, G. C., & G. R. GRADWELL. 1960. Key factors in population studies. J. Anim. Ecol. 29: 399–401.

VARLEY, G. C., & G. R. GRADWELL. 1970. Recent advances in insect population dynamics. Ann. Rev. Entomol. 15: 1–24.

VARLEY, G. C., G. R. GRADWELL, & M. P. HASSELL. 1975. Insect Population Ecology. Blackwell Scientific Publ., Oxford.

VAURIE, C. 1950. Notes on Asiatic nuthatches and creepers. Amer. Mus. Novitates 1472: 1–39.

VEGIS, A. 1964. Dormancy in higher plants. Ann. Rev. Plant Physiol. 15: 185–215.

VEPSALAINEN, K. 1971. The role of gradually changing daylength in determination of winglength, alary polymorphism and diapause in a *Gerris odontogaster* (Zett.) population (Gerridae, Heteroptera) in South Finland. Ann. Acad. Sci. Fenn. Ser. A IV 183: 1–25.

VEPSALAINEN, K. 1973. The distribution and habitats of *Gerris* Fabr. species (Heteroptera, Gerridae) in Finland. Ann. Zool. Fenn. 10: 419–444.

VEPSALAINEN, K. 1974. Determination of wing length and diapause in water-striders (*Gerris* Fabr., Heteroptera). Hereditas 77: 163–176.

VEPSALAINEN, K. 1974. The life cycles and wing lengths of Finnish *Gerris* Fabr. species (Heteroptera, Gerridae). Acta Zool. Fenn. 141: 1–73.

VEPSALAINEN, K. 1974. The winglengths, reproductive stages and habitats of Hungarian *Gerris* Fabr. species (Heteroptera, Gerridae). Ann. Acad. Sci. Fenn. Ser. A IV 202: 1–18.

VERGHESE, M. W., & A. W. NORDSKOG. 1968. Correlated responses in reproductive fitness to selection in chickens. Genet. Res. 11: 221–238.

VERHULST, P. F. 1838. Notice sur la loi que la population suit dans son accroissement. Corresp. Math. Phys. 10: 113–121.

VERNER, J. 1964. Evolution of polygamy in the long-billed marsh wren. Evolution 18: 252–261.

VERNER, J., & M. F. WILLSON. 1966. The influence of habitats on mating systems of North American passerine birds. Ecology 47: 143–147.

VERNER, J., & M. F. WILLSON. 1969. Mating systems, sexual dimorphism, and the role of male North American passerine birds in the nesting cycle. Ornithol. Monogr. No. 9.

VESEY-FITZGERALD, D. F. 1960. Grazing succession among East African game animals. J. Mammol. 41: 161–172.

VIA, S. 1984. The quantitative genetics of polyphagy in an insect herbivore. I. Genotype-environment interactions in larval performance on different host plant species. Evolution 38: 881–895.

VIETMEYER, N. D. 1986. Lesser-known plants of potential use in agriculture and forestry. Science 232: 1379–1384.

VINCENT, T. L, & J. S. BROWN. 1988. The evolution of ESS theory. Ann. Rev. Ecol. Syst. 19: 423–443.

VINCE-PRUE, D. 1975. Photoperiodism in Plants. McGraw-Hill, New York.

VINEGAR, A. , V. H. HUTCHINSON, & H. G. DOWLING. 1970. Metabolism, energetics, and thermoregulation during brooding of snakes of the genus *Python* (Reptilia, Boidae). Zoologica 55: 19–50.

VITOUSEK, P. M. 1982. Nutrient cycling and nutrient use efficiency. Amer. Nat. 119–553–572.

VITOUSEK, P. M. 1984. Litterfall, nutrient cycling, and nutrient limitation in tropical forests. Ecology 65: 285–298.

VITOUSEK, P. M., & P. S. WHITE. 1981. Process studies in succession. Pp. 267–276 in D. C. West, H. H. Shugart, and D. B. Botkin (Eds.), Forest Succession. Concepts and Application. Springer-Verlag, New York.

VOGEL, S. 1970. Convective cooling at low airspeeds and the shape of broad leaves. J. Exp. Bot. 21: 91–101.

VOGEL, S. 1981. Life in Moving Fluids. The Physical Biology of Flow. Princeton Univ. Press, Princeton, New Jersey.

VOGEL, S. 1988. Life's Devices. Princeton Univ. Press, Princeton, New Jersey.

VOGL, R. J. 1973. Ecology of knobcone pine in the Santa Ana Mountains of California. Ecol. Monogr. 43: 125–143.

VOGL, R. J., & P. K. SCHORR. 1972. Fire and manzanita chaparral in the San Jacinto Mountains, California. Ecology 53: 1179–1188.

VOGT, K. A., C. C. GRIER, C. E. MEIER, & R. L. EDMUNDS. 1982. Mycorrhizal role in net primary production and nutrient cycling in *Abies amabilis* ecosystems in western Washington. Ecology 63: 370–380.

VOGT, K. A., C. C. GRIER, & D. J. VOGT. 1986. Production, turnover, and nutrient dynamics of above- and belowground detritus of world forests. Adv. Ecol. Res. 15: 303–337.

VOLTERRA, V. 1926. Variazioni e fluttuazioni del numero d'individui in specie animali conviventi. Mem. Acad. Lincei 2: 31–113.

VOLTERRA, V. 1926. Variations and fluctuations of the numbers of individuals in animal species living together. Reprinted 1931, in R. N. Chapman, Animal Ecology, McGraw-Hill, New York.

WAAGE, J., & D. GREATHEAD. (Eds.). 1986. Insect Parasitoids. Academic Press, London.

WADE, M. J. 1977. Experimental study of group selection. Evolution 31: 134–153.

WADE, M. J. 1978. A critical review of the models of group selection. Quart. Rev. Biol. 53: 101–114.

WAGGONER, P. E., & G. R. STEPHENS. 1970. Transition probabilities for a forest. Nature 255: 1160–1161.

WAGNER, F. H., & L. C. STODDART. 1972. Influence of coyote predation on black-tailed jackrabbit populations in Utah. J. Wildl. Mgmt. 36: 329–343.

WAISEL, Y. 1972. Biology of Halophytes. Academic Press, New York.

WAKE, D. B. 1982. Functional and evolutionary morphology. Perspect. Biol. Med. 25: 603–620.

WALDMAN, B. 1988. The ecology of kin recognition. Ann. Rev. Ecol. Syst. 19: 543–571.

WALKER, B. H. 1981. Is succession a viable concept in African savanna ecosystems? Pp. 430–447 in D. C. West, H. H. Shugart, and D. B. Botkin (Eds.), Forest Succession. Concepts and Application. Springer-Verlag, New York.

WALKER, R. B. 1954. The ecology of serpentine soils, II: factors affecting plant growth on serpentine soils. Ecology 35: 259–266.

WALLACE, A. R. 1876. The Geographical Distribution of Animals, Vol. 1 and 2. Reprinted 1962, Hafner, New York.

WALLACE, A. R. 1878. Tropical Nature and Other Essays. Macmillan, New York and London.

WALOFF, Z. 1966. The upsurges and recessions of the desert locust plague: an historical survey. Anti-Locust Mem. 8: 1–111.

WALSH, J. 1986. Return of the locust: a cloud over Africa. Science 234: 17–19.

WANGENSTEEN, O. D., H. RAHN, R. R. BURTON, & A. H. SMITH. 1974. Respiratory gas exchange of high altitude adapted chick embryos. Respir. Physiol. 21: 61–70.

WARD, P., & A. ZAHAVI. 1973. The importance of certain assemblages of birds as "information-centres" for food-finding. Ibis 115: 517–534.

WARD, R. C. 1967. Principles of Hydrology. McGraw-Hill, New York and London.

WARING, R. H. 1983. Estimating forest growth and efficiency in relation to canopy leaf area. Adv. Ecol. Res. 13: 327–354.

WARING, R. H., & J. MAJOR. 1964. Some vegetation of the California coastal region in relation to gradients of moisture, nutrients, light, and temperature. Ecol. Monogr. 34: 167–215.

WARNER, R. R. 1975. The adaptive significance of sequential hermaphroditism in animals. Amer. Nat. 109: 61–82.

WARREN, W. G., & C. L. BATCHELER. 1979. The density of spatial patterns: robust estimation through distance methods. Pp. 247–270 in R. M. Cormack and J. K. Ord (Eds.), Spatial and Temporal Analysis in Ecology. Intern. Co-op. Publ. House, Fairland, Maryland.

WASHBURN, S. L., & I. DEVORE. 1961. The social life of baboons. Sci. Amer. 204: 62–71.

WATSON, A., & D. JENKINS. 1968. Experiments on population control by territorial behaviour in red grouse. J. Anim. Ecol. 37: 395–614.

WATSON, A., & R. MOSS. 1980. Advances in our understanding of the population dynamics of red grouse from a recent fluctuation in numbers. Ardea 68: 103–111.

WATSON, A., & R. PARR. 1981. Hormone implants affecting territory size and aggressive and sexual behaviour in red grouse. Ornis Scand. 12: 55–61.

WATSON, I. A. 1970. Changes in virulence and population shifts in plant pathogens. Ann. Rev. Phytopathol. 8: 209–230.

WATSON, J. G. 1928. The mangrove swamps of the Malay Peninsula. Malay. For. Rec. 6

WATT, A. S. 1947. Pattern and process in the plant community. J. Ecol. 35: 1–22.

WATT, K. E. F. 1968. Ecology and Resource Management. McGraw-Hill, New York.

WAY, M. J. 1963. Mutualism between ants and honeydew-producing homoptera. Ann. Rev. Entomol. 8: 307–344.

WEBB, W., S. SZAREK, W. LAUENROTH, R. KINERSON, & M. SMITH. 1978. Primary productivity and water use in native forest, grassland, and desert ecosystems. Ecology 59: 1239–1247.

WECKER, S. C. 1963. The role of early experience in habitat selection by the prairie deer mouse, *Peromyscus maniculatus bairdii.* Ecol. Monogr. 33: 307–325.

WECKER, S. C. 1964. Habitat selection. Sci. Amer. 211: 109–116.

WEEKS, H. P., JR., & C. M. KIRKPATRICK. 1976. Adaptations of white-tailed deer to naturally occurring sodium deficiencies. J. Wildl. Mgmt. 40: 610–625.

WEINBERG, G. M. 1975. An Introduction to General Systems Thinking. Wiley, New York.

WEINER, J. 1988. Variation in the performance of individuals in plant populations. Pp. 59–81 in A. J. Davy, M. J. Hutchings, and A. R. Watkinson (Eds.), Plant Population Ecology. Blackwell, Oxford.

WEISSBURG, M. 1986. Risky business: on the ecological relevance of risk-sensitive foraging. Oikos 46: 261–262.

WEISSKOPF, V. F. 1968. How light interacts with matter. Sci. Amer. 219: 60–71.

WELCH, H. 1968. Relationships between assimilation efficiencies and growth efficiencies for aquatic consumers. Ecology 49: 755–759.

WELLER, D. E. 1987. A reevaluation of the -3/2 power rule of plant self-thinning. Ecol. Monogr. 57: 23–43.

WELLINGTON, W. G. 1960. Qualitative changes in natural populations during changes in abundance. Canad. J. Zool. 38: 289–314.

WELLS, H. 1979. Self-fertilization: advantageous or deleterious? Evolution 33: 252–255.

WERNER, E. E. 1977. Species packing and niche complementarity in three sunfishes. Amer. Nat. 111: 553–578.

WERNER, E. E. 1986. Amphibian metamorphosis: growth rate, predation risk, and the optimal size at transformation. Amer. Nat. 128: 319–341.

WERNER, E. E., J. F. GILLIAM, D. J. HALL, & G. G. MITTELBACH. 1983. An experimental test of the effects of predation on habitat use in fish. Ecology 54: 1540–1548.

WERNER, E. E., & D. J. HALL. 1976. Niche shifts in sunfishes: experimental evidence and significance. Science 191: 404–406.

WERNER, E. E., & D. J. HALL. 1977. Competition and habitat shift in two sunfishes (Centrarchidae). Ecology 58: 869–876.

WERNER, E. E., & D. J. HALL. 1979. Foraging efficiency and habitat switching in competing sunfishes. Ecology 60: 256–264.

WERNER, P. A., & W. J. PLATT. 1976. Ecological relationships of co-occurring goldenrods (*Solidago:* Compositae). Amer. Nat. 110. 959–971.

WERREN, J. H. 1980. Sex ratio adaptations to local mate competition in a parasitic wasp. Science 208: 1157–1159.

WEST, G. C. 1972. Seasonal differences in resting metabolic rate of Alaskan ptarmigan. Comp. Biochem. Physiol. 42A: 867–876.

WEST EBERHARD, M. J. 1975. The evolution of social behavior by kin selection. Quart. Rev. Biol. 50: 1–33.

WEST EBERHARD, M. J. 1981. Intragroup selection and the evolution of insect societies. Pp. 3–17 in R. D. Alexander and D. W. Tinkle (Eds.), Natural Selection and Social Behavior. Chiron, New York.

WESTHEIMER, F. H. 1987. Why nature chose phosphates. Science 235: 1173–1178.

WESTOBY, M. 1988. Comparing Australian ecosystems to those elsewhere. BioScience 38: 549–556.

WHEELER, W. M. 1923. Social Life Among the Insects. Harcourt, Brace, New York.

WHEELER, W. M. 1928. The Social Insects, Their Origin and Evolution. Harcourt, Brace, New York.

WHEELER, W. M. 1933. Colony Founding Among Ants. Harvard Univ. Press, Cambridge, Massachusetts.

WHITE, C. D. 1971. Vegetation-Soil Chemistry Correlations in Serpentine Ecosystems. Ph.D. diss., Univ. Oregon, Eugene.

WHITHAM, T. G., & S. MOPPER. 1985. Chronic herbivory: impacts on architecture and sex expression of pinyon pine. Science 228: 1089–1091.

WHITTAKER, R. H. 1952. A study of summer foliage insect communities in the Great Smoky Mountains. Ecol. Monogr. 22: 1–44.

WHITTAKER, R. H. 1953. A consideration of climax theory: the climax as a population and pattern. Ecol. Monogr. 23: 41–78.

WHITTAKER, R. H. 1954. The ecology of serpentine soils, I: introduction. Ecology 35: 258–259.

WHITTAKER, R. H. 1956. Vegetation of the Great Smoky Mountains. Ecol. Monogr. 26: 1–80.

WHITTAKER, R. H. 1960. Vegetation of the Siskiyou Mountains, Oregon and California. Ecol. Monogr. 30: 279–338.

WHITTAKER, R. H. 1965. Dominance and diversity in land plant communities. Science 147: 250–260.

WHITTAKER, R. H. 1967. Gradient analysis of vegetation. Biol. Rev. 42: 207–264.

WHITTAKER, R. H. 1972. Evolution and measurement of species diversity. Taxon 21: 213–251.

WHITTAKER, R. H. 1975. Communities and Ecosystems (2nd ed.). Macmillan, New York.

WHITTAKER, R. H., & P. P. FEENY. 1971. Allelochemics: chemical interactions between species. Science 171: 757–770.

WHITTAKER, R. H., & G. E. LIKENS. 1973. Primary production: the biosphere and man. Human Ecol. 1: 357–369.

WHITTAKER, R. H., & W. A. NIERING. 1965. Vegetation of the Santa Catalina Mountains, Arizona: a gradient analysis of the south slope. Ecology 46: 429–452.

WHITTAKER, R. H., & G. M. WOODWELL. 1968. Dimension and production relations of trees and shrubs in the Brookhaven Forest, New York. J. Ecol. 56: 1–25.

WHITTAKER, R. H., & G. M. WOODWELL. 1969. Structure, production and diversity of the oak-pine forest at Brookhaven, New York. J. Ecol. 57: 155–174.

WICKLER, W. 1968. Mimicry in Plants and Animals. World Univ. Library, London.

WIEBE, W. J. 1975. Nitrogen fixation in a coral reef community. Science 188: 257–259.

WIEBES, J. T. 1979. Co-evolution of figs and their insect pollinators. Ann. Rev. Ecol. Syst. 10: 1–12.

WIEGERT, R. G. 1988. The past, present, and future of ecological energetics. Pp. 29–55 in L. R. Pomeroy and J. J. Albert (Eds.), Concepts of Ecosystem Ecology. A Comparative View. Springer-Verlag, New York.

WIEGERT, R. G., & D. F. OWEN. 1971. Trophic structure, available resources and population density in terrestrial vs. aquatic ecosystems. J. Theoret. Biol. 30: 69–81.

WIENS, D. 1984. Ovule survivorship, brood size, life history, breeding systems, and reproductive success in plants. Oecologia 64: 47–53.

WIENS, J. A. 1966. On group selection and Wynne-Edwards' hypothesis. Amer. Sci. 273–287.

WIENS, J. A. 1973. Pattern and process in grassland bird communities. Ecol. Monogr. 43: 237–270.

WIENS, J. A. 1977. On competition and variable environments. Amer. Sci. 65: 590–597.

WIKLUND, C. 1981. Generalist vs. specialist oviposition behaviour in *Papilio machaon* (Lepidoptera) and functional aspects of the hierarchy of oviposition preferences. Oikos 36: 163–170.

WILBUR, H. M. 1972. Competition, predation, and the structure of the *Ambystoma-Rana sylvatica* community. Ecology 53: 3–21.

WILBUR, H. M., & R. A. ALFORD. 1985. Priority effects in experimental pond communities: responses of *Hyla* to *Bufo* and *Rana.* Ecology 66: 1106–1114.

WILBUR, H. M., P. J. MORIN, & R. N. HARRIS. 1983. Salamander predation and the structure of experimental communities: anuran responses. Ecology 64: 1423–1429.

WILBUR, H. M., D. W. TINKLE, & J. P. COLLINS. 1974. Environmental certainty, trophic level, and resource availability in life history evolution. Amer. Nat. 108: 805–817.

WILDE, S. A. 1968. Mycorrhizae and tree nutrition. BioScience 18: 482–484.

WILEY, R. H. 1974. Evolution of social organization and life-history patterns among grouse. Quart. Rev. Biol. 49: 201–227.

WILKINSON, G. S. 1984. Reciprocal food sharing in the vampire bat. Nature 308: 181–184.

WILLIAMS, C. B. 1964. Patterns in the Balance of Nature and Related Problems in Quantitative Ecology. Academic Press, New York.

WILLIAMS, G. C. 1957. Pleiotropy, natural selection, and the evolution of senescence. Evolution 11: 398–411.

WILLIAMS, G. C. 1966. Natural selection, the costs of reproduction, and a refinement of Lack's principle. Amer. Nat. 100: 687–690.

WILLIAMS, G. C. 1966. Adaptation and Natural Selection. Princeton Univ. Press, Princeton, New Jersey.

WILLIAMS, G. C. 1975. Sex and Evolution. Princeton Univ. Press, Princeton, New Jersey.

WILLIAMS, G. C. 1979. The question of adaptive sex ratio in outcrossed vertebrates. Proc. Roy. Soc. Lond. B 205: 567–580.

WILLIAMS, G. C., & J. B. MITTON. 1973. Why reproduce sexually? J. Theoret. Biol. 39: 545–554.

WILLIAMS, G. C., & D. C. WILLIAMS. 1957. Natural selection of individually harmful social adaptations among sibs with special reference to social insects. Evolution 11: 32–39.

WILLIAMS, P. H. 1975. Genetics of resistance in plants. Genetics 79: 409–419.

WILLIAMSON, G. B. 1975. Pattern and seral composition in an old-growth beech-maple forest. Ecology 56: 727–731.

WILLIS, J. C. 1922. Age and Area. A Study in Geographical Distribution and Origin in Species. Cambridge Univ. Press, Cambridge.

WILLSON, M. F. 1983. Plant Reproductive Ecology. Wiley, New York.

WILLSON, M. F., & N. BURLEY. 1983. Mate Choice in Plants: Tactics, Mechanisms, and Consequences. Princeton Univ. Press, Princeton, New Jersey.

WILSON, D. S. 1980. The Natural Selection of Populations and Communities. Benjamin/Cummings, Menlo Park, California.

WILSON, D. S. 1983. The group selection controversy: history and current status. Ann. Rev. Ecol. Syst. 14: 159–187.

WILSON, E. O. 1961. Nature of the taxon cycle in the Melanesian ant fauna. Amer. Nat. 95: 169–193.

WILSON, E. O. 1971. The Insect Societies. Belknap Press, Cambridge, Massachusetts.

WILSON, E. O. 1973. Group selection and its significance for ecology. BioScience 23: 631–638.

WILSON, E. O. (Ed.). 1988. BioDiveristy. National Academy Press, Washington, D.C.

WILSON, J. W. 1974. Analytical zoogeography of North American mammals. Evolution 28: 124–140.

WIMPENNY, R. S. 1966. The Plankton of the Sea. Faber and Faber, London.

WINGFIELD, J. C. 1984. Androgens and mating systems: testosterone-induced polygyny in normally monogamous birds. Auk 101: 665–671.

WIT, C. T. DE. 1960. On competition. Versl. Landbouwk. Onderz. 66: 1–82.

WITKAMP, M. 1966. Decomposition of leaf litter in relation to environment, microflora, and microbial respiration. Ecology 47: 194–201.

WITKAMP, M., & J. VAN DER DRIFT. 1961. Breakdown of forest litter in relation to environmental factors. Plant and Soil 15: 295–311.

WITTENBERGER, J. F. 1978. The evolution of mating systems in grouse. Condor 80: 126–137.

WITMAN, J. D. 1985. Refuges, biological disturbance, and rocky subtidal community structure in New England. Ecol. Monogr. 55: 421–445.

WITMAN, J. D. 1987. Subtidal coexistence: storms, grazing, mutualism, and the zonation of kelps and mussels. Ecol. Monogr. 57: 167–187.

WITTER, J. A., H. M. KULMAN, & A. C. HODSON. 1975. Life tables for the forest tent caterpillar. Ann. Entomol. Soc. Amer. 65: 25–31.

WOLANSKI, E., & W. M. HAMMER. 1988. Topographically controlled fronts in the ocean and their biological significance. Science 241: 177–181.

WOLFE, J. A. 1975. Some aspects of plant geography of the Northern Hemisphere during the late Cretaceous and Tertiary. Ann. Missouri Bot. Gard. 62: 264–279.

WOLFE, J. A. 1979. Temperature parameters of humid to mesic forests of eastern Asia and relation to forests of other regions of the Northern Hemisphere and Australasia. Geol. Surv. Prof. Paper 1106: 1–37.

WOLFE, J. A. 1981. Vacariance biogeography of angiosperms in relation to paleobotanical data. Pp. 413–427 in G. Nelson and D. E. Rosen (Eds.), Vicariance Biogeography: a critique. Columbia Univ. Press, New York.

WOLIN, M. J. 1979. The rumen fermentation: a model for microbial interactions in anaerobic systems. Adv. Microbial Ecol. 3: 49–77.

WOLK, P. 1973. Physiology and cytological chemistry of blue-green algae. Bacteriol. Rev. 37: 32–101.

WOODWELL, G. M., J. E. HOBBIE, R. A. HOUGHTON, J. M. MELILLO, B. MOORE, B. J. PETERSON, & G. R. SHAVER. 1983. Global deforestation: contribution to atmospheric carbon dioxide. Science 222: 1081–1086.

WOOLFENDEN, G. E., & J. W. FITZPATRICK. 1984. The Florida Scrub Jay. Demography of a Cooperative-breeding Bird. Princeton Univ. Press, Princeton, New Jersey.

WORTHINGTON, E. B. (Ed.). 1975. The Evolution of IBP. Cambridge Univ. Press, London.

WRIGHT, H. E., JR. 1964. Aspects of the early postglacial forest succession in the Great Lakes Region. Ecology 45: 439–448.

WRIGHT, H. E., JR. 1968. The roles of pine and spruce in the forest history of Minnesota and adjacent areas. Ecology 49: 937–955.

WRIGHT, S. 1931. Evolution in Mendelian populations. Genetics 16: 97–159.

WRIGHT, S. 1943. Isolation by distance. Genetics 28: 114–138.

WRIGHT, S. 1945. Tempo and mode in evolution: a critical review. Ecology 26: 415–419.

WRIGHT, S. 1946. Isolation by distance under diverse systems of mating. Genetics 31: 39–59.

WRIGHT, S. 1969. Evolution and the Genetics of Populations. Vol. 2. The Theory of Gene Frequencies. Univ. Chicago Press, Chicago.

WRIGHT, S. J. 1980. Density compensation in island avifaunas. Oecologia (Berl.) 45: 385–389.

WRIGHT, S. J. 1988. Patterns of abundance and the form of the species-area relation. Amer. Nat. 131: 401–411.

WUNDERLE, J. M., M. S. CASTRO, & N. FETCHER. 1988. Risk averse foraging by bananaquits on negative energy budgets. Behav. Ecol. Sociobiol. 21: 249–255.

WYATT, R. 1983. Pollinator-plant interactions and the evolution of breeding systems. Pp. 51–95 in L. Real (Ed.), Pollination Biology. Academic Press, New York.

WYNNE-EDWARDS, V. C. 1962. Animal Dispersion in Relation to Social Behaviour. Oliver and Boyd, Edinburgh.

WYNNE-EDWARDS, V. C. 1963. Intergroup selection in the evolution of social systems. Nature 200: 623–628.

WYNNE-EDWARDS, V. C. 1986. Evolution Through Group Selection. Blackwell Scientific, Palo Alto, California.

YAMPOLSKY, E., & H. YAMPOLSKY. 1922. Distribution of sex forms in phanerogamic flora. Bibl. Genet. 3: 1–62.

YANCEY, P. H., M. E. CLARK, S. C. HAND, R. D. BOWLUS, & G. N. SOMERO. 1982. Living with water stress: evolution of osmolyte systems. Science 217: 1214–1222.

YANCEY, P. H., & G. N. SOMERO. 1980. Methylamine osmoregulatory solutes of elasmobranch fishes counteract urea inhibition of enzymes. J. Exp. Zool. 212: 205–213.

YANG, R. S. H., & F. E. GUTHRIE. 1969. Physiological responses of insects to nicotine. Ann. Entomol. Soc. Amer. 62: 141–146.

YEATON, R. I., & M. L. CODY. 1979. Distribution of cacti along environmental gradients in the Sonoran and Mojave deserts. J. Ecol. 67: 529–541.

YEATON, R. I., R. W. YEATON, J. P. WAGGONER III, & J. E. HORENSTEIN. 1985. The ecology of *Yucca* (Agavaceae) over an environmental gradient in the Mohave Desert: distribution and interspecific interactions. J. Arid Envir. 8: 33–44.

YODA, K., T. KIRA, H. OGAWA, & K. HOZUMI. 1963. Self thinning in overcrowded pure stands under cultivated and natural conditions. J. Biol., Osaka City Univ. 14: 107–129.

YOUNG, E. C. 1965. Flight muscle polymorphism in British Corixidae: ecological observations. J. Anim. Ecol. 34: 353–390.

YULE, G. U. 1924. A mathematical theory of evolution based on the conclusions of Dr. J. C. Willis F. R. S. Phil. Trans. Roy. Soc. Lond. B 213: 21–87.

ZAHAVI, A. 1975. Mate selection—a selection for a handicap. J. Theoret. Biol. 53: 205–214.

ZAHAVI, A. 1977. The cost of honesty (further remarks on the handicap principle). J. Theoret. Biol. 67: 603–605.

ZAHAVI, A. 1977. Reliability in communication systems and the evolution of alturism. Pp. 253–259 in B. Stonehouse and C. M. Perrins (Eds.), Evolutionary Ecology. Macmillan, London.

ZARET, T. M. 1980. Predation and Freshwater Communities. Yale Univ. Press, New Haven, Connecticut.

ZARET, T. M., & R. T. PAINE. 1973. Species introduction in a tropical lake. Science 182: 449–455.

Index

Note on the cover illustration

The illustration on the cover is a slightly cropped reproduction from the painting *Exotic Landscape* by Henri Rousseau. It was painted in 1910, the last year of Rousseau's life, and is reproduced with the generous permission of the Norton Simon Foundation.

Henri Rousseau was born in 1844 in the northwest of France. He spent 24 years as a municipal toll collector in the suburbs of Paris. Too poor to enroll in an art school, he became a diligent Sunday painter. His work was unknown to the public until 1886. That year he exhibited a painting in the Salon of the Independent Artists, a group that eventually included Cézanne, Toulouse-Lautrec, Renoir, Gauguin, van Gogh, and Seurat. Described by all who knew him as a man of childlike simplicity, he nonetheless became friends with most of the figures prominent in the literary and artistic life of Paris. He was also a man of stubborn perseverence and faith in his own work, qualities that led him to retire on a tiny pension in 1893. He chose to live in desperate poverty for the rest of his life so that he might fulfill his passionate wish to show the world what he saw.

How he came to see it was misunderstood for a number of years. Guillaume Apollinaire, a French poet and a champion of Rousseau's work, wrote that Rousseau was sent as a young soldier to Mexico and that his memory of the tropics was expressed in his jungle paintings. In fact, Rousseau's army service never took him more than 20 miles from his birthplace in France. He found his tropics in the Paris botanical garden and zoo, in a book (published by a Parisian department store) called *Wild Beasts: Approximately 200 Amusing Illustrations Drawn from the Life of Animals,* and in the many contemporary illustrated accounts of explorations and colonial expeditions.

Like medieval tapestries, Rousseau's canvases are filled with detail, and, as in our observation of the natural world, we can absorb only a little at a time. Rousseau appears to have taken a profuse joy in every plant and animal he painted. He said of himself that when he saw strange plants from exotic countries, he felt as if he had stepped into a dream. Yet despite the strong note of fantasy in his painting, his depiction of the natural world reflects reality. Maurice Raynal, a friend of Rousseau's who wrote about him and other Parisian painters of the time, quotes a Brazilian painter who had traveled up the Amazon. The man observed that the jungle was so thick it appeared to have no depth, just as though painted in a single plane on a canvas. That, of course, is exactly how the jungle appears in Rousseau's paintings.

In his autobiography, written in the third person, Rousseau says that he made his debut in art after many disappointments, alone and without any master but nature. He goes on to say that he continued to improve himself more and more in the original style he had adopted. So he did, to our great fortune and everlasting amazement. As much as any artist, Rousseau conveys a profound love for the natural world.

Note on the author

Robert E. Ricklefs is a native of northern California, where he received his undergraduate education in biology at Stanford University. He pursued graduate studies at the University of Pennsylvania, to which he returned as a faculty member after postdoctoral work at the Smithsonian Tropical Research Institute in Panama. He is now Professor of Biology.

Recipient of a Guggenheim Fellowship and the Brewster Medal of the American Ornithologists' Union for ecological and evolutionary studies of reproduction in birds, Professor Ricklefs has also conducted research on the ecological relationships of seabirds to the marine environment; on historical biogeography and ecology of island populations; and on the organization of biological communities. In addition to *Ecology,* he is the author of *The Economy of Nature.*

Conversion Factors

Length

1 meter (m) = 39.4 inches (in)
1 meter = 3.28 feet (ft)
1 kilometer (km) = 3281 feet
1 kilometer = 0.621 miles (mi)
1 micron (μ) = 10^{-6} meters
1 inch = 2.54 centimeters (cm)
1 foot = 30.5 centimeters
1 mile = 1609 meters
1 Angstrom unit (Å) = 10^{-10} meters
1 millimicron (mμ) = 10^{-9} meters

Area

1 square centimeter (cm^2) = 0.155 square inches (in^2)
1 square meter (m^2) = 10.76 square feet (ft^2)
1 hectare (ha) = 2.47 acres (A)
1 hectare = 10,000 square meters
1 hectare = 0.01 square kilometer (km^2)
1 square kilometer = 0.386 square miles
1 square mile = 2.59 square kilometers
1 square inch = 6.45 square centimeters
1 square foot = 929 square centimeters
1 square yard (yd^2) = 0.836 square meters
1 acre = 0.407 hectares

Mass

1 gram (g) = 15.43 grains (gr)
1 kilogram (kg) = 35.3 ounces
1 kilogram = 2.205 pounds (lb)
1 metric ton (t) = 2204.6 pounds
1 ounce (oz) = 28.35 grams
1 pound = 453.6 grams
1 short ton = 907 kilograms

Time

1 year (yr) = 8760 hours (hr)
1 day = 86,400 seconds (s)

Volume

1 cubic centimeter (cc or cm^3) = 0.061 cubic inches (in^3)
1 cubic inch = 16.4 cubic centimeters
1 liter = 1,000 cubic centimeters
1 liter = 33.8 U.S. fluid ounces (oz)
1 liter = 1.057 U.S. quarts (qt)
1 liter = 0.264 U.S. gallons (gal)
1 U.S. gallon = 3.79 liters
1 Brit. gallon = 4.55 liters
1 cubic foot (ft^3) = 28.3 liters (l)
1 milliliter (ml) = 1 cubic centimeter
1 U.S. fluid ounce = 29.57 milliliters
1 Brit. fluid ounce = 28.4 milliliters
1 quart = 0.946 liters

Velocity

1 meter per second (m s^{-1}) = 2.24 miles per hour (mi hr^{-1})
1 foot per second (ft s^{-1}) = 1.097 kilometers per hour
1 kilometer per hour = 0.278 meters per second
1 mile per hour = 0.447 meters per second
1 mile per hour = 1.467 feet per second

Energy

1 joule = 0.239 calories (cal)
1 calorie = 4.184 joules
1 kilowatt-hour (kWh) = 860 kilocalories
1 kilowatt-hour = 3600 kilojoules
1 British thermal unit (Btu) = 252.0 calories
1 British thermal unit = 1054 joules
1 kilocalorie (kcal) = 1,000 calories

Power

1 kilowatt (kW) = 0.239 kilocalories per second
1 kilowatt = 860 kilocalories per hour
1 horsepower (hp) = 746 watts
1 horsepower = 15,397 kilocalories per day
1 horsepower = 641.5 kilocalories per hour

Energy per unit area

1 calorie per square centimeter = 3.69 British thermal units per square foot
1 British thermal unit per square foot = 0.271 calories per square centimeter
1 calorie per square centimeter = 10 kilocalories per square meter

Power per unit area

1 kilocalorie per square meter per minute = 52.56 kilocalories per hectare per year
1 footcandle (fc) = 1.30 calories per square foot per hour at 555 mμ wavelength
1 footcandle = 10.76 lux
1 lux (lx) = 1.30 calories per square meter per hour at 555 mμ wavelength

Metabolic energy equivalents

1 gram of carbohydrate = 4.2 kilocalories
1 gram of protein = 4.2 kilocalories
1 gram of fat = 9.5 kilocalories

Miscellaneous

1 gram per square meter = 0.1 kilograms per hectare
1 gram per square meter = 8.97 pounds per acre
1 kilogram per square meter = 4.485 short tons per acre
1 metric ton per hectare = 0.446 short tons per acre